FOUNDATIONS OF PARASITOLOGY

FIFTH EDITION

Juveniles of *Anisakis* sp. on the stomach of a bocaccio (*Sebastes paucispinus*), a commercially important marine fish. The normal definitive hosts of these nematodes are marine mammals, but they cause severe pain when people eat them inadvertently with raw fish (see p. 428).

Photograph courtesy of J. Sakanari.

GERALD D. SCHMIDT & LARRY S. ROBERTS'

FOUNDATIONS OF
PARASITOLOGY

FIFTH EDITION

Larry S. Roberts
University of Miami

John Janovy, Jr.
University of Nebraska–Lincoln

WCB
Wm. C. Brown Publishers

Dubuque, IA   Bogota   Boston   Buenos Aires   Caracas   Chicago
Guilford, CT   London   Madrid   Mexico City   Sydney   Toronto

Book Team

Editor *Margaret J. Kemp*
Developmental Editor *Kathleen R. Loewenberg*
Production Editor *Sue Dillon and Cathy Smith*
Designer *Lu Ann Schrandt*
Art Editor *Tina Flanagan*
Photo Editor *Lori Hancock*
Permissions Coordinator *Patricia Barth*

Wm. C. Brown Publishers

President and Chief Executive Officer *Beverly Kolz*
Vice President, Publisher *Kevin Kane*
Vice President, Director of Sales and Marketing *Virginia S. Moffat*
Vice President, Director of Production *Colleen A. Yonda*
National Sales Manager *Douglas J. DiNardo*
Marketing Manager *Thomas C. Lyon*
Advertising Manager *Janelle Keeffer*
Production Editorial Manager *Renée Menne*
Publishing Services Manager *Karen J. Slaght*
Royalty/Permissions Manager *Connie Allendorf*

A Times Mirror Company

Copyedited by Sarah Lane

Cover photo courtesy of Kevin R. Kazacos and Samuel D. Royer.

Library of Congress Catalog Card Number: 94–72939

ISBN 0–697–26071–2

Printed in the United States of America by Times Mirror Higher Education Group, Inc.,
2460 Kerper Boulevard, Dubuque, IA 52001

10 9 8 7 6 5 4 3 2

PREFACE

It has been too long since the last edition, partly as a result of the tragic death of our friend and colleague, Gerald D. Schmidt. He taught us all much, and we greatly miss him.

One result of the long interval between editions is that this version is the most thoroughly revised of any yet published. A new coauthor has come on board, bringing new insights and a fresh approach. We do believe, however, that we have retained the essential qualities of the text that students and professors liked in the first four editions. The reception accorded *Foundations of Parasitology* has been most gratifying. Your comments and suggestions are always welcome. Keep them coming.

THE SCOPE OF THIS BOOK

This textbook is designed especially for upper-division courses in general parasitology. It emphasizes principles, illustrating them with primary sections on the biology, physiology, morphology, and ecology of the major parasites of humans and domestic animals. We have found that these are of most interest to the majority of students. Other parasites are included as well, where they are of unusual biological interest.

The first three chapters delineate important definitions and principles in various areas, including evolution, ecology, immunology, and pathology. Chapters on specific groups follow, beginning with protistans and ending with arthropods. With the exception of the first three chapters, however, presentations in each chapter are not predicated on the students having first studied groups that come earlier in the book. Therefore, the order can vary as the teacher desires. Different instructors emphasize different materials, so there is more in the book than can be covered adequately in a single semester. As always we have strived for readability, the words being enhanced by photographs, drawings, electron micrographs, and tables.

NEW TO THIS EDITION

The book you are holding represents a milestone in the life of this text. This edition integrates a wealth of new discoveries and literature. Many areas of parasitology are theaters of intense research effort and fruitful results. Chapters covering such areas required extensive updating and rewriting. Addition of so much material compelled us to prune out an equal amount of text and illustrations so as not to increase the book length, but we hope that we have been judicious in our reshaping. Previous users of this text will notice many smaller changes, in addition to the points we mention in the paragraphs to follow, compared with the fourth edition. One of the most obvious of these is the trenchant quotation at the beginning of each chapter. Well, maybe some of them are not so trenchant. Nevertheless, we hope these observations of pioneering researchers, as well as references to literature and even pop culture, will broaden your view of parasitology.

In this edition, coverage of parasite evolution (primarily phylogenetic systematics) and ecology required expansion, and basic definitions have been transferred to Chapter 1. Propelled in large measure by modern molecular methods, immunologists continue their torrent of discoveries. The 1980s saw enormous increases in our understanding of the role and mechanisms of cytokine function and witnessed our realization of the importance of immunopathology in parasitic diseases. Thus, Chapter 3 has been almost completely recast. We hope that instructors will emphasize the concepts in these first three chapters; later chapters harken back to them frequently.

The "Form and Function" chapters on protistan parasites, trematodes, cestodes, nematodes, and arthropods have again been updated and rewritten significantly to provide a stronger base of knowledge with which to investigate each group further. When we could find published cladograms, we have added them to show phylogenetic relationships of some of the major groups.

Chapter 5 on the Kinetoplastida includes more recent discoveries on the molecular biology of trypanosomes and new material on the still-unfolding mystery of immune reactions in Chagas' disease and leishmaniasis. Other protistan chapters address the exploding body of knowledge about opportunistic parasitic infections in immunocompromised persons and the amazing diversity of coccidians as revealed by the active systematic research on these parasites.

Intense scrutiny of malaria continues, reflecting its widespread importance as a human disease, and Chapter 9 has been revised accordingly. Note particularly the expanded table comparing *Plasmodium* spp., new methods of diagnosis, the role of cytokines in pathogenesis and immunity, progress toward vaccines, and drug action and resistance.

The chapter on Monogenea has been moved to a position preceding the cestode chapters, reflecting the closer relationship of monogenes and cestodes as indicated by cladistic analysis.

In Chapter 16, the sections on pathogenesis, diagnosis and treatment, and control of schistosomes have been almost completely rewritten. A table illustrating the major groups of *Schistosoma* spp. has been added. The chapter on echinostomes has been extensively rewritten as warranted by the growing use of these worms in experimental studies of host-parasite relationships.

The chapters on nematodes have been revised in many ways. Some noteworthy changes include emphasis on intracellular parasitism in *Trichinella;* recognition of five species of *Trichinella;* the nasofrontal migration of *Strongyloides;* haplodiploidy in Oxyurida; more evidence for distinctiveness of *Ascaris lumbricoides* and *A. suum;* expanded coverage of visceral larva migrans and *Anisakis* spp.; evidence that *Ancylostoma caninum* is sometimes a parasite of humans; sections on horse strongyles, *Syngamus,* and *Oesophagostomum;* coverages of immunity, immunopathology, immunotolerance, and chemotherapy of lymphatic filariasis; the revolutionary role of ivermectin in treatment of onchocerciasis; expanded coverage of *Dirofilaria immitis;* and the potential eradication of *Dracunculus medinensis.*

Among the numerous revisions of Chapter 34 on parasitic Crustacea, we have added the following: a brief description of parasitic copepods that are so highly modified they cannot even be assigned to orders; the concept of cryptogonochorism as applied to rhizocephalans; and a description of the bizarre little tantulocaridans with a diagram of their proposed life cycle. The arthropod chapters in general include many new illustrations, all of them quite instructive and many of them truly beautiful. Chapter 40, the acarines, has been expanded by additions of sections on Lyme disease, immunity, and pheromonal control of tick behavior.

INSTRUCTIVE DESIGN

Students using the fifth edition of *Foundations of Parasitology* are guided to a clear understanding of the topic through our careful use of study aids. Essential terms, many of which are defined in a complete glossary, are **boldfaced** in the text to provide emphasis and ease in reviewing. New to this edition, and in response to student requests, we are providing pronunciation guides for glossary entries. Numbered references at the end of each chapter make supporting data and further study easily accessible. Clear labeling makes all illustrations approachable and self-explanatory to the student.

We have been fortunate indeed to have William C. Ober and Claire W. Garrison draw many new illustrations for this edition, including many new life cycle diagrams. Their artistic skills and knowledge of biology have enhanced the other zoology texts coauthored by Larry Roberts. Bill and Claire bring to their work as illustrators a unique perspective resulting from their earlier careers as physician and nurse, respectively.

ACKNOWLEDGMENTS

We are indebted to the numerous students and colleagues who have commented on previous editions. We especially wish to thank the following individuals who reviewed certain chapters or the entire text:
Martin L. Adamson, *University of British Columbia*
Daniel R. Brooks, *University of Toronto*
George E. Cain, *University of Iowa*
Janine N. Caira, *University of Connecticut*
Richard E. Clopton, *Texas A & M University*
Dickson D. Despommier, *Columbia University*
Sherwin S. Desser, *University of Toronto*
Michael B. Hildreth, *South Dakota University*
Hadar Isseroff, *SUNY College at Buffalo*
Raymond E. Kuhn, *Wake Forest University*
Mary Louise Nagel Leida, *Morningside College*
Brent B. Nickol, *University of Nebraska*
David F. Oetinger, *Kentucky Wesleyan College*
Wayne Price, *University of Tampa*
Dennis J. Richardson, *University of Nebraska*
J. Teague Self, *University of Oklahoma*
Jon A. Yates, *Oakland University*

We are grateful to Cara Stanko who provided library and office assistance and to Claudette Sterling-Baines of Miami-Dade Community College for her assistance with certain permissions.

We thank the dedicated and conscientious staff of Wm. C. Brown Publishers, especially Marge Kemp, Project Editor; Kathy Loewenberg, Developmental Editor; Cathy Smith and Sue Dillon, Production Editors; Lori Hancock, Photo Editor; Tina Flanagan, Art Editor; Patricia Barth, Permissions Coordinator; and Lu Ann Schrandt, Designer. They have done a marvelous job in facilitating the transition of this book from one publisher to another. The exacting task of copyediting the manuscript was accomplished with unfailing grace and good humor by Sarah Lane.

Larry S. Roberts
John Janovy, Jr.

I want to add a few words of welcome to the new coauthor of *Foundations of Parasitology*. Although John and I have known each other for years, this is our first opportunity to collaborate. He is an excellent writer and a biologist with broad expertise. I count myself and the readers of this book as very fortunate to have him join us. It has been a joy to work with him in preparation of this revision.

LSR

CONTENTS

Chapter 1

INTRODUCTION TO PARASITOLOGY

So, naturalist observe, a flea

Hath smaller fleas that on him prey,

And these have smaller fleas to bite 'em.

And so proceed ad infinitum.

J. Swift

Few persons realize that there are far more kinds of parasitic than nonparasitic organisms in the world. Even if we exclude the viruses and rickettsias, which are all parasitic, and the many kinds of parasitic bacteria and fungi, the parasites are still in the majority. The bodies of "free-living" plants and animals obviously represent a rich environment that has been colonized innumerable times throughout evolutionary history.

In general the parasitic way of life is highly successful because it evolved independently in nearly every phylum of animals, from protistan phyla to arthropods and chordates, as well as in many plant groups. Organisms that are not parasites are usually hosts. Humans, for example, can be infected with more than a hundred kinds of flagellates, amebas, ciliates, worms, lice, fleas, ticks, and mites. It is unusual to examine a domestic or wild animal without finding at least one species of parasite on or within it. Even animals reared under strict laboratory conditions are commonly infected with protozoa and other parasites. Often the parasites themselves are the hosts of other parasites.

The relationships between parasites and hosts are typically quite intimate, biochemically speaking, and it is a fascinating, often compelling, task to explain just why a species of parasite is restricted to one or a few host species. It is no wonder that the science of **parasitology** has developed out of efforts to understand parasites and their relationships with their hosts.

RELATIONSHIP OF PARASITOLOGY TO OTHER SCIENCES

The first and most obvious stage in the development of parasitology was the discovery of parasites themselves. **Descriptive parasitology** probably began in prehistory. **Taxonomy** as a formal science, however, started with Linnaeus's publication of the tenth edition of *Systema Naturae* in 1758. Linnaeus himself is credited with the description of the sheep liver fluke *Fasciola hepatica,* and through the next 100 years, many common parasites, as well as their developmental stages, were described. The discovery and description of

new parasite species continues today, just as does the description of new species in almost every group of plants and animals. Although biologists have a massive "catalog" of the planet's flora and fauna, this list is far from complete. Indeed, based on the rate of new published descriptions, scientists estimate that humans are destroying species faster than they are discovering them, especially in the tropics. There is every reason to believe this generality applies to parasites as well as to butterflies.

Today **systematists** rely on published species descriptions, as well as on studies of DNA, proteins, ecological niches, and geographical distribution, to develop **phylogenies** (phylogeny, singular), or evolutionary histories, of parasites. **Epidemiologists** may need to understand sociological factors, climate, local traditions, and global economics, as well as pharmacology, pathology, biochemistry, and clinical medicine, to devise schemes for controlling parasitic infections.

When people became aware that parasites were troublesome and even serious agents of disease, they began an ongoing effort to heal the infected and eliminate the parasites. Curiosity about routes of infection led to studies of parasite **life cycles;** thus, it became generally understood in the last part of the nineteenth century that certain animals—for example, ticks and mosquitoes—could serve as **vectors** that transmitted parasites to humans and their domestic animals. As more and more life cycles became known, parasitologists quickly realized the importance of understanding these seemingly complex series of ecological and embryological events. It is naive to try to control an infection without knowledge of how the infectious agent, in this case the parasite, reproduces and gets from one host to another.

Parasite biology does not differ fundamentally from the biology of free-living organisms, and parasite systems have provided outstanding models in studies of basic biological phenomena. In the nineteenth century van Beneden described meiosis and Boveri demonstrated the continuity of chromosomes, both in parasitic nematodes. In the twentieth century refined techniques in physics and chemistry applied to parasites have added further to our understanding of basic biological principles and mechanisms. For example, Keilin discovered cytochrome and the electron transport system during his investigations of parasitic worms and insects.[20] Today biochemical techniques are widely used in studies of parasite metabolism, immunology, serology, and chemotherapy. The advent of the electron microscope has resulted in many new discoveries at the subcellular level. Molecular biology and recombinant DNA techniques have contributed new diagnostic techniques and new knowledge of relationships between parasites, and they offer much hope in the development of new vaccines. Certain parasitic protozoa (for example, trypanosomes) today serve as models for some of the most exciting research in molecular genetics and gene expression.[4,9,24]

Historically centered on animal parasites of humans and domestic animals, the discipline of parasitology usually does not include a host of other parasitic organisms, such as viruses, bacteria, fungi, and nematode parasites of plants. Thus, parasitology has evolved separately from virology, bacteriology, mycology, and plant nematology. Medical entomology, too, has branched off as a separate discipline, but it remains a subject of paramount importance to the parasitologist, who must understand the relationships between arthropods and the parasites they harbor and disperse.

PARASITOLOGY AND HUMAN WELFARE

Humans have suffered greatly through the centuries because of parasites. Fleas and bacteria conspired to destroy a third of the European population in the seventeenth century, and malaria, schistosomiasis, and African sleeping sickness have sent untold millions to their graves. Even today, after successful campaigns against yellow fever, malaria, and hookworm infections in many parts of the world, parasitic diseases in association with nutritional deficiencies are the primary killers of humans. Recent summaries of the worldwide prevalence of selected parasitic diseases show that there are more than enough existing infections for every living person to have one, were they evenly distributed:[11,16,26,30,31]

Disease category	Human infections	Deaths per year
All helminths	4.5 billion	
Ascaris	1000 million	20 thousand
Hookworms	900 million	50–60 thousand
Trichuris	750 million	
Filarial worms	657 million	20–50+ thousand
Schistosomes	200 million	0.5–1.0 million
Malaria	489 million	1–2 million

These, of course, are only a few of the many kinds of parasites that infect humans. In addition to causing many deaths, they complicate and contribute to other illnesses. The majority of the more serious infections occur in the so-called tropical zones of the earth, so most dwellers within temperate regions are unaware of the magnitude of the problem. Money for research on tropical infections is very scarce because pharmaceutical companies are reluctant to spend money to develop drugs for treating people who cannot pay for them, and the less-developed countries have many other urgent financial problems. In 1980, $209 for each known case of cancer was spent for cancer research in the United States, $8 per case was spent on research on cardiovascular disease, and 4½ cents per case was spent worldwide for schistosomiasis research.[30]

The notion held by the average person that humans in the United States are free of worms is largely an illusion—an illusion created by the fact that the topic is rarely discussed because of our attitudes that worms are not the sort of thing that refined people talk about, the apparent reluctance of the media to disseminate such information, and the fact that poor people are the ones most seriously affected. Some estimates place the number of children in the United States infected with worms at about 55 million, although this is a gross underestimation if one includes such parasites as pinworms (*Enterobius vermicularis*).

However, the public is becoming more conscious of some other parasites. Some protozoa, such as *Pneumocystis*, *Toxoplasma*, and *Cryptosporidium*, are among the most common opportunistic infections among patients with acquired immunodeficiency syndrome (AIDS). Lyme disease, which is a syndrome produced by infection with a bacterium that can lead to chronic, disabling arthritis, is transmitted by ticks and is by far the most common arthropod-borne disease in North America.[2]

Even though there are many "native-born" parasite infections in the United States, many "tropical" diseases are imported within infected humans coming from endemic areas. After all, one can travel halfway around the world in a day or two. Many thousands of immigrants who are infected with schistosomes, malaria organisms, hookworms, and other parasites—some of which are communicable—currently live in the United States. It is estimated that about 100,000 cases of *Schistosoma mansoni* (Chapter 16) in the continental United States originated in Puerto Rico. Service personnel returning from abroad often bring parasite infections with them. In 1992, 302 of 917 U.S. Peace Corps volunteers in Malawi tested positive for *Schistosoma* infection.[5] There are documented cases of viable filariasis and *Strongyloides* 40 or more years after the initial infection![3,25] A traveler may become infected during a short layover in an airport, and many pathogens find their way into the United States as stowaways on or in imported products. Travel agents and tourist bureaus are reluctant to volunteer information on how to avoid the tropical diseases that a tourist is likely to encounter since bringing up the topic might lose the customer.[12] Small wonder, then, that "exotic" diseases confront the general practitioner with more and more frequency. One family physician claims to have treated virtually every major parasitic disease of humans during the years of his practice in Amherst, Massachusetts. A survey of intestinal parasites of 776 Southeast Asian immigrants in New Mexico revealed 20 different species of parasites, some of which are not common in the United States.[29]

There are other, much less obvious, ways in which parasites affect all of us, even those in comparatively parasite-free areas. Primary among these is malnutrition: 500 million people in the world have protein-energy malnutrition, and 350 million have iron-deficiency anemia.[30] Malnutrition is exacerbated both by population increase and by environmental degradation. From 2 billion in 1930, the population of the earth doubled to 4 billion in 1976, passed 5.5 billion in 1992, and is expected to exceed 10 billion in 2025.[14] Meanwhile, environmental degradation such as erosion continues to decrease the available supply of cropland. The increasing scarcity of resources contributes to violent conflict in the world.[15] The contributions of parasites to malnutrition are important but are underestimated because of underreporting.[30] Hospitals usually list what appears to be the most obvious cause of death, but most patients have multiple infections that have contributed to their disease state.

Even where food is being produced, it is not always used efficiently. Considerable caloric energy is wasted by fevers caused by parasitic infections. Heat production of the human body increases about 7.2% for each degree rise in Fahrenheit. A single, acute day of fever caused by malaria requires approximately 5000 calories, or an energy demand equivalent to two days of hard manual labor. To extrapolate, in a population with an average diet of 2200 calories per day, if 33% had malaria, 90% had a worm burden, and 8% had active tuberculosis (conditions that are repeatedly observed), there would be an energy demand equivalent to 7500 tons of rice per month per million people in addition to normal requirements. That is a waste of 25% to 30% of the total energy yield from grain production in many societies.[27]

Humans create many of their own disease conditions because of high population density and subsequent environmental pollution. Despite great progress in extending water supplies and sewage disposal programs in less-developed countries, not more than 10% to 15% of the world population is thus served.[17] Population shifts from rural to urban areas commonly overload the water and sewage capabilities of even major cities. Usually an adequate water supply has first priority, with sewage disposal running a poor second (Fig. 1.1). When one recalls that most parasite infections are caused by ingesting food or water contaminated with human feces, it is easy to understand why 15 million children die of intestinal infections every year.

Parasites are also responsible for staggering financial loss. Malaria, for example, is usually a chronic, debilitating, periodically disabling disease. In situations where it is prevalent the number of hours lost from productive labor multiplied by the number of malaria sufferers yields a figure that can be charged as loss in the manufacture of goods, in the production of crops, or in the earning of a gross national product. On the basis of estimates this figure is about $2 billion annually. Nations that import goods from countries infected with malaria, schistosomiasis, hookworm, and many other parasitic diseases pay more for these products than they would had the products been produced without the burden of disease. Plant parasites further diminish the productive capacities of all countries.

National and international efforts to increase productivity and standard of living in less-developed countries sometimes inadvertently increase parasitic disease. Schistosomiasis in Egypt increased after construction of the Aswan High Dam on the Nile River (p. 243). Smaller dams for drainage and agriculture have promoted transmission of schistosomiasis, onchocerciasis, dracunculiasis, and malaria.[28] The World Bank loaned Brazil funds to pave highways into the Amazon region to settle poor urban workers for farming, despite contrary advice from their own agricultural experts.[21] This produced an increase in malaria and spread the disease to new foci when the migrants returned to the cities after their farms failed (p. 149).[23]

An important role of parasitologists, together with that of members of other medical disciplines, is to help achieve a lower death rate. However, it is imperative that a lower death rate be matched with a concurrent lower birthrate and higher quality of life. If not, we are faced with the "parasitologist's dilemma"—that of sharply increasing a population that cannot be supported by the resources of the country. Dr. George

FIGURE 1.1

Nightsoil is a logical use of human feces and urine. Here it is applied to a vegetable garden, a technique practiced in much of the world. Although sometimes controlled by government regulations, it still serves as a significant means for distribution of eggs of some helminths and certain protozoan cysts.

Courtesy of Robert E. Kuntz.

Harrar, president of the Rockefeller Foundation, observed, "It would be a melancholy paradox if all the extraordinary social and technical advances that have been made were to bring us to the point where society's sole preoccupation would of necessity become survival rather than fulfillment." Harrar's paradox is already a fact for half the world. Parasitologists have a unique opportunity to break the deadly cycle by contributing to the global eradication of communicable diseases while making possible more efficient use of the earth's resources.

PARASITES OF DOMESTIC AND WILD ANIMALS

Both domestic and wild animals are subject to a wide variety of parasites that demand the attention of the parasitologist. Although wild animals are usually infected with several species of parasites, they seldom suffer as a result of massive deaths, or **epizootics,** because of the normal dispersal and territorialism of most species. However, domesticated animals are usually confined, often in great numbers, to pastures or pens year after year, so the parasite eggs, larvae, and cysts become extremely dense in the soil and the burden of adult parasites within each host becomes devastating. For example, the protozoa known as the coccidia thrive under crowded conditions; they may cause up to 100% mortality in poultry flocks, 28% reduction in wool in sheep, and 15% reduction in weight of lambs.[27] Agriculturists are forced to expend much money and energy in combating the phalanx of parasites that attack their animals. Infections in poultry are controlled by the costly method of prophylactic drug administration in feed. Unfortunately, the coccidia have become resistant to one drug after another.[6] Many other examples can be given, some of which are discussed later in this book. Thanks to the continuing efforts of parasitologists around the world, the identifications and life cycles of most parasites of domestic animals are well-known. This knowledge, in turn, exposes weaknesses in the biology of these pests and suggests possible methods of control. Similarly, studies of the biochemistry of organisms continue to suggest modes of action for chemotherapeutic agents.

Less can be done to control parasites of wild animals. Although it is true that most wild animals tolerate their parasite burdens fairly well, the animals will succumb when crowded and suffering from malnutrition, just as will domestic animals and humans. For example, the range of the big horn sheep in Colorado has been reduced to a few small areas in the high mountains. The sheep are unable to stray from these areas because of human pressure. Consequently, lungworms have so increased in numbers that in some herds no lambs survive the first year of life. These herds seem destined for quick extinction unless a means for control of the parasites can be found in the near future.

A curious and tragic circumstance has resulted in the destruction of many large wild animals in Africa. These animals are heavily infected with species of *Trypanosoma,* a flagellate protozoan of the blood. The wild animals tolerate infection well but function as **reservoirs** of infection for domestic animals, which quickly succumb to trypanosomiasis. One means of control employed is the complete destruction of the wild animal reservoirs themselves. Hence, the parasites of these animals are the indirect cause of their death. It is hoped that this parasitological quandary will be solved in time to save the magnificent wild animals.

Still another important aspect of animal parasitology is the transmission to humans of parasites normally found in wild and domestic animals. The resultant disease is called a **zoonosis.** Many zoonoses are rare and cause little harm, but some are more common and of prime importance to public health. An example is trichinosis, a serious disease caused by minute nematodes, *Trichinella* spp. (Chapter 23). These worms exist in **sylvatic** cycles that involve rodents and carnivores and in an **urban** or **domestic** cycle chiefly among rats and swine. People become infected when they enter any cycle, such as by eating undercooked bear or pork. Another zoonosis is echinococcosis, or hydatid disease, in which humans accidentally become infected with juvenile tapeworms when they ingest eggs from dog feces (Chapter 21). *Toxoplasma gondii,* which is normally a parasite of felines and rodents, is now known to cause many human birth defects (Chapter 8).

We recognize new zoonoses from time to time. Lyme disease, mentioned before, was long present in deer and white-footed mice, but frequent transmission to humans began only in the 1970s.[2] It is the obligation of the parasitologist to identify, understand, and suggest means of control of such diseases. The first step is always the proper identification and description of existing parasites so that other workers can recognize and refer to them by name in their work. Thousands of species of parasites of wild animals are still unknown and will occupy the energies of taxonomists for many years to come.

Aside from their roles as causative agents of disease, parasites provide us with an almost unlimited supply of fascinating, and challenging, problems in ecology and evolution (Chapter 2). Presence of a parasite species with a complex life cycle demonstrates unequivocally that intermediate hosts occupy an area and that an ecological relationship exists between hosts and parasites. Parasites also may be one of the factors, along with predation and abiotic events, that function to regulate host populations. And finally, virtually every species of animal is parasitized by at least one other species. Thus, much of the overall diversity found in any ecosystem can be attributed to parasitism.

CAREERS IN PARASITOLOGY

There is an area within parasitology to interest every biologist. The field is large and has so many approaches and subdivisions that anyone who is interested in biological research can find a lifetime career in parasitology.[1] It is a satisfying career because each bit of progress made, however small, contributes to our knowledge of life and to the eventual conquering of disease. As in all scientific endeavors, every major breakthrough depends on many small contributions made, usually independently, by individuals around the world. Previously little-known parasites suddenly became life-threatening infections in AIDS patients. Had their identifications and life cycles been better understood, much expense and time would have been saved in recognizing this complex disease.

The training required to prepare a parasitologist is rigorous. Modern researchers in parasitology are well-grounded in physics, chemistry, and mathematics, as well as biology from the subcellular through the organismal and populational levels. They must be firmly grounded in medical entomology, histology, and basic pathology. Depending on their interests, they may require advanced work in physical chemistry, immunology, molecular biology, genetics, and systematics. Most parasitologists hold Ph.D.s or other doctoral degrees, but contributions have been made by persons with master's or bachelor's degrees. Such intense training is understandable, since parasitologists must be familiar with the principles and practices that apply to over a million species of animals; in addition they need thorough knowledge of their fields of specialty. Once they have received their basic training, parasitologists continue to learn during the rest of their lives. Even after retirement, many remain active in research for the sheer joy of it. Parasitology indeed has something for everyone.

SOME BASIC DEFINITIONS

The science of parasitology is largely a study of **symbiosis,** or, literally, "living together." Although some authors restrict the term *symbiosis* to relationships wherein both partners benefit, we prefer to use the term in a wider sense, as originally proposed by the German scholar A. de Bary in 1879: *Any two organisms living in close association, commonly one living in or on the body of the other, are symbiotic, as contrasted with "free living."* Usually the **symbionts** are of different species, but not necessarily.

FIGURE 1.2

Gooseneck barnacles (*Poecilasma kaempferi*) growing on the legs and carapace of a crab (*Neolithodes grimaldi*). This is an example of phoresis since the two species are merely "traveling together." However, the relationship could grade into commensalism; some advantages probably accrue to the barnacles.

From R. Williams and J. Moyse, "Occurrence, distribution, and orientation of *Poecilasma kaempferi* Darwin (Cirripedia: Pedunculata) epizoic on *Neolithodes grimaldi* Milne-Edwards and Bouvier (Decapoda: Anomura) in the northeast Atlantic," in *J. Crust. Biol.* 8:177–186. Copyright © 1988.

Symbiotic relationships can be further characterized by specifying the nature of the interactions between the participants. It is always a somewhat arbitrary act, of course, for people to assign definitions to relationships between organisms. But animal species participate in a wide variety of symbiotic relationships, so parasitologists have a need to communicate about these interactions and thus, have coined a number of terms to describe them.

Interactions of Symbionts

• Phoresis

Phoresis exists when two symbionts are merely "traveling together," and there is no physiological or biochemical dependence on the part of either participant. Usually one **phoront** is smaller than the other and is mechanically carried about by its larger companion (Fig. 1.2). Examples are bacteria on the legs of a fly or fungous spores on the feet of a beetle.

• Mutualism

Mutualism describes a relationship in which both partners benefit from the association. Mutualism is usually obligatory, since in most cases physiological dependence has evolved to such a degree that one mutual cannot survive without the other. Termites and their intestinal protistan fauna are an excellent example of mutualism. Termites cannot digest cellulose because they cannot synthesize and secrete the enzyme cellulase. The myriad flagellates in a termite's intestine, however, synthesize cellulase and consequently digest the wood eaten by the host. The termite uses molecules excreted as a by-product of the flagellates'

metabolism. If we kill the flagellates by exposing termites to high temperature or high oxygen concentration, then the termites die even though they continue to eat wood.

An astonishing variety of mutualistic associations can be found among animals, bacteria, fungi, algae, and plants. Blood-sucking leeches cannot digest blood, for example, but their intestinal bacteria, species that are restricted to leech guts, do the digestion for their hosts. What is apparently an evolutionary origin of a case of mutualism has been discovered in a common free-living ameba, *Amoeba proteus*. A strain of *A. proteus* became infected with a parasitic bacteria, and over a period of several years the amebas became unable to survive without the bacteria.[19] As in the termites, when the symbionts are killed by elevated temperature, the amebas (of this particular strain) die unless reinfected by microinjection.[22] Although the nature of this relationship is not known, we presume it is metabolic; that is, the partners exchange needed molecules. As is the case with many such relationships, exploration of the basis for the mutualism would make an interesting doctoral dissertation project!

Mutualistic interactions are not restricted to physiological ones. For example, **cleaning symbiosis** is a behavioral phenomenon and occurs between certain crustaceans and small fish—the cleaners—and larger marine fish. Cleaners often establish stations that the large fish visit periodically, and the cleaners remove ectoparasites, injured tissues, fungi, and other organisms. Some evidence exists that such associations may be in fact obligatory; when all cleaners are carefully removed from a particular area of reef, for example, all the other fish leave too. You can find other examples of mutualistic and related associations in the texts edited by Henry[13] and Cheng.[7]

• Commensalism

In **commensalism** one partner benefits from the association, but the host is neither helped nor harmed. The term means "eating at the same table," and many commensal relationships involve feeding on food "wasted" or otherwise not consumed by the host. Pilot fish (*Naucrates*) and remoras (Echeneidae) are often cited as examples of commensals. A remora is a slender fish whose dorsal fin is modified into an adhesive organ that it attaches to large fish, turtles, and even submarines! The remora gets free rides and scraps, but it does not harm the host or rob it of food. Some remoras, however, are mutuals, since they clean the host of parasitic copepods (Chapter 34).[8]

Commensalism may be **facultative,** in the sense that the commensal may not be required to participate in an association to survive. Stalked ciliates of the genus *Vorticella* are frequently found on small crustaceans, but they survive equally well on sticks in the same pond. Related forms, however, such as *Epistylis* spp., are evidently **obligate** commensals since they are not found except on other organisms, especially crustaceans.

Humans harbor several species of commensal protistans—for example, *Entamoeba gingivalis*. This ameba lives in the mouth where it feeds on bacteria, food particles, and dead epithelial cells but never harms healthy tissues. It has no cyst or other resistant stage in its life cycle. In the laboratory beginning parasitology students are usually asked to learn to distinguish between commensal and parasitic amebas. When you are studying prepared slide specimens, the task is not too difficult. But when faced with real live animals, sometimes the distinctions are not so clear. Adult tapeworms are universally regarded as parasites, yet in some cases they have no known ill effects on their host.[18]

• Parasitism

Parasitism is a relationship in which one of the participants, the parasite, either harms its host or in some sense lives at the expense of the host. Parasites may cause mechanical injury, such as by boring a hole into the host or digging into its skin or other tissues, by stimulating a damaging inflammatory or immune response, or simply by robbing the host of nutrition. Most parasites inflict a combination of these conditions on their hosts.

If a parasite lives on the surface of its host, it is called an **ectoparasite;** if internal, it is an **endoparasite.** Most parasites are **obligate parasites;** that is, they cannot complete their life cycle without spending at least part of the time in a parasitic relationship. However, many obligate parasites have free-living stages outside any host, including some periods of time in the external environment within a protective eggshell or cyst. **Facultative parasites** are not normally parasitic but can become so when they are accidentally eaten or enter a wound or other body orifice. Two examples

are certain free-living amebas, such as *Naegleria* (p. 108), and free-living nematodes belonging to the genus *Micronema*.[10] Infection of humans with either of these is extremely serious and usually fatal.

When a parasite enters or attaches to the body of a species of host different from its normal one, it is called an **accidental,** or **incidental, parasite.** For instance, it is common for nematodes, normally parasitic in insects, to live for a short time in the intestines of birds or for a rodent flea to bite a dog or human. Accidental parasites usually do not survive in the wrong host, but in some cases they can be extremely pathogenic (see *Baylisascaris, Toxocara* in Chapter 26). Parasitism is usually the result of a long, shared evolutionary history between parasite and host species. Accidental parasitism puts both host and parasite into environmental conditions to which neither is well-adapted; it is not surprising that the result may be serious harm to either or both participants.

Some parasites live their entire adult lives within or on their hosts and may be called **permanent parasites,** whereas **temporary,** or **intermittent, parasites,** such as mosquitoes or bedbugs, only feed on the host and then leave. Temporary parasites are often referred to as **micropredators,** in recognition of the fact that they usually "prey" on several different hosts (or the same host at several discrete times).

Predation and parasitism are conceptually similar in that both the parasite and the predator live at the expense of the host or prey. The parasite, however, normally does not kill its host, is small relative to the size of the host, has only one host (or one host at each stage in its life cycle), and is symbiotic. The predator kills its prey, is large relative to the prey, has numerous prey, and is not symbiotic. **Parasitoids,** however, are insects, typically flies or wasps (orders Diptera and Hymenoptera, Chapters 38 and 39, respectively), whose immature stages feed on their hosts' bodies, usually other insects, but finally kill the hosts. Parasitoids resemble predators in this regard, but they only require a single host individual.

Hosts

Parasitologists differentiate between various types of hosts based on the role the host plays in the life cycle of the parasite. A **definitive host** is one in which the parasite reaches sexual maturity. Sexual reproduction has not been clearly shown in some parasites, such as amebas and trypanosomes, and in these cases we arbitrarily consider the definitive host the one most important to humans. An **intermediate host** is one that is required for parasite development, but one in which the parasite does not reach sexual maturity. Definitive hosts are often but not necessarily vertebrates; the malarial parasites, *Plasmodium* spp., reach sexual maturity and undergo fertilization in the mosquito, which by definition is, therefore, the definitive host, while vertebrates are the intermediate hosts (see Chapter 9).

A **paratenic** or **transport host** is one in which the parasite does not undergo any development but in which it remains alive and infective to another host. Paratenic hosts may bridge an ecological gap between the intermediate and definitive hosts. For example, owls may be parasitized by thorny headed worms (Chapter 31), which undergo development to infective stages in insects that pick up the worm eggs from owl feces. Large owls rarely if ever eat insects, but shrews eat them regularly, sometimes accumulating large numbers of juvenile worms encysted in their mesenteries. Owls catch shrews, sometimes becoming heavily infected with the worms. In this case the shrew is the transport host between the insect intermediate and the owl definitive host.

Most parasites develop only in a restricted range of host species. That is, parasites exhibit varying degrees of **host specificity,** some infecting only a single host species, others infecting a number of related species, and a few being capable of infecting many host species. The so-called pork tapeworm, *Taenia solium,* apparently can mature only in humans, so it has absolute host specificity. The nematode *Trichinella spiralis* seems to be able to mature in almost any warm-blooded vertebrate.

Any animal that harbors an infection that can be transmitted to humans is called a **reservoir host,** even if the animal is a normal host of the parasite. Examples are rats and wild carnivores with *Trichinella* spp., dogs with *Leishmania* spp., and armadillos with *Trypanosoma cruzi,* the causative agent of Chagas' disease (see Chapter 5).

Finally, many parasites host other parasites, a condition known as **hyperparasitism.** Examples are *Plasmodium* spp. in mosquitoes, a tapeworm juvenile in a flea, and the many insects whose larvae parasitize other parasitic insect larvae.

Although the vast majority of parasites are different species from their hosts, exceptions do occur. In *Trichosomoides* a nematode parasite of rats, the male lives its mature life within the uterus of the female worm, obtaining its nourishment from her tissues. This is also the case with *Gyrinicola japonicus,* a nematode parasite of frogs. An even stranger relationship has evolved in some species of anglerfish in which the male bites the skin of the female and sucks her blood and tissue fluids for nourishment. Eventually they grow together, and he shares her bloodstream! And, of course, the term *parasite* is used occasionally to condemn our fellow humans who seem to take more than they return to society.

Definitions are always arbitrary, and when we construct pigeonholes to receive descriptions of phenomena in the real world, we expect to find situations that defy easy assignment to one of our precise categories. Many symbiotic associations cannot be classified with certainty as to the effects of the symbiont on the host. An apparent case of commensalism may have damaging effects on the host that humans have not even thought about, much less observed. Conversely, a case of assumed parasitism may, on closer inspection, turn out to be commensalism. Nevertheless, definitions are necessary to communication, so we accept that sometimes they are not exactly precise because they are almost always useful. The terms defined in this chapter convey large amounts of inferred information about ecology, evolutionary history, biochemistry, and development.

References

1. Anonymous. 1989. *Careers in parasitology, medical zoology, tropical medicine.* Lawrence, Kans.: American Society of Parasitologists.

2. Barbour, A. G., and D. Fish. 1993. The biological and social phenomenon of Lyme disease. *Science* 260:1610–16.

3. Behnke, J. M. 1987. Evasion of immunity by nematode parasites causing chronic infections. In Baker, J. R., and R. Muller, eds. *Advances in parasitology* 26. London: Academic Press.

4. Borst, P. 1991. Molecular genetics of antigenic variation. In Ash, C., and R. B. Gallagher, eds. *Immunoparasitol. Today* A29–A33. Cambridge: Elsevier Trends Journals.

5. Centers for Disease Control. 1993. Schistosomiasis in U.S. Peace Corps volunteers—Malawi, 1992. *Morb. Mortal. Weekly Rep.* 42:570.

6. Chapman, H. D. 1993. Resistance to anticoccidial drugs in fowl. *Parasitol. Today* 9:159–62.

7. Cheng, T. C. 1971. *Aspects of the biology of symbiosis.* Baltimore: University Park Press.

8. Cressey, R. F., and E. A. Lachner. 1970. The parasitic copepod diet and life history of diskfishes (Echeneidae). *Copeia* No. 2:310–18.

9. Cross, G. A. M. 1990. Cellular and genetic aspects of antigenic variation in trypanosomes. *Ann. Rev. Immunol.* 8:83–110.

10. Gardiner, C. H., D. S. Koh, and T. A. Cardella. 1981. *Micronema* in man: Third fatal infection. *Am. J. Trop. Med. Hyg.* 30:586–89.

11. Gibbons, A. 1992. Researchers fret over neglect of 600 million patients. *Science* 256:1135.

12. Grimes, P. 1980. Travelers are warned of increasing danger of malaria. Spread of the disease is blamed on ignorance and the failure of travel agents and tourist bureaus to caution clients. *New York Times* (11 May, 15XX).

13. Henry, S. M., ed. 1966, 1967. *Symbiosis* 1 and 2. New York: Academic Press, Inc.

14. Hickman, C. P. Jr., L. S. Roberts, and A. Larson. 1993. *Integrated principles of zoology,* 9th ed. St. Louis: Mosby.

15. Homer-Dixon, T. F., J. H. Boutwell, and G. W. Rathjens. 1993. Environmental change and violent conflict. *Sci. Am.* 268:38–45 (Feb.).

16. Hopkins, D. R. 1992. Homing in on helminths. *Am. J. Trop. Med. Hyg.* 46:626–34.

17. Howard, L. M. 1971. The relevance of parasitology to the growth of nations. *J. Parasitol.* 57(sect. 2, part 5):143–47.

18. Insler, G. D., and L. S. Roberts. 1976. *Hymenolepis diminuta:* Lack of pathogenicity in the healthy rat host. *Exp. Parasitol.* 39:351–57.

19. Jeon, K. W. 1980. Symbiosis of bacteria with Amoeba. In Cook, C. B., P. Pappas, and E. Rudolph, eds. *Cellular interactions in symbiotic and parasitic relationships.* Columbus: Ohio State University Press, 245–62.

20. Keilin, D. 1925. On cytochrome, a respiratory pigment, common to animals, yeast and higher plants. *Proc. R. Soc. Lond. (Biol.)* 98:312–39.

21. Kolata, G. 1987. Anthropologists turn advocates for the Brazilian Indians. *Science* 236:1183–87.

22. Lorch, I. J., and K. W. Jeon. 1980. Resuscitation of amebae deprived of essential symbiotes: Micrurgical studies. *J. Protozool.* 27:423–26.

23. Marques, A. C. 1987. Human migration and the spread of malaria in Brazil. *Parasitol. Today* 3:166–70.

24. Mestel, R. 1991. Parasites help decode genes. *Miami Herald* (25 August, 3J).

25. Pearson, R. D., and R. L. Guerrant. 1991. Intestinal nematodes that migrate through skin and lung. In Strickland, G. T., ed. *Hunter's tropical medicine,* 7th ed.. Philadelphia: W. B. Saunders Company.

26. Peters, W. 1978. Medical aspects—comments and discussion II. In Taylor, A. E. R., and R. Muller, eds. *The relevance of parasitology to human welfare today. Symposia of the British Society of Parasitology* 16. Oxford: Blackwell Scientific Publications Ltd.

27. Pollack, H. 1968. *Disease as a factor in the world food problem.* Arlington, Va.: Institute for Defense Analysis.

28. Ripert, C. L., and C. P. Raccurt. 1987. The impact of small dams on parasitic diseases in Cameroon. *Parasitol. Today* 3:287–89.

29. Skeels, M. R., et al. 1982. Intestinal parasitosis among Southeast Asian immigrants in New Mexico. *Am. J. Public Health* 72:57–59.

30. Stephenson, L. S. 1987. *The impact of helminth infections on human nutrition. Schistosomes and soil-transmitted helminths.* London: Taylor and Francis.

31. Stürchler, D. 1989. How much malaria is there worldwide? *Parasitol. Today* 5:39–40.

Additional References

Ahmadjian, V., and S. Paracer. 1986. *Symbiosis. An introduction to biological associations.* Hanover and London: University Press of New England. Short text but includes consideration of symbioses not usually covered in parasitology courses such as bacterial, viral, fungal, and algal.

Cox, F. E. G., ed. 1993. *Modern parasitology.* Oxford: Blackwell Scientific Publications. Excellent for further reading in epidemiology, biochemistry, molecular biology, physiology, immunology, chemotherapy, and control.

Gallagher, R. B., J. Marx, and P. J. Hines. 1994. Progress in parasitology. *Science* 264:1827. This is the lead editorial in an issue of the journal *Science* featuring parasitology news and research.

Grove, D. I. 1990. *A history of human helminthology.* Oxon, U.K.: CAB International. This book details the history of scientific discovery for each helminth infecting humans.

Hyde, J. E. 1990. *Molecular parasitology.* New York: Van Nostrand Reinhold.

Wyler, D. J., ed. 1990. *Modern parasite biology. Cellular, immunological, and molecular aspects.* New York: W. H. Freeman and Co.

Chapter 2

BASIC PRINCIPLES AND CONCEPTS: PARASITE ECOLOGY AND EVOLUTION

The host is an island invaded by strangers with different needs, different food requirements, different locations within which to raise their progeny.

W. Taliaferro

Ecology is the study of relationships between organisms and their environments. A parasite's environment is primarily the host, of course, but transmission stages such as spores, eggs, and often juveniles must also survive abiotic conditions. A host usually represents a rich and highly regulated supply of nutrients. Most animal body fluids have a wide array of dissolved proteins, amino acids, carbohydrates, and nucleic acid precursors; virtually all animals have mechanisms for maintaining the chemical makeup and osmotic balance of their body fluids; and vertebrates such as birds and mammals control body temperature as well. We should expect parasites to exhibit traits that allow them to exploit such environments but at the same time deny them access to others. A flea cannot live in the white-footed mouse's habitat, for example, unless the mouse itself is living there too.

But hosts are relatively small patches within the vast matrix that is their own habitat; that is, they are "islands" in the true sense of Taliaferro's quote as well as in the metaphorical sense. Thus, suitable parasite environments are dispersed, in addition to being rich and regulated. For example, there is an enormous volume of water in a lake compared to the volume of fish in that same lake. This seemingly trivial observation, however, points to a major problem for monogenean flatworms (p. 281) that must live on these fish: Unless the worm's reproductive stages are able to keep finding fish to infect, the parasite is likely to become locally extinct. Indeed, many parasite control strategies are based on reducing the probability of host and parasite encounter, and conversely, many parasites possess traits that function to increase the probability of finding a host.

The basic situations just described are reflected in a number of parasite characteristics that can be interpreted within an ecological context. These characteristics include feeding adaptations, transmission mechanisms, population structures, and membership in "communities" or assemblages of parasite species within a single host. And ultimately, because the host is the parasite's environment, we should expect evolutionary changes in hosts to be accompanied by parallel, perhaps adaptive, changes in their parasites.

PARASITE ECOLOGY

The Parasite's Ecological Niche

• Trophic Relationships

Parasites always live at a higher trophic level than their hosts.[22] Thus, all parasites are at least primary consumers, and those infecting top predators such as hawks, owls, and carnivores live quite high on a typical food pyramid. Trophic relationships are direct and obvious for parasites that eat host tissue and fluids, such as hookworms (p. 405), frog lung flukes (p. 264), and ticks. But parasite use of the host can also be somewhat indirect. For example, all free-living animals spend significant amounts of energy regulating their internal milieus and producing their own offspring. Thus, parasites can be thought of as "preying" on the homeostatic mechanisms and reproductive efforts of organisms at lower trophic levels.

All parasites are heterotrophic, requiring their energy and carbon in the form of existing complex organic molecules and their nitrogen as a mixture of amino acids.[21] In this respect parasites are no different from other kinds of animals. A parasite's feeding devices, however, may differ considerably from those seen in most free-living animals. For example, tapeworms (phylum Platyhelminthes, Chapter 20) and spiny-headed worms (phylum Acanthocephala, Chapter 31) have no digestive tracts and absorb sugars and amino acids directly across their body walls. These worms feed through uptake sites on the plasma membrane. The anticoagulant of tick saliva (Chapter 40) is an integral part of the parasite's feeding mechanism, another illustration of an adaptation to a characteristic of the host—in this case, the clotting property of blood.

• Dimensions on Resources

Ecologists are able to quantify ecological niches by measuring some aspect of the habitat—such as tree height, water temperature, and seed sizes—and then observing the range of values over which a species occurs or uses. This range is called a species' **dimension on a resource**.[14,18] A parasite's niche consists of dimensions on resources provided by the living body of another species, as well as on abiotic resources encountered by transmission stages such as eggs, cysts, spores, and young. But while it is easy to describe a parasite's niche in very general terms (eye, gut, lung), it is often quite difficult to quantify that niche—that is, measure dimensions on resources—in the same way you could for many free-living animals.

Host intestinal length is one example of an easily measured resource, and much research has been done on the distribution of cestodes, nematodes, and acanthocephalans within

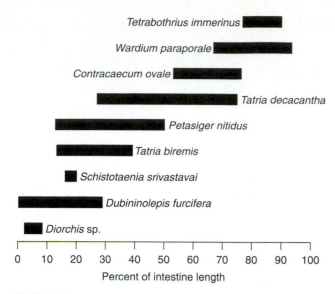

Tetrabothrius immerinus
Wardium paraporale
Contracaecum ovale
Tatria decacantha
Petasiger nitidus
Tatria biremis
Schistotaenia srivastavai
Dubininolepis furcifera
Diorchis sp.

Percent of intestine length

FIGURE 2.1

Distribution of intestinal nematodes, trematodes, and tapeworms in an aquatic bird (eared grebe). The horizontal bars are the average position +/− one standard deviation, in terms of the relative distance from the stomach.

Redrawn by John Janovy, Jr. from T. M. Stock and J. C. Holmes, "Functional relationships and microhabitat distributions of enteric helminths and grebes (Podicipedidae): the evidence for interactive communities" in *J. Parasitol.* 74:214–227. Copyright © 1988. Reprinted with permission of publisher.

the intestine. Intestinal worms usually occur within a particular region of the gut, although that distribution is sometimes influenced by host diet, physiological condition, and the presence of other helminths (Fig. 2.1). In addition, subtle differences occur in oxygen and carbon dioxide tension, pH, and other chemical and physical factors between the mucosa and the center of the lumen. Such differences occur even between the top of a villus and its base, making at least two different habitats available for colonization by parasites of suitable sizes. In a study of parasites in the turtle *Testudo graeca,* for example, Schad found eight species of the nematode genus *Tachygonetria* living in the large intestine.[19] He concluded that even when two parasite species are found in the same region of the intestine, we cannot state conclusively that they use exactly the same resources and thus occupy identical niches.

Digestive systems also vary greatly between species (compare the stomachs of humans and cows) and even between life-cycle stages of a host, as in the case of tadpoles compared to adult toads and frogs. Tadpoles are typically herbivorous: Some are filter feeders; others graze on algae and detritus. Adult anurans, however, are carnivorous. Metamorphosis from tadpole into adult involves loss of intestinal epithelium and significant reduction in intestinal length. Metamorphosis in beetles involves loss of larval gut tissue. Some protozoan parasites of beetles, such as the four species of gregarine parasites in the common mealworm *Tenebrio molitor,* are specific not only to their host species, but also to the life-cycle stage.[6] In this case the parasites "see" the larva as an environment distinct from that of the adult beetle. But even within a single life-cycle stage of a single host species, an intestine may offer many more places, or ways, for parasites to exist than might be suspected. Parasitologists have discovered over 40 species of tapeworms and tens of thousands of individual parasites in scaup ducks alone.[4]

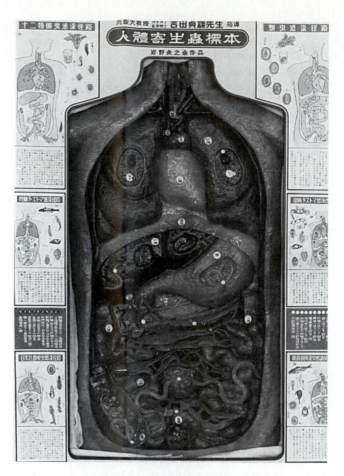

FIGURE 2.2

Pre-World War II wax model of a dissected human, showing various internal parasites in their "ecological niches."

Photograph by John Janovy, Jr.

• Infection Sites

Although most endoparasites of vertebrates live in the digestive system, adult parasites are found in virtually all parts of the body, and juvenile stages often undergo elaborate migrations through the body before arriving at their definitive sites. Parasites that inhabit the lumen of the intestine or other hollow organs are said to be **coelozoic,** while those living within tissues are called **histozoic.** Parasites are generally adapted to, and restricted to, particular sites within or upon a host (Fig. 2.2). Examples of this phenomenon include malarial parasites living inside red blood cells (p. 140), filarial nematodes that congregate in the heart or beneath the skin (pp. 452, 457), bird mites that occur only on flight feathers, and monogeneans found in the urinary bladders of frogs (p. 292). Such observations lead to hypotheses about the evolutionary forces that may have contributed to this circumstance. It is still a matter of some controversy whether parasites compete with one another for resources provided by the host. Some hosts are very heavily infected in nature, with up to several thousand individual worms of a dozen species. It is difficult to imagine that, under these circumstances, some competition is not occurring.

On the other hand, hosts may provide many more microenvironments than we realize. Consider the human eye, an organ not as obviously suited to infection by parasites as the intestine. The retina may be infected by the apicomplexan *Toxoplasma gondii* and juveniles of the filarial nematode *Onchocerca volvulus;* the chamber may harbor the bladder worm metacestodes of the tapeworms *Taenia solium, T. crassiceps, T. multiceps,* or *Echinococcus granulosus;* the conjuctiva may host another wandering filarial nematode, *Loa loa,* and the orbit may be the home of nematodes of the genus *Thelazia.* Parasitologically, the vertebrate body can be considered a mass of habitats that have been colonized by a great diversity of species. It has been said that if a host were infected with all the parasites capable of infecting it and the host tissues were then removed to leave only the parasites, the host could still be recognized!

Parasite Populations

• Quantitative Descriptors

Numbers of parasites are of major interest to epidemiologists, public health workers, ecologists, and evolutionary biologists. A scientist assessing the impact of parasitic diseases on a human population must know who is infected, whether the infections are distributed equally among all age groups and both sexes, and whether certain individuals have unusually high numbers of parasites. An evolutionary biologist is also interested in parasite numbers, since hosts are parasites' environments, and interactions between organisms and their environments are generally considered the driving force behind evolutionary change.

All of the parasites of a particular species, occurring within the body of a single host individual, are called an **infrapopulation** (see Margolis et al.[15] for a discussion of ecological terms in parasitology). The term **intensity** is also used for this quantity; parasitologists often find it useful to average the number of parasites in *infected* hosts, a value called **mean intensity.** A **suprapopulation** is all of the parasites of a single species, regardless of developmental stages (eggs, larvae, juveniles, adults), that occur within the hosts in an ecosystem. It should be obvious that while infrapopulations are relatively easy to count, the size of suprapopulations must be estimated, often by indirect methods or inference.

Not all hosts are always infected. The decimal (or percentage) fraction of hosts that are infected at a given time is called the **prevalence.** Occasionally the term **incidence** is used incorrectly to refer to prevalence; incidence is the number of new infections per unit of time divided by the number of uninfected individuals at the beginning of the measured time. In wild animal populations prevalence is easily measured, but incidence is more difficult to measure unless the parasitic infection can be diagnosed without destroying the host, and the host species is one whose individuals can be marked, released, and recaptured.

The **relative density** is the average number of parasites per host in a sample of hosts and is equal to the **arithmetic mean.** A sample of 10 mice with a total of 75 pinworms would have a relative density or mean of 7.5 worms per host. However, these 75 worms *could* all be in one mouse (in which case the prevalence would be 0.10) or distributed among all the mice (prevalence = 1.00). By pretending to be a veterinarian seeking to rid these mice of their worms, but having only a limited supply of expensive antihelminthic drugs,

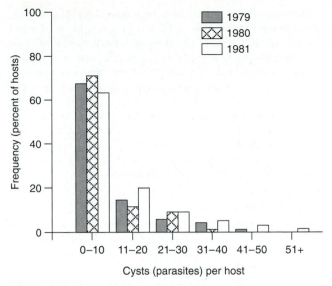

FIGURE 2.3

Population "structure" of the trematode *Uvulifer ambloplitis* (larvae) in bluegill sunfish in North Carolina over a three-year period. Most of the fish are uninfected, while most of the parasites are in the relatively few heavily infected fish, those with more than 25 larval cysts. These frequency distributions match those predicted by the mathematical model (equation) known as the *negative binomial.*

Redrawn by John Janovy, Jr. from D. A. Lemly and G. W. Esch, "Population biology of the trematode *Uvulifer ambloplitis* (Hughes, 1927) in juvenile bluegill sunfish, *Lepomis macrochirus,* and largemouth bass, *Micropterus salmoides,*" in *J. Parasitol.* Copyright © 1984. Reprinted with permission of publisher.

you can see immediately why parasite population structure is of major interest to scientists and clinicians. You do not want to waste your medicine by giving it to noninfected rodents!

• Macro- and Microparasites

Large parasites that do not multiply (in the life-cycle stage of interest) in or on the host are called **macroparasites.** Examples of macroparasites are tapeworms, adult trematodes, most nematodes, acanthocephalans, and arthropods such as ticks and fleas. Macroparasites often, if not typically, occur in **aggregated populations.** That is, *most of the parasites are in a relatively few hosts of a species while the majority of the host species individuals are either uninfected or lightly infected* (Fig. 2.3). This generality was recognized by H. D. Crofton in the early 1970s;[7] Crofton claimed that such population structure was so characteristic of parasites that it should be included in the definition of parasitism. Crofton also offered several explanations for the origin of this aggregation; as a result, he inspired a massive amount of both theoretical and empirical work on parasite population biology.

Frequency distributions found to have greater than expected numbers of observations at the tails when tested against a normal (bell-shaped) distribution also are said to be **contagious** or **overdispersed.** These terms generally describe aggregated distributions, such as those of many parasite populations, with the properties already indicated. Thus, the three verbal descriptions of populations (aggregated, contagious, overdispersed) are sometimes used interchangeably in the parasitology literature.

Small parasites that multiply within the host are called **microparasites,** and these include bacteria, rickettsia, and protozoan infections such as those caused by malarial parasites (genus *Plasmodium,* p. 137), trypanosomes (p. 55), and amebas (p. 99). Whereas in the case of macroparasites one can generally assume that one parasite reflects a single encounter between host and infective stage, that assumption is not valid for microparasites. Thus, a population ecologist must use different methods for microparasites than for macroparasites when attempting to determine the mechanisms responsible for disease transmission in a host species' population; although the most fundamental questions—who is infected, who is immune, and who is at risk—remain the same regardless of the parasite species involved.

• Population Structure

Parasite **population structure** is a critical piece of information for those seeking to control infections. Population structure is often described by the relative density (mean), variance (a statistical parameter whose value is related to the shape of a frequency distribution), and *curve of best fit.* The latter is really an equation that generates a theoretical frequency distribution of the parasites among hosts (Fig. 2.3). A graph can be constructed by plotting parasite per host classes along the X-axis and numbers of hosts that fall into these classes on the Y-axis. The result is a frequency distribution that describes the parasite's population structure. Figure 2.3 is an example of such a graph; it illustrates Crofton's general principle that most of the host individuals are uninfected, or only lightly infected, while most of the parasites are in a few host individuals. This principle, of course, applies to humans as well as to other species including plants.

Parasitologists can quickly form mental pictures of a parasite population from the three pieces of information (mean, variance, curve of best fit). Almost any student could equally quickly form a mental picture of the grade distribution in a large biology class, simply from knowing the class average (relative density or mean), variance (varies according to numbers of As and Fs), and curve of best fit (usually bell-shaped, or normal, in biology classes but steeply falling in parasitized animal populations).

There is some evidence that certain individuals within host populations are either genetically or behaviorally predisposed to heavy infections.[8] In studies conducted in both Kenya and Burma, individuals who were heavily infected with *Ascaris* before treatment of an entire village were most likely also to be heavily infected one or two years later.

• Multiple Species Infections

A single host individual can be infected with a number of parasite species; that is, it can contain a **parasite species assemblage.** These assemblages can be extraordinarily rich, as illustrated by the intestinal parasites of some endothermic (warm-blooded) vertebrates. In a series of typical studies by Holmes and his colleagues, 26 species of intestinal helminths were reported from a sample of 31 eared grebes, and 52 species, with "slightly less than one million individuals," were found in 45 scaup ducks.[4,20] Mammals such as coyotes and black bears may also be heavily and frequently infected.[16]

There is some evidence that parasites interfere with one another in various ways, especially in heavy intestinal infections. This evidence has led workers to postulate a variety of types of parasite communities, from interactive ones in which competition may occur to noninteractive ones, usually with few species, in which there appears to be little if any competition.[10]

Parasite Reproduction

The chance that an individual parasite's offspring will find a new host is very slim. Low individual reproductive success is considered an evolutionary force leading to high reproductive potential in parasites. The inverse relationship between numbers of offspring and the probability of individual success is known for a wide variety of both plants and animals. Among animals, parental care is one of the factors that tends to increase the chance of an offspring surviving.[18] Parasites, on the other hand, exhibit little parental care, although **viviparity,** or live birth, such as occurs in some nematodes and monogeneans, can be considered a more "caring" approach than indiscriminate scattering of eggs. But no amount of parental care can counter the fact that hosts are indeed islands embedded in a large abiotic matrix. Thus, the high reproductive potential of parasites represents a heavy energy investment to counteract the low probability of individual offspring success.

Parasites exhibit a variety of mechanisms that function to increase the reproductive potential of those individuals who succeed at finding a host. These mechanisms often take the form of **asexual reproduction** and **hermaphroditism.** Asexual reproduction often occurs in the larval or sexually immature stages as either **polyembryony** (see Fig. 15.19) or **internal budding** (see Fig. 21.22). Hermaphroditism is the occurrence of both sexes in a single individual. It eliminates the necessity of finding an individual of the opposite sex for fertilization—all individuals are of both sexes. Reproductive encounters result in two fertilized female systems, and some parasites are capable of self-fertilization. The specific manifestations of asexual reproduction and hermaphroditism, however, differ depending on the group of parasites.

Schizogony, or **multiple fission,** is asexual reproduction characteristic of some parasitic protozoa (see Fig. 8.6, Chapter 9, Plates 1 and 3). In schizogony the nucleus divides numerous times before cytokinesis (cytoplasmic division) occurs, resulting in the simultaneous production of many daughter cells. Just as important as the number of daughter cells is the fact that these cells are all in the same ontogenetic state; thus, all offspring are physiologically capable of infecting a new host or entering the next stage of the life cycle. A more detailed discussion of the role of schizogony in parasites' life cycles is found in the chapters on Apicomplexa and Myxozoa (p. 163). Simple **binary fission** is also asexual reproduction. It is common among familiar protozoa such as *Paramecium* as well as some amebas (phylum Sarcomastigophora). But as with any process in which numbers double regularly, rapid fission can easily result in millions of offspring in only a few days.

Trematodes and some tapeworms reproduce asexually as immatures. The juveniles (**metacestodes**) of several tapeworm species are capable of external or internal budding of more metacestodes. The **cysticercus** juvenile of *Taenia crassiceps,* for instance, can bud off as many as a hundred small bladder worms while in the abdominal cavity of a mouse intermediate host. Each new metacestode develops a scolex and neck, and when the mouse is eaten by a carnivore, each scolex develops into an adult tapeworm. The **hydatid** metacestode of *Echinococcus granulosus* is capable of budding off hundreds of thousands of new scolices within itself (see Figs. 21.22 through 21.25). When such a packet of immature worms is eaten by a dog, vast numbers of adult cestodes are produced.

Perhaps the most remarkable asexual reproduction in all zoology is found among the trematodes, a large and successful group of parasites commonly called *flukes.* These animals produce a series of embryo generations within the body of the prior generation, a process that brings to mind the nursery rhyme "Going to St. Ives": *As I was going to St. Ives/I met a man with seven wives./Each wife had seven sacks./Each sack had seven cats./Each cat had seven kits* and so on. Trematode eggs hatch into **miracidia,** which enter a first intermediate host, always a mollusc, and become saclike **sporocysts.** Sporocysts may give rise to **daughter sporocysts,** which, in turn, may each produce a generation of **redia.** These then become filled with **daughter redia,** which finally produce **cercariae** (see Chapter 15 for details). Although there are many variations on this general theme (some eggs must be eaten by the first intermediate host and not all species produce redia), by the time all cercariae from a single successful egg are accounted for, the one miracidium has been responsible for an astonishing number of trematodes. And many flukes give birth to thousands of eggs each day. The biotic potential of these worms makes them the masters of reproduction! On the other hand, this staggering reproductive effort also tells us something about the chances of a single trematode egg reaching adulthood.

With hermaphroditism the parasite has solved the problem of finding a mate. Many tapeworms and trematodes can fertilize their own eggs (through selfing, p. 310); this method, although not likely to produce any unusual genetic recombinations, guarantees offspring. Tapeworms also undergo a continuous asexual production of segments (**strobilization**) from an undifferentiated region immediately behind the attachment organ (**scolex**). These segments, called **proglottids,** are each the reproductive equivalent of a hermaphroditic worm, at least in the vast majority of tapeworm species, since each contains both male and female reproductive organs. And each fertilized female system in each proglottid eventually becomes filled with eggs containing larvae. The result of this combination of asexual reproduction, hermaphroditism, and self-fertilization is a veritable tapeworm egg factory. Whale tapeworms of the genus *Hexagonoporus,* for example, are 100-foot reproductive monsters consisting of about 45,000 proglottids, each with 5 to 14 sets of male and female systems. However, there are not many whales and the ocean is truly a vast space, so perhaps this massive investment of energy in reproduction is the minimum necessary to ensure survival of a parasite whose ancestors colonized whales.

A second way of increasing reproductive potential is through production of vast numbers of eggs. A common rat tapeworm, *Hymenolepis diminuta,* for example, produces up to 250,000 eggs a day for the life of its host. During a period of slightly over a year, a single tapeworm can thus generate a hundred million eggs. If all these eggs reached maturity in new hosts, they would represent more than 20 tons of tapeworm tissue. Female nematodes are also sometimes prodigious egg layers; a single *Ascaris* can produce more than 200,000 eggs a day for several months, and over the course of their lifetimes, members of the filarial genus *Wuchereria* may release several million young into the host's blood. Such high reproductive potential, of course, ensures that such parasites will become medical and veterinary problems when host populations are crowded and transmission conditions are favorable.

Adaptations for Transmission

There are numerous examples of parasite attributes that increase the species' chances of encountering new hosts. These attributes often influence an intermediate host in some way, making it more susceptible to predation by a definitive host. Trematodes of the genus *Leucochloridium,* for example, infect land snails as first intermediate hosts and insectivorous birds as definitive hosts. The sporocysts of *Leucochloridium* are elongated and have pigmented bands (brown or green). These sporocysts move into the snail's tentacles and pulsate, looking for all the world like caterpillars. The bird who snaps up this "caterpillar," however, gets an intestine filled with dozens of infective metacercaria.

The immature stages of some thorny-headed worms (phylum Acanthocephala, Chapter 31) infect freshwater crustaceans of the order Amphipoda (side-swimmers). Some acanthocephalan juveniles appear as conspicuous white or orange spots in the hemocoel of the transluscent amphipods, making the infected ones stand out from the uninfected. Within the genus *Acanthocephalus* infection results in loss of body pigment in the isopod intermediate host, while in *Polymorphus paradoxus,* a parasite of ducks, not only is the juvenile orange, but infection alters an amphipod's behavior so that it becomes positively phototactic and swims closer to the water surface than it would otherwise. Ducks prey on these infected, behaviorally altered amphipods selectively.[2] *Acanthocephalus dirus* eggs also exhibit a feature that might be interpreted as an adaptation for transmission. The eggs possess filaments that get tangled up in the algae that the intermediate host eats.

The most notorious of behavior-changing infections involves trematodes of the genus *Dicrocoelium,* which infect large herbivores such as sheep (p. 263). The second intermediate hosts of *Dicrocoelium* are ants; metacercaria lodge in the ants' brains, making the insects move to the tops of grass blades where their likelihood of being accidently ingested by a definitive host is greatly increased. And among the arthropods, parasite behavior itself often promotes infection, as in the case of nymphal ticks, which climb up on vegetation, thus increasing the chances of encountering a passing host.

Epidemiology

Epidemiology is the study of factors affecting the transmission and distribution of a disease. In the case of infectious diseases one of the most important of these factors is the **vector**, which may be a biting or blood-sucking arthropod, a snail, or even a dust particle. Vectors are vehicles by which infections are transmitted from one host to another, although the term tends to be used most often to describe vehicles for which the hosts are of economic or personal interest to humans, such as ourselves and our domestic animals. Some of the most medically important vectors are anopheline mosquitoes, which transmit malarial parasites (Chapter 9), and snails of certain genera, which carry infective larval blood flukes, or schistosomes (Chapter 16). Malaria and schistosomiasis are still among the most serious human diseases, infecting nearly 700,000,000 people, mostly in developing countries (p. 2).

Vector biology usually must be well understood before disease control measures can become effective. It is standard practice in malaria control efforts, for example, to eliminate mosquito breeding habitat, namely standing water. Unfortunately, in some areas of the world, the only drinking and bathing water available is also the breeding ground of mosquitoes. Agricultural practices have sometimes added to disease control problems, as in Egypt, where irrigation ditches are ideal environments for the snails that serve as intermediate hosts for schistosomes.

Between World Wars I and II it was noted by the Russian school of Pavlovsky that certain parasitic infections occur in some ecosystems but not in others. The components of these ecosystems can be categorized so that they can be recognized wherever they occur. Thus, each disease has a natural focus, or **nidus**—namely, the set of ecological conditions under which it can be predicted to occur. Discovery of this *natural nidality of infection* was a landmark in the history of parasitology because it enabled epidemiologists to recognize "landscapes" where certain diseases could be expected to exist or, equally, where they might be effectively controlled. Such **landscape epidemiology** requires thorough knowledge of all factors that influence transmission, such as climate, plant and animal population densities, geological conditions, and human activities within the nidus. This holistic approach is best applied to **zoonoses,** which are diseases of animals that are transmissable to humans. However, the principles of landscape epidemiology can be applied equally well to pinworm epidemics in day-care centers, whipworm infections in mental institutions, and *Giardia lamblia* outbreaks at posh resorts.

• Mathematical Models

The construction of **mathematical models** to describe parasite population and assemblage dynamics can be valuable to predict both population behavior over periods of time and the outcome of various control strategies and to bring to light questions that may never have been considered by field workers (Fig. 2.4). Nowadays mathematical models are mostly computer programs that use equations, or algorithms, to mimic (or attempt to mimic) natural processes. The advantage of a computer model is that it can be manipulated in many ways not always possible with nature. Furthermore, these manipulations—carried out in seconds, minutes, or hours—can generate theoretical parasite population behaviors corresponding to a period of many years. The disadvantage of models, of course, is that they are made of electronic signals, not real animals, and the signals infect one another inside an environment made of silicon chips, not in a tropical village, a suburban elementary school, a mountain stream, or a rocky intertidal zone.

Models have been of most use to parasitologists by generating predictions, which, in turn, tend to stimulate research. (Scientists are forever trying to falsify their own as well as others' predictions!) Anderson's models, for example, suggested that parasites causing little harm or stress to a host are the most effective regulators of host populations.[1] Anderson's equations tell us that a harmful parasite causes the death of too many parasites (which can live only in hosts), an assertion that supports the prevalent dogma that the most successful parasite is one that causes little harm to its host, even though the dogma never addressed the issue of host population control in a numerical way.

Models can also be of great practical value, especially when their predictions are counterintuitive, violating our "default" perceptions of particular host/parasite relationships. In the Philippines *Schistosoma japonicum* parasitizes not only humans, but also dogs, pigs, and field rats as definitive hosts (see also Chapter 16). In a particularly illustrative study Hairston, using quantitative models, suggested that rats alone could support the suprapopulation of *S. japonicum* because of the high rate of contact between rats and the snail intermediate host.[12,13] Thus, even if all humans were cured at once, only a few years would be required for the parasite infrapopulations in humans to reach the previous level. The Hairston prediction illustrates beautifully the conflict that can occur when science runs into deeply entrenched feelings about our fellow humans. If you are a physician in an endemic area of the Philippines and you are trying to relieve the human population of a parasite burden but are in possession of only limited resources, do you spend your time treating humans or trying to kill rats?

Mathematical models have also shown us that the parasitologist's business is inextricably linked to a broad knowledge of the biology of many animals. Gettinby, for example, predicted that control of *Fasciola hepatica,* an economically important trematode parasite of sheep and cattle, would be more effective if it involved killing the snail intermediate hosts than if it involved treating definitive hosts with antihelminthic.[11] Current models of the transmission of Lyme disease inevitably take into account the biology of ticks as well as that of rodents, deer, and human beings. And the breadth of knowledge required to model a parasitic infection is not reduced when one considers only the definitive host. Ratcliffe's models of the nematode *Hemonchis contortus* in sheep, for example, require the input of numerous physiological parameters obtained only by immunological and hematological study of the host.[17]

PARASITE EVOLUTION

It is a fundamental assumption of evolutionary biology that organisms exist within defined sets of ecological conditions; in other words, there are species-specific and species-defined

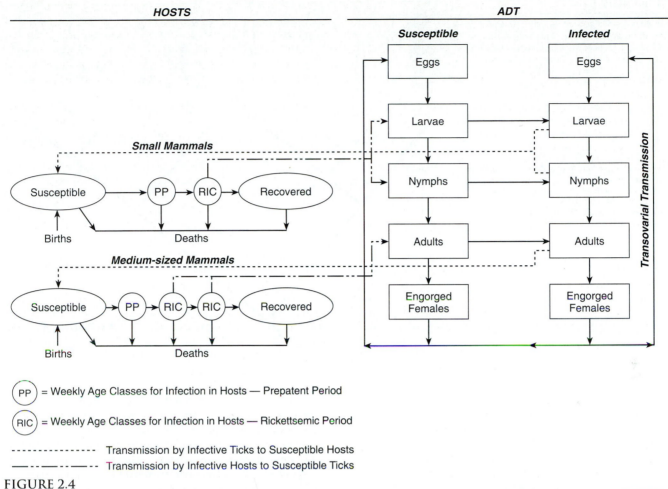

FIGURE 2.4

An example of a parasitological model, in this case a system for predicting the transmission dynamics of a rickettsial disease, rocky mountain spotted fever. **ADT** is the American dog tick, which is a major vector of the disease. The arrows represent the flow of organisms from one category into another. The rates and directions of change can be measured, relationships between boxes can be established mathematically, and the model can be used to predict the effects of various control strategies.

From L. M. Cooksey et al., "Computer simulation of Rocky Mountain spotted fever transmission by the American dog tick (Acari: Ixodidae)" in *J. Med. Ent.* 27:671–680. Copyright © 1990 The Entomological Society of America. Reprinted by permission.

ecological niches. Neo-Darwinism (the merger of Mendelian and population genetics with classical Darwinian natural selection) asserts that organisms adapt to their niches through changes in the genetic makeup of their populations and that mutation is the ultimate source of the necessary genetic variation. Furthermore, competition for limited resources is assumed to drive evolutionary change by reducing the relative fitness of those phenotypes (thus genotypes) least able to compete for these resources.

Commonly cited mechanisms of evolutionary change include **selection, genetic drift,** and the **founder effect.** In selection some genetic variants within a species' population are more successful than others in producing surviving offspring; thus, the population as a whole changes genetically. The relative success is determined "by nature" through competition or other selective pressures—hence, the term *natural selection*. In genetic drift isolated populations change genetically simply by chance because all the possible interbreeding combinations that could occur do not occur. Genetic drift operates most strongly in small populations. The term *founder effect* refers to colonization of an environment by a few individuals, resulting in a gene pool that contains but a small fraction of the variation present in the parent population.

Do these principles apply to parasitic relationships in which the environment is the host and the evolving organism is the parasite? Parasitologists have spent an enormous amount of effort trying to answer this question and in general have concluded that while adaptation has evidently functioned to produce certain structures and relationships, Neo-Darwinian natural selection cannot explain all of our observations. For example, it is not at all clear that resources provided by hosts are always limiting, and other selective pressures such as host defenses may influence parasite fitness. Thus, the essential Darwinian condition of competition has been difficult to demonstrate, at least in a large number of cases. To demonstrate genetic drift one must be able to follow the genetic makeup of a small population for several generations, a requisite that parasitologists have found quite

FIGURE 2.5

A cladogram depicting the phylogenetic relationships of several genera of parasitic Crustacea (copepods). The numbers indicate the apomorphies (see Table 2.1) that are shared by members of a clade. In general, parasites on the two main branches are found, on the left, in elasmobranchs (sharks, skates, rays) and, on the right, in bony fish. The resemblance to classical evolutionary trees is only superficial.

From M. Dojiri and G. B. Deets, *"Norkus cladocephalus,* new genus, new species (Siphonostomatoida: Sphyriidae), a copepod parasitic on an elasmobranch from southern California waters, with a phylogenetic analysis of the Sphyriidae," in *J. Crust. Biol.* 8: 679–687. Copyright © 1988 The Crustacean Society. Reprinted by permission.

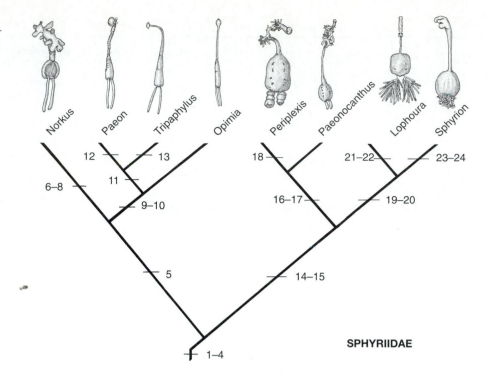

difficult to achieve for a number of technical reasons. And in theory, parasite populations should exhibit a strong founder effect because hosts are "islands" not easily occupied by a large fraction of a parasite's gene pool. But again, studies to demonstrate founder effects are exceedingly difficult technically, they are time and resource consuming, and they have not been conducted with a wide variety of parasite species.

The overriding concern among modern parasitologists studying evolution is the pattern of association between parasites, hosts, and the ecological and geographical distributions of each.[3] In general there are two factors that influence these patterns: descent and colonization. In other words, a parasite may be associated with a host because the two share a long evolutionary history (descent), or the host and parasite may be associated because the parasite has colonized the host in a manner analogous to the colonization of an island. (Because the parasite switches from its ancestral host to a new one, colonization also is called **host switching.**) In colonization geographical distributions of both hosts and parasites are influenced by global events such as continental drift. To explain patterns of host/parasite association one therefore must discover whether the patterns are a product of descent, colonization, physical separation of populations, or some combination of the three.

Advances in mathematical theory and computer design have aided parasitologists immeasurably by enabling them to use **cladistics,** also called **phylogenetic systematics** or **phyletics,** to infer evolutionary histories of hosts, parasites, and the geographical locations both occupy. In cladistic methodology a species' characters are evaluated and then classified as either **plesiomorphic** or **apomorphic.** This evaluation (polarization) is carried out by comparing the group of interest, or **ingroup,** which is usually a set of species or taxa hypothesized to be closely related, to an **outgroup,** a taxon that is related to the ingroup. Plesiomorphic

characters are ones shared by both ingroup and outgroup, while apomorphic characters—evolutionary novelties—occur only in the ingroup. The characters can be morphological, biochemical, ecological, and geographical (Table 2.1). The ingroup members are then evaluated in terms of shared apomorphic characters (**synapomorphies**). The idea is to establish a tree that has only bifurcations, with taxa at the end of two branches sharing more synapomorphies than they share with other groups. Evaluated characters are coded and entered into a data matrix, typically with the taxa as rows and the characters as columns (Table 2.2). This matrix is used to calculate the cladogram, usually by computer.

In practice the phyletic study of parasite evolution requires a fairly large amount of information about both the host and parasite, ideally including structure, habitat, behavior, geographical distribution, and biochemical makeup. The software used in these studies generates **cladograms** (Fig. 2.5), which only superficially resemble the evolutionary trees so often seen in older texts. A cladogram indicates only closest relatives based on numbers of shared derived traits, the latter taken as evidence of common ancestry. Purists assert that such cladograms are in fact hypotheses—that is, statements that may be falsified by additional research. Evolutionary trees, on the other hand, have sometimes been presented as answers to questions about from where plant and animal species come.

The study from which Figure 2.5 was taken provides an excellent illustration of the application of phylogenetics to problems of parasite evolution. The organisms involved are parasitic copepods (see also Chapter 34). The hosts represent several orders of cartilaginous and bony fishes (Elasmobranchii and Teleostii). Parasitic copepods are often quite bizarre structurally, evidently as a result of their extreme adaptations to parasitism (Fig. 2.6). In this particular study

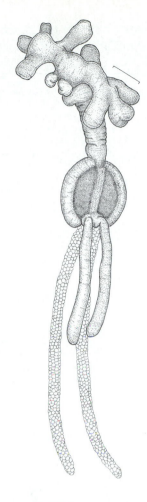

FIGURE 2.6

Norkus cladocephalus, a copepod parasitic on the shovelnose guitarfish, *Rhinobatos productus,* from southern California. Bar equals 2.0 mm.

From M. Dojiri and G. B. Deets, *"Norkus cladocephalus,* new genus, new species (Siphonostomatoida: Sphyriidae), a copepod parasitic on an elasmobranch from southern California waters, with a phylogenetic analysis of the Sphyriidae," in *J. Crust. Biol.* 8:679–687. Copyright © 1988 The Crustacean Society. Reprinted by permission.

the hosts represented sharks and rays on the one hand and diverse groups of bony fishes on the other. The authors, Dojiri and Deets,[9] were seeking to develop a hypothesis about the evolutionary relationships of these copepods, which belonged to the family Sphyriidae, and they were particularly interested in whether the parasites seemed confined to particular evolutionary lines of their hosts or seemed to have colonized a host taxon repeatedly.

Dojiri and Deets chose as an outgroup members of the genus *Ommatokoita* from the family Lernaeopodidae, which is related to the Sphyriidae. The ingroup genera of parasites, all members of the Sphyriidae, are indicated at the ends of the branches in Figure 2.5. The characters that were used are listed in Table 2.1. *Rotation of body* means that the head is twisted with respect to the thorax. The rest of the characters can be understood with references to Chapter 34, which concerns parasitic crustaceans. Table 2.2 shows the coded data matrix derived by the authors from polarization of these characters: 0 indicates the character is plesiomorphic; 1 indicates the

character is apomorphic. The cladogram itself was generated by software called PAUP for Phylogenetic Analysis Using Parsimony (D. Swofford, Illinois Natural History Survey).

The phylogenetic relationships of the hosts were also of interest to Dojiri and Deets.[9] If the authors substituted host orders for their parasites, the resulting phylogeny more or less matched that of the host orders as suggested by the ichthyologists (Fig. 2.7). That is, the taxa of parasites evidently evolved along with the taxa of their hosts. Those parasites on the left-hand branch are found on elasmobranchs, while those on the right are found on teleosts.

In constructing the ecological cladogram Dojiri and Deets were not successful, however (Fig. 2.8). Although the sphyriid genera could be split into two major groups corresponding to both infection site and habitat of the hosts, the outgroup members were so diverse in their parasitic habits that the apomorphic condition could not be determined.

This particular paper illustrates the general techniques used, the types of questions that parasitologists ask, and some of the problems that can arise in the study of parasite evolution. The overall approach has been used repeatedly in recent years, and a rich literature has developed involving many groups of hosts and parasites. Brooks and McLennan's *Parascript*[3] provides a summary of the history, major questions, bibliography, and existing database on parasite evolution. Case studies involving diverse groups of hosts and parasites are analyzed in detail and the results reveal some fascinating and ancient relationships between parasites, hosts, and global geological events. For example, some turtle blood flukes apparently enjoyed a long coevolutionary relationship with their hosts (since the Mesozoic), while others seem to have diversified following the breakup of Pangaea. In an even more intriguing conclusion parasites of freshwater stingrays, now characteristic of South American rivers draining into the Atlantic Ocean, actually originated from the Pacific Ocean prior to the separation of Africa and South America.[3] The literature of parasite evolutionary biology suggests that many young scientists, struggling with large problems and massive data sets, eventually come to see the interactions between hosts, parasites, and global changes in climate and geography over geological time scales as part of a single grand picture of life on earth.

Systematics and Taxonomy of Parasites

In general the world's invertebrates, of which parasites make up a sizeable fraction, are not nearly as well-known as are the vertebrates. Thousands of new species of protozoa, helminths, and arthropods are described every year. Indeed, with a not unreasonable amount of work, almost anyone, undergraduate biology majors included, can find and describe a new species. Then the finder can pick the species name and be immortalized, along with the name, in the parasitological literature. The monogenetic trematode *Salsuginus thalkeni,* for example, is named for a landowner who gave parasitologists permission to use his property for research. *Actinocephalus carrilynnae,* a protozoan parasite of damselflies, was named in "honor" of a little sister by an older sister who continually threatened to "name a parasite after" her. Occasionally one hears parasitologists talk about naming parasites after politicians. But in a more sober and dignified vein, a number of

Table 2.1 *Characters used in phylogenetic analysis of the parasitic copepods of Figure 2.5. Characters are given as plesiomorphic (0)/apomorphic (1); numbers in parentheses refer to the characters mapped onto the cladogram in Figure 2.5.*

A. Body condition: nonrotated/rotated (1)

B. Egg sac length: moderate/elongate (2)

C. Neck length: moderate/elongate (3)

D. Posterior processes: diminutive/prominent (4)

E. Posterior processes: tubercular/single, long, cylindrical (5)

F. Posterior processes: tubercular/single, short, cylindrical (16)

G. Posterior processes: tubercular/single, short, transversely constricted (18)

H. Posterior processes: tubercular/complex, branching (19)

I. Posterior processes: tubercular, branching, grapelike (23)

J. Posterior processes: tubercular/multiple cylinders (21)

K. Genital complex: ovoid/discoid (6)

L. Genital complex: ovoid/gradually expanding (9)

M. Neck holdfast: absent/collar (7)

N. Neck holdfast: absent/multiple planes (17)

O. Neck holdfast: absent/single plane (22)

P. Cephalothorax: cylindrical/antlerlike (8)

Q. Cephalothorax: cylindrical/lateral aliform expansions (24)

R. Cephalothorax: cylindrical/bulbous (10)

S. Cephalothorax: cylindrical/bulbous with protuberances (11)

T. Cephalothorax: cylindrical/lateral hornlike projections (13)

U. Cephalothorax: cylindrical/paired rounded protuberances (12)

V. Gut diverticula: simple/anastomosing (20)

W. First antenna: digitiform/reduced lobe (14)

X. Second antenna: biramous/reduced lobe (15)

Table 2.2 *Character data matrix for the parasitic copepods of the family Sphyriidae. Entries are the plesiomorphic/apomorphic codes (0 or 1) of the characters as indicated in Table 2.1.*

	A	B	C	D	E	F	G	H	I	J	K	L	M	N	O	P	Q	R	S	T	U	V	W	X
Outgroup	0	0	0	0	0	0	0	0	0	0	0	0	0	0	0	0	0	0	0	0	0	0	0	0
Norkus	1	1	1	1	1	0	0	0	0	0	0	1	0	0	1	0	0	0	0	0	0	0	0	0
Paegon	1	1	1	1	1	0	0	0	0	0	0	1	0	0	0	0	0	1	1	0	1	0	0	0
Tripaphylus	1	1	1	1	1	0	0	0	0	0	0	1	0	0	0	0	0	1	1	1	0	0	0	0
Opimia	1	1	1	1	1	0	0	0	0	0	0	1	0	0	0	0	0	1	0	0	0	0	0	0
Periplexis	1	1	1	1	0	1	1	0	0	0	0	0	1	0	0	0	0	0	0	0	0	0	1	1
Paeonocanthus	1	1	1	1	0	1	0	0	0	0	0	0	1	0	0	0	0	0	0	0	0	0	1	1
Lophoura	1	1	1	1	0	0	0	1	0	1	0	0	0	0	1	0	0	0	0	0	0	1	1	1
Sphyrion	1	1	1	1	0	0	0	1	1	0	0	0	0	0	0	0	1	0	0	0	0	1	1	1

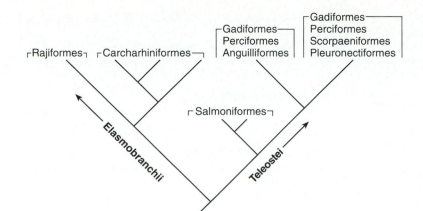

FIGURE 2.7

A phylogeny of the hosts of parasitic copepods of the family *Sphyriidae* as revealed by the parasites.

From M. Dojiri and G. B. Deets, *"Norkus cladocephalus,* new genus, new species (Siphonostomatoida: Sphyriidae), a copepod parasitic on an elasmobranch from southern California waters, with a phylogenetic analysis of the Sphyriidae," in *J. Crust. Biol.* 8: 679–687. Copyright © 1988 The Crustacean Society. Reprinted by permission.

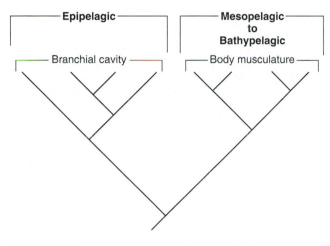

FIGURE 2.8

An ecological summary cladogram of the host habitats and infection sites of sphyriid copepods.

From M. Dojiri and G. B. Deets, *"Norkus cladocephalus,* new genus, new species (Siphonostomatoida: Sphyriidae), a copepod parasitic on an elasmobranch from southern California waters, with a phylogenetic analysis of the Sphyriidae," in *J. Crust. Biol.* 8:679–687. Copyright © 1988 The Crustacean Society. Reprinted by permission.

trematodes and tapeworms were named by Edward Adrian Wilson and Robert Leiper in honor of members of the ill-fated Robert F. Scott expedition to the South Pole who died on the trip.[5] Campbell's story of that expedition is a compelling one, worth reading by any person who feels that scientific names are just biologists' way of separating the Latin scholars from the rest of humanity.

Taxonomy, or the science of classification, is as vibrant an area of biology today as it was a hundred years ago. A good part of the activity is due to the new techniques, especially biochemical ones, that have been adopted by taxonomists. Parasitologists are constantly evaluating the criteria they use to make taxonomic decisions, trying to decipher the relationship between biochemical and morphological characteristics of species, and reexamining the genus, family, and order groupings of the animals they study.

Why is this work important? Taxonomy is a basic subdiscipline of biology. A scientific name carries with it a massive amount of information, some implied, some explicit, and all of value to ecologists, immunologists, epidemiologists, and evolutionary biologists. For example, a doctor or veterinarian cannot make a decision about treatment without knowing what kind of parasite is infecting the patient. And an epidemiologist, looking for ways to control malaria or filariasis, is stumped if unable to tell a mosquito from a housefly.

Taxonomic criteria vary from parasite group to group. In the case of arthropods skeletal morphology is still of primary importance. The classification of Platyhelminthes is based to a large extent on reproductive organs—primarily their numbers, sizes, and relative positions in the body. Nematode taxonomists must focus on reproductive structures, including those at the posterior ends of the male, but the arrangements of sensory papillae and other cuticular features, especially those around the mouth, are considered also. Protozoan taxonomic characters include cyst morphology (amoebas, coccidia), number and arrangement of flagella, and biochemical properties. Members of the genus *Leishmania,* for example, are "typed" using isozyme patterns and DNA buoyant density measurements.

For their study molecular biologists and biochemists often can obtain identified material from the American Type Culture Collection in Rockville, Maryland, a living museum of microorganisms, including many species and strains of parasites. This collection was started to alleviate the taxonomic problems associated with research on microorganisms. It is not only easier but more advisable for experimental biologists to obtain described and documented organisms from such a collection than it is for them to do the taxonomy themselves. A parasitological ecologist, however, must be prepared to identify animals and to describe them if necessary. Thus, the researcher quickly becomes familiar with the massive body of literature, some of it published in obscure and foreign journals, that has accumulated since Linnaeus first described the sheep liver fluke *Fasciola hepatica.*

References

1. Anderson, R. M. 1978. The regulation of host population growth by parasitic species. *Parasitology* 76:119–57.
2. Bethel, W. M, and J. C. Holmes. 1977. Increased vulnerability of amphipods to predation owing to altered behavior induced by larval acanthocephalans. *Can. J. Zool.* 55:110–15.

3. Brooks, D. R., and D. A. McLennan. 1993. *Parascript: Parasites and the language of evolution.* Washington, D.C.: Smithsonian Institution Press. 429p.

4. Bush, A. O., and J. C. Holmes. 1986. Intestinal helminths of lesser scaup ducks: Patterns of association. *Can. J. Zool.* 64:132–52.

5. Campbell, W. C. 1988. Heather and ice: An excursion into historical parasitology. *J. Parasitol.* 74:1–12.

6. Clopton, R. E., J. Janovy Jr., and T. J. Percival. 1992. Host stadium specificity in the gregarine assemblage parasitizing *Tenebrio molitor. J. Parasitol.* 78:334–37.

7. Crofton, H. D. 1971. A quantitative approach to parasitism. *Parasitology.* 62:179–93.

8. Crompton, D. W. T., M. C. Nesheim, and Z. S. Pawlowski. 1985. *Ascariasis and its public health importance.* London: Taylor and Francis.

9. Dojiri, M., and G. B. Deets. 1988. *Norkus cladocephalus,* new genus, new species (Siphonostomatoidida: Sphyriidae), a copepod parasitic on an elasmobranch from southern California waters, with a phylogenetic analysis of the Sphyriidae. *J. Crust. Biol.* 8:679–87.

10. Esch, G. W., A. O. Bush, and J. M. Aho, eds. 1990. *Parasite communities: Patterns and process.* New York: Chapman and Hall.

11. Gettinby, G. 1974. Assessment of the effectiveness of control techniques for liver fluke infection. In Usher, M. B., and M. H. Williamson, eds. *Ecological stability.* London: Chapman & Hall Ltd.

12. Hairston, N. G. 1962. Population ecology and epidemiological problems. In Wolstenholme, G. E. W., and M. O'Connor, eds. *Bilharziasis. Ciba Foundation Symposium.* London: J. & A. Churchill Ltd.

13. Hairston, N. G. 1965. On the mathematical analysis of schistosome populations. *Bull. WHO* 33:45–62.

14. Hutchinson, G. E. 1957. Concluding remarks. *Cold Spring Harbor Symposium on Quantitative Biology* 22:415–27.

15. Margolis, L., G. W. Esch, J. C. Holmes, A. M. Kuris, and G. A. Schad. 1982. The use of ecological terms in parasitology (report of an ad hoc committee of The American Society of Parasitologists). *J. Parasitol.* 68:131–33.

16. Pence, D. B., and L. A. Windberg. 1984. Population dynamics across selected habitat variables of the helminth community in coyotes. *J. Parasitol.* 70:735–46.

17. Ratcliffe, L. H., H. M. Taylor, J. H. Whitlock, and W. R. Lynn. 1969. Systems analysis of a host-parasite interaction. *Parasitology* 59:641–61.

18. Ricklefs, R. E. 1979. *Ecology,* 2d ed. New York: Chiron Press, Inc.

19. Schad, G. A. 1963. Niche diversification in a parasitic species flock. *Nature* 198:404–6.

20. Stock, T. M., and J. C. Holmes. 1988. Functional relationships and microhabitat distribution of enteric helminths of grebes (Podicipedidae): The evidence for interactive communities. *J. Parasitol.* 74:214–27.

21. Von Brand, T. 1973. *Biochemistry of parasites.* New York: Academic Press.

22. Whitfield, P. J. 1979. *The biology of parasitism: An introduction to the study of associating organisms.* Baltimore: University Park Press.

Additional References

Admadjian, V., and S. Paracer. 1986. *Symbiosis. An introduction to biological associations.* Hanover, N.H.: University Press of New England.

Brooks, D. R. 1989. The phylogeny of the Cercomeria (Platyhelminthes: Rhabdocoela) and general evolutionary principles. *J. Parasitol.* 75:606–16.

Brooks, D. R., and D. McLennan. 1991. *Phylogeny, ecology, and behavior: A research program in comparative biology.* Chicago: University of Chicago Press.

Campbell, W. C., and R. S. Rew, eds. 1986. *Chemotherapy of parasitic diseases.* New York: Plenum.

Croll, N. A., and E. Chadirian. 1981. Wormy persons: Contributions to the nature and patterns of overdispersion with *Ascaris lumbricoides, Ancylostoma duodenale, Necator americanus,* and *Trichiuris trichiura. Trop. Geogr. Med.* 33:241–48.

Futuyma, D. J., and M. Slatkin, eds. 1983. *Coevolution.* Sunderland, Mass.: Sinauer Associates, Inc.

Gillett, J. D. 1985. The behavior of *Homo sapiens,* the forgotten factor in the transmission of tropical disease. *Trans. Roy. Soc. Trop. Med. Hyg.* 79:12–20.

Keymer, A. E., and A. F. G. Slater. 1987. Helminth fecundity: Density dependence or statistical illusion? *Parasitol. Today* 3:56–58.

MacInnis, A. J. 1976. How parasites find hosts: Some thoughts on the inception of host-parasite integration. In Kennedy, C. R., ed. *Ecological aspects of parasitology.* Amsterdam: North-Holland Publishing Co.

Pavlovsky, E. N. 1966. Natural nidality of transmissible diseases, with special reference to the landscape epidemiology of zooanthroponoses (English translation). Urbana, Ill.: University of Illinois Press.

Price, P. W. 1980. Evolutionary biology of parasites. *Monographs in population biology* 15. Princeton, N.J.: Princeton University Press.

Schmidt, G. D., ed. 1969. *Problems in systematics of parasites.* Baltimore: University Park Press.

Smith, T. 1963. *Parasitism and disease.* New York: Hafner Publishing Co. A classic, originally published in 1934.

Soulsby, E. J. L., ed. 1976. *Pathophysiology of parasitic infection.* New York: Academic Press, Inc.

Trager, W. 1986. *Living together. The biology of animal parasitism.* New York: Plenum.

Chapter 3

BASIC PRINCIPLES AND CONCEPTS II: IMMUNOLOGY AND PATHOLOGY

. . . in parasitic conditions, there often is limited pathology directly attributable to the organism. Most morbidity is related to the immunoinflammatory response of the host to the parasite.

S. Michael Phillips[31]

A traditional view of host-parasite interaction asserts that as a symbiont becomes progressively more specialized, it increasingly limits its potential host species; that is, it increases its host specificity. A vital component in this process is the habitat (host), which is a dynamic, living, and evolving partner in the relationship. The host reacts to the presence of the symbiont, mounting a defense against the foreign invader, and the successful symbiont must evolve strategies to evade the host defenses. Parasitologists have come to recognize that not only is host specificity determined in great degree by which host individuals can mount an effective defense and which parasites can evade that defense, but also much of the disease caused by parasites is directly related to host defense mechanisms. This chapter will explore the host-parasite relationship by introducing concepts related to host defenses, the evasion of host defenses by parasites, and how parasites cause disease in their hosts.

SUSCEPTIBILITY AND RESISTANCE

A host is **susceptible** to a parasite if the host cannot eliminate the parasite before the parasite can become established. The host is **resistant** if its physiological status prevents the establishment and survival of the parasite. Corresponding terms from the viewpoint of the parasite would be **infective** and **noninfective.**

These terms deal only with the success or failure of infection, not with the mechanisms producing the result. The mechanisms that increase resistance (and correspondingly reduce the susceptibility and infectivity) may involve either attributes of the host not related to active defense mechanisms or specific defense mechanisms mounted by the host in response to a foreign invader. Furthermore, the terms are relative, not absolute; for example, one individual organism may be more or less resistant than another.

The term **immunity** has been, on the one hand, often used as synonymous with resistance and, on the other hand, associated with the sensitive and specific immune response exhibited by vertebrates. However, because many invertebrates can be immune to infection with various agents, a more general yet concise statement is that an animal demonstrates immunity if it possesses tissues capable of recognizing and protecting the animal against nonself invaders.[18] The type of immunity shown by most invertebrates is **innate.** In addition to having innate immunity, vertebrates develop **acquired immunity,** which is specific to the particular nonself material, requires time for its development, and occurs more quickly and vigorously on secondary response.

DEFENSE MECHANISMS

Phagocytosis

Most animals have one or more innate mechanisms to protect themselves against invasion of foreign bodies or infectious agents. These may be coincidental attributes of certain structures (for example, a tough skin or high stomach acidity), or they may be characteristics evolved as adaptations for defense. For defense against an invader the cells in an animal must recognize when a substance does not belong; they must recognize *nonself.* **Phagocytosis** illustrates nonself recognition, and it occurs in almost all metazoa and is a feeding mechanism in protozoa. A cell that has this ability is a **phagocyte.** Phagocytosis is a process of engulfment of the invading particle within an invagination of the phagocyte's cell membrane. The invagination becomes pinched off, and the particle becomes enclosed within an intracellular vacuole. **Lysosomes** empty digestive enzymes (**lysozymes**) into the vacuole to destroy the particle. Lysosomes of many phagocytes also contain, in addition to the digestive enzymes, enzymes that catalyze production of cytotoxic **reactive oxygen intermediates (ROIs)** and **reactive nitrogen intermediates (RNIs).** Examples of ROIs are superoxide radical (O_2^-), hydrogen peroxide (H_2O_2), singlet oxygen ($^1O_2$), and hydroxyl radical (OH•). RNIs include nitric oxide (NO) and its oxidized forms, nitrite (NO_2^-) and nitrate (NO_3^-).[32]

In vertebrates there are **fixed** and **mobile** phagocytes. The fixed phagocytes taken together form the **reticuloendothelial (RE) system** and are spread through a variety of tissues. The RE system filters out and destroys particles and spent red blood cells from the blood that passes throughout the organs where the cells are located. Cells of the RE system of humans include the **Kupffer cells** in sinusoids of the liver; the phagocytic reticular cells of lymphatic tissue, myeloid tissue, and spleen; and the "dust cells" of the lungs. **Macrophages** in the lymph nodes help remove foreign particles from the lymph.

Table 3.1   *Some invertebrate leukocytes and their functions*

Group	Cell types and functions	Phagocytosis	Encapsulation	Transplantation reaction
Sponges	Archaeocytes (wandering cells that differentiate into other cell types and can act as phagocytes)	+	+	?*
Cnidarians	Amebocytes: "lymphocytes"	+		?
Nemertines	Agranular leukocytes; granular macrophagelike cells	+		+
Annelids	Basophilic amebocytes (accumulate as "brown bodies"); acidophilic granulocytes	+	+	+
Sipunculids	Several types	+	+	+
Insects	Several types, depending on family; e.g., plasmatocytes, granulocytes, spherule cells, coagulocytes—blood clotting	+	+	+
Crustaceans	Granular phagocytes; refractile cells that lyse and release contents	+	+	+
Molluscs	Amebocytes	+	+	+
Echinoderms	Amebocytes, spherule cells, pigment cells, vibratile cells—blood clotting	+	+	+
Tunicates	Many types, including phagocytes; "lymphocytes"	+	+	+

From A. M. Lackie, *Parasitology,* 80: 392–412. Copyright © 1980 Cambridge Universtiy Press, Cambridge, England. Reprinted by permission. (See Lackie's article for references)

*Transplantation reactions occur, but the extent to which the leukocytes are involved is unknown.

The most numerous of the mobile or circulating phagocytes are the **polymorphonuclear leukocytes,** a name that refers to the highly variable shape of their nucleus. Another name for these leukocytes is **granulocytes,** a term that alludes to the many small granules that can be seen in their cytoplasm, especially when stained with a Romanovsky type of stain. According to the staining properties of the granules, granulocytes are further subdivided into **neutrophils, eosinophils,** and **basophils.** Neutrophils are the most abundant of the white cells in the blood, and they provide the first line of phagocytic defense in an infection. Eosinophils in normal blood account for about 2.0% to 5.0% of the total leukocytes. Basophils are the least numerous at about 0.5%. A high **eosinophilia** (eosinophil count in the blood) is often associated with allergic diseases and parasitic infections.

Another circulating white cell important in phagocytosis is the **monocyte.** The cell is not considered a granulocyte, though its cytoplasm may contain fine granules. When monocytes move into the tissue, they differentiate into macrophages, which are active phagocytes. Macrophages also have important roles in the specific immune response of vertebrates.

Immunity in Invertebrates

Our presentation will be brief; for further information consult the book edited by Maramorosch and Shope[24] and especially the review by Lackie.[18]

Many invertebrates have specialized cells that function as itinerant troubleshooters within the body, acting to engulf or wall off foreign material (Table 3.1) and repair wounds. The cells are variously known as amebocytes, hemocytes, coelomocytes, and so on, depending on the animals in which they occur. If the foreign particle is small, it is engulfed by phagocytosis; but if it is larger than about 10 µm, it is usually encapsulated. Arthropods can wall off the foreign object also by deposition of melanin around it, either from the cells of the capsule or by precipitation from the **hemolymph** (blood).

One of the principal tests of the ability of invertebrate tissues to recognize nonself is the grafting of a piece of tissue from another individual of the same species (**allograft**) or a different species (**xenograft**) onto the host. If the graft grows in place with no host response, the host tissue is treating it as self, but if cell response and rejection of the graft occur, the host exhibits immune recognition. Most invertebrates tested in this way reject xenografts; and sponges, cnidarians, annelids, sipunculids, echinoderms, and tunicates can reject allografts.[18] Interestingly, nemertines, arthropods, and molluscs apparently do not reject allografts. It is curious that some animals with quite simple body organization (Porifera and Cnidaria) can reject allografts; Lackie[18] suggested that this may be an adaptation to avoid loss of integrity of the individual sponge or colony under conditions of crowding, with attendant danger of overgrowth or fusion with other individuals.

Lysozymes are released from the hemocytes of molluscs during phagocytosis and encapsulation,[6,14] and bactericidal substances occur in the body fluids of a variety of invertebrates.[18] Substances that can agglutinate vertebrate erythrocytes in vitro have been found in many invertebrates, although

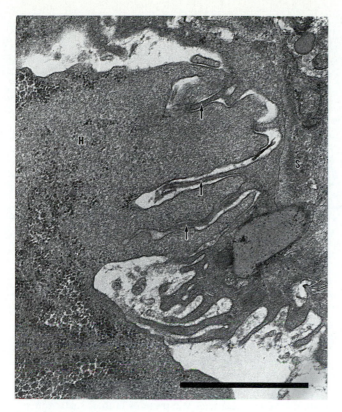

FIGURE 3.1

H, a hemocyte from a schistosome-resistant strain of *Biophalaria glabrata* attacks **S,** a *Schistosoma mansoni* larva under in vitro conditions. Note the hemocyte processes apparently engaged in phagocytosis of portions of the larval tegument (arrows). (Scale bar = 1 µm.)

From Van der Knapp, W.P.W., and E. S. Loker, "Immune mechanisms in trematode-snail interactions," *Parasitol. Today,* 6:175–182, 1990.

the function of these lectinlike molecules in vivo is unclear. A number of reports suggest that such molecules may act as opsonins, enhancing or facilitating phagocytosis, in molluscs and arthropods. Bacterial infection in some insects stimulates production of antibacterial proteins.[36]

Generally, invertebrates do not seem able to acquire a specific immune response. They will either respond at first exposure to a foreign material or not at all; repeated exposure does not elicit immunity. However, contact with infectious organisms can bring the defense systems of snails into enhanced levels of readiness that last for up to two months or more.[3,39] The susceptibility of the snail *Biomphalaria glabrata* to infection with the trematode *Schistosoma mansoni* has been studied, and susceptibility or resistance depends heavily on the genotype of the snail.[33] Excretory/secretory products of the trematode stimulate motility of hemocytes from resistant snails but inhibit motility of hemocytes from susceptible hosts. Hemocytes from resistant snails encapsulate the trematode larva and apparently kill it with superoxide and H_2O_2 and then destroy it by phagocytosis (Fig. 3.1).[39]

Acquired Immune Response in Vertebrates

Vertebrates have a specialized system of nonself recognition that results in increased resistance to *specific* foreign substances or invaders on repeated exposures. Investigations on the mechanisms involved are currently intense, and our knowledge of them is increasing rapidly. For a concise account of general immunology consult Abbas et al.[1] Reviews of various aspects of parasite immunology are in Englund and Sher,[10] Paul,[29] Sher et al.,[34] Wang,[43] and Warren.[44]

The immune response is stimulated by the specific foreign substance called an **antigen,** and, circularly, an antigen is any substance that will stimulate an immune response. Antigens may be any of a variety of substances with a molecular weight of over 3000. They are most commonly proteins and are usually (but not always) foreign to the host. There are two arms of the immune response, known as **humoral** and **cellular.** Humoral immunity is based on **antibodies,** which are mostly dissolved in and circulate in the blood, whereas cellular immunity is associated with cell surfaces. There is extensive communication and interaction among the cells of the two responses.

• Basis of Self and Nonself Recognition

It has been known for many years that nonself recognition is very specific. If tissue from one individual were transplanted into another individual in the same species, the graft would grow for a time and then die as immunity against it arose. Tissue grafts would only grow successfully if they were between identical twins or between individuals of highly inbred strains of animals. The molecular basis for this nonself recognition involves certain proteins imbedded in the cell surface. These proteins are coded by certain genes, now known as the **major histocompatibility complex (MHC).** The MHC proteins are among the most variable known, and unrelated individuals almost always have different genes. There are two types of MHC proteins: class I and class II. Class I proteins are found on the surface of virtually all cells, whereas class II MHC proteins are found only on certain cells participating in the immune responses, such as certain lymphocytes and macrophages.

• Antibodies

Antibodies are proteins called **immunoglobulins.** The basic antibody molecule consists of four polypeptide strands: two identical light chains and two identical heavy chains held together in a Y-shape by disulfide bonds and hydrogen bonds (Fig. 3.2). The amino acid sequence toward the ends of the Y varies in both the heavy and light chains, according to the specific antibody molecule (the **variable region**), and this variation determines with which antigen the antibody can bind. Each of the ends of the Y forms a cleft that acts as the antigen-binding site (see Fig. 3.2), and the specificity of the molecule depends on the shape of the cleft and the properties of the chemical groups that line its walls. The remainder of the antibody is known as the **constant region,** although it also varies to some extent. The variable end of the antibody molecule is referred to as **Fab,** for antigen-binding fragment, and

FIGURE 3.2

Antibody molecule is composed of two shorter polypeptide chains (light chains) and two longer chains (heavy chains) held together by covalent disulfide bonds. The light chains may be either of two types: kappa or lambda. The class of antibody is determined by the type of heavy chain: mu (**IgM**), gamma (**IgG**), alpha (**IgA**), delta (**IgD**), or epsilon (**IgE**). The constant portion of each chain does not vary for a given type or class, and the variable portion varies with the specificity of the antibody. Antigen-binding sites are in clefts formed in the variable portions of the heavy and light chains. IgM normally occurs as a pentamer, five of the structures illustrated being bound together by another chain. IgA may occur as a monomer, dimer, or trimer.

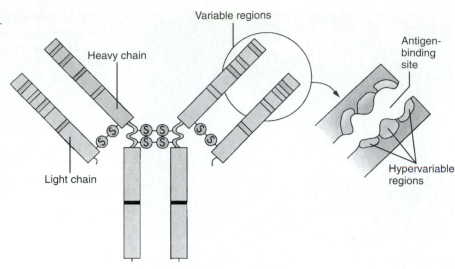

the constant end is known as the **Fc,** for crystallizable fragment (see Fig. 3.2). The constant region of the light chains can be either of two types: kappa or lambda. The heavy chains may be any of five types: mu, gamma, alpha, delta, or epsilon. The latter type determines the **class** of the antibodies, referred to as **IgM, IgG** (now familiar to many people as *gamma globulin*), **IgA, IgD,** and **IgE,** respectively. The class of the antibody determines the role of the antibody in the immune response (for example, the antibody may be secreted or held on a cell surface) but not the antigen it recognizes.

• Cells of the Immune Response

Some of the cells that are important in the immune response, such as PMNs and macrophages, were mentioned previously. A number of others are mostly encompassed by a category of leukocytes known as **lymphocytes. B lymphocytes (B cells)** have antibody molecules in their surface and give rise to cells that actively secrete antibodies into the blood. **T lymphocytes (T cells)** have surface receptors that bind antigens, but the receptors are somewhat different in structure from antibodies. There are a vast number of different kinds of B cells, each bearing molecules of antibody on its surface that will bind with one particular antigen, even though that antigen may have never been present in the body previously. There are probably an equally great number of different T cells with receptors for specific antigens.

Lymphocytes are **activated** when they are stimulated to move from their recognition phase, in which they simply bind with particular antigens, to an activation phase. In this phase they proliferate and differentiate into cells that function to eliminate the antigens. We also speak of activation of effector cells, such as macrophages, when they are stimulated to carry out their protective function.

Subsets of T Cells.   Communication between cells in the immune response, regulation of the response, and certain effector functions are carried out by different kinds of T cells. Although morphologically similar, subsets of T cells can be distinguished by characteristic proteins in their surface membranes. For example, cells with the protein CD4 (for **c**luster of **d**ifferentiation) are CD4+ and those with CD8 are described as CD8+. Until recently immunologists believed that certain CD4+ cells (T helper or T_h) activate immune responses, and certain CD8+ cells (T suppressors) downregulate such responses. Present evidence now suggests a more complicated web of interactions (Fig. 3.3). Some T_h cells (designated T_h1) activate cell-mediated immunity while suppressing the humoral response, and others (called T_h2) activate humoral and suppress cell-mediated immunity. Because each category apparently includes some CD8+ cells, some researchers suggest that the abbreviations T_h1 and T_h2 be further abbreviated to T1 and T2.[9]

Cytotoxic T lymphocytes (CTLs) are CD8+ cells that kill target cells expressing certain antigens. The CTL binds tightly to the target cell and secretes a protein that causes pores to form in the cell membrane. The target cell then lyses.[47]

Other Cells.   **Natural killer cells (NK)** are lymphocytelike cells that can kill virus-infected and tumor cells in the absence of antibody. **Mast cells** are basophil-like cells found in the dermis and other tissues. Their surfaces bear receptors for the Fc portions of IgE and IgG.

• Cytokines

The 1980s saw rapid advances in our knowledge of how cells of immunity communicate with each other. They do this by means of protein hormones called **cytokines.** Cytokines can produce their effects on the same cells that produce them, on

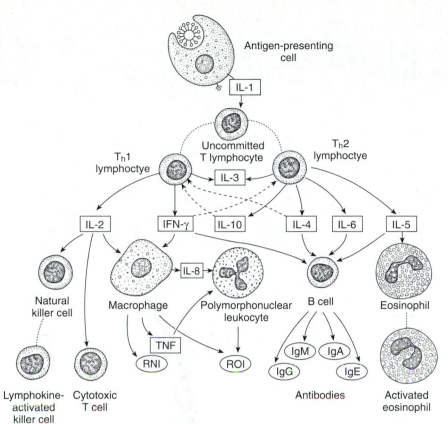

FIGURE 3.3

Major pathways involved in the immune response to parasitic infections as mediated by cytokines. **Solid arrows** indicate positive signals and **broken arrows** indicate inhibitory signals. **Broken lines without arrows** indicate path of cellular activation. **IFN-γ,** inteferon-γ; **Ig,** immunoglobulin; **IL,** interleukin; **TNF,** tumor necrosis factor; **T$_h$1,** helper CD4$^+$ and CD8$^+$ cells that stimulate cell-mediated response; **T$_h$2,** helper CD4$^+$ and CD8$^+$ cells that stimulate humoral response; **RNI,** reactive nitrogen intermediates; **ROI,** reactive oxygen intermediates.

Drawn by William Ober and Claire Garrison from F. E. G. Cox and E. Y. Liew, "T-cell subsets and cytokines in parasitic infections," in *Parasitol. Today,* 8:371–374, 1992.

cells nearby, or on cells distant in the body from those that produced the cytokine. Some cytokines important in immune responses follow:

1. **Interleukin-1 (IL-1).** The interleukins were originally so-called because they are synthesized by leukocytes and have their effect on leukocytes. We now know that some other kinds of cells can produce interleukins, and interleukins produced by leukocytes can affect other kinds of cells. IL-1 is produced by activated macrophages and mediates the host inflammatory response. It also activates T cells and B cells.

2. **Interleukin-2 (IL-2).** IL-2 is produced by CD4$^+$ cells and to a lesser extent by CD8$^+$ cells. It is a major growth factor for T and B cells, and it enhances the cytolytic activity of natural killer cells, causing them to become **lymphocyte-activated killer (LAK) cells.**

3. **Interleukin-3 (IL-3).** IL-3 is produced by CD4$^+$ cells and is a multilineage colony-stimulating factor. It promotes growth and differentiation of all cell types in the bone marrow.

4. **Interleukin-4 (IL-4).** IL-4 is produced mostly by T$_h$2 CD4$^+$ cells. It is a growth factor for B cells, some CD4$^+$ T cells, and mast cells, but it suppresses T$_h$1 differentiation.

5. **Interleukin-5 (IL-5).** IL-5 is produced by certain CD4$^+$ cells and stimulates activation of eosinophils so that they can kill some helminths. It also acts with IL-2 and IL-4 to stimulate growth and differentiation of B cells.

6. **Interleukin-6 (IL-6).** IL-6 is produced by macrophages, endothelial cells, fibroblasts, and T$_h$2 cells. It is an important growth factor for B cells late in the sequence of B-cell differentiation.

7. **Interleukin-8 (IL-8).** IL-8 is one of a family of low molecular weight inflammatory cytokines derived from antigen-activated T cells, activated macrophages, endothelial cells, fibroblasts, and platelets. IL-8 is an activating and chemotactic factor for neutrophils and, to a lesser extent, for other PMNs.

8. **Interleukin-10 (IL-10).** IL-10 is derived from T$_h$2 CD4$^+$ cells, and it inhibits T$_h$1, CD8$^+$, NK, and macrophage cytokine synthesis.

9. **Transforming growth factor-β (TGF-β).** TGF-β is produced by macrophages, lymphocytes, and other cells. It inhibits lymphocyte proliferation, CTL and LAK cell generation, as well as macrophage cytokine production.

10. **Interferon-γ (IFN-γ).** IFN-γ is produced by some CD4$^+$ and almost all CD8$^+$ cells. It is a strong macrophage-activating factor, causes a variety of cells to express class

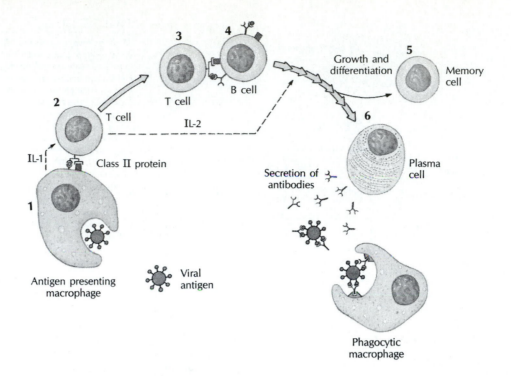

FIGURE 3.4

Humoral immune response. **1,** macrophage consumes antigen, partially digests it, displays epitope on its surface, along with class II MHC protein, and secretes interleukin-1 (**IL-1**); **2,** T helper cell, stimulated by **IL-1,** recognizes epitope and class II protein on macrophage, is activated, and secretes interleukin-2 (**IL-2**); **3,** T helper then activates B cell, which carries antigen and class II protein on its surface; **4,** activated B cell multiplies, finally producing many plasma cells that secrete antibody; **5,** some of B-cell progeny become memory cells; **6,** antibody produced by plasma cells binds to antigen and stimulates macrophages to consume antigen (opsonization).

From Cleveland P. Hickman, Jr. et al., *Integrated Principles of Zoology,* 9th edition, Copyright © 1993 by Mosby-Yearbook, Inc. Reprinted by permission of Times Mirror Higher Education Group, Inc., Dubuque, Iowa. All Rights Reserved.

II MHC molecules, promotes differentiation of T and B cells, activates neutrophils and NK cells, and activates endothelial cells to allow lymphocytes to pass through walls of vessels.

11. **Tumor necrosis factor (TNF).** Activated macrophages secrete most TNF, which is a major mediator of inflammation. In low concentrations TNF activates endothelial cells, activates PMNs, and stimulates macrophages and cytokine production (including IL-1, IL-6, and TNF itself). In higher concentrations TNF causes increased synthesis of prostaglandins in the hypothalamus, resulting in fever.

• Generation of a Humoral Response

When an antigen is introduced into the body, it binds to a specific antibody on the surface of the appropriate B cell, but this is usually insufficient to activate the B cell to multiply. Some of the antigen is taken up by **antigen-presenting cells (APCs),** such as macrophages,[38] that partially digest the antigen. The APCs then each incorporate portions of the antigen into their own cell surface (Figs. 3.3 and 3.4). That portion of the antigen presented on the surface of the macrophages or other APCs is called the **epitope** (or **determinant**). The macrophages also secrete IL-1, which activates appropriate T_h2 cells. These T cells recognize the epitopes on the surface of the macrophages in conjunction with the class II MHC protein. (Both the class II protein and the epitope must be present; neither is effective alone.) The T_h2 cells then secrete IL-4, IL-5, and IL-6, which activate the B cell that has the same epitope and a class II MHC protein on its surface. The B cell multiples rapidly and produces many **plasma cells,** which secrete large quantities of antibody for a period of time and then die. Thus, if we measure the concentration of the antibody (**titer**) soon after the antigen is injected, we can detect little or none. The titer then rises rapidly as the plasma cells secrete antibody, and it may decrease somewhat as they die and the antibody is degraded (Fig. 3.5). However, if we give another dose of antigen (the **challenge**), there is no lag, and the antibody titer rises quickly to a higher level than after the first dose. This is the **secondary** or **anamnestic response,** and it occurs because some of the activated B cells gave rise to long-lived **memory cells.** There are

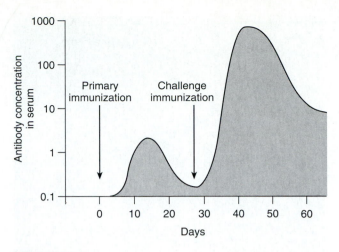

FIGURE 3.5

Typical immunoglobulin response after primary and challenge immunizations. Secondary response is result of large numbers of memory cells produced after primary B-cell activation.

From Cleveland P. Hickman, Jr. et al., *Integrated Principles of Zoology*, 8th edition, Copyright © 1988 by Times Mirror/Mosby College Publishing. Reprinted by permission of Times Mirror Higher Education Group, Inc., Dubuque, Iowa. All Rights Reserved.

many more memory cells present in the body than the original B lymphocyte with the appropriate antibody on its surface, and they rapidly multiply to produce additional plasma cells.

Functions of Antibody in Host Defense.   Antibodies can mediate destruction of an invader (antigen) in a number of ways. A foreign particle, for example, becomes coated with antibody molecules as their Fab regions become bound to it. Macrophages recognize the projecting Fc regions and are stimulated to engulf the particle. This is the process of **opsonization.**

Another important process, particularly in the destruction of bacterial cells, is the interaction with **complement.** Complement is a series of 12 enzymes that are activated by bound antibody. When activated they actually punch holes in the bacterial cell surface. Complement may also play a role in opsonization.

Antibody bound to the surface of an invader may trigger contact killing of the invader by host cells in what is known as **antibody-dependent, cell-mediated cytotoxicity (ADCC).** Eosinophils activated by IL-5 can be effector cells in ADCC. They, as well as lymphoid cells and neutrophils, can destroy bloodstream forms of *Trypanosoma cruzi* in the presence of antibody against this organism.[16,17] The parasites are phagocytized, and the granules in the eosinophils and neutrophils fuse with the phagosome and kill the *T. cruzi* with H_2O_2.[28,41] In the presence of antibody, neutrophils and eosinophils kill newborn juveniles of *Trichinella spiralis* by release of reactive oxygen intermediates, but adults and muscle-stage juveniles are much more resistant to ADCC,

apparently because they secrete antioxidant enzymes.[4] *Schistosoma mansoni* schistosomula (juveniles) are also killed by reactive oxygen intermediates released by neutrophils and eosinophils in the presence of antibody and complement.[4]

• The Cell-Mediated Response

Some immune responses involve little, if any, antibody and depend on the action of cells only. In cell-mediated immunity (CMI) the epitope of the antigen is also presented by macrophages, but the T_h1 arm of the immune response is activated and the T_h2 arm suppressed. Effector cells are macrophages, PMNs, cytotoxic T cells, and activated natural killer cells. The specific interaction of lymphocyte and antigen that generates a CMI greatly influences subsequent events in the nonspecific response we call **inflammation.** The immune response elicited by at least several parasitic worms is not directly responsible for their expulsion from the host. Rather, the expulsion is caused by the inhospitable chemical conditions in the ensuing inflammation that has been enhanced by the specific CMI.[19]

Like humoral immunity, CMI shows a secondary response due to large numbers of memory T cells produced from the original activation. For example, a second tissue graft (challenge) between the same donor and host will be rejected much more quickly than the first.

• Acquired Immune Deficiency Syndrome (AIDS)

AIDS is an extremely serious disease in which the ability to mount an immune response is crippled. The first case was recognized in 1981, and by the end of 1991, over 200,000 individuals, of which almost 138,000 had died, had contracted AIDS in the United States alone.[5] AIDS patients are continuously plagued by infections with agents (often parasites) that cause insignificant problems in persons with normal immune responses. The disease is caused by a virus that preferentially invades and destroys CD4+ lymphocytes. The CD4 surface protein is the major surface receptor for the virus.[45] Normally, CD4+ cells make up 60% to 80% of the T-cell population; in AIDS they can become too rare to be detected.[20] T_h1 cells are relatively more depleted than T_h2 cells, which upsets the balance of immunoregulation and results in persistent B-cell activation.[45]

• Inflammation

Inflammation is a vital process in the mobilization of the body defenses against an invading organism or other tissue damage and in the repair of damage thereafter. The course of events in the inflammatory process is greatly influenced by the prior immunizing experience of the body with the invader and by the duration of the invader's presence or its persistence in the body. The processes by which the invader is actually destroyed, however, are themselves nonspecific. Manifestations of inflammation are **delayed type hypersensitivity (DTH)** and **immediate hypersensitivity,** depending on whether the response is cell mediated or antibody mediated.

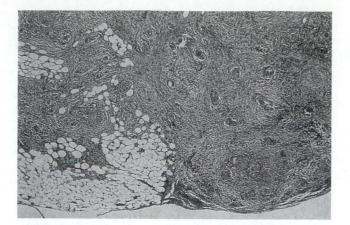

FIGURE 3.6

Granulomatous reaction around eggs of *Schistosoma mansoni* in mesenteries.

Courtesy of H. Zaiman. From H. Zaiman, editor. *A pictorial presentation of parasites.*

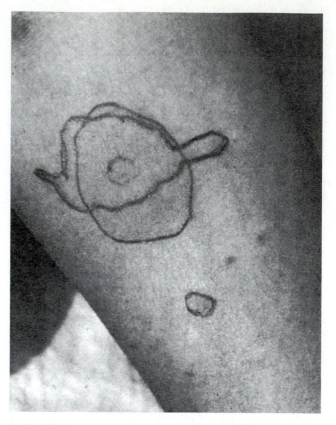

FIGURE 3.7

Immediate hypersensitivity reaction. An intradermal test for schistosomiasis. Limits of swelling are outlined. The two small responses are controls. The immediate (15-minute) response is the irregular outline at upper left, and the larger late-stage reaction has a smooth edge.

From I. G. Kagan and S. E. Maddison, in G. T. Strickland, ed., *Hunter's Tropical Medicine,* 7th ed. © 1991 W. B. Saunders Co.

The DTH reaction is a type of CMI in which the ultimate effectors are activated macrophages. The term *delayed type hypersensitivity* is derived from the fact that a period of 24 hours or more elapses between the time of antigen introduction and the response to it in an immunized subject. This is because the T_h1 cells with the receptors in their surface for that particular antigen require some time to arrive at the antigen site, recognize the epitopes displayed by the APCs (probably tissue macrophages), and become activated and secrete IL-2, TNF, and IFN-γ. TNF causes the endothelial cells of the blood vessels to express on their surface certain molecules to which leukocytes adhere: first neutrophils and then lymphocytes and monocytes. It also causes the endothelium to secrete inflammatory cytokines such as IL-8, which increase the mobility of leukocytes and facilitate their passage through the endothelium. The TNF and IFN-γ also cause the endothelial cells to change shape, favoring leakage of macromolecules and passage of cells. Escape of fibrinogen from the blood vessels leads to conversion of fibrinogen to fibrin, and the area becomes swollen and firm.

As the monocytes pass out of the blood vessel, they become activated macrophages, which are the main effector cells of the DTH. They phagocytize particulate antigen, secrete mediators that promote local inflammation, and secrete cytokines and growth factors that promote healing. If the antigen is not destroyed and removed, its chronic presence leads to deposition of fibrous connective tissue or **fibrosis.** Nodules of inflammatory tissue called **granulomas** may accumulate around persistent antigen and are found in numerous parasitic infections (Fig. 3.6).

Immediate hypersensitivity is quite important in some parasitic infections.[21] This reaction involves degranulation of mast cells in the area. Their surfaces bear receptors for the Fc portions of antibody, especially IgE. Occupation of these sites by antigen-specific antibodies enhances degranulation of the mast cells when the Fab portions bind the particular antigen. There is a rapid release of several mediators, such as histamine, that cause dilation of local blood vessels and increased vascular permeability. Escape of blood plasma into the surrounding tissue causes swelling **(wheal),** and engorgement of vessels with blood produces the characteristic **flare** (Fig. 3.7). Although the wheal and flare of many immediate hypersensitivity reactions resolves in about an hour, some elicit a **late phase reaction** at two to four hours. PMNs and monocytes accumulate, and the reaction resembles DTH. Immediate hypersensitivity in humans is the basis for allergies and asthma, and some workers believe that the allergic response originally evolved to help the body ward off parasites.[22] Systemic immediate hypersensitivity is **anaphylaxis,** which may be fatal if not treated rapidly.

Some degree of cell death **(necrosis)** always occurs, but necrosis may not be prominent in minor inflammation. If the necrotic debris is confined within a localized area, the pus (spent leukocytes and tissue fluid) may increase in hydrostatic pressure, forming an **abscess.** An area of inflammation that opens out to a skin or mucous surface is an **ulcer.**

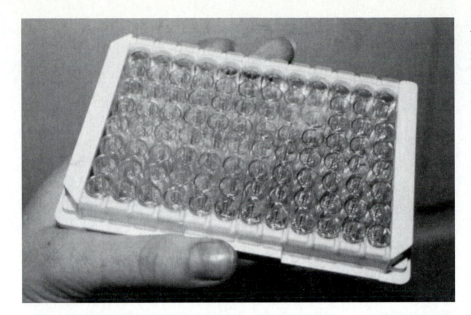

FIGURE 3.8

A microplate for an ELISA test. In addition to positive and negative controls, wells are available on the plate for testing several individuals. Positive controls are wells in which antibody is known to be present, and negative controls omit the enzyme-linked anti-Ig.

Photograph by Larry S. Roberts.

Innate Immunity in Vertebrates

Under this heading one may place the "accidents" of a host's structural and physiological characteristics that reduce its susceptibility to certain parasites as well as "natural" defense adaptations. Examples of the first category include physical barriers, such as a thick, cornified epidermis or other protective external covering; the ability to repair damaged tissues rapidly; and high acidity in the stomach. Such mechanisms are not always easy to distinguish from some "natural" defense adaptations, and the distinction may not be useful. Natural defense adaptations include the nonspecific physiological characteristics possessed by a wide range of hosts, evolved as adaptations for defense and effected on first encounter with a wide range of invading organisms. A variety of parasiticidal substances are present in animal body secretions. In fact, at least one such substance that we previously thought was nonspecific we now know is a class of antibody, IgA. IgA can cross cellular barriers easily. It seems to be an important protective agent in the mucus of the intestinal epithelium, and it is present in mucus in the respiratory tract, in tears, in saliva, and in sweat. Other substances in normal human milk can kill intestinal protozoa such as *Giardia lamblia* and *Entamoeba histolytica,* and these substances may be important in protection of infants against these and other infections.[11] The nature of the substances is not yet known, but they seem to be neither IgA nor another antibody.

Some species of mammals are susceptible to infection with *Schistosoma mansoni,* and others are partially or completely resistant. Without mediation by antibody, macrophages of more resistant species (rats, guinea pigs, rabbits) kill schistosomules in vitro, but macrophages of susceptible species do not.[23]

IMMUNODIAGNOSIS

Although we diagnose many parasitic infections most easily by finding the parasites themselves or their products, such as eggs, in other infections the organisms may be difficult to demonstrate. Thus, numerous tests have been developed that take advantage of the immune response of the patient. Space does not permit a thorough treatment here, but every parasitology student should be aware of these extremely valuable diagnostic tools. You should also be aware of some difficulties. For example, false positives may arise when two related agents have antigens in common or similar enough to cross-react with antibodies raised against the other. This is the case with skin tests, in which a small amount of antigen is injected into the skin of the patient. Many parasitic infections produce immediate or delayed type hypersensitivity reactions, which are easily observed (Fig. 3.7).[13]

The **enzyme-linked immunosorbent assay (ELISA)** and its variants have become quite popular in recent years. They are good diagnostic tests and serve as powerful research tools. They are simple to perform and usually do not require sophisticated equipment. In the assay a small quantity of antigen is adsorbed to the bottom of a small cup in a plastic microplate (Fig. 3.8). Next, a portion of the serum to be tested is added to the cup (Fig. 3.9). The serum is removed, and the cup is rinsed several times. If the serum had contained antibodies to the antigen, they would have bound to the antigen and would not have been removed by the rinses. A solution containing antibodies to human Ig (anti-Ig) is added. The anti-Ig must be prepared beforehand and linked covalently to an enzyme. The enzyme can be any one of several whose reaction product is colored. This solution is then removed from the cup, the cup is rinsed again, and the substrate for the enzymatic reaction is added. If the tested serum contains antibodies against the antigen, the anti-Ig will be bound to them, and the enzymatic reaction will occur, producing a color.

A variation on the ELISA just described is the "sandwich" ELISA, which can detect parasite antigen, rather than host antibody. In this case, antibody to the antigen in question is adsorbed to the plastic cup, the unknown is added, the cup is rinsed, and additional antibody linked to an enzyme is added. Formation of a color indicates a positive result. Adaptations of the sandwich ELISA use a "dipstick" of acetate plastic, with the antibody adsorbed to a film of nitrocellulose.[2] This method can detect antigens of intestinal parasites in the patient's feces and is very convenient to use in the field.

PATHOGENESIS OF PARASITIC INFECTIONS

The pathogenic effects of a parasitic infection may be so subtle as to be unrecognizable, or they may be strikingly obvious. An apparently healthy animal may be host to hundreds of parasitic worms and yet show no obvious signs of distress. On the other hand, another host may be so anemic, unthrifty, and stunted that parasites are undoubtedly the reason for its sad state. The pathogenic effects of parasites are many and varied, but for the sake of convenience they can be discussed under the headings of trauma, nutrition robbing, and interactions of the host immune responses.

Physical trauma, or destruction of cells, tissues, or organs by mechanical or chemical means, is common in parasite infections. When an *Ascaris* or hookworm juvenile penetrates a lung capillary to enter an air space, it damages the blood vessel and causes hemorrhage and possible infection by bacteria that may have been inhaled. The hookworm, after completing its migration to the small intestine, feeds by biting deeply into the mucosa and sucking blood and causing anemia in heavy infections. The dysentery ameba *Entamoeba histolytica* digests away the mucosa of the large intestine, forming ulcers and abscessed pockets that can cause severe disease. These are but a few examples of known physical trauma caused by parasites. Many are discussed in later chapters in conjunction with the particular parasite involved.

A less obvious but often pernicious pathogenic situation is diversion of the host's nutritive substances. Although most tapeworms absorb so little food in proportion to the amount eaten by the host that the host still manages very well, the broad fish tapeworm *Diphyllobothrium latum* has such strong affinity for vitamin B_{12} that it absorbs large amounts from the intestinal wall and contents of its host. Since B_{12} is necessary for erythrocyte production, a severe anemia may result. The large nematode *Ascaris lumbricoides* inhabits the small intestine—often in large numbers—and consumes a good deal of food the host intends for itself. One study showed 21% greater weight gain in children treated for *Ascaris* infection compared with those whose infections went untreated.[46] Other studies showed that removal of the nematode *Trichuris trichiura* resulted in significant improvements in long- and

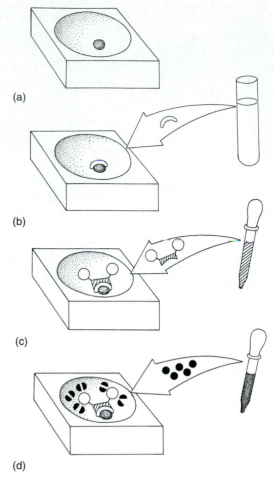

(a)

(b)

(c)

(d)

FIGURE 3.9

Sequence of steps in performance of an ELISA test. (*a*) Known antigen is adsorbed to bottom of microplate well; (*b*) serum from patient is added and then well is rinsed; (*c*) enzyme-linked antibody against human immunoglobulin is added and then well is rinsed again; (*d*) enzyme substrate is added. If colored products of enzyme reaction are observed, this indicates presence of bound anti-Ig, which, in turn, indicates presence of antibody against antigen. Thus, test is positive.

Drawing by William Ober and Claire Garrison.

short-term memory and much higher rate of growth in children.[8,27] The most important forms of malnutrition are aggravated by infection with these and other helminths.[35] We should note that helminths could contribute to malnutrition by decreasing host nutrient intake, increasing nutrient excretion, and/or decreasing nutrient utilization.[35] The tiny protozoan *Giardia* robs its host in a different way. It is concave on its ventral surface and applies this suction cup to the surface of an intestinal epithelial cell. When many of these parasites are present, they cover so much intestinal absorptive surface that they interfere with the host's absorption of nutrients. The unused nutrients then pass uselessly through the intestine and are wasted.

In recent years we have come to realize that a great many, perhaps the most serious and most pervasive, pathogeneses are actually caused by the host's own defense system: the immune response and inflammation.[31] A number of cases previously thought due to toxins released by the parasite are now understood as caused by the host's reaction to parasite products. For example, the protozoan *Trypanosoma cruzi* develops clusters of cells in the smooth and cardiac muscle cells of its host, and when the parasites degenerate—sometimes years later—the inflammatory response damages the supporting cells of the nerve ganglia that control peristalsis and heart contraction. Parasite antigens on the host's own cells, particularly in the endocardium, cause autoimmune reactions, and the host's cells are attacked as foreign by the immune system.[37] Some of the large amount of antigen-antibody complex formed in infections with the African trypanosomes (*T. brucei rhodesiense* and *T. b. gambiense*) adsorbs to the host's red blood cells, activating complement and causing lysis with resulting anemia.[40]

The flow of blood carries many of the eggs laid by schistosomes to the liver where they lodge, leaking antigen and causing a chronic DTH reaction. The formation of granulomas around the eggs eventually impedes blood flow through the liver, resulting in cirrhosis and portal hypertension.[31] Adults of the filarial nematode *Onchocerca volvulus* live in the dermis of humans. They release live juveniles, many of which wander into the eyes, including the cornea. Each generating juvenile in the cornea becomes a focus of inflammation, and over time sclerosing keratitis (hardening inflammation of the cornea) and other complications cause permanent blindness.[15] Today there are villages in Africa and Central America where the majority of adults are blind because of this parasite.

These and many other diseases to be discussed in context are examples of the immune response gone wrong. We could scarcely do without the defenses of our immune system, but some manifestations of the immune response are responsible for much of the pathogenesis hosts suffer.

ACCOMMODATION AND TOLERANCE IN THE HOST-PARASITE RELATIONSHIP

Successful parasites have had to evolve one or more tactics to avoid the defenses of a given host. Otherwise the host simply would not be susceptible. In recent years we have found that parasites display an astonishing array of such tactics (Table 3.2). We will describe only a few examples; for many others and more information see Warren,[44] Goodenough,[12] and the volume introduced by Mitchell.[25]

The location of the parasite may provide some protection against host defenses. The lumen of the intestine is one such site. Although IgA is secreted into the intestine, IgA is not a very potent effector molecule against worms, and complement and phagocytic cells are normally not found in the intestine. However, the rat nematode *Nippostrongylus braziliensis* can be expelled because inflammation and an immediate hypersensitivity reaction change the permeability of the mucosa and apparently allow IgG to leak into the lumen.[42] Many other intestinal parasites, not provoking such inflammation, are relatively long-lived. Numerous parasites, such as juvenile tapeworms (cestodes) in various tissues, achieve protection from the host response by the development of a cyst wall. Others may be shielded by their location within a host cell. Recognition of the infected cell by the host's cell-mediated effector systems is precluded if no parasite antigens are present in the outer membrane of the infected cell, as seems to be the case in liver cells infected with malaria parasites.

Parasites that are constantly or frequently bathed in blood would seem particularly vulnerable to the range of host defenses, but they have evolved fascinating mechanisms for evasion. The protozoa causing African trypanosomiasis display a "moving target"—that is, a continuing succession of variant antigenic types—so that just as the host mounts an antibody response to one, another type proliferates. We will describe this phenomenon further in Chapter 5. Other important mechanisms of evasion are present in these infections as well. Antibody and cell-mediated responses are suppressed, apparently by some substance secreted by the trypanosomes. Suppression may be achieved by polyclonal B-cell activation early in the infection; many subtypes of B cells are stimulated to divide, leading to the production of nonspecific IgG and autoantibodies.[40] There is a suppression of IL-2 secretion and expression of IL-2 receptors, and T cells become refractory to normal signals.[40]

Visceral leishmaniasis, caused by other protozoa, shows a kind of immunosuppression by misdirection of the immune response. The organisms initially infect macrophages near the site of the infection and then invade cells of the reticuloendothelial system throughout the body. The CMI arm of the immune response is necessary to control proliferation of the protozoa; patients with positive DTH reaction to leishmanial antigens successfully resolve the infection. In other patients, however, there is a strong humoral response, and the CMI is suppressed.[30] In these patients continued reproduction of the parasites eventually leads to death (if untreated).

In addition to immunosuppression, polyclonal lymphocyte activation, and other mechanisms, the blood fluke *Schistosoma* actually adsorbs many host antigens so that the host immune system "sees" only self, not recognizing the parasite as foreign.[26] For example, if adult worms are removed from mice and transferred surgically to monkeys, the worms stop producing eggs for a time but then recover and resume normal egg production. However, if the worms from mice are transferred to a monkey that has been previously immunized against mouse red blood cells, the worms are promptly destroyed. Interestingly, several antischistosomal drugs compromise the effectiveness of the worms' immune evasion. Praziquantel, for example, at concentrations too low to be directly lethal to the schistosomes, allows immunological destruction. The drug apparently alters the architecture of the tegumental surface, exposing epitopes to the immune system that are normally sequestered beneath host antigens.[26]

Table 3.2   *Mechanisms favoring immune evasion in human parasitic infections*

Disease	Parasite antigens							Modification of host immune responsiveness			
	Anatomical seclusion	Stage specificity	Antigenic variation	Shedding and renewal	Antigenic disguise	Antibody cleavage	Complement consumption	Modified leukocyte function	Immuno-suppression	Polyclonal lymphocyte activation	Circulating immune complexes
Amebiasis				+							
Giardiasis	+										
African trypanosomiasis	+		+						+	+	+
South American trypanosomiasis				+		+			+		
Leishmaniasis	+			+					+	+	+
Toxoplasmosis	+	+							+		
Malaria	+		+	+				+	+	+	+
Babesiosis			+		+				+	+	
Schistosomiasis				+	+	+	+	+	+	+	+
Fascioliasis				+	+			+			
Filariasis	+								+		
Onchocerciasis							+		+		
Trichinosis	+	+		+					+		
Cestode infections									+		
Nematode infections		+							+		

From S. Cohen, *Immunology of parasitic infections*, 2d ed. pp. 138–161. Copyright © 1982 Blackwell Scientific Publications, Ltd. Reprinted by permission. (See Cohen's article for references)

References

1. Abbas, A. K., A. H. Lichtman, and J. S. Pober. 1991, *Cellular and molecular immunology.* Philadelphia: W. B. Saunders Company.

2. Allan, J. C., F. Mencos, J. Garcia-Noval, E. Sarti, A. Flisser, Y. Wand, D. Liu, and P. S. Craig. 1993. Dipstick dot ELISA for the detection of *Taenia* coproantigens in humans. *Parasitology* 107:79–85.

3. Amen, R. I., and M. De Jong-Brink. 1992. *Trichobilharzia ocellata* infections in its snail host *Lymnaea stagnalis:* an *in vitro* study showing direct and indirect effects on the snail internal defence system, via the host central nervous system. *Parasitology* 105:409–16.

4. Callahan, H. L., R. K. Crouch, and E. R. James. 1988. Helminth anti-oxidant enzymes: a protective mechanism against host oxidants? *Parasitol. Today* 4:218–25.

5. Centers for Disease Control. 1992. Summary of notifiable diseases, United States, 1991. *Morb. Mortal. Weekly Rep.* 40(53):1–63.

6. Cheng, T. C., M. J. Chorney, and T. P. Yoshino. 1977. Lysosome-like activity in the hemolymph of *Biomphalaria glabrata* challenged with bacteria. *J. Invertebr. Pathol.* 29:170–74.

7. Cohen, S. 1982. Survival of parasites in the immunocompetent host. In Cohen, S., and K. S. Warren, eds. *Immunology of parasitic infections,* 2d ed. Oxford: Blackwell Scientific Publications Ltd., 138–61.

8. Cooper, E. S., C. A. M. Whyte-Alleng, and J. S. Finzi-Smith. 1992. Intestinal nematode infections in children: The pathophysiological price paid. *Parasitology* 104:S91–S103.

9. Cox, F. E. G., and E. Y. Liew. 1992. T-cell subsets and cytokines in parasitic infections. *Parasitol. Today* 8:371–74.

10. Englund, P. T., and Sher, A. 1988. *The biology of parasitism.* New York: Alan R. Liss, Inc.

11. Gillen, F. D., D. S. Reiner, and Ch. -S. Wang. 1983. Human milk kills parasitic intestinal protozoa. *Science* 221:1290–91.

12. Goodenough, U. W. 1991. Deception by pathogens. *Am. Sci.* 79:344–55.

13. Kagan, I. G., and S. E. Maddison. 1991. Parasitic immunodiagnosis. In Strickland, G. T., ed. *Hunter's tropical medicine,* 7th ed. Philadelphia: W. B. Saunders Company.

14. Kassim, O. O., and C. S. Richards. 1978. *Biomphalaria glabrata:* Lysozyme activities in the hemolymph, digestive gland, and headfoot of the intermediate host of *Schistosoma mansoni. Exp. Parasitol.* 46:218–24.

15. Kazura, J. W., T. B. Nutman, and B. M. Greene. 1993. Filariasis. In Warren, K. S. ed. *Immunology and molecular biology of parasitic infections,* 3d ed. Boston: Blackwell Scientific Publications.

16. Kierszenbaum, F. 1979. Antibody-dependent killing of bloodstream forms of *Trypanosoma cruzi* by human peripheral blood leukocytes. *Am. J. Trop. Med. Hyg.* 28:965–68.

17. Kierszenbaum, F., S. J. Ackerman, and G. J. Gleich. 1981. Destruction of bloodstream forms of *Trypanosoma cruzi* by eosinophil granule major basic protein. *Am. J. Trop. Med. Hyg.* 30:775–79.

18. Lackie, A. M. 1980. Invertebrate immunity. *Parasitology* 80:393–412.

19. Larsh, J. E. Jr., and N. F. Weatherly. 1975. Cell-mediated immunity against certain parasitic worms. In Dawes, B., ed. *Advances in parasitology* 13. New York: Academic Press, Inc., 183–222.

20. Laurence, J. 1985. The immune system in AIDS. *Sci. Am.* 253(6):84–93 (Dec.).

21. Lee, T. D. G., M. Swieter, and A. D. Befus. 1986. Mast cell responses to helminth infection. *Parasitol. Today* 2:186–91.

22. Lichtenstein, L. M. 1993. Allergy and the immune system. *Sci. Am.* 269(3):116–24 (Sept.).

23. Mahmoud, A. A. F. 1993. Mononuclear phagocytes and resistance to parasitic infections. In Warren, K. S. ed. *Immunology and molecular biology of parasitic infections,* 3d ed. Boston: Blackwell Scientific Publications.

24. Maramorosch, K., and R. E. Shope, eds. 1975. *Invertebrate immunity. Mechanisms of invertebrate vector-parasite relations.* New York: Academic Press, Inc.

25. Mitchell, G. F. 1991. Co-evolution of parasites and adaptive immune responses. In Ash, C., and R. B. Gallagher, eds. *Immunoparasitol. Today* 7:A2–A5. Cambridge: Elsevier Trends Journals.

26. Newport, G. R., and D. G. Colley, 1993. Schistosomiasis. In Warren, K. S. ed. *Immunology and molecular biology of parasitic infections,* 3d ed. Boston: Blackwell Scientific Publications.

27. Nokes, C., S. M. Grantham-McGregor, A. W. Sawyer, E. S. Cooper, B. A. Robinson, and D. A. P. Bundy. 1992. Moderate to heavy infections of *Trichuris trichiura* affect cognitive function in Jamaican school children. *Parasitology* 104:539–47.

28. Okabe, K., T. L. Kipnis, V. L. G. Calish, and W. D. daSilva. 1980. Cell-mediated cytotoxicity to *Trypanosoma cruzi.* I. Antibody-dependent cell-mediated cytotoxicity to trypomastigote bloodstream forms. *Clin. Immunol. Immunopathol.* 16:344–53.

29. Paul, W. E. 1993. Infectious diseases and the immune system. *Sci. Am.* 269(3):90–97 (Sept.).

30. Pearson, R. D., G. Cox, S. M. B. Jeronimo, J. Castracane, J. S. Drew, T. Evans, and J. E. de Alencar. 1992. Visceral leishmaniasis: a model for infection-induced cachexia. *Am. J. Trop. Med. Hyg.* 47(suppl.):8–15.

31. Phillips, M. 1993. Mechanisms of immunopathology in parasitic infections. In Warren, K. S. ed. *Immunology and molecular biology of parasitic infections,* 3d ed. Boston: Blackwell Scientific Publications.

32. Prescott, L. M., J. P. Harley, and D. A. Klein. 1993. *Microbiology,* 2d ed. Dubuque, Iowa: Wm. C. Brown Publishers.

33. Richards, C. S., M. Knight, and F. A. Lewis. 1992. Genetics of *Biomphalaria glabrata* and its effect on the outcome of *Schistosoma mansoni* infection. *Parasitol. Today* 8:171–74.

34. Sher, A., R. T. Gazzinelli, I. P. Oswald, M. Lerici, M. Kullberg, E. J. Pearce, J. A. Berzofsky, T. R. Mosmann, S. L. James, J. C. Morse III, and G. M. Shearer. 1992. Role of T-cell derived cytokines in the downregulation of immune responses in parasitic and retroviral infection. *Immunol. Rev.* 127:183–204.

35. Stephenson, L. S. 1987. *The impact of helminth infections on human nutrition. Schistosomes and soil-transmitted helminths.* London: Taylor and Francis.

36. Sun, S. -C., I. Lindström, H. G. Boman, I. Faye, and O. Schmidt. 1990. Hemolin: An insect-immune protein belonging to the immunoglobulin superfamily. *Science* 250:1729–32.

37. Tarleton, R. L. 1993. Pathology of American trypanosomiasis. In Warren, K. S. ed. *Immunology and molecular biology of parasitic infections,* 3d ed. Boston: Blackwell Scientific Publications.

38. Unanue, E. R., and P. M. Allen. 1987. The basis of the immunoregulatory role of macrophages and other accessory cells. *Science* 236:551–57.

39. van der Knaap, W. P. W., and E. S. Loker. 1990. Immune mechanisms in trematode-snail interactions. *Parasitol. Today* 6:175–82.

40. Vickerman, K., P. J. Myler, and K. D. Stuart. 1993. African trypanosomiasis. In Warren, K. S. ed. *Immunology and molecular biology of parasitic infections,* 3d ed. Boston: Blackwell Scientific Publications.

41. Villalta, F., and F. Kierszenbaum. 1984. Role of inflammatory cells in Chagas' disease. Uptake and mechanisms of destruction of intracellular (amastigote) forms of *Trypanosoma cruzi* by human eosinophils. *J. Immunol.* 132:2053.

42. Wakelin, D., W. Harnett, and R. M. E. Parkhouse. 1993. Nematodes. In Warren, K. S. ed. *Immunology and molecular biology of parasitic infections,* 3d ed. Boston: Blackwell Scientific Publications.

43. Wang, C. C., ed. 1991. *Molecular and immunological aspects of parasitism.* Washington, D.C.: American Association for the Advancement of Science.

44. Warren, K. S., ed. 1993. *Immunology and molecular biology of parasitic infections,* 3d ed. Boston: Blackwell Scientific Publications.

45. Weiss, R. A. 1993. How does HIV cause AIDS? *Science* 260:1273–79.

46. Willett, W. C., W. L. Kilama, and C. M. Kihamia. 1979. *Ascaris* and growth rates: A randomized trial of treatment *Am. J. Public Health* 69:987–91.

47. Young, J. D., and Z. A. Cohn, 1988. How killer cells kill. *Sci. Am.* 258 (1):38–44 (Jan).

Additional References

Abbas, A. K., A. H. Lichtman, and J. S. Pober, 1994. *Cellular and molecular immunology,* 2d ed. Philadelphia: W. B. Saunders Co. Concise, very helpful reference for modern immunology.

Engelhard, V. H. 1994. How cells process antigens. *Sci. Am.* 271(2):54–61 (Aug.). This article focuses on the roles of the MHC proteins.

Chapter 4

PARASITIC PROTISTANS: FORM, FUNCTION, AND CLASSIFICATION

My excrement being so thin, I was at divers times persuaded to examine it; and each time I kept in mind what food I had eaten, and what drink I had drunk, and what I found afterwards. I have sometimes seen animalcules a-moving very prettily. . . .

A. van Leeuwenhoek (November 4, 1681)

Because of the small size of most of them, protozoa were not detected until Antony van Leeuwenhoek developed his microscopes in the seventeenth century. He recounted his discoveries to the Royal Society of London in a series of letters covering a period between 1674 and 1716. Among his observations were oocysts of a parasite in the livers of rabbits, the species known today as *Eimeria stiedai.* Another 154 years passed before the second sporozoan was found, when in 1828 Delfour described gregarines from the intestines of beetles. Leeuwenhoek also found *Giardia lamblia* in his own diarrheic stools, and he found *Opalina* and *Nyctotherus* in the intestines of frogs. By the middle of the eighteenth century other parasitic protozoa were being reported at a rapid rate, and such discoveries have continued unabated to the present. At least 45,000 species of protozoa have been described to date, many of which are parasitic. Parasitic protozoa still kill, mutilate, and debilitate more people in the world than any other group of disease organisms. For this reason, studies on protozoa occupy a prominent place in the history of parasitology.

The word *Protozoa* was once a phylum name. Today, however, the term is used, almost colloquially, as a common noun that refers to a number of phyla. This evolution in the use and meaning of *protozoa* has come about largely as a result of electron microscopy. Ultrastructural research and the accompanying life cycle, genetic, and biochemical work have all shown that organisms once thought to be basically similar are in fact highly diverse and are organized structurally along a number of distinct lines. Thus, most modern texts list at least seven phyla of protozoa.

FORM AND FUNCTION

Protozoa consist of a single cell, although many species contain more than one nucleus during all or portions of their life cycles, and certain stages, such as spores, may be built from more than one cell. By the middle of the nineteenth century,

many genera of protozoa had been described, and their enormous structural diversity, complexity, and even beauty were widely recognized.[15] Early electron microscopists, as had the light microscopists, found the protozoa fascinating subjects, and soon after World War II, researchers recognized that the group was a large, heterogenous assemblage whose members did not all conform to a single body plan (Fig. 4.1). In 1980, a committee of the Society of Protozoologists revised the classification, recognizing seven phyla, and further revisions were recommended by a similar group of experts in 1985.[16,19]

Studies of the ultrastructure of protozoa have shown, however, that regardless of how elaborate or elaborately arranged they are, most protozoan organelles do not differ in any basic way from those of metazoan cells. Indeed, Pitelka[25] concluded "that the fine structure of protozoa is directly and inescapably comparable with that of cells of multicellular organisms," and the "morphologist has to start out by admitting that protozoa are, at the least, cells." On the other hand, to the student who first encounters them, protozoan structures can seem to be bizarre, often multiple versions of the familiar mitochondria, microtubules, flagella, and membranes studied in introductory biology class. If it seems as if the words *may, usually, typically,* and *often* occur more frequently in this chapter than in others, then such use is a reflection of the enormous diversity, even at the subcellular level, found in protozoa.

Nucleus and Cytoplasm

Like all cells, the bodies of protozoa are covered by a trilaminar **plasma membrane,** which should be thought of as the fluid mosaic described in virtually all introductory biology texts. Many protozoa have more than one such membrane as part of their **pellicle.** The additional membranes may be present as **alveoli,** or sacs, which in some ciliates are enlarged, producing ridges and craters on the cell surface. In addition, protozoa may possess a thick **glycocalyx,** or **glycoprotein surface coat,** which, in the case of parasitic forms, has immunological importance (see Chapter 5). Other membrane proteins may serve as binding sites that function during uptake of intracellular parasites by their host cells.

Pellicular microtubules, or fibrils, may course just beneath the plasma membrane, the number and arrangement of such tubules being typical of a group. The pellicle may be thrown into more or less permanent folds, supported by microtubules, as in the gregarine parasites of insects (Fig. 4.2). Or, such microtubules may underlie a flexible membrane, as in kinetoplastid

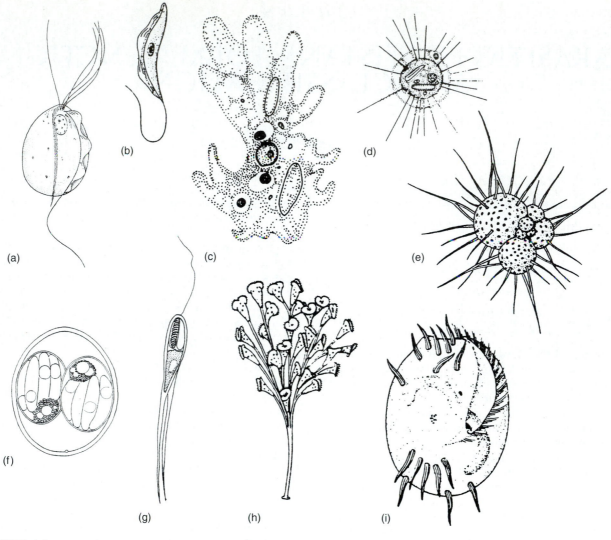

FIGURE 4.1

Representative protozoa showing some of the diversity exhibited by members of the various phyla. They are not drawn to scale, nor are they all the same life-cycle stages. (*a*) *Pentatrichomonas hominis,* 8–20 μm long, a harmless commensal of the human digestive tract; (*b*) *Trypanosoma,* 15–30 μm, from the bloodstream of vertebrates (both *a* and *b* have undulating membranes); (*c*) free-living *Amoeba* sp., 100–150 μm, showing lobopodia; (*d*) *Actinosphaerium,* 200 μm (many species are much smaller), with actinopodia; (*e*) *Globigerina,* a marine foraminiferan up to 800 μm, with filopodia; (*f*) an oocyst of *Levineia canis,* (35–42) × (27–33) μm, a coccidian parasite of dogs; (*g*) *Henneguya* sp., spore, 65 μm, a myxozoan parasite of fish (extruded filament is not a flagellum; see Chapter 10); (*h*) *Epistylis* colony, individuals 50–60 μm, colony up to 2 mm tall, an obligate ectocommensal ciliate of aquatic invertebrates; (*i*) *Euplotes* sp., 100–170 μm, a free-living ciliate with ventral cirri and prominent oral membranes.

(*b, e, h,* and *i*) From Cleveland P. Hickman, Jr. et al.: *Integrated Principles of Zoology,* 9th edition,. Copyright © 1993 by Mosby-Yearbook, Inc. Reprinted by permission of Times Mirror Higher Education Group, Inc., Dubuque, Iowa. All Rights Reserved. (*c* and *d*) Source: R. Kudo, *Protozoology,* 5th ed. 1966, Charles C. Thomas Publishers, Springfield, Il. (*f*) From N. D. Levine and V. Ivens, "*Isopora* species in the dog," in *J. Parasitol.* 51:859–864. Copyright © 1965. Rerpinted by permission of the publisher.

flagellates (Fig. 5.2, p. 54). The structural elaboration of membranes, through folding and addition of electron-dense materials, also occurs in tissue-dwelling cysts of parasitic protozoa. Adjoining membranes may have an electron-dense or fibrous connection between them, such as that between the body and **undulating membrane** of trypanosomes and trichomonads (Fig. 4.1, [*a, b*], Figs. 5.2, and 6.12).

Protozoa possess a great diversity of membranous or-ganelles in their cytoplasm. **Mitochondria,** the organelles that bear the enzymes of oxidative phosphorylation and the

tricarboxylic acid cycle, often have tubular rather than lamellar cristae in protozoa, although the cristae may be ab-sent altogether. Mitochondria may be present as a single, large body, as in some flagellates, or arranged as elongated, sausage-shaped structures, as in the pellicular ridges of some ciliates. The **Golgi apparatus** is quite elaborate in some flagellates, occurring as large and/or multiple **parabasal bodies** in association with kinetosomes, the "basal bodies" of flagella (Fig. 4.3), and is present, although not always as prominent, in amebas and ciliates. The Golgi

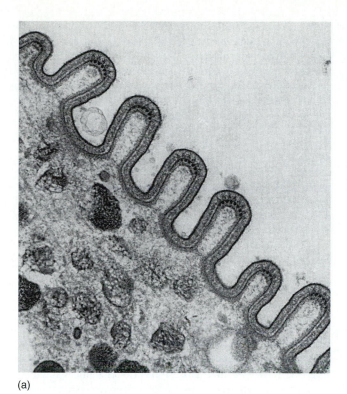

(a)

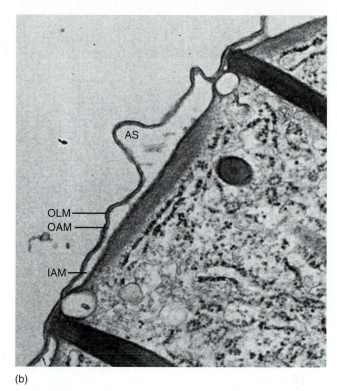

(b)

FIGURE 4.2

Plasma membranes and their modifications in protozoa. (*a*) Epicytic folds of a gregarine parasite of damselflies; (*b*) membranes of *Ichthyophthirius multifiliis,* parasite of fishes; the dark elongate bodies perpendicular to the membranes are mucocysts (**AS,** alveolar sac; **OLM,** outer limiting membrane; **OAM,** outer alveolar membrane; **IAM,** inner alveolar membrane).

(*a*) Courtesy of Tami Percival,; (*b*) From G. B. Chapman and R. C. Kern, "Ultrastructural aspects of the somatic cortex and contractile vacuole of the ciliate *Ichthyophthirius multifiliis* Fouquet," in *J. Protozool.* 30:481–490. Copyright © 1983 The Society of Protozoologists.

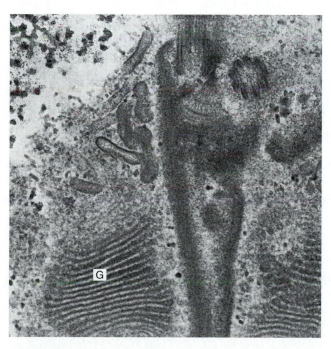

FIGURE 4.3

G, large Golgi apparatus as a "parabasal body" in a flagellate, *Proteromonas* sp., from a lizard.

Courtesy of Kit Lee.

can play diverse roles in the lives of protozoa—for example, as the source of skeletal plates in some amebas and polar filaments in microsporidian parasites.

Protozoa also possess a wide variety of membrane-bound bodies with diverse functions. **Microbodies** are usually, but not always, spherical structures with a dense, granular matrix.[8] In most animal and many plant cells the microbodies contain oxidases and catalase. The oxidases reduce oxygen to hydrogen peroxide, and the catalase decomposes the hydrogen peroxide to water and oxygen. Thus, the microbodies in these cells are called **peroxisomes** because of their biochemical activity. Peroxisomes are found in many aerobic protozoa,[22] protozoa in which oxygen is a terminal electron acceptor in their metabolism. In at least some anaerobic protozoa, such as the parasitic *Trichomonas* spp., the microbodies produce molecular hydrogen and are called **hydrogenosomes** (Figs. 4.4 and 4.5). Microbodies may also contain enzymes of the **glyoxylate cycle,** which functions in the synthesis of carbohydrate from fat. **Glycosomes** are microbodies of Kinetoplastida and contain most of the glycolytic enzymes (which in other eukaryotic cells are found in the cytosol).[21,24]

Other, more unusual, membrane-bound organelles in protozoa include **extrusomes,** which lie beneath the cell membrane and upon proper stimulus, fuse with that membrane, releasing their contents to the exterior. Extrusomes may release toxic substances (**toxosomes**) or function in food capture, (**kinetocysts**),

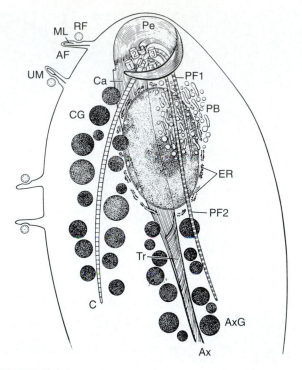

FIGURE 4.4

A composite schematic diagram of a trichomonad flagellate seen from a dorsal and slightly right view. **AF,** accessory filament; **AxG,** paraxostylar granules (hydrogenosomes); **C,** costa; **Ca,** capitulum of the axostyle; **CG,** paracostal granules (hydrogenosomes); **ER,** endoplasmic reticulum; **ML,** marginal lamella; **Pe,** pelta; **PB,** parabasal body; **PF1** and **PF2,** parabasal filaments; **R,** kinetosome of recurrent flagellum; **RF,** recurrent flagellum; **Tr,** trunk of the axostyle; **UM,** undulating membrane; **1** to **4,** kinetosomes of the anterior flagella.

From C. F. T. Mattern et al., "The mastigont system of *Trichomonas gallinae* (Rivolta) as revealed by electron microscopy," in *J. Protozool.* 14:320–339. Copyright © 1967 The Society of Protozoologists. Reprinted by permission.

to paralyze prey **(haptocysts),** or in protection **(trichocysts).** Not all extrusomes, however, have obvious functions; the dark (electron-dense), elongated bodies perpendicular to the cell membrane in Figure 4.2*b* are **mucocysts** of the parasitic ciliate *Ichthyophthirius multifiliis.* Mucocysts are thought to provide a coating that protects the cell against osmotic shock.[4]

The cytoplasmic matrix consists of very small granules and filaments suspended in a low-density medium with the physical properties of a colloid. Central and peripheral zones of cytoplasm can often be distinguished as the **endoplasm** and the **ectoplasm.** The endoplasm is in the **sol state** of the colloid, and it bears the nucleus, mitochondria, Golgi bodies, and so on. The ectoplasm is often in the **gel state;** it appears more transparent under the light microscope than the colloidal sol, and its physical state gives structural rigidity to the protozoan's body. The bases of the flagella or cilia and their associated fibrillar structures, which may be very complex, are embedded in the ectoplasm.

Together with metazoa, fungi, and plants, protozoa are described as **eukaryotes;** that is, their genetic material—**deoxyribonucleic acid, (DNA)**—is carried on well-defined **chromosomes** combined with basic proteins called **his-**

tones, and the chromosomes are contained within a membrane-bound **nucleus.** At the light level, protozoan nuclei are typically oval, discoid, or round, and they are usually **vesicular,** with an irregular distribution of chromatin material and "clear" areas in the nuclear sap. But in the ciliates, which contain at least one macronucleus and one micronucleus, the former may be dense, elongated, chainlike, or branched. In electron micrographs, the nucleoplasm appears finely granular, with aggregations of denser chromatin. The chromosomes may remain as recognizable bodies throughout the cell cycle. Nucleoli are usually present but they typically disappear during nuclear division. **Endosomes,** conspicuous internal bodies, are nucleoli, although they do not disappear during mitosis. Parasitic amebas, trypanosomes, and phytoflagellates have endosomes.

The **nuclear envelope** is similar to that of most eukaryotic cells, consisting of two trilaminar membranes that fuse in the region of pores, but the envelope may be thickened by a fibrous layer or have strange honeycomblike tubes on the outer or inner face. The nuclear envelope may or may not persist during mitosis, again depending on the species, and mitotic spindles can be intra- or extranuclear.

Locomotory Organelles

Protozoa move by three basic types of organelles: flagella, pseudopodia, and cilia. Some amebas possess both flagella and pseudopods, although transformation from flagellated to ameboid cell occurs in response to environmental conditions and is a recognized life-cycle event. Flagella may also occur in large numbers and in rows, thus superficially resembling cilia. The bases of true cilia, possessed only by members of the phylum Ciliophora, are more or less connected by a complex fibrous network.

Flagella are slender, whiplike structures each composed of a central **axoneme** and an outer sheath that is a continuation of the cell membrane (Fig. 4.6). The axoneme consists of nine peripheral and one central pair of microtubules (the nine plus two arrangement found in cilia and flagella throughout the animal kingdom, with a few exceptions). Central microtubules are singlets, but peripheral ones are often doublets, or even doublets "with arms." The central two microtubules are bilateral, and the peripheral ones can thus be numbered with reference to a plane perpendicular to the line between the central pair. The axoneme arises from a **kinetosome,** which is ultrastructurally virtually indistinguishable from the centrioles of other eukaryotic cells, being made up of the nine peripheral elements, typically microtubule triplets arranged in a cartwheel manner. Kinetosomes may lie at the bottom of **flagellar pockets** or **reservoirs** of differing depths, depending on the species.

The entire unit—flagellum, kinetosome, and associated organelles—is called a **mastigont.** Kinetosomes are more or less fixed in position relative to other organelles; thus, flagella may be directed anteriorly, laterally, or posteriorly, independent of their movements. Most flagellates have more than one flagellum, and these may be inserted into the body at different angles. The flagellum may also be bent back along, and loosely attached to, the body, forming a finlike **undulating membrane.** Flagellar movements are generally

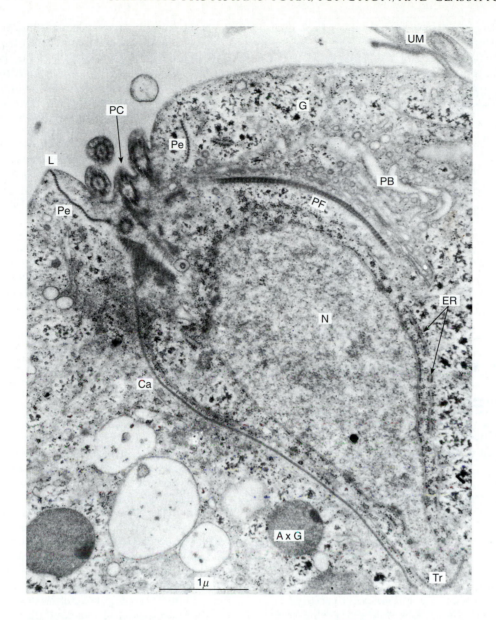

FIGURE 4.5

Section through the anterior portion of *Trichomonas gallinae* seen in dorsal and slightly left view. The capitulum (**Ca**) of the axostyle and a few paraxostylar granules (hydrogenosomes) (**AxG**) are located ventral to the nucleus (**N**); the parabasal body (**PB**) and its accompanying filament (**PF**) are dorsal and to the right of the nucleus. The pelta (**Pe**) extends to the extreme anterior end of the organism, terminating at the cell membrane in the area of the periflagellar lip (**L**). The proximal segments of the flagella are seen within the periflagellar canal (**PC**). The undulating membrane (**UM**) with its recurrent flagellum is located dorsally. Near the lower right corner note the proximal part of the axostylar trunk (**Tr**). Two additional noteworthy features of this micrograph are the absence of mitochondria and the presence of large numbers of dense granules presumed to be glycogen (**G**). **ER,** endoplasmic reticulum. (× 32,600.)

helical waves that begin at either the base or tip, pushing fluids along the axis of the flagellum. The resulting body movement may be fast or slow, forward, backward, lateral, or spiral. In some cases, such as with trichomonad parasites, movement is highly characteristic and recognized instantly by most parasitologists who have previously studied these flagellates in fresh intestinal contents.

In some flagellate groups, the base of the axoneme terminates in a complicated **root system** that varies greatly in complexity in different flagellates (Figs. 4.4 and 4.5). This mastigont system may include a prominent, striated rod, the **costa,** that courses from one of the kinetosomes along the margin of the organism just beneath the recurrent flagellum and undulating membrane. A tubelike **axostyle,** formed by a

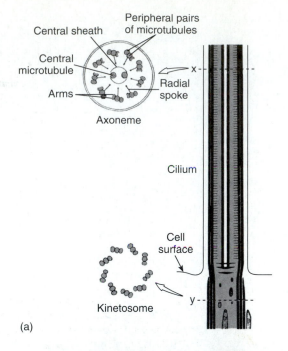

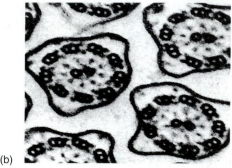

FIGURE 4.6

(*a*) General structure of a cilium or flagellum, showing a section through the axoneme within the cell membrane and a section through the kinetosome. The nine pairs of microtubules plus the central pair make up the axoneme. The central pair ends at about the level of the cell surface in a basal plate (axosome). The peripheral microtubules continue inward for a short distance to comprise two of each of the triplets in the kinetosome. (*b*) Electron micrograph of a section through several flagella, corresponding to section at *x* in (*a*). (×133,000)

(*a*) From Cleveland P. Hickman Jr. and Larry S. Roberts, *Biology of Animals*, 6th edition. Copyright © 1994 Wm.. C. Brown Communications, Inc., Dubuque, Iowa. Reprinted by permission of Times Mirror Higher Education Group, Inc., Dubuque, Iowa. All Rights Reserved. (*b*) Courtesy of L. R. Gibbons.

sheet of microtubules, may run from the area of the kinetosomes to the posterior end, where it may protrude. A Golgi body may be present; if a periodic fibril, the **parabasal filament,** runs from the Golgi body to contact a kinetosome, the Golgi body is referred to as a **parabasal body.** A fibril running from a kinetosome to a point near the surface of the nuclear membrane is called a **rhizoplast,** and the entire complex of organelles of the mastigont and nucleus may be referred to as the **karyomastigont.**

In the order Kinetoplastida, including the trypanosomes (Chapter 5), a dark-staining body called the **kinetoplast** is found near the kinetosome (see Fig. 5.1). The kinetoplast is actually a disc made of DNA circles, located within a single large mitochondrion, and having different genetic properties from the nucleus. Kinetoplastids also have a **paraxial crystalline rod** that lies alongside the axoneme, within the flagellum. And finally, many free-living flagellates possess fine fringes or hairlike **mastigonemes** on their flagella, making them look like motile test-tube brushes in the electron microscope. Students interested in evolutionary biology might find their life's work in trying to explain the origin of all this subcellular complexity; those intrigued by cellular function will discover an equally challenging task.

Cilia are structurally similar to flagella, with a kinetosome and an axoneme composed of two central and nine peripheral microtubules. But beneath the cell membrane, the similarity ends. Ciliate pellicular structure has become almost a subdiscipline of cellular anatomy as a result of the complexity and diversity in microtubular organelles revealed by electron microscopy.

Body cilia are arranged in rows, known as **kineties,** which in turn are composed of **kinetids,** the basic units of ciliate pellicular organization. The pellicle of *Dexiotricha,* a ciliate from an Illinois pig wallow, is simple enough to serve as an introduction to this organization (Fig. 4.7). A kinetid consists of the kinetosome, a small membranous pocket, the **parasomal sac,** and a number of fibers or sheets, made from microtubules, that extend in various directions from the kinetosome. A tapering banded fiber, the kinetodesma (plural kinetodesmata), arises from the clockwise side of each kinetosome (when viewed from the anterior end of the cell), courses anteriorly, and joins a similar fiber from the adjoining cilium in the same row. The resulting compound fiber of kinetodesmata is called a **kinetodesmose.** Flat sheets of microtubules, the **postciliary microtubules,** run posteriorly from each kinetosome, and similarly constructed bands, the **transverse microtubules,** lie perpendicular to the kineties. **Monokinetids** contain a single kinetosome and associated fibers; **dikinetids** contain a pair of kinetosomes; and so on.

The kinetosomes and associated fibrils constitute the **infraciliature.** Ciliates differ significantly in the structure of their infraciliature, and such differences are of major taxonomic importance.

The **oral ciliature** can be amazingly complex and is an outstanding example of the protozoan property just given, namely the elaboration of familiar organelles (see *Euplotes* in Fig. 4.1). **Oral membranes** are actually **polykinetids**—that is, fields or rows of cilia and their kinetosomes linked by electron dense fibrous networks. The **adoral zone of membranelles** is a series of such oral membranes located to the "left" of or counterclockwise from the side of the oral area of the more complex ciliates. Polykinetids may also be found on the body as **cirri** (singular cirrus), tufts of cilia that function together, usually in locomotion along a substrate. A group of kinetosomes forming a tuft of ciliary organelles in the aboral region of peritrich ciliates is called the **scopula.** It is involved in stalk formation.

Cilia beat with a powerful backstroke, pushing the surrounding fluid posteriorly, in metachronal waves.

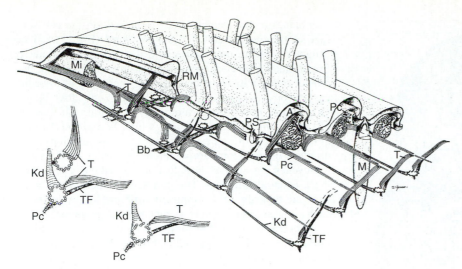

FIGURE 4.7

A diagram of the structure of a ciliate cortex (*Dexiotricha* sp.), reconstructed from electron micrographs, illustrating the relationships between the various elements of the ciliate cortex. **A,** alveolar sac; **BP,** a basal filamentous bundle of fibers; **Kd,** kinetodesmata; **M,** mucocyst; **Mi,** sausage-shaped mitochondrion; **Pc,** postciliary microtubular ribbons; **PS,** parasomal sac; **RM,** a single microtubule running through a pellicular ridge; **T,** transverse microtubule ribbon; **TF,** transverse fiber.

From R. K. Peck, "Cortical ultrastructure of the scuticociliates *Dexiotricha media* and *Dexiotricha colpidiopsis* (Hymenostomata)," in *J. Protozool.* 24:122–134. Copyright © 1977 The Society of Protozoologists. Reprinted with permission of the publisher.

Membranelles have their own beat cycles that are usually independent of the somatic ciliature. Much of the wonder that ciliates seem to produce in students comes from the action of polykinetids; for example, the "walking" motion of cirri tends to make the organism look as if it is behaving in a rather purposeful way. When ciliates divide, the ciliature is usually reorganized according to a sequence of events that is species specific. The reorganization of polykinetids is a complex process, but this "embryological development" is the basis for much of the class level taxonomy in the phylum Ciliophora.

The mechanism by which flagella and cilia move requires ATP and involves the interaction of the arms of each microtubule pair (Fig. 4.6) with the neighboring pair of microtubules. This causes one member of a pair to slide lengthwise relative to the other microtubule in the pair (sliding microtubule model). For a more complete explanation, refer to Preston et al.,[26] Satir,[28] and Stebbings and Hyams.[30]

Pseudopodia are temporary extensions of the cell membrane and are found in the Sarcodina (and other organisms). Pseudopodia function in locomotion and feeding. In some amebas, movement is by flow of the entire body, with no definite extensions. Such amebas are called **limax forms** (see Fig. 7.9), after the slug genus *Limax*. Four types of pseudopodia are found among the rest of the Sarcodina, three of which are illustrated in Figure 4.1. Most of the amebas have **lobopodia,** which are finger-shaped, round-tipped pseudopodia that usually contain both ectoplasm and endoplasm. All parasitic and commensal amebas of humans have this kind of pseudopodium. **Filopodia** are slender, sharp-pointed organelles, composed only of ectoplasm. They are not branched like **rhizopodia,** which branch extensively and fuse together to form netlike meshes. **Axopodia** are like filopodia but each contains a slender axial filament composed of microtubules that extends into the interior of the cell.

The shapes of pseudopods as well as the shapes of **uroids,** or membranous extensions at the (temporarily!) posterior end of the cell, are both taxonomic characters in amebas. Movement by means of pseudopodia is a complex form of protoplasmic streaming. Current evidence suggests that the mechanism, like most other forms of cell movement (except in cilia and flagella), involves the interaction of microfilaments of actin with myosin. Nachmias[23] gives a good review of this mechanism, as do Preston et al.[26] Bailey et al.[2] experimentally demonstrate the role of the cytoskeleton in movement and endocytosis in the parasitic *Entamoeba.* Although protoplasmic streaming is well-studied, the genetic mechanisms, or other cell properties, that determine pseudopod shape are not known. These animals obviously have some characteristics that function to produce extensions of plasma membrane that are indeed temporary but are also consistent enough in structure so that they may be used in identification and classification. Pseudopod formation is no less wondrous than polykinetid function.

In most sporozoan species, the merozoites, ookinetes, and sporozoites appear to glide through fluids with no subcellular motion whatever.[20] Gregarines (p. 114), for example, exhibit a variety of slow, sometimes almost snakelike movements, depending on the species and the kind of fresh tissue preparation that is examined. Electron microscope studies reveal longitudinal pellicular ridges (**epicytic folds**) on these cells, which often appear to have been fixed in the process of forming an undulatory wave. Subpellicular microtubules are found in the folds, and it has been proposed that these fibers function in the gliding locomotion. Experimental work, however, reveals that contact with a substrate is essential to gregarine movement and suggests that mucous secretion may also play a role in locomotion.[20] Parasitologists know sporozoans move; they are just not sure *how.*

Reproduction and Life Cycles

Reproduction in protozoa may be either asexual or sexual, although many species alternate types in their life cycles or perform one or the other reproductive functions in response to environmental conditions. Most often, **asexual reproduction** is by **binary fission,** in which one individual divides into two. The plane of fission is random in Sarcodina, longitudinal in flagellates (between kinetosomes or

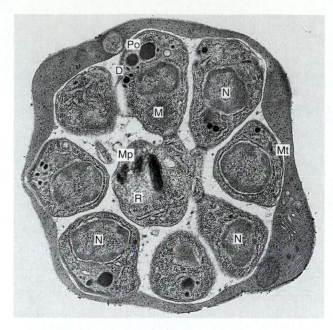

FIGURE 4.8

A late stage in the development of *Plasmodium cathemerium* within the host erythrocyte. The segmentation has been almost completed, and paired organelles (**Po**), dense bodies (**D**), nucleus (**N**), mitochondrion (**M**), pellicular complex with microtubules (**Mt**), and ribosomes are observed in the new merozoites. A residual body (**R**) surrounded by a rim of cytoplasm of the mother schizont contains a cluster of malarial pigment (**Mp**) granules. (× 30,000.)

From M. Aikawa, "The fine structure of the erythrocytic stages of three avian malarial parasites, *Plasmodium fallax, P. lophurae,* and *P. cathemerium,*" in *Am. J. Trop. Med. Hyg.* 15:449–471. Copyright © 1966.

flagellar rows—that is, **symmetrogenic**), and transverse in ciliates (across kineties, or **homothetogenic**). The sequence of division is (1) kinetosome(s), (2) kinetoplast (if present), (3) nucleus, and (4) cytokinesis. Nuclear division during asexual reproduction of protozoa is by mitosis, except in the macronuclei of ciliates, which are highly polyploid and divide amitotically. However, patterns of mitosis are much more diverse among the protozoa than among the metazoa. An inventory of these patterns is beyond the scope of this book, but examples include nuclear membranes that persist through mitosis, spindle fibers that form within the nuclear membrane, missing centrioles, and chromosomes that may not go through a well-defined cycle of condensation and decondensation. Nevertheless, the essential features of mitosis—replication of the chromosomes and regular distribution of the daughter chromosomes to the daughter nuclei—are always present.

Multiple fission (merogony, schizogony) occurs in some Sarcodina and in the Sporozoea. In this type of division the nucleus and other essential organelles divide repeatedly before cytokinesis. Thus, a large number of daughter cells are produced almost simultaneously and are, theoretically, in the same or similar physiological condition. During schiz-ogony the cell is called a **schizont, meront,** or **segmenter.** The daughter nuclei in the schizont arrange themselves peripherally, and the membranes of the daughter cells form beneath the cell surface of the mother cell, bulging outward (Fig. 4.8). The daughter cells are **merozoites,** and they finally break away from a small residual mass of protoplasm remaining from the mother cell to initiate another phase of merogony or to begin gametogony. Schizogony to produce merozoites may be referred to as **merogony.** Another type of multiple fission often recognized is **sporogony,** which is multiple fission after the union of gametes (see Sexual Reproduction). The products of merogony are additional parasites of the same life-cycle stage, such as those that invade red blood cells during a malarial infection. The products of sporogony, however, are typically of a completely different life-cycle stage, such as the sporozoites in resistant spores of gregarines.

Several forms of **budding** can be distinguished. **Plasmotomy,** sometimes regarded as budding, is a phenomenon in which a multinucleate individual divides into two or more smaller, but still multinucleate daughter cells. Plasmotomy itself is not accompanied by mitosis. **External budding** is found among some ciliates, such as the Suctorida. Here nuclear division is followed by unequal cytokinesis, resulting in a smaller daughter cell, which then swims away from the sessile parent and subsequently settles, metamorphoses, and grows to its adult size. **Internal budding,** or **endopolyogeny,** differs from schizogony only in the location of the formation of daughter cells. In this process the daughter cells begin forming within their cell membranes, distributed throughout the cytoplasm of the mother cell rather than at the periphery. The process occurs in some stages of the schizonts of the Eimeriina. **Endodyogeny** is endopolyogeny in which only two daughter cells are formed (Fig. 4.9). The protozoa, it seems, are as varied and elaborate in their asexual reproduction as they are in their structure.

Sexual reproduction involves reductional division in meiosis, resulting in a change from diploidy to haploidy, with a subsequent union of two cells to restore diploidy. The cells that join to restore diploidy are the **gametes,** and the process of producing the gametes is **gametogony.** Cells responsible for the production of gametes are **gamonts** (Fig. 4.10). Reproduction may be **amphimictic,** involving the union of gametes from two parents, or **automictic,** in which one parent gives rise to both gametes. The uniting gametes may be entire cells or only nuclei. When the gametes are whole cells, the union is called **syngamy.**

In syngamy the gametes may be outwardly similar (**isogametes**) or dissimilar (**anisogametes**). Although isogametes look similar, they will fuse only with isogametes of another "mating type" defined by proteins of the glycocalyx. Anisogametes often differ in cytoplasmic contents, in size (sometimes markedly), and in surface proteins. The larger, more quiescent of the pair is the **macrogamete;** the smaller, more active gamete is the **microgamete.** It is tempting to call these forms *female* and *male,* respectively, although it is debatable whether gender, in the commonly used sense, can or even should be distinguished in protozoa. Fusion of the microgamete and macrogamete produces the **zygote,** which is often a resting stage that overwinters or forms spores that enable survival between hosts.

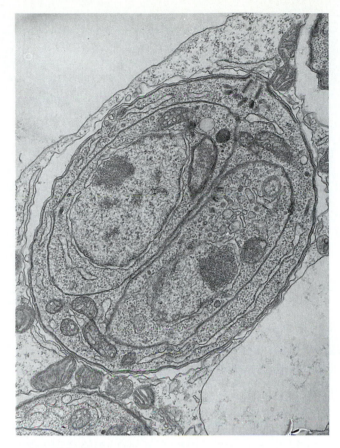

FIGURE 4.9

Toxoplasma gondii exhibiting two daughter cells in a mother cell, formed by endodyogeny.

From E. Vivier and A. Petiprez, "Le complexe membranaire superficiel et son evolution lors de l'elaboration des individus-fils chez *Toxoplasma gondii*," in *J. Cell Biol.* 28:355–373. Copyright © The Rockefeller University Press.

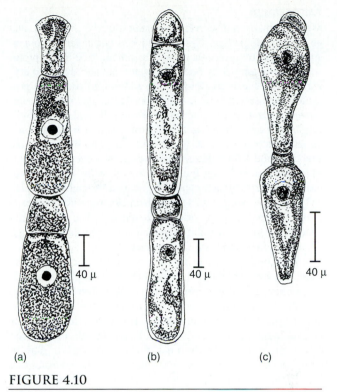

(a)   (b)   (c)

FIGURE 4.10

Paired gamonts of protozoan parasites from the yellow mealworm, *Tenebrio molitor*. (*a*) *Gregarina cuneata;* (*b*) *G. polymorpha;* (*c*) *G. steini.*

From R. E. Clopton et al,. "*Gregarina niphandtodes* n. sp. (Apicomplexa: Eugregarinorida) from adult *Tenebrio molitor* with oocyst descriptions of other gregarine parasites of the yellow mealworm," in *J. Protozool.* 38: 472–479. Copyright © 1991 the Society of Protozoologists. Reprinted with permission of the publisher.

Conjugation, in which only nuclei unite, is found only among the ciliates, whereas syngamy occurs in all other groups in which sexual reproduction is found. Two individuals ready for conjugation unite, and their pellicles fuse at the point of contact. The macronucleus in each disintegrates and the micronuclei undergo meiotic divisions into haploid **pronuclei.** A **migratory pronucleus** from each conjugant passes into the other to fuse with a **stationary pronucleus,** restoring the diploid condition. The cells separate, and subsequent nuclear divisions produce one or more macronuclei. The exconjugants, which are now genetic recombinants, then actively reproduce by fission.

The details of conjugation, including relationships of the conjugants, extent of cytoplasmic sharing, and fate of the exconjugants, vary widely among ciliates. Under natural conditions conjugating pairs are seen occasionally, especially when environmental conditions deteriorate. Clone cultures, descended from single individuals, can be prepared in the lab and stressed to produce cells that are ready to conjugate and will do so en masse when mixed (the mating type reaction). This technique has been useful in the study of mating speci-

ficity, genetics, and surface protein function in ciliates. Variations of conjugation are **cytogamy,** in which two individuals fuse but do not exchange pronuclei, with two pronuclei in each cell rejoining to restore diploidy, and **autogamy,** in which haploid pronuclei from the same cell fuse but there is no cytoplasmic fusion with another individual.

In the majority of the protozoa, including all of the Sporozoea, meiosis occurs in the first division of the zygote (**zygotic meiosis**),[11] and all other stages are haploid. **Intermediary meiosis,** which occurs only in the Foraminiferida among the protozoa but which is widespread in plants, exhibits a regular alternation of haploid and diploid generations.

The infective stages of many parasitic protozoa are protected by **cysts.** In sporozoans, the infective organism is called a **sporozoite,** whether it be inside a cyst, as in the coccidia (see Chapter 8, Figs. 8.1 and 8.3), or in a mosquito's salivary glands, as in the case of malarial parasites (see Chapter 9, Fig. 9.1). The growing stage of a protozoan, following excystment, is called the **trophont,** and other seemingly bizarre stages of obscure symbiotic species, such as those parasitic on sea anemones and various crabs, include encysted dividers (**tomonts**) and encysted, nondividing, stalked forms (**phoronts**).

Encystment

Many protozoa can secrete a resistant covering, the cyst, and go into a resting stage. Cyst formation is particularly common among parasitic protozoa, as well as among free-living protozoa found in temporary bodies of water that are subject to drying or other harsh conditions.[31] In addition to providing protection against unfavorable conditions, cysts may serve as sites for reorganization and nuclear division, followed by multiplication after excystation. In a few forms, such as *Ichthyophthirius,* a ciliate parasite of fish, the cyst falls from the host to the substrate and sticks there until excystation occurs (Chapter 11). Cellulose has been found in the cyst walls of some amebas, and others contain chitin.[1] Cysts can be highly complex and layered structures, as seen with an electron microscope, as in the filamentous cysts of *Giardia* species.[9] The outer layers generally also react with immunodiagnostic reagents, although not always in a highly specific manner.[12]

The conditions favoring encystment are not fully understood, but they are thought in most cases to involve some adverse change in the environment, such as food deficiency, desiccation, increased tonicity, decreased oxygen concentration, or pH or temperature change. It is vitally important for parasitologists to understand the elusive factors that induce cyst formation within the host, the role that cysts play in completion of the parasite life cycle, and the sociological factors that work to disseminate cysts. For example, human amebiasis, caused by *Entamoeba histolytica,* is spread by persons who often have no clinical symptoms but who pass cysts in their feces (Chapter 7).

During encystment the cyst wall is secreted, and some food reserves, such as starch or glycogen, are stored. Projecting portions of locomotor organelles are partially or wholly resorbed, and certain other structures, such as contractile vacuoles, may be dedifferentiated. During the process or following soon thereafter, one or more nuclear divisions give the cyst more nuclei than the trophozoite. In the flagellates and amebas, cytokinesis occurs in a characteristic division pattern after excystation. In Sporozoea the cystic form is the **oocyst,** which is formed after gamete union and in which multiple fission **(sporogony)** occurs with cytokinesis to produce sporozoites. In the Eimeriina the oocyst containing the sporozoites serves as the resistant stage for transferral to a new host, whereas in the Haemosporina (containing the causative agent of malaria, *Plasmodium* spp.) the oocyst merely serves as a developmental capsule for the sporozoites within the insect host.

In species in which the cyst is a resistant stage, a return of favorable conditions stimulates excystation. In parasitic forms some degree of specificity in the requisite stimuli provides that excystation will not take place except in the presence of conditions found in the host gut. Mechanisms for excystation may include absorption of water with consequent swelling of the cyst, secretion of lytic enzymes by the protozoan, and action of host digestive enzymes on the cyst wall. Excystation must include reactivation of enzyme pathways that were "turned off" during the resting stage, internal reorganization, and redifferentiation of cytoplasmic and locomotor organelles.

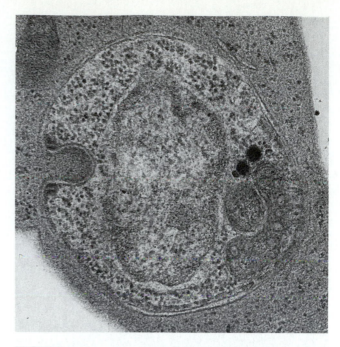

FIGURE 4.11

A uninucleate trophozoite of *Plasmodium cathemerium* ingesting host cell cytoplasm through a cytostome (micropore). (× 52,000.)

From M. Aikawa et al., "Feeding mechanisms of avian malarial parasites," in *J. Cell Biol.* 28:355–373. Copyright © 1966.

Feeding and Metabolism

Some protozoa are **holophytic (photoautotrophic)** and synthesize carbohydrates in chloroplasts, the organelles of "typical" plants. Such plantlike protists are often considered algae and claimed by the phycologists, but a few participate in symbiotic relationships of interest to parasitologists. Zooxanthellae (order Dinoflagellida) are very important mutuals living in the cells of reef-forming corals and other invertebrates (including some other protozoa), contributing significant amounts of carbohydrates to their hosts. Students interested in the biochemistry or evolution of symbiosis can find a fertile field in the obligate relationships between animals and their algal symbionts.

The animal-like protists are all **heterotrophic,** requiring their energy in the form of complex carbon molecules and their nitrogen in the form of a mixture of preformed amino acids. Most protozoa, including many parasitic ones, are also particle feeders—that is, grazers and predators. Their mouth openings may be temporary, as in amebas, or permanent **cytostomes,** as in most ciliates. A submicroscopic micropore is present in *Eimeria* and *Plasmodium* and, in certain stages, is involved in taking in nutrients (Fig. 4.11). Particulate food passes into a food vacuole, which is a digestive organelle that forms around any food thus ingested. Indigestible material is voided either through a temporary opening or through a permanent **cytopyge,** which is found in many ciliates. **Pinocytosis** is an important activity in many protozoa, as is **phagocytosis.** Both pinocytosis and phagocytosis are examples of

endocytosis, differing only in that pinocytosis deals with droplets of fluid, whereas phagocytosis is the process of internalizing particulate matter.

Like most other eukaryotic cells, protozoa generally carry out the many reactions of glycolysis, the Krebs (citric acid) cycle, the pentose-phosphate shunt, electron transport, transaminations, lipid oxidations and syntheses, nucleic acid metabolism, and the multitude of other metabolic events that make biochemical pathways look like printed circuits of high-tech electronic equipment. ATP is the most common form of immediately usable energy, although a few parasites use inorganic pyrophosphate in a similar role. Polysaccharides, especially glycogen or some related molecule, function as deep energy storage. Genes are transcribed in the nucleus and polypeptides are synthesized on ribosomes, as in other cells. Virtually every introductory biology text, as well as more advanced biochemistry texts and references, include chapters, with diagrams, of the major catabolic and anabolic pathways, and you should consult these references if you feel a need to refresh your memory for the generalities of cellular biochemistry.

Comparative biochemical studies reveal that the details of protozoan metabolism are as varied as are the details of protozoan sex. Some important biological factors to consider are that many parasites occupy environments in which the oxygen supply is quite limited, and even in many cases in which oxygen is not limited, neither is glucose. Therefore there is no energy advantage to completely oxidizing glucose. Organisms, including many protozoan parasites, that are adapted to such environments often derive all their energy from glycolysis. The partially oxidized products can be excreted as waste. The complete Krebs cycle and cytochrome system then become so much excess metabolic machinery, at least in terms of energy production. However, the problem of reoxidation of the net accumulation of reduced NAD remains, since even without the need for the energy obtainable in subsequent electron transfer, the oxidized compounds must be available for continuous functioning of glycolysis. In some parasites the electrons are transferred to the pyruvate, and the resulting ethanol or lactate is excreted; however, many organisms excrete such compounds as succinate, acetate, and short-chain fatty acids as end products of glycolysis. Some metabolic solutions to the problem of NAD oxidation will be mentioned in subsequent chapters.

Plantlike protists such as *Euglena* species often use photosynthesis and heterotrophic oxidative assimilation interchangeably, depending on the environment and available substrates. Metabolic flexibility is also a feature of obligate heterotroph protozoa. For example, the Krebs cycle requires a continuous supply of the 4-carbon molecule oxaloacetate, one of the cycle's own end products, as an acceptor of 2-carbon units during the formation of citric acid. Krebs cycle intermediates are routinely taken out of circulation and used in synthetic reactions such as transaminations. Thus, an alternate source of oxaloacetate is required, which in many protozoans is the **glyoxylate cycle,** a metabolic pathway especially important in those species that rely heavily on ethanol, fatty acids, and acetate for their energy and carbon skeletons. The

glyoxylate cycle uses two acetyl-CoA molecules to make a single oxaloacetate molecule; the enzymes for this cycle are found in the glyoxysomes (peroxisomes).

Protozoa also may utilize a variety of hydrogen acceptors in the final oxidations coupled with ATP production. In aerobic metabolism of most animals, this final acceptor is molecular oxygen. Under anaerobic conditions, protozoa may produce lactic acid or ethanol by using pyruvate as a hydrogen acceptor. *Loxodes* sp., a ciliate, evidently uses NO_3^- as a terminal hydrogen acceptor in the mitochondria and contains an enzyme more typical of bacteria than eukaryotes to carry out the feat.[10] In the parasitic protozoa without mitochondria—*Trichomonas vaginalis, T. foetus, Giardia lamblia,* and *Entamoeba histolytica*—the final acceptor can be pyruvate, a key molecule in carbohydrate metabolism, in which case the end product is lactate or ethanol. These protozoa take up molecular oxygen, but the availability of oxygen makes little or no difference in their energy metabolism.

Odd and parasite-specific metabolic pathways are, of course, inviting targets for chemotherapy. Some of the more effective antimalarial drugs interfere with the parasites' ability to metabolize 1-carbon units during nucleic acid synthesis. Intracellular stages of the flagellate genus *Leishmania* do not build their nucleic acid precursors but instead salvage them from their host cells. Allopuranol, a purine analog, cannot be metabolized by the parasites but can be taken up from the host cell and used to build nucleic acids that do not function properly in the parasite. Needless to say, parasitologists leave few metabolic pathways unexplored in their efforts to find ways of treating diseases.

Many parasitic protozoa are intracellular. In some, entry into the host cell is by phagocytosis of the parasite. An example is *Leishmania donovani,* which is eaten by free-roaming macrophages and reticuloendothelial cells. The host cell forms a membrane-bound **parasitophorus vacuole** around the parasite, but instead of killing the parasite with digestive enzymes, as might be expected from a macrophage, the host cell provides it with nutrients. Members of the important genera *Babesia, Eimeria, Plasmodium,* and *Toxoplasma* are all intracellular at least at some stages in their lives, and uptake is by active invasion of host cells by motile infective stages, probably aided by digestive secretions.[27] A different mode of entry into a host cell is employed by members of the phylum Microspora (Chapter 10). The cyst stage of these parasites contains a coiled, hollow filament that apparently is under great pressure. When eaten by the host, which is usually an arthropod, the tubule is forcibly extruded from the cyst and penetrates an adjoining host cell. The organism within the spore (**sporoplasm**) crawls through the tube and enters its host. In this case the membrane of the parasite is in direct contact with the cytoplasm of the host, with no vacuole being formed around it.

Excretion and Osmoregulation

Most protozoa appear to be **ammonotelic;** that is, they excrete most of their nitrogen as ammonia, most of which readily diffuses directly through the cell membrane into the surrounding medium. Other, sometimes unidentified waste products are also produced, at least by intracellular parasites. After these substances are secreted they are accumulated within the host cell and, on the death of the infected cell, have toxic effects on the host. Carbon dioxide, lactate, pyruvate, and short-chain fatty acids are also common waste products.

Contractile vacuoles are probably more involved with osmoregulation than with excretion per se. Because free-living, freshwater protozoa are hypertonic to their environment, they imbibe water continuously by osmosis. The action of contractile vacuoles effectively pumps out the water. Marine species and most parasites do not form these vacuoles, probably because they are more isotonic to their environment. However, *Balantidium* (Chapter 11) contains contractile vacuoles.

Endosymbionts

Just as many protozoa live symbiotically in the bodies of larger animals, many organisms live within the bodies of protozoa. Zooxanthellae were mentioned earlier. Many biologists now believe that chloroplasts and mitochondria and perhaps flagella arose from prokaryotes that came to live inside other cells. Corliss[6] proposed that any such structure or organism be referred to as a **xenosome,** which is any body or constituent organelle that contains DNA, is bounded by at least one membrane, lives within a cell, and is capable of reproducing itself. The term *xenosome* implies that "the symbiont once functioned as a free-living organism outside its present residence." In addition to the previous examples, zoochlorellae (green algal cells in protozoa and some multicellular animals), a variety of prokaryotes in protozoa, many intracellular protozoan parasites of multicellular animals, many hyperparasites of parasitic protozoa, and even nuclei of eukaryotic cells would be considered xenosomes. Nonetheless, endosymbionts living within protozoa are numerous and have been discussed by several authors.[3,16,18,27] Their contribution to and interaction with the metabolism of their hosts undoubtedly varies, but in many cases is poorly understood.

CLASSIFICATION OF THE PROTOZOAN PHYLA

All of the following groups (plus some others) may be properly included in the kingdom Protista. This classification, emphasizing the groups with parasitic members, is based on that of Levine et al.[19] and Lee et al.[17] However, the true phylogenetic relationships of protozoan groups have yet to be elucidated and some, if not many, of the following taxa may be paraphyletic.[5,7] Most of the terminology used in the sections to follow has been covered already in this chapter; some terms will be defined in upcoming chapters.

PHYLUM SARCOMASTIGOPHORA

Single type of nucleus, except in some Foraminiferida; sexuality, when present, essentially syngamy; flagella, pseudopodia, or both types of locomotor organelles.

Subphylum Mastigophora

One or more flagella typically present in trophozoites; asexual reproduction basically by symmetrogenic binary fission; sexual reproduction known in some groups.

Class Phytomastigophorea

Typically with chloroplasts; if chloroplasts lacking, relationship to pigmented forms clearly evident; mostly free-living.

Order Dinoflagellida

Two flagella, typically one transverse and one trailing; body usually grooved transversely and longitudinally, forming a girdle and sulcus, each containing a flagellum; chromatophores usually yellow or dark brown, occasionally green or blue-green; nucleus unique among eukaryotes in having chromosomes that lack or have low levels of histones; mitosis intranuclear; flagellates, coccoid unicells, colonies, and simple filaments; sexual reproduction present; few parasites of invertebrates; one or more species (*Zooxanthella microadriatica*) very important mutuals in tissues of various marine invertebrates, especially cnidarians.[7]

Other Orders

Cryptomonadida, Euglenida, Chrysomonadida, Heteromonadida, Raphidomonadida, Prymnesiida, Volvocida, Prasinomonadida, Silicoflagellida.

Class Zoomastigophorea

Chloroplasts absent; one to many flagella; ameboid forms, with or without flagella in some groups; sexuality known in few groups.

Order Proteromonadida

One or two pairs of unequal flagella without paraxial rods; single mitochondrion, distant from kinetosomes, curling around nucleus, not extending length of body, without kinetoplast; Golgi apparatus encircling band-shaped rhizoplast passing from kinetosomes near surface of nucleus to mitochondrion; cysts present; all species parasitic. *Karotomorpha, Proteromonas.*

Order Retortamonadida

Two to six flagella, one turned posteriorly and associated with ventrally located cytostomal area bordered by fibril; mitochondria and Golgi apparatus absent; intranuclear division spindle; cysts present. *Chilomastix, Retortamonas.*

Order Trichomonadida

Typically at least some kinetosomes associated with rootlet filaments characteristic of trichomonads, parabasal body present; mitochondria absent; division spindle extranuclear; karyomastigonts with four to six flagella, but only one flagellum in one genus and no flagella in another; pelta and noncontractile axostyle in each mastigont, except for one genus; hydrogenosomes present; no sexual reproduction; true cysts rare; all parasitic. *Dientamoeba, Histomonas, Monocercomonas, Trichomonas.*

Order Oxymonadida

One or more karyomastigonts, each containing four flagella typically arranged in two pairs in motile stages; one to many axostyles per organism; mitochondria and Golgi apparatus absent; division spindle intranuclear; cysts in some; sexuality in some; all parasitic. *Monocercomonoides, Oxymonas.*

Order Diplomonadida

One or two karyomastigonts; individual mastigonts with one to four flagella, typically one of them recurrent and associated with cytostome or with organelles forming cell axis; mitochondria and Golgi apparatus absent; intranuclear division spindle; cysts present; free-living or parasitic.

Suborder Enteromonadina

Single karyomastigont containing one to four flagella; one recurrent flagellum in genera with more than single flagellum; frequent transitory forms with two karyomastigonts; all parasitic. *Enteromonas, Trimitus.*

Suborder Diplomonadina

Two karyomastigonts; body with twofold rotational symmetry; each mastigont with four flagella, one recurrent; with variety of microtubular bands; free-living or parasitic. *Giardia, Hexamita.*

Order Hypermastigida

Mastigont system with numerous flagella and multiple parabasal bodies; flagella-bearing kinetosomes distributed in complete or partial circle, in plate or plates, or in longitudinal or spiral rows meeting in a centralized structure; many with microtubule sheets or peltoaxostylar lamellae; one nucleus per cell; mitochondria absent; division spindle extranuclear; cysts in some; sexuality in some; all parasitic.

Suborder Lophomonadina

Extranuclear organelles arranged in one system; typically all old structures resorbed in division and new organelles formed in daughter cells. *Lophomonas, Microjoenia.*

Suborder Trichonymphina

Body divided into anterior rostral and posterior postrostral regions; two or, occasionally, four mastigont systems; typically equal separation of mastigont systems in division, with total or partial retention of old structures when new systems are formed. *Barbulanympha, Trichonympha.*

Suborder Spirotrichonymphina

Flagellar bands begin at the anterior end and spiral in helical coil around body. *Spirotrichonympha.*

Order Kinetoplastida

One or two flagella arising from pocket; flagella typically with paraxial rod that parallels the axoneme; single mitochondrion (nonfunctional in some forms) extending length of body as tube, hoop, or network of branching tubes, usually with conspicuous DNA-containing kinetoplast located near flagellar kinetosomes; Golgi apparatus typically in region of flagellar pocket, not connected to kinetosomes and flagella; majority parasitic, some free-living.

Suborder Bodonina

Typically two unequal flagella, one directed anteriorly, the other posteriorly; no undulating membrane; kinetoplastic DNA in several discrete bodies in some, dispersed throughout mitochondrion in some; free-living and parasitic. *Bodo, Cryptobia, Rhynchomonas.*

Suborder Trypanosomatina

Single flagellum either free or attached to body by undulating membrane; kinetoplast relatively small and compact; all parasitic. *Blastocrithidia, Leptomonas, Herpetomonas, Crithidia, Leishmania, Trypanosoma.*

Other Orders

Choanoflagellida, Cercomonadida, Rhizomastigida, Ebriida.

Subphylum Opalinata

Numerous flagella (also called cilia or undulipodia) in oblique rows over entire body and originating in an anterior field of kinetosomes (falx); some fibrils associated with kinetosomes; cytostome absent; binary fission generally symmetrogenic; known life cycles involve syngamy with anisogamous flagellated gametes; all parasitic.

Class Opalinatea

With characters of the subphylum.

Order Opalinida

With characters of the class. *Opalina, Protoopalina, Cepedea.*

Subphylum Sarcodina

Pseudopodia, or locomotive protoplasmic flow without discrete pseudopodia; flagella, when present, usually restricted to developmental or other temporary stages; body naked or with external or internal test or skeleton; asexual reproduction by fission; sexuality, if present, associated with flagellated or, more rarely, ameboid gametes; most free-living.

SUPERCLASS RHIZOPODA

Locomotion by lobopodia, filopodia, or reticulopodia or by protoplasmic flow without production of discrete pseudopodia.

Class Lobosea

Pseudopodia lobose or more or less filiform but produced from broader hyaline lobe; usually uninucleate; no sporangia or similar fruiting bodies.

Subclass Gymnamoebia

Without test.

Order Amoebida

Typically uninucleate; mitochondria typically present; no flagellate stage; usually asexual.

Suborder Tubulina

Body branched or unbranched cylinder. *Entamoeba, Iodamoeba, Endolimax.*

Suborder Acanthopodina

Pseudopods more or less finely tipped, sometimes filiform, often branched and hyaline, produced from a broad hyaline lobe; cysts common. *Acanthamoeba.*

Other Suborders

Thecina, Flabellina, Conopodina.

Order Schizopyrenida

Body with shape of monopodial cylinder, usually moving with more or less eruptive, hyaline, hemispherical bulges; typically uninucleate; temporary flagellate stages in most species. *Naegleria*.

Other Order

Pelobiontida.

Other Subclass

Testacealobosia.

Class Mycetozoea

Uninuclear, multinucleate, or plasmodial; sporulation either by differentiation of single ameboid cells or by aggregates of amebas that form pseudomultinucleate plasmodia, into fruiting body, or by differentiation of spores from a truly multinucleate plasmodium.

Order Plasmodiophorida

Obligate intracellular parasites of plants, with minute plasmodia; zoospores produced in sporangia and bearing anterior pair of unequal flagella. *Plasmodiophora*.

Other Classes

Acarpomyxea, Filosea, Granuloreticulosea.

OTHER SUPERCLASS

Actinopoda.

Classes

Acantharea, Polycystinea, Phaeodarea, Heliozoea.

PHYLUM LABYRINTHOMORPHA

Trophic stage as ectoplasmic network with spindle-shaped or spherical, nonameboid cells; in some genera ameboid cells move within network by gliding; with sagenogenetosome (unique cell-surface organelle, associated with ectoplasmic network); saprozoic and parasitic on algae; mostly marine and estuarine.

Class Labyrinthulea

With characters of the phylum.

Order Labyrinthulida

With characters of the class. *Labyrinthula*.

PHYLUM APICOMPLEXA

Apical complex (generally consisting of polar ring, micronemes, rhoptries, subpellicular tubules, and conoid) present at some stage; micropore(s) usually present; cilia and flagella absent except for flagellated microgametes in some groups; sexuality by syngamy; all parasitic.

Class Perkinsasida

Conoid forming incomplete cone; "zoospores" (sporozoites?) flagellated, with anterior vacuole; no sexual reproduction; homoxenous.

Order Perkinsorida

With characters of the class. *Perkinsus*.

Class Sporozoasida

Conoid, if present, forming complete cone; reproduction generally both sexual and asexual; oocysts generally containing infective sporozoites resulting from sporogony; locomotion of mature organisms by body flexion, gliding, or undulation of longitudinal ridges; flagella present only in microgametes of some groups; pseudopods ordinarily absent but if present used for feeding, not locomotion.

Subclass Gregarinasina

Mature gamonts large, extracellular; mucron or epimerite in mature organism; mucron formed from conoid; generally syzygy of gamonts; gametes usually isogamous or nearly so; zygotes forming oocysts within gametocysts; in digestive tract or body cavity of invertebrates; generally homoxenous.

Order Archigregarinorida

Life cycle usually with merogony, gametogony, sporogony; in annelids, sipunculids, hemichordates, or ascidians. *Exoschizon, Selenidioides*.

Order Eugregarinorida

Merogony absent; gametogony and sporogony present; typically parasites of arthropods and annelids.

Suborder Blastogregarinorina

Gametogony by gamonts while still attached to intestine; no syzygy; gametocysts absent; gamont of single compartment with mucron, without definite protomerite and deutomerite; in polychaete annelids. *Siedleckia*.

Suborder Aseptatorina

Gametocysts present; gamont of single compartment, without definite protomerite and deutomerite but with mucron in some species; syzygy present. *Lecudina, Lankesteria, Monocystis, Selenidium, Diplocystis*.

Suborder Septatorina

Gametocysts present; gamont divided into protomerite and deutomerite by septum; with epimerite; in alimentary canal of invertebrates, especially arthropods. *Gregarina, Didymophyes, Leidyana, Actinocephalus, Stylocephalus, Acanthospora, Menospora*.

Order Neogregarinorida

Merogony (presumably acquired secondarily); in Malpighian tubules, intestine, hemocoel, or fat tissues of insects. *Gigaductus, Farinocystis (= Triboliocystis)*.

Subclass Coccidiasina

Gamonts ordinarily present; mature gamonts small, typically intracellular, without mucron or epimerite; syzygy generally absent; life cycle characteristically consisting of merogony, gametogony, and sporogony; most species in vertebrates.

Order Agamococcidiorida

Merogony and gametogony absent. *Rhytidocystis*.

Order Protococcidiorida

Merogony absent; in invertebrates. *Eleutheroschizon, Grellia*.

Order Eucoccidiorida

Merogony present; in vertebrates and invertebrates.

Suborder Adeleorina

Macrogamete and microgamont usually associated in syzygy during development; microgamont producing one to four microgametes; sporozoites enclosed in envelope. *Adelea, Haemogregarina, Klossiella*.

Suborder Eimeriorina

Macrogamete and microgamont developing independently; no syzygy; microgamont typically producing many microgametes; zygote not motile; sporozoites typically enclosed in sporocyst within oocyst; homoxenous or heteroxenous. *Aggregata, Eimeria, Isospora, Sarcocystis, Toxoplasma.*

Suborder Haemospororina

Macrogamete and microgamont developing independently; no syzygy; conoid ordinarily absent; microgamont producing eight flagellated microgametes; zygote motile (ookinete); sporozoites naked, with three-membraned wall; heteroxenous, with merogony in vertebrates and sporogony in invertebrates; transmitted by bloodsucking insects. *Haemoproteus, Leucocytozoon, Plasmodium.*

Subclass Piroplasmasina

Piriform, round, rod-shaped, or ameboid; conoid absent; no oocysts, spores, or pseudocysts; flagella absent; usually without subpellicular microtubules, with polar ring and rhoptries; asexual and probably sexual reproduction; parasitic in erythrocytes and sometimes also in other circulating and fixed cells; heteroxenous, with merogony in vertebrates and sporogony in invertebrates; sporozoites with single-membraned wall; known vectors are ticks. *Babesia, Theileria.*

PHYLUM MICROSPORA

Unicellular spores, each with imperforate wall, containing one uninucleate or dinucleate sporoplasm and a polar filament; sporoplasm injected into host cells through extruded polar filament; without mitochondria; intracellular parasites in nearly all major animal groups.

Class Metchnikovellidea

Spherical to lenticular spores, with filament extruding laterally; polaroplast and posterior vacuole absent; sporulation sequence with dimorphism, occurring either in parasitophorous vacuole or in thick-walled cyst; hyperparasites of gregarines in annelids. *Amphiacantha, Metchnikovella.*

Class Microsporididea

Spores elongated, oval, or tubular, with complex extrusion apparatus of Golgi origin, often including polaroplast and posterior vacuole; spore wall with three layers; sporocyst present or absent; often dimorphic in sporulation sequence.

Order Pleistophoridida

Sporulation sequence occurring within more or less persistent intracellular (in host cell) sporocyst (pansporoblastic membrane); often dimorphic, with another sporulation sequence not involving such membrane; spores uninucleate or binucleate, depending on thickness of wall and type of development or transmission; number of spores from sporoblasts variable. *Encephalitozoon, Glugea, Pleistophora, Thelohania, Amblyospora.*

Order Nosematidida

Spores diplokaryotic, with paired nuclei dividing synchronously. *Nosema.*

PHYLUM ASCETOSPORA

Spore multicellular (or unicellular?); with one or more sporoplasms (ameboid cells); without polar capsule or polar filaments; all parasitic in invertebrates.

Class Stellatosporea

Haplosporosomes (membrane-bound bodies that appear dark in electron micrographs) present; spore with one or more sporoplasms.

Order Occlusosporida

Spore with more than one sporoplasm; spore wall entire. *Marteilia.*

Order Balanosporida

Spore with one sporoplasm; spore wall with orifice. *Haplosporidium, Urosporidium.*

Class Paramyxea

Spore bicellular, consisting of parietal cell and one sporoplasm; spore without orifice.

Order Paramyxida

With characters of the class. *Paramyxa.*

PHYLUM MYXOZOA

Spores of multicellular origin, with one or more polar capsules (enclosing polar filaments) and sporoplasms; with one, two, or three (rarely more) valves; all parasitic.

Class Myxosporea

Spore with one or two sporoplasms and one to six (typically two) polar capsules; spore membrane generally with two, occasionally up to six, valves; trophozoite stage well-developed, main site of proliferation; coelozoic or histozoic in ectothermic vertebrates, mainly fishes.

Order Bivalvulida

Spore wall with two valves.

Suborder Bipolarina

Spores with polar capsules at opposite ends of spore or with widely divergent polar capsules located in sutural plane or sutural zone. *Myxidium, Sphaeromyxa.*

Suborder Eurysporina

Spores with two to four polar capsules at one pole in plane perpendicular to sutural plane. *Ceratomyxa, Sphaerospora.*

Suborder Platysporina

Spores with two polar capsules at one pole in sutural plane; spores bilaterally symmetrical, unless with single polar capsule. *Henneguya, Myxobolus* (syn. *Myxosoma*), *Thelohanellus.*

Order Multivalvulida

Spore wall with three or more valves. *Hexacapsula, Kudoa.*

Class Actinosporea

Spores with three polar capsules; membrane with three valves; several to many sporoplasms; trophozoite stage reduced, proliferation mainly during sporogenesis; in invertebrates, especially annelids.

Order Actinomyxida

With characters of the class. *Triactinomyxon* (but see Chapter 10!).

PHYLUM CILIOPHORA

Simple cilia or compound ciliary organelles typical in at least one stage of life cycle; with subpellicular infraciliature present even when cilia absent; two types of nuclei, with rare exception; binary fission transverse, basically homothetogenic, but budding and multiple fission also occur; sexuality involving conjugation, autogamy, and cytogamy; contractile vacuole typically present; most species free-living but many commensal and some parasitic. (Numerous taxa with commensal and some with parasitic species will not be characterized here because they are not covered in the text, and the morphological terminology is beyond the scope of this book.)

Subphylum Postciliodesmatophora

Somatic ciliature dikinetids with postciliodesmata or overlapping postciliary microtubular ribbons; cortical alveolar system poorly developed.

Class Karyorelictea

Long, wormlike, flattened; many with one barren surface; two to many macronuclei.

Class Spirotrichea

Somatic ciliature dikinetids with anterior or both kinetosomes ciliated or with polykinetids, with well-developed overlapping postciliar ribbons; generally with conspicuous oral and/or preoral ciliature with serial polykinetids.

Subclass Heterotrichia

Generally large to very large forms, often highly contractile, sometimes pigmented; body kineties, when present, as dikinetids; left serial of oral polykinetids usually conspicuous; macronucleus oval or, often, beaded; parasitic and free-living species.

Order Clevelandellida

Somatic ciliature well-developed, sometimes separated into distinct areas by well-defined suture lines; several specialized unique fibers associated with kinetosomes; macronuclear karyophore (region of cytoplasm apparently supporting nucleus) and/or conspicuous dorsoanterior sucker characteristic of many species; endoparasitic in digestive tract of insects and other arthropods or lower vertebrates, occasionally in oligochaetes or molluscs. *Clevelandella, Nyctotherus.*

Other Orders

Heterotrichida, Plagiotomida, Armophorida, Phacodiniida, Licnophorida, Odontostomatida.

Other Subclasses

Choreotrichia, Stichotrichia.

Subphylum Rhabdophora

Somatic monokinetids with a crown of dikinetids around cytostome; oral polykinetids usually small and limited to two rows; proter retains oral apparatus during cytokinesis.

Class Prostomatea

Body monokinetids usually with radial transverse ribbon; cytostome apical to subapical.

Orders

Prostomatida, Prorodontida.

Class Litostomatea

Body monokinetids with tangential transverse ribbon and non-overlapping laterally directed kinetodesmal fibrils; simple oral cilia usually not as polykinetids.

Subclass Trichostomatia

Order Vestibuliferida

Apical or near apical densely ciliated vestibulum commonly present; no polykinetids; free-living or parasitic, especially in digestive tract of vertebrates and invertebrates. *Balantidium, Isotricha, Sonderia.*

Order Entodiniomorphida

Somatic ciliature in form of unique ciliary tufts or bands—otherwise body naked; pellicle generally firm, sometimes drawn out into processes; oral area often with retractile cilia and serial polykinetids; skeletal plates in many species; commensals in mammalian herbivores, including anthropoid apes.

Suborder Blepharocorythina

Somatic ciliature markedly reduced; oral ciliature inconspicuous; in herbivorous mammals, especially equids. *Blepharocorys, Ochoterenaia.*

Suborder Entodiniomorphina

Somatic ciliature as tufts, bands, or girdles; oral cilia usually as distinct polykinetids; pellicle rigid, firm, often spiny; in vertebrates, especially artiodactyls and perissodactyls. *Entodinium, Ophryoscolex.*

Other Orders

Haptorida, Pleurostomatida, Pharyngophorida.

Subphylum Cyrtophora

Somatic ciliature with overlapping transverse microtubular ribbons; most species with oral dikinetids and polykinetids; well-developed cortical alveolar system; regression and redevelopment of oral apparatus in proter during cytokinesis.

Class Oligohymenophorea

Somatic monokinetids with forwardly directed, distinctly overlapping fibrils, divergent postciliary ribbons, and radial transverse ribbons; oral apparatus generally well-defined, in buccal cavity, with distinct paroral dikinetid and one to many polykinetids.

Subclass Hymenostomatia

Body ciliation often uniform and heavy; buccal cavity, when present, ventral; sessile forms, stalks, and colony formation relatively rare; freshwater forms predominant.

Order Hymenostomatida

Buccal cavity well-defined; oral area on ventral surface, usually in anterior half of body.

Suborder Ophryoglenina

Large, primarily freshwater, histophagous forms; life cycle with cyst stage; several species causing white spot disease in marine and freshwater fishes. *Ichthyophthirius, Ophryoglena.*

Other Suborder

Tetrahymenina.

Other Order

Scuticociliatida.

Subclass Peritrichia

Oral ciliary field prominent, covering apical end of body, bordered by a dikinetid file and polykinetid that originate in an infundibulum; paroral membrane and adoral membranelles present; somatic ciliature reduced to temporary posterior circlet of locomotor cilia; many stalked and sedentary, others mobile, all with aboral scopula; conjugation total, involving fusion of microconjugants and macroconjugants.

Order Sessilida

Mature trophonts usually sessile, attached, with stalk; some obligate ectosymbionts on aquatic invertebrates. *Epistylis, Rhabdostyla.*

Order Mobilida

Mobile forms, usually conical or cylindrical (or discoidal and orally-aborally flattened), with permanently ciliated trochal band (ciliary girdle); complex thigmotactic apparatus at aboral end, often with highly distinctive denticulate ring; all ectoparasites or endoparasites of freshwater or marine vertebrates and invertebrates. *Trichodina, Urceolaria.*

References

1. Arroyo-Begovich, A., A. Cárabez-Trejo, and J. Ruiz-Herrera. 1980. Identification of the structural component in the cyst wall of *Entamoeba invadens. J. Parasitol.* 66:735–41.

2. Bailey, G. B., D. B. Day, and N. E. McCoomer. 1992. *Entamoeba* motility: Dynamics of cytoplasmic streaming, locomotion and translocation of surface-bound particles, and organization of the actin cytoskeleton in *Entameoba invadens. J. Protozool.* 39:267–72.

3. Cavalier-Smith, T., and J. J. Lee. 1985. Protozoa as hosts for endosymbioses and the conversion of symbionts into organelles. *J. Protozool.* 32:376–79.

4. Chapman, G. B., and R. C. Kern. 1983. Ultrastructural aspects of the somatic cortex and contractile vacuole of the ciliate, *Ichthyophthirius multifiliis* Fouquet. *J. Protozool.* 30:481–90.

5. Corliss, J. O. 1981. What are the taxonomic and evolutionary relationships of the protozoa to the other Protista? *Biosystems* 14:445–59.

6. Corliss, J. O. 1985. Concept, definition, prevalence, and host-interactions of xenosomes (cytoplasmic and nuclear endosymbionts). *J. Protozool.* 32:373–76.

7. Dawes, C. J. 1981. *Marine botany.* New York: John Wiley & Sons, Inc.

8. de Duve, C. 1983. Microbodies in the living cell. *Sci. Am.* 248(5):74–84 (May).

9. Erlandsen, S. L., W. J. Bemrick, and J. Pawley. 1989. High-resolution electron microscope evidence for the filamentous structure of the cyst wall in *Giardia muris* and *Giardia duodenalis. J. Parasitol.* 75:787–97.

10. Finlay, B. J. 1985. Nitrate respiration by protozoa (*Loxodes* spp.) in the hypolimnetic nitrite maximum of a productive freshwater pond. *Freshwater Biol.* 15:333–46.

11. Grell, K. G. 1973. *Protozoology.* New York: Springer-Verlag.

12. Januschka, M. M., E. L. Erlandsen, W. J. Bemrick, D. G. Schupp, and D. E. Feely. 1988. A comparison of *Giardia microti* and *Spironucleus muris* cysts in the vole: An immunocytochemical, light, and electron microscope study. *J. Parasitol.* 74:452–58.

13. Kloetzel, J. A. 1991. Identification and properties of *plateins,* major proteins in the cortical alveolar plates of *Euplotes. J. Protozool.* 38:392–401.

14. Kloetzel, J. A., B. F. Hill, and T. Kosaka. 1992. Genetic variants of plateins (alveolar plate proteins) among and within species of *Euplotes. J. Protozool.* 39:92–101.

15. Kudo, R. R. 1966. *Protozoology.* New York: C. Thomas.

16. Lee, J. J., S. H. Hutner, and E. C. Bovee. 1985. *An illustrated guide to the protozoa.* Lawrence, Kans.: Society of Protozoologists.

17. Lee, J. J., M. J. Lee, and D. S. Weis. 1985. Possible adaptive value of endosymbionts to their protozoan hosts. *J. Protozool.* 32:380–82.

18. Lee, J. J., A. T. Soldo, W. Reisser, M. J. Lee, K. W. Jeon, and H.-D. Gortz. 1985. The extent of algal and bacterial endosymbioses in protozoa. *J. Protozool.* 32:391–403.

19. Levine, N. D., et al. 1980. A newly revised classification of the protozoa. *J. Protozool.* 27:37–58.

20. MacKenzie, C., and M. H. Walker. 1983. Substrate contact, mucus, and eugregarine gliding. *J. Protozool.* 30:3–8.

21. Michels, P. A. M., and F. R. Opperdoes. 1991. The evolutionary origin of glycosomes. *Parasitol. Today* 7:105–9.

22. Müller, M. 1975. Biochemistry of protozoan microbodies: Peroxisomes, α-glycerophosphate oxidase bodies, hydrogenosomes. *Ann. Rev. Microbiol.* 29:467–83.

23. Nachmias, V. T. 1984. *Microfilaments. Carolina biology reader, no. 130.* Burlington, N.C.: Carolina Biological Supply Co.

24. Opperdoes, F. R., and P. Borst. 1977. Localization of nine glycolytic enzymes in a microbody-like organelle in *Trypanosoma brucei:* The glycosome. *FEBS Letters* 80:360–64.

25. Pitelka, D. R. 1963. *Electron-microscopic structure of Protozoa.* Elmsford, N.Y.: Pergamon Press.

26. Preston, T. M., C. A. King, and J. S. Hyams. 1990. *The cytoskeleton and cell motility.* London: Blackie and Son Limited.

27. Reisser, W., R. Meier, H.-D. Gortz, and K. W. Jeon. 1985. Establishment, maintenance, and integration mechanisms of endosymbionts in protozoa. *J. Protozool.* 32:383–90.

28. Satir, P. 1983. *Cilia and related organelles. Carolina biology reader,* no. 123. Burlington, N.C.: Carolina Biological Supply Co.

29. Sheffield, H. J., and M. L. Melton. 1968. The fine structure and reproduction of *Toxoplasma gondii. J. Parasitol.* 54:209–26.

30. Stebbings, H., and J. S. Hyams. 1979. *Cell motility.* London: Longman Group Limited.

31. van Wagtendonk, W. J. 1955. Encystment and excystment of Protozoa. In Hutner, S. H., and A. Lwoff, eds. *Biochemistry and physiology of Protozoa* 2. New York: Academic Press, Inc.

Additional References

Corliss, J. O. 1981. What are the taxonomic and evolutionary relationships of the protozoa to the Protista? *Biosystems* 14:445–59.

Hyman, L. H. 1940. *The invertebrates, vol. 1. Protozoa through Ctenophora.* New York: McGraw-Hill Book Co. An excellent reference to general aspects of the Protozoa.

Jahn, T. L., E. C. Bovee, and F. F. Jahn. 1979. *How to know the Protozoa,* 2d ed. Dubuque, Iowa: Wm. C. Brown Publishers. Identification keys to the common Protozoa.

Kreier, J. P., and J. R. Baker. 1987. *Parasitic protozoa.* Boston: Allen and Unwin.

Lee, J. J., S. H. Hutner, and E. C. Bovee, eds. 1986. *An illustrated guide to the protozoa.* Lawrence, Kans.: Society of Protozoologists.

Levine, N. D. 1973. *Protozoan parasites of domestic animals and man,* 2d ed. Minneapolis: Burgess Publishing Co. A very useful reference to the parasites indicated by the title.

Margulis, L. 1981. *Symbiosis in cell evolution.* San Francisco: W. H. Freeman and Co., Publishers. Presents the case for the symbiotic origin of the eukaryotes.

Scholtyseck, E. 1979. *Fine structure of parasitic protozoa.* Berlin: Springer-Verlag. Atlas of electron micrographs accompanied by labeled diagrams. Heavy on Apicomplexa.

Sleigh, M. A. 1989. *Protozoa and other protists.* New York: Edward Arnold.

Stossel, T. P. 1994. The machinery of cell crawling. *Sci. Am.* 71(3):54–63 (Sept.).

Trager, W. 1986. *Living together. The biology of animal parasitism.* New York: Plenum Press.

Chapter 5

KINETOPLASTIDA: TRYPANOSOMES AND THEIR KIN

They were talking about me in terminal terms.

Page Faegre, teacher (Berkeley, Calif.)

We got down an atlas and figured it out. It's really

very exciting to see something that improbable.

Dr. David Miller, who helped diagnose Page Faegre's case of kala-azar acquired on a trip to the Mediterranean coast

The protozoan order Kinetoplastida contains a group of species that parasitize everything from humans to plants. Members of this group are characterized by a single large mitochondrion containing a body that stains darkly in histological preparations. This body is known as the **kinetoplast,** from which the order name Kinetoplastida is derived. The kinetoplast is located beside the kinetosome at the base of the flagellum, and along with the nearby parts of the mitochondrion, it remains in a more or less established relationship with the kinetosome throughout the parasite's life cycle (Fig. 5.1). The kinetoplast is actually a disc-shaped, DNA-containing organelle within the mitochondrion. Kinetoplast DNA (kDNA) is organized into a network of linked circles, quite unlike the organization of DNA in a chromosome.[7] There are up to 20,000 tiny circles (minicircles) and 20 to 50 larger circles (maxicircles) in the kinetoplast network. Most electron micrographs show no physical connection between the kinetosome and kinetoplast, and the nature of their established association is unknown.

In addition to their distinctive mitochondrial structure, kinetoplastids have a cytoskeleton consisting of microtubules arranged at regular intervals beneath the plasma membrane (Figs. 5.1 and 5.2). Other characteristics include a sizeable flagellar pocket, sometimes elongated; a latticelike crystalline **paraxial rod,** alongside the axoneme, that has short projections that connect it to the axonemal microtubules of the flagellum;[23] an **undulating membrane** (depending on the species); and occasionally a prominent **glycocalyx,** or surface coat, visible in electron micrographs.

Kinetoplastid genera differ considerably in their host distribution, life cycles, and medical and veterinary importance. Three families are recognized: Bodonidae (coprozoic and free-living), Cryptobiidae (parasites of fish and invertebrates), and Trypanosomatidae, some members of which are important human and veterinary pathogens. These parasites provide fascinating challenges for the parasitologist, ranging from extraordinarily difficult control problems to dramatic

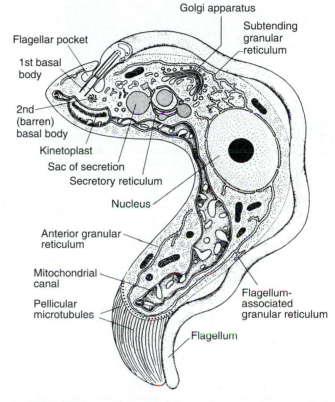

Golgi apparatus

Subtending granular reticulum

Flagellar pocket

1st basal body

2nd (barren) basal body

Kinetoplast

Sac of secretion

Secretory reticulum

Nucleus

Anterior granular reticulum

Mitochondrial canal

Pellicular microtubules

Flagellum-associated granular reticulum

Flagellum

FIGURE 5.1

Diagram to show principal structures revealed by the electron microscope in the bloodstream trypomastigote form of the salivarian trypanosome, *Trypanosoma congolense.* It is shown cut in sagittal sections, except for most of the shaft of the flagellum and the anterior extremity of the body.

From K. Vickerman, "The fine structure of *Trypanosoma congolense* in its bloodstream phase," in *J. Protozool.* 16:54–69. Copyright © 1969 The Society of Protozoologists. Reprinted with permission of the publisher.

pathological effects such as the erosion of facial features. Some have been popular research organisms because of their ease of culture; others defied the taxonomists until molecular biology revealed their enzyme and DNA characteristics; and still others present us with such a diverse clinical picture that we have yet to dissect out the effects of parasite traits, human genetic makeup, and environmental factors such as diet, from the parasites' overall public health impact.

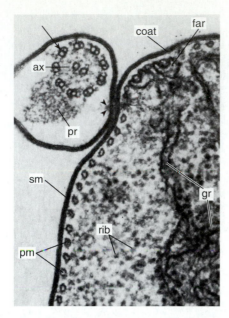

FIGURE 5.2

Trypanosoma congolense. Transverse section of shaft of flagellum and adjacent pellicle in region of attachment. Both flagellum and body surface have a limiting unit membrane **(sm)** covered by a thick coating **(coat)** of dense material. The axoneme **(ax)** of the flagellum shows the partition **(arrow)** dividing one of the tubules of each doublet; alongside the axoneme lies the paraxial rod **(pr).** Pellicular microtubules **(pm)** underlie the surface membrane of the body, and a diverticulum **(far)** of the granular reticulum **(gr)** is always found embracing three or four of these microtubules close to the flagellum. Note the fibrous condensations (*arrowheads*) on either side of the opposed surface membranes, apparently "riveting" the flagellum to the body. A row of these "rivets" replaces a microtubule along the line of adherence. **rib,** ribosomes. (× 66,000.)

From K. Vickerman, "The fine structure of *Trypanosoma congolense* in its bloodstream phase," in *J. Protozool.* 16:54–69. Copyright © 1969 The Society of Protozoologists. Reprinted with permission of the publisher.

FAMILY TRYPANOSOMATIDAE

All species of Trypanosomatidae have a single nucleus and are either elongated with a single flagellum or rounded with a very short, nonprotruding flagellum. Most members of the family are **heteroxenous:** During one stage of their lives they live in the blood and/or fixed tissues of all classes of vertebrates, and during other stages they live in the intestines of bloodsucking invertebrates. In addition, laboratory culture media for these parasites usually must contain blood. Thus, we call them **hemoflagellates.**

Sexual phenomena have not been observed directly in these organisms, but there is a considerable amount of indirect evidence for sexuality.[12,49,96,97,104] In experimental work, parental stocks and hybrids can be identified using isoenzyme markers or fragment length distributions following enzymatic digestion of DNA. "Mating" within the tsetse fly vector—that is, recombination of karyotype or isoenzyme phenotypes—has been demonstrated, although genetic recombination evidently is not obligatory in the life cycle. Segregation of allelic marker genes suggests meiosis, but largely because of the parasites' small size and nonpredictable "mating," meiosis such as seen in larger eukaryote gametocytes

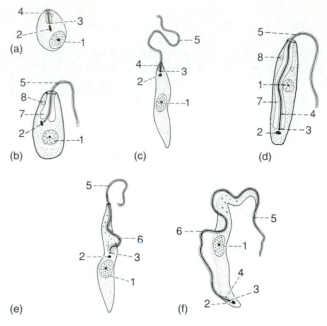

FIGURE 5.3

Genera of Trypanosomatidae. (*a*) *Leishmania* (amastigote form); (*b*) *Crithidia* (choanomastigote); (*c*) *Leptomonas* (promastigote); (*d*) *Herpetomonas* (opisthomastigote); (*e*) *Blastocrithidia* (epimastigote); (*f*) *Trypanosoma* (trypomastigote); **1,** nucleus; **2,** kinetoplast; **3,** kinetosome; **4** and **5,** axoneme and flagellum; **6,** undulating membrane; **7,** flagellar pocket; **8,** contractile vacuole.

From O. W. Olsen, *Animal Parasites, Their Biology and Life Cycles,* 3d ed. Copyright © 1974 Dover Publications, Inc., New York, NY. Reprinted by permission.

has not been observed. Furthermore, at least experimentally, trypanosomatids may take up foreign DNA.[8] Thus, there exists a variety of mechanisms by which strains of these parasites may come to vary genetically, adding, no doubt, to the often confusing clinical picture seen in infections.

The family Trypanosomatidae probably originally parasitized the digestive tract of insects and, possibly, annelids, and some species are still **monoxenous**—that is, parasitic only within a single arthropod host.[110] Most trypanosomatids pass through different morphological stages, depending on the phase of their life cycle and the host they are parasitizing. In the past these stages were named after the genera they most resembled—for example, **leptomonad** for a stage resembling *Leptomonas*—but currently a nomenclature referring to kinetosome and nucleus positions prevails (Fig. 5.3).

The **trypomastigote** stage is characteristic of bloodstream forms of the genus *Trypanosoma* as well as the infective **metacyclic** stages in the tsetse fly vector. In trypomastigotes, the kinetoplast and kinetosome are near the posterior end of the body, and the flagellum runs along the surface, usually continuing as a free whip anterior to the body. The flagellar membrane is closely apposed to the body surface, and when the flagellum beats, this area of the pellicle is pulled up into a fold; the fold and the flagellum constitute the undulating membrane. A second, "barren" kinetosome, without a flagellum, is usually found near the flagellar kinetosome.

In the typical bloodstream form of trypomastigote, a simple mitochondrion with or without tubular cristae runs anteriorly from the kinetoplast. In the insect stage of the organism, the mitochondrion is much larger and more complex, with lamellar cristae. At the base of the flagellum and surrounding the kinetosome is a **flagellar pocket** or reservoir. A system of **pellicular microtubules** spirals around the body just beneath the cell membrane (Figs. 5.1 and 5.2). Rough endoplasmic reticulum is well-developed, and a Golgi body lies between the nucleus and kinetosome.

Other trypanosomatid body forms differ in shape, position of the kinetosome and kinetoplast, development of the flagellum, or shape of the flagellar pocket (Fig. 5.3). A spheroid **amastigote** occurs in the life cycle of some species and is definitive in the genus *Leishmania.* The flagellum is very short, projecting only slightly beyond the flagellar pocket. In the **promastigote** stage the elongated body has the flagellum extending forward as a functional organelle. The kinetosome and kinetoplast are located in front of the nucleus, near the anterior end of the body. The promastigote form is found in the life cycles of several species while they are in their insect hosts. It is the mature form in the genus *Leptomonas.* If the flagellum emerges through a wide, collar-like process, the type is termed a **choanomastigote,** which is found in some species of *Crithidia.*

The **epimastigote** form occurs in some life cycles. Here the kinetoplast and kinetosome are still located between the nucleus and the anterior end, but a short undulating membrane lies along the proximal part of the flagellum. The genera *Crithidia* and *Blastocrithidia,* both parasites of insects, exhibit this form during their life cycles. Finally, the **paramastigote** and **opisthomastigote** forms are found in *Herpetomonas,* a widespread group of insect parasites. In paramastigotes, the kinetosome and kinetoplast are beside the nucleus; in opisthomastigotes, these organelles are located between the nucleus and posterior end, but there is no undulating membrane, the flagellum piercing a long reservoir that passes through the entire length of the body and opens at the anterior end. In the genus *Herpetomonas,* reproduction occurs only in the promastigote form, with the other body forms appearing after populations have reached their peak, such as in culture. Despite their apparent structural simplicity, the trypanosomatids are actually quite diverse, with most of their variability manifested in ultrastructural features and in internal distribution of organelles, differences in life cycles, and host specificity.

Trypanosomatid life cycles vary with respect to the host species, vectors, behavior of the parasites in the vectors and in vertebrate hosts, and life-cycle stages in which reproduction occurs. *Leptomonas* species exhibit the simplest cycle in which an insect is the sole host, multiplication is by promastigotes in the gut, and transmission occurs by way of an ingested amastigotelike cyst. *Leishmania* species undergo multiplication as promastigotes in blood-sucking insects (sand flies; see further on), but they are injected into a vertebrate host when the sand fly feeds, and they undergo additional multiplication, as amastigotes, in a variety of tissues. Members of the genus *Trypanosoma* exhibit the greatest diversity of forms during their life cycles, changing into multiplying epimastigotes in the insect vector's midgut and then into infective trypomastigotes (**metacyclic** forms) in either the hindgut or foregut (depending on the species). Metacyclic

trypomastigotes, depending on the species of parasite, are either passed in the feces to contaminate a wound or injected with saliva during feeding. Tsetse flies of the genus *Glossina* (Fig. 5.4) serve as vectors for the medically important *Trypanosoma brucei,* but fleas, horse flies, true bugs (order Hemiptera), and bats also function as vectors, depending on the parasite species.

Members of the genus *Leishmania* also occupy two strikingly different environments: the insect vector gut and the interior of a vertebrate host cell, typically a macrophage. In the vector (flies of the family Psychodidae, subfamily Phlebotominae) or in culture at 25°C, the parasites are promastigotes and divide rapidly. But in the vertebrate host, promastigotes are phagocytized by macrophages. Within the phagocytic (**parasitophorous**) vacuoles, the promastigotes transform into amastigotes. Although they continue to multiply, they do so at a much slower rate than in culture. Whether inside the phlebotomine gut or in the parasitophorous vacuoles, the parasites are living inside organs, or organelles, that usually function to digest foreign objects.

As in the case of trypanosomes, *Leishmania* species exhibit ultrastructural and metabolic changes as they pass from one life-cycle stage to another. Loss of external flagellum, change from elongated to round body form, rearrangement of the subpellicular microtubules, reduction in oxygen consumption, and activation of metabolic pathways that function to use host cell nucleic acid precursors all accompany transformation from extracellular promastigote to intracellular amastigote.[83] However, clean preparations of intracellular amastigotes, uncontaminated with host cell debris, have proven difficult to obtain, and the search for methods of culturing or harvesting amastigotes has occupied the efforts of many researchers. Heat shocking of promastigotes has proven an effective technique for producing amastigotes, and recent developments have allowed the continuous culture of amastigotes at elevated temperatures.[26]

Biochemical studies on trypanosomatids have not been confined to the medically important species.[67] A number of monoxenous parasites, such as *Crithidia fasciculata* and *Leptomonas costoris,* have been used as "models" for the study of biochemical pathways and gene expression, mainly because of their ease of culture and the fact that many will grow on chemically defined media. The costs and time involved in searching for unique metabolic pathways are greatly reduced by using such parasites as research organisms. The pathways studied are often ones that might be targeted by chemotherapeutic agents in the related pathogenic forms.

Genus *Trypanosoma*

All trypanosomes (except *T. equiperdum*) are **heteroxenous** or at least are transmitted by animal vectors. Various species pass through amastigote, promastigote, epimastigote, and/or trypomastigote stages, with the other forms developing in the invertebrate hosts. Much research has been conducted on this genus because of its extreme importance to the health of humans and domestic animals. Reviews are available dealing with various aspects of the group: host susceptibility,[72] physiology and morphology,[106,107] chemotherapy,[113] taxonomy,[42,76]

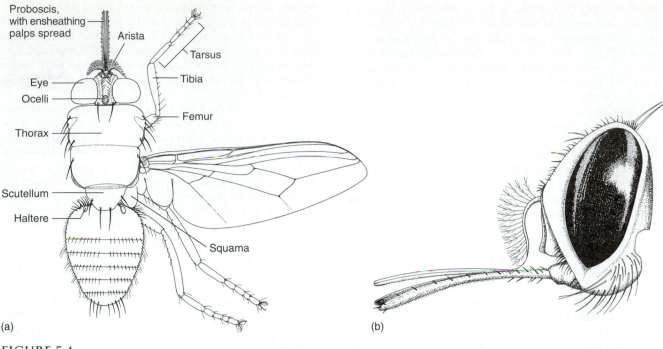

(a)

(b)

FIGURE 5.4

A tsetse fly of the genus *Glossina,* with general anatomical features labelled. (*a*) Dorsal view of *Glossina* showing general anatomical features; (*b*) head, lateral view showing proboscis and palps.

Drawn by J. Janovy, Jr. from a University of Nebraska State Museum specimen (provided by B. C. Ratcliffe, Curator of Entomology).

immunology,[1,6,92,99,108,109] evolution,[33,44] and vector relationships.[68] It is possible that some species now known only from invertebrates may also be parasites of vertebrates.[58]

Members of the genus *Trypanosoma* are parasites of all classes of vertebrates. Most live in the blood and tissue fluids, but some important ones, such as *T. cruzi,* occupy intracellular habitats as well. Although other means of transmission exist, the majority are transmitted by blood-feeding invertebrates.

A few species are responsible for misery and privation of enormous proportions. In Africa alone 4.5 million square miles, an area larger than the United States, are incapable of supporting a cattle industry—not because the land is poor, since it is much like the grasslands of the American West, but because domestic livestock are killed by trypanosomes that are nonfatal parasites of native grazing animals. Thus, semiarid lands that otherwise could support agronomy are denied to millions of persons who most need the protein affordable by the rich soil. More directly affected by trypanosomes are millions of people in South America who have never known a day of good health because of trypanosome infections.

Trypanosomes are divided into two broad groups, or "sections," based on characteristics of their development in the insect hosts. If a species develops in the anterior portions of the digestive tract, it is said to undergo **anterior station** development and is relegated to the section Salivaria, which contains several subgenera. When a species develops in the hindgut of its invertebrate host, it is said to undergo **posterior station** development and is placed in the section Stercoraria. Other developmental and morphological criteria separate the two sections and further aid in placement in the proper section of species that do not require development in an intermediate host (*T. equiperdum, T. equinum*).[42] Classification of the various species into subgenera is based on their physiology, morphology, and biology.

• Section Salivaria

Trypanosoma (Trypanozoon) brucei. The three subspecies of *Trypanosoma brucei*—*T. b. brucei, T. b. gambiense,* and *T. b. rhodesiense*—are morphologically indistinguishable but traditionally have been treated as separate species. They vary in infectivity for different species of hosts and produce somewhat different pathological syndromes. *T. b. brucei* is considered like the ancestral form, with the other two being subspecies.[76] However, more recent biochemical studies of kDNA nucleotide sequences and isoenzyme variations suggest that *T. b. gambiense* is more likely a strain of *T. brucei* than a distinct species or subspecies. Regardless of their taxonomic status, *T. brucei* type typanosomes are widely distributed in tropical Africa between latitude 15° N and 25° S, corresponding in distribution with their vectors, tsetse flies (*Glossina* spp.) (Figs. 5.4 and 5.5).

Trypanosoma brucei brucei is fundamentally a parasite of the bloodstream of native antelopes and other African ruminants, causing a disease called **nagana.** Unfortunately, the

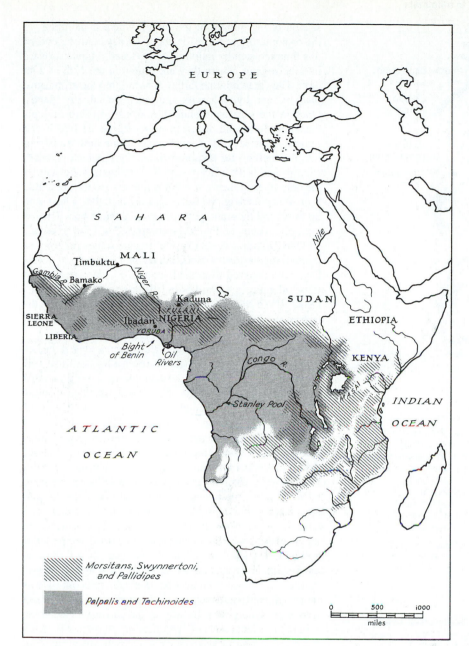

FIGURE 5.5

Distribution of five *Glossina* species across Africa.

From John J. McKelvey, Jr., *Man Against Tsetse: Struggle for Africa*. Maps drawn by John Morris. Copyright © 1973 by Cornell University. Used by permission of the publisher, Cornell University Press.

parasite also infects introduced livestock, including sheep, goats, oxen, horses, camels, pigs, dogs, donkeys, and mules. It is pathogenic to these animals, as well as to several native species. Humans, however, are not susceptible.

Trypanosoma brucei gambiense and *T. b. rhodesiense* are the etiological agents of African sleeping sickness. There are physiological differences between them, and they differ in pathogenesis, growth rate, and biology. As early as 1917 a German researcher named Taute showed that the nagana trypanosome would not infect humans. He repeatedly inoculated himself and native "volunteers" with nagana-ridden blood; none of them acquired sleeping sickness. His work was largely discounted by British experts for reasons that can only be guessed (maybe they could not believe a scientist would be curious enough about a parasite's behavior to try to infect himself with it, but this has happened more than once in the history of parasitology).

Trypanosoma brucei gambiense causes a chronic form of sleeping sickness. It is found in west central and central Africa,

In mammals

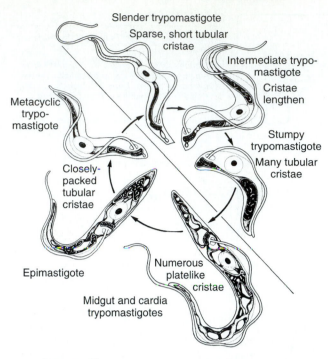

Slender trypomastigote

Sparse, short tubular cristae

Intermediate trypo-mastigote

Cristae lengthen

Metacyclic trypo-mastigote

Stumpy trypomastigote

Many tubular cristae

Closely-packed tubular cristae

Epimastigote

Numerous platelike cristae

Midgut and cardia trypomastigotes

In tsetse flies

FIGURE 5.6

Diagram to show changes in form and structure of the mitochondrion of *Trypanosoma brucei* throughout its life cycle. The slender bloodstream form lacks a functional Krebs cycle and cytochrome chain. Stumpy forms have a partially functional Krebs cycle but still lack cytochromes. The glycerophosphate oxidase system functions in terminal respiration of bloodstream forms. The fly gut forms have a fully functional mitochondrion with active Krebs cycle and cytochrome chain. Cytochrome oxidase may be associated with the distinctive platelike cristae of these forms. Reversion to tubular cristae in the salivary gland stages may therefore indicate loss of this electron transfer system.

From K. Vickerman, "Morphological and physiological considerations of extracellular blood protozoa," in *Ecology and Physiology of Parasites*. Edited by A. M. Fallis. Copyright © 1971 University of Toronto Press. Reprinted by permission.

whereas *T. b. rhodesiense* occurs in central and east central Africa and causes a more acute type of infection. Native game animals serve as reservoirs for Rhodesian trypanosomiasis but apparently not for Gambian trypanosomiasis.

- **Morphology and Life History.** *Trypanosoma brucei* in natural infections tends to be **pleomorphic** (polymorphic) in its vertebrate host, ranging from long, slender trypomastigotes with a long free flagellum through intermediate forms to short, stumpy individuals with no free flagellum (Fig. 5.6). The small kinetoplast is usually very near the posterior end, and the undulating membrane is conspicuous.

The insect vectors of *T. b. brucei* and *T. b. rhodesiense* are *Glossina morsitans, G. pallidipes,* and *G. swynnertoni,* whereas those of *T. b. gambiense* are *G. palpalis* and *G. tachinoides* (Fig. 5.5). At least 90% of the flies are refractive to infection. When sucked up by a susceptible fly along with a blood meal, *T. brucei* locates in the posterior

section of the midgut of the insect, where it multiplies in the trypomastigote form for about 10 days. At the end of this time the slender individuals produced migrate forward into the foregut, where they are found on the 12th to 20th days. They then migrate farther forward into the esophagus, pharynx, and hypopharynx and enter the salivary glands. Once in the salivary glands they transform into the epimastigote form and attach to host cells or lie free in the lumen. After several asexual generations they transform into the **metacyclic trypomastigote** form, which is small, stumpy, and lacks a free flagellum. The metacyclic stage is the only stage in the vector that is infective to the vertebrate host. When feeding, the tsetse fly may inoculate a host with up to several thousand protozoa with a single bite. The entire cycle within the fly can be completed in 15 to 35 days.

Once within a vertebrate, the trypanosomes multiply as trypomastigotes in the blood and lymph. Amastigote forms have been reported from the liver and spleen of experimentally infected mice and from the myocardium of monkeys.[75,114] In the chronic form of the disease many of the trypanosomes invade the central nervous system, multiply, and enter the intercellular spaces within the brain.

Biochemical, ultrastructural, and immunological studies have added greatly to our understanding of trypanosomatids.[11,67,106] Of considerable value is the fact that essentially pure preparations of certain morphological stages can be obtained. For example, when *Trypanosoma brucei* is passed by syringe from one vertebrate host to another, the strain tends to become monomorphic after a period of time, the population consisting only of slender trypomastigotes that are no longer infective to tsetse flies and cannot be cultivated in vitro. Their morphology and metabolism correspond to the slender trypomastigote in natural infections. In contrast, when *T. brucei* is placed in certain in vitro culture systems, its morphology and metabolism revert to that found in the fly midgut, with the kinetoplast farther from the posterior end and close to the nucleus. It has been found that the monomorphic, syringe-passed strain depends entirely on glycolysis for its energy production, degrading glucose only as far as pyruvate and having no tricarboxylic acid cycle or oxidative phosphorylation by way of the classical cytochrome system. The reduced NAD produced in glycolysis is reoxidized by a nonphosphorylating glycerophosphate oxidase system, which, although it requires oxygen, is not sensitive to cyanide. This respiratory system is inhibited by **suramin,** an antitrypanosomal drug, and it is apparently localized in membrane-bound microbodies called alpha-glycerophosphate oxidase bodies, or **glycosomes.**[40,70] The long, slender trypomastigote is very active, and it consumes substantial quantities of both glucose and oxygen in its inefficient energy production. Evidently, the blood and lymph have such a plentiful supply of both glucose and oxygen that no selective value is attached to the conservation of either.

The situation is quite different when the trypanosome finds itself in a blood clot in its vector's midgut. In this case the parasite completely degrades glucose via glycolysis, the tricarboxylic acid cycle, and the cyanide-sensitive cytochrome system. The oxygen and glucose consumption of the midgut (or culture) form is only 1/10 that of the bloodstream form. Culture forms have two oxidase systems with a high affinity for oxygen, one containing cytochrome aa_3, which accounts for more than 50% of the respiration, and the other apparently containing cytochrome o, accounting for 25% to 35% of oxygen consumption.[40] The glycerophosphate oxidase system is also present in the culture forms, but its activity now is sensitive to mitochondrial inhibitors.[70]

Ultrastructural observations on the mitochondria in the respective forms (bloodstream vs. culture or vector) correlate beautifully with the biochemical findings. The long, slender trypomastigote has a single, simple mitochondrion extending anteriorly from its kinetoplast, and the cristae are few, short, and tubular. The midgut stage has an elaborate mitochondrion extending both posteriorly and anteriorly from the kinetoplast, and the cristae are numerous and platelike. The curious movement of the kinetoplast away from the posterior end in the midgut trypomastigote and anterior to the nucleus in the epimastigote can now be understood as reflecting the elaboration of the posterior section of mitochondrion, which "pushes" the kinetosome forward. Furthermore, the short, stumpy form is the only one infective to the tsetse fly, and the intermediate form is transitional from the long, slender noninfective form (Fig. 5.6). Correspondingly, electron microscopy has shown that this transition is marked by increasing elaboration of the mitochondrion; synthesis of mitochondrial enzymes has been shown by cytochemical means. Similarly the metacyclic form has a mitochondrion much like the bloodstream form.

- *Pathogenesis.* In their vertebrate hosts these trypanosomes live in the blood, lymph nodes and spleen, and cerebrospinal fluid. They do not invade or live within cells but inhabit connective tissue spaces within various organs and the reticular tissue spaces of the spleen and lymph nodes. They are particularly abundant in the lymph vessels and intercellular spaces in the brain.

The clinical course in *T. b. brucei* infections depends on the susceptibility of the host species. Horses, mules, donkeys, some ruminants, and dogs suffer acutely, and they may die in 15 days, although sometimes survive for up to four months. Symptoms include anemia, edema, watery eyes and nose, and fever. Within a few days the animals become emaciated, uncoordinated, and paralyzed, and they die shortly afterward. Blindness as the result of the infection is common in dogs. Cattle are somewhat more refractory to the disease, often surviving for several months after onset of symptoms. Swine usually recover from the infection.

In human infections with *T. b. rhodesiense* and *T. b. gambiense* a small sore often develops at the site of inoculation of metacyclic trypansomes. This lesion disappears after one to two weeks. After the protozoa gain entrance to the blood and lymph channels, they reproduce rapidly, producing a parasitemia and invading nearly all organs of the body. *Trypansoma b. rhodesiense* rarely invades the nervous system as does *T. b. gambiense* but usually causes a more rapid course toward death. The lymph nodes become swollen and congested, especially in the neck, groin, and legs. Swollen nodes at the base of the skull were recognized by slave traders as signs of certain death, and slaves who developed them were routinely thrown overboard by slavers bound for the Caribbean markets. Today such swollen lymph nodes are called **Winterbottom's sign,** named after the British officer who first described the symptom. The symptoms of illness usually are more marked in newcomers than in people native to endemic areas.

Intermittent periods of fever accompany the early stages of the disease, and the number of trypanosomes in the circulating blood greatly increases at these times. As previously noted, the successive parasite populations represent different antigenic types. With fever there is an increase in swelling of lymph nodes, generalized pain, headache, weakness, and cramps. Infection of *T. b. rhodesiense* causes rapid weight loss and heart involvement. Death may occur within a few months after infection, but *T. b. rhodesiense* causes no somnambulism or other protracted nervous disorders found with *T. b. gambiense* because the host usually dies before these can develop.

When the trypanosomes of *T. b. gambiense* invade the central nervous system, they initiate the chronic, sleeping-sickness stage of infection. Increasing apathy, a disinclination to work, and mental dullness accompany disturbances of coordination. Tremor of the tongue, hands, and trunk is common, and paralysis or convulsions usually follow. Sleepiness increases, with the patient falling asleep even while eating or standing. Finally, coma and death ensue. Actually, death may result from any one of a number of related causes, including malnutrition, pneumonia, heart failure, other parasitic infections, or a severe fall.

The mechanism of pathogenesis is unclear. In the acute infection of small mammals, in which death occurs rapidly at a time of a high level of parasitemia, mortality probably is a result of overall disruption of normal physiological processes.[62] In the case of mammals, including humans, present evidence suggests that pathogenesis may be caused in part by the antigenic "performance" of the trypanosomes and the host's immune reactions. Because of the repeated changes in the surface antigens of the parasites and the fact that these surface antigens are being released into the blood almost constantly (exoantigens), the immune system of the host is greatly stimulated, and huge amounts of immunoglobulins are produced. Some of the trypanosome antigens can adsorb to the surface of some host cells, and binding of the specific antibody to the adsorbed antigen, in conjunction with complement, leads to lysis of the *host's* cells. Lysis of red blood cells by this mechanism may account for the anemia of trypanosomiasis.

It has been reported that the free tyrosine levels in the brain of voles (*Microtus*) infected with *T. brucei* were only about 50% of the level in controls, possibly caused by nutritional requirements of the trypanosomes. Resulting interference with protein and neurotransmitter synthesis in the brain may help account for neurological symptoms.[74] Subcurative treatment may increase the severity of central nervous system pathology. Experimental work with mice suggests that in such cases the host may produce autoantibodies against its own proteins such as myelin, possibly in response to epitopes shared between parasites and host.[47]

• ***Immunology.*** The trypanosomatids present the parasitologist with a number of especially challenging immunological problems. In trypanosomes, for example, the clinical course characteristic of the infection varies according to the host infected, but in certain hosts (guinea pig, dog, and rabbit), repeated remissions alternate with very high levels of parasitemia. That is, periods with few trypanosomes (and disease symptoms) evident are followed by a large increase in parasite population. This cycle tends to repeat itself until the host dies.[88]

The parasites have evolved an amazing mechanism for escaping obliteration by the host's defenses, namely the successive dominance of each of a series of variant antigenic types (VATs) over time. The remissions seem to result from the generation of protective antibodies by the host that destroy the homologous trypanosomes. But each time the host's antibodies are almost successful in eliminating the infection, the trypanosomes elude destruction by expressing the new surface glycoprotein, thus becoming a new VAT, and then rapidly multiplying! Furthermore, the only apparent limit to the number of VATs that can develop in a clone strain of trypanosomes during an untreated infection is the host's life span; the maximal number shown so far for a trypanosome stock is 101.[11]

The means by which the parasites achieve this succession of antigenic types is a fascinating story of gene expression, and modern molecular biology has done much to explain this phenomenon.[10,11,38,39,78,107] The antigen recognized by the host's immune system is a **variant-specific surface glycoprotein (VSG)** released through the flagellar reservoir of the trypanosome and completely covering the organism as a surface coat. Each *T. brucei* individual possesses approximately 1000 genes coding for VSGs; but only one VSG gene is expressed at any time. The others are transcriptionally silent.[27]

There are evidently three mechanisms of genetic rearrangement that result in expression of a gene.[10,118] In one of these the gene is duplicated and transposed to another position near the end (telomere) of a chromosome, replacing the resident gene. The transposed DNA segment becomes the expression-linked copy; it is located downstream from a promoter on the chromosome and is thus transcribed. A second mechanism involves duplicative transposition of an unexpressed telomeric VSG gene to a second telomeric site where it will be expressed. In a third mechanism inactive telomeric genes become activated and expressed but without duplication. Although these mechanisms involve telomeric positions, some evidence suggests that there may be two or more expression sites for VSG genes.[71]

Expression of the genes occurs in an imprecisely predictable order; that is, expression of a given VSG more commonly occurs after the expression of another, particular VSG but not invariably. Thus, the VSG genes in a population of trypanosomes are heterogeneous at any time in a chronic infection, but there is a single VAT that is predominant in the blood and against which the host mounts its antibody defense. Other dominant VATs can be found in such places as the brain and liver.[93] Adding to the complexity of the system, trypanosomes can lose VSG genes and add new genes to their repertoire by a mechanism of segmental gene conversion (nonreciprocal crossing over).[79,80]

When the trypanosome is ingested by *Glossina*, it loses its VSG surface coat. This implies that there is yet another mechanism for activating and deactivating VSG genes. Expression resumes when the trypanosomes reach the metacyclic state and are then able to infect the mammalian host. A much smaller number of VATs characterizes the metacyclic trypanosomes; only eight VATs make up 60% to 80% of the population in one *T. b. rhodesiense* clone.[103] Although the first VSG gene expressed after infection of the mammalian host is one of the metacyclic VATs, within a few days the VSG found on the trypanosome that had been ingested by the fly will be expressed. This reexpression of the ingested VAT is referred to as **anamnestic** expression. While the promastigotes are without a surface coat in the tsetse fly (a period of several weeks), they are vulnerable to antibodies in subsequent blood meals of the vector. This points to a possible means of protection on cattle ranches if stock have a level of immunity, since the flies have little else to eat.

Although the VAT story is best known for *T. brucei,* antigen switching is also found in at least some other trypanosomes, such as *T. vivax.*[5,30] Through studies of immunology and VSG gene expression, parasitologists have kept alive the line of investigation that began early in this century in an attempt to explain the course of trypanosome infections. Now we know that surface proteins are parasite counter-defenses against host defenses in several groups of parasites (pp. 85, 103, 148, and 226).

• ***Diagnosis and Treatment.*** Demonstration of the parasite in the blood, bone marrow, or cerebrospinal fluid establishes diagnosis. For people in endemic areas in which early infections are usually symptomless, a serological test is available.[3]

Arsenical drugs historically have been used in the treatment of African trypanosomiases, but these drugs have severe drawbacks.[34] They cause eye damage and are best administered intravenously; furthermore, trypanosomes rapidly become tolerant to them. Other drugs (suramin, pentamidine, and Berenil) have been developed in recent years and have proved to be satisfactory in most early cases. Prognosis, however, is poor if the nervous system has become involved. A combination of Berenil and certain nitroimidazoles is effective experimentally in cases with nervous system involvement.[50]

More recently it has been shown that difluoromethylornithine (DFMO) is very efficacious in the treatment of African trypanosomiasis, especially in brain infections. It is the drug of choice at this writing.[90]

Nutrition of the vertebrate host can also affect the course of the disease. Adequate dietary lipid limits the infectivity of *T. brucei* in rats and probably also protects people against African sleeping sickness.[77]

• *Epidemiology and Control.* The most important factors influencing transmission are (1) reservoir hosts and (2) the presence of suitable vectors and the environments necessary for these vectors to reproduce. Tsetse flies occupy 4.5 million square miles of Africa, making much of that area impractical for human habitation (Fig. 5.5). *Glossina* vectors of *T. b. brucei* and *T. b. rhodesiense* occur in open country, pupating in dry, friable earth. Vectors of *T. b. gambiense* are riverine flies, breeding in shady, moist areas along rivers. Trypanosomes of the brucei group do not occur throughout the entire range of tsetse flies, and not all species of *Glossina* are vectors for them. Therefore transmission varies locally, depending on coincidence of the trypanosome and the proper fly species. Furthermore, there is an inheritance of susceptibility to trypanosomes in tsetse flies.[64]

Control of trypanosomiasis brucei is conducted along several lines, most of which involve the vectors. Tsetse flies are larviparous, and they deposit their young on the soil under brush. Because of this, and because the adults rest in bushes at certain heights above the ground and no higher, brush removal and trimming are very successful means of control. When wide belts of land are thus cleared, the flies seldom cross them and can be more easily controlled. However, this method is expensive and must be followed up every year to remove new growth.

Elimination of the wild game reservoirs has been proposed and practiced in some regions, stimulating an outcry among conservation-minded people all over the world.

Programs have been established in which people simply sit and catch flies that try to bite them. Because the flies feed only during the day, some farmers graze their livestock at night, moving them into enclosures during the day and protecting them from flies with switches.

The most satisfactory means of control is by spraying insecticides from aircraft. DDT and benzene hexachloride are inexpensive and highly effective for this purpose. *Glossina pallidipes* was eradicated from Zululand in this manner at a cost of about $0.40 per acre. However, the possibilities of harmful side effects of DDT must be carefully weighed against the benefits gained by its use.

So far, 80 years of tsetse eradication have had little impact on tsetse distribution. An experienced African field researcher, noting the flies' behavioral tendency to follow large, dark, elongate objects (such as cattle), suggested (half in jest to one of the authors) spending all the tsetse control money on vehicles that would then be driven across the continent and into Lake Victoria, drowning the insects that had followed them! A more rational approach has been the development and use of trypanotolerant cattle stocks, which may offer a practical alternative to control of the vectors or trypanosomes.[72] Breeds of cattle that have survived in trypanosome-infested regions of Africa show great promise as trypanotolerant animals. Breeding as few as three generations of trypanotolerant bulls to a trypanosensitive stock will produce a relatively trypanotolerant genotype.[25] However, breeding for tolerance to trypanosomes does not necessarily improve meat or milk production, nor does it convey resistance to other diseases.

The political situation in some African countries, often involving decreased cooperation between adjacent nations and tribes, may contribute to new epidemics. At the same time increased mobility of the human population contributes to wider and faster spread of the disease. Trypanosomiasis is an excellent illustration of the way complex interactions between parasite biology, ecology, economics, and social factors can make disease control extraordinarily difficult.

Trypanosoma (Nannomonas) congolense and *Trypanosoma (Duttonella) vivax.*

Nagana also is caused by *Trypanosoma congolense,* which is similar to *T. brucei* but lacks a free flagellum.[105] It occurs in South Africa, where it is the most common trypanosome of large mammals. The life cycle, pathogenesis, and treatment are as for *T. brucei.* The vascular damage reported in chronic *T. congolense* infections may be a result of the propensity of the trypanosomes to attach to the walls of small blood vessels by their anterior ends.[4]

Trypanosoma vivax is also found in the tsetse fly belt of Africa and has spread to the western hemisphere and Mauritius. Very similar to *T. brucei,* it causes a like disease in the same hosts. In the New World, transmission is mechanical and involves tabanid flies. Pathogenesis and control are as for *T. brucei.* Also, the changes in the mitochondrion through the life cycle in *T. vivax* and *T. congolense* are similar to those in *T. brucei.* However, these two species appear to retain some mitochondrial function in the bloodstream form. A good review of *T. vivax* was given by Gardiner and Wilson.[31]

Trypanosoma (Trypanozoon) evansi and *Trypanosoma (Trypanozoon) equinum.*

Trypanosoma evansi causes a widespread disease of camels, horses, elephants, deer, and many other mammals. The disease goes by many different names in different languages and countries but is most often called **surra**. *Trypanosoma evansi* probably was originally a parasite of camels.[41] Today it is distributed throughout the northern half of Africa, Asia Minor, southern Russia, India, southwestern Asia, Indonesia, Philippines, and Central and South America. The Spaniards introduced the disease to the western hemisphere by way of infected horses in the sixteenth century.

This trypanosome is morphologically indistinguishable from *T. brucei.* Typically it is 15 to 34 μm long. Most are slender in shape, but stumpy forms occasionally appear. However, the biology of *T. evansi* is quite different from that of *T. brucei.* The life cycle does not involve *Glossina* spp. or development within an arthropod vector. In most areas contaminated mouthparts of horseflies (*Tabanus* spp.) mechanically transmit the disease, but *Stomoxys, Lyparosia,* and *Haemotopota* spp. can also transmit it. In South America, vampire bats are common vectors of the disease, known there as **murrina**.[43] The disease is most severe in horses, elephants, and dogs, with nearly 100%

fatalities in untreated cases. It is less pathogenic to cattle and buffalo, which may be asymptomatic for months. In camels surra is serious but tends to remain chronic. Pathogenesis, symptoms, and treatment are the same as for *T. brucei.*

Trypanosoma evansi probably originated from *T. brucei,* when camels were brought into the tsetse fly belt. Subsequently, the organism apparently lost its requirement for development within an insect.

Trypanosoma (Trypanozoon) equinum occurs in South America, where it causes **mal de caderas,** a disease in horses similar to surra. *Trypanosoma equinum* is similar to *T. evansi* except that it appears to lack a kinetoplast. Actually a vestigial kinetoplast can be seen in electron micrographs, but it does not function in activation of the mitochondrion; this condition is known as **dyskinetoplasty.** *Trypanosoma brucei* and *T. evansi* can be rendered dyskinetoplastic with certain drugs, and the character is inherited as a mutation. Such altered organisms can survive as bloodstream parasites but no longer can infect flies. *Trypanosoma equinum* also is transmitted mechanically by tabanid flies. It most likely evolved from *T. evansi.* Pathogenesis, symptoms, and treatment are as for *T. evansi.*

Trypanosoma (Trypanozoon) equiperdum.   Another trypanosome, *T. equiperdum,* also morphologically indistinguishable from *T. brucei,* causes a venereal disease called **dourine** in horses and donkeys. The organisms are transmitted during coitus, and no arthropod vector is known. No doubt this trypanosome also originated from *T. brucei.*

The disease is found in Africa, Asia, southern and eastern Europe, Russia, and Mexico. It was once common in western Europe and North America but has been eradicated from these areas.

Dourine exhibits three stages. In the first the genitalia become edematous, with a discharge from the urethra and vagina. Areas of the penis or vulva may become depigmented. In the second stage a prominent rash appears on the sides of the body, remaining for three or four days. The third stage produces paralysis, first of the neck and nostrils and then of the hind body; the paralysis finally becomes general. Dourine is usually fatal unless treated.

Diagnosis depends on finding trypanosomes in the blood, genital secretions, or fluids from the large urticarious patches of the skin during the second stage. A complement fixation test is very reliable and was used by U.S. Department of Agriculture (USDA) personnel to ferret out infective horses during their successful campaign to eradicate the disease in the United States. All horses now entering the United States must be tested for dourine before being admitted.

• Section Stercoraria

Trypanosoma (Schizotrypanum) cruzi.   *Trypanosoma cruzi* carries the unusual distinction of having been discovered and studied several years before it was known to cause a disease. In 1910 a 40-year-old Brazilian, Carlos Chagas, dissected a number of conenosed bugs (Hemiptera, family Reduviidae, subfamily Triatominae) and found their hindguts swarming with trypanosomes of the epimastigote type. The biology and habits of triatomines are discussed in Chapter 36.

Chagas sent a number of the bugs to the Oswaldo Cruz Institute, where they were allowed to feed on marmosets and guinea pigs. Trypanosomes appeared in the blood of the animals within a month. Chagas thought the parasites went through a type of schizogony in the lungs, so he named them *Schizotrypanum cruzi.* The name *Schizotrypanum* still is employed by some workers, although most prefer to use it as a subgenus of *Trypanosoma.* By 1916 Chagas demonstrated that an acute, febrile disease, common in children throughout the range of conenosed bugs, was always accompanied by the trypanosome. Unfortunately, he thought that goiter and cretinism also were caused by this parasite. When these hypotheses were disproved, suspicion was cast on the rest of his work. Also, Chagas maintained to near the end of his life that transmission of the disease, which now bears his name, occurred through the bite of the insect. It was not until the early 1930s that Chagas' disease was proved to be transmitted by way of the feces of the conenosed bug.

Trypanosoma cruzi is distributed throughout most of South and Central America, where it infects 12 to 19 million persons. Another 35 million are exposed to infection.[116] In some surveys in Brazil it has been reported that 30% of all adults die of *T. cruzi* infection. Many kinds of wild and domestic mammals serve as reservoirs. Animals that live in proximity to humans, such as dogs, cats, opossums, armadillos, and wood rats, are particularly important in the epidemiology of Chagas' disease.

In the United States *T. cruzi* has been found in Maryland, Georgia, Florida, Texas, Arizona, New Mexico, California, Alabama, and Louisiana. Fourteen species of infected mammals have been found in the United States.[51] The first indigenous infection in a human in the United States was reported in 1955.[115] Since then several cases have been reported in Arizona, mainly on Indian reservations. Several North American strains have been isolated. They are morphologically indistinguishable from any other *T. cruzi,* but they seem to be much less pathogenic. It is possible that this infection in humans is more widespread in the United States than is now known; surveys using immunological tests have shown 0.8% positive cases among a random sample of 500 people from the lower Rio Grande Valley of Texas.[13]

- *Morphology.* The trypomastigote form is found in the circulating blood. It is slender, 16 to 20 μm long, and its posterior end is pointed. The free flagellum is moderately long, and the undulating membrane is narrow, with only two or three undulations at a time along its length. The kinetoplast is subterminal and is the largest of any trypanosome; it sometimes causes the body to bulge around it. The protozoan commonly dies in a question mark shape, the appearance it retains in stained smears (Fig. 5.7).

  Amastigotes develop in muscles and other tissues. They are spheroid, 1.5 to 4.0 μm wide, and occur in clusters composed of many organisms. Intermediate forms are easily found in smears of infected tissues.

- *Biology.* When reduviid bugs feed (Fig. 5.8), they often defecate on the skin of their host. Their feces may contain metacylic trypanosomes, which gain entry into the body of the vertebrate host through the bite, through scratched skin, or, most often, through mucous membranes that are rubbed with fingers contaminated with the insect's feces.

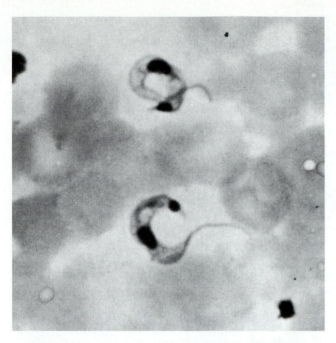

FIGURE 5.7

Trypanosoma cruzi: trypomastigote form in a blood film.

Courtesy of Ann Arbor Biological Center.

Also reservoir mammals can become infected by eating infected insects.[117] Although trypomastigotes are abundant in the blood in early infections, they do not reproduce until they have entered a cell and have transformed into amastigotes. The cells most frequently invaded are reticuloendothelial cells of the spleen, liver, and lymphatics and cells in cardiac, smooth, and skeletal muscles. The nervous system, skin, gonads, intestinal mucosa, bone marrow, and placenta also are infected in some cases. There is some evidence that the trypanosomes can actively penetrate host cells, but they may also enter through phagocytosis by the host macrophages.

The undulating membrane and flagellum disappear soon after the parasite enters a host cell. Repeated binary fission produces so many amastigotes that the host cell soon is killed and lyses. When released, the protozoa attack other cells. Cystlike pockets of parasites, called **pseudocysts,** form in muscle cells (Fig. 5.9). Intermediate forms (promastigotes and epimastigotes) can be seen in the interstitial spaces. Some of these complete the metamorphosis into trypomastigotes and find their way into the blood. *Trypanosoma cruzi* is a "partial aerobic fermenter"; some of the glucose carbon it consumes is degraded completely to carbon dioxide, but it excretes a substantial portion as succinate and acetate.[14] The oxygen consumption of blood and intracellular forms is the same as that of culture forms, and bloodstream forms apparently have a Krebs cycle and classical cytochrome system.[24,37] Thus, *T. cruzi* differs from the African trypanosomes in terms of the spectrum of metabolic properties displayed at various life-cycle stages.

Trypomastigotes that are ingested by triatomine bugs pass through to the posterior portion of the insect's midgut, where they become short epimastigotes. These multiply by longitudinal fission to become long, slender epimastigotes. Short metacyclic trypomastigotes appear in the insect's rectum 8 to 10 days after infection. These pass with the feces and can infect a mammal if rubbed into a mucous membrane or wound in the skin. The first-generation amastigotes in the insect's stomach group together to form aggregated masses.[12] These fuse and may represent a primitive form of sexual reproduction, although Tibayrenc and coworkers dispute this interpretation.[101]

- *Pathogenesis.* Entrance of the metacyclic trypanosomes into cells in the subcutaneous tissue produces an acute local inflammatory reaction. Within one to two weeks after infection, they spread to the regional lymph nodes and begin to multiply in the cells that phagocytose them. The intracellular amastigote undergoes repeated divisions to form large numbers of parasites, producing the so-called pseudocyst. After a few days some of the organisms retransform into trypomastigotes and burst out of the pseudocyst, destroying the cell that contains them. A generalized parasitemia occurs then, and parasites will invade almost every type of tissue in the body, although they show a particular preference for muscle and nerve cells (Fig. 5.10).

Reversion to amastigote, pseudocyst formation, retransformation to trypomastigote, and pseudocyst rupture are repeated in the newly invaded cells; then the process begins again. Rupture of the pseudocyst is accompanied by an acute, local inflammatory response, with degeneration and necrosis (cell or tissue death) of nerve cells in the vicinity, especially ganglion cells. This is the most important pathological change in Chagas' disease, and it appears to be the indirect result of parasitism of supporting cells, such as glial cells and macrophages, rather than of invasion of neurons themselves.[98]

Chagas' disease manifests *acute* and *chronic* phases. The acute phase is initiated by inoculation into the wound of the trypanosomes from the bug's feces. The local inflammation produces a small red nodule, known as a **chagoma,** with swelling of the regional lymph nodes. In about 50% of the cases the trypanosomes enter through the conjunctiva of the eye, causing edema of the eyelid and conjunctiva and swelling of the preauricular lymph node. This symptom is known as **Romaña's sign.** As the acute phase progresses, pseudocysts may be found in almost any organ of the body, although the intensity of attack varies from one patient to another. The heart muscle usually is invaded, with up to 80% of the cardiac ganglion cells being lost. Symptoms of the acute phase include anemia, loss of strength, nervous disorders, chills, muscle and bone pain, and varying degrees of heart failure. Death may ensue three to four weeks after infection.

The acute stage is most common and severe among children less than five years old. The chronic stage, however, is most often seen in adults. Its spectrum of symptoms is primarily the result of central and peripheral nervous dysfunction, which may last for many years. Some patients may be virtually asymptomatic and then suddenly

FIGURE 5.8

Life cycle of *Trypanosoma cruzi.*

From K. M. G. Adam, J. Paul, and V. Zaman, *Medical and Veterinary Protozoology. An Illustrated Guide,* © 1971. Churchill Livingstone. Edinburgh. Reprinted with permission of the publisher.

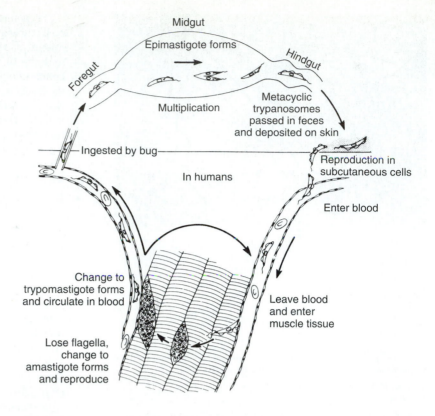

FIGURE 5.9

Trypanosoma cruzi pseudocyst in cardiac muscle. (× 780.)

Courtesy of S. S. Desser.

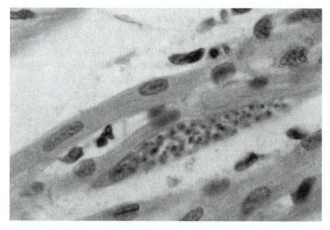

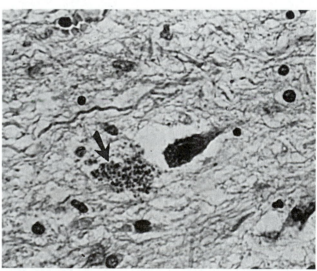

FIGURE 5.10

Pseudocyst of *Trypanosoma cruzi* in brain tissue.

AFIP neg. no. 67-5313.

succumb to heart failure. Chagas' disease accounts for about 70% of cardiac deaths in young adults in endemic areas. Part of the inefficiency in heart function is caused by loss of muscle tone resulting from the destroyed nerve ganglia (Fig. 5.11). The heart itself becomes greatly enlarged and flabby.

Most infections last the life of the individual, but severity of the acute infection may determine course of the chronic disease, possibly by affecting the type of immune response involved.[100] Autoimmunity is still a con-

troversial explanation for pathology of *T. cruzi* infections, but it is clear that the immune system functions to produce disease as well as to protect against parasites. Tarleton[100] reviews an ingenious set of experiments involving heart transplants in mice. Syngenic hearts (same genetic makeup) are accepted, but allogenic hearts (different genetic makeup) are rejected. Mice with chronic

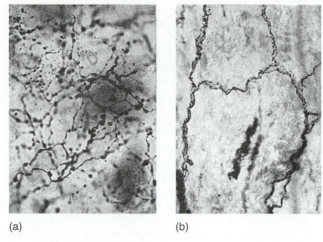

(a)   (b)

FIGURE 5.11

Diaphanised tricuspid valves with zinc-osmium impregnation of nerve fibers (dark lines). (*a*) Normal heart; (*b*) Chagas' cardiopathy with marked reduction of nerve fibers.

From M. S. R. Hutt, F. Koberle, and K. Salfelder, in H. Spencer, editor, *Tropical Pathology,* 1973 Springer Verlag.

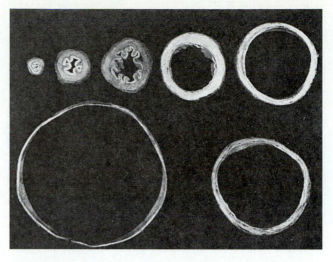

FIGURE 5.12

Different stages of chagasic esophagopathy beginning with a normal organ, passing through hypertrophy and dilatation to the final megaesophagus.

From M. S. R. Hutt, F. Koberle, and K. Salfelder, in H. Spencer, editor, *Tropical Pathology,* 1973 Springer Verlag.

T. cruzi infections, however, reject syngenic hearts, a reaction that can be countered by depleting recipients' CD4$^+$ cells. Host and parasite genetic makeup, sex, age, prior infection, and a variety of other factors influence disease development, and the relationship between these factors is still unresolved.

In some regions of South America it is common for the autonomic ganglia of the esophagus or colon to be destroyed. This ruins the tonus of the muscularis, resulting in deranged peristalsis and gradual flabbiness of the organ, which may become huge in diameter and unable to pass materials within it. This advanced condition is called **megaesophagus** or **megacolon,** depending on the organ involved (Fig. 5.12). Advanced megaesophagus may be fatal when the patient can no longer swallow. It has been experimentally demonstrated that testis tubules and epididymis atrophy in chronic cases.[28]

- *Immunology. Trypanosoma cruzi,* spending most of its vertebrate host phase as an intracellular parasite, presents us with some immunological phenomena quite different from those of the bloodstream trypanosomes. As in the case of the leishmanial parasites, host reactions to *T. cruzi* infections are largely cellular, especially during the acute phase. It is still not clear, however, how these cellular reactions are involved in control of the disease.

As is also the case with *Leishmania* research, our most detailed information on immunology comes from a study of infections in mice. The overriding early event of an experimental *T. cruzi* infection is immunosuppression, which may be responsible also for the pathological effects of the parasite.[99] But mice also evidently kill vast numbers of injected trypomastigotes, which means that some mechanism(s) is at work to help protect the host regardless of the level of prior exposure. Production of the cytokine interleukin-2 (IL-2) is suppressed during the acute phase, an event that in turn affects T-cell growth. Low levels of IL-2 are matched by those of the corresponding mRNA, so regulation of the cytokine is probably at the level of transcription.[73] Production of other cytokines is not generally suppressed, however, and IFN-γ levels are elevated.

Participation of the various T-cell subpopulations in regulating *T. cruzi* infections is still not completely understood, and not everyone agrees that they participate in protective immunity. In addition, these subpopulations may play different roles depending on the phase of the infection. For example, both CD4$^+$ and CD8$^+$ T cells are involved in the initial immune responses in mice, but after the infections have progressed far enough, these cells can be experimentally depleted by use of antibodies without affecting the outcome of the infection.[99] Chronic infections (sometimes called *post-acute* to distinguish them from the long lasting ones resulting in cardiac and gut pathology) are controlled mainly by humoral responses, and in some mouse/parasite strain combinations, circulating IgG is protective. Antibodies are of both the protective and nonprotective kinds; however, the former fix complement and lyse living trypomastigotes. Nonprotective antibodies remain in the blood even after drug cure, and they can be used in serological tests for chronic or past infections.

Some of the elements of the human pathological response are present also in mice, and their presence suggests that autoimmunity is a feature of the disease,[81] although there is disagreement among parasitologists on this point. For example, infection results in both a strong blast transformation response (mitogenesis) in lymphocytes in general (polyclonal activation) and in elevated levels of circulating immunoglobulins.[84] Much of this immunoglobulin does not "recognize" parasite antigens, and the lymphocyte populations stimulated by infection may contain T- and B-cell clones that are autoreactive. Infected cardiac muscle cells eventually rupture, releasing amastigotes and provoking an inflammatory response with infiltrations of lymphocytes and macrophages. This process can lead eventually to fibrosis and loss of cardiac muscle's ability to conduct impulses. But it still is not clear whether autoantibodies are involved in this pathology. The exchange of views on this subject by Kierszenbaum and Hudson[46,52] is an excellent illustration of a gentlemanly debate over a very complex subject. And of course parasite biology is a paragon of complexity.

- **Diagnosis and Treatment.** Diagnosis usually is by demonstration of trypanosomes in blood, cerebrospinal fluid, fixed tissues, or lymph. Trypomastigotes are most abundant in peripheral blood during periods of fever; they may be difficult to find at other times or in cases of chronic infection. In these other cases blood can be inoculated into guinea pigs, mice, or other suitable hosts, and the animals in turn can be examined by heart smear or spleen impression. Another method that is widely employed is **xenodiagnosis.** Laboratory-reared triatomines are allowed to feed on the patient; then after a suitable period of time (10 to 30 days) they are examined for intestinal flagellates. This technique can detect cases in which trypanosomes in the blood are too few to be found by ordinary examination of blood films.

  Complement fixation or other immunodiagnostic tests are extremely effective in demonstrating chronic cases, although they may give false positive reactions if the patient is infected with *Leishmania* or another species of trypanosome. In experiments with infected opossums, *Didelphis marsupialis,* an indirect fluorescent antibody test (IFAT) was the most sensitive test for *T. cruzi,* followed by xenodiagnosis.[48] Both flagellar and cytoplasmic *T. cruzi* proteins have been cloned in *Escherichia coli* and used in ELISAs; these assays were much more specific to *T. cruzi,* as well as more sensitive, than those using crude antigen preparations.[53] Furthermore, the use of recombinant technology reduces the overall costs of diagnostic reagent production. Dot-immunobinding assays using antigen bound to nitrocellulose paper offer the advantage of requiring very small amounts of fluid and because they need no expensive equipment show promise for use under field conditions.[18]

  Diagnostic methods based on detection of parasite DNA using polymerase chain reaction (PCR) techniques have also been developed, but so far they have not come into general use because of the problem of false negatives.[53] Also, PCR is still not quick and cheap enough for the screening of large numbers of samples.

  Unlike the other trypanosomes of humans, *T. cruzi* does not respond well to chemotherapy. The most effective drugs kill only the extracellular protozoa, but the intracellular forms defy the best efforts at eradication. This seems to be because the reproductive stages, inside living host cells, are shielded from the drugs. The lives and strength of millions of Latin American people depend on the discovery of a drug or vaccine that is effective against *T. cruzi.* One hope is the drug ketoconazole, which completely cured 78.5% of otherwise fatally infected mice.[66]

- **Epidemiology.** The principal vectors of *T. cruzi* in Brazil are *Panstrongylus megistus, Triatoma sordida,* and *T. brasiliensis.* In Uruguay, Chile, and Argentina, *Triatoma infestans* is the primary culprit. *Rhodnius prolixus* is the main vector in northern South America and *Triatoma dimidiata* in Central America, whereas species of the *Triatoma protracta* group serve as vectors in Mexico (Fig. 5.13). Several other species of triatomines have been found naturally infected throughout this range. Natural infections in *Triatoma sanguisuga* have been found in the United States. The insects can become infected as nymphs or adults. Triatomines can infect themselves when they feed on each other, presumably by sucking the contents of the intestine. Ticks, sheep keds, and bedbugs have been experimentally infected, but no evidence that they serve as natural vectors has been found. Natural mammalian reservoirs of infection have been mentioned, but domestic dogs and cats probably are the most important to human health. Because the bugs hide by day, primitive or poor-quality housing favors their presence. Thatched roofs, cracked walls, and trash-filled rooms are ideal for the breeding and survival of the insects. Misery compounds itself.

  Transmission from human to human during coitus or through breast milk may be possible, although this has yet to be documented. *Trypanosoma cruzi* can and does cross the placental barrier from mother to fetus. Newborn infants with advanced cases of Chagas' disease, including megaesophagus, have been described in Chile. In some villages in Mexico people believe that triatomines are aphrodisiacs; therefore, they are eaten, and the trypanosomes gain access through the oral mucosa.[89] The age of the victim is important in the epidemiology of Chagas' disease, and most new infections are in children less than two years old. The acute phase is most often fatal in this age group. Finally, the hazard of transmission by blood transfusion from donors with cryptic infection should not be underestimated. The frequency of this mode of transmission now ranks second only to natural vector transmission in endemic areas.[87]

Trypanosoma (Herpetosoma) rangeli. *Trypanosoma rangeli* first was found, as was *T. cruzi,* in a triatomine bug in South America. *Rhodnius prolixus* is the most common vector, but *Triatoma dimidiata* and other species will also serve. Development is in the hindgut, and the epimastigote stages that result are from 32 to more than 100 µm long. The kinetoplast is minute, and the species can thereby be differentiated from *T. cruzi,* with which it often coexists.

Trypanosoma rangeli is common in dogs, cats, and humans in Venezuela, Guatemala, Chile, El Salvador, and Colombia. It has been found in monkeys, anteaters, opossums, and humans in Colombia and Panama. The trypomastigotes,

FIGURE 5.13

Distribution of Chagas' disease in humans and of its four principal vectors.

AFIP neg. no. 65-5015.

Rhodnius prolixus

Triatoma dimidiata

Triatoma infestans

Panstrongylus megistus

● Human infections

26 to 36 μm long, are larger than those of *T. cruzi.* The undulating membrane is large and has many curves. The nucleus is preequatorial, and the kinetoplast is subterminal.

The method of transmission is unclear. Although development is by posterior station, transmissions both by fecal contamination and by feeding inoculation have been reported.[102] *Trypanosoma rangeli* multiplies by binary fission in the mammalian host's blood. No intracellular stage is known, and the organism is apparently not pathogenic in humans. However, infections with *T. rangeli* or mixed infections with *T. rangeli* and *T. cruzi* are potential problems for diagnosis.[36] Conventional immunofluorescence and ELISA assays, reinforced by immunoprecipitation and Western blot analysis, diagnoses either or both infections.

Trypanosoma (Herpetosoma) lewisi. *Trypanosoma lewisi* (Fig. 5.14) is a cosmopolitan parasite of *Rattus* spp. Other rodents, including white-footed mice, deer mice, and kangaroo rats in the United States, are infected with lewisi-like trypanosomes, but it is not completely clear whether these are the same species as found in *Rattus* or a form more closely related to *T. musculi,* a species restricted to mice. The vector of *T. lewisi* is the northern rat flea, *Nosopsyllus fasciatus,* in which the parasite develops inside cells of the posterior midgut. Metacyclic trypomastigotes appear in large numbers in the rectum of the insect, infecting rats that eat fleas or their feces. The parasite seems to be nonpathogenic in most cases, perhaps even promoting the growth of its rat host.[59] However, infection may contribute to abortion and arthritis.

Much research has been conducted on this species because of the ease of maintaining it in the laboratory rat. One fascinating subject of this research is the "ablastin" phenomenon.[106]

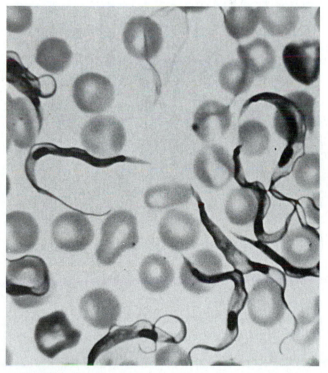

FIGURE 5.14

Trypanosoma lewisi trypomastigotes in the blood of a rat.

Courtesy of Turtox/Cambosco.

Ablastin is an antibody that arises during the course of an infection. After a rat is infected by the metacyclic trypomastigotes the parasite begins reproducing in the epimastigote form in the visceral blood capillaries. After about five days the trypanosome appears in the peripheral blood as a rather "fat" form, and shortly thereafter a crisis occurs in which most of the trypanosomes are killed by a trypanocidal antibody. A small population of slender trypomastigotes remains; they are infective for the flea but do not reproduce further while in the rat. After a few weeks the host produces another trypanocidal antibody, which clears the remaining trypanosomes, and the infection is cured. The slender trypomastigotes are sometimes known as "adults." Their reproduction is inhibited by the ablastin. Ablastin is a globulin with many characteristics of a typical antibody, but its action inhibits reproduction. Nucleic acid and protein synthesis by the trypanosome are inhibited, as is uptake of nucleic acid precursors. However, it is still not clear how this antibody functions.[1]

Trypanosoma (Megatrypanum) theileri.

Trypanosoma theileri is a cosmopolitan parasite of cattle. The vectors are horseflies of the genera *Tabanus* and *Haematopota.* The trypanosome reproduces in the fly gut as an epimastigote.

The size of *T. theileri* varies with the strain—from 12 to 46 μm, 60 to 70 μm, and even up to 120 μm in length. The posterior end is pointed, and the kinetoplast is considerably anterior to it. Both trypomastigote and epimastigote forms can be found in the blood. Reproduction in the vertebrate host is in the epimastigote form and apparently occurs extracellularly in the lymphatics.

Trypanosoma theileri is usually nonpathogenic, but under conditions of stress it may become quite virulent. When cattle are stressed by immunization against another disease, undergo physical trauma, or become pregnant, the parasite may cause serious disease.

This parasite is rarely found in routine blood films. Detection usually depends on in vitro cultivation from blood samples. In fact, during tissue culture of bovine blood or cells, *T. theileri* is the most commonly found contaminant. Strong evidence points to transplacental transmission. In the United States a similar trypanosome is also common in deer and elk.

Other *Trypanosoma* Species.

Other species of *Trypanosoma* are common in other classes of vertebrates—for example, *T. percae* in perch, *T. granulosum* in eels, *T. rotatorium* in frogs, *T. avium* in birds, and incompletely known species in turtles and crocodiles. Trypanosomes are commonly found in a variety of marine fishes.

Genus *Leishmania*

Like the trypanosomes, the leishmanias are heteroxenous. Part of their life cycle is spent in the gut of a fly, where they assume the form of a promastigote; the remainder of their life cycle is completed in vertebrate tissues, where only the amastigote form is found. Traditionally the amastigote is also known as a **Leishman-Donovan (L-D)**

body. The vertebrate hosts of *Leishmania* spp. are primarily mammals. Although nearly a dozen species have been reported from lizards, those species now are considered members of the genus *Sauroleishmania,* based on their biochemical and immunological characteristics.[16,86] The mammals most commonly infected with *Leishmania* are humans, dogs, and several species of rodents. The parasites cause a complex of diseases called **leishmaniasis.** In some cases, especially with some Old World cutaneous infections, leishmaniasis is a zoonosis, with a wild mammal (for example, a gerbil) reservoir.

Species in humans are widely distributed (Fig. 5.15). It is likely that the transport of slaves to the western world from Africa through the Middle East and Asia spread *Leishmania* species into previously uncontaminated areas, where they now apparently are evolving rapidly into new strains. As is the case with virtually all infectious diseases, the popularity of air travel generates the potential for quick spread of leishmanial parasites (as Page Faegre discovered on her Mediterranean vacation). It is not always easy to estimate the numbers of people infected or at risk of acquiring a parasitic disease, especially on a global basis. Recent attempts to estimate the severity of leishmaniasis arrived at a figure of 400,000 new cases annually in 67 countries.[2] This figure is probably low, however, because to increase the accuracy of their study, the researchers tried to eliminate cases with no access to medical facilities, misdiagnosed cases, cases seen in clinics but not officially reported, and infections that remain subclinical. Accurate public health records are not always easy to compile in developed nations, much less in those with less than ideal health care delivery systems.

The intermediate hosts and vectors of leishmaniasis are **sand flies** (Fig. 5.16), small blood-sucking insects in the family Psychodidae, subfamily Phlebotominae (see p. 565). There are over 600 species of sand flies divided into five genera: *Phlebotomus* and *Sergentomyia* in the Old World and *Lutzomyia, Brumptomyia,* and *Warileya* in the New World. When they suck the blood of an infected animal, they ingest amastigote forms. These pass to the midgut or hindgut, where they transform into promastigotes and multiply by binary fission. The parasites attach to the walls of the fly's gut and replicate. By the fourth or fifth day after feeding, promastigotes move forward to the esophagus and pharynx. When promastigotes begin to clog up the esophagus, the feeding sand fly pumps its esophageal contents in and out to clear the obstruction, thereby inoculating promastigotes into the skin of a luckless victim. Transmission also can occur when infected sand flies are crushed into the skin or mucous membrane.

All amastigotes in vertebrate tissues look similar (Figs. 5.3 and 5.17). They are spheroid to ovoid, usually 2.5 to 5.0 μm wide, although some are smaller. They are among the smallest nucleated cells known. In stained preparations only the nucleus and a very large kinetoplast can be seen, and the cytoplasm appears vacuolated. Exceptionally, a short axoneme is visible within the cytoplasm under the light microscope.

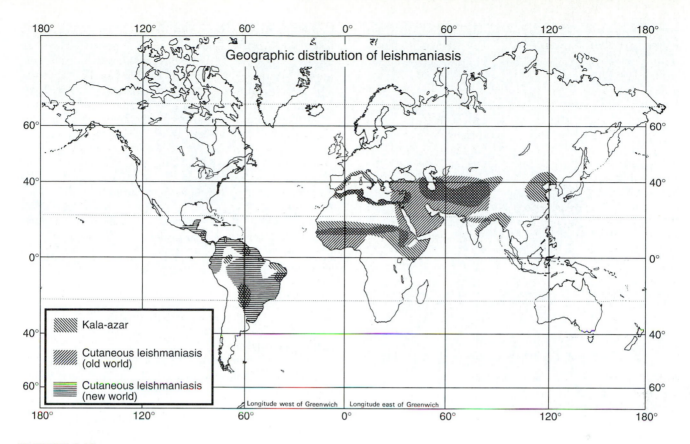

FIGURE 5.15

Geographical distribution of leishmaniasis.

AFIP neg. no. 68-1805-2.

Although all *Leishmania* spp. exhibit similar morphology, they differ clinically, biologically, and serologically.[112] Even so, these characteristics often overlap, so distinctions between species are not always clear-cut. Leishmaniases that normally are visceral may become dermal; dermal forms can become mucocutaneous; and an immunodiagnostic test derived from the antigens of one species may give positive reactions in the presence of other species of *Leishmania* or even *Trypanosoma*.

Leishmania species present us with some of the most baffling problems in immunology, many of which must be solved in order to diagnose, treat, or prevent infections. First, in the vertebrate body, the parasites live inside macrophages, the very cells that in most cases function to kill invading organisms. Second, within macrophages, the parasites reside in phagolysosomes, the compartments that normally function directly to digest foreign particles. Third, *Leishmania* species differ markedly among themselves in terms of clinical manifestations, producing infections that range from self-healing although scarring ones that convey lifetime immunity to fatal visceral involvements or to extremely disfiguring afflictions

that erode facial features. Fourth, the contributions of human host genetic makeup and nutritional state to the course of infection have yet to be completely described. And fifth, drug treatment may precipitate a subsequent clinical manifestion quite different from that of the original infection, such as the post-kala-azar dermal leishmanoid. Needless to say, parasitologists have been fascinated, and fully occupied, with leishmanial parasites for a long time; the complexity and diversity of host/parasite interactions have led to the nickname "leishmaniac" for many such scientists.

As a result of the difficulty in species definition within the genus, several schemes of classification have been proposed, nearly all of them gaining some acceptance. In recent years, however, many strains and species have been "typed," or characterized biochemically, through the study of their isoenzymes and kDNA. Reference centers for the identification and typing of isolates are located in several places in the world, but biochemical identification techniques are often expensive and time-consuming. The older published literature is sometimes confusing because some researchers referred to several species and others considered the same

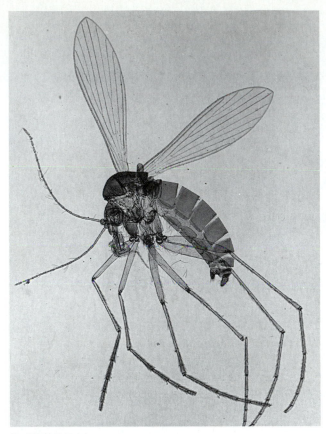

FIGURE 5.16

The sand fly *Phlebotomus,* a vector of *Leishmania* spp. Sand flies are about 3 mm long.

Courtesy of Jay Georgi.

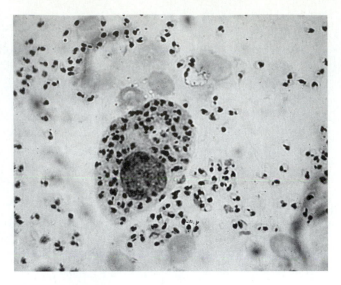

FIGURE 5.17

Spleen smear showing numerous intracellular and extracellular amastigotes of *Leishmania donovani.*

AFIP neg. no. 55-17580.

organisms as a single, widespread species with slightly different clinical manifestations but similar or identical immunological properties among the species.

The difficulty of identifying species of *Leishmania* extends to the forms in the sand fly. Identification of parasites in their vectors is often critical if vector control is part of an overall disease control strategy. It does not help to focus an attack on one species of vector if it is not carrying the parasite or to be fooled into trying to eliminate a species that carries a parasite morphologically identical to those found in but not infective for humans. Attempts to solve this problem through modern biochemical methods such as tests based on hybridization of known kDNA with that of parasites have been partially successful.[91] The most promising of these new approaches involves use of PCR to amplify kinetoplast DNA from samples and parasite species-specific probes for use in dot hybridization tests.[20,63,95] For example, there is evidence that such tools can be used to distinguish *L. donovani* strains producing the disfiguring post-kala-azar dermal leishmanoid from those that do not.[20]

Treatment of leishmanial infections varies according to the clinical manifestations. In earlier years trivalent antimonials were the only drugs available, but they were so toxic as to be downright dangerous. Today pentavalent antimonials are the drugs of choice, but they also are toxic and usually must be administered under the care of a physician. Two pentavalent preparations are available: Pentostam and Glu-

cantime; only Pentostam is available in the United States, through the Centers for Disease Control parasite drug service. Drug resistance has been reported in some strains of some species.[35] Furthermore, relapses and post-kala-azar dermal leishmanoid may follow insufficient treatment.

Some of the most creative, although still largely experimental, approaches to treatment involve turning a liability into an asset (so to speak)—the liability being the fact that in visceral infections the parasites are located within macrophages. Macrophages, however, will eat foreign particles, so that when injected drugs are bound to artificial particles—liposomes or colloidal particles—the efficacy of the compounds is greatly increased.[19,22]

It has long been known that the tropical forests are a rich source of plant molecules with potential medicinal uses. The rapid disappearance of those forests has led to renewed interest in natural plant products, including those that may be effective against leishmanial infections. So far, antileishmanial activity has been found in a number of plant species, including those from the families Apocynaceae (dogbanes), Gentianaceae (gentians), and Euphorbiaceae.[29,94]

Species of flagellate that develop in the sand fly's midgut before moving anteriad are placed in the subgenus *Leishmania.* Those that develop in the hindgut first are placed in the subgenus *Viannia.* Species in subgenus *Leishmania* include both Old World and New World visceral and cutaneous species; those in *Viannia* are New World cutaneous forms including some of the most disfiguring species.[55] Species and subspecies of *Leishmania* currently recognized are listed in Table 5.1; most authors now refer to biochemically related groups as *species complexes.* The taxa are in general agreement with species separation according to criteria based on kDNA buoyant densities and (especially) isozyme patterns, mainly of glycolytic and Krebs cycle enzymes as well as transaminases.[16,55,86] Of those species complexes we will consider the five most important to human welfare: *L. tropica, L. major, L. donovani, L. mexicana,* and *L. braziliensis.*

Table 5.1 *Species and subspecies of* Leishmania

Parasite	Locality
Subgenus *Leishmania* (Ross, 1903)	
L. donovani phenetic complex	
L. donovani (Laveran and Mesnil, 1903)	India, China, Bangladesh
L. archibaldi (Castellani and Chalmers, 1919)	Sudan, Ethiopia
L. infantum phenetic complex	
L. infantum (Nicolle, 1908)	North central Asia, northwest China, Middle East, southern Europe, northwest Africa
L. chagasi (Cunha and Chagas, 1937)	South and Central America
L. tropica phenetic complex	
L. tropica (Wright, 1903)	Urban areas of Middle East and India
L. killicki (Rioux, Lanotte, and Pratlong, 1986)	Tunisia
L. major phenetic complex	
L. major	Africa, Middle East, Soviet Asia
L. gerbilli phenetic complex	
L. gerbilli (Wang, Qu, and Guan, 1973)	China, Mongolia
L. arabica phenetic complex	
L. arabica (Peters, Elbihari, and Evans, 1986)	Saudi Arabia
L. aethiopica phenetic complex	
L. aethiopica (Bray, Ashford, and Bray, 1973)	Ethiopia, Kenya
L. mexicana phenetic complex	
L. mexicana (Biagi, 1953)	Mexico, Belize, Guatemala, south central United States
L. amazonensis (Lainson and Shaw, 1972)	Amazon Basin, Brazil
L. venezuelensis (Bonfante-Garrido, 1980)	Venezuela
L. enrietti phenetic complex	
L. enrietti (Muniz and Medina, 1948)	Brazil
L. hertigi phenetic complex	
L. hertigi (Herrer, 1971)	Panama, Costa Rica
L. deanei (Lainson and Shaw, 1977)	Brazil
Subgenus *Viannia* (Lainson and Shaw, 1987)	
L. braziliensis phenetic complex	
L. braziliensis (Viannia, 1911)	Brazil
L. peruviana (Velez, 1913)	Western Andes
L. guyanensis phenetic complex	
L. guyanensis (Floch, 1954)	French Guiana, Guyana, Surinam
L. panamensis (Lainson and Shaw, 1972)	Panama, Costa Rica

In other classifications, subspecies of *L. mexicana* have been recognized, and these names—*L. mexicana aristedesi, L. m. garnhami,* and *L. m. pifanoi*—do appear in the literature, with the subspecific name sometimes used as a specific epithet. The groupings in the table are those of Rioux et al.[86] and are based on extensive isozyme and cladistic analysis. The term *phenetic complex* refers to zymodemes revealed by cluster analysis. The struggle for a uniformly accepted taxonomic scheme for the leishmanias is likely to continue into the next edition of this book.

Leishmania tropica **and** ***L. major.*** *Leishmania tropica* and *L. major* produce cutaneous ulcers variously known as **oriental sore, cutaneous leishmaniasis, Jericho boil, Aleppo boil,** and **Delhi boil.** They are found in west-central Africa, the Middle East, and Asia Minor into India. These two species have similar life cycles and clinical symptoms; however, *L. tropica* and *L. major* are found in different localities and have different reservoir and intermediate hosts, and the lesions they cause are somewhat different, although in humans the lesions may vary in severity according to age and other factors. The two species can be differentiated biochemically.

• ***Morphology and Life Cycle.*** The appearance of the amastigotes or *L. tropica* and *L. major* is similar to that of the other leishmanias of humans (Figs. 5.3 and 5.17). Sand flies of the genus *Phlebotomus* are the intermediate hosts and vectors. When the fly takes a blood meal containing amastigotes, the parasites multiply in the midgut and then move to the pharynx; they are then inoculated into the next mammalian victim. There they multiply in the reticuloendothelial system and lymphoid cells of the skin. Few amastigotes are found except in the immediate vicinity of the site of infection, so the sand flies must feed there to become infected.

- **Pathogenesis.** The incubation period lasts from a few days to several months. The first symptom of infection is a small, red papule at the site of the bite. This may disappear in a few weeks, but usually it develops a thin crust that hides a spreading ulcer underneath. Two or more ulcers may coalesce to form a large sore (Fig. 5.18). In uncomplicated cases the ulcer will heal within two months to a year, leaving a depressed, unpigmented scar. It is common, however, for secondary infection to occur, including, for example, yaws (a disfiguring disease caused by a spirochete) and myiasis (infection with fly maggots, p. 587).

  Leishmania tropica is found in more densely populated areas. Its lesion is dry, persists for months before ulcerating, and has numerous amastigotes within it. By contrast, *L. major* is found in sparsely inhabited regions. Its papule ulcerates quickly, is of short duration, and contains few amastigotes.

  Most species and subspecies of *Leishmania* can produce cutaneous lesions. There is an astonishing variety of forms of such lesions, ranging from tiny sores to massive, diffused ulcers. Some even have been misdiagnosed as leprosy or tuberculosis. Diagnosis, then, is difficult at times, especially when two species occur in the same locality.

- **Immunology.** In the case of Old World cutaneous leishmaniasis, protective immunity following medical treatment seems to be absolute, and immunity as a result of the natural course of the disease is 97% to 98% effective. Recognizing this, some native peoples deliberately inoculate their children on a part of their body normally hidden by clothing. This practice prevents the child from later developing a disfiguring scar on an exposed part of the body. In Lebanon and Russia, attempts at mass vaccination with promastigotes show promising results. Control of Old World cutaneous leishmaniasis, however, ultimately depends on eliminating the sand flies and reducing the population of rodent reservoirs.

  The gene that controls susceptibility to visceral *L. donovani* infection in mice (the *Lsh* gene) has no effect on resistance or susceptibility to the cutaneous species *L. major* and *L. mexicana*.[9] Instead, the severity of cutaneous infections is influenced by another gene, *Scl-1*, which is nonallelic to *Lsh* and controls healer and nonhealer phenotypes, and a third gene, *Scl-2*, in DBA/2 mice, which exerts a "no growth" lesion phenotype that mimics certain clinical pictures in humans.[9] In mouse strains resistant to infection with *L. major*, the T_h1 arm of the immune response (p. 25) is activated, with production of IFN-γ and a delayed type hypersensitivity reaction.[60] However, in susceptible mouse strains activation of the T_h2 arm stimulates production of IL-4, hyperglobulinemia, and elevated IgE levels.[60] The T cells that respond to infection in both cases are those of the lymph node draining the infection site.

  Without the extensive use of inbred mouse strains of known genetic makeup, progress toward the understanding

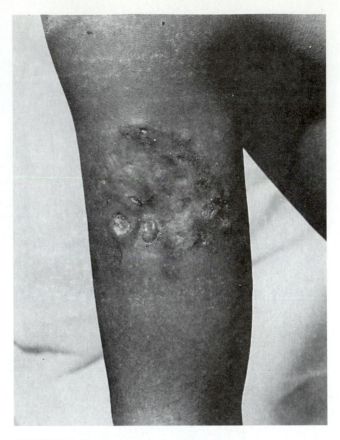

FIGURE 5.18

Oriental sore: a complicated case with several lesions.

AFIP neg. no. A-43418-1.

of leishmanial infections would be greatly slowed. Mice can be obtained with a variety of genotypes that affect their immune reactions to parasites. Furthermore, in such animals, antibodies that neutralize various cytokines can be used as "probes" to neutralize these molecules to follow the resultant course of infection.[60] For example, anti-INF-γ antibody given to protectively immunized (against *L. major*) C3H mice can reduce the levels of immunity, resulting in a disseminated infection. Conversely, nonhealing BALB/c mice can be converted into healers by administration of anti-IL-4 antibody. In both cases, however, the treatment must be given within a week or 10 days of infection. But the interactions of T cells and cytokines within these mice is not a simple matter, for administration of the respective cytokine molecules themselves does not affect the outcome of the experimental infections. The mouse, typically interpreted to be the house mouse *Mus musculus*, runs through our folklore, poetry, nursery rhymes, and popular cultures bringing us much delight. But *M. musculus* also has played a crucial role in the development of our understanding of disease processes. Parasitologists, especially, owe a great deal to this lowly rodent.

- *Diagnosis.* Diagnosis of infection is greatly facilitated by finding amastigotes. Scrapings from the side or edge of the ulcer, smeared on a slide and stained with Wright's or Giemsa's stain, will show the parasites in endothelial cells and monocytes, even though they cannot be found in the circulating blood. Cultures should be made in case amastigotes go undetected.

Leishmania donovani.   In 1900, Sir William Leishman discovered *L. donovani* in spleen smears of a soldier who died of a fever at Dum-Dum, India. The disease was known locally as **Dum-Dum fever** or **kala-azar.** Leishman published his observations in 1903, the year that Charles Donovan found the same parasite in a spleen biopsy. The scientific name honors these men, as does the common name of the amastigote forms, Leishman-Donovan (L-D) bodies. The Indian Kala-azar Commission (1931 to 1934) demonstrated the transmission of *L. donovani* by *Phlebotomus* spp.

- *Morphology and Life Cycle.* *Leishmania donovani* amastigotes cannot be differentiated from other species of *Leishmania* on the basis of morphology as seen in the light microscope; the rounded or ovoid bodies measure 2 to 3 μm, with a large nucleus and kinetoplast. They live within cells of the reticuloendothelial system of the viscera, including spleen, liver, mesenteric lymph nodes, intestine, and bone marrow. Amastigotes have been found in nearly every tissue and fluid of the body.

  The life cycle is similar to that of *L. tropica* except that *L. donovani* is primarily a visceral infection. When a sand fly of the genus *Phlebotomus* ingests amastigotes along with a blood meal, the parasites lodge in the midgut and begin to multiply. They transform into slender promastigotes and quickly block the gut of the insect. Soon they can be seen in the esophagus, pharynx, and buccal cavity, where they are injected into a new host with the fly's bite. Not all strains of *L. donovani* are adapted to all species and strains of *Phlebotomus.* Once in the mammalian host, the parasite is immediately engulfed by a macrophage, in which it divides by binary fission, eventually killing the host cell. Escaping the dead macrophage, the protozoa are engulfed by other macrophages, which they also kill; by this means they eventually severely damage the RE system, a system that plays a critical role in host defense. Interestingly, amastigotes engulfed by neutrophils are killed.[17] Eosinophils also kill the *L. donovani* they consume, but in untreated cases, these polymorphonuclear leucocytes have little or no effect on the eventual outcome of the disease.

- *Pathogenesis.* Clinically, *L. donovani* infections may range from asymptomatic to progressive, fully developed kala-azar. The incubation period in humans may be as short as 10 days or as long as a year, but usually is two to four months. The disease usually begins slowly with low-grade fever and malaise and is followed by progressive wasting and anemia, protrusion of the abdomen from enlarged liver and spleen (Fig. 5.19), and finally death (in untreated cases) in two to three years. In some cases the symptoms may be more acute in onset, with chills, fever up to 104°F, and vomiting; death may occur within 6 to 12 months. Accompanying symptoms are edema, especially of the face, bleeding of the mucous membranes, breathing difficulty, and diarrhea. The immediate cause of death often is the invasion of secondary pathogens that the body is unable to combat. A certain proportion of cases, especially in India, recover spontaneously.

  Visceral leishmaniasis may be viewed essentially as a disease of the reticuloendothelial system. The phagocytic cells, which are so important in defending the host against invasion, are themselves the habitat of the parasites. Blood-forming organs, such as the spleen and bone marrow, undergo compensatory production of macrophages and other phagocytes (hyperplasia) to the detriment of red cell production. Thus, the spleen and the liver become greatly enlarged (hepatosplenomegaly) (Fig. 5.20), while the patient becomes severely anemic and emaciated.

  Experiments with mice and hamsters provide most of our insight into the disease mechanisms of visceral leishmaniasis. Our general picture of pathology as a manifestation of complex interactions between a variety of host cell types is derived mainly from these studies.[81]

  A skin condition known as **post-kala-azar dermal leishmanoid** develops in some cases (Fig. 5.21).[69] It is rare in the Mediterranean and Latin American areas but develops in 5% to 10% of cases in India. The condition usually becomes apparent about one to two years after inadequate treatment for kala-azar. It is marked by reddish, depigmented nodules that sometimes become quite disfiguring.

- *Immunology.* The immunology of *Leishmania donovani* has not been studied to quite the extent that *L. major* has, at least in mice, but experiments have shown that experimental animals respond differently to the visceral parasites than they do to the cutaneous species. *Leishmania donovani* possesses membrane lipophosphoglycans that may inhibit gene expression in macrophages. The inhibition is of protein kinase dependent expression, such as that involved in macrophage activation by TNF and IFN-γ.[61]

  Patients with symptomatic kala-azar do not develop T_h1 responses against *L. donovani*, and their macrophages do not secrete IFN-γ or IL-2 in the presence of leishmanial antigens. However, these patients regularly have high titers of antileishmanial antibodies;[15,82] that is, their T_h2 arm is activated and the T_h1 arm is downregulated. There is an intricate interplay between host immune response and progression of visceral leishmaniasis, and the outcome of this potentially deadly contest is likely influenced by host genotype. Clinical visceral leishmaniasis may not develop for some time, even years, after infection. Asymptomatic infections, of which there are many, probably result from early activation of the T_h1 arm of the immune response, while kala-azar results from the proliferation of nonprotective T_h2 cells.

 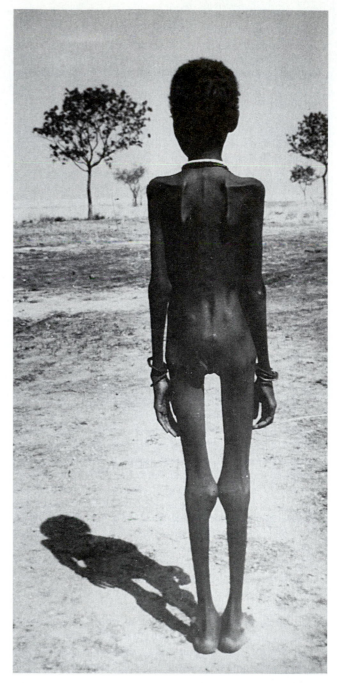

FIGURE 5.19

Advanced kala-azar. Boy, about six years old, from Sudan, showing extreme hepatosplenomegaly and emaciation typical of advanced kala-azar.

From H. Hoogstraal and D. Heyneman, "Leishmaniasis in the Sudan Republic 30. Final epidemiologic report," in *Am. J. Trop. Med. Hyg.* 18:1091–1210. Copyright © 1969.

The relationship between host genetic makeup and the immunological response to leishmanial infection has been studied extensively in mice. Susceptibility of mice to *Leishmania donovani* infection is controlled by a single gene, *Lsh,* which is inherited in Mendelian fashion and expressed in visceral macrophages.

• **Diagnosis and Treatment** As in *L. tropica,* diagnosis of *L. donovani* depends on finding L-D bodies in tissues or secretions. Spleen punctures, blood or nasal smears, bone marrow, and other tissues should be examined for the characteristic parasites, and cultures from these and other organs should be attempted. Immunodiagnostic tests are

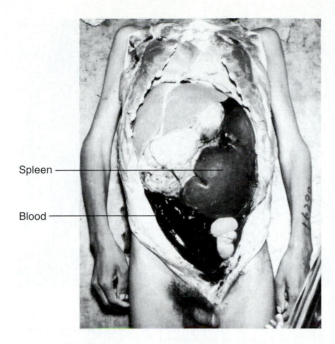

Spleen

Blood

FIGURE 5.20

A patient with kala-azar who died of hemorrhage after a spleen biopsy. Note the greatly enlarged spleen. (The dark matter in the lower abdominal cavity is blood.)

AFIP neg. no. A-45364.

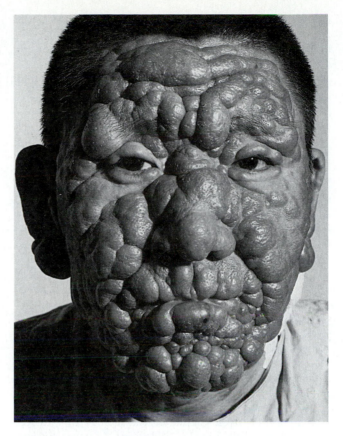

FIGURE 5.21

Post-kala-azar dermal leishmanoid. This patient responded very well to treatment, regaining a nearly normal appearance.

Courtesy of Robert E. Kuntz.

sensitive but cannot differentiate between species of Leishmania or between current and cured cases. The tests most frequently used are the enzyme-linked immunosorbent assay (ELISA) and the indirect fluorescent antibody test (IFA). Until recently the main limitation of these tests was a crossreacting positive response to Trypanosoma cruzi. This problem has been resolved by improved techniques of antigen purification, special antigen polymerization, and antigen selection by use of monoclonal antibodies.[65] Other diseases that might have symptoms similar to kala-azar are typhoid and paratyphoid fevers, malaria, syphilis, tuberculosis, dysentery, and relapsing fevers. Each must be eliminated in the diagnosis of kala-azar.

Treatment consists of injections of various antimony compounds, as previously described for *L. tropica,* and good nursing care.

* *Epidemiology and Control.* Transmission of visceral leishmaniasis is related to the activities of humans and the biology of sand flies. Control of sand flies and reservoir hosts is required in endemic areas. *Phlebotomus* spp. exist mainly at altitudes under 2000 feet, most commonly in flat plains areas. Even in desert areas such as in the Sudan, the flies rest in cracks in the parched earth and under rocks, which offer protection. In such conditions the flies are active only during certain hours of the day. For humans to become infected, they must be in sand fly areas at these times.

A wide variety of animals can be infected experimentally, although dogs are the main important reservoir in most areas. Canine infection is less common in India, where it is believed that a fly-to-human relationship is maintained.

Age of the victim is a factor in the course of the disease, and fatal outcome is most frequent in infants and small children. Males are more often infected than are females, most likely as the result of more exposure to sand flies. Poor nutrition, concomitant infection with other pathogens, and other stress factors predispose the patient to lethal consequences.

Leishmania braziliensis. *Leishmania braziliensis* produces a disease in humans variously known as **espundia, uta,** or **mucocutaneous leishmaniasis.** It is found throughout the vast area between central Mexico and northern Argentina, although its range does not extend into the high mountains, except for the south slope of the Andes. Clinically similar cases have been reported in northwest Africa, due to *L. donovani.* The clinical manifestations of the disease vary along its range, which has led to confusion regarding the identity of the organisms responsible. Several species names have been proposed for different clinical and serological types (see Table 5.1). Once again, it appears that the parasite is rapidly evolving into groups that are adapting to local populations of humans and flies. Morphologically, *L. braziliensis* cannot be differentiated from *L. tropica, L. mexicana,* or *L. donovani.* An interesting historical account of this disease, with evidence of its pre-Columbian existence in South America is given by Hoeppli.[45]

- *Life Cycle and Pathogenesis.* The life cycle and methods of reproduction of *L. braziliensis* are identical to those of *L. donovani* and *L. tropica* except that the promastigotes reproduce in the hindgut of the sand fly, with several species of *Lutzomyia* serving as vectors. Inoculation of promastigotes by the bite of a sand fly causes a small, red papule on the skin. This becomes an itchy, ulcerated vesicle in one to four weeks and is similar at this stage to oriental sore. This primary lesion heals within 6 to 15 months. The parasite never causes a visceral disease but often develops a secondary lesion on some region of the body.

  In Venezuela and Paraguay the lesions more often appear as flat, ulcerated plaques that remain open and oozing. The disease is called **pian bois** in that area. Sloths and anteaters are the primary reservoirs of pian bois in northern Brazil.[56]

  In the more southerly range of *L. braziliensis* the parasites have a tendency to metastasize, or spread directly from the primary lesion to mucocutaneous zones. The secondary lesion may appear before the primary has healed, or it may be many years (up to 30) before secondary symptoms appear.[111]

  The secondary lesion often involves the nasal system and buccal mucosa, causing degeneration of the cartilaginous and soft tissues (Fig. 5.22). Necrosis and secondary bacterial infection are common. **Espundia** and **uta** are the names applied to these conditions. The ulceration may involve the lips, palate, and pharynx, leading to great deformity. Invasion of the infection into the larynx and trachea destroys the voice. Rarely the genitalia may become infected. The condition may last for many years, and death may result from secondary infection or respiratory complications. A similar condition is known to occur in Ethiopia.

- *Diagnosis and Treatment.* Diagnosis is established by finding L-D bodies in affected tissues. Espundialike conditions are also caused by tuberculosis, leprosy, syphilis, and various fungal and viral diseases, and these must be differentiated in diagnosis. Skin tests are available for diagnosis of occult infections. Culturing the parasite in vitro is also a valuable technique when L-D bodies cannot be demonstrated in routine microscope preparation.

  Treatment is similar to that for kala-azar and tropical sore: antimonial compounds applied on the lesions or injected intravenously or intramuscularly. Secondary bacterial infections should be treated with antibiotics. Mucocutaneous lesions are particularly refractory to treatment and require extensive chemotherapy. Relapse is common, but once cured, a person usually has lifelong immunity. However, if the infection is not cured but merely becomes occult, there may be a relapse with onset of espundia many years later. Because this is primarily a sylvatic disease, there is little opportunity for its control.

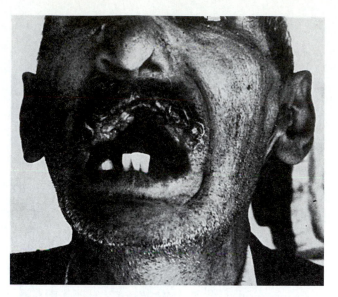

FIGURE 5.22

Espundia of two years' development after 24 years' delay in onset. The upper lip, gum, and palate are destroyed.

From B. C. Walton et al., "Onset of espundia after many years of occult infection with *Leishmania braziliensis,*" in *Am. J. Trop. Med. Hyg.* 22:696–698, Copyright © 1973.

Leishmania mexicana. This parasite is found in northern Central America, Mexico, Texas, and possibly the Dominican Republic and Trinidad. Primarily a cutaneous form, it infects several thousand persons a year, especially agricultural or forest laborers. The disease is on the increase due to increased clearing of forests and concomitant expansion of farmland. Three clinical manifestations are found: cutaneous, nasopharyngeal mucosal, and visceral, although some records probably are due to *L. braziliensis.* Traditionally the cutaneous form of disease has been called **chiclero ulcer** because it is so common in "chicleros," forest-dwelling people who glean a living by harvesting the gum of chicle trees. In Belize, an English-speaking country, it is called **bay sore.**

- *Life Cycle and Pathogenesis.* As in the other species of *Leishmania,* sand flies are the vectors of *L. mexicana.* Several species of *Lutzomyia* are involved. The disease is a zoonosis and the main reservoirs are rodents. The most important reservoirs are those that live or travel at ground level. Obviously, arboreal reservoirs are less efficient sources of infection to humans. No domestic reservoir is known for chiclero ulcer.

  Cutaneous leishmaniasis due to the *L. mexicana* complex usually heals spontaneously in a few months except when the lesions are in the ear. Ear cartilage is poorly vascularized so immune responses are weak. Chronic lesions are known with a duration of up to 40 years. Considerable mutilation may result. Mucocutaneous and visceral manifestations are rare.

At least eight cases of autochthonous infections of *L. mexicana* in Texas are known in humans, with another on the ear of a cat. Visceral cases in dogs in Oklahoma have been discovered; histories of these dogs suggest that canine leishmaniasis (due to what has been called the OKD or Oklahoma Dog strain) may have become endemic in the United States.

- *Diagnosis and Treatment.* The diagnosis and treatment of *L. mexicana* is the same as for *L. tropica*.

Genus *Leptomonas*

Leptomonas spp. are parasitic in invertebrates and are of no medical importance. *Leptomonas* is variously a promastigote and an intracellular amastigote throughout its monoxenous life cycle. Species are found in molluscs, nematodes, insects, and other protozoa.

Genus *Herpetomonas*

Members of the genus *Herpetomonas* also are characteristically monoxenous in insects. They pass through amastigote, promastigote, opisthomastigote, and possibly epimastigote stages in their life cycles. In the opisthomastigote the flagellum arises from a reservoir that runs the entire length of the body.

Genus *Crithidia*

Crithidia spp. are tiny (4 to 10 μm) choanoflagellates of insects. They are often clustered together against the intestine of their host. They can assume the amastigote form and are monoxenous.

Genus *Blastocrithidia*

Blastocrithidia organisms are monoxenous insect parasites, usually found as epimastigotes and amastigotes in the intestines of their hosts. Species are common in water striders (family Gerridae).

Genus *Phytomonas*

Phytomonas is a parasite of milkweeds and related plants. It passes through promastigote and amastigote phases in the intestines of certain beetles and appears as promastigotes in the sap (latex) of its plant hosts.

References

1. Albright, J. W., and J. F. Albright. 1991. Rodent trypanosomes: Their conflict with the immune system of the host. *Parasitol. Today* 7:137–40.

2. Ashford, R. W., P. Desjeux, and P. deRaadt. 1992. Estimation of population at risk of infection and number of cases of leishmaniasis. *Parasitol. Today* 8:104–5.

3. Bailey, N. M. 1967. Recent development in the screening of populations for human trypanosomiasis and their possible application in other immunizing diseases. *East Afr. Med. J.* 44:475–81.

4. Banks, K. L. 1978. Binding of *Trypanosoma congolense* to the walls of small blood vessels. *J. Protozool.* 25:241–45.

5. Barry, J. D. 1986. Antigenic variation during *Trypanosoma vivax* infections of different host species. *Parasitology* 92:51–65.

6. Barry, J. D., and C. M. R. Turner. 1991. The dynamics of antigenic variation and growth of African trypanosomes. *Parasitol. Today* 7:207–11.

7. Battaglia, P. A., M. del Bue, M. Ottaviano, and M. Ponzi. 1983. A puzzle genome: Kinetoplast DNA. In Guardiola, J., L. Luzzatto, and W. Trager, ed. *Molecular biology of parasites.* New York: Raven Press, 107–24.

8. Bellofatto, V. 1990. The new trypanosomatid genetics. *Parasitol. Today* 6:299–302.

9. Blackwell, J. M. 1992. Leishmaniasis epidemiology: All down to the DNA. *Parasitology* 104(S):S19–S34.

10. Borst, P. 1991. Molecular genetics of antigenic variation. *Immunoparasitol. Today* 7:A29–A33.

11. Borst, P., and G. A. M. Cross. 1982. Molecular basis for trypanosome antigenic variation. *Cell* 29:291–303.

12. Brener, Z. 1972. A new aspect of *Trypanosoma cruzi* life cycle in the intermediate host. *J. Protozool.* 19:23–27.

13. Burkholder, J. E., T. C. Allison, and V. P. Kelly. 1980. *Trypanosoma cruzi* (Chagas) (Protozoa: Kinetoplastida) in invertebrate, reservoir, and human hosts of the lower Rio Grande Valley of Texas. *J. Parasitol.* 66:305–11.

14. Cannata, J. J. B., A. C. C. Frasch, M. A. Cataldi de Flombaum, E. L. Segura, and J. J. Cazzulo. 1979. Two forms of "malic" enzyme with different regulatory properties in *Trypanosoma cruzi. Biochem. J.* 184:409–19.

15. Carvalho, E. M., R. Badaro, S. G. Reed, T. C. Jones, and W. D. Johnson Jr. 1985. Absence of gamma interferon and interleukin-2 production during active visceral leishmaniasis. *J. Clinical Invest.* 76:2066–69.

16. Chance, M. C. 1985. The biochemical and immunological taxonomy of *Leishmania.* In Chang, K.-P., and R. S. Bray, eds. *Human parasitic diseases, vol. 1. Leishmaniasis.* New York: Elsevier, 93–110.

17. Chang, K. -P. 1981. Leishmanicidal mechanisms of human polymorphonuclear phagocytes. *Am. J. Trop. Med. Hyg* 30:322–33.

18. Corral, R. S., A. Orn, and S. Grinstein. 1992. Detection of soluble exoantigens of *Trypanosoma cruzi* by a dot-immunobinding assay. *Am. J. Trop. Med. Hyg.* 46:31–38.

19. Croft, S. L., R. N. Davidson, and E. A. Thornton. 1991. Liposomal amphotericin B in the treatment of visceral leishmaniasis. *J. Antimicrobial Chemotherapy.* 28(Suppl. B):111–18.

20. Das Gupta, S., D. K. Ghosh, and H. K. Majumber. 1991. A cloned kinetoplast DNA mini-circle fragment from a *Leishmania* spp. specific for post-kala-azar dermal leishmaniasis strains. *Parasitology* 102:187–92.

21. De Bruijn, M. H. L., and D. C. Barker. 1992. Diagnosis of New World leishmaniasis: Specific detection of species of the *Leishmania braziliensis* complex by amplification of kinetoplast DNA. *Acta Tropica* 52:45–58.

22. Deniau, M., R. Durand, C. Bories, M. Paual, A. Astier, P. Couvreur, and R. Houin. 1993. Étude *in vitro* de médicaments leishmanicides vectorisés. *Ann. Parasitol. Hum. Comp.* 68:34–37.

23. De Souza, W., and T. Souto-Padrón. 1980. The paraxial structure of the flagellum of Trypanosomatidae. *J. Parasitol.* 66:229–35.

24. Docampo, R., F. S. Cruz, W. Leon, and G. A. Schmunis. 1979. Acetate oxidation by bloodstream forms of *Trypanosoma cruzi. J. Protozool.* 26:301–3.

25. Dolan, R. B. 1987. Genetics and trypanotolerance. *Parasitol. Today* 3:137–43.

26. Doyle, P. S., J. C. Engel, P. F. P. Pimenta, P. Pinto da Silva, and D. M. Dwyer. 1991. *Leishmania donovani:* Long-term culture of axenic amastigotes at 37° C. *Parasitology* 73:326–34.

27. Esser, K. M., and M. J. Schoenbechler. 1985. Expression of two variant surface glycoproteins on individual African trypanosomes during antigen switching. *Science* 229:190–93.

28. Ferreira, A. L., and M. A. Rossi. 1973. Pathology of the testis and epididymis in the late phase of experimental Chagas' disease. *Am. J. Trop. Med.* 22:699–704.

29. Fournet, A., A. Angelo, V. Munoz, F. Roblot, R. Hocquemiller, and A. Cave. 1992. Biological and chemical studies of *Pera benensis,* a Bolivian plant used in folk medicine as a treatment of cutaneous leishmaniasis. *J. Ethnopharmacology* 37:159–64.

30. Gardiner, P. R., T. W. Pearson, M. W. Clarke, and L. M. Mutharia. 1987. Identification and isolation of a variant surface glycoprotein from *Trypanosoma vivax. Science* 235:774–77.

31. Gardiner, P. R., and A. J. Wilson. 1987. *Trypanosoma (Duttonella) vivax. Parasitol. Today* 3:49–52.

32. Gibson, W. C. 1986. Will the real *Trypanosoma b. gambiense* please stand up. *Parasitol. Today* 2:255–57.

33. Gibson, W. C., T. F. de C. Marshall, and D. G. Godfrey. 1980. Numerical analysis of enzyme polymorphism: A new approach to the epidemiology and taxonomy of trypanosomes of the subgenus *Trypanozoon.* In Lumsden, W. H. R., R. Muller, and J. R. Baker, eds. *Advances in parasitology* 18. New York: Academic Press, Inc., 175–246.

34. Goodwin, L. G. 1964. The chemotherapy of trypanosomiasis. In Hutner, S. M., ed. *The biochemistry and physiology of protozoa* 3. New York: Academic Press, Inc., 495–524.

35. Grogl, M., T. N. Thomason, and E. D. Franke. 1992. Drug resistance in leishmaniasis: Its implication in systemic chemotherapy of cutaneous and mucocutaneous disease. *Am. J. Trop. Med. Hyg.* 47:117–26.

36. Guhl, F., L. Hudson, C. J. Marinkelle, C. A. Jaramillo, and D. Bridge. 1987. Clinical *Trypanosoma rangeli* infection as a complication of Chagas' disease. *Parasitology* 94:475–84.

37. Gutteridge, W. E., B. Cover, and M. Gaborak. 1978. Isolation of blood and intracellular forms of *Trypanosoma cruzi* from rats and other rodents and preliminary studies of their metabolism. *Parasitology* 76:159–76.

38. Hajduk, S. L., C. R. Cameron, J. D. Barry, and K. Vickerman. 1981. Antigenic variation in cyclically transmitted *Trypanosoma brucei.* Variable antigen type composition of metacyclic trypanosome populations from the salivary glands of *Glossina morsitans. Parasitology* 83:595–607.

39. Hajduk, S. L., and K. Vickerman. 1981. Antigenic variation in cyclically transmitted *Trypanosoma brucei.* Variable antigen type composition of the first parasitaemia in mice bitten by trypanosome-infected *Glossina morsitans. Parasitology* 83:609–21.

40. Hill, G. C. 1976. Characterization of the electron transport systems present during the life cycle of African trypanosomes. In Van den Bossche, H., ed. *Biochemistry of parasites and host-parasite relationships.* Amsterdam: North-Holland Publishing Co., 31–50.

41. Hoare, C. A. 1956. Morphological and taxonomic studies on the mammalian trypanosomes. VIII. Revision of *Trypanosoma evansi. Parasitology* 46:130–72.

42. Hoare, C. A. 1964. Morphological and taxonomic studies on mammalian trypanosomes. X. Revision of the systematics. *J. Protozool.* 11:200–7.

43. Hoare, C. A. 1965. Vampire bats as vectors and hosts of equine and bovine typanosomes. *Acta Tropica* 22:204–16.

44. Hoare, C. A. 1967. Evolutionary trends in mammalian trypanosomes. In Dawes, B., ed. *Advances in parasitology* 5. New York: Academic Press, Inc., 47–91.

45. Hoeppli, R. 1969. *Parasitic diseases in Africa and the Western Hemisphere. Early documentation and transmission by the slave trade.* Basel: Verlag für Recht and Gesellshaft AG.

46. Hudson, L. 1985. Autoimmune phenomena in chronic chagasic cardiopathy. *Parasitol. Today* 1:6–7.

47. Hunter, C. A., F. W. Jennings, J. F. Tierney, M. Murray, and P. G. E. Kennedy. 1992. Correlations of autoantibody titres with central nervous system pathology in experimental African trypanosomiasis. *J. Neuroimmunol.* 41:143–48.

48. Jansen, A. M., P. L. Moriearty, B. G. Castro, and M. P. Deane. 1985. *Trypanosoma cruzi* in the opossum *Didelphis marsupialis:* An indirect fluorescent antibody test for the diagnosis and follow-up of natural and experimental infections. *Trans. R. Soc. Trop. Med. Hyg.* 79:474–77.

49. Jenni, L. 1990. Sexual stages in trypanosomes and implications. *Ann. Parasitol. Hum. Comp.* 65(S):19–21.

50. Jennings, F. W., G. M. Urquhart, P. K. Murray, and B. M. Miller. 1980. 'Berenil' and nitroimidazole combinations in the treatment of *Trypanosoma brucei* infection with central nervous system involvement. *Int. J. Parasitol.* 10:27–32.

51. Kagan, I., L. Norman, and D. S. Allain. 1966. Studies on *Trypanosoma cruzi* isolated in the United States: A review. *Rev. Biol. Trop.* 14:55–73.

52. Kierzenbaum, F. 1985. Is there autoimmunity in Chagas' disease? *Parasitol. Today* 1:4–6.

53. Krieger, M. A., E. Almeida, W. Oelemann, J. J. LaFaille, J. B. Pereira, H. Krieger, M. R. Carvalho, and S. Goldenberg. 1992. Use of recombinant antigens for the accurate immunodiagnosis of Chagas' disease. *Am. J. Trop. Med. Hyg.* 46:427–34.

54. Lainson, R. 1982. Leishmanial parasites of mammals in relation to human disease. In Edwards, M. A., and U. McDonnell, eds. *Symposium of the Zoological Society of London* 50. London: Academic Press, Inc., 137–79.

55. Lainson, R., and J. J. Shaw. 1987. Evolution, classification, and geographical distribution. In Peters, W., and R. Killick-Kendrick, eds. *The leishmaniases in biology and medicine* 1. New York: Academic Press, Inc., 1–120.

56. Lainson, R., J. J. Shaw., and M. Póvoa. 1981. The importance of edentates (sloths and anteaters) as primary reservoirs of *Leishmania braziliensis guyanensis,* causative agent of "pian-bois" in north Brazil. *Trans. R. Soc. Trop. Med. Hyg.* 75:611–12.

57. Laurent, M., E. Pays, E. Delinte, E. Magnus, N. Van Meirvenne, and M. Steinert. 1984. Evolution of a trypanosome surface antigen gene repertoire linked to non-duplicative gene activation. *Nature* 308:370–73.

58. Levine, N. D. 1973. Protozoan parasites of domestic animals and of man, 3d ed. Minneapolis: Burgess Publishing Co.

59. Lincicome, D. R., R. N. Rossan, and W. C. Jones. 1963. Growth of rats infected with *Trypanosoma lewisi. Exp. Parasitol.* 14:54–65.

60. Locksley, R. M., and P. Scott. 1991. Helper T-cell subsets in mouse leishmaniasis: Induction, expansion and effector function. *Immunoparasitol. Today* 7:A58–A61.

61. Locksley, R. M., and J. A. Louis. 1992. Immunology of leishmaniasis. *Current Opinions in Immunology* 4:413–18.

62. Lumsden, W. H. R. 1971. Pathobiology of trypanosomiasis. In Gaafar, S. M., ed. *Pathology of parasitic diseases.* Lafayette, Ind.: Purdue Research Foundation, 1–14.

63. Massamba, N. N., and J. J. Mutinga. 1992. Recombinant kinetoplast DNA (kDNA) probe for identifying *Leishmania tropica*. *Acta Tropica* 52:1–15.

64. Maudlin, I. 1985. Inheritance of susceptibility to trypanosomes in tsetse flies. *Parasitol. Today* 1:59–60.

65. Mauel, J., and R. Behin. 1982. Leishmaniasis: Immunity, immunopathology and immunodiagnosis. In Cohen, S., and K. S. Warren, eds. *Immunology of parasitic infections.* Oxford: Blackwell Scientific Publications Ltd., 299–355.

66. McCabe, R. E., J. S. Remington, and F. G. Araujo. 1987. Ketoconazole promotes parasitological cure of mice infected with *Trypanosoma cruzi. Trans. R. Soc. Trop. Med. Hyg.* 81:613–15.

67. Michels, P. A. M., and F. R. Opperdoes. 1991. The evolutionary origin of glycosomes. *Parasitol. Today* 7:105–9.

68. Molyneux, D. H. 1977. Vector relationships in the Trypanosomatidae. In Dawes, B., ed. *Advances in parasitology* 15. New York: Academic Press, Inc., 1–82.

69. Morgan, F. M., R. H. Watten, and R. E. Kuntz. 1962. Post-kala-azar dermal leishmaniasis. A case report from Taiwan (Formosa). *J. Formosa Med. Assoc.* 61:282–91.

70. Müller, M. 1975. Biochemistry of protozoan microbodies: Peroxisomes, α-glycerophosphate oxidase bodies, hydrogenosomes. *Ann. Rev. Microbiol.* 29:467–83.

71. Murphy, W. J., S. T. Brentano, A. C. Rice-Ficht, D. M. Dorfman, and J. E. Donelson. 1984. DNA rearrangements of the variable surface antigen genes of the trypanosomes. *J. Protozool.* 31:65–73.

72. Murray, M., W. I. Morrison, and D. D. Whitelaw. 1982. Host susceptibility to African trypanosomiasis: Trypanotolerance. In Baker, J. R., and R. Muller, eds. *Advances in parasitology* 21. New York: Academic Press, Inc. 1–68.

73. Nabors, G. S., and R. L. Tarleton. 1991. Differential control of interferon-gamma and IL-2 production during *Trypanosoma cruzi* infection. *J. Immunol.* 146:3591–98.

74. Newport, G. R., and C. R. Page III. 1977. Free amino acids in brain, liver, and skeletal muscle tissue of voles infected with *Trypanosoma brucei gambiense. J. Parasitol.* 63:1060–65.

75. Noble, E. R. 1955. The morphology and life cycles of trypanosomes. *Quart. Rev. Biol.* 30:1–28.

76. Ormerod, W. E. 1967. Taxonomy of the sleeping sickness trypanosomes. *J. Parasitol.* 53:824–30.

77. Ormerod, W. E. 1985. How do lipids affect African trypanosomes? *Parasitol. Today* 1:86–87.

78. Pays, E., M. L'heureux, and M. Steinert. 1982. Structure and expressions of a *Trypanosoma brucei gambiense* variant specific antigen gene. *Nucleic Acids Res.* 10:3149–63.

79. Pays, E., S. Van Assel, M. Laurent, M. Darville, T. Vervoort, N. Van Meirvenne, and M. Steinert. 1983. Gene conversion as a mechanism for antigenic variation in trypanosomes. *Cell* 34:371–81.

80. Pays, E., M. -F. Delauw, S. Van Assel, M. Laurent, T. Vervoort, N. Van Meirvenne, and M. Steinert. 1983. Modifications of a *Trypanosoma b. brucei* antigen gene repertoire by different DNA recombinational mechanisms. *Cell* 35:721–31.

81. Pearson, R. D. 1993. Pathology of leishmaniasis. In Warren, K. S., ed. *Immunology and molecular biology of parasitic infections.* Boston: Blackwell Scientific Publications.

82. Pearson, R. D., T. Evans, D. A. Wheeler, T. G. Naidu, J. E. de Alencar, and J. S. Davis IV. 1986. Humoral factors during South American visceral leishmaniasis. *Ann. Trop. Med. Parasitol.* 80:465–68.

83. Peters, W., and R. Killick-Kendrick, eds. 1987. *The leishmaniases in biology and medicine.* New York: Academic Press, Inc.

84. Piuvezam, M., D. M. Russo, J. M. Burns Jr., Y. A. W. Skeiky, K. H. Grabstein, and S. G. Reed. 1993. Characterization of responses of normal human T cells to *Trypanosoma cruzi* antigens. *J. Immunol.* 150:916–24.

85. Rickman, W. J., and H. W. Cox. 1983. Trypanosome antigen-antibody complexes and immunoconglutinin interactions in African trypanosomiasis. *Int. J. Parasitol.* 13:389–92.

86. Rioux, J. A., G. Lanotte, E. Serres, F. Pratlong, P. Bastien, and J. Perieres. 1990. Taxonomy of *Leishmania.* Use of isozymes. Suggestions for a new classification. *Ann. Parasitol. Hum. Comp.* 65:111–25.

87. Rohwedder, R. 1965. Chagas' infection in blood donors and the possibilities of its transmission by means of transfusion. *Bull. Chil. Parasitol.* 24:88–93.

88. Ross, R., and D. Thompson. 1910. A case of sleeping sickness studied by precise enumerative methods: Regular periodical increase of the parasites disclosed. *Proc. R. Soc. Lond. (Biol.)* 82:411–15.

89. Salazar-Schettino, P. M. 1983. Customs which predispose to Chagas' disease and cysticercosis in Mexico. *Am. J. Trop. Med. Hyg.* 32:1179–80.

90. Schechter, P. J., and A. Sjoerdsma. 1986. Difluoromethylornithine in the treatment of African trypanosomiasis. *Parasitol. Today* 2:223–24.

91. Schoone, G. J., G. J. J. M. van Eys, G. S. Ligthart, F. E. Taub, J. Zaal, Y. Mebrahtu, and P. Lawyer. 1991. Detection and identification of *Leishmania* parasites by *in situ* hybridization with total and recombinant DNA probes. *Exp. Parasitol.* 73:345–53.

92. Scott, M. T., and D. Snary. 1982. American trypanosomiasis (Chagas' disease). In Cohen, S., and K. S. Warren, eds. *Immunology of parasitic infections.* Oxford: Blackwell Scientific Publications Ltd., 261–98.

93. Seed, J. R., R. Edwards, and J. Sechelski. 1984. The ecology of antigenic variation. *J. Protozool.* 31:48–53.

94. Singha, U. K., P. Y. Guru, A. B. Sen, and J. S. Tandon. 1993. Antileishmanial activity of traditional plants against *Leishmania donovani* in golden hamsters. *Int. J. Pharmacognosy* 30:289–95.

95. Smyth, A. J., A. Ghosh, M. Q. Hassan, D. Basu, M. H. L. De Bruijn, S. Adhya, K. K. Mallik, and D. C. Barker. 1992. Rapid and sensitive detection of *Leishmania* kinetoplast DNA from spleen and blood samples of kala-azar patients. *Parasitology* 105:183–92.

96. Sternberg, J., C. M. R. Turner, J. M. Wells, L. C. Ranford-Cartwright, R. W. F. LePage, and A. Tait. 1989. Gene exchange in African trypanosomes: Frequency and allelic segregation. *Molecular and Biochem. Parasitol.* 34:269–79.

97. Tait, A. 1983. Sexual processes in the Kinetoplastida. Symposia of the British Society for Parasitology, 20. *Parasitology* 86(4):29–57.

98. Tanowitz, H. B., C. Brosnan, D. Guastamacchio, G. Baron, C. Raventos-Suarez, M. Bornstein, and M. Wittner. 1982. Infection of organotypic cultures of spinal cord and dorsal root ganglia with *Trypanosoma cruzi. Am. J. Trop. Med. Hyg.* 31:1090–97.

99. Tarleton, R. 1991. Regulation of immunity in *Trypanosoma cruzi* infection. *Exp. Parasitol.* 73:106–9.

100. Tarleton, R. 1993. Pathology of American trypanosomiasis. In Warren, K. S., ed. *Immunology and molecular biology of parasitic infections.* Boston: Blackwell Scientific Publications, 64–86.

101. Tibayrenc, M., L. Echalar, J. P. Dujardin, O. Poch, and P. Desjeux. 1984. The microdistribution of isoenzymic strains of *Trypanosoma cruzi* in southern Bolivia; new isoenzyme profiles and further arguments against Mendelian sexuality. *Trans. R. Soc. Trop. Med. Hyg.* 78:519–25.

102. Tobie, E. J. 1965. Biological factors influencing transmission of *Trypanosoma rangeli* by *Rhodnius prolixus. J. Parasitol.* 51:837–41.

103. Turner, C. M. R., J. D. Barry, and K. Vickerman. 1986. Independent expression of the metacyclic and bloodstream variable antigen repertoires of *Trypanosoma brucei rhodesiense. Parasitology* 92:67–73.

104. Turner, C. M. R., J. Sternberg, N. Buchanan, E. Smith, G. Hide, and A. Tait. 1990. Evidence that the mechanism of gene exchange in *Trypanosoma brucei* involves meiosis and syngamy. *Parasitology* 101:377–86.

105. Vickerman, K. 1969. The fine structure of *Trypanosoma congolense* in its bloodstream phase. *J. Protozool.* 16:54–69.

106. Vickerman, K. 1971. Morphological and physiological considerations of extracellular blood protozoa. In Fallis, A. M., ed. *Ecology and physiology of parasites.* Toronto: University of Toronto Press, 58–91.

107. Vickerman, K. 1974. Antigenic variation in African trypanosomes. In *Parasites in the immunized host: Mechanisms of survival* (Ciba Foundation Symposium 25, new series). Amsterdam: Elsevier, 53–80.

108. Vickerman, K. 1978. Antigenic variation in trypanosomes. *Nature* 273:613–17.

109. Vickerman, K., and J. D. Barry. 1982. African trypanosomiasis. In Cohen, S., and K. S. Warren, eds. *Immunology of parasitic infections.* Oxford: Blackwell Scientific Publications Ltd., 204–60.

110. Wallace, F. G. 1966. The trypanosomatid parasites of insects and arachnids. *Exp. Parasitol.* 18:124–93.

111. Walton, B. C., L. V. Chinel, and O. Eguia y E. 1973. Onset of espundia after many years of occult infection with *Leishmania braziliensis. Am. J. Trop. Med. Hyg.* 22:696–98.

112. Williams, P., and M. de Vasconcellos Coelho. 1978. Taxonomy and transmission of *Leishmania.* In Lumsden, W. H. R., R. Muller, and J. R. Baker, eds. *Advances in parasitology* 16. New York: Academic Press, Inc., 1–42.

113. Williamson, J. 1962. Chemotherapy and chemoprophylaxis in African trypanosomiasis. *Exp. Parasitol.* 12:274–367.

114. Woo, P. T. K., and M. A. Soltys. 1970. Animals as reservoir hosts of human trypanosomes. *J. Wildl. Dis.* 6:313–22.

115. Woody, N. C., and H. B. Woody. 1955. American trypanosomiasis (Chagas' disease). First indigenous case in the United States. *J. A. M. A.* 159:476–77.

116. World Health Organization. 1960. *Chagas' disease. Report of a study group. Technical Report Series no. 202.* Geneva.

117. Yaeger, R. G. 1971. Transmission of *Trypanosoma cruzi* infection to opossums via the oral route. *J. Parasitol.* 57:1375–76.

118. Young, J. R., J. S. Shah, G. Matthyssens, and R. O. Williams. 1983. Relationship between multiple copies of a *T. brucei* variable surface glycoprotein gene whose expression is not controlled by duplication. *Cell* 32:1149–59.

Additional References

Adler, S. 1964. Leishmania. In Dawes, B., ed. *Advances in parasitology* 2. New York: Academic Press, Inc., 1–34. An advanced treatise on the subject. Recommended reading for all who are interested in the Trypanosomatidae.

Barker, D. C. 1987. DNA diagnosis of human leishmaniasis. *Parasitol. Today* 3:177–84.

Blackwell, J., et al. 1986. Molecular biology of *Leishmania. Parasitol. Today* 2:45–53.

Foster, W. D. 1965. *A history of parasitology.* Edinburgh: E. & S. Livingstone, Chapter 10, "The Trypanosomes," is a very interesting account of the history of knowledge about this group.

Hoogstraal, H., and D. Heyneman. 1969. Leishmaniasis in the Sudan Republic. 30. Final epidemiological report. *Am. J. Trop. Med. Hyg.* 18:1089–1210. An extensive account of the aspects of leishmaniasis by two men who have an unashamed love for humanity. It should be required reading for all students of parasitology, and it stands by itself as an example of what scientific writing should be.

Jordan, A. M. 1985. Tsetse eradication plans for southern Africa. *Parasitol. Today* 1:121–23.

Marsden, P. D. 1985. Clinical presentations of *Leishmania braziliensis braziliensis. Parasitol. Today* 1:129–33. An outstanding review of the subject with excellent illustrations.

Vickerman, K. 1985. Leishmaniasis—the first centenary. *Parasitol. Today* 1:149, 172.

Chapter 6

OTHER FLAGELLATE PROTOZOA

Perhaps, science will have replaced the art when the addition of totally defined nutrients, the removal of metabolic wastes, monitoring of physical and chemical conditions of growth, and harvesting of the crop have become automated.

Louis Diamond, on the challenge of "separating a protozoan from its habitat in the wild and inducing it to take up a new existence in the culture tube."

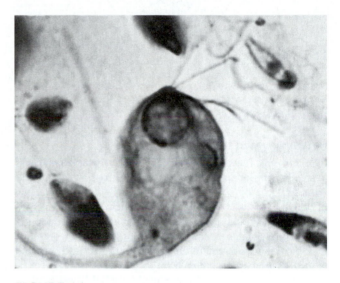

FIGURE 6.1

Trophozoite of *Chilomastix caulleryi*, which is similar morphologically to *C. mesnili*. Note the four flagella and the cytostomal fibrils. It is 6 to 24 μm long.

Photograph by Larry S. Roberts.

Although the Kinetoplastida includes some exceedingly important parasites whose economic impact is quite severe and whose pathology is dramatic, several other groups of flagellated protozoa also have members that are parasitic. These flagellates are likely to be found in every kind of animal from cockroaches to humans. A few of them are structurally complex, and some, such as species of *Giardia,* because of their biochemical characteristics have become favorites of the evolutionary biologists. But space limitations prevent us from covering all of these parasites in detail. Consequently, representative species are drawn from four orders.

ORDER RETORTAMONADIDA

Family Retortamonadidae

Two species in the family Retortamonadidae are commonly found in humans. Although they are apparently harmless commensals, they are worthy of note because they easily can be mistaken for highly pathogenic species.

• *Chilomastix mesnili*

Chilomastix mesnili (see Fig. 6.1) infects about 3.5% of the population of the United States and 6% of the world population.[4] It lives in the cecum and colon of humans, chimpanzees, orangutans, monkeys, and pigs. Other species are known in other mammals, birds, reptiles, amphibians, fish, leeches, and insects.

The living trophozoite is pyriform, with the posterior end drawn out into a blunt point, and is 6 to 24 μm by 3 to 10 μm. A longitudinal **spiral groove** occurs in the surface of the middle of the body, but this is usually visible only on living specimens. A sunken **cytostomal groove** is prominent near the anterior end. Along each side of the cytostome runs a cytoplasmic **cytostomal fibril,** presumably strengthening the

lips of the cytostome. The cytostome leads into the cytopharynx, where endocytosis takes place. Four flagella, one longer than the others, emerge from kinetosomes on the anterior end of the body, and the kinetosomes are interconnected by microfibrillar material.[8] One of the flagella is very short and delicate, curving back into the cytostome, where it undulates. The large nucleus is near the front end.

A cyst stage occurs, especially in formed stools (Fig. 6.2). A typical cyst is thick walled, 6.5 to 10.0 μm long, and pear or lemon shaped. It has a single nucleus and retains all the cytoplasmic organelles, including cytostomal fibrils, kinetosomes, and axonemes.

Transmission is by ingestion of cysts, since trophozoites cannot survive stomach acid. Fecal contamination of drinking water is the most important means of transmission.

Chilomastix mesnili usually is considered nonpathogenic, although Mueller[46] suggested that it might cause a watery stool in some instances. But because *C. mesnili* often co-occurs with other parasites that are pathogenic, it may simply be guilty by association;[16] in such cases, the flagellates are basically confirming what the presence of *Giardia* tells about local sanitary conditions and personal hygiene.

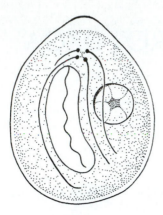

FIGURE 6.2

Cyst of *Chilomastix mesnili* from a human stool, showing the characteristic lemon or pear shape. Also visible are the large, irregular karyosome and the cytostomal fibrils.

Drawing by William Ober.

• *Retortamonas intestinalis*

Retortamonas intestinalis (Fig. 6.3) is a tiny protozoan that is basically similar to *C. mesnili,* but the trophozoite is only 4 μm to 9 μm long. Furthermore, it has only two flagella, one of which extends anteriorly, and the other of which emerges from the cytostomal groove and trails posteriorly. The living trophozoite usually extends into a blunt point at its posterior end, but it bends to round up in fixed specimens. The ovoid to pear-shaped cysts contain a single nucleus.

Like *C. mesnili,* this species is probably a harmless commensal. It lives in the cecum and large intestine of monkeys and chimpanzees as well as of humans, and apparently it is not a common symbiont anywhere in the world. Other members of the genus *Retortamonas* have been reported from crickets, cockroaches, guinea pigs, and toads, including the cane toad, *Bufo marinus,* imported into Australia where it evidently acquired *R. dobelli* from local anurans.[17]

ORDER DIPLOMONADIDA

Family Hexamitidae

Members of the Hexamitidae are easily recognized because they have two equal nuclei lying side by side. There are several species in five genera, most of which are parasitic in vertebrates or invertebrates. One species, *Giardia lamblia,* is a parasite of humans and will serve to illustrate the genus *Giardia. Hexamita meleagridis* is an example of a related species in domestic animals.

• Genus *Giardia*

Members of the genus *Giardia* have come to occupy a prominent place in both the parasitological and evolutionary biology literature in recent years. The lack of mitochondria has been interpreted as a primitive trait, and phylogenetic analysis of ribosomal RNA has been used to place *Giardia* near the point of divergence between pro- and eukaryotes,[31] resulting in use of the term *missing link* to describe this diplomonad's evolutionary position. Not all researchers agree with this assessment, however, and cladistic analysis of

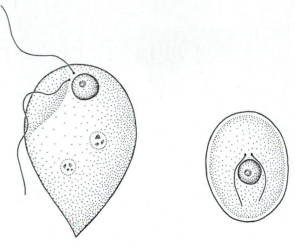

FIGURE 6.3

Retortamonas intestinalis trophozoite and cyst.

Drawing by William Ober.

Giardia's structural features suggests the parasites are actually derived from more recent parasitic ancestors.[58] Regardless of the outcome of this discussion, *Giardia* species will remain of interest to parasitologists and nonparasitologists alike because of their widespread occurrence and fairly frequent infections in people from all nations and socioeconomic levels.

More than 40 species of *Giardia* have been described, but only five are now considered valid:[64] *G. lamblia* (= *intestinalis* = *duodenalis*) and *G. muris* from mammals, *G. ardeae* and *G. psittaci* from birds, and *G. agilis* from amphibians.

Giardia lamblia. *Giardia lamblia* was first discovered in 1681 by Leeuwenhoek, who found it in his own stools. The taxonomy of the species was confused in the nineteenth century, and that confusion is still not completely resolved. Most researchers now refer to the parasites from humans as *Giardia lamblia,* at least in published papers, although *G. intestinalis* and *G. duodenalis* are still used as synonyms.[58, 64] The species is cosmopolitan in distribution but occurs most commonly in warm climates, and children are especially susceptible. *Giardia lamblia* is the most common flagellate of the human digestive tract.

- *Morphology.* The trophozoite (Figs. 6.4 and 6.5) is 12 to 15 μm long, rounded at the anterior end and pointed at the posterior end. The organism is dorsoventrally flattened and convex on the dorsal surface. The flattened ventral surface bears a concave, bilobed **adhesive disc** (Figs. 6.6 and 6.7). The so-called adhesive disc actually is a rigid structure, reinforced by microtubules and fibrous ribbons, surrounded by a flexible, apparently contractile, striated rim of cytoplasm. Application of this flexible rim to a host intestinal cell, working in conjunction with the **ventral flagella,** found in the **ventral groove,** is responsible for the organism's remarkable ability to adhere to the host cell (Fig. 6.8). The pair of ventral flagella, as well as three more pairs of flagella, arises from kinetosomes located between the anterior portions of the two nuclei (Fig. 6.5). The axonemes of all

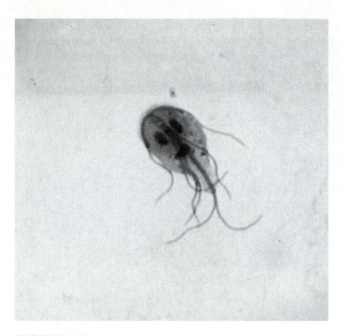

FIGURE 6.4

Giardia lamblia trophozoite in a human stool. It is 12 to 15 μm long.

Courtesy of Sherwin Desser.

flagella course through the cytoplasm for some distance before emerging from the cell body; those of the anterior flagella actually cross and emerge laterally from the adhesive disc area on the side opposite their respective kinetosomes.

A pair of large, curved, transverse, dark-staining **median bodies** lies behind the adhesive disc. These bodies are unique to the genus *Giardia*. Various authors have regarded them as parabasal bodies, kinetoplasts, or chromatoid bodies, but ultrastructural studies have shown they are none of these.[15, 24] Their function is obscure, although it has been suggested that may help support the posterior end of the organism, or they may be involved in its energy metabolism. There is no axostyle; the structure so described by previous authors is formed by the intracytoplasmic axonemes of the ventral flagella and associated groups of microtubules. Interestingly, there are no mitochondria, smooth endoplasmic reticulum, Golgi bodies, or lysosomes.[24]

The overall effect of the two nuclei behind the lobes of the adhesive disc and the median bodies is that of a wry little face that seems to be peering back at the observer.

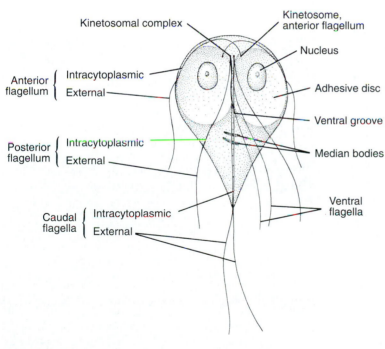

FIGURE 6.5

Diagram of *Giardia lamblia*.

Drawing by William Ober.

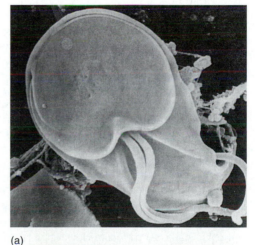

(a)

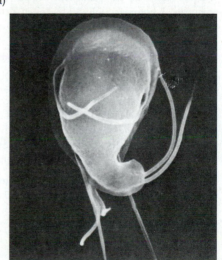

(b)

FIGURE 6.6

Scanning electron micrograph of *Giardia*. (*a*) The ventral view shows the flat adhesive disc and the relationship of the ventral and posterior flagella and ventral groove, but the caudal flagella curve around to the other side in this photograph. (*b*) The dorsal view shows these flagella, as well as the anterior flagella. The organism is 12 to 15 μm long.

Courtesy of Dennis Feely.

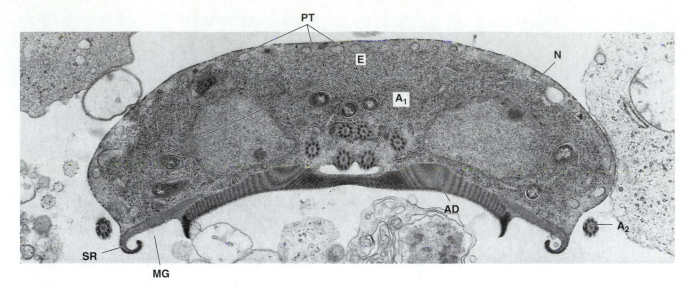

FIGURE 6.7

Transmission electron micrograph of a transverse section of a *Giardia muris* trophozoite found in the small bowel of an infected mouse. The marginal groove is the space between the striated rim of cytoplasm and the lateral ridge of the adhesive disc. The beginning of the ventral groove can be seen dorsal to the central area of the adhesive disc. This specimen bears endosymbionts, which are apparently bacteria. **PT,** peripheral tubules; **E,** endosymbionts; **N,** nucleus; **A₁,** axonemes of posterior, ventral, and caudal flagella; **A₂,** axoneme of anterior flagellum; **AD,** adhesive disc; **MG,** marginal groove; **SR,** striated rim of cytoplasm. (× 15,350.)

From P. C. Nemanic et al., "Ultrastructural observations on giardiasis in a mouse model. II. Endosymbiosis and organelle distribution in *Giardia muris* and *Giardia lamblia,*" in *J. Infect. Dis.* 140:222–228. Copyright © 1979 University of Chicago.

- **Life Cycle.** *Giardia lamblia* lives in the duodenum, jejunum, and upper ileum of humans, with the adhesive disc fitting over the surface of an epithelial cell. In severe infections the free surface of nearly every cell is covered by a parasite. The protozoa can swim rapidly using their flagella.

  Trophozoites divide by binary fission. First the nuclei divide and then the locomotor apparatus and the sucking disc and finally the cytoplasm follow suit. Enormous numbers of flagellates can build up rapidly in this way. It has been calculated that a single diarrheic stool can contain 14 billion parasites, whereas a stool in a moderate infection may contain 300 million cysts.[14] Obviously one infected individual can spread around a lot of misery.

  In the small intestine and in watery stools only the trophic stage can be found. However, as the feces enter the colon and begin to dehydrate, the parasites become encysted. First the flagella shorten and no longer project. The cytoplasm condenses and secretes a thick, hyaline cyst wall. The ovoid cysts (Fig. 6.9) are 8 to 12 µm by 7 to 10 µm in size. Newly formed cysts have two nuclei, but older ones have four. Soon the sucking disc and the locomotor apparatus are doubled, and the twinned flagellates are ready to emerge. When swallowed by the host, they pass safely through the stomach and excyst in the duodenum, immediately completing the division of the cytoplasm. The flagella grow out, and the parasites are once again at home.

- **Metabolism.** *Giardia lamblia* is an aerotolerant anaerobe.[36, 66] As mentioned earlier, these protozoa have no mitochondria. The tricarboxylic acid cycle and cytochrome system are absent, but the organisms avidly consume oxygen when it is present. Glucose is apparently

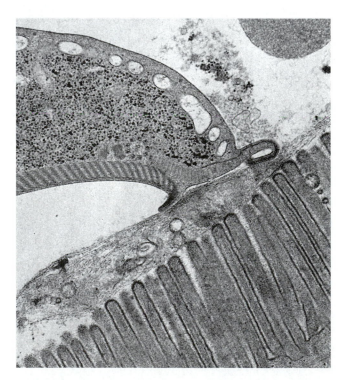

FIGURE 6.8

Periphery of *Giardia muris* in contact with the mucous stream covering the microvilli of a duodenal epithelial cell. It appears that the peripheral flange of striated cytoplasm is the grasping organelle of the ventral surface. (× 33,000.)

From D. S. Friend, "The fine structure of *Giardia muris,*" in *J. Cell Biol.* 29:317–332. Copyright © The Rockefeller University Press.

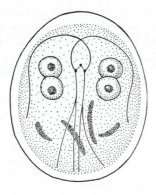

FIGURE 6.9

Cyst of *Giardia lamblia* in the human stool. It is 8 to 12 μm long. The karyosomes of all four cyst nuclei, as well as several intracytoplasmic axonemes, are visible.

Drawing by William Ober.

the primary substrate for respiration, and the parasites store glycogen. But the organisms also multiply and generally produce the same metabolites when glucose is absent or present in low concentrations.[56] The principal end products are ethanol, acetate, and CO_2, both aerobically and anaerobically. In the absence of oxygen, reducing equivalents are transferred to acetaldehyde to produce ethanol. When oxygen is present, the flagellates produce more acetate and less ethanol. All their energy is produced by substrate-level phosphorylation via a flavin, iron-sulfur, protein-mediated fermentative pathway.[36] This pathway is blocked by the flavoantagonists quinacrine and chloroquine.[66]

- **Pathogenesis.** *Giardia* strains differ in their pathogenicity and response to treatment.[63] Many cases of infection show no evidence of disease. Apparently some people are more sensitive to the presence of *G. lamblia* than are others, and considerable evidence suggests that some protective immunity can be acquired. In other individuals, there is a marked increase of mucus production, diarrhea (sometimes incapacitating), dehydration, intestinal pain, flatulence, and weight loss. The stool is fatty but never contains blood. The parasite does not lyse host cells but appears to feed on mucous secretions. A dense coating of flagellates on the intestinal epithelium interferes with the absorption of fats and other nutrients, which probably triggers the onset of disease. The gallbladder may become infected, which can cause jaundice and colic. The disease is not fatal but can be intensely discomforting.

As in the case of trypanosomes, *Giardia lamblia* exhibits antigenic variation, with up to about 180 different antigens being expressed over 6 to 12 generations, depending on the strain.[48, 49] Experimental work has shown that infections are controlled mainly by humoral responses, the major antigens being cysteine-rich surface proteins, which are the same ones that vary antigenically during the course of the infection.[1] Not surprisingly, there also is evidence that these variable surface proteins are related to both infectivity and virulence.[62]

- **Diagnosis and Treatment.** Recognition of trophozoits or cysts in stained fecal smears is adequate for diagnosis. However, an otherwise benign infection with the flagellates may coexist with a peptic ulcer, enteritis, tumor, or strongyloidiasis, any of which could actually be causing the symptoms. In a small percentage of cases cysts are not passed or are passed sporadically. Duodenal aspiration may be necessary for diagnosis by demonstrating trophozoites.

A variety of immunodiagnostic methods, relying on detection of serum antibodies or antigens in feces, are in use although not all of them distinguish between current and past infections.[29] Efforts are being made to develop diagnostic methods based on newer molecular techniques, and such methods may help in cases where cysts are passed in very low numbers. PCR-based techniques can detect a single cyst and also distinguish between species and strains of differing pathogenicity.[38]

Treatment with quinacrine or metronidazole (Flagyl) usually effects complete cure within a few days; both are recommended drugs.[13] All members of a family should be treated simultaneously to avoid reinfection of the others.

- **Epidemiology.** Giardiasis is highly contagious. If one member of a family catches it, others will usually become infected. Transmission depends on the swallowing of mature cysts. Prevention, therefore, depends on a high level of sanitation.

A summary of surveys of 134,966 people throughout the world showed that the prevalence of the infection ranged from 2.4% to 67.5%.[4] In 1984, 26,560 cases of giardiasis were reported in the United States.[12] The Communicable Disease Center in Atlanta, in its 1989–90 summary of waterborne disease outbreaks, indicated that "*Giardia lamblia* was the most frequently identified etiologic agent . . . for the 11th and 12th consecutive years."[13]

Outbreaks continue to flare up in the United States, often without regard for the affluence of the people involved. For instance, an epidemic occurred in Aspen, Colorado, during the 1965–66 ski season, with at least 11% of 1094 skiers infected.[45] Of the permanent population, 5% remained infected after the epidemic.[25] Faunal surveys in watersheds that were known sources of infections to people have shown that several animals, including beavers, dogs, cats, and sheep, serve as reservoirs of infection.[33, 65] Beavers, in particular, are epidemiologically significant in human giardiasis. After hiking for miles in the wild on a hot day, a person is easily tempted to fill a canteen and drink from a crystal-clear beaver pond. Many infections have been acquired in just that way, including some in parasitologists' relatives. In 1980 numerous cases of giardiasis were diagnosed in the resort village of Estes Park, Colorado. Surprisingly, all were in one half of the town, with the other half remaining parasite free. Each half is served with water from a different river. Both rivers have beavers in abundance, but the municipal water filtration system had broken down for one source but not the other. In nearby Rocky Mountain National Park, Monzingo and Hibler[44] found a prevalence of 20% to 60% in beavers from one valley. Muskrats were also infected.

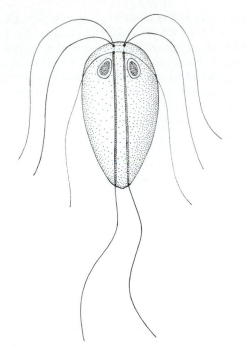

FIGURE 6.10

Diagram of a trophozoite of *Hexamita meleagridis.* It is 6 to 12 μm long.

Drawing by William Ober.

But resorts are certainly not the only places where people can pick up giardiasis. In 1990, an outbreak among Wisconsin insurance company office workers was traced to an employee cafeteria where raw sliced vegetables had been prepared by an infected food handler.[43] Day-care centers also can become foci of transmission.

There have been several reports of a late summer peak in transmission, although the exact reasons for this increased seasonal risk remain somewhat of a mystery.[22]

Hexamita meleagridis. *Hexamita meleagridis* is a parasite of the small intestine of young galliform birds, including turkey, quail, pheasant, partridge, and peafowl. It occurs in the United States, Great Britain, and South America, although it probably is common elsewhere. In the United States *H. meleagridis* causes at least millions of dollars in loss to the turkey industry every year.

Morphologically, *Hexamita* is quite similar to *Giardia,* being elongated, with two nuclei and four pairs of flagella (Fig. 6.10). However, it is smaller, has no sucking disc, possesses karyosomes two-thirds the size of the nuclei, and has no median bodies. As in *Giardia* spp., the kinetosomes are grouped anterior to and between the nuclei, but three pairs of axonemes emerge anteriorly, whereas one pair courses intracytoplasmically. The intracytoplasmic axonemes run posteriorly along granular lines and emerge to become the posterior flagella.

The life cycle is essentially the same as for *Giardia* spp., except that birds rather than mammals are the normal hosts. Hexamitosis is mainly a disease of young animals. Symptomless adults are reservoirs of infection. Mortality in a flock may range from 7% to 80% in very young birds. Survivors are somewhat immune but commonly are stunted in size. They become a ready source of infection for new broods. No

completely satisfactory treatment is available, but prevention in domestic flocks is possible by proper management and sanitation. Separation of chicks from adult birds is mandatory.

Chickens and turkeys are not the only commercially important animals vulnerable to infection. *Hexamita salmonis* is a pathogen of salmon, producing ascites and inflammation of liver and kidneys.[32] *Hexamita* species also have been found in wild frogs and implicated in health problems of cultured oysters.[41, 42]

ORDER TRICHOMONADIDA

Family Trichomonadidae

The many members of this family are rather similar to one another in structure (Figs. 6.11 through 6.14). They are easily recognized because they have an anterior tuft of flagella, a stout median rod (the **axostyle**), and an **undulating membrane** along the recurrent flagellum. They are found in intestinal or reproductive tracts of vertebrates and invertebrates, with one group occurring exclusively in the gut of termites. Unlike other protozoa covered in this chapter, most members of this order do not form cysts. Three species are common in humans, and one is of extreme importance in domestic ruminants. These will serve to illustrate the order.

The three trichomonads of humans, *Trichomonas tenax, T. vaginalis,* and *Pentatrichomonas hominis,* are similar enough morphologically to have been considered conspecific by many taxonomists. More recently, the differences between *P. hominis* and the other two have been recognized. As currently defined, the genus *Trichomonas* contains only three species: *T. tenax, T. vaginalis,* and a species found in birds, *T. gallinae,* which is more like *T. tenax* than is *T. vaginalis.*[26]

Trichomonas tenax. *Trichomonas tenax* (Fig. 6.11) was first discovered by O. F. Müller in 1773, when he examined an aqueous culture of tartar from teeth. *Trichomonas tenax* is now known to have worldwide distribution.

- *Morphology.* Like all species of *Trichomonas, T. tenax* has only a trophic stage. It is an oblong cell 5 to 16 μm long by 2 to 15 μm wide, with size varying according to strain. There are four anterior free flagella, with a fifth flagellum curving back along the margin of an undulating membrane and ending posterior to the middle of the body.[40, 53] The recurrent flagellum is not enclosed by the undulating membrane but is closely associated with it in a shallow groove. A densely staining lamellar structure (**accessory filament**) courses within the undulating membrane along its length. A **costa** arises in the kinetosome complex and runs superficially beneath and generally parallel to the serpentine path of the undulating membrane. The costa distinguishes the Trichomonadidae from other families in its order. It is a rodlike structure with complex cross-striations, which probably serve as a strong, flexible support in the region of the undulating membrane.

A parabasal body (Golgi body) lies near the nucleus, with the parabasal filament running from the kinetosome complex, through or very near the parabasal body, and ending in the posterior portion of the body. A small, "minor" parabasal filament, which is inconspicuous in light microscope preparations, has been shown in other trichomonads, and it probably is present in *T. tenax* as well. The tubelike axostyle extends from the area of the

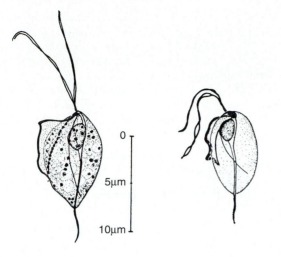

FIGURE 6.11

Typical trophozoites of *Trichomonas tenax*.

From B. M. Honigberg and J. J. Lee, "Structure and division of *Trichomonas tenax* (O. F. Müller)," in *Am. J. Hyg.* 69:177–201. Copyright © 1959.

kinetosomes posteriorly to protrude from the end of the body (covered by cell membrane). The axostylar tube is formed by a sheet of microtubules, and its anterior, middle, and posterior parts are known as **capitulum, trunk,** and **caudal tip,** respectively. Toward the capitulum, the tubular trunk opens out to curve around the nucleus, and the microtubules of the capitulum slightly overlap the curving, collarlike **pelta.** The pelta also comprises a sheet of microtubules and appears to function in supporting the "periflagellar canal," a shallow depression in the anterior end from which all the flagella emerge. A cytostome is not present. *Trichomonas tenax* has concentrations of microbodies traditionally called **paracostal granules** along its costa, and other species of *Trichomonas* have **paraxostylar granules** along the axostyle. These bodies are now called **hydrogenosomes** on the basis of their biochemical characteristics. The metabolic functions of hydrogenosomes are discussed later in the chapter.

- **Biology.** *Trichomonas tenax* can live only in the mouth and, apparently, cannot survive passage through the digestive tract. Transmission, then, is direct, usually through kissing or common use of eating or drinking utensils. Trophozoites divide by binary fission. They are harmless commensals, feeding on microorganisms and cellular debris. They are most abundant between the teeth and gums and in pus pockets, tooth cavities, and crypts of the tonsils, but they also have been found in the lungs and trachea. *Trichomonas tenax* is resistant to changes in temperature and will live for several hours in drinking water. Thus, the "communal dipper" may be a route of infection in some situations. Although good oral hygiene is said to decrease or eliminate the infection, in one survey 15.7% of patients in a clinical practice in New York were positive, and none had oral hygiene rated as poor.[7]

Trichomonas vaginalis.   This species (Figs. 6.12 and 6.13) was first found by Donné in 1836 in purulent vaginal secretions and in secretions from the male urogenital tract. In 1837 he named it *Trichomonas vaginalis,* thereby creating the genus. It is a cosmopolitan species, found in the reproductive tracts of both men and women the world over. Donné thought the organism was covered with hairs, which is what prompted the generic name (Greek *thrix,* hair).

- **Morphology.** *Trichomonas vaginalis* is very similar to *T. tenax* but differs in the following ways: It is somewhat larger, 7 to 32 µm long by 5 to 12 µm wide; its undulating membrane is shorter; and there are more granules along the axostyle and costa. In living and appropriately fixed and stained specimens, the constancy in presence and arrangement of the hydrogenosomes is the best criterion for distinguishing *T. vaginalis* from other *Trichomonas* spp.[27] *Trichomonas vaginalis* frequently produces pseudopodia.

- **Biology.** *Trichomonas vaginalis* lives in the vagina and urethra of women and in the prostate, seminal vesicles, and urethra of men. It is transmitted primarily by sexual intercourse,[30] although it has been found in newborn infants. Its presence occasionally in very young children, including virginal females, suggests that the infection can be contracted from soiled washcloths, towels, and clothing. Viable cultures of the organism have been obtained from damp cloth as long as 24 hours after inoculation. The acidity of the normal vagina (pH 4.0 to 4.5) ordinarily discourages infection, but once established, the organism itself causes a shift toward alkalinity (pH 5 to 6), which further encourages its growth.

- **Metabolism.** Like *Giardia,* trichomonads are aerotolerant anaerobes, degrading carbohydrates incompletely to short-chain organic acids (principally acetate and lactate) and carbon dioxide, regardless of whether oxygen is present.[37] Unlike *Giardia,* however, trichomonads produce molecular hydrogen in the absence of oxygen. These reactions take place in the hydrogenosomes, hence the name of the organelle. Hydrogenosomes are analogous to mitochondria (which are absent in trichomonads) in other eukaryotes; but their distinctness is shown by their morphology, the absence of DNA, and the absence of cardiolipin, which is present in the membranes of mitochondria.[50, 61] Hydrogenosomes are surrounded by two, closely apposed 6 nm membranes.[5] Similar organelles have now been reported in certain rumen ciliates.[47]

  Pyruvate is produced in the cytoplasm by glycolysis. Part of the pyruvate is reduced to lactate by lactic dehydrogenase and excreted. Part of the pyruvate enters the hydrogenosomes where it is oxidatively decarboxylated, and the electrons are accepted by **ferredoxin.**[39] Under anaerobic conditions, the electrons are then transferred to protons by a hydrogenase to form molecular hydrogen. When oxygen is present, it apparently accepts the electrons and, along with H^+, forms water. The oxidation of pyruvate to acetate is coupled to substrate-level generation of ATP; therefore, the hydrogenosomes participate in energy production in the cell. The drug metronidazole is reduced by ferredoxin to form toxic products, thus explaining the effectiveness of this drug in chemotherapy for trichomoniasis. Both metronidazole-sensitive and resistant strains occur, however, and the resistant strains show higher glucose uptake rates, lower hydrogenase activity, and lower H_2 formation than do sensitive strains.[20]

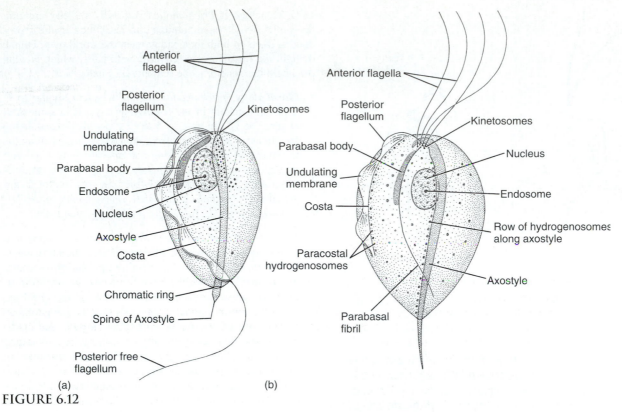

FIGURE 6.12

Morphology of trichomonads: (*a*) *Tritrichomonas foetus;* (*b*) *Trichomonas vaginalis.* The hydrogenosomes are not always in a definite row.

Drawn by William Ober from D. H. Wenrich and M. A. Emmerson, "Studies on the morphology of *Tritrichomonas foetus* (Riedmuller) from american cows," in *J. Morphol.* 55:195, 1933.

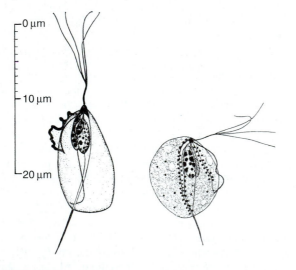

FIGURE 6.13

Typical trophozoites of *Trichomonas vaginalis.*

From B. M. Honigberg and V. M. King, "Structure of *Trichomonas vaginalis* Donné," in *J. Parasitol.* 50:345–364. Copyright ©1964. Reprinted by permission of the publisher.

Recent studies also have shown that trichomonads lack some enzymes necessary to synthesize complex phospholipids and thus must obtain some membrane components from their environment.[2] Such observations suggest additional potential metabolic targets for drug action.

• *Pathogenesis.* Most strains are of such low pathogenicity that the infected person is virtually asymptomatic. However, other strains cause an intense inflammation, with itching and a copious white discharge **(leukorrhea)** that is swarming with trichomonads. They feed on bacteria, leukocytes, and cell exudates and are themselves ingested by monocytes. Like all mastigophorans, *T. vaginalis* divides by longitudinal fission, and, like other trichomonads, it does not form cysts.

A few days after infection there is a degeneration of the vaginal epithelium followed by leukocytic infiltration. The vaginal secretions become abundant and white or greenish, and the tissues become intensely inflamed. An acute infection will usually become chronic, with a lessening of symptoms, but will occasionally flare up again. It should be noted, however, that leukorrhea is not symptomatic of trichomoniasis; indeed, at least half of patients even with severe leukorrhea are negative for *T. vaginalis.*[23] In men the infection is usually asymptomatic, although there may be an irritating urethritis or prostatitis.

Diagnosis depends on recognizing the trichomonad in a secretion or from an in vitro culture made from a vaginal irrigation. Cultivation is recommended to detect low numbers of organisms.[23] Culture of parasitic protozoa is often time-consuming and laborious (see Diamond's epigraph), but plastic envelope methods have been developed for *T. vaginalis,* using dry ingredients that have a long shelf life and are reconstituted with water immediately before use.[3] Dot-blot DNA hybridization assays have also been developed for *T. vaginalis,* and in clinical trials these arrangement is referred to as "four-plus-one," since the fifth flagellum originates and beats independently of the others.[26] A recurrent (sixth) flagellum is aligned alongside the undulating membrane, as in *T. tenax* and *T. vaginalis*, but in contrast to these two species, the recurrent flagellum in *P. hominis* continues as a long, free flagellum past the posterior end of the body. Axostyle, pelta, parabasal body, "major" and "minor" parabasal filaments, costa, and paracostal hydrogenosomes are present. Paraxostylar hydrogenosomes are absent.

- **Biology.** *Pentatrichomonas hominis* lives in the large intestine and cecum, where it divides by binary fission, often building up incredible numbers. It feeds on bacteria and debris, probably taking them in with active pseudopodia. The organism often is present in routine examinations of diarrheic stools, but there is no indication that it contributes to this or other disease conditions. In formed stools the flagellates are rounded and dormant but not encysted. They are difficult to identify at this stage because they do not move, and the structures normally characteristic for the species cannot be distinguished.

  The organism apparently can survive acidic conditions of the stomach, and transmission occurs by contamination. Filth flies can serve as mechanical vectors. High prevalence is correlated with unsanitary conditions. Diagnosis depends on identification of the animal in fecal preparations, and prevention depends on personal and community sanitation. The organism cannot establish in the mouth or urogenital tract.

Tritrichomonas foetus. *Tritrichomonas foetus* (Fig. 6.12) is responsible for a serious genital infection in cattle, zebu, and possibly other large mammals. Probably the third leading cause of abortion in cattle (after brucellosis and leptospirosis), *T. foetus* is especially common in Europe and the United States. The USDA estimated losses from *T. foetus* in the United States between 1951 and 1960 at $8.04 million.

- **Morphology.** The cell is spindle to pear shaped, 10 to 25 μm long by 3 to 15 μm wide. There are three anterior flagella, and the fourth, the recurrent flagellum, extends free from the posterior end of the body about the length of the anterior flagella. The mastigont system is generally similar in organization to the trichomonads described previously, but it is even more complex than the others and will not be described here.[28] The costa is prominent and, although similar in position and function to those of the other trichomonads, differs in ultrastructural detail, resembling a parabasal filament in this respect. The structure of the undulating membrane is curious, consisting of two parts. The proximal part is a foldlike differentiation of the dorsal body surface, and the distal part, which contains the axoneme of the recurrent flagellum, courses along the rim of the proximal part with no obvious physical connection to it. The thick axostyle protrudes from the posterior end of the body. Numerous paraxostylar hydrogenosomes are present in the posterior part of the organism, just anterior to the point of the axostyle, and these are apparent in the light microscope preparations as the "chromatic ring."

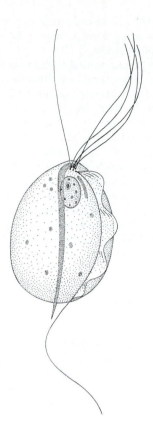

FIGURE 6.14

Pentatrichomonas hominis trophozoite. It ranges from 8 to 20 μm long.

Drawing by William Ober.

assays were more effective than microscopic examination. However, cross-reactions were observed with *Pentatrichomonas hominis*.[54]

Oral drugs, such as metronidazole, usually cure infection in about five days. Some apparently recalcitrant cases may be caused by reinfection by the sexual partner. Suppositories and douches are useful in promoting an acid pH of the vagina. Sexual partners should be treated simultaneously to avoid reinfection. *Trichomonas vaginalis* has been shown to survive cryopreservation of human semen, suggesting that infections could be contracted through artificial insemination.[57]

Pentatrichomonas hominis. The third trichomonad of humans (Fig. 6.14) is a harmless commensal of the intestinal tract. It was first found by Davaine, who named it *Cercomonas hominis* in 1860. Traditionally, it has been called *Trichomonas hominis,* but because most specimens actually bear five anterior flagella, the organism has been assigned to the genus *Pentatrichomonas*. Next to *Giardia intestinalis* and *Chilomastix mesnili,* this is the most common intestinal flagellate of humans. It is also known in other primates and in various domestic animals. The prevalence among 13,517 persons examined in the United States was 0.6%.[4]

- **Morphology.** This species is superficially similar to *T. tenax* and *T. vaginalis* but differs from them in several respects. Its size is 8 μm to 20 μm by 3 μm to 14 μm. Five anterior flagella are present in most specimens, although individuals with fewer flagella are sometimes found. The

- **Biology.** These trichomonads live in the preputial cavity of the bull, although the testes, epididymis, and seminal vesicles also may be infected. In the cow the flagellates first infect the vagina, causing a vaginitis, and then move into the uterus. After establishing in the uterus, they may disappear from the vagina or remain there as a low-grade infection. Bovine genital trichomoniasis is a venereal disease transmitted by coitus, although transmission by artificial insemination is possible. Trichomonads multiply by longitudinal fission and form no cyst.

- **Pathogenesis.** The most characteristic sign of bovine trichomoniasis is early abortion, which usually happens 1 to 16 weeks after insemination. Because of the small size of the fetus, an owner may not notice that the cow has aborted and therefore may believe she did not conceive. The cow may recover spontaneously if all fetal membranes are passed after abortion; however, if they remain, she usually develops chronic endometritis, which may cause permanent sterility. The parasites release extracellular proteases that have the capacity to digest proteins, including immunoglobulins, that might otherwise function in host defense.[60] Normal gestation and delivery occasionally occur with an infected animal.

  Pathogenesis is not observable in bulls, but an infected bull is worthless as a breeding animal; unless treated, it usually remains infected permanently. Treatment is expensive, difficult, and not always effective. Because of the immense prices paid for top-quality bulls, the loss of a single animal may bankrupt the breeder.

- **Epizootiology.** It has not been determined whether *T. foetus* naturally infects animals other than cattle and closely related species. Experimental infections have been established in rabbits, guinea pigs, hamsters, dogs, goats, sheep, and pigs. Trichomonads similar to *T. foetus* occur as natural infections in pigs and horses. Whether these can be transmitted to cattle by contamination is not known.

  Trichomonads can survive freezing in semen ampules, although some media are more detrimental than are others. This precludes use of semen from infected bulls for artificial insemination.

- **Diagnosis, Treatment, and Control.** Direct identification of protozoa from smears or culture remains the only sure means of diagnosis, although a mucus agglutination test is available. In light infections a direct smear of mucus or exudate is sufficient. Smears can be obtained from amniotic or allantoic fluid, vaginal or uterine exudates, placenta, fetal tissues or fluids, or preputial washings from bulls. Flagellates fluctuate in numbers in bulls; in cows they are most numerous in the vagina two or three weeks after infection.

  No satisfactory treatment is known for cows, but the infection is usually self-limiting in them, with subsequent, partial immunity. Bulls can be treated if the condition has not spread to the inner genital tubes and testes. Treatment is usually attempted only on exceptionally valuable animals, since it is a tedious, expensive task. Preputial infection is treated by massaging antitrichomonal salves or ointments into the penis, after it has been let down by nerve block or by injection of a tranquilizer into the penis

retractor muscles. Repeated treatment is usually necessary. Systemic drugs show promise of becoming the standard method of treatment.

Control of bovine genital trichomoniasis depends on proper herd management. Cows that have been infected should be bred only by artificial insemination to avoid infecting new bulls. Bulls should be examined before purchase, with a wary eye for infection in the resident herd. Unless they are extremely valuable, infected bulls should be killed. Like any venereal disease, trichomoniasis can be controlled and eventually eliminated with proper treatment and reporting, but the disease is likely to remain a problem for some time. Vaccines have been developed, some of which employ parasite surface proteins involved in attachment of the flagellates to vaginal epithelial cells.[6] Field trials of a polyvalent vaccine showed that 62.5% of the vaccinated heifers bred to infected bulls produced calves as compared to 31.5% of the controls.[34] These vaccines are most effective when used in conjunction with other control measures, including replacement of older bulls with younger ones.

Family Monocercomonadidae

The Monocercomonadidae show affinities with the Sarcodina because pseudopodia are well-developed, an undulating membrane is absent, and flagella tend to be reduced. Most species are parasites of insects, but three genera infect domestic animals. One of these is economically important and has evolved a unique mode of transmission: in the egg of a nematode.

Histomonas meleagridis. *Histomonas meleagridis,* (Fig. 6.15), a cosmopolitan parasite of gallinaceous fowl, including chickens, turkeys, peafowl, and pheasant, causes a severe disease known variously as **blackhead, infectious enterohepatitis,** and **histomoniasis.** The disease is more virulent in some species of host than others; chickens show disease less often than do turkeys, for example. The USDA estimated that the economic loss in the United States as the result of histomoniasis in chickens and turkeys between 1951 and 1960 amounted to $9.3 million.

The taxonomic history of *H. meleagridis* has been very confused because of its polymorphism in different situations. At various times it has been confused with amebas, coccidia, fungi, and *Trichomonas* spp. Even the disease that it causes has been attributed to different organisms, from amebas to viruses. Today much is known about the organism, and its biology and pathogenesis are less mysterious than they once were.

- **Morphology.** *Histomonas meleagridis* is pleomorphic; its stages change size and shape in response to environmental factors. There is no cyst in the life cycle, only various trophic stages. When they are found in the lumen of the cecum (which is rare) or in culture, the stages are ameboid, 5 to 30 μm in diameter, and almost always with only one flagellum. However, there are usually four kinetosomes, the basic number for trichomonads, although this condition has been attributed to duplication of the kinetic apparatus in preparation for mitosis.[55] The nucleus is vesicular and often has a distinct endosome. One can usually discern a clear ectoplasm and a granular endoplasm.

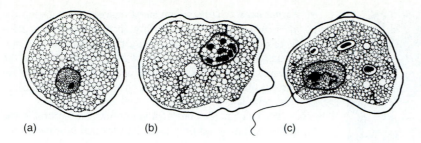

FIGURE 6.15

Examples of *Histomonas meleagridis*. (*a*) Tissue type of *H. meleagridis* in fresh preparation from liver lesion; viewed with phase contrast. (*b*) *H. meleagridis* in transitional stage in lumen of the cecum. Pseudopodia have been formed, and the distribution of chromatin suggests that binary fission is approaching. However, the flagellum has not yet appeared. (*c*) An organism in same cecal preparation as (*b*) but this one completely adapted as a lumen dweller.

From E. E. Lund, "*Histomonas*," in *Advances in Veterinary Science and Comparative Medicine,* edited by C. A. Brandly and C. E. Cornelius. Copyright © 1963 Academic Press, Inc., New York, NY. Reprinted with permission of the publisher.

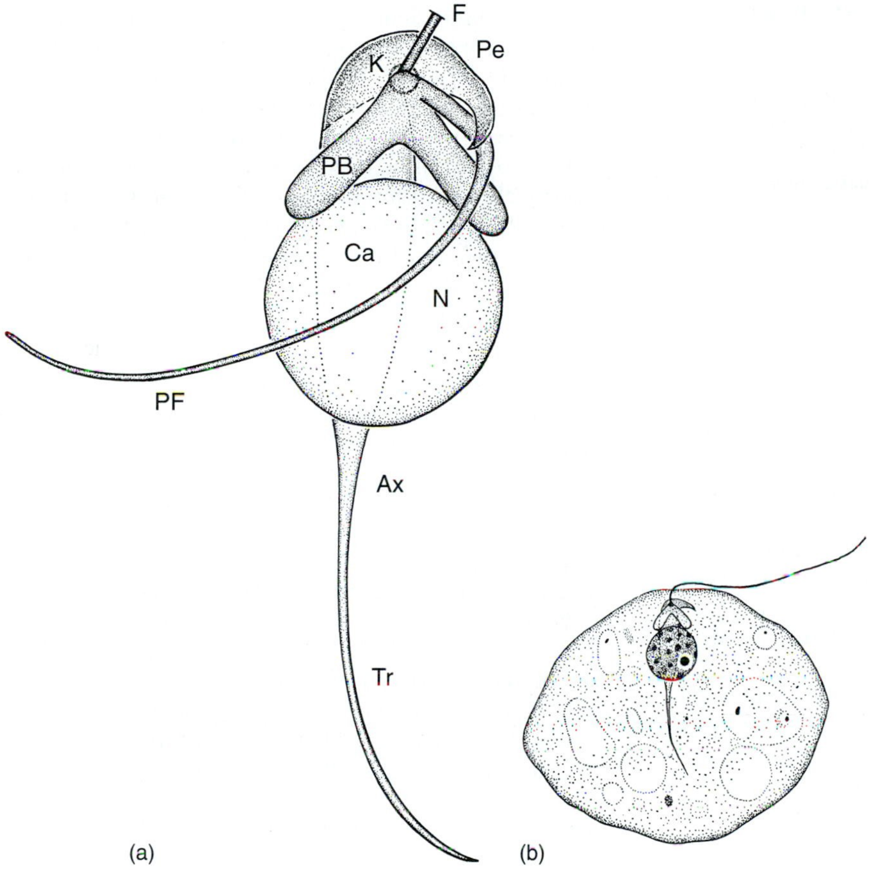

FIGURE 6.16

Histomonas meleagridis. (*a*) Composite, schematic diagram of the mastigont system and nucleus as seen from a dorsal and somewhat right view. (*b*) Composite diagram of an organism, with the mastigont system seen in the same view as in (*a*). The flagellum arises from the kinetosomal complex just anterior to the V-shaped parabasal body. The cytoplasm appears highly vacuolated and contains ingested bacteria and rice starch. **Ax,** axostyle; **Ca,** capitulum; **F,** flagellum; **K,** kinetosomal complex; **N,** nucleus; **Pe,** pelta; **PB,** parabasal body; **PF,** parabasal fibril; **Tr,** trunk of axostyle. (× 4270.)

From B. M. Honigberg and C. J. Bennett, "Light microscope observations on structure and division of *Histomonas meleagridis*," *J. Protozool.* 18:687–697. Copyright © 1971. The Society of Protozoologists. Reprinted with permission of the publisher.

Food vacuoles may contain host blood cells, bacteria, or starch granules. Electron microscope studies have revealed a pelta, a V-shaped parabasal body, a parabasal filament, and a structure resembling an axostyle (Fig. 6.16). These cannot be seen with light microscopy, but their presence supports placement of *Histomonas* spp. in the order Trichomonadida. No mitochondria have been observed. The forms within the tissues have no flagella, although kinetosomes are present near the nucleus.

- ***Biology and Epidemiology.*** Like other flagellates, *H. meleagridis* divides by binary fission. No cysts or sexual stages occur in the life cycle. Trophozoites are fragile

and cannot long survive in the external environment or the host's stomach acids. Certain factors can, and sometimes do, conspire to allow infection by trophozoites. If trophozoites are eaten with certain foods that raise the stomach pH, they may survive to initiate a new infection. This can be the means of an epizootic in a dense flock of birds.

The most important, and by far the most interesting, mode of transmission is within the egg of the cecal nematode, *Heterakis gallinarum*. Since the protozoan undergoes development and multiplication in the nematode, the worm can be considered a true intermediate host.[35] After being ingested by the worm, the flagellates enter

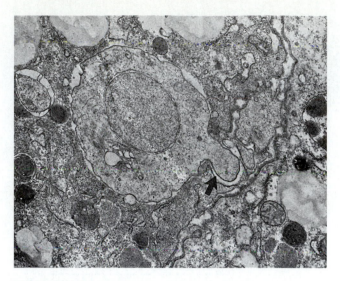

FIGURE 6.17

Electron micrograph of section through the growth zone of the ovary of *Heterakis gallinarum* to show *Histomonas meleagridis* in the process of entering an oocyte (*arrow*). (× 13,800.)

From D. L. Lee, in A. M. Fallis, editor, *Ecology and Physiology of Parasites*, 1971 University of Toronto Press, Toronto.

the nematode's intestinal cells, multiply, and then break out into the pseudocoel and invade the germinative area of the nematode's ovary. There they feed and multiply extracellularly, move down the ovary with the developing oogonia, and then penetrate the oocytes (Fig. 6.17). Feeding and multiplication continue in the oocytes and newly formed eggs. Passing out of the mother worm and out of the bird with its feces, the parasite divides rapidly, invading the tissues of the juvenile nematode, especially those of the digestive and reproductive systems. Interestingly, *H. meleagridis* also parasitizes the reproductive system of the male nematodes.[35] Presumably, it could be transmitted to the female during copulation, thus constituting a venereal infection of nematodes!

Infected nematode eggs can survive for at least two years in the soil. If the worm eggs are eaten by an appropriate bird, they hatch in the intestine, and the juvenile *Heterakis* passes down into the cecum, where *Histomonas* is free to leave its temporary host to begin residence in a more permanent one.

Earthworms are important paratenic hosts of both the *Heterakis* and its contained *Histomonas*. When eaten by an earthworm, the nematode eggs will hatch, releasing second-stage juveniles that become dormant in the earthworm's tissues. When the earthworm is eaten by a gallinaceous fowl, the *Heterakis* juveniles are released, and the bird becomes infected by two kinds of parasites at once. Earthworms can serve to maintain the parasites in the soil for long periods of time. Chickens are the most important reservoirs of infection because they are less often affected by *Histomonas* than are turkeys. Because *Heterakis* eggs and infected earthworms can survive for such long periods in the soil, it is almost impossible to raise uninfected turkeys in the same yards in which chickens have lived.

- **Pathogenesis.** Turkeys are most susceptible between the ages of 3 and 12 weeks, although they can become infected as adults. In very young poults, losses may approach 100% of the flock. Chickens are less prone to the disease, but outbreaks among young birds have been reported. Quails and partridges show varying degrees of susceptibility.

The principal lesions of histomoniasis are found in the cecum and liver. At first, pinpoint ulcers are formed in the cecum. These may enlarge until nearly the entire mucosa is involved. The ceca often become filled with cheesy, foul-smelling plugs that adhere to the cecal walls. Complete perforation of the cecum, with peritonitis and adhesions, can occur. The ceca are usually enlarged and inflamed. Liver lesions are rounded, with whitish or greenish areas of necrosis. Their size varies, and they penetrate deep into the parenchyma.

Infected birds show signs of droopiness, ruffled feathers, and hanging wings and tail. Yellowish diarrhea usually occurs. The skin of the head turns black in some cases, giving the disease the name blackhead; however, other diseases can cause this symptom.

Histomonas meleagridis by itself is incapable of causing blackhead but does so only in the presence of intestinal bacteria of several species, particularly *Escherichia coli* and *Clostridium perfringens*. Birds that survive are immune for life. A related histomonad, *Parahistomonas wenrichi,* also is transmitted by *Heterakis* but is not pathogenic.

- **Diagnosis, Treatment, and Control.** Cecal and liver lesions are diagnostic. Scrapings of these organs will reveal histomonads, thereby distinguishing the disease from coccidiosis. Several types of drugs are used in prevention and treatment, including nitrofurans, nitroimidazoles, and phenylarsonic acid derivatives. These successfully inhibit, suppress, or cure the disease, but some have undesirable side effects, such as delaying sexual maturity of the bird. Treatment of birds with nematocides, such as mebendazole, cambendazole, and levamisole, to eliminate *Heterakis* is effective in preventing future outbreaks, since *H. meleagridis* cannot survive in the soil by itself.

Control depends on effective management techniques, such as rearing young birds on hardware cloth above the ground, keeping young birds on dry ground, and controlling *Heterakis*. Pasture rotation of *Heterakis*-free flocks is also successful.

Dientamoeba fragilis. *Dientamoeba fragilis* (Fig. 6.18) has traditionally been considered a member of the ameba family Endamoebidae, but its differences from other members of this family have long been recognized. For example, a large proportion of individuals have two nuclei, the nuclear structure is rather unlike other Endamoebidae, an extranuclear spindle is present during division, and cysts are not formed. More than 55 years ago Dobell believed that *D. fragilis* was closely related to the ameboflagellate *Histomonas*.[18] On the basis of ultrastructural and immunological evidence, Camp and coworkers placed the genus *Dientamoeba* in a subfamily of the Monocercomonadidae in the flagellate order Trichomonadida.[10] This change seems to

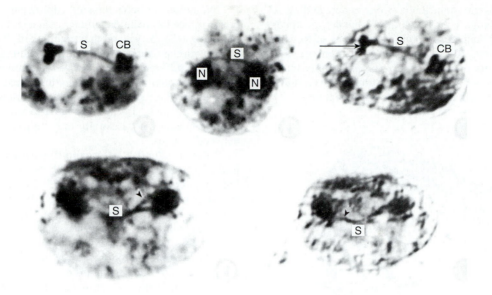

FIGURE 6.18

Dientamoeba fragilis: photomicrographs of binucleate organisms. Four chromatin bodies (**CB**) can be resolved within the telophase nucleus of the organism, shown in the first and third figures. The extranuclear spindle (**S**) extends between the nuclei (**N**) in all figures. Note the branching of the spindle (*arrowheads*) near the nucleus in the fourth and fifth figures. (Bouin's fixative: first, second, and fourth figures—bright field [× 4950]; third and fifth figures—Nomarski differential interference [× 3650].)

From R. R. Camp et al., "Study of *Dientamoeba fragilis* Jepps & Dobell. I. Electronmicroscopic observations of the binucleate stages. II. Taxonomic position and revision of the genus," in *J. Protozool.* 21:69–82. Copyright © 1974 The Society of Protozoologists.

reflect the phylogenetic relationship of the organism rather than the fact that it moves by pseudopodia instead of flagella, and we will use it in this text. *Dientamoeba fragilis,* infecting about 4% of the human population, is the only species known in the genus.

- *Morphology.* Only trophozoites are known in this species; cysts are not formed. The trophozoites (Fig. 6.18) are very delicate and disintegrate rapidly in feces or water. They are 6 to 12 µm in diameter, and the ectoplasm is somewhat differentiated from the endoplasm. A single, broad pseudopodium usually is present. The food vacuoles contain bacteria, yeasts, starch granules, and cellular debris. About 60% of the amebas contain two nuclei, which are connected to each other by a filament, observable by light microscopy; the rest have only one nucleus. By electron microscopy one can discern that the filament connecting the nuclei is a division spindle composed of microtubules; the binucleate individuals are, in reality, in an arrested telophase. The endosome is eccentric, sometimes fragmented or peripheral in the nucleus, and concentrations of chromatin are usually apparent. A filament and Golgi apparatus are present, which are reminiscent of the parabasal fibers and parabasal bodies found in *Histomonas* and trichomonads. There are no kinetosomes or centrioles.

- *Biology. Dientamoeba fragilis* lives in the large intestine, especially in the cecal area. It feeds mainly on debris and traditionally has been considered a harmless commensal. However, a study of 43,029 people in Ontario showed a high percentage of intestinal problems in those infected with *D. fragilis.*[67] Symptoms included diarrhea, abdominal pain, anal pruritus, abnormal stools, and other indications of abdominal distress. It is possible that *D. fragilis* is responsible for many such cases of unknown etiology, especially in small children. However, because *Dientamoeba fragilis* infections often co-occur with other species, it is not always easy to determine which, or which combination, of the parasites is responsible for the most damage. In one study of 414 excised appendices, for example, Cerva et al.[11] found pinworms, ascaris eggs, *Endolimax nana, Entamoeba coli,* and *Giardia* cysts, in addition to *D. fragilis.*

The mode of transmission is unknown, since the parasite does not form cysts, and it cannot survive the upper digestive tract. The organism survives transmission in the eggs of a parasitic nematode, as does its relative, *Histomonas meleagridis.* Small, ameboid organisms resembling *D. fragilis* have been found in the eggs of the common human pinworm, *Enterobius vermicularis,* and there is overwhelming epidemiological evidence that the nematode may be the vector of the protistan.[9,67]

SUBPHYLUM OPALINATA

Order Opalinida

- **Family Opalinidae**

There are about 150 species of opalinids, most of which live in the intestines of amphibians. They are of no economic or medical importance but are of zoological interest because of their peculiar morphology and the fact that their reproductive cycles apparently are controlled by host hormones.[21] Also study of opalinids may contribute to our understanding of amphibian zoogeography and evolution.[19] Finally, opalinids are commonly encountered in routine dissections of frogs in teaching laboratories; the large size, great numbers, and graceful movements of these protozoans inevitably make them exciting finds for students who never gave much thought to the animals that might live in a frog rectum.

Numerous oblique rows of cilia occur over the entire body surface of opalinids, giving them a strong resemblance to ciliates (Fig. 6.19), and they traditionally have been classified with the Ciliophora. However, opalinids have several important differences from ciliates and are now placed as a separate subphylum in the phylum Sarcomastigophora. In contrast to ciliates, for example, opalinids possess only one type of nucleus, they reproduce sexually by anisogamous syngamy, and asexually they undergo binary fission between kineties.

Ultrastructurally, all opalinid genera exhibit cortical folds, corticular ribbons of microtubules, mitochondria with long tubular cristae, and pinocytotic vesicles budding from the bases of the cortical folds (Fig. 6.20). The genera differ, however, in other ultrastructural features such as the presence or absence of fibrous tracts alongside the kinetosomes.[51]

Adult opalinids reproduce asexually by binary fission in the rectum of frogs and toads during the summer, fall, and winter. In the spring, which is their host's breeding season, they accelerate divisions and produce small, precystic forms, which then form cysts and pass out with the feces of the host. When the cysts are eaten by tadpoles, male and female gametes excyst and fuse to form the zygote, which resumes asexual reproduction. The exact chemical identity of the compound(s) that stimulates encystment is not known, but present evidence indicates that it is one or more breakdown products of steroid hormones excreted in the frog's urine. This is an interesting example of a physiological adaptation to ensure the production of infective stages at the time and place of new host availability. The effectiveness of the adaptation is attested to by the prevalence of opalinids in frogs and toads.

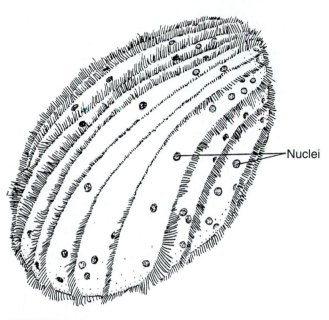

FIGURE 6.19

Opalina sp. from the rectum of a frog. Note the numerous nuclei.

Drawing by Ian Grant.

FIGURE 6.20

The cortex of an opalinid, *Protoopalina australis,* as reconstructed from electron micrographs. **A,** kinetosomal arms; **C,** interkinetosomal connectives; **F,** apical fibers; **H,** transitional helix; **R,** cortical ribbons of microtubules; **S,** kinetosomal shelves; **TD,** transitional disc.

From D. J. Patterson and Ben L. J. Delvinquier, "The fine structure of the cortex of the protist *Protoopalina australis* (Slopalinida, Opalindae) from *Litoria nasuta* and *Litoria inermis* (Amphibia: Anura: Hylidae) in Queensland, Australia," in *J. Protozool.* 37:449–455. Copyright © 1990. The Society of Protozoologists. Reprinted with permission of the publisher.

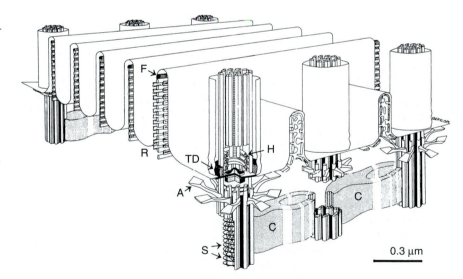

In addition to members of the genus *Opalina,* species of amphibians may also be infected with opalinid species of the genera *Protoopalina, Cepedea,* and *Zelleriella,* although *infected* is a rather strange word for a group of nonpathogenic symbionts so closely, commonly, and inextricably tied to their hosts. A curious symbiosis is found in *Zelleriella opisthocarya,* a parasite of toads, and *Entamoeba* sp., in which more than 200 cysts of the ameba may occur in one opalinid.[59]

References

1. Adam, R. D. 1991. The biology of *Giardia* spp. *Microbiol. Rev.* 55:706–32.

2. Beach, D. H., G. G. Holz Jr., B. N. Singh, and D. G. Lindmark. 1992. Phospholipid metabolism of cultured *Trichomonas vaginalis* and *Tritrichomonas foetus. Molecular and Biochem. Parasitol.* 44:97–108.

3. Beal, C., R. Goldsmith, M. Kotby, M. Sherif, A. El-Tagi, A. Farid, S. Zakarie, and J. Eapen. 1992. The plastic envelope method, a simplified technique for culture diagnosis of trichomoniasis. *J. Clinical Microbiol.* 30:2265–68.

4. Belding, D. L. 1965. *Textbook of clinical parasitology,* 3d ed. New York: Appleton-Century-Crofts, Inc.

5. Benchimol, M., and W. De Souza. 1983. Fine structure and cytochemistry of the hydrogenosome of *Tritrichomonas foetus. J. Protozool.* 30:422–25.

6. Bondurant, R. H., R. R. Corbeil, and L. B. Corbeil. 1993. Immunization of virgin cows with surface antigen TF1.17 of *Trichomonas foetus. Infection and Immunity* 61:1385–94.

7. Brooks, B., and F. L. Schuster. 1984. Oral protozoa: Survey, isolation, and ultrastructure of *Trichomonas tenax* from clinical practice. *Trans. Am. Microsc. Soc.* 103:376–82.

8. Brugerolle, G. 1973. Etude ultrastructural du trophozoite et du kyste chez le genre *Chilomastix* Alexeieff, 1910 (Zoomastigophorea, Retortamonadida Grassé). *J. Protozool.* 20:574–85.

9. Burrows, R. B., and M. A. Swerdlow. 1956. *Enterobius vermicularis* as a probable vector of *Dientamoeba fragilis. Am. J. Trop. Med. Hyg.* 5:258–65.

10. Camp, R. R., C. F. T. Mattern, and B. M. Honigberg. 1974. Study of *Dientamoeba fragilis* Jepps and Dobell. I. Electronmicroscopic observations of the binucleate stages. II. Taxonomic position and revision of the genus. *J. Protozool.* 21:69–82.

11. Cerva, L., M. Schrottenbaum, and V. Kliment. 1991. Intestinal parasites: A study of human appendices. *Folia Parasitologica* 38:5–9.

12. Centers for Disease Control. 1986. *Morbidity and Mortality Weekly Report. Annual Summary 1984.* 33(54):1–135.

13. Centers for Disease Control. 1991. *CDC surveillance summaries, December, 1991. Morbidity and Mortality Weekly Report.* 40(No. SS-3):1–21.

14. Chandler, A. C., and C. P. Read. 1961. *Introduction to parasitology,* 10th ed. New York: John Wiley & Sons, Inc.

15. Cheissin, E. M. 1964. Ultrastructure of *Lamblia duodenalis.* I. Body surface, sucking disc, and median bodies. *J. Protozool.* 11:91–98.

16. Chunge, R. N., N. Nagelkerke, P. N. Karumba, N. Kaleli, M. Wamwea, N. Mutiso, E. O. Andala, J. Gachoya, R. Kiarie, and S. N. Kinoti. 1991. Longitudinal study of young children in Kenya: Intestinal parasitic infection with special reference to *Giardia lamblia,* its prevalence, incidence and duration, and its association with diarrhoea and with other parasites. *Acta Tropica* 50:39–50.

17. Delvinquier, B. L., and W. J. Freeland. 1989. Protozoan parasites of the cane toad, *Bufo marinus,* in Australia. *Aust. J. Zool.* 36:301–16.

18. Dobell, C. 1940. Researches on the intestinal protozoa of monkeys and man. X. The life history of *Dientamoeba fragilis*—observations, experiments and speculations. *Parasitology* 32:417–59.

19. Earl, P. R. 1979. Notes on the taxonomy of the opalinids (Protozoa), including remarks on continental drift. *Trans. Am. Microsc. Soc.* 98:549–57.

20. Ellis, J. E., D. Cole, and D. Lloyd. 1992. Influence of oxygen on the fermentative metabolism of metronidazole-sensitive and resistant strains of *Trichomonas vaginalis. Molecular and Biochem. Parasitol.* 56:79–88.

21. El Mofty, M. M., and I. A. Sadek, 1973. The mechanism of action of adrenaline in the induction of sexual reproduction (encystation) in *Opalina sudafricana* parasitic in *Bufo regularis. Int. J. Parasitol.* 3:425–31.

22. Flanagan, P. A. 1992. *Giardia* diagnosis, clinical course and epidemiology: A review. *Epidemiology and Infection* 109:1–22.

23. Fouts, A. C., and S. J. Kraus. 1980. *Trichomonas vaginalis:* Reevaluation of its clinical presentation and laboratory diagnosis. *J. Infect. Dis.* 141:137–43.

24. Friend, D. S. 1966. The fine structure of *Giardia muris. J. Cell Biol.* 29:317–32.

25. Gleason, N. N., M. S. Horwitz, L. H. Newton, and G. T. Moore. 1970. A stool survey for enteric organisms in Aspen, Colorado. *Am. J. Trop. Med. Hyg.* 19:480–84.

26. Honigberg, B. M. 1963. Evolutionary and systematic relationships in the flagellate order Trichomonadida Kirby. *J. Protozool.* 10:20–63.

27. Honigberg, B. M., and V. M. King. 1964. Structure of *Trichomonas vaginalis* Donné. *J. Parasitol.* 50:345–64.

28. Honigberg, B. M., C. F. T. Mattern, and W. A. Daniel. 1971. Fine structure of the mastigont system in *Tritrichomonas foetus* (Riedmüller). *J. Protozool.* 18:183–98.

29. Isaac-Renton, J. L. 1991. Immunological methods of diagnosis in giardiasis: An overview. *Ann. Clin. Lab. Sci.* 21:116–22.

30. Jírovic, O. 1965. Neuere Forschungen über *Trichomonas vaginalis* und vaginale Trichomonosis. *Angew. Parasitol.* 6:202–10.

31. Kabnick, K. S., and D. A. Peattie. 1991. *Giardia:* A missing link between prokaryotes and eukaryotes. *Am. Sci.* 79:34–43.

32. Kent, M. L., J. Ellis, J. W. Fournie, S. C. Dawe, J. W. Bagshaw, and D. J. Whitaker. 1992. Systemic hexamitid (Protozoa: Diplomonadida) infection in seawater pen-reared chinook salmon *Oncorhynchus tshawytscha. Dis. Aquatic Organisms* 14:81–89.

33. Kirkpatrick, C. E., and J. P. Farrell. 1984. Feline giardiasis: Observations on natural and induced infections. *Am. J. Vet. Res.* 45:2182–88.

34. Kvasnicka, W. G., D. Hanks, J. C. Huang, M. R. Hall, D. Sandblom, H. J. Chu, L. Chavez, and W. M. Acree. 1992. Clinical evaluation of the efficacy of inoculating cattle with a vaccine containing *Trichomonas foetus. Am. J. Vet. Res.* 53:2023–27.

35. Lee, D. L. 1971. Helminths as vectors of micro-organisms. In Fallis, A. M., ed. *Ecology and physiology of parasites.* Toronto: University of Toronto Press, 104–22.

36. Lindmark, D. G. 1980. Energy metabolism of the anaerobic protozoon *Giardia lamblia. Molecular and Biochem. Parasitol.* 1:1–12.

37. Mack, S. R., and M. Müller. 1980. End products of carbohydrate metabolism in *Trichomonas vaginalis. Comp. Biochem. Physiol.* 67B:213–16.

38. Mahbubani, M. H., A. K. Bej, M. H. Perlin, F. W. Schaefer III, W. Jakuowski, and R. M. Atlas. 1992. Differentiation of *Giardia duodenalis* from other *Giardia* spp. by using polymerase chain reaction and gene probes. *J. Clinical Microbiol.* 30:74–78.

39. Marczak, R., T. E. Gorrell, and M. Müller. 1983. Hydrogenosomal ferredoxin of the anaerobic protozoon, *Tritrichomonas foetus. J. Biol. Chem.* 258:12427–33.

40. Mattern, C. F. T., B. M. Honigberg, and W. A. Daniel. 1967. The mastigont system of *Trichomonas gallinae* (Rivolta) as revealed by electron microscopy. *J. Protozool.* 14:320–39.

41. McAllister, C. T. 1991. Protozoan, helminth, and arthropod parasites of the spotted chorus frog, *Pseudacris clarkii* (Anura: Hylidae), from north-central Texas (USA). *J. Helminthol. Soc. Washington* 58:51–56.

42. Meyers, T. R., S. Short, and W. Eaton. 1990. Summer mortalities and incidental parasitisms of cultured Pacific oysters in Alaska (USA). *J. Aquatic Animal Health* 2:172–76.

43. Mintz, E. D., M. Hudson-Wragg, M. L. Cartter, and J. L. Hadler. 1993. Foodborne giardiasis in a corporate office setting. *J. Infect. Dis.* 167:250–53.

44. Monzingo, D. L. Jr., and C. P. Hibler. 1987. Prevalence of *Giardia* sp. in a beaver colony and the resulting environmental contamination. *J. Wildl. Dis.* 23:576–85.

45. Moore, G. T., W. M. Cross, D. McGuire, et al. 1970. Epidemic giardiasis at a ski resort. *N. Engl. J. Med.* 281:402–7.

46. Mueller, J. F. 1959. Is *Chilomastix* a pathogen? *J. Parasitol.* 45:170.

47. Müller, M. 1985. Search for cell organelles in protozoa. *J. Protozool.* 32:559–63.

48. Nash, T. 1992. Surface antigen variability and variation in *Giardia lamblia. Parasitol. Today* 8:229–34.

49. Nash, T. E., S. M. Banks, D. W. Alling, J. W. Merritt, and J. T. Conrad. 1990. Frequency of variant antigens in *Giardia lamblia. Exp. Parasitol.* 71:415–21.

50. Paltauf, F., and J. G. Meingassner. 1982. The absence of cardiolipin in hydrogenosomes of *Trichomonas vaginalis* and *Tritrichomonas foetus. J. Parasitol.* 68:949–50.

51. Patterson, D. J., and B. L. J. Delvinquier. 1990. The fine structure of the cortex of the protist *Protoopalina australis* (Slopalinida, Opalinidae) from *Litoria nasuta* and *Litoria inermis* (Amphibia: Anura: Hylidae) in Queensland, Australia. *J. Protozool.* 37:449–55.

52. Peterson, K. M., and J. F. Alderete. 1984. Selective acquisition of plasma proteins by *Trichomonas vaginalis* and human lipoproteins as a growth requirement for this species. *Molecular and Biochem. Parasitol.* 12:37–48.

53. Poirier, T. P., S. C. Holt, and B. M. Honigberg. 1990. Fine structure of the mastigont system in *Trichomonas tenax* (Zoomastigophorea: Trichomonadida). *Trans. Am. Microsc. Soc.* 109:342–51.

54. Rubino, S., R. Muresu, P. Rappelli, P. L. Fiori, P. Rizzu, G. Erre, and P. Cappuccinelli. 1991. Molecular probe for identification of *Trichomonas vaginalis* DNA. *J. Clinical Microbiol.* 29:702–6.

55. Schuster, F. L. 1968. Ultrastructure of *Histomonas meleagridis* (Smith) Tyzzer, a parasitic amebo-flagellate. *J. Parasitol.* 54:725–37.

56. Schofield, P. J., M. R. Edwards, and P. Kranz. 1991. Glucose metabolism in *Giardia intestinalis. Molecular and Biochem. Parasitol.* 45:39–48.

57. Sherman, J. K., T. L. Hostetler, K. McHenry, and J. J. Daly. 1991. Cryosurvival of *Trichomonas vaginalis* during cryopreservation of human semen. *Cryobiology* 28:246–50.

58. Siddall, M. E., H. Hong, and S. S. Desser. 1992. Phylogenetic analysis of the Diplomonadida (Wenyon, 1926) Brugerolle, 1975: Evidence for heterochrony in protozoa and against *Giardia lamblia* as a "missing link." *J. Protozool.* 39:361–67.

59. Stabler, R. M., and T. Chen. 1936. Observations on an *Endamoeba* parasitizing opalinid ciliates. *Biol. Bull.* 70:56–71.

60. Talbot, J. A., K. Nielsen, and L. B. Corbeil. 1991. Cleavage of proteins of reproductive secretions by extracellular proteinases of *Tritrichomonas foetus. Can. J. Microbiol.* 37:384–90.

61. Turner, G., and M. Müller. 1983. Failure to detect extranuclear DNA in *Trichomonas vaginalis* and *Tritrichomonas foetus. J. Parasitol.* 69:234–36.

62. Udezulu, I. A., G. S. Visvesvara, D. M. Moss, and G. J. Leitch. 1992. Isolation of two *Giardia lamblia* (WB strain) clones with distinct surface protein and antigenic profiles and differing infectivity and virulence. *Infection and Immunity* 60:2274–80.

63. Upcroft, J. A., A. Healey, D. G. Murray, P. F. L. Boreham, and P. Upcroft. 1992. A gene associated with cell division and drug resistance in *Giardia duodenalis. Parasitology* 104:397–405.

64. Van Keulen, H., R. R. Gutell, M. A. Gates, S. R. Campbell, S. L. Erlandsen, E. L. Jarroll, J. Kulda, and E. A. Meyer. 1993. Unique phylogenetic position of Diplomonadida based on the complete small subunit ribosomal RNA sequence of *Giardia ardeae, Giardia muris, Giardia duodenalis* and *Hexamita* sp. *FASEB J.* 7:223–31.

65. Wallis, P. M., J. M. Buchanan-Mappin, G. M. Fauber, and M. Belosevic. 1984. Reservoirs of *Giardia* spp. in southwestern Alberta. *J. Wildl. Dis.* 20:279–83.

66. Weinbach, E. C., C. E. Claggett, D. B. Keister, L. S. Diamond, and H. Kon. 1980. Respiratory metabolism of *Giardia lamblia. J. Parasitol.* 66:347–50.

67. Yang, J., and T. Scholten. 1977. *Dientamoeba fragilis:* A review with notes on its epidemiology, pathogenicity, mode of transmission, and diagnosis. *Am. J. Trop. Med. Hyg.* 26:16–22.

Additional References

Honigberg, B. M. 1978. Trichomonads of importance in human medicine. In Kreier, J. P., ed. *Parasitic protozoa* 3. New York: Academic Press, Inc.

Kulda, J., and E. Nohynkova. 1978. Flagellates of the human intestine and intestines of other species. In Kreier, J. P., ed. *Parasitic protozoa* 3. New York: Academic Press, Inc.

McDougald, I. R., and W. M. Reid. 1978. *Histomonas meleagridis* and its relatives. In Kreier, J. P., ed. *Parasitic protozoa* 3. New York: Academic Press, Inc.

Meyer, E. A., and S. Radulescu. 1979. *Giardia* and giardiasis. In Lumsden, W. H. R., ed. *Advances in parasitology* 17. New York: Academic Press, Inc.

Nadler, S. A., and B. M. Honigberg. 1988. Genetic differentiation and biochemical polymorphism among trichomonads. *J. Parasitol.* 74:797–804.

Schorr, M. S., R. Altig, and W. J. Diehl. 1990. Populational changes of the enteric protozoans *Opalina* spp. and *Nyctotherus cordiformis* during the ontogeny of anuran tadpoles. *J. Protozool.* 37:479–81.

Wolfe, M. S. 1992. Giardiasis. *Clinical Microbiol. Reviews* 5:93–100.

Chapter 7

SUBPHYLUM SARCODINA: AMEBAS

. . . the second species may be . . . the same as found some years ago by . . . Jessionek and Kiolemenglolou in the kidneys, liver, and lungs of an aborted syphilitic fetus, and by Smith and Weidman in the kidneys, lungs, and liver of a non-syphilitic new-born infant and in the lungs of a syphilitic infant one month old, to which these last writers have applied the name of Endameoba mortinatalium.

A. J. Smith and M. T. Barrett

Students of biology are introduced to amebas (subclass Gymnamoebia of the class Lobosea) early in their careers. Most are left with the impression that amebas are harmless, microscopic creatures that spend their lives aimlessly wandering about in mud, water, and soil, occasionally catching a luckless ciliate for food and unemotionally reproducing by binary fission. Actually this is a pretty fair account of most amebas. However, a few species are parasites of other organisms, and one or two are responsible for much misery and death of humans. Still others are commensals, which must be recognized to differentiate them from the pathogenic species.

The subphylum Sarcodina probably appeared early in eukaryote evolutionary history, and some molecular evidence suggests its members are ancestral to the other protists.[26] However, as traditionally conceived, the group is likely polyphyletic. Structural characters that suggest evolutionary lines include permanent cytostomes and both flagellate and ameboid stages, such as found in the flagellate *Tetramitus*. The life cycle of *Naegleria* species also includes flagellate and ameboid stages (Chapter 4), but no permanent cytostome is found in this genus. *Vahlkampfia* has no flagellate stage, but its ameboid stage is like that of *Naegleria*. At least one important parasite, *Entamoeba histolytica*, lacks mitochondria and on the basis of its RNA is thought to have diverged from the eukaryotic line prior to the latter's acquisition of mitochondria and subsequent diversification.[13] Of the many families of amebas, only the Endamoebidae has species of great medical or economic importance. Two other families, Schizopyrinidae and Hartmannelidae, have species that can become facultatively parasitic in humans.

ORDER AMOEBIDA

Family Endamoebidae

Species in the Endamoebidae are parasites or commensals of the digestive systems of arthropods and vertebrates. The genera and species are differentiated on the basis of nuclear structure. Three genera contain known parasites or commensals of humans and domestic animals: *Entamoeba*, *Endolimax*, and *Iodamoeba*.

• Genus *Entamoeba*

Species of *Entamoeba* possess a nucleus that is vesicular and that has a small endosome at or near the center. Chromatin granules are arranged around the periphery of the nucleus and in some species also around the endosome. The cytoplasm contains a variety of food vacuoles, often containing particles of food being digested, usually bacteria or starch grains.[20] On the ultrastructural level, the outer membrane possesses a "fuzzy coat," and the cytoplasm contains numerous vesicles, sometimes considered exocytotic because of their accumulation at the uroid (temporary posterior end).[23] Golgi bodies and mitochondria apparently are absent. Curious, small **helical bodies** can be seen widely distributed in the cytoplasm of some trophozoites. These bodies are 0.3 to 1 μm in length, contain up to 40 distinct ribonucleoproteins, and become crystallized into the **chromatoidal bodies** or **bars** following encystment.[23,25] (Figs. 7.1 and 7.2). These bodies stain darkly with basic dyes and have been known by parasitologists for many years. The chromatoidal bars may be blunt rods or splinter shaped, according to species, and in some species they are noticeable only in young cysts. As the cyst ages, the bars apparently are disassembled and disappear. *Entamoeba histolytica* is also sometimes infected with viruses.[23]

Species of *Entamoeba* occur in both vertebrate and invertebrate hosts. Four species are common in humans (*E. histolytica, E. hartmanni, E. coli,* and *E. gingivalis*) and will be considered here in some detail. *Entamoeba polecki* is mentioned in passing.

Entamoeba histolytica. Dysentery, both bacterial and amebic, has long been known as a handmaiden of war, often inflicting more casualties than bullets and bombs. Accounts of epidemics of dysentery accompany nearly every thorough account of war, from antiquity to the prison camp horrors of World War II and Vietnam. Captain James Cook's first voyage met with amebic disaster in Batavia, Java, and modern tourists, too, often find themselves similarly afflicted on visiting foreign ports.

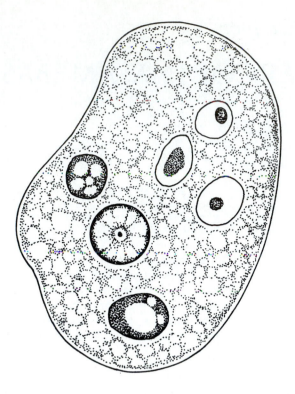

FIGURE 7.1

Entamoeba histolytica trophozoite and cyst.

Drawing by Jeanne Robertson.

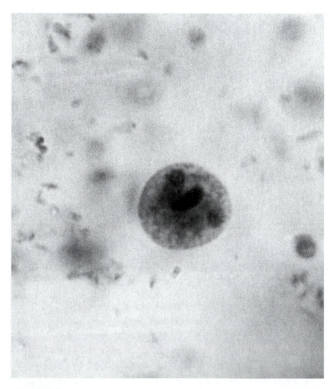

FIGURE 7.2

Young cyst of *Entamoeba histolytica,* containing two nuclei and a prominent chromatoidal bar. Usually, such a cyst is 10 to 20 µm wide.

Photograph by Larry S. Roberts.

Entamoeba histolytica (Fig. 7.3), the ameba responsible for such misery, is the third most common cause of parasitic death in the world. Close to 500 million people are infected at any one time, with up to 100,000 deaths per year. These numbers may increase as urban migration and deteriorating economies of some developing countries result in unhygienic conditions. In addition, high rates of infection exist in certain high-risk groups, such as people who practice anilingus, where infections have reached epidemic levels.

The history of acquired knowledge of the parasite *E. histolytica* is rampant with confusion and false conclusions. Foster[12] provides the interesting account: The ameba was first discovered in 1873 by a clinical assistant, D. F. Lösch, in St. Petersburg (now Leningrad), Russia. The patient, a young peasant with bloody dysentery, was passing large numbers of amebas in his stools. Many of these, Lösch observed, contained erythrocytes in their food vacuoles. He successfully infected a dog by injecting amebas from his patient into the dog's rectum. On dissection Lösch found the dog's colonic mucosa riddled with ulcers that contained amebas. His human patient soon died, and at autopsy Lösch found identical ulcers in the intestinal mucosa. Despite these clear-cut observations, Lösch concluded that the ulcers were caused by some other agent and that the amebas merely interfered with their healing. Nearly 40 years passed before it was generally accepted that an intestinal ameba can cause disease.

A major cause of the 40-year delay was human ignorance about the fact that several species of amebas are found in the

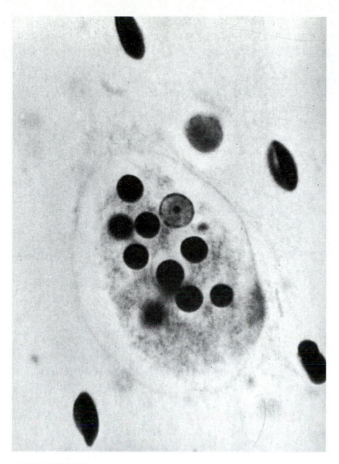

FIGURE 7.3

Trophozoite of *Entamoeba histolytica* with several erythrocytes in food vacuoles.

From M. Kenney and L. K. Eveland, "Transformation *in vivo* of a large race of *Entamoeba histolytica* into a small race," in *Bull. N. Y. Acad. Med.* 57:234–239. Copyright © 1981.

human intestine. Once this situation was recognized and nonpathogenic species were delineated, only one species complex remained that appeared to cause disease, and only occasionally at that. Schaudinn named this group *Entamoeba histolytica* in 1903[33] although the epithet *coli* was already applied to it by Lösch (as *Amoeba coli*). Schaudinn applied the epithet to a nonpathogenic species that he named *Entamoeba coli*.

Through the years it became obvious that *E. histolytica* occurs in two sizes. The smaller-sized amebas have trophozoites 12 to 15 μm in diameter and cysts 5 to 9 μm wide. This form is encountered in about a third of those who harbor amebas and is not associated with disease. The larger form has trophozoites 20 to 30 μm in diameter and cysts 10 to 20 μm wide. The larger form may actually consist of two races, one sometimes pathogenic and the other always a commensal.

The small, nonpathogenic type is considered here as a separate species called *E. hartmanni*. Its life cycle, general morphology, and overall appearance, with the exception of size, are identical to those of *E. histolytica*. The task of proper identification is placed on the diagnostician, whose diagnosis may save the life of the patient or add the burden of unnecessary medication. A third species, *E. moshkovskii*,

is identical in morphology to *E. histolytica,* but it is not a symbiont. It dwells in sewage and is often mistaken for a parasite of humans. Indeed, it may be a strain that recently derived from one of the symbionts of humans.

There are several pathogenic strains of *E. histolytica*. These strains (zymodemes) can be distinguished from the nonpathogenic ones by isozyme analysis. Some authors have proposed that *E. histolytica* be broken into two species, the other one being *E. dispar,* distinguishable not only by biochemical methods, but also by clinical manifestations.[3,10]

- ***Morphology and Life Cycle.*** Several successive stages occur in the life cycle of *E. histolytica:* the **trophozoite, precyst, cyst, metacyst,** and **metacystic trophozoite.** Although the diameter of most trophozoites (Figs. 7.1 and 7.3) falls into the range of 20 to 30 μm, occasional specimens are as small as 10 μm or as large as 60 μm. In the intestine and in freshly passed, unformed stools, the parasites actively crawl about, their short, blunt pseudopodia rapidly extending and withdrawing. They also have filopodia, which are usually not discernible by light microscopy.[21] The clear ectoplasm is rather thin but is clearly differentiated from the granular endoplasm. The nucleus is difficult to discern in living specimens, but nuclear morphology may be distinguished after fixing and staining with iron-hematoxylin. The nucleus is spherical and is about one-sixth to one-fifth the diameter of the cell. A prominent endosome is located in the center of the nucleus, and delicate, achromatic fibrils radiate from it to the inner surface of the nuclear membrane. Chromatin is absent from a wide area surrounding the endosome but is concentrated in granules or plaques on the inner surface of the nuclear membrane. This gives the appearance of a dark circle with a bull's-eye in the center. The nuclear membrane itself is quite thin.

Food vacuoles are common in the cytoplasm of active trophozoites and may contain host erythrocytes in samples from diarrheic stools (Fig. 7.3). Granules typical for all amebas are numerous in the endoplasm. Chromatoidal bars are not found in this stage.

In a normal, asymptomatic infection, the amebas are carried out in formed stools. As the fecal matter passes posteriorly and becomes dehydrated, the ameba is stimulated to encyst. Cysts are neither found in the stools of patients with dysentery nor formed by the amebas when they have invaded the tissues of the host. Trophozoites passed in stools are unable to encyst. At the onset of encystment the trophozoite disgorges any undigested food it may contain and condenses into a sphere, called the **precyst.** A precyst is so rich in glycogen that a large glycogen vacuole may occupy most of the cytoplasm in the young cyst. The chromatoidal bars that form typically are rounded at the ends. The bars may be short and thick, thin and curved, spherical, or very irregular in shape, but they do not have the splinterlike appearance found in *E. coli.*

The precyst rapidly secretes a thin, tough hyaline **cyst wall** around itself to form a **cyst.** The cyst may be somewhat ovoid or elongate, but it usually is spheroid. It is commonly 10 to 20 μm wide but may be as small as 5 μm. The young cyst has only a single nucleus, but this rapidly divides twice to form two- and four-nucleus

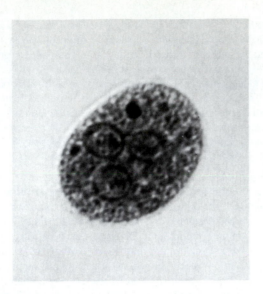

FIGURE 7.4

Three of the four nuclei are in focus in the metacyst of *Entamoeba histolytica,* and two small chromatoidal bodies can be seen.

Photograph by Larry S. Roberts.

stages (Fig. 7.4). As the nuclear division proceeds and the cyst matures, the glycogen vacuole and chromatoidal bodies disappear. In semiformed stools one can find precysts and cysts with one to four nuclei, but quadrinucleate cysts **(metacysts)** are most common in formed stools (Figs. 7.1 and 7.4). This stage can survive outside the host and can infect a new one. After excysting in the small intestine, both the cytoplasm and nuclei divide to form eight small amebulae, or **metacystic trophozoites.** These are basically similar to mature trophozoites except in size.

• *Biology.* Trophozoites may live and multiply indefinitely within the crypts of the mucosa of the large intestine, apparently feeding on starches and mucous secretions and interacting metabolically with enteric bacteria. However, such trophozoites commonly initiate tissue invasion when they hydrolyze mucosal cells and absorb the predigested product. At this stage they no longer require the presence of bacteria to meet their nutritional requirements. It has been shown that under optimal conditions of pH and ionic concentrations, nonvirulent *E. histolytica* becomes virulent when it contains a certain ratio of starch to cholesterol.[35] It has been suggested that the tiny cytoplasmic extensions from the surface (seen in some electron micrographs) could be "triggers" for "surface-active lysosomes" and function in cytolysis of host cells, although there is no evidence for these hypotheses at present.[20,23] Lushbaugh and Pittman[21] showed that these "triggers" were actually filopodia, and they speculated that any or all of the following could be functions of the filopodia: (1) endocytosis or pinocytosis, (2) exocytosis, (3) attachment to the substrate, (4) penetration of tissue, (5) release of cytotoxic substances, or (6) contact cytolysis of host cells. Both pathogenic and nonpathogenic strains can possess proteolytic enzymes that presumably would make tissue invasion possible.

Virulence of a particular strain can be attenuated by in vitro cultivation and sometimes restored by passage through certain experimental hosts. Some evidence exists that viral infection of the amebas may affect virulence.[22] The complex of factors involved in the environmental conditions in the host are even more difficult to untangle because the conditions mutually interact. The oxidation-reduction potential and the pH of the gut contents influence invasiveness, but these conditions are determined largely by the bacterial flora, which is in turn influenced by the host's diet and perhaps even its overall nutritional state. One reason newcomers to areas of endemicity suffer more than the local population may be the differences in their bacterial flora.

Invasive amebas erode ulcers into the intestinal wall, eventually reaching the submucosa and underlying blood vessels. From there, they may travel with the blood to other sites in the body, such as the liver, lungs, or skin. Although these endogenous forms are active, healthy amebas that multiply rapidly, they are on a dead-end course. They cannot leave the host and infect others and so perish with their luckless benefactor.

Mature cysts in the large intestine, on the other hand, leave the host in great numbers. The host that produces such cysts is usually asymptomatic or only mildly afflicted. Cysts of *E. histolytica* can remain viable and infective in a moist, cool environment for at least 12 days, and in water they can live up to 30 days; however, they are rapidly killed by putrefaction, desiccation, and temperatures below 5° C and above 40° C. They can withstand passage through the intestines of flies and cockroaches. The cysts are resistant to levels of chlorine normally used for water purification.

When swallowed, the cyst passes through the stomach unharmed and shows no activity while in an acidic environment. When it reaches the alkaline medium of the small intestine, the metacyst begins to move within the cyst wall, which rapidly weakens and tears. The quadrinucleate ameba emerges and divides into amebulae that are swept downward into the cecum. This is the first opportunity of the organism to colonize, and its success depends on one or more metacystic trophozoites making contact with the mucosa. Obviously chances for establishment are improved when large numbers of cysts are swallowed.

The biochemistry and metabolism of *E. histolytica* has been reviewed by McLaughlin and Aley.[23] The amebas possess several hydrolytic enzymes, including phosphatases, glycosidases, proteinases, and an RNAse. The major metabolic end products are CO_2, ethanol, and acetate, whose proportions vary with the extent to which the animals are deprived of oxygen. Although once thought to be a strict anaerobe, we now know that *E. histolytica* is more of a metabolic opportunist and able to utilize oxygen when it is present in the environment. Glucose, from external sources or stored glycogen, is metabolized via the Embden-Meyerhof pathway exclusively, and fructose phosphate is phosphorylated, prior to lysis, by enzymatic reactions unique to these amebas. Pyruvate is converted mostly to ethanol, even in the presence of oxygen, via Coenzyme-A, and a pyruvate oxidase similar to the one found in the trichomonads (see p. 87). The terminal electron transfers are

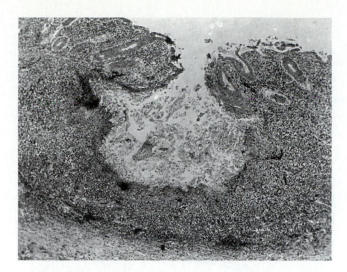

FIGURE 7.5

Typical flask-shaped amebic ulcer of the colon. Extensive tissue destruction has resulted from invasion by *Entamoeba histolytica*.

AFIP neg. no. N–44718.

accomplished with ferredoxin-like iron/sulfur proteins, a trait that may contribute to the efficacy of metronidazole in treatment.[23]

Like all parasites, *E. histolytica* also releases materials into its environment. Some of these compounds—that is, "excretory/secretory products" such as proteins—may alter macrophage metabolism and thus reduce the effectiveness of host immune response, especially in the case of invasive amoebiasis.[38]

- *Pathogenesis.* To quote Elsdon-Dew, "Were one tenth, nay, one hundredth, of the alleged carriers of this parasite to suffer even in minor degree, then the ameba would rank as the major scourge of mankind."[11] Obviously not every infected person shows symptoms of disease.

Entamoeba histolytica is almost unique among the amebas of humans in its ability to hydrolyze host tissues. The trophozoites have active cysteine proteases on their surfaces, and some evidence suggests that the pathogenic strains possess an additional unique form of this enzyme.[2,29] Other studies, however, have failed to make a clear association between phagocytic ability, proteinase activity, and pathogenicity.[24]

The **intestinal lesion** (Figs. 7.5 and 7.6) usually develops initially in the cecum, appendix, or upper colon and then spreads the length of the colon. The number of parasites builds up in the ulcer, increasing the speed of mucosal destruction. The muscularis mucosae is somewhat of a barrier to further progress, and pockets of amebas form, communicating with the lumen of the intestine through a slender, ductlike ulcer. The lesion may stop at the basement membrane or at the muscularis mucosae and then begin eroding laterally, causing broad, shallow areas of necrosis. The tissues may heal nearly as fast as they are destroyed, or the entire mucosa may become pocked. These early lesions usually are not complicated by bacterial invasion, and there is little cellular response by the host. In older lesions the amebas, assisted by bacteria, may break through the muscularis mucosae, infiltrate the submucosa, and even penetrate the muscle layers and serosa. This enables trophozoites to be carried by blood and lymph to ectopic sites throughout the body where secondary lesions then form. A high percentage of deaths results from perforated colons with concomitant peritonitis. Surgical repair of perforation is difficult because a heavily ulcerated colon becomes very delicate.

Sometimes a granulomatous mass, called an **ameboma,** forms in the wall of the intestine and may obstruct the bowel. It is the result of cellular responses to a chronic ulcer and often still contains active trophozoites. The condition is rare except in Central and South America.

Secondary lesions have been found in nearly every organ of the body (Fig. 7.6), but the liver is most commonly affected (about 5% of all cases). Regardless of the secondary site, the initial infection is an intestinal abscess, even though it may go undetected. **Hepatic amebiasis** results when trophozoites enter the mesenteric venules and travel to the liver through the hepatoportal system. They digest their way through the portal capillaries and enter the sinusoids, where they begin to form abscesses. The lesions thus produced may remain pinpoint size, or they may continue to grow, sometimes reaching the size of a grapefruit. The center of the abscess is filled with necrotic fluid, a median zone consists of liver stroma, and the outer zone consists of liver tissue being attacked by amebas, although it is bacteriologically sterile. The abscess may rupture, pouring debris and organisms into the body cavity, where they attack other organs.

Pulmonary amebiasis is the next most common secondary lesion. It usually develops by metastasis from a hepatic lesion but may originate independently. Most cases originate when a liver abscess ruptures through the diaphragm. Other ectopic sites occasionally encountered are the brain, skin, and penis (possibly acquired venereally). Rare ectopic sites include kidneys, adrenals, spleen, male and female genitalia, and pericardium. As a rule all ectopic abscesses are bacteriologically sterile.

- *Symptoms.* Symptoms of infection vary greatly between cases. The strain of *E. histolytica* present, the host's natural or acquired resistance to that strain, and the host's physical and emotional condition when challenged all affect the course of the disease in any individual. When conditions are appropriate, a highly pathogenic strain can cause a sudden onset of severe disease. This usually is the case with waterborne epidemics. More commonly the disease develops slowly, with intermittent diarrhea, cramps, vomiting, and general malaise. Infection in the cecal area may mimic the symptoms of appendicitis. Some patients tolerate intestinal amebiasis for years with no sign of colitis (although they are passing cysts) and then suddenly succumb to ectopic lesions. Depending on the number and distribution of intestinal lesions, the patient might experience pain in the entire abdomen, fulminating diarrhea, dehydration, and loss of blood. Amebic diarrhea is marked by bouts of abdominal discomfort with four to six loose stools per day but little fever.

FIGURE 7.6

Major pathology of amebiasis. Invasion of the intestinal mucosa occurs most commonly in the cecum and next most commonly in the rectosigmoid area. Small lesions develop into large, flask-shaped ulcers with ragged edges (*a*). Passage of trophozoites via the hepatic portal circulation (*arrows, b*) may result in liver abscess formation. Metastasis through the diaphragm may produce secondary abscess formation in the lungs. Trophozoites carried in the bloodstream may cause foci of infection anywhere in the body.

From J. Walter Beck and J. E. Davies, *Medical Parasitology.* Copyright © 1976 Mosby Yearbook, St. Louis, MO. Reprinted by permission.

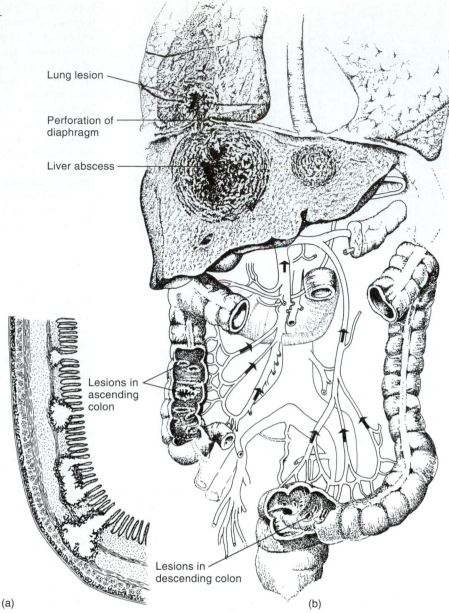

Lung lesion

Perforation of diaphragm

Liver abscess

Lesions in ascending colon

Lesions in descending colon

(a)

(b)

Acute amebic dysentery is a less common condition, but the sufferer from this affliction can best be described as miserable. The onset may be sudden after an incubation period of 8 to 10 days or after a long period as an asymptomatic cyst passer. In acute onset there may be headache, fever, severe abdominal cramps, and sometimes prolonged, ineffective straining at stool. An average of 15 to 20 stools, consisting of liquid feces flecked with bloody mucus, are passed per day. Death may occur from peritonitis, resulting from gut perforation, or from cardiac failure and exhaustion. Bacterial involvement may lead to extensive scarring of the intestinal wall, with subsequent loss of peristalsis. Symptoms arising from ectopic lesions are typical for any lesion of the affected organ.

• ***Diagnosis and Treatment.*** Demonstration of trophozoites or cysts is necessary for the accurate diagnosis of *E. histolytica.* Examination of stool samples is the most effective means of diagnosis of gut infection. A direct smear examined either as a wet mount or fixed and stained will usually reveal heavy infections. Even so, repeated examinations may be necessary; one of us found abundant trophozoites in the stool of a hospital patient after negative findings on three previous days. Lighter infections of cyst passers may be detected with concentration techniques, such as zinc sulfate flotation.

However, a large proportion of patients with extraintestinal amebiasis have no concurrent intestinal infection; diagnosis in such cases must occur, therefore, primarily by clinical and immunological means. X-ray examination and other means of scanning the liver may be useful in diagnosing abscesses. Immunological diagnosis is promising but has yet to be perfected. There is evidence that the surface cysteine proteinases elicit an antibody response in the host,[27] and such response may be of diagnostic value. Modern molecular techniques (polymerase chain reaction, PCR) are also being employed in efforts to detect *E. histolytica* DNA in stool samples and fluid withdrawn from liver abscesses.[1,36] Fluorescent antibody technique has some value in diagnosis but cannot differentiate *E. histolytica* from *E. hartmanni.* Serological procedures that have been adopted widely are the hemagglutination test and agar gel diffusion. Many other diseases can easily be confused with amebiasis; on the hospital chart of the patient who tested negative on three days, a dozen possible explanations *other than amebiasis* for his persistent diarrhea had been listed. Hence demonstration of the organism is nearly mandatory in diagnosis.

Several drugs have a high level of efficacy against colonic amebiasis. Most fall into the categories of arsanilic acid derivatives, iodochlorhydroxyquinolines, and other synthetic and natural chemicals. Antibiotics, particularly tetracycline, are useful as bactericidal adjuvants. These drugs are not as effective in ectopic infections, for which chloroquine phosphate and niridazole show promise of efficacy. Metronidazole (a 5-nitroimidazole derivative) has become the preferred drug in treatment of amebiasis. It is low in toxicity and is effective against both extraintestinal and colonic infections, as well as cysts. However, metronidazole has been reported as being mutagenic in bacteria and carcinogenic in mice at doses not much higher than those given for the treatment of amebiasis. Furthermore, patients must be warned that the drug cannot be taken with alcohol because of its Antabuselike effect. Finally, its efficacy may not be as high as originally reported. Tetracycline in combination with diiodohydroxyquin results in a high rate of cures. Two other 5-nitroimidazole derivatives, ornidazole and tinidazole, can cure amebic liver abscess with a single dose.[17]

- *Epidemiology. Entamoeba histolytica* is found throughout the world. Approximately 500 million persons, of whom about 100 million suffer acute or chronic effects of the disease, are infected. Although clinical amebiasis is most prevalent in tropical and subtropical areas, the parasite is well established from Alaska to the southern tip of Argentina. The prevalence of infection varies widely, depending on local conditions, from less than 1% in Canada and Alaska to 5% in the contiguous United States to 40% in many tropical areas. The prevalence in the United States may be much higher among particular groups, such as persons in mental hospitals or orphanages. Age influences the prevalence of infection: Children less than five years old have a *lower* rate than other age groups. In the United States the greatest prevalence occurs in the age group 26 to 30. The higher prevalence in the tropics results from lower standards of sanitation and the greater

longevity of cysts in a favorable environment. The onset of the disease in persons who travel from temperate regions to endemic tropical areas may be partly the result of lessened resistance from the stress of travel and unaccustomed heat, in addition to the change in bacterial flora in the gut, as mentioned previously. All races are equally susceptible.

Between 1977 and 1978 amebiasis was recognized as a sexually transmitted disease of increasing prevalence in New York City and a major health problem, particularly among gay men. In a study of 126 gay male volunteers who participated in a gay men's health project, 39.7% were infected with *E. histolytica* and 18.3% with *Giardia lamblia,* both fecal-borne organisms.[16] The authors believed that if multiple stools were examined, the figure of 39.7% might have increased at least to 50%. Clearly the primary mode of infection in these cases was oral to anal contact, and certainly the situation is not restricted to New York City. Thus a "new" health problem was discovered that probably has been fairly common for thousands of years.

The manner of disposal of human wastes in a given area is the most important factor in the epidemiology of this organism. Transmission depends heavily on contaminated food and water. Filth flies, particularly *Musca domestica,* and cockroaches also are important mechanical vectors of cysts. Their sticky, bristly appendages can easily carry cysts from a fresh stool to the dinner table, and the habit of the housefly of vomiting and defecating while feeding is an important means of transmission. Polluted water supplies, such as wells, ditches, and springs, are common sources of infection. There have been instances of careless plumbing, in which sanitary drains were connected to freshwater pipes with resultant epidemics. Carriers (cyst passers) handling food can infect the rest of their family groups or even hundreds of people if they work in restaurants. The use of human feces as fertilizer in Asia, Europe, and South America contributes heavily to transmission. Although humans are the most important reservoir of this disease, dogs, pigs, and monkeys are also implicated. A bizarre event occurred in Colorado in 1980, when an epidemic of amebiasis was caused by colonic irrigation with a contaminated enema machine in a chiropractic clinic. Ten patients had to have colectomies; seven of them died.[5]

Entamoeba polecki. *Entamoeba polecki* is usually a parasite of pigs and monkeys, although on rare occasions it occurs in humans. It is generally nonpathogenic in humans, but symptomatic cases may be difficult to treat.[32] It can be distinguished from *E. histolytica* by several morphological criteria, including the fact that cysts of *E. polecki* have just one nucleus, with only about 1% of cysts ever reaching the binucleate stage. Uninucleate cysts of *E. histolytica* are infrequent.

Entamoeba coli. *Entamoeba coli* often coexists with *E. histolytica* and in the living trophozoite stage is difficult to differentiate from it. Unlike *E. histolytica,* however, *E. coli* is a commensal that never lyses its host's tissues. It feeds on bacteria,

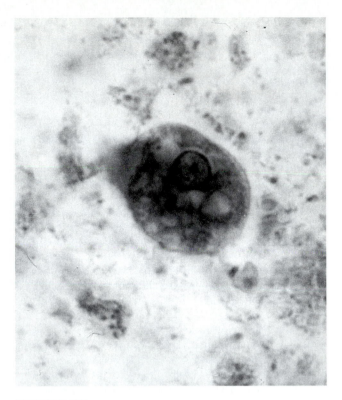

FIGURE 7.7

Trophozoite of *Entamoeba coli,* a commensal in the human digestive tract. Note the characteristic, eccentrically located endosome. The size is usually 20 to 30 μm.

Courtesy of Sherwin Desser.

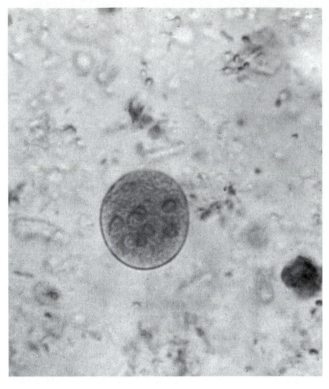

FIGURE 7.8

Metacyst of *Entamoeba coli,* showing eight nuclei. The size is 10 to 33 μm.

Courtesy of David Oetinger.

other protozoa, yeasts, and occasionally blood cells that may be casually available to it. The diagnostician must identify this species correctly; if it is incorrectly diagnosed as *E. histolytica,* the patient may be submitted to unnecessary drug therapy.

Entamoeba coli is more common than *E. histolytica,* partly because of its superior ability to survive in putrefaction.

- **Morphology.** The trophozoite of *E. coli* (Fig. 7.7) is 15 to 50 μm (usually 20 to 30 μm) in diameter and is superficially identical to that of *E. histolytica.* However, their nuclei differ. The endosome of *E. coli* is usually eccentrically placed, whereas that of *E. histolytica* is central. Also, the chromatin lining the nuclear membrane is ordinarily coarser, with larger granules, than that of *E. histolytica.* The food vacuoles of *E. coli* are more likely to contain bacteria and other intestinal symbionts than are those of *E. histolytica,* although both may ingest available blood cells.

  Encystment follows the same pattern as for *E. histolytica.* A precyst is formed; a cyst wall is then rapidly secreted. The young cyst usually has a dense mass of chromatoidal bars that are splinter shaped, rather than blunt as in *E. histolytica.* As the cyst matures, the nucleus divides repeatedly to form eight nuclei (Fig. 7.8). Rarely, as many as 16 nuclei may be produced. The cysts vary in diameter from 10 to 33 μm.

- **Biology.** Infection and migration to the large intestine in the case of *E. coli* are identical to those in the case of *E. histolytica.* The octanucleate metacyst produces 8 to 16 metacystic trophozoites, which first colonize the cecum and then the general colon. Infection is by contamination; in some areas of the world it nearly reaches 100%. Obviously, this widespread infection is a reflection of the level of the sanitation and water treatment. Because *E. coli* is a commensal, no treatment is required. However, infection with this ameba indicates that opportunities exist for ingestion of *E. histolytica.*

Entamoeba gingivalis. *Entamoeba gingivalis* was the first ameba of humans to be described. It is present in all populations, dwelling only in the mouth. Like *E. coli,* it is a commensal and is of interest to parasitologists as another example of niche location and speciation.

- **Morphology.** Only the trophozoite has been found, and encystment probably does not occur. The trophozoite (Fig. 7.9) is 10 to 20 μm (exceptionally 5 to 35 μm) in diameter and is quite transparent in life. It moves rather quickly, by means of numerous blunt pseudopodia. The spheroid nucleus is 2 to 4 μm in diameter and has a small, nearly central endosome. As in all members of this genus, the chromatin is concentrated on the inner surface of the nuclear membrane. Food vacuoles are numerous and contain cellular debris, bacteria, and, occasionally, blood cells.

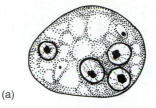
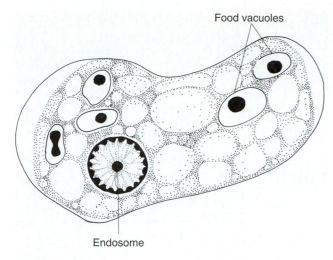

Food vacuoles

Endosome

FIGURE 7.9

Entamoeba gingivalis trophozoite. The size usually is 10 to 20 μm.

Drawing by Ian Grant.

- **Biology.** *Entamoeba gingivalis* lives on the surface of the teeth and gums, in the gingival pockets near the base of the teeth, and sometimes in the crypts of the tonsils. The organisms often are abundant in cases of gum or tonsil disease, but no evidence shows that they cause these conditions. More likely, the protozoa multiply rapidly with the increased abundance of food. They even seem to fare well on dentures if the devices are not kept clean. The commensal also infects other primates, dogs, and cats.

  Because no cyst is formed, transmission must be direct from one person to another, by kissing, by droplet spray, or by sharing eating utensils. Up to 95% of persons with unhygienic mouths may be infected, and up to 50% of persons with healthy mouths may harbor this ameba.[14]

- **Genus *Endolimax***

Members of the genus *Endolimax* live in both vertebrates and invertebrates. These amebas are small, each with a vesicular nucleus. The endosome is comparatively large and irregular and is attached to the nuclear membrane by achromatic threads. Encystment occurs in the life cycle.

Endolimax nana.   *Endolimax nana* lives in the large intestine of humans, mainly at the level of the cecum, and feeds on bacteria. Like *E. coli,* it is a commensal.

- **Morphology.** The trophozoite of this tiny ameba (Fig. 7.10) measures 6 to 15 μm in diameter, but it is usually less than 10 μm. The ectoplasm is a thin layer surrounding the granular endoplasm. The pseudopodia are short and blunt and the ameba moves very slowly, two characteristics from which its name "dwarf internal slug," is derived. The nucleus is small and contains a large centrally or eccentrically located endosome. The marginal chromatin is a thin layer. Large glycogen vacuoles are often present, and food vacuoles contain bacteria, plant cells, and debris.

  Encystment follows the same pattern as in *E. coli* and *E. histolytica.* The precyst secretes a cyst wall, and the young cyst thus formed includes glycogen granules and,

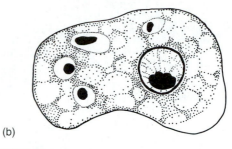

(a)

(b)

FIGURE 7.10

Endolimax nana. (*a*) cyst; (*b*) trophozoite. Note the large karyosome and absence of chromatin granules on the nuclear membrane.

Drawing by Ian Grant.

occasionally, small curved chromatoidal bars. The mature cyst (Fig. 7.10) is 5 to 14 μm in diameter and contains four nuclei.

- **Biology.** As with other cyst-forming amebas that infect humans, the mature cyst must be swallowed. The metacyst excysts in the small intestine, and colonization begins in the upper large intestine. Incidence of infection parallels that of *E. coli* and reflects the degree of sanitation practiced within a community. The cyst is more susceptible to putrefaction and desiccation than is that of *E. coli.* Although the protozoan is not a pathogen, its presence indicates that opportunities exist for infection by disease-causing organisms.

- **Genus *Iodamoeba***

Iodamoeba buetschlii.   The genus *Iodamoeba* has only one species, and it infects humans, other primates, and pigs. Its distribution is worldwide. *Iodamoeba buetschlii* is the most common ameba of swine, which probably are its original host. The prevalence of *I. buetschlii* in humans is 4% to 8%, considerably lower than that of *E. coli* or *E. nana.*

- **Morphology.** The trophozoite (Fig. 7.11) is usually 9 to 14 μm long but may range from 4 to 20 μm. It moves slowly by means of short, blunt pseudopodia. The ectoplasm is not clearly demarcated from the granular endoplasm. The nucleus is relatively large and vesicular, containing a large endosome that is surrounded by lightly staining granules about midway between it and the nuclear membrane. Achromatic strands extend between the endosome and the nuclear membrane, which has no peripheral granules. Food vacuoles usually contain bacteria and yeasts.

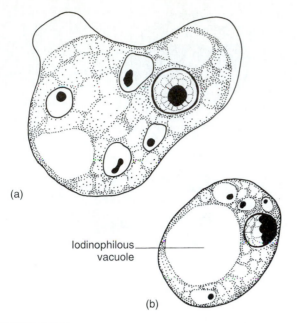

FIGURE 7.11

Iodamoeba buetschlii. (*a*) Trophozoite; (*b*) cysts. Note the persistence of glycogen mass in cyst, and the large eccentric karyosome.

Drawing by Ian Grant.

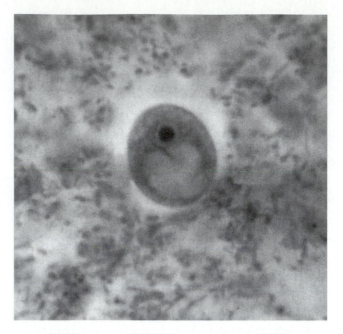

FIGURE 7.12

Iodamoeba buetschlii cyst in human feces. Note the large iodinophilous vacuole. The size is 6 to 15 μm.

Courtesy of James Jensen.

The precyst is usually oblong and contains no undigested food. It secretes the cyst wall that also is usually oblong, measuring 6 to 15 μm long. The mature cyst (Figs. 7.11 and 7.12) nearly always has only one nucleus. A large conspicuous glycogen vacuole stains deeply with iodine, hence the generic name.

- **Biology.** *Iodamoeba buetschlii* lives in the large intestine, mainly in the cecal areas, where it feeds on intestinal flora. Infection spreads by contamination, since mature cysts must be swallowed to induce infection. It is possible that humans become infected through pig feces, as well as human feces. A few reports of *I. buetschlii* causing ectopic abscesses like those of *E. histolytica* probably were actually misidentifications of *Naegleria fowleri.*

ORDER SCHIZOPYRENIDA

Family Vahlkamphidae

Amebas of the order Schizopyrenida are aerobic inhabitants of soil and water and mainly are bacteriophagous. They show affinities with the Mastigophora, since they possess a flagellated stage and an ameboid form. Binary fission seems to take place only in the ameboid form; thus these are diphasic amebas, with the ameboid stage predominating over the flagellated stage. Although the several genera and species in this family live in stagnant water, soil, sewage, and the like, a few are able to become facultative parasites in vertebrates. There has been confusion in the taxonomy of the genera *Naegleria, Hartmannella,* and *Acanthamoeba,* with reports

of all three as facultative parasites of humans. We take the view that *Naegleria* is a member of this family, whereas the other two genera belong in Hartmannellidae.[19]

Soil amebas have evidently been on earth for a long time. These amebas are among the exceedingly few shell- and skeletonless sarcodines represented in the fossil record. Cysts virtually identical to those of *N. gruberi* have been found in Cretaceous amber from Kansas.[39] Research on RNA has shown that some *Naegleria* species are as distantly related as mammals and frogs, but structurally they are very similar. Thus evolutionary divergence has occurred at the molecular level without being matched by structural diversity.

Naegleria fowleri. *Naegleria fowleri* (Fig. 7.13) is also known in some of the literature as *N. aerobia.* It is the major cause of a disease called **primary amebic meningoencephalitis (PAM).** Other known species, *N. gruberi, N. lovaniensis,* and *N. australiensis,* apparently are harmless.

The flagellated stage of *N. fowleri* bears two long flagella at one end, is rather elongated, and does not form pseudopodia; the ameboid stage usually has a single blunt pseudopodium. Pointed tips of the pseudopodia are visible with the scanning electron microscope (Fig. 7.14).

Transformation of the ameboid form to the flagellated form is quite rapid; once the flagella develop, the organism can swim rapidly, like a mastigophoran. The nucleus is vesicular and has a large endosome and peripheral granules. Dark polar masses are formed at mitosis, and Feulgen-negative **interzonal bodies** are present during late stages of nuclear division. A contractile vacuole is conspicuous in free-living forms. Food vacuoles contain bacteria in free-living stages but are filled with

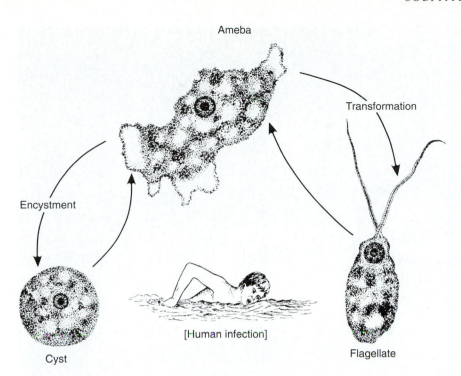

Ameba

Transformation

Encystment

[Human infection]

Cyst

Flagellate

FIGURE 7.13

Life cycle of *Naegleria fowleri.*

From T. D. John, "Amebas," in *McGraw-Hill Yearbook of Science and Technology 1986.* Copyright © 1985, McGraw-Hill, Inc., New York, NY. Reprinted by permission.

host cell debris in parasitic forms. Suckerlike structures called **amebastomes** are present; at least in culture forms, the amebastomes function in phagocytosis (Fig. 7.14).[15] The cyst has a single nucleus.

- *Primary Amebic Meningoencephalitis (PAM).* This is an acute, fulminant, rapidly fatal illness usually affecting children and young adults who have been exposed to water harboring free-living *N. fowleri.* Most cases are contracted in lakes or swimming pools. Probably the flagellated trophozoites are forced deep into the nasal passages when the victim dives into the water. One well-documented case involved ritual washing before prayers by a Nigerian Muslim farmer.[18] The washing occurred five times a day and included sniffing water up the nose. After entrance to the nasal passages, the amebas migrate along the olfactory nerves, through the cribriform plate, and into the cranium. Death from brain destruction is rapid, and few cures have been reported. *Naegleria fowleri* was isolated and cultured from many fatal cases. These amebas kill a variety of laboratory animals when injected intranasally, intravenously, or intracerebrally.[7] They do not form cysts in the host. They have even been isolated from bottled mineral water in Mexico.[30] As of 1962, only 92 cases of PAM had been reported. Since then, over 100 cases have been recorded in widely separated parts of the world, including the United States, Czechoslovakia, Mexico, Africa, New Zealand, and Australia. Undoubtedly, many cases remain undiagnosed. *Naegleria* amebas proliferate rapidly as water temperature rises, so thermal pools that are contaminated by rainwater runoff are particularly at risk (Fig. 7.15). Although these amebas are ubiquitous, the risk of acquiring this infection fortunately is small.

- *Treatment.* Unfortunately most cases of PAM are diagnosed at autopsy. The disease is so rare and its course of brain destruction so rapid that only seldom has it been diagnosed in time for treatment to be attempted. Amphotericin B kills *Naegleria* in vitro and has been used successfully in at least two human cases. Recently *N. fowleri* was shown to be sensitive to qinghaosu (p. 150) in vitro. The lack of toxicity of this drug makes it potentially useful for therapy of PAM.[8] In nature, *N. fowleri* interacts with various organisms in the soil, including, evidently, bacteria that produce substances that in turn kill the amebas.[9] Such substances may hold promise as treatment in the future.

Family Hartmannellidae

A single genus, *Acanthamoeba,* is a facultative parasite of humans in much the same manner as *Naegleria.* The species *A. culbertsoni, A. polyphaga, A. hatchetti, A. castellanii* and *A. rhysodes* have been identified in human tissues. Some of these have been reported as species of *Hartmannella* but that genus is not pathogenic.[40] Biology of the free-living forms is similar to that of *Naegleria* except that flagella are not known to be produced and *Acanthamoeba* cannot tolerate water as hot as can *Naegleria. Acanthamoeba* spp. usually cause chronic infection of the skin or central nervous system in immunocompromised persons, although immunocompetent victims may suffer corneal ulcers and keratitis.

Live trophozoites of *Acanthamoeba* are easily differentiated from *Naegleria,* since they have small spiky acanthopodia and move very slowly. By contrast, *Naegleria* has a single, blunt lobopodium and moves rapidly (more than two body lengths per minute).

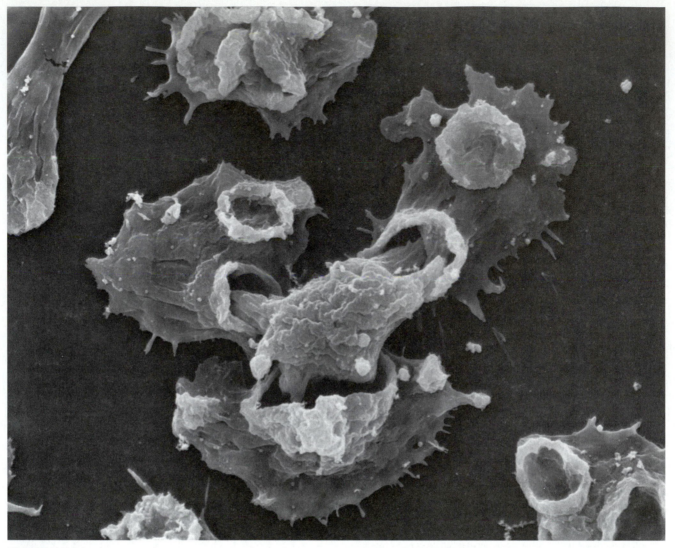

FIGURE 7.14

Three *Naegleria fowleri* from axenic culture, attacking and beginning to devour or engulf a fourth, presumably dead, ameba with their amebastomes. (× 2160.)

From D. T. John et al., "Sucker-like structures on the pathogenic amoeba *Naegleria fowleri*," in *Appl. Envir. Microbiol.* 47:12–14. Copyright © 1984.

Acanthamoeba species are causal agents of keratitis (corneal inflammation and opacity), with contact lenses sometimes considered to be contributing factors. Twenty-four cases of *Acanthamoeba* keratitis were reported to Centers for Disease Control in a nine-month period between 1985 and 1986.[6] In two patients the infected eye was enucleated; 12 patients underwent corneal transplantation. Twenty-two of the patients were initially diagnosed as having corneal herpes simplex virus infections. Most wore contact lenses and were found to use homemade saline washes containing live *Acanthamoeba*. Other recorded cases usually involve some trauma to the cornea before exposure to parasites. Indeed, experimental work with animal systems has shown the corneal abrasion was a necessary condition for keratitis resulting from use of contaminated contact lenses.[37] Public swimming pools have been implicated as sources of infection,[31] but when one notes that *Acanthamoeba* is the most common ameba in fresh water and soil, it is surprising that more infections do not occur.

Treatment is difficult, although some cases have been treated successfully with ketoconazole, miconazole, and propamidine isethionate.[6] Some studies have shown that bacterial contamination in contact lens cleaning solution enhances amoeba multiplication, possibly increasing the chances of eye infection.[4] Other studies have suggested that certain species of bacteria, such as *Pseudomonas aeruginosa*, produce toxins that are lethal to *Acanthamoeba* species,[28] and still other research has shown that the amoebas are chemically attracted to bacteria, even those species that produce toxins.[34] Clearly we have much to learn about *Acanthamoeba* species.

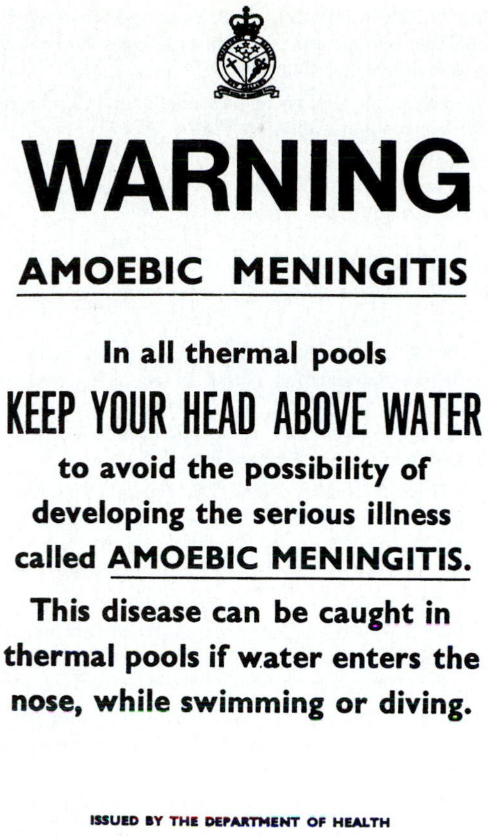

WARNING

AMOEBIC MENINGITIS

In all thermal pools

KEEP YOUR HEAD ABOVE WATER

**to avoid the possibility of
developing the serious illness
called AMOEBIC MENINGITIS.**

**This disease can be caught in
thermal pools if water enters the
nose, while swimming or diving.**

ISSUED BY THE DEPARTMENT OF HEALTH

FIGURE 7.15

Warning encountered in Rotorua, New Zealand, where several cases of primary amebic meningoencephalitis have been contracted in hot pools.
Courtesy of New Zealand Department of Health.

References

1. Acuna-Soto, R., J. Samuelson, P. De-Girolami, L. Zarate, F. Millan-Velasco, G. Schoolnick, and D. Wirth. 1993. Application of the polymerase chain reaction to the epidemiology of pathogenic and nonpathogenic *Entamoeba histolytica*. *Am. J. Trop. Med. Hyg.* 48:58–70.

2. Avila, E. E., and J. Calderon. 1993. *Entamoeba histolytica* trophozoites: A surface-associated cysteine protease. *Exp. Parasitol.* 76:232–41.

3. Blanc, D. S. 1992. Determination of taxonomic status of pathogenic and nonpathogenic *Entamoeba histolytica* zymodemes using isozyme analysis. *J. Protozool.* 39:471–79.

4. Bottone, E. J., R. M. Madayag, and M. N. Qureshi. 1992. *Acanthamoeba* keratitis: Synergy between amebic and bacterial cocontaminants in contact lens care systems as a prelude to infection. *J. Clinical Microbiol.* 30:2447–50.

5. Centers for Disease Control. 1981. Amebiasis associated with colonic irrigation—Colorado. *Morb. Mortal. Weekly Rep.* 30:101–2.

6. Centers for Disease Control. 1986. *Acanthamoeba* keratitis associated with contact lenses—United States. *Morb. Mortal. Weekly Rep.* 35:405–8.

7. Chang, S. H. 1974. Etiological, pathological, epidemiological, and diagnostical consideration of primary amoebic meningoencephalitis. *CRC Crit. Rev. Microbiol.* 3:135–59.

8. Cooke, D. W., G. J. Lalliger, and D. T. Durack. 1987. In vitro sensitivity of *Naegleria fowleri* to qinghaosu and dihydroqinghaosu. *J. Parasitol.* 73:411–13.

9. Cordovilla, P., E. Valdivia, A. Gonzalez-Segura, A. Galvez, M. Martinez-Bueno, and M. Maqueda. 1993. Antagonistic action of the bacterium *Bacillus licheniformis* M-4 toward the amoeba *Naegleria fowleri*. *J. Eukaryotic Microbiol.* 40:323–28.

10. Diamond, L. S., and C. G. Clark. 1993. A redescription of *Entamoeba histolytica* Schaudinn, 1903 (emended Walker, 1911) separating it from *Entamoeba dispar* Brumpt, 1925. *J. Eukaryotic Microbiol.* 40:340–44.

11. Elsdon-Dew, R. 1964. Amoebiasis. *Exp. Parasitol.* 15:87–96.

12. Foster, W. D. 1965. A history of parasitology. Edinburgh: E. & S. Livingstone.

13. Hasegawa, M., T. Hashimoto, J. Adachi, N. Iwabe, and T. Miyata. 1993. Early branchings in the evolution of eukaryotes: Ancient divergence of *Entamoeba* that lacks mitochondria revealed by protein sequence data. *J. Mol. Evol.* 36:380–88.

14. Jaskoski, B. J. 1963. Incidence of oral Protozoa. *Trans. Am. Microsc. Soc.* 82:418–20.

15. John, D. T., T. B. Cole Jr., and R. A. Bruner. 1985. Amebastomes of *Naegleria fowleri*. *J. Protozool.* 32:12–19.

16. Kean, B. H., D. C. William, and S. K. Luminais. 1979. Epidemic of amebiasis and giardiasis in a biased population. *Br. J. Vener. Dis.* 55:375–78.

17. Lasserre, R., et al. 1983. Single-day drug treatment of amebic liver abscess. *Am. J. Trop. Med. Hyg.* 32:723–26.

18. Lawande, R. V., J. T. Macfarlane, W. R. C. Weir, and C. Awunor-Renner. 1980. A case of primary amebic meningoencephalitis in a Nigerian farmer. *Am. J. Trop. Med. Hyg.* 29:21–25.

19. Lee, J. J., S. H. Hutner, and E. C. Bovee. 1985. *An illustrated guide to the protozoa*. Lawrence, Kans.: Society of Protozoologists.

20. Ludvik, J., and A. C. Shipstone. 1970. The ultrastructure of *Entamoeba histolytica*. *Bull. WHO* 43:301–8.

21. Lushbaugh, W. B., and F. E. Pittman. 1979. Microscopic observations on the filopodia of *Entamoeba histolytica*. *J. Protozool.* 26:186–95.

22. Mattern, C. F. T., D. B. Keister, and L. S. Diamond. 1979. Experimental amebiasis. IV. Amebal viruses and virulence of *Entamoeba histolytica*. *Am. J. Trop. Med. Hyg.* 28:653–57.

23. McLaughlin, J., and S. Aley. 1985. The biochemistry and functional morphology of the *Entamoeba*. *J. Protozool.* 32:221–40.

24. Montfort, I., A. Olivos, and R. Pérez-Tamayo. 1993. Phagocytosis and proteinase activity are not related to pathogenicity of *Entamoeba histolytica*. *Parasitol. Res.* 79:160–62.

25. Morgan, R. S., and B. G. Uzman. 1966. Nature of the packing of ribosomes within chromatoid bodies. *Science* 152:214–16.

26. Nanney, D. L., D. O. Mobley, R. M. Preparata, E. B. Meyer, and E. M. Simon. 1991. Eukaryotic origins: String analysis of 5S ribosomal RNA sequences from some relevant organisms. *J. Mol. Evol.* 32:316–27.

27. Osorio, L. M., T. Pico, and A. Luaces. 1992. Circulating antibodies to histolysain, the major cysteine proteinase of *Entamoeba histolytica*, in amoebic liver abscess patients. *Parasitology* 105:207–10.

28. Qureshi, M. N., A. A. Perez, R. M. Madayag, and E. J. Bottone. 1993. Inhibition of *Acanthamoeba* species by *Pseudomonas aeruginosa:* Rationale for their selective exclusion in corneal ulcers and contact lens care systems. *J. Clinical Microbiol.* 31:1908–10.

29. Reed, S., J. Bouvier, A. S. Pollack, J. C. Engel, M. Brown, K. Hirata, X. Que, A. Eakin, P. Hagblom, F. Gillin, and J. H. McKerrow. 1993. Cloning of a virulence factor of *Entamoeba histolytica:* Pathogenic strains possess a unique cysteine proteinase gene. *J. Clinical Invest.* 91:1532–40.

30. Rivera, F., et al. 1981. Bottled mineral waters polluted by protozoa in Mexico. *J. Protozool.* 28:54–56.

31. Rivera, F., E. Ramirez, P. Bonilla, A. Calderon, E. Gallegos, S. Rodriguez, R. Ortiz, B. Zaldivar, P. Ramirez, and A. Duran. 1993. Pathogenic and free-living amoebae isolated from swimming pools and physiotherapy tubs in Mexico. *Environmental Res.* 62:43–52.

32. Salaki, J. S., J. L. Shirey, and G. T. Strickland. 1979. Successful treatment of *Entamoeba polecki* infection. *Am. J. Trop. Med. Hyg.* 28:190–93.

33. Schaudinn, F. 1903. Untersuchungen über die Fortpflanzung einiger Rhizopoden. *Arb. Kaiserl. Gesundheitsamte* 19:547–76.

34. Schuster, F. L., M. Rahman, and S. Griffith. 1993. Chemotactic responses of *Acanthamoeba castellanii* to bacteria, bacterial components, and chemotactic peptides. *Trans. Am. Microsc. Soc.* 112:43–61.

35. Sharma, R. 1959. Effect of cholesterol on the growth and virulence of *Entamoeba histolytica. Trans. R. Soc. Trop. Med. Hyg.* 53:278–81.

36. Tachibana, H., S. Kobayashi, E. Okuzawa, and G. Masuda. 1992. Detection of pathogenic *Entamoeba histolytica* DNA in liver abscess fluid by polymerase chain reaction. *Int. J. Parasitol.* 22:1193–96.

37. Van Klink, F., H. Alizadeh, Y. He, J. A. Mellon, R. E. Silvany, J. P. McCulley, and J. Y. Niederkorn. 1993. The role of contact lenses, trauma, and Langerhans cells in a Chinese hamster model of *Acanthamoeba* keratitis. *Investigative Ophthalmology & Visual Science* 34:1937–44.

38. Wang, W., and K. Chadee. 1992. *Entamoeba histolytica* alters arachidonic acid metabolism in macrophages in vitro and in vivo. *Immunology* 76:242–50.

39. Waggoner, B. M. 1993. *Naegleria*-like cysts in Cretaceous amber from central Kansas. *J. Eukaryotic Microbiol.* 40:97–100.

40. Warhurst, D. C. 1985. Pathogenic free-living amoebae. *Parasitol. Today* 1:24–28.

Additional References

Band, R. N., et al. 1983. Symposium—the biology of small amoebae. *J. Protozool.* 30:192–214.

Chang, S. L. 1971. Small, free-living amebas: Cultivation, quantitation, identification, classification, pathogenesis, and resistance. In Cheng, T. C., ed. *Current topics in comparative pathobiology* 1. New York: Academic Press, Inc., 202–54. A review of the facultatively parasitic amebas.

Connor, D. H., R. C. Neafie, and W. M. Meyers. 1976. Amebiasis. In Binford, C. H., and D. H. Connor, eds. *Pathology of tropical and extraordinary diseases.* Washington, D.C.: Armed Forces Institute of Pathology.

Culbertson, C. G. 1976. Amebic meningoencephalitides. In Binford, C. H., and D. H. Connor, eds. *Pathology of tropical and extraordinary diseases.* Washington, D.C.: Armed Forces Institute of Pathology.

Hoare, C. A. 1958. The enigma of host-parasite relations in amebiasis. *Rice Inst. Pamphlet* 45:23–35. Very interesting reading.

Lösch, F. A. 1875. Massive development of amebas in the large intestine. Kean, B. H., and K. E. Mott, trans., 1975. *Am. J. Trop. Med. Hyg.* 24:383–92.

Martinez-Paloma, A. 1987. The pathogenesis of amoebiasis. *Parasitol. Today* 3:111–18.

Neal, R. A. 1983. Experimental amoebiasis and the development of anti-amoebic compounds. *Parasitology* 86:175–91.

Chapter 8

PHYLUM APICOMPLEXA: GREGARINES, COCCIDIA, AND RELATED ORGANISMS

I began the study of the gregarines of insects in 1942, but I lost many data and manuscripts by the fire caused by the atomic bomb dropped on Hiroshima. After the second World War, I came back to my work, and ressumed [sic] some parts of my previous study.

Kinichiro Obata

The phylum Apicomplexa contains organisms that possess a certain combination of structures, the **apical complex,** distinguishable only with the electron microscope.[36] The structures that make up the apical complex are described in detail next, but they typically include a polar ring, micronemes, rhoptries, subpellicular tubules, micropore(s) (cytostome), and a conoid. All apicomplexans are parasitic. Members of the Apicomplexa have a single type of nucleus and no cilia or flagella, except for the flagellated microgametes in some groups. The phylum contains two classes, Perkinsea and Sporozoea. Perkinsea consists of two species of *Perkinsus,* parasites of oysters and abalones, and will not be considered further here. Sporozoea is divided into three subclasses: Gregarinia, Coccidia, and Piroplasmia.

Included in the Apicomplexa are an astonishing array of parasites, some of which are of major veterinary and medical importance. For example, members of the coccidian genus *Eimeria* cause a variety of intestinal diseases in poultry and cattle, and members of the genus *Plasmodium* (see Chapter 9) cause malaria, one of humankind's most persistent and prevalent public health problems. From an evolutionary perspective, the genus *Gregarina* may be one of the most speciose of the animal kingdom. Nearly a thousand species of *Gregarina* have been described, mostly from insects, but only a tiny fraction of all insect species have been studied.

The Apical Complex

The ultrastructure of the sporozoites and merozoites in the subclass Sporozoea is typical of the Apicomplexa.[38] These banana-shaped organisms are somewhat more attenuated at the anterior, apical complex end (Fig. 8.1) than at the posterior end, which often contains crystalline bodies or granules. A constant feature of the apical complex is one or two **polar rings,** electron-dense structures immediately beneath the cell membrane, which encircle the anterior tip. A **conoid** is found in members of the suborder Eimeriina. This structure is a truncated cone of spirally arranged fibrillar structures just within the polar rings. **Subpellicular microtubules** radiate from the polar rings and run posteriorly, parallel to the axis of the body. These organelles probably serve as structural elements and may be involved with locomotive function. Two to several elongated, electron-dense bodies, the **rhoptries,** extend to the cell membrane within the polar rings (and conoid, if present). Rhoptries probably participate in penetration of host cells through release of enzymes (Fig. 8.1*b*). Smaller, more convoluted elongated bodies, the **micronemes,** also extend posteriorly from the apical complex. The ducts of the micronemes apparently run anteriorly into the rhoptries or join a common duct system with the rhoptries to lead to the cell surface at the apex. The contents of the rhoptries and micronemes seem similar in electron micrographs, and this material is released during entry into the host cell. Host recognition and invasion by various apicomplexan stages are reviewed by Sinden.[66]

Along the side of the organism are one or more **micropores,** which function in ingestion of food material during the intracellular life of the parasite. The edges of the micropore are marked by two concentric, electron-dense rings, located immediately beneath the cell membrane. As host cytoplasm or other food matter within the parasitophorous vacuole is pulled through the rings, the parasite's cell membrane invaginates accordingly and finally pinches off to form a membrane-bound food vacuole. With the exception of the micropore, the structures described previously dedifferentiate and disappear after the sporozoite or merozoite penetrates the host cell to become a trophozoite. Members of the phyla Microspora and Myxozoa, which formerly were included in the subphylum Sporozoa, lack these structures at all stages of their life cycles.

CLASS SPOROZOEA

All classes of vertebrates and most invertebrates harbor Sporozoea. Most Sporozoea produce a resistant spore, or oocyst (containing sporozoites), which survives the elements between hosts (Figs. 8.2 and 8.3). In some, however, the spore wall has been eliminated, and the development of the sporozoites is completed within an invertebrate vector (Fig. 8.4).

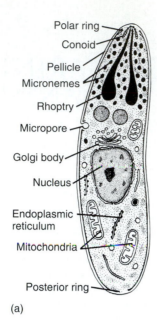

(a)

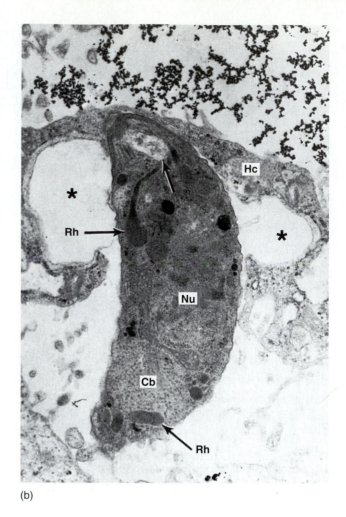

(b)

FIGURE 8.1

(*a*) An apicomplexan sporozoite or merozoite, illustrating the apical complex and other structures typical of this life-cycle stage. (*b*) Electron micrograph of a sporozoite of *Hammondia heydorni* penetrating a cultured cell. *Arrow* indicates empty rhoptry; *asterisks* show host cell vacuoles formed at point of sporozoite entry. **Hc,** host cell; **Nu,** parasite nucleus; **Rh,** rhoptry; **Cb,** crystalline body.

(*a*) From Cleveland P. Hickman, Jr. et al., *Integrated Principles of Zoology,* 9th edition. Copyright © 1993 by Mosby-Year Book Inc. Reprinted by permission of Times Mirror Higher Education Group, Inc., Dubuque, Iowa. All Rights Reserved. (*b*) From C. A. Speer and J. P. Dubey, "Ultrastructure of sporozoites and zoites of *Hammondia heydorni,*" in *J. Protozool.* 36:488–493. Copyright © 1989 by the Society of Protozoologists.

Locomotor organelles are not as obvious as they are in the other phyla of protozoa. Pseudopodia are found only in some tiny, intracellular forms; flagella occur only on gametes of a few species, and a very few have cilialike appendages. Various species have suckerlike depressions, knobs, hooks, myonemes, and/or internal fibrils that aid in limited locomotion. The myonemes and fibrils form tiny waves of contraction across the body surfaces; these can propel the animal slowly through a liquid medium.

Both asexual and sexual reproduction are known in many Sporozoea. Asexual reproduction is either by binary or multiple fission or by endopolyogeny. Sexual reproduction is by isogamous or anisogamous fusion; in many cases this stage marks the onset of spore formation. Insofar as is known, meiosis is postzygotic, so that most of the life-cycle stages are haploid.

Subclass Gregarinia

Members of the large group of Sporozoea known as Gregarinia parasitize only invertebrates, mainly arthropods and especially insects. Because gregarines are widespread, common, and may be large in size, they are often used as instructional materials in elementary and advanced zoology laboratories. Among the most commonly encountered gregarines are members of the genus *Monocystis,* found in earthworm seminal vesicles, and species of *Gregarina,* which parasitize mealworms. Both of these representative gregarines are discussed in some detail next.

In **acephaline** gregarines the body consists of a single unit that may have an anterior anchoring device, the **mucron.** In the **cephaline** species the body is divided by a septum into an anterior **protomerite** and a posterior **deutomerite** that contains the nucleus. Sometimes the protomerite bears an anterior anchoring device, the **epimerite.** Mucrons and epimerites are considered modified conoids. Multiple fission, or **schizogony** (or **merogony,** p. 42), occurs in a few families of gregarines (in the orders Archigregarinida and Neogregarinida). Most gregarines (order Eugregarinida) have no schizogony but undergo multiple fission, within cysts, during gametogenesis. Spore production follows zygote formation. In both the cephalines and acephalines the host becomes infected by swallowing spores. Most species parasitize the body cavity, intestine, or reproductive system of their hosts. Gregarines range in size

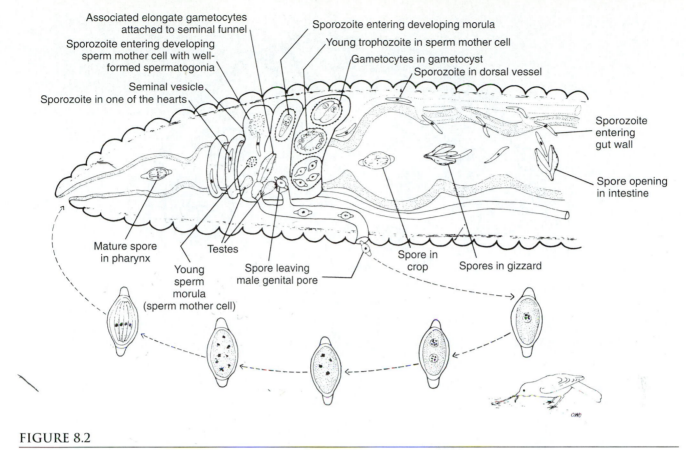

FIGURE 8.2

Life cycle of *Monocystis lumbrici,* an aseptate gregarine of earthworms.

From O.W. Olsen, *Animal Parasites: Their Life Cycles and Ecology.* Copyright © 1974 Dover Publications, Inc., New York, NY. Reprinted by permission.

from only a few micrometers in diameter to at least 10 mm long. Some are so large that nineteenth century zoologists placed them among the worms!

- ## Order Eugregarinida

- ## Suborder Aseptatina

Monocystis lumbrici. *Monocystis lumbrici* (Fig. 8.2) lives in the seminal vesicle of *Lumbricus terrestris* and related earthworms. Worms become infected when they ingest spores, each containing several sporozoites. Spores hatch in the gizzard, and the released sporozoites penetrate the intestinal wall, enter the dorsal vessel, and move forward to the hearts. They then leave the circulatory system and penetrate the seminal vesicles, where they enter into the sperm-forming cells (blastophores) in the vesicle wall.

After a short period of growth, during which the parasites destroy the developing spermatocytes, the sporozoites enter the lumen of the vesicle where they become mature trophozoites, or **sporadins (gamonts),** measuring about 200 μm long by 65 μm wide. Gamonts attach to cells in the region of the sperm tunnel and undergo **syzygy,** in which two or more sporadins connect with one another in tandem. The anterior organism is called the **primite** and the posterior one

the **satellite.** The differences between the two are unclear, but they exhibit different staining reactions.

After syzygy the gamonts flatten against each other and surround themselves with a common cyst envelope, forming the **gametocyst.** Each gamont then undergoes numerous nuclear divisions. The many small nuclei move to the periphery of the cytoplasm and, taking a small portion of the cytoplasm with them, bud off to become gametes. Some of the cytoplasm of each gamont remains as a **residual body.** The gametes from each of the two gamonts are morphologically distinguishable and are thus anisogametes. The fusion of a pair of gametes to form a zygote is followed by the secretion of the **spore** or **oocyst membrane** around that zygote and three cell divisions (sporogony) to form eight sporozoites. Thus each gametocyst now contains many oocysts, and the new host may become infected by eating a gametocyst or, if that body ruptures, an oocyst. As in other sporozoeans, meiosis is zygotic: Only the zygote is diploid, and reductional division in sporogony returns the sporozoites to the haploid condition. Gametocysts or oocysts pass from the host through the sperm duct to be ingested by other worms. The frequency with which one encounters infected worms reveals that this convoluted life history is no barrier to transmission.

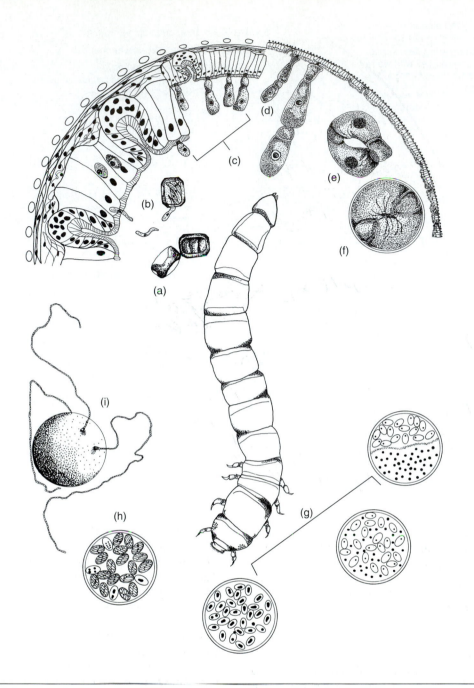

FIGURE 8.3

Life cycle of *Gregarina cuneata* in the yellow mealworm. (*a*) Spores (oocysts); (*b*) exsporulation in the insect's midgut and penetration of epithelial cell by sporozoite; (*c*) growth of the trophont; (*d*) pairing of gamonts; (*e*) syzygy; (*f*) secretion of the gametocyst wall; (*g*) gametogenesis and fertilization; (*h*) division of zygote into sporozoites; (*i*) dehiscence of the gametocyst, with spore chain formation. *Center, Tenebrio molitor* larva.

Drawing by Richard Clopton.

• Suborder Septatina

Gregarina cuneata.   *Gregarina cuneata* (Fig. 8.3) is a common parasite of the mealworm *Tenebrio molitor* and usually infects colonies of beetles maintained in the laboratory. The sporadins are cylindrical, up to 380 μm long by 105 μm wide, each with a small, conical epimerite that is inserted into a host cell. The sporadins undergo syzygy, having detached from the host intestinal epithelium, leaving the epimerite behind. The protomerite is spatula shaped, and the deutomerite is cylindrical, rounded posteriorly. In syzygy the satellite differs structurally from the primite (Fig. 8.3), suggesting that the two gamont mating types are established early following exsporulation. A gametocyst wall is secreted; gametogenesis, fertilization, and oocyst production (sporulation) occur much as in *Monocystis*. The gametocysts pass out with the feces of the host; spores are extruded through tubes and in long chains, in a process called **dehiscence.**

Tenebrio molitor plays host to at least four species of the genus *Gregarina,* the larval stage is parasitized by, in addition to *G. cuneata, G. polymorpha* and *G. steini,* while the adult is parasitized by *G. niphandrodes.*

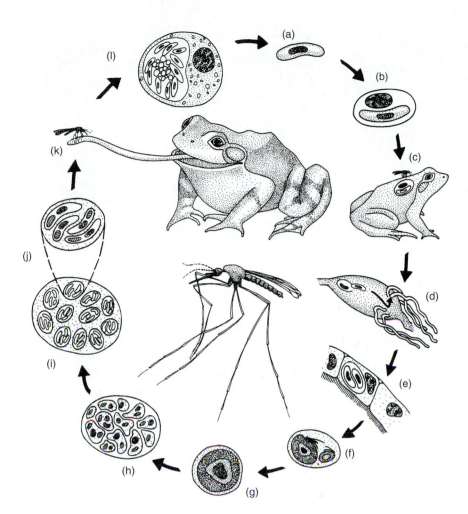

FIGURE 8.4

Life cycle of *Hepatozoon algonquinensis*. (*a, b*) Sausage-shaped gamont in frog erythrocyte; (*c*) mosquito feeding on frog blood; (*d,e*) gamonts escaping erythrocytes and entering Malpighian tubule cells; (*f*) gametogenesis and fertilization; (*g–j*) development of zygote into sporoblasts and, finally, sporocysts with four sporozoites each; (*k*) frog eats mosquito and gets infected; (*l*) parasites in frog's liver cells dividing prior to entering the erythrocytes.

From S. S. Desser et al., "The life history, ultrastructure, and experimental transmission of *Hepatozoon algonquinensis* n. sp., an apicomplexan parasite of the bullfrog, *Rana catesbeiana* and the mosquito, *Culex territans* in Algonquin Park, Ontario," in *J. Parasitol.* 81:212–222. Copyright © 1995. Reprinted with permission of the publisher.

These species differ in size, in body proportions, and in the details of their spore structures. If all the nearly 250,000 described species of beetles have as many species of *Gregarina* as the common yellow mealworm, for example, there may be nearly a million species of this genus parasitizing beetles alone!

Subclass Coccidia

In contrast to the gregarines, Coccidia are small, with a prevalent intracellular reproduction and no epimerite or mucron. Some species are monoxenous, whereas others require two hosts to complete their life cycles. Coccidia live in the digestive tract epithelium, liver, kidneys, blood cells, and other tissues of vertebrates and invertebrates.

The typical coccidian life cycle has three major phases: merogony, gametogony, and sporogony. The infective stage is a rod- or banana-shaped sporozoite that enters a host cell and begins to develop. The organism becomes an ameboid trophozoite that multiplies by merogony to form more rod- or banana-shaped **merozoites,** which then escape from the host cell. These enter other cells to initiate further merogony or transform into gamonts (**gametogony**). Gamonts produce "male" **microgametocytes** or "female" **macrogametocytes.** Most species are thus anisogamous, and the macrogametocyte develops directly into a comparatively large, rounded macrogamete, which is an ovoid body filled with globules of a refractile material and has a central nucleus. The microgametocyte undergoes multiple fission to form tiny, biflagellated microgametes.

Fertilization produces a zygote. Multiple fission of the zygote (sporogony) produces the sporozoite-filled oocyst. In homoxenous life cycles all stages occur in a single host, although the oocyst matures ("sporulates," i.e. completes spore development) in the oxygen-rich, lower-temperature environment outside a host. The sporozoites are then released when the sporulated oocyst is eaten by another host.

In some heteroxenous life cycles, merogony and a part of gametogony occur in a vertebrate host. Sporogony, however, occurs in an invertebrate, and the sporozoites are transmitted by the bite of the invertebrate. In other heteroxenous life cycles, sporozoites are infective to a vertebrate intermediate host, which in turn produces zoites that are infective to the carnivorous vertebrate host.

- **Order Eucoccidiorida**

- **Suborder Adeleorina**

Family Haemogregarinidae

- ***Hepatozoon algonquinensis.*** *Hepatozoon algonquinensis* (Fig. 8.4) is a parasite of bullfrogs, *Rana catesbeiana,* and a mosquito, *Culex territans.*[14a] Similar species have

FIGURE 8.5

(a) Structure of sporulated *Eimeria* oocyst;
(b) sporulated *Isospora* oocyst.

(a) From N. D. Levine, *Protozoan parasites of domestic animals and of man,* 2d ed. Copyright © 1961 Burgess Publishing Co., Minneapolis. Reprinted with permission of the publisher.
(b) From Hoa, et al., "A new coccidian parasite (Apicomplexa: Eimeriorina) from *Anolis distichus* (Sauria: Polychridae) in the Dominican Republic," in *J. Parasitol.* 78:784–785. Copyright © 1992. Reprinted with permission of the publisher.

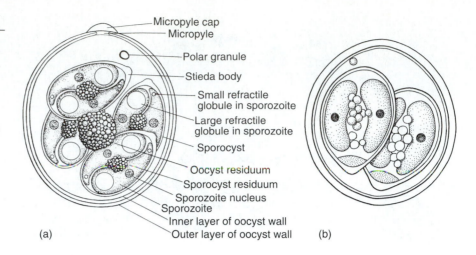

Micropyle cap
Micropyle
Polar granule
Stieda body
Small refractile globule in sporozoite
Large refractile globule in sporozoite
Sporocyst
Oocyst residuum
Sporocyst residuum
Sporozoite nucleus
Sporozoite
Inner layer of oocyst wall
Outer layer of oocyst wall

(a)   (b)

been described under the genus *Haemogregarina* from frogs in Europe, Asia, and Africa, but these parasites cannot be distinguished from *Hepatozoon* species from the intraerythrocytic stages alone, and careful life cycle studies, especially of the stages in the vector, have not always been done.

Gamonts occur in the red blood cells (Fig. 8.4*a,b*). Mosquitoes ingest gamonts in infected erythrocytes with the blood meal. The parasites escape from the red blood cells and enter cells of the Malpighian tubules (8.4*d,e*). Micro- and macrogamonts end up in the same cell and undergo gametogenesis and fertilization (Fig. 8.4*f*). The zygotes segment to form sporoblasts, which then transform into sporocysts, each with four sporozoites (Fig. 8.4*h–j*). Frogs get infected when they eat a mosquito containing mature spores (Fig. 8.4*k*). The life cycle can be completed in about 40 days, half that time spent during development in the mosquito. In Algonquin Park in Canada, where bullfrogs and mosquitoes feed on one another in large numbers, *H. algonquinensis* seems to have taken excellent advantage of this reciprocal trophic relationship and infects nearly a third of the frogs.

• Suborder Eimeriina

In this suborder the macrogamete and microgamete develop independently without syzygy. The microgametocyte produces many microgametes, and the sporozoites are enclosed in a sporocyst. This is a very large group with many families and thousands of species. A few species are found in humans, but most parasitize domestic animals, making this suborder of Eucoccida one of the most important groups of parasites in agriculture.

Family Eimeriidae. This family contains numerous genera, of which several are of medical and veterinary importance. Levine[39] characterized the family as follows: The organisms develop in the host cell proper, and the gamonts and meronts (schizonts) do not have an attachment organelle. Oocysts contain zero, one, two, four, or more sporocysts, each with one or more sporozoites. Merogony

and gametogony occur within a host; sporogony typically, although not necessarily, occurs outside. Microgametes have two or three flagella, and particular species may be monoxenous or heteroxenous.

The taxonomy of the Eimeriidae is an area of active research interest, with new species being described annually from all classes of vertebrates. The size and shape of the oocyst and its contents, the presence or absence of several of the internal structures described next, and the texture of the outer wall are all important taxonomic characters. Identification of a coccidian usually can be accomplished by examining its oocyst, which is remarkably constant in its characters in a given species. However, coccidian taxonomy has also undergone a rather remarkable upheaval in the past 20 years, largely as a result of life-cycle studies involving parasites with very similar oocysts[37] (see Table 8.1).

A typical oocyst (*Eimeria* sp.) is diagrammed in Figure 8.5*a*. The oocyst wall is of two layers, and with the electron microscope a membrane surrounding the outer wall may be distinguished.[62] In many species there is a tiny opening at one end of the oocyst, the **micropyle,** and this may be covered by the **micropylar cap.** A refractile **polar granule** may lie somewhere within the oocyst. The oocyst wall (and probably the sporocyst wall, too) is of a resistant material that helps the organism to survive harsh conditions in the external environment. Wilson and Fairbairn showed that the wall was composed of a chitinlike substance but not chitin, since it does not contain N-acetyl-glucosamine.[76] The chitinlike substance is probably found in the inner layer of the wall because that layer is resistant to sodium hypochlorite. Figure 8.5*b* shows an oocyst of an *Isospora* species. Comparison of the two oocysts of Figure 8.5 reveals the major difference between the two genera, namely the number of sporocysts contained within the sporulated oocyst.

Most species form sporocysts, which contain the sporozoites, within the oocyst. During sporogony to form the sporozoites, the cytoplasmic material not incorporated into the sporozoites forms the **oocyst residuum.** In like manner some material may be left over within the sporocysts to become the **sporocyst residuum.** However, it appears that the sporocyst residuum is more than a depository for waste. It contains a

Table 8.1 COMPARISON OF HOMOXENOUS AND HETEROXENOUS COCCIDIA WITH 4 SPOROZOITES IN EACH OF 2 SPOROCYSTS PER OOCYST*

| | Eimeriidae | | Sarcocystidae | | | | | |
| | | | Cystoisosporinae | Toxoplasmatinae | | | Sarcocystinae | |
Genus	*Isospora*	*Atoxoplasma*	*Cystoisospora*	*Toxoplasma*	*Hammondia*	*Besnoitia*	*Sarcocystis*	*Frenkelia*
Definitive host	Specific	Specific	Specific	Variable	Specific	Carnivores	Variable	Variable
Proliferative stages in gut	Present	Present, also in liver, spleen, lung	Present	Present	Present	Present	Absent	Absent
Zygote formed in:	Epithelium (exceptions)	Epithelium	Epithelium	Epithelium	Epithelium	Epithelium	Lamina propria cell	Lamina propria cell
Oocysts shed	Unsporulated	Unsporulated	Unsporulated	Unsporulated	Unsporulated	Unsporulated?	Sporulated, usually as sporocysts	Sporulated, usually as sporocysts
Prepatent or patent period	Days	Days	Days	Days	Days	Days	Weeks or months	Weeks or months
Intermediate host	None	None	Many hosts	Many hosts	Many hosts	Many hosts	Specific (except in birds)	Variable
Tissue cysts	None	None	Monozoic	Polyzoic	Polyzoic	Polyzoic	Polyzoic	Polyzoic
Cystozoites	None	None	Bradyzoite	Bradyzoite	Bradyzoite	Bradyzoite	Metrocytes bradyzoite	Metrocytes bradyzoite
Location of cyst	—	—	Lymphoid muscle	Many cells	Striated muscle	Many tissues	Striated muscle	Neurons
Cyst wall	—	—	Thin, within cell	Thin, within cell	Thin, within cell	Thick, surrounds cell	Thin or with spines of taxonomic value	Thin, within cell
Host cell nucleus	—	—	±	±	±	Undergoes hypertrophy and hyperplasia	±	Enlarged
Infectivity								
Oocyst to:								
definitive host	Yes	Yes	Yes	Yes	No	No	No	No
intermediate host	None	None	Yes	Yes	Yes	Yes	Yes	Yes
Tissue cysts to:								
definitive host	—	—	Yes	Yes	Yes	Yes	Yes	Yes
intermediate host	—	—	No	Yes	No	Variable	No	No
Infectivity of tissue cysts	—	—	Immediate	Immediate	Immediate	Immediate	After weeks/ months	After weeks/ months

*The genus *Arthrocystis*, with spherical zoites and cysts articulated like bamboo, has not been studied experimentally. Some believe it to be a stage of *Leucocytozoon.*

From J. K. Frenkel, et al. "Beyond the oocyst: Over the molehills and mountains of Coccidialand," in *Parasitol. Today* 3:250–251. Copyright © 1987. Elsevier Publications: Cambridge U. K. Reprinted with permission of publisher and author.

FIGURE 8.6

Life cycle of the chicken coccidian *Eimeria tenella*. A sporozoite (*1*) enters an intestinal epithelial cell (*2*), rounds up, grows, and becomes a first-generation schizont (*3*). This produces a large number of first-generation merozoites (*4*), which break out of the host cell (*5*), enter new intestinal epithelial cells (*6*), round up, grow, and become second-generation schizonts (*7, 8*). These produce a large number of second-generation merozoites (*9, 10*), which break out of the host cell (*11*). Some enter new host intestinal epithelial cells and round up to become third-generation schizonts (*12, 13*), which produce third-generation merozoites (*14*). The third-generation merozoites (*15*) and the great majority of second-generation merozoites (*11*) enter new host intestinal epithelial cells. Some become microgametocytes (*16, 17*), which produce a large number of microgametes (*18*). Others turn into macrogametes (*19, 20*). The macrogametes are fertilized by the microgametes and become zygotes (*21*), which lay down a heavy wall around themselves and turn into young oocysts. These break out of the host cell and pass out in the feces (*22*). The oocysts then sporulate. The sporont throws off a polar body and forms four sporoblasts (*23*), each of which forms a sporocyst containing two sporozoites (*24*). When the sporulated oocyst (*24*) is ingested by a chicken, the sporozoites are released (*1*).

From N. D. Levine, *Protozoan parasites of domestic animals and of man,* 2d edition. Copyright © 1961 by Burgess Publishing Co., Minneapolis. Reprinted by permission.

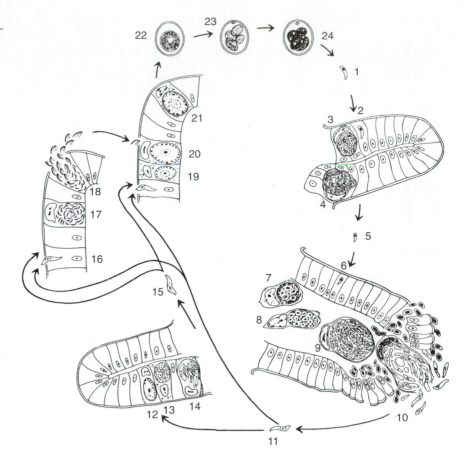

large amount of lipid that seems to be an important source of energy for the sporozoites during their sojourn outside a host.[76] The sporocyst wall consists of a thin outer granular layer surrounded by two membranes and a thick, fibrous inner layer. At one end of the sporocyst, a small gap in the inner layer is plugged with a homogeneous **Stieda** body. In some species additional plug material underlies the Stieda body and is designated the **substiedal body.** When the sporocysts reach the intestine of a new host or are treated in vitro with trypsin and bile salt, the Stieda body is digested, the substiedal body pops out, and the sporozoites wriggle through the small opening thus created.[62] In addition to the apical complex, nucleus, and so on, the sporozoites themselves may contain one or more prominent **refractile bodies** of unknown function.

Eimeria spp. are often restricted to a certain host species, but a given species of *Eimeria* also may be limited to certain organ systems, narrow zones in that system, specific kinds of cells in that zone, and even specific locations within the cells.[52] One species may be found only at the tips of the intestinal villi, another in the crypts at the bases of the villi, and a third in the interior of the villi, all in the same host. Some species develop below the nucleus of the host cell, others above it, and a few within it. Most coccidia inhabit the digestive tract, but a few are found in other organs, such as the liver and kidneys.

The number of species of coccidia is staggering. Levine and Ivens[44] recognized 204 species of *Eimeria* in rodents, but they estimated that there must be at least 2700 species of *Eimeria* in rodents alone. It has been calculated that *Eimeria* has been described from only 1.2% of the chordate species and 5.7% of the mammals in the world.[37] If all chordates were examined, Levine estimated that 34,000 species of *Eimeria* would be found—3500 of them in mammals—and if all animals were examined, a total of 45,000 species would be found of this single genus.[34] One never knows where scientists are likely to discover new *Eimeria* species; recent descriptions have come from hosts as disparate as marine fish, tropical lizards, Argentine rodents, and even domestic animals whose parasites one would expect had been studied extensively for decades.

- *Eimeria tenella.* *Eimeria tenella* (Fig. 8.6) lives in the epithelium of the intestinal ceca of chickens, where it destroys tissues, causing a high mortality rate in young birds. This and related species are of such consequence that all commercial feeds for young chickens now contain anticoccidial agents ("coccidiostats").

• *Biology and Course of Infection.* Chickens become infected when they swallow food or water that is contaminated with sporulated oocysts. The micropyle ruptures in the bird's gizzard, and activated sporozoites escape the sporocyst in the small intestine. Once in a cecum the sporozoites enter the cells of the surface epithelium and pass through the basement membrane into the lamina propria.

There they are engulfed by macrophages that carry them to the glands of Lieberkuhn. They then escape the macrophages and enter into a glandular epithelial cell of the crypt, where they locate between the nucleus and the basement membrane.

Within the epithelial cell the sporozoite becomes a trophozoite, feeding on the host cell and enlarging to become a meront. During merogony the meront separates into about 900 first-generation merozoites, each about 2 to 4 μm long. These break out into the lumen of the cecum about two and one-half to three days after infection, destroying the host cell. Surviving first-generation merozoites enter other cecal epithelial cells to initiate the second endogenous generation. Merozoites develop into meronts that live between the nuclei and free borders of host cells. A great many merozoites will form meronts in the lamina propria under the basement membrane.

About 200 to 350 second-generation merozoites, each about 16 μm long, are then formed by merogony. These rupture the host cell and enter the lumen of the cecum about five days after infection. Some of these merozoites enter new cells to initiate a third generation of merogony below the nucleus, producing 4 to 30 third-generation merozoites, each about 7 μm long. Many merozoites are engulfed and digested by macrophages during these cycles of merogony.

Some of the second-generation merozoites enter new epithelial cells in the cecum to begin gametogony. Most develop into macrogametocytes. Both male and female gamonts lie between the host cell nucleus and the basement membrane. Microgametocytes bud to form many slender, biflagellated microgametes that leave the host cell and enter cells containing macrogametes, where fertilization takes place.

Macrogametes have many granules of two types. Immediately after fertilization these granules pass peripherally toward the surface of the zygote, flatten out, and coalesce to form first the outer and then the inner layer of the oocyst wall. This coalescence takes place within the cell membrane of the zygote, and the membrane becomes the covering of the outer wall. Oocysts are then released from the host cells, move with the cecal contents into the large intestine, and pass out of the body with the feces. Oocysts appear in the feces within six days after infection. Oocysts are passed for several days because not all second-generation merozoites reenter host cells at the same time; furthermore, oocysts often remain in the lumen of the cecum for some time before moving to the large intestine.

Freshly passed oocysts each contain a single cell, the **sporont.** Sporogony (often called **sporulation**), or development of the sporont into sporocysts and sporozoites, is exogenous (occurs outside the host). The sporont is diploid, and the first division is reductional, a polar body being expelled. The haploid number of chromosomes is two. The sporont divides into four **sporoblasts,** each of which forms a sporocyst containing two sporozoites. Sporulation takes two days at summertime temperatures, whereupon the oocysts are infective.

Although the organisms can survive anaerobic conditions, as might be found in freshly passed feces, the metabolism of sporulation is an aerobic process and will not proceed in the absence of oxygen.[76] Oxygen consumption is high at first but falls steadily as sporulation is completed. The organisms have large amounts of glycogen, which is rapidly consumed, and measurements of the respiratory quotient indicate that they depend primarily on carbohydrate oxidation for energy during sporoblast formation and then change over to lipid for energy as sporulation is completed. Thus the biochemistry suggests an interesting developmental control in metabolism: First a rapid burst of energy fuels sporulation, and then a shift to a low level of maintenance metabolism conserves resources until a new host is reached.

Eimeria infections are self-limiting; that is, asexual reproduction does not continue indefinitely. If the chicken survives through oocyst release, it recovers. It may become reinfected, but a primary infection usually imparts some degree of protective immunity to the host.

The number of oocysts produced in any infection can be astounding. Theoretically one oocyst of *E. tenella,* containing eight sporozoites, can produce 2.52 million second-generation merozoites, most of which will become macrogametes and thereby oocysts. However, many merozoites and sporozoites are discharged with the feces before they can penetrate host cells, and many are destroyed by host defenses. A complete replacement of cecal epithelium normally occurs about every two days, so any merozoite or sporozoite that invades a cell that is about to be sloughed is out of luck. Young chickens are more susceptible to infection and discharge more oocysts than do older birds.

• *Pathogenesis and Economic Importance.* Cecal coccidiosis is a serious disease that causes a bloody diarrhea, sloughing of epithelium, and commonly death of the host.[47] Emergence of merozoites destroys tissues and cells. The large schizonts, especially when packed close together, disrupt the delicate capillaries that service the epithelium, further altering normal physiology of the tissues and also causing hemorrhage (Fig. 8.7). A hard core of the clotted blood and cell debris often plugs up the cecum, causing necrosis of that organ (Fig. 8.8). Birds that are not killed outright by the infection become listless and are susceptible to predation and other diseases. The USDA estimated that loss to poultry farmers in the United States alone in the mid-1980s was $80 million, counting the extra cost of medicated feeds and added labor. Annual broiler production in the United States is about 4.2 trillion birds.[48] Worldwide expenditures for coccidiostats are estimated to be $250 to $300 million annually.[55]

Many useful drugs are available as prophylaxes against coccidiosis. However, once infection is established, there is no effective chemotherapy. Therefore a coccidiostat must be administered continuously in food or water to prevent an outbreak of disease. These compounds affect the schizont primarily, so the host can still build up an immunity in response to invading sporozoites. Some success has been obtained in producing coccidia-resistant chickens, but these efforts are still in the research phase.[60] Vaccination methods using attenuated parasite strains or irradiated oocysts also show some promise in experimental studies.

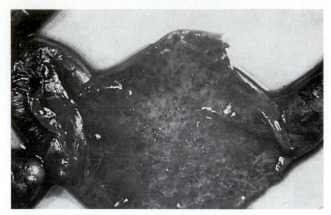

FIGURE 8.7

Cecum of chicken, opened to show patches of hemorrhage caused by *Eimeria tenella*.

Courtesy of James Jensen.

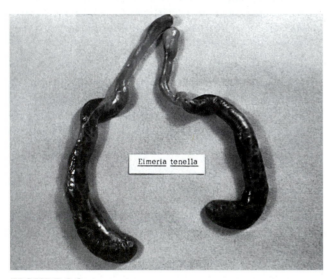

FIGURE 8.8

Ceca of chicken infected by *Eimeria tenella*. Note distention caused by clotted blood and debris and dark color from hemorrhage.

Courtesy of James Jensen.

- *Other Eimeria Species.* Levine[38] summarized the species that parasitize domestic animals, some of the most common of which are *E. auburnensis* and *E. bovis* in cattle, *E. ovina* in sheep, *E. debliecki* and *E. porci* in pigs, *E. stiedai* in rabbits, *E. necatrix* and *E. acervulina* in chickens, *E. meleagridis* in turkeys, and *E. anatis* in ducks. All have life cycles similar to that of *E. tenella* but differ in details of their courses of infection and effects on the host. Since the late 1980s additional species have been described from stingrays, marine fish, freshwater fish, frogs, lizards, snakes, turtles, doves, llamas, gazelles, and manatees. Clearly members of the genus *Eimeria* know few limits, evolutionarily speaking, on the types of hosts they colonize.

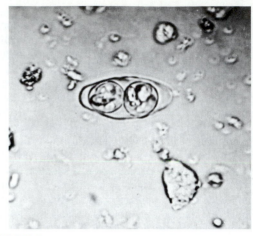

FIGURE 8.9

Isospora belli oocyst. It averages 35 by 9 µm.

From J. W. Beck and J. E Davies. *Medical Parasitology*, 3d. ed. The C. V. Mosby Co.

- *Isospora Group.* Oocysts of *Isospora* species contain two sporocysts, each with four sporozoites. Oocysts of the genera *Toxoplasma, Sarcocystis, Levineia, Besnoitia, Frenkelia,* and *Arthrocystis* have similarly constructed oocysts, but these parasites are heteroxenous, with vertebrate intermediate hosts. For this reason they are placed in the family Sarcocystidae, discussed next.

    For an enlightening discussion of the taxonomic problems surrounding *Isospora* and the sarcocystid genera, see the lively exchanges between Baker,[6] Frenkel et al.,[24] and Levine.[43] It is only within the past two decades that we have begun to understand the taxonomic position of many of these parasites. Taxonomic difficulties (and naivetes!) resulted from ignorance of the life cycles, and the confusion was compounded through description of different life-cycle stages as different species (an event not unknown in other groups of parasites).

    Isospora contains far fewer species than *Eimeria,* most of them in birds, but it includes a human parasite, *I. belli* (Fig. 8.9). Infection with *I. belli* is rather uncommon, and most cases have been reported from the tropics. It can cause severe disease with fever, malaise, persistent diarrhea, and even death, especially in AIDS patients.[16,18,22] Species of *Atoxoplasma* are also parasites of birds. They produce *Isospora*-like oocysts and have long been confused with that genus.[40] *Atoxoplasma* has a monoxenous life cycle with merogony in blood and intestinal cells and gametogony in intestinal cells of the same host. Sporogony is outside the host; infection occurs through swallowing the oocyst.

    Table 8.1 summarizes the present state of our knowledge about the major genera of Eimeriidae and Sarcocystidae that possess two sporocysts per oocyst.

Family Sarcocystidae. Members of this family differ from those of the Eimeriidae principally in having heteroxenous life cycles (Table 8.1). Asexual development occurs in vertebrate intermediate hosts, whereas other vertebrates, mainly carnivorous mammals and birds, are definitive hosts. The oocyst contains two sporocysts, each with four sporozoites. There are well over a hundred described species of *Sarcocystis,* and new species, as well as new genera, of Sarcocystidae are being discovered regularly.[6,42] Their classification has been changing rapidly as new information becomes available. One genus, *Toxoplasma,* is of importance to humans. Others are of veterinary importance.

The **basic life cycle** found throughout this family is as follows: Oocysts from a definitive host sporulate and are swallowed by an intermediate host. Sporozoites released from the cysts infect various tissues and rapidly undergo endodyogeny to form merozoites, also known as **tachyzoites.** These can infect other tissues in the body, such as muscles, fibroblasts, the liver, and nerves. Asexual reproduction in these tissues is much slower than in the original site, and the parasite develops large, cystlike accumulations of merozoites that are now called **bradyzoites.** The cyst itself is called a **zoitocyst,** or simply a **tissue cyst.** A definitive host is infected when it eats meat containing bradyzoites or, rarely, tachyzoites or in some cases when it swallows a sporulated oocyst. Tachyzoites and bradyzoites have antigenic differences.[49]

- *Toxoplasma gondii.*   Like *Trypanosoma cruzi, Toxoplasma gondii* (Figs. 8.10 to 8.13) was discovered before it was known to cause disease in humans. It was first discovered in 1908 in a desert rodent, the gondi, in a colony maintained in the Pasteur Institute in Tunis. Since then the parasite has been found in almost every country of the world in many species of carnivores, insectivores, rodents, pigs, herbivores, primates, and other mammals, as well as in birds. We now realize that it is cosmopolitan in the human population and can cause disease. The importance of the organism as a human pathogen has stimulated a huge amount of research in recent years. A 1963 bibliography on the subject contained 3706 references, and Jacobs[30] reported more than 2000 references for the years 1967 through 1972. Between 1968 and 1975 an additional 12,500 references were added. Since the mid-1980s *Toxoplasma gondii* has joined a number of other parasites recognized as complicating factors for immunosuppressed patients. This once obscure protozoan parasite of an obscure African rodent has become one of many new exciting subjects whose importance has been revealed by research.

• *Biology. Toxoplasma* is an intracellular parasite of many kinds of tissues, including muscle and intestinal epithelium. In heavy acute infections the organism can be found free in the blood and peritoneal exudate. It may inhabit the nucleus of the host cell but usually lives in the cytoplasm. The life cycle includes intestinal-epithelial (**enteroepithelial**) and **extraintestinal** stages in domestic cats and other felines, but extraintestinal stages only in other hosts. Sexual reproduction of *Toxoplasma* occurs while in the cat, and only asexual reproduction is known while in other hosts.

Extraintestinal stages begin when a cat or other host ingests bradyzoites. Ingested tachyzoites or sporocysts also sometimes are infective. Intrauterine infection is possible (see discussion of pathogenesis). The oocyst is 10 to 13 μm by 9 to 11 μm and is basically similar in appearance to those of isosporan species (Fig. 8.11). There is no oocyst residuum or polar granule, and the sporocysts have a sporocyst residuum but no Stieda body. The sporozoites escape from the sporocysts and the oocyst in the small intestine. In cats some of the sporozoites enter epithelial cells and remain to initiate the enteroepithelial cycle, whereas others penetrate through the mucosa to begin development in the lamina propria, mesenteric lymph nodes and other distant organs, and white blood cells. In hosts other than cats there is no enteroepithelial development; the sporozoite enters a host cell and begins multiplying by endodyogeny. These rapidly dividing cells in acute infections are called tachyzoites (Fig. 8.12). Eight to 32 tachyzoites accumulate within the host cell's parasitophorous vacuole before the cell disintegrates, releasing the parasites to infect new cells. These accumulations of tachyzoites in a cell are called **groups.** Tachyzoites apparently are relatively less resistant to stomach secretions; therefore they are less important sources of infection than are other stages.

As infection becomes chronic, the zoites that affect brain, heart, and skeletal muscles multiply much more slowly than in the acute phase. They are now called bradyzoites, and they accumulate in large numbers within a host cell. They become surrounded by a tough wall and are called zoitocysts or tissue cysts (Fig. 8.13). Cysts may persist for months or even years after infection, particularly in nervous tissue. Cyst formation coincides with the time of development of immunity to new infection, which is usually permanent. If immunity wanes, released bradyzoites can boost the immunity to its prior level. This protection against superinfection by the presence of the infectious agent in the body is called **premunition.** Immunity to *Toxoplasma* is of both the antibody and cell-mediated types; the latter is more important. The tough, thin cyst wall, except when the cyst breaks down, effectively separates the parasite from the host, and the cyst does not elicit an inflammatory reaction. The cyst wall and its bradyzoites develop intracellularly, but they may eventually become extracellular because of distention and rupture of the host cell. Bradyzoites are resistant to digestion by pepsin and trypsin, and when eaten they can infect a new host.

Enteroepithelial stages are initiated when a cat ingests zoitocysts containing bradyzoites, oocysts containing sporozoites, or occasionally tachyzoites. Another possible means of epithelial infection is by migration of extraintestinal zoites into the intestinal lining within the cat. Once inside an epithelial cell of the small intestine or colon, the parasite becomes a trophozoite that grows and prepares for merogony. At least five different strains have been studied well enough to allow characterization of the enteroepithelial stages.[23] These strains differ in duration

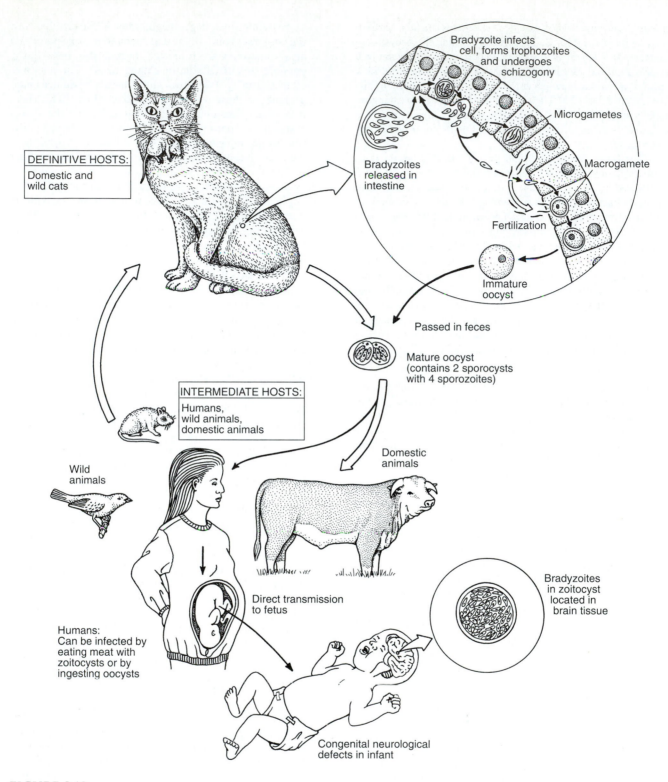

FIGURE 8.10

Life cycle and transmission of *Toxoplasma gondii.*

Drawing by William Ober and Claire Garrison.

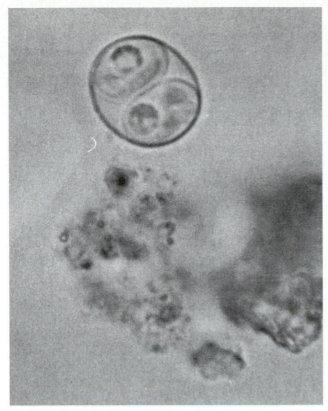

FIGURE 8.11

Oocyst of *Toxoplasma gondii* from cat feces. It is 10 to 13 by 9 to 11 μm. Courtesy of Harley Sheffield.

FIGURE 8.12

Tachyzoites *Toxoplasma gondii*. They are about 7 by 12 μm.

of stages, number of merozoites produced, shape, and other details. Basically, from 2 to 40 merozoites are produced by merogony, endopolyogeny, or endodyogeny, and these initiate subsequent asexual stages. The number of merogonous cycles is variable, but gametocytes are produced within 3 to 15 days after cyst-induced infection. Gametocytes develop throughout the small intestine but are more common in the ileum. From 2% to 4% of the gametocytes are male; each produces about 12 microgametes. Oocysts appear in the cat's feces from three to five days after infection by cysts, with peak production occurring between days five and eight. Oocysts require oxygen for sporulation; they sporulate in one to five days. The extraintestinal development can proceed simultaneously with the enteroepithelial in the cat. Ingested bradyzoites penetrate the intestinal wall and multiply as tachyzoites in the lamina propria. They may disseminate widely in the extraintestinal tissues of the cat within a few hours of infection.[16]

One final note of interest is that in the Pasteur Institute in Tunis in 1908, when gondis were brought in from the field and died, the source of their infection was never established. However, it is known that at the time a cat had been roaming the laboratory.[30]

• *Pathogenesis.* In view of the fact that antibody to *Toxoplasma* is widely prevalent in humans throughout the world yet clinical toxoplasmosis is less common, it is clear that most infections are asymptomatic or mild. Several factors influence this phenomenon: the virulence of the strain of *Toxoplasma,* the susceptibility of the individual host and of the host species, the age of the host, and the degree of acquired immunity of the host. Pigs are more susceptible than cattle; white mice are more susceptible than white rats; chickens are more susceptible than most carnivores. The reasons for natural resistance or susceptibility to infection are not known. Occasionally, circumstances conspire to make a mild case important, as when Martina Navratilova lost the U.S. Open tennis championship and $500,000 in 1982 when she had toxoplasmosis.

About 13% of the world population is infected. In 1976 the global prevalence of toxoplasmosis was estimated at over 500 million.[29] In countries like France, where raw meat is popular, the prevalence may be higher; in Paris, for example, it may be as high as 50%. In the United States, about 3500 infants are born each year with severe infections, costing $300 million annually in care and treatment.[61]

Tachyzoites proliferate in many tissues and this rapid reproduction tends to kill host cells at a faster rate than

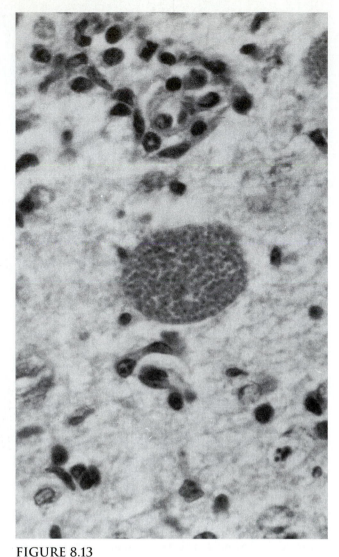

FIGURE 8.13

Zoitocyst of *Toxoplasma gondii* in the brain of a mouse.

Courtesy of Sherwin Desser.

does the normal turnover of such cells. Enteroepithelial cells, on the other hand, normally live only a few days, especially at the tips of the villi. Therefore, the extraintestinal stages, particularly in sites such as the retina or brain, tend to cause more serious lesions than do those in the intestinal epithelium.

Since there seems to be an age resistance, infections of adults or weaned juveniles are asymptomatic, although exceptions occur. Asymptomatic infections can suddenly become fulminating if immunosuppressive drugs such as corticosteroids are employed for other conditions. Symptomatic infections can be classified as acute, subacute, and chronic.

In most **acute infections** the intestine is the first site of infection. Cats infected by oocysts usually show little disease beyond loss of individual epithelial cells, and these are rapidly replaced. Actually, oocysts probably are of little importance in infecting cats. In massive infections, however, intestinal lesions can kill kittens in two to three weeks. The first extraintestinal sites to be infected in both cats and other hosts, including humans, are the mesenteric lymph nodes and the parenchyma of the liver. These sites, too, experience rapid regeneration of cells and perform an effective preliminary screening of the parasites. The most common symptom of acute toxoplasmosis is painful, swollen lymph glands in the cervical, supraclavicular, and inguinal regions. This symptom may be associated with fever, headache, muscle pain, anemia, and sometimes lung complications. This syndrome can be mistaken easily for the flu. Acute infection can, although rarely does, cause death. If immunity develops slowly, the condition can be prolonged and is then called *subacute.*

In **subacute infections** pathogenic conditions are extended. Tachyzoites continue to destroy cells, causing extensive lesions in the lung, liver, heart, brain, and eyes. Damage may be more extensive in the central nervous system than in unrelated organs because of lower immunocompetence in these tissues.

Chronic infection results when immunity builds up sufficiently to depress tachyzoite proliferation. This coincides with the formation of zoitocysts. These cysts can remain intact for years and produce no obvious clinical effect. Occasionally, a cyst wall will break down, releasing bradyzoites; most of these are killed by host reactions, although some may form new cysts. Death of the bradyzoites elicits an intense hypersensitive inflammatory reaction, the area of which, in the brain, is gradually replaced by nodules of glial cells. If many such nodules are formed, the host may develop symptoms of chronic encephalitis, with spastic paralysis in some cases. Chronic active or relapsing infections of retinal cells by tachyzoites causes blind spots and extensive infection of the central, macular area, which may lead to blindness. Cysts and cyst rupture in the retina can also lead to blindness. Other kinds of extensive pathological conditions such as myocarditis, with permanent heart damage and with pneumonia, can occur in chronic toxoplasmosis.

In the immunocompetent person *T. gondii* ordinarily is kept at bay by cell-mediated immunity. When an infected person becomes immunosuppressed, the organism will disseminate rapidly, which may lead to ocular toxoplasmosis and to fatal CNS disorders such as encephalitis. Any long-term steroid therapy, such as is given to some cancer patients, can result in disseminated toxoplasmosis. Currently, *T. gondii* is a serious opportunistic infection in AIDS. Death usually results from cyst rupture with continued multiplication of tachyzoites.

Another tragic form of this disease is **congenital toxoplasmosis.** If a mother contracts acute toxoplasmosis at the time of her child's conception or during pregnancy, the organisms will often infect her developing fetus. Fortunately, most neonatal infections are asymptomatic, but a significant number cause death or disability to newborns. It is generally assumed that *Toxoplasma* crosses the placental barrier from the mother's blood.

The transmission rate to the fetus from a maternal infection is about 45%. Of those infected, about 60% will be subclinical, 9% may die, and 30% may suffer severe damage such as hydrocephalus, intracerebral calcification, retinochoroiditis, and mental retardation. However, even subclinical cases may develop into ocular toxoplasmosis later in life.

Stillbirths and spontaneous abortions may result from fetal infection with *Toxoplasma* in humans and other animals. Sheep seem to be particularly susceptible, and *Toxoplasma*-caused abortions in this host often reach epidemic proportions. Congenital toxoplasmosis probably accounts for half of all ovine abortions in England and New Zealand.[9]

In a study of more than 25,000 pregnant women in France, no case of congenital toxoplasmosis was found whenever maternal infection occurred before pregnancy.[12] However, of 118 cases of maternal infection near the time of or during pregnancy, there were nine abortions or neonatal deaths without confirmation by examination of the fetus, 39 cases of acute congenital toxoplasmosis with two deaths, and 28 cases of subclinical infection. The remainder of fetuses were free of infection. Maternal infection in the first three months of pregnancy results in more extensive pathogenesis, but transmission to the fetus is more frequent if the maternal infection occurs in the third trimester.

In cases of twins one may have severe symptoms and the other no overt evidence of infection. In children who survive infection there is often congenital damage to the brain, manifested as mental retardation and retinochoroiditis. Thus toxoplasmosis is a major cause of human birth defects, probably causing more congenital abnormalities in the United States than rubella, herpes, and syphilis combined.

• *Diagnosis and Treatment.* Specific diagnosis in humans is based on one or more laboratory tests. Demonstration of the organism at necropsy or biopsy is definitive. Intraperitoneal inoculation of a biopsy of lymph node, liver, or spleen into mice is useful and accurate, as is culture of parasites in fibroblast cells in vitro. Demonstration of specific antibody, using an enzyme-linked, immunosorbent assay (ELISA), is also employed, and molecular methods are currently used in the preparation of antigen reagents. In recent years attempts have been made to develop diagnostic techniques that rely on nucleic acid probes to detect small amounts of parasite DNA. These probes are made using PCR (polymerase chain reaction) amplification of parasite ribosomal DNA.[26,54] Such techniques allow for the detection of single organisms in tissue samples (0.1 pg *T. gondii* DNA).

Pyrimethamine and sulfonamides given together are drugs widely used against *Toxoplasma*. They act synergistically by blocking the pathway involving *p*-aminobenzoic acid and the folic-folinic acid cycle, respectively. Possible side effects of this treatment are thrombocytopenia and/or leukopenia, but these can be avoided by administration of folinic acid and yeast to the patient. Vertebrates can employ presynthesized folinic acid, whereas *Toxoplasma* cannot. Experimental chemotherapy may involve additional drugs in combination with the aforementioned compounds.[2]

• *Epidemiology.* In the United States the prevalence of chronic, asymptomatic toxoplasmosis is age related,

increasing 0.5% to 1.0% per year of age.[33] Although clinical toxoplasmosis usually affects only scattered individuals, small epidemics occur from time to time. For example, in the spring of 1968, several Cornell University Medical College students were infected simultaneously by wolfing down undercooked hamburgers between classes.[31] In 1969, at a university in São Paulo, Brazil, 110 persons were diagnosed with acute toxoplasmosis in a three-month period. Most admitted to eating undercooked meat. Therefore, raw meat seems to be an important source of infection. One large sheep or pig might conceivably be the source of an epidemic at any time. Considering the custom of backyard cooking and Americans' fondness for rare beef, many cases of toxoplasmosis thus may be acquired every day.

Although beef is certainly a potential source of infection, pork and lamb are much more likely to be contaminated. Freezing at $-14°$ C for even a few hours apparently will kill most cysts. To avoid a multitude of parasites, persons who insist on eating undercooked meat would do well to see that it has been hard frozen.

Feral and domestic cats will continue to be a source of infection of humans. Stray cats lead to problems of several kinds and are reservoirs of several diseases; efforts should be made to keep their numbers down. A more difficult problem to resolve is the household pet, the tabby that spends most of its time in a close relationship with its owners. Any cat, no matter how well fed and protected, may be passing oocysts of *Toxoplasma*, although for only a few days after infection. The possibilities are particularly alarming if someone in the house becomes pregnant. Certainly, a woman who knows she is pregnant should never empty the litterbox or clean up after the cat's occasional indiscretion. (Emptying the box every two days should help, but since cysts require one to three days to sporulate, it is better to have someone else do the job.) Having a cat tested for antibodies is impractical, for their presence does not correlate with shedding of oocysts. Also, because children's sandboxes become litter boxes for neighborhood cats, they should have tightly fitting covers. Covers also will protect children from larva migrans from hookworms and ascaridoid juveniles. Any soil reservoir of oocysts is a most important source of infection of humans.

Filth flies and cockroaches are capable of carrying *Toxoplasma* oocysts from cat feces to the dinner table.[74] Earthworms may serve to move oocysts from where cats have buried them to ground surface.

Toxoplasma tachyzoites have been isolated in humans from nasal, vaginal, and eye secretions; milk; saliva; urine; seminal fluid; and feces. The role of any of these in spreading infection is unknown, but it seems reasonable that any or all may be involved. Whole blood or leukocyte transfusions and organ transplants are also potential sources of serious infection, given that recipients may be immunodeficient because of disease or treatment.

• **Sarcocystis Species and Related Parasites.**   *Sarcocystis* spp. have been known from their zoitocysts in muscle of reptiles, birds, and mammals since the late nineteeth century; but the

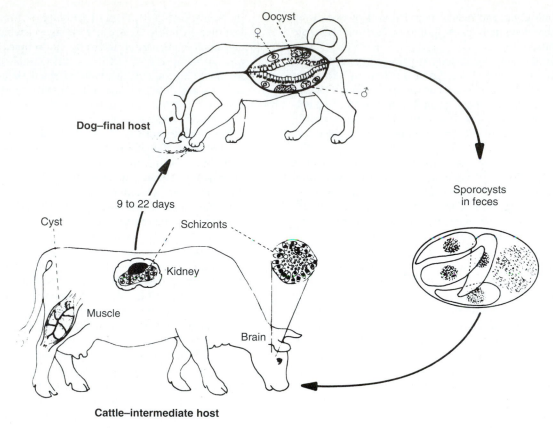

Oocyst

Dog–final host

9 to 22 days

Cyst

Schizonts

Sporocysts in feces

Kidney

Muscle

Brain

Cattle–intermediate host

FIGURE 8.14

Life cycle of *Sarcocystis cruzi* of cattle with the dog exemplifying the definitive host. Dogs, wolves, coyotes, raccoons, and foxes shed sporulated oocysts or sporocysts in their feces after eating infected bovine musculature. Cattle become infected by ingesting sporocysts from the feces of carnivores. Generalized infection occurs in bovine tissues, and schizonts are formed in many tissues, especially in the kidneys and brain. After schizogonic cycles, cysts are formed in the musculature in two months. Current evidence indicates that canines become infected by ingesting only mature cysts. Sporocysts are noninfectious to definitive hosts.

From J. P. Dubey, "A review of *Sarcocystis* of domestic animals and of other coccidia of cats and dogs," in *Am. Vet. Med. Assoc.* 169:1061–1078. Copyright © 1976. Reprinted with permission of the publisher.

life cycle remained obscure until 1972 when it was discovered that the bradyzoites would lead to the development of coccidian gametes in cell culture and of oocysts after being fed to cats.[20,63] Since then, it has been found that some species of what was called *Isospora* were in fact stages of *Sarcocystis* in their definitive hosts (for example, *S. bigemina* and *S. hominis*),[39] and what had been considered single species of *Sarcocystis* from particular hosts comprised several species in each. For example, oocysts of the three following species cannot be distinguished morphologically: *Sarcocystis cruzi* (syn. *S. bovicanis*), *S. tenella* (syn. *S. ovicanis*), and *S. meischeriana* (syn. *S. suicanis*). For a review of the taxonomy of the genus see Levine.[42]

Sarcocystis spp. are obligately heteroxenous, with a herbivorous intermediate host—such as various species of reptiles, birds, small rodents, and hoofed animals—and a carnivorous definitive host (Fig. 8.14). When sporozoites are released from sporocysts consumed by

the intermediate host, they penetrate the intestinal epithelium, are distributed through the body, and invade the endothelial cells of blood vessels in many tissues. There they undergo merogony, and additional merogonous generations may ensue. Zoitocysts (tissue cysts) (Fig. 8.15) then form in skeletal and cardiac muscle and occasionally the brain. The cysts are also known as **sarcocysts** or **Miescher's tubules,** and some species are large enough to be seen by the unaided eye. They usually have internal septa and compartments. They are elongated, cylindroid, or spindle shaped, but they may be irregularly shaped. They lie within a muscle fiber, in the same plane as the muscle bundle. The overall size varies, reaching 1 cm in diameter in some cases, but they usually are 1 to 2 mm in diameter and 1 cm or less long.

The structure of the cyst wall varies among "species" and in different stages of development of the parasite. In some cases the outer wall is smooth; in others it has an

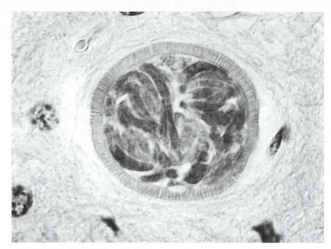

FIGURE 8.15

Cross section of zoitocyst of *Sarcocystis tenella* in muscle of experimentally infected sheep. (× 6000.)

From J. P. Dubey et al., "Development of sheep-canid cycle of *Sarcocystis tenella*," in *Canad. J. Zool.* 60:2464–2477. Copyright © 1982.

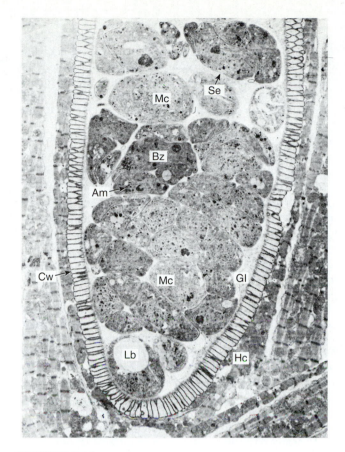

FIGURE 8.16

Transmission electron micrograph of *S. tenella* sarcocyst with fully formed wall composed of cytophaneres (**CW**); note indistinct granular septum (**Se**), bradyzoites (**Bz**), metrocytes (**Mc**), amylopectin (**Am**), and lipid bodies (**Lb**).
(× 6000.)

From J. P. Dubey et al., "Development of sheep-canid cycle of *Sarcocystis tenella*," in *Canad. J. Zool.* 60:2464–2477. Copyright © 1982.

outer layer of fibers, the **cytophaneres,** which radiate out into the muscle (Fig. 8.16). The origin of the cyst wall is controversial: Some authors conclude that it is of host origin; others maintain that it is of parasite origin. It may well be derived from both sources. Two distinct regions can be distinguished in the cyst. The peripheal region is occupied by globular **metrocytes.** After several divisions the metrocytes give rise to the more elongated bradyzoites. The bradyzoites resemble typical coccidian merozoites except that they have a larger number of micronemes; the metrocytes are also structurally similar but lack rhoptries and micronemes. Only bradyzoites are infective to definitive hosts.

When the zoitocyst is consumed by the definitive host, its wall is digested away, and the bradyzoites penetrate the lamina propria of the small intestine. There they undergo gamogony without an intervening merogonic generation. The male gametes penetrate the female gametes, and the oocyst sporulates in the lamina propria. The oocyst wall is thin and is usually broken during passage through the intestine; thus sporocysts rather than oocysts are normally passed in the feces. Sporocysts can infect the intermediate hosts but not the definitive hosts.

Humans have been named the definitive hosts for some species *(S. hominis, S. suihominis),* but zoitocysts of several unidentified species occasionally are found in human muscle.[8,56] Discovery of a generalized infection in dogs indicates that carnivores may develop infections typical of those found in intermediate hosts.[17] *Sarcocystis* species are globally distributed, being found in animals as diverse as Florida whitetail deer and the endangered Hierro giant lizard *Gallotia simonyi* on the Canary Islands.[5,7] It is likely

that many *Sarcocystis* species await discovery, especially in sylvatic prey-predator systems. Illustrations of the latter include parasites that utilize small owls and deer mice, kingsnakes and voles, and opossums and birds.[11,19,46] More than 50% of adult swine, cattle, and sheep probably are infected with *Sarcocystis* spp.[16] Some of these parasites are nonpathogenic (Table 8.1), but some may cause serious symptoms, which may include loss of appetite, fever, lameness, anemia, weight loss, and abortion in pregnant animals. Heavily infected animals may die. Flies may act as transport hosts.[51] In vitro cultivation may lead the way to immunization and more accurate diagnosis.[68]

- **Besnoitia Species.** *Besnoitia* species are parasites of vertebrates. At least some species are facultatively heteroxenous, and they produce characteristic cysts with a very thick wall, mostly in the connective tissue of their intermediate hosts. The cyst wall contains several flattened giant host cell nuclei. The natural intermediate hosts of *Besnoitia besnoiti* are apparently cattle and wild hoofed animals in Africa, the Mediterranean countries, China, and the former Soviet Union. The definitive hosts possibly are cats. The organisms can cause considerable economic losses in cattle, and death may occur in severe infections. In the intermediate host, infection proceeds in two distinct phases: acute and chronic. Acute infections produce weakness, fever, and swelling of the lymph nodes. Chronic infections result in various skin problems and, in bulls, infertility.

Family Cryptosporidiidae. This family contains the single genus *Cryptosporidium,* parasites of the brush borders of epithelia of many species of mammals, birds, reptiles, and fishes. About 19 species names have been proposed, but Levine[41] considers valid only *C. muris* in mammals, *C. meleagridis* in birds, *C. crotali* in reptiles, and *C. nasorum* in fishes, although most of the human isolates are now referred to as *C. parvum.* Lack of host specificity is one of the major characteristics that sets *Cryptosporidium* apart from many of the other coccidia. For an excellent review of this organism see Tzipori.[70] Until recently cryptosporidiosis was considered to be an infection in animals other than humans. From 1907, when *Cryptosporidium* was first described by Tyzzer, until 1975, 15 reports describing *Cryptosporidium* infection in eight species of animals were published. From 1989 to 1993 nearly 400 papers have been referenced in *Biological Abstracts*, most of them dealing with infections in humans or domestic animals or with research efforts to improve diagnosis and treatment. Infections have been reported in guinea pigs,[73] turkeys,[28] chickens,[15] calves,[58,72] lambs,[71] and a wide variety of pet and zoo animals.[45] Now *Cryptosporidium* is known as an opportunistic parasite of humans, especially young children, both those who are immunodeficient and those who are immunocompetent. Cryptosporidiosis commonly occurs in patients with AIDS, and it is an important contributory factor in the deaths of some infected patients who have AIDS.[75]

These coccidians are very small (2 to 6 μm), living in the brush border or just under the free-surface membrane of the host gastrointestinal or respiratory epithelial cell (Fig. 8.17). Oocysts are seen only in the feces, and diagnosis is made using formalin-ethyl acetate and hypertonic sodium chloride flotation followed by Ziehl-Neelsen staining methods (fuchsin followed by methylene blue) or by use of Giemsa, nigrosin, merbronime, or light-green. Methods for large-scale purification of oocysts and sporozoites, using differential centrifugation and sucrose or percoll gradients, have been described, especially for use in research.[4,32] Examination by Nomarski or phase-contrast is often preferred over typical light microscopy.

- **Biology.** The spherical oocysts (Fig. 8.18*a*) are 4 to 5 μm wide, are highly refractile, and contain one to eight prominent granules, usually in a small cluster near the margin of the cell. Sporocysts are absent. Each oocyst contains four slender, fusiform sporozoites (Fig. 8.18*b*). Oocysts generally live a long time in water, including seawater, but they do not survive drying.

  When swallowed, the sporozoites excyst in the intestine and invade either epithelial cells of the respiratory system or the intestine (from the ileum to the colon). The meronts are about 7 μm wide and produce eight banana-shaped merozoites and a small residuum. Microgamonts produce 16 rod-shaped, nonflagellated microgametes that are 1.5 to 2.0 μm long. Oocysts are passed as early as five days after infection. Virulence may be strain specific; calves experimentally infected with various human isolates developed infections that were significantly different in their severity.[59]

- **Pathogenesis and Treatment.** In patients with AIDS, the parasites cause profuse, watery diarrhea lasting for several months. Bowel-movement frequency ranges from 6 to 25 per day, and the maximal stool volume ranges from 1 to 17 liters per day. Despite intensive trials in humans and animals, no effective drug treatment for cryptosporidiosis has been found, although paromomycin has been used with mixed success.[3] One study failed to demonstrate that antigens in human breast milk reduced the severity of infection, but experiments using animal models have suggested that oral treatments with monoclonal antibodies and hyperimmune colostrum were effective.[21,57,69] The infection is much less severe in immunocompetent patients, with no symptoms in some and with a self-limiting diarrhea and abdominal cramps lasting from 1 to 10 days in others.

- **Epidemiology.** Infection is by fecal-oral contamination. A number of animals can serve as reservoirs of infection. Current and his coworkers experimentally infected kittens, puppies, and goats with oocysts from an immunodeficient person.[14] They also infected calves and mice using oocysts from infected calves and humans. Finally, they diagnosed 12 infected immunocompetent persons who worked closely with calves that were infected with *Cryptosporidium.* Thus cryptosporidiosis should be considered a zoonosis and in fact may be a fairly common cause of short-term diarrhea in the population at large. The zoonotic potential of *Cryptosporidium* is illustrated by surveys on cattle. In one study Anderson[1] examined nearly a hundred thousand cattle and discovered that 65% of the dairies and 80% of the feedlots had infected animals, and although the overall prevalence was low (less than 5% by state), in some pens 31% of the cattle were passing oocysts. In other studies swimming pools have been implicated as a periodical source of infection.[67]

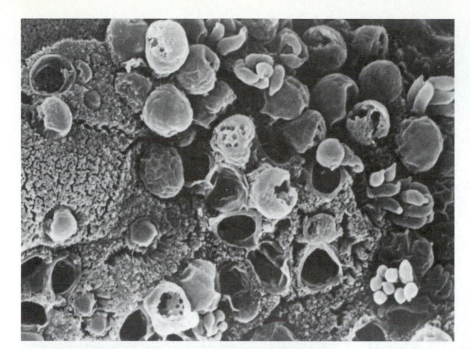

FIGURE 8.17

Oocysts of *Cryptosporidium* in various stages of development in intestinal epithelium. The slender, elongated bodies are emerging sporozoites.

Courtesy of S. Tzipori, Royal Children's Hospital, Australia.

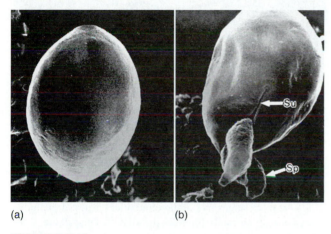

(a)   (b)

FIGURE 8.18

Cryptosporidium parvum (*a*) oocyst; (*b*) three sporozoites (**Sp**) emerging from a suture (**Su**) in a oocyst obtained from a calf and excysted experimentally in vitro.

From D. W. Reduker et al., "Ultrastructure of *Cryptosporidium parvum* oocysts and excysting sporozoites as revealed by high resolution scanning electron microscopy." in *J. Protozool.* 32:708–711. Copyright © 1985 by the Society of Protozoologists.

Cryptosporidium infections dramatically illustrate the manner in which discovery of a medical problem can suddenly focus attention on organisms previously thought obscure and rare. Recognition of opportunistic parasites as a cause of disease in persons with AIDS led to interest in the distribution of such parasites in the immunocompetent and nonsymptomatic population. We now know from a large number of studies that cryptosporidiosis is a serious problem especially in the warmer parts of the world, and it may be one of the three most common causative agents of chronic diarrhea in humans.[13]

Pneumocystis carinii. *Pneumocystis carinii* is a parasite whose taxonomic position is still undetermined nearly a century after its discovery. It is considered a fungus by some workers and a protozoan by others, but we include it here because of its importance as one of a group of "opportunistic parasites" that often cause severe pathology in immunodeficient hosts. *Pneumocystis carinii* produces chitin, has ultrastructural properties similar to some fungi, and stains with certain fungal stains, such as methenamine silver. On the other hand, the species has other structural characteristics resembling *Toxoplasma* and *Plasmodium* and is sensitive to antiprotozoal agents such as pentamidine, trimethoprim-sulfamethoxazole, isethionate, pyrimethamine, and sulfadiazine.[25,77] Its membrane properties, including formation of small pseudopodia, resemble those of the amebas, although in *P. carinii* these pseudopodia interdigitate with membranes of the host's cells, enabling the parasites to remain attached.

Pneumocystis carinii causes interstitial plasma cell pneumonia, especially in the immunosuppressed host, either human or rodent. It occurs commonly in humans of all age groups and is particularly important in the elderly and in infants and children with primary disorders of immune deficiency. It is also a critical problem in patients receiving

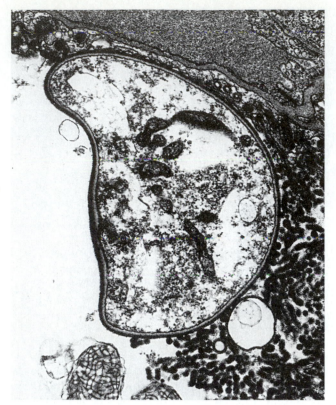

FIGURE 8.19

Transmission electron micrograph of a precyst *Pneumocystis carinii.* Note tightly packed outer layer of cell membrane.

From K. Yoneda et al., *"Pneumocystis carinii:* Freeze-fracture study of stages of the organism," in *Exp. Parasitol.* 53:68–76. Copyright © 1982.

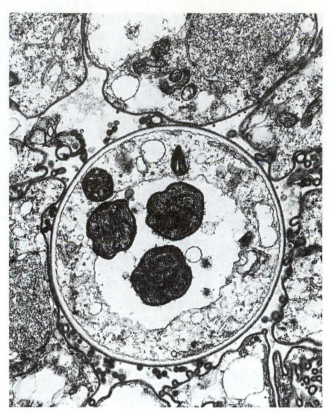

FIGURE 8.20

Transmission electron micrograph of a *Pneumocystis carinii* cyst. Note the intracystic bodies.

From K. Yoneda et al., *"Pneumocystis carinii:* Freeze-fracture study of stages of the organism," in *Exp. Parasitol.* 53:68–76. Copyright © 1982.

cytotoxic or immunosuppressive drugs for lymphoreticular cancers, organ transplantation, and a variety of other disorders. Persons with AIDS are particularly susceptible; 85% of such patients eventually present infections,[10] and pneumocystis pneumonia is a major cause of death in that segment of the population. In properly prepared laboratory rats the organism appears to respond identically as in humans; thus a laboratory model for experimentation is available, although in vitro systems are under development and other host species such as pigs have been used in recent research.[64] In nature the organism is widespread in mammals. Many human infections may be acquired from pets.

- *Morphology and Biology.* *Pneumocystis carinii* assumes three major morphological forms while in the lung: the **trophozoite,** the **precyst,** and the **cyst** stages. The trophozoite is pleomorphic, 1 to 5 μm wide, and has small filopodia that form pockets in the membranes of interstitial cells. The precyst (Fig. 8.19) is oval, with few filopodia, has a clump of mitochondria in its center, and has a nucleus that is more evident in electron micrographs than in light-level preparations. The precyst nucleus undergoes three divisions, after which the nuclei become "delimited" by membranes. The mature cyst (Fig. 8.20) is spherical,

has a thick chitinous membrane, and contains eight "intracystic bodies," which are the infective young trophozoites. Collapsed cysts are crescent shaped (Fig. 8.21).

All three stages live in the interstitial tissues of the lungs and are not normally found in the alveoli. Their mode of reproduction is unknown, although there is evidence of both sexual and asexual reproduction, and some authors consider the initial divisions of precystic nuclei to be meiotic.[53] Occasionally, the parasites break into air spaces, especially in virulent cases seen in immunologically deficient patients, for they are abundant in pulmonary exudate. Human-to-human transmission probably is by aerosol droplets and direct contact. Congenital infection is possible, as *P. carinii* has been found in stillborn infants, newborn germ-free rats, and three-day-old children.[53]

- *Pathogenesis.* In the infected lung the alveolar epithelium becomes partly desquamated, and the alveoli fill with foamy exudate containing parasites. The disease, **interstitial plasma cell pneumonitis,** has a rapid onset associated with fever, cough, rapid breathing, and cyanosis (blue skin around the mouth and eyes). Death is caused by asphyxia. The mortality rate is virtually

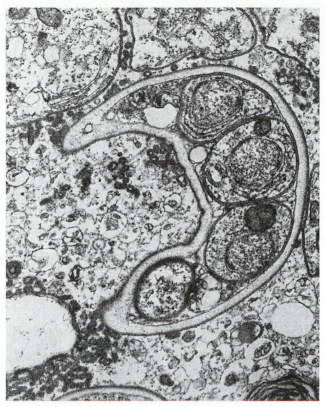

FIGURE 8.21

Transmission electron micrograph of a crescent-shaped cyst of *Pneumocystis carinii*. Note intracystic bodies.

From K. Yoneda et al., *"Pneumocystis carinii:* Freeze-fracture study of stages of the organism," in *Exp. Parasitol.* 53:68–76. Copyright © 1982.

100% in untreated patients. In some cases the parasites disseminate to the spleen, lymph nodes, bone marrow, and even the eyes. Gutierrez[27] gives an excellent detailed discussion of both the clinical manifestations and pathology of *P. carinii.*

- **Diagnosis.**   Infections with *P. carinii* are suspected in any patient with appropriate risk factors and who presents clinical symptoms consistent with the disease, but positive diagnosis is possible only by demonstrating the organisms with special staining. Noninvasive diagnostic procedures such as sputum examination are effective in about half the cases. Lung biopsy or bronchial lavage yield infected material most often. Toluidine blue or methenamine silver stains are apparently reliable, and the Gram-Weigert stain is accurate in demonstrating the cysts. Although a number of other methods—such as ELISA, immunofluorescence assay, and DNA amplification techniques—are being developed, none as yet has proven quicker, easier, and more reliable than the classical staining.

- **Treatment.**   Even with treatment the mortality rate is high in patients with this disease because the immune system is deficient to begin with. The treatment of choice is the combination of trimethoprim-sulfamethoxazole (Bactrim). Equally effective is pentamidine isethionate, delivered as an inhalant spray, but pentamidine is toxic to the patient, too, and treatment must be monitored carefully for dangerous side effects.

References

1. Anderson, B. C. 1991. Prevalence of *Cryptosporidium muris*-like oocysts among cattle populations of the United States: Preliminary report. *J. Protozool.* 38:14–15S.

2. Araujo, F. G., T. Lin, and J. S. Remington. 1993. The activity of atovaquone (566C80) in murine toxoplasmosis is markedly augmented when used in combination with pyrimethamine or sulfadiazine. *J. Infect. Dis.* 167:494–97.

3. Armitage, K., T. Flanigan, J. Carey, I. Frank, R. MacGregor, P. Ross, R. Goodgame, and U. Turner. 1992. Treatment of cryptosporidiosis with paromomycin. A report of five cases. *Arch. Intern. Med.* 152:2497–9.

4. Arrowood, M. J., and C. R. Sterling. 1987. Isolation of *Cryptosporidium* oocysts and sporozoites using discontinuous sucrose and isopycnic percoll gradients. *J. Parasitol.* 73:314–19.

5. Atkinson, C. T., S. D. Wright, S. R. Telford Jr., G. S. McLaughlin, D. J. Forrester, M. E. Roelke, and J. W. McCown. 1993. Morphology, prevalence, and distribution of *Sarcocystis* spp. in white-tailed deer (*Odocoileus virginianus*) from Florida. *J. Wildl. Dis.* 29:73–84.

6. Baker, J. R. 1987. The *Toxoplasma* tangle. *Parasitol. Today* 3:103–5.

7. Bannert, B. 1992. *Sarcocystis simonyi* new species (Apicomplexa: Sarcocystidae) from the endangered Hierro giant lizard *Gallotia simonyi* (Reptilia: Lacertidae). *Parasitol. Res.* 78:142–45.

8. Beaver, P. C., R. K. Gadgil, and P. Morera, 1979. *Sarcocystis* in man: A review and report of five cases. *Am. J. Trop. Med. Hyg.* 28:819–44.

9. Beverly, J. K. A., W. A. Watson, and J. B. Spence. 1971. The pathology of the foetus in ovine abortion due to toxoplasmosis. *Vet. Rec.* 88:174–78.

10. Boylan, C., and W. Current. 1991. An improved rat model of *Pneumocystis carinii* pneumonia induced infections in *Pneumocystis*-free animals. *J. Protozool.* 38:138–40S.

11. Clubb, S. L., and J. K. Frenkel. 1992. *Sarcocystis falcatula* of opossums: Transmission by cockroaches with fatal pulmonary disease in psittacine birds. *J. Parasitol.* 78:116–24.

12. Couvreur, J. 1971. Prospective study in pregnant women with a special reference to the outcome of the foetus. In Hentsch, D., ed. *Toxoplasmosis*. Bern: Hans Huber Medical Publisher, 119–35.

13. Current, W. L., and B. L. Blagburn. 1991. *Cryptosporidium* and microsporidia: Some closing comments. *J. Protozool.* 38:244–45S.

14. Current, W. L., N. C. Reese, J. V. Ernst, W. S. Bailey, M. B. Heyman, and W. M. Weinstein. 1983. Human cryptosporidiosis in immunocompetent and immunodeficient persons. Studies of an outbreak and experimental transmission. *N. Eng. J. Med.* 308:1252–7.

14a. Desser, S. S., H. Hong, and D. S. Martin. 1995. The life history, ultrastructure, and experimental transmission of *Hepatozoon algonquinensis* n. sp., an apicomplexan parasite of the bullfrog, *Rana catesbeiana* and the mosquito, *Culex territans* in Algonquin Park, Ontario. *J. Parasitol.* 81:212–222.

15. Dhillon, A. S., H. L. Thacker, V. Dietzel, and R. W. Winterfield. 1981. Respiratory cryptosporidiosis in broiler chickens. *Avian Dis.* 25:747–51.

16. Dubey, J. P. 1977. *Toxoplasma, Hammondia, Besnoitia, Sarcocystis,* and other tissue cyst-forming coccidia of man and animals. In Kreier, J. P., ed. *Parasitic protozoa, vol. III.* New York: Academic Press, Inc.

17. Dubey, J. P., and C. A. Speer. 1991. *Sarcocystis canis,* new species (Apicomplexa: Sarcocystidae), the etiologic agent of generalized coccidiosis in dogs. *J. Parasitol.* 77: 522–27.

18. Dubey, J. P., C. A. Speer, and R. Fayer. 1990. Cryptosporidiosis of man and animals. Boca Raton, Fla.: CRC Press.

19. Espinosa, R. H., M. C. Sterner, J. A. Blixt, and R. J. Cawthorn. 1988. Description of a species of *Sarcocystis* (Apicomplexa: Sarcocystidae), a parasite of the northern saw-whet owl, *Aegolius acadicus,* and experimental transmission to deer mice *Peromyscus maniculatus. Can. J. Zool.* 66: 2118–21.

20. Fayer, R. 1972. Gametogony of *Sarcocystis* sp. in cell culture. *Science* 175:65–67.

21. Flanigan, T., R. Marshall, D. Redman, C. Kaetzel, and B. Ungar. 1991. In vitro screening of therapeutic agents against *Cryptosporidium:* Hyperimmune cow colostrum is highly inhibitory. *J. Protozool.* 38:225–27S.

22. Forthal, D. N., and S. S. Guest. 1984. *Isospora belli* enteritis in three homosexual men. *Am. J. Trop. Med. Hyg.* 33:1060–4.

23. Frenkel, J. K. 1973. Toxoplasmosis: Parasite life cycles, pathology, and immunology. In Hammond, D. M., and P. L. Long, eds. *The Coccidia.* Eimeria, Isospora, Toxoplasma, *and related genera.* Baltimore: University Park Press, 343–410.

24. Frenkel, J. K., H. Mehlorn, and A. O. Heydorn. 1987. Beyond the oocyst: Over the molehills and mountains of coccidialand. *Parasitol. Today* 3:250–51.

25. Gajdusek, D. C. 1957. *Pneumocystis carinii*—etiologic agent of interstitial plasma cell pneumonia of premature and young infants. *Pediatrics* 19:543–45.

26. Guay, J. M., D. Dubois, M. J. Morency, S. Gagnon, J. Mercier, and R. C. Levesque. 1993. Detection of the pathogenic parasite *Toxoplasma gondii* by specific amplification of ribosomal sequences using comultiplex polymerase chain reaction. *J. Clinical Microbiol.* 31:203–7.

27. Gutierrez, Y. 1990. *Diagnostic pathology of parasitic infections with clinical correlations.* Philadelphia, Pa.: Lea and Febiger.

28. Hoerr, F. J., F. M. Ranck, and T. F. Hastings. 1978. Respiratory cryptosporidiosis in turkeys. *J. Am. Vet. Med. Assoc.* 173:1591–3.

29. Hughes, H. P. A. 1985. Toxoplasmosis—a neglected disease. *Parasitol. Today* 1:41–44.

30. Jacobs, L. 1973. New knowledge of *Toxoplasma* and toxoplasmosis. In Dawes, B., ed. *Advances in parasitology,* 11. New York: Academic Press, Inc., 631–69.

31. Kean, B. H., A. C. Kimball, and W. N. Christenson. 1969. An epidemic of acute toxoplasmosis. *J.A.M.A.* 208:1002–4.

32. Kilani, R. T., and L. Sekla. 1987. Purification of *Cryptosporidium* oocysts and sporozoites by cesium chloride and percoll reagents. *Am. J. Trop. Med. Hyg.* 36:505–8.

33. Krick, J. A., and J. S. Remington. 1978. Toxoplasmosis in the adult. *N. Eng. J. Med.* 298:550–53.

34. Levine, N. D. 1962. Protozoology today. *J. Protozool.* 9:1–6.

35. Levine, N. D. 1963. Coccidiosis. *Ann. Rev. Microbiol.* 17:179–98.

36. Levine, N. D. 1970. Taxonomy of the Sporozoa. *J. Parasitol.* 56(sect. II, part 1):208–9.

37. Levine, N. D. 1973. Introduction, history and taxonomy. In Hammond, D. M., and P. L. Long, eds. *The Coccidia.* Eimeria, Isospora, Toxoplasma, *and related genera.* Baltimore: University Park Press, 1–22.

38. Levine, N. D. 1973. *Protozoan parasites of domestic animals and of man,* 2d ed. Minneapolis: Burgess Publishing Co.

39. Levine, N. D. 1977. Nomenclature of *Sarcocystis* in the ox and sheep and of fecal coccidia of the dog and cat. *J. Parasitol.* 63:36–51.

40. Levine, N. D. 1982. The genus *Atoxoplasma* (Protozoa, Apicomplexa). *J. Parasitol.* 68:719–23.

41. Levine, N. D. 1984. Taxonomy and review of the coccidian genus *Cryptosporidium* (Protozoa, Apicomplexa). *J. Protozool.* 31:94–98.

42. Levine, N. D. 1986. The taxonomy of *Sarcocystis* (Protozoa, Apicomplexa) species. *J. Parasitol.* 72:372–82.

43. Levine, N. D. 1987. Whatever became of *Isospora bigemina? Parasitol. Today* 3:101–3.

44. Levine, N. D., and V. Ivens. 1965. The coccidian parasites (Protozoa, Sporozoa) of rodents. *Ill. Biol. Monogr.* 33. Urbana, Ill.: University of Illinois Press.

45. Lindsey, D. S., B. L. Blagburn, F. J. Hoerr, and P. C. Smith. 1991. Cryptosporidiosis in zoo and pet birds. *J. Protozool.* 38:180–81S.

46. Lindsay, D. S., S. J. Upton, B. L. Blagburn, M. Toivio-Kinnucha, J. P. Dubey, C. T. McAllister, and S. E. Trauth. 1992. Demonstration that *Sarcocystis montanaensis* has a speckled kingsnake-prairie vole life cycle. *J. Helm. Soc. Wash.* 59:9–15.

47. Long, P. L. 1973. Pathology and pathogenicity of coccidial infections. In Hammond, D. L., and P. L. Long, eds. *The Coccidia.* Eimeria, Isospora, Toxoplasma, *and related genera.* Baltimore: University Park Press.

48. Long, P. L., and T. K. Jeffers. 1986. Control of chicken coccidiosis. *Parasitol Today.* 2:236–40.

49. Lunde, M. N., and L. Jacobs. 1983. Antigenic differences between endozoites and cystozoites of *Toxoplasma gondii. J. Parasitol.* 69:806–8.

50. Maddison, S. E., et al. 1979. Lymphocyte proliferative responsiveness in 31 patients after an outbreak of toxoplasmosis. *Am. J. Trop. Med. Hyg.* 28:955–61.

51. Markus, M. B. 1980. Flies as natural transport hosts of *Sarcocystis* and other coccidia. *J. Parasitol.* 66:361–62.

52. Marquardt, W. C. 1973. Host and site specificity in the Coccidia. In Hammond, D. M., and P. L. Long, eds. *The Coccidia.* Eimeria, Isospora, Toxoplasma, *and related genera.* Baltimore: University Park Press, 23–43.

53. Matsumoto, Y., and Y. Yoshida. 1986. Advances in *Pneumocystis* biology. *Parasitol. Today* 2:137–42.

54. MacPherson, J. M., and A. A. Gajadhar. 1993. Sensitive and specific polymerase chain reaction detection of *Toxoplasma gondii* for veterinary and medical diagnosis. *Can. J. Vet. Res.* 57:45–48.

55. Parry, S., M. E. J. Barratt, S. Jones, S. McKee, and J. D. Murray. 1992. Modelling coccidial infection in chickens: Emphasis on vaccination by in-feed delivery of oocysts. *J. Theoret. Biol.* 157:407–25.

56. Pathmanathan, R., and S. P. Kan. 1992. Three cases of human *Sarcocystis* infection with a review of human muscular sarcocystosis in Malaysia. *Trop. Geogr. Med.* 44:102–8.

57. Perryman, L. E., and J. M. Bjorneby. 1991. Immunotherapy of cryptosporidiosis in immunodeficient animal models. *J. Protozool.* 38:98–100S.

58. Polenz, J., H. W. Moon, N. F. Cheville, and W. J. Bemrick. 1978. Cryptosporidiosis as a probable factor in neonatal diarrhea of calves. *J. Am. Vet. Med. Assoc.* 172:452–57.

59. Pozio, E., M.A.G. Morales, F. M. Barbieri, and G. La-Rosa. 1992. *Cryptosporidium:* Different behaviour in calves of isolates of human origin. *Trans. R. Soc. Trop. Med. Hyg.* 86:636–38.

60. Quist, K. L., R. L. Taylor Jr., L. W. Johnson, and R. G. Strout. 1993. Comparative development of *Eimeria tenella* in primary chick kidney cell cultures derived from coccidia-resistant and -susceptible chickens. *Poultry Sci.* 72:82–87.

61. Remington, J. S., and G. Desmonts. 1976. In Remington, J. S., and J. O. Klein, eds. Infectious diseases of the fetus and newborn infant. Philadelphia: W. B. Saunders Company, 191–332.

62. Roberts, W. L., C. A. Speer, and D. M. Hammond. 1970. Electron and light microscope studies of the oocyst walls, sporocysts, and excysting sporozoites of *Eimeria callospermophili* and *E. larimerensis. J. Parasitol.* 56:918–26.

63. Rommel, M., A. O. Heydorn, and F. Gruber. 1972. Beiträge zum Lebenzyklus der Sarkosporidien. I. Die Sporozyste von *S. tenella* in der Fäzes der Katze. *Berl. Münch. Tierärztl. Wochenschr.* 85:101–5.

64. Settnes, O. P., V. Bille-Hansen, S. E. Jorsal, and S. A. Henriksen. 1991. The piglet as a potential model of *Pneumocystis carinii* pneumonia. *J. Protozool.* 38:140–1S.

65. Siddall, M. E., and S. S. Desser. 1990. Gametogenesis and sporogonic development of *Haemogregarina balli* (Apicomplexa: Adeleina: Haemogregarinidae) in the leech *Placobdella ornata. J. Protozool.* 37:511–20.

66. Sinden, R. E. 1985. A cell biologist's view of host cell recognition and invasion by malarial parasites. *Trans. R. Soc. Trop. Med. Hyg.* 79:598–605.

67. Sorvillo, F. J., K. Fujioka, B. Nahlen, M. P. Tormey, R. Kebabjian, and L. Mascola. 1992. Swimming associated cryptosporidiosis. *Amer. J. Public Health* 82:742–44.

68. Speer, C. A., and D. E. Burgess. 1987. *In vitro* cultivation of *Sarcocystis* merozoites. *Parasitol. Today* 3:2–3.

69. Sterling, C. R., R. H. Gilman, N. A. Sinclair, V. Cama, R. Castillo, and F. Diaz. 1991. The role of breast milk in protecting urban Peruvian children against cryptosporidiosis. *J. Protozool.* 38:23–25S.

70. Tzipori, S. 1985. *Cryptosporidium:* Notes on epidemiology and pathogenesis. *Parasitol. Today* 1:159–65.

71. Tzipori, S., K. W. Angus, E. W. Gray, I. Campbell, and F. Allen. 1981. Diarrhea in lambs experimentally infected with *Cryptosporidium* isolated from calves. *Am. J. Vet. Res.* 42:1400–4.

72. Tzipori, S., I. Campbell, D. Sherwood, and D. R. Snodgrass. 1980. An outbreak of calf diarrhea attributed to cryptosporidial infection. *Vet. Rec.* 107:579–80.

73. Vetterling, J. M., H. R. Jervis, T. G. Merrill, and H. Sprinz. 1971. *Cryptosporidium wrairi* sp.n. from the guinea pig *Cavia porcellus,* with an emendation of the genus. *J. Protozool.* 18:243–47.

74. Wallace, G. D. 1971. Experimental transmission of *Toxoplasma gondii* by filth flies. *Am. J. Trop. Med. Hyg.* 20:411–13.

75. Whiteside, M. E., J. S. Barkin, R. G. May, S. D. Weiss, M. A. Fischl, and C. L. MacLeod. 1984. Enteric coccidiosis among patients with the acquired immunodeficiency syndrome. *Am. J. Trop. Med. Hyg.* 33:1065–72.

76. Wilson, P. A. G., and D. Fairbairn. 1961. Biochemistry of sporulation in oocysts of *Eimeria acervulina. J. Protozool.* 8:410–16.

77. Yoneda, K., P. D. Walzer, C. S. Richey, and M. G. Birk. 1982. *Pneumocystis carinii:* Freeze-fracture study of stages of the organism. *Exp. Parasitol.* 53:68–76.

Additional References

Brandberg, L. L., S. B. Goldberg, and W. C. Breidenbach. 1970. Human coccidiosis—a possible cause of malabsorption. The life cycle in small-bowel mucosal biopsies as a diagnostic feature. *N. Eng. J. Med.* 283:1306–13. A study of six cases and a review of previous reports. Endogenous stages are illustrated for the first time.

Desmonts, G., and J. Couveur. 1974. Congenital toxoplasmosis. *N. Eng. J. Med.* 290:1110–16. A study of 378 pregnancies.

Feldman, H. A. 1974. Congenital toxoplasmosis, at long last . . . *N. Eng. J. Med.* 290:1138–40. A short summary of the discovery of congenital toxoplasmosis.

Long, P. L., ed. 1982. *The biology of the Coccidia.* Baltimore: University Park Press.

Ma, P. 1987. Protozoa and acquired immune deficiency syndrome (AIDS). *ATCC Quart. Newsletter* 7:1–2, 7.

Markus, M. B. 1978. *Sarcocystis* and sarcocystosis in domestic animals and man. *Adv. Vet. Sci. Comp. Med.* 22:159–93.

Ryley, J. F. 1980. Recent developments in coccidian biology: Where do we go from here? *Parasitology* 80:189–209.

Tenter, A. M., P. R. Baverstock, and A. M. Johnson. 1992. Phylogenetic relationships of *Sarcocystis* species from sheep, goats, cattle and mice based on ribosomal RNA sequences. *Int. J. Parasitol.* 22:503–13.

Chapter 9

PHYLUM APICOMPLEXA: MALARIA ORGANISMS AND PIROPLASMS

Parasitic elements are found in the blood of patients who are ill with malaria. Up to now, these elements were thought incorrectly to be pigmented leukocytes. The presence of these parasites in the blood probably is the principal cause of malaria.

Charles Louis Alphonse Laveran, 1880

SUBORDER HAEMOSPORINA

This suborder contains the family Plasmodiidae, including the genera *Plasmodium, Haemoproteus,* and *Leucocytozoon,* which are the malaria and malarialike organisms. When in host cells, *Plasmodium* and *Haemoproteus* usually produce a pigment called **hemozoin** from host hemoglobin, distinguishing them from the closely related *Leucocytozoon.* The ultrastructure of these parasites is basically similar to the coccidia, except that they lack conoids. Syzygy is absent, and the macrogametocyte and microgametocyte develop independently. The microgametocyte produces about eight flagellated gametes. The zygote is motile and is called an **ookinete;** the sporozoites are not enclosed within sporocysts. They are heteroxenous, with merozoites produced in the vertebrate host and sporozoites developing in the invertebrate host. It is possible that these parasites evolved from the coccidia of vertebrates rather than of invertebrates, with mites or other bloodsuckers initiating the cycle in arthropods.

Although most species of Haemosporina are parasites of wild animals and appear to cause little harm in most cases, a few cause diseases that are among the worst scourges of humanity. Indeed, malaria has played an important part in the rise and fall of nations and has killed untold millions the world over. John F. Kennedy said in 1962,[49]

For centuries, malaria has outranked warfare as a source of human suffering. Over the past generation it has killed millions of human beings and sapped the strength of hundreds of millions more. It continues to be a heavy drag on man's efforts to advance his agriculture and industry.

Despite the combined efforts of 102 countries to eradicate malaria, it remains the most important disease in the world today in terms of lives lost and economic burden. Progress has been made, however. In some countries, such as the United States, eradication of endemic malaria is complete. Between 1948 and 1965 the number of cases was cut from a worldwide total of 350 million to fewer than 100 million. However, more recent estimates put the worldwide prevalence as high as 489 million.[101] Development of resistance in the parasite to antimalarial drugs and in the vector to insecticides deserves much of the credit for the increase in prevalence.[61]

About 1472 million persons live in malarious areas of the world. This unprotected population lives in countries without the administrative, financial, and human resources necessary for control.

Genus *Plasmodium*

History is, after all, a review of past experiences which influence present events.

Elvio H. Sadun

Malaria has been known since antiquity; recognizable descriptions of the disease were recorded in various Egyptian papyri. The Ebers papyrus (3550 B.P.) mentions fevers, splenomegaly, and the use of oil of the Balamites tree as a mosquito repellent. Hieroglyphs on the walls of the ancient Temple of Denderah in Egypt describe an intermittent fever following the flooding of the Nile.[37] Hippocrates studied medicine in Egypt and clearly described quotidian, tertian, and quartan fevers with splenomegaly. He believed that bile was the cause of the fevers. Greek states built beautiful cities in the lowlands only to see them devastated by the disease, and wealthy Greeks and Romans traditionally summered in the highlands to escape the heat, mosquitoes, and mysterious fevers. Herodotus (c. 2500–2424 B.C.) states that Egyptian fishermen slept with their nets arranged around their beds so that mosquitoes could not reach them. Homer also noted that malaria is most prevalent in the later summer: In the *Iliad* (XXII, 31) we read, " . . . like that star which comes on in the autumn . . . , the star they give the name of Orion's dog which is brightest among the stars, and yet is wrought as a sign of evil and brings on the great fever for unfortunate mortals." Medieval England saw crusaders falter and fail as they encountered malaria. As had happened before and has happened since, malaria killed more warriors than did warfare.

When Europeans imported slaves and returned their colonial armies to their continent, they brought malaria with them, increasing the concentration of the disease with devastating results.

Throughout history a connection between swamps and fevers has been recognized. It was commonly concluded that the disease was contracted by breathing "bad air" or *mal aria.* This belief flourished until near the end of the nineteenth century. Another name for the disease, *paludism* (marsh disease), is still in common use in the world.

There has been much speculation as to whether malaria existed in the Western Hemisphere before the Spanish conquest. It seems inconceivable that the great Olmec and Mayan civilizations could have developed in highly malarious regions. The Spanish conquistadores made no mention of fevers during the early years of the conquest, and in fact they holidayed in Guayaquil and the coastal area near Veracruz, regions that soon after became very unhealthy because of malaria. Balboa did not mention any encounters with malaria while traversing the Isthmus of Panama. It therefore seems likely that malaria was introduced into the New World by the Spaniards and their African slaves. However, some evidence that Africans reached South America during pre-Columbian times suggests that, while improbable, it is not impossible that malaria existed in localized areas of the continent before the Spanish conquest,[44] could have been brought from Oceania or from Asia by way of the Bering Strait, or could have been introduced by the Vikings.

No progress was made in the etiology of malaria until 1847, when Meckel observed black pigment granules in the blood and spleen of a patient who died of the disease. He even stated that the granules lay within protoplasmic masses. Was he the first to actually see the parasite? In 1879 Afanasiev suggested that the granules caused the disease.

During the next 30 years physicians and scientists of high stature searched diligently for the cause of the disease and its means of transmission to people. Two obscure army medical officers, working in their spare time, under primitive and difficult circumstances, were to make these cardinal discoveries.

Most research was directed toward finding an infective organism in water or in the air. Many false hopes were generated when a previously unknown ameba or fungus was discovered. When Edwin Klebs (German) and the equally prestigious Corrado Tommasi-Crudelli (Italian) declared *Bacillus malariae* the causative organism, few doubted the truth of their momentous discovery.

Meanwhile, in North Africa, far from academic circles, a young French Army physician named Charles Louis Alphonse Laveran decided that the mysterious pigment in his malarious patients would be a good starting point for further research. He observed the pigment not only free in the plasma but also within leukocytes, and he saw clear bodies within erythrocytes. As the hyaline bodies of irregular shape grew, he saw the erythrocytes grow pale and pigment form within them. He little doubted the parasitic nature of the organisms he saw. Then, on November 6, 1880, he witnessed one of the most dramatic events in protozoology: the formation of male gametes by the process of exflagellation. He quickly wrote of his discovery, reporting on November 23, 1880, to the Academy of Medicine in Paris, where much

skepticism followed his report. Most scientists were loath to abandon the Klebs/Tommasi-Crudelli bacillus in favor of a protozoan that an army physician claimed to have discovered in Algeria. His "organisms" were assumed to be degenerating blood cells. In addition to the prestige of Klebs and Tommasi-Crudelli, other factors influenced this skeptical attitude.[97] Opportunities to study the malarial fevers in the academic medical centers of Europe were limited, and there were real limitations and difficulties in interpreting microscopic observations. Technical progress in microscopy was rapid in the decade of 1880–1890, however, and Ettore Marchiafava (a favorite student of Tommasi-Crudelli) and Angelo Celli (his longtime collaborator), who originally favored the bacillus hypothesis, became convinced that Laveran was correct. More strong support came in 1885, when Camillo Golgi differentiated between species of *Plasmodium* and demonstrated the synchronism of the parasite in relation to paroxysm.

Laveran had accurately described the male and female gametes, the trophozoite, and the schizont while working with a poor, low-power microscope and with unstained preparations. By 1890 several scientists in different parts of the world verified his findings. In Russia in 1891, Romanovsky developed a new method of staining blood smears based on methylene blue and eosin. A hundred years later, modifications of his stain remain in wide use.

The mode of transmission of malaria was, however, still unknown. Although ideas were rampant ("bad night air" was still a popular candidate), few were as well thought out as that of Patrick Manson, who favored the hypothesis of transmission by mosquitoes. True, he was conditioned by the proof of mosquitoes as vectors of filariasis, which gave him some insight. Surgeon-Major Ronald Ross was 38 years old when he met Manson for the first time, while on leave from the Indian Medical Service. Finding in Ross a man who was interested in malaria and who could test his ideas for him, Manson lost no time in convincing him that malaria was caused by a protozoan parasite. For the next several years, in India, Ross worked during every spare minute, searching for the mosquito stages of malaria that he was certain existed. Dissecting mosquitoes at random and also after allowing them to feed on malarious patients, he found many parasites, but none of them proved to be what he searched for. During this time he had a steady correspondence with Manson, who encouraged him and brought his discourses to the learned societies of England. Ross left a wonderful record of his moods of excitement, frustration, disappointment, and triumph. His journals also contain long quotations from Manson's letters written to him at that time.

Ross's first significant observation was that exflagellation normally occurs in the stomach of a mosquito, rather than in the blood as was then thought. At this time he was posted to Bangalore to help fight a cholera epidemic, the first in a series of frustrating interruptions by superiors who had no concept of the importance of the work Ross was doing in his spare time. Returning from Bangalore, he continued the search for further development of the parasite within the mosquito. Failing in this effort, he concluded that he had been working with the wrong kinds of mosquitoes (*Culex* and *Stegomyia*). He tried other kinds and was led astray time after time by gregarines and other mosquito parasites, each of which had to be eliminated as possible malaria organisms by laborious experimentation. After two years of work, which his superior officers ignored as harmless lunacy, he seemed to have reached an impasse. He was eligible for retirement soon but was determined to try "one more desperate effort to solve the Great Problem." He toiled far into the nights, dissecting mosquitoes, in a hot little office. He could not use the overhead fan lest it blow his mosquitoes away, so while he worked, swarms of gnats and mosquitoes avenged themselves "for the death of their friends." At last, late in the night of August 16, 1897, he dissected some "dapple-winged" mosquitoes (*Anopheles* spp.) that had fed on a malaria patient, and he found some pigmented, spherical bodies in the walls of the insects' stomachs. The next day he dissected his last remaining specimen and found the spheroid cells had grown. They were most certainly the malaria parasites! That night he penned in a notebook,

This day relenting God

  Hath placed within my hand

A wondrous thing, and God

  Be praised. At this command,

Seeking His secret deeds

  With tears and toiling breath,

I find thy cunning seeds,

  Oh million-murdering Death.

He reported his discovery to Manson and immediately set about breeding the correct kind of mosquito in preparation for the first step of transmitting the disease from the insect to humans. Unfortunately, he was immediately posted to Bombay, where he could do no further research on human malaria. Nevertheless, he found similar organisms (*Plasmodium relictum*) in birds. He repeated his feeding experiments with mosquitoes and found similar parasites when they fed on infected birds. He also found that the spheroid bodies ruptured, releasing thousands of tiny bodies that dispersed throughout the insect's body, including into the salivary glands.

Through Manson, Ross reported to the world how malaria was transmitted by mosquitoes. It remained only for a single experiment to prove the transmission to humans. Ross never did it. The authorities were so impressed with his work they ordered Ross to work out the biology of kala-azar in another part of India. This transfer seems to have broken his spirit, for he never really tried again to finish the study of malaria. The concentration had made him ill, his eyes were bothering him, and his microscope had rusted tight from his sweat. Anyway, he was a physician, not a zoologist, and he was most interested in learning how to prevent the disease, as opposed to determining the finer points of the parasite's biology. This he considered done, and he retired from the Army. He was awarded the Nobel Prize in Medicine in 1902 and was knighted in 1911. He died in 1932 after a distinguished postarmy career in education and research.

Unfortunately, the history of malariology is tarnished by strife and bitterness. Several persons who were working on the life cycle of the parasite claimed credit for the discovery that pointed to the means of control for malaria. Italian, German, and American scientists all made important contributions to the solution of the problem of malaria transmission. Several of them, including Ross, spent a good portion of their lives quibbling about priorities in the discoveries. Manson-Bahr[60] and Harrison[40] give fascinating accounts of the personalities of the men who conquered the life cycle of malaria. Credit for completing study of the life cycle should go to Amigo Bignami and Giovanni Grassi, who experimentally transmitted the malaria parasite from mosquito to human in 1898.

Although medical scientists thought they knew the life cycle of malaria after Ross's work, they knew nothing of the stages in the liver. In the early twentieth century they thought that the cycle progressed from the blood to the mosquito and back to the blood. This concept gained support from the published work of Fritz Schaudin, who claimed to have seen sporozoites penetrating red blood cells and transforming into trophozoites. Schaudin's work remained unchallenged until World War I, when a fact began to emerge that could not be explained by the direct cycle between mosquito and blood. Quinine was a well-known antimalarial drug, but it had effect only on the erythrocytic forms. Soldiers treated with the drug were apparently cured; that is, no parasites could be found in their blood. However, when the treatment stopped and the patients moved to a nonmalarious area, parasites returned to their blood at certain time intervals.

In 1917 Julius von Wagner-Jauregg discovered that the high fevers of malaria could be used to treat neurosyphilis. From this work two additional facts emerged. When a patient was infected by injection with parasitized blood, the incubation period could be shortened or lengthened by changing the number of parasites injected. However, the number of bites—whether 1 or 200—of infected mosquitoes did not alter the incubation period.

In 1938 S. P. James and P. Tate discovered the exoerythrocytic stages of *P. gallinaceum*. After this discovery large-scale work began to find the exoerythrocytic stages of human malaria parasites. Finally, in 1948 H. C . Shortt and P. C. C. Garnham demonstrated the exoerythrocytic stages of *P. cynomolgi* in monkeys and *P. vivax* in humans.[93]

These historical notes should not be concluded without mention of a man who applied these early discoveries for the immense benefit of his country and humanity: William C. Gorgas. Gorgas was the medical officer placed in charge of the Sanitation Department of the Canal Zone when the United States undertook the construction of the Panama Canal. Were it not for his mosquito control measures, malaria and yellow fever would have defeated American attempts to build the Canal, just as they had defeated the French. In July 1906 the malaria rate in the Canal Zone was 1263 hospital admissions per 1000 population![86] Gorgas's work reduced the rate to 76 hospital admissions per 1000 in 1913, saving his country $80 million and the lives of 71,000 fellow humans. Gorgas was a hero: The President made him Surgeon General, Congress promoted him, Oxford University made him an honorary Doctor of Science, and the King of England knighted him. Sir William Osler stated, "There is nothing to match the work of Gorgas in the history of human achievement." It is a sad commentary on our cultural memory that the name of Gorgas is now known by so few, while

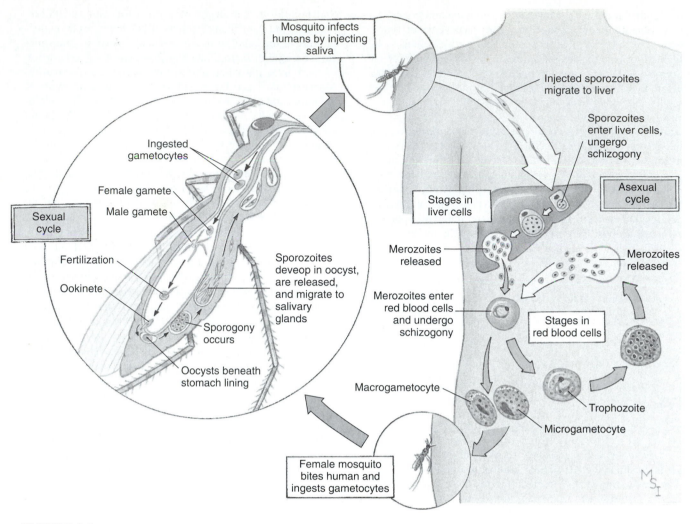

FIGURE 9.1

Life cycle of *Plasmodium vivax*. (*a*) Sexual cycle produces sporozoites in body of mosquito. Meiosis occurs just after zygote formation (zygotic meiosis). (*b*) Sporozoites infect a human and reproduce asexually, first in liver cells and then in red blood cells.

From Cleveland P. Hickman, Jr. et al., *Integrated Principles of Zoology,* 9th edition. Copyright © 1993 by Mosby-Year Book Inc. Reprinted by permission of Times Mirror Higher Education Group, Inc., Dubuque, Iowa. All Rights Reserved.

so many easily remember the names of generals and tyrants who caused great bloodshed.

For students interested in more details about humanity's fight against malaria, we highly recommend Harrison's book.[40]

• Life Cycle and General Morphology

Following is a general account of the development and structure of malaria parasites (Fig. 9.1), without reference to particular species. Specific morphological details for each species are in Table 9.1 *Plasmodium* spp. require two types of hosts: an invertebrate (mosquito) and a vertebrate (reptile, bird, or mammal). Technically the invertebrate can be considered the definitive host because sexual reproduction occurs there. Asexual reproduction takes place in the tissues of a vertebrate, which thus can be called the intermediate host. However, the gametocytes actually form in the blood of the vertebrate, and fertilization occurs while still in this medium in the stomach of the mosquito. By this

reasoning the vertebrate is the definitive host.[21] Also, *Plasmodium* spp. were probably derived from an ancestral coccidian whose asexual and sexual reproduction took place in the same (presumably vertebrate) host.

Vertebrate Phases.   When an infected mosquito takes blood from a vertebrate, she injects saliva containing tiny, elongated sporozoites into the bloodstream. The sporozoite basically is similar in morphology to that of *Eimeria* and other coccidia. It is about 10 to 15 μm long by 1 μm in diameter and has a pellicle composed of a thin outer membrane, a doubled inner membrane, and a layer of subpellicular microtubules. There are three polar rings. The rhoptries are long, extending to the midportion of the organism, and much of the rest of the anterior cytoplasm is taken up by the micronemes. An apparently nonfunctional cytostome is present, and there is a mitochondrion in the posterior end of the sporozoite.[3]

After being injected into the bloodstream, the sporozoites disappear from the circulating blood within an hour. Their

Table 9.1

SOME CHARACTERISTICS OF *PLASMODIUM* SPP. IN HUMANS

Stage or period	P. vivax	P. falciparum	P. ovale	P. malariae
Early trophozoite	About ⅓ diameter of red cell; chromatin dot heavy; vacuole prominent	About ⅕ diameter of red cell; chromatin dot small; two dots frequent; marginal forms frequent	Like *P. vivax* and *P. malariae*	Single, heavy chromatin dot; cytoplasmic circle often smaller, thicker, heavier than in *P. vivax;* vacuole fills in early
Growing trophozoite	Pseudopodia common; one or more food vacuoles	This stage not usually seen in peripheral blood	Compact, little vacuolation	Cytoplasm usually compact; little or no vacuole; sometimes in a band form across the red cell
Late trophozoite	Large mass of chromatin; fine brown hemozoin; almost fills red cell	This stage not usually seen in peripheral blood		Chromatin often elongated, less definite in outline than in *P. vivax;* cytoplasm dense, rounded, oval, or band shape; almost fills red cell
Hemozoin	Short delicate rods, irregularly scattered; yellowish brown	Granular, tendency to coalesce; coarse in gametocytes	Hemozoin lighter than *P. malariae;* similar to *P. vivax*	Granules rounded; larger, darker than in *P. vivax;* tendency to peripheral arrangement
Appearance of erythrocyte	Larger than normal, often oddly shaped; Schüffner's dots at all stages but young rings; multiple infection occasional	Normal size; Maurer's spots common in cells with later trophozoites (not usually seen in peripheral blood)	Schüffner's dots often present in ring and later stages; red cell larger than normal, oval, often with irregular edge	About normal or slightly smaller; stippling rarely seen; multiple infection rare
Schizont	12 to 24 merozoites; hemozoin in one or two clumps; almost fills red cell	8 to 24 or more merozoites; rare in peripheral blood	4 to 16 but usually 8 merozoites	6 to 12 but usually 8 or 10 merozoites in a rosette or cluster arrangement; often found in peripheral blood
Microgametocytes (usually smaller and fewer than macrogametocytes)	Rounded or oval; almost fill red cell; dark hemozoin throughout cytoplasm; chromatin diffuse, in large mass, pink; small amount of light blue cytoplasm; no vacuoles	Crescent shaped; length about 1.5 times diameter of red cell; chromatin diffuse, pink; hemozoin granules in central portion; cytoplasm pale blue	Like *P. vivax* but somewhat smaller; mature macrogametocyte fills infected cell; microgametocytes smaller	Like *P. vivax,* but smaller; pigment more conspicuous
Macrogametocytes	As in microgametocytes except cytoplasm stains darker blue; chromatin more compact, dark red	Size and shape about as in microgametocytes; chromatin more compact, red; cytoplasm darker; hemozoin concentrated		Pigment abundant; round, dark brown granules; coarser than *P. vivax*
EE cycle	8 days	5½ to 6 days	9 days	13 days
Prepatent pd, minimum	11 to 13 days	9 to 10 days	10 to 14 days	15 to 16 days
Schizogonic cycle	48 hours	36 to 48 hours, usually 48	about 48 hours	72 hours
Development in mosquito	10 days at 25 to 30° C	10 to 12 days at 27° C	14 days at 27° C	25 to 28 days at 22 to 24° C

immediate fate was a great mystery until the mid-1940s, when it was shown that within one or two days they enter the parenchyma of the liver or other internal organ, depending on the species of *Plasmodium.* Where they are the first 24 hours still is unknown. A protein covering the surface of the sporozoite (circumsporozoite protein) bears a ligand (molecule that specifically and noncovalently binds to another molecule) that binds to receptors on the basolateral domain of the hepatocyte

cell membrane.[16] That is why the sporozoites enter liver cells and not other cells in the body. Extrusion of proteins from the rhoptries facilitates penetration into host cells by a mechanism that is still unclear.[79] Entry into the hepatocyte initiates a series of asexual reproductions known as the **preerythrocytic cycle** or **primary exoerythrocytic schizogony,** often abbreviated as the **PE** or **EE** stage. Once within a hepatic cell, the parasite metamorphoses into a feeding trophozoite. The

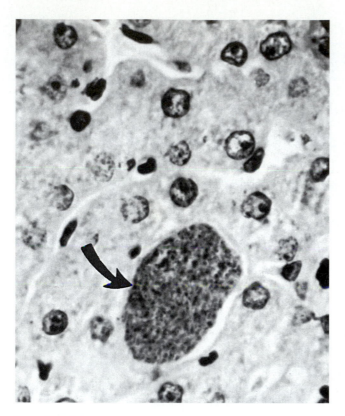

FIGURE 9.2

Preerythrocytic schizont of *Plasmodium* (*arrow*) in liver tissue.
Courtesy of Peter Diffley.

organelles of the apical complex disappear, and the trophozoite feeds on the cytoplasm of the host cell by way of the cytostome and, in the species in mammals, by pinocytosis.

After about a week, depending on the species, the trophozoite is mature and begins schizogony. Numerous daughter nuclei are first formed, transforming the parasite into a schizont (Fig. 9.2), also known as a **cryptozoite.** During the nuclear divisions the nuclear membranes persist, and the microtubular spindle fibers are formed within the nucleus. The mitochondrion becomes larger during the growth of the trophozoite, forms buds, and then breaks up into many mitochondria. Elements of the apical complex form subjacent to the outer membrane, and schizogony proceeds as previously described. The merozoites thus formed after cytokinesis are referred to in the EE stage as **metacryptozoites.** The merozoites are much shorter than sporozoites—2.5 μm long by 1.5 μm in diameter—and have small, teardrop-shaped rhoptries and small, oval micronemes.

What happens next has been a subject of lively debate. For many years it was believed that the merozoites entered new hepatocytes to form new schizonts and then merozoites, at least in the species of *Plasmodium* that are capable of causing a relapse.[93] However, as early as 1913 it was

postulated that some sporozoites become dormant for an indefinite time after entering the body.[9] Such dormant cells, now called **hypnozoites,** have now been demonstrated.[53,54] They are discussed under relapse in malaria (p. 149).

Eventually merozoites leave liver cells to penetrate erythrocytes in the blood, initiating the **erythrocytic cycle.** Some of the merozoites may be phagocytized by Kupffer cells in the liver, which may be an important host defense mechanism.[102] On entry into an erythrocyte, the merozoite again transforms into a trophozoite. The host cytoplasm ingested by the trophozoite forms a large food vacuole, giving the young *Plasmodium* the appearance of a ring of cytoplasm with the nucleus conspicuously displayed at one edge (Plate 1, *1* and *2*). The distinctiveness of the "signet-ring stage" is accentuated by the Romanovsky stains: The parasite cytoplasm is blue, and the nucleus is red. As the trophozoite grows (Plate 1, *3* to *15*), its food vacuoles become less noticeable by light microscopy, but pigment granules of **hemozoin** in the vacuoles become apparent. Hemozoin is an end product of the parasite's digestion of the host's hemoglobin. It contains 65% protein, 16% ferriprotoporphyrin-IX (hematin), 6% carbohydrate, and traces of lipid and nucleic acid.[34] The protein is mostly host globin.

The parasite rapidly develops into a schizont (Plate 1, *16* to *20*). The stage in the erythrocytic schizogony at which the cytoplasm is coalescing around the individual nuclei, before cytokinesis, is called the **segmenter.** When development of the merozoites is completed, the host cell ruptures, releasing parasite metabolic wastes and residual body, including hemozoin. The metabolic wastes thus released are one factor responsible for the characteristic symptoms of malaria. A great many of the merozoites are ingested and destroyed by reticuloendothelial cells and leukocytes, but, even so, the number of parasitized host cells may become astronomical because erythrocytic schizogony takes only from one to four days, depending on the species. Hemozoin has a toxic effect on macrophages, depressing their effectiveness as phagocytes.[105]

After an indeterminate number of asexual generations,[94] some merozoites enter erythrocytes and become **macrogamonts (macrogametocytes)** and **microgamonts (microgametocytes)** (Plate 1, *21* to *24*). The size and shape of these cells are characteristic for each species (Table 9.1); they also contain hemozoin. Unless they are ingested by a mosquito, gametocytes soon die and are phagocytized by the reticuloendothelial system.

Invertebrate Stages.   When erythrocytes containing gametocytes are imbibed by an unsuitable mosquito, they are digested along with the blood. However, if a susceptible mosquito is the diner, the gametocytes develop into gametes. Suitable hosts for the *Plasmodium* spp. of humans are a wide variety of *Anopheles* spp. (see Fig. 38.1). After release from its enclosing erythrocyte, the macrogametocyte matures to the macrogamete in a process involving little obvious change other than a shift of the nucleus toward the periphery. In contrast, the microgametocyte displays a rather astonishing transformation, **exflagellation.** As the microgametocyte becomes extracellular,

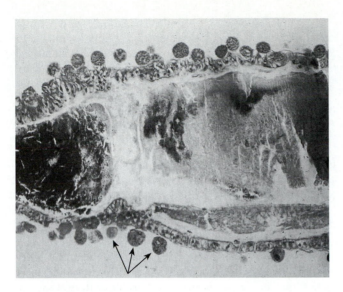

FIGURE 9.3

Longitudinal section of a mosquito intestine, with numerous oocysts (*arrows*) of *Plasmodium* sp.

From H. Zaiman, editor. *A Pictorial Presentation of Parasites.*

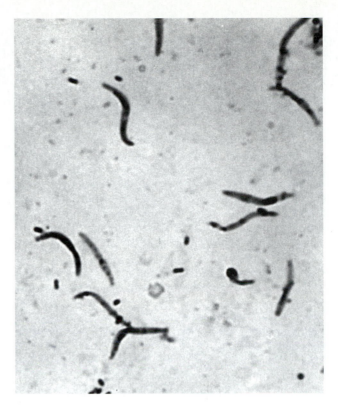

FIGURE 9.4

Plasmodium sporozoites.

Courtesy of Peter Diffley.

within 10 to 12 minutes its nucleus divides repeatedly to form six to eight daughter nuclei, each of which is associated with the elements of a developing axoneme. The doubled outer membrane of the microgametocyte becomes interrupted; the flagellar buds with their associated nuclei move peripherally between the interruptions and then continue outward covered by the outer membrane of the gametocyte. These break free and are the microgametes. The stimulus for exflagellation is an increase in pH caused by escape of dissolved carbon dioxide from the blood.[72] The life span of the microgametes is short since they contain little more than the nuclear chromatin and the flagellum covered by a membrane. The microgamete swims about until it finds a macrogamete, which it penetrates and fertilizes. The resultant diploid zygote quickly elongates to become a motile **ookinete.** The ookinete is reminiscent of a sporozoite and merozoite in morphology. It is 10 to 12 μm in length and has polar rings and subpellicular microtubules but no rhoptries or micronemes.

The ookinete penetrates the peritrophic membrane in the mosquito's gut and migrates intracellularly or intercellularly (or perhaps both)[67,103] to the hemocoel side of the gut. There it begins its transformation into an oocyst. The oocyst (Fig. 9.3) is covered by an electron-dense capsule and soon extends out into the insect's hemocoel. The initial division of its nucleus is reductional; meiosis takes place immediately after zygote formation as in other Sporozoea.[95,96] The oocyst reorganizes internally into a number of haploid nucleated masses called **sporoblasts,** and the cytoplasm contains many ribosomes, endoplasmic reticulum, mitochondria, and other inclusions. The sporoblasts, in turn, divide repeatedly to

form thousands of sporozoites[82] (Fig 9.4). These break out of the oocyst into the hemocoel and migrate throughout the mosquito's body. On contacting the salivary gland, sporozoites enter its channels and can be injected into a new host at the next feeding.

Sporozoite development takes from 10 days to two weeks, depending on the species of *Plasmodium* and the temperature. Once infected, a mosquito remains infective for life, capable of transmitting malaria to every susceptible vertebrate it bites. *Anopheles* spp. that are good vectors for human malaria live long enough to feed on human blood repeatedly.[41]

Plasmodium sometimes is transmitted by means other than the bite of a mosquito. The blood cycle may be initiated by blood transfusion, by syringe-passed infection among drug addicts, in laboratory accidents,[43] or, rarely, by congenital infection.

• Classification of *Plasmodium*

The genus *Plasmodium* was divided by Garnham[27] into nine subgenera, of which three occur in mammals, four in birds, and two in lizards. Most *Plasmodium* spp. are parasites of birds; others occur in such animals as rodents, primates, and reptiles. Some species, such as the rodent parasite *Plasmodium berghei* and the chicken parasite *P. gallinaceum,* are very useful in laboratory studies of immunity, physiology, and so forth.

Still other species, normally parasitic in nonhuman primates, occasionally infect humans as zoonoses or can be acquired by humans when infected experimentally.

There are four species of *Plasmodium* normally parasitic in humans: *P. falciparum, P. vivax, P. malariae,* and *P. ovale.* Molecular analysis of the small subunit rRNA gene of three of these species suggests they are members of separate phylogenetic lineages, each more closely related to *Plasmodium* of other animal parasites than to each other.[110] *Plasmodium falciparum* is apparently most closely related to *P. gallinaceum* and *P. lophurae* from birds, *P. malariae* seems to form a lineage of its own, and *P. vivax* seems most closely related to several species from other primates. Insufficient data are available to place *P. ovale* in a lineage. Ultrastructural evidence also supports the relationship of *P. falciparum* to the avian plasmodia.[94]

Plasmodium vivax. *Plasmodium vivax* (Plates 1 and 2) is the cause of **benign tertian malaria,** also known as **vivax malaria** or **tertian ague.** When early Italian investigators noted the actively motile trophozoites of the organism within host corpuscles, they nicknamed it *vivace,* foreshadowing the Latin name *vivax,* which later was accepted as its epithet. The designation *tertian* is based on the fact that fever paroxysms typically recur every 48 hours (Table 9.1); the name is derived from the ancient Roman custom of calling the day of an event the first day, 48 hours later hence being the third.

The species flourishes best in temperate zones, rarely as far north as Manchuria, Siberia, Norway, and Sweden and as far south as Argentina and South Africa. Because malaria eradication campaigns have been so successful in many of the temperate areas of the world, however, the disease has practically disappeared from them. Most vivax malaria today is found in Asia; about 40% of malaria among United States military personnel in Vietnam resulted from *P. vivax.*[13] It is common in North Africa but drops off in tropical Africa to very low levels, partly because of a natural resistance of black persons to infection with this species (see Duffy blood groups). About 43% of malaria in the world is caused by *P. vivax.*

Sporozoites that are 10 to 14 μm long invade cells of the liver parenchyma within one or two days after injection with the mosquito's saliva. By the seventh day the exoerythrocytic schizont is an oval body about 40 μm long that has blue-staining cytoplasm, a few large vacuoles, and lightly staining nuclei. On attaining maturity, the schizonts' vacuoles disappear, and about 10,000 merozoites are produced. The fate of these merozoites is a subject of debate. Certainly many of them are killed outright by host defenses. Others invade erythrocytes to initiate the erythrocytic stages of development. Still others may possibly reinfect hepatic cells (see relapse of malaria p. 149).

Relapses up to eight years after initial infection are characteristic of vivax malaria. The patient is in normal health during the intervening periods of latency. The relapses are believed to result from genetic differences in the original sporozoites; that is, some give rise to tissue schizonts that take much longer to mature.[20] However, occurrence of relapses may also be related to the immune state of the host (see discussion of immunity later in the chapter).

Plasmodium vivax merozoites invade only young erythrocytes, the reticulocytes, and apparently are unable to penetrate mature red cells. Merozoites can only penetrate erythrocytes with mediated receptor sites, such sites being genetically determined.[39] Known as the **Duffy blood groups,** two codominant alleles, Fy^a and Fy^b are recognized by their different antigens. A third allele, *Fy,* has no corresponding antigen. The *Fy/Fy* genotype is common in black Africans and Americans (40% or more) and rare in white people (about 0.1%). Fy^a and Fy^b are receptors for *P. vivax* and *P. knowlesi;*[70] hence *Fy/Fy* is refractory to infection. This explains the natural resistance of many black people to vivax malaria. The Duffy negative genotype may represent the original, rather than the mutant, condition in tropical Africa.[63]

Soon after invasion of the erythrocyte and formation of the ring stage, the cytoplasm becomes actively ameboid, throwing out pseudopodia in all directions and fully justifying the name *vivax.* As the trophozoite grows, the red cell enlarges, loses its pink color, and develops a peculiar stippling known as **Schüffner's dots** (Plate 1, *5*). These dots are visible by light microscopy after Romanovsky staining. With electron microscopy they can be seen as small surface invaginations **(caveolae),** surrounded by small vesicles.[91] The trophozoite occupies about two-thirds of the red cell after 24 hours. The vacuole disappears, the organism becomes more sluggish, and hemozoin granules accumulate as the trophozoite grows. By 36 to 42 hours after infection, nuclear division begins and is repeated several times, yielding 12 to 24 nuclei in the mature schizont. Once schizogony begins, the hemozoin granules accumulate in two or three masses in the parasite, ultimately to be left in the **residual body** (see Fig. 4.8) and engulfed by the host's reticuloendothelial system. The rounded merozoites, about 1.5 μm in diameter, immediately attack new erythrocytes. Erythrocytic schizogony takes somewhat less than 48 hours, although early in the disease there are usually two populations, each maturing on alternate days, resulting in a daily, or **quotidian,** periodicity (refer to discussion of pathology later in the chapter).

Some merozoites develop into gametocytes rather than into schizonts. The factors determining the fate of a given merozoite are not known, but since the gametocytes have been found as early as the first day of parasitemia in rare instances, it may be possible for exoerythrocytic merozoites to produce gametocytes. Some differences in appearance of the micro- and macrogametocytes are given in Table 9.1. The mature macrogametocyte fills most of the enlarged erythrocyte and measures about 10 μm wide. The mature microgametocyte is smaller than the macrogametocyte and usually does not fill the erythrocyte.

Gametocytes take four days to mature, twice the time required for schizont maturation. Macrogametocytes often outnumber microgametocytes by two to one. A single host cell may contain both a gametocyte and a schizont.

Formation of zygote, ookinete, and oocyst are as described previously. The oocyst may reach a size of 50 μm and produce up to 10,000 sporozoites. If the ambient temperature is too high or too low, the oocyst blackens with pigment and degenerates, a phenomenon noted by Ross. Too many developing oocysts kill the mosquito before the sporozoites complete development.

Plasmodium falciparum.   Malaria known as **malignant tertian, subtertian,** or **estivoautumnal (E-A)** is caused by *P. falciparum* (Plates 3 and 4), the most virulent of *Plasmodium* spp. in humans. It was nearly cosmopolitan at one time, with a concentration in the tropics and subtropics. It still extends into the temperate zone in some areas, although it has been eradicated in the United States, the Balkans, and around the Mediterranean. Nevertheless, falciparum malaria reigns supreme as the greatest killer of humanity in the tropical zones of the world today, accounting for about 50% of all malaria cases.

Among the many cases studied by Laveran, persons suffering from "malignant tertian malaria" interested him the most. He had long noticed a distinct darkening of the gray matter of the brain and abundant pigment in other tissues of his deceased patients. When in 1880 he saw crescent-shaped bodies in the blood and watched them exflagellate, he knew he had found living parasites. The confusion that surrounded the correct name for this species continued until 1954, when the International Commission of Zoological Nomenclature validated the epithet *falciparum.*

Malignant tertian malaria is usually blamed for the decline of the ancient Greek civilization, the halting of Alexander the Great's progress to the East, and the disintegration of some of the Crusades. In more modern times, the Macedonian campaign of World War I was destroyed by falciparum malaria, and the disease caused more deaths than did battles in some theaters of World War II.

As in other species, the exoerythrocytic schizont of *P. falciparum* grows in liver cells. It is more irregularly shaped than that of *P. vivax,* with projections extending in all directions by the fifth day. The schizont ruptures in about five and one-half days, releasing about 30,000 merozoites. There seems to be no second exoerythrocytic cycle, and true relapses do not occur. However, recrudescences of the disease may follow remissions of up to a year, occasionally up to two or three years, after initial infection, apparently because small populations of the parasites remain in the red cells.

Merozoites can invade erythrocytes of any age, including reticulocytes; therefore, falciparum malaria usually has much higher levels of parasitemia than the other types. Soon after invasion of the erythrocyte, the trophozoite produces a protein that is deposited beneath and within the erythrocyte surface membrane, and surface deformations called "knobs" appear on the surface.[8] In addition, novel antigens, some of which may be derived from the parasite, appear on the cell surface.[71,92] One or more of these proteins bind to certain glycoproteins on postcapillary venular endothelium. This causes sequestration of infected erythrocytes in the vascular endothelium of deep tissues,[1,2] including the brain, spleen, and bone marrow. Gametocyte-infected erythrocytes have no knobs and do not stick to the endothelium.[91] Hence one usually observes only early ring stages and/or gametocytes in blood smears from patients with falciparum malaria. If schizogony is well synchronized, parasites may be practically absent from peripheral blood toward the end of the 48-hour cycle. The possible adaptive benefit to *P. falciparum* is obscure and still controversial.[8] The knobs themselves do not seem to be necessary for cytoadherence because rare knobless phenotypes of *P. falciparum* can cytoadhere.[83] Infected erythrocytes can also bind to uninfected red cells, forming rosettes, which may also play a role in clogging venules.[106]

The early ring-stage trophozoite is the smallest of any *Plasmodium* spp. of humans: about 1.2 μm. Other diagnostic aids for ring stages of *P. falciparum* are given in Table 9.1 and illustrated in Plate 3. The frequency of multiple infections in the same cell has led some parasitologists to believe that the ring stages divide and that the binucleate rings are division stages. As it grows, the protozoan extends wispy pseudopodia, but it is never as active as *P. vivax.* The infected erythrocyte develops irregular blotches known as **Maurer's clefts** (Plate 3, 9). These are much larger than the fine Schüffner's dots found in *P. vivax* infections. They are caused by extension of the parasitophorous vacuole within the host cytoplasm.[52]

The mature schizont is less symmetrical than those of the other species infecting humans. It develops 8 to 32 merozoites, with 16 being the usual number. In contrast to the usual situation, schizonts may be fairly common in peripheral blood in some geographical areas. This may reflect strain differences. The erythrocytic cycle takes 48 hours, but the periodicity is not as marked as in *P. vivax,* and it may vary considerably with the strain of parasite. Extremely high levels of parasitemia may occur, with more than 65% of the erythrocytes containing parasites; a density of 25% is usually fatal. Two or three parasites per milliliter of blood may be sufficient to cause disease symptoms.

In *P. vivax* the gametocytes may appear in the peripheral blood almost at the same time as the trophozoites, but in *P. falciparum* the sexual stages require nearly 10 days to develop, and then they appear in large numbers. They develop in the blood spaces of the spleen and bone marrow—first assuming bizarre, irregular shapes, then becoming round, and finally changing into the crescent shape so distinctive of the species (Plate 3, 26 and 27). The hemozoin granules cluster around the nucleus in the micro- and macrogametocyte. This differs from *P. vivax,* in which pigment is diffuse throughout the cytoplasm (Table 9.1).

Plasmodium malariae.   **Quartan malaria,** with paroxysms every 72 hours, is caused by *P. malariae* (Plates 5 and 6). It was recognized by the early Greeks because the timing of the fevers differed from that of the tertian malaria parasites. Although Laveran saw and even illustrated the characteristic schizonts of this parasite, he refused to believe it was different from *P. falciparum.* In 1885 Golgi differentiated the tertian and quartan fevers and gave an accurate description of what is now known as *P. malariae.*

Plasmodium malariae is a cosmopolitan parasite but does not have a continuous distribution anywhere. It is common in many regions of tropical Africa, Myanmar (formerly Burma), India, Sri Lanka, Malaya, Java, New Guinea, and Europe. It is also distributed in the New World, including Guadeloupe, Guyana, Brazil, Panama, and at one time the United States. The peculiar distribution of this parasite has never been satisfactorily explained. It may be the only species of human malaria organism that also regularly lives in wild animals. Chimpanzees are infected at about the same rate as humans but are unimportant as reservoirs, since they do not live side by side with people. Some workers believe that *P. brasilianum* is really *P. malariae* in New World monkeys.[55] As noted before, present molecular evidence suggests that *P. malariae* is alone in its evolutionary lineage. This species accounts for about 7% of malaria cases in the world.

Exoerythrocytic schizogony is completed in 13 to 16 days. Erythrocytic forms build up slowly in the blood; the characteristic symptoms of the disease may appear before it is possible to find the parasites in blood smears. The ring forms are less ameboid than those of *P. vivax,* and the cytoplasm is somewhat thicker. Rings often retain their shape for as long as 48 hours, finally transforming into an elongated "band form," which begins to collect pigment along one edge (Plate 3, *6* and *10*). The nucleus divides into 6 to 12 merozoites at 72 hours. The segmenter is strikingly symmetrical and is called a *rosette* or *daisy-head* (Plate 5, *20*). Parasitemia levels are characteristically low, with one parasite per 20,000 red cells representing a high figure for this species. This low density is accounted for by the fact that the merozoites apparently can invade only aging erythrocytes, which are soon to be removed from circulation by the normal process of blood destruction.

Gametocytes probably develop in the internal organs, since immature forms are rare in peripheral blood. They are slow to develop in sporozoite-induced infections. Recrudescences of quartan malaria can occur up to 53 years after initial infection.[27] Because it can live in the blood so long, it is the most important cause of transfusion malaria.

Plasmodium ovale.   This species causes **ovale, or mild tertian, malaria** and is the rarest of the four malaria parasites of humans. It is confined mainly to the tropics, although it has been reported from Europe and the United States. Although common on the west coast of Africa, which may be its original home, the species is scarce in central Africa and present but not abundant in eastern Africa. It is known also in India, the Philippine Islands, New Guinea, and Vietnam. *Plasmodium ovale* is difficult to diagnose because of its similarity to *P. vivax* (Table 9.1).

The youngest ring stage has a large, round nucleus and a rather small vacuole that disappears early. The mature schizont is oval or spheroid and is about half the size of the host cell. Eight merozoites are usually formed, but there is a range of 4 to 16. Schüffner's dots appear early in the infected blood cells. They are very numerous and larger than those in *P. vivax* infections and stain a brighter red color. As in *P. vivax,* the Schüffner's dots are due to caveolae.

Gametocytes of *P. ovale* take longer to appear in the blood than do those of other species. They are numerous enough three weeks after infection to infect mosquitoes regularly.

• Malaria: The Disease

Certain disease aspects of *Plasmodium* spp. have been mentioned in the preceding pages; following is a brief consideration of the subject, particularly in relation to pathogenesis and public health. We urge you to consult other references for more complete treatment.[12,100,113]

Diagnosis.   Diagnosis depends to some extent on the clinical manifestations of the disease, but most important is demonstration of the parasites in stained smears of peripheral blood. Technical details can be found in many texts and laboratory manuals of medical parasitology. Some useful criteria for differential diagnosis are summarized in Table 9.1.

Because diagnosis by this conventional method requires training and time and results in very low parasitemias being easily missed, some newer techniques can be very helpful. An inexpensive method of visualizing the parasites after staining with a fluorescent dye is simple, very sensitive, and adaptable for field conditions.[48] A DNA probe specific for the detection of *P. falciparum* is sensitive and suitable for field conditions,[7] and a refined ELISA test, also suitable for field use, has been described.[87]

Pathogenesis.   Most of the major clinical manifestations of malaria may be attributed to two general factors: (1) the host inflammatory response, which produces the characteristic chills and fever, as well as other related phenomena; and (2) anemia, arising from the enormous destruction of red blood cells. Severity of the disease is correlated with the species producing it: Falciparum malaria is most serious and vivax and ovale the least dangerous.

Fever is a common, nonspecific reaction of the body to infection, functioning at least in part to increase the rate of metabolic reactions important in host defenses. Fever in malaria is correlated with the maturation of a generation of merozoites and the rupture of the red blood cells that contain them. It is widely believed that fever is stimulated by the waste products of the parasites, released when the erythrocytes lyse. Release of these malarial antigens into the circulation apparently triggers a burst of tumor necrosis factor, or TNF (p. 26), from activated macrophages.[18,68,113] Induction of fever is among the effects of overproduction of TNF, and TNF toxicity can account for most or all of the symptoms described in the next few paragraphs.[18]

A few days before the first paroxysm, the patient may feel malaise, muscle pain, headache, loss of appetite, and slight fever; or the first paroxysm may occur abruptly, without any prior symptoms. A typical attack of benign tertian or quartan malaria begins with a feeling of intense cold as the hypothalamus, the body's thermostat, is activated, and the temperature then rises rapidly to 104° to 106° F. The teeth chatter, and the bed may rattle from the victim's shivering. Nausea and vomiting are usual. The hot stage begins within one-half to one hour later, with intense headache and feeling of intense heat. Often a mild delirium stage lasts for several hours. As copious perspiration signals the end of the hot stage, the temperature drops back to normal within two to three hours, and the entire paroxysm is over within eight to twelve hours. The person may sleep for a while after an episode and feel fairly well until the next paroxysm. The foregoing time periods for the stages are usually somewhat shorter in quartan malaria, and the paroxysms recur every 72 hours. In vivax malaria the periodicity is often quotidian early in the infection, since two populations of merozoites usually mature on alternate days. "Double" and "triple" quartan infections also are known. Only after one

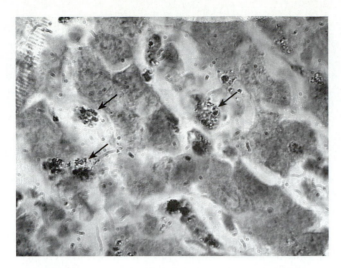

FIGURE 9.5

Section of liver tissue with numerous deposits of malarial pigment (*arrows*).

Photograph by Larry S. Roberts.

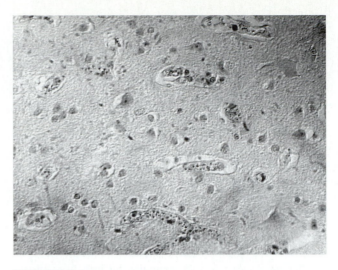

FIGURE 9.6

Section of cerebral tissue, demonstrating capillaries filled with erythrocytes infected by *Plasmodium falciparum.* Infected red cells are marked by pigment; the parasites themselves are transparent.

Photograph by Larry S. Roberts.

or more groups drop out does the fever become tertian or quartan, and the patient experiences the classical good and bad days.

Because the synchrony in falciparum malaria is much less marked, the onset is often more gradual, and the hot stage is extended. The fever episodes may be continuous or fluctuating, but the patient does not feel well between paroxysms, as in vivax and quartan malaria. In cases in which some synchrony develops, each episode lasts 20 to 36 hours, rather than 8 to 12, and is accompanied by much nausea, vomiting, and delirium. Concurrent infections with *P. vivax* and *P. falciparum* are not uncommon.

The main causes of the anemia are destruction of both parasitized and nonparasitized erythrocytes, inability of the body to recycle the iron bound in the insoluble hemozoin, and an inadequate erythropoietic response of the bone marrow. Why such large numbers of nonparasitized red cells are destroyed is still not understood, but some evidence has indicated complement-mediated, autoimmune hemolysis. For reasons yet unclear, however, the spleen in acute malaria removes substantial numbers of unparasitized red cells from the blood, an effect that may persist beyond the time of parasite clearance.[113] Both the splenic removal of red cells and the defective bone marrow response may be due in part to TNF toxicity.[18] Destruction of erythrocytes leads to an increase in blood bilirubin, a breakdown product of hemoglobin. When excretion cannot keep up with formation of bilirubin, jaundice yellows the skin. The hemozoin is taken up by circulating leukocytes and is deposited in the reticuloendothelial system. In severe cases the viscera, especially the liver, spleen, and brain, become blackish or slaty as the result of pigment deposition (Fig. 9.5, Plate 8). After ingesting hemozoin, macrophages suffer impairment in phagocytic ability.[105]

Falciparum malaria is always serious, and sometimes severe complications are produced. The most common of these is **cerebral malaria,** which may account for 10% of falciparum malaria admitted to the hospital and 80% of such deaths.[114] Cerebral malaria may be gradual in onset, but it is commonly sudden; a progressive headache may be followed by a coma, an uncontrollable rise in temperature to above 108° F, and psychotic symptoms or convulsions, especially in children. Death may ensue within a matter of hours. Initial stages of cerebral malaria are easily mistaken for a variety of other conditions, including acute alcoholism, usually with disastrous consequences. Another grave and usually fatal complication of severe falciparum malaria is **pulmonary edema,** which in some cases may be a result of overadministration of intravenous fluids.[113] Difficulty in breathing increases and death may ensue in a few hours.[114] A combination of other severe manifestations leads to a condition known as **algid malaria,** which is a rapid development of shock. There is a circulatory collapse with markedly low blood pressure. The skin is cold and clammy; peripheral veins are constricted. Algid malaria is sometimes associated with a bacterial infection of the blood (**septicemia**) with toxemia and massive gastrointestinal hemorrhage.

These severe complications traditionally have been blamed on a "plugging" of the capillaries in the affected organs by clots (Fig. 9.6), leading to circulatory stasis and hypoxia. More recently, some investigators have concluded that symptoms of cerebral malaria are due to stagnant hypoxia caused by adherence of the parasitized erythrocytes to the endothelium of cerebral venules and capillaries.[108,109] The symptoms of algid malaria may result from circulatory stasis in the gastrointestinal tract due to the same mechanism. However, residual neurological deficit is rare in patients who recover from cerebral malaria, a surprising result if the condition is due to hypoxia of brain tissues. Therefore, Clark and his coworkers have proposed that the excess TNF causes release of nitric oxide (NO) from vascular endothelium and vascular smooth muscle in the brain.[19] The NO would diffuse into the surrounding neural tissue and disrupt neuronal signalling.

Blackwater fever is another grave condition associated with falciparum malaria, but its clinical picture is distinct from the foregoing. It is an acute, massive lysis of erythrocytes, marked by high levels of free hemoglobin and breakdown products of hemoglobin in the blood and urine and by renal insufficiency. Because of the presence of hemoglobin and its products in the urine, the fluid is quite dark, hence the name of the condition. A prostrating fever, jaundice, and persistent vomiting occur. Renal failure is usually the immediate cause of death. Damage to the kidney is now thought to result from renal anoxia, reducing efficiency of the glomerular filtration and tubular resorption. The massive hemolysis is not directly attributable to the parasites; the organisms frequently cannot be demonstrated. However, the condition is almost always associated with areas of *P. falciparum* hyperendemicity, found in persons with prior falciparum malaria and very frequently with irregular or inadequate treatment for the infection. Inadequate suppressive or therapeutic doses of quinine most often have been implicated, but many cases have been reported following treatment with quinacrine and pamaquine, and the condition can occur in persons who have not been treated at all. It is now believed that blackwater fever is an autoimmune phenomenon and is triggered by some stimulus that results in release of large amounts of antibodies, which act as hemolysins, into the circulation. Mortality is 20% to 50%. The incidence of blackwater fever has declined in recent years, perhaps due to the use of drugs other than quinine for prophylaxis.

Hypoglycemia (reduced concentration of blood glucose) is a common symptom in falciparum malaria. It is usually found in women with uncomplicated or severe malaria who are pregnant or have recently delivered as well as in other cases of severe falciparum malaria.[114] Coma produced by hypoglycemia has commonly been misdiagnosed as cerebral malaria. This condition is usually associated with quinine treatment. The pancreatic islet cells are stimulated by quinine to increase insulin secretion, thus lowering blood glucose.[108] This effect may also be due to excessive TNF.[18]

Immunity and Resistance.   Despite the fact that much of the disease results from the inflammatory and immune responses of the host, host defenses are vital in limiting the infection. One vivax segmenter producing 24 merozoites every 48 hours would give rise to 4.59 billion parasites within 14 days, and the host would soon be destroyed if the organisms continued reproducing unchecked.[55] The development of some protective immunity is evident in malaria, and we will consider only briefly some of the practical effects.

Relapses and recrudescences may be due to lowered antibody titers or increased ability of the parasite to deal with the antibody, but they also may depend on genetic differences in sporozoite populations. Symptoms in a relapse are usually less severe than those in the primary attack, but the level of parasitemia is higher. After the primary attack and between relapses, the patient may have a **tolerance** to the effects of the organisms, remaining asymptomatic, while in fact having as high a circulating parasitemia level as during the primary attack. Such tolerant carriers are very important

in the epidemiology of the disease. Tolerance may be related to loss of reactivity to TNF; humans can become refractory to TNF on continued exposure.[18]

Protective immunity to malaria is primarily a premunition—that is, a resistance to superinfection. It is effective only as long as a small, residual population of parasites is present; if the person is completely cured, susceptibility returns. Thus, in highly endemic areas, infants are protected by maternal antibodies, and young children are at greatest risk after weaning. The immunity of children who survive a first attack will be continuously stimulated by the bites of infected mosquitoes as long as the children live in the malarious area. Nonimmune adults are highly susceptible. Immunity is species specific and to some degree strain specific so that a person may risk a new infection by migrating from one malarious area to another. Falciparum malaria is unmitigated in its severity to a person who is immune to vivax malaria.

The protective mechanisms by immune effectors against the parasites remain unclear. In vitro binding of specific antibodies to surface proteins of sporozoites and merozoites can prevent penetration of host cells, and there is some evidence for an antibody-dependent, cell-mediated cytotoxicity (ADCC).[68] At least part of sporozoite-induced immunity depends on killing of infected liver cells by cytotoxic T lymphocytes.[24] However, the immune response is inefficient: Malaria induces a polyclonal B-cell activation, with dramatic synthesis of especially IgG and IgM, only 6% to 11% of which is specifically against malarial antigens.[68] It is probable that an important mechanism for *Plasmodium* to evade the host defense system involves exposure of the host to a large repertoire of antigenic epitopes.[24] Analogous systems of immune evasion are shown in trypanosomes (p. 60) and schistosomes (p. 31).

Black persons are much less susceptible to vivax malaria than are whites, and falciparum malaria in blacks is somewhat less severe. The genetic basis for this phenomenon is explained by the inheritance of Duffy blood groups (p. 144). Other factors that can contribute to genetic resistance are certain heritable anemias: sickle cell, favism, and thalassemia. Although these conditions are of negative selective value in themselves, they have been selected for in certain populations because they confer resistance to falciparum malaria. The most well-known of these is **sickle-cell anemia.** In persons homozygous for this trait a glutamic acid residue in the amino acid sequence of hemoglobin is replaced by a valine, interfering with the conformation of the hemoglobin and oxygen-carrying capacity of the erythrocytes. Persons with sickle-cell anemia usually die before the age of 30. In heterozygotes some of the hemoglobin is normal, and these persons can live relatively normal lives, but the presence of the abnormal hemoglobin inhibits growth and development of *P. falciparum* in their erythrocytes. The selective pressure of malaria in Africa has led to maintenance of this otherwise undesirable gene in the population. This legacy has unfortunate consequences when the people are no longer threatened by malaria, as in the United States, where 1 in 10 Americans of African ancestry is heterozygous for the sickle-cell gene, and 1 in 400 is homozygous.

Relapse in Malarial Infections.   Since the advent of an antimalarial drug (quinine) in the sixteenth century, it has been noted that some persons who have been treated and seemingly recovered relapse back into the disease weeks, months, or even years after the apparent cure.[36] Coatney contributed an interesting history of the phenomenon.[20] Malarial relapse engendered much speculation and research for many years. The discovery of preerythrocytic schizogony in the liver by Shortt and Garnham in 1948 seemed to have solved the mystery. It appeared most reasonable to assume that preerythrocytic merozoites simply reinfected other hepatocytes, with subsequent reinvasion of red blood cells. This would explain why relapse occurred after erythrocytic forms were eliminated by erythrocytic schizontocides, such as quinine and chloroquine.

However, not all species of *Plasmodium* cause relapse. Among the parasites of primates, only *P. vivax* and *P. ovale* of humans and *P. cynomolgi, P. fieldi,* and *P. simiovale* of simians cause true relapse. If preerythrocytic merozoites reinvade hepatocytes, then relapse should occur in all species.

In species that undergo relapse, there are two populations of exoerythrocytic forms.[53,54] One develops rapidly into schizonts, as previously described, but the other remains dormant as **hypnozoites** ("sleeping animalcules"). *P.vivax, P. ovale,* and *P. cynomolgi* have hypnozoites, but they have not been found in any species that does not cause relapse. How long the hypnozoite can remain capable of initiating schizogony and what triggers it to do so are unknown. Primaquine is an effective hypnozoiticide.

Malariologists long thought that *P. malariae,* a dangerous species in humans, also exhibited relapse, but we now know that this species can remain in the blood for years, possibly for the lifetime of the host, without showing signs of disease and then suddenly initiate a clinical condition. This is more correctly known as a **recrudescence,** since preerythrocytic stages are not involved. The danger of transmission of this parasite through blood transfusion is evident. Because there are no hypnozoites, treatment of *P. malariae* with primaquine is unnecessary.

Epidemiology, Control, and Treatment.   In light of the prevalence and seriousness of the disease, epidemiology and control are extremely important, and thorough consideration is far beyond the scope of this book. Some aspects of these subjects have been touched on in the preceding pages, and the following will give you additional insight into the problem involved (see also Chapter 38; Strickland;[100] and Bruce-Chwatt[12]).

In addition to natural or biological transmission, discussed next, human to human transmission can spread malaria. Accidental transmission can occur by blood transfusion and by the sharing of needles by IV drug users. Although rare, infection of the newborn from an infected mother also occurs.[12] Neurosyphilis was formerly treated by deliberate infection with malaria. (A great deal of knowledge about malaria was gained during these treatments, but we still do not understand why infection with malaria alleviated the symptoms of the terrible disease of neurosyphilis.)[17]

A variety of interrelated factors contributes to the level of natural transmission of the disease in a given area (Fig. 9.7). One must thoroughly study and understand all these factors before undertaking a malaria control program with any hope of success. Following (modified from Strickland[100]) are the most important: (1) Reservoir—the prevalence of the infection in humans, including persons with symptomatic disease and tolerant individuals, and in some cases other primates, with high enough levels of parasitemia to infect mosquitoes. (2) Vector—suitability of the local anophelines as hosts; their breeding, flight, and resting behavior; feeding preferences; and abundance. (3) New hosts—availability of nonimmune hosts. (4) Local climatic conditions. (5) Local geographical and hydrographical conditions and human activities that determine availability of mosquito breeding areas.

Of the approximately 390 species of *Anopheles,* some are more suitable hosts for *Plasmodium* than are others. Of those which are good hosts, some prefer animal blood other than human; therefore, transmission may be influenced by the proximity in which humans live to other animals. The preferred breeding and resting places are very important. Some species breed only in fresh water; some prefer brackish water; some like standing water around human habitations, such as puddles or trash that collects water, such as bottles and broken coconut shells. Water, vegetation, and amount of shade are important, as are whether the species enters dwellings and rests there after feeding and whether the species flies some distance from breeding areas. *Anopheles* spp. show an astonishing variety of such preferences; two specific examples can be cited for illustration. *Anopheles darlingi* is the most dangerous vector in South America, extending from Venezuela to southern Brazil, breeding in shady, fresh water among debris and vegetation. It invades houses and prefers human blood. *Anopheles bellator* is an important vector in cocoa-growing areas of Trinidad and coastal states of southern Brazil, breeding in partial shade in the "vases" of epiphytic bromeliads (plants that grow attached to trees and collect water in the center of their leaf rosettes). It prefers humans but enters dwellings only occasionally and then returns to the forest. The importance of thorough investigation of such factors is demonstrated by cases in which swamps have been flooded with seawater to destroy the breeding habitat of the species, only to create extensive breeding areas for a brackish water species that turned out to be just as effective a vector.

Sometimes deliberate government policy exacerbates transmission of malaria. During the 1970s and 1980s Brazil was attempting to open up the Amazon regions to farming. This caused a great influx of nonimmune people from the cities into the newly cleared land, and malaria incidence jumped from 6 in 1000 per year in 1971 to 30 in 1000 per year in 1986.[23] Because the soils of the former rain forest were poorly suited for agriculture, return of the failed farmers (often infected with malaria) to their old homes was inevitable. This created many new foci of transmission outside the Amazon region.[23]

Valuable actions in mosquito control include destruction of breeding places when possible or practical, introduction of

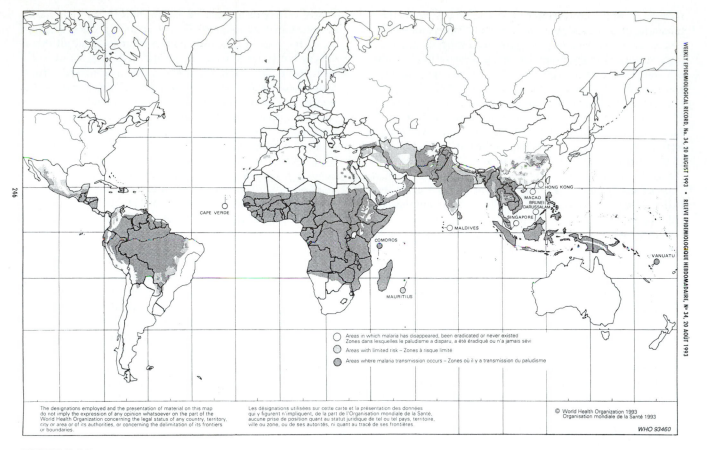

WEEKLY EPIDEMIOLOGICAL RECORD, No. 34, 20 AUGUST 1993 • RELEVÉ EPIDEMIOLOGIQUE HEBDOMADAIRE, N° 34, 20 AOÛT 1993

Legend on map:

○ Areas in which malaria has disappeared, been eradicated or never existed
Zones dans lesquelles le paludisme a disparu, a été éradiqué ou n'a jamais sévi

◔ Areas with limited risk – Zones à risque limité

● Areas where malaria transmission occurs – Zones où il y a transmission du paludisme

The designations employed and the presentation of material on this map do not imply the expression of any opinion whatsoever on the part of the World Health Organization concerning the legal status of any country, territory, city or area or of its authorities, or concerning the delimitation of its frontiers or boundaries.

Les désignations utilisées sur cette carte et la présentation des données qui y figurent n'impliquent, de la part de l'Organisation mondiale de la Santé, aucune prise de position quant au statut juridique de tel ou tel pays, territoire, ville ou zone, ou de ses autorités, ni quant au tracé de ses frontières.

© World Health Organization 1993
Organisation mondiale de la Santé 1993

WHO 93460

FIGURE 9.7

Areas of risk for malaria transmission, 1991.

Reproduced by permission of the World Health Organization, from: "World malaria situation 1991," Parts I and II. *Weekly Epidemiological Record,* 68:245–252/253–260, 1993.

mosquito predators such as the mosquito-eating fish *Gambusia affinis,* and judicious use of insecticides. The efficacy and economy of DDT have been a boon to such efforts in underdeveloped countries. Although we now are aware of the environmental dangers of DDT, these dangers may be preferable and minor compared with the miseries of malaria. Unfortunately, reports of DDT-resistant strains of *Anopheles* are increasing, and this phase of the battle is becoming more difficult. For exterminating susceptible *Anopheles* spp. that enter dwellings and rest there after feeding, spraying the insides of houses with residual insecticides can be effective and cheap, without incurring any environmental penalty. Unfortunately, some *Anopheles* rest in houses only briefly before or after feeding, and sufficient quantities of DDT are difficult to obtain on the world market.[23]

Appropriate drug treatment of persons with the disease, as well as prophylactic drug treatment of newcomers to malarious areas, is an integral part of malaria control. Centuries ago the Chinese used extracts of certain plants, such as *chang shan* and *shun qi* (the roots and leaves of *Dichroa febrifuga,* family Saxifragaceae) and *qing hao* (the annual *Artemisia annua,* family Compositae) (Fig. 9.8), that had antimalarial properties.[50,85] In the meantime, Europeans were medically powerless and depended on absurd and superstitious remedies.

FIGURE 9.8

Artemisia annua, the source of the antimalarial drug qinghaosu, growing in the herb garden of the College of Traditional Medicine, Guangzhou, China.

Photograph by Larry S. Roberts.

Extracts of bark from Peruvian trees were used with varying success to treat malaria, but alkaloids from the bark of the Peruvian tree *Cinchona ledgeriana* proved dependable and effective. The most widely used of these alkaloids had been **quinine,** discovered in the sixteenth century. The alkaloid of *D. febrifuga,* febrifugine, is now considered too toxic for human use, but the terpene from *A. annua,* called *qinghaosu,* recently has been "rediscovered" and promises to be a valuable drug.[50]

Only two synthetic antimalarials were discovered before World War II. Japanese capture of cinchona plantations early in the war created a severe quinine shortage in the United States, stimulating a burst of investigation that produced a number of effective drugs. The most important of these was **chloroquine.** Subsequently a number of valuable drugs have been developed, including **primaquine, mefloquine, pyrimethamine, proguanil,** sulfonamides such as **sulfadoxine,** and antibiotics such as **tetracycline.** Only primaquine is effective against all stages of all species; the others vary in efficacy according to stages and species, with the erythrocytic stages being most susceptible. The drugs of choice are chloroquine and primaquine for *P. vivax* and *P. ovale* malarias and chloroquine alone for *P. malariae* infections. Chloroquine is still recommended for strains of *P. falciparum* sensitive to that drug.[29]

Resistance of *P. falciparum* to chloroquine has now spread through Asia, Africa, and South America,[78] and resistance to other drugs is often present. A combination of sulfadoxine and pyrimethamine (Fansidar) was used for chloroquine-resistant falciparum malaria, but Fansidar-resistant *P. falciparum* is now present in a number of areas. For multidrug resistant *P. falciparum,* mefloquine (Lariam) is still effective (although there are reports of mefloquine resistance).[64,112] Resistance to qinghaosu has not been reported in the field, but resistant strains have been produced in the laboratory.[45] Current recommendations for travelers in malarious regions are to take prophylactic doses of mefloquine regularly.[15] For those unable to tolerate mefloquine, the U.S. Centers for Disease Control recommends prophylaxis with doxycycline (a tetracycline derivative).

Because of the ominous and dangerous multidrug resistance in various strains of *P. falciparum,* it is clear that the search for satisfactory malaria treatments must continue; perhaps the answer lies in the development of vaccines. This area is the subject of intensive investigation, and much progress has been made.[5,33,69,73,74] The current thrust has been made possible by two major advances: the development of methods whereby *P. falciparum* could be cultured in vitro,[104] thus making a large supply of organisms available, and the development of recombinant DNA technology. However, difficulties have been numerous. Different stages of the parasite have different antigens on their surface; therefore, antibodies against merozoites will not affect sporozoites or gametocytes. Also, as we already noted, the immunity produced in natural infections is only partial. A sporozoite continually sloughs off its outer coat and restores it with newly synthesized protein, and it has multiple epitopes.[33] Thus, sporozoites evade the host immune response by producing new binding sites and producing "decoys" in the form of sloughed molecules.

Most effort for a vaccine has centered on the sporozoites, since killing sporozoites would prevent initial infection. The gene for a surface protein in sporozoites (circumsporozoite protein, or CS protein) of *P. falciparum* was cloned in bacteria.[73] The protein produced was used to immunize volunteers with mixed results;[6] therefore, merozoite surface proteins are also under investigation.[25] Patarroyo and his colleagues[77] have reported good results in immunizing volunteers with a synthetic vaccine based on peptides from both sporozoite and merozoite coats. Additional trials of Patarroyo's vaccine are ongoing.[31,62,99] Human volunteers have been immunized against *P. falciparum* with radiation-attenuated sporozoites.[42] Because CS vaccine immunization is predominantly antibody mediated, while that with the attenuated sporozoites is cell mediated, strategies designed to induce stronger cell-mediated immunity may be required for a successful vaccine.[18,74]

In the 1950s many parasitologists thought that an expenditure of effort and money could achieve the eradication of malaria from large areas of the globe: Its scourge would rest only in history. Such views were naively optimistic. Not only did we not anticipate insecticide-resistant *Anopheles* spp., drug-resistant *Plasmodium,* and animal reservoirs, but also we took insufficient account of the enormous logistical problems of control in wilderness areas; neither did we consider the disruptive effects of wars and political upheavals on control programs. Malaria will be with us for a long time, probably as long as there are people.

Metabolism, Drug Action, and Drug Resistance.

- *Energy Metabolism.* Zhang[116] likened the metabolism of *Plasmodium* spp. to that of cancer cells. They derive energy primarily by the degradation of glucose to lactate, even though oxygen is available, and a complete Krebs cycle has not been demonstrated. Species from birds have cristate mitochondria; they nevertheless depend heavily on glycolysis for energy, converting four to six molecules of glucose to lactate for every one they oxidize completely. Most mammalian species have acristate mitochondria. Interestingly, the sporogonic stages of these organisms in the mosquito possess prominent, cristate mitochondria, reflecting perhaps a developmental change in metabolic pattern analogous to that observed in trypanosomes (p. 58). Treatment of the host with qinghaosu leads to swelling of the mitochondria of *P. inui* (a mammalian species with prominent mitochondria) within two and one-half hours.[47] Host mitochondria are unaffected. Similar reactions have been observed after primaquine treatment, leading to the suggestion that these drugs act via inhibition of mitochondrial metabolic reactions.

  The erythrocytic forms of *Plasmodium* act as facultative anaerobes, consuming oxygen when it is available. They probably use oxygen for biosynthetic purposes, especially synthesis of nucleic acids. Also, a branched electron transport system, analogous to that suggested for some helminths (p. 317), has been proposed,[90] but a classical cytochrome system has not been demonstrated. A limiting factor may be the parasite's inability to synthesize coenzyme A, which it must obtain from its host; this cofactor is necessary to introduce two-carbon fragments into the Krebs cycle. Supplies of CoA in the mammalian erythrocyte may be even more limited and may impose restrictions on any CoA-dependent reaction.

Both bird and mammal plasmodia fix carbon dioxide into phosphoenolpyruvate, as do numerous other parasites (see Fig. 20.29). In plasmodia the carbon dioxide-fixation reaction can be catalyzed by either phosphoenolpyruvate carboxykinase or phosphoenolpyruvate carboxylase. Chloroquine and quinine inhibit both enzymes, possibly accounting for some of the antimalarial activity of these drugs. The significance of the carbon dioxide fixation is not clearly understood; it may be to reoxidize NADH produced in glycolysis, or its reactions may function to maintain levels of intermediates for use in other cycles.

Activity of the pentose phosphate pathway is low in plasmodia, and the parasites apparently lack the first enzyme in the pathway, glucose 6-phosphate dehydrogenase (G6PDH). Because an important function of the pathway is to furnish reducing power in the form of NADH, it has been suggested that *Plasmodium* gets NADH from its host. Persons with a genetic deficiency in erythrocytic G6PDH are more resistant to malaria. Ingestion of various substances such as the antimalarial drug primaquine or the broad bean *Vicia favia* can bring on a hemolytic crisis of varying severity.[59] Such genes are relatively frequent in black people and some Mediterranean whites. Over 5% of Southeast Asian refugees entering the United States have had a G6PDH deficiency.[89] People with a deficiency in G6PDH are protected to some extent against *P. falciparum*. However, presence of the deficiency should be determined before treatment with primaquine to avoid a hemolytic crisis.[89]

- **Protein Degradation.** Some 25 different proteases have been described from various species of *Plasmodium*.[11] They are vital in maturation and release of merozoites from red cells, invasion of cells, and digestion of hemoglobin in the food vacuole.

Plasmodia depend heavily on host hemoglobin as a source of nutrition. They ingest a portion of host cytosol via the cytostome, and the vesicle thus formed migrates to and joins the central food vacuole, where the hemoglobin is rapidly degraded.[115] One of the products of hemoglobin digestion is ferriprotoporphyrin IX (FP, hematin), but FP inhibits several of the plasmodial proteases.[22] Therefore, the parasites sequester FP in the hemozoin pigment complex. Chloroquine is a dibasic amine (a weak base) and increases the pH in the food vacuole and prevents the digestion of hemoglobin. It binds to FP and prevents sequestration of the FP into inert hemozoin. Chloroquine is not effective against *Plasmodium* stages that do not form hemozoin.[22] Mefloquine also affects the food vacuoles,[46] and quinine may act by a similar mechanism.[57]

In chloroquine-sensitive strains of *P. falciparum*, chloroquine accumulates in the food vacuole, but in chloroquine-resistant strains, the drug moves out again just as rapidly. The rapid-efflux phenotype is apparently due to a mutation at a single genetic locus that spread rapidly from one or two foci in Southeast Asia and South America.[111] The mechanism of the efflux is still controversial. It is reversed (and chloroquine sensitivity regained) by verapamil and other Ca^+ channel blockers.[32] Some scientists believe that interaction with a permease on the lysosomal membrane may be involved.[107]

Resistance to *P. falciparum* by persons homozygous and heterozygous for sickle-cell hemoglobin (HbS) may involve several mechanisms, partly involving feeding and digestion by the protozoa. The parasite develops normally in cells with HbS until those cells are sequestered in the tissues.[26] Kept in this low oxygen environment for several hours, the cells have more of a tendency to sickle than do cells that pass through at a normal rate. When sickling occurs, HbS forms filamentous aggregates. The filamentous aggregates actually pierce the *Plasmodium*, apparently releasing digestive enzymes that lyse both parasite and host cell. Furthermore, K^+ leaks out of the sickled cell, depriving the parasite of this ion. Sickled cells also may block capillaries, further decreasing local oxygen concentration. Furthermore, sickling denatures hemoglobin and releases ferriprotoporphyrin IX, which has a membrane toxicity;[75] the effect of FP on plasmodial proteases was already mentioned.

- **Synthetic Metabolism.** As a specialized parasite, *Plasmodium* depends on its host cell for a variety of molecules other than the strictly nutritional ones. Specific requirements for maintenance of the parasites free of host cells are pyruvate, malate, NAD, ATP, CoA, and folinic acid. The inability of the organisms to synthesize CoA has been mentioned. They are unable to synthesize the purine ring de novo, thus requiring an exogenous source of purines for DNA and RNA synthesis. The purine source seems to be hypoxanthine salvaged from the normal purine catabolism of the host cell.[58]

Although plasmodia have cytoplasmic ribosomes of the eukaryotic type, several antibiotics that specifically inhibit prokaryotic (and mitochondrial) protein synthesis, such as tetracycline and tetracycline derivatives, have a considerable antimalarial potency. Tetracycline inhibits protein synthesis in *P. falciparum* as well as growth in vitro.[10]

Tetrahydrofolate is a cofactor that is very important in the transfer of one-carbon groups in various biosynthetic pathways in both prokaryotes and eukaryotes. Mammals require a precursor form, **folic acid,** as a vitamin, and dietary deficiency in this vitamin inhibits growth and produces various forms of anemia, particularly because of impaired synthesis of purines and the pyrimidine thymine. In contrast, *Plasmodium* (in common with bacteria) synthesizes tetrahydrofolate from simpler precursors, including *p*-aminobenzoic acid, glutamic acid, and a pteridine (Fig. 9.9); the organisms are apparently unable to assimilate folic acid. Analogs of *p*-aminobenzoic acid such as **sulfones** and **sulfonamides** block its incorporation, and some of these (for example, sulfadoxine and dapsone) are effective antimalarials. In both the mammalian pathway and the plasmodial-bacterial pathway an intermediate product is dihydrofolate, which must be reduced to

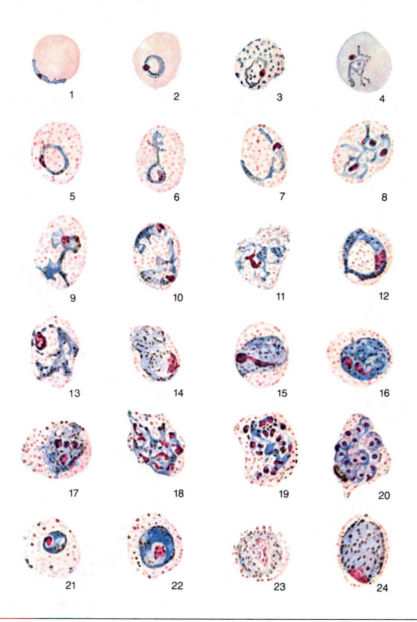

PLATE 1

Plasmodium vivax. **1,** normal-sized red cell with marginal ring-form trophozoite. **2,** young signet-ring-form trophozoite in macrocyte. **3,** slightly older ring-form trophozoite in red cell showing basophilic stippling. **4,** polychromatophilic red cell containing young tertian parasite with pseudopodia. **5,** ring-form trophozoite showing pigment in cytoplasm, in enlarged cell containing Schüffner's stippling (*dots*). (Schüffner's stippling does not appear in all cells containing growing and older forms of *P.vivax,* as would be indicated by these pictures, but it can be found with any stage from fairly young ring form onward.) **6, 7,** very tenuous medium trophozoite forms. **8,** three ameboid trophozoites with fused cytoplasm. **9, 11–13,** older ameboid trophozoites in process of development. **10,** two ameboid trophozoites in one cell. **14,** mature trophozoite. **15,** mature trophozoite with chromatin apparently in process of division. **16–19,** schizonts showing progressive steps in division (presegmenting schizonts). **20,** mature schizont. **21, 22,** developing gametocytes. **23,** mature microgametocyte. **24,** mature macrogametocyte.

From A. Wilcox, *Manual for the Microscopial Diagnosis of Malaria in Man.* Public Health Service, Publication No. 796. U. S. Government Printing Office, Washington, D. C., 1960.

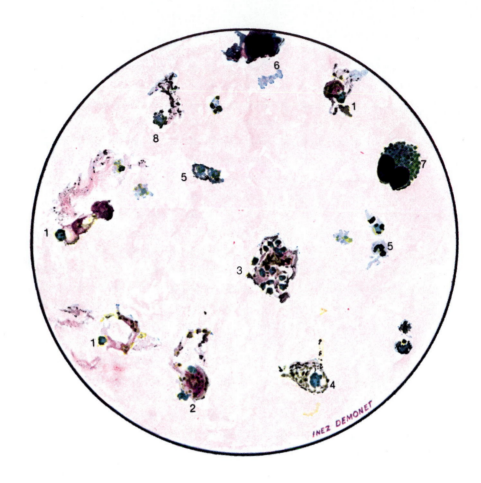

PLATE 2

Plasmodium vivax in thick smear. **1,** Ameboid trophozoites, **2,** Schizont—two divisions of chromatin. **3,** Mature schizont. **4,** Microgametocyte. **5,** Blood platelets. **6,** Nucleus of neutrophil. **7,** Eosinophil. **8,** Blood platelet associated with cellular remains of young erythrocytes.

From Wilcox, A. 1960. Manual for the microscopical diagnosis of malaria in man. U.S. Government Printing Office, Washington, D.C.

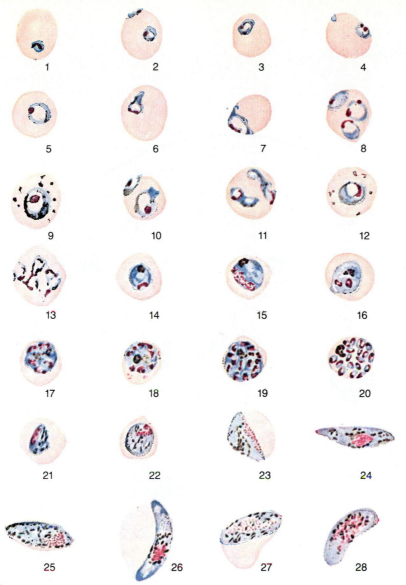

PLATE 3

Plasmodium falciparum. **1,** Very young ring-form trophozoite. **2,** Double infection of single cell with young trophozoites, one a marginal form, the other a signet-ring form. **3** and **4,** Young trophozoites showing double chromatin dots. **5** to **7,** Developing trophozoite forms. **8,** Three medium trophozoites in one cell. **9,** Trophozoite showing pigment, in cell containing Maurer's dots. **10** and **11,** Two trophozoites in each of two cells, showing variations of forms that parasites may assume. **12,** Almost mature trophozoite showing haze of pigment throughout cytoplasm. Maurer's dots in cell. **13,** Estivoautumnal slender forms. **14,** Mature trophozoite showing clumped pigment. **15,** Parasite in process of initial chromatin division. **16** to **19,** Various phases of development of schizont (presegmenting schizonts). **20,** Mature schizont. **21** to **24,** Successive forms in development of gametocyte—usually not found in peripheral circulation. **25,** Immature macrogametocyte. **26,** Mature macrogametocyte. **27,** Immature microgametocyte. **28,** Mature microgametocyte.

From Wilcox, A. 1960. Manual for the microscopical diagnosis of malaria in man. U.S. Government Printing Office, Washington, D.C.

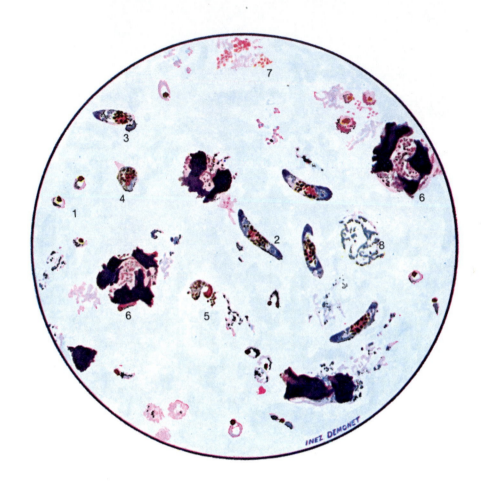

PLATE 4

Plasmodium falciparum in thick film. **1,** small trophozoites; **2,** gametocytes—normal; **3,** slightly distorted gametocyte; **4,** "rounded-up" gametocyte; **5,** disintegrated gametocyte; **6,** nucleus of leukocyte; **7,** blood platelets; **8,** cellular remains of young erythrocyte.

From A. Wilcox, U. S. Government Printing Office, Washington, D. C., 1960.

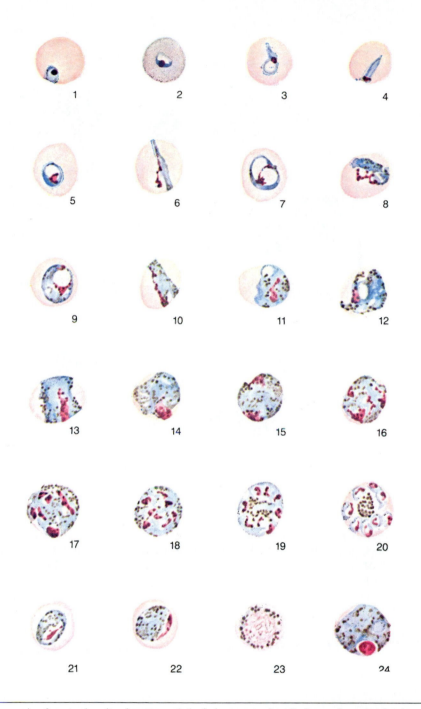

PLATE 5

Plasmodium malariae. **1,** young ring-form trophozoite of quartan malaria. **2–4,** young trophozoite forms of parasite showing gradual increase of chromatin and cytoplasm. **5,** developing ring-form trophozoite showing pigment granule. **6,** early band-form trophozoite—elongate chromatin, some pigment apparent. **7–12,** some forms that developing trophozoite of quartan may take. **13, 14,** mature trophozoites—one a band form. **15–19,** phases in development of schizont (presegmenting schizonts). **20,** mature schizont. **21,** immature microgametocyte. **22,** immature macrogametocyte. **23,** mature microgametocyte. **24,** mature macrogametocyte.

From A. Wilcox, U. S. Government Printing Office, Washington, D. C., 1960.

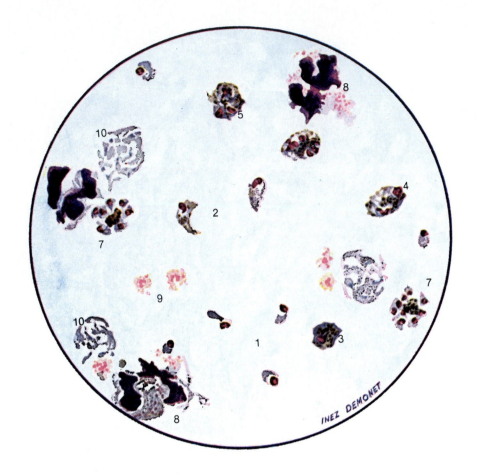

PLATE 6

Plasmodium malariae in thick smear. **1,** Small trophozoites. **2,** Growing trophozoites. **3,** Mature trophozoites. **4** to **6,** Schizonts (presegmenting) with varying numbers of divisions of chromatin. **7,** Mature schizonts. **8,** Nucleus of leukocyte. **9,** Blood platelets. **10,** Cellular remains of young erythrocytes.

From Wilcox, A. 1960. Manual for the microscopical diagnosis of malaria in man. U.S. Government Printing Office, Washington, D.C.

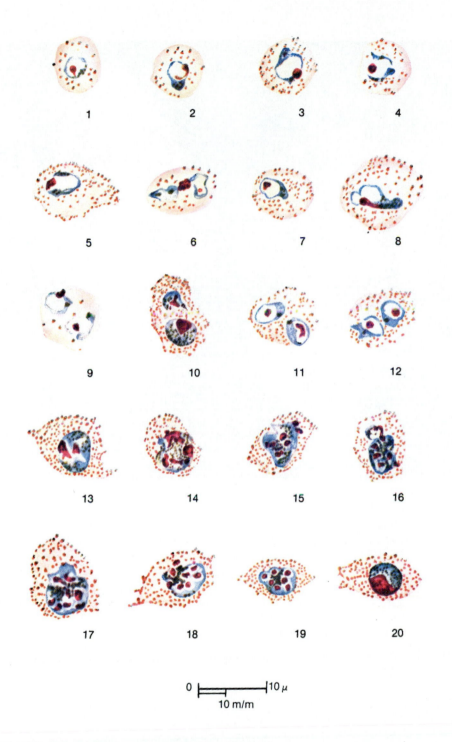

PLATE 7

Plasmodium ovale. **1,** young ring-shaped trophozoite; **2–5,** older ring-shaped trophozoites; **6–8,** older ameboid trophozoites; **9, 11, 12,** doubly infected cells, trophozoites; **10,** doubly infected cell, young gametocytes; **13,** first stage of schizont; **14–19,** schizonts, progressive stages; **20,** mature gametocyte.

From A. Wilcox, U. S. Government Printing Office, Washington, D. C., 1960.

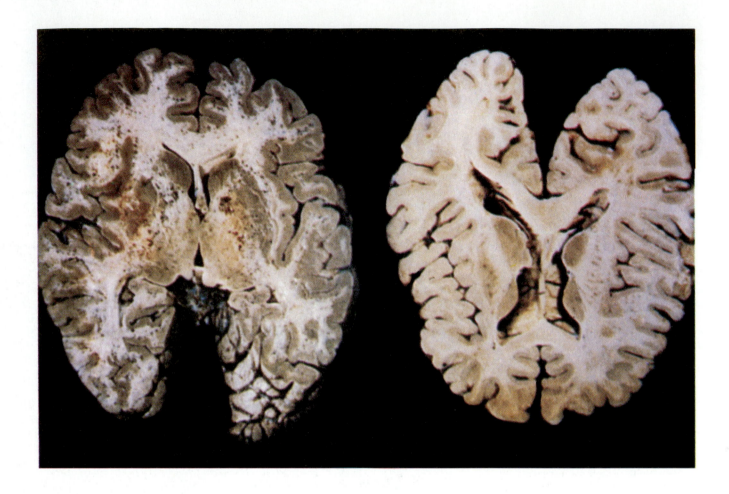

PLATE 8

Cut section of brain from cerebral malaria victim (*left*) compared with normal brain (*right*). The cortex shows slate-gray color from hemozoin, and tiny hemorrhages (petechiae) around blood vessels in white matter can be seen. The degree to which there is swelling from fluid in the nervous tissue (edema) can be observed by comparing to ventricles and sulci of the normal brain.

From G. Toro González, G. Román-Campos, and L. Navarro de Roman, *Neurologia Tropical: Aspectos Neuropatológicos de la Medicina Tropical,* 1983 Editorial Printer Columbiana Ltda. Colombía.

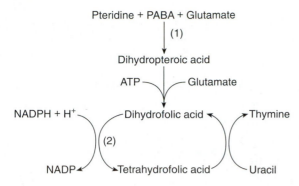

FIGURE 9.9

Metabolism of folate in *Plasmodium*. **1,** site of action of PABA analogs, such as sulfadoxine, which inhibit the synthesis of dihydropteroic acid from PABA and pteridine. **2,** site of action of pyrimethamine, which inhibits synthesis of tetrahydrofolic acid from dihydrofolic acid, which prevents the synthesis of thymine required for DNA synthesis.

Source: D. L. Looker, J. J. Marr, and R. L. Stotish, *Chemotherapy of Parasitic Diseases,* 1986, Plenum Press, New York, NY.

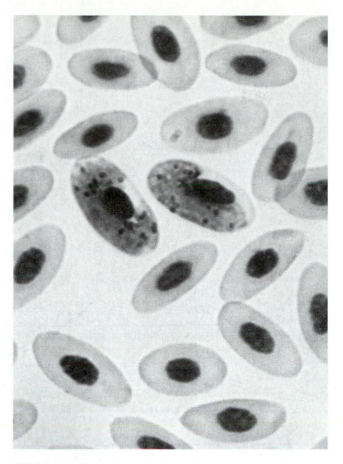

FIGURE 9.10

Haemoproteus gametocytes in blood of a mourning dove. They are about 14 μm long.

Courtesy of Sherwin Desser.

tetrahydrofolate by the enzyme **dihydrofolate reductase.** Also, this enzyme is necessary for tetrahydrofolate regeneration from dihydrofolate, which is produced in a vital reaction for which tetrahydrofolate is a cofactor: thymidylic acid synthesis. Thus, the enzyme is vital to both parasite and host, but fortunately the dihydrofolate reductases from the two sources vary in several respects. These differences include affinity for certain inhibitors.[57] A concentration of pyrimethamine and trimethoprim required to produce 50% inhibition of the mammalian enzyme is more than 1000 times that yielding 50% inhibition of the plasmodial one.

Resistance to pyrimethamine is due to point mutations, changing one or another of the amino acids in the parasite's dihydrofolate reductase. These changes reduce the binding of the protein with pyrimethamine, but they do not affect its enzymatic function.[111]

Genus *Haemoproteus*

Protozoa belonging to the genus *Haemoproteus* are primarily parasites of birds and reptiles and have their sexual phases in insects other than mosquitoes. Exoerythrocytic schizogony occurs in endothelial cells; merozoites enter erythrocytes to become pigmented gametocytes in the circulating blood (Fig. 9.10).

Haemoproteus columbae is a cosmopolitan parasite of pigeons. The definitive hosts and vectors of this parasite are several species of ectoparasitic flies in the family Hippoboscidae (Chapter 38), which inject sporozoites with their bite. Exoerythrocytic schizogony takes about 25 days in the capillary endothelium of the lungs, producing thousands of merozoites from each schizont. Merozoites presumably can develop directly from a schizont, or the schizont can break into numerous multinucleate "cytomeres." In the latter case the host endothelial cell breaks down, releasing the cytomeres, which usually lodge in the capillary lumen, where they grow, become branched, and rupture, producing many thousands of merozoites. A few of these may attack other endothelial cells, but most enter erythrocytes and develop into gametocytes. At first they resemble ring stages of *Plasmodium,* but they grow into mature microgametocytes or macrogametocytes in five or six days. Multiple infections of young forms in a single red blood cell are common, but one rarely finds more than one mature parasite per cell.

Mature macrogametocytes are 14 μm long and grow in a curve around the nucleus. Their granular cytoplasm stains a deep blue color and contains about 14 small, dark-brown pigment granules. The nucleus is small. Microgametocytes are 13 μm long, are less curved, have lighter-colored cytoplasm, and have six to eight pigment granules. The nucleus is diffuse.

Exflagellation in the stomach of the fly produces four to eight microgametes. The ookinete is like that of *Plasmodium* except there is a mass of pigment at its posterior end. It penetrates the intestinal epithelium and encysts between the muscle layers. The oocyst grows to maturity within nine days, when it measures 40 μm in diameter. Myriad sporozoites are released when the oocyst ruptures. Many of the sporozoites reach the salivary glands by the following day. Flies remain infected throughout the winter and can transmit infection to young squabs the following spring.

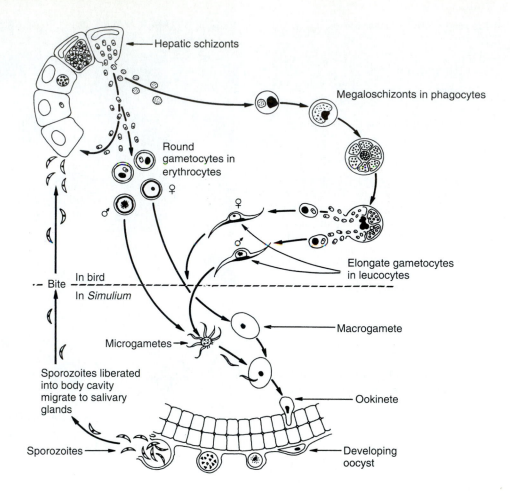

FIGURE 9.11

Life cycle of *Leucocytozoon simondi.*

From K. M. G. Adam, J. Paul, and V. Zaman, *Medical and veterinary protozoology. An illustrated guide.* 1971. Churchill Livingstone, Edinburgh.

Pathogenesis in pigeons is slight, and infected birds usually show no signs of disease. Exceptionally, birds appear restless and lose their appetite, and their lungs may become congested. Some anemia may result from loss of functioning erythrocytes, and the spleen and liver may be enlarged and dark with pigment.

More than 80 species of *Haemoproteus* have been named from birds, mainly Columbiformes. The actual number may be much less than that, since life cycles of most of them are unknown. *Culicoides* spp. are vectors of *H. meleagridis* of turkeys in Florida.[56]

Of the related genera, *Hepatocystis* spp. parasitize African and oriental monkeys, lemurs, bats, squirrels, and chevrotains; *Nycteria* and *Polychromophilus* are in bats; *Simondia* occurs in turtles; *Haemocystidium* lives in lizards; and *Parahaemoproteus* is common in a wide variety of birds.

Genus *Leucocytozoon*

Species of *Leucocytozoon* (Fig. 9.11) are parasites of birds. Schizogony is in fixed tissues, gametogony is in both leukocytes and immature erythrocytes, and sporogony occurs in insects other than mosquitoes. Pigment is absent from all phases of the life cycles. A related genus, *Akiba,* with only one species, occurs in chickens. *Leucocytozoon* has about 60 species in various birds. These are the most important blood protozoa of birds, since they are pathogenic in both domestic and wild hosts.

Leucocytozoon simondi is a circumboreal parasite of ducks, geese, and swans. The definitive hosts and vectors are black flies, family Simuliidae (see Fig. 38.7b). Sporozoites injected when the black fly feeds enter hepatocytes of the avian host where they develop into small schizonts (11 to 18 μm). They produce merozoites in four to six days. Merozoites that enter red blood cells become round gametocytes (Fig. 9.12). If, however, the merozoite is ingested by a macrophage in the brain, heart, liver, kidney, lymphoid tissues, or other organ, it develops into a huge megaloschizont 100 to 200 μm in diameter. The large form is more abundant than the small hepatic schizont.

The megaloschizont divides internally into primary cytomeres, which, in turn, multiply in the same manner. Successive cytomeres become smaller and finally multiply by

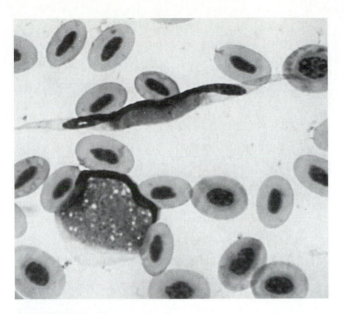

FIGURE 9.12

Avian blood cells infected with elongate and round gametocytes of *Leucocytozoon simondi.* The elongate form is up to 22 μm long.

Courtesy of Sherwin Desser.

schizogony into merozoites. Up to a million merozoites may be released from a single megaloschizont.

Merozoites penetrate leukocytes or developing erythrocytes to become elongated gametocytes (Fig. 9.12). Gametocytes of both sexes are 12 to 14 μm in diameter in fixed smears and may reach 22 μm in living cells. The macrogametocyte has a discrete, red-staining nucleus. The male cell is pale staining and has a diffuse nucleus that takes up most of the space within the cell. The diffuse nucleus of the macrogametocyte has large numbers of ribosomes.[51] As the gametocytes mature, they cause their host cells to become elongated and spindle shaped (Fig. 9.12).

Exflagellation begins only three minutes after the organism is eaten by the fly. A typical ookinete entering an intestinal cell becomes a mature oocyst within five days. Only 20 to 30 sporozoites form and slowly leave the oocyst. Rather than entering the salivary glands of the vector, they enter the proboscis directly and are transmitted by contamination or are washed in by saliva.

Leucocytozoon simondi is highly pathogenic for ducks and geese, especially young birds. The death rate in ducklings may reach 85%; older ducks are more resistant, and the disease runs a slower course in them, but they still may succumb. Anemia is a prominent symptom of leukocytozoonosis, as are elevated numbers of leukocytes. The liver enlarges and becomes necrotic, and the spleen may increase to as much as 20 times the normal size. *Leucocytozoon simondi* probably kills the host by destroying vital tissues, such as brain and heart. An outstanding feature of an outbreak of leukocytozoonosis is the suddenness of its onset. A flock of ducklings may appear normal in the morning, become ill in the afternoon, and be dead by the next morning. Birds that survive are prone to relapses but as the result of premunition are generally immune to reinfection.

Another species of importance is *L. smithi,* which can devastate domestic and wild turkey flocks. Its life cycle is similar to that of *L. simondi.*

SUBCLASS PIROPLASMEA

Members of the subclass Piroplasmea are small parasites of ticks and mammals. They do not produce spores, flagella, cilia, or true pseudopodia; their locomotion, when necessary, is accomplished by body flexion or gliding. No stages produce intracellular pigment. Asexual reproduction is in the erythrocytes or other blood cells of mammals by binary fission or schizogony. Sexual reproduction occurs, at least in some species.[65] The components of the apical complex are reduced but warrant placement in the phylum Apicomplexa.

The single order Piroplasmida contains the two families Babesiidae and Theileriidae, both of which are of considerable veterinary importance.

Family Babesiidae

Babesiids are usually described from their stages in the red blood cells of vertebrates. They are pyriform, round, or oval parasites of erythrocytes, lymphocytes, histiocytes, erythroblasts, or other blood cells of mammals and of various tissues of ticks. The apical complex is reduced to a polar ring, rhoptries, micronemes, and subpellicular microtubules. A cytostome is present in at least some species. Schizogony occurs in ticks. By far the most important species in America is *Babesia bigemina,* the causative agent of **babesiosis,** or **Texas red-water fever,** in cattle.

• *Babesia bigemina*

By 1890 the entire southeastern United States was plagued by a disease of cattle, variously called Texas cattle fever, red-water fever, or hemoglobinuria. Infected cattle usually had red-colored urine resulting from massive destruction of erythrocytes, and they often died within a week after symptoms first appeared. The death rate was much lower in cattle that had been reared in an enzootic area than in northern animals that were brought south. Also, it was noticed that when southern herds were driven or shipped north and penned with northern animals, the latter rapidly succumbed to the disease. The cause of red-water fever and its mode of dissemination were a mystery when Theobald Smith and Frank Kilbourne began their investigations in the early 1880s. In a series of intelligent, painstaking experiments, they showed that the tick *Boophilus annulatus* (Fig. 9.13) was the vector and alternate host of a tiny protozoan parasite that inhabited the red blood cells of cattle and killed these relatively immense animals.[98] This not only pointed the way to an effective means of control but was also the first demonstration that a protozoan parasite could develop in and be transmitted by an arthropod. This book is replete with other examples of this phenomenon, as we have already seen.

Babesia bigemina infects a wide variety of ruminants, such as deer, water buffalo, and zebu, in addition to cattle.

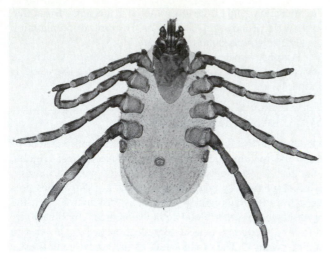

FIGURE 9.13

Boophilus annulatus, the vector of *Babesia bigemina.*
Courtesy of Jay Georgi.

When in the erythrocyte of the vertebrate host, the parasite is pear shaped, round, or, occasionally, irregularly shaped, and it is 4.0 μm long by 1.5 μm wide. The organisms usually are seen in pairs within the erythrocyte (hence the name *bigemina,* the twins) and are often united at the pointed tips (Fig. 9.14). At the light microscope level, they appear to be undergoing binary fission, but the electron microscope has revealed that the process is a kind of binary schizogony, a budding analogous to that occurring in the Haemosporina, with redifferentiation of the apical complex and merozoite formation.[3]

Biology. The infective stage of *Babesia* in the tick is a sporozoite. It is about 2 μm long and is pyriform, spherical, or ovoid. After completing development, sporozoites in the salivary glands of the tick are injected with its bite. There is no exoerythrocytic schizogony in the vertebrate. The parasites immediately enter erythrocytes, where they become trophozoites and escape from the parasitophorous vacuole.[4] They undergo binary fission and ultimately kill their host cell. The merozoites attack other red blood cells, building up an immense population in a short time. This asexual cycle continues indefinitely or until the host succumbs. Erythrocytic phases are reduced or apparently absent in resistant hosts. Some of the intraerythrocytic parasites do not develop further and are destined to become gametocytes, called **ray bodies,** when ingested by a tick.[65]

Ticks of the genus *Boophilus* transmit *Babesia bigemina;* distribution of the tick limits distribution of babesiosis. *Boophilus annulatus* is the vector in the Americas. It is a one-host tick, feeding, maturing, and mating on a single host. After engorging and mating, the female tick drops to the ground, lays her eggs, and dies. The larval, six-legged ticks that hatch from the eggs climb onto vegetation and attach to animals that brush by the plants.

One would think that a one-host tick would be a poor vector—if they do not feed on successive hosts, how can they transmit pathogens from one animal to another? This question was answered when it was discovered that the protozoan infects the developing eggs in the ovary of the tick, a phenomenon called **transovarian transmission.**

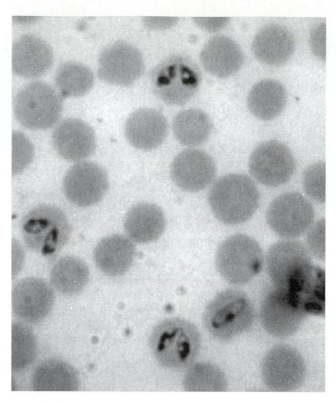

FIGURE 9.14

Babesia bigemina trophozoites in the erythrocytes of a cow.
Courtesy of Warren Buss.

After ingestion by a feeding tick, the parasites are freed by digestion from their dead host cells, and they develop into ray bodies. These are bizarrely shaped stages that have a thornlike process and several stiff, flagellalike protrusions (Fig. 9.15).[65] Fusion of two ray bodies forms a zygote, which becomes a primary **kinete.** The primary kinetes leave the intestine and penetrate various cells, such as hemocytes, muscles, malpighian tubule cells, and ovarian cells including oocytes. They enlarge and become polymorphic, dividing by multiple fission into a number of **cytomeres,** which differentiate into new kinetes.[65] Some of the secondary kinetes migrate to the salivary glands, penetrate the gland cells, and become polymorphic. They stimulate the host cells and nuclei to hypertrophy. When the host begins to feed, the parasites rapidly undergo multiple fission to produce enormous numbers of sporozoites about 2 to 3 μm long by 1 to 2 μm wide. They enter the channels of the salivary glands and are injected into the vertebrate host by the feeding tick.[81]

Although this is the life cycle as it occurs in a one-host tick, two- and three-host ticks serve as hosts and vectors of *B. bigemina* in other parts of the world. With these ticks transovarian transmission is not required and may not occur. All instars of such ticks can transmit the disease.

Pathogenesis. *Babesia bigemina* is unusual in that the disease it causes is more severe in adult cattle than in calves. Calves less than a year old are seldom seriously affected, but the mortality rate in acute cases in untreated adult cattle is as high as 50% to 90%. The incubation period is 8 to 15 days, but an acutely ill animal may die only four to eight days after

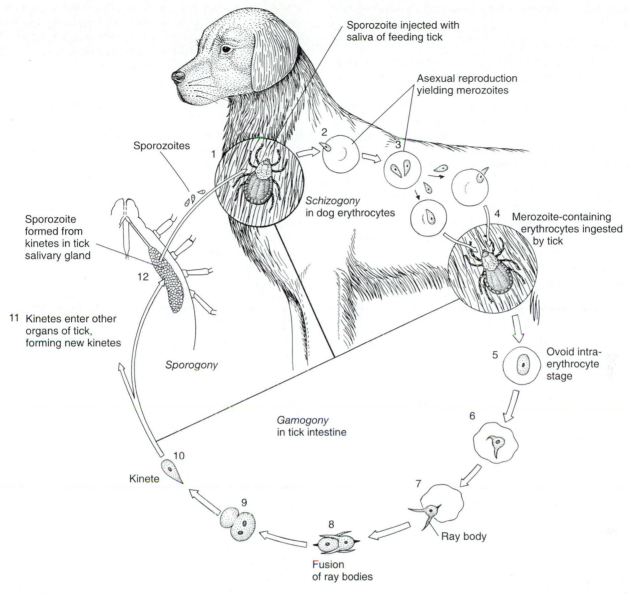

Sporozoite injected with
saliva of feeding tick

Asexual reproduction
yielding merozoites

Sporozoites

Schizogony
in dog erythrocytes

Merozoite-containing
erythrocytes ingested
by tick

Sporozoite
formed from
kinetes in tick
salivary gland

11 Kinetes enter other
organs of tick,
forming new kinetes

Sporogony

Ovoid intra-
erythrocyte
stage

Gamogony
in tick intestine

Kinete

Ray body

Fusion
of ray bodies

FIGURE 9.15

Life cycle of *Babesia canis*. **1,** sporozoite injected with saliva of feeding tick. **2, 3,** asexual reproduction in red blood cells of vertebrate host by binary fission, yielding merozoites. **4,** merozoites in erythrocytes are digested in tick. **5, 6,** gametocytes form protrusions after ingestion by tick and become ray bodies. **7–9,** two ray bodies fuse to form zygote. **10,** zygote becomes motile kinete. **11,** kinetes leave intestine, enter other cells, and form new kinetes. **12,** kinetes that enter cells of salivary gland give rise to thousands of small sporozoites.

Drawing by William Ober and Claire Garrison.

infection. The first symptom is a sudden rise in temperature to 106° to 108° F; this may persist for a week or more. Infected animals rapidly become dull and listless and lose their appetite. Up to 75% of the erythrocytes may be destroyed in fatal cases, but even in milder infections so many erythrocytes are destroyed that a severe anemia results. Mechanisms for clearance of hemoglobin and its breakdown products are overloaded, producing jaundice, and much excess hemoglobin is excreted by the kidneys, giving the urine the red color mentioned earlier. Chronically infected animals remain thin, weak, and out of condition for several weeks before recovering. Levine described damage to internal organs.[55]

Cattle that recover are usually immune for life with a sterile immunity or, more commonly, premunition. There are strain differences in the degree of immunity obtained; furthermore, little cross-reaction occurs between *B. bigemina* and other species of *Babesia*.

For unknown reasons drugs that are effective against trypanosomes are also effective against *Babesia* spp. A number of chemotherapeutic agents are available, some allowing recovery but leaving latent infection, others effecting a complete cure. It should be remembered that elimination of all parasites also eliminates premunition.

Infection can be prevented by tick control, the means by which red-water fever was eliminated from the United States. Regular dipping of cattle in a tickicide effectively eliminates the tick vector, especially if it is a one-host species. Another method that has been used is artificial premunizing of young animals with a mild strain before shipping them to enzootic areas.

• *Babesia microti*

Prior to 1969 *Babesia* infections in humans were rare. There were two infections with *B. divergens,* a cattle parasite, and some others were caused by species normally parasitic in rodents. In several of the cases, three of which were fatal, the patients had been splenectomized some time before infection, and it was believed that the disabling of the immune system by splenectomy rendered the humans susceptible. However, human infection in a nonsplenectomized patient was reported from Nantucket Island off the coast of Massachusetts in 1969. Since then, additional cases have been reported from Nantucket Island, Martha's Vineyard, Shelter Island near Long Island, New York, eastern Long Island itself, and Connecticut.[14,35,76,88] Currently about 10 to 15 clinical cases per year are being reported.[80] These have all been infections with *B. microti,* a parasite of meadow voles and other rodents that can also infect pets. The vector is *Ixodes dammini,* whose adults feed on deer. Deer are refractory to infection with *B. microti,* and the infection is transmitted among rodents and to humans by nymphs of *I. dammini* and among rodents by *I. muris,* which does not feed on humans.[84] It is unclear why this formerly rare infection has now become almost common. However, as with Lyme disease (p. 605), the explanation probably lies in the increased contact of humans with ticks and the reservoir hosts.

• Other Species of Babesiidae

Cattle seem particularly suitable as hosts to piroplasms. Other species of *Babesia* in cattle are *B. bovis* in Europe, Russia, and Africa; *B. berbera* in Russia, north Africa, and the Middle East; *B. divergens* in western and central Europe; *B. argentina* in South America, Central America, and Australia; and *B. major* in north Africa, Europe, and Russia. Several other species are known from deer, sheep, goats, dogs, cats, and other mammals, as well as birds. Their biology, pathogenesis, and control are generally the same as for *B. bigemina.*

Family Theileriidae

Like the Babesiidae, the members of this family lack a conoid. The rhoptries, micronemes, subpellicular tubules, and polar ring are well demonstrated in the tick stages. The Theileriidae parasitize blood cells of mammals, and the vectors are hard ticks of the family Ixodidae. Gamogony occurs in the gut of the nymphal tick, resulting in the formation of **kinetes,** which are very similar to the ookinetes of Haemosporina.[65] The kinetes grow in the gut cells of the tick for a time and then leave and penetrate the cells of the salivary glands, where sporogony takes place. Several members of this family infect cattle, sheep, and goats, causing a disease called **theileriosis,** which results in heavy losses in Africa, Asia, and southern Europe.

• *Theileria parva*

Theileria parva causes a disease called East Coast fever in cattle, zebu, and Cape buffalo. It has been one of the most important diseases of cattle in southern, eastern, and central Africa, although it has been eliminated from most of southern Africa. After Romanovsky staining, the forms within erythrocytes have blue cytoplasm and a red nucleus in one end. At least 80% of them are rod shaped, about 1.5 to 2.0 µm by 0.5 to 1.0 µm in size. Oval and ring- or comma-shaped forms are also found.

Biology. East Coast fever, like red-water fever, is a disease of ticks and cattle, flourishing in both. The principal vector is the brown cattle tick *Rhipicephalus appendiculatus,* a three-host species. Other ticks, including one- and two-host ticks, can also serve as hosts for this parasite.

When the tick feeds, it injects sporozoites present in its salivary glands into the next host;[30] there they enter T and B lymphocytes within lymphoid tissue, escape from the phagolysosome into the cytosol,[4] grow, and undergo schizogony. Schizonts, called **Koch's blue bodies,** can be seen in circulating lymphocytes within three days after infection. Two types of schizonts are recognized. The first generation in lymph cells comprises **macroschizonts** and produces about 90 macromerozoites, each 2.0 to 2.5 µm in diameter. Some of these enter other lymph cells, especially in fixed tissues, and initiate further generations of macroschizonts. Others enter lymphocytes and become **microschizonts,** producing 80 to 90 micromerozoites, each 0.7 to 1.0 µm wide. If microschizonts rupture while in lymphoid tissues, the micromerozoites enter new lymph cells, maintaining the lymphatic infection. However, if they rupture in the circulating blood, the micromerozoites enter erythrocytes to become the "piroplasms" typical of the disease. Apparently the parasites do not multiply in erythrocytes.

Ticks of all instars can acquire infection when they feed on blood containing piroplasms. However, because three-host ticks drop off the host to molt immediately after feeding, only the nymph and adult are infective to cattle. Transovarian transmission does not occur, as it does in *Babesia* spp.

Ingested erythrocytes are digested, releasing the piroplasms that undergo gamogony, differentiating into ray bodies. Fusion of ray bodies produces kinetes, as described before.[65]

Pathogenesis. As in babesiosis, calves are more resistant to *T. parva* than are adult cattle. Nevertheless, *T. parva* is highly pathogenic: Strains with low pathogenicity kill around 23% of infected cattle, whereas highly pathogenic strains kill 90% to 100%. Symptoms such as high fever first appear 8 to 15 days after infection. Other signs are nasal discharge, runny eyes, swollen lymph nodes, weakness, emaciation, and diarrhea. Hematuria and anemia are unusual, although blood is often present in feces.

Animals that recover from theileriosis are immune from further infection, without premunition. The immunity is cell mediated, including destruction of infected lymphocytes by activated cytotoxic T cells and natural killer cells.[39] Diagnosis depends on finding the parasites in blood or lymph smears. No cheap and effective drug is currently available.[39] Control depends on tick control and quarantine rules.

Other species of *Theileria* are *T. annulata, T. mutans, T. hirei, T. ovis,* and *T. camelensis,* all parasites of ruminants. Other genera in the family are *Haematoxenus* in cattle and zebu and *Cytauxzoon* in antelope, both in Africa.

References

1. Aikawa, M., J. R. Rabbege, I. Udeinya, and L. H. Miller. 1983. Electron microscopy of knobs in *Plasmodium falciparum*-infected erythrocytes. *J. Parasitol.* 69:434–37.

2. Aikawa, M., I. J. Udeinya, J. Rabbege, M. Dayan, J. H. Leech, R. J. Howard, and L. H. Miller. 1985. Structural alteration of the membrane of erythrocytes infected with *Plasmodium falciparum*. *J. Protozool.* 32:424–29.

3. Aikawa, M., and C. R. Sterling. 1974. *Intracellular parasitic Protozoa*. New York: Academic Press, Inc.

4. Andrews, N. W., and P. Webster. 1991. Phagolysosomal escape by intracellular pathogens. *Parasitol. Today.* 7:335–40.

5. Ash, C. 1991. First impressions of the malaria vaccine. *Parasitol. Today* 7:63–64.

6. Ballou, W. R., J. A. Sherwood, F. A. Neva, D. M. Gordon, R. A. Wirtz, G. F. Wasserman, C. L. Diggs, S. L. Hoffman, M. R. Hollingdale, W. T. Hockmeyer, I. Schneider, J. F. Young, P. Reeve, and J. D. Chulay. 1987. Safety and efficacy of a recombinant DNA *Plasmodium falciparum* sporozoite vaccine. *The Lancet* (6 June 1987):1277–81.

7. Barker, R. H. Jr., L. Suebsaeng, W. Rooney, and D. F. Wirth. 1989. Detection of *Plasmodium falciparum* infection in human patients: A comparison of the DNA probe method to microscopic diagnosis. *Am. J. Trop. Med. Hyg.* 41:266–72.

8. Berendt, A. R., D. J. P. Ferguson, and C. I. Newbold. 1990. Sequestration in *Plasmodium falciparum* malaria: Sticky cells and sticky problems. *Parasitol. Today* 6:247–54.

9. Bignami, A. 1913. Concerning the pathogenesis of relapses in malarial fevers. *South. Med. J.* 6:79–88.

10. Blum, J. J., A. Yayon, S. Friedman, and H. Ginsberg. 1984. Effects of mitochondrial protein synthesis inhibitors on the incorporation of isoleucine into *Plasmodium falciparum* in vitro. *J. Protozool.* 31:475–79.

11. Breton, C. B., and L. H. Pereira da Silva. 1993. Malaria proteases and red blood cell invasion. *Parasitol. Today* 9:92–96.

12. Bruce-Chwatt, L. J. 1980. *Essential malariology*. London: William Heinemann Medical Books Ltd.

13. Canfield, C. J. 1972. Malaria in U.S. military personnel 1965–1971. In Sadun, E. H., ed. Basic research in malaria. *Proc. Helm. Soc. Wash.* (special issue) 39:15–18.

14. Centers for Disease Control. 1989. Babesiosis—Connecticut. *Morb. Mortal. Weekly Rep.* 38:649–50.

15. Centers for Disease Control. 1990. Recommendations for the prevention of malaria among travelers. *Morb. Mortal. Weekly Rep.* 39:1–10.

16. Cerami, C., U. Frevert, P. Sinnis, B. Takacs, P. Clavijo, M. J. Santos, and V. Nussenzweig. 1992. The basolateral domain of the hepatocyte plasma membrane bears receptors for the circumsporozoite protein of Plasmodium falciparum sporozoites. *Cell* 70:1021–33.

17. Chernin, E. 1984. The malariatherapy of neurosyphilis. *J. Parasitol.* 70:611–17.

18. Clark, I. A. 1987. Cell-mediated immunity in protection and pathology of malaria. *Parasitol. Today* 3:300–305.

19. Clark, I. A., K. A. Rockett, and W. B. Cowden. 1991. Proposed link between cytokines, nitric oxide and human cerebral malaria. *Parasitol. Today* 7:205–7.

20. Coatney, G. R. 1976. Relapse in malaria—an enigma. *J. Parasitol.* 62:3–9.

21. Corradetti, A. 1950. Ospite definitivo e ospite intermedio di parassiti della malaria. *Riv. Parasitol.* 11:89.

22. Cowman, A. F., and S. J. Foote. 1990. Chemotherapy and drug resistance in malaria. *Int. J. Parasitol.* 20:503–13.

23. Cruz Marques, A. 1987. Human migration and the spread of malaria in Brazil. *Parasitol. Today* 3:166–70.

24. Day, K. P., and K. Marsh. 1991. Naturally acquired immunity to *Plasmodium falciparum*. *Parasitol. Today* 7:A68–A71.

25. Diggs, C. L., W. R. Ballou, and L. H. Miller. 1993. The major merozoite surface protein as a malaria vaccine target. *Parasitol. Today* 9:300–302.

26. Friedman, M. J., and W. Trager. 1981. The biochemistry of resistance to malaria. *Sci. Am.* 244:154–64 (Mar.).

27. Garnham, P. C. C. 1966. *Malaria parasites and other Haemosporidia* Oxford: Blackwell Scientific Publications Ltd.

28. Garnham, P. C. C. 1977. The continuing mystery of relapses in malaria. *Protozool. Abstr.* 1:1–12.

29. Geary, T. G., and J. B. Jensen. 1986. Protozoan infections of man: Malaria. In Campbell, W. C., and R. S. Rew, eds. *Chemotherapy of parasitic diseases*. New York: Plenum Press.

30. Gettinby, G., and W. Byrom. 1989. The dynamics of East Coast fever. *Parasitol. Today* 5:68–73.

31. Gibbs, W. W. 1993. Back to basics. Mapping malaria's genome may help produce a vaccine. *Sci. Am.* 268:134–36.

32. Ginsberg, H. 1988. Effect of calcium antagonists on malaria susceptibility to chloroquine. *Parasitol. Today* 4:209–11.

33. Godson, G. N. 1985. Molecular approaches to malaria vaccines. *Sci. Am.* 252:52–59 (May).

34. Goldie, P., E. F. Roth Jr., J. Oppenheim, and J. P. Vanderberg. 1990. Biochemical characterization of *Plasmodium falciparum* hemozoin. *Am. J. Trop. Med. Hyg.* 43:584–96.

35. Grunwaldt, E. 1977. Babesiosis on Shelter Island. *N.Y. State J. Med.* 77:1320–21.

36. Guazzi, M., and S. Grazi. 1963. Consideratione sa un caso di malaria quartana recidivante dopo se anni di latenza. *Riv. Malar.* 42:55–59.

37. Halawani, A., and A. A. Shawarby. 1957. Malaria in Egypt. *J. Egypt. Med. Assoc.* 40:753–92.

38. Hall, A. P., and C. J. Canfield. 1972. Resistant falciparum malaria in Vietnam: Its rarity in Negro soldiers. In Sadun, E. H., ed. *Basic research in malaria, Proc. Helm. Soc. Wash.* (special issue) 39:66–70.

39. Hall, F. R. 1988. Antigens and immunity in *Theileria annulata*. *Parasitol. Today* 4:257–61.

40. Harrison, G. 1978. *Mosquitoes, malaria and man: A history of the hostilities since 1880*. New York: E. P. Dutton & Co., Inc.

41. Harwood, R. F., and M. T. James. 1979. Entomology in human and animal health, 7th ed. New York: Macmillan Publishing Co., Inc.

42. Herrington, D., J. Davis, E., Nardin, M. Beier, J. Cortese, H. Eddy, G. Losonsky, M. Hollingdale, M. Sztein, M. Levine, R. S. Nussenzweig, D. Clyde, and R. Edelman. 1991. Successful immunization of humans with irradiated malaria sporozoites: Humoral and cellular responses of the protected individuals. *Am. J. Trop. Med. Hyg.* 45:539–47.

43. Herwaldt, B. L., and D. D. Juranek. 1993. Laboratory-acquired malaria, leishmaniasis, trypanosomiasis, and toxoplasmosis. *Am. J. Trop. Med. Hyg.* 48:313–23.

44. Hoeppli, R. 1969. *Parasitic diseases in Africa and the western hemisphere. Early documentation and transmission by the slave trade.* Basel: Verlag für Recht und Gesellschaft AG.

45. Inselberg, J. 1985. Induction and isolation of artemisinine-resistant mutants of *Plasmodium falciparum. Am. J. Trop. Med. Hyg.* 34:417–18.

46. Jacobs, G. H., M. Aikawa, W. K. Milhous, and J. R. Rabbege. 1987. An ultrastructural study of the effects of mefloquine on malaria parasites. *Am. J. Trop. Med. Hyg.* 36:9–14.

47. Jiang, J. -B., G. Jacobs, D. -S. Liang, and M. Aikawa. 1985. Qinghaosu-induced changes in the morphology of *Plasmodium inui. Am. J. Trop. Med. Hyg.* 34:424–28.

48. Kawamoto, F., and P. F. Billingsley. 1992. Rapid diagnosis of malaria by fluorescence microscopy. *Parasitol. Today* 8:69–71.

49. Kennedy, J. F. 1962. *Message of first day of issue of U.S. malaria eradication stamp.*

50. Klayman, D. L. 1985. *Qinghaosu* (artemisinin): An antimalarial drug from China. *Science* 228:1049–55.

51. Kocan, A. A., and K. M. Kocan. 1978. The fine structure of elongate gametocytes of *Leucocytozoon ziemanni* (Laveran). *J. Parasitol.* 64:1057–59.

52. Kreier, J. P., and J. R. Baker. 1987. *Parasitic protozoa.* Boston: Allen and Unwin.

53. Krotoski, W. A., W. E. Collins, R. S. Bray, P. C. C. Garnham, F. B. Cogswell, R. W. Gwadz, R. Killick-Kendrick, R. Wolf, R. Sinden, L. C. Koontz, and P. S. Stanfill. 1982. Demonstration of hypnozoites in sporozoite-transmitted *Plasmodium vivax* infection. *Am. J. Trop. Med. Hyg.* 31:1291–93.

54. Krotoski, W. A., P. C. C. Garnham, R. S. Bray, D. M. Krotoski, R. Killick-Kendrick, C. C. Draper, G. A. T. Targett, and M. W. Guy. 1982. Observations on early and late post-sporozoite tissue stages in primate malaria. I. Discovery of a new latent form of *Plasmodium cynomolgi* (the hypnozoite), and failure to detect hepatic forms within the first 24 hours after infection. *Am. J. Trop. Med. Hyg.* 31:24–35.

55. Levine, N. D. 1973. Protozoan parasites of domestic animals and of man, 2d ed. Minneapolis: Burgess Publishing Co.

56. Long, P. L., W. L. Current, and G. P. Noblet. 1987. Parasites of the Christmas turkey. *Parasitol. Today* 3:360–66.

57. Looker, D. L., J. J. Marr, and R. L. Stotish. 1986. Modes of action of antiprotozoal agents. In Campbell, W. C., and R. S. Rew, eds. *Chemotherapy of parasitic diseases.* New York: Plenum Press.

58. Manandhar, M. S. P., and K. van Dyke. 1975. Detailed purine salvage metabolism in and outside the free malarial parasite. *Exp. Parasitol.* 37:138–46.

59. Mange, A. P., and E. J. Mange. 1990. Genetics: Human aspects, 2d ed. Sunderland, Mass.: Sinauer Associates, Inc., Publishers.

60. Manson-Bahr, P. 1963. The story of malaria: The drama and the actors. *Int. Rev. Trop. Med.* 2:329–90.

61. Marshall, E. 1991. Malaria parasite gaining ground against science. *Science* 254:190.

62. Marshall, E. 1992. Malaria vaccine on trial at last? *Science* 255:1063–64.

63. Mathews, H. M., and J. C. Armstrong. 1981. Duffy blood types and vivax malaria in Ethiopia. *Am. J. Trop. Med. Hyg.* 30:299–303.

64. Meek, S. R., E. B. Doberstyn, B. A. Gaüzère, C. Thanapanich, E. Nordlander, and S. Phuphaisan. 1986. Treatment of falciparum malaria with quinine and tetracycline or combined mefloquine/sulfadoxine/pyrimethamine on the Thai-Kampuchean border. *Am. J. Trop. Med. Hyg.* 35:246–50.

65. Mehlhorn, H., and E. Schein. 1984. The piroplasms: Life cycle and sexual stages. In Baker, J. R., and R. Muller, eds. *Advances in parasitology* 23. London: Academic Press.

66. Mehlhorn, H., E. Schein, and M. Warnecke. 1979. Electron-microscopic studies on *Theileria ovis* Rodhain, 1916: Development of kinetes in the gut of the vector tick, *Rhipicephalus evertsi evertsi* Neumann, 1897, and their transformation within the cells of the salivary glands. *J. Protozool.* 26:377–85.

67. Meis, J. F. G. M., G. Pool, G. J. van Gemert, A. H. W. Lensen, T. Ponnudurai, and J. H. E. T. Meuwissen. 1989. *Plasmodium falciparum* ookinetes migrate intercellularly through *Anopheles stephensi* midgut epithelium. *Parasitol. Res.* 76:13–19.

68. Melancon-Kaplan, J., J. M. Burns Jr., A. B. Vaidya, H. K. Webster, and W. P. Weidanz. 1993. Malaria. In Warren, K. S., ed. *Immunology and molecular biology of parasitic infections.* Boston: Blackwell Scientific Publications.

69. Miller, L. H., R. J. Howard, R. Carter, M. F. Good, V. Nussenzweig, and R. S. Nussenzweig. 1986. Research toward malaria vaccines. *Science* 234:1349–56.

70. Miller, L. H., S. J. Mason, D. F. Clyde, and M. H. McGinnis. 1976. The resistance factor to *Plasmodium vivax* in blacks. The Duffy bloodgroup genotype FyFy. *N. Eng. J. Med.* 295:302–4.

71. Newbold, C. I., and K. Marsh. 1990. Antigens on the *Plasmodium falciparum* infected erythrocyte surface are parasite derived: A reply. *Parasitol. Today* 6:320–22.

72. Nijhout, M. M., and R. Carter. 1978. Gamete development in malaria parasites: Bicarbonate-dependent stimulation by pH *in vitro. Parasitology* 76:39–53.

73. Nussenzweig, V., and R. S. Nussenzweig. 1986. Development of a sporozoite malaria vaccine. *Am. J. Trop. Med. Hyg.* 35:678–88.

74. Nussenzweig, V., and R. S. Nussenzweig. 1990. Sporozoite malaria vaccine. Where do we stand? *Ann. Parasitol. Hum. Comp.* 65:49–52.

75. Orjih, A. U., R. Chevli, and C. D. Fitch. 1985. Toxic heme in sickle cells: An explanation for death of malaria parasites. *Am. J. Trop. Med. Hyg.* 34:223–27.

76. Parry, M. F., M. Fox, S. A. Burka, and W. J. Richar. 1977. *Babesia microti* infection in man. *J.A.M.A.* 238:1282–83.

77. Patarroyo, M. E., P. J. Amador, A. Moreno, F. Guzman, P. Romero, R. Tason, A. Franco, L. A. Murillo, G. Ponton, and G. Trujillo. 1988. A synthetic vaccine protects humans against challenge with asexual blood stages of *Plasmodium falciparum. Nature* 332:158–61.

78. Payne, D. 1987. Spread of chloroquine resistance in *Plasmodium falciparum. Parasitol. Today* 3:241–46.

79. Perkins, M. E. 1992. Rhoptry organelles of apicomplexan parasites. *Parasitol. Today* 8:28–32.

80. Piesman, J. 1987. Emerging tick-borne diseases in temperate climates. *Parasitol. Today* 3:197–99.

81. Riek, R. F. 1964. The life cycle of *Babesia bigemina* (Smith and Kilbourne, 1893) in the tick vector *Boophilus microplus* (Canastrini). *Aust. J. Agr. Res.* 15:802–21.

82. Rosenberg, R., and J. Rungsiwongse. 1991. The number of sporozoites produced by individual malaria oocysts. *Am. J. Trop. Med. Hyg.* 45:574–77.

83. Ruangjirachuporn, W., B. A. Afzelius, S. Paulie, M. Wahlgren, K. Berzins, and P. Perlmann. 1991. Cytoadherence of knobby and knobless *Plasmodium falciparum*-infected erythrocytes. *Parasitology* 102:325–34.

84. Ruebush, T. K. II, D. D. Juranek, A. Spielman, J. Piesman, and G. R. Healy. 1981. Epidemiology of human babesiosis on Nantucket Island. *Am. J. Trop. Med. Hyg.* 30:937–41.

85. Russell, P. F. 1955. *Man's mastery of malaria.* London: Oxford University Press.

86. Russell, P. F., L. S. West, and R. D. Manwell. 1946. *Practical malariology.* Philadelphia: W. B. Saunders Co.

87. Sato, K., S. Kano, H. Yamaguchi, F. M. Omer, S. H. Safi, A. A. El Gaddal, and M. Suzuki. 1990. An ABC-ELISA for malaria serology in the field. *Am. J. Trop. Med. Hyg.* 42:24–27.

88. Scharfman, W. B., and E. G. Taft. 1977. Nantucket fever. An additional case of babesiosis. *J.A.M.A.* 238:1281–82.

89. Schwartz, I. K., W. Chin, J. Newman, and J. M. Roberts. 1984. Glucose-6-phosphate dehydrogenase deficiency in Southeast Asian refugees entering the United States. *Am. J. Trop. Med. Hyg.* 33:182–84.

90. Sherman, I. W. 1979. Biochemistry of *Plasmodium* (malarial parasites). *Microbiol. Rev.* 43:453–95.

91. Sherman, I. W. 1985. Membrane structure and function of malaria parasites and the infected erythrocyte. *Parasitology* 91:609–45.

92. Sherman, I. W., and E. Winograd. 1990. Antigens on the *Plasmodium falciparum* infected erythrocyte surface are not parasite derived. *Parasitol. Today* 6:317–20.

93. Shortt, H. C., and P. C. C. Garnham. 1948. Demonstration of a persisting exoerythrocytic cycle in *Plasmodium cynomolgi* and its bearing on the production of relapses. *Br. Med. J.* 1:1225–32.

94. Sinden, R. E. 1983. Sexual development of malarial parasites. In Baker, J. R., and R. Muller, eds. *Advances in parasitology* 22. London: Academic Press, 153–216.

95. Sinden, R. E., and R. H. Hartley. 1985. Identification of the meiotic division of malarial parasites. *J. Protozool.* 32:742–44.

96. Sinden, R. E., R. H. Hartley, and L. Winger. 1985. The development of *Plasmodium* ookinetes *in vitro:* An ultrastructural study including a description of meiotic division. *Parasitology* 91:227–44.

97. Smith, D. C., and L. B. Sanford. 1985. Laveran's germ: The reception and use of a medical discovery. *Am. J. Trop. Med. Hyg.* 34:2–20.

98. Smith, T., and F. L. Kilbourne. 1893. Investigations into the nature, causation, and prevention of Texas or southern cattle fever. *U.S. Dept. Agr. Bur. Anim. Indust. Bull.* 1.

99. Speck, M. 1993. Breakthrough in malaria research. Miami Herald (June 27:1A, 8A).

100. Strickland, G. T. 1991. Malaria. In Strickland, G. T., ed. *Hunter's tropical medicine,* 7th ed. Philadelphia: W. B. Saunders Co.

101. Stürchler, D. 1989. How much malaria is there worldwide? *Parasitol. Today* 5:39–40.

102. Terzakis, J. A., J. P. Vanderberg, D. Foley, and S. Shustak. 1979. Exoerythrocytic merozoites of *Plasmodium berghei* in rat hepatic Kupffer cells. *J. Protozool.* 26:385–89.

103. Torii, M., K. -I. Nakamura, K. P. Sieber, L. H. Miller, and M. Aikawa. 1992. Penetration of the mosquito (*Aedes aegypti*) midgut wall by the ookinetes of *Plasmodium gallinaceum. J. Protozool.* 39:449–54.

104. Trager, W., and J. B. Jensen. 1976. Human malaria parasites in continuous culture. *Science* 193:673–75.

105. Turrini, F., E. Schwarzer, and P. Arese. 1993. The involvement of hemozoin toxicity in depression of cellular immunity. *Parasitol. Today* 9:297–300.

106. Wahlgren, M., J. Carlson, R. Udomsangpetch, and P. Perlmann. 1989. Why do *Plasmodium falciparum*-infected erythrocytes form spontaneous erythrocyte rosettes? *Parasitol. Today* 5:183–85.

107. Warhurst, D. C. 1988. Mechanism of chloroquine resistance in malaria. *Parasitol. Today* 4:211–13.

108. Warrell, D. A. 1987. Pathophysiology of severe falciparum malaria in man. Symposia of the British Society of Parasitology, vol. 24. *Parasitology* 94:S53–S76.

109. Warrell, D. A., S. Looareesuwan, R. E. Phillips, N. J. White, M. J. Warrell, H. M. Chapel, S. Areekul, and S. Tharavanij. 1986. Function of the blood-cerebrospinal fluid barrier in human cerebral malaria: Rejection of the permeability hypothesis. *Am. J. Trop. Med. Hyg.* 35:882–89.

110. Waters, A. P., D. G. Higgins, and T. F. McCutchan. 1993. The phylogeny of malaria: A useful study. *Parasitol. Today* 9:246–50.

111. Wellems, T. E. 1991. Molecular genetics of drug resistance in *Plasmodium falciparum* malaria. *Parasitol. Today* 7:110–12.

112. White, N. J. 1988. The treatment of falciparum malaria. *Parasitol. Today* 4:10–14.

113. White, N. J., and M. Ho. 1992. The pathophysiology of malaria. In Baker, J. R., and R. Muller, eds. *Advances in parasitology* 31. London: Academic Press, 83–173.

114. World Health Organization Malaria Action Programme. 1986. Severe and complicated malaria. *Trans. R. Soc. Trop. Med. Hyg.* 80(suppl.):1–50.

115. Yayon, A. R., Timberg, S. Friedman, and H. Ginsberg. 1984. Effects of chloroquine on the feeding mechanism of the intraerythrocytic human malarial parasite *Plasmodium falciparum. J. Protozool.* 31:367–72.

116. Zhang, Y. 1987. Malaria: An intra-erythrocytic neoplasm? *Parasitol. Today* 3:190–92.

Chapter 10

PHYLA MYXOZOA AND MICROSPORA: PROTOZOA WITH POLAR FILAMENTS

. . . an enigma wrapped in a puzzle.

Arthur Koestler

Members of these two phyla formerly were placed in a class (Cnidosporidea) of the Sporozoa because they form spores. However, it is now recognized that they are quite different from the apicomplexans and, indeed, bear little if any relationship to each other. In fact, some workers do not consider the Myxozoa protozoa because their spores are composed of more than one cell. In both groups the polar filaments are tubelike and held coiled within the spores. When the spores encounter the proper environment, typically the digestive system of a host, the polar filaments are expelled. In the Myxozoa the filaments lie within **polar capsules** and apparently serve an anchoring function after expulsion. In the Microspora the polar filament pierces the intestinal epithelium of the host, and the amebalike **sporoplasm** passes through the filament into the host cell. The Myxozoa attack lower vertebrates, and a few are known from invertebrates.[26] Microsporans are mostly parasites of invertebrates, especially insects, but some are found in lower vertebrates and rarely in humans. Life cycles have not been worked out for many of these parasites, but recent research suggests myxozoans and some microsporans require a second host. They probably are a major natural control of some insect populations.

PHYLUM MYXOZOA

In the phylum Myxozoa the spores are of multicellular origin and are surrounded by two, three, or rarely more valves of various shapes (Fig. 10.1). They have one or more polar capsules and are mostly parasites of fish. A few are reported from amphibians and reptiles, but none is known from birds or mammals. From one to four polar capsules can be found at one end of the spore, except in the suborder Bipolarina where one capsule is located at each end of the spore. Next to the polar capsules is an ameboid sporoplasm that is infective to the host (Fig. 10.2). Some species have large vacuoles in the sporoplasm that stain readily with iodine and are therefore called **iodinophilous vacuoles.** The valves join at a **sutural plane** that is either twisted or straight. The valves may bear various markings and often are extended as pointed processes at the "posterior" end. More than 1200 species in 46 genera are described in this phylum. Most are host and tissue specific.

Family Myxosomatidae

Of the many families of myxozoans few are more striking in appearance and importance than the Myxosomatidae. Fish parasites, they have two or four polar capsules in the spore stage, and their sporoplasm lacks iodinophilous vacuoles.[12] One species, *Myxobolus cerebralis,* is of circumboreal importance to salmonid fish, including trout.

Elucidation of the life histories of myxozoans has been one of the more interesting parasitological developments in recent years. An increasing number of reports indicate that myxozoans require annelids as intermediate hosts.[16,35] In the case of *M. cerebralis,* Markiw and Wolf[25] showed that tubificid oligochaetes were required intermediate hosts and, more remarkably, that the parasite stages infective for fish were identical to members of the genus *Triactinomyxon* (Fig. 10.3), which had formerly been placed in a separate class (Actinosporea) of the phylum! Triactinomyxons liberated into the water from the worms are infective for the fish, which develop typical *M. cerebralis* infections.[24,34]

The triactinomyxon stage has its own complex life cycle in the worm, involving sexual reproduction and the production of spores with three valves (contrasted with two in spores that develop in fish).[22] The initial stages of spore development occur in cysts in the intestinal epithelium. Sporogenesis begins when two cells envelop two others (Fig. 10.4). The enclosed cells then undergo a series of divisions and arrangements into their relative positions in the finished spore, with the three flattened valve cells surrounding the sporoplasm (Fig. 10.5) and capsulogenic cells (those that will develop the polar filaments). The shaft and long hooks of the triactinomyxon are inflated to their final form after release of the spore from the worm. In addition to *M. cerebralis,* a number of other species of myxozoans have likewise been transmitted to various fishes by way of triactinomyxons.[7,8]

There have been no detailed studies of the fate of infective spores once they encounter the vertebrate host. When exposed to a variety of stimuli (pressure, acid), myxosporean polar capsules shoot out their filaments in a manner analogous

FIGURE 10.1

Spores of representative genera of myxosporans.
(*a*) *Myxobilatus;* (*b*) *Ceratomyxa;* (*c*) *Myxidium,*
with a polar capsule at each end; (*d*) *Henneguya;*
(*e*) *Thelohanellus,* with a single polar capsule;
(*f*) *Chloromyxum,* with four polar capsules.

Drawings by Ian Grant.

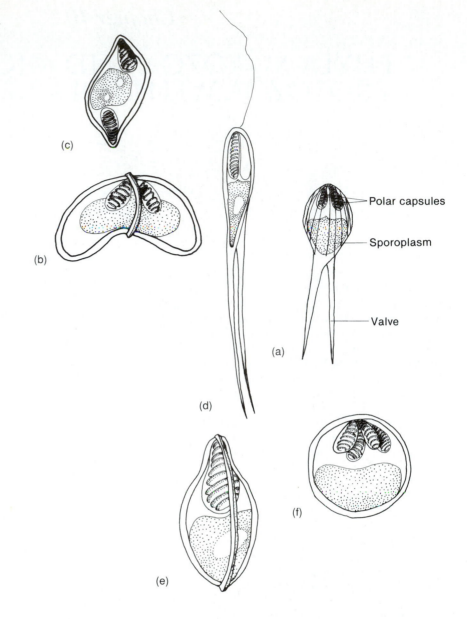

to eversion of the finger of a glove. This event presumably initiates the infection, exporulation occurring in the tissues. Once the sporoplasm leaves the valves and migrates (?) to its characteristic infection site, it begins to grow, the nuclei dividing repeatedly.[6] The multinucleate trophozoite often grows until it is visible to the unaided eye (Fig. 10.6)—some species can reach a size of several millimeters—feeding from the surrounding tissue by pinocytosis[5] (Fig. 10.7).

During the course of growth and nuclear divisions, two types of nuclei can be distinguished, **generative** and **somatic** (Fig. 10.8). As development proceeds, a certain amount of cytoplasm becomes segregated around each generative nucleus to form a separate cell within the trophozoite. These cells will produce the spores; hence they are called **sporoblasts.** Because in most species each will give rise to more

than one spore, they also are called **pansporoblasts.** Each pansporoblast in *M. cerebralis* will produce two spores. The generative nucleus for each spore will divide four times, one of the daughter nuclei of each division remaining generative and the other becoming somatic. The first somatic daughter nucleus will form the outer envelope of the spore; the second will divide again to give rise to the valvogenic cells; and the third nucleus will divide to produce the nuclei of the polar cells.[6] Thus, the spore of the Myxozoa is of multicellular origin.

• Genus *Myxobolus*

Myxobolus species (some of which were formerly placed in the genus *Myxosoma*) have ovoid or teardrop-shaped spores with a distinct sutural line (Fig. 10.9) and two polar capsules.

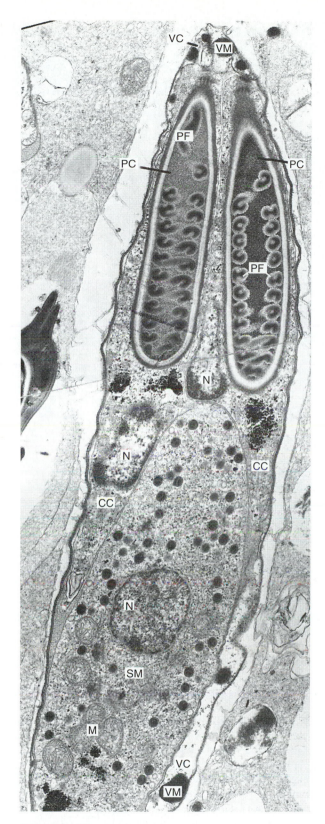

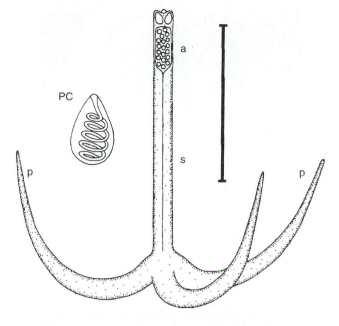

FIGURE 10.3

A *Triactinomyxon* spore with enlargement of polar capsule (**PC**), anterior part of the spore with the sporoplasm and PC (**a**), stylus (**s**), and projections (**p**). (Bar = 100 μm.)

From J. Lom and I. Dyková, "Fine structure of *Triactinomyxon* early stages and sporogony: Myxosporean and actinospoorean features compared," in *J. Protozool.* 39:16–27. Copyright © 1992 by the Society of Protozoologists.

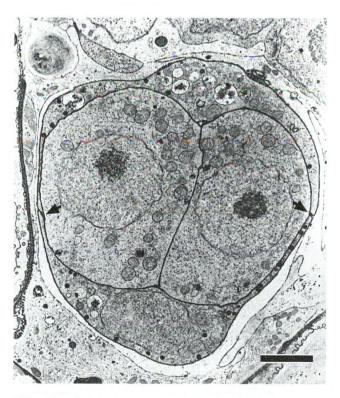

FIGURE 10.2

Longitudinal section of a *Henneguya adiposa* spore. **CC,** capsulogenic cells with nuclei (**N**) and polar capsules (**PC**); **PF,** polar filament; **SM,** sporoplasm; **M,** mitochondria; **VM,** accumulation of valve-forming material in the valvogenic cells (**VC**).

From W. L. Current, "*Henneguya adiposa* Minchew (Myxosporida) in the channel catfish: Ultrastructure of the plasmodium wall and sporogenesis," in *J. Protozool.* 26:209–217. Copyright © 1979 by the Society of Protozoologists..

FIGURE 10.4

Beginnings of spore formation in *Triactinomyxon*. Two outer cells (contact points indicated by *arrows*) envelop two inner cells. (Bar = 3 μm.)

From J. Lom and I. Dyková, "Fine structure of *Triactinomyxon* early stages and sporogony: Myxosporean and actinospoorean features compared," in *J. Protozool.* 39:16–27. Copyright © 1992 by the Society of Protozoologists.

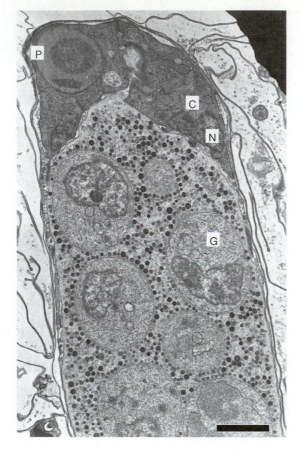

FIGURE 10.5

Anterior part of triactinomyxon with two capsulogenic cells (**C**) containing capsule primordium (**P**) and nucleus (**N**). **G,** germinal cells of the sporoplasm. (Bar = 2 µm.)

From J. Lom and I. Dyková, "Fine structure of *Triactinomyxon* early stages and sporogony: Myxosporean and actinospoorean features compared," in *J. Protozool.* 39:16–27. Copyright © 1992 by the Society of Protozoologists.

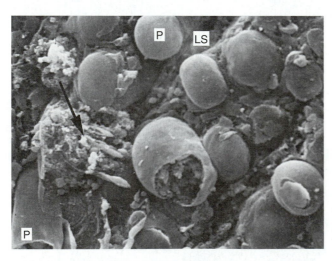

FIGURE 10.6

Scanning electron micrograph of the interior of a channel catfish gill infected with *Henneguya exilis*. **P,** parasite cysts (plasmodia); *arrows* point to broken cyst with spores protruding; **LS,** lamellar sinuses.

From W. L. Current and J. Janovy Jr., "Comparative study of ultrastructure of interlamellar and intralamellar types of *Henneguya exilis* Kudo from channel catfish," in *J. Protozool.* 25:56–65. Copyright © 1978 by the Society of Protozoologists.

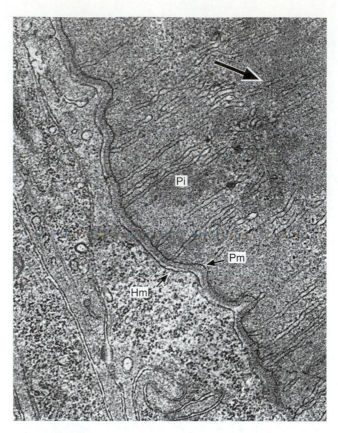

FIGURE 10.7

Transmission electron micrograph of the *Myxobolus (Myxosoma) funduli* cyst wall. **Pi,** zone of pinocytic canals; **Pm,** parasite cyst membrane; **Hm,** host cell membranes; arrow, pinocytic vesicle at end of canal.

From W. L. Current et al., "*Myxosoma funduli* Kudo (Myxosporida) in *Fundulus kansae:* Ultrastructure of the plasmodium wall and of sporogenesis, in *J. Protozool.* 26:574–583. Copyright © 1979 by the Society of Protozoologists.

A wide variety of fish, especially minnows, are infected with *Myxobolus* sp. Infections occur in several tissues: skin, gills, and various internal organs.

Myxobolus cerebralis. *Myxobolus cerebralis* causes in salmonids **whirling disease,** so called because fish with the disease swim in circles when disturbed or feeding. The parasite apparently was formerly endemic in the brown trout from central Europe to southeast Asia, and it causes no symptoms in that host. The disease was first noticed in 1900 after the introduction of the rainbow trout to Europe. Since then it has spread to other localities in Europe, including Sweden and Scotland; to the United States; to South Africa; and to New Zealand.[11] Whirling disease results in a high mortality rate in very young fish and causes corresponding economic loss, especially in hatchery-reared brook and rainbow trout. If a fish survives, damage to the cranium and vertebrae can cause crippling and malformation.

- *Morphology.* The mature spore of *M. cerebralis* is broadly oval, with thick sutural ridges on the edges of the valves. It measures 7.4 to 9.7 µm long by 7 to 10 µm wide. The entire spore is covered with a mucoidlike envelope. There are two polar capsules at the anterior end, each with a filament twisted into five or six coils. During

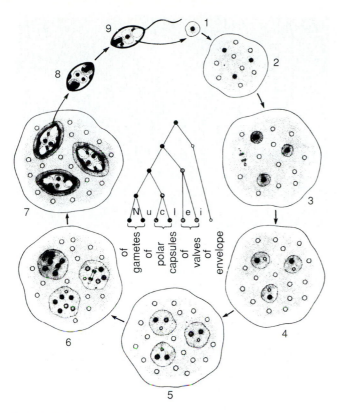

FIGURE 10.8

Diagram of the development of a myxosporidian in the vertebrate host. **1,** uninucleate amebula. **2,** multinucleate plasmodium: differentiation into generative (**dark**) and somatic (**light**) nuclei. **3,** segregation of the sporoblasts. **4–6,** different stages of nuclear multiplication in the sporoblasts, corresponding to the divisional sequence shown in the middle of the diagram (envelope nucleus, **white;** valve nuclei, **crosshatched;** nuclei of the polar capsule, **dotted;** gamete nuclei [germ line!], **black). 7,** plasmodium with spores. **8,** single spore with binucleate amebula. **9,** Single spore with uninucleate amebula and a polar capsule with discharged polar filament.

From Karl G. Grell, *Protozoology.* Copyright © 1973 Springer-Verlag GmbH & Co., KG., Heidelberg. Reprinted with permission.

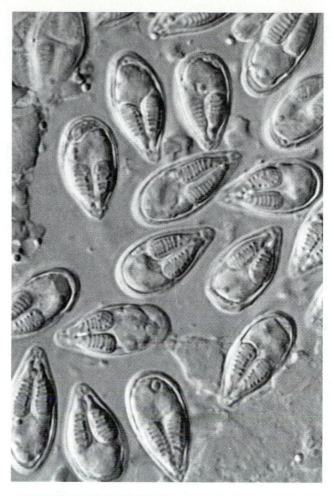

FIGURE 10.9

Spores from a *Myxobolus* species parasitizing a minnow, as seen in a fresh squash preparation of a cyst.

Courtesy of W. L. Current.

development each polar capsule lies within a polar cell that also contains a nucleus, and the nuclei of the two valvogenic cells may be seen lying adjacent to the inner surface of each valve. The sporoplasm contains two nuclei (presumably haploid), numerous ribosomes, mitochondria, and other typical organelles.[21]

- *Biology.* Following encounter with the fish, the triactinomyxon exsporulates, and the sporoplasm migrates to the spine and head cartilages; it begins growing, and its nuclei divide, as discussed previously, forming cavities in the surrounding tissue. The cavities within the cartilage become packed with trophozoites and spores by eight months after infection. Spores may live in the fish for three or more years. How they escape into the water is speculative, but it seems reasonable to assume that, when the host is devoured by a larger fish or other piscivorous predator, such as a kingfisher or heron, the spores are released by digestion of their former home. The crippling

effect of the parasite can make the host especially vulnerable to predation. The feces of birds that have been fed fish infected with *M. cerebralis* can carry the disease organism and subsequently infect fish held in previously uncontaminated water.[31]

- *Pathogenesis.* The main pathogenic effects of this disease are damage to the cartilage in the axial skeleton of young fish, consequent interference with function of adjacent neural structures, and subsequent granuloma formation in healing of the lesions. Invasion of the cartilaginous capsule of the auditory-equilibrium organ behind the eye interferes with coordinated swimming; thus, when the fish is disturbed or tries to feed, it begins to whirl frantically, as if chasing its tail. It may become so exhausted by this futile activity that it sinks to the bottom and lies on its side until it regains strength. Predation most likely occurs at this stage.[13] Often the cartilage of the spine is invaded, especially posterior to the 26th vertebra. Function of the sympathetic nerves controlling the melanocytes is impaired, and the posterior part of the fish becomes very dark, producing the "black

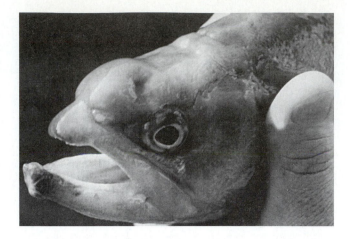

FIGURE 10.10

Axial skeleton deformities in living rainbow trout that have recovered from whirling disease (*Myxobolus cerebralis*). (*a*) Note bulging eyes, shortened operculum, and both dorsoventral and lateral curvature of the spinal column (lordosis and scoliosis). (*b*) Note gaping, underslung jaw and grotesque cranial granuloma.

Photograph by Larry S. Roberts.

tail." If the fish survives, granulomatous tissue infiltration of the skeleton may produce permanent deformities: misshapen head, permanently open or twisted lower jaw, or severe spinal curvature (scoliosis) (Fig. 10.10).

- *Epizootiology and Prevention.* It seems clear that in ponds in which infected fish are held, spores can accumulate, whether by release from dead and decomposing fish, passage through predators, or some kind of escape from the tissue of infected living fish. Severity of an outbreak depends on the degree of contamination of a pond, and light infections cause little or no overt disease. Spores are resistant to drying and freezing, surviving for a long period of time, up to 18 days, at $-20°$ C.[14]

No effective treatment for infected fish is known, and such fish should be destroyed by burial or incineration. Great care must be exercised to avoid transferring spores to uncontaminated hatcheries or streams, either by live fish that might be carriers or by feeding possibly contaminated food materials, for example, tubificids, to hatchery fish. Earthen and concrete ponds in which infected fish have been held can be disinfected by draining and treating with calcium cyanamide or quicklime.

- *Extrasporogonic Phases of the Life Cycle.* Several species of Myxozoa, including *Sphaerospora renicola* in the commercially important carp, have an asexually proliferative phase in the blood of the host. This stage only increases the number of parasites; it does not develop directly into spores. A second extrasporogonic phase invades the swim bladder of carp fry, causing swimbladder inflammation that results in high mortality or growth retardation. Some small plasmodia (ameboid forms) reach the renal tubules where they either produce spores (seasonally) or are destroyed by host reactions.[20]

Some other species, belonging to different genera and families, also commonly occur in fish, amphibians, and reptiles. For general reviews and keys see Hoffman,[12] Hoffman et al.,[15] Landsberg and Lom,[18] and Lom.[20]

PHYLUM MICROSPORA

The phylum Microspora includes fewer than 1000 described species of intracellular parasites of invertebrates and some vertebrates.[30] These species have been found in protozoa, platyhelminths, nematodes, bryozoa, rotifers, annelids, all classes of arthropods, fish, amphibians, reptiles, birds, and some mammals. A few species are known to infect humans. A number are pathogenic, and several are of economic importance. The spores are unicellular and have a single sporoplasm; the spore walls are complete, without suture lines, pores, or other openings. They have a simple or complex extrusion apparatus with polar tube and polar cap. The extrusion apparatus of the class Microsporea is described next; the class Rudimicrosporea has a simple extrusion apparatus without a polaroplast and posterior vacuole. The rudimicrosporeans are all hyperparasites of gregarines.

The Microspora formerly included the class Haplosporea. The haplosporideans and some newly described and lesser known forms are now placed in the phylum Ascetospora.[19] The spores are often, or always, multicellular and complex, with one or more sporoplasms but without polar capsules or filaments. They are parasites of a variety of invertebrates; *Haplosporidium* spp. (formerly *Minchinia* spp.) are pathogens in the economically important oyster *Crassostrea virginica.*

For techniques of study see Canning and Lom.[4]

Class Microsporea

The spore is the most conspicuous and morphologically distinctive stage in the life cycle of microsporeans. Spores are ovoid, spheroid, or cylindroid. The spore wall is trilaminar, consisting of an outer, dense exospore, an electron-lucent middle layer (endospore), and a thin membrane surrounding the cytoplasmic contents (Fig. 10.11). Some species have two to five layers of exospore. The wall is dense and refractile; its resistant properties contribute greatly to the survival of the spore. Spores are usually about 3 to 6 μm in length, and little structure can be discerned under the light microscope other

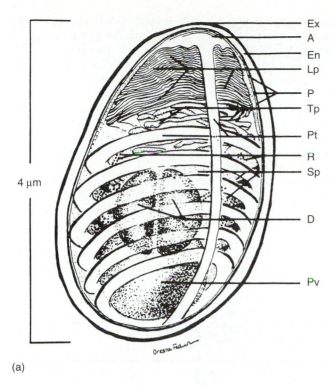

4 µm

Ex
A
En
Lp
P
Tp
Pt
R
Sp
D
Pv

(a)

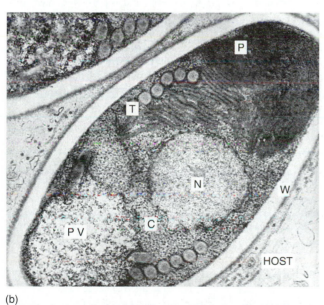

(b)

FIGURE 10.11

(a) Diagram of the internal structure of a microsporidian spore. **Ex,** electron dense exospore; **A,** filament anchoring disc; **Lp,** lamellar polaroplast; **Tp,** tubular polaroplast; **Pt,** polar tubule; **D,** diplokaryon nuclei; **Pv,** posterior vacuole. (b) *Nosema lophii* spore displaying polaroplast (**P**), nucleus (**N**), ribosome-rich cytoplasm (**C**), polar tube (**T**), posterior vacuole (**PV**), and wall (**W**).

(a) From A. Cali, "General microsporidian features and recent findings on AIDS isolates," in *J. Protozool.* 38:625–630. Copyright © 1991. The Society of Protozoologists. Reprinted with the permission of the publisher. (b) From E. Weidner, "Ultrastructural study of microspor.idian invasion into cells," in *Z. Parasitenkd.* 40:227–242. Copyright © 1972.

than an apparent vacuole at one or both ends. The smallest known spore is *Encephalitozoon* from mammals (2.5 × 1.5 µm), while the largest is *Mrazekia piscicola* from cod (20 × 6 µm). It has long been known that microsporeans possess a **polar tube,** or **filament,** because the structure can be stimulated to extrude artificially.

Possession of the spore filament was formerly the basis for uniting the microsporans with the myxozoans in the Cnidospora. However, use of the electron microscope has shown that the polar filaments of the two groups are basically dissimilar. There is no polar capsule in the microsporans, and neither is the polar filament formed by a separate capsulogenic cell. At the ultrastructural level we can see a small **polar cap** or **sac** covering the attached end of the filament and just overlying the **polaroplast** (the apparent anterior vacuole, which is actually a stack of sacs and tubules) (Fig. 10.11).

The ameboid sporoplasm surrounds the extrusion apparatus, with its nucleus and most of its cytoplasm lying within the coils of the filament. A **posterior vacuole** may be found at the end opposite the polaroplast. The cytoplasm of the sporoplasm has many free ribosomes and some endoplasmic reticulum but no mitochondria, peroxisomes, or typical Golgi membranes. The membrane and matrix of the polar cap are continuous, with a highly pleated membrane comprising the polaroplast. This in turn is continuous with the anchoring disc or base of the polar filament.[33] When extrusion of the polar filament is stimulated in the host, a permeability change in the polar cap apparently allows water to enter the spore, and the filament is expelled explosively, simultaneously turning "inside out." The stacked membrane in the polaroplast unfolds as the filament discharges, and the membrane contributes to the expelled filament so that it is much longer than when it is coiled within the spore. The force with which the filament discharges causes it to penetrate any cell in its path, and the sporoplasm flows through the tubular filament, thereby gaining access to its host cell. The end of the filament within the host cell expands to enclose the sporoplasm and becomes the parasite's new outer membrane.

The nuclei of the intracellular trophozoite divides repeatedly, and the organism becomes a large, multinucleate plasmodium. Finally cytokinesis takes place, and the process may then be repeated. In some species the nuclei may be associated in pairs (**diplokarya),** but such association apparently is not involved with sexual reproduction. The multiple fission of the trophozoites (or meronts) is usually regarded as schizogony, but the process may not be strictly analogous to the schizogony found in the Apicomplexa.

Sporogenesis occurs when the nuclear divisions of the monokaryotic or dikaryotic trophozoites give rise to nuclei destined to become spore nuclei. In a number of genera, *not* including *Nosema,* the nuclear division preceding sporogony is meiotic (reductional), giving rise to haploid spores.[10,23] In these genera the spores are not directly infective to new hosts, leading to the suggestion that there is an alternate (intermediate?) host in which restoration of the diploid condition occurs. For example, see Canning and Hollister[3] for *Amblyospora* in copepods and mosquitoes. Sexual reproduction seems to be restricted to plasmogamy, not karyogamy.

During sporogony the organism becomes a multinucleate, sporogonial plasmodium. This can occur either by internal segregation of cytoplasm around the nuclei to become sporont-determinate areas or by the formation of an envelope at the sporont surface and subsequent separation from developing sporoblasts, leaving a vacuolar space.[27] The spores then differentiate and mature within the pansporoblast. In each sporoblast, there forms a mass of tubules, which becomes the polar tube and polaroplast.[17] Mitochondria are not present at any stage. A xenoma (combination parasite and hypertrophied cell) of considerable size develops in some species.[4]

• Family Nosematidae

The genera of Nosematidae are separated on the basis of the number of spores produced by each sporoblast mother cell during the life cycle (from 1 to 16 or more).

Nosema apis.   *Nosema apis* is a common parasite of honeybees in many parts of the world, causing much loss annually to beekeepers. It infects the epithelial cells in the midgut of the insect. Infected bees lose strength, become listless, and die. Although the ovaries of the queen are not directly infected, they degenerate when her intestinal epithelium is damaged, an example of parasitic castration. The disease is variously known as **nosema disease, spring dwindling, bee dysentery, bee sickness,** and **May sickness.**

The spore of *N. apis* is oval, measuring 4 to 6 μm long by 2 to 4 μm wide. The extended filament is 250 to 400 μm long. Infected bees defecate spores that are infective to other bees. Swallowed spores enter the midgut and lodge on the peritrophic membrane. Extruded filaments pierce the peritrophic membrane and intestinal epithelial cells, and the sporoplasm enters an epithelial cell. The entire process is accomplished within 30 minutes. Sporogony takes place in the second multiple fission generation, and the spores rupture the host cell to be passed with the feces. The entire life history in the bee is completed in four to seven days. Destruction of the intestinal epithelium may kill the host.

Other *Nosema* Species.   Besides *N. apis,* a few other of the many species in this genus are important to humans. However, many additional ones may be important biological controls of insect populations.

Nosema bombycis is a parasite of silk moth larvae, flourishing in the unnaturally crowded conditions of silkworm culture. The parasite affects nearly all tissues of the insect's body, including the intestinal epithelium. Parasitized larvae show brown or black spots on their bodies, giving them a peppered appearance. There is a high rate of mortality. Pasteur devoted considerable effort in 1870 to understanding and controlling this disease and is credited with saving the silk industry in the French colonies. *Nosema bombycis* also was one of the first "germs" proved to cause disease. Its life cycle is basically similar to that of *N. apis* and can be completed in four days.

Ameson michaelis is a parasite of the economically important blue crab, *Callinectes sapidus,* which it may kill 15 to 29 days after infection.[32] *Ameson michaelis* undergoes schizogony, probably in the hemocytes, and then undergoes sporogony in the striated muscle, which is extensively dam-

aged. Species of *Glugea, Pleistophora, Nosema,* and other genera parasitize fish, including several economically important groups. Serious epizootics have been reported.

Because of the pathological effects, microsporidians are also being studied as biological control agents. *Nosema algerae* infection, for example, reduces the number of malarial oocysts formed in *Anopheles* mosquitoes.[28] *Nosema whitei* is pathological to *Tribolium* spp. (flour beetles) species, which in addition to being favorites of the experimental ecologists, are among the many stored grain pests that significantly reduce global food supplies.[1,9]

Encephalitozoon cuniculi occurs in laboratory mice and rabbits, and it is also known in monkeys, dogs, rats, birds, guinea pigs, and other mammals, including humans, usually in the brain.[29] It may be transmitted in body exudates or transplacentally. Although damage is usually minimal, the infection can be fatal. It is the most extensively studied of all Microsporidia. At various early times it was thought to cause rabies and polio.

Microsporidian spores must be exceedingly common in the environment, and consequently these parasites are candidates for opportunistic infections in immunodeficient people.[3] For example, high levels of anti–*E. cuniculi* antibodies are common in immunodeficient patients but low in uncompromised people. Other microsporidian species have been isolated from AIDS patients and others who were unable to rally their lymphocytic defenses. Documented cases have been reported for species of the genera *Pleistophora, Nosema, Enterocytozoon* (*E. bieneusi*), "*Microsporidium*" (a catchall genus for microsporidian parasites of unknown affinities), and *Nosema connori* in humans.[2] There are no effective drugs for the treatment of microsporidiosis.

References

1. Bass, L. K., and E. Armstrong. 1992. *Nosema whitei*: Effects on oocyte development and maturation in *Tribolium castaneum. J. Invertebr. Pathol.* 59:115–23.

2. Cali, A. 1991. General microsporidian features and recent findings on AIDS isolates. *J. Protozool.* 38:625–30.

3. Canning, E. U., and W. S. Hollister. 1987. Microsporidia of mammals—widespread pathogens or opportunistic curiosities? *Parasitol. Today* 3:267–73.

4. Canning, E. U., and J. Lom. 1986. *The microsporidia of vertebrates.* London: Academic Press.

5. Current, W. L., and J. Janovy Jr. 1976. Ultrastructure of interlamellar *Henneguya exilis* in the channel catfish. *J. Parasitol.* 62:975–81.

6. Current, W. L., J. Janovy Jr., and S. A. Knight. 1979. *Myxosoma funduli* Kudo (Myxosporida) in *Fundulus kansae:* Ultrastructure of the plasmodium wall and of sporogenesis. *J. Protozool.* 26:574–83.

7. El-Matbouli, M., and R. W. Hoffmann. 1989. Experimental transmission of two *Myxobolus* spp. developing bisporogeny via tubificid worms. *Parasitol. Res.* 75:461–64.

8. El-Matbouli, M., T. Fischer-Scherl, and R. W. Hoffmann. 1992. Transmission of *Hoferellus carassi* Achmerov, 1960, to goldfish *Carassius auratus* via an aquatic oligochaete. *Bull. European Assoc. Fish Pathology* 12:54–56.

9. Ghosh, S., and K. Sara. 1992. Experimental infection of *Nosema* sp. Ghosh in an unnatural host *Lesioderma sericorne*. *J. Protozool.* 39:410–12.

10. Hazard, E. I., T. G. Andreadis, D. J. Joslyn, and E. A. Ellis. 1979. Meiosis and its implications in the life cycles of *Amblyospora* and *Parathelohania* (Microspora). *J. Parasitol.* 65:117–22.

11. Hewitt, G. C., and R. W. Little. 1972. Whirling disease in New Zealand trout caused by *Myxosoma cerebralis* (Hofer, 1903) (Protozoa; Myxosporida). *N. Z. J. Mar. Freshwater Res.* 6:1–10.

12. Hoffman, G. L. 1967. *Parasites of North American freshwater fishes.* Berkeley: University of California Press.

13. Hoffman, G. L., C. E. Dunbar, and A. Bradford. 1969. Whirling disease of trouts caused by *Myxosoma cerebralis* in the United States. *U.S. Department of Interior, Fish and Wildlife Service, Special Scientific Report, Fisheries No. 427* (1962 report issued with addendum, 1969).

14. Hoffman, G. L., and R. E. Putz. 1969. Host susceptibility and the effect of aging, freezing, heat, and chemicals on spores of *Myxosoma cerebralis*. *Progressive Fish-Culturist* 31:35–37.

15. Hoffman, G. L., R. E. Putz, and C. E. Dunbar. 1965. Studies on *Myxosoma cartilaginis* n. sp. (Protozoa: Myxosporidea) of centrarchid fish and a synopsis of the *Myxosoma* of North American freshwater fishes. *J. Protozool.* 12:319–32.

16. Kent, M. L., D. J. Whitaker, and L. Margolis. 1992. Transmission of *Myxobolus arcticus* Pugachev and Khokhlov, 1979, a myxosporean parasite of Pacific salmon, via a triactinomyxon from the aquatic oligochaete *Stylodrilus heringianus* (Lumbriculidae). *Can. J. Zool.* 71:1207–11.

17. Krinsky, W. L., and S. F. Hayes. 1978. Fine structure of the sporogonic stages of *Nosema parkeri*. *J. Protozool.* 25:177–86.

18. Landsberg, J. H., and J. Lom. 1991. Taxonomy of the genera of the *Myxobolus* and *Myxosoma* group (Myxobolidae: Myxosporea), current listing of species and revision of synonyms. *Systematic Parasitol.* 18:165–86.

19. Levine, N. D. et al. 1980. A newly revised classification of the Protozoa. *J. Protozool.* 27:37–58.

20. Lom, J. 1987. Myxosporea: A new look at long-known parasites of fish. *Parasitol. Today* 3:327–32.

21. Lom, J., and P. dePuytorac. 1965. Studies on the myxosporidean ultrastructure and polar capsule development. *Protistologica* 1:53–65.

22. Lom, J., and I. Dykova. 1992. Fine structure of *Triactinomyxon* early stages and sporogony: Myxosporean and actinosporean features compared. *J. Protozool.* 39:16–27.

23. Loubès, C. 1979. Recherches sur la méiose chez les microsporidies: Conséquences sur les cycles biologiques. *J. Protozool.* 26:200–208.

24. Markiw, M. W. 1986. Salmonid whirling disease: Dynamics of experimental production of the infective stage—the triactinomyxon spore. *Can. J. Fish. Aquatic Sci.* 43:521–26.

25. Markiw, M. E., and K. Wolf. 1983. *Myxosoma cerebralis* (Myxozoa: Myxosporea) etiologic agent of salmonid whirling disease requires tubificid worm (Annelida: Oligochaeta) in its life cycle. *J. Protozool.* 30:561–64.

26. Overstreet, R. M. 1976. *Fabespora vermicola* sp. n., the first myxosporidan from a platyhelminth. *J. Parasitol.* 62:680–84.

27. Overstreet, R. M., and E. Weidner. 1974. Differentiation of microsporidian spore-tails in *Inodosporus spraguei* gen. et sp. n. *Z. Parasitenkd.* 44:169–86.

28. Schenker, W., W. A. Maier, and H. M. Seitz. 1992. The effects of *Nosema algerae* on the development of *Plasmodium yoelii nigeriensis* in *Anopheles stephensi*. *Parasitol. Res.* 78:56–59.

29. Shadduck, J. A., W. T. Watson, S. P. Pakes, and A. Cali. 1979. Animal infectivity of *Encephalitozoon cuniculi*. *J. Parasitol.* 65:123–29.

30. Sprague, V. 1982. Microspora. In Parker, S. P., ed. *Synopsis and classification of living organisms* 1. New York: McGraw-Hill Book Co., 589–94.

31. Taylor, R. L., and M. Lott. 1978. Transmission of salmonid whirling disease by birds fed trout infected with *Myxosoma cerebralis*. *J. Protozool.* 25:105–6.

32. Weidner, E. 1970. Ultrastructural study of microsporidian development. 1. *Nosema* sp. Sprague, 1965, in *Callinectes sapidus* Rathbun. *Z. Zellforsch.* 105:33–54.

33. Weidner, E. 1972. Ultrastructural study of microsporidian invasion into cells. *Z. Parasitenkd.* 40:227–42.

34. Wolf, K., and M. E. Markiw. 1984. Biology contravenes taxonomy in the Myxozoa: New discoveries show alternation of invertebrate and vertebrate hosts. *Science* 225:1449–52.

35. Yokoyama, H., K. Ogawa, and H. Wakabayashi. 1991. A new collection method of actinosporeans: A probable infective stage of myxosporeans to fishes from tubificids and experimental infection of goldfish with the actinosporean, *Raabeia* sp. *Fish Pathology* 26:133–38.

Additional References

The Cali (1991) reference is from a series of papers, all published in vol. 38 of the *Journal of Protozoology*, from a symposium on microsporidiosis in AIDS patients.

Sprague, V. 1982. Ascetospora. In Parker, S. P., ed. *Synopsis and classification of living organisms* 1. New York: McGraw-Hill Book Co., 599–601.

Sprague, V. 1982. Myxozoa. In Parker, S. P., ed. *Synopsis and classification of living organisms* 1. New York: McGraw-Hill Book Co., 595–97.

Chapter 11

PHYLUM CILIOPHORA: CILIATED PROTISTAN PARASITES

Nearly all the endoparasitic protozoa offer problems in the question of their transmission, which seem opposed to a simple solution.

Justus F. Mueller, discussing the life cycle of *Trichodina renicola,* a parasite of fish kidneys

The possession of simple cilia or compound ciliary organelles in at least one stage of their life cycle is the most conspicuous feature of the Ciliophora (see Chapter 4). A compound subpellicular infraciliature is universally present even when cilia are absent. Most species have one or more macronuclei and micronuclei, and fission is homothetogenic. Some species exhibit sexual reproduction involving conjugation, autogamy, and cytogamy. Although each cilium has a kinetosome, centrioles functioning as such are absent. Most ciliates are free-living, but many are commensals of vertebrates and invertebrates, and a few are parasitic.

The phylum Ciliophora probably has undergone more revision in the past two decades, and especially more extensive evaluation of higher taxonomic criteria, than any other group of organisms. Today ciliate taxonomy at virtually all levels depends on cortical structure, position and arrangement of kinetosomes, and ontogeny of the ciliary distribution patterns during cell division. The primary investigative tools are various techniques for staining the cortical structures with silver.

The following examples represent the most common and widely recognized ciliate commensals and parasites.

CLASS KINETOFRAGMINOPHOREA

Subclass Vestibulifera; Order Trichostomatida; Family Balantidiidae

The class is named for structures typically found in the oral region of its members: kinetofragments. These are patches or short files of somatic (body vs. strictly oral) kinetids, only some of which may bear cilia. As now conceived, the class includes some very primitive ciliates, as well as some very specialized ones. The subclass Vestibulifera is composed of those that have a **vestibulum** at the apical or near-apical end of the body. The vestibulum is a depression or invaginated area that leads directly to the cytostome; it is lined with cilia predominantly somatic in nature and origin. In the order Trichostomatida the somatic ciliature is typically distributed uniformly over the body, and there is little reorganization of

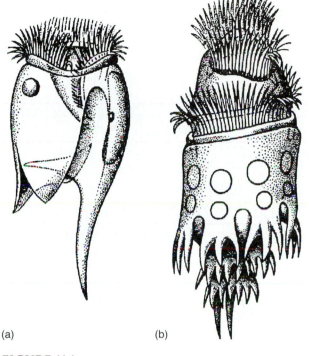

(a)   (b)

FIGURE 11.1

Examples of rumen ciliates. (*a*) *Entodinium caudatum*; (*b*) *Ophryoscolex purkinjei.*

From Karl G. Grell, *Protozoology.* Copyright © 1973 Springer-Verlag GmbH & Co. KG., Heidelberg, Germany. Reprinted with permission.

the somatic kineties at the level of the vestibulum. The curiously appearing entodiniomorphids (order Entodiniomorphida) have unique tufts of cilia on an otherwise naked body and a generally firm pellicle. They are commensals in mammalian herbivores, especially ruminants (Fig. 11.1). The remaining order (Colpodida) has a highly reorganized vestibular ciliature; members of the genus *Colpoda* are relatively large and are commonly the first to appear in large numbers in an infusion culture (weeds and water in a jar). In our consideration of this ciliate class we will discuss in detail only the trichostomatid species *Balantidium coli,* which is an important parasite of humans.

The family Balantidiidae has the single genus *Balantidium,* species of which are found in the intestines of crustaceans, insects, fish, amphibians, and mammals. The vestibulum leading into the cytostome is at the anterior end, and a cytopyge is present at the posterior tip (Fig. 11.2).

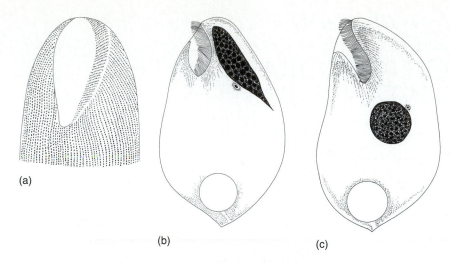

FIGURE 11.2

(*a*) Vestibule infraciliature of *Balantidium*, showing how vestibular infraciliature is actually a continuation of body kineties.
(*b,c*) *Balantidium* species from cockroaches;
(*b*) *B. praenucleatum* from *Blatta orientalis*;
(*c*) *B. ovatum* from *Blatta americana*.

From E. Fauré-Fremiet, "La position systematique du genre *Balantidium*," *J. Protozool.* 2:54–58. Copyright © 1955. Reprinted with permission of the publisher.

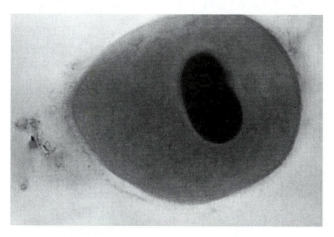

FIGURE 11.3

Trophozoite of *Balantidium coli.* Trophozoites range from 30 to 150 μm long by 25 to 120 μm wide.

Courtesy of James Jensen.

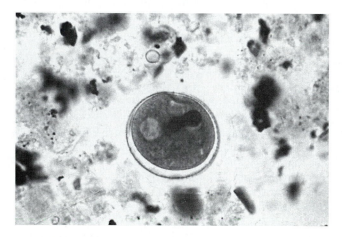

FIGURE 11.4

Encysted form of *Balantidium coli.* Cysts are 40 to 60 μm in diameter.

Balantidium coli. *Balantidium coli* (Fig. 11.3) is the largest protozoan parasite of humans. It is most common in tropical zones but is present throughout the temperate climes as well. The epidemiology and effects on the host are similar to those of *Entamoeba histolytica*. The organism appears to be primarily a parasite of pigs, with strains adapted to various other hosts.

* ***Morphology.*** Trophozoites of *B. coli* are oblong, spheroid, or more slender, 30 to 150 μm long by 25 to 120 μm wide (Fig. 11.3). Encysted stages (Fig. 11.4), which are most commonly found in stools, are spheroid or ovoid, measuring 40 to 60 μm in diameter. The macronucleus is a large, sausage-shaped structure. The single micronucleus is much smaller and often hidden from view by the macronucleus. There are two contractile vacuoles, one near the middle of the body and the other near the posterior end. The cytostome is at the anterior end. Food vac-

uoles contain erythrocytes, cell fragments, starch granules, and fecal and other debris. Living trophozoites and cysts are yellowish or greenish.

* ***Biology.*** The ciliate *B. coli* lives in the cecum and colon of humans, pigs, guinea pigs, rats, and many other mammals. It is not readily transmissible from one species of host to another, since it seems to require a period of time to adjust to the symbiotic flora of a new host. However, when adapted to a host species, the protozoan flourishes and can become a serious pathogen, particularly in humans. In animals other than primates the organism is unable to initiate a lesion by itself, but it can become a secondary invader if the mucosa is breached by other means.

The trophozoite multiplies by transverse fission. Conjugation has been observed in culture but may occur only rarely, if at all, in nature. Encystment is instigated by dehydration of feces as they pass posteriorly in the rectum. These protozoa can encyst after being passed in stools—an important factor in the epidemiology of the disease.

Infection occurs when the cyst is ingested, usually in contaminated food or water. Unencysted trophozoites may live up to 10 days and may possibly be infective if eaten, although this is unlikely under normal circumstances. Because *B. coli* is destroyed by a pH lower than 5, infection is most likely to occur in malnourished persons with low stomach acidity.

- *Pathogenesis.* Under ordinary conditions the trophozoite feeds much like a *Paramecium,* ingesting particles with the vestibulum and cytostome. However, sometimes it appears that the organisms can produce proteolytic enzymes that digest away the intestinal epithelium of the host. Production of hyaluronidase has been detected, and this enzyme could help enlarge the ulcer. The ulcer usually is flask shaped, like an amebic ulcer, with a narrow neck leading into an undermining saclike cavity in the submucosa. The colonic ulceration produces lymphocytic infiltration with few polymorphonuclear leukocytes, and hemorrhage and secondary bacterial invasion may follow. Fulminating cases may produce necrosis and sloughing of the overlying mucosa and occasionally perforation of the large intestine or appendix, as in amebic dysentery. Death often follows at this stage. Secondary foci, such as the liver or lung, may become infected.[4] Urogenital organs are sometimes attacked after contamination, and vaginal, uterine, and bladder infections have been discovered.

- *Epidemiology.* Balantidiasis in humans is most common in the Philippines but can be found almost anywhere in the world, especially among those who are in close contact with swine. Generally the disease is considered rare and occurs in less than 1% of the human population. Higher infection rates have been reported among institutionalized persons. However, in pigs the infection rate may be quite high; in a typical survey of pigs brought to slaughter in Japan, the prevalence was 100%.[6] Primates other than humans sometimes are infected and may represent a reservoir of infection to humans, although the reverse is probably more likely. The ability of the ciliates to encyst after being passed increases the number of potential infections from a single reservoir host, and cysts can remain alive for weeks in pig feces, if the feces do not dry out. The pig is probably the usual source of infection for humans, but the relationship is not clear. The protozoa in swine are essentially nonpathogenic and are considered by some a separate species, *B. suis.* There may be strains of *B. coli* that vary in their adaptability to humans. Infections often disappear spontaneously in healthy persons, or they can become symptomless, making the person a carrier.

- *Treatment and Control.* Several drugs are used to combat infections of *B. coli,* including carbarsone, diiodohydroxyquin, and tetracycline. Prevention and control measures are similar to those for *Entamoeba histolytica,* except that particular care should be taken by those who work with pigs.

Other species of *Balantidium* are *B. praenuleatum,* common in the intestines of American and oriental cockroaches, *B. duodeni* in frogs, *B. caviae* in guinea pigs, and *B. procypri* and *B. zebrascopi* in fishes.

CLASS OLIGOHYMENOPHOREA

Members of this class have a buccal cavity bearing a well-defined oral ciliary apparatus. However, the oral ciliary apparatus may be rather inconspicuous. It is composed of only three or four specialized membranelles.

Subclass Hymenostomatia; Order Hymenostomatida; Family Ophryoglenidae

The body ciliature in members of this subclass is often uniform and heavy, and conspicuous kinetodesmata are regularly present. The Hymenostomatida have a well-defined buccal cavity on their ventral surface. Most species are small, but *Ichthyophthirius multifiliis* is a very large one, and cysts on infected fish are often visible to the unaided eye.

The family Ophryoglenidae contains one genus of parasites, most species of which are unimportant to humans. One species, however, is a common pest in freshwater aquaria and in fish farming, causing much economic loss.

Ichthyophthirius multifiliis. *Ichthyophthirius multifiliis* (Fig. 11.5) causes in aquarium and wild freshwater fish a common disease known as **ick** to many fish culturists. It attacks the epidermis, cornea, and gill filaments.

- *Morphology.* Adult trophozoites are as large in diameter as 1 mm. The macronucleus is a large, horseshoe-shaped body that encircles the tiny micronucleus. Each of several contractile vacuoles has its own micropore in the pellicle. A permanent cytopyge is located at the posterior end of the animal.

- *Biology.* Mature trophozoites form pustules in the skin of their fish hosts (Fig. 11.5). They are set free and swim feebly about when the pustules rupture, finally settling on the bottom of their environment or on vegetation. Within an hour the ciliate secretes a thick, gelatinous cyst about itself and begins a series of transverse fissions, producing up to 1000 infective cells. The daughter trophozoites, or **tomites,** also termed **theronts** or **swarmers,** represent the infective stage and can survive about 96 hours without a host. The tomite is about 40 by 15 µm. Its narrowed anterior end carries a characteristic long filament that emerges from a conical depression in the pellicle.[5] Apparently the parasite burrows into the fish's skin with its pointed end and filament. There it becomes a trophozoite within three days, ingesting debris of host cells and forming a pustule that reaches over 1 mm in diameter.

- *Pathogenesis.* Grayish pustules form wherever the parasites colonize in the skin (Fig. 11.6). Epidermal cells combat the irritation by producing much mucus, but many die and are sloughed. When many parasites attack the gill filaments, they so interfere with gas exchange that the fish may die. Catfish that recover show protective immunity for up to eight months, which shows promise for the development of vaccines.[1,9] Research with carp has shown that the immune response is of a cellular nature, with macrophages accumulating at the site of epidermal infections in immunized fish, while in naive fish the cellular response to theronts consisted of a diffuse infiltration of neutrophils.[3]

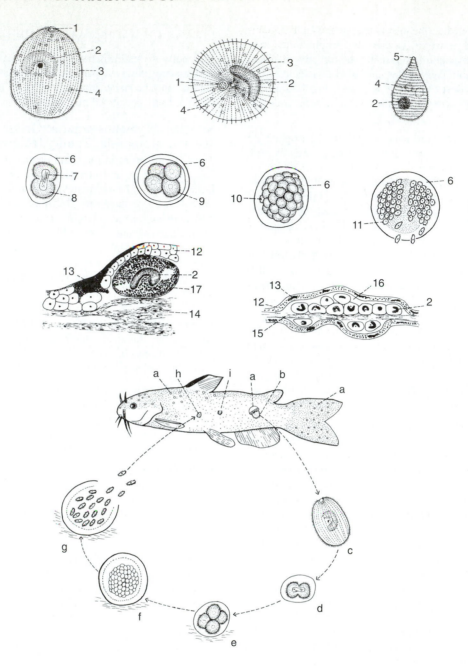

FIGURE 11.5

Life cycle of *Ichthyophthirius multifiliis.* (*a*) Fully developed trophozoite from pustule. (*b*) Anterior end of fully developed trophozoite. (*c*) Tomite from cyst. (*d, e*) First and second divisions of encysted trophozoite. (*f*) Later stage of cystic multiplication. (*g*) Cyst filled with tomites, some of which are escaping into water. (*h*) Section of skin of fish, showing full-grown trophozoite embedded in it. (*i*) Section of tail of carp, showing ciliates developing in pustule. (*j*) Infected bullhead (*Ameiurus melas*). **1,** cytostome; **2,** macronucleus with nearby micronucleus; **3,** longitudinal rows of cilia; **4,** contractile vacuoles; **5,** boring or penetrating apparatus; **6,** cyst; **7,** dividing of macronucleus; **8,** two daughter cells formed by first division; **9,** four daughter cells formed by second division in cyst; **10,** numerous daughter cells; **11,** tomites; **12,** epidermis of fish skin; **13,** pigment cell in epidermis; **14,** dermis; **15,** cartilaginous skeleton of tail of carp; **16,** pustule containing trophozoites; **17,** trophozoite under skin; **a,** pustules; **b,** trophozoite escaping from pustule into water; **c,** trophozoite free in water; **d,** encysted trophozoite on bottom of pond in first division, showing two daughter cells; **e,** cyst in second division with four daughter cells; **f,** cyst with many daughter cells; **g,** ruptured cyst liberating tomites; **h,** tomite attached to skin; **i,** tomite partially embedded in skin.

From O. W. Olsen, *Animal Parasites: Their Life Cycles and Ecology.* Copyright © 1974 Dover Publications, Inc., New York, NY. Reprinted by permission.

FIGURE 11.6

Sunfish infected with *Ichthyophthirius multifiliis.* Note the light-colored pustules in the skin.

From G. Hoffman, in J. Kreier, editor, *Protozoa of Medical and Vetinary Interest,* © 1977 Academic Press, Inc.

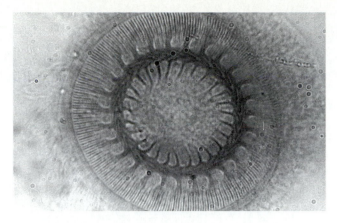

FIGURE 11.7

Trichodina sp. from the gill of a fish. *Trichodina* are 35 to 60 μm in diameter, with a height of 25 to 55 μm.

Courtesy of Warren Buss.

Species of fish, as well as populations of a single species, may differ significantly in their susceptibility to *I. multifiliis.* Susceptibility also can vary according to the time a species has been under domestication (from wild populations), the most recently isolated stocks being least resistant.[2]

Aquarium fish can be treated successfully with very dilute concentrations of formaldehyde, malachite green, or methylene blue. There are also commercial preparations, available in most pet stores, that usually are quite effective. Food with malachite green has been developed and been shown to be effective in the control of ick.[7] *Ichthyophthirius multifiliis* is an exceedingly common and widespread parasite in nature, but in the confines of an aquarium its populations can explode. One of the surest ways to infect an expensive carnivorous ornamental pet fish is to feed it wild caught minnows.

Subclass Peritrichia; Order Peritrichida; Family Trichodinidae

Although the oral ciliary field is prominent in members of this subclass, the somatic ciliature is much reduced. There is a temporary posterior circlet of locomotor cilia, and many are stalked and sessile. All possess an aboral **scopula,** a structure at the aboral pole composed of a field of kinetosomes with immobile cilia. It functions as a holdfast or may be involved in the formation of the stalk.

Species in the family Trichodinidae lack stalks and are mobile. The oral-aboral axis is shortened, with a prominent basal disc usually at the aboral pole. A protoplasmic fringe, or velum, lies on the margin of the basal disc, and a circle of strong cilia lies underneath. A second circle of cilia, above the disc, cannot always be found. The biology of the group is poorly known. The family contains seven genera, with *Trichodina* being a typical example.[8]

***Trichodina* Species.**   Members of this genus parasitize a wide variety of aquatic invertebrates, fish, and amphibians.

The basal disc contains a ring of sclerotized "teeth" that aid the parasite in attaching to its host (Fig. 11.7). The number, arrangement, and shapes of these teeth are useful taxonomic characters. The buccal ciliary spiral makes more than one, but fewer than two, complete turns. Species of *Trichodina* may cause some damage to the gills of fish, but most produce little pathogenic effect and are of interest only as beautiful examples of highly evolved protozoa with incredibly specialized organelles. Typical examples are *T. californica* on the gills of salmon, *T. pediculus* on *Hydra,* and *T. urinicola* in the urinary bladder of amphibians.

CLASS POLYHYMENOPHOREA

The Polyhymenophorea have a well-developed, conspicuous system of membranelles in and around their buccal cavity **(adoral zone of membranelles or AZM).** The body ciliature may be reduced, or the cilia may be joined into compound organelles called **cirri.**

Order Heterotrichida; Family Plagiotomidae

Somatic ciliature is sparse in most species in this order but, when present, is usually uniform. Buccal ciliature is conspicuous, with the AZM typically composed of one to many membranelles or undulating membranes that wind clockwise to the cytosome. Most species are quite large.

The Plagiotomidae are robust parasites of the intestine of vertebrates and invertebrates. The entire body has tiny cilia arranged in longitudinal rows. A single undulating membrane extends from the anterior end to deep within the cytopharynx.

The most common genus is *Nyctotherus.* These ciliates (Fig. 11.8) are ovoid to kidney shaped, with the cytostome on one side. The anterior half contains a massive macronucleus, with a small micronucleus nearby. The genus has numerous species, some of which are useful in routine laboratory exercises. Common species are *N. ovalis* in cockroaches and *N. cordiformis* in the colon of frogs and toads.

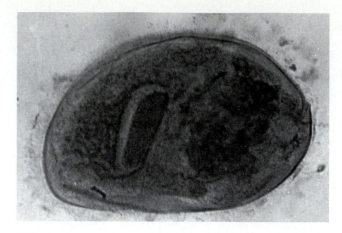

FIGURE 11.8

Nyctotherus cordiformis trophozoite from the colon of a frog. These protozoa range from 60 to 200 μm in length.

Courtesy of Warren Buss.

References

1. Burkart, M. A., T. G. Clark, and H. W. Dickerson. 1990. Immunization of channel catfish, *Ictalurus punctatus* Rafinesque, against *Ichthyophthirius multifiliis* (Fouquet): Killed versus live vaccines. *J. Fish Diseases* 13:401–10.

2. Clayton, G. M., and D. J. Price. 1992. Interspecific and intraspecific variation in resistance to ichthyophthiriasis among poeciliid and goodeid fishes. *J. Fish Biol.* 40:445–53.

3. Cross, M. L., and R. A. Matthews. 1992. Ichthyophthiriasis in carp, *Cyprinus carpio* L.: Fate of parasites in immunized fish. *J. Fish Diseases* 15:497–505.

4. Dorfman, S., O. Rangel, and L. G. Bravo. 1984. Balantidiasis: Report of a fatal case with appendicular and pulmonary involvement. *Trans. Soc. Trop. Med. Hyg.* 78:833–34.

5. McCartny, J. B., G. W. Fortner, and M. F. Hansen. 1985. Scanning electron microscopic studies of the life cycle of *Ichthyophthirius multifiliis. J. Parasitol.* 71:218–26.

6. Nakauchi, K. 1990. A survey on the prevalence rate of *Balantidium coli* in pigs in Japan. *Japanese J. Parasitol.* 39:351–55.

7. Schmahl, G., S. Ruider, H. Mehlorn, H. Schmidt, and G. Ritter. 1992. Treatment of fish parasites: 9. Effects of a medicated food containing malachite green on *Ichthyophthirius multifiliis* Fouquet, 1876 (Hymenostomatida, Ciliophora) in ornamental fish. *Parasitol. Res.* 78:183–92.

8. Van As, J. G., and L. Basson. 1987. Host specificity of trichodinid ectoparasites of freshwater fish. *Parasitol. Today* 3:88–90.

9. Woo, P. T. K. 1987. Immune response of fish to parasitic protozoa. *Parasitol. Today* 3:186–88.

Additional References

Bykhovskaya-Pavlovskaya, I. E. 1962. *Key to the parasites of freshwater fish.* Israel Program for Scientific Translations, Jerusalem (1964), trans. Moscow: Academy of Science. An outstanding reference to ciliate parasites.

Corliss, J. O. 1979. *The ciliated protozoa. Characterization, classification and guide to the literature.* Oxford: Pergamon Press. Advanced treatise but essential to serious students of ciliates.

Hoffman, G. L. 1967. *Parasites of North American freshwater fishes.* Berkeley: University of California Press. Ciliates of North American fish are listed in this useful reference work.

Levine, N. D. 1973. *Protozoan parasites of domestic animals and of man,* 2d ed. Minneapolis: Burgess Publishing Co.

Chapter 12

PHYLUM MESOZOA: PIONEERS OR DEGENERATES?

It has proved to be a docile animal, easily anesthetized and with vascular and renal systems superbly available for preparative surgery . . . catheters have been inserted wherever needed.

A. W. Martin, describing the renal system of a 15 kg *Octopus dofleini*

Mesozoa are tiny, ciliated animals that parasitize marine invertebrates. Their affinities with other phyla are obscure, chiefly because of the simplicity of their structure and their unusual biology. Digestive, circulatory, nervous, and excretory systems are lacking. Basically, a mesozoan's body is made of two layers of cells, but these are not homologous with the endoderm and ectoderm of diploblastic animals.

Two distinct groups traditionally are placed in the phylum Mesozoa: the classes Rhombozoa and Orthonectida. However, these two groups are so different in morphology and life cycles that some authors feel they should be placed in separate phyla.[1] Since no one has thus far formally proposed the separation, we will include both within the phylum Mesozoa, although recognizing the possible artificiality of the scheme.

CLASS RHOMBOZOA

Rhombozoans are parasites of the renal organs of cephalopods, either lying free in the kidney sac or attached to the renal appendages of the vena cava. Partial life cycles are known for a few species, but certain details are lacking in all cases. Interesting histories of the group are presented by Stunkard.[10,11]

Order Dicyemida

The most prominent developmental stages in the cephalopod are the **nematogens** and the **rhombogens** (Fig. 12.1). Their bodies are composed of a **polar cap,** or **calotte,** and a **trunk.** The calotte is made up of two tiers of cells, usually with four or five cells in each. The anterior tier is called the **propolar;** the posterior is called the **metapolar.** The cells in the two tiers may be arranged opposite or alternate to each

other, depending on the genus. The trunk comprises relatively large axial cells surrounded by a single layer of ciliated, somatic cells. The axial cells give rise to new individuals, as in the following description.

The earliest known stage in the cephalopod is a ciliated larva, the **larval stem nematogen.** The axial cells of the larval stem nematogen each contain a **vegetative nucleus** and a **germinative nucleus.** The germinative nucleus becomes an **agamete.** The stem nematogen grows larger while agametes continue to divide, becoming aggregates of cells, in a process much like the asexual, internal reproduction (germ balls) found in miracidia, sporocysts, and rediae of digenetic trematodes (Chapter 15). The animal is now called an **adult stem nematogen.**

Within the axial cell, agametes develop into vermiform embryos (Fig. 12.2) of **primary nematogens** that escape the body of the stem nematogen and attach to the kidney tissues of the host. Agametes within the axial cell of the primary nematogen produce many generations of identical vermiform embryos that develop into primary nematogens, building up a massive infection in the cephalopod. When the host becomes sexually mature, the production of primary nematogens ceases. Instead, the vermiform embryos form stages that become primary rhombogens, similar to nematogens in cell number and distribution but with a different method of reproduction and with lipoprotein- and glycogen-filled somatic cells (Fig. 12.3). These cells may become so engorged that they swell out, and the animal appears lumpy. Some primary nematogens metamorphose directly into rhombogens, and rhombogens derived by this route are referred to as **secondary rhombogens.**

Both primary and secondary rhombogens produce, in the axial cell, agametes that divide to become nonciliated **infusorigens.** An infusorigen is a mass of reproductive cells that represents either a hermaphroditic sexual stage or a hermaphroditic gonad.[9] It remains within the axial cell and produces male and female gametes, which fuse in fertilization. The zygotes detach from the infusorigen, and each then divides to become a hollow, ciliated ovoid stage called an **infusoriform larva,** which is the most complex stage in the life cycle.[9] This microscopic larva consists of a fixed number of cells of several different types. In some species that number is 37,[2] which is small enough for the complete cell lineages to be described (Fig. 12.4).

FIGURE 12.1

Dicyemennea antarcticensis. (*a*) Entire nematogens. (*b*) Vermiform embryos within axial cells of nematogens. (Scales in micrometers.)

From R. B. Short and F. G. Hochberg, Jr., "A new species of *Dicyemennea* (Mesozoa: Dicyemidae) from near the Antarctic peninsula," in *J. Parasitol.* 56:517–522. Copyright © 1970. Reprinted with permission of the publisher.

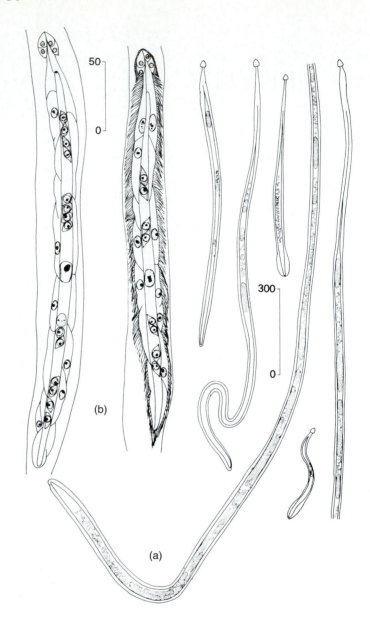

The infusoriform larva escapes from the axial cell and parent rhombogen and leaves the host. It is the only stage known that can survive in seawater. Subsequent to leaving the host, the fate of the larva is unknown because attempts to infect new hosts with it have failed. It is possible that an alternate or intermediate host exists in the life cycle.

Order Heterocyemida

Although they are also parasites of cephalopods, the heterocyemids differ in morphology from the dicyemids. The nematogens of heterocyemids have no cilia or calotte and are covered by a syncytial external layer. Rhombogens are much like nematogens, and they produce infusorigens and infusoriform larvae, as in the dicyemids.

CLASS ORTHONECTIDA

The Orthonectida are quite different from the Rhombozoa in their biology and morphology. The 17 known species parasitize marine invertebrates, including brittle stars, nemerteans, annelids, turbellarians, and molluscs. Complete life cycles are known for some.

Morphology and Biology

The best known orthonectid is *Rhopalura ophiocomae,* a parasite of brittle stars along the coast of Europe (Figs. 12.5 and 12.6). Both sexual and asexual stages exist in the life cycle.

A **plasmodium stage** lives in the tissues and spaces of the gonads and genitorespiratory bursae of the ophiuroid

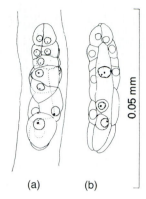

(a)   (b)

FIGURE 12.2

Primary nematogens developing within adult stem nematogens of *Dicyema typoides,* mature. (*a*) Somatic cell outlines and nuclei. (*b*) Optical section.

From R. B. Short, "*Dicyema typoides* sp. n. (Mesozoa: Dicyemidae) from the northern Gulf of Mexico," in *J. Parasitol.* 50:646–651. Copyright © 1964. Reprinted with permission of the publisher.

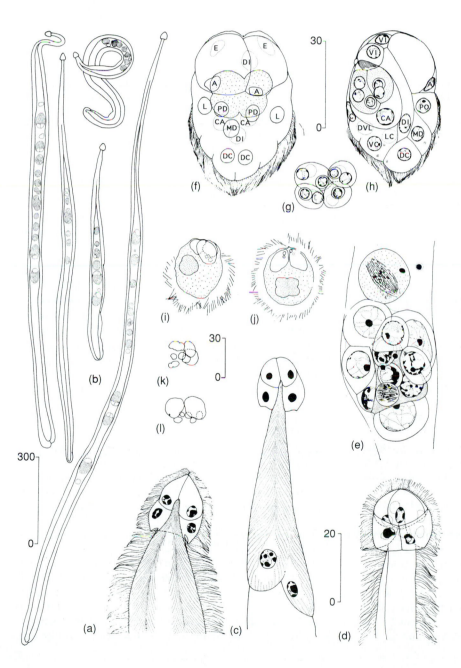

FIGURE 12.3

Dicyemennea antarcticensis life stages (continued from Fig. 12.1). (Scale between *c* and *d* also applies to *a*. Scale between *f* and *h* also applies to *e* and *g*. Scale to right of *k* also applies to *i, j,* and *l.*) (*a*) Nematogen, anterior end. (*b*) Entire rhombogens. (*c, d*) Rhombogens, anterior ends. (*e*) Infusorigen. (*f–j*) Infusoriform larvae. Abbreviations denoting cells (in *f,* only nuclei of cells are shown): **A**, apical; **CA**, capsule; **C**, couvercle; **DC**, dorsal caudal; **DI**, dorsal interior; **E**, enveloping; **L**, lateral; **LC**, lateral caudal; **MD**, median dorsal; **PD**, paired dorsal; **VI**, ventral internal; **V1**, first ventral. **F**, dorsal view, position of urn cells stippled. **G**, urn cells. **H**, side view, optical section. **I, J**, views to show relative sizes of refringent bodies and urn cells. **K, L**, refringent bodies.

From R. B. Short and F. G Hochberg, Jr., "A new species of *Dicyemennea* (Mesozoa: Dicyemidae) from near the Antartic peninsula," in *J. Parasitol.* 56:517–522. Copyright © 1970. Reprinted with permission of the publisher.

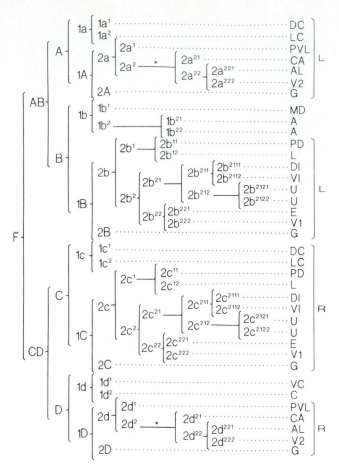

FIGURE 12.4

Cell lineage in the infusiform embryo of *Dicyema japonicum*. **F,** fertilized egg; **L,** left side of embryo; **R,** right side. Cells are **A,** apical; **AL,** anterior lateral; **C,** couvercle; **CA,** capsule; **DC,** dorsal caudal; **DI,** dorsal internal; **E,** enveloping; **G,** germinal; **L,** lateral; **LC,** lateral caudal; **MD,** dorsal median; **PVL,** posterior ventral lateral; **U,** urn; **V1, V2,** first and second ventral cells.

From Hidetaka Furuya, et al., "Development of the infusoritform embryo of *Dicyema japonicum* (Mesozoa: Dicyemidae)," in *Biol. Bull.* 183:248–257. Copyright © 1992. Reprinted with permission of the author.

Amphipholis squamata. It may spread into the aboral side of the central disc, around the digestive system, and into the arms. Developing host ova degenerate, with ultimate castration, but male gonads usually are unaffected.[4] The multinucleate plasmodia are usually male or female but are sometimes hermaphroditic. Some of the nuclei are vegetative, whereas others are agametes that divide to form balls of cells called **morulas.** Each morula differentiates into an adult male or female, with a ciliated somatoderm of **jacket cells** and numerous internal cells that become gametes. Monoecious plasmodia that produce both male and female offspring may represent the fusion of two separate, younger plasmodia. Male ciliated forms are elongated and 90 to 130 μm long. Constrictions around the body divide it into a conical cap, a middle portion, and a terminal portion. A genital pore, through which sperm escape, is located in one of the constrictions. Jacket cells are arranged in rings around the body; the number of rings and their arrangement are of taxonomic importance.

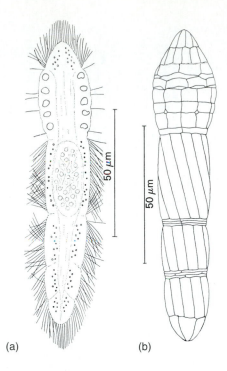

(a)   (b)

FIGURE 12.5

Rhopalura ophiocomae, representing adult stages of an orthonectid mesozoan, male. (*a*) Living individual, as seen in optical section, showing distribution of cilia, lipid inclusions, crystal-like inclusions of the second superficial division of the body, and testis. (*b*) Boundaries of jacket cells, at the surface; silver nitrate impregnation.

From E. N. Kozloff, "Morphology of the orthonectid *Rhopalura ophiocomae,*" in *J. Parasitol.* 55:171–195. Copyright © 1969. Reprinted with permission of the publisher.

There are two types of females in this species. One type is elongated, 235 to 260 μm long and 65 to 80 μm wide, whereas the other is ovoid, 125 to 140 μm long and 65 to 70 μm wide. Otherwise, the two forms are similar to each other and differ from the male in lacking constrictions that divide the body into zones. The female genital pore is located at about midbody. The oocytes are tightly packed in the center of the body.

Males and females emerge from the plasmodia and escape from the ophiuroid into the sea. There, tailed sperm somehow transfer into females, where they fertilize the ova. Within 24 hours of fertilization, the zygote has developed into a multicellular, ciliated larva that is born through the genital pore of its mother and enters the genital opening of a new host.

It is not known whether a plasmodium is derived from an entire ciliated larva or from certain of its cells or whether one larva can propagate more than one plasmodium.

PHYLOGENETIC POSITION

The phylogenetic position of the Mesozoa is most obscure. Early taxonomists placed them between protozoa and sponges because of their cilia, small size, and simple cellularity. Certainly their structure and life cycles are no more complex than those of some protozoa. A good argument has been made for considering rhombozoans to be primitive or

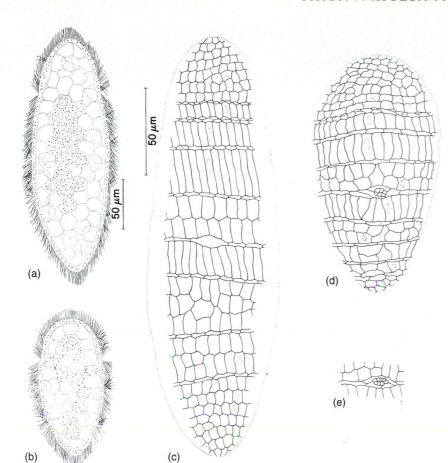

(a)

(b)

(c)

(d)

(e)

FIGURE 12.6

Adult stages of *Rhopalura ophiocomae* (continued from Fig. 12.5), female. (*a*) Living specimen of elongated type, as seen in optical section. (*b*) Living specimen of ovoid type, as seen in optical section. (*c*) Boundaries of jacket cells of elongated type; silver nitrate impregnation. (The cells surrounding the genital pore have been omitted because they were not distinct; approximate proportions of nuclei of representative cells are based on specimens impregnated with Protargol.) (*d*) Cell boundaries of ovoid type; silver nitrate impregnation. (*e*) Genital pore of ovoid type; silver nitrate impregnation.

From E. N. Kozloff, "Morphology of the orthonectid *Rhopalura ophiocornae*," in *J. Parasitol.* 55:171–195. Copyright © 1969. Reprinted with permission of the publisher.

degenerate Platyhelminthes. The ciliated larva is similar to a miracidium (p. 216) in some ways, and the internal reproduction by agametes in nematogens and rhombogens parallels similar processes in germinal sacs of digenetic trematodes.

Biochemical studies have shown that mesozoan DNA has the exceptionally low GC content of 40%,[5] which is closer to that of ciliates than to most metazoans. Although GC content does not prove a particular relationship, it does preclude the existence of many nucleotide sequences homologous with those of species with higher contents.[5] Other biochemical studies have concluded that mesozoans diverged from the main line of metazoan evolution at about the same time as ciliates and euglenoids, which would suggest that they are not degenerate flatworms (which evidently originated much later).[3]

PHYSIOLOGY AND HOST-PARASITE RELATIONSHIPS

What little is known of the physiology of the Mesozoa was reviewed by McConnaughey,[6] and most of that concerns the rhombozoans, based on the early observations of Nouvel.[7] Good ultrastructural studies of both rhombozoans and orthonectids are available.[4,8,9]

Most rhombozoans attach themselves loosely to the lining of the cephalopod kidney by their anterior cilia. They are easily dislodged and can swim about freely in their host's urine. The relationship appears to be entirely commensalistic; no pathogenic consequences of the infection can be discerned. However, a few species have morphological adaptations for gripping the renal cell surface more firmly, and, on the parasites' dislodgement, the renal tissue shows an eroded appearance.

The ruffle membrane surface of the nematogens and rhombogens (Fig. 12.7) evidently is an elaboration to facilitate uptake of nutrients. Ridley[8] showed that the membranes could fuse at various points and form endocytotic vesicles, and "transmembranosis" was suggested by uptake of ferritin. The peripheral cells of infusoriform larvae do not have ruffle membranes but do have microvilli.[9] Clearly the nutritive substances of nematogens and rhombogens must be derived largely or entirely from the host's urine, whereas the infusoriform must live for a period on stored food molecules. Oxygen is very low or absent in the cephalopod's urine, and the nematogens and rhombogens apparently are obligate anaerobes. The organisms live longer in vitro under nitrogen, or even in the presence of cyanide, than when maintained in urine under air or in the absence of cyanide. The infusoriform can live anaerobically only until its glycogen supply is consumed. Adult orthonectids, on the other hand, require aerobic conditions.

A plethora of questions remains, including those enumerated by McConnaughey:[6]

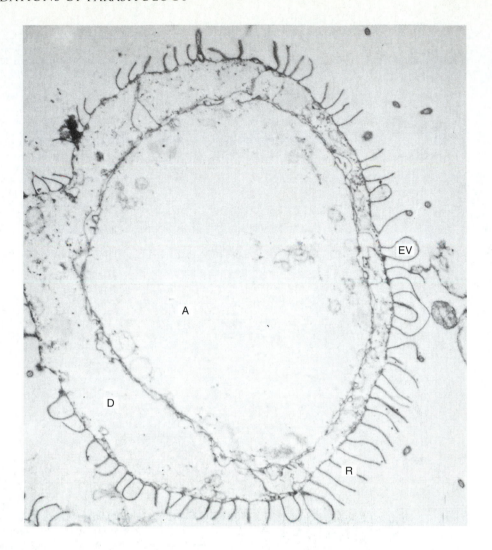

FIGURE 12.7

Nematogen of *Dicyema aegira,* transverse section through somatic cells and axial cells. Note scarcity of organelles, imparting hyaline appearance to cells. Ruffles on somatic cells fused distally at several locations around periphery, forming large endocytotic vesicles. (× 8500.) **A,** axial cell; **D,** somatic cell; **EV,** endocytotic vesicle; **R,** ruffle membrane.

From R. K. Ridley, "Electron microscopic studies on dicyemid Mesozoa. I. Vermiform stages," in *J. Parasitol.* 54:975–998. Copyright © 1968.

1. What are the actual nutritional requirements of the Mesozoa, and to what extent are these met by the host's urine?

2. What are the factors responsible for differentiation of the nematogens from rhombogens, and do these include host hormones?

3. What is responsible for the host specificity, and how do the invasive stages find and invade young specimens of the correct host species?

4. Why is cephalopod urine usually sterile except for the Mesozoa, in spite of the fact that it is a good culture medium?

CLASSIFICATION OF PHYLUM MESOZOA

Class Rhombozoa
Order Dicyemida

Family Dicyemidae

- Genera: *Dicyema, Pseudicyema, Pleodicyema, Dicyemennea*

Order Heterocyemida

Family Conocyemidae

- Genera: *Conocyema, Microcyema*

Class Orthonectida

Order Orthonectida

Family Rhopaluridae

- Genera: *Rhopalura, Stoecharthrum*

Family Pelmatosphaeridae

- Genus: *Pelmatosphaera*

References

1. Dodson, E. O. 1956. A note on the systematic position of the Mesozoa. *Systematic Zool.* 5:37–40.
2. Furuya, H., K. Tsuneki, and Y. Koshida. 1992. Development of the infusiform embryo of *Dicyema japonicum* (Mesozoa: Dicyemidae). *Biol. Bull.* 183:248–57.
3. Hori, H., and S. Osawa. 1987. Origin and evolution of organisms as deduced from 5S ribosomal RNA sequences. *Molecular Biol. and Evolution* 4:445–72.
4. Kozloff, E. N. 1969. Morphology of the orthonectid *Rhopalura ophiocomae. J. Parasitol.* 55:171–95.
5. Lapan, E. A., and H. J. Morowitz. 1974. Characterization of mesozoan DNA. *Exp. Cell Res.* 83:143–51.
6. McConnaughey, B. H. 1968. The Mesozoa. In Florkin, M., and B. T. Scheer, eds. *Chemical zoology, vol. 2. Porifera, Coelenterata, and Platyhelminthes.* New York: Academic Press, Inc., 537–70.
7. Nouvel, H. 1933. Recherches sur la cytologie, la physiologie et la biologie des dicyemides. *Ann. Inst. Oceanogr.* 13:163–255.
8. Ridley, R. K. 1968. Electron microscopic studies on dicyemid Mesozoa. I. Vermiform stages. *J. Parasitol.* 54:975–98.
9. Ridley, R. K. 1969. Electron microscopic studies on dicyemid Mesozoa. II. Infusorigen and infusoriform stages. *J. Parasitol.* 55:779–93.
10. Stunkard, H. W. 1954. The life history and systematic relations of the Mesozoa. *Q. Rev. Biol.* 29:230–44.
11. Stunkard, H. W. 1972. Clarification of taxonomy in the Mesozoa. *Syst. Zool.* 21:210–14.

Additional References

Grassé, P. P., and M. Caullery. 1961. Embranchement des mésozoaires. In Grassé, P., ed. *Traité de zoologie: Anatomie, systématique, biologie, vol. 4. Platheminthes, Mésozoaires, Acanthocéphales, Némertiens.* Paris: Masson & Cie, 693–729.

Stunkard, H. W. 1982. Mesozoa. In Parker, S. P., ed. *Synopsis and classification of living organisms* 1. New York: McGraw-Hill Book Co., 853–55.

Chapter 13

INTRODUCTION TO THE PHYLUM PLATYHELMINTHES

There's no god dare wrong a worm.

Ralph Waldo Emerson

The Platyhelminthes, or flatworms, are so called because most are dorsoventrally flattened. They are usually leaf shaped or oval, but some, such as tapeworms and terrestrial planarians, are extremely elongated. Flatworms range in size from nearly microscopic to over 200 feet in length. These worms lack a coelom, but do possess a well-developed mesoderm, which becomes parenchyma, reproductive organs, and musculature in the adult animal. Traditionally the phylum has contained four classes. The "free-living" flatworms were included in the class Turbellaria, which is no longer recognized, but we will use the term *turbellaria* as a common noun to refer to those generally free-living or ectocommensal platyhelminths that typically have ciliated epidermis in the adult.

Platyhelminthes also are bilaterally symmetrical and thus have a definite anterior end, with associated sensory and motor nerve elements. This nervous system is surprisingly elaborate in many species and helps enable them to invade a wide variety of ecological niches, including lakes and streams, moist terrestrial environments, and ocean sediments from pole to pole. The bodies of other kinds of animals have proven quite hospitable to flatworms, and in fact most platyhelminths are parasitic. But flatworms can even serve as hosts for other flatworms; some cercariae (free-swimming transmission stages of trematodes) can and do penetrate planarians and encyst, becoming infective stages (metacercariae) for the next host in a complex life cycle.[10]

A peculiarity of platyhelminth physiology is their apparent inability to synthesize fatty acids and sterols *de novo,* which may explain why flatworms are most often symbiotic with other organisms, either as commensals or parasites.[16] The free-living acoel turbellarians, sometimes considered illustrative of ancestral flatworms, also seem to lack this ability, indicating that the parasites may not have lost it secondarily as a response to parasitism. Being soft bodied, the Platyhelminthes have left a relatively poor fossil record, but some evidence suggests they've been on Earth for eons. Fossil tracks from a slab of Italian lower Permian grey-green siltstone have been interpreted as those of a land planarian.[1]

The **tegument** varies in structure among the major taxonomic groups. Generally speaking, the turbellaria and some free-living stages of Cestoidea and Trematoda have a ciliated epithelium, which in some cases is their primary mode of locomotion. This epithelium is very thin, being formed of a single layer of cells, and contains many glandular cells and ducts from subepithelial glands. Sensory nerve endings are abundant in the epithelium. In some flatworms cells that produce adhesive secretions are paired with those that produce releasing secretions; the combination is known as a **duo-gland adhesive system.**

The Trematoda and Cestoidea have lost the external cilia except in certain larval stages. During metamorphosis of these parasitic forms, the larval epidermis is replaced by a syncyticial adult tegument, the nuclei of which are in cell bodies (**cytons**) located beneath a superficial muscle layer. Thus, the name *Neodermata* ("new skin") has been used in some classifications to distinguish such worms from the free-living species that retain the ciliated epithelium as adults.

Embedded in the tegument in most free-living turbellarians and in the trematode *Rhabdiopoeus* are numerous rodlike bodies called **rhabdites.** Their function is not always clear, but various authors have attributed lubrication, adhesion, and predator repellancy to them; they are generally absent in symbiotic turbellaria.

Most of the body of a flatworm is made up of **parenchyma,** a loosely arranged mass of fibers and cells of several types. Some of these cells are secretory, others store food or waste products, and still others have huge mitochondria and function in regeneration. The internal organs are so intimately embedded in the parenchyma that dissecting them out is nearly impossible. The bulk of the parenchyma probably is composed of myocytons.

Muscle fibers course through the parenchyma. Contractile portions of the muscle fibers are rarely striated and are usually arranged in one or two longitudinal layers near the body surface. Circular and dorsoventral fibers also occur.

The **nervous system** of acoel turbellarians (Fig. 13.1) includes central and peripheral components, the central nervous system consisting of ganglia around the statocyst, and the peripheral portion consisting of networks supplying the epithelium, muscles, and sensory structures.[2] In the larger and more structurally complex turbellarians and in the trematodes and cestodes, the nerve system is a "ladder type," with paired ganglia near the anterior end, nerves running anteriorly toward sensory or holdfast organs, and longitudinal nerve trunks extending posteriorly to near the end of the body. The number of trunks varies, but most trunks are lateral and are connected by transverse commissures. Sensory

FIGURE 13.1

Amphiscolops, an aceol from Bermuda. (*a*) Whole worm; (*b*) dorsal portion of the nervous system; **1,** frontal organ; **2,** eyes; **3,** statocyst; **4,** mouth; **5,** penis; **6,** brain.

Source: L.H. Hyman, *The Invertebrates,* vol. 2. 1951, McGraw-Hill Book Company, Inc., New York, NY. (Courtesy of the American Museum of Natural History.)

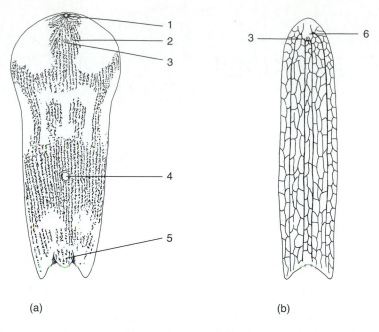

(a)   (b)

elements are abundant, especially in the turbellaria, and may be distributed in a variety of patterns depending on the species. Tactile cells, chemoreceptors, eyespots, and statocysts have been found. The nervous system of turbellarians has attracted the interest of a number of recent researchers, primarily because this system may hold clues to the evolutionary origin of bilateral symmetry in animals. Most of this work is done at the electron microscope level, and as might be expected, has shown that the diversity of neuron types and synaptic junctions is greater than that expected of seemingly primitive animals.[2]

The **digestive system** is typically a blind sac, although acoels and a few trematodes (*Anenterotrema, Austromicrophallus*) have only a mouth but no permanent gut, food being digested by individual cells of the parenchyma. Most flatworms have a mouth near the anterior end, and many turbellarians and most trematodes have a muscular **pharynx,** behind the mouth, with which they suck in food. In the familiar planarians as well as in some other free-living flatworms, the mouth is located midventrally and the pharynx can be extended outwards. The gut varies from a simple sac to a highly branched tube, but only rarely does the flatworm have an anus. Digestion is primarily extracellular, with phagocytosis by intestinal epithelium (gastrodermis), which may contain both secretory and phagocytic cells.[4] Undigested wastes are eliminated through the mouth. A digestive system is completely absent in all life-cycle stages of cestodes.

The functional unit of most flatworms' **excretory system** is the **flame cell,** or **protonephridium** (see Fig. 20.15). This is a single cell with a tuft of flagella that extends into a delicate tubule, which may consist of another cell interdigitating with the first.[11] As is the case with the nervous system, ultrastructural studies aimed partly at uncovering characters of evolutionary significance have shown that the structure of platyhelminth excretory systems is far more complex than originally thought. We now know that such systems contain at least three types of flame cells and as many kinds of tubule cells.[11] Excess water, which may contain soluble nitrogenous wastes, is forced into the tubule, which joins with other tubules, eventually to be eliminated through one or more excretory pores. The filtration occurs through minute slits formed by **rods,** or extensions of the cell, collectively called the **weir** (Old English *wer,* a fence placed in a stream to catch fish). In parasitic flatworms the weir is formed by rods from both the terminal flagellated cell (the **cyrtocyte**) and a tubule cell and is thus referred to as a **two-cell weir.** Because the excreta are mainly excess water, the system is often referred to as an **osmoregulatory system,** with excretion of other wastes considered a secondary function. Some species have an excretory bladder just inside the pore.

The **reproductive systems** follow a common basic pattern in all Platyhelminthes. However, extreme variations of this basic pattern are found among various groups. Most species are monoecious, but a few are dioecious. Because the reproductive organs are so important in identification of parasites and therefore are considered in great detail for each group, we will not discuss them here. Most hermaphrodites can fertilize their own eggs, but cross-fertilization occurs in many. Some turbellarians and cestodes practice **hypodermic impregnation,** which is sperm transfer through piercing the body wall with a male organ, the **cirrus,** and injecting sperm into the parenchyma of the recipient. How the sperm find their way into the female system is not known. Most worms, however, deposit sperm directly into the female tract. The young are usually born within egg membranes, but a few species are viviparous or ovoviviparous. In parasitic species the egg yolk is supplied by cells other than the ovum, and the egg is thus **ectolecithal.** Asexual reproduction is also common in trematodes and a few cestodes.

PLATYHELMINTH SYSTEMATICS

Platyhelminth systematics has been one of the most active areas of parasitological research in recent years, when some of the older classification schemes have been reconsidered in light of evidence from electron microscopy and as a result of the application of cladistic techniques. The traditional class Turbellaria is considered paraphyletic by some authors,[9] and is not even mentioned in the more recent treatments of platyhelminth taxonomy.[6]

On the other hand, a great deal of useful and basic platyhelminth information is found in the older literature organized according to the traditional classifications. Historically, the phylum has included four classes: Turbellaria, Monogenea, Trematoda (Digenea), and Cestoda, corresponding to the "free-living" flatworms, ectoparasitic single host worms, endoparasitic flukes with two or more hosts (one a mollusc), and the tapeworms, respectively. Nevertheless, free-living and parasitic are not valid taxonomic characters, and other features such as the posterior larval attachment organ (cercomer), the nature of the egg yolk, and replacement of larval integument are evidently important characters shared by groups formerly considered only distantly related (such as tapeworms and monogeneans). Other important characters include structure of the pharynx and excretory organs, especially the flame cells.

In Figure 13.4 on page 191 you will find a number of anatomical features and terms listed as synapomorphies. These terms are defined in the appropriate chapters or in the anatomical sections pertaining to the worms involved in this chapter.

The classification that follows is modified from that of Brooks and McLennan[6] and Ehlers.[9] It is based largely on cladistic analysis of the phylum using a variety of characters, many of which are ultrastructural. We have attempted to preserve enough of the traditional nomenclature to enable students to gain access to the older literature but at the same time understand the general directions that platyhelminth systematics is now taking. Whatever else it may accomplish, modern evolutionary biology has clearly demonstrated that we still have a great deal to learn about animals we have been studying for a long time.

Not all scientists agree upon the taxon names or hierarchical levels. Among the most detailed classifications is that of Brooks and McLennan,[6] which utilizes more taxonomic levels than typically encountered in literature used by undergraduates. Thus, you will find superclasses, subsuperclasses, infraclasses, cohorts, and subcohorts, in addition to the familiar classes and orders, when you go searching in the modern platyhelminth taxonomic literature. You should interpret this complexity as part of the human struggle to resolve knotty problems. The parasites involved do not give up their secrets easily, nor do they always fall into convenient categories constructed by humans.

According to Brooks and McLennan,[6] the subphylum Catenulida is a sister group of the "true" Platyhelminthes. The Catenulida includes a number of delightful little worms whose ease of culture and asexual reproductive habits have made them favorite experimental animals for regeneration studies (Fig. 13.2). The major structural feature dividing the catenulid platyhelminthes from the rest is the presence of a frontal organ in the latter. A frontal organ is a terminal or subterminal pit with mucoid gland cells and sometimes cilia. The catenulids lack this organ, although some species have lateral pits. The remaining platyhelminths (Euplatyhelminthes) also possess dense epidermal ciliature (three to six cilia per μm^2)

compared to the catenulids, which have about a tenth that many per unit area. The acoel flatworms, typically small species without an intestine (digesting food intracellularly and in temporary cavities), lack protonephridia. The remaining groups, including all the parasitic ones (Figs. 13.3 and 13.4), have protonephridial flame cells with more than two, and sometimes more than a hundred, flagella.[9]

CLASSIFICATION OF THE PHYLUM PLATYHELMINTHES (WITH EMPHASIS ON THE COMMENSAL AND PARASITIC REPRESENTATIVES)

PHYLUM PLATYHELMINTHES

Subphylum Catenulida
Lack a frontal organ and have monociliated epidermal cells.

Subphylum Euplatyhelminthes
With a frontal organ, high density of epidermal cilia, and multiflagellated flame cells (when present).

SUPERCLASS ACOELOMORPHA
Reduction and loss of protonephridia and a modified (or missing) gut.

SUPERCLASS RHABDITOPHORA
With lamellated rhabdites, a duo-gland adhesive system, and multiflagellated flame cells.

SUPERCLASS CERCOMERIA
With a doliiform pharynx, saccate gut, and posterior adhesive organ. Lack locomotor cilia in the adult.

Class Temnocephalidea
With cephalic tentacles. (This group may not be correctly placed.)

NOTE: The following taxa share some important characteristics, primarily loss of larval ciliated epidermis and a syncytial adult epidermis. In addition, their eggs are ectolecithal.

Class Udonellidea
With a secondary protonephridial system made of pores and canals. The older literature places this group with the Monogenea. Ectoparasitic on ectoparasitic arthropods, mainly crustaceans.[7]

Class Cercomeridea
With an oral sucker. The uterus has lateral coiling. The adult intestine is bifurcate.

Subclass Trematoda
Posterior adhesive organ a sucker; male genital pore opening into an atrium; adults with pharynx near the oral sucker.

Infraclass Aspidobothrea
With specialized microvilli and microtubules in neodermis. Posterior sucker is divided into compartments.
***Major order
Aspidobothriiformes.***

FIGURE 13.2

Some representative Catenulida.
(*a*) *Stenostomum tenuicauda* showing unpaired protonephridia and sites of asexual division (zooid ciliated pits). (*b*) *Catenula lemnae,* also in the process of sexual reproduction.
(*c*) *Rhynchoscolex* sp. **1,** ciliated pits (not frontal organs); **2,** mouth; **3,** pharynx; **4,** protonephridium; **5,** intestine; **6,** ciliated pits of zooids; **7,** nephridiopore; **8,** fission lines of zooid formation.

Source: L.H. Hyman, *The Invertebrates,* vol. 2. 1951, McGraw-Hill Book Company, Inc., New York, NY. (Courtesy of the American Museum of Natural History.)

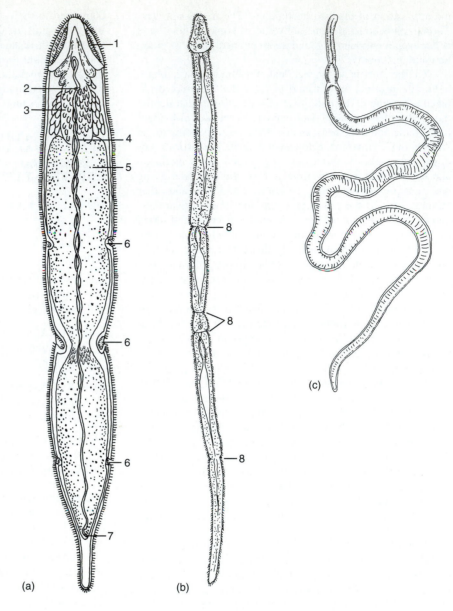

(a)   (b)

(c)

FIGURE 13.3

Major divisions of the phylum Platyhelminthes. Frontal organs are considered an apomorphy possessed by all groups except the Catenulida. All parasitic flatworms have ectolecithal eggs.

Modified from U. Ehlers, "Comments on a phylogenetic system of the Platyhelminthes," in *Hydrobiologia* 132:1–12.

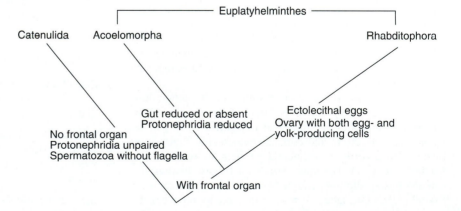

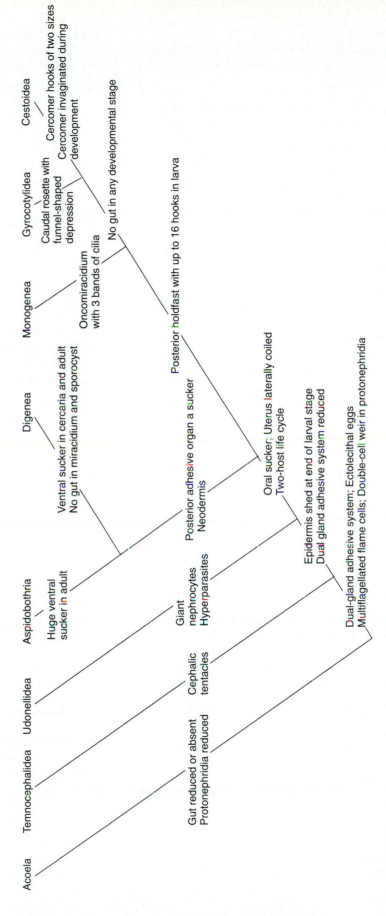

FIGURE 13.4

Cladogram depicting relationships of the parasitic Platyhelminthes. Larval characters establish the relationship between monogeneans and tapeworms and separate the former from the Digenea and Aspidobothria with which they were associated in older classifications.

Based on several sources.

Infraclass Digenea
First larval stage a miracidium; life cycle with one or more sporocyst generations and cercarial stage; gut development is paedomorphic.

Important orders
Paramphistomiformes, Echinostomiformes, Hemiuriformes, Strigeiformes, Opisthorchiformes, Plagiorchiformes.

Subclass Cercomeromorphae
Posterior adhesive organ (cercomer) with hooks larval cercomer with 16 hooks.

Infraclass Monogenea
Oncomiracidium (larva) with three ciliary bands; adults with a single testis; all ectoparasitic.

Important orders
Monocotyliformes, Gyrodactyliformes, Dactylogyriformes, Polystomatiformes, Diclybothriiformes.

Infraclass Cestodaria
Intestine lacking; cercomer paedomorphic and somewhat reduced in size; oral sucker and pharynx vestigal; larval cercomer with 10 hooks.

Cohort Gyrocotylidea
Rosette with funnel at the posterior end; body margins crenulate.

Cohort Cestoidea
Male and female genital pores close to each other and cercomer totally invaginated during development; cercomer hooks of two sizes.

Subcohort Amphilinidea
Genital pores at posterior end; uterus N-shaped.

Subcohort Eucestoda
Adults polyzoic; six-hooked larval cercomer lost during ontogeny; life cycles with more than one host.

Orders
Pseudophyllidea, Caryophyllidea, Spathebothriidea, Cyclophyllidea, Proteocephalata, Tetraphyllidea, Trypanorhyncha.

CLASSIFICATION OF THE PLATYHELMINTHES AS FOUND IN EARLIER LITERATURE

Class Turbellaria
Mostly free-living worms in terrestrial, freshwater, and marine environments; some commensals or parasites of invertebrates, especially of echinoderms and molluscs.

Class Monogenea
All parasitic, mainly on the skin or gills of fish; although mostly ectoparasites, a few living within the stomodaeum, proctodaeum, or their diverticula.

Class Trematoda
All parasitic, mainly in the digestive tract, of all classes of vertebrates.

Subclass Digenea
At least two hosts in life cycle, first almost always a mollusc; perhaps most diversification in bony marine fish, although many species in all other groups of vertebrates.

Subclass Aspidogastrea
Most with only one host, a mollusc; a few mature in turtles or fishes with mollusc or lobster intermediate host.

Subclass Didymozoidea
Tissue-dwelling parasites of fish; no complete life cycle known, but intermediate host may not be required.

Class Cestoidea
All parasitic, common in all classes of vertebrates except agnathan classes; intermediate host required for almost all species.

THE TURBELLARIANS

Most turbellarians are free-living predators, but many of the orders contain species that maintain varying degrees and types of symbiosis. Of these, most are symbionts of echinoderms, but others are found on or in sipunculids, arthropods, annelids, molluscs, coelenterates, other turbellarians, and fish. At least 27 families have symbiotic species. A considerable degree of host specificity is manifested by these worms. Most symbionts are commensals; a few are true parasites and several degrees of these relationships are known. Although it is tempting to array these in a series of ectocommensals, endocommensals, ectoparasites, and so on to postulate how parasitism evolved in this phylum, it is clear that most individual cases are the end results of their particular situations and have not given rise to succeedingly complex associations.

The Acoels
The acoels are entirely marine and are from one to several millimeters long. They exhibit several primitive characteristics, including the absence of an excretory system, pharynx, and permanent gut, and many have no rhabdites. Most are free-living, feeding on algae, protozoa, bacteria, and various other microscopic organisms. A temporary gut with a syncytial lining appears whenever food is ingested, and digestion occurs in vacuoles within it. After digestion is completed the gut disappears.

Few species have adopted a symbiotic existence, and it is difficult to decide which, if any, are true parasites. *Ectocotyla paguri* is the only ectocommensal known. It lives on hermit crabs, but nothing is known of its biology or feeding habits. Several species of acoels live in the intestines of Echinoidea and Holothuroidea. It is not known if any are parasites, but because no apparent harm comes to the hosts, these acoels are usually considered endocommensals.

The Rhabditophorans
Many of the orders of turbellarians contain mainly free-living species and would be placed on the cladogram in Figure 13.4 on branches between the Acoela and the neoophoran flatworms. Apomorphies include unflagellated sperm, the presence of duo-gland adhesive organs, the branching structure of the gut, and the fate of the blastomeres during development. Space limitations prevent us from presenting a detailed taxonomy of these groups, but Meglitsch and Schram[15] and Ehlers[9] provide informative reviews. Meglitsch and

Schram[15] give class status to the Rhabditophora and subclass status to the Macrostomida and Neoophora. They include within their Neoophora the orders Seriata, Typhloplanoida, and Dalyelloida. Most symbiotic turbellarians belong to one of these orders. Again most seem to be commensals, but a few are definitely parasitic. Dalyelloids are small, like acoels, but they have a permanent, straight gut and a complex, bulbous pharynx. Most are predators of small invertebrates. Of the four suborders in the order, three have symbiotic species.

Fecampia erythrocephala (order Dalyelloida) lives in the hemocoel of decapod crustaceans. During their development in the host, the young worms lose their eyes, mouth, and pharynx, and they absorb their nutrients from the host's blood. When sexually mature they mate and leave the host. After cementing itself to a substrate, the flatworm shrinks until all internal tissues vanish, leaving only a bottle-shaped cocoon made of the degenerated epidermis. Each cocoon contains two eggs and several vitelline cells that produce two ciliated, motile juveniles. These swim about until contacting a crustacean.[3] Their mode of entry into a host is not known.

The host is not killed by the parasite but does suffer adverse effects of the hepatopancreas and ovaries. Because of its fertility, this parasite presents a high risk for culture of prawns in Atlantic and Mediterranean marine areas. *Fecampia* may illustrate a hypothetical stage in the origin of the Digenea.

Kronborgia amphipodicola is very unusual among the turbellarians because it is dioecious.[8] Furthermore, there is pronounced sexual dimorphism: The males are 4 to 5 mm long, whereas the females are 20 to 30 mm long and can stretch to 45 mm. Both sexes lack eyes and digestive systems at all stages of their life cycles. They mature in the hemocoel of the tube-dwelling amphipod *Amphiscela macrocephala,* with the male near the anterior end and the female filling the rest of the available space. On reaching sexual maturity, the worms burrow out of the posterior end of the host, which becomes paralyzed and quickly dies; as if to add insult to injury, before the host is killed it is castrated. After emergence from the amphipod, the female worm quickly secretes a cocoon around herself and attaches the elongated cocoon to the wall of the burrow, from which it protrudes 2 to 3 cm into the open water. The male enters the cocoon, crawls down to the female, and inseminates her. He then leaves the cocoon and dies. The female produces thousands of capsules, each with two eggs and some vitelline cells, and then also dies. A ciliated larva hatches from each egg and eventually encysts on the cuticle of another amphipod. While in the cyst, the larva bores a hole through the host's body wall and enters the hemocoel to begin its parasitic existence.

The tegumental ultrastructure of *K. amphipodicola* has been studied.[14] The lateral membranes of the epidermal cells break down, and the epidermis thus becomes syncytial. Although short microvilli are not unusual on the outer surface of epithelial cells of free-living turbellaria, the microvilli of *K. amphipodicola* are quite long and constitute an adaptation for increasing surface area to absorb nutrients. Subepidermal gland cells with long processes extending to the surface

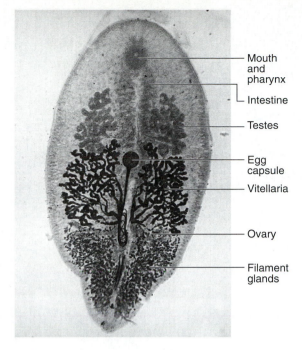

FIGURE 13.5

Syndesmis sp., a turbellarian from the intestine of a sea urchin. It is about 2.5 to 3.0 mm long.

Courtesy of Warren Buss.

probably function in the escape of the worm from its host and in construction of the cocoon.

Members of the dalyelloid family Umagillidae live in the digestive tract or coelom of Holothuroidea and Echinoidea. Crinoidea and Sipunculida also are infected. Traditionally considered harmless commensals, we now know that some species consume host intestinal cells as well as commensal ciliated protozoa.[12] For example, *Syndesmis franciscanus* and *Syndesmis dendrastrorum* ingest host intestinal tissue along with intestinal contents, whereas *Syndesmis echinorum* subsists entirely on host intestinal tissue.[18]

Syndesmis spp. (Fig. 13.5) and *Syndisyrinx* spp. are found in the intestines of sea urchins and therefore are available to nearly any college laboratory with preserved or living sea urchins in its stock. Very little is known of their biology, but they appear to be excellent subjects for study. About 50 species have been described in this family.

Syndesmis franciscanus inhabits sea urchins of the genus *Strongylocentrotus* along the northwest coast of North America. It produces an egg capsule about every one and one-half days; these capsules are released one at a time into the intestine of the host and pass to the outside with feces. Each capsule contains two to eight oocytes and several hundred vitelline cells. Embryogenesis requires about two months. The worms hatch when eaten by a suitable host and mature with no further migration.[19]

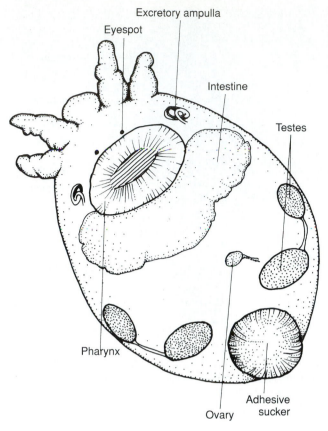

FIGURE 13.6

Temnocephala tasmanica, an ectoparasite of Australian crayfish.
Drawing by Ian Grant.

Molluscs also play host to turbellarians. Bivalves, especially, have invaded the New World as a result of commerce, and as other animals are inclined to do when they travel, have brought along their symbionts. Thus, dalyelloid flatworms (along with a whole community of protozoans), interpreted to be natural occupants of the Manila clam (*Tapes philippinarum*), have been found in the intestinal lumen of these bivalves accidently introduced into Canada.[5]

Temnocephalida

The Temnocephalida are given class status by Brooks and McLennan.[6] Most are ectocommensals on crustaceans in South and Central America, Australia, New Zealand, Madagascar, Sri Lanka, and India; a few are known from Europe. A few species occur on turtles, molluscs, and freshwater hydromedusae. Probably they are much more widespread but have gone undiscovered or unrecognized as the result of the paucity of trained specialists.

Temnocephalids are small and flattened, with tentacles at the anterior end and a weak, adhesive sucker at the posterior end (Fig. 13.6). They have leechlike movements, alternately attaching with the tentacles and posterior sucker. The tegument is syncytial with varying numbers of cilia, or at least ciliated receptors, depending on the species.[21] Scanning electron microscope studies have shown the tegument to be structurally complex, with folds, microvilli, and occasionally scales. (Fig. 13.7)[13,21] Rhabdites are located mainly at the anterior end, and mucous glands are most numerous around the posterior sucker.

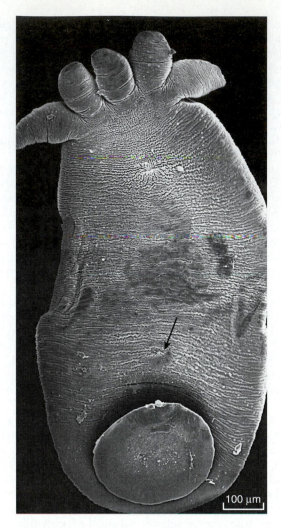

FIGURE 13.7

Scanning electron micrograph of *Temnocephala dendyi,* showing structure of the epidermis. *Arrow* indicates location of the gonopore.
From Williams, *Australian Journal of Zoology,* 26:217–224, 1978.

The biology of temnocephalids is simple, as far as it is known. Eggs are laid in capsules and attached to the exoskeleton of the host. Each hatches as an immature adult and matures with no further ado. What happens to those that are lost at ecdysis of the host is unknown; the fate of the adults at that time is also unknown. It is possible that a free-living stage is present in the life cycle of these worms but has yet to be found.

The pattern of nutrition apparently does not differ from that of free-living flatworms, with protozoa, bacteria, rotifers, nematodes, and other microscopic creatures serving as food. Cannibalism has been established. The host serves only as a substrate for attachment.

The Alloeocoels

Alloeocoels are turbellarians with an irregular gut. Most are marine, but a few inhabit brackish or fresh water, and a few are terrestrial. Several are commensal on snails, clams, and crustaceans, but *Ichthyophaga subcutanea* is clearly a parasite of marine teleost fish. It lives in cysts under the skin in the branchial and anal regions of its host and apparently ingests

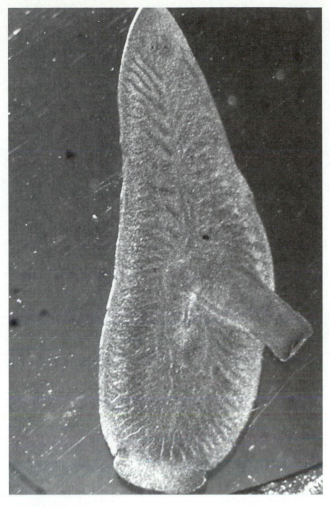

FIGURE 13.8

Bdelloura candida, a triclad turbellarian from the gills of a horseshoe crab. Note the eyespots and the huge midventral pharynx. Overall length may reach 20 mm.

Courtesy of Warren Buss.

blood. Morphologically it has nonparasite features, such as eyes and a ciliated epithelium.[20]

Monocelis sp. lives within the valves of intertidal barnacles and snails during low tide but returns to the open water when the tide is in. This may illustrate a case of incipient endosymbiosis.

The Tricladids

Tricladids are large worms, up to 50 cm in length, that occupy marine, freshwater, and terrestrial habitats. They are easily recognized by their tripartite intestine. Nearly all are free-living predators, feeding on small invertebrates and sucking the contents out of larger ones by means of their eversible pharynges.

Three genera, *Bdelloura, Syncoelidium,* and *Ectoplana,* live on the book gills of horseshoe crabs, *Limulus polyphemus.* Of these, *Bdelloura candida* (Fig. 13.8) is the most common. It has a large adhesive disc at its posterior end and well-developed eyespots. Apparently it feeds on particles of food torn apart by the gnathobases of its host and washed back to the gill area. No evidence of harm to its host has been detected. It lays its eggs in capsules on the book gill

lamellae. The tricladids may migrate from one horseshoe crab to another during copulation of their hosts, a sort of marine, verminous venereal disease! The biology and physiology of these worms would surely prove to be a rewarding area of research.

The Polycladids

The polycladids have a complex gut with many radiating branches. Except for one freshwater species, they are all marine. No parasites are known in this group, and the few reported "commensals" are suspect of even that degree of symbiosis. Although some species are found together with hermit crabs, they are also found in empty shells. Others, such as the "oyster leech," *Stylochus frontalis,* live between the valves of oysters and are predators on the original owner, devouring large pieces of it at a time.[17]

The truly parasitic turbellarians show structural changes expected with their specialized way of life: losses of ciliated epidermis, eyes, mucous glands, and rhabdites. The various commensals, however, show few or no specializations over their free-living brethren. The prevalence of rhabdocoels in echinoderms may simply be a result of the diverse fauna of ciliated, protozoan commensals in the latter, which offer rich pickings for the former. The origin of trematodes and cestodes from acoel ancestors, which became adapted to endocommensalism within molluscs and crustaceans, is not difficult to visualize.

References

1. Alessandrello, A., G. Pinna, and G. Teruzzi. 1988. Land planarian locomotion trail from the lower Permian of Lombardian Pre-Alps. *Atti Della Societa Italiana di Scienze Naturali e del Museo Civico d Storia Naturale di Milano* 129:139–45.

2. Bedini, C., and A. Lanfranchi. 1991. The central and peripheral nervous system of Acoela (Platyhelminthes). An electron microscope study. *Acta Zoologica* 72:101–6.

3. Bellon-Humbert, C. 1983. *Fecampia erythrocephala* Giard (Turbellaria Neorhabdocoela), a parasite of the prawn *Palaemon serratus* Pannart: The adult phase. *Aquaculture* 31:117–40.

4. Bogitsh, B. J. 1993. A comparative review of the flatworm gut with emphasis on the Rhabdocoela and Neodermata. *Trans. Am. Microsc. Soc.* 112:1–9.

5. Bower, S. M., J. Blackbourn, and G. R. Meyer. 1992. Parasite and symbiont fauna of Japanese littlenecks, *Tapes philippinarum* (Adams and Reeve, 1850), in British Columbia. *J. Shellfish Res.* 11:13–19.

6. Brooks, D. R., and D. A. McLennan. 1993. *Parascript: Parasites and the language of evolution.* Washington, D.C.: Smithsonian Institution Press, 429p.

7. Ching, H. L., and B. J. Leighton. 1993. The presence of *Udonella ophiodontis* in Washington and of *U. caligorum* in British Columbia. *J. Helm. Soc. Wash.* 60:137–40.

8. Christiansen, A. P., and B. Kanneworff. 1965. Life history and biology of *Kronborgia amphipodicola* Christiansen and Kanneworff (Turbellaria, Neorhabdocoela). *Ophelia* 2:237–51.

9. Ehlers, U. 1986. Comments on the phylogenetic system of the Platyhelminthes. *Hydrobiologia* 132:1–12.

10. Fried, B., and L. C. Rosa-Brunet. 1991. Exposure of *Dugesia tigrina* (Turbellaria) to cercariae of *Echinostoma trivolvis* and *Echinostoma caproni* (Trematoda). *J. Parasitol.* 77:113–16.

11. Hertel, L. 1993. Excretion and osmoregulation in the flatworms. *Trans. Am. Microsc. Soc.* 112:10–17.

12. Jennings, J. B., and D. F. Mettrick. 1968. Observations on the ecology, morphology and nutrition of the rhabdocoel turbellarian *Syndesmis franciscana* (Lehman, 1946) in Jamaica. *Caribb. J. Sci.* 8:57–69.

13. Jennings, J. B., L. R. G. Cannona, and A. J. Hick. 1992. The nature and origin of the epidermal scales of *Notodactylus handschini*—an unusual temnocephalid turbellarian ectosymbiotic on crayfish from northern Queensland. *Biol. Bull.* 182:117–28.

14. Lee, D. L. 1972. The structure of the helminth cuticle. In Dawes, B., ed. *Advances in parasitology* 10. New York: Academic Press, Inc., 347–79.

15. Meglitsch, P. A., and F. R. Schram. 1991. Invertebrate zoology. New York: Oxford University Press.

16. Meyer, F., and H. Meyer. 1972. Loss of fatty acid biosynthesis in flatworms. In Van den Bossche, H. ed. *Comparative biochemistry of parasites.* New York: Academic Press, Inc., 383–93.

17. Pearse, A. S., and G. W. Wharton. 1938. The oyster "leech" *Stylochus inimicus* Palomi, associated with oysters on the coasts of Florida. *Ecol. Monogr.* 8:605–55.

18. Shinn, G. L. 1981. The diet of three species of umagillid neorhabdocoel turbellarians inhabiting the intestine of echinoids. *Hydrobiologia* 84:155–62.

19. Shinn, G. L. 1983. The life history of *Syndisyrinx franciscanus,* a symbiotic turbellarian from the intestine of echinoids, with observations on the mechanism of hatching. *Ophelia* 22:57–79.

20. Syriamiatnikova, I. P. 1949. A new turbellarian of fish, *Ichthyophagia subcutanea* n. g. n. sp. *C. R. Akad. Nauk.* 68:805–8.

21. Williams, J. B. 1978. Studies on the epidermis of *Temnocephala* III. Scanning electron microscope study of the epidermal surface of *Temnocephala dendyi. Aust. J. Zool.* 26:217–24.

Additional References

The entire volume 132 of the journal *Hydrobiologia* is devoted to the newer information on phylogeny, development, reproduction, regeneration, and ecology; this publication is entitled "Advances in the biology of turbellarians and related platyhelminthes" and will be a primary reference for several years.

Baer, J. F. 1961. Classe des Temnocéphales. In Grassé, P., ed. *Traité de zoologie: Anatomie, systématique, biologie, vol. 4, part I. Plathelminthes, Mésozoaires, Acanthocéphales, Némertiens.* Paris: Masson & Cie, 213–41.

Brooks, D. R. 1989. The phylogeny of the Cercomeria (Platyhelminthes: Rhabdocoela) and general evolutionary principles. *J. Parasitol.* 75:606–16.

Jennings, J. B. 1971. Parasitism and commensalism in the Turbellaria. In Dawes, B., ed. *Advances in parasitology* 9. New York: Academic Press, Inc., 1–32. A most readable account of the subject. Recommended for all parasitologists.

Chapter 14

TREMATODA: ASPIDOBOTHREA

There's a sucker born every minute.

Phineas Taylor Barnum (attributed)

The Aspidobothrea constitute a small group of Digenea-like worms, that in most species, have established a loosely parasitic relationship with molluscs, but some are facultative or obligate parasites of fishes or turtles.[2] Two other names have often been used for this group: Aspidocotylea and Aspidogastrea. Although most of the literature has accumulated under the name Aspidogastrea, Aspidobothrea undoubtedly has priority.

By any name, this group of organisms has attracted perhaps less than its fair share of attention because it contains no species of medical or known economic importance. Nevertheless these innocuous little worms are of considerable biological interest: They seem to represent a step between free-living and parasitic organisms. And like many seemingly unimportant parasites, their structures and lives have been remarkable enough to capture the interest of many well-known parasitologists at some time in their careers. A comprehensive review of Aspidobothrea has been presented by Rohde;[5] the literature cited section of that review reads like a "Who's Who" of 20th century helminthology, although few of these famous people spent very much time on the Aspidobothrea.

FORM AND FUNCTION

Body Form

Externally aspidobothreans exhibit three basic types of anatomy, corresponding to the three families that have been established for them. The Aspidogastridae (Fig. 14.1) have a huge **ventral sucker,** extending most of the length of the body. This sucker (also known as an **opisthaptor** or **Baer's disc**) has muscular **septa** in longitudinal and transverse rows, dividing it into shallow depressions called **alveoli** or **loculi** (Fig. 14.2).[3] The number, shape, and arrangement of these loculi are of considerable taxonomic importance. Hooks or other sclerotized structures are never present. Between the marginal loculi usually are **marginal bodies,** which are secretory organs, or short tentacles, also presumably secretory in nature. Exceptionally both are absent.

In the Stichocotylidae (see Fig. 14.11 on p. 203) occurs a longitudinal series of individual suckers instead of a single complex of loculi, whereas in Rugogastridae the ventral holdfast is made up of transverse ridges called **rugae** (see Schell[10]).

The marginal bodies are round to oval organs and are connected to each other by fine ducts. They consist of gland cells, storage chambers (ampullae), and secretory ducts (Fig. 14.3). In some species the ampulla empties through a muscular papilla, and the terminal duct can be protruded or retracted.[8] Although a sensory function has been suggested for the marginal bodies, no indication exists that their function is other than secretory. The tentacles of *Lophotaspis* (Fig. 14.4) are probably modified marginal organs.

The **longitudinal septum** is a peculiar morphological characteristic of the Aspidogastrea. It is a horizontal layer of connective tissue and muscle in the anterior part of the body, projecting like a shelf and dividing the body into dorsal and ventral compartments. The function of the septum is not known, but it might be correlated with pressures exerted by contraction of the giant ventral sucker.

Tegument

The tegument of aspidobothreans seems to be basically similar to that of other groups of parasitic flatworms, although that conclusion is based largely on the study of one species (*Multicotyle purvisi*). The tegument is syncytial and has an outer stratum of *distal cytoplasm,* containing mitochondria and numerous vesicles of various types. The tegumental nuclei are in **cytons** internal to the superficial muscle layer and connected to the distal cytoplasm by internuncial processes. The cytons are rich in Golgi complexes. A mucoid layer of variable thickness is found on the outer surface membrane, and in some areas the surface membrane has riblike elevations to support the thick mucoid layer.

Digestive System

The digestive tract is simple. In some species the **mouth** is funnel-like, whereas in others it is surrounded by a muscular sucker or several muscular lobes. At the base of the mouth funnel is a spheroid **pharynx,** a powerful muscular pump. The **intestine,** or **cecum,** is a single, simple sac that usually extends to near the posterior end of the body. Its epithelial cells bear a complex reticulum of lamellae on their luminal

FIGURE 14.1

Examples of the family Aspidogastridae. (*a*) *Lobatostomum ringens*. (*b*) *Cotylogaster michaelis*. (*c*) *Lophotaspis vallei*. (*d*) *Cotylaspis insignis*.

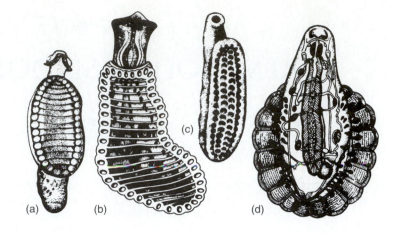

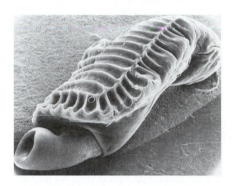

FIGURE 14.2

Scanning electron micrograph: ventral view of *Cotylogaster occidentalis*. The cup-shaped buccal funnel is at left, and the neck is semiretracted into the body by a fold. Note the smooth tegument and prominent alveoli in the ventral haptor.

From H. S. Ip, S. S. Desser, and I. Weller, "*Cotylogaster occidentalis* (Trematoda: Aspidogastrea): Scanning electron microscopic observations of sense organs and associated surface structures," in *Trans. Am. Microsc. Soc.* 101:253–261. Copyright © 1982.

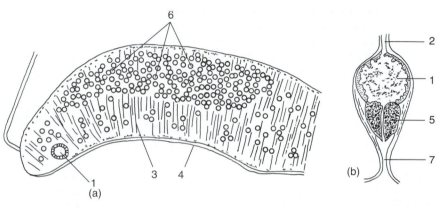

FIGURE 14.3

(*a*) Diagram of a section through a marginal alveolus of *Lobatostoma manteri*. (*b*) Diagram through one of the marginal bodies. **1,** ampulla of marginal body; **2,** duct of marginal gland; **3,** dorsoventral muscles; **4,** longitudinal muscles; **5,** muscular papilla; **6,** nuclei of the marginal gland; **7,** terminal duct.

From K. Rohde and N. Watson, "Ultrastructure of the marginal glands of *Lobatostoma manteri* (Trematoda, Aspidogastrea)," in *Zool. Anz.* 223:301–310. Copyright © 1989. Reprinted with permission of the publisher and author.

surface, presumably vastly increasing the absorptive surface. A layer of muscles, usually of both circular and longitudinal fibers, surrounds the cecum.

Osmoregulatory System

This system consists of numerous **flame cell protonephridia** connected to capillaries feeding into larger excretory ducts and eventually into an **excretory bladder** near the posterior end of the body. The flame cells are peculiar in that their flagellar membranes continue beyond the tips of the flagella and anchor apically in the cytoplasm of the flame cell. Lateral or nonterminal flagellar flames have been reported in a number of species. The small capillaries have numerous microvilli projecting into their lumina, and the larger capillaries and excretory ducts are abundantly provided with lamellar projections of their

surface membranes, thus suggesting secretory-absorptive function. The **excretory pore** is dorsosubterminal or terminal and usually single.

Nervous System

The nervous system of aspidobothreans is very complex for a parasitic flatworm, reminiscent of a condition more typical of free-living forms. As in many turbellaria, there is a complex set of anterior nerves called the **cerebral commissure** and a modified ladder type of peripheral system. A wide variety of sensory receptors has been observed, mostly around the mouth and on the margins of the ventral disc. In a specimen of *Multicotyle purvisi* 6.1 mm long, Rohde[5] counted 360 dorsal and 260 ventral receptors in the prepharyngeal region and 140 in the oral cavity, not counting free nerve endings below the tegument. Three

FIGURE 14.4

Lophotaspis interiora from an alligator snapping turtle.

From H.B. Ward and S.H. Hopkins, "A new North American aspidogastrid, *Lophotaspis interiora*," in *J. Parasitol.* 18:69–78. Copyright © 1931.

types of "ciliated sense organs" (sensilla) have been described on the body of *Cotylogaster occidentalis,*[4] and Rohde and Watson[9] distinguished nine types of receptors, most of them with cilia, in *Lobatostoma manteri,* a parasite of snails and fish of Australia's Great Barrier Reef. These receptors differed in the structure of their cilia and in the presence of a rootlet fiber (Fig. 14.5).

A complex system of connectives and commissures occurs in the ventral disc and walls of the alveoli, indicating a high degree of neuromuscular coordination. The septum, intestine, pharynx, prepharynx, cirrus pouch, uterus, and genital and excretory openings are all innervated by plexuses. Some cells in the nervous system are positive for paraldehyde-fuchsin stain, indicating possible neurosecretory function.

Reproductive Systems

The male reproductive system of Aspidobothrea is similar to that of the Digenea (see Figs. 15.2 and 15.5). One, two, or many **testes** are present, located posterior to the ovary (Fig. 14.6). The **vas deferens** expands to form an **external seminal vesicle** before it enters the **cirrus pouch** to become the **ejaculatory duct.** A cirrus pouch is absent in some species. The **cirrus** is unarmed and opens through the genital pore into a common genital atrium, located on the midventral surface just anterior to the leading margin of the ventral disc. The axonemes in the spermatozoon filament have the 9 plus 1 structure, as is the case in other platyhelminth sperm (Fig. 14.7).

The female reproductive system consists of an ovary, vitelline cells, uterus, and associated ducts. The **ovary** (Fig. 14.6) is lobated or smooth and empties its products into an **oviduct.** The oviduct is peculiar among the Platyhelminthes in that its lumen is divided into many tiny chambers by septa, and the lining along much of its length is ciliated (Fig. 14.8). Each septum has a small hole in it through which the oocytes pass. The oviduct empties onto the **ootype,** which is surrounded by Mehlis' gland cells (Fig. 14.9). A short tube leading from the ootype and ending blindly in the parenchyma or, in a few cases, connecting with the excretory canal is called **Laurer's canal** and probably represents a vestigial vagina.

Vitelline follicles occur in two lateral fields, each of which has a main **vitelline duct** that fuses with that from the other field to form a small **vitelline reservoir,** which in turn opens into the ootype. Finally, a **uterus** extends from the ootype to course toward the genital atrium, usually with a posterior loop and anterior, distal stem. The distal end of the uterus has powerful muscles in its walls and is called the **metraterm.** This propels the eggs out of the system.

Some aspidobothreans are apparently self-fertilizing, with the cirrus depositing sperm in the terminal end of the uterus, which serves as a vagina. However, self-fertilization does not occur in *Lobatostoma manteri,* and unfertilized eggs do not develop beyond the blastula stage.[6]

DEVELOPMENT

As in other platyhelminths with separate vitellaria, the eggs of aspidobothreans are ectolecithal; that is, most of the embryo's yolk supply is derived from separate cells packaged with the zygote inside the eggshell. Some species' eggs are completely embryonated when they pass from the parent and hatch within a matter of hours, whereas others require three to four weeks of embryonation in the external environment. The larvae (**cotylocidia**) (Fig. 14.10) hatching from the egg in most species have a number of ciliary tufts that are effective in swimming. These larvae possess a mouth, pharynx, simple gut, and a prominent posterior-ventral disc without alveoli; there are no hooks. As the worm develops in its host, alveoli begin to form, tier by tier, in the anterior part of the ventral disc. The original cup of the disc remains apparent for some time behind the new ventral sucker and then disappears forever.

Larval ultrastructure has been studied in *M. purvisi* and in *C. occidentalis.*[1,5] The tegument pattern is similar to that of the adult, with the distal cytoplasmic, syncytial layer at the surface and with internal cytons. Between the ciliary tufts and covering most of the body the tegument surface in *M. purvisi* bears unique filiform structures called **microfila.** These have one central filament and about 9 to 12 peripheral filaments, differing from microvilli in that they do not have a cytoplasmic core. Their function is unknown, but it has been suggested that they help the larva to float.[5] In contrast, the tegument of *C. occidentalis* bears short microvilli with an external glycocalyx coat.[1]

Most aspidobothreans have a direct life cycle, requiring no intermediate host. Those parasitic in vertebrates appear to require an intermediate host; no case is known in which the

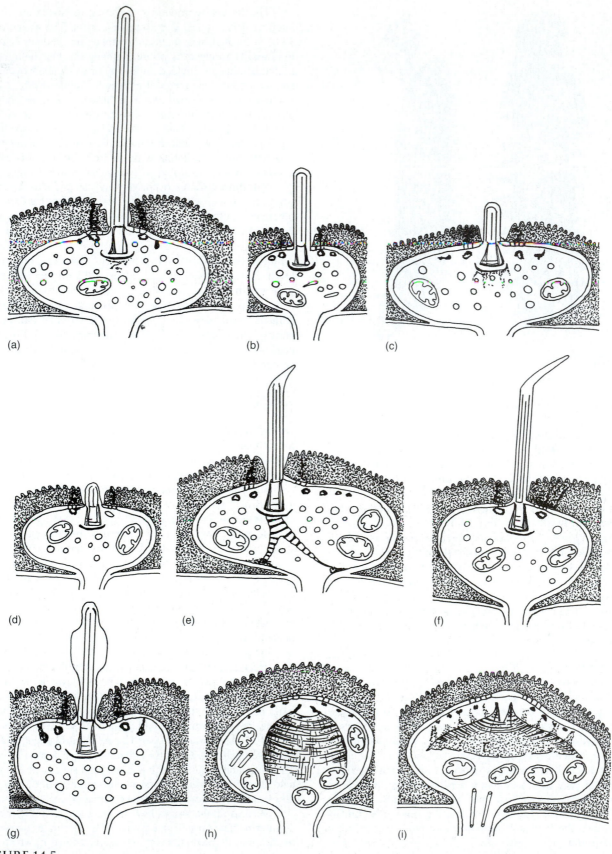

FIGURE 14.5

Diagrams of the various types of receptors of *Lobatostoma manteri:* (*a*) Uniciliate receptors with long cilium; (*b*) receptor with cilium of medium length; (*c*) short cilium receptor from posterior body surface; (*d*) short cilium receptor from anterior body surface; (*e*) receptor with ciliary rootlet; (*f*) cilium with bent tip; (*g*) inflated cilium receptor; (*h*) nonciliated receptor with large rootlet; (*i*) nonciliate disclike receptor.

From K. Rohde and N. Watson, "Sense receptors of *Lobatostoma manteri* (Trematoda, Aspidogastrea)," in *Int. J. Parasitol.* 22:35–42. Copyright © 1992. Reprinted with permission of the publisher and author.

(a)

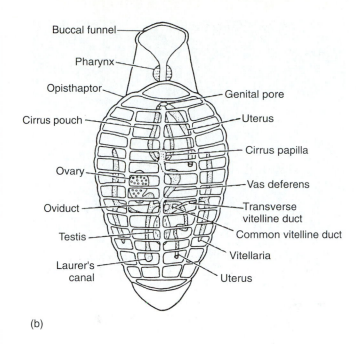

(b)

FIGURE 14.6

Aspidogaster conchicola, a common parasite of freshwater clams. (*a*) Lateral view, showing general body form. (*b*) Ventral view. The gut and the portion of the uterus between the descending limb and the terminal portion have been omitted in (*b*).

(*a*) Courtesy of Warren Buss. (*b*) From O.W. Olsen, *Animal Parasites: Their Life Cycles and Ecology.* Copyright © 1974 Dover Publications, Inc., New York, NY. Reprinted by permission.

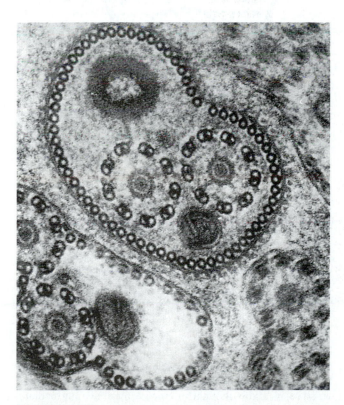

FIGURE 14.7

Cross section of sperm filament of *Aspidogaster conchicola.*

Courtesy of Ronald P. Hathaway.

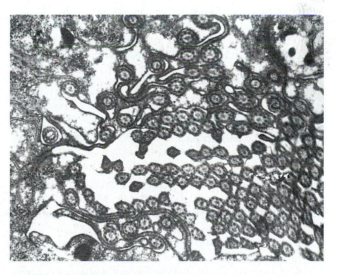

FIGURE 14.8

Section of proximal oviduct of *Aspidogaster conchicola,* showing the cilia that line much of its length.

Courtesy of Ronald P. Hathaway.

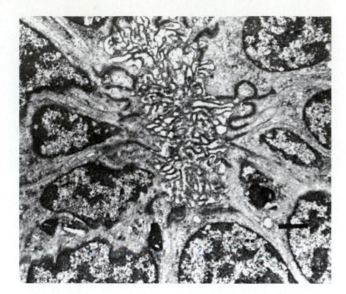

FIGURE 14.9

Ootype of *Aspidogaster conchicola* surrounded by Mehlis's gland cells. Courtesy of Ronald P. Hathaway.

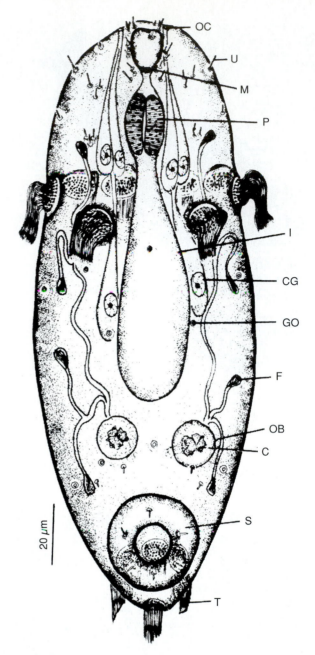

FIGURE 14.10

Composite drawing of cotylocidium of *Cotylogaster occidentalis.* **C,** Concretion; **CG,** cephalic gland; **GO,** opening of gobletlike gland cells; **F,** flame cell; **I,** intestine; **M,** mouth; **OB,** osmoregulatory bladder; **OC,** opening of cephalic gland; **P,** pharynx; **S,** sucker; **T,** tuft of cilia; **U,** uniciliated sensory structure.

From D.W. Fredericksen, "The fine structure and phylogenetic position of the cotylocidium larva of *Cotylogaster occidentalis* Nickerson 1902 (Trematoda: Aspidogastridae)," in *J. Parasitol.* 64:961–976. Copyright ©1978. Reprinted with permission of the publisher.

free larva is directly infective to vertebrates. Individuals can be removed from their definitive hosts and are capable of surviving for several days in water or saline, suggesting that they are rather generalized physiologically and not highly specialized for parasitism. Furthermore if they are eaten by a fish or turtle, they can live for a considerable length of time in this new host. Therefore it is not uncommon to find an aspidobothrean in the intestine of a fish, although it normally parasitizes a mollusc. Some species have so little host specificity that they can mature both in clams and in fish, although those in fish may be larger and produce more progeny.[1] Others apparently will not mature in a mollusc and need a fish final host.[2] *Lobatostoma manteri* preadults develop in any of several species of snails, but must reach the intestine of a snail-eating fish (*Trachinotus blochi*) to mature.[6,7]

The following life cycles illustrate the biology of two families in the subclass.

Aspidogaster conchicola

This common representative of the Aspidogastridae (see Fig. 14.6) is most often found in the pericardial cavity of freshwater clams in Europe, Africa, and North America, although it is known from other molluscs, fishes, and turtles. The adult is 2.5 to 3.0 mm long by 1.0 mm wide; it is oval and has a long, mobile "neck" with a buccal funnel at its end. The loculi on the ventral sucker are arrayed in four longitudinal rows, totaling 64 to 66.

When the eggs hatch within the host mollusc, the young can develop without further migration. If the egg or cotylocidium leaves the mollusc and is drawn into the incurrent siphon of the same or another clam, it can reach the nephridiopore and migrate through the kidney into the pericardium.

The cotylocidium is 13 to 17 μm long at hatching, lacks external cilia, and bears a simple posterior sucker without loculi. Growth and metamorphosis are rapid.

Lophotaspis vallei, also in Aspidogastridae, may use a marine snail as intermediate host. Mature forms have been found in marine turtles, but it is possible that they normally mature in molluscs.

Stichocotyle nephropsis

This parasite (Fig. 14.11) lives in the bile ducts of rays in the Atlantic Ocean. It has been found in lobsters and other crustaceans and is thought to employ them as intermediate hosts. The adult is slender, 115 mm long, and has 24 to 30 separate suckers along its ventral surface. This is the only species in Stichocotylidae and the only aspidogastrean

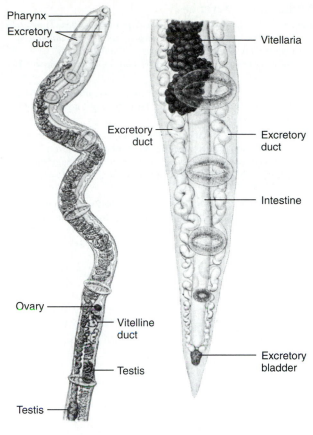

FIGURE 14.11

Stichocotyle nephropsis from the bile ducts of rays.

From T. Odhner, "*Stichocotyle nephropsis* J. T. Cunningham ein aberranter Trematode der Digenenfamilie Aspidogastridae," in *K. Svenska Vetensk. Acad. Handl.* 45:3–16. Copyright © 1910.

found in crustaceans. It is also possible that the crustacean is not a normal host in the life cycle of this parasite and that the occurrence of *S. nephropsis* in these animals is accidental. It is also possible that this worm does not belong in Aspidobothrea.

PHYLOGENETIC CONSIDERATIONS

Superficially the Aspidobothrea appear to be a link between the Monogenea and Digenea. Their anatomy is rather digenean, whereas their biology is suggestive of the Monogenea. This tiny group of worms displays sufficient individuality that it becomes apparent that the Aspidobothrea are distinct and separate from both groups. Aspidobothreans differ morphologically from both Digenea and Monogenea in that the ventral sucker develops as a new structure, unrelated to any homologue in larvae or adults of the other two groups. The frontal septum of aspidobothreans is not found in either of the other groups; neither is the septate, ciliated oviduct. However, the predominance of mollusc hosts and the presence of Laurer's canal and a highly developed nervous system are suggestive of Digenea. The simple, saclike cecum and undemanding physiological requirements are more in keeping with free-living turbellaria. Rohde[6] believes that an organism such as *Lobatostoma manteri,* with its mollusc intermediate host and fish definitive host, probably lies close to the ancestral protodigenean stock.

Modern phylogenetic analysis places the aspidobothreans as a sister group to the Digenea (see Chapter 13).

CLASSIFICATION OF THE ASPIDOBOTHREA

Trematodes with single, large, ventral sucker subdivided by septa into numerous, shallow loculi or with one ventral row of individual suckers; no sclerotized armature on any species; mouth with or without sucker, sometimes lobated; pharynx well-developed; intestine with a single or double median sac; testis single, double, or numerous; cirrus pouch present or absent; genital pores median, in front of sucker; ovary single, pretesticular; vagina absent, Laurer's canal sometimes present; vitellaria follicular, usually lateral but occasionally otherwise; eggs lacking polar prolongations; excretory pores on or near posterior end; development direct, without intermediate host; parasites of molluscs, fishes, and turtles.

Order Aspidobothriiformes

Family Aspidogastridae
   Body oval or elongate; ventral sucker with numerous shallow loculi; one or two testes present; vitellaria follicular, lateral; parasites of molluscs, fishes, or turtles; cosmopolitan.

Subfamilies
   Aspidogasterinae, Cotylaspidinae, Rohdellinae.

Order Stichocotylida

Family Stichocotylidae
   Body elongated, slender; ventral surface with longitudinal row of separate suckers; two testes present; vitellaria tubular, unpaired; parasites of Batoidea.

Family Rugogastridae
   Body elongated; most of ventral and lateral body surface has transverse rugae; musculature of buccal funnel weakly developed; pharynx, prepharynx, and esophagus present; two ceca; testes multiple; ovary pretesticular; Laurer's canal present, seminal receptacle absent; vitellaria distributed along ceca, uterus ventral to testes; eggs operculate; parasites of Holocephali.

Family Multicalycidae
   Body elongated; holdfast composed of fused suckers, otherwise similar to Rugogastridae.

References
1. Fredericksen, D. W. 1978. The fine structure and phylogenetic position of the cotylocidium larva of *Cotylogaster occidentalis* Nickerson 1902 (Trematoda: Aspidogastridae). *J. Parasitol.* 64:961–76.

2. Hendrix, S. S., and R. M. Overstreet. 1977. Marine aspidogastrids (Trematoda) from fishes in the northern Gulf of Mexico. *J. Parasitol.* 63:810–17.

3. Ip, H. S., S. S. Desser, and I. Weller. 1982. *Cotylogaster occidentalis* (Trematoda: Aspidogastrea): Scanning electron microscopic observations of sense organs and associated surface structures. *Trans. Am. Microsc. Soc.* 100:253–61.

4. Ip, H. S., and S. S. Desser. 1984. Transmission electron microscopy of the tegumentary sense organs of *Cotylogaster occidentalis* (Trematoda: Aspidogastrea). *J. Parasitol.* 70:563–75.

5. Rohde, K. 1972. The Aspidogastrea, especially *Multicotyle purvisi* Dawes, 1941. In Dawes, B., ed. *Advances in parasitology* 10. New York: Academic Press, Inc., 77–151.

6. Rohde, K. 1973. Structure and development of *Lobatostoma manteri* sp. nov. (Trematoda: Aspidogastrea) from the Great Barrier Reef, Australia. *Parasitology* 66:63–83.

7. Rohde, K. 1975. Early development and pathogenesis of *Lobatostoma manteri* Rohde (Trematoda: Aspidogastrea). *Int. J. Parasitol.* 5:597–607.

8. Rohde, K., and N. Watson. 1989. Ultrastructure of the marginal glands of *Lobatostoma manteri* (Trematoda, Aspidogastrea). *Zoo. Anz.* 223:301–10.

9. Rohde, K., and N. Watson. 1992. Sense receptors of larval *Lobatostoma manteri* (Trematoda, Aspidogastrea). *Int. J. Parasitol.* 22:35–42.

10. Schell, S. C. 1973. *Rugogaster hydrolagi* gen. et sp. n. (Trematoda: Aspidobothrea: Rugogastridae fam. n.) from the ratfish, *Hydrolagus colliei* (Lay and Bennett, 1839). *J. Parasitol.* 59:803–5.

Additional References

Baer, J. G., and C. Joyeux. 1961. Classe des trematodes (Trematoda Rudolphi). In Grassé, P., ed. *Traité de zoologie: Anatomie, systématique, biologie, vol. 4, part I. Plathelminthes, Mésozoaires, Acanthocéphales, Némertiens.* Paris: Masson & Cie, 561–70.

Dollfus, R. P. 1958. Trematodes. Sous-classe Aspidogastrea. *Ann. Parasitol.* 33:305–95. A detailed summary of knowledge of this group to 1958.

Gibson, D. I., and S. Chinabut. 1984. *Rohdella siamensis* gen. et sp. nov. (Aspidogastridae: Rohdellinae subfam. nov.) from freshwater fishes in Thailand, with a reorganization of the classification of the subclass Aspidogastrea. *Parasitology* 88:383–93.

Yamaguti, S. 1963. *Systema Helminthum* 4. New York: Interscience Publishers. A most useful taxonomic treatment of the group.

Chapter 15

TREMATODA: FORM, FUNCTION, AND CLASSIFICATION OF DIGENEANS

The life cycle of the great majority of digeneans displays a remarkable and highly characteristic alternation of asexual and sexual reproductive phases. . . .

P. J. Whitfield

The digenetic trematodes, or flukes, are among the most common and abundant of parasitic worms, second only to nematodes in their distribution. They are parasites of all classes of vertebrates, especially marine fishes, and some species, as adults or juveniles, inhabit nearly every organ of the vertebrate body. Their development occurs in at least two hosts. The first is a mollusc or, very rarely, an annelid. Many species include a second and even a third intermediate host in their life cycles. Several species cause economic losses to society through infections of domestic animals, and others are medically important parasites of humans. Because of their importance, the Digenea have stimulated vast amounts of research, and the literature on the group is immense. We will summarize the morphology and biology of the group, illustrating it with some of the more important species.

Trematode development will be considered in detail later (p. 215), but a "typical" life cycle is as follows: A ciliated, free-swimming larva, the **miracidium,** hatches from its shell and penetrates the first intermediate host, usually a snail. At the time of penetration or soon after, the larva discards its ciliated epithelium and metamorphoses into a rather simple, saclike form, the **sporocyst.** Within the sporocyst a number of embryos develop asexually to become **rediae.** The redia is somewhat more differentiated than the sporocyst, possessing, for example, a pharynx and a gut, neither of which are present in a miracidium or sporocyst. Additional embryos develop within the redia, and these become **cercariae.** The cercaria emerges from the snail and usually has a tail to aid in swimming. Although many species require further development as **metacercariae** before they are infective to the definitive host, the cercariae are properly considered juveniles; they have organs that will develop into the adult digestive tract and suckers, and genital primordia are often present. The fully developed, encysted metacercaria is infective to the definitive host and develops there into the adult trematode.

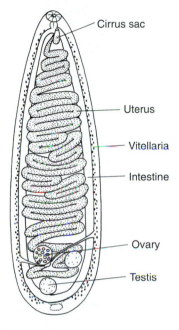

FIGURE 15.1

Cyclocoelum lanceolatum, a common monostome fluke from the air sacs of shore birds.

Drawing by William Ober.

FORM AND FUNCTION

Body Form

Flukes exhibit a great variety of shapes and sizes, as well as variations in internal anatomy. They range from the tiny *Levinseniella minuta,* only 0.16 mm long, to the giant *Fascioloides magna,* which reaches 5.7 cm in length and 2.5 cm in width.

Most flukes are dorsoventrally flattened and oval in shape, but some are as thick as they are wide. Some species are filiform, round or even wider than they are long. Flukes usually possess a powerful oral sucker that surrounds the mouth, and most also have a midventral acetabulum or ventral sucker. The words **distome, monostome,** and **amphistome** are sometimes used as descriptive terms, although they formerly had taxonomic significance, and of course they refer to suckers, not mouths (Gr., *stoma,* mouth). If a worm has only an oral sucker, it is called a monostome

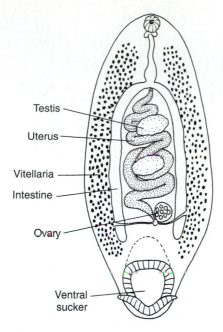

FIGURE 15.2

Zygocotyle lunata, an amphistome fluke from ducks.
Drawing by William Ober.

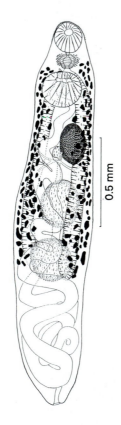

FIGURE 15.3

Alloglossidium hirudicola, a distome trematode from leeches.

From G.D. Schmidt and K. Chaloupka, *Alloglossidium hirudicola* sp. n., a
neotenic trematode (Plagiorchiidae) from leeches, *Haemopis* sp.," in *Int. J.
Parasitol.* 55:1185–1186. Copyright © 1969. Reprinted with permission of
the publisher.

FIGURE 15.4

Bunodera sacculata from yellow perch. Note the muscular lappets on
the oral sucker.

From H. J. Van Cleave and J. F. Mueller, "Parasites of Oneida Lake fishes,
Part 1. Descriptions of new genera and new species," in *Roosevelt Wildl.
Ann.* 3:9–71. Copyright © 1932. Reprinted with permission of the publisher.

(Fig. 15.1); with an oral sucker and an acetabulum at the
posterior end of the body, it is an amphistome (Fig. 15.2);
and if the acetabulum is elsewhere on the ventral surface,
the worm is a distome (Fig. 15.3). The oral sucker may
have muscular lappets, as in *Bunodera* (Fig. 15.4), or there
may be an anterior adhesive organ with tentacles, as in *Bucephalus* (Fig. 15.5). *Rhopalias* spp., parasites of American
opossums, have a spiny, retractable proboscis on each side
of the oral sucker (Fig. 15.6). In species of Hemiuridae the
posterior part of the body telescopes into the anterior portion. Some workers use additional terms to describe body
forms of digeneans, such as **holostome** (p. 234, Fig. 16.1),
schistosome (p. 237), and **echinostome** (p. 251).

Tegument

The tegument of trematodes, like that of cestodes, traditionally was considered a nonliving, secreted cuticle; but as in
cestodes, the electron microscope reveals that the body covering of trematodes is a living, complex tissue. In common
with the Monogenea and the Cestoidea, digenetic trematodes have a "sunken" epidermis; that is, there is a distal,
anucleate layer (**distal cytoplasm**). The cell bodies containing the nuclei (**cytons**) lie beneath a superficial layer of
muscles, connected to the distal cytoplasm by way of channels (**internuncial processes**) (Fig. 15.7). Because the distal
cytoplasm is continuous, with no intervening cell membranes, the tegument is **syncytial.** Although this is the same
general organization found in the cestodes, trematode tegument differs in many details, and striking differences in
structure may occur in the same individual from one region

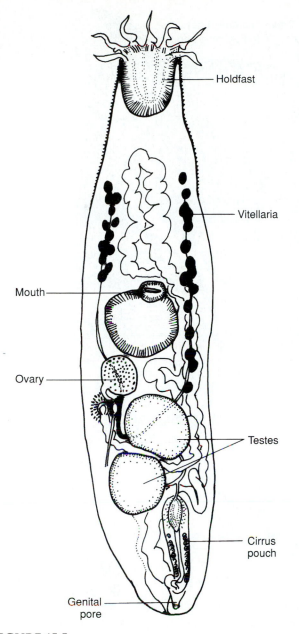

FIGURE 15.5

Bucephalus polymorphus from European and Asian fishes.

Source: From S. Yamaguti, *Synopsis of Digenetic Trematodes of Vertebrates.*
© 1971 Keigaku Publishing Company, Tokyo.

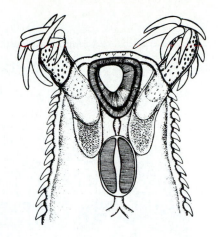

FIGURE 15.6

Rhopalias coronatus, a parasite of American opossums. Retractable
proboscides are located on each side of the mouth.

Source: From S. Yamaguti, *Synopsis of Digenetic Trematodes of Vertebrates,*
Volume 1. Copyright © 1971 Keigaku Publishing Company, Tokyo.
Reprinted with permission of the publisher.

of the body to another. Ornamentation such as **spines** is
often present in certain areas of the trematode's body and
may be discernible with the light microscope. The oral and
ventral suckers of *Schistosoma mansoni* are densely beset
with spines, and much of the male's dorsum bears **bosses**
(rounded, elevated areas, Fig. 15.8) with 50 to 250
spines.[80] Papillae, many with craterlike sensory openings,
are interspersed (Fig. 15.8). Bosses are absent from the
male's gynecophoral canal (see p. 237) and from the female, but the females have many *anteriorly directed* spines
on their posterior ends (Fig 15.9). The spines consist of

crystalline actin;[24] their bases lie above the basement
membrane of the distal cytoplasm, and their apices project
above the surface, although generally they are covered by
the outer plasma membrane.

The distal cytoplasm usually contains vesicular inclusions, more or less dense, and the tegument of the same
worm sometimes may bear several recognizable types. The
function of the vesicles is unclear, although in some cases
they contribute to the outer surface. The surface membrane
of *S. mansoni* is continuously renewed by multilaminar vesicles moving outward through the distal cytoplasm, perhaps
to replace membrane damaged by host antibodies.[59] The
outer layers with host antibody adsorbed to them are indeed
shed by the worm.[65] In *Megalodiscus* the contents of some
vesicles seem to be emptied to the outside.[7]

Vesicles of the distal cytoplasm are Golgi derived. They
usually pass outward from the cytons through the internuncial processes, although Golgi bodies occasionally occur in
the distal cytoplasm as well.

Mitochondria occur in the distal cytoplasm in most
species examined, although not in *Megalodiscus* and some
paramphistomes.[7,10,37]

The outer surface of adult trematodes is not modified for
absorption by the elaboration of microvilluslike microtriches,
as in the cestodes, but some structural features that increase
surface area occur. The tegument of some trematodes is penetrated by many deep pits (Fig. 15.10) and channels.[73,107]

Miracidia of *Fasciola* and *Schistosoma* (at least) are
covered by ciliated epithelial cells with nuclei as is typical
for such cells.[121] The epithelial cells are interrupted by
"intercellular ridges," extensions of cells whose perikarya
lie beneath the superficial muscle layer and that bear no
cilia, although some microvilli may be present (Fig.
15.11). On loss of the ciliated epithelium, metamorphosis to the sporocyst involves a spreading of the distal
cytoplasm over the worm's surface, although whether

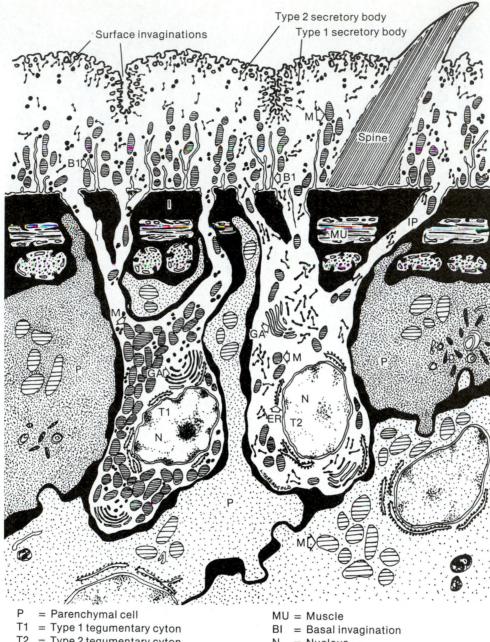

P = Parenchymal cell
T1 = Type 1 tegumentary cyton
T2 = Type 2 tegumentary cyton
GA = Golgi complex
I = Interstitial material (connective tissue)
IP = Internunical process

MU = Muscle
BI = Basal invagination
N = Nucleus
ER = Granular endoplasmic reticulum
M = Mitochondria

FIGURE 15.7

Diagram of the tegument of *Fasciola hepatica* at the ultrastructural level.
Drawing by L. T. Threadgold.

this comes from the intercellular ridges is unclear. Well-developed microvilli are present on the surface of both the sporocyst and redia. The *luminal* surface of the tegumental cells in the redia may be thrown into a large number of flattened sheets that extend to other cells in the body wall and to the cercarial embryos contained in the lumen. Nutritive molecules, such as glucose, and molecules up to the size of horseradish peroxidase can pass through the tegument to the developing cercariae.[7]

The early embryos of the cercariae are covered with a primary epidermis below which a definitive epithelium forms. The nuclei of this secondary epithelium sink into the parenchyma, and the final form of the cercarial tegument results in an organization similar to that of the adult. Cystogenic cells in the parenchyma begin to secrete cyst material, which passes into the distal cytoplasm of the tegument. The metacercarial cyst wall forms when the cercarial tegument sloughs off, and the cyst material it contains undergoes chemical

FIGURE 15.8

Tegument of *Schistosoma mansoni:* dorsal region in distal third of male, showing spines in interbossal spaces and papillae with openings. (× 780.)

From F. H. Miller, G. S. Tulloch, and R. E. Kuntz, "Scanning electron microscopy of integumental surface of *Schistosoma mansoni*," in *J. Parasitol.* 58:693–698. Copyright © 1972.

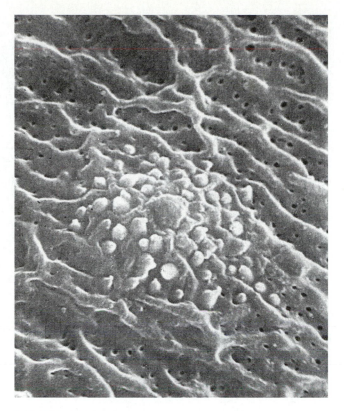

FIGURE 15.10

A tubercle and spines on the tegument of a male *Schistosoma mansoni*. Note the many pits in the surface. (× 6500.)

From D. J. Hockley, "Ultrastructure of the tegument of *Schistosoma*," in *Advances in Parasitology,* vol. 11. Edited by B. Dawes. Copyright © 1973 Academic Press LTD, London, England. Reprinted with permission.

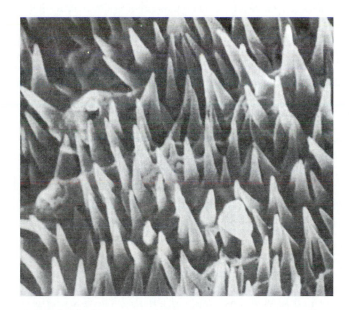

FIGURE 15.9

Schistosoma mansoni: dense covering of forward-projecting spines at posterior tip of female. (× 1700.)

From F. H. Miller, G. S. Tulloch, and R. E. Kuntz, "Scanning electron microscopy of integumental surface of *Schistosoma mansoni*," in *J. Parasitol.* 58:693–698. Copyright © 1972.

and/or physical changes to envelop the worm in its cyst. The cystogenic cells in the parenchyma then flow toward the surface, their nuclei being retained beneath the superficial muscles, and a thin layer of cytoplasm spreads over the organism to become the definitive adult tegument. A contrasting mode of cercarial tegument formation has been reported for *S. mansoni.*[57] The definitive tegument, including its nuclei, forms beneath the primitive tegument and is syncytial.

Subsequently the nuclei of the primitive tegument degenerate and are lost, as processes from subtegumental cells grow outward and join the distal cytoplasmic layer (Fig. 15.12). The end result is the same.

The tegument is variously interrupted by cytoplasmic projections of gland cells, by openings of excretory pores, and by nerve endings. Both miracidia and cercariae may have penetration glands that open at the anterior, and the adults of some species have prominent glandular organs opening to the exterior.

Muscular System

The muscles that occur most consistently throughout the Digenea are the superficial muscle layers (formerly called *subcuticular*), and these usually comprise circular, longitudinal, and diagonal layers enveloping the rest of the body like a sheath below the distal cytoplasm of the tegument. The degree of muscularization varies considerably in the group; some species have a rather feeble musculature, some are very robust and strong, and other examples fit all conditions in between. The deep musculature found in cestodes is generally absent in trematodes. Muscles are often more prominent in the anterior parts of the body, and strands connecting the dorsal to the ventral superficial muscles are usually found in the lateral areas. The fibers are smooth, and the nuclei occur in cytons called *myoblasts* connected to the fiber bundles and located in various sites around the body, often in syncytial clusters. It is likely that the "parenchymal" cells are in fact

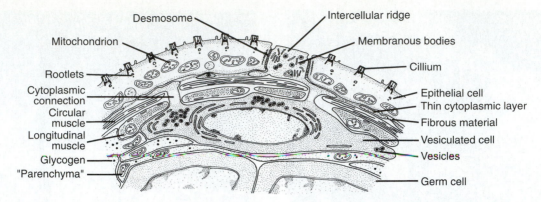

FIGURE 15.11

Miracidium of *Fasciola hepatica*. Line drawing reconstruction of transverse segment of body wall in region of the germ cell cavity.

From R. A. Wilson, "Fine structure of the tegument of the miracidium of *Fasciola hepatica* L.," in *J. Parasitol.* 55:124–133. Copyright © 1969. Reprinted with permission of the publisher.

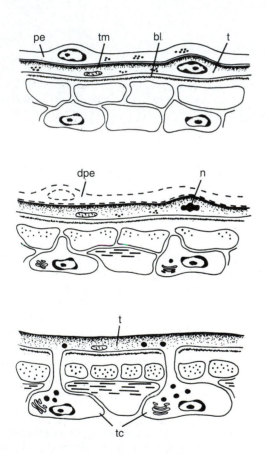

FIGURE 15.12

Diagram summarizing three stages in the formation of the tegument during cercarial development. *Top,* germ ball covered with a primitive epithelium **(pe)** and the tegument **(t)**, which has a thickened outer membrane **(tm)** and an underlying basal lamina **(bl)**. *Center,* young cercaria with degenerating primitive epithelium **(dpe)** and degenerating tegumental nucleus **(n)**. *Bottom,* cercaria nearly ready to emerge from the sporocyst; the primitive epithelium has been lost, and the tegument **(t)** is connected to nucleated tegumental cytons **(tc)**.

From D. J. Hockley, "Ultrastructure of the tegument of *Schistosoma*," in *Advances in Parasitology,* vol. 11. Edited by B. Dawes. Copyright © 1973 Academic Press LTD, London, England. Reprinted with permission.

cytons of muscle cells, as has been found in cestodes.[77] The suckers and pharynx are supplied with radial muscle fibers, often very strongly developed. A network of fibers may surround the intestinal ceca, helping to fill and empty these structures.

Nervous System

The organization of the nervous system is the typical platyhelminth ladder type.[89] A pair of cerebral ganglia is connected by a commissure (connecting band of nerve tissue) in the anterior part of the animal. From this, several nerves issue anteriorly, and three main pairs of trunks—dorsal, lateral, and ventral—supply the posterior parts of the body. The ventral nerves are usually best developed, and a variable number of commissures link these and the other longitudinal nerves. Branches provide motor and sensory endings to muscles and tegument. The anterior end, especially the oral sucker, is well supplied with sensory endings.

Sensory endings in the Digenea are an interesting array of types, particularly in the miracidia and cercariae. The adults require no orientation to such stimuli as light and gravity, and in the few forms in which the ultrastructure has been studied, only one type of sensory ending has been described.[58] This is a bulbous nerve ending in the tegument that has a short, modified cilium projecting from it, and the cilium is enclosed throughout its length by a thin layer of tegument. The general structure is similar to sense organs described in cestodes (see Fig. 20.13). These structures generally have been regarded as tangoreceptors (receptors sensitive to touch) in trematodes.

Cercariae and miracidia show more variety in sense organs, doubtlessly related to the adaptive value of finding a host quickly.[11] Uniciliated bulbous endings are found on the anterior portion of the cercariae of *S. mansoni,* similar to but smaller than those on the adult. The tegumentary sheath opens at the ciliary apex. In addition, a bulbous type with a long (7 μm) unsheathed cilium is widely distributed over the

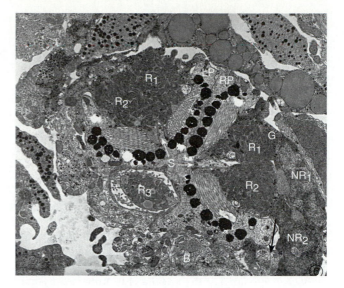

FIGURE 15.14

Ultrastructure of eyespots in a miracidium of *Fasciola hepatica*. Nearly frontal section in dorsal aspect. *Arrow* indicates a junction between a lateral retinular cell and an end-bulb of an axon from the brain. Retinular cells have closely packed mitochondria with rhabdomeric microvilli adjacent to pigment cells. **B,** brain; **G,** glycogen; **LP,** left pigment cell; **RP,** right pigment cell; **NR_1,** nucleus of anterior lateral retinular cell; **NR_2,** nucleus of posterior lateral retinular cell; **R_1,** anterior lateral retinular cell; **R_2,** posterior lateral retinular cell; **R_3,** median or posterior (5th) retinular cell; **S,** septum. (× 7000.)

From H. Isseroff and R. M. Cable, "Fine structure of photoreceptors in larval trematodes. A comparative study," in *Z. Zellforsch.* 86: 511–534. Copyright © 1968.

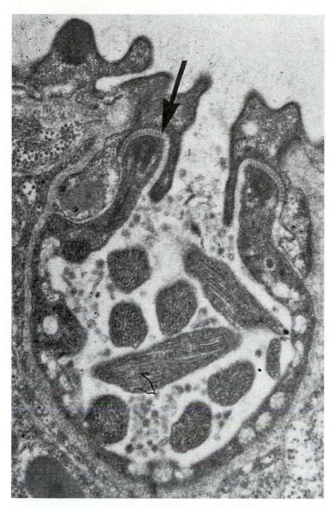

FIGURE 15.13

Multiciliated pit in anterior body tegument of cercaria of *Schistosoma mansoni. Arrow* indicates septate desmosome. (× 36,000.)

From G. P. Morris, "The fine structure of the tegument and associated structures of the cercaria of *Schistosoma mansoni,*" in *Z. Parasitenkd.* 36:115–31. Copyright © 1971.

body of the cercaria, and its lateral areas bear small, flask-shaped endings containing five or six cilia and opening to the outside through a 0.2 μm pore (Fig. 15.13). This latter type is probably a chemosensory ending.

Another apparent chemoreceptor described in the miracidium of *D. spathaceum* consists of two dorsal papillae between the first series of ciliated plates. Each papilla consists of a nerve ending and has radiating from it a number of modified cilia, which are parallel to the surface of the miracidium. These sensory endings are strikingly similar to the olfactory receptors of the vertebrate nasal epithelium!

Eyespots are present in many species of miracidia and in some cercariae. Although also present in some adult trematodes, eyespots are apparently functionless in adults. The structure of eyespots in miracidia is generally similar to that of such organs found in turbellaria and some Annelida.

The eyespots consist of one or two cup-shaped pigment cells surrounding the parallel rhabdomeric microvilli of one or more retinular cells (Fig. 15.14). The mitochondria of the retinular cells are packed in a mass near the rhabdomere. Because the rhabdomeres are the photoreceptors, the cup shape of the pigment cells allows the organism to distinguish light direction. Interestingly, some miracidia do not have eyespots yet can orient with respect to light. Some cells in the miracidia of *Diplostomum spathaceum* and *S. mansoni* have large vacuoles; into these vacuoles project a number of cilia, each of which has a conspicuous membrane evagination. These membranes, which are stacked in a lamellar fashion, might be photoreceptors, thus providing, in the case of *S. mansoni,* a means of light sensitivity for a miracidium without eyespots.[11]

An important excitatory neurotransmitter is 5-hydroxy-tryptamine, and acetyl choline is apparently the major inhibitor of neuromuscular transmission.[47]

A large number of neuropeptides have been found distributed through the nervous system of trematodes and other flat-worms.[53] Although they probably serve as messenger systems that regulate and control a variety of bodily processes, their specific functions remain obscure. There is evidence that some neuropeptides help coordinate complex muscular activities involved in the formation of "eggs" in the oogenotop (p. 214).

Excretion and Osmoregulation

In her concise review, Hertel[54] paraphrased Beklemishev,[4] who said that excretion included (1) removal of waste products of metabolism; (2) regulation of internal osmotic pressure; (3) regulation of internal ionic composition; and (4) removal of unnecessary or harmful substances. Thus, by this definition excretion includes osmoregulation, and this is an important function in the so-called excretory systems of many flatworms. Removal of metabolic wastes takes place by diffusion across the tegument and epithelial lining of the gut and by exocytosis of vesicles, in addition to that portion of removal that may be attributed to the excretory system.

The excretory system of Digenea is based on the flame bulb **protonephridium** (a unit of an excretory system closed at the proximal end and opening to the exterior at the distal end by way of a pore). The **flame bulb** (Fig. 20.15), or cell, is flask shaped and contains a tuft of fused flagella to provide the motive force for the fluid in the system. In the Trematoda and Cercomeromorphae the flagella are surrounded by rodlike extensions of the flame cell, which interdigitate with rodlike extensions of the proximal tubule cell.[54] These interdigitations form the latticelike weir (p. 188). The weir is the filtering apparatus. A thin membrane usually extends between the rods, and beating of the flagella creates a pressure gradient that draws fluid through the weir and into the collecting tubule. Fingerlike **leptotriches** sometimes extend from both the internal and external surfaces of the weir. They appear to increase filtration, possibly by keeping surrounding cells away from the weir and keeping the wall of the weir away from the tuft of flagella.

The ductules of the flame cells join collecting ducts, those on each side eventually feeding into an **excretory bladder** in the adult that opens to the outside with a single pore. In digeneans the pore is almost always located near the posterior end of the worm. In some trematodes the walls of the collecting ducts are supplied with microvilli, indicating that some transfer of substances, absorption or secretion, is probably occurring.[103] That the system has osmoregulatory function may be inferred from the fact that among free-living Platyhelminthes, freshwater forms have much better-developed protonephridial systems than do marine flatworms. When the two free-swimming stages of digeneans occur in fresh water, they require an efficient water-pumping system.

The primary nitrogenous excretory product of trematodes is apparently ammonia, although excretion of uric acid and urea has been reported. The proportion of ammonia excretion that takes place through the tegument, ceca, or excretory system is not known.

Acquisition of Nutrients and Digestion

Feeding and digestion in trematodes vary with nutrient type and habitat within the host.[52] For example, two lung flukes of frogs, *Haematoloechus medioplexus* and *Haplometra cylindracea,* feed predominately on blood from the capillaries. Both species draw a plug of tissue into their oral sucker and then erode the host tissue by a pumping action of the strong, muscular pharynx.[52] Other trematode species characteristically found in the intestine, urinary bladder, rectum, and bile ducts feed more or less by the same mechanism, although their food may consist of less blood and more mucus and tissue from the wall of their habitat, and it may even include gut contents. In species without a pharynx that feed by this mechanism, the walls of the esophagus are quite muscular, and this apparently serves the function of the pharynx. In contrast *S. mansoni,* living in the blood vessels of the hepatoportal system and immersed in its semifluid blood food, has no necessity to breach host tissues, and not surprisingly, this species has neither pharynx nor muscular esophagus.

Digestion in most species studied is predominately extracellular in the ceca, but in *Fasciola hepatica* it occurs by a combination of intracellular and extracellular processes. One of the frog lung flukes, *Haplometra cylindracea,* has pearshaped gland cells in its anterior end, and a nonspecific esterase is secreted from these cells through the tegument of the oral sucker, beginning the digestive process even before the food is drawn into the ceca.

Those trematodes that feed on blood cope in various ways with the iron component of the hemoglobin molecule. In *F. hepatica,* in which the final digestion of hemoglobin is intracellular, the iron is expelled through the excretory system and tegument. The fate of the iron in *H. cylindracea* is unclear, but apparently it is stored within the worm, tightly bound to protein. The extracellular digestion in *H. medioplexus* and *S. mansoni* produces insoluble end products within the cecal lumen, and these are periodically regurgitated.

In *S. mansoni* the end products are a heterogeneous population of molecules, but the worms digest and incorporate some of both the globin and heme moieties of hemoglobin.[42] The ceca of trematodes apparently do not bear any gland cells, but the gastrodermal cells themselves may in certain species secrete some digestive enzymes: Proteases, a dipeptidase, an aminopeptidase, lipases, acid and alkaline phosphatases, and esterases have been detected.[100,106,124] A protease from *S. mansoni* has marked substrate specificity for hemoglobin and an optimal pH of 3.9 to 4.5. The schistosome gut is an acidic compartment containing acid hydrolases, and it reacts similarly to lysosomotropic agents and to a secondary lysosome.[9] Expression and sequence of mRNA for one schistosome hemoglobinase have been studied.[40] Enzyme activity and mRNA are low in young schistosomules but increase to high levels coincident with the onset of feeding on red blood cells.

All species investigated have microvilli on the gastrodermal cells, although these vary from short (1 to 15 μm) and irregular to long (10 to 20 μm) and are organized into a definite brush border, according to species. It is clear, however, that the structures interpreted as microvilli in several species at the light microscope level are in fact flattened platelike or lamelloid processes projecting into the lumen[33,45] (Fig. 15.15). Digitiform processes more like typical microvilli are borne by the gut cells of some species. *Fasciola hepatica* and *Echinostoma hortense* have distal and/or marginal filamentous extensions on the lamellae. In all cases the absorptive surface area is vastly greater than it would be if the cell surface were flat. Within the gut cells of both *Gorgodera amplicava* and *Haematoloechus medioplexus* are abundant rough endoplasmic reticulum, many mitochondria, and frequent Golgi bodies and membrane-bound, vesicular inclusions. High activity of acid phosphatase is found in the vesicles of *H. medioplexus* and *Paragonimus kellicotti,* and after incubation in ferritin the

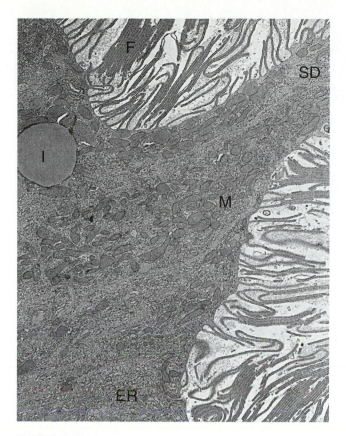

FIGURE 15.15

Apical portion of the cecal epithelium of *Paragonimus kellicotti*. The apical surface has numerous folds (**F**) extending into the lumen. The cecal epithelial cells are joined by septate desmosomes (**SD**). The cytoplasm contains a well-developed granular endoplasmic reticulum (**ER**) and numerous mitochondria (**M**). An inclusion (**I**) is indicated.

From S. C. Dike, "Acid phosphatase activity and ferritin incorporation in the ceca of digenetic trematodes," in *J. Parasitol.* 55:111–123. Copyright © 1969.

material is found within them. No evidence of "transmembranosis" has been found, but the vesicles may be lysosomes that would function in degradation of nutritive materials after phagocytosis.

It is not surprising that trematodes can absorb small molecules through the tegument. The uptake of nutrients through the surface of sporocysts and rediae has already been mentioned. In the few species examined glucose was absorbed through the tegument and not by way of the gut, although it has not always been clear whether the worms might not have been able to absorb this hexose by the intestinal route had they been feeding normally.[88] However, in one report (*Philophthalmus megalurus*), the trematodes must have been feeding in vitro, since they absorbed the amino acids tyrosine and leucine only through the gut, whereas they absorbed glucose mostly through the tegument.[83] Thymidine was absorbed by *P. megalurus* through *both* routes. *Gorgoderina* can absorb tyrosine, thymidine, adenosine, and glucose through its tegument, whereas *Haematoloechus* can absorb glucose via its tegument but arginine only by its gut.[86,88] Also, is it known that *S. mansoni* takes in glucose only through its tegument,[95] and male *S. mansoni* somehow passes glucose to the female lying in the gynecophoral canal.[28] Schistosomes absorb glucose both

by diffusion and by a carrier-mediated system,[26] and even brief exposure to a glucose-free medium disrupts uptake of a variety of other small molecules.[27]

Megalodiscus temperatus cannot absorb glucose or galactose across its tegument,[104] and this species, as well as several other paramphistomes, has no mitochondria in the tegumental cytoplasm.[10,37] This could be a reason the tegument of these trematodes may have little or no absorptive capacity.

Reproductive Systems

Most trematodes are hermaphroditic (important exceptions are the schistosomes), and some are capable of self-fertilization. Others, however, require cross-fertilization to produce viable progeny. Some species inseminate themselves readily; others will do so if there is only one worm present, but they always seem to cross-inseminate when there are two or more in the host.[84] A few instances are known in which adult trematodes can reproduce parthenogenetically.[119]

• Male Reproductive System

The male reproductive system (Fig. 15.16) usually includes two testes, although some species have from one testis to several dozen. Their shape varies from round to highly branched, according to species. Each testis has a vas efferens that connects with the other to form a vas deferens. This courses toward the genital pore, which is usually found within a shallow genital atrium. The genital atrium is most often on the midventral surface, anterior to the acetabulum, but it can be found nearly anywhere, including at the posterior end, beside the mouth, or even dorsal to the mouth in some species. Before reaching the genital pore, the vas deferens usually enters a muscular cirrus pouch where it may expand into an **internal seminal vesicle** for sperm storage. Constricting again, the duct forms a thin ejaculatory duct, which extends the rest of the length of the cirrus pouch and forms at its distal end a muscular cirrus. The cirrus is the male copulatory organ. It can be invaginated into the cirrus pouch and evaginated for transfer of sperm to the female system. The cirrus may be naked or covered with spines of different sizes. The ejaculatory duct is usually surrounded by numerous unicellular **prostate gland cells.** At this point a muscular dilation may form a **pars prostatica.**

Much variation in these terminal organs occurs among families, genera, and species. The cirrus pouch and prostate gland may be absent, with the vas deferens expanded into a powerful seminal vesicle that opens through the genital pore, as in *Clonorchis*. The vas deferens may expand into an **external seminal vesicle** before continuing into the cirrus pouch. Other, more specialized modifications are described and illustrated by Yamaguti.[123]

• Female Reproductive System

The single ovary in the female reproductive tract (Fig. 15.16) is usually round or oval, but it may be lobated or even branched. The short oviduct is provided with a proximal sphincter, the **ovicapt,** that controls the passage of ova. The oviduct and most of the rest of the female ducts are ciliated. A seminal receptacle forms as an outpocketing of the wall of the oviduct. It may be large or small, but it is almost always present. At the base of the seminal receptacle there often

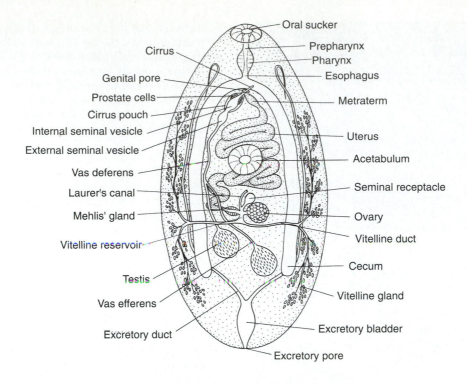

FIGURE 15.16

Diagrammatic representation of digenetic fluke, showing male and female reproductive systems.
Drawing by William Ober.

arises a slender tube, Laurer's canal, which ends blindly in the parenchyma or opens through the tegument. Laurer's canal is probably a vestigial vagina that no longer functions as such (with a few possible exceptions), but it may serve to store sperm in some species.

Unlike other animals but in common with the Cercomeromorphae and some free-living flatworms, the yolk is not stored in the female gamete (**endolecithal**) but is contributed by separate cells called vitelline cells. Such a system is described as **ectolecithal.** The vitelline cells are produced in follicular vitelline glands, usually arranged in two lateral fields and connected by ductules to the main right and left vitelline ducts. These ducts carry the vitelline cells to a single, median vitelline reservoir, from which extends the common vitelline duct joining the oviduct. The anatomical distribution of vitelline glands tends to be constant within a species and so is an important taxonomic character. After the junction with the common vitelline duct the oviduct expands slightly to form the ootype. Numerous unicellular Mehlis' glands surround the ootype and deposit their products into it by means of tiny ducts.

The structural complex just described (Fig. 15.17), as well as the upper uterus, is called the *egg-forming apparatus,* or **oogenotop.**[49] Beyond the ootype, the female duct expands to form the uterus, which extends to the female genital pore. The uterus may be short and fairly straight, or it may be long

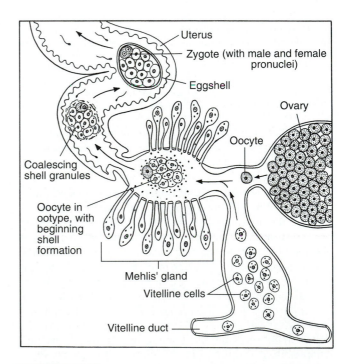

FIGURE 15.17

Schematic representation of the oogenotop of a digenetic trematode.
Drawing by William Ober and Claire Garrison.

and coiled or folded. The distal end of the uterus is often quite muscular and is called the **metraterm.** The metraterm functions as ovijector and as a vagina. The female genital pore opens near the male pore, usually together with it in the genital atrium. In some species, such as in the Heterophyidae, the genital atrium is surrounded by a muscular sucker called a **gonotyl.**

At the time the germ cells leave the ovary, they have not completed meiosis and thus are not ova at all but oocytes. Meiosis is completed after sperm penetration. The first meiotic division may reach pachytene or diplotene, at which point meiotic activity arrests, and the chromosomes return to a diffuse state. After sperm penetration the chromosomes quickly reappear as bivalents and proceed into the first meiotic metaphase. The two meiotic divisions occur, with extrusion of polar bodies, and only then do the male and female pronuclei fuse.[18,66]

As the oocyte leaves the ovary and proceeds down the oviduct, it becomes associated with several vitelline cells and a sperm emerging from the seminal receptacle. These all come together in the area of the ootype, and there are contributions from the cells of Mehlis' gland as well. It was long thought that Mehlis' gland contributed the shell material; the organ was called a *shell gland* in older texts. However, we now know that the bulk of the shell material is contributed by the vitelline cells, and the function of the Mehlis' gland remains obscure. In at least some species two distinct types of secretions are released—mucoid dense bodies and membranous bodies.[8] The mucoid dense bodies may serve as an adhesive mediating coalescence of vitelline globules to form the shell, or they may serve as a lubricant for the various components in the ootype. The membranous bodies aggregate to surround the oocyte, two or three vitelline cells, and some spermatozoa. Globules released from the vitelline cells coalesce against the membranous aggregate; therefore, the aggregate forms a kind of template for the shell material before stabilization. The results of Moczon and coworkers[81] are consistent with the idea that the Mehlis' gland secretions serve as an eggshell template, but there may be other functions as well.

"Stabilization" of structural proteins (e.g., sclerotin, keratin, and resilin) to impart qualities of physical strength and intertness occurs by crosslinkage to amino acid moieties in adjacent protein chains. Most trematode eggshells appear to be stabilized primarily by the quinone-tanning process of sclerotization (see Fig. 33.4).[25,118] In some trematodes keratin or elastin may be the major structural proteins.

DEVELOPMENT

At least two hosts serve in the life cycle of a typical digenetic fluke. One is a vertebrate (with a few exceptions) in which sexual reproduction occurs, and the other is usually a mollusc in which one or more generations are produced by an unusual type of asexual reproduction. A few species have asexual generations that develop in annelids.

This alternation of sexual and asexual generations in different hosts is one of the most striking biological phenomena. The variability and complexity of life cycles and ontogeny have stimulated the imaginations of zoologists for more than 100 years, creating a huge amount of literature on the subject. Even so, many mysteries remain, and research on questions of trematode life cycles remains active.

As many as six recognizably different body forms may develop during the life cycle of a single species of trematode (see p. 222 for summary). In a given species certain stages may be repeated during ontogeny, whereas stages found in other species may be absent. So many variations occur that few generalizations are possible. Therefore we will first examine each form separately and then illustrate the subject with a few examples.

Embryogenesis

Apart from the fact that the embryo produced by the sexual adult begins with a union of female and male gametes, the early embryogenesis of progeny produced asexually and sexually is basically similar. The first cleavage produces a **somatic cell** and a **propagatory cell** (stem cell), which are cytologically distinguishable. Daughter cells of the somatic cell will contribute to the body tissues of the embryo, whether miracidium, sporocyst, redia, or cercaria. Further divisions of the propagatory cell each may produce another somatic cell and another propagatory cell, but at some point, propagatory cell divisions produce only more propagatory cells. Each of these will become an additional embryo in the miracidium, sporocyst, or redia. In the developing cercaria the propagatory cells become the gonad primordia. Thus, the propagatory cells are the germinal cells in the asexually reproducing forms, and they give rise to the germ cells in the sexual adult. As noted previously, the miracidium metamorphoses into the sporocyst; however, if a sporocyst stage is absent in a particular species, redial embryos develop in the miracidium to be released after penetration of the intermediate host. The youngest embryos developing in a given stage are usually seen in the posterior portions of its body and are often referred to as **germ balls.**

The nature of this asexual reproduction has long been controversial; different workers have argued that it represented alternatively **budding, polyembryony,** or **parthenogenesis.** The view of early zoologists—that it is an example of metagenesis (an alternation of generations in which the asexual generation reproduces by budding)—was discarded when it was realized that the specific reproductive cells (the propagatory cells) are kept segregated in the germinal sacs. The most widely held opinion has been that the process is one of sequential polyembryony;[29,30] that is, multiple embryos are produced from the same zygote with no intervening gamete production as, for example, in monozygotic twins in humans. Whitfield and Evans[119] reviewed the evidence for parthenogenesis and found it insubstantial. They felt that the asexual reproduction in Digenea "most probably represents a budding process in which the development of the buds is initiated by the division of diploid totipotent (propagatory) cells."

Larval and Juvenile Development

• Egg (Shelled Embryo)

The structure referred to as an *egg* of trematodes is not an ovum but the developing (or developed) embryo enclosed by its shell, or capsule. The egg capsule of most flukes has an operculum at one end, through which the larva eventually will escape. It is not clear how the **operculum** is formed, but it appears that the embryo presses pseudopodiumlike processes against the inner surface of the shell while it is being formed, thereby forming a circular groove. An operculum is absent in the eggshell of blood flukes. Considerable variation exists in the shape and size of fluke eggs, as well as in the thickness and coloration of the capsules.

In many species the egg contains a fully developed miracidium by the time it leaves the parent; in others development has advanced to only a few cell divisions by that time. In *Heronimus* the miracidium hatches while still in the uterus. For eggs that embryonate in the external environment, certain factors influence embryonation. Water is necessary, since the eggs desiccate rapidly in dry conditions. Development is stimulated by high oxygen tension, although eggs can remain viable for long periods under conditions of low oxygen. Eggs of *Fasciola hepatica* will not develop outside a pH range of 4.2 to 9.0.[97] Temperature is critical, as would be expected. Thus, *F. hepatica* requires 23 weeks to develop at 10°C, whereas it takes only eight days at 30°C. However, above 30°C development again slows and completely stops at 37°C. Eggs are killed rapidly at freezing. Light may be a factor influencing development in some species, but this has not been thoroughly investigated.

Eggs of many species hatch freely in water, whereas others hatch only when eaten by a suitable intermediate host. Light and osmotic pressure are important in stimulation of hatching for species that hatch in water, and osmotic pressure, carbon dioxide tension, and probably host enzymes initiate hatching in those which must be eaten. The eggs of *Transversotrema patialense* hatch spontaneously and show what appears to be a circadian rhythm, hatching at the same time of day whether kept under constant light or a 12 hour light/12 hour dark cycle.[16] The time of hatching is correlated with the time the snail intermediate host is nearby. Light is also required for hatching of *Echinostoma caproni* eggs, which likewise show a circadian hatching pattern.[3]

The miracidium of *F. hepatica* within its capsule is surrounded by a thin **vitelline membrane,** which also encloses a padlike **viscous cushion** between the anterior end of the miracidium and the operculum.[120] Light stimulates hatching activity. Apparently the miracidium releases some factor that alters the permeability of the membrane enclosing the viscous cushion. The latter structure contains a mucopolysaccharide that becomes hydrated and greatly expands the volume of the viscous cushion. The considerable increase in pressure within the capsule causes the operculum to pop open, remaining attached at one point, and the miracidium rapidly escapes, propelled by its cilia. The nonoperculated capsules of *Schistosoma* are fully embryonated when passed from the host, and they hatch spontaneously in fresh water. The miracidia release substantial quantities of leucine aminopeptidase, and this enzyme probably helps digest the capsule from the inside.[122] Unlike leucine aminopeptidases

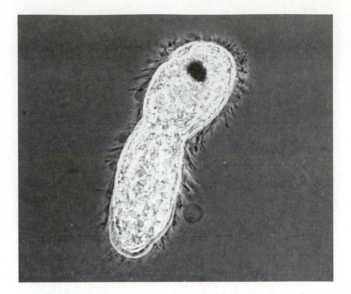

FIGURE 15.18

Miracidium of *Alaria* sp.
Courtesy of Jay Georgi.

from other sources, the enzyme produced by schistosome miracidia is inhibited by NaCl, which would prevent hatching while in the body of the host.

• Miracidium

The typical miracidium (Fig. 15.18) is a tiny, ciliated organism that could easily be mistaken for a protozoan by the casual observer. It is piriform, with a retractable **apical papilla** at the anterior end. The apical papilla has no cilia but bears five pairs of duct openings from glands and two pairs of sensory nerve endings (Fig. 15.19). The gland ducts connect with **penetration glands** inside the body. A prominent **apical gland** can be seen in the anterior third of the body. This probably secretes histolytic enzymes. An apical stylet is present on some species, and spines are found on others. The sensory nerve endings connect with nerve cell bodies that in turn communicate with a large ganglion. Miracidia have a variety of sensory organs and endings, including adaptations for photoreception, chemoreception, tangoreception, and statoreception.

The outer surface of a miracidium is covered by flat, ciliated epidermal cells, the number and shape of which are constant for a species. Underlying the surface are longitudinal and circular muscle fibers. Cilia are restricted to protruding **ciliated bars** in the genus *Leucochloridiomorpha* (Brachylaimidae) and family Bucephalidae, and they are absent altogether in the families Azygiidae and Hemiuridae. One or two pairs of protonephridia are connected to a pair of posterolateral excretory pores.

In the posterior half of the miracidium are found propagatory cells, or germ balls (embryos), which will be carried into the sporocyst stage to initiate further individuals.

Free-swimming miracidia are very active, swimming at a rate of about 2 mm per second, and they must find a suitable molluscan host rapidly, since they can survive as free-living organisms for only a few hours. In many cases the mucus produced by the snail is a powerful attractant for miracidia.[21]

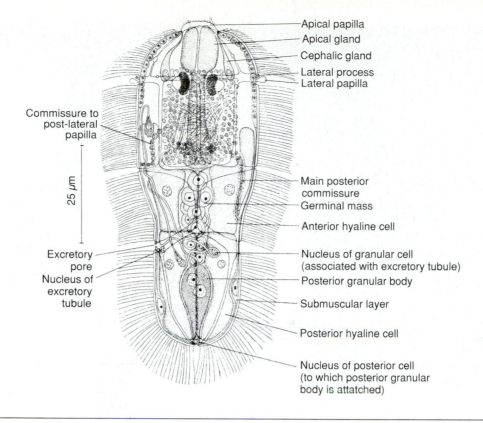

Apical papilla
Apical gland
Cephalic gland
Lateral process
Lateral papilla

Commissure to
post-lateral
papilla

25 μm

Main posterior
commissure
Germinal mass

Anterior hyaline cell

Nucleus of granular cell
(associated with excretory tubule)

Posterior granular body

Excretory
pore
Nucleus of
excretory
tubule

Submuscular layer

Posterior hyaline cell

Nucleus of posterior cell
(to which posterior granular
body is attatched)

FIGURE 15.19

Miracidium of *Neodiplostomum intermedium,* dorsal view.

From J. C. Pearson, "Observation on the morphology and life cycle of *Neodiplostomum intermedium,*" in *Parasitology* 51:133–172. Copyright © 1961. Reprinted with permission of the publisher.

On contacting a proper mollusc, the miracidium attaches to it with the apical papilla, which actively contracts and extends, in an augerlike motion. Cytolysis of snail tissues can be seen as the miracidium embeds itself deeper and deeper. As penetration proceeds, the miracidium loses its ciliated epithelium, although this may be delayed until penetration is complete. The miracidium takes about 30 minutes to complete penetration and begin the next phase of its life cycle as the sporocyst.

Miracidia of many species will not hatch until they are eaten by the appropriate snail, after which they penetrate the snail's gut.

• Sporocyst

Metamorphosis of the miracidium into the sporocyst involves extensive changes. In addition to the loss of the ciliated epithelial cells, there is a formation of the new tegument with its microvilli.[31] The sporocysts retain the subtegumental muscle layer and protonephridia of the miracidium, but all other miracidial structures generally disappear. The sporocyst has no mouth or digestive system; it absorbs nutrients from the host tissue, with which it is in intimate contact, and the entire structure serves only to nurture the developing embryos. The sporocyst (or other stage with embryos developing within it—that is, the miracidium or redia) may be referred to as a **germinal sac.** Often sporocysts grow near the site of penetration, such as foot, antenna, or gill, but they may be found in any tissue, depending on the species, and sometimes they may become very slender and extended, branched, or highly ramified.

The embryos in the sporocyst may develop into another sporocyst generation, **daughter sporocysts;** into a different form of germinal sac, the redia; or directly into cercariae (Fig. 15.20).

Leucochloridium paradoxum has a specialized sporocyst with a fascinating adaptation that evidently enhances transfer to its bird definitive host.[94] The sporocyst is divided into three parts: a central body located in the snail's hepatopancreas where the embryos are produced, a broodsac lying in the head-foot of the snail and entering its tentacles, and a tube connecting the broodsac to the central body.[91] The embryos pass from the central body through the tube to the broodsac, where they mature into cercariae. The sporocyst within the snail's tentacles causes the tentacles to enlarge, become brightly colored, and pulsate rapidly. The effect is analogous to a flashing neon sign attracting the birds to "Dine here!"

• Redia

Rediae (Fig. 15.21) burst their way out of the sporocyst or leave through a terminal birth pore and usually migrate to the hepatopancreas or gonad of the molluscan host. They are commonly elongated and blunt at the posterior end and may have one or more stumpy appendages called **procrusculi.** More active than most sporocysts, they crawl about within their host. They have a rudimentary but functional digestive system, consisting of a mouth, muscular pharynx, and short, unbranched gut. Rediae pump food into their gut by means of the pharyngeal muscles, as previously described in adults. They not only feed on host tissue but also can prey on sporocysts of their

FIGURE 15.20

Ruptured sporocyst releasing furcocercous cercariae.
Courtesy of James Jenson.

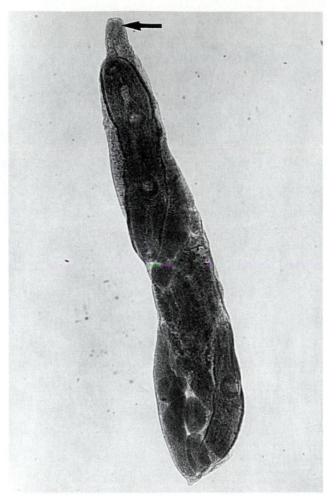

FIGURE 15.21

A redia. Note the large, muscular pharynx (*arrow*) just inside the mouth.
Courtesy of Warren Buss.

own or other species.[75] The luminal surface of their gut is greatly amplified by flattened, lamelloid or ribbonlike processes.[70] The gut cells are apparently capable of phagocytosis. The outer surface of the tegument also functions in absorption of food, and it is provided with microvilli or lamelloid processes.

The embryos in the redia develop into daughter rediae or into the next stage, the cercaria, which emerges through a birth pore near the pharynx. The epithelial lining of the birth pore in *Cryptocotyle lingua* (and probably other species) is highly folded so that it can withstand the extreme distortion produced by the exit of a cercaria.[61] It appears that rediae must reach a certain population density before they stop producing more rediae and begin producing cercariae: Young rediae have been transplanted from one snail to another through more than 40 generations without cercariae being developed.[35] This is an interesting parallel to certain free-living invertebrates that reproduce parthenogenetically only as long as certain environmental conditions are maintained.[19]

• Cercaria

The cercaria represents the juvenile stage of the vertebrate-inhabiting adult. There are many varieties of cercariae, and most have specializations that enable them to survive a brief free-living existence and make themselves available to their definitive or second intermediate hosts (Fig. 15.22). Most have tails that aid them in swimming, but many have only rudimentary tails or none at all; these cercariae can only creep about, or they may remain within the sporocyst or redia that produced them until they are eaten by the next host.

The structure of a cercaria is easily studied, and cercarial morphology often has been considered a more reliable indication of phylogenetic relationships among families than has the morphology of adults. Cercariae are widely distributed, abundant, and easily found; hence they have attracted much attention from zoologists. The name *Cercaria* can be used properly in a generic sense for a species in which the adult form is unknown, as is done with the term *Microfilaria* among some nematodes.

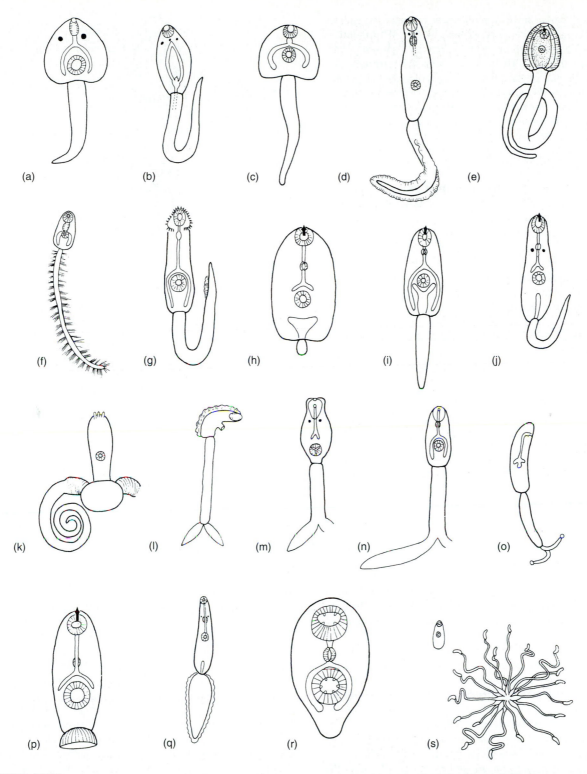

FIGURE 15.22

A few of the many types of cercariae. (*a*) Amphistome cercaria; (*b*) monostome cercaria; (*c*) gymnocephalous cercaria; (*d*) gymnocephalous cercaria of pleurolophocercous type; (*e*) cystophorous cercaria; (*f*) trichocercous cercaria; (*g*) echinostome cercaria; (*h*) microcercous cercaria; (*i*) xiphidiocercaria; (*j*) ophthalmoxiphidiocercaria; (*k–o*) furocercous types of cercariae: (*k*) gasterostome cercaria; (*l*) lophocercous cercaria; (*m*) apharyngeate furcocercous cercaria; (*n*) pharyngeate furcocercous cercaria; (*o*) apharyngeate monostome furcocercous cercaria without oral sucker; (*p*) cotylocercous cercaria; (*q*) rhopalocercous cercaria; (*r*) cercariae; (*s*) rattenkönig, or rat-king, cercariae.

Most cercariae have a mouth near the anterior end, although it is midventral in the Bucephalidae. The mouth is usually surrounded by an oral sucker, and a prepharynx, muscular pharynx, and forked intestine are normally present. Each branch of the intestine is simple, even those that are ramified in the adult. Many cercariae have various glands opening near the anterior margin, often called *penetration glands* because of their assumed function. Cercariae of most trematodes probably have glands that serve several functions; schistosome cercariae have no less than four distinguishable types[109]:

1. **Escape glands.** So-called because their contents are expelled during emergence of the cercaria from the snail, but their function is not known.
2. **Head gland.** Secretion is emitted into the matrix of the tegument and is thought to function in the postpenetration adjustment of the schistosomule.
3. **Postacetabular glands.** Produce mucus, help cercariae adhere to surfaces, and have other possible functions.
4. **Preacetabular glands.** Secretion contains calcium and a variety of enzymes including a protease. The function of these glands seems most important in actual penetration of host skin.

Secretory **cystogenic cells** are particularly prominent in cercariae that will encyst on vegetation or other objects.

Many morphological variations exist in cercariae that are constant within a species (or larger taxon); thus, certain descriptive terms are of value in categorizing the different varieties. Some of the more commonly used terms are **xiphidiocercaria** (have a stylet in anterior margin of oral sucker), **ophthalmocercaria** (with eyespots), **cercariaeum** (without a tail), **microcercous cercaria** (with a small, knoblike tail), and **furcocercous cercaria** (with a forked tail).

The excretory system is well-developed in the cercaria. In some cercariae the excretory vesicle empties through one or two pores in the tail. Because the number and arrangement of the protonephridia are constant for a species, these are important taxonomic characters. Each flame cell has a tiny **capillary duct** that joins with others to form an **accessory duct.** The accessory ducts join the **anterior** or **posterior collecting ducts,** whose junction forms a **common collecting duct** on each side (Fig. 15.23). When the common collecting ducts extend to the region of the midbody and then fuse with the excretory vesicle, the cercaria is called **mesostomate.** If the tubules extend to near the anterior end and then pass posteriorly to join the vesicle, the cercaria is known as **stenostomate.** The number and arrangement of flame cells can be expressed conveniently by the **flame cell formula.** For example, 2[(3 + 3) (3 + 3)] means that both sides of the cercaria (2) have three flame cells on each of two accessory tubules (3 + 3) on the anterior collecting tubule, plus the same arrangement on the posterior collecting tubule (3 + 3). The flame cell formula for the cercaria in Figure 15.23 would be 2[(3 + 3 + 3) (3 + 3 + 3)].

Mature cercariae emerge from the mollusc and begin to seek their next host. Many remarkable adaptations can be found among them that enable them to do this. Most cer-

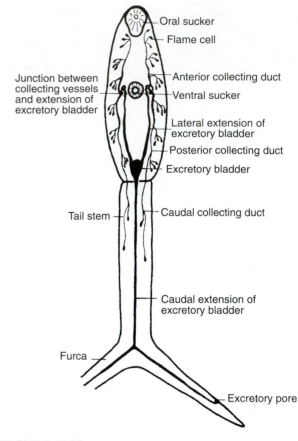

FIGURE 15.23

Diagrammatic representation of the excretory system of a fork-tailed cercaria. The caudal flame cells are absent from the tail in nonfurcate forms.

From D. A. Erasmus, *The Biology of Trematodes.* Copyright © 1972 Crane-Russak, & Co., New York, NY. Reproduced with permission. All rights reserved.

cariae are active swimmers, of course, and rely on chance to place them in contact with an appropriate organism. Some species are photopositive, dispersing themselves as they swim toward the surface of the water, but then become photonegative, returning to the bottom where the next host is. Some opisthorchiform cercariae remain quiescent on the bottom until a fish swims over them; the resulting shadow activates them to swim upward. Some plagiorchiform cercariae cease swimming when in a current; hence, when drawn over the gills of a crustacean host, they can attach and penetrate rather than swim on. Large, pigmented azygiids and bivesiculids are enticing to fish, which eat them and become infected. Some cercariae float; some unite in clusters; some creep at the bottom. Cercariae of *Schistosoma mansoni,* which directly penetrate the warm-blooded definitive host, concentrate in a thermal gradient near a heat source (34°C).[23] In the case of certain cystophorous hemiurid cercariae, the body is withdrawn into the tail, which becomes a complex injection device.[78] The second intermediate host of the trematode is a copepod crustacean, which attempts to eat the caudal cyst bearing the cercaria. When the narrow end of the cyst is broken by the mandibles of the copepod, a delivery tube is rapidly everted into the mouth of the copepod, piercing its midgut, and then the cercaria slips through the delivery tube into the hemocoel of the crustacean! These and many more adaptations help the trematode reach its next host.

• Mesocercaria

Species of the strigeiform genus *Alaria* have a unique larval form, the **mesocercaria,** which is intermediate between the cercaria and metacercaria (see Fig. 16.3).

• Metacercaria

Between the cercaria and the adult is a quiescent stage, the metacercaria, although this stage is absent in the families of blood flukes. The metacercaria is usually encysted, but in *Brachycoelium, Halipegus, Panopistus,* and others it is not. Most metacercariae are found in or on an intermediate host, but some (Fasciolidae, Notocotylidae, and Paramphistomidae) encyst on aquatic vegetation, sticks, and rocks or even freely in the water.

The cercaria's first step in encysting is to cast off its tail. Cyst formation is most elaborate in metacercariae encysting on inanimate objects or vegetation. The cystogenic cells of *F. hepatica* are of four types, each with the precursors of a different cyst layer. Metacercariae encysting in intermediate hosts have thinner and simpler cyst walls, with some components contributed by the host.

Development that occurs in the metacercaria varies widely according to species, from those in which a metacercaria is absent (*Schistosoma*) to those in which the gonads mature and viable eggs are produced (*Proterometra*). Often some amount of development is necessary in the metacercaria before the organism is infective for the definitive host. We can arrange metacercariae in three broad groups on this basis:[34]

1. Species whose metacercariae encyst in the open, on vegetation and inanimate objects, such as *Fasciola* spp. Members of this group can infect the definitive host almost immediately after encystment, in some cases within only a few hours, with no growth occurring.

2. Species that do not grow in the intermediate host but that require at least several days of physiological development to infect the definitive host, such as Echinostomatidae.

3. Species whose metacercariae undergo growth and metamorphosis before they enter their resting stage in the second intermediate host and usually require a period of weeks for this development, such as Diplostomidae.

These developmental groups are correlated with the longevity of the metacercariae: Those in group 1 must live on stored food and can survive the shortest time before reaching the definitive host, whereas those in groups 2 and 3 obtain some nutrient from their intermediate hosts and so can remain viable for the longest periods—in one case up to seven years.

After the required development the metacercaria goes into the quiescent stage and remains in readiness to excyst on reaching the definitive host. *Zoogonus lasius,* a typical example of group 2, has a high rate of metabolism for the first few days after infecting its second intermediate host, a nereid polychaete, and then drops to a low level, only to return to a high rate on excystation.[116] Metacercariae of *Bucephalus haimeanus* remain active in the liver of their fish host and increase threefold in size. They take up nutrients from degenerating liver cells, including large molecules, by pinocytosis.[55] Metacercariae of *Clinostomum marginatum* take up glucose both by facilitated diffusion and by active transport.[114]

The metacercarial stage has a high selective value for most trematodes. It can provide a means for transmission to a definitive host that does not feed on the first intermediate host or is not in the environment of the mollusc, and it can permit survival over unfavorable periods, such as a season when the definitive host is absent.

Development in the Definitive Host

Once the cercaria or metacercaria has reached its definitive host, it matures in a variety of ways: either by penetration (if a cercaria) or by excystation (if a metacercaria) and then by migration, growth, and morphogenesis to reach gamete production. If the species does not have a metacercaria and the cercaria penetrates the definitive host directly, as in the schistosomes, the most extensive growth, differentiation, and migration will be necessary. At the other extreme some species acquire adult characters while in the metacercarial stage, the gonads may be almost mature, or some eggs may even be present in the uterus; and little more than excystation is needed before the production of progeny (*Bucephalopsis, Coitocaecum, Transversotrema*). A very few species (*Proctoeces maculatus*[1] and *Proterometra dickermani*) reach sexual reproduction in the mollusc and apparently do not have a vertebrate definitive host. Others may mature in another invertebrate; for example, several species of Macroderoididae mature in leeches (Fig. 15.3)[115] and *Allocorrigia filiformis* matures in the antennal gland of a crayfish.[113] These are probably examples of neoteny.

Normally development in the definitive host begins with excystation of the metacercaria, and species with the heaviest, most complex cysts, such as those with cysts on vegetation (for example, *Fasciola hepatica*), seem to require the most complex stimuli for excystation. The outer cyst of *F. hepatica* is largely removed by digestive enzymes, but escape from the inner cyst requires the presence of a temperature of about 39°C, a low oxidation-reduction potential, carbon dioxide, and bile. Comparable conditions produce maximal excystment of *Himasthla quissetensis*.[67] This combination of conditions is not likely to be present anywhere but in the intestine of an endothermic vertebrate; like the conditions required for the hatching or exsheathment of some nematodes, the requirements constitute an adaptation that avoids premature escape from the protective coverings. Such an adaptation is less important to metacercariae that are not subjected to the widely varying physical conditions of the external environment, such as those encysted within a second intermediate host. These have thinner cysts and excyst on treatment with digestive enzymes. A number of species require the presence of a bile salt(s) or excyst more rapidly in its presence.[6,41,44,108] Some metacercariae may release enzymes that assist in excystment.[62]

After excystation in the intestine, a more or less extensive migration is necessary if the final site is in some other organ. The main sites of such parasites are the liver, lungs, and circulatory system. Probably the most common way to reach the liver is by way of the bile duct (*Dicrocoelium dendriticum*), but *F. hepatica* burrows through the gut wall into the peritoneal cavity and finally, wandering through the tissues, reaches the liver. *Clonorchis sinensis* usually penetrates the gut wall and is carried to the liver by the hepatoportal system. *Paragonimus westermani* penetrates the gut wall, undergoes a developmental phase of about a week in the abdominal wall, and then reenters the abdominal cavity and makes its way through the diaphragm to the lungs.

Trematode Transitions

A remarkable physiological aspect of trematode life cycles is the sequence of totally different habitats in which the various stages must survive, with physiological adjustments that must often be made extremely rapidly. As the egg passes from the vertebrate, it must be able to withstand the rigors of the external environment in fresh water or seawater, if only for a period of hours, before it can reach haven in the mollusc. There conditions are quite different from both the water and the vertebrate. The trematode's physiological capacities must again be readjusted on escape from the intermediate host and again on reaching the second intermediate or definitive host. Environmental change may be somewhat less dramatic if the second intermediate host is a vertebrate, but often it is an invertebrate. Although the adjustments must be extensive, the nature of these physiological adjustments made by trematodes during their life cycles has been little investigated, the most studied trematode in this respect being *Schistosoma*.[22]

Penetration of the definitive host is a hazardous phase of the life cycle of schistosomes, and it requires an enormous amount of energy. The hazards include a combination of the dramatic changes in the physical environment, in the physical and chemical nature of the host skin through which the schistosome must penetrate, and in host defense mechanisms. Depending on the host species, losses at this barrier may be as high as 50%, and the glycogen content of the newly penetrated schistosomules (**schistosomule** is the name given the young developing worm) is only 6% of that found in cercariae. Among the most severe physical conditions the organism must survive is the sequence of changes in ambient osmotic pressure. The osmotic pressure of fresh water is considerably below that in the snail, and that in the vertebrate is twice as great as in the mollusc. Assuming that the osmotic pressure of the cercarial tissues approximates that in the snail, the trematode must avoid taking up water after it leaves the snail and avoid a serious water loss after it penetrates the vertebrate. Aside from the possible role of the osmoregulatory organs (protonephridia), there appear to be major changes in the character, and probably permeability, of the cercarial surface. The cercarial surface is coated with a fibrillar layer, or glycocalyx, which is lost on penetration of the vertebrate, and with it is lost the ability to survive in fresh water; 90% of schistosomules recovered from mouse skin 30 minutes after penetration die rapidly if returned to fresh water. That chemical changes have occurred is indicated by the fact that the schistosomule surface is much less easily dissolved by a number of chemical reagents, including 8 M urea, than that of the cercaria. The antigenic determinants of the schistosomule as compared with the cercaria are changed as well. When cercariae are incubated in immune serum, a thick envelope called the CHR (*cercarienhüllenreaktion*) forms around them, but schistosomules do not give this reaction.

In several cases cercarial attraction to the next host is mediated by substances different from those that stimulate actual penetration. *Opisthorchis viverrini* is attracted to its fish host by glycosaminoglycans, but it penetrates in response to proteins.[85] These signals differ from those attracting *Acanthostomum brauni* to its fish host (certain glycoproteins) and stimulating it to penetrate (fatty acids and a protein in the mucus on the fish surface). Schistosome cercariae are apparently attracted to host skin by the amino acid arginine,[50] whereas the most important stimulus for actual penetration is the skin lipid film, specifically essential fatty acids, such as linoleic and linolenic acids, and certain nonessential fatty acids.[99] Human skin surface lipid applied to the walls of their glass container will cause cercariae to attempt to penetrate it, lose their tails, evacuate their preacetabular glands, and become intolerant to water. The presence of the penetration-stimulating substances causes loss of osmotic protection and a reduction of the CHR, even in cercariae free in the water.[51] Successful penetration and transformation have been correlated with cercarial production of eicosanoids, such as leukotrienes and prostaglandins (fatty acid derivatives with potent pharmacological activity).[98] These eicosanoids may enable the schistosomules to evade the host defenses by inhibiting superoxide production by neutrophils.[82]

After penetration the tegument of developing schistosomules undergoes a remarkable morphogenesis. Within 30 minutes numerous subtegumental cells have connected with the distal cytoplasm and are passing abundant "laminated bodies" into it. These bodies each have *two* trilaminar limiting membranes (hence heptalaminar). The bodies move to the surface of the tegument to become the new tegumental outer membrane; the old cercarial outer membrane, along with its remaining glycocalyx, is cast off. The schistosomule outer membrane is almost entirely heptalaminar three hours after penetration. During the next two weeks the main changes in the tegument are a considerable increase in thickness and the development of many invaginations and deep pits. The pits increase the surface area fourfold between 7 and 14 days after penetration. It is likely that this represents an adaptation for nutrient absorption through the tegument.

Summary of Life Cycle

In summary the basic pattern of the life cycle of digenetic trematodes is egg → miracidium → sporocyst → redia → cercaria → metacercaria → adult.

The student should learn this pattern well, because it is the theme on which to base the variations. The most common variations are (1) more than one generation of sporocysts or

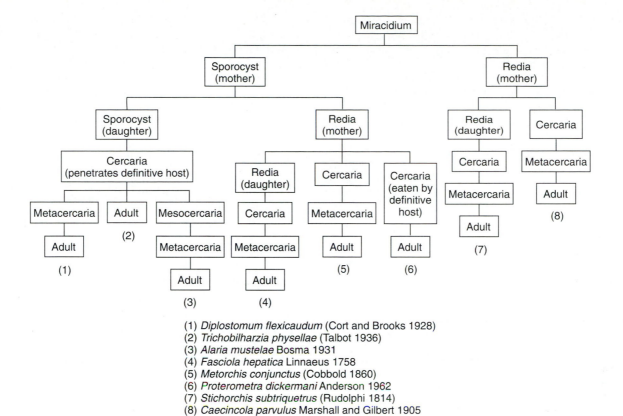

(1) *Diplostomum flexicaudum* (Cort and Brooks 1928)
(2) *Trichobilharzia physellae* (Talbot 1936)
(3) *Alaria mustelae* Bosma 1931
(4) *Fasciola hepatica* Linnaeus 1758
(5) *Metorchis conjunctus* (Cobbold 1860)
(6) *Proterometra dickermani* Anderson 1962
(7) *Stichorchis subtriquetrus* (Rudolphi 1814)
(8) *Caecincola parvulus* Marshall and Gilbert 1905

FIGURE 15.24

Some life cycles of digenetic trematodes.

Modified from S. C. Schell, *How to Know the Trematodes.* © 1970 Wm. C. Brown Publishers, Dubuque, Iowa.

rediae, (2) deletion of either sporocyst or redial generations, and (3) deletion of metacercaria. Much less common are cases in which miracidia are produced by sporocysts and forms with adult morphology in the mollusc that produce cercariae (these in turn lose their tails and produce another generation of cercariae).[63] Figure 15.24 shows some possible life cycles.

METABOLISM

Energy Metabolism

Metabolism of trematodes has been accorded considerable study.[13,76,101,106,110] Compared with that of certain vertebrates, however, trematode metabolism is meagerly known, and the metabolism of larval stages has received scant attention. Furthermore, because of size, availability, and/or medical importance, the metabolism of *Schistosoma* and *Fasciola* has been much more thoroughly investigated than that of other species.

The overall scheme of nutrient catabolism is surprisingly similar in adult trematodes, cestodes, and even nematodes. Their main sources of energy are from the degradation of carbohydrate from glycogen and glucose. They are facultative anaerobes, and even in the presence of oxygen they excrete large amounts of short-chain acid end products (Fig.

15.25).[13,56] In other words the energy potential in the glucose molecule is far from completely harvested. Why the worms should excrete such reduced compounds, from which so much additional energy could be derived, is not at all obvious. In the past most investigators concluded that since the parasites had what was, for practical purposes, an inexhaustible food supply, they simply did not need to catabolize the glucose molecule completely. Current workers, unsatisfied with that conclusion, have maintained that "there is a payoff somewhere."[13] Some possibilities follow.

The "usual" glycolytic pathway produces pyruvate, which in most aerobic organisms is then decarboxylated and condensed with coenzyme A to form acetyl CoA. The acetyl CoA enters the Krebs cycle, reactions that release carbon dioxide and electrons. The electrons are passed through a series of oxidation-reduction reactions (electron transport pathway) where energy is reaped in the generation of ATP, and the final electron acceptor is oxygen. In the absence of oxygen, pyruvate becomes the final electron acceptor and is reduced to the end product of anaerobic glycolysis, lactate. Glycolysis yields only $1/18$ the amount of energy resulting from full oxidation of the glucose molecule through the Krebs cycle and electron transport system. Although adults of many parasitic worms do excrete lactate (sometimes almost exclusively), many have a further series of reactions that derive

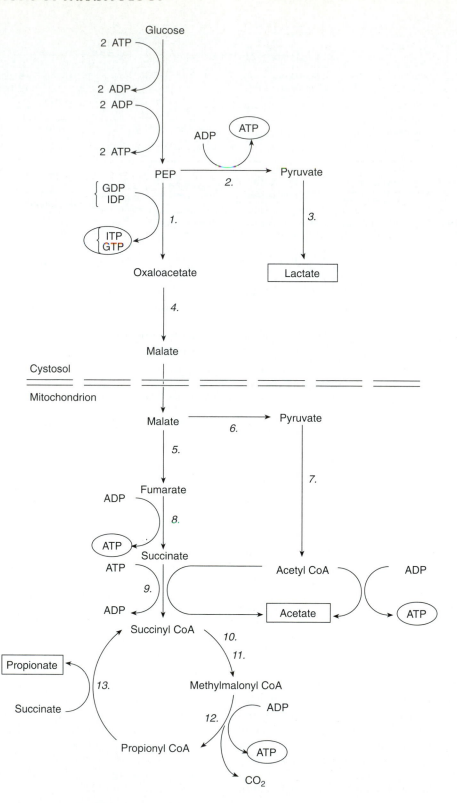

FIGURE 15.25

Possible overall pathway for energy metabolism of *Fasciola hepatica.* Compounds in boxes represent end products; circled compounds are net energy derived in phosphate bonds. **1,** PEP carboxykinase; **2,** pyruvate kinase; **3,** lactate dehydrogenase; **4,** malate dehydrogenase; **5,** fumarate hydratase; **6,** malate dehydrogenase (decarboxylating); **7,** pyruvate dehydrogenase; **8,** fumarate reductase; **9,** succinyl CoA synthetase; **10,** methylmalonyl CoA mutase; **11,** methylmalonyl CoA racemase; **12,** propionyl CoA carboxylase; **13,** acyl CoA transferase.

Source: C. M. Lloyd, "Energy Metabolism and its Regulation in the Adult Liver Fluke *Fasciola hepatica,*" in *Parasitology,* 93:217–248, 1986.

some additional energy (Fig. 15.25).[68,69,92] Carbon dioxide is fixed into phosphoenolpyruvate (PEP), which is reduced to malate and then passed into the mitochondria where these reactions take place. The end products are varying amounts of acetate, succinate, and propionate, which are excreted. (Additional reduced end products are excreted by some worms; for example see p. 377.)

Several trematode species can catabolize certain of the Krebs cycle intermediates and they have various enzymes in that cycle. In fact all enzymes necessary for a functional citric acid cycle are present in *F. hepatica*,[76] but the levels of aconitase and isocitrate dehydrogenase activities are quite low. There is evidence for a functional Krebs cycle in schistosomes,[24] and some investigators believe that aerobic respiration contributes significantly to the energy requirements of schistosomes; it is, after all, 18 times more efficient.[13] Cheah and Prichard[20] concluded that *F. hepatica* has a branched electron transport system: one branch being the classical mammalian type, with cytochrome a_3 as terminal electron acceptor and the other with cytochrome o. Cheah and Prichard suggested that the branched chain is an adaptation of large parasites to low environmental oxygen. However, NADH–cytochrome c oxidoreductase, succinate–cytochrome c oxidoreductase, NADH oxidase, and cytochrome c–oxygen oxidoreductase were all present in the trematode. Barrett[2] concluded that parasitic helminths generally are capable of oxidative phosphorylation.

Whether or not these organisms oxidize any glucose completely, they nevertheless excrete copious amounts of the short-chain acids. Bryant[13] speculated that, like some microorganisms, the worms might obtain an energetic advantage by pumping out acids. The microorganisms establish a proton gradient across their cell membranes and use the energy of that gradient to generate ATP. Could the mitochondria of parasitic helminths retain that ability? Alternatively, considering that the "waste" molecules of the worm remain quite usable by the host, the explanation may lie in the continuing evolution of the host-parasite relationship and physiological interactions with the host.

Although little is known of the lipid metabolism, we have no evidence that lipids are used as energy sources or energy storage compounds. However, trematodes may contain considerable lipid, and sizable quantities may be excreted. *Fasciola hepatica* excretes about 2% of its net weight per day as polar and neutral lipids (including cholesterol and its esters), and this is mainly by way of its excretory system.[17]

Consequently it is not surprising that digeneans contain large amounts of stored glycogen: 9% to 30% of dry weight, according to species. Amounts in female *S. mansoni* are unusually low: only about 3.5% of dry weight. Although the glycogen content of cestodes may range higher than 30%, it is still surprising that trematodes, even tissue-dwelling species, store so much, since the availability of their food should not be subject to the vagaries of their host's feeding schedule, as it is with cestodes. In cases in which measurements have been performed, a large proportion of the trematode's glycogen is consumed under starvation conditions in vitro. In fact the maintenance of a high glycogen concentration in the worms may be of critical importance.

A pentose phosphate pathway may function in schistosomes, but critical enzymes for a glyoxylate cycle have not been found. In contrast the pentose cycle in *Fasciola* appears to be minimal, but enzymes necessary for the glyoxylate path are all present.[76] Schistosomes apparently have all the necessary enzymes for gluconeogenesis, but no one has been able to demonstrate that the process occurs.[112]

The astonishing ability of trematodes to survive the radical changes in environment requires important adjustments in their energy metabolism. Clearly the ability to derive every possible ATP from every glucose unit would be of great selective value to the free-swimming miracidium or cercaria that does not feed. Thus, the miracidia and the cercariae are obligate aerobes in all species investigated so far, and they are killed by short exposures to anaerobiosis. The cercariae of *S. mansoni* oxidize pyruvate rapidly; carbon dioxide is produced from all three of the pyruvate carbons. The miracidium may have a functional Krebs cycle within the egg. Study of the sporocysts is difficult because of the problem of separating host from parasite tissue; however, the same drugs that kill sporocysts within the snail affect the adults, and sporocysts (or the cercariae within) produce lactate under aerobic conditions. Therefore the metabolism of the sporocysts is apparently like that of the adults. Use of inhibitors suggests that a functional Krebs cycle is present in cercariae; they exhibit a Pasteur effect, and cytochromes a/a_3, b, and c are all present. Immediately after penetration schistosome energy metabolism seems to undergo a major adjustment, and the adjustment occurs entirely within the "head" of the cercaria and not the discarded tail.[60] The ability to use pyruvate drops dramatically, and the schisotosomules produce lactate aerobically.[24] Developmental studies on the metabolism of other trematodes would be very interesting, but few have been reported. The oxygen consumption of adult *Gynaecotyla adunca,* an intestinal parasite of fish and birds, drops sharply 24 hours after excystation and then even more after 48 and 72 hours.[116] Juvenile *F. hepatica,* living in the liver parenchyma, have a cyanide-sensitive respiration but are facultative anaerobes, so they seem to be in transition from the aerobic cercariae to the anaerobic adults.[111]

Transamination ability appears limited, but the α-ketoglutarate–glutamate transaminase reaction is active.[32,117]

Ammonia and urea are both important end products in degradation of nitrogenous compounds in *Fasciola* and *Schistosoma* spp., and both worms excrete several amino acids as well. A full complement of the enzymes necessary for the ornithine-urea cycle is not present in *Fasciola;* the urea produced must be by other pathways.[64]

• Effect of Drugs on Energy Metabolism

Niridazole, an antischistosomal drug, causes glycogen depletion in the schistosome, and its mode of action is very interesting. The glucose moieties in glycogen are mobilized for glycolysis by the action of glycogen phosphorylase, as in other systems, and the extent of the mobilization is controlled by how much of the enzyme is in the physiologically active a form. Niridazole inhibits the conversion of phosphorylase a to the inactive b form; thus, the phosphorolysis of glycogen is uncontrolled, the glycogen stores of the worm are depleted, and it is finally killed if the niridazole concentration is maintained.[15] As with any good chemotherapeutic agent, the corresponding host enzyme is not affected.

Organic trivalent antimonials, traditional antischistosomal drugs, inhibit a critical enzyme in glycolysis, phosphofructokinase (PFK). The PFK of the schistosomes is much more sensitive to the antimonials than is the corresponding host enzyme.[14] However, there is evidence that the action of PFK does not fully account for the effect of these drugs.[5,24] Antimonials have severe side effects on the host and have now been replaced by other compounds.

Synthetic Metabolism

Stimulated by the search for chemotherapeutic agents, researchers have studied purine and pyrimidine metabolism in *Schistosoma. Schistosoma mansoni* cannot synthesize purines de novo, but they are capable of de novo pyrimidine synthesis.[39] However, the worms probably depend on salvage pathways for supplies of both types of bases. Kurelec[71] showed that *F. hepatica* and *Paramphistomum cervi* could not synthesize carbamyl phosphate and concluded that they depended on their hosts for both pyrimidines and arginine. In light of the high arginine requirement of *S. mansoni,* it would seem probable that the situation is the same in that species.

The requirement of schistosomes for arginine is so high, in fact, that they reduce the level of serum arginine to almost zero in mice with severe infections. The worms more rapidly take up arginine than histidine, tryptophan, or methionine. Proline from the host is also rapidly consumed by the male schistosome through both the gut and the tegument, but only a little is absorbed by the tegument of the female.[102] Interestingly the proline consumed is concentrated in the ventral arms of the gynecophoral canal, the region of contact with the female. Glycogen concentrations in male and female schistosomes fluctuate in a parallel manner.[74] This and the foregoing suggest that the embrace of the male has a nutritive as well as sexual function.

We indicated earlier that trematodes excrete lipids; this seems a rather profligate custom, since it appears that they cannot synthesize their own complex lipids. The inability to synthesize fatty acids or sterols de novo may be a characteristic shared by trematodes and free-living flatworms.[79] Both *S. mansoni* and the planarian *Dugesia dorotocephala* can synthesize their complex lipids provided that they are supplied with a source of long-chain fatty acids. So far as known, no parasitic flatworm can synthesize long-chain fatty acids or sterols de novo.[43]

Biochemistry of the Tegument

Recognition that the tegument of schistosomes represents their barrier of defense against the host has led in recent years to much investigation of its structure and chemistry. This research has shown that the tegumental surface is active and complex. We have already mentioned the heptalaminate structure of the tegument in schistosomes (p. 222). The vesicles and granules in the distal cytoplasm appear to replace the outer membranes continuously, and turnover is quite rapid.[24] A variety of carbohydrates, including mannose, glucose, galactose, *N*-acetyl glucosamine, *N*-acetyl galactosamine, and sialic acid, are exposed on the surface. There are receptors for both host antigens (for example, blood group antigens) and antibodies, including IgG, IgA, and IgM. At least some of the antibodies that bind to the schistosome surface are not antischistosome antibodies; therefore these antibodies are apparently binding to Fc receptors. A number of enzymes have been detected, including alkaline phosphatase, ATPase, alkaline phosphodiesterase, glycerol triphosphatase, glucose-6-phosphatase, and a protease that can cleave the IgG bound there. Phospholipids and lipoproteins bound to the tegument may modulate the host immune response.[46]

• Praziquantel and the Tegument

It appears that the aggregate of host molecules bound on the tegument, plus the rapid turnover of tegumental membrane, effectively shields the schistosome from the host's immune defenses. It would follow that chemicals that could disrupt the integrity of the membrane could allow host immune effectors to recognize the worm. Such an action appears to be important in the effect of praziquantel, a drug that is highly effective against many flatworms. In addition, praziquantel affects permeability to calcium ions, allowing a rapid influx and resulting in a muscular tetany. Some evidence suggests that both the effects on Ca++ metabolism and on tegumental structure are necessary for the lethal effect of the drug.[5] Praziquantel is not effective against *Fasciola hepatica,* possibly because the tegument of *Fasciola* is much thicker than that of *Schistosoma.* Also, tetanic contraction of *Fasciola* in vitro requires 100 times greater concentration of praziquantel than that required for *Schistosoma.*[5]

PHYLOGENY OF DIGENETIC TREMATODES

Numerous schemes have been suggested for the origin of the Digenea. Various authors have derived the ancestral form from Monogenea, Aspidobothrea, and even insects.[90,105] However, most authorities today believe that trematodes share an ancestor with some of the free-living flatworms, probably rhabdocoels.[38,48] Whatever the ancestral digenean, any system of their phylogeny must rationalize the evolution of their complex life cycles in terms of natural selection, a most perplexing task.

Digeneans display much more host specificity to their molluscan hosts than to their vertebrate hosts. This suggests that they may have established themselves as parasites of molluscs first and then added a vertebrate host as a later adaptation. It is not difficult to imagine a small, rhabdocoel-like worm, invading the mantle cavity of a mollusc and feeding on its tissues. In fact the main hosts of known endocommensal rhabdocoels are molluscs and echinoderms. The stage of the protodigenean in the mollusc was probably a developmental one, with a free-living, sexually reproducing adult. There are several reasons for believing this. First, a free reproductive stage would be of selective value in dispersion and transferral to new hosts. Precisely this lifestyle is shown by *Fecampia,* a rhabdocoel symbiont of various marine crustaceans. Second, the possession of a cercarial stage is surprisingly ubiquitous among digeneans, and most of these are adapted for swimming. Those without tails show evidence that the structure has been secondarily lost.

If one grants that the present adult represents the ancestral adult, which was free living, it is clear that additional (asexual) multiplication in the mollusc would have been advantageous also, and the alternation of the two reproductive generations could have been established. It is likely that such free-living adults would often be eaten by fish, and individuals in the population that could survive and maintain themselves in the fish's digestive tract for a period of time would have selective advantage in extending their reproductive life. *Fecampia,* for example, dies after depositing its eggs.

Further evidence that the protodigenean was originally free living as an adult is demonstrated by the fluke still having to leave the snail to infect the next host. With few exceptions, it is incapable of infecting a definitive host while still in its first intermediate host, even if eaten. In most cases when a life cycle requires that the fluke be eaten within a mollusc, it has left its first host and penetrated a second to become infective. Some workers, however, feel that the protodigenean adult was a parasite of molluscs, as in modern aspidobothreans.[48,96]

It is likely that the miracidium represents the larval form of the fluke's ancestor; all digeneans still have them, even though they are not now all free swimming.

With the basic two-host cycle with two reproducing generations established in the protodigeneans, it is less difficult to visualize how further elaborations of the life cycle could have been selected for. The ecological value of the metacercaria, already mentioned, may have had some protective function as well.

We can assume that the digenean adaptation to vertebrate hosts has occurred relatively recently. Digenetic trematodes are very common in members of all classes of vertebrates except Chondrichthyes; extremely few species of digeneans are found in sharks and rays. The urea in the tissue of most elasmobranchs, which plays such an important role in their osmoregulation, is quite toxic to the flukes on which it has been tested. It is supposed that elasmobranchs did not have digenean parasites when that particular osmoregulatory adaptation was evolved and that the urea has since proved a barrier to invasion of the elasmobranch habitat by flukes. The situation is quite the opposite with the cestodes; sharks and rays have a rich tapeworm fauna, and their cestodes either tolerate the urea or degrade it.[93]

Cladistic analysis yields the relationships among the groups of Digenea shown in Figure 15.26.

CLASSIFICATION OF SUBCLASS DIGENEA
The most widely accepted system of classification of the higher taxa has been based on that of LaRue,[72] using characteristics of the excretory bladder (superorders Anepitheliocystidia and Epitheliocystidia). Due to doubts cast upon this system by ultrastructural studies,[48] we will not here divide the Digenea at the superorder level. Diagnoses are adapted from the synapomorphies and diagnoses listed by Brooks and McLennan.[12]

Subclass Digenea
With the characteristics of the Trematoda (p. 189); in addition with primitive character states: first larval stage a miracidium; miracidium with a single pair of flame cells; saclike sporocyst stage in snail host following miracidium; cercarial stage developing in snail host following mother sporocyst; cercariae with tail; cercariae with primary excretory pore at posterior end of tail; cercarial excretory ducts stenostomate; cercarial intestine bifurcate; gut development pedomorphic (gut does not appear until redial or cercarial stage).

Order Heronimiformes
With symmetrically branched sporocysts; eggs hatching in utero; ventral sucker degenerating in adults.

Family
Heronimidae

Order Paramphistomiformes
Cercariae with two eyespots; redial stage with appendages; cercariae leaving snail and encysting in the open ("on" something, either animal, vegetable, or mineral); pharynx in adults at junction of esophagus and cecal bifurcation.

Families
Gyliauchenidae, Paramphistomidae, Microscaphidiidae, Pronocephalidae, Notocotylidae

Order Echinostomatiformes
Redial stage with appendages; cercariae encyst in the open; primary excretory pore in anterior half of cercarial tail; ventral sucker in cercariae midventral; secondary excretory pore terminal; acetabulum in adult midventral; adult body with spines; no eyespots in cercariae; rediae with collars; uterus extending from ovary to preacetabular.

Families
Cyclocoelidae, Psilostomidae, Fasciolidae, Philophthalmidae, Echinostomidae, Rhopaliasidae

Order Haploporiformes
Cercariae with two eyespots; cercariae encyst in the open; primary excretory pore in anterior half of cercarial tail; ventral sucker in cercariae midventral; secondary excretory pore terminal; acetabulum in adult midventral; adult body with spines; rediae without appendages; hermaphroditic duct present; uterus extending from ovary anteriorly to halfway between bifurcation and pharynx.

Families
Haploporidae, Haplosplanchnidae, Megaperidae

Order Transversotrematiformes
Cercariae with two eyespots; primary excretory pore in anterior half of cercarial tail; furcocercous cercariae; body transversely elongated; rediae with appendages.

Family
Transversotrematidae

Order Hemiuriformes
Cercariae with two eyespots; primary excretory port in anterior half of cercarial tail; ventral sucker in cercariae midventral; secondary excretory pore terminal; acetabulum in adult midventral; adult body with spines; rediae without appendages; furcocercous cercariae; cystophorous cercaria.

Families
Vivesiculidae, Ptychogonimidae, Azygiidae, Hirudinellidae, Bathycotylidae, Hemiuridae, Accacoeliidae, Syncoeliidae

Order Strigeiformes
Cercariae with two eyespots; acetabulum in adult midventral; adult body with spines; rediae without appendages; cercariae encyst in second intermediate host; mesostomate excretory

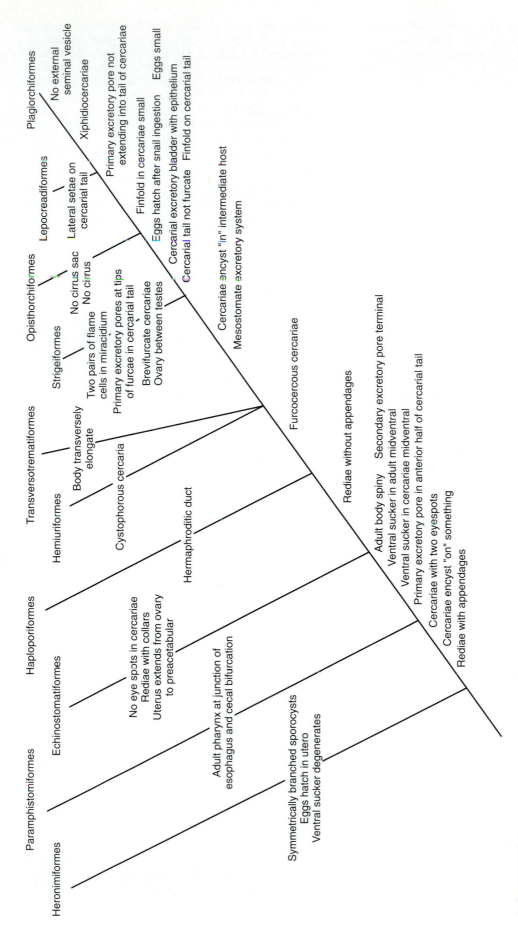

FIGURE 15.26

A hypothetical phylogenetic tree for the orders of Digenea.
Modified from *Parascript: Parasites and the Language of Evolution* Washington, DC: Smithsonian Institution Press, page 211, by permission of the publisher. Copyright © 1993.

system; two pairs of flame cells in miracidium; no secondary excretory pore in cercariae; brevifurcate cercariae; primary excretory pores at tips of furcae in cercarial tail; ovary between testes; genital pore midhindbody; uterus extending anteriorly from ovary to near acetabulum then posteriorly to genital pore.

Families

Clinostomidae, Sanguinocolidae, Spirorchidae, Schistosomatidae, Gymnophallidae, Fellodistomidae, Brachylaemidae, Bucephalidae, Liolopidae, Cyathocotylidae, Proterodiplostomidae, Neodiplostomidae, Bolbophoridae, Diplostomidae, Strigeidae

Order Opisthorchiformes

Cercariae with two eyespots; secondary excretory pore in cercariae terminal; acetabulum in adult midventral; adult body with spines; rediae without appendages, cercariae encyst in second intermediate host; mesostome excretory system; cercarial tail not furcate; cercarial excretory bladder lined with epithelium; seminal receptacle present; primary finfold present on cercarial tail; eggs small, generally less than 40 µm; eggs ingested and hatch in molluscan host; no cirrus sac; no cirrus.

Families

Opisthorchiidae, Cryptogonimidae, Heterophyidae

Order Lepocreadiiformes

Cercariae with two eyespots; secondary excretory pore in cercariae terminal; acetabulum in adult midventral; adult body with spines; rediae without appendages, cercariae encyst in second intermediate host; mesostome excretory system; cercarial tail not furcate; cercarial excretory bladder lined with epithelium; seminal receptacle present; primary excretory vesicle in cercariae extending a short distance into tail; dorsoventral finfold on cercarial tail; eggs small, generally less than 40 µm; eggs hatch in molluscan host; primary excretory pore not extending into tail of cercariae; finfold in cercariae small; lateral setae on cercarial tail.

Families

Deropristidae, Homalometridae, Lepocreadiidae

Order Plagiorchiformes

Secondary excretory pore in cercariae terminal; acetabulum in adult midventral; adult body with spines; rediae without appendages; cercariae encyst in second intermediate host; mesostome excretory system; cercarial tail not furcate; cercarial excretory bladder lined with epithelium; seminal receptacle present; primary excretory vesicle in cercariae extending a short distance into tail; dorsoventral finfold present on cercarial tail; eggs small, generally less than 40 µm; primary excretory pore not extending into tail of cercariae; finfold in cercariae small; xiphidiocercariae; no external seminal vesicle.

Families

Allocreadiidae, Acanthocolpidae, Campulidae, Troglotrematidae, Renicolidae, Macroderoididae, Opecoelidae, Zoogonidae, Lissorchiidac, Microphallidae, Lecithodendriidae, Prosthogonimidae, Plagiorchiidae, Dicrocoeliidae, Brachycoeliidae, Cephalogonimidae, Gorgoderidae, Auridistomidae, Rhytidodidae, Telorchiidae, Ochetosomatidae, Urotrematidae, Pleorchiidae, Pachypsolidae, Calycodidae, Haematoloechidae

References

1. Aitken-Ander, P., and N. L. Levin. 1985. Occurrence of adult and developmental stages of *Protoeces maculatus* (Trematoda: Digenea) in the gastropod *Crepidula convexa*. *Trans. Am. Microsc. Soc.* 104:250–60.

2. Barrett, J. 1976. Bioenergetics in helminths. In Van den Bossche, H., ed. *Biochemistry of parasites and host-parasite relationships.* Amsterdam: Elsevier/North Holland Biomedical Press.

3. Behrens, A. C., and P. M. Nollen. 1993. Hatching of *Echinostoma caproni* miracidia from eggs derived from adults grown in hamsters and mice. *Parasitol. Res.* 79:28–32.

4. Beklemishev, W. N. 1969. Origin and development of the excretory apparatus. In Kabata, Z., ed. *Principles of comparative anatomy of invertebrates* 2. Edinburgh: Oliver & Boyd Ltd., 327–52.

5. Bennett, J. L., and D. P. Thompson. 1986. Mode of action of antitrematodal agents. In Campbell, W. C., and R. S. Rew, eds. *Chemotherapy of parasitic diseases.* New York: Plenum Press, 427–43.

6. Bock, D. 1986. *In vitro* excystment of the metacercaria of *Plagiorchis species 1* (Trematoda, Plagiorchiidae). *Int. J. Parasitol.* 16:641–45.

7. Bogitsh, B. J. 1968. Cytochemical and ultrastructural observation on the tegument of the trematode *Megalodiscus temperatus. Trans. Am. Microsc. Soc.* 87:477–86.

8. Bogitsh, B. J. 1987. Further observations on eggshell formation in *Haematoloechus medioplexus* (Trematoda: Digenea). *Trans. Am. Microsc. Soc.* 106:373–78.

9. Bogitsh, B. J., and G. R. Davenport. 1991. The in vitro effects of various lysosomotropic agents on the gut of *Schistosoma mansoni* schistosomula. *J. Parasitol.* 77:187–93.

10. Brennan, G. P., R. E. B. Hanna, and W. A. Nizami. 1991. Ultrastructural and histochemical observations on the tegument of *Gasterodiscoides hominis* (Paramphistoma: Digenea). *Int. J. Parasitol.* 21:897–905.

11. Brooker, B. E. 1972. The sense organs of trematode miracidia. In Canning, E. U., and C. A. Wright, eds. *Behavioural aspects of parasite transmission. Linnean Society of London.* London: Academic Press, 171–80.

12. Brooks, D. R., and D. A. McLennan. 1993. *Parascript. Parasites and the language of evolution.* Washington, D.C.: Smithsonian Institution Press.

13. Bryant, C. 1993. Organic acid excretion by helminths. *Parasitol. Today* 9:58–60.

14. Bueding, E., and J. Fisher. 1966. Factors affecting the inhibition of phosphofructokinase activity of *Schistosoma mansoni* by trivalent antimonials. *Biochem. Pharmacol.* 15:1197–211.

15. Bueding, E., and J. Fisher. 1970. Biochemical effects of niridazole on *Schistosoma mansoni. Molecular Pharmacol.* 6:532–39.

16. Bundy, D. A. P. 1981. Periodicity in the hatching of digenean eggs: A possible circadian rhythm in the life cycle of *Transversotrema patialense. Parasitology* 83:13–22.

17. Burren, C. H., I. Ehrlich, and P. Johnson. 1967. Excretion of lipids by the liver fluke (*Fasciola hepatica* L). *Lipids* 2:353–56.

18. Burton, P. R. 1960. Gametogenesis and fertilization in the frog lung fluke, *Haematoloechus medioplexus* Stafford (Trematoda: Plagiorchiidae). *J. Morphol.* 107:92–122.

19. Cable, R. M. 1971. Parthenogenesis in parasitic helminths. *Am. Zool.* 11:267–72.

20. Cheah, K. S., and R. K. Prichard. 1975. The electron transport systems of *Fasciola hepatica* mitochondria. *Int. J. Parasitol.* 5:183–86.

21. Christensen, N. O. 1980. A review of the influence of host- and parasite-related factors and environmental conditions on the host-finding capacity of the trematode miracidium. *Acta Tropica* 37:303–18.

22. Clegg, J. A. 1972. The schistosome surface in relation to parasitism. In Taylor, A. E. R., and R. Muller, eds. *Functional aspects of parasite surfaces.* Symposia of the British Society of Parasitology, vol. 10. Oxford: Blackwell Scientific Publications Ltd., 23–40.

23. Cohen, L. M., H. Neimark, and L K. Eveland. 1980. *Schistosoma mansoni:* Response of cercariae to a thermal gradient. *J. Parasitol.* 66:362–64.

24. Coles, G. C. 1984. Recent advances in schistosome biochemistry. *Parasitology* 89:603–37.

25. Cordingley, J. S. 1987. Trematode eggshells: Novel protein biopolymers. *Parasitol. Today* 3:341–44.

26. Cornford, E. M., W. D. Bocash, and W. H. Oldendorf. 1981. Transintegumental glucose uptake in *Schistosomatium douthitti. J. Parasitol.* 67:24–30.

27. Cornford, E. M., and A. M. Fitzpatrick. 1992. Glucose-induced modulation of nutrient influx in *Schistosoma mansoni. J. Parasitol.* 78:266–70.

28. Cornford, E. M., and M. E. Huot. 1981. Glucose transfer from male to female schistosomes. *Science* 213:1269–71.

29. Cort, W. W. 1944. The germ cell cycle in the digenetic trematodes. *Q. Rev. Biol.* 19:275–84.

30. Cort, W. W., D. J. Ameel, and A. Van der Woude. 1954. Parasitological reviews—germinal development in the sporocysts and rediae of the digenetic trematodes. *Exp. Parasitol.* 3:185–225.

31. Daniel, B. E., T. M. Preston, and V. R. Southgate. 1992. The *in vitro* transformation of the miracidium to the mother sporocyst of *Schistosoma margrebowiei;* changes in the parasite surface and implications for interactions with snail plasma factors. *Parasitology* 104:41–49.

32. Daugherty, J. W. 1952. Intermediary protein metabolism in helminths. I. Transaminase reactions in *Fasciola hepatica. Exp. Parasitol.* 1:331–38.

33. Dike, S. C. 1969. Acid phosphatase activity and ferritin incorporation in the ceca of digenetic trematodes. *J. Parasitol.* 55:111–23.

34. Dönges, J. 1969. Entwicklungs- und Lebensdauer von Metacercarien. *Z. Parasitenkd.* 31:340–66.

35. Dönges, J. 1970. Transplantation of rediae—a device for solving special problems in trematodology. *J. Parasitol.* 54:82–83.

36. Dunn, T. S., R. E. B. Hanna, and W. A. Nizami. 1987. Sensory receptors of the miracidium of *Gigantocotyle explanatum* (Trematoda: Paramphistomidae). *Int. J. Parasitol.* 17:1131–40.

37. Dunn, T. S., R. E. B. Hanna, and W. A. Nizami. 1987. Ultrastructural and cytochemical observations on the tegument of three species of paramphistomes (Platyhelminthes: Digenea) from the Indian water buffalo, *Bubalus bubalis. Int. J. Parasitol.* 17:1153–61.

38. Ehlers, U. 1986. Comments on a phylogenetic system of the Platyhelminthes. *Hydrobiologia* 132:1–12.

39. El Kouni, M. H., and F. N. M. Naguib. 1990. Pyrimidine salvage pathways in adult *Schistosoma mansoni. Int. J. Parasitol.* 20:37–44.

40. El Meanawy, M. A., T. Aji, N. F. B. Phillips, R. E. Davis, R. A. Salata, I. Malhotra, D. McClain, M. Aikawa, and A. H. Davis. 1990. Definition of the complete *Schistosoma mansoni* hemoglobinase mRNA sequence and gene expression in developing parasites. *Am. J. Trop. Med. Hyg.* 43:67–78.

41. Fashuyi, S. A. 1986. Excystment of the metacercaria of the trematode *Mesocoelium monodi. Int. J. Parasitol.* 16:237–39.

42. Foster, L. A., and B. J. Bogitsh. 1986. Utilization of the heme moiety of hemoglobin by *Schistosoma mansoni* schistosomules *in vitro. J. Parasitol.* 72:669–76.

43. Frayha, G. J., and J. D. Smyth. 1983. Lipid metabolism in parasitic helminths. In Baker, J. R., and R. Muller, eds. *Advances in parasitology* 22. London: Academic Press, 309–87.

44. Fried, B., and G. B. Ramundo. 1987. Excystation and cultivation *in vitro* and *in vivo* of *Cyathocotyle bushiensis* (Trematoda) metacercariae. *J. Parasitol.* 73:541–45.

45. Fujino, T., and Y. Ishii. 1979. Comparative ultrastructural topography of the gut epithelia of some trematodes. *Int. J. Parasitol.* 9:435–48.

46. Furlong, S. T. 1991. Unique roles for lipids in *Schistosoma mansoni. Parasitol. Today* 7:59–62.

47. Geary, T. G., R. D. Klein, L. Vanover, J. W. Bowman, and D. P. Thompson. 1992. The nervous systems of helminths as targets for drugs. *J. Parasitol.* 78:215–30.

48. Gibson, D. I. 1987. Questions on digenean systematics and evolution. *Parasitology* 95:429–60.

49. Gönnert, R. 1962. Histologische Untersuchungen über den Feinbau der Eibildungsstatte (Oogenotop) von *Fasciola hepatica. Z. Parasitenkd.* 21:457–92.

50. Granzer, M., and W. Haas. 1986. The chemical stimuli of human skin surface for the attachment response of *Schistosoma mansoni* cercariae. *Int. J. Parasitol.* 16:575–79.

51. Haas, W., and R. Schmitt. 1982. Characterization of chemical stimuli for the penetration of *Schistosoma mansoni* cercariae. II. Conditions and mode of action *Z. Parasitenkd.* 66:309–19.

52. Halton, D. W. 1967. Observations on the nutrition of digenetic trematodes. *Parasitology* 57:639–60.

53. Halton, D. W., C. Shaw, A. G. Maule, C. F. Johnston, and I. Fairweather. 1992. Peptidergic messengers: A new perspective of the nervous system of parasitic platyhelminths. *J. Parasitol.* 78:179–93.

54. Hertel, L. A. 1993. Excretion and osmoregulation in the flatworms. *Trans. Am. Microsc. Soc.* 112:10–17.

55. Higgins, J. C. 1979. The role of the tegument of the metacercarial stage of *Bucephalus haimeanus* (Lacaze-Duthiers, 1854) in the absorption of particulate material and small molecules in solution. *Parasitology* 78:99–106.

56. Hochachka, P. W., and T. Mustafa. 1972. Invertebrate facultative anaerobiosis. *Science* 178:1056–60.

57. Hockley, D. J. 1972. *Schistosoma mansoni:* The development of the cercarial tegument. *Parasitology* 64:245–52.

58. Hockley, D. J. 1973. Ultrastructure of the tegument of *Schistosoma.* In Dawes, B., ed. *Advances in parasitology* 2. New York: Academic Press, Inc., 233–305.

59. Hockley, D. J., and D. J. McLaren. 1973. *Schistosoma mansoni:* Changes in the outer membrane of the tegument during development from cercaria to adult worm. *Int. J. Parasitol.* 3:13–25.

60. Horemans, A. M. C., A. G. M. Tielens, and S. G. van den Bergh. 1991. The transition from an aerobic to an anaerobic energy metabolism in transforming *Schistosoma mansoni* cercariae occurs exclusively in the head. *Parasitology* 102:259–65.

61. Irwin, S. W. B., L. T. Threadgold, and N. M. Howard. 1978. *Cryptocotyle lingua* (Creplin) (Digenea: Heterophyidae): Observations on the morphology of the redia, with special reference to the birth papilla and release of cercariae. *Parasitology* 76:193–99.

62. Irwin, S. W. B., G. McShane, and D. H. Saville. 1989. A study of metacercarial excystment in *Parapronocephalum symmetricum* (Trematoda: Notocotylidae). *Parasitol. Res.* 76:45–49.

63. James, B. L. 1964. The life cycle of *Parvatrema homoeotecnum* sp. nov. (Trematoda; Digenea) and a review of the family Gymnophallidae Morozov., 1955. *Parasitology* 54:1–41.

64. Janssens, P. A., and C. Bryant. 1969. The ornithine-urea cycle in some parasitic helminths. *Comp. Biochem. Physiol.* 30:261–72.

65. Kemp. W. M., P. R. Brown, S. C. Merritt, and R. E. Miller. 1980. Tegument-associated antigen modulation by adult male *Schistosoma mansoni. J. Immunol.* 124:806–11.

66. Khalil, G. M., and R. M. Cable. 1968. Germinal development in *Philophthalmus megalurus* (Cort, 1914) (Trematoda: Digenea). *Z. Parasitenkd.* 31:211–31.

67. Kirschner, K., and W. J. Bacha Jr. 1980. Excystment of *Himasthla quissetensis* (Trematoda: Echinostomatidae) metacercariae in vitro. *J. Parasitol.* 66:263–67.

68. Köhler, P., C. Bryant, and C. A. Behm. 1978. ATP synthesis in a succinate decarboxylase system from *Fasciola hepatica* mitochondria. *Int. J. Parasitol.* 8:399–404.

69. Köhler, P., and D. F. Stahel. 1972. Metabolic end products of anaerobic carbohydrate metabolism of *Dicrocoelium dendriticum* (Trematoda). *Comp. Biochem. Physiol.* 43B:733–41.

70. Krupa, P. L., A. K. Bal, and G. H. Cousineau. 1967. Ultrastructure of the redia of *Cryptocotyle lingua. J. Parasitol.* 53:725–34.

71. Kurelec, B. 1972. Lack of carbamyl phosphate synthesis in some parasitic platyhelminths. *Comp. Biochem. Physiol.* 43B:769–80.

72. LaRue, G. R. 1957. The classification of digenetic Trematoda: A review and a new system. *Exp. Parasitol.* 6:306–49.

73. Leitch, B., A. J. Probert, and N. W. Runham. 1984. The ultrastructure of the tegument of adult *Schistosoma haematobium. Parasitology* 89:71–78.

74. Lennox, R. W., and E. L. Schiller, 1972. Changes in dry weight and glycogen content as criteria for measuring the postcercarial growth and development of *Schistosoma mansoni. J. Parasitol.* 58:489–94.

75. Lie, K. J., D. Heyneman, and N. Kostanian. 1975. Failure of *Echinostoma lindoense* to reinfect snails already harboring that species. *Int. J. Parasitol.* 5:483–86.

76. Lloyd, G. M. 1986. Energy metabolism and its regulation in the adult liver fluke *Fasciola hepatica. Parasitology* 93:217–48.

77. Lumsden, R. D., and R. Specian. 1980. The morphology, histology, and fine structure of the adult stage of the cyclophyllidean tapeworm *Hymenolepis diminuta.* In Arai, H. P., ed. *Biology of the tapeworm Hymenolepis diminuta.* New York: Academic Press, Inc., 157–280.

78. Matthews, B. F. 1981. *Cercaria vaullegeardi* Pelseneer, 1906 (Digenea: Hemiuridae); the infection mechanism. *Parasitology* 83:587–93.

79. Meyer, F., H. Meyer, and E. Bueding. 1970. Lipid metabolism in the parasitic and free-living flatworms, *Schistosoma mansoni* and *Dugesia dorotocephala. Biochem. Biophys. Acta* 210:257–66.

80. Miller, F. H. Jr., G. S. Tulloch, and R. E. Kuntz. 1972. Scanning electron microscopy of integumental surface of *Schistosoma mansoni. J. Parasitol.* 58:693–98.

81. Moczon, T., Z. Swiderski, and H. Huggel. 1992. *Schistosoma mansoni:* The chemical nature of the secretions produced by the Mehlis' gland and ootype as revealed by cytochemical studies. *Int. J. Parasitol.* 22:65–73.

82. Nevhutalu, P. A., B. Salafsky, W. Haas, and T. Conway. 1993. *Schistosoma mansoni* and *Trichobilharzia ocellata:* Comparison of secreted cercarial eicosanoids. *J. Parasitol.* 79:130–33.

83. Nollen, P. M. 1968. Uptake and incorporation of glucose, tyrosine, leucine, and thymidine by adult *Philophthalmus megalurus* (Cort, 1914) (Trematoda), as determined by autoradiography. *J. Parasitol.* 54:295–304.

84. Nollen, P. M. 1983. Patterns of sexual reproduction among parasitic platyhelminths. Symposia of the British Society for Parasitology, vol. 20. *Parasitology* 86(4):99–120.

85. Ostrowski de Nuñez, M., and W. Haas. 1991. Penetration stimuli of fish skin for *Acanthostomum brauni* cercariae. *Parasitology* 102:101–4.

86. Pappas, P. W. 1971. *Haematoloechus medioplexus:* Uptake, localization, and fate of tritiated arginine. *Exp. Parasitol.* 30:102–19.

87. Pappas, P. W., and C. P. Read. 1975. Membrane transport in helminth parasites: A review. *Exp. Parasitol.* 37:469–530.

88. Parkening, T. A., and A. D. Johnson. 1969. Glucose uptake in *Haematoloechus medioplexus* and *Gorgoderina* trematodes. *Exp. Parasitol.* 25:358–67.

89. Pax, R. A., and J. L. Bennett. 1992. Neurobiology of parasitic flatworms: How much "neuro" in the biology? *J. Parasitol.* 78:194–205.

90. Pigulevskii, S. V. 1958. On the question of the phylogeny of flatworms. *Rabot. Gelymintol. 80-Let. Skrjabin* 265–70.

91. Pojmanska, T., and K. Machaj. 1991. Differentiation of the ultrastructure of the body wall of the sporocyst of *Leucochloridium paradoxum. Int. J. Parasitol.* 21:651–59.

92. Prichard, R. K., and P. J. Schofield. 1968. The glycolytic pathway in adult liver fluke, *Fasciola hepatica. Comp. Biochem. Physiol.* 24:697–710.

93. Read, C. P., L. T. Douglas, and J. E. Simmons Jr. 1959. Urea and osmotic properties of tapeworms from elasmobranchs. *Exp. Parasitol.* 8:58–75.

94. Rennie, J. 1992. Living together. *Sci. Am.* 266(1):122–33.

95. Rogers, S. H., and E. Bueding. 1975. Anatomical localization of glucose uptake by *Schistosoma mansoni* adults. *Int. J. Parasitol.* 5:369–71.

96. Rohde, K. 1971. Phylogenetic origin of trematodes. *Parasitol. Schrift.* 21:17–27.

97. Rowcliffe, S. A., and C. B. Ollerenshaw. 1960. Observations on the bionomics of the egg of *Fasciola hepatica*. *Ann. Trop. Med. Parasitol.* 54:172–81.

98. Salafsky, B., and A. Fusco. 1987. Eicosanoids as immunomodulators of penetration by schistosome cercariae. *Parasitol. Today* 3:279–281.

99. Salafsky, B., Y. -S. Wang, A. C. Frusco, and J. Antonacci. 1984. The role of essential fatty acids and prostaglandins in cercarial penetration (*Schistosoma mansoni*). *J. Parasitol.* 70:656–60.

100. Sauer, M. C. V., and A. W. Senft. 1972. Properties of a proteolytic enzyme from *Schistosoma mansoni*. *Comp. Biochem. Physiol.* 42B:205–20.

101. Saz, H. J. 1990. Helminths: Primary models for comparative biochemistry. *Parasitol. Today* 6:92–93.

102. Senft, A. W. 1968. Studies in proline metabolism by *Schistosoma mansoni*. I. Radioautography following in vitro exposure to radioproline C^{14}. *Comp. Biochem. Physiol.* 27:251–61.

103. Senft. A. W., D. E. Philpott, and A. H. Pelofsky. 1961. Electron microscope observations of the integument, flame cells and gut of *Schistosoma mansoni*. *J. Parasitol.* 47:217–29.

104. Shannon, W. Jr., and B. J. Bogitsh. 1971. *Megalodiscus temperatus:* Comparative radioautography of glucose-^{3}H and galactose-^{3}H incorporation. *Exp. Parasitol.* 29:309–19.

105. Sinitsin, D. 1931. Studien über die Phylogenie der Trematoden. IV. The life histories of *Plagioporus silicus* and *Plagioporus virens,* with special reference to the origin of Digenea. *Z. Wiss. Zool.* 138:409–56.

106. Smyth, J. D., and D. W. Halton. 1983. *The physiology of trematodes,* 2d ed. Cambridge: Cambridge University Press.

107. Sobhon. P., E. S. Upatham, and D. J. McLaren. 1984. Topography and ultrastructure of the tegument of adult *Schistosoma mekongi*. *Parasitology* 89:511–21.

108. Spellman, S. J., and A. D. Johnson. 1987. *In vitro* excystment of the black spot trematode *Uvulifer ambloplitis* (Trematoda: Diplostomatidae). *Int. J. Parasitol.* 17:897–902.

109. Stirewalt, M. A. 1974. *Schistosoma mansoni:* Cercaria to schistosomule. In Dawes, B., ed. *Advances in parasitology* 12. New York: Academic Press, Inc., 115–82.

110. Thompson, S. N. 1991. Applications of nuclear magnetic resonance in parasitology. *J. Parasitol.* 77:1–20.

111. Tielens, A. G. M., P. van der Meer, and S. S. van den Bergh. 1981. The aerobic energy metabolism of the juvenile *Fasciola hepatica. Molecular and Biochem. Parasitol.* 3:205–14.

112. Tielens. A. G. M., P. van der Meer, and J. M. van den Heuvel. 1991. The enigmatic presence of all gluconeogenic enzymes in *Schistosoma mansoni* adults. *Parasitology* 102:267–76.

113. Turner, H. M. 1984. Orientation and pathology of *Allocorrigia filiformis* (Trematoda: Dicroeliidae) from the antennal glands of the crayfish *Procambarus clarkii*. *Trans. Am. Microsc. Soc.* 103:434–37.

114. Uglem, G. L., and O. R. Larson. 1987. Facilitated diffusion and active transport systems for glucose in metacercariae of *Clinostomum marginatum* (Digenea). *Int. J. Parasitol.* 17:847–50.

115. Vande Vusse, F. J., T. D. Fish, and M. P. Neumann. 1981. Adult Digenea from upper midwest hirudinid leeches. *J. Parasitol.* 67:717–20.

116. Vernberg, W. B., and F. J. Vernberg. 1971. Respiratory metabolism of a trematode metacercaria and its host. In Cheng. T. C., ed. *Aspects of the biology of symbiosis.* Baltimore: University Park Press, 91–102.

117. Watts, S. D. M. 1970. Transamination in homogenates of rediae of *Cryptocotyle lingua* and of sporocysts of *Cercaria emasculans* Pelseneer, 1900. *Parasitology* 61:499–504.

118. Wharton, D. A. 1983. The production and functional morphology of helminth egg-shells. Symposia of the British Society for Parasitology, vol. 20. *Parasitology* 86(4):85–97.

119. Whitfield, P. J., and N. A. Evans. 1983. Parthenogenesis and asexual multiplication among parasitic platyhelminths. *Parasitology* 86:121–60.

120. Wilson. R. A. 1968. The hatching mechanism of the egg of *Fasciola hepatica* L. *Parasitology* 58:79–89.

121. Wilson, R. A. 1969. Fine structure of the tegument of the miracidium of *Fasciola hepatica* L. *J. Parasitol.* 55:124–33.

122. Xu, Y., and M. H. Dresden. 1986. Leucine aminopeptidase and hatching of *Schistosoma mansoni* eggs. *J. Parasitol.* 72:507–11.

123. Yamaguti, S. 1971. *Synopsis of digenetic trematodes of vertebrates* l. Tokyo: Keigaku Publishing Co.

124. Yamasaki, H., E. Kominami, and T. Aoki. 1992. Immunocytochemical localization of a cysteine protease in adult worms of the liver fluke *Fasciola* sp. *Parasitol. Res.* 78:574–80.

Additional References

Baer, J. G., and C. Joyeux. 1961. Classe des Trématodes (Trématoda Rudolphi). In Grassé, P. P., ed. *Traité de zoologie: Anatomie, systématique, biologie, vol. 4, part I. Plathelminthes, Mésozoaires, Acanthocéphales, Némertiens.* Paris: Masson & Cie, 561–692. A well-illustrated overview of trematodes.

Barrett, J. 1981. *Biochemistry of parasitic helminths.* Baltimore: University Park Press.

Cable, R. M. 1972. Behaviour of digenetic trematodes. *Zool. J. Linn. Soc.* 51(suppl. 1):1–18.

Conn, D. B. 1991. *Atlas of invertebrate reproduction and development.* New York: John Wiley & Sons, Inc.

Dawes, B. 1946. *The Trematoda, with special reference to British and other European forms.* Cambridge: Cambridge University Press. A classic reference work, of value to all interested in trematodes.

Hyman, L. H. 1951. *The invertebrates, vol. 2. Platyhelminthes and Rhynchocoela. The acoelomate Bilateria.* New York: McGraw-Hill Book Co. A standard reference to all aspects of Trematoda.

Schell, S. C. 1985. *Handbook of trematodes of North America north of Mexico.* Moscow, Idaho: University Press of Idaho.

Trager, W. 1986. *Living together. The biology of animal parasitism.* New York: Plenum Press.

Chapter 16

DIGENEANS: STRIGEIFORMES

. . . though fully appreciating professor Looss's vast erudition, we must not forget that without the complement of good judgment, it is quite easy to strain learning into absurdity.

L. W. Sambon, on Looss's insistence that there was only one species of schistosome[9]

Of the several superfamilies in this order, only two, Strigeoidea and Schistosomatoidea, are of much economic or medical significance. The latter, however, contains some of the most important disease agents of humans.

SUPERFAMILY STRIGEOIDEA

Strigeoidea are bizarre in appearance, with their bodies divided into two portions (Fig. 16.1). The anterior portion usually is spoon or cup shaped, with accessory **pseudosuckers** on each side of the oral sucker. Behind the acetabulum is a spongy, padlike organ referred to as the **adhesive** or **tribocytic organ.** This structure secretes proteolytic enzymes that digest host mucosa, probably functioning both as an accessory holdfast and as a digestive-absorptive organ. The hindbody contains most of the reproductive organs, although vitelline follicles often extend into the forebody. The genital pore is located at the posterior end.

Most strigeoids are quite small and are found commonly in the digestive tracts of fish-eating vertebrates. Their cercariae are easily recognized by the fact that they have both a pharynx and a forked tail. No adult strigeoids are known to parasitize humans, but they are so ubiquitous and their biology so interesting that we will briefly consider a few species.

Family Diplostomidae

• *Alaria americana*

The genus *Alaria* contains several very similar species, all of which mature in the small intestines of carnivorous mammals. *Alaria americana* is found in various species of Canidae in northern North America. They are about 2.5 to 4.0 mm long, with the forebody longer than the hindbody. The forebody has a pair of ventral flaps that are narrowest at the anterior end (Fig. 16.2a). A pointed process flanks each side of the oral sucker. The tribocytic organ is relatively large and elongated and has a ventral depression in its center.

The life cycles of *Alaria* spp. are remarkable in that the worms may require four hosts before they can develop to maturity (Fig. 16.2). The eggs are unembryonated when laid, and they hatch in about two weeks. The miracidium swims actively and will attack and penetrate any of several species of planorbid snails.[33] Mother sporocysts develop in the renal veins and produce daughter sporocysts in about two weeks. Daughter sporocysts migrate to the digestive gland and need about a year to mature and begin producing cercariae. The furcocercous cercaria leaves the snail during daylight hours and swims to the surface, where it hangs upside down. Occasionally it sinks a short distance and then returns to the surface. If a tadpole swims by, the resulting water currents stimulate the cercaria to swim after it. If it contacts the tadpole, the cercaria will quickly attack, drop its tail, penetrate the skin, and begin wandering within the amphibian. The trematode remains viable if the tadpole undergoes metamorphosis. In about two weeks the cercaria has transformed into a **mesocercaria,** an unencysted form between a cercaria and a metacercaria. It is then infective to the next host, which may be the definitive host or a paratenic host. If a canid eats an infected tadpole or adult frog, the mesocercariae are freed by digestion, penetrate into the coelom, and then move to the diaphragm and lungs. After about five weeks in the lungs, the mesocercaria has transformed into a **diplostomulum metacercaria** (Fig. 16.2). Diplostomula migrate up the trachea and then to the intestine, where they mature in about a month.

Tadpoles, however, are not always available to terrestrial canids and, furthermore, are distasteful to all but the hungriest carnivores. This ecological barrier is overcome when a water snake eats the infected tadpole or frog and thereby becomes a paratenic host. A snake (or other animal) can accumulate large numbers of mesocercariae in its tissues, rendering a heavy infection to the definitive host when the animal is eaten. The mesocercariae then migrate, develop into diplostomula, and mature in the intestine, as do those from tadpoles. Life cycles of other species of *Alaria* are similar.

Mature *Alaria* spp. are quite pathogenic, causing severe enteritis that often kills the definitive hosts in severe infections. Also, the mesocercariae are pathogenic, especially

233

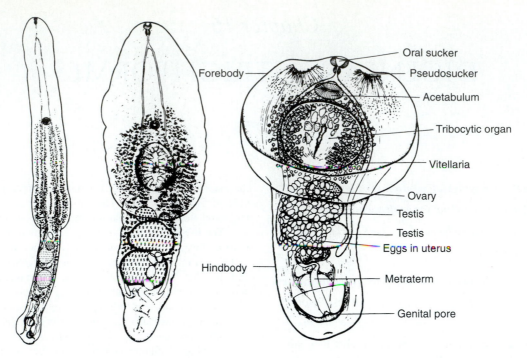

FIGURE 16.1

Typical strigeoid trematodes, illustrating body forms. (*a*) *Mesodiplostomum gladiolum* Dubois 1936. (*b*) *Pseudoneodiplostomum thomasi* (Dollfus 1935). (*c*) *Proalarioides serpentis* (Yamaguti 1933).

From S. Yamaguti, *Synopsis of Digenetic Trematodes of Vertebrates*, Volume 1. Copyright © 1971 Keigaku Publishing Co., Tokyo. Reprinted with permission of the publisher.

when accumulated in large numbers. Figure 16.3 is from a fatal infection of mesocercariae in a human.

Shoop and Corkum[39] demonstrated that a related species, *Alaria marcianae,* can be transmitted to a juvenile definitive host through the milk of its mother. In this species the parasite matures normally in the intestine of the adult male and nonlactating females but remains as diplostomula disseminated throughout the tissues of lactating females. In an experimental infection a single cat infected 21 of her offspring via milk over the course of five litters and still harbored infective juveniles after three years. Primates can also transmit this worm by transmammary means.[40]

• *Uvulifer ambloplitis*

Several species of strigeoid trematodes cause black spots in the skin of fish; one such species is *Uvulifer ambloplitis,* a parasite of kingfishers, fish-eating birds that are widely distributed across the United States. The spoon-shaped forebody of the parasite is separated from the longer hindbody by a slender constriction. Adults are 1.8 to 2.3 mm long.

The eggs, which are unembryonated when laid, hatch in about three weeks. The miracidium will penetrate snails of the genus *Helisoma* and will transform into mother sporocysts that retain the eyespots of the first larva. Daughter sporocysts invade the digestive gland and produce cercariae in about six weeks. The cercariae escape from the tissues of the snail and rise to the surface of the water, where they are sensitive to the passing of fish. If they contact a fish, especially a centrarchid or percid, they drop their tails and penetrate the skin. Once inside the dermis,

the flukes metamorphose into **neascus metacercariae** and secrete a delicate, hyaline cyst wall around themselves. A neascus is similar to a diplostomulum except that the forebody is spoon shaped, without anterolateral "points." The fish host responds to the neascus by laying down layers of melanin granules. The result is a conspicuous black spot indicating the presence of a metacercaria (Fig. 16.4). When such fish are heavily infected, they are often discarded as diseased by people who catch them. Kingfishers become infected when they eat such a fish. The flukes mature in 27 to 30 days.

Other, related flukes also cause black spots in a wide variety of fish and have similar life cycles. When a neascus larva is encountered for which the adult genus is unknown, it is proper to refer it to the genus *Neascus*. This is also true for *Diplostomulum, Tetracotyle,* and *Cercaria*. In fact new species can be named in these genera; of course, when the adult form becomes known, the species reverts to the proper genus.

Family Strigeidae

• *Cotylurus flabelliformis*

This is a common parasite of wild and domestic ducks in North America. Adult flukes are 0.5 to 1.0 mm long. The forebody is cup shaped, with the acetablum and tribocytic organ located at its depths. The hindbody is short and stout and is curved dorsally.

Adult worms live in the small intestines of ducks. Eggs passed in the feces hatch in about three weeks, and the

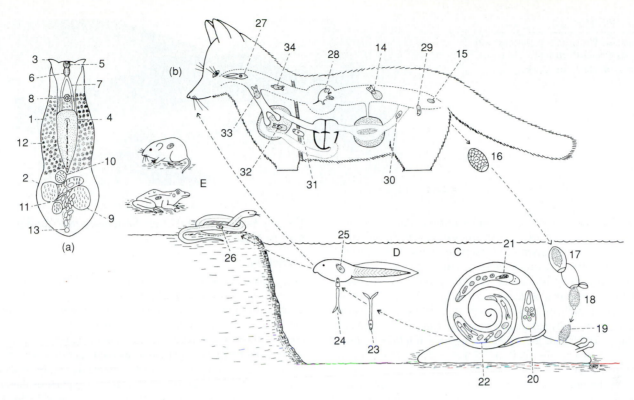

FIGURE 16.2

Life cycle of *Alaria americana*. (*a*) Ventral view of adult fluke. (*b*) Fox definitive host. (*c*) Planorbid snail (*Helisoma*), first intermediate host. (*d*) Tadpole (*Rana, Bufo*), second intermediate host. (*e*) Paratenic hosts (snakes, frogs, mice). **1,** forebody; **2,** hindbody; **3,** lappet or pseudosucker; **4,** holdfast (tribocytic) organ; **5,** oral sucker; **6,** pharynx; **7,** cecum; **8,** ventral sucker; **9,** testes; **10,** ovary; **11,** eggs in uterus; **12,** vitelline glands; **13,** common genital pore; **14,** adult fluke in small intestine; **15,** egg passing out of body in feces; **16,** unembryonated egg; **17,** embryonated egg; **18,** egg hatching; **19,** miracidium penetrating *Helisoma* snail; **20,** young mother sporocyst; **21,** mature mother sporocyst; **22,** daughter sporocyst; **23,** cercaria free in water; **24,** cercaria penetrating tadpole, casting tail as it enters; **25,** mesocercaria; **26,** mesocercaria in snake, frog, and mouse paratenic hosts; **27,** infection of definitive host by swallowing tadpole, second intermediate host; **28,** infection of definitive host by swallowing infected paratenic host; **29,** mesocercariae migrate through gut wall into coelom; **30,** mesocercariae enter hepatoportal vein, but it has not been shown that they reach the lungs by way of the blood; **31,** mesocercariae pass through the diaphragm and penetrate the lungs; **32,** in the lungs the mesocercariae transform to a diplostomulum stage; **33,** diplostomulae migrate up trachea; **34,** diplostomulae are swallowed, go to small intestine, and develop to maturity (**14**) in five to six weeks.

From O. W. Olsen, *Animal Parasites: Their Life cycles and Ecology.* Copyright © 1974 Dover Publications, Inc., New York, NY.

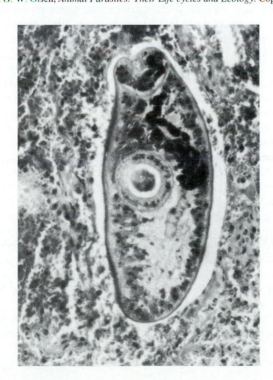

FIGURE 16.3

Mesocercaria of *Alaria americana* in human lung biopsy. The case proved to be fatal, with nearly every organ of the body infected, presumably as a result of eating undercooked frogs' legs.

From R. Freeman, et al, *American Journal of Tropical Medical Hygeine,* 25:803–807, 1976.

From R. S. Freeman et al., "Fatal human infection with mesocercariae of the trematode *Alaria americana*," in *Am. J. Trop. Med. Hyg.* 25:803–807. Copyright © 1976.

FIGURE 16.4

A minnow, *Pimephales* sp., infected with neascus-type metacercariae, "black spot."
Courtesy of John S. Mackiewicz.

miracidia attack snails of the family Lymnaeidae (*Lymnaea, Stagnicola*). There are two sporocyst generations. After about six weeks the sporocysts begin to release furcocercous cercariae. If the cercariae contact a snail of the same family Lymnaeidae, they penetrate, migrate to the ovotestis, and transform into **tetracotyle metacercariae.** The tetracotyle is similar to a diplostomulum except that it has an extensive system of excretory canals that often are filled with excretory products. The canals are called the **reserve bladder system.** When a duck eats the snail, the flukes excyst and mature in about one week.

On the other hand, if the cercaria enters a snail of the families Planorbidae or Physidae, it will attack sporocysts or rediae of other species of flukes already present and will develop into a tetracotyle metacercaria within them. These too will mature in about a week if a duck eats them.

Strigeoid trematodes, then, exhibit complex life cycles that involve several unrelated hosts. Their adaptability seems amazing when one considers the differences in environments provided by snails, pond water, fish, amphibians, reptiles, and birds or mammals. It is also remarkable that a parasite that may require more than a year to complete its juvenile development can become sexually mature in its definitive host in a week or less and die a few days later. Such is the pattern in the life cycles of the strigeoids.[34]

SUPERFAMILY SCHISTOSOMATOIDEA

Flukes of the superfamily Schistosomatoidea are peculiar in that they have no second intermediate host in their life cycles and also in that they mature in the blood vascular system of their definitive hosts. Most species are dioecious. Popiel[36] discusses the tantalizing puzzle of the possible selective advantage in being dioecious in a class of worms that is overwhelmingly monoecious. They are parasites of fishes, turtles, birds, and mammals throughout the world. Several species are parasites of humans, causing misery and death wherever they are distributed.[45] The families Sanguinicolidae and Spirorchidae parasitize fish and turtles. The family Schistosomatidae, however, includes species that are among the most dreaded parasites of humans. To date, 10 species and two varieties of schistosomes have been reported in Africa. The taxonomy of this group and the validity of some species of parasites have been subjects of controversy for years; the final decision on species recognition awaits further studies.

Family Schistosomatidae: *Schistosoma* Species and Schistosomiasis

• History

Three species of schistosomes are of vast medical significance: *Schistosoma haematobium, S. mansoni,* and *S. japonicum*—all parasites of humans since antiquity. Bloody urine was a well-recognized disease symptom in northern Africa in ancient times. At least 50 references to this condition have been found in surviving Egyptian papyri, and calcified eggs of *S. haematobium* have been found in Egyptian mummies dating from about 1200 B.C. Hulse[20] presented a well-reasoned hypothesis that the curse that Joshua placed on Jericho can be explained by the introduction of *S. haematobium* into the communal well by the invaders. The removal of the curse occurred after the abandonment of Jericho and subsequent droughts eliminated the snail host, *Bulinus truncatus.* Today Jericho (Ariha, Jordan) is well known for its fertile lands and healthy, well-nourished people.

The first Europeans to record contact with *S. haematobium* were surgeons with Napoleon's army in Egypt (1799–1801). They reported that **hematuria** (bloody urine) was prevalent among the troops, although the cause, of course, was unknown. Nothing further was learned about **schistosomiasis haematobia** for more than 50 years, until a young German parasitologist, Theodor Bilharz, discovered the worm that caused it. He announced his discovery in letters to his former teacher, Von Siebold, naming the parasite *Distomum haematobium.*[14] (Tragically, Bilharz died of typhus at the age of 37.) During the next few years, it was discovered that 30% to 40% of the population in Egypt bore infections of *S. haematobium,* and the worm was even found in an ape dying in London. The peculiar morphology of the worm made it clear that it could not be included in the genus *Distomum,* so in 1858 Weinland proposed the name *Schistosoma.* Three months later Cobbold named it *Bilharzia,* after its discoverer. This latter name became widely accepted throughout the world, and the parasite was even given the nickname "Bill Harris" by British soldiers serving in Europe during World War I. Today, however, the strict rules of zoological nomenclature decree that *Schistosoma* has priority and is thus the current name for the parasite. Even so, health officers in many parts of the world erect signs next to ponds and streams that warn prospective bathers of the dangers of "bilharzia." Nonetheless, *Schistosoma* is an apt name, referring to the "split body" (gynecophoral canal) of the male.

While information was accumulating on the biology of *S. haematobium,* some investigators began to doubt whether it

Table 16.1

COMPARATIVE MORPHOLOGY OF THE THREE PRIMARY SPECIES OF HUMAN SCHISTOSOMES

Characteristic	*S. haematobium*	*S. mansoni*	*S. japonicum*
Tegumental papillae	Small tubercles	Large papillae with spines	Smooth
Size			
Male			
Length	10 to 15 mm	10 to 15 mm	12 to 20 mm
Width	0.8 to 1.0 mm	0.8 to 1.0 mm	0.50 to 0.55 mm
Female			
Length	ca. 20 mm	ca. 20 mm	ca. 26 mm
Width	ca. 0.25 mm	ca. 0.25 mm	ca. 0.3 mm
Number of testes	4 to 5	6 to 9	7
Position of ovary	Near midbody	In anterior half	Posterior to midbody
Uterus	With 20 to 100 eggs at one time; average 50	Short; few eggs at one time	Long; may contain up to 300 eggs; average 50
Vitellaria	Few follicles, posterior to ovary	Few follicles, posterior to ovary	In lateral fields, posterior quarter of body
Egg	Elliptical, with sharp terminal spine; 112 to 170 μm × 40 to 70 μm	Elliptical, with sharp lateral spine; 114 to 175 μm × 45 to 70 μm	Oval to almost spherical; rudimentary lateral spine; 70 to 100 μm × 50 to 70 μm

was a single species or whether two or more species were being confused. The problem was confounded by the observation in some patients of eggs with terminal spines in both urine and feces. Whenever eggs with lateral spines were noticed, they were ignored as "abnormal." In 1905 Sir Patrick Manson decided that intestinal and vesicular (urinary bladder) schistosomiasis usually were distinct diseases, caused by distinct species of worms. He reached this conclusion when he examined a man from the West Indies who had never been to Africa and who passed laterally-spined eggs in his feces but none at all in his urine.[27] Sambon argued in favor of the two-species concept in 1907, and he named the parasites producing laterally-spined eggs *Schistosoma mansoni*. (Japanese zoologists had already detected still another species by this time, but their reports were generally unknown to Europeans.) However, the eminent German parasitologist Looss disagreed and brought the full sway of his reputation and dialectic against the notion and even stated that he had seen a female worm with both kinds of eggs in its uterus. Sambon, undaunted, replied that until Professor Looss could "show me an actual specimen, I am bound to place the worm capable of producing the two kinds of eggs with the phoenix, the chimaera and other mythical monsters."[38]

The question was finally resolved by Leiper[25] in 1915. He first visited Japan to acquaint himself with the work of Miyairi and Suzuki on *S. japonicum*. Then, working in Egypt, he discovered that cercariae emerging from the snail *Bulinus* could infect the vesicular veins of various mammals, and they always produced eggs with terminal spines. Those emerging from a different snail, *Biomphalaria*, infected the intestinal veins and produced laterally-spined eggs. It was soon determined that *S. mansoni* had a broad distribution in the world, having been widely scattered by the slave trade. It is now widespread in Africa and the Middle East and is the

only blood fluke of humans in the New World, with the possible exception of a small focus of *S. haematobium* in Surinam.[26a] The original endemic area of *S. mansoni* was probably the Great Lakes region of central Africa.

While Cobbold, Weinland, Bancroft, Sambon, and others were wrestling with the problem of *S. haematobium* and *S. mansoni*, Japanese researchers were investigating a similar disease in their country. For years physicians in the provinces of Hiroshima, Saga, and Yamanachi had recognized an endemic disease characterized by an enlarged liver and spleen, ascites, and diarrhea. At autopsy they noted eggs of an unknown helminth in various organs, especially in the liver. In 1904 Professor Katsurada of Okayama recognized that the larvae in these eggs resembled those of *S. haematobium*. Because he was unable to make a postmortem examination of an infected person, he began examining local dogs and cats, in hopes that they were reservoirs for the parasite. He soon found adult worms containing eggs identical to those from humans and named them *Schistosoma japonicum*. The experimental elucidation of the life cycle by various Japanese researchers was a milestone in the history of parasitology and formed the basis for Leiper's work on blood flukes in Egypt. The distribution of *S. japonicum* is limited to Japan, China, Taiwan, the Philippines, and Southeast Asia.

In more recent years other species of *Schistosoma* parasitic in humans have been distinguished (p. 246), and *Schistosoma* spp. seem to be evolving actively.[21]

• Morphology

Although *Schistosoma* spp. are generally similar structurally, several differences in detail are listed in Table 16.1. Considerable sexual dimorphism exists in the genus, the males being shorter and stouter than the females (Fig. 16.5). The males have a ventral, longitudinal groove, the **gynecophoral**

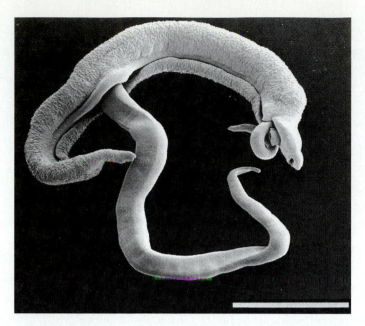

FIGURE 16.5

Scanning electron micrograph of male and female *Schistosoma mansoni.*
The female is lying in the gynecophoral groove in the ventral surface of
the male. (Bar = 2mm.)

Courtesy of D. W. Halton.

canal, where the female normally resides. The mouth is surrounded by a strong oral sucker, and the acetabulum is near
the anterior end. There is no pharynx. The paired intestinal
ceca converge and fuse at about the midpoint of the worm
and then continue as a single gut to the posterior end. The
male possesses five to nine testes, according to species, each
of which has a delicate vas efferens, and these combine to
form the vas deferens. The latter dilates to become the seminal vesicle, which opens ventrally through the genital pore
immediately behind the ventral sucker. Cirrus pouch, cirrus,
and prostate cells are absent.

The suckers of the females are smaller and not so muscular
as those of the males, and the tegumental tubercles (Fig. 16.5), if
any, are confined to the ends of the female. The ovary is anterior
or posterior to or at the middle of the body, and the uterus is correspondingly short or long, depending on the species.

• Biology

Adult worms live in the veins that drain certain organs of
their host's abdomen (Fig. 16.6), and the three main species
have distinct preferences: *S. haematobium* lives principally
in veins of the urinary bladder plexus; *S. mansoni* prefers the
portal veins draining the large intestine; and *S. japonicum* is
more concentrated in the veins of the small intestine. The female worm is usually in the gynecophoral canal of the male
worm, where copulation takes place, and there are other
physiological reasons for this habitus.[3] The worms work
their way "upstream" into smaller veins, where the female
may leave the gynecophoral canal, reaching still smaller
venules where she deposits eggs. The eggs (Fig. 16.7) must
then traverse the wall of the venule, some intervening tissue,
and the gut or bladder mucosa before they are in a position to

be expelled from the host. The mechanism by which this "escape" is achieved is not self-evident and has been the subject
of much speculation. The spines on the eggs were traditionally credited with contributing to the expulsion, but the feat
is also accomplished by *S. japonicum* and other Oriental
schistosomes that have only the most rudimentary spines. It
appears that the worms enlist the aid of the host. According
to File,[13] the endothelial cells lining the venule actively
move over the schistosome eggs to exclude them from the
lumen (Fig. 16.8). Damian[10] and Doenhoff et al.[12] postulated
that the worm then exploits the host immune response to
transport its egg to the lumen of the gut or the bladder. The
extravasated egg stimulates a granuloma to form around it
(Fig. 16.15). The granuloma, consisting of motile cells (such
as eosinophils, plasma cells, and macrophages), then moves
to the intestinal or bladder lumen, carrying the egg with it.
Once in the lumen the cells of the granuloma disperse, and
the egg is excreted with the feces or urine. In any case about
two-thirds of the eggs do not make it, and large numbers
build up in the gut or bladder wall, particularly in chronic
cases in which the wall is toughened by a great amount of fibrous tissue. Of course many eggs are never expelled from
the venules but are swept away by the blood, eventually to
lodge in the liver or capillary beds of other organs. By the
time the eggs reach the outside by way of the urine or feces,
they are completely embryonated and hatch when exposed to
the lower osmolarity of fresh water.

The mechanism of hatching is poorly understood. The
first indication of hatching is activation of the cilia on the
miracidium. This increases until the miracidium is a veritable
spinning ball. Then, suddenly, an osmotically induced vent
opens on the side of the egg, and the miracidium emerges
(Fig. 16.9). The miracidium usually contracts a few times,
completely clearing the shell, after which it rapidly swims
away. However, some eggs do not hatch, no matter how active the miracidium becomes, and others hatch before the
larva becomes activated.

The miracidium is a typical one and swims ceaselessly
during its short life. If hatching from an old egg, it will live
only one to two hours; in optimal conditions it will survive
for five to six hours. Although the miracidia of schistosomes
do not have eyespots, they apparently have photoreceptors,
and they are positively phototropic.[7] When miracidia enter
the vicinity of a snail host, they are stimulated to swim more
rapidly and change direction much more frequently, thus increasing their chances of encountering the host. Following
are the most important snails (Fig. 16.6):

1. For *S. haematobium,* several species of *Bulinus* and
   Physopsis, possibly also *Planorbarius.*

2. For *S. mansoni, Biomphalaria alexandrina* in northern
   Africa, Saudia Arabia, and Yemen; *B. sudanica, B.*
   rupellii, B. pfeifferi, and others in the genus in other parts
   of Africa; *B. glabrata* in the Western Hemisphere; and
   Tropicorbis centrimetralis in Brazil.

3. For *S. japonicum,* several species of *Oncomelania.*

After penetration of the snail the miracidium sheds its epithelium and begins development into a mother sporocyst, usually near its point of entrance. After about two weeks the
mother sporocyst, which has four protonephridia, gives birth

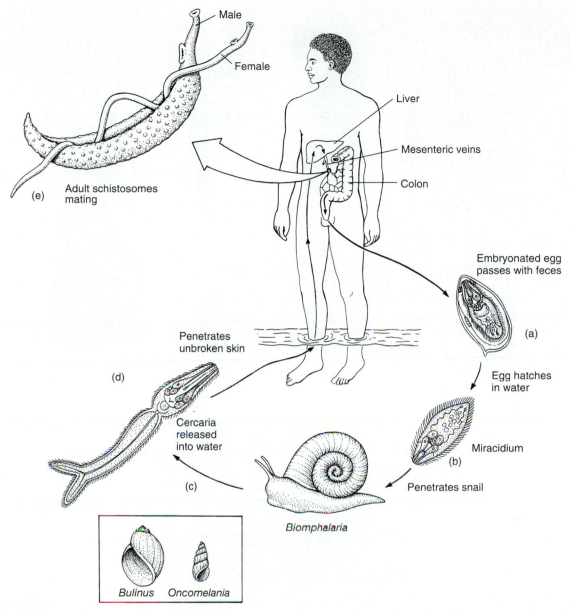

FIGURE 16.6

Life cycle of *Schistosoma mansoni*. Main intermediate hosts of *S. mansoni* are species of *Biomphalaria*. Those of *S. haematobium* are *Bulinus* spp. (*left, in box*) and of *S. japonicum* are *Oncomelania* spp. (*right, in box*). (*a*) Shelled miracidium is passed in feces. (*b*) Miracidium hatches spontaneously and penetrates *Biomphalaria*. (*c*) Two sporocyst generations in snail. (*d*) Cercaria leaves snail and penetrates skin of definitive host. (*e*) Adult schistosomes in portal venules of intestine.

Drawing by William Ober and Claire Garrison.

to daughter sporocysts, which usually migrate to other organs of the snail, if there is room. The mother sporocyst continues producing daughter sporocysts for up to six to seven weeks.[30] There is no redial generation.

The furcocercous cercariae (Fig. 16.6) start to emerge from the daughter sporocysts and the snail host about four weeks after initial penetration by the miracidium. The cercaria has a body 175 to 240 μm long by 55 to 100 μm wide and a tail 175 to 250 μm long by 35 to 50 μm wide, bearing a pair of furci 60 to 100 μm long. The oral sucker is absent, being replaced by a head organ composed of penetration glands, and the ventral sucker is small and covered with

minute spines. Four types of glands open through bundles of ducts at the anterior margin of the head (p. 220).

There is no second intermediate host in the life cycle. The cercariae alternately swim to the surface of the water and slowly sink toward the bottom, continuing to live this way for one to three days. If they come into contact with the skin of a prospective host, such as a human, they attach and creep about for a time as if seeking a suitable place to penetrate (Fig. 16.10). They are attracted to secretions of the skin, showing a strongly positive response to arginine. Upon stimulation by arginine the cercaria begin to produce arginine themselves from postacetabular glands. This may attract

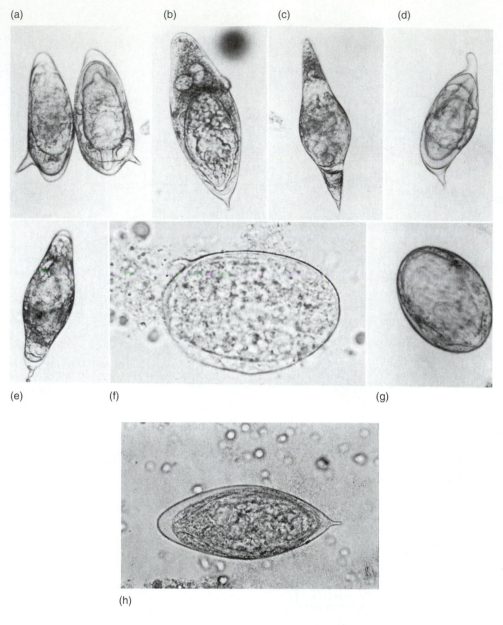

(a) (b) (c) (d)

(e) (f) (g)

(h)

FIGURE 16.7

Eggs of schistosome flukes. (*a*) *Schistosoma mansoni;* (*b*) *Schistosoma intercalatum;* (*c*) *Schistosoma bovis;* (*d*) *Schistosoma rodhaini;* (*e*) *Schistosoma mattheei;* (*f*) *Schistosoma japonicum;* (*g*) *Schistosomatium douthitti;* (*h*) *Schistosoma haematobium.*

Courtesy of Robert E. Kuntz and Jerry A. Moore.

FIGURE 16.8

Scanning electron micrograph of endothelial cells and eggs of *Schistosoma japonicum* in vitro. The eggs have just been expelled by a female worm, and the endothelial cells are moving over them.

Courtesy of S. K. File.

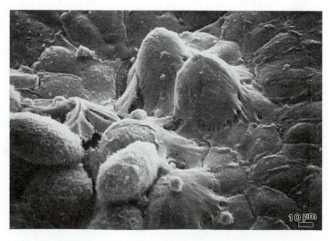

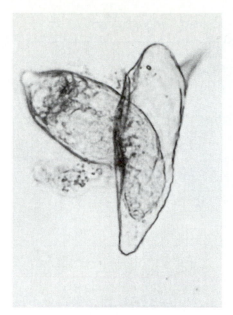

FIGURE 16.9

Miracidium of *Schistosoma mansoni* escaping
from its eggshell.
Courtesy of Robert E. Kuntz.

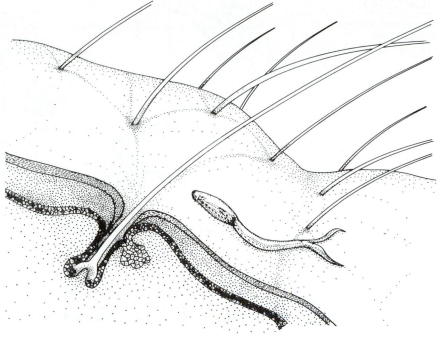

FIGURE 16.10

Diagram of a cercaria of *Schistosoma* sp. in exploring position on skin.
Drawing by Ian Grant.

other cercariae in the neighborhood.[15] They require only half
an hour or less to completely penetrate the epidermis, and
they can disappear through the surface in 10 to 30 seconds.
Penetration is accompanied by a vigorous wiggling, together
with secretion of the products of the head organ. The tail
drops off in the process. The worms are somewhat smaller
now that the penetration glands have emptied their contents.
Within 24 hours the **schistosomules** (little schistosomes)
enter the peripheral circulation and are swept off to the heart.
Some of the schistosomules may migrate through the lymphatics to the thoracic duct and from there to the subclavian
veins and heart. Leaving the right side of the heart, the small
worms wriggle their way through the pulmonary capillaries
to gain access to the left heart and systemic circulation. It appears that only the schistosomules that enter the mesenteric
arteries, traverse the intestinal capillary bed, and reach the
liver by the hepatoportal system can continue to grow. After
undergoing a period of about three weeks of development in
the liver sinusoids, the young worms migrate to the walls of
the gut or bladder (according to species), copulate, and begin
producing eggs. The entire prepatent period is about five to
eight weeks. Adult schistosomes may live 20 to 30 years.[23]

Unpaired female worms do not become sexually mature
and have the appearance of starving. Their esophageal musculature is weak and thin, they produce little of at least some digestive enzymes, and they ingest about one-fourth as many
erythrocytes as paired females. A growth-stimulating function
may result from the muscular action of the clasping male, which
helps the immature female pump blood into her intestine.[3,17]

• Epidemiology

Human waste in water containing intermediate hosts of the
Schistosoma worm is the single most important epidemiological factor in schistosomiasis, and the availability of
suitable species of snail host will determine the endemicity
of the particular species of *Schistosoma*. The latter is well-
illustrated by the fact that although both *S. mansoni* and *S.
haematobium* are widespread in Africa, only *S. mansoni*
became established in the New World by the slave trade, almost certainly because snails suitable for only that species
were present there (Fig. 16.11). Survival of these parasites
depends on human insistence on polluting water with their
organic wastes. Hygienic waste disposal is sufficient to eliminate schistosomiasis as a disease of humans (Fig. 16.12).
Tradition, at once the salvation and the bane of culture,
prompts people to use the local waterway for sewage disposal instead of foul-smelling outhouses (Figs. 16.13 and
16.14). A bridge across a small stream becomes a convenient
toilet; a grove of mango trees over a rivulet is a haven for
children who bombard the area with their feces. Especially
vulnerable to infection are farmers who wade in their irrigation water, fishermen who wade in their lakes and streams,
children who play in any contaminated body of water, and
people who wash clothes in streams. A focus of infection in
Brazil was a series of ditches in which watercress was grown
for food. In some Moslem countries the religious requirement of ablution—that is, washing the anal or urethral orifices after urination or defecation—is an important factor in
transmission. Not only is the convenient water source used to

FIGURE 16.11

Geographical distribution of
schistosomiasis.
AFIP neg. no. 68-4866-3.

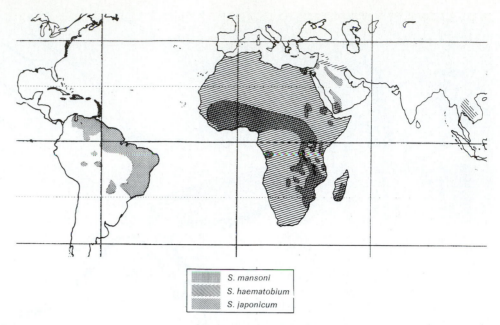

	S. mansoni
	S. haematobium
	S. japonicum

Geographic distribution of schistosomaisis

FIGURE 16.12

Government-encouraged pit latrines in the Philippines serve as a means
to prevent schistosomiasis.

Courtesy of Robert E. Kuntz.

FIGURE 16.13

Typical native dwelling in a schistosomiasis area of the Philippines.
Oncomelania, the snail host of *S. japonicum,* is found in the stream near
the house, which is also likely to be contaminated by human feces.

Courtesy of Robert E. Kuntz.

FIGURE 16.14

Slow-running streams and protecting tropical vegetation provide ideal habitats for *Biomphalaria,* a snail host for *S. mansoni* in Puerto Rico. Photograph by Larry S. Roberts.

perform ablution likely to be a contaminated river or canal, but it is likely to be near the spot chosen for the deposition of additional feces and urine, ensuring further contamination.

Clearly the economic and education level of the population will influence the transmission of the disease, and age and sex are important factors as well. Males usually show the highest rates of infection and the most intense infections, and the most hazardous age is the second decade of life. This appears to reflect occupational and recreational differences, rather than sex or age resistance to infection. In Surinam, where both sexes work in the fields, the highest prevalence occurs in adults of both sexes. Certain other factors, such as immunity and the cessation of egg release in chronic infections, must be considered when a population is surveyed and transmission is studied. The fact that buildup of granulation tissue in the gut or bladder wall prevents release of eggs into the feces or urine will mask infection unless immunodiagnostic or biopsy methods are used. Furthermore, even though infected persons may be gravely disabled, they are removed from the cycle of transmission.

During the course of the infection some protective immunity to superinfection is elicited, either by repeated exposures to the cercariae or by the presence of the adults, although the adult worms themselves are not affected by the immune response (p. 226). The resistance to reinfection becomes apparent only after many years of exposure and

occurs earlier in locations where levels of infection are high.[18] For an excellent review of immunity in human schistosomiasis see Newport and Colley.[29]

It is of utmost importance to recognize that, by extending snail habitats, agricultural projects intended to increase food production in underdeveloped countries have, in many cases, created more misery than they have alleviated.[8,46] A $10 million irrigation project in southern Zimbabwe had to be abandoned 10 years after it was started because of schistosomiasis.[32] The Aswan High Dam in Egypt, much acclaimed in its inception, may have its benefits canceled by the increase in disease it has caused. Restraint of the wide fluctuations in the water level of the Nile, although making possible four crops per year by perennial irrigation, has also created conditions vastly more congenial to snails.[43] Before the dam construction, perennial irrigation was already practiced in the Nile delta region, and the prevalence of schistosomiasis was about 60%; in the 500 miles of river valley between Cairo and Aswan, where the river was subject to annual floods, the prevalence was only about 5%. Four years after the dam was completed the prevalence of *S. haematobium* ranged from 19% to 75%, with an average of 35%, between Cairo and Aswan, or an average sevenfold increase! In the area above the dam, prevalence was very low before its construction; in 1972, 76% of the fishermen examined in the impounded area were infected. A 1982 study showed a continued increase in prevalence in six villages of upper Egypt.[24]

The cost, in terms of productivity, of *S. haematobium* to a resident of affected areas may well equal the worker's per capita income. Even oil production in some countries of the Middle East may be directly, albeit slightly, affected by the presence of schistosomes in oil refinery employees.

The roles of reservoir hosts and of strains of the parasite have some importance as epidemiological factors, depending on the species. Members of no less than seven mammalian orders have been successfully infected experimentally with *S. mansoni;* however, certain monkeys and a variety of rodents are probably important natural reservoir hosts in Africa and tropical America. *Schistosoma haematobium* is more host specific than is *S. mansoni,* and it is thought that no natural reservoir hosts exist for it. The opposite is true of *S. japonicum,* which seems to be the least host specific. It can develop in dogs, cats, horses, swine, cattle, caribou, rodents, and deer; but there seems to be more than one race of this worm, and the susceptibility of a given host varies. For example, *S. japonicum* is widely prevalent in rats in Taiwan, but it is rare in humans there. The more recent descriptions of *S. mekongi* and *S. malayanum* as distinct from *S. japonicum* suggest that the traditional *S. japonicum* may be a complex of cryptic species.

• Pathogenesis

Schistosomiasis is unusual among parasitic infections in that the pathogenesis is almost entirely due to the eggs and not to the adult worms. We mentioned before that the eggs traverse the gut or bladder wall surrounded by a granuloma (Fig. 16.15) which disperses upon reaching the lumen. Clearly the granulomas do not disperse from the eggs that do not reach the lumen, remaining in place and leaking antigens over a considerable length of time. Thus, the primary lesion in

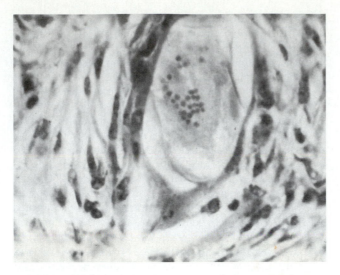

FIGURE 16.15

Egg of *Schistosoma mansoni* in granuloma in intestinal wall.
Courtesy of David F. Oetinger.

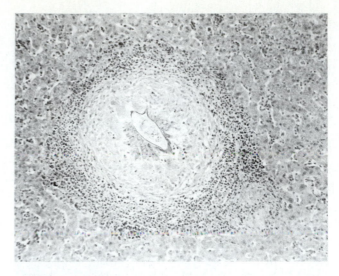

FIGURE 16.16

Egg of *Schistosoma mansoni* in granuloma. Note leukocytic infiltration around the granuloma.
AFIP neg. no. 64-6532.

schistosomiasis is a delayed type hypersensitivity (DTH) reaction around the egg (Fig. 16.16). However, the progress and outcome of the disease are a result of a complex interplay of immunopathology involving both the T_h1 and T_h2 arms of the immune response.

Schistosomiasis is often divided into three phases: migratory, acute, and chronic. The **migratory phase** encompasses the time from penetration until maturity and egg production; it is often symptomless. Penetration of the cercariae may produce a dermatitis if the patient's immune system has been sensitized by earlier experiences of cercarial penetration. Schistosome dermatitis is usually much more severe when caused by bird schistosomes (p. 247), probably because the cercariae are killed.

The **acute phase** is sometimes called **Katayama fever** and occurs when the schistosomes begin producing eggs about 4 to 10 weeks after initial infection. By this time the body has had considerable exposure to various schistosome antigens, sufficient to mount a humoral response, but advent of egg production substantially increases the amount of antigen release. The change in antigen-antibody ratio leads to formation of large immune complexes that must be cleared by cells of the RE system. The syndrome is marked by chills and fever, fatigue, headache, malaise, muscle aches, lymphadenopathy, and gastrointestinal discomfort.[29] There is a high eosinophilia in the peripheral blood, and the granulomas around the eggs contain large numbers of eosinophils, as well as neutrophils and macrophages. Macrophages in the early acute granuloma secrete predominately IL-1, and then TNF increases after one to two weeks. Chronic granulomas are dominated by macrophages, lymphocytes, fibroblasts, and multinucleated giant cells.[42] These become small fibrous granulomas, or **pseudotubercles,** so-called because of their resemblance to the localized nodules of tissue reaction (tubercles) in tuberculosis. Many eggs are carried by the hepatic portal circulation back up into the liver, where they stimulate granuloma formation (Fig. 16.16), and some may be carried to the lungs or other tissues.

In the **chronic phase** patients indigenous to endemic areas are commonly asymptomatic,[42] or with intestinal schistosomiasis, they may show mild, chronic, bloody diarrhea with mild abdominal pain and lethargy. With schistosomiasis haematobia, there may be pain on urination and blood in the urine. Affected people usually accept these conditions as normal and only seek medical assistance with heavy infections or when more serious complications develop. In most cases the immune responses of the patient are modulated so that the granulomatous reactions do not become too severe. For example, although the macrophages secrete fibroblast growth factors, which mediate fibrosis, they also secrete collagenases that digest the collagen fibers. A balance between collagen synthesis and degradation in most cases prevents progression to serious hepatic fibrosis.

In about 8% of cases of infection with *S. japonicum* and *S. mansoni,* development of egg granulomas and fibrosis in the liver seriously impedes portal blood flow. As the eggs accumulate and the fibrotic reactions in the liver continue, a periportal cirrhosis and portal hypertension ensue. A marked enlargement of the spleen (splenomegaly) occurs, partly because of eggs lodged in it and partly because of the chronic passive congestion of the liver. Ascites (accumulation of fluid in the abdominal cavity) is common at this stage (Fig. 16.17). Some eggs pass the liver, lodging in the lungs, nervous system, or other organs and produce pseudotubercles there.

The pathological changes due to *S. japonicum* tend to involve the small intestine more extensively than those due to *S. mansoni.* Frequently fibrous nodules containing nests of eggs occur on the serosal and peritoneal surfaces. Eggs of *S. japonicum* seem to reach the brain more often than do those of the other species: 60% of all neurological disease in schistosomiasis and almost all brain lesions are due to *S. japonicum.* Schistosomiasis japonica is the most grave of the three, and the prognosis is unfavorable in heavy infections without early treatment.[42]

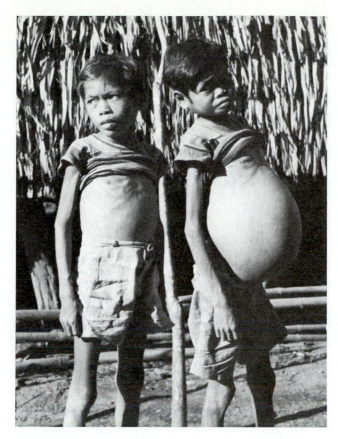

FIGURE 16.17

Ascites in advanced schistosomiasis japonica, Leyte, Philippines (*right*). This is an example of dwarfing caused by schistosomiasis. The male on the left is 13 years old; the one on the right is 24 years old.

Courtesy of Robert E. Kuntz.

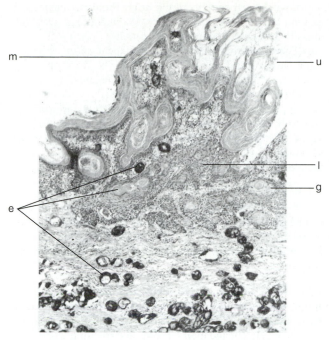

FIGURE 16.18

Schistosomiasis of the urinary bladder. In this case, many eggs (**e**) of *S. haematobium* can be seen in all the layers of the bladder. Many of the eggs are calcified. The epithelium has undergone squamous metaplasia (**m**). Note also leukocytic infiltration (**l**), granulosis (**g**), and ulceration (**u**).

AFIP neg. no. 65-6779.

Infection with *S. haematobium* is considered the least serious. Since the adults of this species live in the venules of the urinary bladder, the chief symptoms are associated with the urinary system. The onset of hematuria is usually gradual and becomes marked as the disease develops and the bladder wall becomes more ulcerated. Pain is most intense at the end of urination.[42] Changes in the bladder wall (Fig. 16.18) are associated with the DTH reactions around the eggs—that is, pseudotubercles, fibrous infiltration, thickening of the muscularis layer, and ulceration. Evidence that *S. haematobium* causes bladder cancer is strong.[5] Chronic, heavy infections lead to genital, ureteral, and kidney involvement and to lesions in other parts of the body, as with other species.

• Diagnosis and Treatment

As in the diagnosis of many other helminths, the demonstration of eggs in the excreta is the most straightforward mode of diagnosis. However, the number of eggs produced per female schistosome, even for *S. japonicum,* is far smaller than for most other helminth parasites of humans; therefore, direct smears must be augmented by concentration techniques and other diagnostic methods, such as biopsy and immunodiagnosis. With concentration techniques, such as gravity or centrifugal sedimentation, more than 90% of the coprologically demonstrable cases can be diagnosed. The Kato technique is

a simple method for discovery and quantification of eggs in 20 to 50 mg stool samples.[35] However, few or no eggs may be passed, particularly in chronic cases. In such cases rectal, liver, or bladder biopsies may be of great value, but these require the services of specialists and the availability of appropriate surgical facilities. Hence, substantial effort has been directed at finding sensitive, accurate, and reliable immunodiagnostic methods.

Serological tests based on detection of antibodies in the patient's blood have several inherent problems:[42] (1) they only become positive some time after infection, (2) they only become negative some time after cure, and (3) they crossreact with other helminth infections. Tests designed to detect schistosome antigens, however, offer potential solutions to these problems. They become positive as soon as antigens are present and become negative quickly after cure. Monoclonal antibodies to one of the best characterized schistosome antigens were prepared by Deelder and his coworkers.[11] In a modified ELISA test they detected the antibody in patients' sera at levels as low as 1 ng/ml, and the antigen concentration was highly correlated with egg output—that is, it provided estimation of worm burden.

Difficulty in treating schistosomiasis has been a major factor contributing to the disease as a world health problem. Until recently the most effective drugs were the organic trivalent antimonials, but these are quite toxic to humans and must be given carefully—in small doses over a period of two to six weeks, depending on the drug. Problems inherent in treating large numbers of people with such drugs over wide areas in developing countries are obvious.

Therefore, much effort has been spent on investigating other, less toxic but effective drugs. The drug of choice is now praziquantel, which is effective against all species of schistosomes in humans. We described its mode of action in Chapter 15 (p. 226). Fortunately, appropriate chemotherapy can lead to reversal and even resolution of much fibrosis and urinary pathology.[1,29] However, long-standing heavy infection can result in irreversible liver or bladder damage.[42]

• Control

Control of schistosomiasis, as with many infectious diseases, can be a multipartite effort, including (1) education of a population to undertake activities to prevent transmission, (2) curing of infected persons, (3) control of the vector, and (4) protective vaccination.

Education.   Although potentially very effective, education is often exceedingly difficult, depending ultimately on the almost intractable task of persuading masses of uneducated, poor people to change their customs and traditions.

Control by Chemotherapy.   In the past curing infected persons was not a practical strategy for control. The development of safe and effective schistosomacides has altered that circumstance. For example, in one study of schoolchildren in the Nile Delta, the prevalence of S. mansoni was lowered from 60.3% to 24.8% one year after treatment with praziquantel, and it was 41.1% three years after treatment.[41] Corresponding data for S. haematobium were 37.6%, 5.5% at one year, and 9.9% after three years. Another survey credited educational efforts by the Egyptian government, as well as praziquantel availability, for decreasing prevalence (in a 17,310-person sample) of S. mansoni from 39% to 23% and of S. haematobium from 5% to 3% in the Nile Delta between 1983 and 1990.[27a]

Vector Control.   Control of snails was the major thrust of control efforts before the advent of safe and effective antihelminthics, and it is still important to support chemotherapy campaigns and to reduce reinfection.[26] Snail control may be undertaken by environmental management, by molluscicides, and by biological agents.

Although draining snail habitats is of value, we pointed out before that many more good habitats are being created by efforts to increase agricultural production. This need not be so. Environmental management measures, such as stream channelization, seepage control, canal lining, and canal relocation with deep burial of snails, can prevent increase in transmission.[26] Such factors helped prevent an increase in prevalence after construction of a lake and irrigation canals in Cameroon between 1979 and 1985.[2] This project included regular cleaning of vegetation from secondary and tertiary canals, use of only the concrete-lined primary canals for bathing and clothes washing, and provision of a clean source of drinking water.

Use of chemical molluscicides has met with some success but problems involved include determination and application of the proper quantity in a given body of water, dilution, effects on other organisms in the environment, and errors in estimating the physical and chemical characteristics of the water. Niclosamide is the only molluscicide currently available. Molluscicidal control of S. japonicum is virtually ineffective because Oncomelania is amphibious and only visits water to lay its eggs.

Some biological control efforts have been successful, mostly the introduction of potential predators and competitors of vector snails.[26] The best-studied are potential competitive species of snails: Marisa cornuarietis, Helisoma duryi, Thiara granifera, and Melanoides tuberculata. In Martinique, for example, M. tuberculata colonized rapidly after its introduction in January 1983, and by October 1984, Biomphalaria glabrata and B. straminea had disappeared and have not been found since. Introduction of M. cornuarietis has successfully controlled Biomphalaria in several habitats in Puerto Rico. Marisa not only competes with Biomphalaria for food, but also preys on the vector snails.[26] A North American crayfish, Procambarus clarkii, was introduced into East Africa for aquaculture in about 1970 and has since then dispersed into all major drainages in Kenya.[19] Procambarus feeds on Biomphalaria, and it reduces populations of the snails significantly.

Snail-eating fish have been cultured and released in infected waters with some success.

Vaccination.   The development of an effective vaccine would have great potential value in the control of schistosomiasis, and this area of research is active. An advantage for investigators searching for an antischistosome vaccine is that the vaccine would not have to be more than 90% effective, as with a vaccine against a virus or bacteria. Because serious symptoms are shown only by patients with large numbers of worms, dramatic reduction in numbers of severe cases would result from only 50% or so reduction in worm burden. Some protection can be conferred by vaccination with irradiated cercariae and/or schistosomules,[4] but for practical use, a long-acting, killed antigen would be more desirable.

The best hope at present lies in use of genetic engineering to incorporate one or more specific worm antigens in the vaccine.[8,28] One example of such an antigen is a protein secreted by the excretory system of S. mansoni and S. japonicum. It is one of a family of enzymes called **glutathione S-transferases (GSTs).** These enzymes are normally found in hepatocytes and erythrocytes and are involved in detoxification reactions. The worm enzyme may help it detoxify products generated by host defense cells at the worm surface, or it may help dissolve hematin in the worm's gut.[28] An injection of worm GSTs confers significant protection in mice, and tests for safety in humans are proceeding.[8]

• Other Schistosoma spp.

In addition to S. mansoni, S. haematobium, and S. japonicum, several other species of schistosomes can infect humans, although their distribution and prevalence are much less than those of the "big three." In Africa hundreds of cases are known in which terminal-spined eggs are recovered from stools only. The worms were named Schistosoma intercalatum, but they are morphologically indistinguishable from S. mattheei, a natural parasite of African ruminants. However, S. mattheei can apparently only infect primates in

Table 16.2

DISTRIBUTION AND HOST SPECIFICITY OF *SCHISTOSOMA* SPP. (modified from Johnston et al.)[21]

Species group	Distribution[a]	Snail host	Mammalian host[b]
Schistosoma haematobium			
S. haematobium	Af & ad	*Bulinus*	Pr
S. intercalatum	Af	*Bulinus*	Pr
S. mattheei	Af	*Bulinus*	Pr, Ar
S. bovis	Af & ad	*Bulinus, Planorbarius*	Ar
S. curassoni	Af	*Bulinus*	Ar
S. margrebowiei	Af	*Bulinus*	Ar
S. leiperi	Af	*Bulinus*	Ar
Schistosoma mansoni			
S. mansoni	Af, SA	*Biomphalaria*	Pr, R
S. rodhaini	Af	*Biomphalaria*	R, C
S. edwardiense	Af	*Biomphalaria*	Ar
S. hippopotamii	Af	?	Ar
Schistosoma japonicum			
S. japonicum	SEA	*Oncomelania*	Pr, Ar, R, C, Pe
S. mekongi	SEA	*Neotricula*	Pr, C
S. sinensium	SEA	*Neotricula*	R
S. malayensis	SEA	*Robertsiella*	Pr, R
Schistosoma indicum			
S. indicum	SEA, SWA	*Indoplanorbis*	Ar
S. spindale	SEA, SWA	*Indoplanorbis*	Ar
S. nasale	SWA	*Indoplanorbis*	Ar
S. incognitum	SEA, SWA	*Lymnea, Radix*	Ar, R, C

[a] Af = Africa; Af & ad = Africa and adjacent regions; SA = South America and Caribbean; SEA = Southeast Asia; SWA = Southwest Asia.
[b] Pr = Primates; Ar = Artiodactyla; R = Rodentia; C = Carnivora; Pe = Perissodactyla.

the presence of a coinfection with *S. haematobium* or *S. mansoni*.[21] In Southeast Asia, careful examination of biology, morphology, and molecular analysis has supported recognition of *S. mekongi* and *S. malayensis* as distinct from *S. japonicum*, which they closely resemble.[16,44]

A number of other species of *Schistosoma* exist, and these differ with respect to morphology, host specificity, and distribution (Table 16.2), but some are quite difficult to distinguish from each other. These can be arranged in four groups: *S. mansoni* group (with lateral-spined eggs), *S. haematobium* group (with terminal-spined eggs), *S. japonicum* group (rounded, minutely-spined, or spineless eggs), and *S. indicum* group (an as yet poorly defined group from India and Southeast Asia).[21] Species in the *S. japonicum* group use snails of the class Prosobranchia as intermediate hosts, and members of all other groups use snails of the class Pulmonata.

Characterization of the groups on the basis of egg shape and spines is convenient but not entirely satisfactory. For example, *S. margrebowiei* in the *S. haematobium* group has round eggs with small spines, and *S. sinensium* in the *S. japonicum* group has lateral-spined eggs. They nevertheless fit into their assigned groups on the basis of other criteria, such as specificity of intermediate hosts. Molecular analysis has so far supported the identity of the species recognized,[22] but within groups, some species are very closely related. *Schistosoma mansoni* and *S. rodhaini* can form hybrids that produce viable and fertile young in the laboratory.[6] *Schistosoma intercalatum* and *S. haematobium* also produce hybrids in the laboratory, and naturally occurring hybrids have been found in certain areas in Africa.[37]

Molecular analysis of at least one species from each of the four groups has suggested that *S. haematobium* and *S. spindale* are sister taxa in a clade with *S. mansoni*.[22] *S. japonicum* was not closely related to that clade.

- ## Schistosome Cercarial Dermatitis ("Swimmer's Itch")

Several species in the genus *Schistosoma* cause a severe rash when their cercariae penetrate the skin of an unsuitable host; hence, *S. spindale* and *S. bovis* are agents of dermatitis in humans throughout the range of the schistosomes.

More importantly, several species of bird schistosomes are distributed throughout the world and cause "swimmer's itch" when their cercariae attack anyone on whose skin the organisms land. The genera *Trichobilharzia, Gigantobilharzia,*

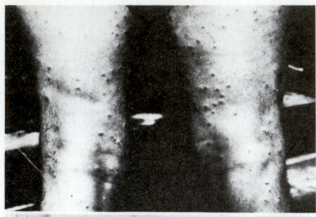

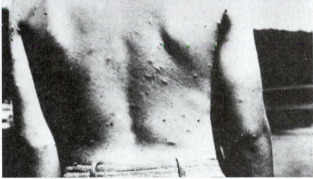

FIGURE 16.19

Cercarial dermatitis, or "swimmer's itch," caused by cercariae of avian blood flukes.

AFIP neg. no. 77203.

Ornithobilharzia, Microbilharzia, and *Heterobilharzia* are the guilty parties.

For the most part the skin reaction is a product of sensitization, with repeated infections causing increasingly severe reactions.[31] When the cercaria penetrates the skin and is unable to complete its migration, the host's defense responses rapidly kill it. At the same time the cercariae release allergens that cause inflammation and, typically, a pus-filled pimple (Fig. 16.19). The reaction may also be general, with an itching rash produced over much of the body. The condition is not a serious threat to health but is a terrific annoyance, much like poison ivy, that interrupts a summer vacation, for instance, or decreases the income of someone who rents lakefront cottages to the summer crowd. In the United States the problem is most serious in the Great Lakes area, but it has been reported from nearly all states.

Control depends mainly on molluscicides, but their usefulness is limited because they threaten sport fishing by poisoning the fish and, of course, because of the other problems we mentioned previously. Ocean beaches are occasionally infested with avian schistosome cercariae, for which no control has yet been devised.

References

1. Andrade, Z. A., T. M. Cox, and A. M. Cheever. 1993. Regression of hepatic lesions after treatment of *Schistosoma mansoni* or *Schistosoma japonicum* infection in mice: A comparative study. *Am. J. Trop. Med. Hyg.* 49:1–9.

2. Audibert, M., R. Josseran, R. Josse, and A. Adjidji. 1990. Irrigation, schistosomiasis, and malaria in the Logone Valley, Cameroon. *Am. J. Trop. Med. Hyg.* 42:550–60.

3. Basch, P. F. 1990. Why do schistosomes have separate sexes? *Parasitol. Today* 6:160–63.

4. Bickle, Q. D., M. G. Taylor, M. J. Doenhoff, and G. S. Nelson. 1979. Immunization of mice with gamma-irradiated intramuscularly injected schistosomula of *Schistosoma mansoni. Parasitology* 79:209–22.

5. Brand, K. G. 1979. Schistosomiasis—cancer: Etiological considerations. *Acta Tropica* 36:203–14.

6. Brémond, P., A. Théron, and D. Rollinson. 1989. Hybrids between *Schistosoma mansoni* and *S. rodhaini:* Characterization by isoelectric focusing of six enzymes. *Parasitol. Res.* 76:138–45.

7. Brooker, B. E. 1972. The sense organs of trematode miracidia. In Canning, E. U., and C. A. Wright, eds. *Behavioural aspects of parasite transmission. Linnean Society of London.* London: Academic Press, 171–80.

8. Cherfas, J. 1991. New hope for vaccine against schistosomiasis. *Science* 251:630–31.

9. Chernin, E. 1983. The curious case of the lateral-spined egg: *Schistosoma mansoni. Trans. R. Soc. Trop. Med. Hyg.* 77:847–50.

10. Damian, R. T. 1987. Presidential address—the exploitation of host immune responses by parasites. *J. Parasitol.* 73:1–13.

11. Deelder, A. M., N. de Jonge, O. C. Boerman, Y. E. Fillié, G. W. Hilberath, J. P. Rotmans, M. J. Gerritse, and D. W. O. L. Schut. 1989. Sensitive determination of circulating anodic antigen in *Schistosoma mansoni* infected individuals by an enzyme-linked immunosorbent assay using monoclonal antibodies. *Am. J. Trop. Med. Hyg.* 40:268–72.

12. Doenhoff, M. J., O. A. Hassounah, and S. B. Lucas. 1985. Does the immunopathology induced by schistosome eggs potentiate parasite survival? *Immunol. Today* 6:203–6.

13. File, S. K. 1994. Personal communication.

14. Foster, W. D. 1965. *A history of parasitology.* Edinburgh: E. & S. Livingstone.

15. Granzer, M., and W. Haas. 1986. The chemical stimuli of human skin surface for the attachment response of *Schistosoma mansoni* cercariae. *Int. J. Parasitol.* 16:575–79.

16. Greer, G. J., C. K. Ow-Yang, and H.-S. Yong. 1988. *Schistosoma malayensis* n. sp.: A *Schistosoma japonicum*-complex schistosome from peninsular Malaysia. *J. Parasitol.* 74:471–80.

17. Gupta, B. C., and P. F. Basch. 1987. The role of *Schistosoma mansoni* males in feeding and development of female worms. *J. Parasitol.* 73:481–86.

18. Hagan, P. 1992. Reinfection, exposure and immunity in human schistosomiasis. *Parasitol. Today.* 8:12–16.

19. Hofkin, B. V., G. M. Mkoji, D. K. Koech, and E. S. Loker. 1991. Control of schistosome-transmitting snails in Kenya by the North American crayfish *Procambarus clarkii. Am. J. Trop. Med. Hyg.* 45:339–44.

20. Hulse, E. V. 1971. Joshua's curse and the abandonment of ancient Jericho: Schistosomiasis as a possible medical explanation. *Med. Hist.* 15:376–86.

21. Johnston, D. A., E. Dias Neto, A. J. G. Simpson, and D. Rollinson. 1993. Opening the can of worms: Molecular analysis of schistosome populations. *Parasitol. Today* 9:286–91.

22. Johnston, D. A., R. A. Kane, and D. Rollinson. 1993. Small subunit (18S) ribosomal RNA gene divergence in the genus *Schistosoma. Parasitology* 107:147–56.

23. Jordan, P. 1985. *Schistosomiasis—the St. Lucia Project.* Cambridge: Cambridge Unviersity Press.

24. King, C. L., F. D. Miller, M. Hussein, R. Rarkat, and A. S. Monto. 1982. Prevalence and intensity of *Schistosoma haematobium* infection in six villages in Upper Egypt. *Am. J. Trop. Med. Hyg.* 31:320–27.

25. Leiper, R. T. 1915. Observations on the mode of spread and prevention of vesicle and intestinal bilharziosis in Egypt, with additions to August, 1916. *Proc. R. Soc. Med.* 9:145–72.

26. Madsen, H. 1990. Biological methods for the control of freshwater snails. *Parasitol. Today* 6:237–41.

26a. Maldonado, J. F. 1967. *Schistosomiasis in America.* Barcelona: Editorial Científico-Médica.

27. Manson, P. 1905. *Lectures on tropical diseases.* London: Constable, 54.

27a. Michelson, M. K., F. A. Azziz, F. M. Gamil, A. A. Wahid, F. O. Richards, D. D. Juranek, M. A. Habib, and H. C. Spencer. 1993. Recent trends in the prevalence and distribution of schistosomiasis in the Nile Delta region. *Am. J. Trop. Med. Hyg.* 49:76–87.

28. Mitchell, G. F. 1989. Glutathione *S*-transferases—potential components of anti-schistosome vaccines? *Parasitol. Today* 5:34–37.

29. Newport, G. R., and D. G. Colley. 1993. Schistosomiasis. In Warren, K. S., ed. *Immunology and molecular biology of parasitic infections,* 3d ed. Boston: Blackwell Scientific Publications, 387–437.

30. Okabe, K. 1964. Biology and epidemiology of *Schistosoma japonicum* and schistosomiasis. In Morishita, K., Y. Komiya, and H. Matsubayashi, eds. *Progress of medical parasitology in Japan,* 1. Tokyo: Meguro Parasitological Museum, 185–218.

31. Olivier, L. J. 1949. Schistosome dermatitis, a sensitization reaction. *Am. J. Hyg.* 49:209–301.

32. Osmundsen, J. A. 1965. Science: Battle is on against a dread crippler. *New York Times* (22 August, 1965), 8E.

33. Pearson, J. C. 1956. Studies on the life cycles and morphology of the larval stages of *Alaria orisaemoides* Augustine and Uribe, 1927 and *Alaria canis* La Rue and Fallis. *Can. J. Zool.* 34:295–387.

34. Pearson, J. C. 1959. Observations on the morphology and life cycle of *Strigea elegans* Chandler and Rausch, 1947 (Trematoda: Strigidae). *J. Parasitol.* 45:155–70, 171–74.

35. Peters, P. A., M. El Alamy, K. S. Warren, and A. A. F. Mahmoud. 1980. Quick Kato smear for field quantificaton of *Schistosoma mansoni* eggs. *Am. J. Trop. Med. Hyg.* 29:217–19.

36. Popiel, I. 1986. The reproductive biology of schistosomes. *Parasitol. Today* 2:10–15.

37. Ratard, R. C., and G. J. Greer. 1991. A new focus of *Schistosoma haematobium/S. intercalatum* hybrid in Cameroon. *Am. J. Trop. Med. Hyg.* 45:332–38.

38. Sambon, L. W. 1909. What is *Schistosoma mansoni* Sambon, 1907? *J. Trop. Med.* 12:1–11.

39. Shoop, W. L., and K. C. Corkum. 1987. Maternal transmission by *Alaria marcianae* (Trematoda) and the concept of amphiparatenesis. *J. Parasitol.* 73:110–13.

40. Shoop, W. L., W. F. Font, and P. F. Malatesta. 1990. Transmammary transmission of mesocercariae of *Alaria marcianae* (Trematoda) in experimentally infected primates. *J. Parasitol.* 76:869–73.

41. Spencer, H. C., E. Ruiz-Tibén, N. S. Mansour, and B. L. Cline. 1990. Evaluation of UNICEF/Arab Republic of Egypt/WHO Schistosomiasis Control project in Beheira Governorate. *Am. J. Trop. Med. Hyg.* 42:441–48.

42. Strickland, G. T., and M. F. Abdel-Wahab. 1991. Schistosomiasis. In G. T. Strickland, ed. *Hunter's tropical medicine,* 7th ed. Philadelphia: W. B. Saunders Co., 781–809.

43. Van der Schalie, H. 1974. Aswan Dam revisited. *Environment* 16(9):18–20, 25–26.

44. Voge, M., D. Bruckner, and J. I. Bruce, 1978. *Schistosoma mekongi* sp. n. from man and animals, compared with four geographic strains of *Schistosoma japonicum. J. Parasitol.* 64:577–84.

45. Wright, W. H. 1968. Schistosomiasis as a world problem. *Bull. N.Y. Acad. Med.* 44:301–2.

46. Wright, W. H. 1972. A consideration of the economic import of schistosomiasis. *Bull. WHO* 197:559–66.

Additional References

Ansari, N., ed. 1973. *Epidemiology and control of schistosomiasis (bilharziasis).* Basel: S. Karger AG. An official publication of the World Health Organization, outlining advances in the area.

Berrie, A. D. 1970. Snail problems in African schistosomiasis. In Dawes, B., ed. *Advances in parasitology* 8. New York: Academic Press, Inc., 43–96.

Bruce, J. I., and S. Sornmani, eds. 1980. *The Mekong schistosome. Malacological Reviews, Suppl. 2.* Whitmore Lake, Mich.

Loker, E. S. 1983. A comparative study on the life-histories of mammalian schistosomes. *Parasitology* 87:343–69.

McCully, R. M., C. N. Barron, and A. W. Cheever. 1976. Schistosomiasis (bilharziasis). In Binford, C. H., and D. H. Connor, eds. *Pathology of tropical and extraordinary diseases, vol. 2, sect. 10.* Washington, D.C.: Armed Forces Institute of Pathology.

Sobhon, P., and E. S. Upatham. 1990. *Snail hosts, life-cycle, and tegumental structure of oriental schistosomes.* Geneva: UNDP/WORLD BANK/WHO Special Program Research Training Tropical Diseases.

Chapter 17

DIGENEANS: ECHINOSTOMATIFORMES

Die Frage nach der Lebens- und

Entwickelungsgeschichte des Distomum hepaticum

hat mich bereits seit vielen Jahren beschäftigt.

Dr. R. Leuckart, Leipzig, 1881

Members of the order Echinostomatiformes often show little resemblance to one another in the adult stages, but studies on their embryology have shown common ancestries through developmental similarities. Often, though by no means always, the tegument bears well-developed scales or spines, particularly near the anterior end. The acetabulum is near the oral sucker. In many cases a second intermediate host is absent, and the metacercaria encysts on underwater vegetation or debris. Most species are parasitic in wild animals, but a few are important as agents of disease in humans, domestic animals, or both.

SUPERFAMILY ECHINOSTOMATOIDEA

Parasites of the superfamily Echinostomatoidea infect all classes of vertebrates and are found in marine, freshwater, and terrestrial environments. Some are among the most common parasites encountered, and a few cause devastating losses to agriculture.

Family Echinostomatidae

Echinostomes are easily recognized by their circumoral collar of peglike spines; hence their name (Fig. 17.1). The spines are arranged either in a single, simple circle or in two circles, one slightly lower than and alternating with the other. The collar is interrupted ventrally and at each end has a group of "corner spines." The size, number, and arrangement of these spines are of considerable taxonomic importance in both cercariae and adults. Echinostomes typically are slender worms with large preequatorial acetabula, pretesticular ovaries, and tandem testes, although exceptions occur. The vitellaria are voluminous and mainly postacetabular. These worms are parasites of the intestine or bile duct of reptiles, birds, and mammals, particularly those frequenting aquatic environments.

• Genus *Echinostoma*

Members of the genus *Echinostoma* (Fig. 17.2) are cosmopolitan, sometimes rather nonhost specific, parasites that are among the most widespread and abundant of all trematodes of warm-blooded, semiaquatic vertebrates. At least 15

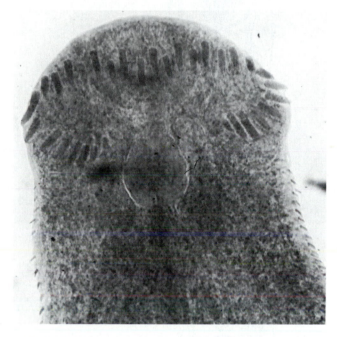

FIGURE 17.1

Anterior end of *Echinostoma* sp., showing the double crown of peglike spines on the circumoral collar.

Courtesy of Warren Buss.

species have been reported from humans, and human echinostomiasis is fairly common in the Orient, particularly in Taiwan and Indonesia.[2,13]

There has been some confusion in the literature regarding the taxonomy of *Echinostoma,* with up to 114 species being described from the genus. However, the bulk of the published research, especially experimental work, involves five species: *E. caproni, E. paransei, E. trivolvus, E. echinatum,* and *E. revolutum,* although the latter is found mostly in the older literature about studies that may actually have used one of the other species. Some of the taxonomic confusion results from the relatively low host specificity for members of the genus. "Preferred" natural definitive hosts for *E. caproni,* for example, include domestic ducks, rats, and Egyptian giant shrews.[13] *Echinostoma* species have proven so useful in the lab, however, that parasitologists are working diligently to solve the taxonomic problems. (See the review by Huffman and Fried.)[13]

Eggs hatch in water and the miracidia penetrate the first intermediate host. Routine inspections of snails of the genera

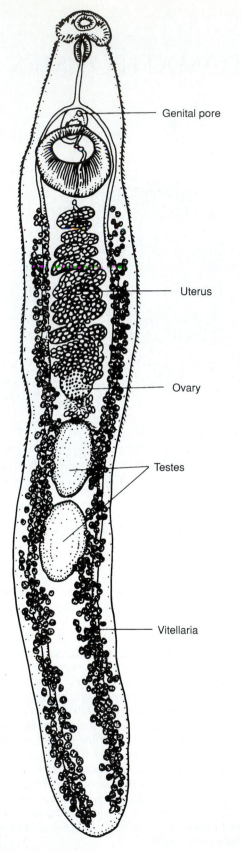

FIGURE 17.2

Echinostoma trivolvus, a common intestinal parasite of aquatic birds and mammals.

Drawing from Ian Grant.

Physa, Lymnaea, Helisoma, Paludina, and *Segmentina*—all common genera—often reveal echinostome infections. Metacercariae occur in molluscs, planaria, fish, and tadpoles. Infection of the definitive host is accomplished when the definitive host eats one of these. Humans are usually infected by eating raw mussels or snails.

Morphologically, worms of the genus *Echinostoma* are easily identified by their circumoral collar spines arranged in two rows. The number of spines ranges from 27 to 51, but the common species used in experimental work belong to a group with 37. The operculate eggs are large, typically about 100 μm by 60 μm, and few of them occur in the uterus at any one time. The genital pore is median and preacetabular, and the cirrus pouch is large, passing dorsal to the voluminous acetabulum. The short uterus has an ascending limb only. Overall size varies considerably among species.

Echinostoma caproni.   Huffman and Fried[13] consider many of the reports of *E. revolutum* actually studies of *E. caproni,* and Christensen et al.[6] consider *E. liei,* described by Jeyarasasingam et al.,[14] synonymous with *E. caproni.* Snails of the genus *Biomphalaria,* especially *B. glabrata* and *B. alexandrina,* are good first intermediate hosts. Daughter rediae migrate into the snail's gonad and digestive gland. Cercariae have a number of options, including migration up a snail's nephridiopore or penetration of a variety of second intermediate hosts such as clams, frogs, and sometimes fish.[13] *Echinostoma caproni* adults survive longer in mice and hamsters than in chickens.

Echinostoma paransei.   *Echinostoma paransei* is similar to the other species in many respects. Lie and Basch[18] reported both *Biomphalaria glabrata* and *Physa rivalis*—that is, snails of two distinct families (Planorbidae and Physidae)—served as first intermediate hosts. Sporocysts develop in the snail's ventricle, but rediae migrate through the tissues to a variety of organs. Cercariae emerge about 25 days after infection, live for six hours, and in experimental situations encyst, as metacercariae, in the very snails they came from. Lie and Basch[18] reported that metacercariae accumulated in the tentacle tips of *B. glabrata* snails, and dead or dying snails had many cysts in their heads. Although adult worms lived for five months in hamsters, infections of over 100 parasites killed the hosts.

Echinostoma trivolvus.   *Echinostoma trivolvus* is distinguished by a rather remarkable list of definitive hosts, including several species of ducks, geese, hawks, owls, doves, flamingos, dogs, cats, guinea pigs, rabbits, pigs, rats, and of course mice.[13] Not all hosts are equal, however; both worm size and number of eggs produced differ, depending on the definitive host. Christensen et al.[6] state that much of the early research on "*E. revolutum*" was actually done on *E. trivolvus.* For anyone who thinks all the world's systematic problems are solved or easily solvable, a journey through these discussions in the echinostome literature would be exceedingly educational.

***Echinostoma revolutum* (syn. *E. audyi*).**   There are numerous reports in the older literature of trematodes identified as *E. revolutum,* but the name is now restricted to worms matching Frolich's 1802 description. Huffman and Fried[13]

The figure labels read: Genital pore, Uterus, Ovary, Testes, Vitellaria.

consider *E. audyi* a synonym of *E. revolutum,* and the former's life cycle description by Lie and Umathevy[19] is a fine illustration of the way these seemingly complex events are discovered and deciphered. In these experiments, the snail *Lymnaea rubiginosa* served as both first and second intermediate hosts, and both rediae and daughter rediae were produced. Cercariae encysted in *L. rubiginosa* as well as in species of other snail genera and in some cases within the redia itself. Lie and Umathevy[19] were able to infect pigeons, ducklings, and sparrows experimentally, but they could not determine the preferred host in nature. Egg production started eight days after infection in the birds, but adult worms evidently died after eight weeks. Worms from pigeons were over 10 mm long, while those from ducklings were less than 7 mm, a difference that might lead a naive parasitologist to suspect he or she was dealing with more than one species.

Euparyphium ilocanum.   This species was formerly placed in the genus *Echinostoma,* but Huffman and Fried now consider it a member of the related genus *Euparyphium.* The eggs of *Eu. ilocanum* were first seen in the stool of a prisoner in Manila in 1907. The organism has since been found commonly throughout the East Indies and China. Tubangui[30] discovered that the Norway rat was an important reservoir of infection. *Euparyphium ilocanum* has 49 to 51 spines, with five or six corner spines at each end. The double row of spines is continuous dorsally, and the testes are deeply lobate.

The biology of *Eu. ilocanum* is similar to that of other echinostomes, with the metacercaria encysting in any freshwater mollusc. Infected snails are eaten raw by native peoples, who thereby become infected. The worms cause inflammation at their sites of attachment within the small intestine. Intestinal pain and diarrhea may develop in severe cases.

Other Echinostomid Species Reported from Humans.   Several species of echinostomes in different genera, which normally parasitize wild animals, have been reported from humans. These include *E. lindoense* in Celebes and possibly Brazil (considered a synonym of *E. echinatum* by Christensen et al.);[6] *E. malayanum* from India (Fig. 17.3), southeast Asia, and the East Indies; *E. cinetorchis* from Japan, Taiwan, and Java; *E. hortense* from Japan and Korea; *E. melis,* which is circumboreal; and *Hypoderaeum conoidum* in Thailand. Others are *Himasthla muehlensi* in New York; *Paryphostomum surfrartyfex* in Asia Minor; and *Echinochasmus perfoliatus* from eastern Europe and from Asia. These are only some of the species of echinostomes of wild animals that have found their ways into humans, but considering the lack of host specificity in this group, other species probably do so fairly often and remain undetected. Thus, when zoonotic infections are found, it immediately becomes necessary to identify the pathogen and determine how the person became infected. Faunal and systematic surveys, sometimes considered a less glamorous sort of science than experimental work, are the only ways to fulfill this need.

• Echinostomatids as Models in Experimental Parasitology

Echinostoma caproni, in nature a parasite of a Madagascar falcon, has proven very useful as an experimental animal. For example, experiments with this species, as well as some other echinostomes, especially *E. paransei,* have shown parasite

FIGURE 17.3

Echinostoma malayanum, a parasite of humans in southern Asia.

From T. Odhner, "Ein zweites Echinostomum aus dem Menschen in Ostasien (*Ech. malayanum* Leiper)," in *Zool. Anz.* 41:577–582. Copyright © 1913.

species specific pathological effects on vertebrate host intestinal epithelium.[12] Other studies have revealed alterations in the fatty acid composition of infected snails[11] and snail strain specific responses to parasites that are correlated with resistance or susceptibility to important trematodes such as *Schistosoma mansoni.* Snail resistance to trematodes is a well-known phenomenon, but in the case of echinostomes experimental work shows that parasite excretory and secretory proteins inhibit a host snail's ability to mount a cellular immune response.[20] And these tractable generalist parasites do not always need a complete definitive host in order to complete their life cycles in the lab. Fried and his coworkers have grown *E. caproni* (as well as several other trematode species) to maturity on the allantoic membranes of chick embryos.[5]

Family Fasciolidae

Members of the family Fasciolidae are large, leaf-shaped parasites of mammals, mainly of herbivores. They have a tegument covered with scalelike spines, and the acetabulum is close to the oral sucker. The testes and ovary are dendritic, and the vitellaria are extensive, filling most of the postacetabular space. There is no second intermediate host in the life cycle; the metacercariae encyst on submerged objects or freely in the water. One important species lives in the intestinal lumen, but most parasitize the liver of mammals.

Fasciola hepatica.   Although *Fasciola hepatica* (Fig. 17.4) is rare in humans in most countries, it has been known as an important parasite of sheep and cattle for hundreds of years. Because of its size and economic importance, it has been the subject of many scientific investigations and is probably the best known of any trematode species. Jean de Brie published

FIGURE 17.4

Fasciola hepatica, the sheep liver fluke.
Courtesy of Turtox/Cambosco.

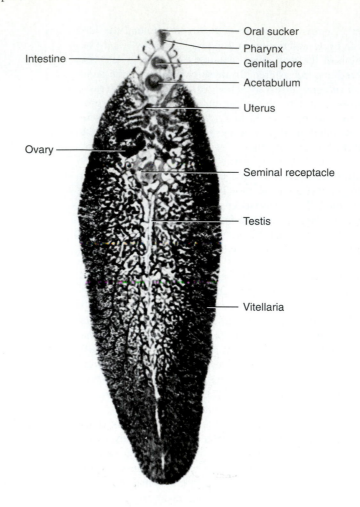

the first record of it in 1379. He was well-acquainted with a disease of sheep called "liver rot," in which the liver of an animal is infected with large, flat worms. In 1668 the great pragmatist Francisco Redi was the first to illustrate this fluke, thereby stimulating other researchers to investigate its biology. Leeuwenhoek was interested in the organism but apparently was distracted by all of the other beings he found with his microscope.

The cercaria and redia of *F. hepatica* were described in 1737 by Jan Swammerdam, a man with a remarkable ability to see and understand microscopic objects with the use of a primitive microscope. Linnaeus gave the worm its name in 1758 but considered it a leech. Pallas, in 1760, was the first to find it in a human. Professor C. L. Nitzsch, in 1816, was the first to recognize the similarity of cercariae and adult liver flukes. Thus, the history of *F. hepatica* parallels the history of trematodology itself in that the discoveries proved to be generally applicable to the biology of digeneans. In 1844 Johannes Steenstrup published a landmark book, *Alternation of Generations,* in which he postulated that trematodes have two generations, one adult and one not.

By the mid-1800s, circumstantial evidence indicated that molluscs were involved in the transmission of *F. hepatica.* In

1880 George Rolleston, professor of anatomy and physiology at Oxford, was convinced that a common slug was the intermediate host of *F. hepatica.* Although he was wrong in this assumption, he recommended that A. P. Thomas undertake an investigation to determine the life cycle of this parasite. Thomas was a 23-year-old demonstrator at the time but took on this formidable task with zeal. He soon found the snail *Lymnaea truncatula* infected with rediae and cercariae that were similar in many regards to *Fasciola.* Then he successfully infected this snail with miracidia and followed their development through the sporocyst, redia, and cercarial stages.

At the same time as this "lowly" Oxford demonstrator was investigating liver rot, fascioliasis was also engrossing the mind of the greatest parasitologist then living, Rudolph Leuckart. After a series of false starts Leuckart traced the development of this parasite through the same species of snail and, as a final irony, published his results 10 days before Thomas published his. Credit is given to both men equally, but one can scarcely refrain from lending sympathy to the young Englishman who elucidated the first trematode life cycle in a truly scientific manner, without the advantages of a large budget and long experience. Even today, there are many trematode species, indeed many parasite species,

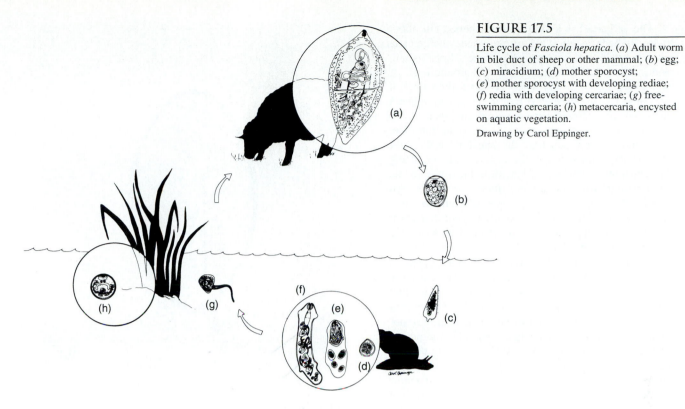

FIGURE 17.5

Life cycle of *Fasciola hepatica*. (*a*) Adult worm in bile duct of sheep or other mammal; (*b*) egg; (*c*) miracidium; (*d*) mother sporocyst; (*e*) mother sporocyst with developing rediae; (*f*) redia with developing cercariae; (*g*) free-swimming cercaria; (*h*) metacercaria, encysted on aquatic vegetation.

Drawing by Carol Eppinger.

whose life cycles are yet to be fully described and might well yield to insight, creative experiments, and hard work, regardless of the funds available for research.

Neither Thomas nor Leuckart determined the mode of infection of the definitive host. This was done by Adolph Lutz, a Brazilian working in Hawaii, who demonstrated between 1892 and 1893 that ruminants become infected by eating juveniles encysted on vegetation. That Lutz was actually working with a different species, *F. gigantica,* is immaterial, since the biology of both species is the same.

- **Morphology.** *Fasciola hepatica* is one of the largest flukes of the world, reaching a length of 30 mm and a width of 13 mm. It is rather leaf shaped, pointed posteriorly and wide anteriorly, although the shape varies somewhat. The oral sucker is small but powerful and is located at the end of a cone-shaped projection at the anterior end. The marked widening of the body at the base of the so-called oral cone gives the worm the appearance of having shoulders. The combination of an oral cone and "shoulders" is an immediate means of identification. The acetabulum is somewhat larger than the oral sucker and is quite anterior, at about the level of the shoulders. The tegument is covered with large, scalelike spines, reminding one of echinostomes, to which they are closely related. The intestinal ceca are highly dendritic (branched) and extend to near the posterior end of the body.

The testes are large and greatly branched, arranged in tandem behind the ovary. The smaller, dendritic ovary lies on the right side, shortly behind the acetabulum, and the uterus is short, coiling between the ovary and the preacetabular cirrus pouch. The vitelline follicles are extensive, filling most of the lateral body and becoming confluent behind the testes. The operculate eggs are 130 to 150 µm by 63 to 90 µm.

- **Biology.** Adult *F. hepatica* (Fig. 17.5) live in the bile passages of the liver of many kinds of mammals, especially ruminants. Humans are occasionally infected. In fact, fascioliasis is one of the major causes of hypereosinophilia in France.[7] The flukes feed on the lining of the biliary ducts. Their eggs are passed out of the liver with the bile and into the intestine to be voided with the feces. If they fall into water, the eggs will complete their development into miracidia and hatch in 9 to 10 days during warm weather. Colder water retards their development. On hatching, the miracidium has about 24 hours in which to find a suitable snail host, which in the United States is *Fossaria modicella* or *Stagncola bulimoides.*

In other parts of the world different but related snails are the important first intermediate hosts. Mother sporocysts produce first-generation rediae, which in turn produce daughter rediae that develop in the snail's digestive gland. Cercariae begin emerging five to seven weeks after infection. If the water in which the snails live dries up, the snails burrow into the mud and survive, still infected, for months at a time. When water is again present, the snails emerge and rapidly shed many cercariae.

The cercaria has a simple, club-shaped tail about twice its body length. Once in the water, the cercaria quickly attaches to any available object, drops its tail, and produces a thick, transparent cyst around itself. If it does not encounter an object within a short time, it will drop its tail and encyst free in the water. When a mammal eats metacercariae encysted on vegetation or in water, juvenile flukes excyst in the small intestine. They immediately penetrate the intestinal wall, enter the coelom, and creep over the viscera until contacting the capsule of the liver. Then they burrow into the liver parenchyma and wander about for almost two months, feeding and growing and finally entering bile ducts.[8] The worms become sexually mature in another month and begin producing eggs. Adult flukes can live as long as 11 years.

Fascioloid trematodes have been used to explore one of parasitology's most obdurate mysteries, namely the mechanisms by which parasites find their infection sites within hosts. Sukhdeo[29] showed that *F. hepatica* has a fixed behavioral pattern cued by a single component of bile, namely glycocholic acid. This molecule elicited emergence from the cyst. Migration to the bile ducts is evidently independent of at least some brain function; developmental studies showed that major portions of the cerebral gangionic complex were not present until after the worms had reached their final infection site.[29]

- *Epidemiology.* Infection begins when metacercaria-infected aquatic vegetation is eaten or when water containing metacercariae is drunk. Humans are often infected by eating watercress. Human infections occur in parts of Europe, northern Africa, Cuba, South America, and other locales. Surprisingly, few cases are known in humans in the United States, although the worm is fairly common in parts of the South and West. Sheep, cattle, and rabbits are the most frequent reservoirs of infection.

Whether or not humans are infected, veterinary fascioliasis is a major economic problem. For example, out of the nearly six million bovines slaughtered for food in Mexico between 1979 and 1987, 424,000 livers were confiscated.[4] And in a recent study in Montana , over 17% of the livers were infected, as determined by inspection during slaughter.[15] Regardless of whether you like liver, that is a lot of high quality food lost to the human population.

- *Pathology.* Little damage is done by juveniles penetrating the intestinal wall and the capsule surrounding the liver (Glisson's capsule), but much necrosis results from the migration of flukes through the parenchyma of the liver. During this time, they feed on liver cells and blood.[8] Anemia sometimes results from heavy infections. There is evidence that this anemia is not caused by hematophagia but instead probably is caused by a chemical released, perhaps proline.[27] Further, it is known that deposition of bile duct collagen is induced by proline from the worm.[31]

Worms in the bile ducts cause inflammation and edema, which in turn stimulate the production of

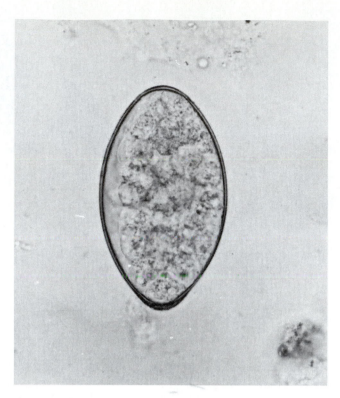

FIGURE 17.6

Egg of *Fasciola hepatica*. Eggs of this species are 130 to 150 μm by 63 to 90 μm.

Courtesy of Jay Georgi.

fibrous tissue in the walls of the ducts (pipestem fibrosis) (see Fig. 18.25). Thus thickened, the ducts can handle less bile and are less responsive to the needs of the liver. Back pressure causes atrophy of the liver parenchyma, with concomitant cirrhosis and possibly jaundice. In heavy infections the gallbladder is damaged, and the walls of the bile ducts are eroded completely through, with the worms then reentering the parenchyma, causing the large abscesses. Migrating juveniles frequently produce ulcers in ectopic locations, such as the eye, brain, skin, and lungs.

Ingestion of raw sheep or goat liver in the Middle East may result in adult worms establishing in the nasopharynx. The resulting respiratory blockage is called **halzoun** in Arabic. Recent research casts doubt on this disease entity, placing the full blame on pentastomid larvae and leeches. At this writing, *F. hepatica* remains suspect.

- *Diagnosis and Treatment.* Whenever liver blockage coincides with a history of watercress consumption, fascioliasis should be suspected. Specific diagnosis depends on finding the eggs (Fig. 17.6) in the stool. A false record can result when the patient has eaten infected liver and *Fasciola* eggs pass through with the feces. Daily examination during a liver-free diet will unmask this false diagnosis. An enzyme-linked immunosorbent assay (ELISA) test also is available. Early diagnosis is

FIGURE 17.7

Fasciola gigantica, a liver fluke. It may be as large as 75 mm long.
Courtesy of Warren Buss.

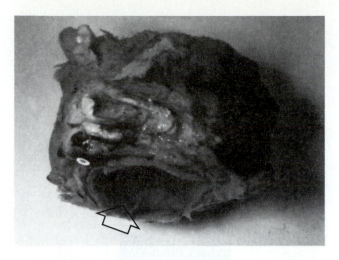

FIGURE 17.8

Large calcareous cyst from the liver of a steer (*arrow* points to its opening).
This cyst contained two *Fascioloides magna* and is about 3 in. wide.
Courtesy of Warren Buss.

important to avoid irreparable damage to the liver.
Prevention in humans depends on eschewing raw
watercress. In domestic animals the problem is
much more difficult to avoid. Snail control is always a possibility, although this is almost impossible in areas of high precipitation, where nearly
every cow track is filled with water and snails.
Reservoir hosts, particularly rabbits, can maintain
infestation of a pasture when pasture rotation is attempted as a control measure. *Fasciola hepatica* is
one of the most important disease agents of domestic stock throughout the world and shows
promise of remaining so for years to come. Losses
are enormous because of mortality, condemned
livers, reduction of milk and meat production, secondary bacterial infection, and expensive anthelmintic treatment.

Molecular techniques have been used in a number
of recent studies to address questions regarding diagnosis, pathogenesis, and evolution in fascioloid
trematodes. For example, use of isoelectric focusing
of soluble proteins has shown that regardless of the
worms' high morphological variability, *F. hepatica*
can be distinguished from *F. gigantica* on the basis
of protein separation banding patterns, but that *F.
hepatica* isolates from Korea and North America are
indistinguishable from each other.[16,17]

Excretory/secretory (ES) antigens are also potentially useful in immunodiagnostic tests. Experimental
work has shown that although monoclonal antibodies
to some ES products are crossreactive with antigens
from other helminths, at least two such antibodies
were specific to *F. hepatica,* and antigens from bile
and feces included a 26 kDa protein stable enough to
use in diagnostic assays.[10,26]

Several drugs are effective in chemotherapy of fascioliasis, both in humans and in domestic animals. One
of these, rafoxanide, apparently acts by uncoupling oxidative phosphorylation in the fluke.[24] Praziquantel is
ineffective against *F. hepatica.*

• Other Fasciolid Trematodes

Fasciola gigantica (Fig. 17.7), a species that is longer and
more slender than but otherwise very similar to *F. hepatica,*
is found in Africa, Asia, and Hawaii, being relatively common in herbivorous mammals, especially cattle, in these
areas. The morphology, biology, and pathology are nearly
identical to those of *F. hepatica,* although different snail
hosts are necessary.

Fasciola jacksoni causes anemia, loss of weight, hy-
poproteinemia, abdominal and submandibular edema, and
sometimes death in Asian elephants.[3]

Fascioloides magna is a giant in a family of large flukes,
reaching nearly 7.5 cm in length and 2.5 cm in width. Formerly strictly an American species, it was first discovered in
Europe, in an American elk from an Italian zoo. Now the
fluke has become established in Europe, mainly in game reserves. It is easily distinguished from *Fasciola* spp. by its
large size and the absence of a cephalic cone and "shoulders."
The life cycle is similar to that of *F. hepatica,* except that the
adults live in the liver parenchyma rather than in the bile
ducts. Because of their large size, they cause extensive damage. Often, but not always, they become encased in a calcareous cyst of host origin (Fig. 17.8). Their excretory system
produces great amounts of melanin, which fills their excretory
canals and also the cyst containing them. The normal hosts

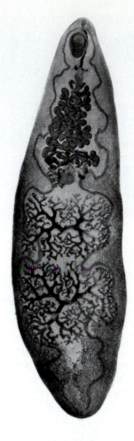

FIGURE 17.9

Fasciolopsis buski, an intestinal fluke in the Fasciolidae. It may reach 75 mm long.

Courtesy of Robert E. Kuntz.

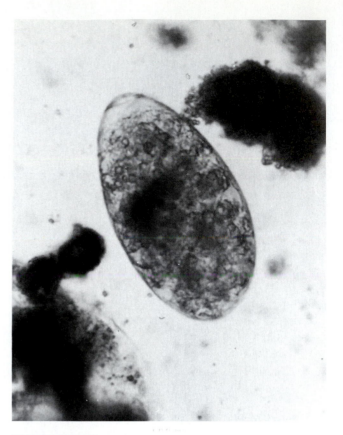

FIGURE 17.10

Egg of *Fasciolopsis buski* in a human stool in Taiwan. Eggs of this species are 130 to 140 μm by 80 to 85 μm.

Courtesy of Robert E. Kuntz.

are probably elk and other Cervidae, but cattle are commonly infected in endemic areas. Human infections have not been found.

In recent years evolutionary biologists have implicated parasites as selective factors influencing reproductive success, often through reduction of a male host's ability to compete for females. Infection with *Fascioloides magna,* for example, is associated with reduced body size and lowered number of antler points in white-tailed deer.[23]

Fasciolopsis buski (Fig. 17.9) is a common parasite of humans and pigs in the Orient. Stoll[28] estimated 10 million human infections in 1947. The number may be greater today. Although basically a typical fasciolid, it is peculiar because it lives in the small intestine of its definitive host rather than in the liver. It is elongated-oval, reaching a length of 20 to 75 mm and a width of up to 20 mm. There is no cephalic cone or "shoulders." The acetabulum is larger than the oral sucker and is located close to it. Another difference from "typical" fasciolids is the presence of unbranched ceca. The dendritic testes are tandem in the posterior half of the worm. The ovary is also branched and lies in the midline anterior to the testes. Vitelline follicles are extensive, filling most of the lateral parenchyma all the way to the caudal end. The uterus is short, with an ascending limb only. The eggs are almost identical to those of *F. hepatica.*

The life cycle of *F. buski* parallels that of *F. hepatica.* Each worm daily produces about 25,000 eggs (Fig. 17.10), which take up to seven weeks to mature and hatch at 27° to

32° C. Several species of snails of the genera *Segmentina* and *Hippeutis* (Planorbidae) serve as intermediate hosts. Cercariae encyst on underwater vegetation, including cultivated water chestnut, water caltrop, lotus, bamboo, and other edible plants (Fig. 17.11). Metacercariae are swallowed when these plants are eaten raw or when they are peeled or cracked with the teeth before eating. The worms excyst in the small intestine, grow, and mature in about three months without further migration. Infection, then, depends on human or pig feces being introduced directly or indirectly into bodies of water in which edible plants grow.[25]

Disease conditions resulting from *F. buski* are immunopathologic, obstructive, and traumatic. Inflammation at the site of attachment provokes excess mucous secretion, which is a typical symptom of infection. Heavy infections block the passage of food and interfere with normal digestive juice secretions. Ulceration, hemorrhage, and abscess of the intestinal wall result from long-standing infections. Chronic diarrhea is symptomatic. Another aspect of disease is a sensitization caused by absorption of the worm's allergenic metabolites. This may eventually cause the death of the patient. Treatment is usually effective in early or lightly infected cases. Late cases do not fare so well. Prevention is easy. Immersion of vegetables in boiling water for a few seconds will kill the metacercariae. Snail control should be attempted whenever it is impractical to prevent the use of night soil as a fertilizer.

FIGURE 17.11

Woman harvesting water caltrop, which serves as a medium for transport of metacercariae of *Fasciolopsis buski* to humans in Taiwan.
Courtesy of Robert E. Kuntz.

FIGURE 17.12

Stichorchis subtriquetrus, a stomach parasite of the American beaver. These parasites are about 10 mm long.
Courtesy of Warren Buss.

SUPERFAMILY PARAMPHISTOMOIDEA

The superfamily Paramphistomoidea contains the "amphistomes," flukes in which the acetabulum is located at or near the posterior end. Usually they are thick, fleshy worms with the genital pore preequatorial and the ovary usually posttesticular. Species are found in fishes, amphibians, reptiles, birds, and mammals. Of several families in this group we will consider three.

Family Paramphistomidae

Members of the family Paramphistomidae parasitize mammals, especially herbivores. Several species in different genera of this family parasitize sheep, goats, cattle, cervids, water buffalo, elephants, and other important animals. One species was found once in humans.

Paramphistomum cervi. The species *Paramphistomum cervi* lives in the rumen of domestic animals throughout most of the world. Adults are almost conical in shape and are pink when living. The testes are slightly lobate.

The life cycle is similar to that of *F. hepatica* and in North America at least, the parasites develop in the same snail hosts. The cercariae are large and pigmented and have eyespots. There is no second intermediate host; the metacercariae encyst on aquatic vegetation. When eaten, the worms excyst in the duodenum, penetrate the mucosa, and migrate anteriorly through the tissues. On reaching the abomasum, or true stomach, they return to the lumen and creep farther forward to the rumen. There they attach among the villi and mature in two to four months.

Paramphistomum cervi is a particularly pathogenic species. Migrating juveniles cause severe enteritis and hemorrhage, often killing the host. Secondary bacterial infection often complicates the problem. No adequate prevention or treatment is known.

Stichorchis subtriquetrus (Fig. 17.12) is a parasite of beavers, occurring throughout their range. Like *P. cervi,* the metacercariae encyst on underwater objects, including sticks that beavers embed in the bottom of their ponds and streams. When they later eat the bark off these sticks, they swallow any attached metacercariae. A peculiarity of this worm's early embryogenesis is the development of a mother redia within the miracidium. This is released in the snail immediately after penetration, and the remainder of the miracidium then disintegrates. Adult worms live in the stomach and reportedly have been causing mortality in beavers in the former Soviet Union.

Family Diplodiscidae

Flukes in the family Diplodiscidae have a pair of posterior diverticula in the oral sucker.

Megalodiscus temperatus. *Megalodiscus temperatus* and other genera and species of amphistomes are common parasites of the rectum and urinary bladder of frogs. They measure up to 6.00 mm long and 2.25 mm wide at the posterior end. The posterior sucker is equal to about the greatest width of the body.

The life cycle of this species is similar to that of other amphistomes in that no second intermediate host is required. Miracidia hatch soon after the eggs reach water and penetrate snails of the genus *Helisoma.* Cercariae have eyespots and swim toward lighted areas. If they contact a frog, they will encyst almost immediately on its skin, especially on its dark spots. Frogs molt the outer layers of their skin regularly and not infrequently will eat the sloughed skin. Metacercariae excyst in the rectum and mature in one to four months. If a tadpole eats a cercaria, the worm will encyst in the stomach and excyst when it reaches the rectum. At metamorphosis, when the amphibian's intestine shortens considerably, the flukes migrate anteriorly as far as the stomach and then to the rectum. These parasites are an easily obtained amphistome for general studies.

Family Gastrodiscidae

The morphology of the family Gastrodiscidae is essentially similar to that of Paramphistomidae and perhaps should not be separate from it. One of its species is a common parasite of humans in restricted areas of the world.

Gastrodiscoides hominis. This typical amphistome is cone shaped, fleshy, and pink. It is an important parasite of humans in Assam, India, and in southeastern Asia and the Philippines, inhabiting the lower small intestine and the upper colon. Rodents and primates are reservoirs.

Adult worms are 5 to 8 mm long by 5 to 14 mm wide at the ventral disc, which occupies about two-thirds of the ventral surface. There is a conspicuous posterior notch in the rim of the ventral sucker.

The complete life cycle of *G. hominis* is unknown, but the planorbid snail *Helicorbus coenosus* serves as an experimental host in India.[9] Presumably, humans are infected by eating uncooked aquatic plants. An adult worm draws a mass of mucosal tissue into the ventral sucker and remains attached for some time, causing a nipplelike projection on the intestinal lining.[1] The most common symptom is mucoid diarrhea. Treatment and prevention have not been well studied.

References

1. Ahluwalia, S. S. 1960. *Gastrodiscoides hominis* (Lewis and McConnell) Leiper 1913. *Indian J. Med. Res.* 48:315–25.

2. Bonne, C., G. Bras, and Lie Kian Joe. 1948. Five human echinostomes in the Malayan Archipelago. *Med. Monandbl.* 23:456–65.

3. Caple, I. W., M. R. Jainudeen, T. D. Buick, and C. Y. Song. 1979. Some clinico-pathologic findings in elephants (*Elephas maximus*) infected with *Fasciola jacksoni*. *J. Wildl. Dis.* 14:110–15.

4. Castellanos-Hurtado, A. A., I. Escutia-Sanchea, and H. Quiroz-Romero. 1992. Frequency of liver flukes in bovines in federal inspection slaughter-houses in Mexico from 1979 to 1987. *Veterinaria México* 23:339–42.

5. Chien, W. Y., and B. Fried. 1992. Cultivation of excysted metacercariae of *Echinostoma caproni* to ovigerous adults in the allantois of the chick embryo. *J. Parasitol.* 78:1019–23.

6. Christensen, N. O., B. Fried, and I. Kanev. 1990. Taxonomy of 37-collar spined *Echinostoma* (Trematoda: Echinostomatidae) in studies on the population regulation in experimental rodent hosts. *Angew. Parasitol.* 31:127–30.

7. Danis, M., J. -P. Nozais, and J. Chandenier. 1985. La fasciolose humaine en France. *L'Action Veterinaire* 907:1–3.

8. Dawes, B. 1961. On the early stages of *Fasciola hepatica* penetrating into the liver of an experimental host, the mouse: A histological picture. *J. Helminthol.*, R. T. Leiper Supplement, 41–52.

9. Dutt, S. C., and H. D. Srivastava. 1966. The intermediate host and the cercaria of *Gastrodiscoides hominis* (Trematoda: Gastrodiscidae). *J. Helminthol.* 40:45–52.

10. El-Bahi, M. M., J. B. Malone, W. J. Todd, and K. L. Schnorr. 1992. Detection of stable diagnostic antigen from bile and feces of *Fasciola hepatica* infected cattle. *Vet. Parasitol.* 45:157–67.

11. Fried, B., J. Sherma, K. S. Rao, and R. G. Ackman. 1993. Fatty acid composition of *Biomphalaria glabrata* (Gastropoda: Planorbidae) experimentally infected with the intramolluscan stages of *Echinostoma caproni* (Trematoda). *Comp. Biochem. Physiol.* (B) 104:595–98.

12. Fujino, T., and B. Fried. 1993. Expulsion of *Echinostoma trivolvis* (Cort, 1914) Kanev, 1985 and retention of *Echinostoma caproni* Richard, 1964 (Trematoda: Echinostomatidae) in C3H mice: Pathological, ultrastructural, and cytochemical effects on the host intestine. *Parasitol. Res.* 79:286–92.

13. Huffman, J., and B. Fried. 1990. *Echinostoma* and echinostomiasis. In Baker, J. R., and R. Muller, eds, *Advances in Parasitology* 29. New York: Academic Press, Inc., 215–69.

14. Jeyarasasingam, U., D. Heyneman, L. Hok-Kan, and N. Mansour. 1972. Life cycle of a new echinostome from Egypt, *Echinostoma liei* sp. nov. (Trematoda: Echinostomatidae). *Parasitology* 65:203–22.

15. Knapp, S. E., A. M. Dunkel, K. Han, and L. A. Zimmerman. 1992. Epizootiology of fascioliasis in Montana. *Vet. Parasitol.* 42:241–46.

16. Lee, C. G., G. L. Zimmerman, and S. H. Wee. 1992. *Fasciola hepatica:* Comparison of flukes from Korea and the United States by isoelectric focusing banding patterns of whole-body protein. *Vet. Parasitol.* 42:311–16.

17. Lee, C. G., and G. L. Zimmerman. 1993. Banding patterns of *Fasciola hepatica* and *Fasciola gigantica* (Trematoda) by isoelectric focusing. *J. Parasitol.* 79:120–23.

18. Lie, K. J., and P. F. Basch. 1967. The life history of *Echinostoma paraensei* sp. n. (Trematoda: Echinostomatidae). *J. Parasitol.* 53:1192–99.

19. Lie, K. J., and T. Umathevy. 1965. Studies on Echinostomatidae (Trematoda) in Malaya VIII. The life history of *Echinostoma audyi* sp. n. *J. Parasitol.* 51:781–88.

20. Loker, E. S., D. F. Cimino, and L. A. Hertel. 1992. Excretory-secretory products of *Echinostoma paraensei* sporocysts mediated interference with *Biomphalaria glabrata* hemocyte functions. *J. Parasitol.* 78:104–15.

21. Manger, P. M. Jr., and B. Fried. 1993. Infectivity, growth and distribution of preovigerous adults of *Echinostoma caproni* in ICR mice. *J. Helminthol.* 67:158–60.

22. Monroy, F., L. A. Hertel, and E. S. Loker. 1992. Carbohydrate-binding plasma proteins from the gastropod *Biomphalaria glabrata:* Strain specificity and the effects of trematode infection. *Dev. Comp. Immunol.* 16:355–66.

23. Mulvey, M., and J. M. Aho. 1993. Parasitism and mate competition: Liver flukes in white-tailed deer. *Oikos* 66:187–92.

24. Prichard, R. K. 1978. The metabolic profile of adult *Fasciola hepatica* obtained from rafoxanide-treated sheep. *Parasitol.* 76:277–88.

25. Sadun, E. H., and C. Maiphoom. 1953. Studies on the epidemiology of the human intestinal fluke, *Fasciolopsis buski* (Lankester) in central Thailand. *Am. J. Trop. Med. Hyg.* 2:1070–84.

26. Solano, M., R. K. Ridley, and H. C. Minocha. 1991. Production and characterization of monoclonal antibodies against excretory-secretory products of *Fasciola hepatica*. *Vet. Parasitol.* 40:227–40.

27. Spengler, R. N., and H. Isseroff. 1981. Fascioliasis: Is the anemia caused by hematophagia? *J. Parasitol.* 67:886–92.

28. Stoll, N. R. 1947. This wormy world. *J. Parasitol.* 33:1–18.

29. Sukhdeo, M. 1992. The behavior of parasitic flatworms in vivo: What is the role of the brain? *J. Parasitol.* 78:231–42.

30. Tubangui, M. A. 1931. Worm parasites of the brown rat, *Rattus norvegicus,* in the Philippine Islands, with special reference to those that may be transmitted to human beings. *Phil. J. Sci.* 46:537–91.

31. Wolf-Spengler, M. L., and H. Isseroff. 1983. Fascioliasis: Bile duct collagen induced by proline from the worm. *J. Parasitol.* 69:290–94.

Additional References

Boray, J. C. 1969. Experimental fascioliasis in Australia. In Dawes, B., ed. *Advances in parasitology* 7. New York: Academic Press, Inc., 96–210. A very interesting general account of fascioliasis, with many experimental approaches. Accent is on special problems of Australia.

Connor, D. H., and R. C. Neafie. 1976. Fasciolopsiasis. In Binford, C. H., and D. H. Connor, eds. *Pathology of tropical and extraordinary diseases,* vol. 2, sect. 10. Washington, D.C.: Armed Forces Institute of Pathology.

Dawes, B., and D. L. Hughes. 1964. Fascioliasis: The invasive stages of *Fasciola hepatica* in mammalian hosts. In Dawes, B., ed. *Advances in parasitology* 2. New York: Academic Press, Inc., 97–168.

Horak, I. G. 1971. Paramphistomiasis of domestic ruminants. In Dawes, B., ed. *Advances in parasitology* 9. New York: Academic Press, Inc., 33–72. A thorough discussion of paramphistomiasis, mainly outside North America.

Kendall, S. B. 1970. Relationships between the species of *Fasciola* and their molluscan hosts. In Dawes, B., ed. *Advances in parasitology* 8. New York: Academic Press, Inc., 251–58. This short article brings the earlier one [1966] by this author up-to-date.

Meyers, W. M., and R. C. Neafie. 1976. Fascioliasis. In Binford, C. H., and D. H. Connor, eds. *Pathology of tropical and extraordinary diseases,* 2, sect. 10. Washington, D.C.: Armed Forces Institute of Pathology.

Reinhard, E. G. 1957. Landmarks of parasitology. I. The discovery of the life cycle of the liver fluke. *Exp. Parasitol.* 6:208–32.

Chapter 18

DIGENEANS: PLAGIORCHIFORMES AND OPISTHORCHIFORMES

Suddenly the respiratory pore opened and a new slimeball was produced with an almost explosive expulsion from the respiratory chamber.

W. H. Krull and C. R. Mapes, describing a snail infected with *Dicrocoelium dendriticum.*

ORDER PLAGIORCHIFORMES

Even more than in some other orders, adults of the Plagiorchiformes often show little resemblance to each other. However, they have many larval and juvenile similarities. The wall of the excretory bladder is epithelial. The cercaria has a simple tail with a dorsal finfold, the primary excretory vesicle extends a short distance into the tail, and an oral stylet is usually present (xiphidiocercariae). In most members of the order, the eggs are small and must be eaten by the snail intermediate host before hatching.

Brooks and McLennan[6] include six suborders in the Plagiorchiformes, of which we will describe some examples from the Plagiorchiata and the Troglotrematata. Members of the Troglotrematata retain the homoplasious character of free-swimming miracidia.

Suborder Plagiorchiata

Members of the Plagiorchiata parasitize fishes, amphibians, reptiles, birds, and mammals and thereby are among the most commonly encountered flukes. They inhabit hosts in marine, freshwater, and terrestrial environments. Most are unimportant parasites of wild animals, but a few are important disease agents of humans and domestic animals. Cercariae possess an oral stylet, and the metacercaria usually encysts in an invertebrate intermediate host. We will discuss only a few of the many families in the suborder.

• Family Dicrocoeliidae

This is one of the three major families of liver flukes that we will consider in this book (see also Fasciolidae and Opisthorchiidae). Some species parasitize the gallbladder, pancreas, or intestine. All are parasites of terrestrial or semiterrestrial vertebrates and use land snails as first intermediate hosts. All dicrocoeliids are medium sized and flattened, with a subterminal oral sucker and a powerful acetabulum in the anterior half of the body. The body is usually pointed at both ends. The ceca are simple. Testes are preequatorial, and the ovary is posttesticular. The voluminous uterus has a descending and ascending limb, commonly filling most of the medullary parenchyma.

Dicrocoeliids are common among a wide variety of familiar vertebrates, but they rarely parasitize humans. One cosmopolitan species is an important parasite of domestic mammals and occasionally of humans.

Dicrocoelium dendriticum. *Dicrocoelium dendriticum* (Fig. 18.1) is common in the bile ducts of sheep, cattle, goats, pigs, and cervids and rarely is found in humans. It is common throughout most of Europe and Asia and has foci in North America and Australia, where it was recently introduced. Ducommun and Pfister[11] found almost 68% of female cattle and over 11% of male cattle in Switzerland infected with liver flukes, most of which were *D. dendriticum.* The trematode is commonly known as the **lancet fluke** because of its bladelike shape.

- ***Morphology.*** *Dicrocoelium dendriticum* is 6 to 10 mm long by 1.5 to 2.5 mm at its greatest width, near the middle. Both ends of the body are pointed. The ventral sucker is larger than the oral sucker and is located near it. The large, lobate testes lie almost in tandem directly behind the acetabulum, and the small ovary lies immediately behind them. Loops of the uterus fill most of the body behind the ovary. The vitellaria are lateral and restricted to the middle third of the body. The operculate eggs are 36 to 45 μm by 22 to 30 μm.

- ***Biology.*** *Dicrocoelium dendriticum* is an interesting example of a trematode that has dispensed with the aquatic environment at all stages of its life cycle. Adult *D. dendriticum* live in the bile ducts within the liver, much like *F. hepatica.* When laid, the eggs contain miracidia and must be eaten by land snails before they will hatch (Fig. 18.2). *Cionella lubrica* appears to be the most important snail host in the United States; other species serve in other lands. On hatching in the snail's intestine, the miracidium penetrates the gut wall and transforms into a mother sporocyst in the digestive gland. Mother sporocysts produce daughter sporocysts, which in turn produce xiphidiocercariae. The fact that these cercariae possess well-developed tails probably indicates a recent aquatic origin. About three months after infection, cercariae may accumulate in the "lung" (mantle cavity) of the snail or on its body surface. The snail surrounds this irritant with

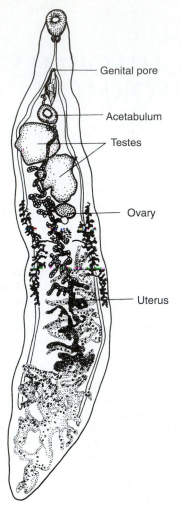

FIGURE 18.1

Dicrocoelium dendriticum, a liver fluke of mammals.
Drawing by Ian Grant.

thick mucus and eventually deposits these cercaria-containing **slime balls** as it crawls along. The slime ball may be expelled from the snail's pneumostome with some force. Slime balls are most abundantly produced during a wet period immediately following a drought. Individual slime balls may contain up to 500 cercariae each. Drying of the slime ball surface retards desiccation of the interior and thereby prolongs the lives of the cercariae within.

Continued development of the fluke depends on its ingestion by an ant, which becomes the second intermediate host. The common brown ant *Formica fusca* is the arthropod host in North America. On eating the delectable slime balls or feeding them to their larvae, the ants become host to metacercariae, most of which encyst in the hemocoel and are then infective to the definitive host. Over 100 metacercariae may occur in a single ant. One or two, however, migrate to the subesophageal ganglion and encyst there.[18] These will not

become infective, but they alter the ant's behavior in a most remarkable way. When the temperature drops in the evening, ants thus infected climb to the tops of grasses and other plants and grasp them firmly in their mandibles, while the uninfected nestmates return to warmer digs. The infected ants remain attached until later the next day when they warm up and seemingly resume normal behavior.[3] This behavioral pattern keeps the infected ants near the tops of vegetation in the active periods of grazing by ruminants in the evening and morning hours but allows them to retreat to cooler places during the hot hours of the day. The parasite thus influences its intermediate host to behave in a manner that increases the probability of passage to the definitive host. A similar phenomenon is found when the metacercariae of the dicrocoeliid *Brachylecithum mosquensis,* a parasite of American robins, encyst near the supraesophageal ganglion of carpenter ants, *Campanotus* spp. Instead of retreating from brightly lighted areas, as is normal for these ants, infected individuals actually seek such places and wander aimlessly or in circles on exposed surfaces. This makes them much more attractive to the bird definitive host.[8]

On being eaten by a definitive host, *D. dendriticum* excysts in the duodenum. It apparently is attracted by bile and quickly migrates upstream to the common bile duct and thence into the liver. The flukes mature in sheep in six or seven weeks and begin producing eggs about a month later. Up to 50,000 *D. dendriticum* have been found in a single sheep.

Pathological conditions of dicrocoeliiasis are basically the same as those for fascioliasis, except that there is no trauma to the gut wall or liver parenchyma resulting from migrating larvae. General biliary dysfunction, with its several symptoms, is typical.

Numerous cases of *D. dendriticum* in humans have been reported. Most of these were false infections. That is, the eggs that were detected in the stool were actually part of a liver repast that the person had enjoyed a few hours earlier. There have been a few genuine infections in humans, however, mainly in Russia, Europe, Asia, and Africa. One human case was reported in New Jersey.[10] Many cases of human infection with a related species, *D. hospes,* have been reported from Africa.[21]

Some benzimidazoles and praziquantel are effective against this trematode in domestic animals, but at a high dose rate that may not be economically feasible.[5] Praziquantel is the drug of choice for humans.[7] Control promises to be difficult in the foreseeable future because of the ubiquity of land snails and ants.

• Family Haematoloechidae

The flukes of the family Haematoloechidae are parasitic in the lungs of frogs and toads. They are of no economic or medical importance to humans, but because of their large size and easy availability, they are often the first live parasites seen by students of biology. Their transparent beauty,

FIGURE 18.2

Life cycle of *Dicrocoelium dendriticum*.
(*a*) Adult, in bile duct of sheep or other plant-eating mammal. (*b*) Egg. (*c*) Miracidium hatching from egg after being eaten by snail. (*d*) Mother sporocyst. (*e*) Daughter sporocyst. (*f*) Cercaria. (*g*) Slime balls containing cercariae. (*h*) Ant, eating slime balls. (*i*) Metacercaria encysted in ant; the definitive host becomes infected when it accidentally eats the ant.

Drawing by Carol Eppinger.

enigmatic location, and fascinating biology have led more than one biology student into a career in parasitology. In fact the parasitology literature abounds with first papers by later famous scientists dealing with medically unimportant organisms. One wonders if these parasites are unimportant if they served to attract serious scientists into the discipline.

Haematoloecus medioplexus (Fig. 18.3) is a typical species among the more than 40 known species that are found in all parts of the world where amphibians occur. It is a flat nonmuscular worm up to 8.0 mm long and 1.2 mm wide. The acetabulum is small and inconspicuous in this and related species. The uterus is voluminous, with a descending limb reaching near the posterior end and then ascending with wide loops to the genital pore near the oral sucker. So many eggs fill the uterus that most internal organs are obscured. However, when living worms are placed in tap water, they will expel most of the eggs and thereby become transparent enough to study.

Adult flukes lay prodigious numbers of eggs, which are carried out of the respiratory tract by ciliary action and thence through the gut to the outside. When swallowed by a scavenging *Planorbula armigera* snail, the miracidium hatches and migrates to the hepatic gland, where it develops into a sporocyst. Cercariae escape the snail by night and live the free life for up to 30 hours. When sucked into the rectal branchial chamber of a dragonfly nymph, the cercaria penetrates the thin cuticle and encysts in nearby tissues. When the insect metamorphoses into an adult, the metacercariae remain in the posterior end of the abdomen, to be eaten, along with the rest of the luckless dragonfly, by a frog or toad. Excystation occurs in the stomach. The little flukes creep through the stomach, up the esophagus, through the glottis, and into the respiratory tree. As many as 75 worms have been found in a single lung, although two or three are average.

The known life cycles of other haematoloechids are quite similar.

• Family Prosthogonimidae

Most prosthogonimids are parasites living in the oviduct, bursa of Fabricius, or gut of birds. They are remarkably transparent, stain well, and make excellent examples for classroom studies of trematode morphology.

Prosthogonimus macrorchis (Fig. 18.4) is about 8 mm long and is found in the oviduct flukes of domestic fowl and various wild birds in North America. It causes considerable damage to the oviduct and can decrease or even prevent egg laying. Many have been found within eggs after being trapped in the membranes formed by the oviduct, presumably giving cooks an unexpected surprise.

The life cycle of *P. macrorchis* is similar to that of *Haematoloechus* spp. When embryonated eggs pass into water, they sink to the bottom. They do not hatch until eaten by a snail (*Amnicola*). Then they burrow into the digestive gland, become sporocysts, and produce short-tailed xiphidiocercariae. When these are sucked into the rectal branchial chamber of a dragonfly nymph, they attach and penetrate into

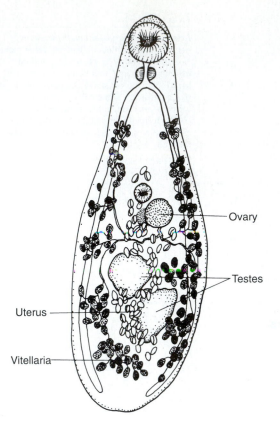

FIGURE 18.3

Haematoloechus medioplexus, common in the lungs of frogs.
Drawing by Ian Grant.

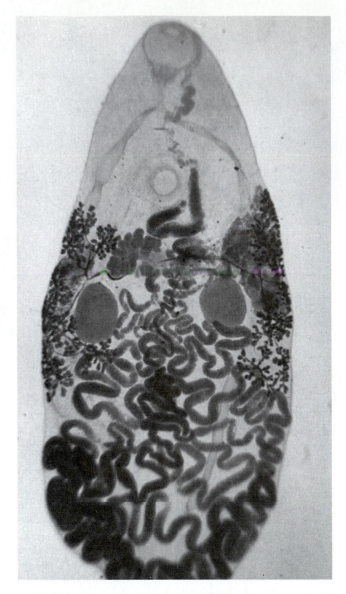

FIGURE 18.4

Prosthogonimus macrorchis, an oviduct fluke of birds.
Courtesy of Warren Buss.

the hemocoel and encyst in the muscles of the body wall, re-maining infective after the insect metamorphoses. When eaten by a bird, they excyst in the intestine, migrate down-stream to the cloaca and into the bursa of Fabricius or the oviduct, and mature in about a week. In male birds the infec-tion is lost when the bursa atrophies. In the United States the prevalence of *P. macrorchis* is declining because chickens are increasingly maintained in confined spaces, no longer able to pursue dragonfly prey. More than 30 species of *Prosthogo-nimus* are known from various areas around the world.

• Some Other Plagiorchiata

Another family of flukes often encountered in wild animals is Plagiorchiidae. They parasitize vertebrate classes from fishes through birds and mammals and are the typical flukes in the animal world. All species use aquatic snails as first in-termediate hosts and insects, such as mayflies and dragon-flies, as usual second intermediate hosts. The literature on plagiorchiids is replete with variations of life cycles and de-scriptions of the many species of worms encountered. Appar-ently this is an evolutionarily plastic group, adaptable to dif-ferent situations. Representative species are *Plagiorchis maculosus,* cosmopolitan in swallows (Fig. 18.5); *P. nobeli* in American blackbirds; and *Neoglyphe* in shrews. Several species of *Plagiorchis* have been reported from humans in

Japan, Java, and the Philippines, but these no doubt were zoonotic infections.[13]

Snakes that feed on amphibians, which bear the meta-cercariae, are often infected by members of the Ochetoso-matidae: *Ochetosoma (= Renifer)* and *Dasymetra* in the mouth and esophagus and *Pneumatophilus* and *Lechriorchis* in the trachea and lung.

Suborder Troglotrematata

• Family Troglotrematidae

The Troglotrematidae are oval, thick flukes with a spiny tegument and dense vitellaria. They are parasites of the lungs, intestine, nasal passages, cranial cavities, and various

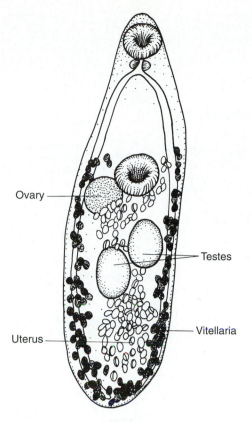

FIGURE 18.5

Plagiorchis maculosus, a common parasite of swallows. Birds are infected when they eat mayflies containing metacercariae.

Drawing by Ian Grant.

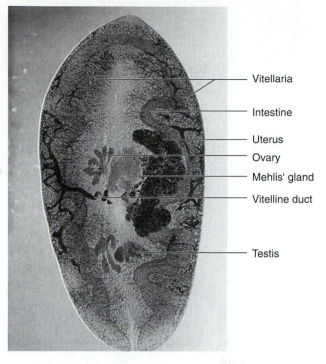

FIGURE 18.6

Adult *Paragonimus westermani.*
Courtesy of Robert E. Kuntz and Jerry A. Moore.

ectopic locations in birds and mammals in many parts of the world. We will illustrate the biology of this interesting group with discussions of two species.

***Paragonimus* spp.**   Paragonimiasis is an excellent example of a zoonosis. About 48 species and subspecies of *Paragonimus* have been described as parasites of carnivorous mammals, but not all of these may be valid. Seven species have been recorded from humans from three main foci: Asia and Oceania (*P. westermani, P. skrjabini, P. miyazakii, and P. heterotremus*).; western, sub-Saharan Africa (*P. africanus* and *P. uterobilateralis*); and South and Central America (*P. mexicanus*).[2,15] Several million people are infected in Asia.[15]

Paragonimus westermani is the most widely prevalent species. It was first described from two Bengal tigers that had died in zoos in Europe in 1878. During the next two years, infections by this worm in humans were found in Formosa. It was very quickly found in the lungs, brain, and viscera of humans in Japan, Korea, and the Philippines. The life cycle was worked out by Kobayashi[22] and Yokagawa.[38]

- ***Morphology.*** Adult worms (Fig. 18.6) are 7.5 to 12.0 mm long and 4 to 6 mm at their greatest width. They are very thick, however, measuring 3.5 to 5.0 mm in the dorsoventral axis. In life they are reddish brown, lending

the worms the overall size, shape, and color of coffee beans. The tegument is densely covered with scalelike spines. The oral and ventral suckers are about equal in size, with the latter placed slightly preequatorially. The excretory bladder extends from the posterior end to near the pharynx. The lobated testes are at the same level, located at the junction of the posterior fourth of the body. A cirrus and cirrus pouch are absent. The genital pore is postacetabular.

The ovary is also lobated and is found to the left of midline, slightly postacetabular. The uterus is tightly coiled into a rosette at the right of the acetabulum and opens into the common genital atrium with the vas deferens. Vitelline follicles are extensive in the lateral fields, from the level of the pharynx to the posterior end. The eggs are ovoid and have a rather flattened operculum set into a rim. They measure 80 to 118 μm by 48 to 60 μm.

Identification of the 30 or so species of *Paragonimus* is difficult, with much emphasis being placed on the characters of the metacercaria and the shape of the tegumental spines.[26] Several nominal species are probably synonyms.

- ***Biology.*** Adult *Paragonimus* (Fig. 18.7) usually live in the lungs, encapsulated in pairs by layers of granuloma (Fig. 18.8). They sometimes occur in many other organs of the body, however (Fig. 18.9). Cross-fertilization normally occurs. The eggs (Fig. 18.10) are often trapped in surrounding tissues and cannot leave the lungs, but those

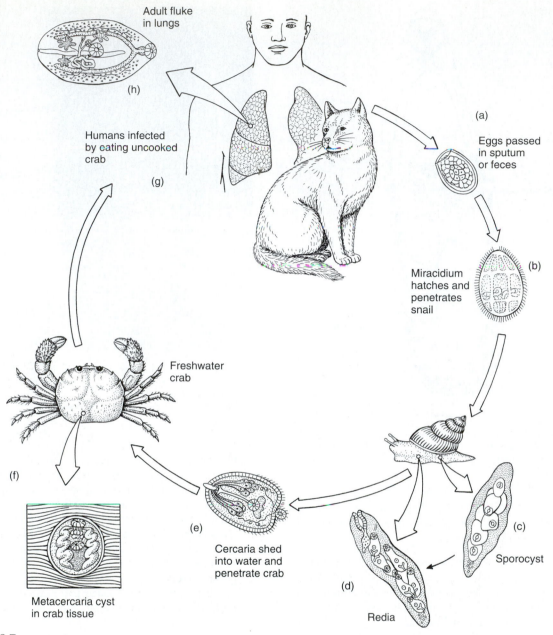

Adult fluke
in lungs

(h)

Humans infected
by eating uncooked
crab

(g)

Eggs passed
in sputum
or feces

(a)

Miracidium
hatches and
penetrates
snail

(b)

Freshwater
crab

Sporocyst

(c)

(f)

Cercaria shed
into water and
penetrate crab

(e)

Redia

(d)

Metacercaria cyst
in crab tissue

FIGURE 18.7

Life cycle of *Paragonimus westermani*. (*a*) Shelled embryo passed in feces or sputum. (*b*) After development miracidium hatches spontaneously and penetrates snail. (*c*) Sporocyst. (*d*) Redia. (*e*) Cercaria shed into water and penetrates crab. (*f*) Metacercarial cyst in tissue of freshwater crab. (*g*) Cats or humans infected by eating uncooked crab. (*h*) Adult fluke in lungs.
Drawing by William Ober and Claire Garrison.

that escape into the air passages are moved up and out by the ciliary epithelium. Most eggs escape before encapsulation is complete. Arriving at the pharynx, they are swallowed and passed through the alimentary canal to be voided with the feces. The larvae require from 16 days to several weeks in water before development of the miracidium is complete. Hatching is spontaneous, and the miracidium must encounter a snail in the family Thieridae if it is to survive. Since these snails usually live in swift-flowing streams, the chances of survival of any miracid-

ium are slight. This is offset by the numbers of eggs produced by the adult. On entering a snail, the miracidium forms a sporocyst that produces rediae, which in turn develop many cercariae. These cercariae (Fig. 18.11) are microcercous, with spined, knoblike tails and minute oral stylets.

After escaping from the snail, cercariae become quite active, creeping over rocks in inchworm fashion, and attack crabs and crayfish of at least 11 species, encysting in the viscera and muscles. A common second intermediate

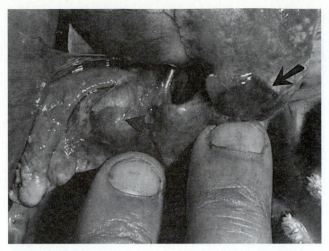

FIGURE 18.8

Lung of a cat with two cysts containing adult *Paragonimus westermani* *(arrows)*.

Courtesy of Robert E. Kuntz.

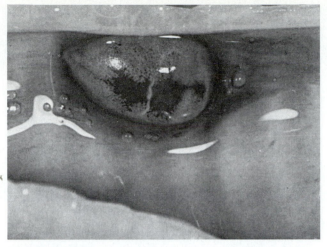

FIGURE 18.9

Adult *Paragonimus westermani* in the trachea of an experimentally infected cat.

Courtesy of Robert E. Kuntz.

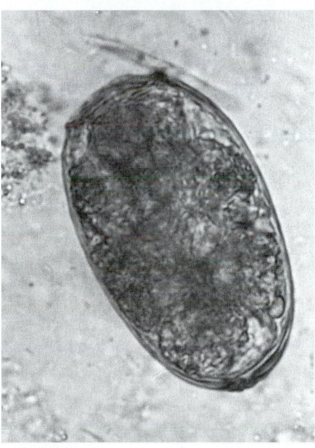

FIGURE 18.10

Egg of *Paragonimus westermani* from the feces of a cat. Eggs average 87 by 50 μm.

Courtesy of Robert E. Kuntz.

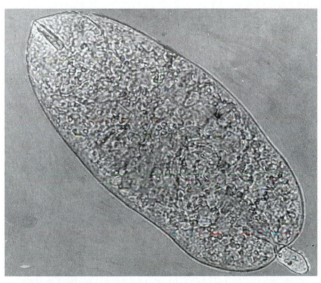

FIGURE 18.11

Microcercous cercaria of *Paragonimus westermani*. It is about 500 μm long.

Courtesy of Robert E. Kuntz.

host in Taiwan is *Eriocheir japonicus* (Fig. 18.12). Some evidence suggests that crustaceans may become infected by eating infected snails.[27] The metacercariae (Fig. 18.13) are pearly white in life and can be identified to species by an expert. When the crustacean is eaten by a proper definitive host, the worms excyst in the duodenum, pierce its wall, and embed themselves in the abdominal wall. Several days later they reenter the coelom, penetrate the diaphragm and pleura, and enter the bronchioles of the

FIGURE 18.12

Eriocheir japonicus, second intermediate host for *Paragonimus westermani* in Taiwan.

Courtesy of Robert E. Kuntz.

lungs. They mature in 8 to 12 weeks. Wandering juveniles may locate in ectopic locations, such as the brain, mesentery, pleura, or skin.

- *Epidemiology.* The natural, definitive, and, therefore, reservoir hosts of *Paragonimus* spp. are several species of carnivores, including felids, canids, viverrids, and mustelids, as well as some rodents and pigs. Humans are probably a lesser source of infective eggs than are other mammals, but like the others, humans become infected when they eat raw or insufficiently cooked crustaceans. Crab collectors in some countries distribute their catch miles from their source, effectively propagating paragonimiasis (Fig. 18.14). Completely raw crab or crayfish is not as commonly eaten in the Orient as that prepared by marination in brine, vinegar, or wine, which coagulates the protein in the muscles, giving it a cooked appearance and taste but not affecting the metacercariae. Exposure commonly is effected by contamination of fingers or cooking utensils during food preparation (Fig. 18.15).[36] It is even possible that persons accidentally become infected when they smash rice-eating crabs in the paddies, splashing themselves with juices that contain metacercariae. Another factor of possible epidemiological significance in some ethnic groups is the medicinal use of juices strained from crushed crabs or crayfish.

- *Pathology.* The early, invasive stages of paragonimiasis cause few or no symptomatic pathological conditions. Once in a lung or an ectopic site, the worm stimulates an inflammatory response that eventually will enshroud it in a capsule of granulation tissue. Such capsules often ulcerate and heal slowly. Eggs in surrounding tissues will themselves become centers of pseudotubercles. Worms in

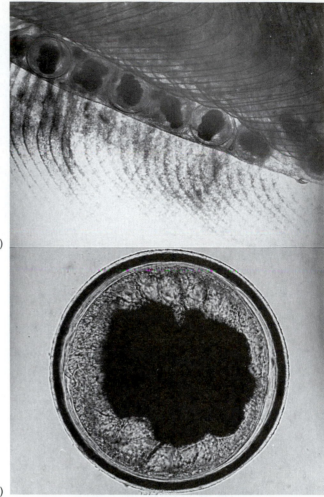

(a)

(b)

FIGURE 18.13

Metacercariae of *Paragonimus westermani.* (*a*) Several metacercariae in a gill filament of a crab. (*b*) A single metacercaria. The opaque mass is characteristic for this genus. The size is 340 to 480 µm.

Courtesy of Robert E. Kuntz.

the spinal cord can cause paralysis, which sometimes is total. Fatal cases of *Paragonimus* in the heart have been recorded. Cerebral cases have the same results as those of cerebral cysticercosis (see p. 334).[25] Pulmonary cases usually cause chest symptoms, with breathing difficulties, chronic cough, and sputum containing blood or brownish streaks (fluke eggs). Fatal cases are rare in pulmonary paragonimiasis.[15]

- *Diagnosis and Treatment.* The only sure diagnosis, aside from surgical discovery of the adult worm, is by finding the highly characteristic eggs in sputum, aspirated pleural fluid, feces, or matter from a *Paragonimus*-caused ulcer. The pulmonary type is easily mistaken for tuberculosis, pneumonia, spirochaetosis, and other such illnesses; and X-ray examination may be incorrectly interpreted.

FIGURE 18.14

Crab collectors stringing crabs for a trip to market in Taiwan. This practice distributes *Paragonimus* far from its source.

Courtesy of Robert E. Kuntz.

FIGURE 18.15

Children "cooking" fresh-caught crabs on an open fire. Such practices contribute to the widespread incidence of infection in eastern Asia.

Courtesy of Robert E. Kuntz.

Cerebral involvement requires differentiation from tumors, cysticercosis, hydatids, encephalitis, and others. Seroimmunological diagnosis is useful and particularly valuable in detecting ectopic infection. The intradermal test is practiced for surveys but must be followed by other assays on persons testing positive, because the dermal reaction persists for long periods after recovery from the disease. An assay that detects worm antigens using a monoclonal antibody is now available.[39]

The drug of choice is praziquantel.[15] Clinical symptoms decrease after five to six years of infection, but worms can live for 10 to 20 years. Infection can be avoided by cooking crustaceans before eating them and by avoiding contamination with their juices.

Paragonimus kellicotti closely resembles *P. westermani*. It has been found in a wide variety of mammals (cat, dog, raccoon, opossum, skunk, mink, muskrat, bobcat, pig, goat, red fox, coyote, weasel) in North America east of the Rocky Mountains, and many details of its life history and pathogenesis are known.[33,34] The first intermediate host is *Pomatiopsis lapidaria*. Crayfish of the common genus *Cambarus* serve as second intermediate hosts, with the metacercariae usually encysting on the heart. Like *P. westermani*, *P. kellicotti* in the definitive hosts is usually found in cysts occupied by pairs of worms in the lung. How the migrating worms find each other is still unknown, but encounter with another worm may be necessary for them both to mature.[32] One case of infection in a human has been reported.

Nanophyetus salmincola. By 1814 people realized that when dogs ate raw salmon in northwestern North America, they were prone to a disease so severe that scarcely 1 in 10 survived.[17] Over 100 years later an association with a minute fluke (Fig. 18.16) was found.[9] Metacercariae were common in the flesh and viscera of salmon. We now know that the disease itself is due to a rickettsia, *Neorickettsia helminthoeca,* which is transmitted to the dogs by the flukes. Infected dogs can be treated effectively with sulfanilamides and antibiotics.

- ***Morphology.*** Adult worms are 0.8 to 2.5 mm long and 0.3 to 0.5 mm wide. The oral sucker is slightly larger than the midventral acetabulum. The testes are side by side in the posterior third of the body. A cirrus pouch is present, but there is no cirrus. The small ovary is lateral to the acetabulum, and the uterus is short, containing only a few eggs at a time.

- ***Biology.*** Adult *N. salmincola* live deeply embedded in crypts in the wall of the small intestine of at least 32 species of mammals, including humans, as well as of fish-eating birds. They produce unembryonated eggs that hatch in water after 87 to 200 days. The snail host in the northwestern United States is *Oxytrema silicula,* an inhabitant of fast-moving streams. Experimental infection of snails in the laboratory has not been accomplished.

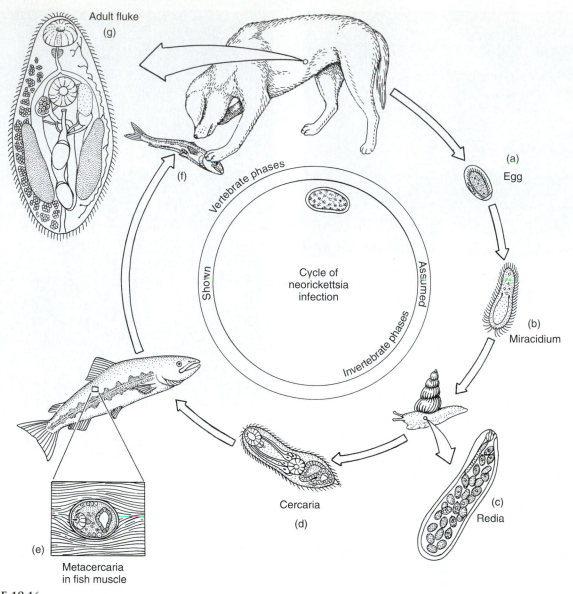

FIGURE 18.16

Life cycle of *Nanophyetus salmincola* and of the neorickettsia it harbors. (*a*) Shelled embryo is passed in feces. (*b*) After development, miracidium hatches spontaneously. (*c*) Redia. (*d*) Cercaria leaves snail and penetrates fish. (*e*) Metacercaria in fish muscle. (*f*) Dog eats raw fish. (*g*) Adult fluke in small intestine of dog.

Based on C. B. Philip, "Canine rickettsiosis in western United States and comparison with a similar disease in the Old World," in *Arch. Inst. Pasteur Tunis.* 36:505–603. Copyright © 1959.
Drawing by William Ober and Claire Garrison.

Sporocysts have not been found, but rediae are well-known, occurring in nearly all tissues of the snail. The xiphidiocercaria is microcercous. It penetrates and encysts in at least 34 species of fish, but salmonid fish are more susceptible than are fish of other families. Metacercariae can be found in nearly any tissue of the fish, but they are most numerous in the kidneys, muscles, and fins. Young fish have a high rate of mortality in heavy infection.[14] A variety of mammals and even two bird species (heron and merganser) can be infected with the trematode, but the raccoon and spotted skunk are clearly the main definitive hosts in nature.[31] The worm has been reported from humans in North America on at least 10 occasions and infects up to 98% of the people in some villages in Siberia.[12]

• **Pathology.** Adult flukes themselves cause surprisingly little disease. Philip[29] noted that the inflammatory changes in the intestine of a dog carrying hundreds of *Nanophyetus* were no more extensive than in animals infected with salmon poisoning disease by injection with lymph node suspensions. Salmon poisoning disease is restricted to dogs, coyotes, and other canids and does not affect humans. Its course in dogs is rapid and severe. After an incubation period of 6 to 10 days, the dog's temperature rises to 40°C to 42°C, often with edematous swelling of the face and discharge of pus from the eyes. The dog exhibits depression, loss of appetite, and increased thirst and then vomiting and diarrhea by four to seven days after the onset of symptoms. The fever usually lasts from four to seven days, and the dog usually dies about 10 days to two weeks after onset; however, those that recover are immune for the rest of their lives.

Much remains to be learned about the biology of this fluke and the rickettsia it harbors. Dogs are extremely susceptible, and when untreated, their mortality is about 90%. The disease can be transmitted experimentally by injection of lymph node preparations from other infected dogs or by injecting eggs (evidence of transovarial transmission in the fluke), metacercariae, or adult flukes and digestive glands from infected snails.[28] The geographical range of the disease coincides with the distribution of the snail host of the fluke, and the proportion of salmonids within this range that are infected is extremely high. In light of these facts and the mortality in dogs, we can assume that there is some reservoir of the rickettsia, but the identity of that reservoir is not at all clear. Raccoons do not seem to be susceptible; after fluke infection or injection with infected lymph nodes, they have a transitory, low-grade fever, but attempts to transmit the disease from them to dogs by way of lymph node preparations were unsuccessful.[29]

ORDER OPISTHORCHIFORMES

These are medium to small flukes, often spinose and with poorly developed musculature. The testes are at or near the posterior end, and a cirrus pouch and cirrus are absent. A seminal receptacle is present, and the metraterm and ejaculatory ducts unite to form a common genital duct. Eggs are embryonated when passed, but hatching occurs only after ingestion by a suitable snail. Adults live in the intestine or biliary system of fishes, reptiles, birds, and mammals. Metacercariae are in fishes.

• Family Opisthorchiidae

Opisthorchiids are delicate, leaf-shaped flukes with weakly developed suckers. Most are exceptionally transparent when prepared for study and so are popular subjects for parasitology classes. Adults are in the biliary system of reptiles, birds, and mammals. Three species, assigned to the genera *Clonorchis* and *Opisthorchis*, are of substantial consequence to humans. Over 30 million people are infected with *Clonorchis* or *Opisthorchis*.[19]

Clonorchis sinensis. *Clonorchis sinensis* was first discovered in the bile passages of a Chinese carpenter in Calcutta in 1875. Other infections were quickly discovered in Hong Kong and Japan. Although some authors have used the name *Opisthorchis sinensis*, Looss erected the genus *Clonorchis* in 1907 on the basis of its branched testes in contrast to the lobed testes of *Opisthorchis*.[23]

Today we know that the *C. sinensis* is widely distributed in Japan, Korea, China, Taiwan, and Vietnam, where it causes untold suffering and economic loss. Reports of this parasite outside the Orient involve infections people acquired while visiting there or by eating frozen, dried, or pickled fish imported from endemic areas. Prevalence of infection among 150 New York City immigrant Chinese was 26%. Four cases were described by Sun.[35]

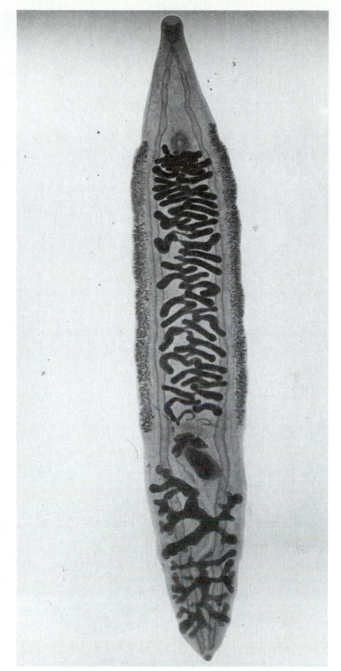

FIGURE 18.17

Chinese liver fluke, *Clonorchis sinensis*. Adults measure 8 to 25 mm long by 1.5 to 5.0 mm wide.

Courtesy of Robert E. Kuntz.

• ***Morphology.*** Adults (Fig. 18.17) measure 8 to 25 mm long by 1.5 to 5.0 mm wide. The tegument lacks spines, and the musculature is weak. The oral sucker is slightly larger than the acetabulum, which is about a fourth of the way from the anterior end.

The male reproductive system consists of two large, branched testes in tandem near the posterior end and a large, serpentine seminal vesicle leading to the

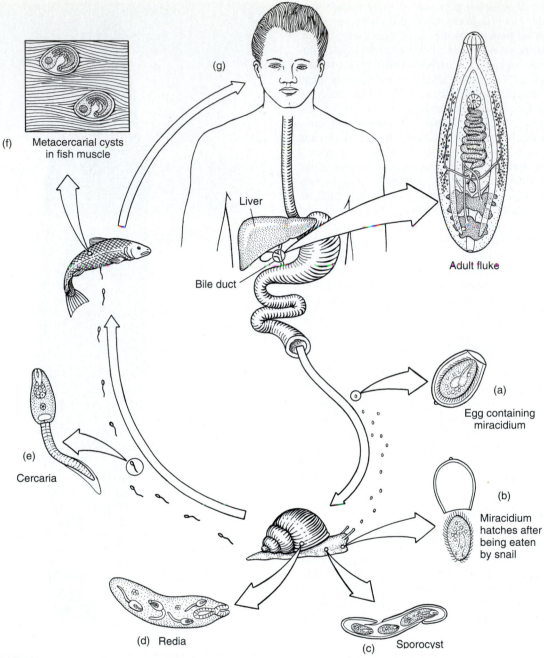

FIGURE 18.18

Life cycle of *Clonorchis sinensis*. (*a*) Egg containing miracidium is passed in feces. (*b*) Miracidium hatches after being eaten by snail. (*c*) Sporocyst. (*d*) Redia. (*e*) Cercaria leaves snail and penetrates fish. (*f*) Metacercarial cysts in fish muscle. (*g*) Human becomes infected by eating raw fish. (*h*) Adult fluke in bile ducts.

Drawing by William Ober and Claire Garrison.

genital pore. A cirrus and cirrus pouch are absent. The pretesticular ovary is relatively small and has three lobes. The seminal receptacle is large and transverse and is located just behind the ovary. The uterus ascends in broad, tightly packed loops and joins the ejaculatory duct to form a short, common genital duct. The genital pore is median, just anterior to the acetabulum. Vitelline follicles are small and dense and are confined to the level of the uterus. Laurer's canal is conspicuous.

- **Biology.** Human liver flukes (Fig. 18.18) mature in the bile ducts and produce up to 4000 eggs per day for at least six months. The mature egg (Fig. 18.19) is yellow-brown, 26 to 30 μm long and 15 to 17 μm wide. The operculum is large and fits into a broad rim of the eggshell. There is usually a small knob or curved spine on the aboperculum end that helps distinguish the eggs of this species. When passed, the egg contains a well-developed miracidium that is rather asymmetrical in its internal organization.

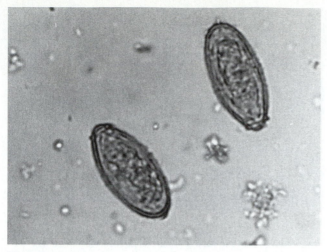

FIGURE 18.19

Eggs of *Clonorchis sinensis* from a human stool. They are 26 to 30 μm long. Note the small knob on the abopercular end.

Courtesy of Robert E. Kuntz and Jerry A. Moore.

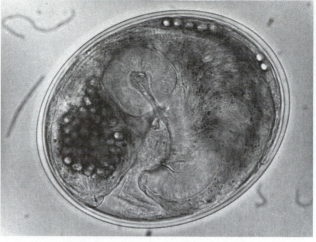

FIGURE 18.20

Encysted metacercaria of *Clonorchis sinensis* from fish muscle. The oral and ventral suckers are clearly seen; the round bodies are excretory corpuscles.

From J. B. Gibson and T. Sun, *Pathology of Protozoal and Helminthic Diseases with Clinical Correlation,* in Marcial-Rojas, editor, © 1971 Williams & Wilkins.

FIGURE 18.21

Grass carp, *Ctenopharyngodon idellus,* a common second intermediate host of *Clonorchis sinensis.* This fish is widely cultivated in eastern Asia.

From J. B. Gibson and T. Sun, *Pathology of Protozoal and Helminthic Diseases with Clinical Correlation,* in Marcial-Rojas, editor, © 1971 Williams & Wilkins.

Hatching of the miracidium will occur only after the egg is eaten by a suitable snail, of which *Parafossarulus manchouricus* is the most common and, therefore, most important first intermediate host throughout east Asia. The miracidium transforms into a sporocyst in the wall of the intestine or in other organs within four hours of infection. The sporocysts produce rediae within 17 days. Each redia produces from 5 to 50 cercariae. The cercaria has a pair of eyespots and is beset with delicate bristles and tiny spines. The entire cercaria is brownish. The tail has dorsal and ventral fins (**pleurolophocercous cercaria**).

The cercaria hangs upside down in the water and slowly sinks to the bottom. When contacting any object, it rapidly swims upward toward the surface and again begins to sink. Even a slight current of water will also cause this reaction. Thus, when a fish swims by, the cercaria is stimulated to react in a way favoring its contact with its next host. On touching the epithelium of the fish, the cercaria attaches with its suckers, casts off its tail, and bores through the skin, coming to rest and encysting under a scale or in a muscle (Fig. 18.20). Nearly a hundred species of fishes, mostly in Cyprinidae (Fig. 18.21), have been found naturally infected with metacercariae of *C. sinensis,* although some species are more susceptible than others. Thousands of metacercariae may accumulate in a single fish, but the number usually is much smaller. Metacercariae will also develop in the crustaceans *Caridina, Macrobrachium,* and *Palaemonetes;* and such metacercariae are infective, at least to guinea pigs.[37] The definitive host is infected when it eats raw or undercooked fish or crustaceans.

Mammals other than humans that have been found infected with adult *C. sinensis* are pigs, dogs, cats, rats, and camels.[24] Experimentally, rabbits and guinea pigs are highly susceptible. Perhaps any fish-eating mammal can become infected. Dogs and cats undoubtedly are important reservoir hosts. Birds may possibly be infected.

The young flukes excyst in the duodenum. The route of migration to the liver is not clear; conflicting reports have been published. It seems probable to us that the juveniles migrate up to the common bile duct to the liver. Young flukes have been found in the liver 10 to 40 hours after infection of experimental animals. The worms mature and begin producing eggs in about a month. The entire life cycle can be completed in three months under ideal conditions. Adult worms can live at least eight years in humans.

- *Epidemiology.* It is easy to see why clonorchiasis is common in countries in which raw fish is considered a delicacy (Fig. 18.22). In some areas the most heavily infected people are wealthy epicures who can afford beautifully cut and arranged slices of raw fish. On the other hand, the poor are also afflicted, since fish is often their only source of animal protein. The prevalence may range from an average of 14% in cities such as Hong Kong to 80% in some

FIGURE 18.22

"Yue-shan chuk," thin slices of raw carp with rice soup, vegetable garnishing, and soy sauce—a Cantonese delicacy.

From J. B. Gibson and T. Sun, *Pathology of Protozoal and Helminthic Diseases with Clinical Correlation,* in Marcial-Rojas, editor, © 1971 Williams & Wilkins.

FIGURE 18.23

Privy over a fish-culture pond in Hong Kong. The Chinese characters on the structure advertise a worm medicine.

From J. B. Gibson and T. Sun, *Pathology of Protozoal and Helminthic Diseases with Clinical Correlation,* in Marcial-Rojas, editor, © 1971 Williams & Wilkins.

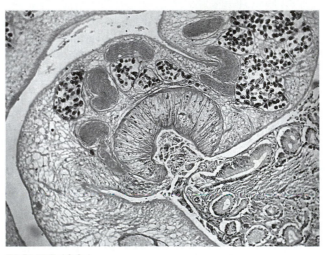

FIGURE 18.24

Adult *Clonorchis sinensis* attached by its ventral sucker to biliary epithelium in a human.

From J. B. Gibson and T. Sun, *Pathology of Protozoal and Helminthic Diseases with Clinical Correlation,* in Marcial-Rojas, editor, © 1971 Williams & Wilkins.

endemic rural areas. Although complete protection is achieved simply by cooking fish, it would be a futile exercise to try to get millions of people to change centuries-old eating habits. In addition, educating these people to cook their fish would not change matters, since fuel is a luxury that many cannot afford.

Fish farming is a mainstay of protein production throughout eastern Asia, in Europe, and increasingly in the United States. More protein in the form of fish can be harvested from an acre of pond than protein in the form of beef, beans, or corn from an acre of the finest farmland. The fastest growing fish are primary consumers of algae and other plants. Fish ponds typically are fertilized with human feces (Fig. 18.23), which increases the growth rate of water plants and thereby that of the fish. Of course this practice abets the life cycle of *C. sinensis.* Where fish farming is not so important, dogs and cats serve as reservoirs of infection, contaminating streams and ponds with their feces.

Metacercariae will withstand certain types of preparation of fish, such as salting, pickling, drying, and smoking. Because of this, people can become infected thousands of miles from an endemic area when they eat imported fish.[4]

- **Pathology.** The basic pathogenesis of *Clonorchis* infection is erosion of the epithelium lining the bile ducts (Fig. 18.24). The ultimate effect depends mainly on the intensity and duration of infection; fortunately worm burdens are usually small. The mean intensity of infection in most endemic areas is 20 to 200 flukes, but as many as 21,000 have been removed at a single autopsy. Chronic defoliation of the biliary epithelium leads to gradual thickening and occlusion of the ducts (Fig. 18.25). Pockets form in

the walls of bile ducts, and complete perforation into surrounding parenchyma may result. Infiltrating eggs become surrounded by granulomas, thereby interfering with liver function.

Ascites nearly always occurs in fatal cases, but its relationship to *Clonorchis* infection is uncertain. Jaundice is found in a small percentage of cases and is probably caused by bile retention when ducts are obstructed. Eggs and sometimes entire worms often become nuclei of gallstones. Cancer of the liver is more prevalent in Japan than elsewhere, and its relationship to clonorchiasis should be investigated.

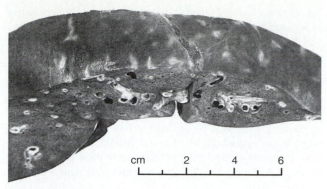

FIGURE 18.25

Severe clonorchiasis with "pipestem fibrosis" in a human. The dilated, thick-walled bile ducts are full of flukes.

From J. B. Gibson and T. Sun, *Pathology of Protozoal and Helminthic Diseases with Clinical Correlation,* in Marcial-Rojas, editor, © 1971 Williams & Wilkins.

- *Diagnosis and Treatment.* Diagnosis is based on the recovery of the characteristic eggs in the feces. Liver abnormalities just described should suggest clonorchiasis in endemic areas, but care must be taken to exclude cancer, hydatid disease, beriberi, amebic abscess, and other types of hepatic disease. Intradermal tests seem promising in making a diagnosis. Praziquantel is the drug of choice.

***Opisthorchis* spp.** *Opisthorchis felineus* is very similar to *C. sinensis* but has a more European distribution. Originally described from a domestic cat in Russia, it is common throughout southern, central, and eastern Europe, Turkey, southern Russia, Vietnam, India, and Japan. Besides parasitizing cats and other carnivores, it parasitizes humans, probably infecting more than a million persons within its range.

   Opisthorchis viverrini infects at least 7 million people in Thailand.[30] The pathology and epidemiology of this species, as well as those of *O. felineus,* are similar to those of *C. sinensis.* A control program in northeast Thailand consisting of education combined with praziquantel treatment has produced encouraging results.[30]

- **Family Heterophyidae**

Heterophyids are tiny, teardrop-shaped flukes, usually maturing in the small intestine of fish-eating birds and mammals. The distal portion of the vas deferens and uterus join to form a **hermaphroditic duct,** which opens into a **genital sac.** The genital sac may bear a muscular sucker, or **gonotyl,** which is greatly modified in different species. In some heterophyids the genital sinus encloses the acetabulum. A cirrus pouch is absent. The tegument is scaly, especially anteriorly. This is a large family with several subfamilies. Several species are important parasites of humans.

Heterophyes heterophyes. *Heterophyes heterophyes* (Fig. 18.26) is a minute fluke that was first discovered in an Egyptian in Cairo in 1851. It is common in northern Africa, Asia Minor, and the Far East, including Korea, China, Japan, Taiwan, and

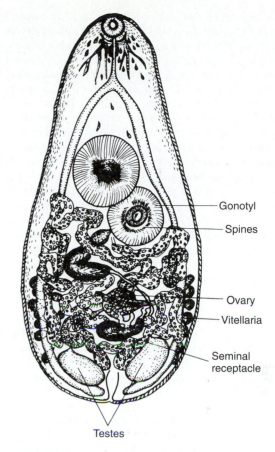

FIGURE 18.26

Heterophyes heterophyes. Its size is 1.0 to 1.7 mm long.
Drawing by Ian Grant.

the Philippines. Because the eggs cannot be differentiated from related species, an accurate estimate of human infections cannot be made.

- *Morphology.* The adults are 1.0 to 1.7 mm long and 0.3 to 0.4 mm at their greatest width. The entire body is covered with slender scales, most numerous near the anterior end. The oral sucker is only about 90 μm in diameter, whereas the acetabulum is around 230 μm wide and is located at the end of the first third of the body. The two oval testes lie side by side near the posterior end of the body. The vas deferens expands to form a sinuous seminal vesicle, which constricts again, becoming a short ejaculatory duct. The ovary is small, medioanterior to the testes, at the beginning of the last fourth of the body. A seminal receptacle and Laurer's canal are present. The uterus coils between the ceca and constricts before joining the ejaculatory duct to form a short common genital duct, which then opens into the genital sinus. The gonotyl is about 150 μm wide and has 60 to 90 toothed spines on its margin. Lateral vitelline follicles are few in number and are confined to the posterior third of the worm. The eggs are 28 to 30 μm by 15 to 17 μm.

- *Biology.* Adult worms live in the small intestine, burrowed between the villi. The egg contains a fully developed miracidium when laid but hatches only when eaten by an appropriate freshwater or brackish-water snail (*Pironella conica* in Egypt, *Cerithidia cingula* in Japan). After penetrating the gut of the snail, the miracidium transforms into a sporocyst that produces rediae. A second-generation redia gives birth to cercariae with eyespots and finned tails (**ophthalmolophocercous cercariae**), which emerge from the snail. Like the cercaria of *Clonorchis sinensis,* that of *H. heterophyes* swims toward the surface of the water and slowly drifts downward. On contacting a fish, it penetrates the epithelium, creeps beneath a scale, and encysts in muscle tissue. Metacercariae are most abundant in various species of mullet, which are exposed when they enter estuaries or brackish-water shorelines. Several thousand metacercariae have been found in a single, small fish. The definitive host becomes infected when it eats raw or undercooked fish.

- *Epidemiology.* For eggs to be available to estuarine and brackish-water snails, pollution must occur in these waters. Therefore, boatmen, fishermen, and others who live by or on the water are often the main sources of infection. Infected fish are distributed widely in fish markets. Other fish-eating mammals, such as cats, foxes, and dogs, serve as reservoirs of infection.

- *Pathology.* Each worm elicits a mild inflammatory reaction at the site of contact with the intestine. Heavy infections, which are common, cause damage to the mucosa and produce intestinal pain and mucous diarrhea. Perforation of the mucosa and submucosa sometimes occurs and allows eggs to enter the blood and lymph vascular systems and to be carried to ectopic sites in the body.[1] The heart is particularly affected, with tissue reactions in the valves and myocardium leading to heart failure. Kean and Breslau[20] reported that 14.6% of cardiac failure in the Philippines resulted from heterophyid myocarditis. Eggs in the brain or spinal cord lead to neurological disorders that are sometimes fatal. Two bizarre cases are known where adult *H. heterophyes* were found in the brains of humans, and in another case an adult worm was found in the myocardium.[1] Such infections are probably more common than previously thought, for experimental infections in laboratory animals often lead to ectopic flukes.[16] In these experiments immature flukes were found inside lymphoid follicles and Peyer's patches. Young flukes migrated from the sinuses in Peyer's patches via the lymphatics to the mesenteric lymph glands, which became enlarged and hyperplastic and contained mature worms.

  Diagnosis is difficult when adult worms are not available. The eggs closely resemble those of several other heterophyids and are not very different from those of *C. sinensis.* Praziquantel is effective in treatment.

Other Heterophyid Parasites of Humans. *Heterophyes katsuradai* is very similar to *H. heterophyes.* It has been found in humans near Kobe, Japan. Infection is acquired by eating raw mullet. *Metagonimus yokagawai* is a very common heterophyid in the Far East, the former Soviet Union, and the Balkan region, where it infects humans. It superficially resembles *H. heterophyes,* but its acetabulum is displaced to the left, where it is fused with the gonotyl. The biology of *M. yokagawai* is identical to that of *H. heterophyes,* except that a different snail host (*Semisulcospira* spp.) is required and the second intermediate hosts are freshwater fish of several species. The definitive host becomes infected when it eats uncooked fish. Various fish-eating mammals are natural reservoirs, and even pelicans have been implicated in this regard. Pathogenesis, diagnosis, and treatment are as for *H. heterophyes.*

Until proved otherwise, all species of Heterophyidae should be considered potential parasites of humans. More than a dozen species have been found infective to date.

References

1. Africa, C. M., W. de Leon, and E. Y. Garcia. 1937. Heterophyidiasis. VI. Two more cases of heart failure associated with the presence of eggs in sclerosed veins. *J. Philippine Isl. Med. Assoc.* 17:605–9.

2. Alarcón de Noya, B., O. Noya G., J. Torres, and C. Botto. 1985. A field study of paragonimiasis in Venezuela. *Am. J. Trop. Med. Hyg.* 34:766–69.

3. Anokhin, I. A. 1966. Daily rhythm in ants infected with metacercariae of *Dicrocoelium lanceatum. Dokl. Akad. Nauk SSSR* 166:757–59.

4. Binford, C. H. 1934. Clonorchiasis in Hawaii. Report of cases in natives of Hawaii. *Public Health Rep.* 49:602–4.

5. Boray, J. C. 1986. Trematode infections of domestic animals. In Campbell, W. C., and R. S. Rew, eds. *Chemotherapy of parasitic diseases.* New York: Plenum Press, 401–25.

6. Brooks, D. R., and D. A. McLennan. 1993. *Parascript. Parasites and the language of evolution.* Washington, D.C.: Smithsonian Institution Press.

7. Bunnag, D., T. Bunnag, and R. Goldsmith. 1991. Liver fluke infections. In Strickland, G. T., ed. *Hunter's tropical medicine,* 7th ed. Philadelphia: W. B. Saunders Company, 818–27.

8. Carney, W. P. 1969. Behavioral and morphological changes in carpenter ants harboring dicrocoeliid metacercariae. *Am. Midland Natur.* 82:605–11.

9. Donham, C. R., B. T. Simms, and F. W. Miller. 1926. So-called salmon poisoning in dogs (progress report). *J. Am. Vet. Med. Assoc.* 68:701–15.

10. Drabick, J. J., J. E. Egan, S. L. Brown, R. G. Vicr, B. M. Sandman, and R. C. Neafie. 1988. Dicroceliasis (lancet fluke disease) in an HIV seropositive man. *JAMA* 259:567–68.

11. Ducommun, D., and K. Pfister. 1991. Prevalence and distribution of *Dicrocoelium dendriticum* and *Fasciola hepatica* infections in cattle in Switzerland. *Parasitol. Res.* 77:364–66.

12. Eastburn, R. L., T. R. Fritsche, and C. A. Terhune Jr. 1987. Human intestinal infection with *Nanophyetus salmincola* from salmonid fishes. *Am. J. Trop. Med. Hyg.* 36:586–91.

13. Faust, E. C., P. F. Russell, and R. C. Jung. 1970. *Craig and Faust's clinical parasitology,* 8th ed. Philadelphia: Lea & Febiger.

14. Gebhardt, G. A., R. E. Millemann, S. E. Knapp, and P. A. Nyberg. 1966. "Salmon poisoning" disease. II. Second intermediate host susceptibility studies. *J. Parasitol.* 52:54–59.

15. Goldsmith, R., D. Bunnag, and T. Bunnag. 1991. Lung fluke infections: paragonimiasis. In Strickland, G. T. (ed). *Hunter's tropical medicine*, 7th ed. Philadelphia: W. B. Saunders Co., 827–31.

16. Hamdy, E. L., and E. Nicola. 1981. On the histopathology of the small intestine in animals experimentally infected with *H. heterophyes. J. Egypt. Med. Assoc.* 63:179–84.

17. Henry's Astoria Journal. 1814. *In the Oregon country under the Union Jack. A reference book of historical documents for Scholars and Historians.* 1962. Montreal: Rayette Radio Ltd.

18. Hohorst, W. 1964. Die Rolle der Ameisen in Entwicklungsangan des Lanzettegels *(Dicrocoelium dendriticum). Z. Parasitenkd.* 22:105–6.

19. Hopkins, D. R. 1992. Homing in on helminths. *Am. J. Trop. Med. Hyg.* 46:626–34.

20. Kean, B. H., and R. C. Breslau. 1964. *Parasites of the human heart.* New York: Grune & Stratton, Inc., 95–103.

21. King, E. V. J. 1971. Human infection with *Dicrocoelium hospes* in Sierra Leone. *J. Parasitol.* 57:989.

22. Kobayashi, H. 1918. Studies on the lung-fluke in Korea. *Mitteil. Med. Hochschule Keijo* 2:95–113.

23. Komiya, Y. 1966. *Clonorchis* and clonorchiasis. In Dawes, B., ed. *Advances in parasitology* 4. New York: Academic Press, Inc., 53–106.

24. Komiya, Y., and N. Suzuki. 1964. Biology of *Clonorchis sinensis.* In Morishita, K., et al., eds. *Progress of medical parasitology in Japan* 1. Tokyo: Meguro Parasitological Museum, 551–645.

25. Madrigal, R. B., B. Rodriquez-Otiz, G. V. Solano, E. M. O. Oband, and P. J. R. Sotela. 1982. Cerebral hemorrhagic lesions produced by *Paragonimus mexicanus. Am. J. Trop. Med. Hyg.* 31:522–26.

26. Miyazaki, I. 1965. Recent studies on *Paragonimus* in Japan with special reference to *P. ohirai* Miyazaki, 1939, *P. iloktsuenensis* Chen, 1940 and *P. miyazakii* Kamo, Nishida, Hatsushika et Tomimura, 1961. In Morishita K., et al., eds. *Progress of medical parasitology in Japan* 2. Tokyo: Meguro Parasitological Museum, 349–54.

27. Noble, G. A. 1963. Experimental infection of crabs with *Paragonimus. J. Parasitol.* 44:352.

28. Nyberg, P. A., S. E. Knapp, and R. E. Millemann. 1967. "Salmon poisoning" disease. IV. Transmission of the disease to dogs by *Nanophyetus salmincola* eggs. *J. Parasitol.* 53:694–99.

29. Philip, C. B. 1955. There's always something new under the "parasitological" sun (the unique story of the helminthborne salmon poisoning disease). *J. Parasitol.* 41:125–48.

30. Saowakontha, S., V. Pipitgool, S. Pariyanonda, S. Tesana, K. Rojsathaporn, and C. Intarakhao. 1993. Field trials on the control of *Opisthorchis viverrini* with an integrated programme in endemic areas of northeast Thailand. *Parasitology* 106:283–88.

31. Schlegel, M. W., S. E. Knapp, and R. E. Millemann. 1968. "Salmon poisoning" disease. V. Definitive hosts of the trematode vector, *Nanophyetus salmincola. J. Parasitol.* 54:770–74.

32. Sogandares-Bernal, F. 1966. Studies on American paragonimiasis. IV. Observations on pairing of adult worms in laboratory infections of cats. *J. Parasitol.* 52:701–3.

33. Sogandares-Bernal, F., and J. R. Seed. 1973. American paragonimiasis. *Curr. Top. Comp. Pathobiol.* 2:1–56.

34. Stromberg, P. C., and J. P. Dubey. 1978. The life cycle of *Paragonimus kellicotti* in cats. *J. Parasitol.* 64:998–1002.

35. Sun, T. 1980. Clonorchiasis: A report of four cases and discussion of unusual manifestations. *Am. J. Trop. Med. Hyg.* 29:1223–27.

36. Suzuki, Z. 1958. Epidemiological studies on paragonimiasis in South Izu District, Shizuoka Prefecture. *Japan Kiseichugaku Zasshi* 7:560–72.

37. Tang, C. C., et al. 1963. Clonorchiasis in South Fukien with special reference to the discovery of crayfishes as second intermediate hosts. *Chinese Med. J.* 82:545–618.

38. Yokagawa, S. 1919. A study of the lung distoma. Third report. *Formosan Endoparasitic Disease Research.*

39. Zhang, Z., Y. Zhang, Z. Shi, K. Sheng, L. Liu, Z. Hu, and W. F. Piessens. 1993. Diagnosis of active *Paragonimus westermani* infections with a monoclonal antibody-based antigen detection assay. *Am. J. Trop. Med. Hyg.* 49:329–34.

Additional References

Dooley, J. R., and R. C. Neafie. 1976. Clonorchiasis and opisthorchiasis. In Binford, C. H., and D. H. Connor, eds. *Pathology of tropical and extraordinary diseases,* vol. 2, sect. 10. Washington, D.C.: Armed Forces Institute of Pathology.

Holmes, J. C., and W. M. Bethel. 1972. Modification of intermediate host behaviour by parasites. In Canning, E. U., and C. A. Wright, eds. *Behavioural aspects of parasite transmission. Linnaean Society of London.* London: Academic Press, 123–49. An outstanding summary of the subject. Should be required reading for all students of parasitology.

Komiya, Y. 1966. *Clonorchis* and clonorchiasis. In Dawes, B., ed. *Advances in parasitology* 4. New York: Academic Press, Inc., 53–106.

Meyers, W. M., and R. C. Neafie. 1976. Paragonimiasis. In Binford, C. H., and D. H. Connor, eds. *Pathology of tropical and extraordinary diseases,* vol. 2, sect. 10. Washington, D.C.: Armed Forces Institute of Pathology.

Millemann, R. E., and S. E. Knapp. 1970. Biology of *Nanophyetus salmincola* and "salmon poisoning" disease. In Dawes, B., ed. *Advances in parasitology* 8. New York: Academic Press, Inc., 1–41.

Travassos, L. 1944. Revisão da família Dicrocoeliidae Odhner, 1910. *Inst. Oswaldo Cruz Monogr.* 2:1–357. The definitive monograph on Dicrocoeliidae.

Yokagawa, M. 1965. *Paragonimus* and paragonimiasis. In Dawes, B., ed. *Advances in parasitology* 3. New York: Academic Press, Inc., 99–158. A complete summary of paragonimiasis. Required reading for all who are interested in the subject.

Chapter 19

MONOGENEA

For oaths are straws, men's faiths are wafer-cakes,

And hold-fast is the only dog, my duck.

—William Shakespeare, *Henry V*

The Monogenoidea are given class status by some authors[4] and infraclass status by others.[5] Regardless of where and how humans classify them, monogeneans are hermaphroditic flatworms that mainly are external parasites of vertebrates, particularly fish. Some species, however, are found internally in diverticula of the stomodeum or proctodeum and also in the ureters of fishes and the bladders of turtles and frogs. A single species is known from mammals: *Oculotrema hippopotami* from the eye of the hippopotamus.[39] Nevertheless, monogeneans are primarily fish parasites, particularly of the gills and external surfaces. Although a few fish deaths have been attributed to monogeneans in nature, the worms are not usually regarded as hazardous to wild populations. However, like copepods and numerous other fish pathogens, monogeneans become a serious threat when fish are crowded together, as in hatcheries and in farming operations.

Despite their remarkable morphology and life cycles and their considerable economic importance, monogeneans have not attracted the attention of large numbers of parasitologists. Probably fewer than half of the existing species have been described.[38] These worms were aligned loosely with digenetic trematodes by early parasitologists. The first comprehensive overview of the group was by Braun[6] in 1889 and 1893. Next, Fuhrmann,[12] in 1928, helped establish the Monogenea as a category separate from the Digenea, although closely allied with it. Bychowsky,[7] in 1937, was apparently the first to propose Monogenea as a separate class, apart from and equal to Digenea. However, this viewpoint is not universally accepted.[36] Nevertheless, modern phylogenetic analysis of the parasitic flatworms reveals that the monogeneans are probably more closely related to the tapeworms than to the trematodes (see Chapter 13).

Monogeneans are often very particular about both the species of host and the site where they live on that host, restricting themselves to extremely narrow niches in many cases. Thus, one species may live only at the base of a gill filament, whereas another is found only at its tip.[34] Furthermore, many species are found on certain gill arches but not on others

within the same fish. It is possible that such niche specificity is influenced by the physical attachment abilities of the highly specialized, posterior attachment organ, the opisthaptor. A similar phenomenon occurs in the case of the scolex of tetraphyllidean cestodes of elasmobranchs. Some monogeneans remain fixed to the original site of attachment and cannot relocate later. Others, especially those on the skin, move about actively, leechlike, relocating at will. Certain species are found only on young fish, whereas others occur only on mature fish. Nutritional requirements of the parasites may play a role in the determination of such host specificity, but in some cases the free-swimming larvae are particularly attracted by mucus produced by the epidermis of their host species.[18]

The life span of monogeneans varies from a few days to several years. Many are incapable of living more than a short time after the death of the host. When sampling a fish population for monogeans, researchers should not transport dead fish to the laboratory in water because the worms may drop off. A recommended procedure is to place the fish on ice until examination or to remove the gills and drop them into a 10% formalin solution while still in the field.

FORM AND FUNCTION

Body Form

Monogeneans are basically bilaterally symmetrical, with partial asymmetry superimposed on a few species, particularly involving the opisthaptor. The body can be subdivided roughly into the following regions: **cephalic region** (anterior to pharynx), **trunk** (body proper), **peduncle** (portion of body tapered posteriorly), and **opisthaptor** (Fig. 19.1).

Most monogeneans are quite small, but a few are large; their sizes range from 0.03 to 20.00 mm long. Marine forms are usually larger than those from fresh water. All are capable of stretching and compressing their bodies, so unless the worms are properly relaxed before fixing, a permanent slide preparation may give a false impression of the true morphology. The dorsal side of the body is usually convex, while the ventral side is concave. The body is usually colorless or gray, but eggs, internal organs, or ingested food may cause it to be red, pink, brown, yellow, or black.

The anterior end of the body bears various adhesive and feeding organs, collectively called the **prohaptor,** which sometimes is associated with compound sense organs.[32]

FIGURE 19.1

Anatomy of an adult specimen of *Entobdella soleae* (ventral view).

From G. C. Kearn, *Ecology and Physiology of Parasites,* © 1971. M. Fallis, editor. University of Toronto Press, Toronto.

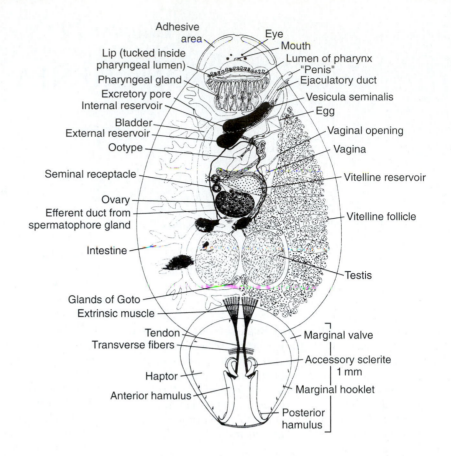

There are two main types of prohaptor: those that are not connected with the mouth funnel and those that are. In the first case (Fig. 19.2), the head end usually is truncated, lobated, or broadly rounded. This group usually bears **cephalic,** or **head glands,** which are unicellular organs that release sticky substances through individual ducts or groups of them. The utility of the substances produced by the head glands for adhesion is clear to anyone who has watched a monogenean with no anterior sucker progress down a fish gill filament in an inchwormlike manner, alternately attaching and releasing the anterior and posterior ends. In one species of *Gyrodactylus* at least three different head gland types have been recognized. Two to eight clusters of such ducts, called **head organs,** are usual. These areas usually bear dense, long microvilli on the tegument, in contrast to the short, scattered microvilli on the remainder of the body. These microvilli may function to spread and mix the secretions of the different types of head glands. Some species in this group have shallow, muscular **bothria,** which serve as suckers, in conjunction with the head gland secretions. Most species have two bothria, but some species have four.

The second type of prohaptor (Fig. 19.3) involves specializations of the mouth and buccal funnel. The simplest types have an **oral sucker** that surrounds the mouth. This may be a slightly muscular anterior rim of the mouth or a powerful circumoral sucker. In members of the order Mazocraeidea, two

buccal organs (buccal suckers) are embedded within the walls of the buccal funnel. Ultrastructural studies of the buccal organs have shown muscular, glandular, and sensory components; and they appear to play some role in the worm's feeding on host blood.[35]

The posterior end of all monogeneans also bears a highly characteristic organ, the opisthaptor (Fig. 19.4). It is clear that, in a group whose primary habitat is the surface and gills of fish, great adaptive value will accrue from an efficient attachment organ that prevents dislodgment by strong water currents, particularly one that will allow the mouth end to "hang downstream" and graze at will. In the Monogenoidea the opisthaptor is such an organ; unsurprisingly, it exhibits great variation within the group, and many forms have been interpreted as adaptations to particular hosts and infection sites.

The opisthaptor may extend for a considerable distance anteriorly along the trunk of the worm or may be confined to the posterior extremity. It may be sharply delineated from the body by a peduncle or may be merely a broad continuation of it. Opisthaptors develop into one or two basic types during ontogeny. The larva that hatches from the egg always has a tiny opisthaptor armed with sclerotized hooks or spines. This is retained in the adults of most species and either expands into the definitive opisthaptor or remains juvenile, while the adult organ develops from other sources near or surrounding it. In this first basic type the muscles expand into a large disc

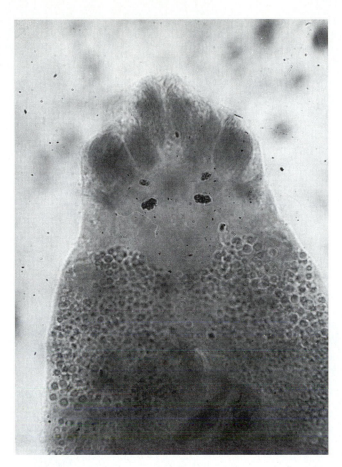

FIGURE 19.2

Prohaptor not connected to mouth funnel.
Courtesy of Warren Buss.

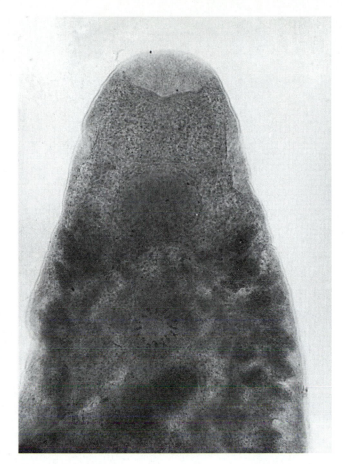

FIGURE 19.3

Prohaptor with an oral sucker.
Courtesy of Warren Buss.

that often has shallow **loculi** or well-developed **suckers,** as well as large hooks called **anchors,** referred to by some workers as **hamuli.** Anchors occur in one to three pairs, usually in the center of the opisthaptor, although they may be displaced to the side or posterior margin of the disc. Hamuli often have **connecting bars,** or **accessory sclerites,** supporting them (Fig. 19.5). The homologies of central hooks and their supporting bars are not always clear, at least at higher taxonomic levels. The protein in the anchors and marginal hooks appears to be keratin.

The tiny hooklets of the larva, and their persistent forms in the adult, are termed **hooks,** as opposed to anchors (Fig. 19.5). Opisthaptor hooks are characterized as being either "marginal" or "central." **Marginal hooks,** which are not always strictly marginal, are usually the tiny hooklets of the larval opisthaptor, some of which may be missing. It takes careful and skillful microscopy to find these.

Rarely, **supplementary discs,** or **compensating discs,** are developed near the base of the opisthaptor (family Diplectanidae). These are accessory to the opisthaptor and are not technically a part of the larval or adult opisthaptor.

They consist of a series of sclerotized lamellae or spines. **Suckers** are found on the ventral surface of the opisthaptors of many species. They range in number from two to eight.

Complex **clamps** are found on many species of monogeneans. On some species the clamp is muscular, whereas on others it is mainly sclerotized. It functions as a pinching mechanism, aiding in adherence to the host. Although many variations of structure occur, all are based on a single, basic type of clamp (Fig. 19.6). The identity of the material of which the clamps are constructed is enigmatic; it is not keratin, chitin, quinone-tanned protein, or collagen.[24] The number of clamps varies from eight to several hundred, distributed symmetrically in some species and asymmetrically in others. Finally the combinations of hooks, suckers, and clamps vary among several families (Fig. 19.4).

Because it has undergone such evolutionary diversification and varies considerably among species, the opisthaptor is an important taxonomic character. Consequently, specialists studying the monogeneans rely heavily on the sclerotized hooks, bars, clamps, and so on, which are relatively easy to study. However, recent research has shown that in a single

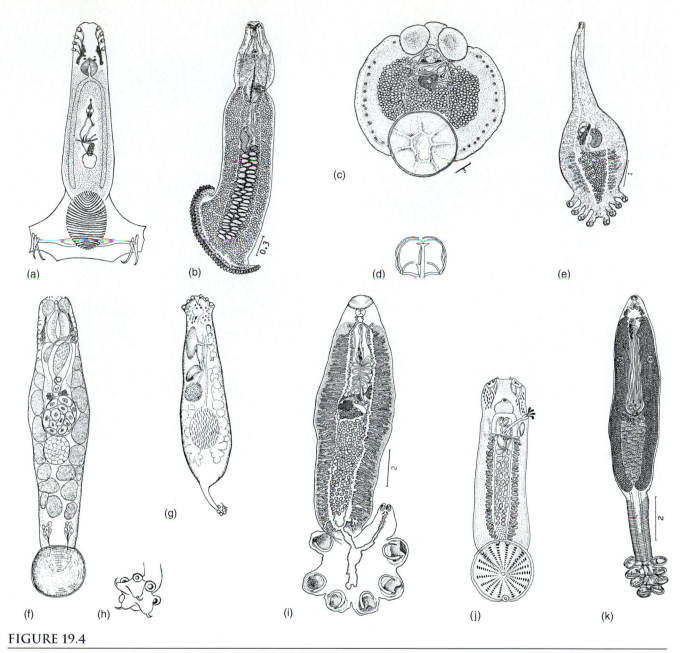

FIGURE 19.4

A variety of monogeneans, showing variations of opisthaptors. (*a*) *Diplectanum aculeatum.* (*b*) *Neoaxine constricta.* (*c*) *Capsala pricei.* (*d*) Sclerites in clamp of *Neoaxine constricta.* (*e*) *Diclidophora merlangi.* (*f*) *Udonella caligorum.* (*g*) *Aviella baikalensis.* (*h*) Opisthaptor of *Aviella baikalensis.* (*i*) *Erpocotyle borealis.* (*j*) *Acanthocotyle lobianchi.* (*k*) *Chimaericola leptogaster.* (Scales in millimeters, where available.)

After various authors in *System Helminthum,* vol. 4. Monogenea and Aspidocotylea, edited by S. Yamaguti. Copyright © 1963 Interscience Publishers, New York, NY. Reprinted by permission of John Wiley & Sons, Inc.

species, sizes of opisthaptor hooks may vary considerably with the season, primarily due to development of worms at different temperatures.[29] But while sizes vary, hamulus shape and proportions remain stable, suggesting that these features are influenced more by genes than by the environment.[11] The sclerotized portions of the male reproductive system also vary significantly between species, and consequently they have been used extensively in taxonomy and phylogenetic analysis.[3,30]

Tegument

As in the digeneans and cestodes, the tegument of monogeneans traditionally has been called a cuticle because light microscopists could discern little structure within it. However, by use of the electron microscope, the "cuticle" has now been recognized as a living tissue, the **tegument.** Its fundamental structure is similar to that of the tegument of digeneans and cestodes, with some noteworthy differences. The

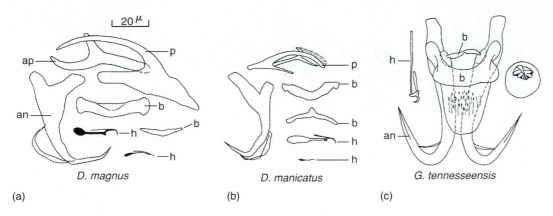

(a) D. magnus   (b) D. manicatus   (c) G. tennesseensis

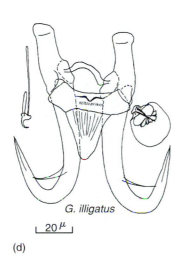

G. illigatus

(d)

FIGURE 19.5

Examples of opisthaptor hooks, anchors, and sclerotized portions of the male reproductive system, as typically seen in papers describing new species from small fishes (not all drawn to the same scale). (*a*) *Dactylogyrus magnus* from a silver chub; (*b*) *D. manicatus* from a striped shiner; (*c*) *Gyrodactylus tennesseensis* from a redbelly dace; (*d*) *G. illigatus* from a silverband shiner. **an,** anchor; **ap,** accessory piece; **b,** bar(s) and shields; **c,** cirrus; **h,** hook(s); **p,** penis. Note that bar shape, anchor (hamulus) shape, hook shape, and structure of the sclerotized parts of the male system all vary among species. Terminal portions of the male reproductive system, as seen at the worm's surface, are to the right of the anchors in (*c*) and (*d*).

(*a, b*) From W. A. Rogers, "Studies on *Dactylogyrus* (Monogenea) with descriptions of 24 new species of *Dactylogyrus*, 5 new species of *Pellucidhaptor*, and the proposal of *Aplodiscus* gen.n," in *J. Parasitol.* 53:501–524. Copyright © 1967. Reprinted with permission of the publisher. (*c, d*) From W. A. Rogers, "Eight new species of *Gyrodactylus* (Monogenea) from the southeastern U.S. with redescriptions of *G. fairporti* Van Cleave, 1921, and *G. cyprini* Diarova, 1964, "in *J. Parasitol.* 54:490–495. Copyright © 1968. Reprinted with permission of the publisher.

surface layer of the tegument is, as in cestodes and digeneans, a syncytial stratum, laden with vesicles of various types and mitochondria, bounded externally by a plasma membrane and glycocalyx (fine filamentous layer on surface) and internally by a membrane and basal lamina. This stratum is the **distal cytoplasm,** and it is connected by trabeculae **(internuncial processes)** to the cell bodies, or **cytons** (perikarya), located internal to a superficial muscle layer. Such is the case in all species studied so far except those of *Gyrodactylus,* in which trabeculae and cytons could not be observed. *Gyrodactylus* spp. are very peculiar in other respects and will be discussed further. Often the tegument of monogeans is supplied with short, scattered microvilli; in some species these are absent, and shallow pits occur.

A curious condition has been reported in certain species. Some areas of the body are without a tegument, and large pieces of the tegument are only loosely connected to the surface. In these areas the basal lamina constitutes the external covering. Rohde[33] contends that the condition is not artifactual but that pieces of the tegument are being secreted into the environment. Rohde also posits that these cases might offer a clue to the adaptive value of the tegumental arrangement of the monogeneans, digeneans, and

cestodes—that is, syncytial distal cytoplasm with internal perikarya. He has suggested that the transfer of an original superficial epithelium into the interior of the body may prevent permanent damage by the hosts. Since the tegument may be subjected to such damaging influences as host secretions, it then easily can be replaced from the cell machinery that is still intact.

Muscular and Nervous Systems

The main musculature, other than that in the opisthaptor, consists of the **superficial muscles,** immediately below the distal cytoplasm of the tegument, arranged in circular, diagonal, and longitudinal layers. The muscles of the opisthaptor in the suckers or inserted on the hooks and accessory sclerites are clearly important in adhesion. We understand the mechanics of their operation in several species, an example of which is *Entobdella soleae* (Fig. 19.7). This species lives on the skin of the sole, and its opisthaptor anchors the animal firmly in its relatively smooth and exposed site on the host.[18] The disc-shaped opisthaptor forms an effective suction cup. Prominent muscles in the peduncle insert on a tendon that passes down to near the ventral surface of the disc, up over a notch in the accessory sclerites, and

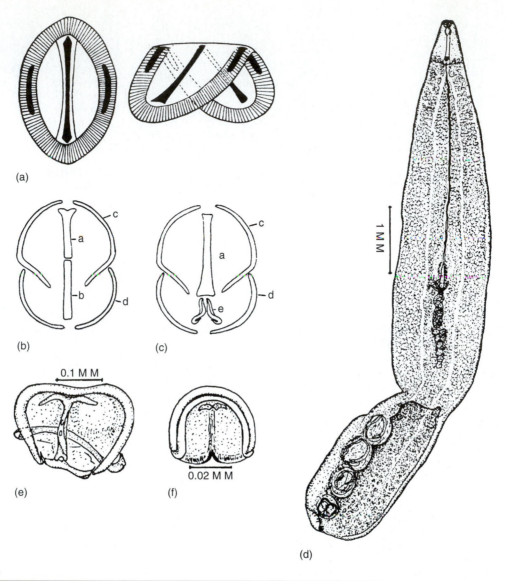

FIGURE 19.6

Monogenean clamps and associated sclerites. (*a*) Diagram of an attaching clamp. On the left, it is fully open; on the right, it is partially closed. The sclerotized parts are black; the musculature is crosshatched. (*b, c*) Types of clamp sclerites from members of the order Mazocraeidea; **a,** mid-sclerite or anterior mid-sclerite; **b,** posterior mid-sclerite; **c,** anterolateral sclerite; **d,** posterolateral sclerite; **e,** accessory sclerite. (*d*) *Grubea pneumatomphori* from the chub mackerel, showing two sizes of clamps. (*e, f*) Large and small clamps from *G. pneumatomphori.*

(*a*) Source: B. E. Bychowsky, *Monogenetic Trematodes, Their Systematics and Phylogeny.* 1961 American Institute of Biological Sciences, Washington, D.C. (*b, c*) From W. A. Boeger and D. C. Kritsky, "Phylogeny and a revised classification of the Monogenoidea Bychowsky, 1937 (Platyhelminthes)," in *Syst. Parasitol.* 26:1–32. 1993. Reprinted with permission of the publisher. (*d, e, f*) From E. W. Price, "North American monogenetic trematodes IX. The families Mazocraeidae and Plectanocotylidae," in *Proc. Biol. Soc. Wash.* 74: 127–156. Copyright © 1961 Biological Society of Washington. Reprinted with permission of the publisher.

then to the proximal end of the large anchors. Contraction of the muscles erects the accessory sclerites so that their distal ends push down against the fish's skin and their proximal ends serve as a prop toward which the proximal ends of the anchors are pulled. This action tends to lift the center area of the opisthaptor, thus reducing pressure and creating suction, at the same time that the distal, pointed ends of the anchors are forced downward to penetrate the host's epidermis.

Studies on neural and neuromuscular conduction have not been carried out, but cholinesterase was demonstrated in the nervous system of *Diclidophora merlangi*; therefore, at least some fibers are probably cholinergic.[16]

The general pattern of the nervous system is the ladder type with **cerebral ganglia** in the anterior and several nerve trunks coursing posteriorly from them. The nerve trunks connect by the ladder commissures, and additional nerves emanate from the cerebral ganglia to connect with the pharyngeal commissure. As would be expected, the adhesive organs of the opisthaptor are well-innervated.

Monogeneans have a fairly wide variety of sense organs. Most have pigmented eyes in the free-swimming larval stage, and in many species the adults have eyes as well. Oncomiracidia of subclass Polyonchoinea usually have four eyespots, which persist in the adult, perhaps somewhat

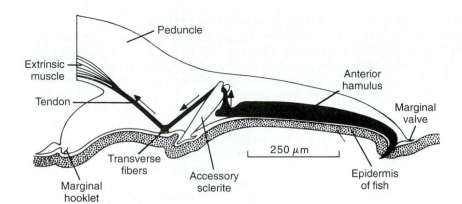

FIGURE 19.7

A diagrammatic parasagittal section through the adhesive organ of *Entobdella soleae.* The *arrows* show the direction of movement of the tendon when the extrinsic muscle contracts. The posterior hamulus (*anchor*) has been omitted.

From G. C. Kearn, *Ecology and Physiology of Parasites,* © 1971. A. M. Fallis, editor. University of Toronto Press, Toronto.

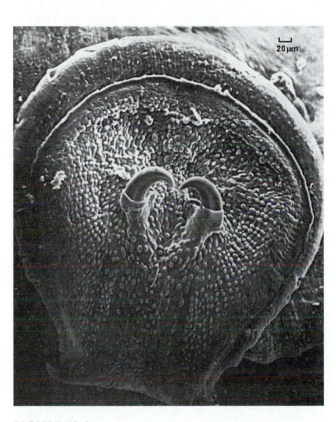

FIGURE 19.8

Scanning electron micrograph of the opisthaptor of *Entobdella soleae,* showing the probable sensory papillae.

From K. M. Lyons, "Scanning and transmission electron microscope studies on the sensory sucker papillae of the fish parasite *Entobdella soleae* (Monogenea)," in *Zeitschr. Zellforsch.* 137:471–480. Copyright © 1973. Springer-Verlag.

reduced; the two larval eyespots of Polystomatoinea, on the other hand, disappear during maturation. These are rhabdomeric eyes similar to those found in turbellaria and some larval Digenea. In addition, what appears to be a nonpigmented ciliary photoreceptor has been found in the larva of *Entobdella,* with counterparts in structures described in larval Digenea. Several different types of ciliary sense organs in the tegument have been described, including single receptors (one modified cilium in a single nerve ending) and compound receptors (consisting both of several associated nerve endings, each with a single cilium, and of one or a few nerves, each with many cilia).[26]

Finally, a very interesting, nonciliated sense organ occurs on the opisthaptor of *Entobdella.* The disc surface of the opisthaptor is covered with more than 800 small papillae (Fig. 19.8), and beneath the tegument of each papilla are packed nerve endings that double over and pile on top of one another. The function of these peculiar organs is probably mechanoreception, perhaps to sense contact with the host or detect local tensions in the opisthaptor. It is not known whether similar organs occur in other monogeneans.[27]

Osmoregulatory System

The excretory system has not been used as a tool for systematics in this group as it has in the Digenea. Typical of the Platyhelminthes, the excretory unit is the **flame cell protonephridium.** A thin-walled capillary leads from this unit to fuse with a succession of ducts leading to two lateral **excretory pores** near the anterior end of the worm. The terminal ducts often each have a contractile bladder at their distal ends.

The fine structure of the excretory system has been studied in two species of *Polystomoides,* and it is generally similar to that of Digenea and Cestoda with minor differences.[33] The internal surface area of the tubules is increased in a manner differing from that in either of the other groups—that is, by strongly reticulated walls. Lateral or nonterminal flames are frequent.

Acquisition of Nutrients

The mouth and buccal funnel often have associated suckers. Behind the buccal funnel a short **prepharynx** is followed by a muscular and glandular **pharynx.** This powerful sucking apparatus draws food into the system. In *Entobdella soleae* the pharynx can be everted and the pharyngeal lips closely applied to the host's skin.[18] The pharyngeal glands secrete a strong protease that erodes the host epidermis, and the worm sucks up the lysed products. Fortunately for the fish, its epidermis is capable of rapid migration and regeneration to close the wound left by the parasite's feeding. Posterior to the pharynx may be an **esophagus,** although it is absent in many species. The esophagus may be simple or have lateral branches and may have unicellular digestive glands opening into it.

In most monogeneans the **intestine** divides into two lateral **crura,** which are often highly branched and may even connect along their length. If the crura join near the posterior end of the body, it is common for a single tube to continue posteriorly for some distance. There is no anus. Digestion and ultrastructure of the gut of several species have been studied.[13,17,33]

Many monogeneans seem to feed mostly on mucus and epithelial cells of their hosts, although feeding on blood has been demonstrated in *Dactylogyrus* spp. and blood appears to be the dominant component of the diet of some families and genera. It was believed formerly that the unusable breakdown product of hemoglobin digestion, hematin, was eliminated in the gut by sloughing off gut cells containing it so that parts of the cecal wall were denuded. However, ultrastructural studies on *Polystomoides* and *Diclidophora* have shown that the cecal epithelium is not discontinuous but that the hematin-containing cells are interspersed with a different kind of cell called the "connecting cell."[15] In *Diclidophora* both the hematin cell and the connecting cell have their luminal surface increased by long, thin lamellae, but in *Polystomoides* only the hematin cell has such lamellae. It appears that the digestion of hemoglobin in *Diclidophora*, at least, is mostly or entirely intracellular: The protein is taken into the cell by pinocytosis and digested within an extensive, intracellular reticular space, and the hematin is subsequently extruded by temporary connections between the reticular system and the gut lumen. Indigestible particles are eliminated through the mouth in all monogeneans. Finally, *Diclidophora* can absorb neutral amino acids through its tegument, suggesting the possibility that direct absorption of low-molecular weight organic compounds could supplement its blood diet.[14]

Male Reproductive System

Monogeneans are hermaphroditic with cross-fertilization usually taking place (Fig. 19.1). This is epitomized by the genus *Diplozoon,* in which individuals completely fuse into pairs with their genital ducts together, the ultimate in "oneness."

Testes usually are round or ovoid, but they may be lobated. Most species have only one testis, but the number varies according to species, and one species has more than 200 per individual. Each testis has a **vas efferens,** which expands or fuses into an ejaculatory duct. There is no trace of a cirrus pouch or eversible cirrus, in the sense of those in cestodes or trematodes. In some cases the ejaculatory duct is simple and terminates within a shallow, sometimes suckerlike, **genital atrium,** which propels sperm into the female system at copulation. In many species the tissues surrounding the terminal ejaculatory duct are thick and muscular, forming a papillalike **penis.** Hooks of consistent size and form for each species commonly arm the distal end of the penis. In many the lining of the distal ejaculatory duct is sclerotized, sometimes for a considerable portion of its length. (We use the term **sclerotized,** although the chemical nature of the stabilized protein is unknown.) A simple, saclike seminal vesicle is present in some species. Unicellular **prostatic glands** are usually present.

In several families still another type of copulatory organ exists: a complex **sclerotized copulatory apparatus** that joins with the ejaculatory duct. This apparatus commonly consists of a penis and **accessory piece.** These components vary widely in structure among species but are similar within a species and, therefore, are important taxonomic characters.[3] The structures are contained in a membranous sac and are controlled by muscles.

Female Reproductive System

The single **ovary (germarium)** of all species of monogenea is usually anterior to the testes (Figs. 19.1 and 19.9). Between species it varies in shape from round or oval to elongated or lobated. The **oviduct** leaves the ovary and courses toward the ootype, receiving the vitelline, vaginal, and genitointestinal ducts along the way. More specifically, the oviduct extends from the ovary to a confluence with the vitelline duct (canal); the remainder is often referred to as the **female sex duct.** A seminal receptacle is present, either as a simple swelling of the oviduct or as a special sac with a separate duct to the oviduct.

The **vitellaria** are abundant, usually extending throughout the parenchyma and often even into the opisthaptor. Despite their many ramifications, the vitellaria consist basically of left and right groups. Each has an efferent duct; they fuse midventrally near the oviduct, forming a small **vitelline reservoir.** Each vitelline follicle consists of a few cells surrounded by a thin, muscular membrane. The vitelline ducts are lined with ciliated epithelium.

There are two basic types of female reproductive systems in monogeans,[4] distinguished by the connections of the vagina(s) and the presence or absence of a curious structure called the **genitointestinal canal,** connecting the female system to the gut (Fig. 19.9). The vagina may be present or not and when present it may be doubled. Vaginal openings are dorsal, ventral, or lateral. The terminal portion is sclerotized in some species; in others the vaginal pore is multiple or surrounded by spines. In those species with a "true" vagina, the vaginal opening and duct lead directly to the oviduct. The second basic type of system is one in which there is a "ductus vaginalis" connecting the vagina to the vitelline canals (Fig. 19.9).

In some species, such as *Entobdella* (Fig. 19.1), the vagina may be much smaller than the penis, and sperm transfer is achieved by deposition of a spermatophore adjacent to the vagina of the mating partner, rather than by direct copulation. *Diclidophora merlangi,* which does not have a vagina, practices a kind of hypodermic impregnation.[28] The suckerlike penis of an individual attaches at a ventrolateral position posterior to the genital openings of its partner, draws up a papilla of tegument into the penis, and breaches the tegument with spines in the penis. Sperm enter and make their way between cells of the partner to the seminal receptacle, a distance of 1 to 2 mm.

The function, if any, of the genitointestinal canal is unknown. Sometimes yolk granules and sperm are observed in the gut, presumably having arrived there through the genitointestinal canal. One hypothesis is that the canal represents a vestige of a mechanism by which eggs are passed into the intestine to be expelled through the mouth. Such a canal occurs in many tubellarians, especially polyclads, but the homology of this canal in the various platyhelminth groups, as well as its function in the polyclads, is obscure.

After being fertilized in the oviduct or ovary itself, the zygote and attendant vitelline cells pass into the ootype, a muscular expansion of the female duct. In the species studied, **Mehlis' gland** around the ootype comprises two cell types, mucous and serous. (The ootype epithelium may also

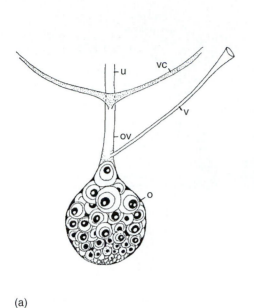

(a)

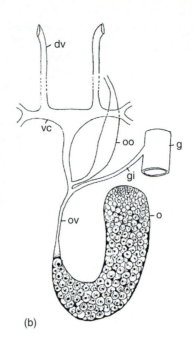

(b)

FIGURE 19.9

Basic types of female reproductive systems in monogeneans. (*a*) Vagina connecting to oviduct ("true" vagina), genitointestinal canal absent; (*b*) vagina connecting to vitelline ducts ("ductus vaginalis"), genitointestinal canal present; **dv**, "ductus vaginalis"; **g**, gut; **gi**, genitointestinal canal; **o**, germarium; **oo**, ootype; **ov**, oviduct; **u**, uterus; **v**, "true" vagina; **vc**, vitelline canal.

From W. A. Boeger and D. C. Kritsky, "Phylogeny and a revised classification of the Monogenoidea Bychowsky, 1937 (Platyhelminthes)," in *Systematic Parasitol.* 26:1–32.

be secretory.) The function of Mehlis' gland is not known. It was formerly thought to contribute shell material, but in the Monogenoidea, as in the Digenea and Cestoda, the shell material seems to come from the vitelline cells. The shape of the egg is apparently determined by the walls of the ootype. In *Entobdella* the tetrahedral egg shape is imparted by four pads in the ootype walls.[18] The eggs of many monogeneans have a filament at one or both ends, also characteristic of a given species. The filament may have an adhesive property that serves to attach the egg to the host or substrate on which it falls after release into the open water. Further observations on egg development of *Entobdella soleae* are given by Kearn.[20]

It is generally believed that the protein in the eggshell is stabilized by a process of quinone tanning to form sclerotin, but some work indicates the stabilization may not be by quinone tanning but by means of dityrosine and disulfide links as in resilin and keratin.[31]

Although many eggs may be produced (*Polystoma* produces one to three eggs every 10 to 15 seconds), they are passed out of the worm fairly rapidly; therefore, not many may be present within the parent at one time. Some species may store a few eggs in the ootype and then pass them to the outside directly through a pore, but in most species the eggs pass from the ootype into a uterus, which courses anteriorly to open into the genital atrium, together with the ejaculatory duct. Hence the uterus, at least in most cases, does not function as a vagina as it does in digeneans.

DEVELOPMENT

The life cycles of a few species have been well studied, but little or nothing is known about most. With the exception of the viviparous Gyrodactylidae, monogeneans usually have a single-host life cycle involving an egg, oncomiracidium, and adult. Some evidence suggests that two species of gastro-

cotylids that parasitize predatory fish do not infect their definitive hosts directly but undergo a period of development on fish preyed on by the parasite's definitive hosts.[22]

Oncomiracidium

The oncomiracidium (Fig. 19.10) hatches from the egg and rather resembles a ciliate protozoan in size and shape. It is elongated and bears three zones of cilia: one in the middle and one at each end. The zones of ciliated epidermal cells are separated by an interciliary, nonnucleate syncytium. It has been shown in *Entobdella* that the nuclei of the interciliary regions are actually extruded during embryogenesis. Subsequently, the cytons of the "presumptive adult" tegument, which are located within the superficial muscle layer, extend processes out to underlay the ciliated cells and join the syncytial interciliary regions. The animal is thus prepared for rapid shedding of the ciliated cells on attachment to the host; the stimulus for this shedding in *Entobdella* is mucus from the host epidermis, and the shedding takes only 30 seconds.

The oncomiracidium has cephalic glands with efferent ducts opening on the anterior margin and as previously noted has one or two pairs of eyes. The digestive tract is well-differentiated, and the excretory pores are already formed. The posterior end always is developed into an attachment organ that bears hook sclerites, and these sclerites are retained in the adult. The larvae swim about until they contact a host; then they attach, lose their ciliated cells, and develop into adults. Oncomiracidia of *Entobdella soleae* swim alternately toward and away from light, apparently performing a search pattern for their host, a bottom-dwelling flatfish.[19] Rates of development into adults are largely unknown.

Inasmuch as oncomiracidia are free swimming and the primary hosts of monogeneans are fishes or other aquatic or amphibious vertebrates, it would seem that infection of new hosts would not be a problem. However, the free-swimming

FIGURE 19.10

The oncomiracidium of *Entobdella soleae* (ventral view).

From G. C. Kearn, *Ecology and Physiology of Parasites,* © 1971. A. M. Fallis, editor. University of Toronto Press, Toronto.

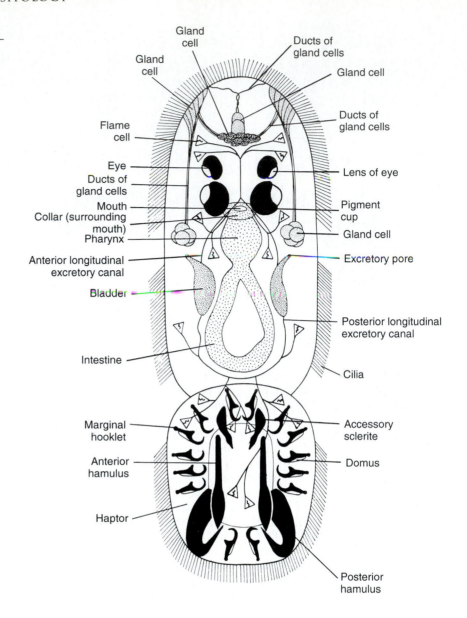

life of an oncomiracidium is short; its potential hosts are widely dispersed most of the time and in any case can swim much faster than the larva. In addition, potential hosts may not even be present in the aquatic habitat except during breeding season. Thus, it is of great selective value for the worm's egg production to be closely related to its host's reproduction, to coincide, for instance, with a time when the host is concentrated in spawning areas. Other features of the host's habits also may enhance chances for infection. Such correlation has been shown in several species, and similar adaptations are probably more widely prevalent.[23] Some of these adaptations will be illustrated in the discussion of life cycles that follow.

• Subclass Polyonchoinea

***Dactylogyrus* Species.**   A large number of species (Fig. 19.11) in this genus have been described, and some of them,

such as *D. vastator, D. anchoratus,* and *D. extensus,* are of great economic importance as pathogens of hatchery fish. *Dactylogyrus* has large anchors on its opisthaptor and lives on the gill filaments of its host. Heavy infections cause loss of blood, erosion of epithelium, and access for secondary bacterial or fungal infections. Irritation to the gills stimulates increased mucus production, which often smothers the fish. Heavy infections may kill the host; massive die-offs are common in the crowded situations of fish culture ponds.

The life cycles and factors influencing the economically important species are reasonably well-known.[8] The main features of the life cycle correspond to the preceding general outline. *Dactylogyrus vastator* on carp shows marked seasonal fluctuations correlated with temperature. Each worm deposits 4 to 10 eggs per 24 hours during the summer, and this rate increases with increasing temperature. The eggs require four to five days with temperatures between 20° and

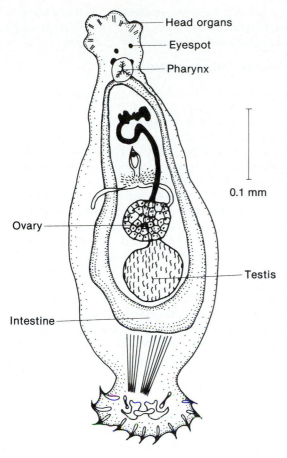

FIGURE 19.11

Dactylogyrus vastator.

Source: B. E. Bychowsky, *Monogenetic Trematodes, Their Systematics and Phylogeny.* Copyright © 1961. American Institute of Biological Sciences, Washington, D.C.

28° C for embryonation, but this rate slows with temperatures down to 4° C, at which point development is completely suppressed. The adult worms are adversely affected by lower temperatures so that the number of parasites on the fish decreases greatly during the winter. The net effect is that the parasite population builds up over the summer, but the eggs deposited toward the end of the season winter over and result in a mass emergence in the spring to infest the young of the year.

***Gyrodactylus* Species.** Despite the (unfortunate) similarity in name and pathologic effects with *Dactylogyrus, Gyrodactylus* spp. (Fig. 19.12) are very different organisms. They are also economically significant as important pests, particularly of trout, bluegills, and goldfish in fish ponds.[10] The family Gyrodactylidae is unusual among the Monogenoidea in that it is viviparous. The young are retained in the uterus until they develop into functional subadults. Inside such a developing juvenile, one can often see a second juvenile developing, with a third juvenile inside of it and a fourth inside the

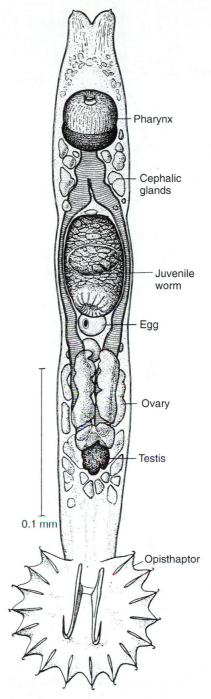

FIGURE 19.12

Gyrodactylus cylindriformis, ventral view.

From J. F. Mueller and H. J. Van Cleave, "Parasites on Oneida Lake fishes. Part II. Descriptions of new species and some general taxonomic considerations, especially concerning the trematode family Heterophyidae," in *Roosevelt Wildl. Ann.* 3:73–154. Copyright © 1932.

third! The exact mechanism of this unique embryogenesis is unknown, but it may be considered a type of sequential polyembryony; as many as four individuals usually result from

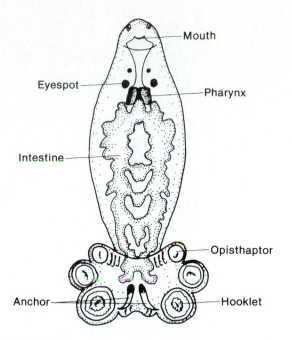

FIGURE 19.13

Polystoma integerrimum, a parasite of Old World frogs.

Source: E. Zeller, "Untersuchungen über die Entwicklung und den Bau des *Polystomum integerrimum* Rud." in *Zeitschr. Wissensch. Zool.* 22:1–28. Copyright © 1872.

one zygote. After birth the young worm begins feeding on its host and gives birth to the juvenile "remaining" inside. Only then can an egg from its own ovary be fertilized to repeat the sequence. Since only a day or so is required for a worm to mature after birth and give birth to another worm already developing within it, massive infection can build up quickly.

Certain other peculiarities of *Gyrodactylus* have been described that may correlate with their unusual embryogenesis. One of these is the possession of an **ovovitellarium,** a fused mass of ova and vitelline cells. Another is the fact that the "distal cytoplasm" of the tegument apparently is not connected to cytons in the parenchyma. The epidermis of the developing embryo has nuclei that are not present in the adult tegument. The tegument of *Gyrodactylus* may be embryologically equivalent to the ciliated epidermal cells on the oncomiracidium of other monogeneans.[25]

Not having an oncomiracidium, *Gyrodactylus* must depend on transmission of the adult or subadult from one host to another. Since these forms appear unable to swim, it is clear that the prospective host must be quite close to the worm's current host for the transferral to take place. However, *Gyrodactylus* infections seem to spread easily in hatcheries, so that the lack of a swimming stage is not a serious barrier to transmission. The worms also sometimes leave a dead fish and remain active for several hours. Experiments have shown that *G. salaris* can leave a dead salmon and in-

fect a live eel. The same studies showed that when infected salmon were removed from a tank, an uninfected eel placed in the same tank the next day became infected.[1]

• Subclass Polystomatoinea

Polystomatidae. *Polystoma integerrimum* (Fig. 19.13) is a parasite of Old World frogs. It is of particular interest because it is known that the worm's reproductive cycle is synchronized with that of its host by means of host hormones, a mechanism to provide a ready supply of hosts to the hatching oncomiracidia. Furthermore, two different types of adults develop: normal and neotenic.

Adult worms live in the urinary bladder of their host. They are dormant during the winter, while the frogs hibernate, but they become active in the spring along with their hosts. When the frog's gonads begin to swell and produce gametes, the worms begin to copulate and produce eggs that are released into the surrounding urine. In the laboratory, maturation and stimulation of worm gamete production can be elicited by injecting the frogs with pituitary extract, although it is not known whether the effect is direct by way of the gonadotropins or stimulated by the injection via the gonadal hormones of the host.[37] In nature the eggs are voided into the water in the frog's spawning area. Depending on the temperature, the oncomiracidia hatch in 20 to 50 days. By then the frog's eggs have developed into tadpoles that will be the next host generation. Tadpoles first have external and then internal gills and breathe by sucking water into the mouth, over the gills, and out the side of the pharynx through slits, much like fish. An oncomiracidium contacting a gill attaches, metamorphoses, and begins producing eggs within 20 to 25 days.

The gill form of *P. integerrimum* has a narrower body than the bladder form, and the opisthaptor is not as sharply set off from the body. The gill form's intestine has fewer lateral branches, and the ovary is a different shape from that of bladder worms. Furthermore, there is no uterus or genitointestinal canal. Some authors consider the gill stage neotenic.[2]

Eggs of *P. integerrimum* hatch in 15 to 20 days, when the water is warm. These larvae also attach to the gills of tadpoles, but by the time the water warms in the spring, the tadpoles are older, and worm maturation is delayed. When the tadpoles begin their metamorphosis, including resorption of the gills, the worms migrate to the bladder. The migration occurs over the ventral skin of the tadpole at night and takes only about a minute.[22] Oddly, even when the metamorphosing tadpole is exposed to newly hatched larvae, the worms go first to the gills and then to the bladder, and this has been taken as evidence that the migration is controlled endogenously and not by stimuli from the host. It takes as long as four to five years for the bladder form to mature and begin egg production.

A similar species, *P. nearcticum,* occurs in tree frogs in the United States. It also has a neotenic gill form but has no slowly developing and then migrating immature gill form.

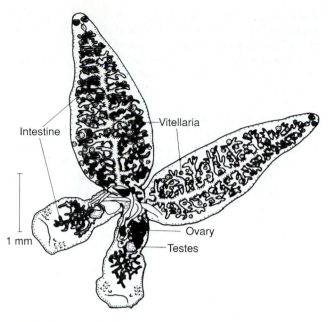

FIGURE 19.14

Diplozoon paradoxum, a parasite of freshwater fishes in Europe and Asia.

Source: B. E. Bychowsky and L. F. Nagibina, in B. E. Bychowsky, *Monogenetic Trematodes, Their Systematics and Phylogeny.* Copyright © 1961. American Institute of Biological Sciences, Washington, D. C.

Oncomiracidia enter the cloaca directly when they contact metamorphosing tadpoles. Larvae of *Protopolystoma xenopi* also directly enter the cloaca of their host, *Xenopus laevis.* Interestingly, *Xenopus* remains in water all year, and reproduction of *Protopolystoma* continues correspondingly.

At the other extreme are *Pseudodiplorchis americanus* and *Neodiplorchis scaphiopodis,* parasites in the urinary bladder of spadefoot toads (*Scaphiopus* spp.) in Arizona. The toads live in one of the hottest and driest areas in the United States and spend about 10 months per year beneath the soil in a state of torpor. They breed during only one to three nights per year in temporary pools formed by the brief desert rains. These conditions present an extraordinary challenge to a parasite that depends on an aqueous environment for transmission, and transmission must be exquisitely coordinated with the activities of the hosts. During the time that the toads are in hibernation, the encapsulated oncomiracidia become fully developed in the uterus of the adult worms.[40] When the toad enters water, the larvae are deposited, pass out with the urine, hatch within seconds, and are fully infective for the next host. The oncomiracidia are much larger than most other oncomiracidia (up to 600 μm), have a much longer free-swimming life (up to 48 hours), and are more resistant to drying (up to one hour). Thus, if they do not reach another host during the first night when the toads are spawning, they have a good chance of succeeding the next night. When the oncomiracidia contact another spadefoot toad as it is floating with the end of its nose above water, they crawl on the toad's skin up *out* of the water and enter its nares. Then they migrate to the lungs, undergo a period of development, and finally migrate to the urinary bladder by passing through the stomach and intestine.

Obviously these worms pass through many relatively hostile environments in a fairly short length of time before reaching their final home in the urinary bladder. Unique vesicles found in their tegument allow such adventurous migrations.[9] The contents of these vesicles are released during migration through the gut, producing a thick, presumably protective, glycocalyx and mucin layer over the worm. The toads feed gluttonously during their brief annual stay above ground, consuming nearly 30% of their body weight in a single feed and replenishing the fat stores needed to sustain hibernation in just two nights of eating. The massive food intake may stimulate the worms to migrate from the lungs and into the bladder. The vesicles that protect them are an adaptation for running the digestive system gauntlet, an event that consumes about 30 minutes of their year-long lives.[9]

• Subclass Oligonchoinea

Diplozoon paradoxum. *Diplozoon paradoxum* (Fig. 19.14) is a common parasite of the gills of species of European cyprinid fish. Like *Dactylogyrus, Diplozoon* exhibits a strong seasonal variation in its reproductive activity. Virtually no gametes are produced during the winter, but gonads begin to function during the spring, reaching a peak during May to June and continuing through the summer. The eggs, which have a long, coiled filament at their ends, can hatch about 10 days after they are deposited, and light intensity and turbulence of the water, as might be caused by host feeding or spawning activity, stimulate hatching. The oncomiracidium bears two clamps on its opisthaptor with which it attaches to a gill filament; it loses its cilia almost immediately. The worm feeds and begins to grow, adding another pair of clamps to the opisthaptor. A small sucker also appears on the ventral surface and a tiny papilla on the dorsal surface, slightly more posterior than the sucker. When this stage (Fig. 19.15) was first discovered, it was thought to represent a new genus, and it was named *Diporpa.* When *Diporpa* was recognized as a juvenile stage of *Diplozoon,* the term **diporpa** was applied to the stage. Curiously, in contrast to *Dactylogyrus* spp., *Diplozoon,* even diporpae, rarely infect young-of-the-year fish.

A diporpa juvenile can live for several months, but it cannot develop further until encountering another diporpa; unless this happens, the diporpa usually perishes by winter. When one diporpa finds another, each attaches its sucker to the dorsal papilla of the other. Thus begins one of the most intimate associations of two individuals in the animal kingdom. The two worms fuse completely, with no trace of partitions separating them. The fusion stimulates maturation. Gonads appear; the male genital duct of one terminates near the

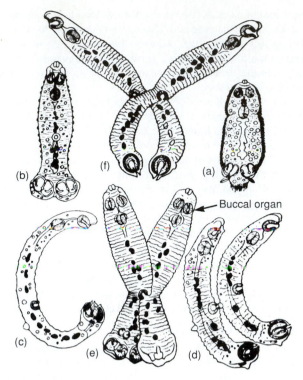

FIGURE 19.15

Development of *Diplozoon paradoxum*. (*a*) Freshly hatched, free-swimming juvenile. (*b*) Diporpa juvenile. (*c–f*) Diporpa juveniles attaching themselves to one another.

From E. Zeller, "Untersuchungen über die Entwicklung des *Diplozoon paradoxum*, in *Zeitschr. Wissensch. Zool.* 22:168–180. Copyright © 1872.

female genital duct of the other, permitting cross-fertilization. Two more pairs of clamps develop in the opisthaptor of each. Adults apparently can live in this state for several years.

PHYLOGENY

Although some modern authors persist in allying the Monogenea with the Digenea and Aspidogastrea as an order or subclass within the Trematoda,[36] others have argued that the Monogenea are more closely related to the Cestoidea,[8,21] and most platyhelminth systematists now consider the monogeneans to be a sister group to the Cestodaria, an infraclass that includes the cohorts Gyrocotylidea, Amphilinidea, and Eucestoda (see Chapter 13 and Brooks and McLennan[5]). The major feature uniting all these taxa is a posterior adhesive organ with hooks.

An extensive and detailed revision of the group has been published by Boeger and Kritsky,[4] and the following classification is based on their work. The characteristics given are the major synapomorphies that distinguish the groups in their cladogram.

CLASSIFICATION OF CLASS MONOGENOIDEA
Hermaphroditic, dorsoventrally flattened, elongated or oval worms with a syncytial tegument; conspicuous posterior adhesive organ present (opisthaptor) that is muscular, some-

times divided into loculi, usually with sclerotized anchors, hooks, and/or clamps, often subdivided into individual suckers or clamps without sclerites; an anterior adhesive organ (prohaptor) usually present, consisting of one or two suckers, grooves, glands, or expanded ducts from deeper glands; eyes, when present, usually of two pairs; mouth near anterior, pharynx usually present; gut usually with two simple or branched stems often anastomosing posteriorly, rarely a single median tube or sac; male genital pore usually in atrium common with female pore; genitointestinal duct present or absent; reproduction oviparous or viviparous; vitelline follicles extensive, usually lateral; two lateral osmoregulatory canals present, each with an expanded vesicle opening dorsally near anterior end; parasites on or in aquatic vertebrates, especially fishes, or rarely on aquatic invertebrates; cosmopolitan.

Subclass Polyonchoinea
Mouth ventral; sperm microtubules lying along one-fourth of cell periphery; male copulatory organ sclerotized, muscular, elongated, with spines absent; 14 marginal and two central hooks on the oncomiracidium.

Order Dactylogyridea
Two pairs of ventral anchors; sperm with one axoneme.

Suborder Dactylogyrinea
Twelve marginal and two central hooks on the oncomiracidium; eight marginal, two central, and four dorsal hooks on the adult.

Families
Dactylogyridae, Pseudomurraytrematidae, Diplectanidae.

Suborder Tetraonchinea
Gut single.

Families
Tetraonchidae, Neotetraonchidae.

Suborder Amphibdellatinea
Eyes absent on the oncomiracidium.

Family
Amphibdellatidae.

Suborder Neodactylodiscinea
Bars absent; two lateral "true" vaginas.

Family
Neodactylodiscidae.

Suborder Calceostomatinea
Twelve marginal and two central hooks on the oncomiracidium; 12 marginal and two central hooks on the adult.

Family
Calceostomatidae.

Order Gyrodactylidea
Two ventral bars and 16 marginal hooks on the oncomiracidium; 16 marginal hooks on the adult; hooks hinged.

Families
Gyrodactylidae, Anoplodiscidae, Bothitrematidae, Tetraonchoididae.

Order Montchadskyellidea
Gut diverticula present; germarium/oviduct looping around right caecum; one midventral "true" vagina.

Family
Montchadskyellidae.

Order Capsalidea
Single genital aperture is marginal.

Families
Acanthocotylidae, Capsalidae, Dionchidae.

Order Monocotylidea
Sperm with two axonemes, one reduced; 14 marginal hooks on both oncomiracidium and adult.

Families
Monocotylidae, Loimoidae

Subclass Polystomatoinea
More than two testes; gastrointestinal canal present; haptoral suckers, with associated hook, present in the adult; three pairs of haptoral suckers; two lateral "ducti vaginalis."

Order Polystomatidea
Egg filaments absent.

Family
Polystomatidae, Sphyranuridae.

Subclass Oligonchoinea
One pair of haptoral suckers; eyes absent on the oncomiracidium; "crochet en fleau" hooklike; one pair of lateral sclerites; four pairs of haptoral suckers; gut diverticula present; mid-sclerite terminates in hook.

Order Mazocraeidea
Two oral suckers present; germarium elongated, inverted U-shaped; egg with two filaments; mid-sclerite flared or truncated; one pair of eyes fused on the oncomiracidium; eyes absent on the adult; two pairs of lateral sclerites.

Suborder Microcotylinea
Germarium elongated, double inverted U-shaped.

Families
Axinidae, Diplasiocotylidae, Heteraxinidae, Microcotylidae, Allopyragraphioridae, Diclidophoridae, Pterinotrematidae, Rhinecotylidae, Pyragraphoridae, Heteromicrocotylidae.

Suborder Hexostomatinea
Spines of male copulatory organ absent; two pairs of eyes in oncomiradicium.

Family
Hexostomatidae.

Suborder Discocotylinea
Haptoral suckers present on oncomiracidium; six marginal hooks on oncomiracidium; anchor absent in all developmental stages.

Families
Discocotylidae, Diplozoidae, Octomacridae.

Suborder Gastrocotylinea
Accessory sclerite parallel to mid-sclerite (see Fig. 19.6).

Families
Anthocotylidae, Pseudodiclidophoridae, Protomocrocotylidae, Allodiscocotylidae, Pseudomazocreaidae, Chauhaneidae, Bychowskycotylidae, Gastrocotylidae, Neothoracocotylidae, Gotocotylidae.

Suborder Mazocraeinea
Posterior mid-sclerite platelike; two pairs of lateral sclerites, posterior pair broken.

Families
Plectanocotylidae, Mazoplectidae, Mazocraeidae.

Order Diclybothriidea
Haptoral suckers present in appendix; lateral sclerites absent.

Families
Diclybothriidae, Hexabothriidae.

Order Chimaericolidea
Fourteen marginal hooks on the oncomiracidium and adult; germarium lobated; eyes absent on the oncomiracidium.

Family
Chimaericolidae.

References

1. Bakke, T. A., P. A. Jansen, and L. P. Hansen. 1990. Experimental transmission of *Gyrodactylus salaris* Malmberg, 1957 (Platyhelminthes, Monogenea) from the Atlantic salmon (*Salmo salar*) to the European eel (*Anguilla anguilla*). *Can. J. Zool.* 69:733–37.

2. Baer, J. G., and L. Euzet. 1961. Classe des monogènes. Monogenoidea Bychowsky. In Grassé, P., ed. *Traité de zoologie: Anatomie, systématique, biologie* 4. Paris: Masson & Cie, 243–325.

3. Beverley-Burton, M., and G. J. Klassen. 1990. New approaches to the systematics of the ancyrocephalid monogenea from Nearctic fishes. *J. Parasitol.* 76:1–21.

4. Boeger, W. A., and D. C. Kritsky. 1993. Phylogeny and a revised classification of the Monogenoidea Bychowsky, 1937 (Platyhelminthes). *Systematic Parasitol.* 26:1–32.

5. Brooks, D. R., and D. A. McLennan. 1993. *Parascript. Parasites and the language of evolution.* Washington, D.C.: Smithsonian Institution Press.

6. Braun, M. 1889–1893. Trematodes Rudolphi 1808. *Bronn's Klassen und Ordnugen des Tierreichs* 4:306–925.

7. Bychowsky, B. E. 1937. Ontogenese und phylogenetische Beziehungen der parasitischen Platyhelminthes. *Izvest. Akad. Nauk: SSSR Seria Biol.* 4:1353–83.

8. Bychowsky, B. E. 1957. *Monogenetic trematodes: Their systematics and phylogeny.* Moscow: Akademii Nauk: SSSR. English trans. 1961. Hargis, W. J., ed. Washington, D.C.: American Institute of Biological Sciences.

9. Cable, J., and R. C. Tinsley. 1992. Unique ultrastructural adaptations of *Pseudodiplorchis americanus* (Polystomatidae: Monogenea) to a sequence of hostile conditions following host infection. *Parasitology* 105:229–41.

10. Cone, D. K., and P. H. Odense. 1984. Pathology of five species of *Gyrodactylus* Nordmann, 1832 (Monogenea). *Can. J. Zool.* 62:1084–88.

11. Ferdig, M. T., M. A. McDowell, J. Janovy Jr., and R. E. Clopton. 1993. Patterns of morphological variation of *Salsuginus yutanensis* (Monogenea: Ancyrocephalidae) over space and time. *J. Parasitol.* 79:744–50.

12. Fuhrmann, O. 1928. Trematoda. Zwite Klasse des Cladus Platyhelminthes. *Kükenthal's and Krumbach's Handbuch der Zoologie* 2:1–140.

13. Halton, D. W. 1974. Hemoglobin absorption in the gut of a monogenetic trematode, *Diclidophora merlangi. J. Parasitol.* 60:59–66.

14. Halton, D. W. 1978. Trans-tegumental absorption of L-alanine and L-leucine by a monogenean, *Diclidophora merlangi. Parasitology* 76:29–37.

15. Halton, D. W., E. Dermott, and G. P. Morris. 1968. Electron microscope studies on *Diclidophora merlangi* (Monogenea: Polyopisthocotylea). I. Ultrastructure of the cecal epithelium. *J. Parasitol.* 54:909–16.

16. Halton, D. W., and G. P. Morris. 1969. Occurrence of cholinesterase and ciliated sensory structures in fish-gill fluke, *Diclidophora merlangi* (Trematoda: Monogenea). *Z. Parasitenkd.* 33:21–30.

17. Jennings, J. B. 1968. Nutrition and digestion. In Florkin, M., and B. T. Scheer, eds. *Chemical zoology, vol. 2, sect. III. Platyhelminthes, Mesozoa.* New York: Academic Press, Inc., 303–26.

18. Kearn, G. C. 1971. The physiology and behaviour of the monogenean skin parasite *Entobdella soleae* in relation to its host (*Solea solea*). In Fallis, A. M., ed. *Ecology and physiology of parasites.* Toronto: University of Toronto Press, 161–87.

19. Kearn, G. C. 1980. Light and gravity responses of the oncomiracidium of *Entobdella soleae* and their role in host location. *Parasitology* 81:71–89.

20. Kearn, G. C. 1985. Observations on egg production in the monogenean *Entobdella soleae. Int. J. Parasitol.* 15:187–94.

21. Llewellyn, J. 1965. The evolution of parasitic platyhelminths. In Taylor, A. E. R., ed. *Evolution of parasites. Third Symposium of the British Society for Parasitology.* Oxford: Blackwell Scientific Publications Ltd., 47–78.

22. Llewellyn, J. 1968. Larvae and larval development of monogeneans. In Dawes, B., ed. *Advances in parasitology* 6. New York: Academic Press, Inc., 373–83.

23. Llewellyn, J. 1972. Behaviour of monogeneans. In Canning, E. U., and C. A. Wright, eds. *Behavioural aspects of parasite transmission. Linnaean Society of London.* London: Academic Press, 19–30.

24. Lyons, K. M. 1966. The chemical nature and evolutionary significance of monogenean attachment sclerites. *Parasitology* 56:63–100.

25. Lyons, K. M. 1970. Fine structure of the outer epidermis of the viviparous monogenean *Gyrodactylus* sp. from the skin of *Gasterosteus aculeatus. J. Parasitol.* 56:1110–17.

26. Lyons, K. M. 1972. Sense organs of monogeneans. In Canning, E. U., and C. A. Wright, eds. *Behavioural aspects of parasite transmission. Linnaean Society of London.* London: Academic Press, 181–99.

27. Lyons, K. M. 1973. The epidermis and sense organs of the Monogenea and some related groups. In Dawes, B., ed. *Advances in parasitology* 11. New York: Academic Press, Inc., 193–232.

28. Macdonald, S., and J. Caley. 1975. Sexual reproduction in the monogenean *Diclidophora merlangi:* Tissue penetration by sperms. *Z. Parasitenkd.* 45:323–34.

29. Mo, T. A. 1991. Seasonal variations of opisthaptoral hard parts of *Gyrodactylus salaris* Malmberg, 1957 (Monogenea: Gyrodactylidae) on parr of Atlantic salmon *Salmo salar* L. in the River Batnfjordselva, Norway. *Systematic Parasitol.* 19:231–40.

30. Murith, D., and M. Beverley-Burton. 1985. *Salsuginus* Beverley-Burton, 1984 (Monogenea: Ancyrocephalidae) from Cyprinodontoidei (Atheriniformes) in North America with descriptions of *Salsuginus angularis* (Mueller, 1934) Beverley-Burton, 1984 from *Fundulus diaphanus* and *Salsuginus heterocliti* n. sp. from *F. heterocliti. Can. J. Zool.* 63:703–14.

31. Ramalingam, K. 1973. Chemical nature of the egg shell in helminths. II. Mode of stabilization of egg shells of monogenetic trematodes. *Exp. Parasitol.* 34:115–22.

32. Rees, J. A., and G. C. Kearn. 1984. The anterior adhesive apparatus and an associated compound sense organ in the skin-parasitic monogenean *Acanthocotyle lobianchi. Z. Parasitenkd.* 70:609–25.

33. Rohde, K. 1975. Fine structure of the Monogenea, especially *Polystomoides* Ward. In Dawes, B., ed. *Advances in parasitology* 13. New York: Academic Press, Inc., 1–33.

34. Rohde, K. 1977. Habitat partitioning in Monogenea of marine fishes. *Z. Parasitenkd.* 53:171–82.

35. Rohde, K. 1979. The buccal organ of some Monogenea Polyopisthocotylea. *Zoologica Scripta* 8:161–70.

36. Schell, S. C. 1985. *Handbook of trematodes of North America north of Mexico.* Moscow, Idaho: University Press of Idaho.

37. Smyth, J. D., and D. W. Halton. 1983. *The physiology of trematodes,* 2d ed. Cambridge: Cambridge University Press.

38. Sproston, N. G. 1946. A synopsis of the monogenetic trematodes. *Trans. Zool. Soc. Lond.* 23:183–600.

39. Thurston, J. P., and R. M. Laws. 1965. *Oculotrema hippopotami* (Trematoda: Monogenea) in Uganda. *Nature* 205:1127.

40. Tinsley, R. C., and C. M. Earle. 1983. Invasion of vertebrate lungs by the polystomatid monogeneans *Pseudodiplorchis americanus* and *Neodiplorchis scaphiopodis. Parasitology* 85:501–17.

41. Tinsley, R. C., and H. C. Jackson. 1986. Intestinal migration in the life cycle of *Pseudodiplorchis americanus* (Monogenea). *Parasitology* 93:451–69.

Additional References

Hargis, W. J. Jr. 1957. The host specificity of monogenetic trematodes. *Exp. Parasitol.* 6:620–25.

Hargis, W. J. Jr., A. R. Lawler, R. Morales-Alamo, and D. E. Zwerner. 1969. *Bibliography of the monogenetic trematode literature of the world.* Gloucester Point, Va.: Virginia Institute of Marine Science Special Scientific Report 55:1–95.

Llewellyn, J. 1963. Larvae and larval development of monogeneans. In Dawes, B., ed. *Advances in parasitology* 1. New York: Academic Press, Inc., 287–326.

Sproston, N. G. 1946. A synopsis of the monogenetic trematodes. *Trans. Zool. Lond.* 25:185–600.

Yamaguti, S. 1963. *Systema Helminthum* 4, New York: Interscience Publishers.

Chapter 20

CESTOIDEA: FORM, FUNCTION, AND CLASSIFICATION OF THE TAPEWORMS

He was as fitted to survive in this modern world as

a tapeworm in an intestine.

—William Golding

Fear and superstition still abound among laypersons, who generally view tapeworms as the lowliest and most degenerate of creatures (Fig. 20.1). Most of the repugnance with which people regard these animals derives from the fact that the tapeworms live in the intestine and are only seen when they are passed with the feces of the host. Furthermore, tapeworms seem to be generated spontaneously, and mystery is nearly always accompanied by fear. Finally, in a few instances their presence initiates disease conditions that traditionally have been difficult to cure. Be that as it may, a scientific approach to cestodology has increased understanding of tapeworms and shown that they are one of the most fascinating groups of organisms in the animal kingdom. Their complex life cycles and intricate host-parasite relationships are rivaled by few known organisms. Observations of cestodes no doubt began in earliest times and extend today into sophisticated laboratories throughout the world.

In ancient times, Hippocrates, Aristotle, and Galen appreciated the animal nature of tapeworms.[28] The Arabs suggested that the segments passed with the feces were a separate species of parasite from tapeworms; they called these segments the cucurbitini, after their similarity to cucumber seeds.[36] Andry, in 1718, was the first to illustrate the scolex of a tapeworm from a human (Fig. 20.2). Three common species in humans, *Taeniarhynchus saginatus,* *Taenia solium,* and *Diphyllobothrium latum,* were confused by all scientists until the brilliant efforts of Küchenmeister, Leuckart, Mehlis, Siebold, and others in the nineteenth century determined both the external and internal anatomy of these and other common species. These researchers also showed conclusively that bladder worms, hydatids, and coenuri were juvenile tapeworms and not separate species or degenerate forms in improper hosts. Although these organisms have been removed from the realms of ignorance and superstition within the past 150 years, much misconception persists.

Sexually mature tapeworms live in the intestine or its diverticula (rarely in the coelom) of all classes of vertebrates. Two forms are known that mature in invertebrates: *Archigetes* spp. (order Caryophyllidea) in the coelom of a freshwater oligochaete and *Cyathocephalus truncatus* (order Spathebothriidea) in the hemocoel of an amphipod.[1]

FORM AND FUNCTION

Strobila

The **strobila** (Fig. 20.3) of cestodes is a structure unique among the Metazoa. Typically it consists of a linear series of sets of reproductive organs of both sexes; each set is referred to as a **genitalium,** and the area around it is a **proglottid** or **proglottis.** Cestodes with multiple proglottids are described as **polyzoic,** but of the Eucestoda, members of the order Caryophyllidea have only one genitalium and so are (**monozoic**) (Fig. 21.11). Some workers advise avoiding the use of the terms *polyzoic* and *monozoic* because such usage implies that polyzoic tapeworms are chains of zoids (individuals), which may not be the case.[55] Nonetheless, in the absence of a better term, we shall retain *polyzoic* to describe tapeworms with multiple genitalia. Polyzoic cestodes may have only a few proglottids, but others may have thousands. Usually there are constrictions between proglottids, and the worms are said to be segmented. However, tissues such as tegument and muscle are continuous between proglottids, and no membranes separate them; some workers have contended, therefore, that the word *segment* may be a misnomer.[55] Because some polyzoic cestodes lack constrictions of any kind between proglottids (order Spathebothriidea), the words *segment* and *proglottid* are not synonymous, although they are often used as such by parasitologists. Finally, *segments* of tapeworms should not be confused with segments or metameres of metameric animals such as annelids and arthropods (p. 297).

In many polyzoic species new proglottids (and segments) are continuously differentiated near the anterior end in a process called **strobilation.** Each segment moves toward the posterior end as a new one takes its place and during the process becomes sexually mature. By the time they approach

FIGURE 20.1

Advertisements of this kind illustrate the low regard most people have for cestodes.

Courtesy SANE, Washington, D.C.

the posterior end of the strobila, the genitalia will have copulated and produced eggs. A given proglottid can copulate with itself, with others in the strobila, or with those in other worms, depending on the species. After the segment contains fully developed eggs or shelled embryos, it is said to be **gravid.**

When a segment reaches the end of the strobila, it often detaches and passes intact out of the host with the feces, as *Taenia* does, or disintegrates en route, releasing the eggs, as *Hymenolepis* does. This process is called **apolysis.** In some species the eggs are released from the gravid segment through a uterine pore, such as in *Diphyllobothrium* spp., or through tears or slits in the segment (as in Trypanorhyncha); the segment only detaches when it is senile or exhausted (**pseudapolysis** or **anapolysis).** In

some forms the segments may be shed while immature and lead an independent existence in the gut while developing to maturity (**hyperapolysis),** as in some Tetraphyllidea. If the posterior margin of a segment overlaps the anterior of the following one, the strobila is said to be **craspedote;** if not, it is **acraspedote** (Fig. 20.4).

Scolex

Most tapeworms bear a "head," or **scolex** (plural **scolices),** at the anterior end that may be equipped with a variety of holdfast organs to maintain the position of the animal in the gut (Fig. 20.5). The scolex may bear suckers, grooves, hooks, spines, glands, tentacles, or combinations of these (Fig. 20.6). However, the scolex can be simple or absent altogether. In some forms the holdfast function of the scolex is

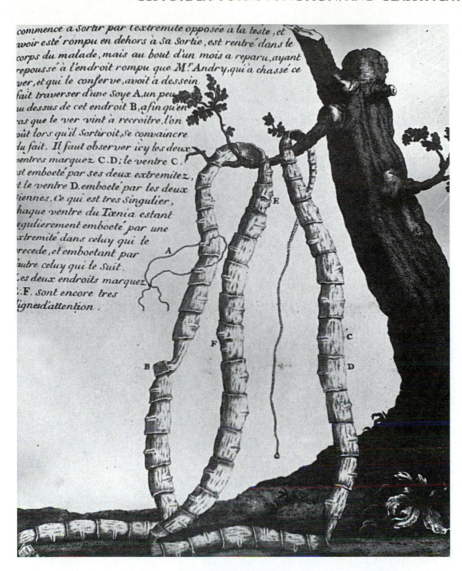

commence a sortir par l'extremite opposée à la teste, et avoir esté rompu en dehors à sa sortie, est rentré dans le corps du malade, mais au bout d'un mois a reparu, ayant repoussé à l'endroit rompu que Mr. Andry, qui a chassé ce ver, et qui le conserve, avoit à dessein fait traverser d'une soye A, un peu au dessus de cet endroit B, afin qu'en cas que le ver vint à recroitre, l'on vit lors qu'il sortiroit, se convaincre du fait. Il faut observer icy les deux ventres marquez C.D: le ventre C. est emboeté par ses deux extremitez, et le ventre D, emboeté par les deux siennes. Ce qui est tres singulier, chaque ventre du Taenia estant regulierement emboeté par une extremité dans celuy qui le precede, et emboetant par l'autre celuy qui le suit. Les deux endroits marquez E.F. sont encore tres dignes d'attention.

FIGURE 20.2

Taeniid tapeworm, probably *T. saginatus,* illustrated for the first time by Andry in 1718.

lost early in life, and the anterior end of the strobila becomes distorted into a **pseudoscolex** to function as a holdfast (Fig. 20.7). Some species penetrate the gut wall of the host to a considerable distance, with the scolex and a portion of the strobila then encapsulated by reacting host tissues.

Suckerlike organs on scolices of tapeworms can be divided into three types: **acetabula, bothria,** and **bothridia.** The acetabulum is more or less cup shaped, circular or oval in outline, with a heavy muscular wall. There are normally four acetabula on a scolex, spaced equally around it. Bothridia usually are in groups of four; are quite muscular, projecting sharply from the scolex; and can have highly mobile, leaflike margins. Bothria are usually two in number, although as many as six may occur, and take the form of shallow pits or longer grooves. They are arranged either in lateral or dorsoventral pairs. Accessory suckers sometimes occur, and most cestodes have a variety of proteinaceous hooks for anchoring the scolex to the host gut. In acetabulate worms the hooks often are arranged in one or more circles anterior to the suckers, borne on a protrusible, dome-shaped area on

the apex of the scolex called the **rostellum.** Both the presence or absence and the shape and arrangements of the hooks are of great taxonomic value. If the rostellum is armed with hooks, it is supplied internally with a heavy muscular pad, which becomes flat and disc shaped when the hooks attach to the host's gut wall. Retraction of the central area of the pad allows withdrawal of the hooks.

Various kinds of gland cells occur in the scolices of a variety of tapeworms, but their function(s) remain(s) enigmatic.[45,46] In some Pseudophyllidea, secretions of the glands may aid in adhesion of the scolex to the host gut mucosa.[34,71] The contents of one type of gland in the pseudophyllidean *Diphyllobothrium dendriticum* are expelled within three days after infection of the definitive host; another gland type remains active and is associated with the nervous system in the scolex.[31]

Hymenolepis diminuta, a cyclophyllidean with an unarmed rostellum, has an invagination of the apical tegument termed an *apical organ* or *anterior canal* (Fig. 20.8).[85] Modified tegumentary cytons (see p. 302) secrete material into

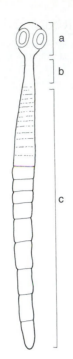

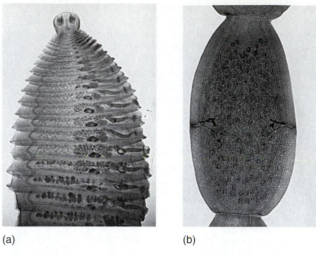

(a)   (b)

FIGURE 20.4

(*a*) Scolex and proglottids of *Paranoplocephala mamillana,* a craspedote cestode. (*b*) A proglottid of *Dipylidium caninum,* an acraspedote species.

Courtesy of Jay Georgi.

FIGURE 20.3

Generalized diagram of a tapeworm, showing scolex (*a*), neck (*b*), and strobila (*c*).

From G. D. Schmidt, *How to Know the Tapeworms.* Copyright © 1970. Reprinted with permission.

FIGURE 20.5

Two types of holdfast organs: (*a*) Bothridea of a tetraphyllidean, *Phyllobothrium* sp., with accessory sucker (*arrow*) (first bar in upper left = 10 μm); (*b*) spiny tentacles and bothridea of a trypanorhynchan, *Callitetrarhynchus gracilis* (× 80).

Courtesy of Frederick H. Whittaker.

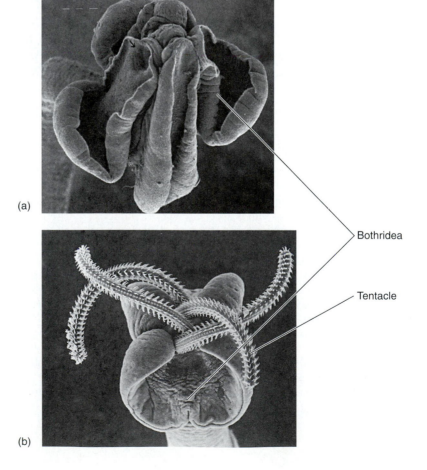

(a)

Bothridea

Tentacle

(b)

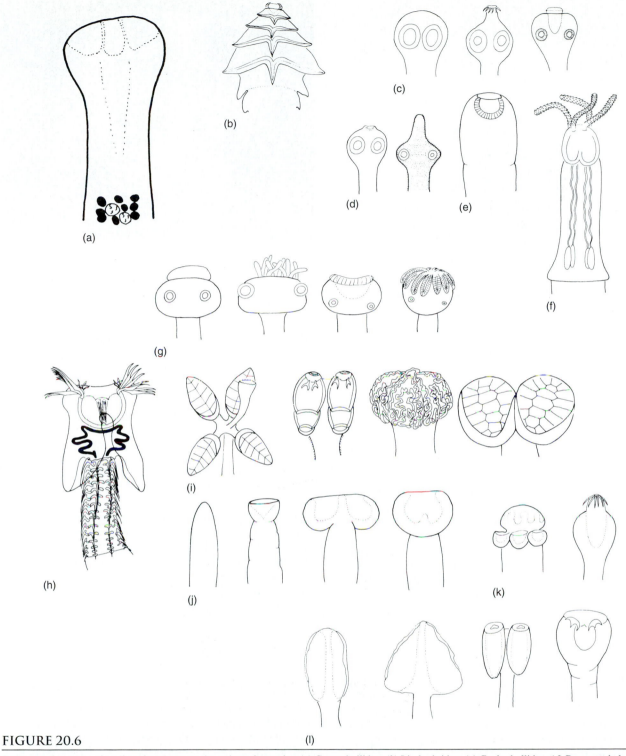

FIGURE 20.6

Representative types of scolices found among the orders of cestodes. (*a*) Caryophyllidea; (*b*) Litobothridea; (*c*) Cyclophyllidea; (*d*) Proteocephalata; (*e*) Nippotaeniidea; (*f*) Trypanorhyncha; (*g*) Lecanicephalidea; (*h*) Diphyllidea; (*i*) Tetraphyllidea; (*j*) Spathebothriidea; (*k*) Aporidea; (*l*) Pseudophyllidea.

From G. D. Schmidt, *How to Know the Tapeworms*. Copyright © 1970. Reprinted with permission.

the apical organ and through the surrounding rostellar tegument.[86] It is possible that these materials play some regulatory role in the development of the worms; there is circumstantial evidence that they are antigenic.[22,38] Apical organs are found in many other cestodes, but they may not be homologous or even structurally similar to that of *H. diminuta*. Apparently similar and homologous structures are found in the Proteocephalata, where, at least in certain cases, their secretion has proteolytic activity and probably functions in penetration (p. 345).[18]

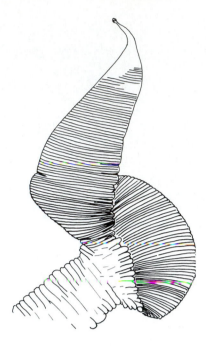

The scolex contains the chief neural ganglia of the worm (as we will describe further), and it bears numerous sensory endings on its anterior surface, probably detecting both physical and chemical stimuli. Such sensory input may allow optimal placement of the scolex and entire strobila with respect to the gut surface and physicochemical gradients within the intestinal milieu.

Commonly, between the scolex and the strobila lies a relatively undifferentiated zone called the **neck,** which may be long or short. It contains stem cells that apparently are responsible for giving rise to new proglottids. In the absence of a neck, similar cells may be present in the posterior portion of the scolex. Praziquantel, a chemotherapeutic agent highly effective against cestodes, preferentially damages the tegument of the neck region and leaves the tegument of segments farther down the strobila unaffected.[7]

Tegument

Cestodes lack any trace of a digestive tract and therefore must absorb all required substances through their external covering. Because of this, the structure and function of the body covering have been of great interest to parasitologists, who have used electron microscopy and radioactive tracers to contribute much to this area of cestodology. Before 1960 the body covering of cestodes and trematodes was commonly referred to as a *cuticle,* but it is now known that it is a living tissue with high metabolic activity, and most parasitologists prefer the term **tegument.**

Tegumental structure is generally similar in all cestodes studied, differing in details according to species. The basic

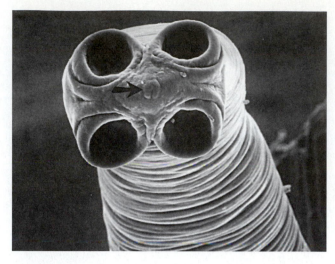

FIGURE 20.8

plan is similar to that of the Trematoda. One major difference is that cestode tegument is covered by minute microvilli called **microtriches** (singular microthrix) (Fig. 20.9). The distal cytoplasm is connected to cytons by internuncial processes that run through the superficial muscle layer (Fig. 20.10). The microtriches are similar in some respects to the microvilli found on gut mucosal cells and other vertebrate and invertebrate transport epithelia, and they completely cover the worm's surface, including the suckers. They have a dense distal portion set off from the base by a multilaminar plate (Fig. 20.11). The distal portion can take one of a variety of forms in different cestode species groups. The cytoplasm of the base is continuous with that of the rest of the tegument, and the entire structure is covered by a plasma membrane.

A layer of carbohydrate-containing macromolecules, the **glycocalyx,** is found on the surface membrane. The microtriches serve to increase the absorptive area of the tegument. A number of phenomena, apparently depending on interaction of certain molecules with the glycocalyx, have been reported: enhancement of host amylase activity; inhibition of host trypsin, chymotrypsin, and pancreatic lipase; absorption of cations; and adsorption of bile salts. Several of these phenomena seem to depend on adsorption of the molecules to the glycocalyx, but present evidence suggests that this is not the case with trypsin inhibition.[79] When incubated in the presence of *H. diminuta,* trypsin seems to undergo a subtle conformational change that decreases its proteolytic activity. The functional value of such phenomena to the worm is uncertain, but interaction with nutrient absorption, protection against digestion by host enzymes, and maintenance of the integrity of the worm's surface membrane may be involved. There is evidence for the presence of a G protein-linked signal transduction system on the surface membrane.[94]

The distal cytoplasm beneath the microtriches contains abundant vesicles and electron-dense bodies, as well as numerous mitochondria. The tegumental nuclei are not found in this

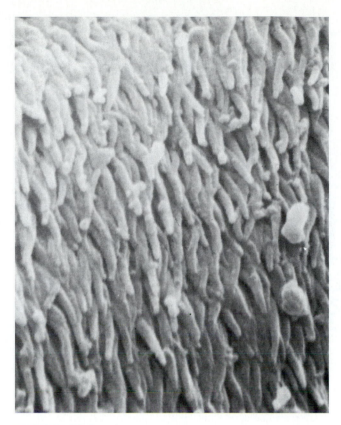

FIGURE 20.9

Posteriorly directed microtriches on the surface of a proglottid of *Hymenolepis diminuta*. (× 44,925.)

From: J. E. Ubelaker et al., "Surface topography of *Hymenolepis diminuta* by scanning electron microscopy," in *J. Parasitol.* 59:667–671. Copyright © 1973.

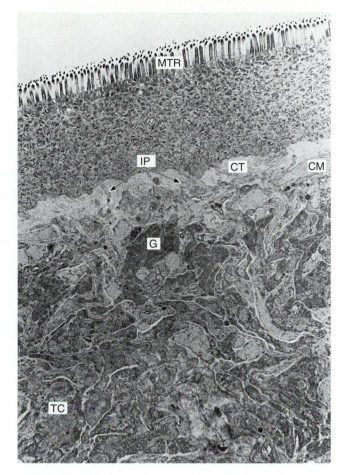

FIGURE 20.10

Longitudinal section through immature proglottid of *Hymenolepis diminuta,* showing nature of tegumentary cortical region. Basal tegumentary cytons (perikarya) (**TC**) are surrounded by glycogen-filled processes (**G**) of cortical myocytons. Internuncial processes (trabeculae) (**IP**) from tegumentary cytons extend through longitudinal and circular (**CM**) muscles, as well as a connective tissue (**CT**) layer (basement lamina), before joining syncytial distal cytoplasm. Microtriches (**MTR**) line free surface of syncytial layer, and discoidal vesicles occupy distal cytoplasm. (× 5900.)

From R. D. Lumsden and R. D. Specian, in H. P. Arai, editor, *Biology of the Rat Tapeworm, Hymenolepis diminuta,* Academic Press, Inc.

layer but lie in the cytons. The vesicles are secreted in the cytons, they are passed to the distal cytoplasm through the internuncial processes, and at least some of them contribute to microthrix and hook formation.[48,58,71] Although each cyton contains but one nucleus, the distal cytoplasm is continuous, with no intervening cell membranes; therefore, the tegument of cestodes is a syncytium. However, there is immunocytochemical evidence that different subpopulations of cytons exist along the strobila, which may reflect functional differences.[37]

Calcareous Corpuscles

The tissues of most cestodes contain curious structures termed **calcareous corpuscles;** they also are found in the excretory canals of some trematodes.[11] They are secreted in the cytoplasm of differentiated calcareous corpuscle cells, which are themselves destroyed in the process. The corpuscles are from 12 to 32 μm in diameter, depending on the species, and consist of inorganic components, principally calcium and magnesium carbonates, along with a hydrated form of calcium phosphate embedded in an organic matrix.[82] The organic matrix is organized into concentric rings and a double outer envelope; it contains protein, lipid, glycogen, mucopolysaccharides, alkaline phosphatase, RNA, and DNA.

The bodies always contain a series of minor inorganic elements, and these, as well as the amount of phosphate, are affected by the diet of the host.

The possible function of the calcareous corpuscles has been the subject of much speculation. For example, mobilization of the inorganic compounds might buffer the tissues of the worm against the large amounts of organic acids produced in its energy metabolism (p. 315). Another suggestion has been that they might provide depots of ions or carbon dioxide for use when such substances are present in insufficient quantity in the environment, such as on initial establishment in the host gut. Still another idea is that they could be an excretory product.[23] Confirmation of these proposals or discovery of the true function of the corpuscles will require the application of a creative scientific mind.

FIGURE 20.11

(a) Sagittal section of tegumental microtriches. Electron-opaque cap (C) separated from the base (B) by multilaminar baseplate (BP). Microfilaments (MF) regularly arranged within base. Tegumental plasmalemma extends over entire length of each microthrix. (× 71,000.) (b) Cross section through bases of tegumental microtriches revealing orderly array of microfilaments (MF) surrounded by accumulation of electron-dense material. (× 120,000.)

From R. D. Lumsden and R. D. Specian, in H. P. Arai, editor, *Biology of the Rat Tapeworm, Hymenolepis diminuta*, Academic Press, Inc.

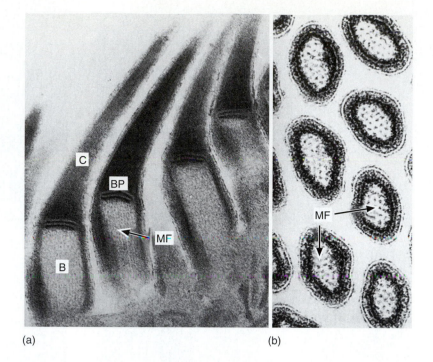

(a)   (b)

Muscular System

The muscle cells of *Hymenolepis diminuta* consist of two portions: the contractile myofibril and the noncontractile myocyton.[49] The contractile portion contains actin and myosin fibrils, and like muscles of other platyhelminths, it is nonstriated and lacks transverse sarcolemmal tubules (T tubules),[47] as might be expected of muscles with slow contraction. The myocytons comprise the bulk of the parenchyma of the worm, and they have been referred to as *parenchymal* cells. Myocytons contain a nucleus, rough endoplasmic reticulum, free ribosomes, a vesicular Golgi apparatus, few mitochondria, and abundant glycogen. Lipid is stored here as well. Although these cytological details are best known for *H. diminuta,* it is highly likely that they pertain to all other cestodes and even trematodes.

The contractile portions of the muscle cells are arranged in discrete bundles in specific regions of the worms. Just internal to the distal cytoplasm are bundles of longitudinal and circular fibers. More powerful musculature lies below the superficial muscles. The longitudinal bundles are usually arranged around a central parenchymal area, which is largely free of contractile elements. There may be a zone of cortical parenchyma, also free of longitudinal fibers. There are numbers of dorsoventral and transverse fibers, and sometimes radial fibers as well. The pattern and relative development of muscle bundles are highly variable in the Cestoidea but constant within a species; therefore, they are often valuable taxonomic characters.

Internal musculature of the scolex is complex, making the scolex extraordinarily mobile. Three distinct muscle types have been found in scolices of trypanorhynchs (p. 347): peripheral myofibers similar to those previously described, tentacle retractor muscles, and tentacle bulb muscles. The bulb muscles are obliquely striated and have numerous motor end plates; motor innervation of the peripheral muscles and retractor muscles has not been found.[95]

Nervous System

The main nerve center of the cestode is in its scolex, and the complexity of ganglia, commissures, and motor and sensory innervation there depends on the number and complexity of other structures on the scolex. Among the simplest are the bothriate cestodes such as *Bothriocephalus,* which have only a pair of lateral cerebral ganglia united by a single ring and a transverse commissure. Arising from the cerebral ganglia is a pair of anterior nerves, supplying the apical region of the scolex; four short posterior nerves; and a pair of lateral nerves that continue posteriorly through the strobila. The bothria are innervated by small branches from the lateral nerves.

In contrast, worms with bothridia or acetabula and hooks, rostellum, and so on may have a substantially more complex system of commissures and connectives in the scolex, with five pairs of longitudinal nerves running posteriorly from the cerebral ganglia through the strobila (Fig. 20.12). In addition to the motor innervation of the scolex, there may be many sensory endings, particularly at the apex of the tegument. Stretch receptors have been described.[70]

As the longitudinal nerves proceed posteriorly, they are connected by intraproglottidal commissures in a ladderlike fashion. Smaller nerves emanate from them to supply the general body musculature and sensory endings. The cirrus and vagina are richly innervated, and sensory endings around the genital pore are more abundant than in other areas of the strobilar tegument.[52,98] Such an arrangement has obvious value.

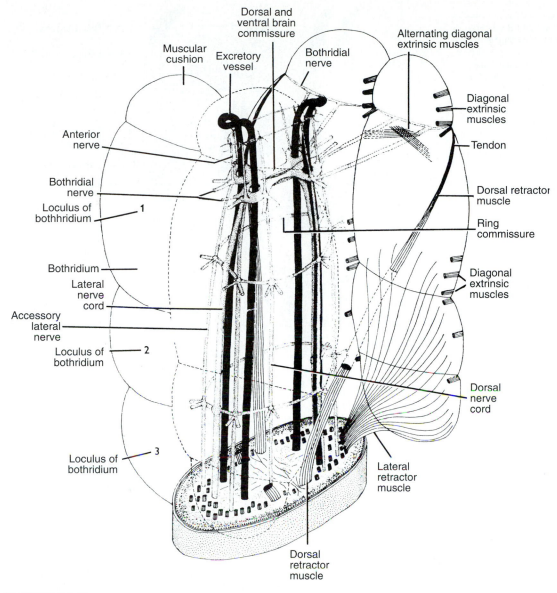

FIGURE 20.12

Acanthobothrium coronatum. Reconstruction of nervous system of scolex. Excretory vessels and some muscles included.

From G. Rees and H. H. Williams, "The functional morphology of the scolex and the genitalia of *Acanthobothrium coronatum* (Rud.) (Cestoda: Tetraphyllidea)," in *Parasitology* 55:617–651. Copyright © 1965. Reprinted with permission of the publisher.

Study of the neuroanatomy of cestodes formerly was difficult because the nerves are unmyelinated and do not stain well with conventional histological stains. However, histochemical techniques that show sites of acetylcholinesterase activity and immunocytological techniques to demonstrate neuropeptides and 5-hydroxytryptamine (serotonin, 5-HT) have permitted elegant studies of tapeworm nervous systems. Serotonin is an important excitatory neurotransmitter, and acetylcholine seems to be the main inhibitory neurotransmitter.[83] Some 20 different neuropep-tides occur in cestodes. There are cholinergic, serotoninergic, and peptidergic components throughout the central and peripheral nervous systems in *Moniezia expansa,*[52] and the "classical" neurotransmitters and peptides may coexist in certain populations of neurons in flatworms.[33] The functions of the plethora of neuropeptides are even more obscure in cestodes than in trematodes (p. 211).

Sensory function probably includes both tactoreception and chemoreception, and cestodes possess at least two, and perhaps more, morphologically distinct types of sensory endings

FIGURE 20.13

Schematic drawing of a longitudinal section through a sensory ending in the tegument of *Echinococcus granulosus.*

From D. J. Morseth, "Observations on the fine structure of the nervous system of *Echinococcus granulosus,*" in *J. Parasitol.* 53:492-500. Copyright © 1967.

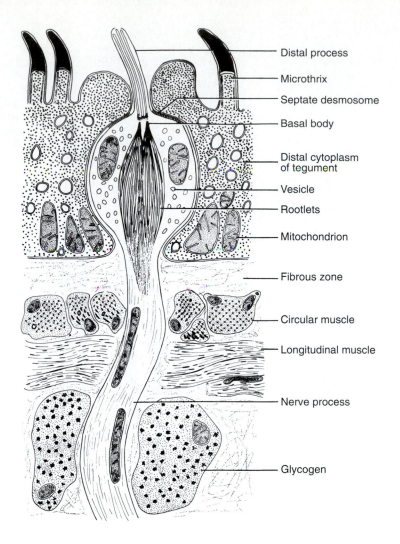

Distal process

Microthrix

Septate desmosome

Basal body

Distal cytoplasm of tegument

Vesicle

Rootlets

Mitochondrion

Fibrous zone

Circular muscle

Longitudinal muscle

Nerve process

Glycogen

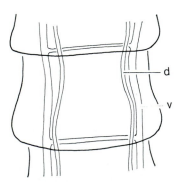

d

v

FIGURE 20.14

Diagram showing the typical arrangement of dorsal (*d*) and ventral (*v*) osmoregulatory canals.

From G. D. Schmidt, *How to Know the Tapeworms.* Copyright © 1970. Reprinted with permission.

in their tegument.[83] One of these has a modified cilium projecting as a terminal process (Fig. 20.13), as is common in such cells in invertebrates.

Excretion and Osmoregulation

In many families of cestodes the main excretory canals run the length of the strobila from the scolex to the posterior end. These are usually in two pairs, one ventrolateral and the other dorsolateral on each side (Fig. 20.14). Most often the dorsal pair is smaller in diameter than is the ventral pair, a useful criterion for determining the dorsal and ventral sides of a tapeworm. The canals may branch and rejoin throughout the strobila or may be independent. Usually a transverse canal joins the ventral canals at the posterior margin of each proglottid. The dorsal and ventral canals unite in the scolex, often with some degree of branching. Posteriorly, the two

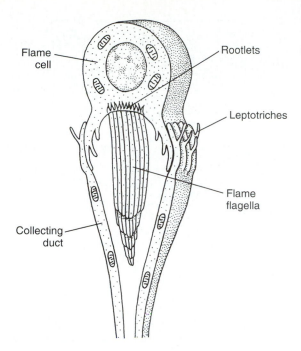

FIGURE 20.15

Diagram of terminal organ of flame cell protonephridium in *Hymenolepis diminuta*. Flame composed of approximately 50 flagella. Collecting duct is syncytial.

Source: M. B. Hildredth, R. D. Lumsden, and R. D. Specian, *Biology of the Rat Tapeworm, Hymenolpis diminuta,* 1980, edited by H. P. Arai, Academic Press, Inc., New York, NY.

pairs of canals merge into an excretory bladder with a single pore to the outside. When the terminal proglottid of a polyzoic species is detached, the canals empty independently at the end of the strobila. Rarely the major canals also empty through short, lateral ducts. In some orders, such as Pseudophyllidea, the canals form a network that lacks major dorsal and ventral ducts.

Embedded throughout the parenchyma are flame cell protonephridia (Fig. 20.15), whose ductules feed into the main canals. The flagella of the flame cell provide motive force to the fluid in the system. Protonephridia of tapeworms show the **weir** construction already described for trematodes (p. 188). In contrast to trematodes, however, the distal tubules of tapeworms are not formed by a single cell but may be syncytial.[35] Furthermore, the excretory ducts are lined with microvilli (Fig. 20.16), thus suggesting that the duct linings serve a transport function. Therefore, functions of the system might include active transport of excretory wastes and ionic regulation of the excretory fluid. Fluid from the excretory canals of *H. diminuta* demonstrated glucose, soluble proteins, lactic acid, urea, and ammonia; lipid was absent.[97]

The principal end products of cestode energy metabolism, short-chain organic acids, are probably excreted through the tegument.

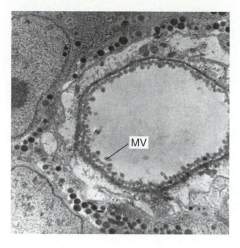

FIGURE 20.16

Low-magnification electron micrograph of excretory duct of *Hymenolepis diminuta* showing beadlike microvilli (**MV**).
Courtesy of H. E. Potswald.

Osmoregulation is another function of the tegumental surface. Although cestodes have been regarded as osmoconformers, with little ability to regulate their body volume in media of differing osmotic concentrations, *Hymenolepis diminuta* can osmoregulate between 210 and 335 mOsm/L in a balanced salt solution if 5 mM glucose is present.[92] The worms rapidly lose water at pH 7.4 and 300 mOsm/L without glucose. Uglem[92] concluded that water balance in *H. diminuta* is closely related to excretory acid concentration and pH of the medium.

Reproductive Systems

Tapeworms are monoecious with the exception of a few rare species from water birds and two from a stingray that are dioecious. Usually each segment has one complete set of both male and female systems, but some genera have two sets of genitalia per segment (Fig. 20.4*b*), and a few species in water birds have one male and two female systems in each segment.

As a segment moves toward the posterior end of the strobila, the reproductive systems mature, sperm are transferred, and oocytes are fertilized. Usually the male organs mature first and produce sperm that are stored until maturation of the ovary; this is called **protandry** or **androgyny.** In a few species the ovary matures first; this is called **protogyny** or **gynandry.** This may be an adaptation that avoids self-fertilization of the same proglottid. Many variations occur in structure, arrangement, and distribution of reproductive organs in tapeworms. These variations are useful at all levels of taxonomy.

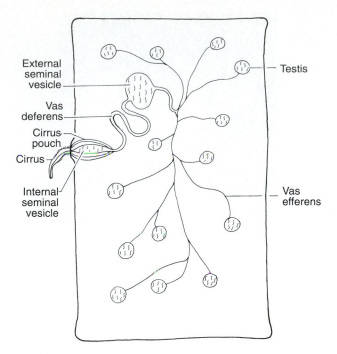

FIGURE 20.17

Diagrammatic representation of a tapeworm male reproductive system.
From G. D. Schmidt, *How to Know the Tapeworms.* Copyright © 1970.
Reprinted with permission.

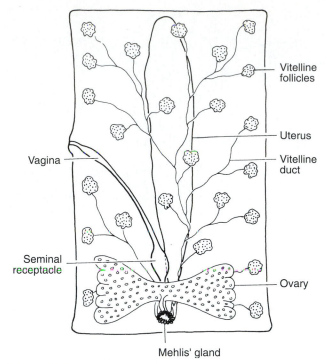

FIGURE 20.18

Diagram of a tapeworm female reproductive system.
From G. D. Schmidt, *How to Know the Tapeworms.* Copyright © 1970.
Reprinted with permission.

• Male Reproductive System

The male reproductive system (Fig. 20.17) consists of one to many testes, each of which has a fine vas efferens. The vasa efferentia unite into a common vas deferens that channels the sperm toward the genital pore. The vas deferens may be a simple duct, or it may have sperm storage capacity in convolutions or in a spheroid external seminal vesicle. As the vas deferens leads into the cirrus pouch, which is a muscular sheath containing the terminal organs of the male system, it may form a convoluted ejaculatory duct or dilate into an internal seminal vesicle. The male copulatory organ is the muscular cirrus, which may or may not bear spines. It can invaginate into the cirrus pouch and evaginate through the cirrus pore.

Commonly the reproductive pores of both sexes open into a common sunken chamber, the genital atrium, which may be simple or equipped with spines, stylets, glands, or accessory pockets. The cirrus pore may open on the margin or somewhere on the flat surface of the segment. If two male systems are present, they open on the segmental margins opposite from one another.

• Female Reproductive System

The female reproductive system (Fig. 20.18) consists of an ovary and associated structures, which are variable in size, shape, and location, depending on the genus. The entire complex is called the **oogenotop.** Vitelline cells, which contribute yolk and shell material to the embryo, may be arranged into a single, compact vitellarium, or they may be scattered as follicles in various patterns. The female gametes are termed **ectolecithal** because they do not produce their own yolk. As oocytes mature, they leave the ovary through a single oviduct, which usually has a controlling sphincter, the **oocapt.**

Oocytes leave the ovary arrested in prophase of meiosis I.[21] Sperm penetration occurs in the proximal oviduct and stimulates resumption and completion of meiotic divisions. One or more cells from the vitelline glands pass through a common vitelline duct, sometimes equipped with a small vitelline reservoir, and join with the zygote. Together they pass into an area of the oviduct known as the ootype. This zone is surrounded by unicellular Mehlis' glands, which appear to secrete a thin membrane around the zygote and its associated vitelline cells.

Shell formation is then completed from within by the vitelline cells and in many cases the cells of the embryo. Eggs of Pseudophyllidea are covered by a thick **capsule** of sclerotin. These capsules are apparently homologous with trematode eggshells and are formed in a similar manner (Fig. 20.19 and p. 214). Some shelled embryos develop in water after passing from the host and usually hatch to release a free-swimming larval stage that is eaten by the aquatic intermediate host.

Shell formation in cestodes in the infracohort Saccouterina (p. 320) is complicated by a variety of layers contributed by embryonic cells.[20,91] These layers include the **coat, embryophore,** and **oncospheral membrane;** the capsule is thin or lacking. Three different types are distinguished (Fig. 20.19): (1) *Dipylidium* type with a thin capsule and an embryophore (as in the cyclophyllidean genera *Dipylidium,*

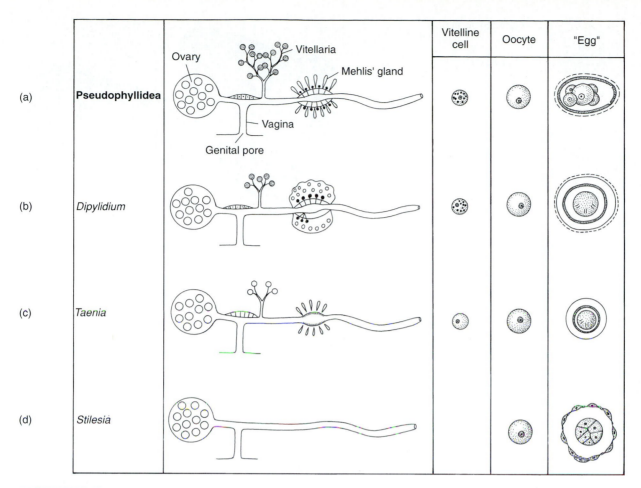

		Vitelline cell	Oocyte	"Egg"
(a)	**Pseudophyllidea**			
(b)	*Dipylidium*			
(c)	*Taenia*			
(d)	*Stilesia*			

FIGURE 20.19

Types of egg-forming systems in cestodes. (*a*) Pseudophyllidean type, found also in the Caryophyllidea. A relatively thick capsule of sclerotin is formed from material from the vitelline cells. (*b–d*) "Oligolecithal" types found in the infracohort Saccouterina. (*b*) *Dipylidium* type, found in some Cyclophyllidea, Tetraphyllidea, and Proteocephalata. A thin coat from the outer envelope is added to the thin capsule. In Tetraphyllidea, two to three vitelline cells contribute to the capsule. (*c*) *Taenia* type with thick embryophore and very thin capsule. (*d*) *Stilesia* type, found in cestodes with no distinct vitellaria, in which cellular covering is apparently laid down by the uterine wall.

Modified from E. Löser, "Die Eibildung bei Cestoden," in *Z. Parasitenkd.* 25:556–580. Copyright © 1965.

Moniezia, and *Hymenolepis,* Proteocephalata, and Tetraphyllidea); (2) *Taenia* type with a very thin capsule but a thick embryophore (as in *Taenia* and *Echinococcus* spp.); and (3) *Stilesia* type, formed by species with no distinct vitellaria, with cellular covering apparently laid down by the uterine wall.[83] In contrast with the pseudophyllidean type, in the *Dipylidium* and *Taenia* types only one or a few vitelline cells associate with the zygote. During early embryogenesis some cells become segregated from the rest of the embryo, fuse, surround the embryo, and form the **outer envelope (OE)** (Fig. 20.20). Other cells become an **inner envelope (IE).** The vitelline cell contributes to the OE. The coat forms within the OE and adds to or replaces the capsule. The embryophore and oncospheral membrane are formed by the IE.

As the zygote and vitelline cells pass through the ootype, the secretions of Mehlis' glands are added. The secretions may cause exocytosis of shell material from the vitelline cells and form the structural support component for the capsule.[91] Leaving the ootype, the developing larva passes into the uterus where embryonation is completed.

The form of the uterus varies considerably among groups. It may be reticulated, lobulated, or circular; it may be a simple sac or a simple or convoluted tube; or it may be replaced by other structures. In some tapeworms the uterus disappears, and the eggs, either singly or in groups, are enclosed within the hyaline **egg capsules** embedded within the parenchyma. In some species one or more fibromuscular structures called **paruterine organs** form, attached to the uterus. In these species the eggs pass from the uterus into the paruterine organ, which assumes the functions of a uterus. The uterus then usually disintegrates.

DEVELOPMENT

Nearly every life cycle known for tapeworms requires two hosts for its completion. One notable exception is *Vampirolepis nana,* a cyclophyllidean parasite of mice and humans, which can complete its juvenile stages within the definitive host (p. 341). (*V. nana* is widely called *Hymenolepis nana,* but the type species of *Hymenolepis,*

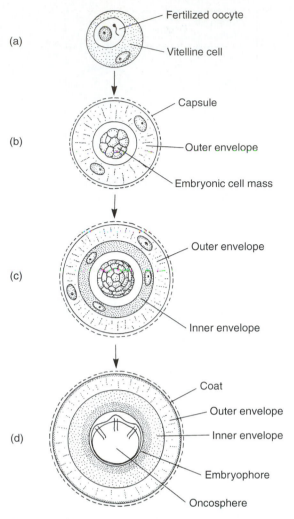

(a) Fertilized oocyte
Vitelline cell

(b) Capsule
Outer envelope
Embryonic cell mass

(c) Outer envelope
Inner envelope

(d) Coat
Outer envelope
Inner envelope
Embryophore
Oncosphere

FIGURE 20.20

Diagram showing formation of embryonic envelopes in Cyclophyllidea. The organization of the envelopes is similar in other cestodes. (*a*) Fertilized oocyte surrounded by vitelline cell and capsule. (*b*) Early phase of development showing formation of outer envelope from vitelline cell and embryonic blastomeres. (*c*) Later phase showing formation of inner envelope from other blastomeres. (*d*) Mature oncosphere with fully developed embryonic envelopes: coat, embryophore of inner envelope, oncospheral membrane (the oncospheral membrane cannot be seen by light microscopy).

Source: K. Rybicka, "The embryonic envelopes in cyclophyllidean cestodes," *Acta Parasitol. Pol.* 13:25–34, 1965.

H. diminuta, has an unarmed rostellum. The rostellar hooks of *V. nana* are ample basis for putting it in a separate genus.)[78] Complete life cycles are known for only a comparatively few species of tapeworms. In fact there are some orders in which not a single life cycle has been determined. Among the life cycles that are known, much variety exists in juvenile forms and patterns of development.

Sexually mature tapeworms live in the intestine or its diverticula or rarely in the coelom of all classes of vertebrates. As stated earlier, two genera are known that can mature in invertebrates. A mature tapeworm may live for a few days or up to many years, depending on the species. During its re-

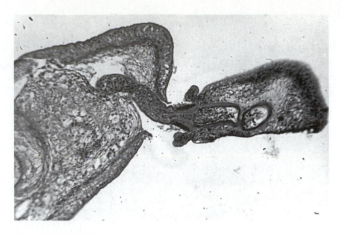

FIGURE 20.21

Hypodermic impregnation of *Dioecotaenia cancellatum.* A cross-section view: the smaller male proglottid (*right*) has pierced the female proglottid (*left*) with its cirrus.

Photograph by Gerald D. Schmidt.

productive life a single worm produces from a few to millions of eggs, each with the potential of developing into the adult form. Because of the great hazards obstructing the course of transmission and development of each worm, mortality is high.

Most tapeworms are hermaphroditic and are capable of fertilizing their own eggs. Sperm transfer is usually from the cirrus to the vagina of another proglottid in the same strobila or between adjacent strobilas, if the opportunity affords. A few species are known in which a vagina is absent; **hypodermic impregnation** has been observed in some of these. In these species the cirrus is forced through the body wall, and the sperm are deposited within the parenchyma (Fig. 20.21). It is not known how they find their way into the seminal receptacle.

A few species of tapeworms are dioecious. In these it is not clear what determines the gender of a given strobila, because it appears that each strobila has the potential of maturing as either male or female. Interaction between two or more strobilas is important in sex determination of dioecious forms. For example, in *Shipleya* (Cyclophyllidea, Dioecocestidae), if a single strobila is present in a host, it is usually female; if two are present, one is nearly always a male. In fact, most of the time the host intestine contains a single female and a single male worm.[80]

Both invertebrates and vertebrates serve as intermediate hosts of tapeworms. Nearly every group of invertebrates has been discovered harboring juvenile cestodes, but the most common are crustaceans, insects, molluscs, mites, and annelids. As a general rule, when a tapeworm occurs in an aquatic definitive host, the juvenile forms are found in aquatic intermediate hosts. A similar assumption can be made for terrestrial hosts.

Vertebrate intermediate and paratenic hosts are found among fishes, amphibians, reptiles, and mammals. Tapeworms found in these hosts normally mature within predators whose diets include the intermediary.

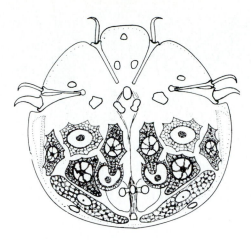

FIGURE 20.22

Diagram of oncosphere of *Hymenolepis diminuta,* dorsal view.

From R. E. Ogren, "The cellular pattern in invasive oncospheres of *Hymenolepis diminuta* as revealed by an enzyme-acetic acid-orcein method," in *Trans. Am. Microsc. Soc.* 86:250–260. Copyright © 1967. Reprinted with permission of the publisher.

Larval and Juvenile Development

Among the life histories that are known, much variety exists in the juvenile forms and details of development, but there seems to be a single basic theme:[30] (1) embryogenesis within the egg to result in a larva, the **oncosphere;** (2) hatching of the oncosphere after or before being eaten by the next host, where it penetrates to a parenteral (extraintestinal) site; (3) metamorphosis of the larva in the parenteral site into a juvenile (**metacestode**) usually with a scolex; and (4) development of the adult from the metacestode in the intestine of the same or another host. The oncospheres of all Eucestoda have three pairs of hooks (Fig. 20.22) and thus also are referred to as **hexacanths.** The free-swimming oncospheres hatching from the egg of some Pseudophyllidea and a few Tetraphyllidea have a ciliated inner envelope and are called **coracidia** (Fig. 20.23).[91] The larvae of gyrocotylideans and amphilinideans have 10 hooks (hence, **decacanths**), are also ciliated, and are called **lycophoras.**

In cestodes with free-swimming larvae the coracidium must be eaten by an intermediate host, usually an arthropod, within a short time. There the coracidium sheds its ciliated IE and actively uses its six hooks to penetrate the gut of its host. In the hemocoel it metamorphoses into a **procercoid** (Fig. 20.24). During this reorganization the oncospheral hooks are relegated to the posterior end in a structure known as the **cercomer.** The procercoid is defined as the stage in which the larval hooks are still present but the definitive holdfast has not developed. It is regarded by some authors as a differentiating metacestode.[30] When the first intermediate host is consumed by the second intermediate host—often a fish—the procercoid penetrates the host gut into the peritoneal cavity and mesenteries and then commonly into the skeletal muscles. Development of the scolex characterizes the **plerocercoid** (Fig. 20.24), and there is commonly strobila formation at this stage, with or without concomitant proglottid formation.

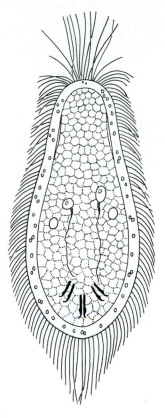

FIGURE 20.23

Coracidium of *Diphyllobothrium erinacei.*

Source: Neveu Lemaie, Traité d'helminthologie médicale et veterinaire, 1936, in *The Zoology of the Tapeworm,* edited by Wardle and McLeod. 1952 University of Minnesota Press.

In the pseudophyllideans *Ligula* and *Schistocephalus,* development as plerocercoids proceeds so far that little growth occurs when these worms reach the definitive host, and the gonads mature within 72 hours and start producing eggs within 36 hours thereafter.[53,60] The Proteocephalata develop a first-stage plerocercoid in the arthropod intermediate host, with no intervening procercoid, and a second-stage plerocercoid in a parenteral site in the second intermediate host. In some species of this family, metacestode development (plerocercoid II) may be completed in the gut of the definitive host, or the metacestodes may develop through a sequence of sites: parenterally in an intermediate host, then parenterally in the definitive host, and finally enterally in the definitive host.[26,27,99] Coracidia, procercoids, and plerocercoids of pseudophyllideans and plerocercoids of proteocephalatans are all plentifully supplied with penetration glands to aid in penetration of, and migration in, host tissues.[18,45]

The life cycles of cyclophyllideans, which are of most concern in this discussion, contrast with the foregoing in that there is neither a procercoid nor a plerocercoid. The larvae are fully developed and infective when they pass from the definitive host, but they do not hatch until eaten by an intermediate host. The oncosphere penetrates the gut of the intermediate host to reach a parenteral site and metamorphoses to

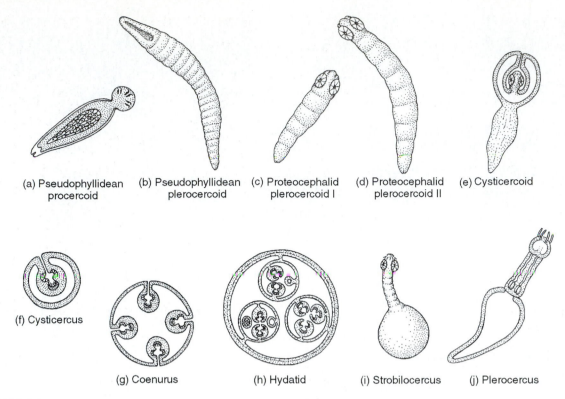

(a) Pseudophyllidean procercoid

(b) Pseudophyllidean plerocercoid

(c) Proteocephalid plerocercoid I

(d) Proteocephalid plerocercoid II

(e) Cysticercoid

(f) Cysticercus

(g) Coenurus

(h) Hydatid

(i) Strobilocercus

(j) Plerocercus

FIGURE 20.24

Types of cestode metacestodes. (*a*) The procercoid can be regarded as a differentiating plerocercoid of pseudophyllideans (*b*). According to some authors, the differentiating plerocercoid of proteocephalatans is a plerocercoid I (*c*), and the infective stage develops into plerocercoid II (*d*). Cysticercoids and cysticerci are metacestodes in the Cyclophyllidea. Cysticercoids (*e*) have an invaginated scolex and a solid body, and the scolices of cysticerci (*f–i*) are both invaginated and introverted into a fluid-filled bladder. The plerocercus (*j*), found in some Trypanorhyncha, is like a plerocercoid with a posterior bladder. Drawings by William Ober and Claire Garrison.

a **cysticercoid** or to a **cysticercus** type of metacestode. The cysticercoid (Figs. 20.24 and 20.25) is a solid-bodied organism with a fully developed scolex invaginated into its body. It is surrounded by cystic layers, and the cercomer, which contains the larval hooks, is outside the cyst. If not displaced mechanically, the cercomer will be digested away, along with parts of the cyst, in the gut of the definitive host. A few cysticercoids have been described that undergo asexual reproduction by budding.

Members of the cyclophyllidean family Taeniidae form a cysticercus metacestode (Figs. 20.24, 21.13, and 21.18), which differs from the cysticercoid in that the scolex is *introverted* as well as invaginated, and the scolex forms on a germinative membrane enclosing a fluid-filled bladder. Several variations from the simple cysticercus in the Taeniidae undergo asexual reproduction by budding (as will be discussed further). They are of considerable medical and veterinary importance.

Numerous other kinds of metacestodes can be distinguished from the typical forms described previously, but they are, for the most part, simply modifications of the types already described:

1. **Sparganum**—a term originally proposed to be applied to any pseudophyllidean plerocercoid of unknown species but now usually used for some plerocercoids of the genus *Diphyllobothrium* (formerly *Spirometra*).

2. **Plerocercus**—a modified plerocercoid found in some Trypanorhyncha, in which the posterior forms a bladder, the **blastocyst,** into which the rest of the body can withdraw (as in *Gilquinia* spp.). Also applied to plerocercoids of proteocephalatans with invaginated scolex.

3. **Strobilocercoid**—a cysticercoid that undergoes some strobilation; found only in *Schistotaenia* spp.

4. **Tetrathyridium**—a fairly large, solid-bodied juvenile that can be regarded as a modified cysticercoid, developing in vertebrates that have ingested the cysticercoid encysted in the invertebrate host. It is known only in the atypical cyclophyllidean *Mesocestoides.*

5. Variations on cysticercus.

   a. **Strobilocercus** (Fig. 20.26)—a simple cysticercus in which some strobilation occurs within the cyst (for example, *Taenia taeniaeformis*).

   b. **Coenurus** (Figs. 20.27 and 21.21)—budding of a few to many scolices (called **protoscolices**) from the germinative membrane of the cyst, each on a simple stalk invaginated into the common bladder (as in *Taenia multiceps*).

   c. **Unilocular hydatid** (Fig. 21.24)—up to several million protoscolices present; occasional sterile specimens. Usually there is an inner, or **endogenous, budding** of **brood cysts,** each with many

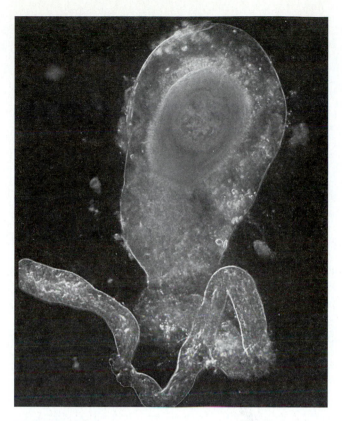

FIGURE 20.25

Fully developed cysticercoid of *Hymenolepis diminuta*.

From M. Voge, in G. D. Schmidt, editor, *Problems in Systematics of Parasites*, © 1969 University Park Press.

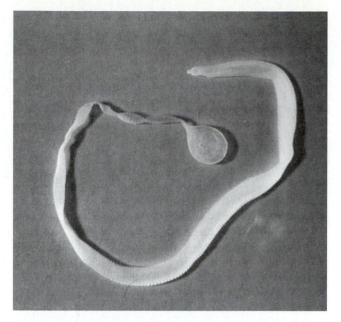

FIGURE 20.26

Strobilocercus from the liver of a rat. Note the small bladder at the posterior end.

Courtesy of James Jensen.

protoscolices inside. **Exogenous budding** rarely occurs, resulting in two more hydatids called **daughter cysts.** This form may grow very large, sometimes containing several quarts of fluid. Occasionally many protoscolices break free and sink to the bottom of the cyst, forming **hydatid sand** (Fig. 21.26), but this is probably rare in the living, normal cyst. This metacestode form is known only for the cyclophyllidean genus *Echinococcus.*

d. **Multilocular** or **alveolar hydatid** (Fig. 21.28)—known only for *Echinococcus multilocularis,* exhibiting extensive exogenous budding, resulting in an infiltration of host tissues by numerous cysts. It forms a single mass with many little pockets that contain protoscolices when in a normal host.

Development in the Definitive Host

As with many other areas of parasitology, generalizations regarding this phase of development may be ill advised because detailed studies of relatively few species are available. However, substantial data have accumulated for the species that have been examined.[75]

When the juvenile tapeworm reaches the small intestine of its definitive host, certain stimuli cause it to excyst, evaginate, or both and begin growth and sexual maturation. In encysted forms action of digestive enzymes in the host's gut may be necessary to at least partially free the organism from its cyst. In *Hymenolepis diminuta* most of the cyst wall may be removed by treatment with pepsin and then with trypsin, but few worms will evaginate and emerge from the cyst unless bile salts are present.[77]

In some pseudophyllideans with a well-developed strobila in the plerocercoid (for example, *Ligula* and *Schistocephalus*), an increase in temperature to that of their definitive host is all that is required for them to mature.[3] The temperature "activation" of such plerocercoids is accompanied by a great increase in the rate of carbohydrate catabolism, excretion of organic acids, and levels of tricarboxylic acid cycle intermediates.[8,44] A burst of neurosecretory activity occurs during activation of *Diphyllobothrium dendriticum* plerocercoids.[32] Contact of the rostellum with a suitable protein substrate is necessary to induce strobilar growth in *Echinococcus.*[84]

As strobilar development begins, subsequent events are influenced by a variety of conditions, including size of the infecting juvenile, species of the worm and host, size and diet of the host, presence of other worms, and the immune and/or inflammatory state of the host intestine.[40] Under optimal conditions certain species have a burst of growth that must surely rival growth rates found anywhere in the animal kingdom. *Hymenolepis diminuta* can increase its weight by up to 1.8 million times within 15 to 16 days.[72] Such rapid growth, accompanied by strictly organized differentiation, makes this worm a fascinating system for the study of development, particularly since the course of the growth may be altered experimentally.

Worm growth is especially sensitive to the composition of the host diet with respect to carbohydrates. The situation is best known for *H. diminuta,* but the findings can be extended to other tapeworms, to some extent at least. *Hymenolepis*

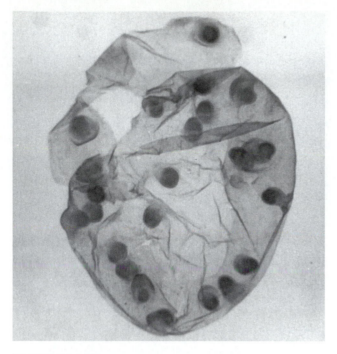

FIGURE 20.27

Coenurus. Each round body in the bladder is an independent protoscolex.

Courtesy of Warren Buss.

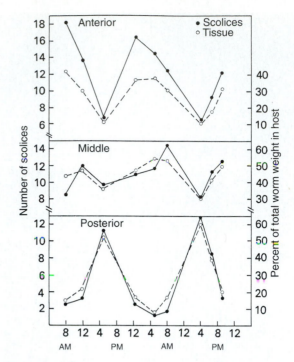

FIGURE 20.28

Distribution of scolices and of wet tissue of *H. diminuta* in the host intestine at various times of the day. **Anterior** refers to the first 10 in., **middle** to the second 10 in., and **posterior** to the remainder of the small intestine. Each point is the mean of determinations from four host animals, representing 110 to 120 worms.

From C. P. Read, "Some physiological and biochemical aspects of host-parasite relations," in *J. Parasitol.* 56:643–652. Copyright © 1970. Reprinted with the permission of the publisher.

diminuta apparently has a high carbohydrate requirement, but it can only absorb glucose and to a lesser degree galactose across its tegument. This is true for other cestodes tested, although some can absorb a limited number of other monosaccharides and disaccharides.[12] For optimal growth the carbohydrate must be supplied in the host diet in the form of a polysaccharide so that the glucose will be released as the digestion proceeds in the host gut. If glucose per se—or a disaccharide containing glucose, such as sucrose—is furnished in the host diet, the worm is placed at a competitive disadvantage for glucose with respect to the gut mucosa, physiological conditions in the gut are altered, or both, such that the worm's growth is substantially restrained.

Another important condition affecting worm growth is the increased presence of other tapeworms in the gut, the so-called **crowding effect.** This may be viewed as an interesting adaptation by which the parasite biomass adjusts to the carrying capacity of the host. Again, evidence exists that, although best known in *H. diminuta,* the crowding effect occurs in at least several other species.[69] Within certain limits, the weight of the individual worms in a given infection is, on the average, inversely proportional to the number of worms present. In consequence the total worm biomass and the number of eggs produced is the same and is maximal for that host, regardless of the number of worms present.

The operational mechanism of the crowding effect is of considerable biological interest as a mode of developmental control. One view was that the individual worms compete for available host dietary carbohydrate. However, the means by which the competition might be translated into lower rates of cell division and cell growth have not been elucidated, and the worms apparently secrete "crowding factors" that influence the development of other worms in the population.[19,41,100]

As the worm approaches maximal size, growth rate decreases, and production of new proglottids is only sufficient to replace those lost by apolysis. Although some species, such as *V. nana,* characteristically become senescent and pass out of the host after a period, others may be limited only by the length of their host's life. *Taeniarhynchus saginatus* may live in a human for more than 30 years, and *H. diminuta* may live as long as the rat it inhabits. In fact Read[68] reported an "immortal" worm that he kept alive for 14 years by periodically removing it from its host, severing the strobila in the region of the germinative area, and then surgically reimplanting the scolex in another rat.

Finally, we should note that some tapeworms manage a surprising degree of mobility within their host's intestine. Cestodes may establish initially in one part of the gut and then move to another as they grow. For example, *D. dendriticum* in rats passes all the way to the large intestine within a few hours of infection, but less than 24 hours later it moves back to the duodenum to start growth.[2] *Hymenolepis diminuta* actually undergoes a diurnal migration in the rat's gut (Fig. 20.28). This migration correlates with the nocturnal feeding habits of rats and can be reversed by giving food to the rat only in the daytime. In fact migration of the worms is apparently mediated by vagal nerve stimulation of gastrointestinal function rather than by the presence of food itself.[56]

METABOLISM

Acquisition of Nutrients

All nutrient molecules must be absorbed across the tegument. The mechanisms of absorption include active transport, mediated diffusion, and simple diffusion.[63] Whether pinocytosis is possible at the cestode surface has been the subject of some dispute,[50] but the plerocercoids of *Schistocephalus* and *Ligula* are capable of this process.[39,90] Cysticerci of *Taenia crassiceps* are capable of pinocytosis, and the process is stimulated by the presence of glucose, yeast extract, or bovine serum albumin in the medium.[88,89]

Glucose is the most important nutrient molecule to fuel energy processes in tapeworms. As noted before, the only carbohydrates that most cestodes can absorb are glucose and galactose, and although some tapeworms can absorb other monosaccharides and disaccharides, we know of none other than glucose and galactose that can actually be metabolized. The primary fate of galactose seems to be incorporation into membranes or other structural components, such as glycocalyx.[59] Galactose can be incorporated into glycogen but does not support net glycogen synthesis.[43] Both glucose and galactose are actively transported and accumulated in the worm against a concentration gradient. Of the two sugars, glucose has been studied more extensively. Glucose influx in a number of species couples to a sodium pump mechanism, that is, the maintenance of a sodium concentration difference across the membrane. The accumulation of glucose, in *H. diminuta* at least, is also sodium dependent. At least two transport sites for glucose are kinetically distinct in the tegument of *H. diminuta*, and the relative proportion of these sites changes during development.[73,87] Fully developed larvae of *H. diminuta* with intact shells absorb very little glucose, but when the shell is removed, as it would be when eaten by the beetle intermediate host, they absorb much larger amounts.[61,62]

Amino acids are also actively transported and accumulated, although less is known about them than about glucose. However, the presence of other amino acids in the ambient medium stimulates efflux of amino acids from the worm; therefore, the worm pool of amino acids rapidly comes to equilibrium with the amino acids in the intestinal milieu.

Purines and pyrimidines are absorbed by facilitated diffusion, and the transport locus is distinct from the amino acid and glucose loci.[51]

The actual mechanism of lipid absorption has not been investigated, but it is likely to be a form of diffusion. Fatty acids, monoglycerides, and sterols are absorbed at a considerably greater rate when they are in a micellar solution with bile salts.[5]

Requirements for external supplies of vitamins are substantiated in only two cases. Investigations of vitamin requirements are difficult, as they often are in parasites, because of limitations in in vitro cultivation techniques, because the worm may be less sensitive than its host to a vitamin-deficient diet, or both. In any case the pathogenesis of vitamin deficiency in the host may have indirect effects on the worm. The necessity for an external supply of a vitamin has been demonstrated unequivocally in only one case—that of pyridoxine in *H. diminuta*.[66,76] By inference we can assume that *Diphyllobothrium latum* has a requirement for vitamin B_{12}, since the worm accumulates unusually large amounts of it.[10] In some cases *D. latum* can compete so successfully with its host for the vitamin that the worm can cause pernicious anemia in persons genetically susceptible to its effects (Chapter 21).

Energy Metabolism

• Glycolysis

The patterns of energy metabolism in cestodes are much like those already described in trematodes (p. 223). In brief, adult cestodes are facultative anaerobes that derive energy from catabolism of glucose and glycogen, but they only oxidize the glucose molecule in part, and they excrete highly reduced end products, such as short chain organic acids (Fig. 20.29).[74]

Because cestodes have very limited ability to degrade fatty and amino acids, their processes of carbohydrate storage and catabolism assume critical importance for energy production. Indeed, juvenile and adult cestodes characteristically store enormous amounts of glycogen, ranging from about 20% to more than 50% of dry weight. Whereas the tissue-dwelling juveniles will be exposed to a reasonably constant glucose concentration maintained by the homeostatic mechanisms of the host, the adults must survive between host feeding periods. The large amount of stored glycogen serves at these times as an effective cushion. *Hymenolepis diminuta* consumes 60% of its glycogen during 24 hours of host starvation and another 20% during the next 24 hours. When glucose is again available, the glycogen stores are rapidly replenished.[24]

As in trematodes, glucose from glycogen or absorbed directly from the host intestine is degraded by classical glycolysis as far as phosphoenolpyruvate (PEP), but at this point there is a branch in the pathway (Fig. 20.29). Either lactate is produced by dephosphorylation of PEP and reduction of pyruvate, or malate is produced by fixation of carbon dioxide to form oxaloacetate, which is then reduced to malate.[74] Both branches thus far are functionally equal because each generates a high-energy phosphate bond and reoxidizes the NADH formed in glycolysis; therefore, the cytoplasmic redox balance is preserved.

However, additional energy is obtained when the malate enters the mitochondria, where part of the malate is metabolized and excreted as acetate or is transaminated and excreted as alanine. The other half of the malate is metabolized and reduced to succinate (Fig. 20.29). Reducing equivalents for the reduction of fumarate are provided by the oxidation of malate. However, in *H. diminuta* and *H. microstoma,* the oxidative decarboxylation of malate is NADP dependent, whereas the fumarate reduction is NAD dependent. Therefore, a hydride ion must be transferred from NADPH to NAD, and this is accomplished by an NADPH:NAD transhydrogenase.[25] The excretion of succinate and acetate produces two more ATPs than if the glucose carbon were excreted solely as lactate. In some cestodes propionate is formed by decarboxylation of succinate, generating additional ATP.[65] An advantage of alanine excretion would be that it is less acidic than lactate.

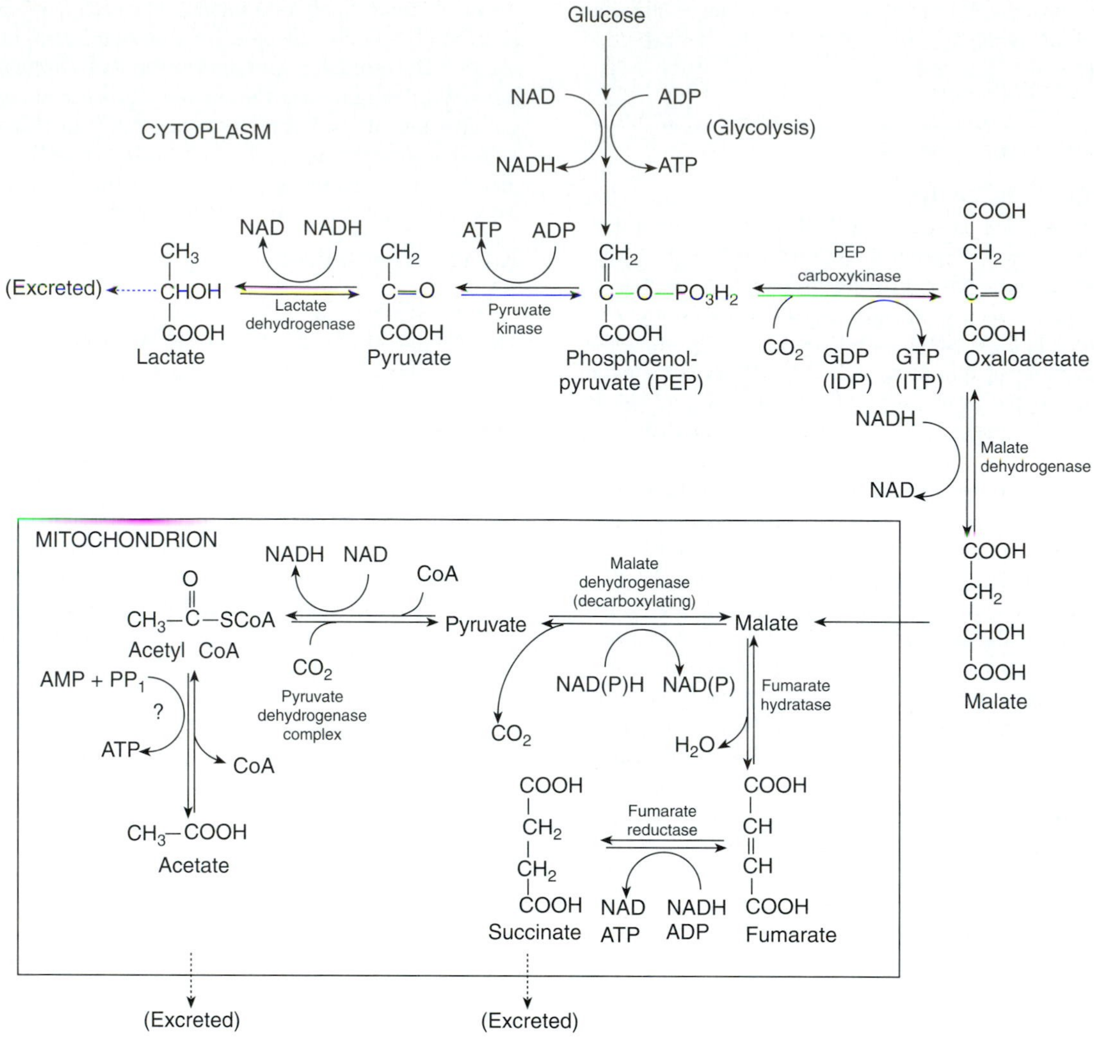

FIGURE 20.29

Reactions forming the major end products of energy metabolism from phosphoenolpyruvate in *Hymenolepis diminuta* (adapted and proposed from various sources). These reactions yield additional ATP above that from classical glycolysis, with a balanced cytoplasmic oxidation-reduction and a balanced mitochondrial oxidation-reduction (ratio of succinate to acetate excreted approximately 2:1). Fumarate reductase usually is referred to as succinate dehydrogenase, but it acts in an opposite direction from mammalian systems; that is, as a fumarate reductase (Watts and Fairbairn).[96] In *Hymenolepis,* the malate dehydrogenase (decarboxylating) reaction is NADP dependent, and the hydride ion is transferred to NAD for the fumarate reductase reaction by a transhydrogenase, thus maintaining the redox balance in the mitochondrion.[25]

Despite the energetic advantage of the mitochondrial reactions and excretion of acetate and succinate, the value of these reactions to the worms remains quite unclear. Some strains of *Hymenolepis diminuta* excrete mostly lactate and little acetate and succinate, and others excrete predominately acetate and succinate.[14] Moreover, within the same strain the proportions of these acids excreted varies according to worm development, part of the strobila, and even the immune status of the host. The acids excreted by the tapeworms can be catabolized by the host to CO_2 and water, and the explanation may lie in still obscure aspects of the host-parasite interaction.[14]

• Krebs Cycle

The Krebs acid cycle is of little or no importance in adult cestodes, but a substantial amount of glucose carbon may flow through the Krebs cycle in certain metacestodes. As much as 40% of the carbohydrate utilized by protoscolices of *Echinococcus multilocularis* and the sheep strain of *E. granulosus* may be channeled into the Krebs cycle, and only 22% of the glycogen catabolized by plerocercoids of *Schistocephalus solidus* is accounted for by excreted acids.[44,54] Activity of the Krebs cycle increases in *S. solidus* when the plerocercoids are activated by an increase in the ambient temperature.[8]

• Electron Transport

Cestodes take up oxygen when it is available, but the function of oxygen is probably not a terminal electron acceptor in an energy-producing series of reactions (for example, oxidative phosphorylation via the "classical" cytochrome system). Although earlier research indicated that some cytochromes might be present in some cestodes, later research failed to confirm that a cytochrome system was operating, and the function of such cytochromes as were present was a mystery. Use of more sensitive techniques now has provided evidence that a classical mammalian type of electron transport system is present in at least some cestodes, that the classical chain is probably of minor importance (Fig. 20.30), and that the major cytochrome system is a so-called *o*-type, similar to that reported in many bacteria.[17] This system may be an adaptation to facultative anaerobiosis. The terminal oxidase can transfer electrons to either fumarate or oxygen, depending on whether conditions are aerobic or anaerobic, and the products are either succinate or hydrogen peroxide, respectively. A peroxidase destroys the hydrogen peroxide before it reaches toxic levels. Under anaerobic conditions, the succinate formed in this pathway would be excreted.

Oxygen consumption increases by 40% when cysticerci of *Taenia solium* are stimulated to evaginate by treatment with trypsin, but evagination is not affected by respiratory poisons such as cyanide.[16] These metacestodes apparently also have a branched electron transport system.

Tapeworms probably do not derive any energy from degradation of lipids or proteins. *H. diminuta* has only a modest capacity for carrying out transaminations and can degrade only four amino acids.[93] The function served by much of the lipid in cestodes remains a mystery, since no one has been able to show that lipids are depleted at all during starvation, even though they may comprise up to 20% of total worm dry weight or more than 30% in the parenchyma of gravid proglottids. *Schistocephalus solidus* has all the enzymes necessary for the β-oxidation sequence of lipids; nevertheless, it appears unable to catabolize them.[6] Lipids in cestodes may represent metabolic end products, since they are relatively nontoxic to store, and the parenchyma of gravid proglottids is discarded during apolysis.

Nitrogenous end products excreted by *H. diminuta* include considerable quantities of ammonia, α-amino nitrogen, and urea.[24]

Synthetic Metabolism

Little need be said of the protein and nucleic acid synthetic abilities of cestodes. It is clear that they can absorb amino acids, purines, pyrimidines, and nucleosides from the intestinal milieu and synthesize their own proteins and nucleic acids (see, for example, Bolla and Roberts[9]). *Moniezia expansa* cannot synthesize carbamyl phosphate; therefore, it depends on its host for both pyrimidine and arginine.

In contrast, capacity for synthesis of lipids appears minimal. The worm can neither synthesize fatty acids de novo from acetyl-CoA nor introduce double bonds into the fatty acids it absorbs.[42] *Hymenolepis diminuta* rapidly hydrolyzes monoglycerides after absorption, and it can then resynthesize triglycerides. It can lengthen the chain of fatty acids provided that the acid already contains 16 or more carbons. Similar observations have been reported for *Diphyllobothrium mansonoides*.[57] There is some evidence that *Vampirolepis microstoma* might be able to synthesize fatty acids de novo.[67]

Finally, *H. diminuta* cannot synthesize cholesterol, a biosynthesis that requires molecular oxygen in other systems.[29]

Hormonal Effects of Metabolites

Certain substances produced by cestodes have effects on their hosts that mimic the host's own hormones. Fish infected with the plerocercoids of *Ligula* are unable to reproduce: Their

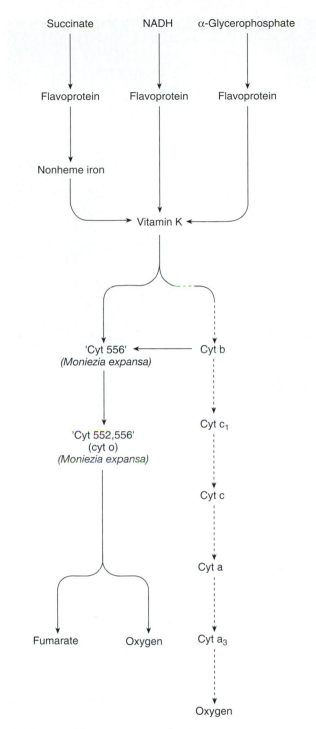

FIGURE 20.30

Branched chain electron transport system with cytochrome *o*, facultatively transporting electrons to fumarate or oxygen. Evidence exists that a similar system operates in *Moniezia, Taenia,* and probably other cestodes. *Solid line,* major pathway; *dotted line,* minor pathway.

Modified from C. Bryant, "Electron transport in parasitic helminths and protozoa," in *Advances in Parasitology,* Vol. 8, edited by B. Dawes. Copyright © 1970 Academic Press, Inc., New York, NY.

FIGURE 20.31

Illustration of growth hormone-like action of *Diphyllobothrium mansonoides* plerocercoids. All rats were hypophysectomized when they weighed 90 g, but the two larger rats received 20 juvenile scolices of *D. mansonoides* approximately one month after the operation. The photograph was taken six months later, and the experimental animals outweigh the controls by three or four times.

From J. F. Mueller, "The biology of *Spirometra*," in *J. Parasitol.* 60:3–14. Copyright © 1974.

gonads do not develop, and there is an apparent suppression of the presumed gonadotropin-producing cells in their pituitary glands. On the possibility that a sex steroid was produced by the worms and that the steroid interfered with gonadotrophin production and hence gonad development, the presence of such compounds was investigated.[4] None was found, and the mechanism of the effect on the host remains enigmatic.

A more surprising case is that of a substance produced by the plerocercoids of *Diphyllobothrium* (= *Spirometra*) *mansonoides*. Referred to as **plerocercoid growth factor (PGF)**, the compound closely resembles human growth factor (hGF) and acts as a growth hormone in several mammalian species (Fig. 20.31).[64] It is definitely not the same compound, although the pituitary recognizes it as such and decreases its own production of growth hormone. Under normal circumstances administration of hGF can have anti-insulin and diabetogenic activities; these properties do not accompany administration of PGF. Growth hormones from other vertebrates (except primates) are not active in humans because of the strict binding specificity of the receptor molecules in humans. However, PGF has the same binding specificity as hGF, and monoclonal antibodies raised against the unique epitope of hGF crossreact with PGF. This and other evidence has suggested the following hypothesis: The gene for PGF is a *human gene* (for hGF) that *has been sequestered by the tapeworm* during its evolution.[64] The possible value of PGF to the tapeworm is a mystery, but there is some evidence of interaction with the immune system of the host. Perhaps suppression of one or more pathways by PGF might indirectly suppress the immune response, thus allowing the worms to evade host defenses while becoming established.[81] Nevertheless, there is significant potential for use of PGF in human endocrinology because of its lack of anti-insulin and diabetogenic activities.

CLASSIFICATION OF CESTOIDEA

The classification of tapeworms is in a state of flux as the phylogenetic systematists strive to construct a system that avoids paraphyly and polyphyly. We are adopting what we believe is the most acceptable system for the higher classification of the Cestoidea currently available.[13] Thus, the cohort Cestoidea belongs to the subclass Cercomeromorphae (possessing a cercomer with hooks) and the infraclass Cestodaria (intestine lacking, cercomer reduced) (p. 319). Figure 20.32 is a cladogram showing relationships among the Cestodaria. Keep in mind that each character listed in the diagnoses and cladograms is apomorphic for the groups but that all members of a group do not necessarily have that character. Just as snakes are tetrapods with no legs, caryophylleids are eucestodans with but one proglottid.

We believe, however, that the arrangement of tapeworm orders proposed by Brooks and McLennan[13] has not yet been sufficiently examined by systematists for use in a general text. Therefore, for this edition we continue to recognize the traditional orders and their respective diagnoses.

Cohort Cestoidea
Male genital pore and vagina proximate; cercomer totally invaginated during ontogeny; hooks on larval cercomer in two size classes (six large and four small); protonephridial ducts lined with microvilli; protonephridia in larvae in posterior end of body; genital pores marginal.

Subcohort Amphilinidea
Uterine pore and genital pores not proximate; male pore and vaginal pore at posterior end; uterus N-shaped; uterine pore proximal to vestigial pharynx; inner longitudinal muscle layer weakly developed; adults parasitic in body cavity of fishes and turtles.

Subcohort Eucestoda
Adults polyzoic; cercomer lost during ontogeny; hooks on larval cercomer reduced to six; medullary portion of proglottids restricted; tegument covered with microtriches; sperm lacking mitochondria.

Infracohort Pseudophylla
Bilaterally symmetrical, bipartite scolices ("difossate" condition), with modifications for attachment consisting of longitudinal flaps (bothria) and their modifications; polylecithal

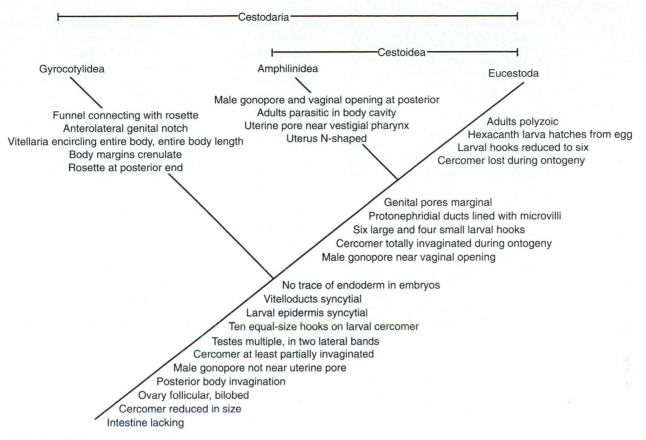

FIGURE 20.32

Cladogram showing hypothetical relationships of groups within the infraclass Cestodaria. Gyrocotylidea and Amphilinidea will be described in Chapter 21. The name Cestodaria is commonly applied to the Gyrocotylidea and Amphilinidea, but it appears that the Amphilinidea and Eucestoda share a common ancestor and form a clade with the Gyrocotylidea as sister group. One primitive character state of the Cestodaria is 10 larval hooks, of which four become reduced in the Cestoidea and disappear in the Eucestoda.

Source: Based on D. R. Brooks and D. A. McLennan, *Parascript: Parasites & the Language of Evolution.* Copyright © 1993 Smithsonian Institution Press, Washington, D.C.

eggs (a large component of vitelline material forming a true shell that is quinone tanned); one embryonic membrane (no embryophore); hexacanth larva hatches from egg, is ingested in water, and is followed by procercoid and plerocercoid stages; oncospheres with unicellular protonephridium.

Order Caryophyllidea
Scolex unspecialized or with shallow grooves or loculi or shallow bothria; monozoic; genital pores midventral; testes numerous; ovary posterior; vitellaria follicular, scattered or lateral; uterus a coiled median tube, opening, often together with vagina, near male pore; parasites of teleost fishes and aquatic annelids.

Families
Caryophyllaeidae, Balanotaeniidae, Lytocestidae, Capingentidae.

Order Spathebothriidea
Scolex feebly developed, undifferentiated or with funnel-shaped apical organ or one or two hollow, cuplike organs; constrictions between proglottids absent; proglottids distinguished internally; genital pores and uterine pore ventral or alternating dorsal and ventral; testes in two lateral bands; ovary dendritic; vitellaria follicular, lateral or scattered; uterus coiled; parasites of teleost fishes.

Families
Cyathocephalidae, Spathebothriidae, Bothrimonidae.

Order Pseudophyllidea
Scolex with two bothria, with or without hook; neck present or absent; strobila variable; proglottids anapolytic; genital pores lateral, dorsal, or ventral; testes numerous; ovary posterior; vitellaria follicular, lateral or cortical, and encircling other organs; uterine pore present, dorsal or ventral; egg usually operculate, containing a coracidium; parasites of fishes, amphibians, reptiles, birds, and mammals.

Families
Amphicotylidae, Bothriocephalidae, Cephalochlamydidae, Diphyllobothriidae, Echinophallidae, Haplobothriidae, Parabothriocephalidae, Ptychobothriidae, Triaenophoridae.

Infracohort Saccouterina

Bilateral saccate uteri lacking permanent pores; oncospheral flame cells absent; oncospheres not ciliated; generally "oligolecithal" eggs (minimal vitelline component, with shell formed by embryo); two embryonic membranes; oncosphere matures in utero; indeterminate growth (continuous budding from a growth zone); generally apolytic.

Order Nippotaeniidea

Scolex with single sucker at apex, otherwise simple; neck short or absent; strobila small; proglottids each with single set of reproductive organs; genital pores lateral; testes anterior; ovary posterior; vitelline gland compact, single, between testes and ovary; osmoregulatory canals reticular; parasites of teleost fishes.

Family
Nippotaeniidae.

Order Lecanicephalidea

Scolex divided into anterior and posterior regions by transverse groove; anterior portion cushionlike or with unarmed tentacles, capable of being withdrawn into posterior portion, forming a large suckerlike organ; posterior portion usually with four suckers; neck present or absent; testes numerous; ovary posterior; vitellaria follicular, lateral or encircling proglottid; uterine pore usually present; parasites of elasmobranchs.

Families
Adelobothriidae, Disculicepitidae, Lecanicephalidae.

Order Trypanorhyncha

Scolex elongated, with two or four bothridia and four eversible (rarely atrophied) tentacles armed with hooks; each tentacle invaginates into internal sheath provided with muscular bulb; neck present or absent; strobila apolytic, anapolytic, or hyperapolytic; genital pores lateral, rarely ventral; testes numerous; ovary posterior; vitellaria as in Pseudophyllidea; uterine pore present or absent; parasites of elasmobranchs.

Families
Dasyrhynchidae, Eutetrarhynchidae, Gilquiniidae, Gymnorhynchidae, Hepatoxylidae, Hornelliellidae, Lacistorhynchidae, Mustelicolidae, Otobothriidae, Paranybeliniidae, Pterobothriidae, Sphyriocephalidae, Tentaculariidae, Mixodigmatidae, Rhinoptericolidae.

Order Aporidea

Scolex with simple suckers or grooves and armed rostellum; constrictions between proglottids absent; proglottids distinguished internally or separate proglottids not evident; genital ducts and pores, cirrus, ootype, and Mehlis' gland absent; hermaphroditic, rarely dioecious; vitelline cells mixed with ovarian cells; parasites of Anseriformes.

Family
Nematoparataeniidae.

Order Tetraphyllidea

Scolex with highly variable bothridia, sometimes also with hooks, spines, or suckers; myzorhynchus present or absent; proglottids commonly hyperapolytic; hermaphroditic, rarely dioecious; genital pores lateral, rarely posterior; testes numerous; ovary posterior; vitellaria follicular, usually medullary in lateral fields; uterine pore present or absent; vagina crosses vas deferens; parasites of elasmobranchs.

Families
Onchobothriidae, Phyllobothriidae, Triloculariidae, Dioecotaeniidae.

Order Diphyllidea

Scolex with armed or unarmed peduncle; two spoon-shaped bothridia present, lined with minute spines, sometimes divided by median, longitudinal ridge; apex of scolex with insignificant apical organ or with large rostellum bearing dorsal and ventral groups of T-shaped hooks; strobila cylindrical, acraspedote; genital pores posterior, midventral; testes numerous, anterior; ovary posterior; vitellaria follicular, lateral, or surrounding other organs; uterine pore absent; uterus tubular or saccular; parasites of elasmobranchs.

Families
Ditrachybothridiidae, Echinobothriidae.

Order Litobothridea

Scolex a single, well-developed apical sucker; anterior proglottids modified, cruciform in cross section; neck absent; strobila dorsoventrally flattened, with numerous segments, each with single set of medullary genitalia; proglottids laciniated and craspedote, apolytic or anapolytic; testes numerous, preovarian; genital pores lateral; ovary with two or four lobes, posterior; vitellaria follicular, encircling medullary parenchyma; parasites of elasmobranchs.

Family
Litobothridae.

Order Proteocephalata

Scolex with four suckers, often with prominent apical organ, occasionally with armed rostellum; neck usually present; genital pores lateral; testes numerous; ovary posterior; vitelline glands follicular, usually lateral, either cortical or medullary; uterine pore present or absent; parasites of fishes, amphibians, and reptiles.

Families
Proteocephalidae, Monticellidae.

Order Cyclophyllidea

Scolex usually with four suckers; rostellum present or absent, armed or unarmed; neck present or absent; strobila usually with distinct segments, monoecious or rarely dioecious; genital pores lateral (ventral in Mesocestoididae); vitelline gland compact, single (double in Mesocestoididae), posterior to ovary (anterior or beneath ovary in Tetrabothriidae); uterine pore absent; parasites of amphibians, reptiles, birds, and mammals.

Families
Amabiliidae, Anoplocephalidae, Catenotaeniidae, Davaineidae, Dilepididae, Dioecocestidae, Diploposthidae, Hymenolepididae, Mesocestoididae, Nematotaeniidae, Progynotaeniidae, Taeniidae, Tetrabothriidae, Triplotaeniidae.

References

1. Amin, O. M. 1978. On the crustacean hosts of larval acanthocephalan and cestode parasites in southwestern Lake Michigan. *J. Parasitol.* 64:842–45.
2. Archer, D. M., and C. A. Hopkins. 1958. Studies on cestode metabolism. III. Growth pattern of *Diphyllobothrium* sp. in a definitive host. *Exp. Parasitol.* 7:125–44.

3. Arme, C. 1966. Histochemical and biochemical studies on some enzymes of *Ligula intestinalis* (Cestoda: Pseudophyllidea). *J. Parasitol.* 52:63–68.

4. Arme, C., D. V. Griffiths, and J. P. Sumpter. 1982. Evidence against the hypothesis that the plerocercoid larva of *Ligula intestinalis* (Cestoda: Pseudophyllidea) produces a sex steroid that interferes with host reproduction. *J. Parasitol.* 68:169–71.

5. Bailey, H. H., and D. Fairbairn. 1968. Lipid metabolism in helminth parasites. V. Absorption of fatty acids and monoglycerides from micellar solution by *Hymenolepis diminuta* (Cestoda). *Comp. Biochem. Physiol.* 26:819–36.

6. Barrett, J., and W. Körting. 1977. Lipid catabolism in the plerocercoids of *Schistocephalus solidus* (Cestoda: Pseudophyllidea). *Int. J. Parasitol.* 7:419–22.

7. Becker, B., H. Mehlhorn, P. Andrews, and H. Thomas. 1981. Ultrastructural investigations on the effect of praziquantel on the tegument of five species of cestodes. *Z. Parasitenkd.* 64:257–69.

8. Beis, I., and J. Barrett. 1979. The content of adenine nucleotides and glycolytic and tricarboxylic acid cycle intermediates in activated and non-activated plerocercoids of *Schistocephalus solidus* (Cestoda: Pseudophyllidea). *Int. J. Parasitol.* 9:465–68.

9. Bolla, R. I., and L. S. Roberts. 1971. Developmental physiology of cestodes. X. The effect of crowding on carbohydrate levels and on RNA, DNA, and protein synthesis in *Hymenolepis diminuta*. *Comp. Biochem. Physiol.* 40A:777–87.

10. von Bonsdorff, B. 1956. *Diphyllobothrium latum* as a cause of pernicious anemia. *Exp. Parasitol.* 5:207–30.

11. von Brand, T. 1973. *Biochemistry of parasites,* 2d ed. New York: Academic Press, Inc.

12. von Brand, T., P. McMahon, E. Gibbs, and H. Higgins. 1964. Aerobic and anaerobic metabolism of larval and adult *Taenia taeniaeformis*. II. Hexose leakage and absorption; tissue glucose and polysaccharides. *Exp. Parasitol.* 15:410–29.

13. Brooks, D. R., and D. A. McLennan. 1993. *Parascript. Parasites and the language of evolution.* Washington, D.C.: Smithsonian Institution Press.

14. Bryant, C. 1993. Organic acid excretion by helminths. *Parasitol. Today* 9:58–60.

15. Bryant, C. 1983. Australian Society for Parasitology. Presidential Address. Intraspecies variations of energy metabolism in parasitic helminths. *Int. J. Parasitol.* 13:327–32.

16. Cervantes-Vazquez, M., D. Correa, M. Merchant, J. J. Hicks, and J. P. Laclette. 1990. Respiratory changes associated with the in vitro evagination of *Taenia solium* cysticerci. *J. Parasitol.* 76:108–12.

17. Cheah, K. S. 1983. Electron-transport systems. In Arme, C., and P. W. Pappas, eds. *The biology of the Eucestoda* 2. London: Academic Press, 421–40.

18. Coggins, J. R. 1980. Tegument and apical end organ fine structure in the metacestode and adult *Proteocephalus ambloplitis*. *Int. J. Parasitol.* 10:409–18.

19. Cook, R. L., and L. S. Roberts. 1991. In vivo effects of putative crowding factors on development of *Hymenolepis diminuta*. *J. Parasitol.* 77:21–25.

20. Davis, R. E., and L. S. Roberts. 1983. Platyhelminthes— Eucestoda. In Adiyodi, K. G., and R. G. Adiyodi, eds. *Reproductive biology of invertebrates, vol. I. Oogenesis, oviposition, and oosorption.* Chicester, England: John Wiley & Sons, 709–33.

21. Davis, R. E., and L. S. Roberts. 1989. Platyhelminthes— Eucestoda. In Adiyodi, K. G., and R. G. Adiyodi, eds. *Reproductive biology of invertebrates, vol. IV, part A. Fertilization, development, and parental care.* New Delhi: Oxford & IBH Publishing Co. Pvt. Ltd., 92–133.

22. Elowni, E. E. 1982. *Hymenolepis diminuta:* the origin of protective antigens. *Exp. Parasitol.* 53:157–63.

23. Etges, F. J., and V. Marinakis. 1991. Formation and excretion of calcareous bodies by the metacestode (tetrathyridium) of *Mesocestoides vogae*. *J. Parasitol.* 77:595–602.

24. Fairbairn, D., G. Wertheim, R. P. Harpur, and E. L. Schiller. 1961. Biochemistry of normal and irradiated strains of *Hymenolepis diminuta*. *Exp. Parasitol.* 11:248–63.

25. Fioravanti, C. F., J. R. McKelvey, and J. M. Reisig. 1992. Energy-linked mitochondrial pyridine nucleotide transhydrogenase of adult *Hymenolepis diminuta*. *J. Parasitol.* 78:774–78.

26. Fischer, H. 1968. The life cycle of *Proteocephalus fluviatilis* Bangham (Cestoda) from the smallmouth bass, *Micropterus dolomieu* Lacepede. *Can. J. Zool.* 46:569–79.

27. Fischer, H., and R. S. Freeman. 1973. The role of plerocercoids in the biology of *Proteocephalus ambloplitis* (Cestoda) maturing in smallmouth bass. *Can. J. Zool.* 51:133–41.

28. Foster, W. D. 1965. *A history of parasitology.* Edinburgh: E. & S. Livingston.

29. Frayha, G. J., and D. Fairbairn. 1969. Lipid metabolism in helminth parasites. VI. Synthesis of 2-*cis*, 6-*trans* farnesol by *Hymenolepis diminuta* (Cestoda). *Comp. Biochem. Physiol.* 28:1115–24.

30. Freeman, R. 1973. Ontogeny of cestodes and its bearing on their phylogeny and systematics. In Dawes, B., ed. *Advances in parasitology* 11. New York: Academic Press, Inc., 481–557.

31. Gustafsson, M. K. S., and B. Vaihela. 1981. Two types of frontal glands in *Diphyllobothrium dendriticum* (Cestoda, Pseudophyllidea) and their fate during the maturation of the worm. *Z. Parasitenkd.* 66:145–54.

32. Gustafsson, M. K. S., and M. C. Wikgren. 1981. Activation of the peptidergic neurosecretory system in *Diphyllobothrium dendriticum* (Cestoda: Pseudophyllidea). *Parasitology* 83:243–47.

33. Halton, D. W., I. Fairweather, C. Shaw, and C. F. Johnston. 1990. Regulatory peptides in parasitic platyhelminths. *Parasitol. Today* 6:284–90.

34. Hayunga, E. G. 1979. The structure and function of the glands of three caryophyllid tapeworms. *Proc. Helm. Soc. Wash.* 46:171–79.

35. Hertel, L. A. 1993. Excretion and osmoregulation in the flatworms. *Trans. Am. Microsc. Soc.* 112:10–17.

36. Hoeppli. R. J. C. 1959. *Parasites and parasitic infections in early medicine and science.* Singapore: University of Malaya Press.

37. Holy, J. M., J. A. Oaks, M. Mika-Grieve, and R. Grieve. 1991. Development and dynamics of regional specialization within the syncytial epidermis of the rat tapeworm, *Hymenolepis diminuta. Parasitol. Res.* 77:161–72.

38. Hopkins, C. A., and I. F. Barr. 1982. The source of antigen in an adult tapeworm. *Int. J. Parasitol.* 12:327–33.

39. Hopkins, C. A., L. M. Law, and L. T. Threadgold. 1978. *Schistocephalus solidus:* Pinocytosis by the plerocercoid tegument. *Exp. Parasitol.* 44:161–72.

40. Howard, R. J., et al. 1978. The effect of concurrent infection with *Trichinella spiralis* on *Hymenolepis microstoma* in mice. *Parasitology* 77:273–79.

41. Insler, G. D., and L. S. Roberts. 1980. Developmental physiology of cestodes. XV. A system for testing possible crowding factors in vitro. *J. Exp. Zool.* 211:45–54.

42. Jacobsen, N. S., and D. Fairbairn. 1967. Lipid metabolism in helminth parasites. III. Biosynthesis and interconversion of fatty acids by *Hymenolepis diminuta* (Cestoda). *J. Parasitol.* 53:355–61.

43. Komuniecki, R., and L. S. Roberts. 1977. Galactose utilization by the rat tapeworm, *Hymenolepis diminuta. Comp. Biochem. Physiol.* 57B:329–33.

44. Körting, W., and J. Barrett. 1977. Carbohydrate catabolism in the plerocercoids of *Schistocephalus solidus* (Cestoda: Pseudophyllidea). *Int. J. Parasitol.* 7:411–17.

45. Kuperman, B. I., and V. G. Davydov. 1982. The fine structure of glands in oncospheres, procercoids and plerocercoids of Pseudophyllidea (Cestoda). *Int. J. Parasitol.* 12:135–44.

46. Kuperman, B. I., and V. G. Davydov. 1982. The fine structure of frontal glands in adult cestodes. *Int. J. Parasitol.* 12:285–93.

47. Lumsden, R. D., and J. Bryam III. 1967. The ultrastructure of cestode muscle. *J. Parasitol.* 53:326–42.

48. Lumsden, R. D., J. A. Oaks, and J. F. Mueller. 1974. Brush border development in the tegument of the tapeworm, *Spirometra mansonoides. J. Parasitol.* 60:209–26.

49. Lumsden, R. D., and R. D. Specian. 1980. The morphology, histology, and fine structure of the adult stage of the cyclophyllidean tapeworm *Hymenolepis diminuta.* In Arai, H. P., ed. *Biology of the rat tapeworm,* Hymenolepis diminuta. New York: Academic Press, Inc., 157–280.

50. Lumsden, R. D., L. T. Threadgold, J. A. Oaks, and C. Arme. 1970. On the permeability of cestodes to colloids: An evaluation of the transmembranosis hypothesis. *Parasitology* 60:185–93.

51. MacInnis, A. J., F. M. Fisher Jr., and C. P. Read. 1965. Membrane transport of purines and pyrimidines in a cestode. *J. Parasitol.* 51:260–67.

52. Maule, A. G., D. W. Halton, C. Shaw, and C. F. Johnston. 1993. The cholinergic, serotoninergic and peptidergic components of the nervous system of *Moniezia expansa* (Cestoda, Cyclophyllidea). *Parasitology* 106:429–40.

53. McCaig, M. L. O., and C. A. Hopkins. 1963. Studies on *Schistocephalus solidus.* II. Establishment and longevity in the definitive host. *Exp. Parasitol.* 13:273–83.

54. McManus, D. P., and J. D. Smyth. 1982. Intermediary carbohydrate metabolism in protoscoleces of *Echinococcus granulosus* (horse and sheep strains) and *E. multilocularis. Parasitology* 84:351–66.

55. Mehlhorn, H., B. Becker, P. Andrews, and H. Thomas. 1981. On the nature of the proglottids of cestodes: A light and electron microscopic study on *Taenia, Hymenolepis,* and *Echinococcus. Z. Parasitenkd.* 65:243–59.

56. Mettrick, D. F., and C. H. Cho. 1981. Effect of electrical vagal stimulation on migration of *Hymenolepis diminuta. J. Parasitol.* 67:386–90.

57. Meyer, F., S. Kimura, and J. F. Mueller. 1966. Lipid metabolism in the larval and adult forms of the tapeworm *Spirometra mansonoides. J. Biol. Chem.* 241:4224–32.

58. Mount, P. M. 1970. Histogenesis of the rostellar hooks of *Taenia crassiceps* (Zeder, 1800) (Cestoda). *J. Parasitol.* 56:947–61.

59. Oaks. J. A., and R. D. Lumsden. 1971. Cytological studies on the absorptive surfaces of cestodes. V. Incorporation of carbohydrate-containing macromolecules into tegument membranes. *J. Parasitol.* 57:1256–68.

60. Orr, T. S. C., and C. A. Hopkins. 1969. Maintenance of *Schistocephalus solidus* in the laboratory with observations on rate of growth of, and proglottid formation in, the plerocercoid. *J. Fish. Res. Bd. Can.* 26:741–52.

61. Pappas, P. W., and G. M. Durka. 1993. A study of carbohydrate metabolism in the eggs (oncospheres) of the tapeworm, *Hymenolepis diminuta,* with special reference to glucose. *Parasitology* 106:201–9.

62. Pappas, P. W., and G. M. Durka. 1993. Temporal changes in glucose and carbohydrate metabolism in the oncospheres of the cyclophyllidean tapeworm, *Hymenolepis diminuta. Parasitology* 106:317–25.

63. Pappas, P. W., and C. P. Read. 1975. Membrane transport in helminth parasites: A review. *Exp. Parasitol.* 37:469–530.

64. Phares, C. K. 1987. Plerocercoid growth factor: A homologue of human growth hormone. *Parasitol. Today* 3:346–49.

65. Pietrzak, S. M., and H. J. Saz. 1981. Succinate decarboxylation to propionate and the associated phosphorylation in *Fasciola hepatica* and *Spirometra mansonoides. Molecular and Biochem. Parasitol.* 3:61–70.

66. Platzer, E. G., and L. S. Roberts. 1969. Developmental physiology of cestodes. V. Effects of vitamin deficient diets and host coprophagy prevention on development of *Hymenolepis diminuta. J. Parasitol.* 55:1143–52.

67. Rath, E. A., and M. Walkey. 1987. Fatty acid and cholesterol synthesis in mice infected with the tapeworm *Hymenolepis microstoma. Parasitology* 95:79–92.

68. Read, C. P. 1967. Longevity of the tapeworm, *Hymenolepis diminuta. J. Parasitol.* 53:1055–56.

69. Read, C. P., and J. E. Simmons Jr. 1963. Biochemistry and physiology of tapeworms. *Physiol. Rev.* 43:263–305.

70. Rees, G. 1966. Nerve cells in *Acanthobothrium coronatum* (Rud.) (Cestoda: Tetraphyllidea). *Parasitology* 56:45–54.

71. Richards, K. S., and C. Arme. 1981. Observations on the microtriches and stages in their development and emergence in *Caryophyllaeus laticeps* (Caryophyllidea: Cestoda). *Int. J. Parasitol.* 11:369–75.

72. Roberts, L. S. 1961. The influence of population density on patterns and physiology of growth in *Hymenolepis diminuta* (Cestoda: Cyclophyllidea) in the definitive host. *Exp. Parasitol.* 11:332–71.

73. Roberts, L. S. 1980. Development of *Hymenolepis diminuta* in its definitive host. In Arai, H. P., ed. *Biology of the rat tapeworm,* Hymenolepis diminuta. New York: Academic Press, Inc., 357–423.

74. Roberts, L. S. 1983. Carbohydrate metabolism. In Arme, C., and P. W. Pappas, eds. *The biology of the Eucestoda* 2. London: Academic Press, 343–90.

75. Roberts, L. S., and R. E. Davis. 1992. Platyhelminthes—Eucestoda. In Adiyodi, K. G., and R. G. Adiyodi, eds. *Reproductive biology of invertebrates, vol. V. Sexual differentiation and behaviour.* Chichester, England: John Wiley & Sons, 75–103.

76. Roberts, L. S., and F. N. Mong. 1973. Developmental physiology of cestodes. XIII. Vitamin B_6 requirement of *Hymenolepis diminuta* during in vitro cultivation. *J. Parasitol.* 59:101–4.

77. Rothman, A. H. 1959. Studies on the excystment of tapeworms. *Exp. Parasitol.* 8:336–64.

78. Schmidt, G. D. 1986. *Handbook of tapeworm identification.* Boca Raton, Fla.: CRC Press.

79. Schroeder, L. L., P. W. Pappas, and G. E. Means. 1981. Trypsin inactivation by intact *Hymenolepis diminuta* (Cestoda): Some characteristics of the inactivated enzyme. *J. Parasitol.* 67:378–85.

80. Self, J. T., and J. S. Pipkin. 1966. Sex distribution of the dioecious cestode *Shipleya intermis* Fuhrmann, 1908 in dowitchers. *J. Parasitol.* 52:45.

81. Sharp, S. E., C. K. Phares, and M. L. Heidrick. 1982. Immunological aspects associated with suppression of hormone levels in rats infected with plerocercoids of *Spirometra mansonoides* (Cestoda). *J. Parasitol.* 68:993–98.

82. Smith, S. A., and K. S. Richards. 1993. Ultrastructure and microanalyses of the calcareous corpuscles of the protoscoleces of *Echinococcus granulosus. Parasitol. Res.* 79:245–50.

83. Smyth, J. D., and D. W. Halton. 1989. *The physiology and biochemistry of cestodes.* Cambridge: Cambridge University Press.

84. Smyth, J. D., H. J. Miller, and A. B. Howkins. 1967. Further analysis of the factors controlling strobilization, differentiation and maturation of *Echinococcus granulosus* in vitro. *Exp. Parasitol.* 21:31–41.

85. Specian, R. D., and R. D. Lumsden. 1980. The microanatomy and fine structure of the rostellum of *Hymenolepis diminuta. Z. Parasitenkd.* 63:71–88.

86. Specian, R. D., and R. D. Lumsden. 1981. Histochemical, cytochemical and autoradiographic studies on the rostellum of *Hymenolepis diminuta. Z. Parasitenkd.* 64:335–45.

87. Starling, J. A. 1975. Tegumental carbohydrate transport in intestinal helminths: Correlation between mechanisms of membrane transport and the biochemical environment of absorptive surfaces. *Trans. Am. Microsc. Soc.* 94:508–23.

88. Threadgold, L. T., and J. Dunn. 1983. *Taenia crassiceps:* Regional variations in ultrastructure and evidence of endocytosis in the cysticerus' tegument. *Exp. Parasitol.* 55:121–31.

89. Threadgold, L. T., and J. Dunn. 1984. *Taenia crassiceps:* Basic mechanisms of endocytosis in the cysticercus. *Exp. Parasitol.* 58:263–69.

90. Threadgold, L. T., and C. A. Hopkins. 1981. *Schistocephalus solidus* and *Ligula intestinalis:* Pinocytosis by the tegument. *Exp. Parasitol.* 51:444–56.

91. Ubelaker, J. E. 1983. The morphology, development and evolution of tapeworm larvae. In Arme, C., and P. W. Pappas, eds. *Biology of the Eucestoda* 1. London: Academic Press.

92. Uglem, G. L. 1991. Water balance and its relation to fermentation acid production in the intestinal parasites *Hymenolepis diminuta* (Cestoda) and *Moniliformis moniliformis* (Acanthocephala). *J. Parasitol.* 77:874–83.

93. Wack, M., R. Komuniecki, and L. S. Roberts. 1983. Amino acid metabolism in the rat tapeworm, *Hymenolepis diminuta. Comp. Biochem. Physiol.* 74B:399–402.

94. Walker, J., and J. Barrett. 1993. Evidence for a G protein system in the tegumental brush border plasma membrane of *Hymenolepis diminuta. Int. J. Parasitol.* 23:281–84.

95. Ward, S. M., G. McKerr, and J. M. Allen. 1986. Structure and ultrastructure of muscle systems within *Grillotia erinaceus* metacestodes (Cestoda: Trypanorhyncha). *Parasitology* 93:587–97.

96. Watts, S. D. M., and D. Fairbairn. 1974. Anaerobic excretion of fermentation acids by *Hymenolepis diminuta* during development in the definitive host. *J. Parasitol.* 60:621–25.

97. Webster, L. A., and R. A. Wilson. 1970. The chemical composition of protonephridial canal fluid from the cestode *Hymenolepis diminuta. Comp. Biochem. Physiol.* 35:201–9.

98. Wilson, V. C. L. C., and E. L. Schiller. 1969. The neuroanatomy of *Hymenolepis diminuta* and *H. nana. J. Parasitol.* 55:261–70.

99. Wooten, R. 1974. Studies on the life history and development of *Proteocephalus percae* (Müller) (Cestoda: Proteocephalida). *J. Helminthol.* 48:269–81.

100. Zavras, E. T., and L. S. Roberts. 1985. Developmental physiology of cestodes: Cyclic nucleotides and the identity of putative crowding factors in *Hymenolepis diminuta. J. Parasitol.* 71:96–105.

Additional References

Arme, C., and P. W. Pappas, eds. 1983. *The biology of the Eucestoda,* 2 vols. London: Academic Press. Summary of cestodology; covers evolution and systematics, ecology, morphology and fine structure, development, biochemistry and physiology, pathology, immunology, and chemotherapy.

Barrett, J. 1981. *Biochemistry of parasitic helminths.* Baltimore: University Park Press.

Fairbairn, D. 1970. Biochemical adaptation and loss of genetic capacity in helminth parasites. *Biol. Rev.* 45:29–72.

Hyman, L. H. 1951. *The invertebrates* 2. New York: McGraw-Hill Book Co. A complete summary of knowledge of cestodes up to 1951.

Khalil, L. F., and A. Jones, eds. 1994. *Keys to the cestode parasites of vertebrates.* Wallingford, Oxon, Eng.: CAB International.

Pax, R. A., and J. L. Bennett. 1992. Neurobiology of parasitic flatworms: How much "neuro" in the biology? *J. Parasitol.* 78:194–205.

Read, C. P. 1959. The role of carbohydrates in the biology of cestodes. VIII. *Exp. Parasitol.* 8:365–82.

Schmidt, G. D. 1986. *Handbook of tapeworm identification.* Boca Raton, Fla.: CRC Press.

Wardle, R. A., and J. A. McLeod. 1952. *The zoology of tapeworms.* New York: Hafner Publishing Co. This monograph is the classic in its field. No student of tapeworms should be without it.

Yamaguti, S. 1959. *Systema helminthum, vol. 2. The cestodes of vertebrates.* New York: Interscience.

Chapter 21

TAPEWORMS

. . . we should all brush up on tapeworms from time to time . . .

Dave Barry

Although most species of cestodes are parasites of wild animals, a few infect humans or domestic animals and so are of particular interest. All tapeworms of humans belong to the orders Pseudophyllidea and Cyclophyllidea.

Many groups of tapeworms cause no medical or economic problem. Still, they are interesting in their own right and deserve at least an introduction. Their diversity of morphology is astonishing, and the study of their many varieties of life cycles is a science in itself. Many opportunities are available for research on these worms. For example, many groups of cestodes exist for which not a single life cycle is known. Following the discussion of Pseudophyllidea, Cyclophyllidea, and brief descriptions of several other orders, we will give very brief accounts of the sister group of the subcohort Eucestoda, the Amphilinidea, and of the sister group of the cohort Cestoidea, the Gyrocotylidea.

ORDER PSEUDOPHYLLIDEA

A pseudophyllidean cestode typically has a scolex with two longitudinal bothria. The bothria may be deep or shallow, smooth or fimbriated, and in some cases they are fused along all or part of their length, forming longitudinal tubes. Proteinaceous hooks accompany the bothria in some species. The genital pores may be lateral or medial, depending on the species. The vitellaria are always follicular and scattered throughout the segment. The testes are numerous. Some species are fairly small, but the largest tapeworms known are in the Pseudophyllidea. For example, *Hexagonoporus* from the sperm whale measures more than 30 meters long. In addition, each segment has 4 to 14 complete sets of genitalia. The worm has up to 45,000 segments. The reproductive capacity of such an animal is staggering. Generally, the life cycles of pseudophyllideans involve crustacean first intermediate hosts and fish second intermediate hosts.

Family Diphyllobothriidae

Diphyllobothrium latum. Usually called the broad fish tapeworm, this cestode is common in fish-eating carnivores, particularly in northern Europe. It appears to exhibit a striking lack of host specificity, occurring in many canines and

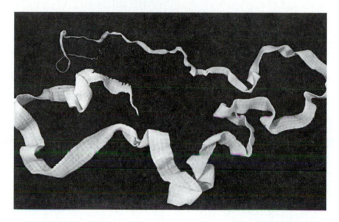

FIGURE 21.1

Diphyllobothrium latum. The scolex is at the tip of the threadlike end at upper left.

Courtesy of Warren Buss.

felines, mustelids, pinnipeds, bears, and humans. Many of these records, however, are misidentifications; in northeastern North America, humans may also become infected with *D. ursi,* a closely related species. Humans seem to be quite suitable as hosts; *D. latum* infects an estimated 9 million people worldwide.[42] It is most abundant in Scandinavia, the Baltic states, and Russia and is present in the Arctic, Great Lakes area, and West Coast of North America.[71] It has also been found in Africa, Japan, South America, Ireland, and Israel, but some of these records may be erroneous. Although it has recently been reported for the first time from Argentina,[69] *Diphyllobothrium* spp. apparently parasitized native South Americans well before the "discovery" of the New World by Columbus.[49]

- ***Morphology.*** The adult worm (Fig. 21.1) may attain a length of 10 meters and shed up to a million eggs a day. The species is anapolytic and characteristically releases long chains of spent segments, usually the first indication that the infected person has a secret guest.

  The scolex (Fig. 21.2) is finger shaped and has dorsal and ventral bothria. Proglottids (Fig. 21.3) usually are wider than long. There are numerous testes and vitelline follicles scattered throughout the proglottid, except for a narrow zone in the center. The male and female genital pores open midventrally. The bilobed ovary is near the rear of the segment. The uterus consists of short loops and extends from the ovary to a midventral uterine pore.

325

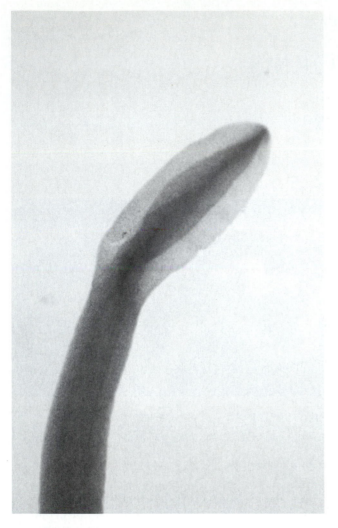

FIGURE 21.2

Scolex of *Diphyllobothrium latum*. It is about 1 mm long.

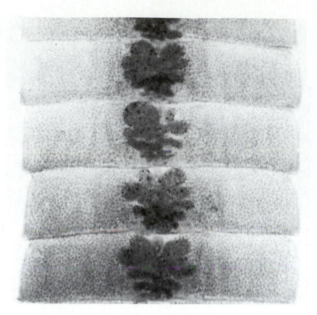

FIGURE 21.3

Gravid proglottids of *Diphyllobothrium latum*, showing the characteristic rosette-shaped uterus.

Courtesy of Larry Jensen.

• **Biology.** The ovoid eggs measure about 60 by 40 μm and have a lidlike operculum at one end and a small knob on the other (Fig. 21.5). When released through the uterine pore, the shelled embryo is at an early stage of development, and it must be deposited in water for development to continue (Fig. 21.4). Completion of development to coracidium takes from eight days to several weeks, depending on the temperature. Emerging through the operculum (Fig. 21.6), the ciliated coracidium swims randomly about, where it may attract the attention of predaceous copepods such as *Diaptomus* and related genera. Soon after being eaten, the coracidium loses its ciliated epithelium and immediately begins to attack the wall of the midgut with its six tiny hooks. Once through the intestine and into the crustacean's hemocoel, it becomes parasitic, absorbing nourishment from the surrounding hemolymph. In about three weeks it increases its length to around 500 μm, becoming an elongated, undifferentiated mass of parenchyma with a cercomer at the posterior end.

It is now a procercoid (Fig. 21.7), incapable of further development until eaten by a suitable second intermediate host—any of several species of freshwater fishes, especially pike and related fish, or any of the salmon family. The cercomer may be lost while still in the copepod or soon after the procercoid enters a fish. Large, predaceous fish eat comparatively few microcrustaceans but can still become infected by eating smaller fish containing plerocercoids, which then migrate into the new host.

When a fish eats the infected copepod, the procercoid is released and bores its way through the intestinal wall and into the body muscles. Here it absorbs nutrients and grows rapidly into a plerocercoid. Mature plerocercoids vary in length from a few millimeters to several centimeters. They are still mainly undifferentiated, but there may be evidence of shallow bothria at the anterior end. Usually plerocercoids are found unencysted and coiled up in the musculature, although they may be encysted in the viscera. They are easily seen as white masses in uncooked fish (Fig. 21.8), but when the flesh is cooked, the worms are seldom noticed. Plerocercoids of other pseudophyllideans, as well as those of proteocephalatans and trypanorhynchans, are also found in fish and are often mistaken for those of *D. latum*. When the plerocercoid is ingested by a suitable host, it survives the digestive fate of its late host and begins a close relationship with a new one. The worms grow rapidly and may begin egg production by 7 to 14 days. Little of the growth may be attributed to the production of new proglottids but is caused by growth in primordia already in the plerocercoid. As much as 70% of the strobila may mature on the same day.[4]

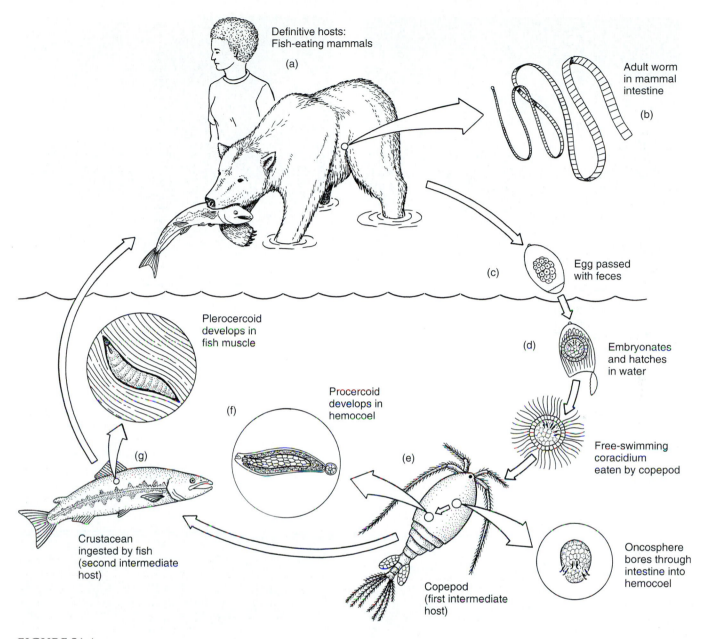

FIGURE 21.4

Life cycle of *Diphyllobothrium latum.* (*a*) Definitive hosts: any of a number of fish-eating mammals. (*b*) Adult worm is in mammal small intestine. (*c*) Shelled embryo passed in feces at an early stage of development. (*d*) Embryogenesis continues in water, and free-swimming coracidium hatches. (*e*) Coracidium eaten by copepod, and oncosphere penetrates intestine into hemocoel. (*f*) Procercoid develops in hemocoel. (*g*) Copepod eaten by fish, where procercoid penetrates into muscle and develops into pleroceroid.

Drawing by William Ober and Claire Garrison.

- **Epidemiology.** Obviously persons become infected when they eat raw or undercooked fish. Hence, infection rates are highest in countries where raw fish is eaten as a matter of course. Communities that dispose of sewage by draining it into lakes or rivers without proper treatment create an opportunity for a massive buildup of *D. latum* in local fish. These fish may be harvested for local consumption or shipped thousands of miles by refrigerated freight to distant markets. There an unsuspecting customer may gain infection in a restaurant or at home by tasting such dishes as gefilte fish during preparation. The higher prevalence of *D. latum* in women is probably due to the higher prevalence of women among the ranks of cooks. The fad in the United States of eating raw salmon as sushimi has led to infections.[71]

- **Pathogenesis.** Many cases of **diphyllobothriasis** are apparently asymptomatic or have poorly defined symptoms

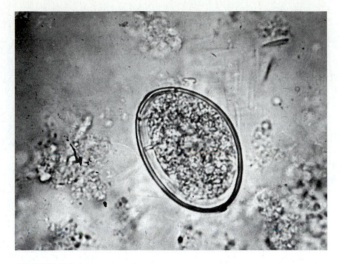

FIGURE 21.5

Egg of *Diphyllobothrium latum* in a human stool; note operculum at upper end and small knob at opposite end. It is 40 to 60 μm long.

Courtesy of David Oetinger.

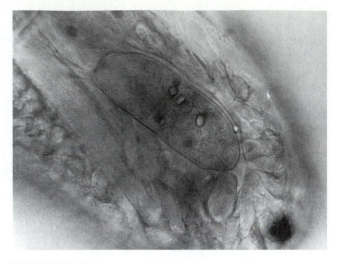

FIGURE 21.7

Procercoid in the hemocoel of a copepod. Note the anterior pit, the posterior cercomer, and the internal calcareous granules.

Courtesy of Justus F. Mueller.

FIGURE 21.6

Diphyllobothriid coracidium emerging from its eggshell. The cilia are moving so fast that they appear as a blur.

Courtesy of Justus F. Mueller.

FIGURE 21.8

Two plerocercoids in the flesh of a perch.

From R. Vik, in Marcial-Rojas, editor, *Pathology of Protozoal and Helminthic Diseases, with Clinical Correlation*, © 1971 Williams & Wilkins.

associated with other tapeworms, such as vague abdominal discomfort, diarrhea, nausea, and weakness. However, in a small number of cases the worm causes a serious megaloblastic anemia; virtually all of these cases are in Finnish people. Almost a fourth of the population of Finland may be infected with *D. latum,* and about 1000 of these will have pernicious anemia.[12] It was thought originally that toxic products of the worm produced the anemia, but we now know that the large amount of vitamin B_{12} absorbed by the cestode, in conjunction with some degree of impairment of the patient's normal absorptive mechanism for vitamin B_{12}, is responsible for the disease. Nyberg[64] reported that an average of 44% of a single oral dose of vitamin B_{12} labeled with cobalt 60 was absorbed by *D. latum* in otherwise healthy patients, but in patients with tapeworm pernicious anemia 80% to 100% of the dose was absorbed by the cestode. The clinical symptoms of tapeworm pernicious anemia are similar in many respects to "classical" pernicious anemia (caused by a failure in intestinal absorption of vitamin B_{12}), except that expulsion of the worm generally brings a rapid remission of the anemia.

- *Diagnosis and Treatment.* Demonstration of the characteristic eggs or proglottids passed with the stool gives positive diagnosis. In the past a variety of drugs has been used against *D. latum* and other tapeworms; aspidium oleoresin (extract of male fern), mepacrine, dichlorophen, and even extracts of fresh pumpkin seeds (*Cucurbita* spp.) have anticestodal properties.[24] However, the drugs of choice are now niclosamide (Yomesan) and praziquantel.[82] The mode of action of niclosamide seems to be an inhibition of an inorganic phosphate—ATP exchange reaction associated with the worm's anaerobic electron transport system. The action of praziquantel is described on p. 226.

Other Pseudophyllideans Found in Humans

Several other species of *Diphyllobothrium* have been reported from humans in different parts of the world. These include *D. chordatum* and *D. pacificum,* parasites of pinnipeds in the northern and southern hemispheres, respectively, and *D. ursi* of bears. Other species of *Diphyllobothrium* reported from humans are probably synonyms of *D. latum.*

Diphyllobothrium erinacei, Digramma brauni, and *Ligula intestinalis* have also been reported from humans, but such occurrences must be rare. *Diplogonoporous grandis (D. balaenopterae)* has been reported numerous times from humans in Japan.[47] A parasite of whales, its plerocercoid occurs in marine fish, the mainstay of the Japanese protein diet.

- ## Sparganosis

With the exception of the forms with scolex armature, it is impossible to distinguish the species of plerocercoids found in humans by examining their morphology. When procercoids of some species are ingested accidentally, usually by swallowing an infected copepod in drinking water, they can

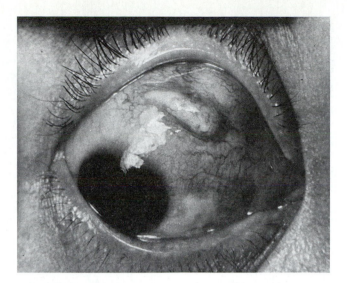

FIGURE 21.9

Right eye of patient with sparganosis. Note the protruding mass in the upper conjunctiva.

From L. T. Wang and J. H. Cross, "Human sparganosis on Taiwan. A report of two cases," in *J. Formosan Med. Assoc.* 73:173–177. Copyright © 1974.

migrate from the gut and develop into plerocercoids, sometimes reaching a length of 14 inches. The infection is called **sparganosis** and may cause severe pathological consequences. Cases have been reported from most countries of the world but are most common in eastern Asia. Yamane, Okada, and Takihara[88] reported a living sparganum that had infected a woman's breast for at least 30 years.

Another means of infection is by ingestion of insufficiently cooked amphibians, reptiles, birds, or even mammals such as pigs.[22] Plerocercoids present in these animals may then infect the person indulging in such delicacies. Many Chinese are infected in this way because of their tradition of eating raw snake to cure a panoply of ills.[50]

A third method of infection results from the east Asian treatment of skin ulcers, inflamed vagina, or inflamed eye (Fig. 21.9) by poulticing the area with a split frog or flesh of a vertebrate that may be infected with spargana. The active worm then crawls into the orbit, vagina, or ulcer and establishes itself. Most cases of sparganosis in eastern Asia are probably caused by *D. erinacei,* a parasite of carnivores.

In North America most spargana are probably *D. mansonoides,* a parasite of cats.[62] It usually does not proliferate, except by occasionally breaking transversely, and may live up to 10 years in a human.[81] The current public awareness of the symptoms of cancer has led to an increase in reported cases of sparganosis in this country. Subdermal lumps are no longer ignored by the average person, and more than one physician has been shocked to find a gleaming, white worm in a lanced nodule. Wild vertebrates are commonly infected with spargana (Fig. 21.10).

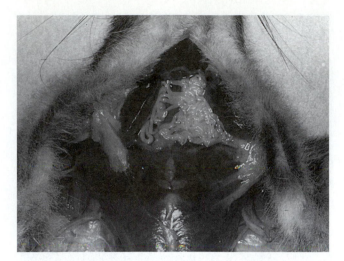

FIGURE 21.10

Spargana in subcutaneous connective tissues of a wild rat in Taiwan.
Courtesy of Robert E. Kuntz.

Rarely a sparganum will be proliferative, splitting longitudinally and budding profusely. Such cases are very serious, since many thousands of worms can result, with the infected organs becoming honeycombed.[62]

Treatment of sparganosis is usually by surgery, but supplementary treatment with praziquantel may be advisable.[82]

ORDER CARYOPHYLLIDEA

Caryophyllideans are intestinal parasites of freshwater fishes, except for a few that mature in the coelom of freshwater oligochaete annelids.[54,56] All are monozoic, showing no trace of internal proglottisation or external segmentation (Fig. 21.11). The scolex is never armed. Usually it is quite simple, bearing shallow depressions (loculi), or it is frilled or entirely smooth (Fig. 21.12). Some species seem to lack a scolex altogether. The anterior end of the worm is very motile, however, and functions well as a holdfast. Some species induce a pocket in the wall of the host's intestine in which one or more worms remain.

Each worm has a single set of male and female reproductive organs. In most the ovary is near the posterior end. Testes fill the median field of the body, and vitelline follicles are mainly lateral. Male and female genital pores open near each other on the midventral surface.

Catfishes, true minnows (Cyprinidae), and suckers are the most common hosts of the Caryophyllidea. *Glaridacris* spp., predominantly *G. catostomi*, are found abundantly in suckers (*Catostomus* spp.) in North America.[55] Intermediate hosts are aquatic annelids. After the oligochaete eats the egg, the oncosphere hatches and penetrates the coelom. There it grows into a procercoid with a prominent cercomer, similar to that of *Diphyllobothrium*. When eaten by a fish, the procercoid loses its cercomer and grows directly into an adult.

The biology and morphology of Caryophyllidea and Pseudophyllidea are similar, the main differences being the absence of a plerocercoid and a strobilated adult in the

FIGURE 21.11

Penarchigetes oklensis, a typical caryophyllidean cestode, from a spotted sucker.

From J. S. Mackiewicz, "*Penarchigetes oklensis* gen. et sp. n. and *Biacetabulum carpiodi* sp. n. (Cestoidea: Caryophyllaeidae) from castostomid fish in North America," in *Proc. Helm. Soc. Wash.* 37:110–118. Copyright © 1969. Reprinted with permission of the publisher.

Caryophyllidea. Also, caryophyllideans use annelids as intermediate hosts, whereas pseudophyllideans employ crustaceans.

It has been suggested that segmented adults once existed but became extinct with their hosts, probably aquatic reptiles. However, this did not happen before the plerocercoid developed neotenically in the fish second intermediate host. If this hypothesis is true, extant caryophyllaeid species actually are **neotenic plerocercoids.** Support is lent to this idea by the existence of several species of *Archigetes*, which become sexually mature *while in the annelid*. The reproductive adult retains the cercomer and infects no additional host, although it can live for some time if eaten by a fish. *Archigetes*, then, appears to be a **neotenic procercoid.**

ORDER SPATHEBOTHRIIDEA

These are peculiar parasites of marine and freshwater teleost fishes. Their most striking characteristic is a complete absence of segmentation, with possession of a typically linear

FIGURE 21.12

Typical scolices of Caryophyllidea.

From G. D. Schmidt, *How to Know the Tapeworms.* Copyright © 1970. Reprinted with permission.

series of internal proglottids. The scolex always lacks armature. It may be totally undifferentiated, as in *Spathebothrium;* it may be a shallow funnel-shaped organ, as in *Cyathocephalus;* or it may consist of one or two powerful cuplike organs (Fig. 20.6*j*). The genital pores are ventral, the testes are in two lateral bands, the ovary is dendritic, and the vitellaria are follicular and lateral or scattered. The uterus is rosettelike and opens ventrally, usually near the vaginal pore.

No life cycles are known. Although these worms are of no known economic importance, they remain an interesting zoological group that should be studied further. *Bothrimonus,* a common genus in North America, has been investigated more fully.[15]

ORDER CYCLOPHYLLIDEA

The Cyclophyllidea and the Proteocephalata both have scolices with four acetabula. Their parenchyma is divided into highly distinct medullary and relatively extensive cortical regions defined by the longitudinal musculature. Brooks and McLennan[13] combined these two orders into the Proteocephaliformes, but we are retaining the traditional separation for the present.

The most obvious morphological feature that characterizes the Cyclophyllidea is a single compact, postovarian vitelline gland. A rostellum, which usually bears an armature of hooks, is commonly present. The genital pores are lateral in all except the family Mesocestoididae, in which they are midventral. The number of testes varies from one to several hundred, depending on the species. Most species are rather small, although some are giants of more than 10 meters in length. Most tapeworms of birds and mammals belong to this order.

Family Taeniidae

The largest cyclophyllideans are in the family Taeniidae, as are the most medically important tapeworms of humans. A remarkable morphological similarity occurs among most species in the family; a striking exception is *Echinococcus,* which is much smaller than cestodes of the other genera. An armed rostellum is present on most species and when present is not retractable. The testes are numerous, and the ovary is a bilobed mass near the posterior margin of the proglottid. The metacestodes are various types of bladderworms (Fig. 21.13), and mammals serve as their intermediate hosts.

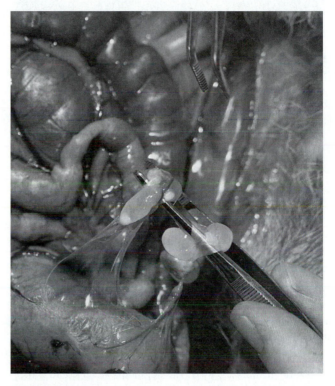

FIGURE 21.13

Cysticerci of *Taenia pisiformis* in the mesenteries of a rabbit.

Courtesy of John Mackiewicz.

Taeniarhynchus saginatus.   Among the Taeniidae, *Taeniarhynchus saginatus* is by far the most common in humans, occurring in nearly all countries where beef is eaten. The beef tapeworm, as it is usually known, lacks a rostellum or any scolex armature (Fig. 21.14). Individuals of this exceptionally large species may attain a length of over 20 meters, but 3 to 5 meters is much more common. Even the smaller specimens may consist of as many as 2000 segments.

- *Morphology.* The scolex, with its four powerful suckers, is followed by a long, slender neck. Mature segments are slightly wider than long, whereas gravid ones are much longer than wide. Usually it is the gravid segment passed

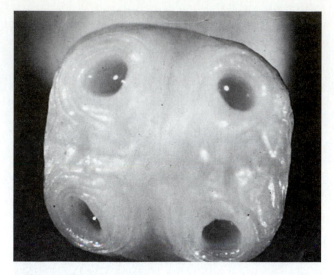

FIGURE 21.14

En face view of the scolex of *Taeniarhynchus saginatus.* Note the absence of a rostellum or armature.

AFIP neg. no. 65-12073-2.

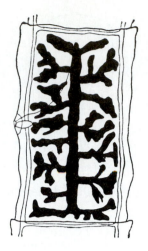

FIGURE 21.15

Gravid segments of *Taenia solium* and *Taeniarhynchus saginatus.* The uterus is characteristic for these genera, consisting of a median stem with lateral branches. *Taenia solium* has 7 to 13 lateral branches, and *Taeniarhynchus saginatus* has 15 to 20 lateral branches. Each segment is around 10 mm long but may be longer or shorter, depending on the degree of contraction.

From G. D. Schmidt, *Essentials of Parasitology,* 5th ed. Copyright © 1992 W. C. Brown Communications, Inc., Dubuque, Iowa. Reprinted by permission of Times Mirror Higher Education Group, Inc., Dubuque, Iowa. All Rights Reserved.

in the feces that is first noticed and taken to a physician for diagnosis. Because the eggs of this species cannot be differentiated from those of *Taenia solium,* the next most common taeniid of humans, accurate diagnosis depends on a critical examination of a gravid uterus. Of course, if the entire worm is passed, the combination of unarmed scolex and taeniid type of proglottid (Fig. 21.15) leads to an unmistakable diagnosis.

The spherical eggs are characteristic of the Taeniidae (Fig. 21.16). A thin, hyaline, outer membrane is usually lost by the time the egg is voided with the feces. The embryophore is very thick and riddled with numerous tiny pores, giving it a striated appearance in optical section. Unfortunately, the egg sizes of several taeniids in humans overlap, making diagnosis of species impossible on this character alone.

- *Biology.* When gravid, the segments detach and either pass out with the feces or migrate out of the anus. Each segment behaves like an individual worm, crawling actively about, as if searching for something. The segments are easily mistaken for trematodes or even nematodes at this stage. As a segment begins to dry up, a rupture occurs along the midventral body wall, allowing eggs to escape. The larvae are fully developed and infective to the intermediate host at this time; they remain viable for many weeks. Cattle are the usual intermediate host, although cysticerci have also been reported from llamas, goats, sheep, giraffes, and even reindeer (perhaps incorrectly).

When eaten by a suitable intermediate host, the egg hatches in the duodenum under the influence of gastric and intestinal secretions. The released hexacanth quickly penetrates the mucosa and enters an intestinal venule, to

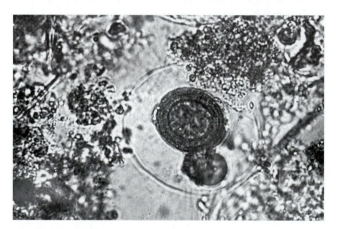

FIGURE 21.16

Taenia egg in human feces. The thin outer membrane is often lost at this stage.

Courtesy of David Oetinger.

be carried throughout the body. Typically it leaves a capillary between muscle cells and enters a muscle fiber, developing into an infective cysticercus in about two months. This metacestode is white, pearly, and about 10 mm at its greatest diameter and contains a single, invaginated scolex. Humans are probably an unsuitable intermediate host, and the few records of *Taeniarhynchus saginatus* cysticerci in humans are most likely misidentifications. Before the beef cysticercus was known to be a juvenile form of *T. saginatus,* it was placed in a separate genus under the name of *Cysticercus bovis.* The disease produced in cattle is thus known as **cysticercosis bovis,** and flesh riddled with the juveniles is called **measly beef.**

A person who eats infected beef, cooked insufficiently to kill the juveniles, becomes infected. The invaginated scolex and neck of the cysticercus evaginate in response to bile salts. The bladder is digested by the host or absorbed by the scolex, and budding begins. Within 2 to 12 weeks the worm will begin shedding gravid proglottids.

- *Epidemiology.* Human infection is highest in areas of the world where beef is a major food and sanitation is deficient. Thus, in several developing nations of Africa and South America, for instance, ample opportunity exists for cattle to eat tapeworm eggs and for people to eat infected flesh. Many persons are content to eat a chunk of meat that is cooked in a campfire, charred on the outside and raw on the inside.

Local custom may have profound effect on infection rates. Hence, in India there may be a high rate of infection among Moslems, whereas Hindus, who do not eat beef, are unaffected. In the United States, federal meat inspection laws and a high degree of sanitation combine to keep the incidence of infection low. However, not all cattle slaughtered in the United States are federally inspected, and standard inspection procedures fail to detect a fourth of infected cattle.[25] One wonders if backyard cookery and the popularity of steak tartare might not contribute to prevalence of taeniiasis.

Despite the high level of sanitation in any country, it still is possible for cattle to be exposed to the eggs of this parasite. One infected person who defecates in a pasture or cattle-feeding area can quickly infect an entire herd. The use of human feces as fertilizer can have the same effect. Shelled larvae can remain viable in liquid manure for 71 days, in untreated sewage for 16 days, and on grass for 159 days.[45] Cattle are coprophagous and often will eat human dung, wherever they find it. In India, where cattle roam at will, it is common for a cow to follow a person into the woods, in hopes of obtaining a fecal meal.[19]

Prevention of human infection is easy; when meat is cooked until it is no longer pink in the center, it is safe to eat, since cysticerci are killed at 56°C. Furthermore, meat is also rendered safe by freezing at −5°C for at least a week.

- *Pathogenesis.* Disease characteristics of *T. saginatus* infection are similar to those of infection by any large tapeworm, except that the avitaminosis B_{12} found in association with *D. latum* is unknown. Most people infected with *T. saginatus* are asymptomatic or have mild to moderate symptoms of dizziness, abdominal pain, diarrhea, headache, localized sensitivity to touch, and nausea. Delirium is rare but does occur. Intestinal obstruction with need for surgical intervention sometimes occurs. Hunger pains, universally accepted by lay people as a symptom of tapeworm infection, are not common, but *loss* of appetite is frequent. The worms release antigens, which sometimes result in allergic reactions. In addition, it is difficult to estimate the psychological effects on an infected person of observing continued migration of proglottids out of the anus.

- *Diagnosis and Treatment.* Identification of taeniid eggs according to species is impossible. Therefore, accurate diagnosis depends on examination of a scolex or a gravid segment. The latter is characterized by having 15 to 20 lateral branches on each side (contrasted with 7 to 13 in *T. solium*). Because these branches tend to fuse in old segments, freshly passed specimens must be obtained for reliable results. Work is proceeding on immunodiagnostic tests for worm antigens passed in the feces (coproantigens) using variants of the ELISA.[1]

Numerous taeniicides have been used in the past. Today niclosamide and praziquantel are the drugs of choice.

Taenia solium. The most dangerous adult tapeworm of humans is the pork tapeworm, *Taenia solium,* because of the possibilities of self-infection with cysticerci. Furthermore, it is possible to infect others in the same household with juveniles of this parasite, often with grave results.

- *Morphology.* The scolex of the adult (Fig. 21.17) bears a typical, nonretractable taeniid rostellum, armed with two circles of 22 to 32 hooks measuring 130 to 180 μm long. Whereas the scolex of *Taeniarhynchus saginatus* is cuboidal and up to 2 mm in diameter, that of *Taenia solium* is spheroid and only half as large. There are reports of strobilas as long as 30 feet, but 6 to 10 feet is much more common. Mature segments are wider than long and are nearly identical to those of *T. saginatus,* differing in number of testes (150 to 200 in *T. solium,* 300 to 400 in *T. saginatus*). Gravid segments are longer than wide and have the typical taeniid uterus, a medial stem with 7 to 13 lateral branches.

- *Biology.* The life cycle of *Taenia solium* (Fig. 21.18) is in most regards like that of *T. saginatus,* except that the intermediate hosts are pigs instead of cattle. Gravid proglottids passed in the feces are laden with eggs infective to swine. When eaten, the oncospheres develop into cysticerci (*Cysticercus cellulosae*) in the muscles and other organs. Blowflies can carry eggs from infected feces to uninfected meat, which is readily eaten by pigs,[51] and pigs feed on human feces where it is available.[35]

A person easily becomes infected when eating a bladderworm along with insufficiently cooked pork. Evaginating by the same process as in *T. saginatus,* the worm attaches to the mucosa of the small intestine and matures in

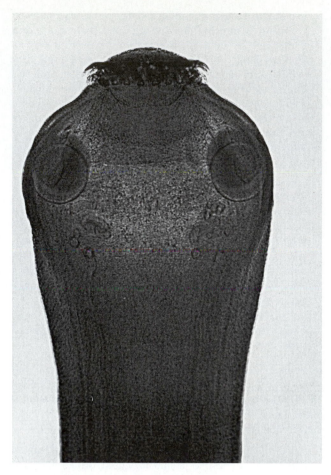

FIGURE 21.17

Scolex of *Taenia solium.* Note the large rostellum with two circles of hooks.

Courtesy of David Oetinger.

5 to 12 weeks. Specimens of *T. solium* can live for as long as 25 years. Pathogenesis caused by the adult worm is similar to that in taeniiasis saginatus.

• *Cysticercosis.* Unlike those of most other species of *Taenia,* the cysticerci of *T. solium* develop readily in humans. Infection occurs when shelled larvae pass through the stomach and hatch in the intestine. Persons who are infected by adult worms may contaminate their households or food with eggs that are accidentally eaten by themselves or others. Possibly, a gravid proglottid may migrate from the lower intestine to the stomach or duodenum, or it may be carried there by reverse peristalsis. Subsequent release and hatching of many eggs at the same time results in a massive infection by cysticerci.

Virtually every organ and tissue of the body may harbor cysticerci. Most commonly they are found in the subcutaneous connective tissues. The second most common site is the eye, followed by the brain (Fig. 21.19), mus-

cles, heart, liver, lungs, and coelom. A fibrous capsule of host origin surrounds the metacestode, except when it develops in the chambers of the eye. The effect of any cysticercus on its host depends on where it is located. In skeletal muscle, skin, or liver, little noticeable pathogenesis usually results, except in massive infection. Ocular cysticercosis may cause irreparable damage to the retina, iris, or choroid. A developing cysticercus in the retina may be mistaken for a malignant tumor, resulting in the unnecessary surgical removal of the eye. Removal of the cysticercus by fairly simple surgery is usually successful.

Cysticerci occur rarely in the spinal cord but commonly in the brain.[10] Symptoms of infection are vague and rarely diagnosed except at autopsy. Radiological diagnosis of neurocysticercosis is practical.[17] Pressure necrosis may cause severe central nervous system malfunction, blindness, paralysis, disequilibrium, obstructive hydrocephalus, or disorientation. Perhaps the most common symptom is epilepsy of sudden onset. When this occurs in an adult with no family or childhood history of epilepsy, cysticercosis should be suspected.[26] Praziquantel is the drug of choice for neurocysticercosis, but it is not used against cysticerci in the brain ventricles or the eye.[35] Praziquantel can also be used against porcine cysticercosis, thus avoiding the need for condemnation of the carcass.[85]

When a cysticercus dies, it elicits a rather severe inflammatory response. Many of them may be rapidly fatal to the host, particularly if the worms are located in the brain. This was observed frequently in former British soldiers; a high proportion of those who served in India became infected. Other types of cellular reaction also occur, usually resulting in eventual calcification of the parasite (Fig. 21.20). If this occurs in the eye, there is little chance of corrective surgery.

Cysticerci occur in three distinct morphological types, of which the most common is the ordinary "cellulose" cysticercus, with an invaginated scolex and a fluid-filled bladder about 0.5 to 1.5 cm in diameter. The "intermediate" form (with a scolex) and the "racemose" (no scolex can be found) are much larger and more dangerous.[66] They can measure up to 20 cm and contain 60 ml of fluid. Up to 13% of patients may have all three types in the brain.[66]

Prevention of cysticercosis depends on early detection and elimination of the adult tapeworm and a high level of personal hygiene. Fecal contamination of food and water must be avoided and the use of untreated sewage on vegetable gardens eschewed. The majority of cases apparently originate from such sources, including contamination by infected food handlers.[29]

Although neurocysticercosis has been considered uncommon, improved brain imaging through computerized axial tomography (CAT) and magnetic resonance imaging (MRI) has demonstrated a higher frequency in the United States than we once thought.[18] Of 138 cases reported from Los Angeles County, California, from 1988

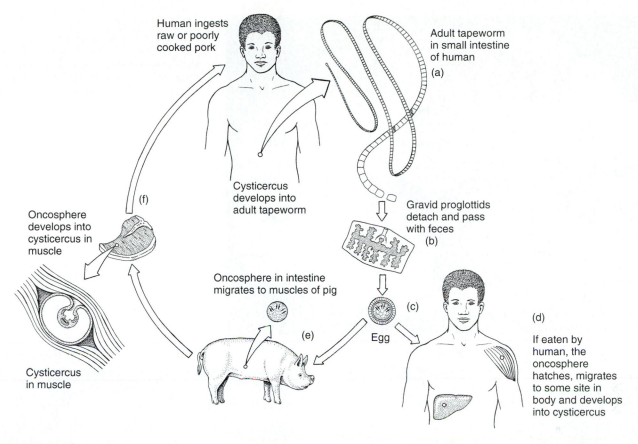

Human ingests
raw or poorly
cooked pork

Adult tapeworm
in small intestine
of human
(a)

Cysticercus
develops into
adult tapeworm

Gravid proglottids
detach and pass
with feces
(b)

(f)

Oncosphere
develops into
cysticercus in
muscle

Oncosphere in intestine
migrates to muscles of pig

(c)

Cysticercus
in muscle

(e)   Egg

(d)

If eaten by
human, the
oncosphere
hatches, migrates
to some site in
body and develops
into cysticercus

FIGURE 21.18

Life cycle of *Taenia solium*. (*a*) Adult tapeworm in the small intestine of a human. (*b*) Gravid proglottids detach from the strobila and migrate out of the anus or pass with feces. (*c*) Shelled oncoshpere. (*d*) If eaten by a human, the oncosphere hatches, migrates to some site in the body, and develops into a cysticercus. (*e*) Cysticerci will also develop if the eggs are eaten by a pig. (*f*) The life cycle is completed when a person eats pork containing live cysticerci.

Drawing by William Ober and Claire Garrison.

to 1990, most were in immigrants from Mexico,[77] where cysticercosis is a major public health problem;[35] however, nine were travel-associated cases, and 10 were infections acquired in the United States. Detection of *T. solium* antigens in cerebrospinal fluid can aid in diagnosis of neurocysticercosis.[20]

Curiously, swine cysticercosis is fairly common, even in countries where the adult is rare. It is evident that the multitudes of eggs produced by even one adult worm overcome great odds to infect pigs and thereby continue the species.

• Other Taeniids of Medical Importance

A taeniid very similar morphologically to *Taeniarhynchus saginatus* has been distinguished in Southeast Asia and China.[33,44] Most authors now refer to this form as Asian *Taenia*. A striking biological difference from classical *T. saginatus* is that cysticerci develop in pigs, primarily in the liver and other viscera, not in the muscles.[31] (Many rural Asians relish raw pig viscera.) The cysticercus of Asian

Taenia has small hooklets on its scolex. Some investigators consider these differences sufficient to designate it a separate species.[30] However, molecular analysis shows that Asian *Taenia* is genetically much closer to *Taeniarhynchus saginatus* than to other species of taeniids and probably should be considered a subspecies or strain of *T. saginatus*.[11]

Taenia multiceps, T. glomeratus, T. brauni, and *T. serialis* are all characterized by a coenurus type of bladderworm (Fig. 21.21). This is similar to an ordinary cysticercus but has many rather than one protoscolex. Such coenuri occasionally occur in humans, particularly in the brain, eye, muscles, or subcutaneous connective tissue, where they often grow to be longer than 40 mm. The resulting pathogenesis is similar to that of cysticercosis. The adults are parasites of carnivores, particularly dogs, with herbivorous mammals serving as intermediate hosts. Accidental infection of humans occurs when the eggs are ingested. Coenuriasis of sheep, caused by *T. multiceps,* causes a characteristic vertigo called gid, or staggers.

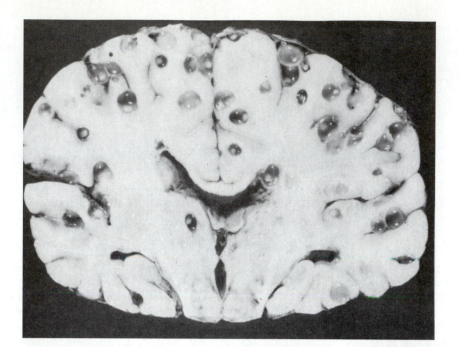

FIGURE 21.19

Human brain containing numerous cysticerci of *Taenia solium.*

From A. Flisser, "Neurocysticercosis in Mexico," in *Parasitol. Today* 4:131–137. Copyright © 1988.

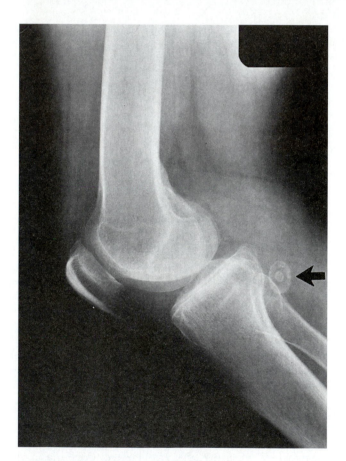

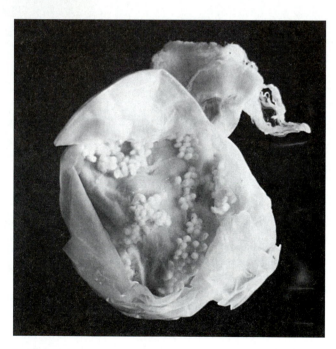

FIGURE 21.21

Coenurus metacestode of *Taenia serialis,* from the muscle of a rabbit, that has been opened to show the numerous protoscolices arising from the germinal epithelium. The cyst is about 4 in. wide.

Courtesy of James Jensen.

FIGURE 21.20

Cysticercus cellulosae: partially calcified cyst (*arrow*) found in a routine x-ray examination of a human leg.

From R. L. Roudabush and G. A. Ide, "*Cysticercus cellulosae* on X-ray," in *J. Parasitol.* 61:512. Copyright © 1975.

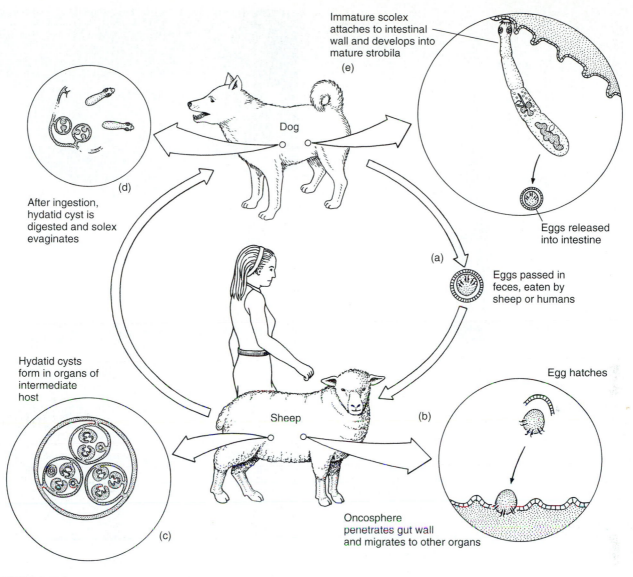

Immature scolex
attaches to intestinal
wall and develops into
mature strobila
(e)

Eggs released
into intestine

After ingestion,
hydatid cyst is
digested and solex
evaginates
(d)

Dog

(a)

Eggs passed in
feces, eaten by
sheep or humans

Egg hatches

Hydatid cysts
form in organs of
intermediate
host

Sheep

(b)

(c)

Oncosphere
penetrates gut wall
and migrates to other organs

FIGURE 21.22

Life cycle of *Echinococcus granulosus.* (*a*) Shelled oncosphere is passed in feces and eaten by hooved animals or humans. (*b*) Oncosphere penetrates gut wall and is carried to liver and other sites by circulation. (*c*) Hydatid cysts form in organs of intermediate host. (*d*) After ingestion by canid, hydatid cyst is digested and scolex evaginates. (*e*) Scolex attaches to intestinal wall and develops strobila.

Drawing by William Ober and Claire Garrison.

Echinococcus granulosus. The genus *Echinococcus* contains the smallest tapeworms in the Taeniidae. However, their juvenile forms are often huge and are capable of infecting humans, resulting in **hydatidosis,** a very serious disease in many parts of the world.

Echinococcus granulosus uses carnivores, particularly dogs and other canines, as definitive hosts (Fig. 21.22). Many mammals may serve as intermediate hosts, but herbivorous species are most likely to become infected by eating the eggs on contaminated herbage.

The adult (Fig. 21.23) lives in the small intestine of the definitive host. It measures 3 to 6 mm long when mature and consists of a typically taeniid scolex, a short neck, and usually

only three proglottids. The nonretractable rostellum bears a double crown of 28 to 50 (usually 30 to 36) hooks. The anteriormost segment is immature; the middle one is usually mature; and the terminal one is gravid. The gravid uterus is an irregular longitudinal sac. The eggs cannot be differentiated from those of other taeniids. The ripe segment detaches and develops a rupture in its wall, releasing the eggs, which are fully capable of infecting an intermediate host.

Hatching and migration of the oncosphere are the same as previously described for *Taeniarhynchus saginatus,* except that the liver and lungs are the usual sites of development. By a very slow process of growth, the oncosphere metamorphoses into a type of bladderworm called a **unilocular hydatid**

FIGURE 21.23

Adult *Echinococcus granulosus* from the intestine of a dog. Adults are only 3 to 8 mm long.

Courtesy of Ann Arbor Biological Center.

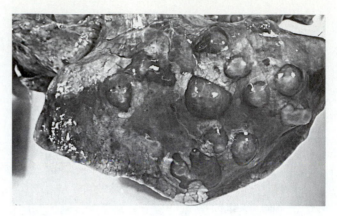

FIGURE 21.24

Several unilocular hydatids in the lung of a sheep. Each hydatid contains many protoscolices.

Courtesy of James Jensen.

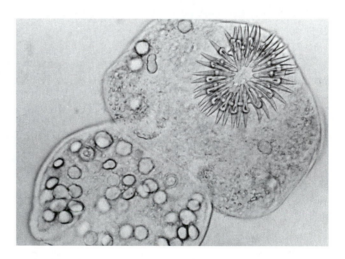

FIGURE 21.25

Protoscolex of *Echinococcus granulosus,* removed from a hydatid cyst.
Courtesy of Sharon File.

(Fig. 21.24). In about five months the hydatid develops a thick outer, laminated, noncellular layer and an inner, thin, nucleated germinal layer. The inner layer eventually produces the protoscolices that are infective to the definitive host. Protoscolices (Fig. 21.25) are usually produced singly into the lumen of the bladder as in a coenurus and also within **brood capsules.** The latter are small cysts, containing 10 to 30 protoscolices, which usually are attached to the germinal layer by a slender stalk; they may break free and float within the hydatid fluid. Similarly, individual scolices and brood capsules may break free and sink to the bottom of the bladder, where they are known as **hydatid sand** (Fig. 21.26) (although it is possible that this happens only in dead cysts). Rarely germinal cells penetrate the laminated layer and form **daughter capsules.** When a carnivore eats the hydatid, the cyst wall is digested away, freeing the protoscolices, which evaginate and attach among the villi of the small intestine. A small percentage of hydatids lack protoscolices and are sterile, being unable to infect a definitive host. The worm matures in about 56 days and may live for 5 to 20 months.

- *Epidemiology.* The life cycle of *E. granulosus* in wild animals may involve a wolf-moose, wolf-reindeer, dingo-wallaby, lion-warthog, or other carnivore-herbivore relationship, which is known as **sylvatic echinococcosis.**

Humans are seldom involved as accidental intermediate hosts in these cases. However, ample opportunities exist for human infection in situations in which domestic herbivores are raised in association with dogs. For example, hydatid disease is a very serious problem in sheep-raising areas of Australia, New Zealand, North and South America, Europe, Asia, and Africa. Similarly, goats, camels, reindeer, and pigs, together with dogs, maintain the cycle in various parts of the world. Dogs are infected when they feed on the offal of butchered animals, and herbivores are infected when they eat herbage contaminated with dog

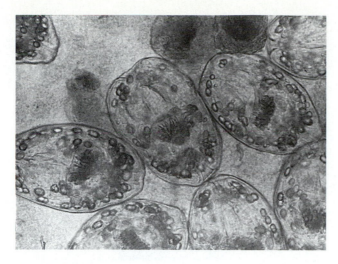

FIGURE 21.26

"Hydatid sand," consisting of unattached protoscolices of *Echinococcus granulosus*.
Courtesy of Robert E. Kuntz.

dung. Humans are infected with hydatids when they accidentally ingest *Echinococcus* eggs, usually as a result of fondling dogs.

The species *E. granulosus* is composed of a number of genetically differing strains.[53,83,84] Strain differences include morphology, development, metabolism, DNA hybridization and restriction site analysis, and intermediate host specificity. Worms of one strain are adapted to one species of intermediate host—for example, cattle, horses, sheep, or pigs—and they do not develop well in other species. The strains have considerable epidemiological significance for humans: Horse and pig strains in Europe probably do not infect humans, but sheep and cattle strains do.

Local traditions may contribute to massive infections. Some tribes of Kenya, for instance, are said to relish dog intestine roasted on a stick over a campfire. Because cleaning of the intestine may involve nothing more than squeezing out its contents, and cooking may entail nothing more than external scorching, these people probably have the highest rate of infection with hydatids in the world.[63] A further complication lies in the lack of burial of the dead by the Turkana people of Kenya. When the corpses are eaten by carnivores, humans become true intermediate hosts of *E. granulosus.*[57]

A different set of circumstances leads to infection in tanners in Lebanon, where dog feces are used as an ingredient of the tanning solution. Scats picked off the street are added to the vats, and any eggs present may infect their handler by contamination.[75]

Sheepherders in the United States and elsewhere risk infection by living closely with their dogs. Surveys of cat-

tle, hogs, and sheep in abattoirs reveal that *E. granulosus* occurs throughout most of the United States, with greatest concentrations in the deep South and far West. Recent outbreaks have been diagnosed in California and Utah. Nevertheless, 98 percent of hydatid infections diagnosed in the United States are imported, the most frequent contributor of cases being Italy.[14]

This disease can be eliminated from an endemic area only by interrupting the life cycle by denying access by dogs to offal, by destroying stray dogs, and by a general education program.[38,60]

- **Pathogenesis.** The effects of a hydatid may not become apparent for many years after infection because of its usual slow growth. Up to 20 years may elapse between infection and overt pathogenesis. If infection occurs early in life, the parasite may be almost as old as its host.[8]

The type and extent of pathological conditions depend on the location of the cyst in the host. As the size of the hydatid increases, it crowds adjacent host tissues and interferes with their normal functions (Fig. 21.27). The results may be very serious. If the parasite is lodged in the nervous system, clinical effects may be manifested relatively early in the infection before much growth occurs. When bone marrow is affected, the growth of the hydatid is restricted by lack of space. Chronic internal pressure caused by the parasite usually causes necrosis of the bone, which becomes thin and fragile; characteristically, the first sign of such an infection is a spontaneous fracture of an arm or leg. When the hydatid grows in an unrestricted location, it may become enormous, containing more than 15 quarts of fluid and millions of protoscolices. Even if it does not occlude a vital organ, it can still cause sudden death if it ruptures. The host is sensitized to *Echinococcus* antigens during its long-standing infection, and sudden release of massive amounts in the hydatid fluid induces an adverse host reaction called anaphylactic shock. Unconsciousness and death are nearly instantaneous in such instances.

- **Diagnosis and Treatment.** When hydatids are found, it is often during X-radiography or ultrasonography. Several immunodiagnostic techniques are available, but these are generally less sensitive than imagery.[6]

Surgery remains the only routine method of treatment and then only when the hydatid is located in an unrestricted location; for treatment of inoperable hydatid, albendazole is recommended.[86] The typical surgical procedure involves incising the surrounding adventitia until the capsule is encountered and aspirating the hydatid fluid with a large syringe. Considerable delicacy is required at this point, since fluid spilled into a body cavity can quickly cause fatal anaphylactic shock. After aspiration of the cyst contents, 10% formalin is injected into the hydatid to kill the germinal layer. This fluid is withdrawn after five minutes, and the entire cyst is then excised. A high rate of surgical success is obtained on ocular hydatidosis.

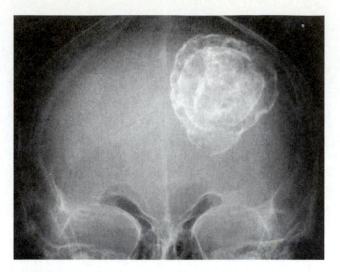

FIGURE 21.27

Partially calcified hydatid cyst in the brain.
AFIP neg. no. 68-2740.

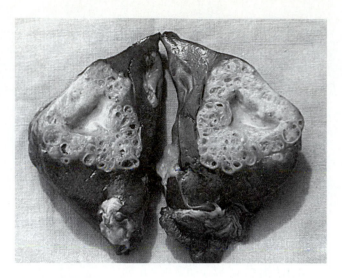

FIGURE 21.28

Alveolar hydatid cyst on the liver of an experimentally infected rhesus monkey.

Courtesy of Robert Rausch.

Echinococcus multilocularis. *Echinococcus multilocularis* is primarily boreal in its distribution. It is known from Europe, Asia, New Zealand, and South and North America. It has been reported from the United States as far south as Nebraska and Iowa and as far east as Ohio.[14,79] Human cases in the United States occur most commonly from Alaska,[87] but a human infection has also been reported from Minnesota.[37] The adult is mainly a parasite of foxes, but dogs, cats, and coyotes may also serve as definitive hosts. The hydatid develops in several species of small rodents such as voles, lemmings, and mice.

The adult is very similar to *E. granulosus,* differing from it in the following characteristics: (1) *E. granulosus* is 3 to 6 mm long, whereas *E. multilocularis* is only 1.2 to 3.7 mm long; (2) the genital pore of *E. granulosus* is about equatorial, but it is preequatorial in *E. multilocularis;* (3) *E. granulosus* has 45 to 65 testes with a few located anterior to the cirrus pouch, but *E. multilocularis* has 15 to 30 testes, all located posterior to the cirrus pouch.

The juvenile form (Fig. 21.28) differs in several respects from that of *E. granulosus.* Instead of developing a thick, laminated layer and growing into large, single cysts, this parasite has a thin outer wall that grows and infiltrates processes into the surrounding host tissues like a cancer. Each process may have several small, fluid-filled pockets containing several protoscolices. In humans and other unnatural hosts the pockets typically lack protoscolices. In natural intermediate hosts the cyst is more regular. In humans pieces of the cyst sometimes break off and metastasize to other parts of the body.[58] Because of its type of construction, this metacestode form is called an **alveolar** or **multilocular hydatid.** Some authorities, especially in the Soviet Union, place this species in a separate genus, *Alveococcus,* because of its unique form.

Human infection with alveolar hydatid is rare because the normal life cycle is sylvatic rather than urban and because humans do not seem to be very good hosts. Although protoscolices may not develop in human hosts, the germinal membrane is still viable.[68] Anyone handling wild foxes may be exposed to infection. Thus, this disease is most common among professional trappers and among handlers of sled dogs, where the dogs catch and eat wild mice as a regular part of their diet. There is some evidence that there are strains that differ in virulence.[52] Some human infections seem to disappear spontaneously, while other seem to march inexorably toward death.

Diagnosis of alveolar hydatid is difficult, particularly because the protoscolices may not be found. Even at necropsy cysts may be mistaken for malignant tumors. As a result of the difficulties of liver surgery, excision is usually practical only when the hydatid is localized near the tip of a lobe of the liver; infections of the hilar area are inoperable. The infiltrative nature of the cyst and its slow rate of growth may advance the disease to an inoperable state before its presence is detected. Praziquantel, the drug that is so effective for most flatworm parasites, may actually *enhance* growth of alveolar hydatids.[58]

Alveolar hydatidosis can be prevented only by avoiding dogs and their feces in endemic regions, by carefully washing all strawberries, cranberries, and the like that may be contaminated by dung, and by regularly worming dogs that may be liable to infection. Because it is a sylvatic disease it is much more difficult to eradicate than *E. granulosus.*[38]

Echinococcus vogeli. This parasite of canids in Central and South America rarely causes hydatidosis in humans. Previously attributed to *E. oligarthrus,* a tapeworm of felids, all known cases are now identified as *E. vogeli.*[23] The most important source of infection for humans is the domestic dog. The suitability of humans as hosts for *E. vogeli* seems to be intermediate between that of *E. granulosus* and *E. multilocularis;* although polycystic in humans, *E. vogeli* produces relatively large, fluid-filled vesicles with numerous protoscolices. Its natural intermediate host is the paca (a rodent).[67]

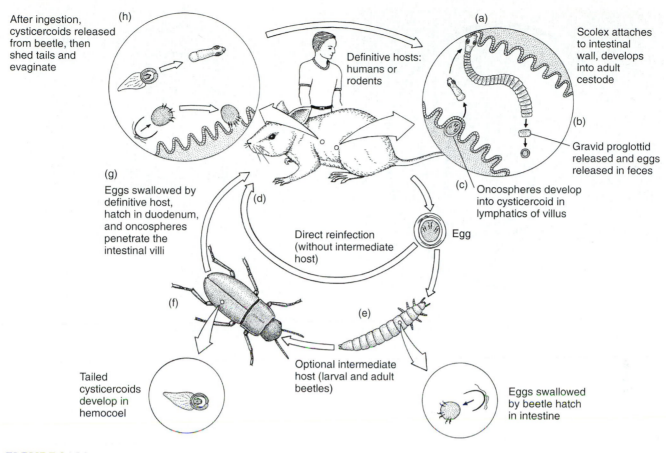

After ingestion, cysticercoids released from beetle, then shed tails and evaginate (h)

Definitive hosts: humans or rodents (a)

Scolex attaches to intestinal wall, develops into adult cestode

Gravid proglottid released and eggs released in feces (b)

(g) Eggs swallowed by definitive host, hatch in duodenum, and oncospheres penetrate the intestinal villi

(c) Oncospheres develop into cysticercoid in lymphatics of villus

(d) Direct reinfection (without intermediate host)

Egg

(f) Tailed cysticercoids develop in hemocoel

Optional intermediate host (larval and adult beetles)

(e) Eggs swallowed by beetle hatch in intestine

FIGURE 21.29

Life cycle of *Vampirolepis nana,* the dwarf tapeworm. (*a*) Adult attached to intestinal wall releases gravid proglottids (*b*) and shelled oncospheres in feces (*c*). (*d*) Direct reinfection when definitive host swallows shelled oncosphere. (*e*) Larval and adult beetles are optional intermediate hosts. (*f*) Cysticercoids develop in hemocoel of beetle. (*g*) When shelled oncospheres are ingested by definitive host, they hatch in the duodenum and penetrate the intestinal villi. (*h*) Ingested cysticercoids shed tails and evaginate, and scolices attach to intestinal wall.
Drawing by William Ober and Claire Garrison.

Family Hymenolepididae

The huge family Hymenolepididae consists of numerous genera with species in birds and mammals. Only two species, *Vampirolepis nana* and *Hymenolepis diminuta,* can infect humans. The family offers considerable taxonomic difficulties because of the large number of species and the immense and far-flung literature that has accumulated. However, their morphology is relatively simple compared with, for example, the Pseudophyllidea, and most species are small, transparent, and easy to study.

The most obvious morphological feature of the group is the small number of testes: usually one to four. The combination of few testes, usually unilateral genital pores, and large external seminal vesicle allows easy recognition of the family. All (except *V. nana*) require arthropod intermediate hosts.

Vampirolepis nana.   Commonly called the dwarf tapeworm, *Vampirolepis nana* (Fig. 21.29) previously known as *Hymenolepis nana,* is a cosmopolitan species that is the most common cestode of humans in the world, especially among children. Rates of infection run from 1% in the southern United States to 9% in Argentina and to 97.3% in Moscow.[48]

As its name implies, this is a small species, seldom exceeding 40 mm long and 1 mm wide. The scolex bears a retractable rostellum armed with a single circle of 20 to 30 hooks. The neck is long and slender, and the segments are wider than long. The genital pores are unilateral, and each mature segment contains three testes. After apolysis the gravid segments disintegrate, releasing the eggs, which measure 30 to 47 μm in diameter. The oncosphere (Fig. 21.30) is covered with a thin, hyaline, outer membrane and an inner, thick membrane with polar thickenings that bear several filaments. The heavy embryophores that give taeniid eggs their characteristic striated appearance are lacking in this and the other families of tapeworms infecting humans.

The life cycle of *V. nana* is unique among tapeworms in that an intermediate host is optional (Fig. 21.29). When eaten by a person or a rodent, the eggs hatch in the duodenum, releasing the oncospheres, which penetrate the mucosa and come to lie in the lymph channels of the villi. Here each develops into a cysticercoid (Fig. 21.31). In five to six days the cysticercoid emerges into the lumen of the small intestine, where it attaches and matures.

This direct life cycle is doubtless a recent modification of the ancestral two-host cycle, found in other species of hymenolepidids, since the cysticercoid of *V. nana* can still develop normally within larval fleas and beetles (Fig. 21.32).

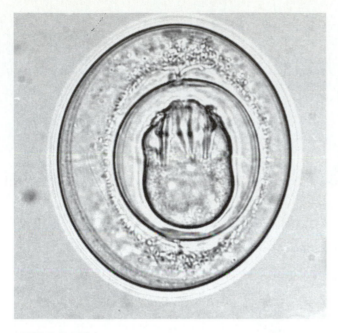

FIGURE 21.30

Egg of *Vampirolepis nana.* Note the polar filaments on the inner membrane and the well-developed oncosphere. Its size is 30 to 47 μm.

Courtesy of Jay Georgi.

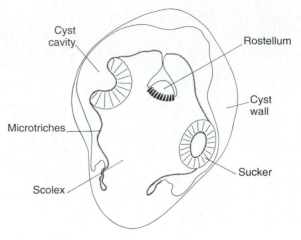

FIGURE 21.31

Vampirolepis nana: diagrammatic representation of a longitudinal section through a cysticercoid from a mouse villus.

From J. Caley, "A comparative study of two alternative larval forms of *Hymenolepis nana,* the dwarf tapeworm, with special reference to the process of encystment," in *Z. Parasitenkd.* 47:218–228. Copyright © 1975. Reprinted with permission of the publisher.

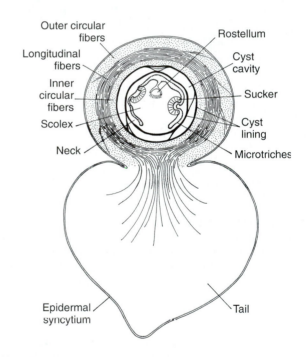

FIGURE 21.32

Diagrammatic representation of a longitudinal section through a cysticercoid of *Vampirolepis nana* from the insect host.

From J. Caley, "A comparative study of two alternative larval forms of *Hymenolepis nana,* the dwarf tapeworm, with special reference to the process of encystment," in *Z. Parasitenkd.* 47:218–228. Copyright © 1975. Reprinted with permission of the publisher.

One reason for the facultative nature of the life cycle is that *V. nana* cysticercoids can develop at higher temperatures than can those of other hymenolepidids. Direct contaminative infection by eggs is probably the most common route in human cases, but accidental ingestion of an infected grain beetle or flea cannot be ruled out.

Besides humans, domestic mice and rats also serve as suitable hosts for *V. nana.*[34] Some authors contend that two subspecies exist: *V. nana nana* in humans and *V. nana fraterna* in murine rodents. Differences do seem to exist in the physiological host-parasite relationships of these two subspecies, since higher rates of infection result from eggs obtained from the same host species than from the other.[65] This is probably an example of **allopatric speciation** in action.

Pathological results of infection by *V. nana* are rare and usually occur only in massive infections. Heavy infections can occur through autoinfection,[41] and the symptoms are similar to those already described for *Taeniarhynchus saginatus* infection. Treatment with niclosamide is efficacious but may have to be repeated in a month to remove the worms that were developing in villi at the time of treatment. Praziquantel acts very rapidly against *V. nana* and *H. diminuta.*[2,9] In vitro the drug produces vacuolization and disruption of the tegument in the neck of the worms but not in more posterior portions of the strobila.

Hymenolepis diminuta. *Hymenolepis diminuta* is a cosmopolitan worm that is primarily a parasite of rats (*Rattus* spp.), but human infections are not uncommon. It is a much larger species than *V. nana* (up to 90 cm) and differs from *V. nana* in lacking hooks on the rostellum. Typical of the genus, it has unilateral genital pores and three testes per proglottid.

The eggs (Fig. 21.33) are easily differentiated from those of *V. nana,* since they are larger and have no polar filaments. It has been demonstrated experimentally that more than 90 species of arthropods can serve as suitable intermediate hosts. Stored-grain beetles (*Tribolium* spp.) are probably

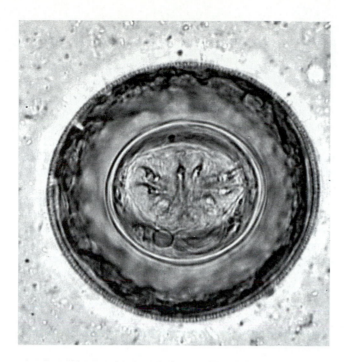

FIGURE 21.33

Egg of *Hymenolepis diminuta*. It is 40 to 50 μm wide.
Courtesy of Jay Georgi.

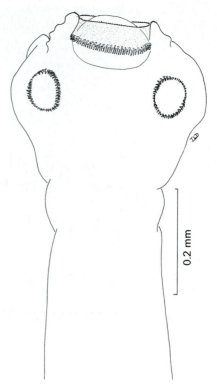

FIGURE 21.34

Scolex of *Raillietina*. The suckers are weak and have a double circle of spines, and the massive rostellum has many hammer-shaped hooks.
Drawing by Thomas Deardorff.

most commonly involved in infections of both rats and humans. A household shared with rats is also likely to have its cereal foods infested with beetles. Treatment is as recommended for *V. nana*.[46]

The ease with which this parasite is maintained in laboratory rats and beetles makes it an ideal model for many types of experimental studies; its physiology, metabolism, development, genetics, and nutrient uptake have been more thoroughly examined than those of any other tapeworm.[3] Research on *H. diminuta* has contributed enormously to our concepts of the world of tapeworms, including our understanding of how they survive, reproduce, and develop and their adaptations for parasitism.

Family Davaineidae

• *Raillietina* Species

The following species of *Raillietina* have been reported from humans: *R. siriraji, R. asiatica, R. garrisoni, R. celebensis,* and *R. demarariensis*. All normally parasitize domestic rats and possibly represent no more than two actual species. The genus is easily recognized by its large rostellum with hundreds of tiny, hammer-shaped hooks and by its spiny suckers (Fig. 21.34). Their life cycle is unknown, but they probably use insects as intermediate hosts. Their epidemiology is probably similar to that of *H. diminuta*.

Raillietina cesticillus is one of the most common poultry cestodes in North America, and a wide variety of grain, dung, and ground beetles serve as intermediate hosts. The genus is very large, with species in many birds and mammals. The closely related *Davainea* is identical to *Raillietina*,

except that its strobila is very short, consisting of only a few proglottids. *Davainea* spp. are found in galliform birds and use terrestrial molluscs as intermediate hosts. Other genera infect a wide variety of hosts, from passeriform birds to scaly anteaters.

Family Dilepididae

Dipylidium caninum.   A cosmopolitan, common parasite of domestic dogs and cats, *Dipylidium caninum* often occurs in children.[61] It is easily recognized because each segment has two sets of male and female reproductive systems and a genital pore on each side (Fig. 21.35). The scolex has a retractable, rather pointed rostellum with several circles of rosethorn-shaped hooks. The uterus disappears early in its development and is replaced by hyaline, noncellular egg capsules, each containing 8 to 15 eggs. Gravid proglottids detach and either wander out of the anus or are passed with feces. They are very active at this stage and are the approximate size and shape of cucumber seeds. As the detached segments begin to desiccate, the egg capsules are released. Fleas are the usual intermediate hosts, although chewing lice have also been implicated. Unlike the adult, a larval flea has simple, chewing mouthparts and feeds on organic matter, which may include *Dipylidium* egg capsules. The resulting cysticercoids survive their host's metamorphosis into the parasitic adult stage, when the flea may be nipped or licked out of the

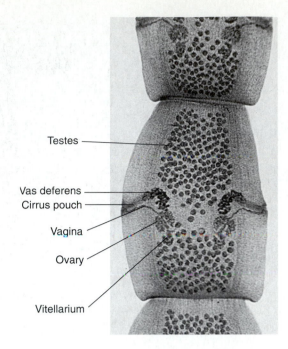

Testes

Vas deferens
Cirrus pouch

Vagina

Ovary

Vitellarium

FIGURE 21.35

Mature segment of *Dipylidium caninum,* the "double-pored tapeworm" of dogs and cats. The two vitelline glands are directly behind the larger ovaries. The smaller spheres are testes.

Courtesy of Ann Arbor Biological Center.

fur of the dog or cat, thereby completing the life cycle. This, by the way, is an example of **hyperparasitism,** since the flea is itself a parasite.

Nearly every reported case of infection of humans has involved a child. Adult humans may be more resistant, or else children may have increased chances of accidentally swallowing a flea. The symptoms and treatment are the same as for *Vampirolepis nana.*[46]

Basically, the only feature separating this family from Hymenolepididae is an increased number of testes, usually more than 12. This family, too, consists of hundreds of species that parasitize birds and mammals. Taxonomic difficulties also attend this family.

Family Anoplocephalidae

• *Moniezia* Species

The numerous species of *Moniezia* use hoofed animals as definitive hosts. The most frequently encountered species in domestic animals are *M. expansa* and *M. benedeni* in sheep, cattle, and goats. These are large tapeworms, up to 6 meters, and their proglottids are much wider than long. They have two sets of reproductive organs in each proglottid, and their scolex is unarmed. They have curious **interproglottidal glands** that open along the junctions between proglottids; the function of the glands is unknown.[76] The eggs of *M. benedeni* are rather square in shape, and those of *M. expansa* are more triangular. The

oncospheres of both are borne within an oddly shaped **pyriform apparatus,** an embryophore with long hook- or hornlike extensions.

Moniezia expansa was the first anoplocephalid for which a life cycle was discovered. Parasitologists had been mystified for many years as to how herbivorous animals could be infected with tapeworms, and small arthropods that might be ingested along with herbage were diligently investigated as potential intermediate hosts. Finally in 1937 Horace Stunkard announced that the intermediate host of *M. expansa* was a minute, free-living mite in the family Oribatidae.[80]

Ba and coworkers[5] sorted *M. expansa* and *M. benedeni* from France and Senegal, Africa, on the basis of the patterns of their interproglottidal glands and then performed isoenzyme analysis on them. They found the French worms very similar genetically to some of the African worms, and these were identified as *M. expansa.* They came from sheep and goats but not from cattle. The other African worms, putatively assigned to either *M. expansa* or *M. benedeni,* seemed to represent at least four distinct species. Thus, species diversity of *Moniezia* in domesticated ruminants may be greater than previously believed.

Bertiella studeri. Normally a parasite of Old World primates, *Bertiella studeri* has been reported many times from humans, especially in southern Asia, the East Indies, and the Philippines. The scolex is unarmed, and the proglottids are much wider than they are long, with the ovary located between the middle of the segment and the cirrus pouch. The egg is characteristic: 45 to 50 µm in diameter, with a bicornuate pyriform apparatus on the inner shell.

Ripe segments are shed in chains of about a dozen at a time. The intermediate hosts are various species of oribatid mites. Accidental ingestion of mites infected with cysticercoids completes the life cycle within primates. No disease has been ascribed to infection by *B. studeri.* Treatment is as for *Vampirolepis nana.*

Bertiella mucronata is similar to *B. studeri* and also has been reported from humans. It appears to be a parasite of New World monkeys, and children may become infected when living with a pet monkey and the ubiquitous oribatid mite. Distinguishing this species from *B. studeri* normally requires a specialist.

Inermicapsifer madagascariensis. *Inermicapsifer madagascariensis* is normally parasitic in African rodents, but it has been reported repeatedly in humans in several parts of the world, including South America and Cuba. Baer[7] concluded that humans are the only definitive host outside Africa.

The scolex is unarmed. The strobila is up to 42 cm long. Mature proglottids are somewhat wider than long. The uterus becomes replaced by egg capsules in ripe segments, each capsule containing 6 to 10 eggs, which do not possess a pyriform apparatus.

The life cycle of this parasite is unknown but undoubtedly involves an arthropod intermediate host. Clinical pathological conditions have not been studied, and treatment is similar to that described for other species.

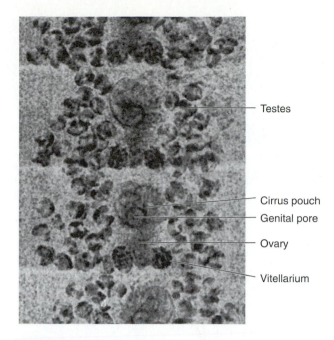

FIGURE 21.36

Mesocestoides sp., a cyclophyllidean cestode with a midventral genital pore and a bilobed vitellarium.

Courtesy of Larry Shults.

Family Mesocestoididae

• *Mesocestoides* Species

Unidentified specimens of the genus *Mesocestoides,* whose definitive hosts are normally various birds and mammals, have occasionally been reported from humans in Denmark, Africa, the United States, Japan, and Korea.[43] The ventrome-dial location of the genital pores is clearly diagnostic of the genus (Fig. 21.36). The complete life cycle is not known for any species in this difficult family, but many have a rodent or reptile intermediate host, in which a cysticercoid type of larva known as a **tetrathyridium** (Fig. 21.37) develops. Nei-ther mammals nor reptiles can be infected directly by eggs, so a first host must be involved. As yet, such a host has not been identified (Fig. 21.38). Pathological conditions and treatment of humans have not been studied.

Mesocestoides is very curious in that it may undergo asexual multiplication in the definitive host (Fig. 21.38)—not by budding, as in coenuri and hydatids, but by longitudi-nal fission of the scolex! An inwardly directed protuberance of the tegument between the suckers, the "apical massif," has morphocytogenetic power.[40] Although we know that this one isolate of *Mesocestoides* reproduces in this astonishing way (and the isolate has been widely propagated and used as an experimental model), the phenomenon may not be typical of the group.[21]

The scolex of *Mesocestoides* has four simple suckers and no rostellum. Each proglottid has a single set of male and female reproductive systems; the genital pores are

FIGURE 21.37

Tetrathyridial metacestodes of *Mesocestoides* sp. in the mesenteries of a baboon, *Papio cyanocephalus.*

Courtesy of Robert E. Kuntz.

median and ventral. Otherwise the morphology is typi-cally cyclophyllidean, with a paruterine organ replacing the uterus in most species.

Mesocestoides spp. are widespread in carnivores throughout most of the world. No complete life cycle is known, since a first intermediate host has never been discov-ered (Fig. 21.38). Rodents and reptiles are the most common second intermediate hosts, in which the parasite develops into a tetrathyridium (Fig. 21.37). When eaten by the preda-tor, the tetrathyridium develops into an adult (Fig. 21.36).

Some specialists consider members of this family a dis-tinct order of tapeworms.

Family Dioecocestidae

Except for *Dioecotaenia* (whose members form a family of tetraphyllideans in rays), the only dioecious tapeworms are found in this family. All are parasites of shorebirds, grebes, or herons. Some species are completely dioecious, whereas others are regionally so. There are wide variations in scolex types, although all have four suckers. In some species, such as *Shipleya inermis* in dowitchers, both sexes have sec-ondary sex organs of the opposite sex. Hence, the male has a uterus and the female a cirrus and cirrus pouch, but each has only an ovary or testes.[74]

ORDER PROTEOCEPHALATA

The proteocephalatans are all parasites of freshwater fishes, amphibians, or reptiles. The scolices (Fig. 20.6*d*) are much like those occurring in the cyclophyllideans, bearing four simple suckers and occasionally armature or a rostellum. The proglottids, however, are much more like those of the Tetra-phyllidea. Genital pores are lateral. The ovary is posterior, and the numerous testes fill most of the region anterior to it. The vitellaria are follicular and are restricted to the lateral margins of the proglottid.

FIGURE 21.38

Developmental sequence of *Mesocestoides corti,* illustrating early developmental stages (**2–5**), tetrathyridium and asexual multiplication in second intermediate host (**6–8**), and asexual multiplication with subsequent formation of adult worms in intestine of definitive host (**9–10**). Not illustrated here is the potential reinvasion of tissues from intestinal lumen of carnivore with continuing asexual multiplication of tetrathyridial stage.

From M. Voge, in *Problems in Systemics of Parasites.* Edited by G. D. Schmidt. Copyright © 1969 University Park Press, Baltimore, MD.

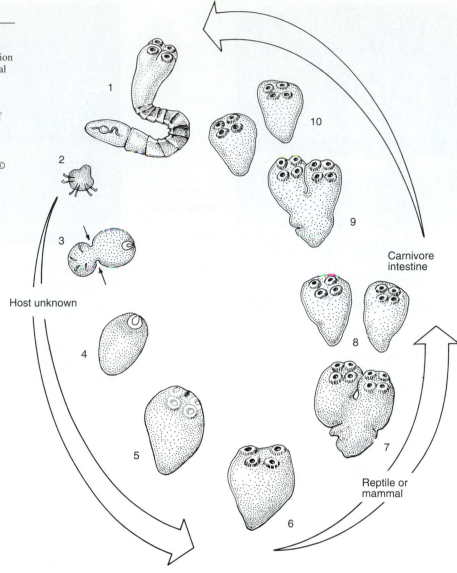

Host unknown

Carnivore intestine

Reptile or mammal

We know complete life cycles for several species. All involve a cyclopoid crustacean intermediate host, in which the worm develops into a procercoid (the plerocercoid I, according to some authors; see p. 311). This metacestode has a well-developed scolex and a cercomer at the posterior end. In some species the procercoid is directly infective to the definitive host; in others it burrows into the viscera for a time before reemerging and maturing in the lumen of the gut. Paratenic hosts are common in proteocephalid life cycles. The boring action of the plerocercoid (as it is now called, since it loses the cercomer as it penetrates the intestinal wall; = plerocercoid II) may be highly pathogenic to the host. For example, *Proteocephalus ambloplitis* in bass in North America sometimes castrates its fish host.

ORDER TETRAPHYLLIDEA

Tetraphyllideans are notable for their astonishing variety of scolex forms (Figs. 20.5*a* and 20.6*i*). Basically there are four bothridia, which may be stalked or sessile, smooth or crenate, or subdivided into loculi or major units. Often there are accessory suckers (Fig. 21.39) and/or hooks (Fig. 21.40) or spines. An apical, stalked, suckerlike organ, the **myzorhynchus,** is present on some. A neck is present or absent. The strobila and proglottids are essentially identical to those of the Lecanicephalidea and Trypanorhyncha, and like members of those orders, adult tetraphyllideans are all parasites of the spiral intestine of elasmobranchs.

As far as is known, the life cycles are also similar. No complete cycle has been discovered, but infective plerocercoids

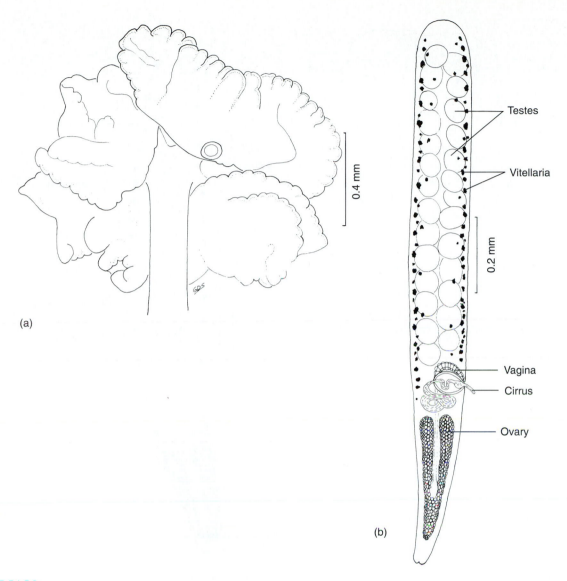

FIGURE 21.39

Phyllobothrium kingae, (*a*) unarmed bothridia with accessory suckers; (*b*) mature proglottid.

From G. D. Schmidt, "*Phyllobothrium kingae* sp. n., a tetraphyllidean cestode from a yellow-spotted stingray in Jamaica," in *Proc. Helm. Soc. Wash.* 45:132–134. Copyright © 1978. Reprinted with permission of the publisher.

are common in molluscs, crustaceans, and fishes. The fishes undoubtedly are paratenic hosts, as may be some of the molluscs and crustaceans. In vitro cultivation of *Acanthobothrium* plerocercoids in the presence of urea, a substance they encounter in their definitive host, causes the scolex to differentiate into the adult condition.[39]

An old but useful monograph on this order is that of Southwell.[78]

ORDER TRYPANORHYNCHA

The scolex (Fig. 20.6*f*) of trypanorhynchans is an extraordinary organ. It usually is elongated with two or four shallow bothridia, which may be covered with minute microtriches. Four eversible tentacles (atrophied in *Apororhynchus*) emerge from the apex of the scolex. The tentacles are armed with an astonishing array of hooks and spines (Fig. 21.41),

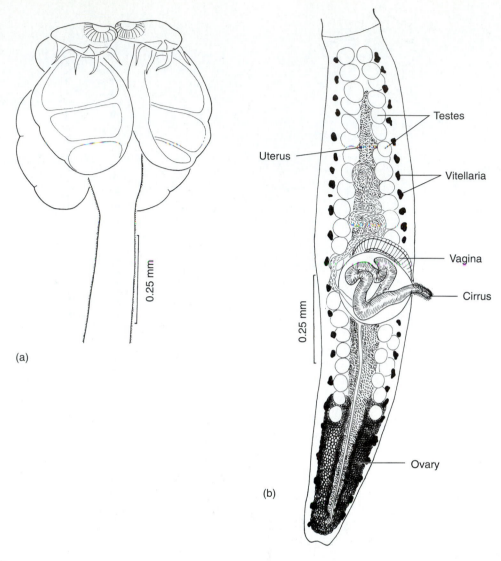

(a)

Testes

Uterus

Vitellaria

Vagina

Cirrus

Ovary

(b)

FIGURE 21.40

Acanthobothrium urolophi, an armed tetraphyllidean. (*a*) Scolex; (*b*) mature proglottid.

From G. D. Schmidt, "*Acanthobothrium urolophi* sp. n., a tetraphyllidean cestode (Oncobothriidae) from an Australian stingaree," in *Proc. Helm. Soc. Wash.* 40:91–93. Copyright © 1973. Reprinted with permission of the publisher.

shaped and arranged differently in each species. Interpretation of the hook arrangement is difficult but must be accomplished before species identification is possible. Each tentacle invaginates into an internal tentacle sheath, provided at its base with a muscular bulb. A retractor muscle originates at the base or front end of the bulb, courses through the tentacle sheath, and inserts inside the tip of the tentacle. When the retractor muscle contracts, it invaginates the tentacle, detaching it from host tissues. When the bulb contracts, it hydraulically evaginates the tentacle, driving it deep into the host's intestinal wall. This process is very similar to that which manipulates the proboscis of an acanthocephalan (Chapter 31).

A neck is present or absent; the strobila varies from hyperapolytic to anapolytic. The proglottids of try-

panorhynchans are morphologically very consistent with those of tetraphyllideans. The single ovary is basically bilobed and posterior. Vitellaria are follicular, cortical, and lateral or circummedullary. The uterus is a simple sac, usually in the anterior two-thirds of the gravid segment. Testes are few to many and medullary, and the cirrus pouch and cirrus often are huge relative to the proglottid. All genital pores are lateral.

Adult trypanorhynchans are all parasites of the spiral intestine of sharks and rays. Infective metacestodes are common in marine molluscs,[16] crustaceans, and fishes. Sakanari and Moser [72] reported experimental infection of copepods with coracidia of *Lacistorhynchus tenuis* which developed into procercoids. These grew into plerocercoids after being

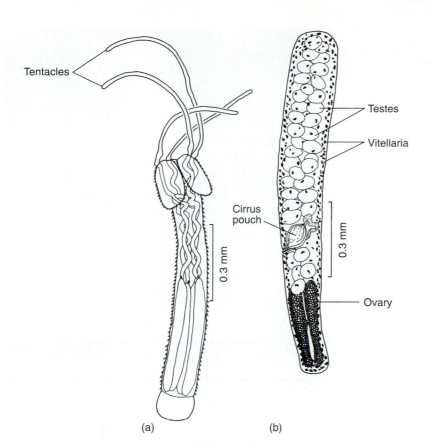

(a)   (b)

FIGURE 21.41

Basic morphology of *Eutetrarhynchus,* a typical trypanorhynch genus. (*a*) Scolex; (*b*) proglottid.

From K. J. Kovacs and G. D. Schmidt, "Two new species of cestode (Trypanorhyncha, Eutetrarhynchidae) from the yellow-spotted stingray, *Urolophus jamaicensis,*" *Proc. Helm. Soc. Wash.* 47:10–14. Copyright © 1980. Reprinted with permission of the publisher.

eaten by mosquitofish, which produced immature adults after being fed to leopard sharks. This life cycle is similar to that of another trypanorhynchan, *Grillotia erinaceus,* but many other members of this order do not have operculated eggs that release ciliated coracidia.[36]

The plerocercoid, sometimes called a plerocercus, may or may not bear a posterior sac, the **blastocyst,** into which the scolex is inverted. Plerocerci may be so plentiful in the flesh of certain fish or shrimps as to make them unpalatable and thereby unsalable. This is one known economic importance of trypanorhynchans. They have never been reported from humans. However, they remain among the most enigmatic and challenging invertebrates for the taxonomist. Dollfus published classical reviews,[27,28] and Schmidt provided a key to families and genera.[73,74]

SUBCOHORT AMPHILINIDEA

The Amphilinidea is the sister group of the Eucestoda. Their body is monozoic and dorsoventrally flattened, with an indistinct, proboscislike holdfast at the anterior end (Fig. 21.42). The genital pores are near the posterior end; the uterine pore is near the anterior end. The ovary is posterior, the vitelline follicles are bilateral, and the testes are preovarian. The uterus is N-shaped or looped. The oncosphere is provided with six large and four small hooks.

Amphilinideans are parasites of the body cavity of fishes and turtles in Asia, Japan, Europe, North America, Sri Lanka, Brazil, Africa, East Indies, and Australia. They are of no medical or economic importance. Rohde and Georgi[70] presented an excellent study on the structure and life history of *Austramphilina.*

COHORT GYROCOTYLIDEA

This is the sister group of the Cestoidea. They are also monozoic, and their anterior end is provided with a small, inversible holdfast organ. The posterior end is a frilled, rosettelike organ, and the lateral margins may be frilled (*Gyrocotyle* Fig. 21.43), or it is a long, simple cylinder, and the lateral margins are smooth (*Gyrocotyloides*). The ovary is posterior; the uterus has extensive lateral loops, terminating in a midventral pore in the anterior half. Testes are anterior. The genital pores are near the anterior end. Gyrocotylideans have a larva with 10 equal-sized hooks. They are parasites of the spiral intestine of Holocephali.

The Gyrocotylidea have traditionally been placed with the Amphilinidea in a subclass Cestodaria of the Class Cestoidea. Present opinion places them as a sister group of the cohort Cestoidea in the infraclass Cestodaria, and the Cestodaria is a sister group of the infraclass Monogenea in the subclass Cercomeromorphae.[13]

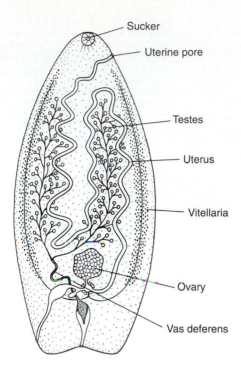

FIGURE 21.42

The monozoic tapeworm *Amphilina foliacea.*

Source: R. A. Wardle and J. A. McLeod, *The Zoology of Tapeworm,* 1952; Hafner Publishing Co., New York, NY.

References

1. Allan, J. C., F. Mencos, J. Garcia-Noval, E. Sarti, A. Flisser, Y. Wang, D. Liu, and P. S. Craig. 1993. Dipstick dot ELISA for the detection of *Taenia* coproantigens in humans. *Parasitology* 107:79–85.

2. Andrews, P., and H. Thomas. 1979. The effect of praziquantel on *Hymenolepis diminuta* in vitro. *Tropenmed. Parasitol.* 30:391–400.

3. Arai, H. P., ed. 1980. *Biology of the tapeworm* Hymenolepis diminuta. New York: Academic Press, Inc.

4. Archer, D. M., and C. A. Hopkins. 1958. Studies on cestode metabolism, III. Growth pattern of *Diphyllobothrium* sp. in a definitive host. *Exp. Parasitol.* 7:125–44.

5. Ba, C. T., W. Q. Wang, F. Renaud, L. Euzet, B. Marchand, and T. De Meeüs. 1993. Diversity and specificity in cestodes of the genus *Moniezia*: Genetic evidence. *Int. J. Parasitol.* 23:853–57.

6. Babba, H., A. Messedi, S. Masmoudi, M. Zribi, R. Grillot, P. Ambroise-Thomas, I. Beyrouti, and Y. Sahnoun. 1994. Diagnosis of human hydatidosis: Comparison between imagery and six serologic techniques. *Am. J. Trop. Med. Hyg.* 50:64–68.

7. Baer, J. G. 1956. The taxonomic position of *Taenia madagascariensis* Davaine, 1870, a tapeworm parasite of man and rodents. *Ann. Trop. Med. Parasitol.* 50:152–56.

8. Barnett, L. 1939. Hydatid disease: Errors in teaching and practice. *Br. Med. J.* 2:593–99.

9. Becker, B., H. Mehlhorn, P. Andrews, and H. Thomas. 1980. Scanning and transmission electron microscope studies on the efficacy of praziquantel on *Hymenolepis nana* (Cestoda) in vitro. *Z. Parasitenkd.* 61:121–33.

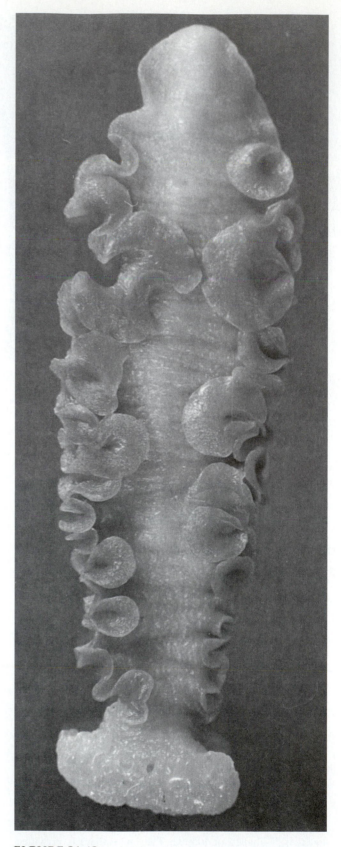

FIGURE 21.43

Gyrocotyle parvispinosa from the ratfish *Hydrolagus colliei.*
Courtesy of Warren Buss.

10. Berman, J. D., P. C. Beaver, A. W. Cheever, and E. A. Quindlen. 1981. Cysticercus of 60-milliliter volume in human brain. *Am. J. Trop. Med. Hyg.* 30:616–19.

11. Bowles, J., and D. P. McManus. 1994. Genetic characterization of the Asian *Taenia,* a newly described taeniid cestode of humans. *Am. J. Trop. Med. Hyg.* 50:33–44.

12. von Bonsdorff, B. 1956. *Diphyllobothrium latum* as a cause of pernicious anemia. *Exp. Parasitol.* 5:207–30.

13. Brooks, D. R., and D. A. McLennan. 1993. *Parascript. Parasites and the language of evolution.* Washington, D.C.: Smithsonian Institution Press.

14. Bryan, R. T., and P. M. Schantz. 1989. Echinococcosis (hydatid disease). *J. Am. Vet. Med. Assoc.* 195:1214–16.

15. Burt, M. D. B., and I. M. Sandeman. 1969. Biology of *Bothrimonus* (=*Diplocotyle*) (Pseudophyllidea: Cestoda). I. History, description, synonymy, and systematics. *J. Fish. Res. Bd. Can.* 26:975–96.

16. Cake, E. W. Jr. 1976. A key to larval cestodes of shallow-water, benthic mollusks of the northern Gulf of Mexico. *Proc. Helm. Soc. Wash.* 43:160–71.

17. Camargo, C. A., and W. H. Marshall. 1987. Radiological diagnosis of neurocysticercosis. *Parasitol. Today* 3:30–31.

18. Centers for Disease Control. 1992. Locally acquired neurocysticercosis—North Carolina, Massachusetts, and South Carolina, 1989–-1991. *Morb. Mortal. Weekly Rep.* 41:1–4.

19. Chandler, A. C., and C. P. Read. 1961. *Introduction to parasitology,* 10th ed. New York: John Wiley & Sons, Inc.

20. Choromanski, L., J. J. Estrada, and R. E. Kuhn. 1990. Detection of antigens of larval *Taenia solium* in the cerebrospinal fluid of patients with the use of HPLC and ELISA. *J. Parasitol.* 76:69–73.

21. Conn, D. B. 1990. The rarity of asexual reproduction among *Mesocestoides* tetrathyridia (Cestoda). *J. Parasitol.* 76:453–55.

22. Corkum, K. C. 1966. Sparganosis in some vertebrates of Louisiana and observations of human infection. *J. Parasitol.* 52:444–48.

23. D'Alessandro, A., R. L. Rausch, C. Cuello, and N. Aristizabal. 1979. *Echinococcus vogeli* in man, with a review of polycystic hydatid disease in Columbia and neighboring countries. *Am. J. Trop. Med. Hyg.* 28:303–17.

24. Davis, A. 1973. *Drug treatment in intestinal helminthiases.* Geneva: World Health Organization.

25. Dewhirst, L. W., J. D. Cramer, and J. J. Sheldon. 1967. An analysis of current inspection procedures for detecting bovine cysticercosis. *J. Am. Vet. Med. Assoc.* 150:412–17.

26. Dixon, H. B. F., and D. W. Smithers. 1934. Epilepsy in cysticercosis (*Taenia solium*). A study of seventy-one cases. *Q. J. Med.* 3:603–16.

27. Dollfus, R. P. 1942. Etudes critiques sur les Tétrarhynques du Museum de Paris. *Arch. Mus. Nat. Hist. Nat.* Ser. 6, 19:1–466.

28. Dollfus, R. P. 1946. Notes diverses sur des Tétrarhynques. *Mem. Mus. Nat. Hist. Nat.* n. s. 22:179–220.

29. Dooley, J. F. 1980. Health precautions in Mexico. Letter to the editor. *J. A. M. A.* 243:1524.

30. Eom, K. S., and H. J. Rim. 1993. Morphologic descriptions of *Taenia asiatica* sp. n. *Korean J. Parasitol.* 31:1–6.

31. Eom, K. S., H. J. Rim, and S. Geerts. 1992. Experimental infection of pigs and cattle with eggs of Asian *Taenia saginata* with special reference to its extrahepatic viscerotropism. *Korean J. Parasitol.* 30:269–75.

32. Etges, F. J. 1991. The proliferative tetrathyridium of *Mesocestoides vogae* sp. n. (Cestoda). *J. Helm. Soc. Wash.* 58:181–85.

33. Fan, P. C. 1988. Taiwan *Taenia* and taeniasis. *Parasitol. Today* 4:86–88.

34. Ferretti, G., F. Gabriele, and C. Palmas. 1981. Development of human and mouse strain of *Hymenolepis nana* in mice. *Int. J. Parasitol.* 11:424–30.

35. Flisser, A. 1988. Neurocysticercosis in Mexico. *Parasitol. Today* 4:131–37.

36. Freeman, R. S. 1973. Ontogeny of cestodes and its bearing on their phylogeny and systematics. In Dawes, B., ed. *Advances in parasitology,* 11. London: Academic Press. 481–557.

37. Gamble, W. G., M. Segal, P. M. Schantz, and R. L. Rausch. 1979. Alveolar hydatid disease in Minnesota. First human case acquired in the contiguous United States. *J. A. M. A.* 241:904–7.

38. Gemmell, M. A., J. R. Lawson, and M. G. Roberts. 1987. Towards global control of cystic and alveolar hydatid diseases. *Parasitol. Today* 3:144–51.

39. Hamilton, K. A., and J. E. Byram. 1974. Tapeworm development: The effects of urea on a larval tetraphyllidean. *J. Parasitol.* 60:20–28.

40. Hess, E. 1980. Ultrastructural study of the tetrathyridium of *Mesocestoides corti* Hoeppli, 1925: Tegument and parenchyma. *Z. Parasitenkd.* 61:135–59.

41. Heyneman, D. 1962. Studies of helminth immunity: I. Comparison between lumenal and tissue phases of infection in the white mouse by *Hymenolepis nana* (Cestoda: Hymenolepididae). *Am. J. Trop. Med. Hyg.* 11:46–63.

42. Hopkins, D. R. 1992. Homing in on helminths. *Am. J. Trop. Med. Hyg.* 46:626–34.

43. Hutchison, W. F., and J. B. Martin. 1980. *Mesocestoides* (Cestoda) in a child in Mississippi treated with paromomycin sulfate (Humantin). *Am. J. Trop. Med. Hyg.* 29:478–79.

44. Ito, A. 1992. Cysticercosis in Asian-Pacific regions. *Parasitol. Today* 8:182–83.

45. Jepsen, A., and H. Roth. 1952. Epizootiology of *Cysticercus bovis*-resistance of the eggs of *Taenia saginata.* Report 14. *Int. Vet. Cong.* 22:43–50.

46. Jones, W. E. 1979. Niclosamide as a treatment for *Hymenolepis diminuta* and *Dipylidium caninum* infection in man. *Am. J. Trop. Med. Hyg.* 28:300–2.

47. Kamegai, S., A. Ichihara, H. Nonobe, M. Machida, and T. Hara. 1968. A case of human infection with the immature worm of *Diplogonoporus grandis. Res. Bull. Meguro Parasitol. Mus.* 2:1–8.

48. Karnaukov, V. K., and A. I. Laskovenko. 1984. Clinical picture and treatment of rare human helminthiases (*Hymenolepis diminuta* and *Dipylidium caninum*). *Meditsinskaya Parazitol. i Parazitornye Bolezni* no. 4, 77–79.

49. Kliks, M. M. 1990. Helminths as heirlooms and souvenirs: A review of New World paleoparasitology. *Parasitol. Today* 6:93–100.

50. Kuntz, R. E. 1963. Snakes of Taiwan. *Q. J. Taiwan Mus.* 16:1–79.

51. Lawson, J. R., and M. A. Gemmell. 1990. Transmission of taeniid tapeworm eggs via blowflies to intermediate hosts. *Parasitology* 100:143–46.

52. Liance, M., S. Bresson-Hadni, D. Vuitton, S. Bretagne, and R. Houin. 1990. Comparison of the viability and developmental characteristics of *Echinococcus multilocularis* isolates from human patients in France. *Int. J. Parasitol.* 20:83–86.

53. Lymbery, A. J. 1992. Interbreeding, monophyly and the genetic yardstick: Species concepts in parasites. *Parasitol. Today* 8:208–11.

54. Mackiewicz, J. S. 1972. Caryophyllidea (Cestoidea): A review. *Exp. Parasitol.* 31:417–512.

55. Mackiewicz, J. S. 1976. *Glaridacris vogei* n. sp. (Cestoidea: Caryophyllidea) from catostomid fishes in western North America. *Trans. Am. Micros. Soc.* 95:92–97.

56. Mackiewicz, J. S. 1982. Caryophyllidea (Cestoidea): Perspectives. *Parasitology* 84:397–417.

57. MacPherson, C. L. 1983. An active intermediate host role for man in the life cycle of *Echinococcus granulosus* in Turkana, Kenya. *Am. J. Trop. Med. Hyg.* 32:397–404.

58. Marchiondo, A. A., R. Ming, F. L. Andersen, J. H. Slusser, and G. A. Conder. 1994. Enhanced larval cyst growth of *Echinococcus multilocularis* in praziquantel-treated jirds (*Meriones unguiculatus*). *Am. J. Trop. Med. Hyg.* 50:120–27.

59. McGreevy, P. B., and G. S. Nelson. 1991. Larval cestode infections. In Strickland, G. T., ed. *Hunter's tropical medicine,* 7th ed. Philadelphia: W. B. Saunders Company, 843–59.

60. McManus, D. P., and J. D. Smyth. 1986. Hydatidosis: Changing concepts in epidemiology and speciation. *Parasitol. Today* 2:163–67.

61. Moore, D. V. 1962. A review of human infections with the common dog tapeworm, *Dipylidium caninum,* in the United States. *Southwestern Vet.* 15:283–88.

62. Mueller, J. F. 1974. The biology of *Spirometra. J. Parasitol.* 60:3–14.

63. Nelson, G. S., and R. L. Rausch. 1963. *Echinococcus* infections in man and animals in Kenya. *Ann. Trop. Med. Parasitol.* 57:136–49.

64. Nyberg, W. 1958. Absorption and excretion of vitamin B_{12} in subjects infected with *Diphyllobothrium latum* and in noninfected subjects following oral administration of radioactive B_{12}. *Acta Haematol.* 19:90–98.

65. Pampiglione, S. 1962. Indagine sulla diffusion dell' imenolepiasi nella Sicilia occidentale. *Parassitologia* 4:49–58.

66. Rabiela, M. T., A. Rivas, and A. Flisser. 1989. Morphological types of *Taenia solium* cysticerci. *Parasitol. Today* 5:357–59.

67. Rausch, R. L., A. D'Allessandro, and V. R. Rausch. 1981. Characteristics of the larval *Echinococcus vogeli* Rausch and Bernstein, 1972 in the natural intermediate host, the paca, *Cuniculus paca* L. (Rodentia: Dasyproctidae). *Am. J. Trop. Med. Hyg.* 30:1043–52.

68. Rausch, R. L., and J. F. Wilson. 1973. Rearing of the adult *Echinococcus multilocularis* Leuckart, 1863, from sterile larvae from man. *Am. J. Trop. Med. Hyg.* 22:357–60.

69. Revenga, J. E. 1993. *Diphyllobothrium dendriticum* and *Diphyllobothrium latum* in fishes from southern Argentina: Association, abundance, distribution, pathological effects, and risk of human infection. *J. Parasitol.* 79:379–83.

70. Rohde, K., and M. Georgi. 1983. Structure and development of *Austramphilina elongata* Johnston, 1931 (Cestodaria: Amphilinidea). *Int. J. Parasitol.* 13:273–87.

71. Ruttenber, A. J., B. G. Weniger, F. Sorvillo, R. A. Murray, and S. L. Ford. 1984. Diphyllobothriasis associated with salmon consumption in Pacific coast states. *Am. J. Trop. Med. Hyg.* 33:455–59.

72. Sakanari, J., and M. Moser. 1985. Infectivity of, and laboratory infection with, an elasmobranch cestode, *Lacistorhynchus tenuis* (van Beneden, 1858). *J. Parasitol.* 71:788–91.

73. Schmidt, G. D. 1970. *How to know the tapeworms.* Dubuque, Iowa: Wm. C. Brown Publishers.

74. Schmidt, G. D. 1986. *Handbook of tapeworm identification.* Boca Raton, Fla.: CRC Press.

75. Schwabe, C. W., and K. A. Daoud. 1961. Epidemiology of echinococcosis in the Middle East. I. Human infection in Lebanon, 1949–1959. *Am. J. Trop. Med. Hyg.* 10:374–81.

76. Smyth, J. D., and D. P. McManus. 1989. *The physiology and biochemistry of cestodes.* Cambridge: Cambridge University Press.

77. Sorvillo, F. J., S. H. Waterman, F. O. Richards, and P. M. Schantz. 1992. Cysticercosis surveillance: Locally acquired and travel-related infections and detection of intestinal tapeworm carriers in Los Angeles County. *Am. J. Trop. Med. Hyg.* 47:365–71.

78. Southwell, T. 1925. A monograph on the Tetraphyllidea. *Liverpool University Press, Liverpool School of Tropical Medicine Memoir* n. s. no. 2, 1–368.

79. Storandt, S. T., and K. R. Kazacos. 1993. *Echinococcus multilocularis* identified in Indiana, Ohio, and east-central Illinois. *J. Parasitol.* 79:301–5.

80. Stunkard, H. W. 1944. How do tapeworms of herbivorous animals complete their life cycles? *Trans. N.Y. Acad. Sci.* 6:108–21.

81. Swartzwelder, J. C., P. C. Beaver, and M. W. Hood. 1964. Sparganosis in southern United States. *Am. J. Trop. Med. Hyg.* 13:43–48.

82. Tanowitz, H. B., and M. Wittner. 1991. Tapeworm infections. In Strickland, G. T. ed. *Hunter's tropical medicine,* 7th ed. Philadelphia: W. B. Saunders Company, 834–59.

83. Thompson, R. C. A., and A. J. Lymbery. 1988. The nature, extent and significance of variation within the genus *Echinococcus.* In Baker, J. R., and R. Muller, eds. *Advances in parasitology* 27. New York: Academic Press, Inc., 209–58.

84. Thompson, R. C. A., and A. J. Lymbery. 1990. *Echinococcus:* Biology and strain variation. *Int. J. Parasitol.* 20:457–70.

85. Torres, A., A. Plancarte, A. N. M. Villalobos, A. S. de Aluja, R. Navarro, and A. Flisser. 1992. Praziquantel treatment of porcine brain and muscle *Taenia solium* cysticercosis. 3. Effect of 1-day treatment. *Parasitol. Res.* 78:161–64.

86. Webbe, G. 1986. Cestode infections of man. In Campbell, W. C., and R. S. Rew, eds. *Chemotherapy of parasitic diseases.* New York: Plenum Press, 457–77.

87. Wilson, J. F., and R. L. Rausch. 1980. Alveolar hydatid disease. A review of clinical features of 33 indigenous cases of *Echinococcus multilocularis* infection in Alaskan Eskimos. *Am. J. Trop. Med. Hyg.* 29:1340–55.

88. Yamane, Y., N. Okada, and M. Takihara. 1975. On a case of long term migration of *Spirometra erinacei* larva in the breast of a woman. *Yonago Acta Medica.* 19:207–13.

Additional References

Andersen, F. L., J. Chai, and F. Liu, eds. 1993. *Compendium on cystic echinococcosis with special reference to the Xinjian Uygur Autonomous Region, The People's Republic of China.* Provo, Utah: Brigham Young University. Contains much basic information, in addition to reports on the Xinjian Uygur Autonomous Region.

Arai, H. P., ed. 1980. *Biology of the tapeworm* Hymenolepis diminuta. New York: Academic Press, Inc.

Arme, C., and P. W. Pappas, eds. 1983. *Biology of the Eucestoda.* London: Academic Press, 2 vols.

Binford, C. H., and D. H. Connor, eds. 1976. *Pathology of tropical and extraordinary diseases, sect. 11. Disease caused by cestodes.* Washington, D.C.: Armed Forces Institute of Pathology.

Hoffman, G. L. 1967. *Parasites of North American freshwater fishes.* Berkeley: University of California Press. Keys to all genera of tapeworms of North American fishes with lists of species and hosts.

Soulsby, E. J. L. 1965. *Textbook of veterinary clinical parasitology 1.* Philadelphia: F. A. Davis Co. A useful reference to tapeworms of veterinary significance.

Spasskaya, L. P. 1966. *Cestodes of birds of SSSR.* Moscow: Akademii Nauk. SSSR. A useful, illustrated survey of tapeworms of northern birds.

Stunkard, H. W. 1962. The organization, ontogeny and orientation of the cestodes. *Q. Rev. Biol.* 37:23–34.

Thompson, R. C. A., ed. 1986. *The biology of* Echinococcus *and hydatid disease.* London: George Allen and Unwin.

Chapter 22

PHYLUM NEMATODA: FORM, FUNCTION, AND CLASSIFICATION

If all the matter in the universe except the

nematodes were swept away, . . . we should find

[our world's] mountains, hills, vales, rivers, lakes

and oceans represented by a thin film of

nematodes.

N. A. Cobb

Nematodes are among the most abundant animals on earth. Of course more species of insects have been described, but when one realizes that many kinds of insects harbor at least one species of parasitic nematode and when one further calculates the number of kinds of nematodes parasitic in the rest of the animal kingdom, there is no contest. There are also many species of nematodes that parasitize plants. Finally, the species of free-living marine, freshwater, and soil-dwelling nematodes probably far outnumber those that are parasitic. Numbers of individuals are often extremely high: 90,000 were found in a single rotting apple; 1074 individuals, representing 36 species, were counted in 6 to 7 ml of mud; and 3 to 9 billion per acre may be found in good farmland in the United States.[76]

Obviously, these animals have a lot going for them. We hope that this chapter will give you some insight into their enormous numbers and diversity in parasitic habits.

Most nematodes are small, inconspicuous, and seemingly unimportant to humans and therefore attract the attention only of specialists. A few, however, cause diseases of great importance to humans and domestic and wild plants and animals. These are the roundworms that are studied by many parasitologists and nematologists.

HISTORICAL ASPECTS

Ancient people were probably familiar with the larger nematodes, which they encountered when they slew game or disemboweled an opponent at war. The earliest records mention the worms or contain recognizable allusions to them. Aristotle discussed the worm we now call *Ascaris lumbricoides,* and the Ebers Papyrus of 1550 B.C. Egypt described clinical hookworm disease, as did Hippocrates, Lucretius, and the ancient Chinese. Moses wrote of a scourge that probably was caused by the guinea worm. The eggs of *Ascaris lumbricoides* and *Trichuris trichiura* were found in the intestine of a 2300-year-old body found preserved in a peat bog in the Orkney Islands.[19]

The Arabians Avicenna and Avenzoar, who kept parasitology (and much other science) alive during the Dark Ages in Europe, studied elephantiasis, differentiating it from leprosy.[43]

Linnaeus, in 1758, placed the roundworms in his class Vermes, along with all other worms and wormlike animals. Goeze, Zeder, and Rudolphi made great advances in recognition of various nematodes, although they still believed the worms arose by spontaneous generation. Further work by Gegenbauer, Huxley, Hatschek, Leuckart, Beneden, Diesing, Linstow, Looss, Railliet, Stossich, and many others established the nematodes as a distinct and important group of animals.

The name Nematoda is a modification of Rudolphi's Nematoidea and was placed in Nemathelminthes, itself first considered a class in phylum Vermes, by Gegenbaur in 1859 and later was elevated to phylum status. Although still employed by some authors today, the name Nemathelminthes is little used. The name Aschelminthes was proposed by Grobben in 1910 as a superphylum to contain several divergent groups of wormlike animals that had in common a pseudocoelomic body cavity. Hyman resurrected the name Aschelminthes for use as a phylum containing the classes Nematoda, Rotifera, Priapulida, Gastrotricha, Nematomorpha, and Kinorhyncha.[55] Today most authorities consider each of these groups as separate phyla, sometimes placing them in the superphylum Aschelminthes to recognize their rather distant phyletic relationship. Attempts by a few students of free-living nematodes to change Nematoda to Nema or Nemata have not been widely accepted.

Curiously studies of nematodes developed along two separate lines, with parasitologists claiming the parasites of vertebrates and nematologists accounting for free-living and plant- and invertebrate-parasitic roundworms. This division of labor is a result of parasitologists' historical concern for parasites of medical and veterinary importance. Exceptions occur, of course; a few individuals, such as Chitwood, Bird, Crites, Inglis, Mawson, and Schuurmanns-Steckhoven, have made significant contributions to both areas. Nevertheless, each discipline publishes in its own journals, uses unique terminology and taxonomic formulas, and to a large extent employs different techniques for handling and preparing specimens for study. Recent trends suggest that the two disciplines are beginning to merge. If so, each school of

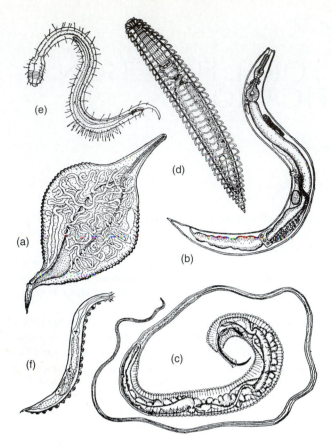

FIGURE 22.1

Variety of form in the nematodes. (*a*) *Tetrameres;* (*b*) *Rhabditis;* (*c*) *Trichuris;* (*d*) *Criconema;* (*e*) *Chaetosoma;* (*f*) *Bunonema.*

From H. D. Crofton, *Nematodes.* Copyright 1966 Hutchinson University Library, London. Reprinted with permission of publisher.

thought will profit from the special knowledge of the other, and the waste engendered by lack of communication will be minimized.

FORM AND FUNCTION

Typical nematodes are elongated, tapered at both ends, are bilaterally symmetrical, and possess a pseudocoel, a body cavity derived from the embryonic blastocoel. There are variations of this basic shape, however (Fig. 22.1). The digestive system is complete, with a mouth at the extreme anterior end and an anus near the posterior tip (Fig. 22.2). The lumen of the pharynx is characteristically **triradiate** (see Fig. 22.20). The body is covered with a noncellular cuticle that is secreted by an underlying hypodermis and is shed four times during ontogeny. The muscles of the body wall are only one layer thick and are distinguished by all being *longitudinally arranged* with no circular layer. The excretory system consists of lateral canals, ventral glands, or both, which open near the anterior end through a ventral excretory pore. Except for some sensory endings of modified cilia, neither cilia nor flagella are present, even in the male gamete. Most nematodes are dioecious and show considerable sexual dimorphism: The

females are usually larger, and the tail of the male is more curled. Some species are hermaphroditic, and others are parthenogenetic. Most are oviparous, but some are ovoviviparous. The female reproductive system opens through a ventral genital pore; the male system opens into a cloaca, together with the digestive system. Adult nematodes vary in size from less than 1 mm, as in the genus *Caenorhabditis,* to more than a meter, as in *Dracunculus.*

A considerable body of knowledge has accumulated on the function and structure (both at the light and electron microscope levels) of nematodes, far beyond our ability to review within the confines of this chapter. Many reviews and literature references are available.[13,34,35,68,87,118]

Body Wall

The body wall of nematodes comprises the cuticle, hypodermis, and body wall musculature. The outermost covering is the **cuticle,** a complex structure of great functional significance to the animals. The cuticle also lines the stomodeum, proctodeum, excretory pore, and vagina. Overlying the cuticle in free-living and parasitic nematodes is a carbohydrate-rich surface coat, 5 to 20 nm in thickness.[14] The surface coat may be important in evasion of the immune response in parasites of animals.

The cuticle itself is basically of three regions—the cortical, the middle (also called homogeneous or matrix), and the inner fibrous layers—that are commonly subdivided (Fig. 22.3). The cortex is covered by a thin layer called the **epicuticle,** about 20 nm thick. This appears in electron micrographs as a trilaminar membrane, but it is not a cell membrane.[70] The main body of the cortex is divided into an **external** and an **internal cortical layer.** The external layer contains a highly resistant protein called **cuticlin,** stabilized by dityrosine crosslinks.[40,88] The inner cortical layer, as well as the other layers in the cuticle, is primarily of collagen, a protein type also abundant in vertebrate connective tissue. Even under the electron microscope one can distinguish little structure in the **middle layer,** although some researchers have reported fine striations.

There are two or three **fibrous layers,** each of parallel strands of collagen, running at an angle of about 75 degrees to the longitudinal axis of the worm. Strands of the second fiber layer run at an angle of about 135 degrees to those of the first (and third, if present) layer, thus forming a latticelike arrangement. The fibrous layers are important components of the hydrostatic skeleton in larger nematodes.[68] The strands themselves are not extensible, but they do allow longitudinal stretching and compression of the overlying cuticle by changes in the angles between the layers. The innermost layer of the cuticle is the **basal lamella,** a layer of fine fibrils that merges with the underlying hypodermis. Ringlike depressions occur in the cuticle, enhancing flexibility of the animal. These striations are more prominent in some species than in others. The cuticle of the parasitic juveniles of mermithids (described later) is quite different in structure from that just described.[9]

Cuticular markings and ornamentations of many types occur in various kinds of nematodes. These include shallow **punctations,** deeper **pores,** and **spines** of varying

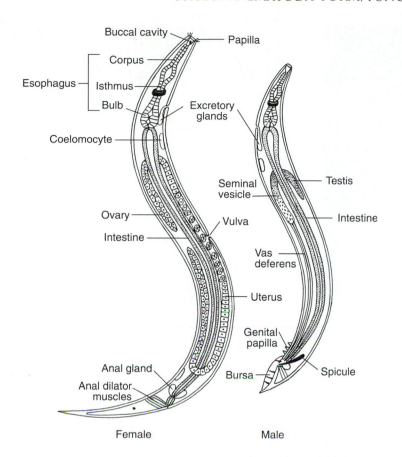

Buccal cavity
Papilla
Corpus
Esophagus
Isthmus
Bulb
Excretory glands
Coelomocyte
Testis
Seminal vesicle
Ovary
Intestine
Vulva
Intestine
Vas deferens
Uterus
Genital papilla
Anal gland
Bursa
Spicule
Anal dilator muscles

Female   Male

FIGURE 22.2

Morphology of a typical nematode male and female.

Drawing by William Ober and Claire Garrison.

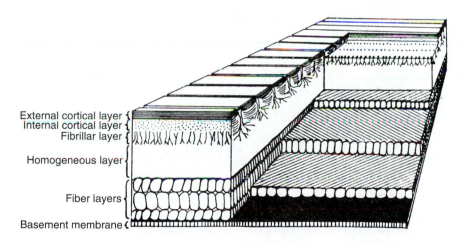

External cortical layer
Internal cortical layer
Fibrillar layer
Homogeneous layer
Fiber layers
Basement membrane

FIGURE 22.3

Diagram showing transverse, longitudinal, and tangential sections of the cuticle of *Ascaris*. The strands of each of the three fiber layers run at an angle of about 75° to the longitudinal axis of the worm, and the strands of the middle layer run about 135° from those of the inner and outer layers.

From A. F. Bird and K. Deutsch, "The structure of the cuticle of *Ascaris lumbricoides* var. *suis*," in *Parasitology* 47:319–329. Copyright © 1957. Reprinted with permission of the publisher.

complexity. Lateral or sublateral cuticular thickenings called **alae** are present in many species. Cervical alae (Fig. 22.4) are found on the anterior part of the body; caudal alae are on the tail ends of some males; and longitudinal alae, when present, extend the entire length of both sexes. The lateral alae may be of value to the animal when it is swimming or lend greater stability on solid substrate, when the nematode is crawling on its side by dorsoventral undulations, as does the juvenile of *Nippostrongylus brasilien-*

sis[65] (Fig. 22.5). Longitudinal ridges occur in many adult trichostrongylids. Cuticular ultrastructure has been studied in *N. brasiliensis;* Lee reported that the ridges are supported by a series of struts or skeletal rods in the middle layer of the cuticle[64] (Fig. 22.6). The struts are held erect by collagenous fibers inserted in the cortical and fibrous layers, but the middle layer itself is fluid filled and contains hemoglobin. The functions of the longitudinal ridges in trichostrongylids are to aid in locomotion, as the worm

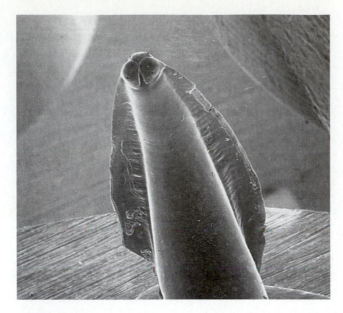

FIGURE 22.4

Scanning electron micrograph of *Toxocara cati*, illustrating cervical alae.
Courtesy of John Ubelaker.

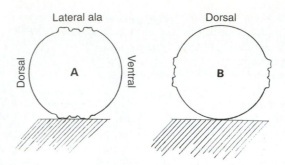

FIGURE 22.5

Outline of a transverse section through a third-stage juvenile to show the
more stable position when it lies on its side (*a*) when moving by two-
dimensional undulatory propulsion, and the less stable position, when it
lies on its ventral surface (*b*).

From D. L. Lee, "*Nippostrongylus brasiliensis:* some aspects of the fine
structure and biology of the infective larva and the adult," in *Nippostrongylus
and Toxoplasma,* edited by A. E. R. Taylor. Copyright © 1969 Blackwell
Science, Ltd., Oxford, UK. Reprinted with permission of the publisher.

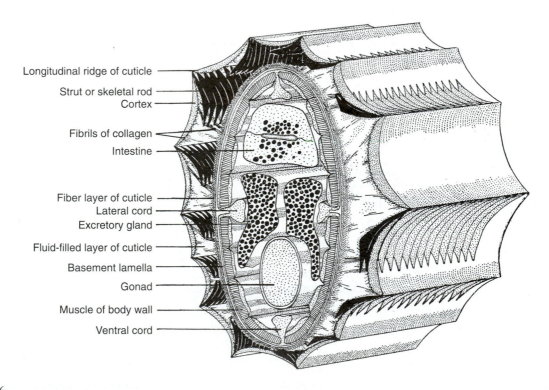

Longitudinal ridge of cuticle
Strut or skeletal rod
Cortex
Fibrils of collagen
Intestine
Fiber layer of cuticle
Lateral cord
Excretory gland
Fluid-filled layer of cuticle
Basement lamella
Gonad
Muscle of body wall
Ventral cord

FIGURE 22.6

Stereogram of a thick section taken from the middle region of an adult *Nippostrongylus brasiliensis* to show the arrangement of the various layers of the
cuticle and other internal organs.

From D. L. Lee, "The cuticle of adult *Nippostrongylus brasiliensis,*" in *Parasitology* 55:173–181. Copyright © 1965. Reprinted with permission of the publisher.

moves between villi with a corkscrew type of motion (Fig. 22.7), and to abrade the microvillar surface, thus helping the animal to obtain food in the absence of biting or piercing mouthparts. The arrangement of these rods is called the **synlophe.**

The **hypodermis** lies just beneath the basal lamella of the cuticle. It is usually syncytial in adult worms, and the nuclei lie in four thickened portions (six to eight in mermithids), the **hypodermal cords,** that project into the pseudocoel. The hypodermal cords run longitudinally and divide the somatic musculature into four quadrants. On the large nematodes, these may be discernible with the unaided eye as pale lines. The dorsal and ventral cords contain longitudinal nerve trunks, whereas the lateral cords contain the lateral canals of the excretory system in most species. Especially in the regions of the cords, the hypodermis contains mitochondria and endoplasmic reticulum. An important function of the hypodermis is secretion of the cuticle, as described in the section on development.

Specialized areas of the hypodermis, the **bacillary bands,** occur in at least two genera of trichuroids, *Trichuris* and *Capillaria.* Gland cells open through pores lateral to the esophagus in *Trichuris* and extend the length of the body in *Capillaria.* Dendritic processes of adjacent nerve cells project into the gland cells.[13] The function of the bacillary bands is unknown.

The hypodermal cords of enoplian nematodes bear structures that apparently function as proprioceptors.[53] These structures were named **metanemes** by Lorenzen.[72]

Musculature

The somatic musculature is technically a part of the body wall, but it is convenient to consider its function along with that of the pseudocoel. Indeed, the somatic musculature and pseudocoels, along with the cuticle, function together as a **hydrostatic skeleton.**

Nematode muscles commonly have a contractile portion and a noncontractile "cell body" or **myocyton.** The **platymyarian** muscle cell is rather ovoid in cross section; contains its contractile fibrils at one side, adjacent to the hypodermis; and has a myocyton of about the same width bulging into the pseudocoel (Fig. 22.8*b*). The myocyton contains the nuclei, large mitochondria with numerous cristae, ribosomes, endoplasmic reticulum, glycogen, and lipid. The **coelomyarian** cell is more spindle shaped, with the contractile portion at the distal end in the shape of a narrow U (Fig. 22.9). The distal end of the U is placed against the hypodermis, the contractile fibrils extend up along its sides, and the space in the middle is tightly packed with mitochondria. In some cases the elongated contractile portion does not sandwich the mitochondria, but these organelles are concentrated in the distal portion of the myocyton close to the contractile fibrils.[129] The myocyton bulges medially into the pseudocoel. It contains the nucleus, some mitochondria, endoplasmic reticulum, a

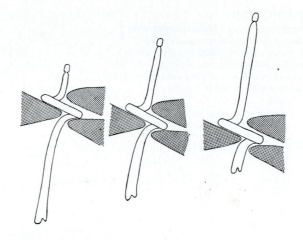

FIGURE 22.7

Locomotion of an adult *Nippostrongylus* removed from the intestine of the host and placed among moist sand grains. It is probable that similar movements are performed by the nematode among the villi of the host intestine.

From D. L. Lee, *The Physiology of Nematodes.* Copyright © 1965 Oliver & Boyd Ltd., Essex, UK. Reprinted with permission of the publisher.

Golgi body, and a large amount of glycogen. An important function of the cyton is as a glycogen storage depot. In the **circomyarian** cell the contractile fibrils at the periphery entirely encircle the myocyton.

The myofilaments seem to be essentially similar in all muscle types and are of two sizes. Contraction apparently occurs in a manner similar to the Hanson-Huxley model for vertebrate striated muscle, with thick filaments containing myosin and thin filaments of actin. The actin filaments slide past the myosin filaments in contraction. The A, H, and I band typical of striated muscle can be distinguished, but Z lines are absent. Thus, the structure of nematode muscle is similar to vertebrate striated muscle and insect flight muscle, except that rows of myofilaments are offset, a condition referred to as "obliquely striated"[92–94] (Fig. 22.10).

Processes run from the myocytons to the nerves instead of from the nerve cytons to the muscles (Fig. 22.11). This pattern of innervation is also known in platyhelminths and some other invertebrate groups.[127] Nematode muscle cells have frequent muscle-muscle connectives, at least in coelomyarian types.[130] These occur most often in the anterior regions of the worms and between the innervation processes of the muscle cells, although they may be between cytons. A higher degree of muscular coordination may result from transmission of nerve impulses between muscle cells that are so connected.

Pseudocoel and Hydrostatic Skeleton

The somatic musculature and the rest of the body wall enclose a fluid-filled cavity, the **pseudocoelom,** or **pseudocoel** (Fig. 22.12). A pseudocoel is derived from a persistent

FIGURE 22.8

Diagrams depicting a typical platymyarian muscle type at different magnifications. (a) Whole transverse section; (b) portion of muscle cells on either side of dorsal nerve cord; (c) two types of muscle filaments as seen at high resolution with the aid of the electron microscope.

From A. F. Bird, *The Structure of Nematodes.* Copyright © 1971 Academic Press, Inc., New York, NY. Reprinted with permission of the publisher and author.

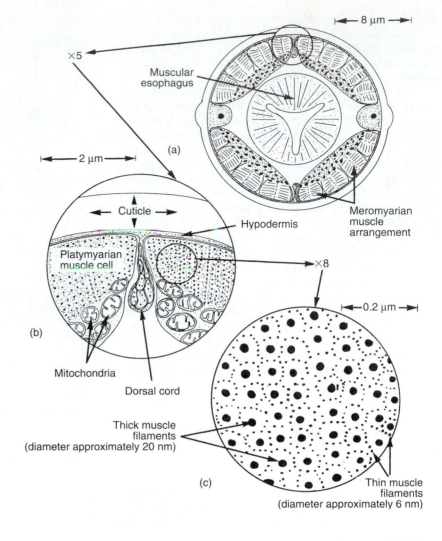

embryonic blastocoel, rather than being a cavity within the endomesoderm; thus, a pseudocoel differs from a true coelom in that it has no peritoneal (mesodermal) lining. The nematode pseudocoel functions as a hydrostatic skeleton. Hydrostatic skeletons are widespread in invertebrates. Their function depends on the enclosure of a volume of noncompressible fluid, the ability of muscle contraction to apply pressure to that fluid, and the transmission of the pressure in all directions in the fluid as the result of its incompressibility. Thus, in a simple case, simultaneous contraction of circular muscles and relaxation of longitudinal muscles cause an animal to become thinner and longer, whereas relaxation of circular muscles and a contraction of longitudinal muscles make an animal shorter and thicker. However, in nematodes the somatic musculature is entirely composed of longitudinal fibers, and the muscles act not against other antagonistic muscles but against stretching and compression of the cuticle.[50]

The mechanism of body movement can be summarized as follows: As the muscles on the ventral or dorsal side of the body contract, they compress the cuticle on that side, and the force of the contraction is transmitted (by the fluid in the pseudocoel) to the other side of the nematode, stretching the cuticle on the opposite side. The compression

and stretching of the cuticle serve to antagonize the muscle and are the forces that return the body to resting position when the muscles relax. The alternation of contraction and relaxation in the dorsal and ventral muscles impels the body into a series of curves in a single, dorsoventral plane, producing the characteristic motion seen in nematode movement.[113] An increase in efficiency of this system can only be achieved by an increase in hydrostatic pressure, and the hydrostatic pressure in the pseudocoel of nematodes is extraordinarily high. In *Ascaris* the pressure can average from 70 to 120 mm Hg and vary up to 210 mm Hg.[48,50] This is an order of magnitude higher than the pressure in the body fluids of animals with hydrostatic skeletons in other phyla. The limitations imposed by this high internal pressure determine many features of nematode morphology and physiology, such as how they eat, defecate, copulate, and lay eggs.

The pseudocoelomic fluid is known as **hemolymph.** In *Ascaris,* and probably in other nematode parasites of animals, it is a clear, pink, almost cell-free, complex solution. Aside from its structural significance, it almost certainly is important in transport of solutes from one tissue to another. These solutes include a variety of electrolytes, proteins, fats, and carbohydrates. Curiously, the fluid has far less chloride

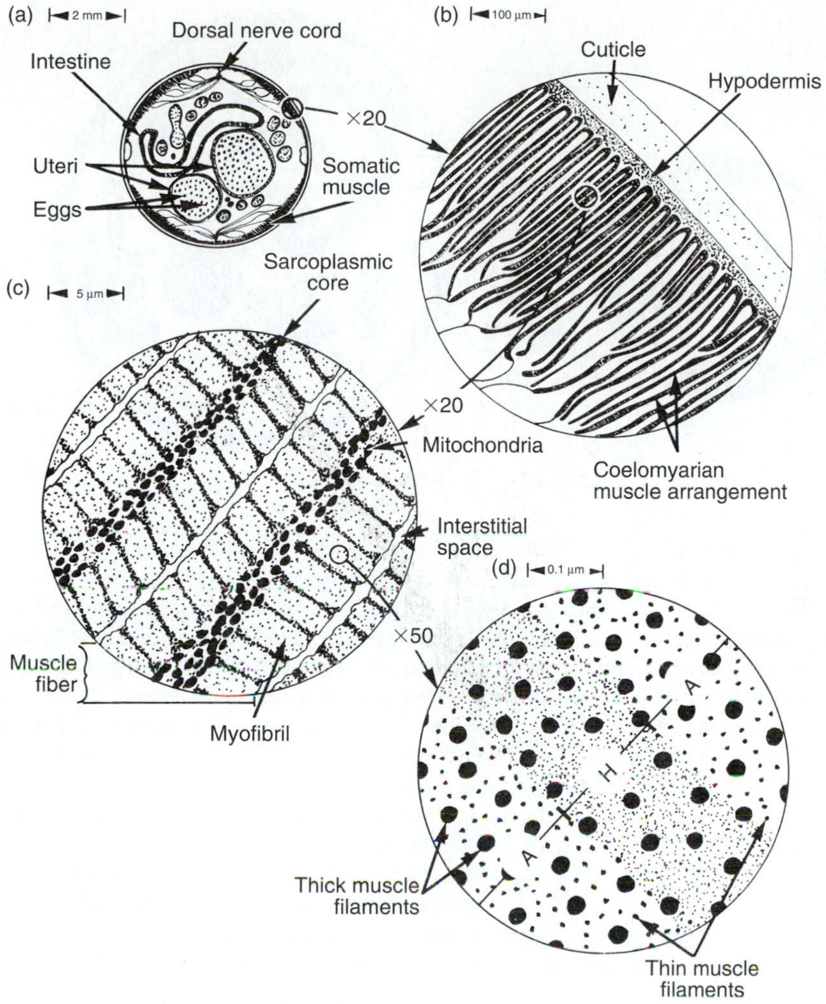

(a) ⊢◄ 2 mm ►⊣
Dorsal nerve cord
Intestine
×20
Uteri
Eggs
Somatic muscle

(b) ⊢◄ 100 μm ►⊣
Cuticle
Hypodermis
×20
Coelomyarian muscle arrangement

Sarcoplasmic core

(c) ⊢◄ 5 μm ►⊣
Mitochondria
Interstitial space
Muscle fiber
Myofibril

(d) ⊢◄ 0.1 μm ►⊣
×50
A
H
A
Thick muscle filaments
Thin muscle filaments

FIGURE 22.9

Diagrams depicting the typical coelomyarian muscle type of *Ascaris lumbricoides* over a wide range of magnifications. (*a*) Whole transverse section; (*b*) part of the muscle quadrant between the dorsal nerve and lateral hypodermal cord; (*c*) fibers of two muscle cells; (*d*) an **H** and two **A** bands and the two types of muscle filaments at high resolution.

From A. F. Bird, *The Structure of Nematodes.* Copyright © 1971 Academic Press, Inc., New York, NY. Reprinted with permission of the publisher and author.

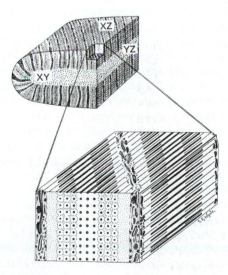

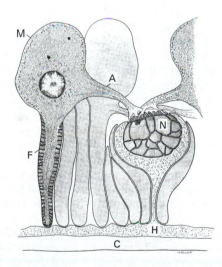

FIGURE 22.10

Diagram of a muscle fiber showing the pattern of striation in three planes. In the **XZ** plane, the myofilaments are staggered with the result that the striations are oblique rather than transverse. A second consequence of the stagger is that the adjacent rows of myofilaments do not reach the **XY** plane in phase, resulting in the appearance of striation in this plane also. The **YZ** plane shows cross-striation.

From J. Rosenbluth, "Ultrastructural organization of obliquely striated muscle fibers in *Ascaris lumbricoides.*" in *J. Cell Biol.* 25:495–515. Copyright © 1965 The Rockefeller University Press, New York, NY. Reprinted with permission.

FIGURE 22.11

Diagram of muscle cells and myoneural junctions in transverse section. The myocyton (**M**), containing the nucleus of the muscle cell, is continuous with the core of the striated fiber (**F**) and with the elongate arm (**A**). The arm subdivides as it approaches the nerve cord (**N**). The individual axons comprising the nerve cord are embedded in a troughlike extension of the hypodermis (**H**), which underlies the animal's cuticle (**C**).

From J. Rosenbluth, "Ultrastructure of somatic cells in *Ascaris lumbricoides.* II. Intermuscular junctions, neuromuscular junctions, and glycogen stores," in *J. Cell Biol.* 26:579–591. Copyright © 1965. The Rockefeller University Press, New York, NY. Reprinted with permission.

FIGURE 22.12

Cross sections of male and female *Ascaris suum.* Space between organs is pseudocoel. (*a*) Male; (*b*) female.

Courtesy of Warren Buss.

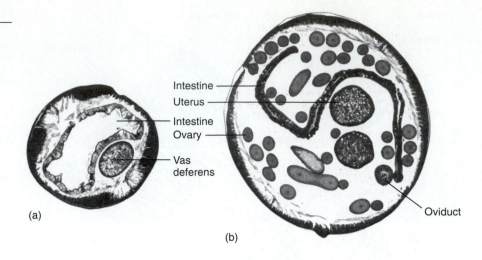

Intestine
Uterus
Intestine
Ovary
Vas deferens

Oviduct

(a)

(b)

than would be required to balance the cations present, and the anion deficiency is made up mostly of volatile and non-volatile organic acids.[37]

A peculiar and unique cell type found in the pseudocoel is the **coelomocyte.** Usually two, four, or six such cells, ovoid or with many branches, lie in the pseudocoel, attached to surrounding tissues. Although often small, in some species they are enormous; in *Ascaris* the coelomocytes are 5 mm by 3 mm by up to 1 mm thick. Their function is still obscure, although they may have a role in the accumulation and storage of vitamin B_{12} and in protein synthetic and secretory function.[17]

Nervous System

• Morphology

The nervous system of nematodes is relatively simple. There are two main concentrations of nerve elements in nematodes, one in the esophageal region and one in the anal area, connected by longitudinal nerve trunks. The most prominent feature of the anterior concentration is the **nerve ring,** or **circumesophageal commissure.** In *Ascaris* the nerve ring comprises eight cells, four of which are nerve cells and four of which are supporting, or **glial,** cells. The ring lies close to the outer wall of the esophagus and is fairly easily seen in most species. Because its location is constant within a species, it is a good taxonomic character. The ring serves as a commissure for the **ventral, lateral,** and **dorsal** cephalic **ganglia** (Fig. 22.13*a*), which are usually paired. The ventral ganglia are largest; the dorsal ganglia are smallest. Emanating from each ganglion posteriorly are the **longitudinal nerve trunks,** which become embedded in the hypodermal cords, and again the ventral nerve is largest. Proceeding anteriorly from the lateral ganglia are two **amphidial nerves.** Six **papillary nerves,** which are derived directly from the nerve ring, innervate the cephalic sensory papillae surrounding the mouth.

The ventral nerve trunk runs posteriorly as a chain of ganglia, the last of which is the **preanal ganglion.** The preanal ganglion gives rise to two branches that proceed dorsally into the pseudocoel to encircle the rectum, thus forming the **rectal commissure,** or **posterior nerve ring.** Other posterior nerves and ganglia are depicted in Figure 22.13*b*. The peripheral nervous system consists of a latticework of nerves that interconnect with fine commissures and supply nerves to sensory endings within the cuticle.

The main sense organs are cephalic and caudal papillae, amphids, phasmids, and in certain free-living species, ocelli. The pattern of sensory papillae on the head of a nematode is a very important taxonomic character. The primitive pattern of lips surrounding the mouth of the ancestral nematodes probably was two lateral, two dorsolateral, and two ventrolateral, each of which was supplied with sensory papillae (Fig. 22.14). In addition to the papillae forming the **inner** and **outer labial circles,** there were four **cephalic papillae,** one located behind the lips in each of the dorsolateral and ventrolateral quadrants. Most parasitic nematodes are modified from this basic form. Labial papillae are often lost or fused together, and cephalic papillae usually are very small. However, some papillae are found on all species, and careful study will reveal all 16 nerve endings on most species, even those that have lost all semblances of lips. The pattern of lips and papillae on nematodes is studied by slicing the anterior tip from the worm with a sharp blade and orienting the end for an en face view on a microscope slide. The sensory endings of the papillae are modified cilia.[13] The papillae are probably tactile receptors.

The **amphids** are a pair of somewhat more complex sensory organs that open on each side of the head at about the same level as the cephalic circle of papillae. They are most conspicuous in marine, free-living forms and usually are reduced in animal parasites. The amphidial opening, which ordinarily is at the tip of a papilla, leads into a deep,

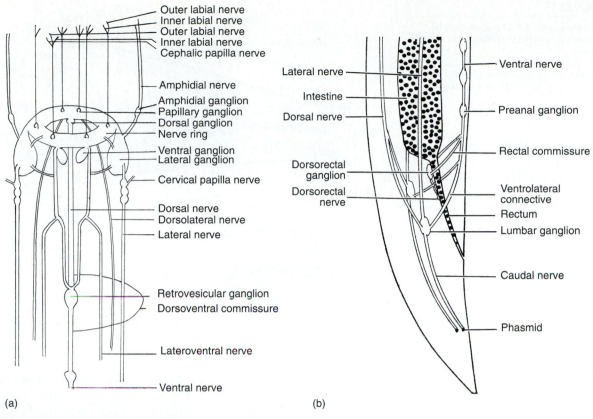

FIGURE 22.13

Diagrammatic representation of the nervous system of a nematode. (*a*) Anterior end; (*b*) posterior end.

From H. D. Crofton, *Nematodes.* Copyright © 1966 Hutchinson University Library, London. Reprinted with permission of the publisher.

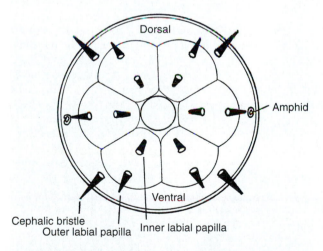

FIGURE 22.14

Diagram of the anterior end of a hypothetical primitive nematode to show the arrangement of the sense organs.

Source: L. A. P. de Coninck, "Les relations de symetrie, regissant la distribution des organes sensibles anterieurs chez les nematodes," in *Ann. Soc. Roy. Zool. Belgique* 81:25–31. Copyright © 1950.

cuticular pit, at the base of which is a nerve bulb with several nerve processes (Fig. 22.15). The sensory endings are modified cilia, up to 23 in one amphid, in contrast to the one to three per papilla. Until modified cilia were discovered in the sense organs of nematodes, it was thought that these worms had no cilia. Of course their structure is rather different from ordinary motile cilia. They have no kinetosomes, and the microtubules usually diverge from the normal 9 + 2 pattern (p. 38), for example, to 9 + 4, 8 + 4, or 1 + 11 + 4. The amphids are considered chemoreceptors[68,114] but may have a secretory function in some species;[111] extracts of hookworm amphids inhibit clotting of vertebrate blood.[116]

Most parasitic nematodes have a pair of cuticular papillae, the **deirids,** or **cervical papillae,** at about the level of the nerve ring, and other sensory papillae are at different levels along the body of many species. **Caudal papillae** (Fig. 22.16) are more elaborate in males, aiding in copulation. The pattern of distribution is an important taxonomic character. These papillae reach maximal development in the order Strongylida, where they help to form a complex copulatory bursa (Chapter 25).

FIGURE 22.15

Scale diagram of part of the tip of the head of
Meloidogyne sp. cut open to reveal one of the
two amphids.

From A. F. Bird, *The Structure of Nematodes.*
Copyright © 1971 Academic Press, Inc., New
York, NY. Reprinted with permission of publisher
and author.

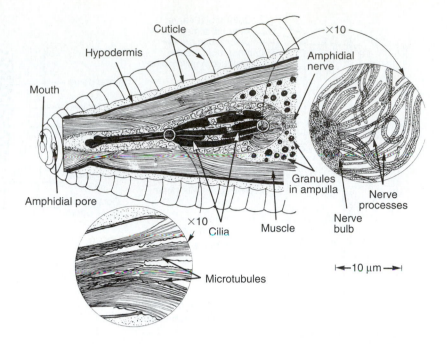

Near the posterior end of many nematodes (subclass
Rhabditia) is a bilateral pair of cuticle-lined organs, the
phasmids (Fig. 22.17). The phasmids are similar in structure
to the amphids except that they have fewer neural endings
and the gland, if present, is smaller.[68] The presence or absence of phasmids has been the criterion to separate the two
classes of nematodes: Adenophorea (= Aphasmidea, without
phasmids) and Secernentea (= Phasmidea, with phasmids).
However, more recent analyses suggest that the class Aphasmidea was paraphyletic (p. 378). Although difficult to see in
some species, in most the phasmids are easily recognized by
their cuticle-lined ducts (Fig. 22.18) that open at the apices
of papillae near the tip of the tail.

• Neurotransmission

The predominant excitatory neurotransmitter in nematodes is
acetylcholine.[68] The muscle cell undergoes spontaneous depolarization in the innervation arm and then generation of action potentials in a repeated or oscillatory manner, somewhat
similar to vertebrate cardiac muscle, in which there is a spontaneous, rhythmic spike production. The rate of firing increases with lowered resting potential and decreases with
higher resting potential. The nerve fibers play primarily a
modulating role, with both excitatory and inhibitory fibers.
Stimulation of excitatory fibers releases acetylcholine at the
neuromuscular junction, depolarizes the muscle membrane,
and increases the rate of spikes (Fig. 22.19). The inhibitory
fibers release gamma-aminobutyric acid (GABA), hyperpolarize the muscle, and decrease the rate of action potentials.

As in the platyhelminths, numerous neuropeptides have
been discovered in nematodes,[20,21,112] and similarly their
functions remain mostly a mystery. One of them hyperpolarizes (and thus relaxes) body wall muscle of *Ascaris* by some
mechanism that does not involve stimulation of GABA

FIGURE 22.16

Ventral view of female *Toxascaris* sp., showing caudal papillae.
Courtesy of Jay Georgi.

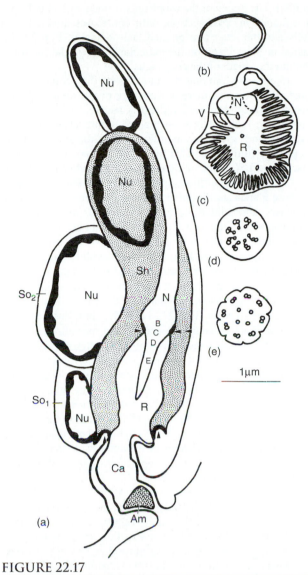

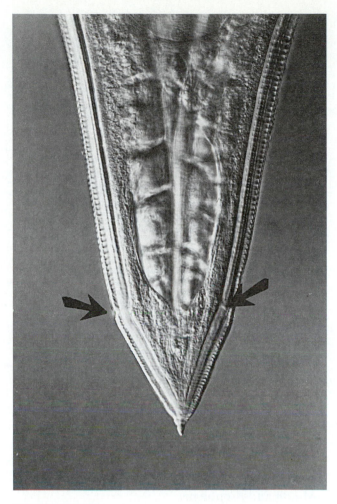

FIGURE 22.18

Ventral view of female *Toxascaris* sp., showing ducts (*arrows*) leading to phasmids.

Courtesy of Jay Georgi.

FIGURE 22.17

Reconstruction of a phasmid in adult male *Meloidodera floridensis*. (*a*) Phasmid from dorsoventral view. Sensory ending is a modified cilium extending into a receptor cavity (**R**). The cavity opens to the outside at the ampulla (**Am**), which is often plugged by an electron-dense material. The neuron (**N**) is surrounded by the cytoplasm of a sheath cell (**Sh**), which in turn is partially or completely surrounded by socket cells (**So**). **Ca,** canal; **Nu,** nucleus; **arrows,** intercellular junctions. (*b–e*) Cross sections at levels indicated by small letters inside neuron in (*a*). **V,** vesicle.

From L. K. Carta and J. G. Baldwin, "Ultrastructure of plasmid development in *Meloidodera floridensis* and *M. charis* (Heteroderinae), in *J. Nematol.* 22:362–385. Copyright © 1990. Reprinted with permission of publisher and author.

release from inhibitory presynaptic terminals.[44] Neuropeptides, along with serotonin and acetylcholine, may function in control and modulation of feeding activities by their effects on muscles of the esophagus.[22]

• Effects of Drugs

The foremost reason that neurobiology of nematodes has elicited so much research interest is that several nematicidal drugs interfere with neural function. For example, piperazine

hyperpolarizes the muscle membrane, effectively paralyzing the worms, so that they pass out of the host.[46] Levamisole and pyrantel mimic the effects of acetylcholine, depolarizing the muscle membrane, resulting in paralysis.

Earlier results suggested that the action of ivermectin was due to stimulation of GABA release and enhancement of its binding to postsynaptic receptors.[28,119] However, it now seems that the action of ivermectin is due to inhibition of esophageal pumping.

The benzimidazoles (mebendazole, parbendazole, fenbendazole) appear to have two modes of action.[97] They inhibit mitochondrial electron transport, especially the fumarate reductase system (see p. 316), thus inhibiting energy metabolism; and they also bind with tubulin, which interferes with microtubule-dependent processes such as acetylcholinesterase secretion, and paralyze the worms.

Digestive System and Acquisition of Nutrients

The digestive system is complete in most nematodes, with mouth, gut, and anus, although in mermithids and a few filariids the anus is atrophied. Cuticle lines the stomodeum

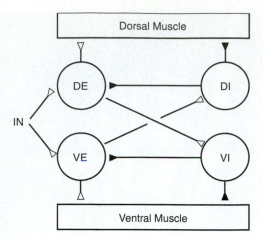

FIGURE 22.19

Diagram illustrating the synaptic relationships of excitatory and inhibitory motor neurons in nematodes. Open triangles are excitatory, and closed triangles are inhibitory synapses. Stimulation of dorsal excitatory motor neuron (**DE**) by interneuron (**IN**) causes dorsal muscle contraction and stimulation of ventral inhibitory motor neuron (**VI**). **VI** inhibits ventral excitatory motor neuron (**VE**) and hyperpolarizes the ventral muscle, thus relaxing it. Stimulation of the **VE** neuron has the opposite effect, causing ventral muscle to contract and dorsal muscle to relax. (Direct recordings have not been made from **VE** neurons, but their action is inferred from recordings made from **DE** neurons.)

Source: A. Stretton et al., "Motor behavior and motor nervous system function in the nematode *Ascaris suum*," in *J. Parasitol.* 78:206–214. Copyright © 1992.

(buccal cavity and esophagus) and proctodeum (rectum), and nematodes shed the cuticular lining of these cavities along with the exterior cuticle when they molt.

• Mouth and Esophagus

The mouth is usually a circular opening surrounded by a maximum of six lips. Few parasitic nematodes possess as many as six lips; in some they fuse in pairs to form three. In many species the lips are absent altogether, whereas in others two lateral lips develop as new structures derived from the inner margin of the mouth. Regardless of the morphology of a given species, it is a variation of the primitive, six-lipped form.

A buccal cavity lies between the mouth and esophagus of most nematodes. The size and shape of this area vary among species and are important taxonomic characters. In some species the cuticular lining is quite thick, forming a rigid structure known as a **buccal capsule;** in others the lining is thin. The cavity may be elongated, reduced, or absent altogether, with a mouth that opens almost immediately into the lumen of the esophagus. Buccal armament is often present in parasitic and predaceous nematodes. The elements arise from the cavity wall or as anterior projections of the esophagus. Some nematodes have such teeth of both types.

Food ingested by a nematode moves into a muscular region of the digestive tract known as the esophagus or pharynx. This is a pumping organ that sucks food into the alimentary canal and forces it into the intestine. This arrangement is necessary because of the high pressure in the surrounding pseudocoel. The esophagus assumes a variety of shapes, depending on the order and species of nematode, and for this reason is an important taxonomic character. It is highly muscular and cylindrical and often has one or more enlargements

(**bulbs**). The lumen of the esophagus is lined with cuticle and is triradiate in cross section, with one radius directed ventrally and the other two pointed laterodorsally (Fig. 22.20). Radial muscles insert on the cuticular lining in the interradii and run the length of the esophagus.

Interspersed among the muscles of the esophagus are three glands (five or more in Enoplea), one in each of the interradial zones. The dorsal gland is usually more extensive than are the ventrolaterals. Each gland is uninucleate or multinucleate and opens independently into the lumen of the esophagus. In the Rhabditea the dorsal one commonly opens near the buccal cavity, and the ducts of the subventral glands open near the base of the esophagus. The secretions produced by these glands are digestive: amylase, proteases, pectinases, chitinases, and cellulases have been detected in them.[59,68] In hookworms the secretions have anticoagulant properties.[116] In some species the glands fuse together near the posterior end of the esophagus, and in some nematodes, such as the Spirurida, the posterior portion of the esophagus is mostly glandular. In some species the glands, especially the dorsal gland, are so extensive that much of their mass lies outside the esophagus proper (Fig. 22.21). In the class Enoplea there are five or more esophageal glands; in the specialized esophagus of the Trichurida and Mermithida the anterior portion is a thin-walled, muscular tube, whereas the posterior portion is a very thin tube surrounded by a column of single glandular cells, the **stichocytes,** the entire structure being referred to as the **stichosome.** The stichocytes communicate with the esophageal lumen by small ducts.[107] The stichosome may be homologous to esophageal glands of other nematodes, having been derived by multiplication of the number of glands.[1]

Rapid contraction of the buccal muscles and anterior esophageal muscles opens the mouth and dilates the anterior end of the esophagus, sucking in food (Fig. 22.22). The high hydrostatic pressure in the pseudocoel surrounding the esophagus closes the mouth and esophageal lumen when the muscles relax. The food passes down the esophagus by the posteriorly progressing wave of muscle contraction opening the lumen for it until the food reaches the intestine. The posterior bulb of many species seems to function as a one-way, nonregurgitation valve for food in the intestine. Thus, the mechanism is a kind of peristalsis in which the force moving the food is not the contraction of circular muscles but the closure of the esophageal lumen by hydrostatic pressure behind the food. The frequency of pumping has been recorded as 2 to 24 per second.[33]

In a few ascaroids (*Contracaecum, Multicaecum, Polycaecum,* and others) one to five posteriorly directed esophageal ceca originate from a short, glandular **ventriculus** between the body of the esophagus and the intestine (Fig. 22.23).

• Intestine

The intestine is a simple, tubelike structure, extending from the esophagus to the proctodeum, and is constructed of a single layer of intestinal cells.

In females a short terminal, cuticle-lined rectum runs between the anus and intestine. In males the rectum receives the products of the reproductive system into its terminal portion and, therefore, is a cloaca. The dorsal wall of the cloaca is usually invaginated into two pouches, the spicule sheaths, that contain the copulatory spicules, to be described along with the

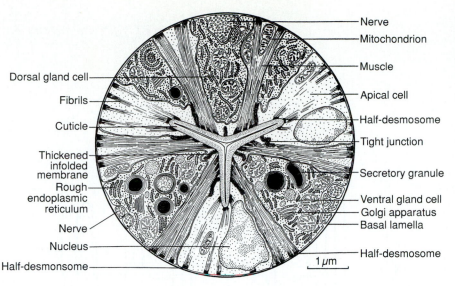

- Nerve
- Mitochondrion
- Muscle
- Apical cell
- Half-desmosome
- Tight junction
- Secretory granule
- Ventral gland cell
- Golgi apparatus
- Basal lamella
- Half-desmosome

Dorsal gland cell
Fibrils
Cuticle
Thickened infolded membrane
Rough endoplasmic reticulum
Nerve
Nucleus
Half-desmonsome

1 µm

FIGURE 22.20

Diagram of a transverse section through the posterior part of the esophagus of *Nippostrongylus brasiliensis* to show the arrangement of various cells, cell membranes, and cellular organelles. Reconstructed from several electron micrographs.

From D. L. Lee, "The ultrastructure of the alimentary tract of the skin-penetrating larva of *Nippostrongylus brasiliensis* (Nematoda)," in *J. Zool., Lond.* 154:9–18. Copyright © 1968. Reprinted with permission of the Oxford University Press, Oxford, UK.

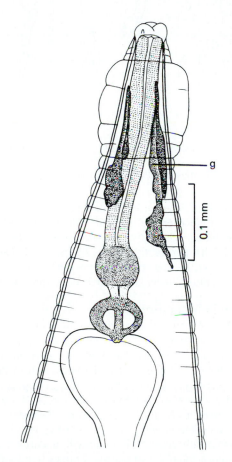

g

0.1 mm

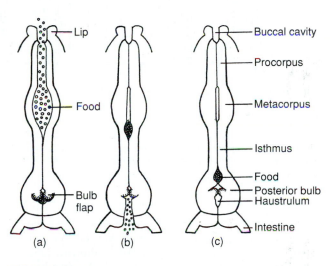

- Lip
- Food
- Bulb flap

- Buccal cavity
- Procorpus
- Metacorpus
- Isthmus
- Food
- Posterior bulb
- Haustrulum
- Intestine

(a)   (b)   (c)

FIGURE 22.22

Diagrams to show the structure and function of the *Rhabditis* type of esophagus during feeding. Food particles, small enough to pass through the buccal cavity, are drawn into the lumen of the metacorpus by sudden dilation of the procorpus and metacorpus (*a*). Closure of the lumen of the esophagus in these regions expels excess water (*b*), and the mass of food particles is passed backward along the isthmus (*b, c*). Food is drawn between the bulb flaps of the posterior bulb by dilation of the haustrulum, which inverts the bulb flaps (*a*) and is passed to the intestine by closure of haustrulum and by dilation, followed by closure of the pharyngeal-intestinal valve (*b*). The bulb flaps contribute to the closure of the valve in the posterior bulb and, when they invert (*a*), also crush food particles.

From D. L. Lee, *The Physiology of Nematodes.* Copyright © 1965 Oliver & Boyd Ltd., Essex, UK. Reprinted with permission of the publisher.

FIGURE 22.21

Syphacia, a rodent pinworm with enlarged esophageal glands (**g**).

From G. D. Schmidt and R. E. Kuntz, "Nematode parasites of Oceanica. IV. Oxyurids of mammals of Palawan, P.I., with descriptions of four new species of *Syphacia,*" in *Parasitology* 58:845–854. Copyright © 1968. Reprinted with permission of the publisher.

reproductive system. The vas deferens opens into the ventral wall of the cloaca.

The intestine is nonmuscular. Its contents are forced posteriorly by the action of the esophagus as it adds more food to the front end of the system and perhaps by locomotor activity of the worm. Internal pressure in the pseudocoel flattens the intestine when empty. Between the dorsal wall of the cloaca and the body wall is a powerful muscle bundle called the

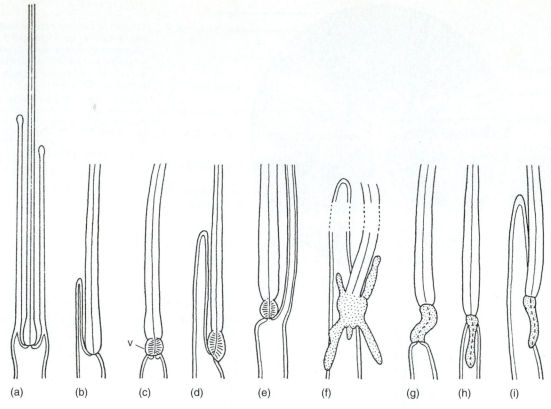

FIGURE 22.23

Variations in esophagi in some ascaroid nematodes: (*a*) *Crossophorus;* (*b*) *Angusticaecum;* (*c*) *Toxocara; v,* ventriculus; (*d*) *Porrocaecum;* (*e*) *Paradujardinia;* (*f*) *Multicaecum;* (*g*) *Anisakis;* (*h*) *Raphidascaris;* (*i*) *Contracaecum.*

Source: G. Hartwich, *CIH Keys to the Nematode Parasites of Vertebrates, no. 2,* 1974; Commonwealth Agricultural Bureaux, Farnham Royal, Bucks, England.

depressor ani. This is a misnomer because, when it contracts, the anus is opened; it is therefore a dilator rather than a depressor. Hydrostatic pressure surrounding the intestine causes defecation when the anus is opened. The hydrostatic pressure expels the feces with some force: When lifted from the saline solution maintaining it, *Ascaris* can project its feces 60 cm.[33]

The wall of the intestine consists of tall, simple columnar cells with prominent brush borders of microvilli (Fig. 22.24).[108] Although several digestive enzymes have been identified in the intestinal lumen, intestinal digestion is probably of minor importance in most forms because of the rapid rate of food movement through the intestine.

The number of intestinal cells varies from about 30 in some free-living species to more than a million in the larger parasitic forms. These cells rest on a basement membrane, which is attached to random extensions of the body wall musculature. It is probable that the intestine serves as the primary means of excretion of nitrogenous waste products, in addition to its function in nutrient absorption. Crofton[33] states that the intestine of *Ascaris lumbricoides* is emptied by defecation every three minutes under experimental conditions. Such a rapid turnover of materials must surely limit the amount of enzymatic action possible in the intestinal lumen but would favor the excretion of water-soluble waste products.

• Food

The food of nematodes parasitic in animals is blood, tissue cells and fluids, intestinal contents, or some combination of these. Some species parasitic in the intestine feed only on tissue and not blood or host ingesta.[6] Hookworms, which feed solely on blood, accumulate granules of zinc sulfide in their intestinal cells, apparently as a waste product.[47] Nematodes feed extravagantly and wastefully, and the thin-walled intestine with its brush border is an efficient absorptive mechanism.

Feeding in Mermithids

Members of the family Mermithidae are unusual in that the adults are free living, but the juveniles are parasitic in invertebrates, primarily insects. The adults do not feed, and at no stage is there a functional gut. The body wall in the adults and the first-stage juveniles has a structure typical of other nematodes, described previously, but the body wall in the parasitic juveniles is greatly modified for absorption of nutrients.[9] The cuticle is very thin, and the hypodermis is thick and metabolically active, with microvilli underlying the cuticle. It is connected by cytoplasmic bridges to a food storage organ, the **trophosome.**[10] During the sometimes long, nonfeeding adult life, the worm apparently lives on nutrients stored in the trophosome. Because mermithids almost always kill their host, they have potential as biological control agents of insect pests; *Romanomermis culicivorax,* for example, is a parasite of mosquitoes (p. 574).[83,86]

Secretory-Excretory System

So-called excretory systems have been observed in all nematodes except the Trichurida and Dioctophymatida,[30] but

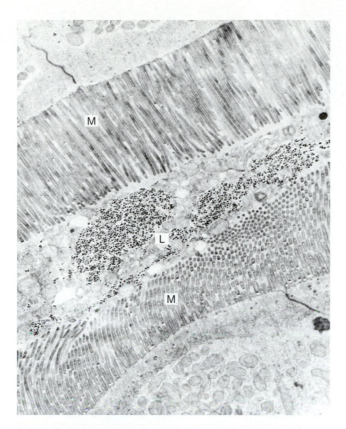

FIGURE 22.24

Cross section of intestine showing microvilli (**M**) of dorsal and ventral sides. Cellular debris fills the lumen (**L**). ($\times$ 10,800.)

From H. G. Sheffield, "Electron microscope studies on the intestinal epithelium of *Ascaris suum*," in *J. Parasitol.* 50:365–379. Copyright © 1964.

an excretory function was assigned to the systems in various nematodes solely on a morphological basis; that is, the systems looked like excretory systems. However, there is little evidence that the "excretory systems" are involved in the elimination of waste; strong evidence exists that most excretion occurs through the intestine.[98] The actual functions of these systems varies according to the species of nematode and its stage of development, but both osmoregulatory and secretory functions have been described. Bird and Bird[13] suggest the term **secretory-excretory system (S-E system),** which we will adopt here.

The presence of an S-E system is apparently primitive and probably evolved first in marine forms. There are no flame cells or nephridia; in fact the nematode S-E system seems to be a neoformation of the Nematoda.[1] The two basic types are **glandular** and **tubular.** The glandular type is typical of free-living groups and may be involved in secretion of enzymes, proteins, or mucoproteins. Adamson[1] regards the glandular type as plesiomorphic. We will confine our discussion to the tubular type, which is characteristic of the Rhabditia (= Phasmidia).

Several varieties of tubular excretory systems occur (Fig. 22.25). Basically two long canals in the lateral hypodermis connect to each other by a transverse canal near the anterior end. This transverse canal opens to the exterior by means of a median, ventral duct and pore, the excretory pore. This pore is conspicuous in most species; its location is fairly constant within a species and, therefore, is a useful taxonomic character.

• Osmoregulation

The ability to osmoregulate varies greatly among nematodes and corresponds generally with the requirements of their habitats. The body fluids of species parasitic in animals may be somewhat different in osmotic pressure from the tissues they inhabit but not dramatically so. For example, *Ascaris* hemolymph is about 80% to 90% of the osmotic pressure of pig intestinal contents.[49] *Ascaris* clearly can control its electrolyte concentrations to some degree: Chloride ion concentration of host intestinal contents varies between 34 and 102 mM, but *Ascaris* hemolymph is fairly constant at around 52 mM. Adults of most parasitic species cannot tolerate media much different in osmotic pressure from their hemolymph; when placed in tap water, they will burst, sometimes within minutes, from addition of the imbibed water to the already high internal pressure. Of course freshwater and terrestrial nematodes, including juveniles of many parasitic species, must withstand (and regulate their internal osmotic constitution in) extremely hypotonic conditions.

Details of water and ion excretion are poorly known. Contractions of S-E canals and the ampulla near the pore have been observed in several species. Contractions of the ampulla in free-living, third-stage juveniles of *Ancylostoma* and *Nippostrongylus* are inversely proportional to the salt concentration of the solution in which they are maintained. Some evidence suggests that osmoregulation by the ampulla is part of a homeostatic mechanism to maintain constant volume so that locomotor activity is not impaired. Nelson and Riddle[78] ablated different portions of the S-E system of the free-living nematode *Caenorhabditis elegans* with a laser microbeam. Destruction of the pore cell, duct cell, or excretory cell led to accumulation of water and death of the worm, but ablation of the gland cell had no effect.

The ultrastructure of the S-E system strongly suggests that the system functions in osmoregulation and perhaps in excretion of waste products and in secretion as well.[71] The surface area of the peripheral cell membrane may be greatly increased by numerous bulbular invaginations, and on the interior the lumen is perforated by drainage ductules, or canaliculi (Fig. 22.26). Filaments that are presumably contractile may surround the lumen of the duct. The hydrostatic pressure in the pseudocoel probably provides filtration pressure to excrete substances through the canals embedded in the hypodermal cords.

• Secretion

The ultrastructure of the gland cells clearly suggests secretory function. Enzymes responsible for exsheathment (shedding the old cuticle at ecdysis) are produced there by various strongyle juveniles. A variety of nematodes excrete substances antigenic for their hosts through the S-E pores. Lee[65,66] suggested that digestive enzymes were secreted by adult *Nippostrongylus* to act in conjunction with the abrading action of the cuticle.

• Excretion

The major nitrogenous waste product of nematodes is ammonia. In normal saline, *Ascaris* excretes 69% of the total nitrogen excreted as ammonia and 7% as urea. Under conditions of osmotic stress, these proportions can be changed to 27%

FIGURE 22.25

Excretory systems. (*a*) Single renette in a dorylaimid; (*b*) two-celled renette in *Rhabdias;* (*c*) larval *Ancylostoma;* (*d*) rhabditoid type; (*e*) oxyuroid type; (*f*) *Ascaris;* (*g*) *Anisakis;* (*h*) *Cephalobus;* (*i*) *Tylenchus.*

From H. D. Crofton, *Nematodes.* Copyright © 1966 Hutchinson University Library, London. Reprinted with permission of the publisher.

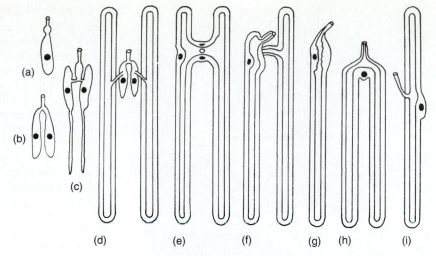

ammonia and 52% urea. Amino acids, peptides, and amines may be excreted by nematodes. Other excretory products include carbon dioxide and a variety of fatty acids. The fatty acids are end products of energy metabolism (p. 376). The role of the S-E system in the elimination of the foregoing substances is not well-established. Juvenile *Nippostrongylus* excrete several primary aliphatic amines through their S-E pore. A large proportion of nitrogenous waste products can be excreted via the intestine and anus by *Ascaris,*[98] and it would seem that the cuticle must play a major role in ammonia excretion in most nematodes.

Reproduction

Most nematodes are dioecious, although a few monoecious species are known. Parthenogenesis also exists in some. Sexual dimorphism usually attends dioecious forms, with females growing larger than the males. Furthermore, males have a more coiled tail and often have associated external features, such as bursae, alae, and papillae. Such dimorphism achieves the ultimate in the Tetrameridae and the plant-parasitic Heteroderidae, where the males have typical nematode anatomy, but the females are little more than swollen bags of uteri.

The gonads of nematodes are solid cords of cells that are continuous with the ducts that lead to the external environment. This allows the reproductive systems to function despite the high pressure of the surrounding pseudocoelom. Were it not continuous, an oocyte would not be able to gain access to the oviduct, which would be squeezed closed by the surrounding pressure. When germ cells proliferate only at the inner end of the gonad, the gonad is described as **telogonic;** but if germ cells proliferate throughout the length of the gonad (as in Trichuroidea), it is **hologonic.**

• Male Reproductive System

Testes and Ducts. Testes are generally paired in the Enoplea and some Rhabditea, and the paired condition is probably plesiomorphic. However, in the Rhabditia there is but a single testis. This organ may be relatively short and uncoiled, but in the larger animal parasites it appears as a long, threadlike structure that is coiled around the intestine and itself at various

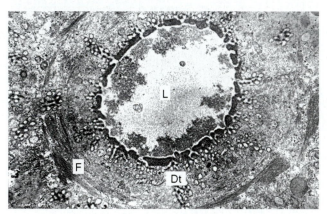

FIGURE 22.26

Transverse section through main excretory canal of *Anisakis,* showing round canal with interrupted dense material lining the lumen (**L**), ramifying drainage tubules (**Dt**), and congregated vesicles surrounding main canal and drainage tubules. Filaments (**F**) appear in circular arrangement around main canal. (× 10,000.)

From H.-F. Lee et al., "Ultrastructure of the excretory system of *Anisakis* larva (Nematoda: Anisakidae)," in *J. Parasitol.* 59:289–298. Copyright © 1973.

levels of the body. There are usually two zones in telogonic testes: the **germinal zone,** incorporating the blind end and in which spermatogonial divisions take place, and the **growth zone.** The end of the growth zone merges with a more tubular structure, the **seminal vesicle,** which is a sperm storage organ. The seminal vesicle merges into the vas deferens, which is usually divided into an anterior, glandular region and a posterior, muscular region, the ejaculatory duct. The ejaculatory duct opens into the cloaca. Some species have a pair of cement glands near the ejaculatory duct that secrete a hard, brown material to plug the vulva after copulation.

Accessory Reproductive Organs. Nearly all nematodes have a pair of sclerotized, acellular, copulatory **spicules** (Fig. 22.27). They originate within dorsal outpocketings of the cloacal wall and are controlled by proximal muscles. Each spicule is surrounded by a fibrous spicule sheath. The spicule structure varies between species but is fairly constant among individuals

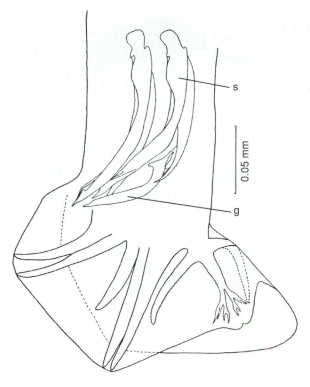

FIGURE 22.27

Bursate nematode *Molineus,* showing complex spicules (**s**) and a gubernaculum (**g**).

From G. D. Schmidt, "*Molineus mustelae* sp. n. (Nematoda: Trichostrongylidae) from the long-tailed weasel in Montana and *M. chabaudi* nom. n. with a key to the species of *Monlineus,*" in *J. Parasitol.* 51:164–168. Copyright © 1965. Reprinted with permission of the publisher.

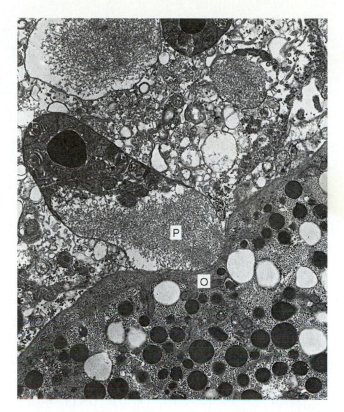

FIGURE 22.28

Angiostrongylus cantonensis sperm in contact with an oocyte (**O**) in the uterus of the female worm. Note large pseudopod (**P**) and continuity between the membrane specialization and the sperm plasma membrane. (× 14,500.)

From W. E. Foor, "Spermatozoan morphology and zygote formation in nematodes," *Biol. Reprod.*" 2 (suppl.): 177–202. Copyright © 1970.

within a species, making the size and morphology of the spicules two of the most important taxonomic characters. A dorsal sclerotization of the cloacal wall, the **gubernaculum,** occurs in many species. It guides the exsertion of the spicules from the cloaca at copulation. In several strongyloid genera an additional ventral sclerotization of the cloaca, the **telamon,** has the same general function as that of the gubernaculum. Both structures are important taxonomic characters. The spicules are inserted into the vulva at copulation. They are not true intromittent organs, since they do not conduct the sperm, but are another adaptation to cope with the high internal hydrostatic pressure. The spicules must hold the vulva open while the ejaculatory muscles overcome the hydrostatic pressure in the female and rapidly inject sperm into her reproductive tract.

Spermatozoa. Nematode spermatozoa are unusual among those studied in the animal kingdom in that they lack a flagellum and acrosome. Furthermore, internal organization of organelles differs markedly from that of all other sperm described in other groups. The sperm are rather diverse between species in cytological characteristics; Foor[42] recognized at least four types, ranging from small, rounded structures to ameboid cells with distinct anterior and posterior cytoplasm to elongated, tadpole-shaped structures, with "head" and "tail." In none of the types is the nucleus bounded by a

nuclear membrane. Some types undergo further morphological development after insemination of the female; for example, the "tadpole" becomes ameboid, and the sperm may not be able to fertilize an ovum until these changes have occurred.[110] Motility is not easy to discern, although clearly the gametes must travel up the female tract. The "tail" of the tadpole-shaped sperm apparently is nonmotile, but the "head" can put out pseudopodia.[67] There is good ultrastructural evidence for pseudopodial movement (Fig. 22.28). Activation of nematode sperm does not seem to be associated with increased levels of cyclic AMP, as it is with flagellated sperm.[105]

• Female Reproductive System

Most female nematodes have two ovaries, although some have from one to more than six. The general pattern of structure of the reproductive system in the female is similar to that in the male, except the gonopore is independent of the digestive system. The pattern is a linear series of structures, with the gonad at the internal or proximal end, followed by developmental, storage, and ejective areas. Female reproductive tracts of most nematodes are telogonic, but some are hologonic.

Ovaries and Oviducts. The ovaries are solid cords of cells that produce gametes and move them distally into the terminal portion of the system. The proximal end of a telogonic ovary

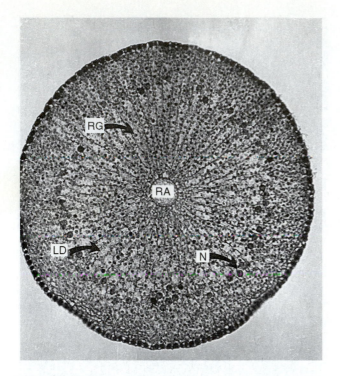

FIGURE 22.29

Ascaris lumbricoides: transverse section through growth zone of ovary. **LD,** lipid droplet; **N,** nucleus; **RA,** rachis; **RG,** refringent granule. (× 440.)

From W. E. Foor, "Ultrastructure aspects of oocyte development and shell formation in *Ascaris lumbricoides,*" in *J. Parasitol.* 53:1245–1261. Copyright © 1967.

is the **germinal zone,** which produces oogonia; the oogonia become oocytes and move into the **growth zone** of the ovary, toward the oviduct. In the large ascarids the oocytes are attached to a central supporting structure, the **rachis.** In *Ascaris* the germinal zone is very short, and most of the 200 to 250 cm length of the ovary comprises oocytes attached to the rachis in a radial manner by cytoplasmic bridges[41] (Fig. 22.29). The oocytes increase in size as they move down the rachis, and when about 3 to 5 cm from the oviduct they become detached from it. In some nematodes the rachis ends at the beginning of the growth zone, and the oocytes pass single file down the growth zone, increasing greatly in volume as the growth zone also increases.[74]

The proximal end of the oviduct in most nematodes is a distinct **spermatheca,** or sperm storage area. As the oocytes enter the oviduct (spermathecal area), sperm penetrate them; only then do the oocytes proceed with meiosis. A polar body is extruded at each of the two meiotic divisions. Concurrent with these events, shell formation occurs, which we will describe in the section on development.

Uterus and Vulva. The wall of the uterus has well-developed circular and diagonal muscle fibers, and these move the developing embryos ("eggs") distally by peristaltic action. The shape of the eggs may be molded by the uterus, and uterine secretory cells may contribute additional material to the eggshells. The distal end of the uterus is usually quite muscular and constitutes an **ovijector.** The ovijectors of the uteri fuse to form a short vagina that opens through a ventral, transverse slit in the body wall, the

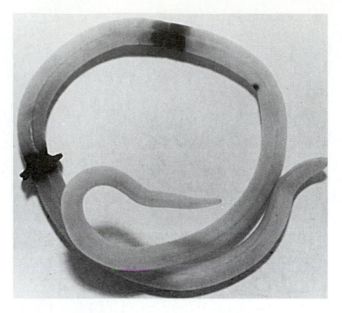

FIGURE 22.30

Female *Ascaris lumbricoides* strangled by a shoe eyelet. This illustrates the tropism of female nematodes to seek the coiled tail of males.

From P. C. Beaver, "*Ascaris* strangled in a shoe-eyelet," in *Am. J. Trop. Med. Hyg.* 13:295–296. Copyright © 1964.

vulva. The vulva may be located anywhere from near the mouth to immediately in front of the anus, depending on the species. The vulva never opens posterior to the anus and only very rarely into the rectum to form a cloaca. The muscles of the vulva act as dilators, and constriction of the circular muscles of the ovijectors both expel the eggs and restrain more proximal, undeveloped eggs from being expelled because of hydrostatic pressure.

• Mating Behavior

Clearly, adult worms of opposite sexes must find each other and copulate for reproduction to occur. Both chemotactic and thigmotactic mechanisms operate in these processes.

Pheromone sex attractants have now been shown for about 40 species of nematodes,[51,73] usually by means of an in vitro assay. For example, male *N. brasiliensis* migrate toward a source of medium in which females have been incubated or that contains an aqueous extract of females. There may be a "medley" of attractants, more complex than hitherto realized. The use of pheromones as biological control agents has some potential.

After a male and female nematode have found each other, thigmotactic responses mediated by the papillae facilitate copulation. The female in some species seeks the coiled posterior end of the male, which she enters. The caudal papillae of the male detect the vulva; this excites a probing response of the spicules, leading to sperm transfer. Females of some species have vulvar papillae. Curiously, if no males are present within a host, females of some species tend to wander, seeking a constriction to squeeze through. This may result in dire consequences to the host if a bile duct, for example, is selected for exploration. Other unexpected results of this behavior have been recorded (Fig. 22.30). Serial copulations of a female with several males have been observed.[3]

DEVELOPMENT

Studies on the development of nematodes have led to fundamental discoveries in zoology. For example, in 1883 van Beneden[12] was the first to elucidate the meiotic process and realize that equal amounts of nuclear material were contributed by sperm and egg after fertilization. Boveri[18] (1899) first demonstrated the genetic continuity of chromosomes and determinate cleavage—that is, embryogenesis in which the fate of the blastomeres is determined very early. Both men based their insights on studies of nematode material.

Not surprisingly, in such a large and diverse phylum as Nematoda, details of development and life history differ greatly among the various groups. However, the general pattern is remarkably conserved. The four juvenile stages and the adult are each separated from the preceding one by an ecdysis, or molting of cuticle. Animal parasitologists traditionally (but incorrectly) refer to the juvenile stages as "larvae." The first-stage juvenile is quite similar in body form to the adult. No real metamorphosis occurs during ontogeny, and with certain possible exceptions, all somatic cells of the adult may be present in the embryo.[33]

Eggshell Formation

Penetration of the oocyte by the sperm initiates the process by which protective layers are produced around the zygote and developing embryo. The fully formed shell in most nematodes consists of three layers: (1) an outer **vitelline layer,** often not detectable by light microscopy; (2) a **chitinous layer;** and (3) an innermost **lipid layer.** A fourth, **proteinaceous layer,** which consists of an acid mucopolysaccharide-tanned protein complex, is contributed by uterine cell secretions in some nematodes (*Ascaris, Thelastoma, Meloidogyne*). Formation of the shell layers is best known in *Ascaris,* but it seems likely that the process is similar in other nematodes.

Immediately after sperm penetration a new plasma membrane forms beneath the original; the old plasma membrane becomes the vitelline layer and separates from the peripheral cytoplasm, and the cytoplasm shrinks back, leaving a clear space within which the chitinous layer forms[41,69] (Fig. 22.31). Refringent bodies, previously dispersed throughout the cytoplasm, migrate to the periphery and extrude their contents, the fusion of which forms the lipid layer.

The so-called chitinous layer is probably supportive or structural in function and also contains protein; the proportion of chitin present varies among groups from great (ascaroids, oxyuroids) to very small (strongyloids). The lipid layer confers resistance to desiccation and to penetration of polar substances. At least in ascarids, the lipid layer is composed of 25% protein and 75% **ascarosides.** Ascarosides are very interesting and unique glycosides (compounds with a sugar and an alcohol joined by a glycosidic bond). In ascarosides the sugar is **ascarylose** (3,6-dideoxy-L-arabinohexose), and the alcohols are a series of secondary monols and diols containing 22 to 37 carbon atoms.[56] The ascarosides render the eggshell virtually impermeable to substances other than gases and lipid solvents; we further discuss the resistance of *Ascaris* eggs is in Chapter 27. We do not know whether ascarosides are present in the lipid layers of nematode eggs

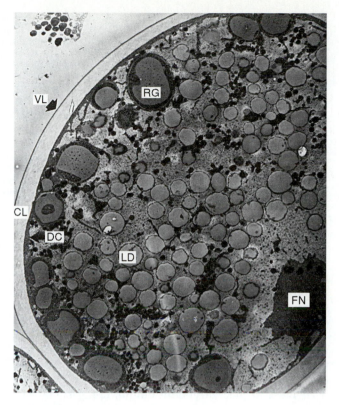

FIGURE 22.31

Low magnification electron micrograph of newly fertilized egg, showing vitelline layer (**VL**); incipient chitinous layer (**CL**); dense, particulate, cortical cytoplasm (**DC**); and numerous lipid droplets (**LD**). Female nucleus (**FN**) lies near surface, and refringent granules (**RG**) have migrated to position just beneath cortical cytoplasm. After extrusion, the contents of the refringent granules will become the lipid layer. (× 3800.)

From W. E. Foor, "Ultrastructural aspects of oocyte development and shell formation in *Ascaris lumbricoides,*" in *J. Parasitol.* 53:1245-1261. Copyright © 1967.

other than ascaroid. Some water can pass across the lipid layer of *Ascaris* and at least some other nematodes, but the embryos can continue to develop despite water stress.[109,126]

Eggshell formation is similar in oxyurids, but there are two uterine layers, and the lipid layer in some (*Syphacia* spp., for example) is thin and of doubtful protective value.[124,125] Oxyurids and some other nematodes also have an operculum, which is a specialized area on the egg to facilitate hatching of the juvenile.[126] Trichurid eggs have an opercular plug at each end.

Embryogenesis

Molecular phenomena in early development have been studied in *Ascaris,* and some data indicate similar phenomena in *Parascaris.*[39] During oogenesis in most animals, there is considerable synthesis of ribosomal and informational RNA (rRNA, mRNA). These are conserved in the unfertilized egg, and little or no RNA synthesis follows fertilization and early cleavage. In contrast there is almost no RNA in mature *Ascaris* oocytes, and fertilization is followed by a burst of rRNA synthesis in the *male* pronucleus.

The determinate cleavage of the nematode embryo is the clearest and best documented example of germinal lineage in

FIGURE 22.32

Cell lineage of nematodes. The two cells produced at the first cleavage of the zygote are P_1 and S_1. The diagram indicates the progeny of these cells and the tissues to which they give rise.

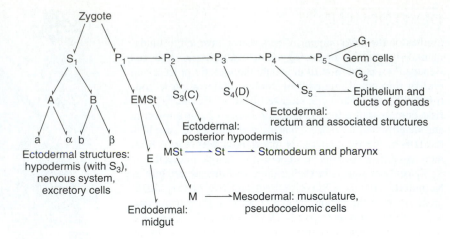

the animal kingdom.[33] Because of the early determination of the fate of each cell (blastomere) in the cleaving embryo, names or letter designations can be given to each blastomere, and the tissues that will develop from each are known (Fig. 22.32). At the first cleavage the zygote produces one cell that will give rise to somatic tissues and one cell whose progeny will comprise more somatic cells and the germinal cells. The early cleavages of nematodes are marked by a very curious phenomenon called **chromatin diminution.** The chromosomes fragment, and only the middle portions are retained, the ends being extruded to the cytoplasm to degenerate. Oddly, during the diminution there is an approximate doubling of the histone/DNA ratio. The explanation for this observation is unclear; the distance between the nucleosomes does not decrease, nor is there a detectable free histone pool.[36]

Since chromatin diminution only occurs in the somatic cells, the germ line can be recognized by its full chromatin complement. Finally, the only cells left with complete chromosomes are G_1 and G_2, which will give rise to the gonads. Interestingly, further differentiation of the nuclei in the various tissues seems to go in both directions with respect to chromatin content. Some, such as muscle and ganglia nuclei, further diminish until DNA can no longer be detected, whereas others, particularly those with protein synthetic activity such as excretory and pharyngeal glands and uterine cells, exhibit polyploidy with respect to DNA content.[115] Clearly there must be a great redundancy of the genes left after chromatin diminution.

Rather typical morula and blastula stages are formed. Gastrulation is by invagination and also by epiboly (movement of the micromeres down over the macromeres).

In the fully formed embryo the nuclei other than the germinal cells cease to divide; thus, all cells of the adult are present at this time. The phenomenon is known as **cell** or **nuclear constancy,** or **eutely,** and it is characteristic of several pseudocoelomate phyla. There are some exceptions—cells of the intestine and hypodermis of the large nematodes divide further—but in most species growth after embryogenesis is a matter of cell enlargement rather than cell division. The number of cells per individual is fairly constant within a species and varies among species.

The timing, site, and physical requirements for embryogenesis vary greatly among species. In some, cleavage will not begin until the egg reaches the external environment and oxygen is available. Others begin (or even complete) embryogenesis before the egg passes from the host, whereas in some the juveniles complete development and hatch within the female nematode (ovoviviparity).

Embryonic Metabolism

Embryonation of *Ascaris* eggs demonstrates a most fascinating sequence of biochemical epigenetic adaptation: adaptive appearance and disappearance of biochemical pathways through ontogeny, based on repression and derepression of genes.[39]

The energy metabolism of adult *Ascaris* is anaerobic, but that of the embryonating egg is obligately aerobic. Dependence on pathways such as glycolysis would not only be wasteful of the limited stored nutrient in the embryos, but also a toxic concentration of acidic end products would soon build up as the result of impermeability of the eggshell. Eggs survive temporary anaerobiosis, but they do not develop unless oxygen is present. They are completely embryonated and infective after 20 days at 30°C, and throughout this time a Krebs cycle and cytochrome *c*-cytochrome oxidase electron transport system are present.

The infective stage is the second-stage juvenile, the worms having undergone one molt in the egg. (A second molt before hatching has been reported, which would make the infective stage the third, as in most other nematodes.)[75] The egg hatches in the host intestine and the juvenile goes through a tissue migration. It breaks out into the lung alveoli, travels up the trachea, and then by being swallowed gains access to the intestine, where it becomes an adult. Cytochrome oxidase is still present in the juveniles recovered from the lungs, and they require oxygen for motility. Oxidase activity disappears from the fourth-stage juvenile in the intestine and is essentially repressed through adult life.

A similar phenomenon has been observed with regard to the enzymes of the glyoxylate cycle.[8] *Ascaris* embryos consume both lipid and carbohydrate reserves during the first 10 days of development and then *resynthesize carbohydrate* (glycogen and trehalose) from fat.[81] *Ascaris* is the only metazoan in which the glyoxylate cycle and its role in the conversion of fat to carbohydrate have been established.[8] Finally, all activity of the two critical enzymes (isocitrate lyase and malate synthase) seems to be repressed in the adult muscle.

Hatching

Hatching of nematodes whose juveniles are free living before becoming parasitic occurs spontaneously.[82] Some plant-parasitic species hatch in the presence of substances from their prospective hosts.[13] A number of animal-parasitic species, however, will hatch only after being swallowed by a prospective host. On reaching the infective stage, such eggs remain dormant until the proper stimulus is applied, and this requirement has the obvious adaptive value of preventing premature hatching. Ascarid eggs require a combination of conditions: temperature about 37° C, a moderately low oxidation-reduction potential (the presence of an oxidizing agent reversibly inhibits hatching),[54] a high carbon dioxide concentration, and a pH of about 7. These conditions are present in the gut of many warm-blooded vertebrates, and indeed *Ascaris* will hatch in a wide variety of mammals and even in some birds, but all four conditions are unlikely to be present simultaneously in the external environment.

The lipid layer is permeable to water, but the fluid around the juvenile has a trehalose concentration of about 0.2 M, resulting in a high osmotic pressure surrounding the juvenile.[31] The first change detectable on application of the stimulus is a rapid change in permeability; trehalose from the perivitelline fluid leaks from the eggs. The increase in water concentration apparently activates the juveniles; Clarke and Perry[32] suggested a mechanism for the permeability change. The lipid layer is now also permeable to chitinase secreted by the juvenile. Esterases and proteinases also are secreted, and these enzymes attack the hard shell, digesting it sufficiently for the worm to force a hole in it and escape.[37]

First-stage juveniles of some nematodes, such as *Trichuris*, possess a stylet on their anterior end, and when the juveniles are activated by the hatching stimulus, they penetrate the operculum (polar plug) with the stylet and emerge from the eggshell.[80]

Growth and Ecdysis

• Molting

Unlike most arthropods, there is growth in body dimensions of nematodes between molts of their cuticle (Fig. 22.33). After the fourth molt in large nematodes such as *Ascaris,* there is considerable increase in size, and the cuticle itself continues to grow after the last ecdysis. The molting process has been studied in several species. First the hypodermis detaches from the basal lamella of the old cuticle and starts to secrete a new one, beginning with the cortical layers. This process may continue until the new cuticle is substantially folded under the old cuticle, to be stretched out later after ecdysis. In some cases the old cuticle up to the cortical layer dissolves and the new cuticle absorbs the resulting solutes. This is particularly important when materials and space are limited, such as in the first molt of *Ascaris*, and less so when

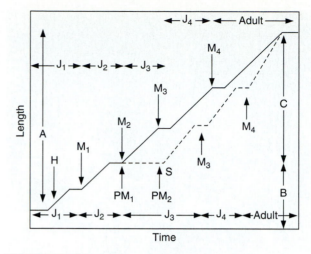

FIGURE 22.33

Idealized form of the basic life cycle of nematodes. The life cycle of a free-living nematode is represented by a solid line. Hatching (**H**) is "spontaneous," and there are four molts (**M₁–M₄**). The broken line represents a life cycle in which a change in environment is necessary to stimulate (**S**) the completion of the second molt (**PM₂**). (**A–C**) are different environments. (**J₁–J₄** are the juvenile stages.)

Modified from W. P. Rogers and R. I. Sommerville, "The infective stage of nematode parasites and its significance in parasitism," in *Advances in Parasitology,* Vol. 1. Edited by B. Dawes. Copyright © 1963 Academic Press, Inc., New York, NY.

there is plenty of food and the old cuticle is very complex in structure, as in the fourth molt of *Nippostrongylus.*[66] Escape from the old cuticle seems to be facilitated by several enzymes. A collagenaselike enzyme attacks the old cuticle.[90]

• Developmental Arrest

A common adaptation in many parasites is a resting stage at some point in their development, enabling them to survive adverse conditions while awaiting access to a new host. Such **developmental arrest** or **hypobiosis**[103] is of particular interest in nematodes, not only because of the variety of stages and situations in which it takes place but also because fundamentally similar processes are demonstrated in some free-living species.[91] An example is *Rhabditis dubia*, which lives in cow dung. It may go through an indefinite number of generations, developing normally, but when unsuitable conditions occur, special third-stage juveniles called **dauer juveniles** are produced. The dauer juveniles develop no further but await access to psychodid flies, to which they attach. When the fly obligingly transports them to a new pile of cow dung, they detach and proceed with development. Another species is *R. coarctata,* which uses dung beetles for transport and in which dauer juveniles are produced every generation. Several other examples could be cited, and a particularly interesting one was reported by Hominick and Aston.[52] The dauer juveniles of *Pelodera strongyloides* attach not to invertebrates but to mice, where they enter the hair follicles of the abdominal skin and molt to fourth-stage juveniles. They will develop no further at the body temperature of the mouse, and the mouse may accumulate hundreds, or even thousands, of nematode juveniles during its life. When the mouse dies and

FIGURE 22.34

Third-stage infective juveniles of *Nippostrongylus brasiliensis,*
illustrating the typical behavior of crawling up on pebbles, blades of
grass, or the like and waving anterior ends to and fro. In this photograph
of living worms, the juveniles have mounted granules of charcoal and
even each other.

Photograph by Larry S. Roberts.

its body cools, the nematodes rapidly emerge and, in the
presence of a food source, molt to the adult stage. The mouse
seems little inconvenienced by its passengers.

A wide variety of parasitic nematodes produce infective
third-stage juveniles that are quite comparable to dauer juveniles. They develop no further until a new host is available,
remaining ensheathed in the second-stage cuticle. They live
on stored food reserves and usually exhibit behavior patterns
that enhance the likelihood of reaching a new host. For example, third-stage juveniles of *Haemonchus* and *Trichostrongylus* migrate out of the fecal mass and onto vegetation that is
eaten by the host. Third-stage juveniles of species that penetrate the host skin, such as hookworms and *Nippostrongylus,*
migrate onto small objects (sand grains, leaves, and others)
and move their anterior ends freely back and forth, in the
same manner as some dauer juveniles (Fig. 22.34).

In both dauer juveniles and infective juveniles, a more
or less specific stimulus is required for resumption of development and completion of the ecdysis of the second-stage
cuticle. Those that penetrate skin usually exsheath in the
processes of penetration, but the stimuli for exsheathment of
swallowed juveniles (*Haemonchus, Trichostrongylus,* and
others) are very similar to those required for hatching of *As-*

caris eggs, including carbon dioxide, temperature, redox potential, and pH. In fact infective eggs (shelled juveniles) are
fundamentally the same as dauer juveniles.[91] For most nematodes tested, carbon dioxide seems to be the most important
stimulus for hatching or exsheathing.[84,85]

Nematodes with intermediate hosts normally undergo hypobiosis at the third stage and remain dormant until they reach
the definitive host. Some species are astonishingly plastic in
their capacities to sustain more than one developmental arrest
in their ontogenies. For example, if some species of hookworms and ascarids infect an unsuitable host, they enter another developmental arrest and lie dormant in the host tissues
until they receive another stimulus to migrate.[77] In several of
these, the older animal is an unsuitable host, and the worms lie
dormant until they are stimulated by the hormones of host
pregnancy. They then migrate to the uterus or mammary
glands and infect the infant by way of the placenta in utero or
the milk after birth.[26] Some species, for example, *Strongyloides ratti,* may not undergo a second developmental arrest at
this stage, but if the lactating female is infected, the juveniles
are somehow diverted from completing their migration to the
adult's intestine and migrate instead to the mammary glands
and infect the suckling young.[128]

In some species adverse environmental conditions, such
as chilling or the onset of the dry season, can predispose infective juveniles to undergo hypobiosis when they reach a definitive host.[2,104] Thus, maturation and production of eggs at a
time when progeny cannot survive are avoided.[5] More examples of adaptational arrests in development will be found in
the nematode life cycles described in the chapters to follow.

METABOLISM

Energy Metabolism

• Adult Nematodes

Probably more is known about nematode metabolism than
about metabolism of any other group of parasitic
helminths,[23–25,99] but most of what we know has been derived from studies on *Ascaris suum. Ascaris* was one of the
first organisms in which cytochrome was demonstrated.[58]
Nevertheless, many questions await resolution.

The overall scheme of energy metabolism of adult
Ascaris—and other such nematode parasites of animals that
have been examined—is basically similar to that of adult
trematodes and cestodes (Figs. 15.25 and 20.29). Although
many nematode parasites live in the presence of substantial
quantities of oxygen, they do not completely oxidize all of
their nutrient molecules. They degrade glucose to phosphoenolpyruvate and fix CO_2 to form oxaloacetate, which is
reduced to malate, and the malate enters the mitochondrion
to undergo further reactions. Additional ATP is derived in
the mitochondrial reactions, and the reduced end products
include lactate, acetate, and succinate. Some of the succinate is decarboxylated to form propionate. Further reactions may occur in the cytoplasm to produce a variety of
other unusual end products, such as α-methylbutyrate and
α-methylvalerate (Fig. 22.35).

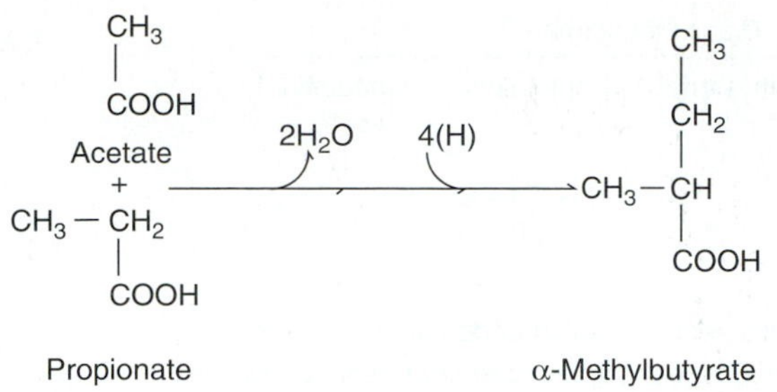

FIGURE 22.35

Mode of α-methylbutyrate formation in *Ascaris* muscle; α-methylvalerate is formed in a similar manner except that the α-carbon of one propionate unit condenses with the carboxyl carbon of another propionate unit.[101,102]

As with the trematodes and cestodes, the adaptive value of this seemingly wasteful scheme is not at all obvious. We can speculate that it may involve some as yet unclear aspects of the host-parasite relationship.[25]

Other electron transport reactions are present in *Ascaris* spp., but their importance and sequence are still not known with certainty. A classical electron transport system is unlikely to be of physiological importance,[60] and oxygen in moderate concentrations is toxic to this nematode. Succinate and malate are oxidized in mitochondrial preparations with hydrogen peroxide as an end product. The toxicity of hydrogen peroxide probably accounts for the nematode's intolerance of oxygen. However, small amounts may be necessary in certain biosynthetic reactions.

Other nematodes have been studied that survive and metabolize carbohydrates in the absence of oxygen for extended periods; examples are *Heterakis gallinarum* and *Trichuris vulpis*. These organisms serve as particularly good examples of how some parasites have solved the metabolic problem of reoxidizing NADH (Chapter 4) in the absence of oxygen as a terminal electron acceptor. Some adults (*Haemonchus contortus*) apparently have a tricarboxylic acid cycle that operates when oxygen is present and an *Ascaris* type of system operative in the absence of oxygen.[122,123] *Rhabdias bufonis*, a parasite in the lungs of frogs, also apparently has alternative systems,[4] in spite of the unlikelihood of its being subjected to anaerobiosis.

Nevertheless, other adult nematodes seem to be obligate aerobes with respect to their energy metabolism, requiring the presence of at least low concentrations of oxygen for survival and motility. Even so, glucose is not oxidized completely to carbon dioxide and water, and substantial quantities of various reduced end products are excreted. Some species apparently have a classical cytochrome system, or alternatively the electrons may be transported by a flavoprotein and terminal flavin oxidase to oxygen, producing hydrogen peroxide.

Nippostrongylus brasiliensis and *Litosomoides carinii* can survive short periods of anaerobiosis but are killed by longer periods (a few hours).[89,100] *Nippostrongylus* has a fluid-filled layer in its cuticle (see Fig. 22.6) that contains hemoglobin, and the hemoglobin loads and unloads oxygen in the living animal.[106] Thus, the worm can exploit areas in the intestine that are quite hypoxic. *Brugia pahangi* produces three times more lactate in the absence of oxygen than in its presence (when glucose is present).[7]

Some evidence suggests that the extent of dependence on aerobic pathways in nematodes is correlated with body diameter; that is, the larger nematodes are more anaerobic, whereas the smaller ones can get to the oxygen near the mucosa and are more aerobic. Fry and Jenkins view the parasitic nematodes as "metabolic opportunists, combining the versatility of an anaerobic and aerobic energy metabolism."[45]

• Juveniles

This discussion so far has referred to adult nematodes. Different stages in the life cycles, such as the aerobic embryos and the anaerobic adults of *Ascaris*, often show dramatic biochemical adaptations in energy metabolism. Developing juveniles change over to anaerobic metabolism during the J_3 stage.[62] Although metabolism of the *Ascaris* testis/seminal vesicle is anaerobic, their mitochondria have elevated levels of Krebs cycle enzymes, possibly for use later during embryonic development.[61] Juveniles of *Trichinella spiralis* in the muscles of their host have more aerobic metabolism than do adult *Ascaris*, but some of their energy is apparently generated anaerobically.[15]

Strongyloides is another interesting example of biochemical epigenetic adaptation.[63] *Strongyloides* has a complex life cycle with free-living adults (males and females) and parasitic adults (parthenogenetic females only). The first three juvenile stages of both types are free living, but those destined to become parasitic undergo developmental arrest at the third stage, until penetration of the host (Chapter 24). All free-living stages are subjected to a selective pressure common to other free-living animals—to use as completely as possible the energetic value in their nutrient molecules—and they have a complete tricarboxylic acid cycle and probably a cytochrome system. In contrast the parasitic females have neither a complete tricarboxylic acid cycle nor cytochrome system, a situation similar to many other intestinal helminths. Juveniles of several other parasitic species have apparently functional tricarboxylic acid cycles,[16,121] although in some species the significance of the cycle may lie in regulation of four-carbon intermediates rather than in energy production. Oddly, the first and second stage juveniles of *Ancylostoma tubaeforme* and *Haemonchus contortus* are apparently anaerobic, and the infective third stage is aerobic.[79]

The normal pathway of fatty acid oxidation is called β-oxidation because the β-carbon of the fatty acetyl-CoA is oxidized, and the two-carbon fragment, acetyl-CoA, is cleaved off to enter the Krebs cycle. It would be expected, therefore, that the presence of β-oxidation enzymes would be correlated with a functional Krebs cycle (though not necessarily so), and in the few cases investigated this is the case. β-oxidation of fatty acids has been found in developing *Ascaris* embryos and in free-living *Strongyloides* juveniles and adults.[63,120] Tissue lipids gradually disappear (are consumed?) from infective eggs or juveniles of several species. Interestingly neither *Ascaris* muscle nor parasitic females of *Strongyloides* can carry out β-oxidation.

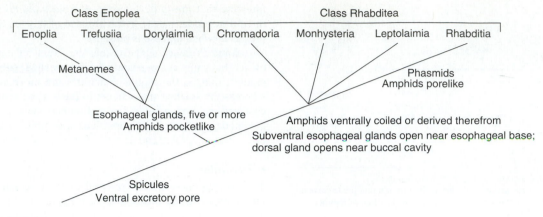

FIGURE 22.36

Hypothesized phylogenetic relationships among the major taxonomic divisions of nematodes. According to this analysis, the conventional Class Adenophorea (Aphasmidea) would be paraphyletic, including the Chromadoria, Monhysteria, and Leptolaimia. Here, these later taxa are grouped with the Rhabditia (= Phasmidea) to form the Class Rhabditea, a sister group of the Enoplea.

Modified from M. Adamson, "Phylogenetic analysis of the higher classification of the Nematoda," in *Canad. J. Zool.* 31:243–255. Copyright © 1987.

Synthetic Metabolism

• Proteins and Nucleic Acids

Synthetic metabolism of nematodes has not been as intensively studied as has energy metabolism, probably because, in contrast to prokaryotes, energy pathways of helminths usually offer the better sites for chemotherapeutic action. However, there are several points of interest. In light of the enormous number of progeny produced by an organism such as *Ascaris*, protein and nucleic acid synthetic ability must be correspondingly great. In this connection the RNA metabolism of fertilized *Ascaris* eggs deserves further comment (see earlier discussion of embryogenesis). The young oocytes have nucleoli and large amounts of cytoplasmic RNA, and these presumably are responsible for the very large amount of yolk protein synthesized in the developing oocyte. By the time the oocyte matures, the nucleoli and most of the cytoplasmic RNA have disappeared.[57] At the same time the sperm contains little or no RNA. Immediately after fertilization, there is a massive ribosomal RNA synthesis in the male pronucleus, along with a smaller amount of messenger RNA, while the female pronucleus is going through its maturation divisions. Kaulenas and Fairbairn suggested, therefore, that the female genome is responsible for the high rate of oocyte production and yolk synthesis, whereas ribosomes provided by the male genome largely support shell formation and cleavage.[57]

Presumably much of the amino acid supply for protein synthesis in the oocyte is furnished by the nearby intestinal absorption, but some amino acids are synthesized in the ovaries as well. The ovaries contain active transaminases, which form amino acids from the corresponding α-keto acids derived from carbohydrate metabolism. In addition the ovaries can condense pyruvate with ammonia to form the amino acid alanine. Some free-living nematodes can synthesize a wide variety of amino acids from a simple substrate, such as acetate. When incubated in a medium containing glycine, glucose, and acetate, *Caenorhabditis briggsae* synthesizes an array of "nonessential" and "essential" amino acids. This was the first metazoan known that could synthesize "essential" amino acids.[96]

As noted in the discussion of the body wall, much collagen is found in the cuticle of *Ascaris*. Collagens are also found in the muscle, intestine, and reproductive organs. Such stabilized proteins are important factors in the resistance and strength of the nematode's cuticle. Collagens are stabilized by bonds between lysine residues in the subunits, and they are unusual in that they contain around 12% proline and 9% hydroxyproline; hydroxyproline is an amino acid rarely found in other proteins. The normal collagen precursor is a polypeptide called protocollagen, and the proline in protocollagen is hydroxylated to hydroxyproline by the enzyme **protocollagen proline hydroxylase** (**PPH** or **proline monoxygenase**). Among other cosubstrates, this enzyme requires molecular oxygen to carry out the hydroxylation of proline. This is an example of a biosynthetic reaction that requires oxygen in an organism that is anaerobic with respect to its energy metabolism. Oxygen concentration greater than 5% even inhibits PPH from *Ascaris* muscle but not the enzyme from the embryos.[27]

• Lipids

At least some nematodes can synthesize polyunsaturated fatty acids de novo but apparently are unable to synthesize sterols de novo.[95] *Ascaris* incorporates acetate into long-chain fatty acids, probably by the malonyl-CoA pathway as found in vertebrates.[11] The nonsugar parts of the ascarosides (the alcohols) in *Ascaris* spp. are synthesized from long-chain fatty acids. This involves a condensation in which the carboxyl carbon of one fatty acid condenses with carbon number 2 of another, with the elimination of a molecule of carbon dioxide.[38] The ascarylose is freely synthesized by *Ascaris* ovaries from glucose or glucose-1-phosphate, and the end product of the synthesis is probably ascarylose-dinucleotide phosphate, which then condenses with the nonsugar moiety to give the ascaroside. *Dirofilaria immitis* can synthesize all classes of complex lipids, including cholesterol.[117]

CLASSIFICATION OF PHYLUM NEMATODA

For the higher classification of the Nematoda, we are following the system proposed by Adamson[1] (Fig. 22.36). Adamson's analysis suggests that the Adenophorea (=Aphasmidia), as traditionally constituted, is paraphyletic (does not include all descendants of the most recent common ancestor) and that the Enoplea and Rhabditea are sister groups. For classification of orders and superfamilies, the following classification is a combination of Chabaud,[29] Adamson,[1] and Adamson (personal communication). Nematode classification is subject to further analysis and thus to further change. We omit detailed diagnoses and orders for strictly free-living groups.

Members of the phylum Nematoda can be distinguished from other pseudocoelomate groups by their possession of spicules and a ventral excretory pore.

Class Enoplea (= Adenophorea; = Aphasmidea)

Amphids generally well-developed (except in parasitic forms), well behind the lips, often pocketlike; caudal and hypodermal glands common; phasmids absent; excretory system lacking lateral canals, formed of single, ventral, glandular cells, or entirely absent; deirids absent; mostly free living, some parasitic on plants and animals.

Subclass Enoplia
Metanemes present.

Subclass Dorylaimia
Five or more esophageal glands; buccal stylet present at some stage in life cycle; free living, plant parasitic or animal parasitic.

Order Trichurida

Anterior end more slender than posterior end; lips and buccal capsule absent or much reduced; esophagus a very slender, capillarylike tube, embedded within one or more rows of large, glandular cells (stichocytes) along its posterior portion; bacillary band present; both sexes with a single gonad; males with one spicule or none; eggs with polar plugs (opercula) except in *Trichinella*; histiotrophic parasites of nearly all organs of all classes of vertebrates.

Families
Anatrichosomatidae, Capillariidae, Cystoopsidae, Trichinellidae, Trichosomoididae, Trichuridae.

Order Dioctophymatida

Stout worms, often very large; anterior cuticle spinose in some species; esophageal glands highly developed, multinucleate; lips and buccal capsule reduced, replaced by muscular oral sucker in Soboliphymatidae; esophagus cylindrical; nerve ring far anterior; anus at posterior end in both sexes; male with bell-shaped muscular copulatory bursa without rays; both sexes with a single gonad; male with single spicule; eggs deeply sculptured or pitted; histiotrophic parasites of birds and mammals.

Families
Dioctophymatidae, Eustrongylidae, Soboliphymatidae.

Order Muspiceida

Alimentary tract reduced or vestigial; male unknown; parasites of skin and deeper tissues of rodents, deer, bats, marsupials, and crows.

Families
Muspiceidae, Robertdollfusiidae.

Order Mermithida

Six or eight hypodermal cords visible in cross section; juvenile stages parasitic in body cavity of various invertebrates.

Families
Mermithidae, Marimermithidae, Echinodermellidae, Tetradonematidae.

Class Rhabditea

Amphids ventrally coiled or derived therefrom; three esophageal glands; subventral glands open near base of esophageal corpus; dorsal gland opens at or near buccal cavity.

Subclasses Chromadoria, Monohysteria, Leptolaimia
No known parasites in these groups.

Subclass Tylenchia
Buccal stylet present; esophagus with median, valved bulb and posterior glandular swelling; excretory canal limited to one side of body; most species parasitic on or in plants, some parasitic in hemocoel of insects (occasionally other arthropods), others essentially free living, often using insects as phoretic hosts.

Subclass Rhabditia (= Phasmidea; = Secernentea)
Amphids generally poorly developed, with small, simple pores near or on the lips; caudal and hypodermal glands absent; phasmids present; excretory system with one or two lateral canals, with or without associated glandular cells; deirids commonly present; free living in soil or fresh water or parasitic in plants or animals.

Order Rhabditida

Tiny to small worms, commonly with six small lips; esophagus muscular, divided into anterior corpus, median isthmus, and posterior bulb; pseudobulb often present between corpus and isthmus; bulb usually absent in parasitic stages; buccal capsule small or absent; tail conical in both sexes, spicules equal, gubernaculum usually present; mostly free living; some parasites in lungs of amphibians and reptiles or in intestine of amphibians, reptiles, birds, and mammals.

Families
Rhabdiasidae, Strongyloididae.

Order Drilonematida

Phasmids suckerlike; cephalic hooks often present; parasites in coelomic cavity of earthworms.

Families
Ungellidae, Homungellidae, Pharyngonematidae, Drilonematidae, Creagrocercidae.

Order Rhigonematida

Complex cuticular modifications present at base of buccal cavity; vagina long and muscular; parasites in posterior gut of Diplopoda.

SUPERFAMILIES
Rhigonematoidea, Ransomnematoidea.

Order Strongylida

Commonly long, slender worms; esophagus usually swollen posteriorly but lacking definite bulb; male with well-developed copulatory bursa supported by sensory rays; ovijector complex, with well-developed sphincters; excretory system with H-shaped tubular arrangement and two subventral glands; first-, second-, and beginning of third-stage juveniles free living or parasitic in invertebrates; usually oviparous; eggs thin shelled, rarely developed beyond morula when laid; parasites of all classes of vertebrates (rare in fishes).

SUPERFAMILIES AND FAMILIES
Diaphanocephaloidea: Diaphanocephalidae; Ancylostomatoidea: Ancylostomidae, Uncinariidae, Globocephalidae;

Strongyloidea: Strongylidae, Cloacinidae, Syngamidae; Trichostrongyloidea: Trichostrongylidae, Amidostomatidae, Strongylacanthidae, Heligmosomidae, Ollulanidae, Dictyocaulidae; Metastrongyloidea: Metastrongylidae, Angiostrongylidae.

Order Ascaridida

Three prominent lips usually present; lateral external labial papillae present; numerous caudal papillae, of which two to three pairs are dorsolateral in position; esophagus variable in structure; life cycle variable, direct or indirect, with invertebrate or vertebrate intermediate hosts.

SUPERFAMILY ASCARIDOIDEA

Lips often prominent; buccal capsule weakly cuticularized and surrounded by esophageal tissue; esophagus consisting of corpus and posterior ventriculus which may be muscular or glandular; eggs thick shelled; life cycle direct or indirect with invertebrate or vertebrate intermediate hosts.

Families
Ascarididae, Crossophoridae, Anisakidae, Acanthocheilidae.

SUPERFAMILY COSMOCERCOIDEA

Esophagus with well-defined corpus, isthmus, and bulb with valve; parasites of the posterior gut of various vertebrate hosts; oviparous with thin-shelled eggs passed in feces to external environment or ovoviviparous with autoinfective cycle (Atractidae).

Families
Cosmocercidae, Atractidae, Kathlaniidae.

SUPERFAMILY HETERAKOIDEA

Lips prominent; esophagus with well-developed posterior bulb with valve; male with prominent preanal sucker surrounded by cuticularized ring; parasites of posterior gut of tetrapods.

Families
Heterakidae, Ascaridiidae.

SUPERFAMILY SUBULUROIDEA

Lips absent; esophagus with well-developed posterior bulb with valve; anterior lobes of esophagus modified to form complex pharyngeal teeth; eggs thick shelled, with first-stage juveniles in utero; life cycle indirect, with development from first- to third-stage juvenile in insect.

Families
Subuluridae, Maupasiidae.

SUPERFAMILY SEURATOIDEA

Artificial grouping of intestinal parasites of various vertebrates; esophagus typically club shaped, without swellings or valves; ventral sucker present in some; life cycles little known, some with aquatic arthropod intermediate hosts.

Families
Seuratidae, Cucullanidae, Quimperiidae.

Order Oxyurida

Medium to small worms often with sharply pointed tails; esophagus with prominent posterior bulb with valve; excretory system X shaped with prominent sinus and vesiculate duct; males with single (or no) spicule and reduced number of caudal papillae; sperm comet shaped; eggs often flattened on one side; haplodiploid (male haploid derived from unfer-

tilized egg, female diploid derived from fertilized egg); parasites of colon or rectum of arthropods (rarely annelids) and vertebrates; life cycles direct.

SUPERFAMILIES AND FAMILIES

Thelastomatoidea (in arthropods and annelids): Thelastomatidae, Travassosinematidae, Pseudonymidae, Hystrignathidae, Protrelloididae; Oxyuroidea (in vertebrates): Oxyuridae, Pharyngodonidae, Heteroxynematidae.

Order Spirurida

Mouth surrounded by six lips, lips absent, or lateral pseudolabia present; buccal capsule often well-developed; esophagus divided into anterior muscular and posterior glandular portions, never with bulb; development from first- to third-stage juvenile in arthropod intermediate host; parasites in intestine and deeper tissues of all vertebrate classes.

Suborder Camallanina
First-stage juveniles with dorsal prominence or tooth; ovoviviparous; development from first to third stage in copepod.

SUPERFAMILY CAMALLANOIDEA

Lips absent; buccal capsule well-developed, often in the form of two opposing valves; anterior muscular and posterior glandular portion of esophagus separated into distinct cellular compartments; male often with caudal alae; female often with single ovary and uterus; parasites in intestine of fish, turtles, and amphibians.

Families
Camallanidae, Oceanicucullanidae.

SUPERFAMILY DRACUNCULOIDEA

Buccal capsule reduced; anterior muscular and posterior glandular portions of esophagus not separated into distinct compartments; female often highly enlarged, filled with first-stage juveniles; parasites of various tissue sites of vertebrates (mostly fish).

Families
Dracunculidae, Philometridae, Phlyctainophoridae, Skrjabillanidae, Anguillicolidae.

Suborder Spirurina
Eggs thick shelled; first-stage juveniles with dorsal cuticular hooks or spines; parasites of intestine and deeper tissues of all vertebrate classes.

SUPERFAMILIES AND FAMILIES

Gnathostomatoidea: Gnathostomatidae; Physalopteroidea: Physalopteridae; Rictularoidea: Rictulariidae; Thelazoidea: Thelaziidae, Rhabdochonidae, Pneumospiruridae; Spiruroidea: Gongylonematidae, Spiruridae, Spirocercidae, Hartertiidae; Habronematoidea: Hedruridae, Habronematidae, Tetrameridae, Cystidicolidae; Acarioidea: Acuariidae; Filaroidea: Filariidae, Onchocercidae; Aproctoidea: Aproctidae, Desmidocercidae; Diplotriaenoidea: Diplotriaenidae.

References

1. Adamson, M. L. 1987. Phylogenetic analysis of the higher classification of the Nematoda. *Can. J. Zool.* 65:1478–1482.

2. Altaif, K. I., and W. H. Issa. 1983. Seasonal fluctuations and hypobiosis of gastro-intestinal nematodes of Awassi lambs in Iraq. *Parasitology* 86:301–10.

3. Anderson, R. V., and H. M. Darling. 1964. Embryology and reproduction of *Ditylenchus destructor* Thorne, with emphasis on gonad development. *Proc. Helm. Soc. Wash.* 31:240–56.

4. Anya, A. O., and G. M. Umezurike. 1978. Respiration and carbohydrate energy metabolism of the lung-dwelling parasite *Rhabdias bufonis* (Nematoda: Rhabdiasoidea). *Parasitology* 76:21–27.

5. Armour, J., and M. Duncan. 1987. Arrested larval development in cattle nematodes. *Parasitol. Today* 3:171–76.

6. Bansemir, A. D., and M. V. K. Sukhdeo. 1994. The food resource of adult *Heligmosomoides polygyrus* in the small intestine. *J. Parasitol.* 80:24–28.

7. Barrett, J., A. H. W. Mendis, and P. E. Butterworth. 1986. Carbohydrate metabolism in *Brugia pahangi* (Nematoda: Filaroidea). *Int. J. Parasitol.* 16:465–69.

8. Barrett, J., C. W. Ward, and D. Fairbairn. 1970. The glyoxylate cycle and the conversion of triglycerides to carbohydrates in developing eggs of *Ascaris lumbricoides. Comp. Biochem. Physiol.* 35:577–86.

9. Batson, B. S. 1979. Body wall of juvenile and adult *Gastromermis boophthorae* (Nematoda: Mermithidae): Ultrastructure and nutritional role. *Int. J. Parasitol.* 9:495–503.

10. Batson, B. S. 1979. Ultrastructure of the trophosome, a food-storage organ in *Gastromermis boophthorae* (Nematoda: Mermithidae). *Int. J. Parasitol.* 9:505–14.

11. Beames, C. G. Jr., B. G. Harris, and F. A. Hopper Jr. 1967. The synthesis of fatty acids from acetate by intact tissue and muscle extract of *Ascaris lumbricoides suum. Comp. Biochem. Physiol.* 20:509–21.

12. van Beneden, E. 1883. Recherches sur la maturation de l'oeuf et la fécondation *(Ascaris megalocephala). Arch. Biol.* 4:265–641.

13. Bird, A. F., and J. Bird. 1991. *The structure of nematodes,* 2d ed. San Diego: Academic Press.

14. Blaxter, M. L., A. P. Page, W. Rudin, and R. M. Maizels. 1992. Nematode surface coats: Actively evading immunity. *Parasitol. Today* 8:243–47.

15. Boczon, K. 1991. Differences between the energy metabolism of *Trichinella spiralis* and *Trichinella pseudospiralis. J. Helm. Soc. Wash.* 58:122–25.

16. Boczon, K., and J. W. Michejda. 1978. Electron transport in mitochondria of *Trichinella spiralis* larvae. *Int. J. Parasitol.* 8:507–13.

17. Bolla, R. I., P. P. Weinstein, and G. D. Cain. 1972. Fine structure of the coelomocyte of adult *Ascaris suum. J. Parasitol.* 58:1025–36.

18. Boveri, T. 1899. *Die Entwicklung von* Ascaris megalocephala *mit besonderer Rücksicht auf die Kernverhältnisse.* Jena, Germany: Festschrift für C. Von Kupffer.

19. Brothwell, D. 1987. *The bog man and the archeology of people.* Cambridge, Mass.: Harvard University Press.

20. Brownlee, D. J. A., I. Fairweather, C. F. Johnston, D. Smart, C. Shaw, and D. W. Halton. 1993. Immunocytochemical demonstration of neuropeptides in the central nervous system of the roundworm, *Ascaris suum* (Nematoda: Ascaroidea). *Parasitology* 106:305–16.

21. Brownlee, D. J. A., I. Fairweather, and C. F. Johnston. 1993. Immunocytochemical demonstration of neuropeptides in the peripheral nervous system of the roundworm *Ascaris suum* (Nematoda, Ascaroidea). *Parasitol. Res.* 79:302–8.

22. Brownlee, D. J. A., I. Fairweather, C. F. Johnston, and C. Shaw. 1994. Immunocytochemical demonstration of peptidergic and serotoninergic components in the enteric nervous system of the roundworm, *Ascaris suum* (Nematoda, Ascaroidea). *Parasitology* 108:89–103.

23. Bryant, C. 1975. Carbon dioxide utilization and the regulation of respiratory metabolic pathways in parasitic helminths. In Dawes, B., ed. *Advances in parasitology* 13. New York: Academic Press, Inc., 36–69.

24. Bryant, C. 1982. Biochemistry. In Cox, F. E. G., ed. *Modern parasitology.* Oxford: Blackwell Scientific Publications Ltd., 84–115.

25. Bryant, C. 1993. Organic acid excretion by helminths. *Parasitol. Today* 9:58–60.

26. Burke, T. M., and E. L. Roberson. 1985. Prenatal and lactational transmission of *Toxocara canis* and *Ancylostoma caninum:* Experimental infection of the bitch before pregnancy. *Int. J. Parasitol.* 15:71–75.

27. Cain, G. D., and D. Fairbairn. 1971. Protocollagen proline hydroxylase and collagen synthesis in developing eggs of *Ascaris lumbricoides. Comp. Biochem. Physiol.* 40B:165–79.

28. Campbell, W. C. 1985. Ivermectin: An update. *Parasitol. Today* 1:10–16.

29. Chabaud, A. G. 1974. Class Nematoda. Keys to subclasses, orders, and superfamilies. In Anderson, R. C., A. G. Chabaud, and S. Willmott, eds. *CIH keys to the nematode parasites of vertebrates,* no. 1. Bucks, Eng.: Commonwealth Agricultural Bureaux, Farnham Royal, 6–17.

30. Chitwood, B. G., and M. B. Chitwood. 1950. *An introduction to nematology.* Baltimore: Monumental Printing Co.

31. Clarke, A. J., and R. N. Perry. 1980. Egg shell permeability and hatching of *Ascaris suum. Parasitology* 80:447–56.

32. Clarke, A. J., and R. N. Perry. 1988. The induction of permeability in egg shells of *Ascaris suum* prior to hatching. *Int. J. Parasitol.* 18:987–90.

33. Crofton, H. D. 1966. *Nematodes.* London: Hutchinson University Library.

34. Croll, N. A., ed. 1976. *The organization of nematodes.* New York: Academic Press, Inc.

35. Croll, N. A., and B. E. Matthews. 1977. *Biology of nematodes.* New York: John Wiley & Sons, Inc.

36. Davis, A. H., and C. E. Carter. 1980. Chromosome diminution in *Ascaris suum. Exp. Cell Res.* 128:59–62.

37. Fairbairn, D. 1960. The physiology and biochemistry of nematodes. In Sasser, J. N., and W. R. Jenkins, eds. *Nematology.* Chapel Hill, N.C.: University of North Carolina Press, 267–96.

38. Fairbairn, D. 1969. Lipid components and metabolism of Acanthocephala and Nematoda. In Florkin, M., and B. T. Scheer, eds. *Chemical zoology,* 3. New York: Academic Press, Inc., 361–78.

39. Fairbairn, D. 1970. Biochemical adaptation and loss of genetic capacity in helminth parasites. *Biol. Rev.* 45:29–72.

40. Fetterer, R. H., M. L. Rhoads, and J. F. Urban Jr. 1993. Synthesis of tyrosine-derived crosslinks in *Ascaris suum* cuticular proteins. *J. Parasitol.* 79:160–66.

41. Foor, W. E. 1967. Ultrastructural aspects of oocyte development and shell formation in *Ascaris lumbricoides. J. Parasitol.* 53:1245–61.

42. Foor, W. E. 1970. Spermatozoan morphology and zygote formation in nematodes. *Biol. Rep.* 2(suppl.):177–202.

43. Foster, W. D. 1965. *A history of parasitology.* Edinburgh: E. & S. Livingstone.

44. Franks, C. J., L. Holden-Dye, R. G. Williams, R. Y. Pang, and R. J. Walker. 1994. A nematode FMRFamide-like peptide, SDPNFLRFamide (PF1), relaxes the dorsal muscle strip preparation of *Ascaris suum. Parasitology* 108:229–36.

45. Fry, M., and D. C. Jenkins. 1984. *Nematoda:* Aerobic respiratory pathways of adult parasitic species. *Exp. Parasitol.* 57:86–92.

46. Geary, T. G., R. D. Klein, L. Vanover, J. W. Bowman, and D. P. Thompson. 1992. The nervous systems of helminths as targets for drugs. *J. Parasitol.* 78:215–30.

47. Gianotti, A. J., D. T. Clark, and J. Dash. 1991. Zinc sulfide in intestinal cell granules of *Ancylostoma caninum* adults. *J. Parasitol.* 77:285–89.

48. Harpur, R. P. 1964. Maintenance of *Ascaris lumbricoides* in vitro. III. Changes in the hydrostatic skeleton. *Comp. Biochem. Physiol.* 13:71–85.

49. Harpur, R. P., and J. S. Popkin, 1965. Osmolality of blood and intestinal contents in the pig, guinea pig, and *Ascaris lumbricoides. Can. J. Biochem.* 43:1157–69.

50. Harris, J. E., and H. D. Crofton. 1957. Structure and function in the nematodes: Internal pressure and cuticular structure in *Ascaris. J. Exp. Biol.* 34:116–30.

51. Haseeb, M. A., and B. Fried. 1988. Chemical communication in helminths. In Baker, J. R., and R. Muller, eds. *Advances in parasitology* 27. London: Academic Press, 169–207.

52. Hominick, W. M., and A. J. Aston. 1981. Association between *Pelodera strongyloides* (Nematoda: Rhabditidae) and wood mice, *Apodemus sylvaticus. Parasitology* 83:67–75.

53. Hope, W. D., and S. L. Gardiner. 1982. Fine structure of a proprioceptor in the body wall of the marine nematode *Deontostoma californicum* Steiner and Albin, 1933 (Enoplida: Leptosomatidae). *Cell Tissue Res.* 225:1–10.

54. Hurley, L. C., and R. I. Sommerville. 1982. Reversible inhibition of hatching of infective eggs of *Ascaris suum* (Nematoda). *Int. J. Parasitol.* 12:463–65.

55. Hyman, L. H. 1951. *The invertebrates: Acanthocephala, Aschelminthes, and Entoprocta, the pseudocoelomate Bilateria* 3. New York: McGraw-Hill Book Co.

56. Jezyk, P. F., and D. Fairbairn. 1967. Metabolism of ascarosides in the ovaries of *Ascaris lumbricoides* (Nematoda). *Comp. Biochem. Physiol.* 23:707–19.

57. Kaulenas, M. S., and D. Fairbairn. 1968. RNA metabolism of fertilized *Ascaris lumbricoides* eggs during uterine development. *Exp. Cell Res.* 52:233–51.

58. Keilin, D. 1925. On cytochrome, a respiratory pigment common to animals, yeasts and higher plants. *Proc. R. Soc. Ser. B* 98:312–39.

59. Knox, D. P., and D. G. Jones. 1990. Studies on the presence and release of proteolytic enzymes (proteinases) in gastro-intestinal nematodes of ruminants. *Int. J. Parasitol.* 20:243–49.

60. Köhler, P., and R. Bachmann. 1980. Mechanisms of respiration and phosphorylation in *Ascaris* muscle mitochondria. *Molecular and Biochem. Parasitol.* 1:75–90.

61. Komuniecki, P. R., J. Johnson, M. Kamhawi, and R. Komuniecki. 1993. Mitochondrial heterogeneity in the parasitic nematode, *Ascaris suum. Exp. Parasitol.* 76:424–37.

62. Komuniecki, P. R., and L. Vanover. 1987. Biochemical changes during the aerobic-anaerobic transition in *Ascaris suum* larvae. *Molecular and Biochem. Parasitol.* 22:241–48.

63. Körting, W., and D. Fairbairn. 1971. Changes in beta-oxidation and related enzymes during the life cycle of *Strongyloides ratti* (Nematoda). *J. Parasitol.* 57:1153–58.

64. Lee, D. L. 1965. The cuticle of adult *Nippostrongylus brasiliensis. Parasitology* 55:173–81.

65. Lee, D. L. 1969. *Nippostrongylus brasiliensis:* Some aspects of the fine structure and biology of the infective larva and the adult. In Taylor, A. E. R., ed. Nippostrongylus *and* Toxoplasma. *Symposia of the British Society of Parasitology* 7. Oxford: Blackwell Scientific Publications Ltd., 3–16.

66. Lee, D. L. 1970. Moulting in nematodes: The formation of the adult cuticle during the final moult of *Nippostrongylus brasiliensis. Tissue Cell* 2:139–53.

67. Lee, D. L., and A. O. Anya. 1967. The structure and development of the spermatozoon of *Aspiculuris tetraptera* (Nematoda). *J. Cell Sci.* 2:537–44.

68. Lee, D. L., and H. J. Atkinson. 1977. *Physiology of nematodes,* 2d ed. New York: Columbia University Press.

69. Lee, D. L., and P. Lestan. 1971. Oogenesis and eggshell formation in *Heterakis gallinarum* (Nematoda). *J. Zool.* 164:189–96.

70. Lee, D. L., K. A. Wright, and R. R. Shivers. 1993. A freeze-fracture study of the cuticle of adult *Nippostrongylus brasiliensis* (Nematoda). *Parasitology* 107:545–52.

71. Lee, H., I. Chen, and R. Lin. 1973. Ultrastructure of the excretory system of *Anisakis* larva (Nematoda: Anisakidae). *J. Parasitol.* 59:289–98.

72. Lorenzen, S. 1978. Discovery of stretch receptor organs in nematodes—structure, arrangement and functional analysis. *Zoologica Scripta* 7:175–78.

73. MacKinnon, B. M. 1987. Sex attractants in nematodes. *Parasitol. Today* 3:156–58.

74. MacKinnon, B. M. 1987. An ultrastructural and histochemical study of oogenesis in the trichostrongylid nematode *Heligmosomoides polygyrus. J. Parasitol.* 73:390–99.

75. Maung, M. 1978. The occurrence of the second moult of *Ascaris lumbricoides* and *Ascaris suum. Int. J. Parasitol.* 8:371–78.

76. Meglitsch, P.A. 1972. *Invertebrate zoology,* 2d ed. New York: Oxford University Press.

77. Michel, J. F. 1974. Arrested development of nematodes and some related phenomena. In Dawes, B., ed. *Advances in parasitology* 12. New York: Academic Press, Inc., 280–366.

78. Nelson, F. K., and D. L. Riddle. 1984. Functional study of the *Caenorhabditis elegans* secretory-excretory system using laser microsurgery. *J. Exp. Zool.* 231:45–56.

79. Onwuliri, C. O. E. 1985. Energy metabolism in the developing larval stages of *Ancylostoma tubaeforme* and *Haemonchus contortus:* Glycolytic and tricarboxylic acid cycle enzymes. *Parasitology* 90:169–77.

80. Panesar, T. S., and N. A. Croll. 1981. The hatching process in *Trichuris muris* (Nematoda: Trichuroidea). *Can. J. Zool.* 59:621–28.

81. Passey, R. F., and D. Fairbairn. 1957. The conversion of fat to carbohydrate during embryonation of *Ascaris* eggs. *Can. J. Biochem. Physiol.* 35:511–25.

82. Perry, R. N., and A. J. Clarke. 1981. Hatching mechanisms of nematodes. *Parasitology* 83:435–49.

83. Petersen, J. J. 1982. Current status of nematodes for the biological control of insects. In Anderson, R. M., and E. U. Canning, eds. Parasites as biological control agents. Symposia of the British Society for Parasitology 19. *Parasitology* 84(4):177–204.

84. Petronijevic, T., W. P. Rogers, and R. I. Sommerville. 1985. Carbonic acid as the host signal for the development of parasitic stages of nematodes. *Int. J. Parasitol.* 15:661–67.

85. Petronijevic, T., W. P. Rogers, and R. I. Sommerville. 1986. Organic and inorganic acids as the stimulus for exsheathment of infective juveniles of nematodes. *Int. J. Parasitol.* 16:163–68.

86. Platzer, E. G. 1980. Nematodes as biological control agents. *California Agric.* 34(3):27.

87. Poinar, G. O. Jr. 1983. *The natural history of nematodes.* Englewood Cliffs, N.J.: Prentice-Hall, Inc.

88. Politz, S. M., and M. Philipp. 1992. *Caenorhabditis elegans* as a model for parasitic nematodes: A focus on the cuticle. *Parasitol. Today* 8:6–12.

89. Roberts, L. S., and D. Fairbairn. 1965. Metabolic studies on adult *Nippostrongylus brasiliensis* (Nematoda: Trichostrongyloidea). *J. Parasitol.* 51:129–38.

90. Rogers, W. P. 1982. Enzymes in the exsheathing fluid of nematodes and their biological significance. *Int. J. Parasitol.* 12:495–502.

91. Rogers, W. P., and R. I. Sommerville. 1963. The infective stage of nematode parasites and its significance in parasitism. In Dawes, B., ed. *Advances in parasitology* 1. New York: Academic Press, Inc., 109–77.

92. Rosenbluth, J. 1965. Ultrastructural organization of obliquely striated muscle fibers in *Ascaris lumbricoides. J. Cell Biol.* 25:495–515.

93. Rosenbluth, J. 1965. Ultrastructure of somatic cells in *Ascaris lumbricoides.* II. Intermuscular junctions, neuromuscular junctions, and glycogen stores. *J. Cell Biol.* 26:579–91.

94. Rosenbluth, J. 1967. Obliquely striated muscle. III. Concentration mechanism of *Ascaris* body muscle. *J. Cell Biol.* 34:15–33.

95. Rothstein, M. 1970. Nematode biochemistry. XI. Biosynthesis of fatty acids by *Caenorhabditis briggsae* and *Panagrellus redivivus. Int. J. Biochem.* 1:422–28.

96. Rothstein, M., and H. Mayoh. 1964. Glycine synthesis and isocitrate lyase in the nematode, *Caenorhabditis briggsae. Biochem. Biophys. Res. Comm.* 14:43–47.

97. Sangster, N. C., R. K. Prichard, and E. Lacey. 1985. Tubulin and benzimidazole-resistance in *Trichostrongylus colubriformis* (Nematoda). *J. Parasitol.* 71:645–51.

98. Savel, J. 1955. Études sur la constitution et le métabolism protéiques d'*Ascaris lumbricoides* Linné, 1758. *Rev. Path. Comp. Hyg. Gen. Comp.* 55:52–121.

99. Saz, H. J. 1990. Helminths: Primary models for comparative biochemistry. *Parasitol. Today* 6:92–93.

100. Saz, D. K., T. P. Bonner, M. Karlin, and H. J. Saz. 1971. Biochemical observations on adult *Nippostrongylus brasiliensis. J. Parasitol.* 57:1159–62.

101. Saz, H. J., and A. Weil. 1960. The mechanism of the formation of α-methylbutyrate from carbohydrate by *Ascaris lumbricoides* muscle. *J. Biol. Chem.* 235:914–18.

102. Saz, H. J., and A. Weil. 1962. Pathway of formation of α-methylvalerate by *Ascaris lumbricoides. J. Biol. Chem.* 237:2053–56.

103. Schad, G. A. 1977. The role of arrested development in the regulation of nematode populations. In Esch, G. W., ed. *Regulation of parasite populations.* New York: Academic Press, Inc., 111–67.

104. Schad, G. A. 1983. Arrested development of *Ancylostoma caninum* in dogs: Influence of photoperiod and temperature on induction of a potential to arrest. In Meerovitch, E., ed. *Aspects of parasitology.* Montreal: Institute of Parasitology, McGill University, 361–91.

105. Sepsenwol, S., M. Nguyen, and T. Braun. 1986. Adenylate cyclase activity is absent in inactive and motile sperm in the nematode parasite, *Ascaris suum. J. Parasitol.* 72:962–64.

106. Sharpe, M. J., and D. L. Lee. 1981. Observations on the structure and function of the haemoglobin from the cuticle of *Nippostrongylus brasiliensis* (Nematoda). *Parasitology* 83:411–24.

107. Sheffield, H. G. 1963. Electron microscopy of the bacillary band and stichosome of *Trichuris muris* and *T. vulpis. J. Parasitol.* 49:998–1009.

108. Sheffield, H. G. 1964. Electron microscope studies on the intestinal epithelium of *Ascaris suum. J. Parasitol.* 50:365–79.

109. Smales, L. R. 1984. The egg shell of *Labiostrongylus eugenii* (Nematoda, Strongyloidea): Structure and function. *Int. J. Parasitol.* 14:231–39.

110. Sommerville, R. I., and P. P. Weinstein. 1964. Reproductive behavior of *Nematospiroides dubius* in vivo and in vitro. *J. Parasitol.* 50:401–9.

111. Stewart, G. R., R. N. Perry, J. Alexander, and D. J. Wright. 1993. A glycoprotein specific to the amphids of *Meloidogyne* species. *Parasitology* 106:405–12.

112. Stretton, A. O. W., C. Cowden, P. Sithigorngul, and R. E. Davis. 1991. Neuropeptides in the nematode *Ascaris suum. Parasitology* 102:S107–S116.

113. Stretton, A., J. Donmoyer, R. Davis, J. Meade, C. Cowden, and P. Sithigorngul. 1992. Motor behavior and motor nervous system function in the nematode *Ascaris suum. J. Parasitol.* 78:206–14.

114. Strote, G., and I. Bonow. 1993. Ultrastructural observations on the nervous system and the sensory organs of the infective stage (L3) of *Onchocerca volvulus* (Nematoda: Filarioidea). *Parasitol. Res.* 79:213–20.

115. Swartz, F. J., M. Henry, and A. Floyd. 1967. Observations on nuclear differentiation in *Ascaris. J. Exp. Zool.* 164:297–307.

116. Thorson, R. E. 1956. The effect of extracts of the amphidial glands, excretory glands, and esophagus of adults of *Ancylostoma caninum* on the coagulation of dog's blood. *J. Parasitol.* 42:26–30.

117. Turner, A. C., and W. F. Hutchison. 1979. Lipid synthesis in the adult dog heartworm, *Dirofilaria immitis. Comp. Biochem. Physiol.* 64B:403–5.

118. Viglierchio, D. R. 1991. *The world of nematodes.* Davis, Calif.: D. R. Viglierchio.

119. Wann, K. T. 1987. The electrophysiology of the somatic muscle cells of *Ascaris suum* and *Ascaridia galli. Parasitology* 94:555–66.

120. Ward, C. W., and D. Fairbairn. 1970. Enzymes of β-oxidation and their function during development of *Ascaris lumbricoides* eggs. *Dev. Biol.* 22:366–87.

121. Ward, C. W., and P. J. Schofield. 1967. Comparative activity and intracellular distribution of tricarboxylic acid cycle enzymes in *Haemonchus contortus* larvae and rat liver. *Comp. Biochem. Physiol.* 23:335–59.

122. Ward, P. F. V., and N. S. Huskisson. 1978. The energy metabolism of adult *Haemonchus contortus,* in vitro. *Parasitology* 77:255–71.

123. Ward, P. F. V., and N. S. Huskisson. 1980. The role of carbon dioxide in the metabolism of adult *Haemonchus contortus,* in vitro. *Parasitology* 80:73–82.

124. Wharton, D. A. 1979. The structure and formation of the egg-shell of *Hammerschmidtiella diesingi* (Hammerschmidt) (Nematoda: Oxyuroidea). *Parasitology* 79:1–12.

125. Wharton, D. A. 1979. The structure and formation of the egg-shell of *Syphacia obvelata* Rudolphi (Nematoda: Oxyurida). *Parasitology* 79:13–28.

126. Wharton, D. A. 1980. Nematode egg shells. *Parasitology* 81:447–63.

127. Willmer, P. 1990. *Invertebrate relationships. Patterns in animal evolution.* Cambridge: Cambridge University Press.

128. Wilson, P. A. G., M. Cameron, and D. S. Scott. 1978. *Strongyloides ratti* in virgin female rats: Studies of oestrous cycle effects and general variability. *Parasitology* 76:221–27.

129. Wright, K. A. 1964. The fine structure of the somatic muscle cells of the nematode *Capillaria hepatica* (Bancroft, 1893). *Can. J. Zool.* 42:483–90.

130. Wright, K. A. 1966. Cytoplasmic bridges and muscle systems in some polymyarian nematodes. *Can. J. Zool.* 44:329–40.

Additional References

Anderson, R. C. 1993. *Nematode parasites of vertebrates. Their transmission and development.* Wallingford, Oxon, Eng.: CAB International. A complete compilation of what is known about nematode life cycles in vertebrates.

Anderson, R. C., A. G. Chabaud, and S. Willmott. 1974–1985. *CIH keys to the nematode parasites of vertebrates.* Bucks, Eng.: Commonwealth Agricultural Bureaux, Farnham Royal. A 10-part up-to-date series of keys for the identification of nematode parasites.

Barrett, J. 1976. Bioenergetics in helminths. In Van den Bossche, H., ed. *Biochemistry of parasites and host-parasite relationships.* Amsterdam: Elsevier/North Holland Biomedical Press.

Bird, A. F., and J. Bird. 1991. *The structure of nematodes,* 2d ed. New York: Academic Press, Inc. This is a very good summary of nematode morphology. Indispensible for students of nematodology.

Bryant, C. 1978. The regulation of respiratory metabolism in parasitic helminths. In Lumsden, W. H. R., R. Muller, and J. R. Baker, eds. *Advances in parasitology* 16. New York: Academic Press, Inc.

Crofton, H. D. 1966. *Nematodes.* London: Hutchinson University Library. A very useful summary of nematode characteristics.

Croll, N. A., ed. 1976. *The organization of nematodes.* New York: Academic Press, Inc.

Croll, N. A., and B. E. Matthews. 1977. *Biology of nematodes.* New York: John Wiley & Sons, Inc.

Grassé, P. P. 1965. *Traité de zoologie: Anatomie, systématique, biologie, vol. 4, parts 2 and 3. Némathelminthes (Nématodes-Gordiacés), rotifères-gastrotriches, kinorinques.* Paris: Masson & Cie. An indispensable reference for serious students of nematodes.

Lee, D. L., and H. J. Atkinson. 1977. *Physiology of nematodes,* 2d ed. New York: Columbia University Press.

Levine, N. D. 1980. *Nematode parasites of domestic animals and of man,* 2d ed. Minneapolis: Burgess Publishing Co. An excellent general reference. The introductory chapter is useful for anatomy; Levine's proposed standard endings for taxa are used throughout.

Chapter 23

NEMATODES: TRICHURIDA AND DIOCTOPHYMATIDA, ENOPLEAN PARASITES

Trichinella spiralis: *the worm that would be virus.*

D. D. Despommier

Some of the most dreaded, disfiguring, and debilitating diseases of humans are caused by nematodes. In addition, agriculture suffers mightily from attacks by these animals. Nematodes normally parasitic in wild animals can occasionally infect humans and domestic animals, causing mystifying diseases. Furthermore, nonparasitic nematodes may accidentally find their way into a vertebrate and become a short-lived, but pathogenic, parasite. Many thousands of nematodes are known to parasitize vertebrates; many still are unknown. A few examples are presented here and in the following chapters, selected for their interest as parasites of humans and as illustrations of parasitism as exemplified by nematodes.

This chapter is devoted to the two orders of the Class Enoplea that are parasites of animals. Most enopleans are nonparasitic worms.

ORDER TRICHURIDA

The Trichurida contains, among others, three genera of medical importance: *Trichuris, Capillaria,* and *Trichinella.*

Family Trichuridae

Whipworms, members of the family Trichuridae, are so called because they are threadlike along most of their body, and then they abruptly become thick at the posterior end, reminiscent of a whip with a handle (Fig. 23.1). The name *Trichocephalus* (thread-head), in widespread use in some countries, was coined when it was realized that the "thread" was the anterior end rather than the tail, but the term *Trichuris* has priority. There are many species in a wide variety of mammalian hosts, and one is a very important parasite of humans.

Eggs of *Trichuris* have been found in fossil feces more than 2000 years old, and the worm has likely been with us for much longer; it probably coevolved with us as a parasite of our nonhuman ancestors.[7]

Trichuris trichiura

- **Morphology.** *Trichuris trichiura* measures 30 to 50 mm long, with males being somewhat smaller than females. The mouth is a simple opening, lacking lips. The buccal cavity is tiny and is provided with a minute spear. The esophagus is very long, occupying about two-thirds of

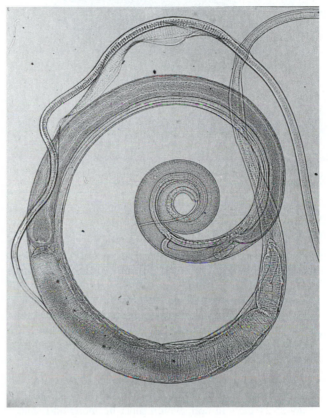

FIGURE 23.1

Male *Trichuris.* Note the slender anterior end and the stout posterior end with a single, terminal spicule.

Courtesy of Jay Georgi.

the body length, and consists of a thin-walled tube surrounded by large, unicellular glands, the **stichocytes.** The entire structure often is referred to as the **stichosome.** The anterior end of the esophagus is somewhat muscular and lacks stichocytes. Both sexes have a single gonad, and the anus is near the tip of the tail. Males have a single spicule that is surrounded by a spiny spicule sheath. The ejaculatory duct joins the intestine anterior to the cloaca. In the female the vulva is near the junction of the esophagus and the intestine. The uterus contains many unembryonated, lemon-shaped eggs, each with a prominent opercular plug at each end (Fig. 23.2).

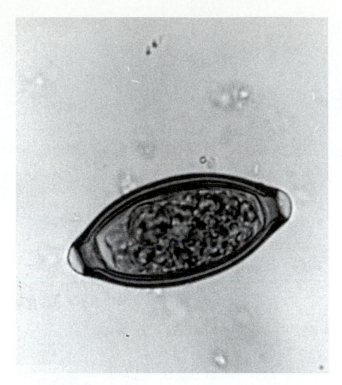

FIGURE 23.2

Egg of *Trichuris trichiura*. It measures 50 to 54 μm by 22 to 23 μm.
Courtesy of Robert E. Kuntz.

The excretory system is absent. The ventral surface of the esophageal regions bears a wide band of minute pores, leading to underlying glandular and nonglandular cells.[43,51] This **bacillary band** is typical of the order. Although the function of the cells in the bacillary band is unknown, their ultrastructure suggests that the gland cells may have a role in osmotic or ion regulation, and the nongland cells may function in cuticle formation and food storage.

- ***Biology.*** Each female worm produces from 3000 to 20,000 eggs per day.[7] Embryonation is completed in about 21 days in soil, which must be moist and shady. When swallowed, the infective juveniles hatch and enter the crypts of Lieberkühn. After penetration of the cells in the base of the crypts, the worms begin to grow and tunnel within the epithelium back toward the luminal surface. They can penetrate the gut mucosa in many places, but according to Bundy and Cooper, there is no evidence at present that worms entering cells other than in the large intestine develop further, nor is there evidence of a duodenal phase and later migration down the gut.[7] As the worms approach maturity, the enlarging posterior portion breaks out of the epithelium and protrudes into the intestinal lumen. The process requires about three months. The slender anterior ends remain embedded in the gut mucosa (Fig. 23.3), and the worms essentially are tissue parasites. Adults live for several years, so large numbers may accumulate in a person, even in areas in which the rate of new infection is low.

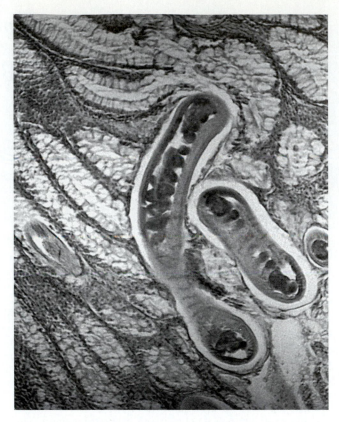

FIGURE 23.3

Section of large intestine with sections of *Trichuris trichiura* embedded in the mucosa. Individual stichocytes are evident.
Courtesy of Robert E. Kuntz.

- ***Epidemiology.*** The two requirements for *T. trichiura* to become a serious health problem are poor standards of sanitation in which human feces are deposited on the soil and the combination of physical conditions that allows the worm's survival and development: a warm climate, high rainfall and humidity, moisture-retaining soil, and dense shade. Although generally coextensive in distribution with *Ascaris, T. trichiura* is more sensitive to the effects of desiccation and direct sunlight. Appropriate physical conditions exist in much of the world, including parts of the southeastern United States, where the prevalence of infection may reach 20% to 25%, mainly in small children. Hopkins[23] estimated the world prevalence at 750 million. Often both *Ascaris* and *Trichuris* infections are present concurrently.[6,41]

Infected eggs are picked up from contaminated soil, and the practice of geophagy (eating soil) contributes to transmission in some localities.[7] Use of nightsoil as fertilizer for vegetables can be an important source of infection, and houseflies can serve as mechanical vectors.

Infections with *T. trichura* in human populations are characteristically overdispersed.[11,50] Furthermore, when individuals are treated for the infection, they tend to be predisposed to picking up large numbers of worms after

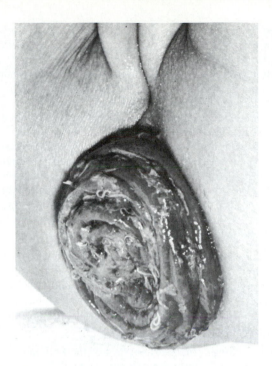

FIGURE 23.4

Prolapse of the rectum caused by whipworm infection.

Courtesy of University of Miami School of Medicine. From Beck, J. W., and J. E. Davies. 1981. Medical parasitology, ed. 3. The C. V. Mosby Co., St. Louis.

treatment, as with *Ascaris* (p. 421). Although there is no evidence that predisposition is genetic, that it exists has important implications for control programs.[11]

- **Pathology.** Fewer than 100 worms rarely cause clinical symptoms, and the majority of infections are symptomless. A heavier burden may result in a variety of conditions, occasionally terminating in death. Small children are particularly prone to heavy infections, which may involve 200 to more than 1000 worms.[15] In intense trichuriasis, dysentery, anemia, and growth retardation are very common, and finger clubbing and rectal prolapse (Fig. 23.4) are frequent. Moderate to heavy infections adversely affect cognitive function in children.[34] Finger (and toe) clubbing, which is an odd thickening of the ends of the digits, is not a significant disability, but it does indicate the systemic nature of the effects of this infection.

    With their anterior ends buried in the mucosa, the worms feed on cell contents and blood, although blood loss by this mechanism is negligible. Trauma to the intestinal epithelium and underlying submucosa, however, can cause a chronic hemorrhage that results in anemia. Pathogenesis is in large measure related to the host inflammatory response, but, there is a striking *absence* of normal markers of cell-mediated immunopathology.[14] Nonetheless, there is an increase in macrophages in the colonic lamina propria and increased TNF (p. 26) concentration both in the mucosa and systemic circulation. There is an increase in degranulating mast cells and a tenfold increase in proportion of lamina propria cells with surface IgE. This and other evidence suggests that inflammation in *Trichuris* colitis may be considered a local, tissue anaphylactic response.[14]

- **Diagnosis and Treatment.** Specific diagnosis depends on demonstrating a worm or egg in the stool. The eggs, with distinctive bipolar plugs, are 50 to 54 μm by 22 to 23 μm and have smooth outer shells. Their structure and formation have been reported by Preston and Jenkins.[39] Clinical symptoms may be confused with those of hookworm, amebiasis, or acute appendicitis. Worms can be dramatically demonstrated by colonoscopy.[7]

    Because of their frequent location in the cecum, appendix, or lower ileum, whipworms are difficult to reach with oral drugs or medicated enemas. Mebendazole and albendazole are effective drugs.[13] Training of children and adults in sanitary disposal of feces and in washing of hands is necessary to prevent reinfection.

• Other *Trichuris* Species

Some 60 to 70 other species of *Trichuris* have been described from a wide variety of mammals. Several species occur in wild and domestic ruminants, of which *T. ovis* is the most important in domestic sheep and cattle. *Trichuris suis,* which is indistinguishable from *T. trichiura,* is found in swine; *T. vulpis* is found in the cecum of dogs, foxes, and coyotes and is common in the United States except in the drier areas. This species occasionally infects humans.[25] *Trichuris muris* occurs often in rats and mice.

Family Capillariidae

Members of the genus *Capillaria* look very much like *Trichuris* spp., except that the transition between the anterior, filiform portion and the posterior, stout portion is gradual, rather than sudden. Other morphological features are similar. A large genus, *Capillaria* includes species that are parasitic in nearly all organs and tissues of all classes of vertebrates.

Capillaria hepatica

- **Biology.** *Capillaria hepatica* is a parasite of the liver, mainly of rodents, but it has been found in a wide variety of mammals, including humans. The female deposits eggs in the liver parenchyma, where they have no means of egress until they are eaten by a predator or until the liver decomposes after death. The eggs cannot embryonate while in the liver, so a new host cannot be infected when it eats an egg-laden liver. The eggs merely pass through the digestive tract of the predator with feces. Embryonation occurs on the soil, and new infection is by contamination. After hatching in the small intestine, the juveniles migrate to the liver, where they mature.

- **Epidemiology.** As with *T. trichiura,* infection occurs when contaminated objects or food is ingested. Choe and coworkers, who reported the first case from Korea, believed that geophagy is especially important.[12] Unlike with whipworm, however, human feces are not the source of contamination; more likely, the feces of carnivores or flesh-eating rodents are involved. Eggs of *C. hepatica*

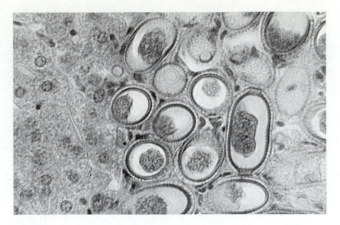

FIGURE 23.5

Eggs of *Capillaria hepatica* in liver. Note the extensive damage to hepatic parenchyma.

Courtesy of Warren Buss.

have been found in several species of earthworms. These transport hosts may facilitate infections in normal definitive hosts.[42]

This nematode has some potential as a biological control agent for rodent populations.[4,45]

- *Pathology.* Wandering of adult *C. hepatica* through the host liver causes loss of liver cells and thereby loss of normal function. Large areas of parenchyma may be replaced by masses of eggs (Fig. 23.5). Rarely, eggs will be carried to the lungs or other organs by the bloodstream.

  Hepatomegaly can become severe, and eggs become encased in granulomatous tissue, with heavy infiltration of eosinophils and other leukocytes.[12]

- *Diagnosis and Treatment.* There are only 26 reported cases of this parasite in humans,[12] partly because of difficulties of diagnosis. Most cases have been determined after death, but liver biopsy has uncovered others. Untreated cases may be fatal. Clinical symptoms resemble numerous liver disorders, especially hepatitis with eosinophilia. Specific diagnosis depends on demonstrating the eggs, which closely resemble those of *Trichuris* except that they measure 51 to 67 μm by 30 to 35 μm and have deep pits in the shells. Treatment for this disease is poorly known, but albendazole is currently the drug of choice.[13]

  Discovery of *C. hepatica* eggs in human feces may indicate the presence of a spurious infection caused by eating an infected liver.

Capillaria philippinensis. *Capillaria philippinensis* was discovered in 1963 as a parasite of humans in the Philippines. In contrast to *C. hepatica*, *C. philippinensis* is an intestinal parasite. Its appearance as a human pathogen was sudden and unexpected. One or two isolated cases were followed by an epidemic in Luzon in 1967 that killed several dozen persons.[16] It has now been reported from Thailand, Iran, Japan, and Egypt.

It is probably a zoonotic disease, but the original animal host remains unknown. *Capillaria philippinensis* has been transmitted experimentally to monkeys, gerbils, *Rattus* spp., and several species of migratory fish-eating birds.[16] Some female worms bear living juveniles, and eggs, juveniles, and adults pass from the definitive host in the feces. When the feces reach water, the eggs embryonate and are eaten by small fishes. After hatching in the fish intestine, the juveniles develop for a few weeks until they become infective for the definitive host. Juveniles liberated in the host intestine are autoinfective, and massive populations can accumulate, causing severe pathology.

- *Morphology.* This parasite is very small; males measure 2.3 to 3.2 mm, and females measure 2.5 to 4.3 mm long. The male has small caudal alae and a spineless spicule sheath. The esophagus of the female is about half as long as the body. The female produces typical *Capillaria*-type eggs that lack pits.

- *Epidemiology.* Intensive surveys of the Philippine fauna have so far failed to identify any reservoir host, but fish-eating birds are prime suspects. Migratory birds are probably the means by which the infection has spread to other Asian countries and even to the Middle East. Because the infective juveniles are in fish intestines, any region where people savor small, whole raw fish may experience new cases.

- *Pathology.* The worms repeatedly penetrate the mucosa of the small intestine and reenter the lumen, especially in the jejunum, leading to progressive degeneration of the epithelium and submucosa. Infected persons usually experience diarrhea and abdominal pain, progressing to weight loss, weakness, malaise, anorexia, and emaciation.[16] Protein and electrolytes, especially potassium, are lost, and there is malabsorption of fats and sugars. Patients die from loss of electrolytes, heart failure, and sometimes secondary bacterial infection.

- *Diagnosis and Treatment.* Both adults and eggs, as well as juveniles, are abundant in feces of heavily infected persons, and at least one of them is necessary for specific diagnosis. The only other intestinal nematode in humans in which the juveniles pass in the feces is *Strongyloides stercoralis* (p. 400), and the presence of the easily distinguished eggs and adults of *C. philippinensis* differentiates these two.

  Mebendazole and albendazole are effective in curing this disease. Control consists of persuading people to refrain from eating small raw fish whole.

• Other *Capillaria* Species

Several species of *Capillaria* are important parasites of domestic animals. *Capillaria aerophila* is a lung parasite of

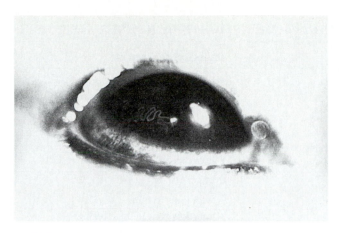

FIGURE 23.6

Anatrichosoma ocularis in the eye of a tree shrew, *Tupaia glis.*

From S. K. File, "*Anatrichosoma ocularis* sp. n. (Nematoda: Trichosomoididae) from the eye of the common tree shrew, *Tupaia glis,*" in *J. Parasitol.* 60:985-988. Copyright © 1974.

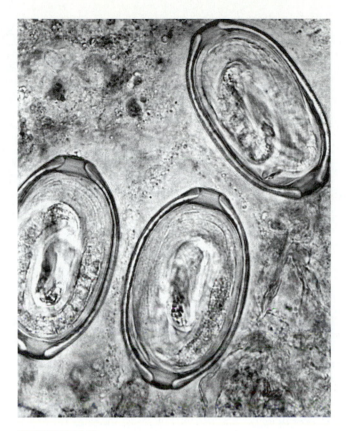

FIGURE 23.7

Eggs of *Anatrichosoma ocularis* from eye secretions.

Courtesy of Sharon K. File.

cats, dogs, and other carnivores and has been reported several times from humans. It is probably the most destructive parasite of commercial fox farms.

Capillaria annulata and *C. caudinflata* infect the esophagus and crop of chickens, turkeys, and several other species of birds. Unlike most species in the genus, these species require an intermediate host (earthworm) in the life cycle.

Few species of terrestrial vertebrates, wild or domestic, are free from at least one species of *Capillaria.*

Family Anatrichosomatidae

• *Anatrichosoma* Species

Species of *Anatrichosoma* are very similar to *Capillaria,* except that they lack a spicule and spicule sheath. They have been reported from the tissues of a wide variety of Asian and African monkeys and gerbils and from the North American opossum. *Anatrichosoma ocularis* (Fig. 23.6) lives in the corneal epithelium of tree shrews, *Tupaia glis.* The eggs of this genus (Fig. 23.7) have the polar plugs characteristic of the order. No species is known to parasitize humans, but the species in monkeys, at least, should be considered as potential zoonoses.

Family Trichinellidae

Trichinella Spp. Curiously, the smallest nematode parasite of humans, which exhibits the most unusual life cycle, is one of the most widespread and clinically important parasites in the world. This parasite was described in 1835 and since 1895 has been known as *Trichinella spiralis.* We have come to realize that *T. spiralis* is actually several sibling species, subspecies, or strains, according to various authors. There are, however, no morphological differences between the different kinds of *Trichinella.* Sufficient evidence has accumulated that we can recognize at least five sibling species (Table 23.1).[37] There are other biological characteristics of these species, in addition to those shown in Table 23.1.[36] Allozyme and ribosomal DNA analysis support these distinctions. *Trichinella* spp. can also be distinguished on the basis of the disease produced in humans,[38] recognition of antigens by monoclonal antibodies,[24] and some repetitive sequences in their DNA amplified by the polymerase chain reaction.[3,47]

In the discussion to follow, we will be referring to *T. spiralis* in the strict sense, except where noted otherwise.

These parasites are responsible for the disease variously known as trichinosis, trichiniasis, or trichinelliasis. It is common in carnivorous mammals, including rodents and humans, primarily on the circumboreal continents. *Trichinella* spp. are less common in tropical regions but are well-known in Mexico, parts of South America, Africa, southern Asia, and the Middle East. Incidence of infection is always higher than suspected because of the vagueness of symptoms, which usually suggest other conditions; more than 50 different diseases have been diagnosed incorrectly as trichinosis.

Table 23.1   BIOLOGICAL AND DISTRIBUTIONAL CHARACTERISTICS OF *TRICHINELLA* SPP.*

Species	Distribution	Biological characters
Trichinella spiralis, strict sense	Cosmopolitan	• high RCI** in rats and pigs • newborn juvenile production/72 hr in vitro > 90 (others < 60) • no freezing resistance
T. nativa	Arctic and subarctic zones, Holarctic Region	• low RCI in rats and pigs • high freezing resistance
T. pseudospiralis	Cosmopolitan	• nurse cell absent • infectious to birds • very low RCI in pigs, high in rats • no freezing resistance
T. nelsoni, strict sense	Tropical Africa	• low RCI in pigs and rats • no freezing resistance
T. britovi	Temperate zone, Palaearctic Region	• low RCI in rats and pigs • low freezing resistance

*Modified from Pozio et al.[37]

**RCI = reproductive capacity index.

- *Morphology.* The males (Fig. 23.8) measure 1.4 to 1.6 mm long and are more slender at the anterior than the posterior end. The anus is nearly terminal and has a large copulatory **pseudobursa** on each side of it (Fig. 23.9). A copulatory spicule is absent. Like other members of the order Trichurida, stichocytes are arranged in a row following a short muscular esophagus. Females are about twice the size of males, also tapering toward the anterior end. The anus is nearly terminal. The vulva is near the middle of the esophagus, which is about a third the length of the body. The single uterus is filled with developing eggs in its posterior portion, whereas the anterior portion contains fully developed, hatching juveniles.

- *Biology.* The biology of this organism is unusual in that the same individual animal serves as both definitive and intermediate host, with the juveniles and the adults located in different organs. Even more bizarre, however, is another unique character: It is the world's largest intracellular parasite! As Dickson Despommier described it, "*Trichinella spiralis:* the worm that would be virus."[17] The adults are intramulticellular parasites in the intestinal epithelium, and the juveniles reside in nurse cells, whose formation they themselves induce, in skeletal muscle.

  When the infective juveniles are swallowed and reach the small intestine of the host, they are released from their nurse cells and enter the intestinal mucosa. Four molts, growth, and copulation occur within the mucosal epithelium within 30 to 32 hours after infection.[17,22] The worms lie directly in the cytoplasm and thread through a serial row of intestinal cells (Fig. 23.10). During her sojourn in the intestinal epithelium, the female gives birth to between hundreds and thousands of juveniles over a period of 4 to 16 weeks. Eventually the spent female dies and passes out of the host. Males can copulate several times but then die shortly after.

  Most juveniles are carried away by the hepatoportal system through the liver and then to the heart, lungs, and the arterial system, which distributes them throughout the body. During this migration, they may be found in literally every kind of tissue and space in the body. When they reach skeletal muscle, they penetrate individual fibers. Then, in another strategy reminiscent of viruses, they subvert and redirect the host cell activities to their own survival. Gene expression of the host cell is altered from that of a contractile fiber to that of a nurse cell, a cell that functions in nourishing the worm. After the nematode enters it, the fiber loses its myofilaments, but its nuclei enlarge (hypertrophy) and the smooth endoplasmic reticulum increases. Mitochondria degenerate, while collagen is overexpressed, and eventually the entire unit becomes encapsulated with colla-

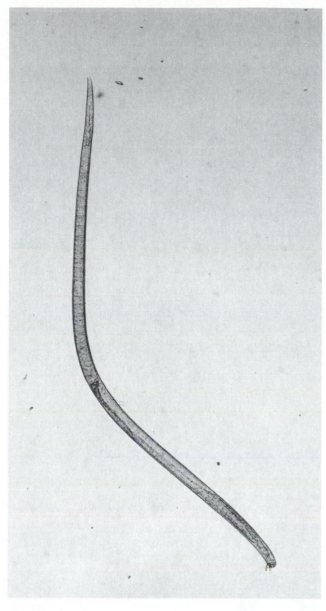

FIGURE 23.8

Male *Trichinella spiralis* (1.4 to 1.6 mm in length) from the intestine of a rat.

Courtesy of Jay Georgi.

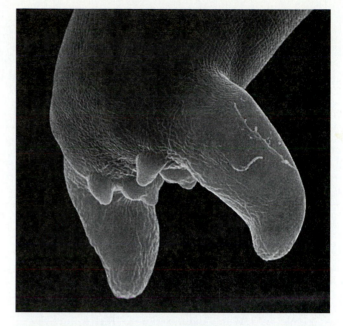

FIGURE 23.9

Scanning electron micrograph of *Trichinella nativa*, posterior end of male, showing copulatory pseudobursae and papillae.

From J. R. Lichtenfels et al., "Comparison of three subspecies of *Trichinella spiralis* by scanning electron microscopy," in *J. Parasitol.* 69:1131–1140. Copyright © 1983.

gen (Fig. 23.11). A network of tiny blood vessels (a **circulatory rete**) (Fig. 23.12) forms around the parasite-nurse-cell complex.[18] The enclosing nurse cell has until recently been referred to as a "cyst" and the worm as "encysted," on the assumption that the "cyst" existed only for containment.

We do not yet understand the mechanism by which this astonishing transformation occurs, but it is under active investigation. Material (antigen) secreted from the stichocytes has been identified in the cytoplasm and hypertrophic nuclei of nurse cells.[19,27] The antigen is present in stichocytes of six-day-old juveniles but not in nurse-cell nuclei at that time. However, it can be found in nurse-cell nuclei by eight days, when nuclear hypertrophy is greatest.[18] These observations led to the hypothesis that secretory-excretory (S-E) substances from the stichocytes mediate alteration of host gene expression. Injection into muscle of S-E material collected from juveniles cultured in vitro can produce similar changes in muscle fibers.[26]

Some muscles are much more heavily invaded than are others, but the reasons for this are not understood. Most susceptible are muscles of the eye and tongue and

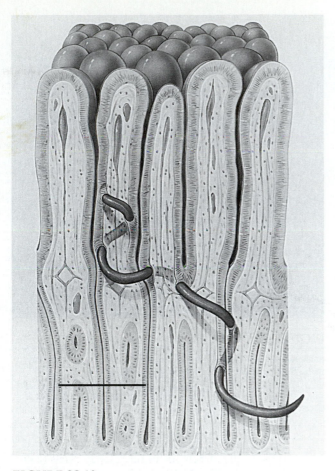

FIGURE 23.10

Conceptual illustration of an adult *T. spiralis* threading its way through the intestinal epithelium. (Scale bar = 400 μm.)

Illustration by J. Karapelou. From D. D. Despommier, "*Trichinella spiralis* and the concept of niche," in *J. Parasitol.* 79:472–482. Copyright © 1993. Reprinted with permission of the publisher and author.

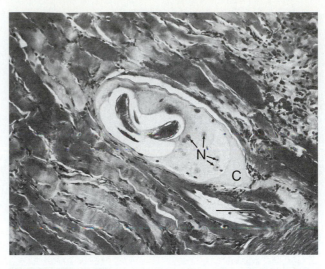

FIGURE 23.11

Juvenile of *Trichinella spiralis* in a muscle nurse cell, the nurse cell-parasite complex. The outer layer is collagen (**C**). Note the hypertrophied nuclei (**N,** *arrows*). (Scale bar = 100 μm.)

From D. D. Despommier, "*Trichinella spiralis:* the worm that would be virus," in *Parasitol. Today.* 6:193–196. Copyright © 1990 Elsevier Trends Journals. Reprinted with permission of the publisher and author.

FIGURE 23.12

Schematic drawing of an intact nurse cell-parasite complex, showing the surrounding circulatory rete.

From D. D. Despommier, "*Trichinella spiralis:* the worm that would be virus," in *Parasitol. Today.* 6:193–196. Copyright © 1990 Elsevier Trends Journals. Reprinted with permission of the publisher and author.

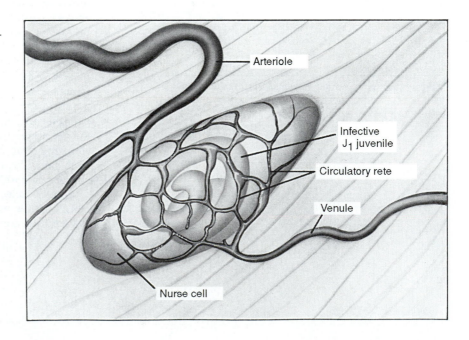

masticatory muscles, then the diaphragm and intercostals, and finally the heavy muscles of the arms and legs. Juveniles of *T. spiralis* invade only slow-twitch fibers, whereas those of *T. pseudospiralis* invade both fast-twitch and slow-twitch fibers.[2] The juveniles absorb their nutrients from the enclosing nurse cell and increase in length to about 1 mm in about four to eight weeks, at which time they are infective to their next host.

The nurse cell becomes gradually thicker during this time, and finally achieves a length of 0.25 to 0.50 mm. The juveniles enter a developmental arrest and can live for months or years while encased. Gradually host reactions begin to calcify the nurse cell and eventually the worms themselves. However, living juveniles in muscle up to 39 years after first infection have been documented.[18]

It was the presence of such calcified granules in human cadavers that led to the discovery of this species in 1835 by James Paget, a medical student in London. Noticing that his subject had gritty particles in its muscles that tended to dull his scalpels, he studied the particles and demonstrated their wormlike nature to his fellow students. He then showed them to the eminent anatomist Richard Owen, who reported on them further and gave them their scientific name. It was another 25 years before it was determined that these minute animals cause disease.

- *Epidemiology.* Trichinosis today may best be considered a zoonotic disease, since humans can scarcely be important in the life cycle of the parasite. Unless an infected person is eaten by a carnivorous predator or becomes a cannibal's supper, both unlikely events these days, the parasites are at a dead end in a human.

We have traditionally considered the life cycle of *Trichinella spiralis* as two epidemiologically distinct types: the urban cycle (involving pigs and rats, around human habitation) and the sylvatic cycle (involving wild animals). Campbell[8] provided a different framework, taking account of the host preferences of the different species of *Trichinella*. He recognized four distinct epidemiological cycles: **domestic, sylvatic—temperate zone, sylvatic—tropical variant,** and **sylvatic—arctic variant** (Fig. 23.13).

Domestic trichinosis involves *Trichinella spiralis*, in the strict sense (Table 23.1). It is epidemiologically most important to humans because of the close relationship among rats, pigs, and people. Infected pork is our most common source of infection. Pigs become infected by eating offal or trichinous meat in garbage or by eating rats, which are ubiquitous in pig farms. Garbage containing raw pork scraps is probably the usual source of infection for pigs, but pigs will greedily devour dead or even live rats when they can catch them. The rats likely maintain their infections by cannibalism, although rat and mouse feces can contain juveniles capable of infecting rats, pigs, or humans.[28,29,35] Juveniles in the muscles of mice can alter the behavior of the rodent, making them more vulnerable to predation.[52] Infection can spread from pig to pig when they nip off and eat each other's tails, a common practice in crowded piggeries.[46] Piggish cannibalism also is involved. The worm can survive and remain infective after the anaerobic digestion of sewage sludge,[21] but freezing at −15°C destroys all parasites.

The importance of cooking pork thoroughly before it is eaten cannot be overstated. A roast or other piece of solid meat is safe when all traces of pink have disappeared. Many persons are careful about this but become careless when cooking sausage, which is equally dangerous. Raw sausage is a delicacy among many peoples of the world, particularly in the areas where trichinosis is a chronic health problem. Even a casual taste to determine proper seasoning can be fatal. Particularly important in transmission is meat processed by "backyard butchers" (individuals who slaughter their own stock) or by very small packing houses. Sausage from such sources is unlikely to be diluted with uninfected meat.

Reported cases (which are likely a tiny fraction of the true number) in the United States declined from over 400 per year in the 1940s to about 30 to 40 per year from 1987 to 1989. In 1990 the incidence jumped to over 100; many of these cases were Southeast Asian immigrants, who are fond of raw pork sausage.[10]

Sylvatic trichinosis in the temperate zone is due to both *T. spiralis* and *T. britovi*. *Trichinella britovi* has some resistance to freezing but not nearly so much as *T. nativa,* which is responsible for sylvatic trichinosis in the arctic. Because sylvatic trichinosis involves wild mammals, humans are infected only when they interject themselves into the sylvatic food chain. Native Americans and others who rely on wild carnivores for food and urban dwellers who return home with the spoils of the hunt are all subject to infection with *Trichinella*. Fatal cases of trichinosis are common among those who eat undercooked or underfrozen bear, wild pig, cat, dog, or walrus meat. Theoretically, any wild mammal may be a source of infection, but of course most rarely find their way to the dinner table. In Alaska, polar bears, black bears, and walrus are common sources of *T. nativa*.[9,31,40] Arctic explorers have been killed by *Trichinella* acquired from uncooked polar bear meat. The cause of death of the three members of the ill-fated André polar expedition of 1897 was determined *50 years later* when *Trichinella* juveniles were found in museum specimens of the polar bear meat that the men had been eating before they died.[49] An outbreak at Barrow, Alaska, in 1980 was due to grizzly bear meat that had been stored frozen.[9]

The cause of sylvatic trichinosis in tropical Africa, *T. nelsoni,* has a low infectivity for pigs and rats, but it can cause intensive infections and even death in humans.[8] The carrion-eating habits of hyenas suggest that they are important in transmission, but any predator feeding on the carcass of another predator could transmit the infection. Humans in this cycle would not be dead-end hosts if they were fed upon by wild predators.

The epidemiological importance of *T. pseudospiralis* has yet to be assessed. Although it has a high reproductive capacity in rats, it has a low capacity in pigs. It may be primarily a parasite of birds. It has no freezing resistance.

Survivors of trichinosis have varying degrees of immunity to further infection, and this immunity in mice can

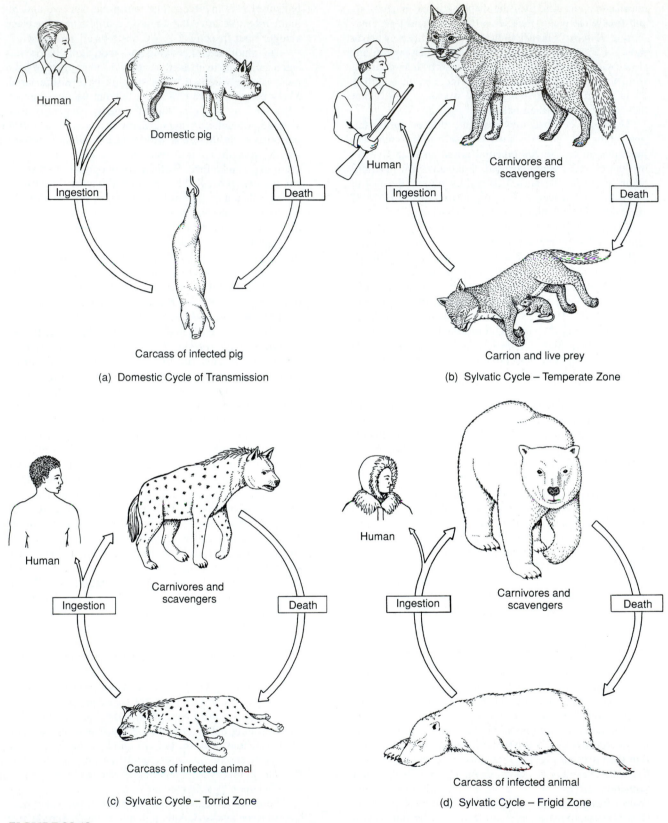

(a) Domestic Cycle of Transmission

(b) Sylvatic Cycle – Temperate Zone

(c) Sylvatic Cycle – Torrid Zone

(d) Sylvatic Cycle – Frigid Zone

FIGURE 23.13

Various life cycles of *Trichinella* spp. These are all essentially zoonoses, with humans becoming infected incidentally and not playing an essential role in the life cycle. (*a*) Domestic cycle, primarily *T. spiralis.* (*b*) Temperate zone sylvatic cycle, *T. spiralis* and *T. britovi.* (*c*) Tropic zone sylvatic cycle, *T. nelsoni.* (*d*) Arctic zone sylvatic cycle, *T. nativa.*

Drawing by William Ober and Claire Garrison from W. C. Campbell, "Trichinosis revisited—Another look at modes of transmission," *Parasitol. Today* 4:83-86. Copyright © 1988.

be passed by a mother to her young by suckling.[20] Suckling rats rapidly expel *Trichinella* if the mother is immune. Her serum antibodies are passed through the milk to the pups.[1]

- **Pathogenesis.** The pathogenesis of *Trichinella* infection can be considered in three successive stages: penetration of adult females into the mucosa, migration of juveniles, and penetration and nurse cell formation.

    First symptoms may appear between 12 hours and two days of ingestion of infected meat. Commonly this phase is clinically inapparent because of low-grade infection or is misdiagnosed because of the vagueness of symptoms. Worms migrating in the intestinal epithelium cause traumatic damage to the host tissues; the host begins to react to their waste products. Inflammation causes symptoms such as nausea, vomiting, sweating, and diarrhea. Respiratory difficulties may occur, and red blotches erupt on the skin in some cases. This period usually terminates with facial edema and fever five to seven days after the first symptoms.

    During migration the newborn juveniles damage blood vessels, resulting in localized edema, particularly in the face and hands. Wandering juveniles may also cause pneumonia, pleurisy, encephalitis, meningitis, nephritis, deafness, peritonitis, brain or eye damage, and subconjunctival or sublingual hemorrhage. Death resulting from myocarditis (inflammation of the heart muscle) may occur at this stage. Although the juveniles do not stay in the heart, they migrate through its muscle, causing local areas of necrosis and infiltration of leukocytes.

    By the tenth day after the first symptoms appear, the juveniles begin penetration of muscle fibers. Attendant symptoms are again varied and vague: intense muscular pain, difficulty in breathing or swallowing, swelling of masseter muscles (occasionally leading to a misdiagnosis of mumps), weakening of pulse and blood pressure, heart damage, and various nervous disorders, including hallucination. Extreme eosinophilia is common but may not be present even in severe cases. Death is usually caused by heart failure, respiratory complications, or kidney malfunction.

- **Diagnosis and Treatment.** Most cases of trichinosis, particularly subclinical cases, go undetected. Routine examinations rarely detect juveniles in feces, blood, milk, or other secretions. Although muscle biopsy is seldom employed, it remains an accurate diagnostic if trichinosis is suspected. Pressing the tissue between glass plates and examining it by low-power microscopy is useful, although digestion of the muscle in artificial gastric enzymes for several hours provides a much more reliable diagnostic technique. Several immunodiagnostic tests are available.[33]

    No really satisfactory treatment for trichinosis is known. Treatment is basically that of relieving the symptoms by use of analgesics and corticosteroids. Purges during the initial symptoms may dislodge the females that have not yet begun penetrating the intestinal epithe-

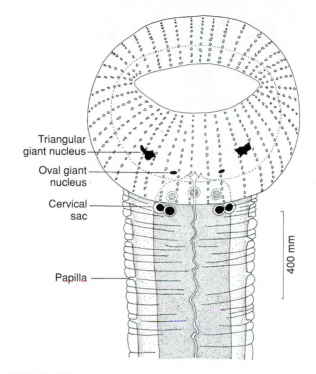

FIGURE 23.14

Soboliphyme baturini from a short-tailed weasel. Note the swollen mouth capsule typical of this genus.

From G. D. Schmidt and J. M. Kinsella, "Contribution to the morphology of *Soboliphyme baturini* Petrow, 1930 (Dioctophymoidea: Nematoda)," in *Trans. Am. Microsc. Soc.* 84:413–415. Copyright © 1965. Reprinted with permission of the publisher.

lium. Thiabendazole has been shown effective in experimental animals, but results in clinical cases have been variable.

Despite immense research, trichinosis remains an important disease of humans, one that has the potential of striking anyone, anywhere. One hopeful note: For unknown reasons the incidence of infection has slowly but steadily declined throughout the world.

ORDER DIOCTOPHYMATIDA

The few members of the order Dioctophymatida are parasites of aquatic birds and terrestrial mammals. There are three families in the order: Soboliphymidae has the single genus *Soboliphyme,* parasites of shrews and mustelid carnivores (Fig. 23.14), Eustrongylidae has the genera *Eustrongylides*[48] and *Hystrichis,* both parasites of wild birds, although infection of humans by juvenile *Eustrongylides* has been reported.[44] Dioctophymatidae is of more concern than the first two and will be considered more fully.

Family Dioctophymatidae

The family Dioctophymatidae has three genera, *Dioctowittius, Mirandonema,* and *Dioctophyme,* each with a single

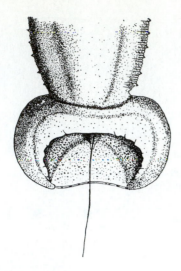

FIGURE 23.15

Posterior end of a male *Dioctophyme renale*. Note the single spicule and the powerful copulatory bursa.

From W. Stefanski, "Quelques précisions sur les caractères spécifiques du strongle géant du chien," in *Ann. Parasitol. Hum. Comp.* 6:93–100. Copyright © 1928. Reprinted with permission of the publisher.

FIGURE 23.16

Egg of *Dioctophyme renale,* showing the corrugated shell and the two-cell embryo. This stage is released by the female worm.

From T. F. Mace and R. C. Anderson, "Development of the giant kidney worm, *Dioctophyma renale* (Goeze, 1782 Nematoda: Dioctophymatoidea)," in *Can. J. Zool.* 53:1552–1568. Copyright © 1975. Reprinted with permission of the publisher.

species. The first is known only from a snake and the second from a Brazilian carnivore, but the third has been reported from a wide variety of mammals, including humans, in many parts of the world.

Dioctophyme renale

- **Morphology.** *Dioctophyme renale* is truly a giant among nematodes, with the males up to 20 cm long and 6 mm wide and the females up to 100 cm long and 12 mm wide. They are blood red and rather blunt at the ends. The male has a conspicuous, bell-shaped copulatory bursa that lacks any supporting rays or papillae (Fig. 23.15). A single, simple spicule is 5 to 6 mm long. The vulva is near the anterior end. Eggs (Fig. 23.16) are lemon shaped, with deep pits in the shells, except at the poles.

- **Biology.** The life cycle of this parasite has been described.[30] The thick-shelled eggs require about 0.5 to 3.0 months in water to embryonate, depending on the temperature. The first-stage juvenile, which bears an oral spear, hatches when eaten by the aquatic oligochaete annelid *Lumbriculus variegatus.* There it penetrates into the ventral blood vessel, where it develops into the third-stage juvenile. When the small annelid is swallowed, the juvenile migrates to a kidney of the new host, where it matures. If the annelid is eaten by any of several species of fish and frogs, the juvenile will encyst in the muscle or viscera, using the fish or frog as a paratenic host.[32] When swallowed by a definitive host, the juvenile penetrates the stomach wall to the submucosa. After about five days it migrates to the liver and remains in the liver parenchyma

for about 50 days; then it migrates to the kidney. Usually the right kidney is invaded, perhaps because of the proximity of the stomach to the right lobes of the liver, from which the worms can migrate directly to the kidney.[30] The worms mature in the kidney, and the eggs are voided from the host in its urine. In the rare human infections the worms usually are in a kidney, but one was a third-stage juvenile in a subcutaneous nodule.[5]

- **Epidemiology.** Probably any species of large mammal can serve as definitive host. Because of their fish diets, mustelids, canids, and bears are particularly susceptible, as are humans. However, even such non-fish-eating mammals as cows, horses, and pigs can become infected when they accidentally ingest an infected annelid. Thorough cooking of fish and drinking of only pure water will prevent infection in people.

- **Pathology.** Pressure necrosis caused by the growing worm, together with its feeding activities, reduces the infected kidney to a thin-walled, ineffective organ (Figs. 23.17 and 23.18). Loss of kidney function is compounded by uremic poisoning. The worms will sometimes penetrate the renal capsule and wander in the coelomic cavity.

- **Diagnosis and Treatment.** The rarity of this parasite makes physicians unlikely to suspect its presence. Demonstration of the characteristic eggs in the urine is the only positive means of diagnosis, aside from surgical discovery of the worm itself. Surgical removal of the worm is the only treatment known.

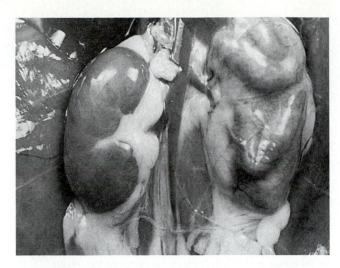

FIGURE 23.17

Ferret dissection: The kidney on the left is normal, whereas that on the right is distended by an adult *Dioctophyme renale*. (Anterior of animal is toward the bottom.)

Arthur E. Woodhead. Courtesy Ann Arbor Biological Center.

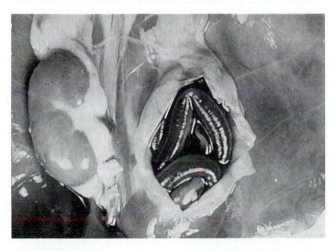

FIGURE 23.18

Specimen in Figure 23.17 with infected kidney opened to reveal the worm. The organ is reduced to a hollow shell.

Arthur E. Woodhead. Courtesy Ann Arbor Biological Center.

References

1. Appleton, J. A., and D. D. McGregor. 1984. Rapid expulsion of *Trichinella spiralis* in suckling rats. *Science* 226:68–72.

2. Bagheri, A., J. E. Ubelaker, G. L. Stewart, and B. Wood. 1986. Muscle fiber selectivity of *Trichinella spiralis* and *Trichinella pseudospiralis*. *J. Parasitol.* 72:277–82.

3. Bandi, C., G. La Rosa, S. Comincini, G. Damiani, and E. Pozio. 1993. Random amplified polymorphic DNA technique for the identification of *Trichinella* species. *Parasitology* 107:419–24.

4. Barker, S. C., G. R. Singleton, and D. M. Spratt. 1991. Can the nematode *Capillaria hepatica* regulate abundance in wild house mice? Results of enclosure experiments in southeastern Australia. *Parasitology* 103:439–49.

5. Beaver, P. C., and J. H. Theis. 1979. Dioctophymatid larval nematode in a subcutaneous nodule from man in California. *Am. J. Trop. Med. Hyg.* 28:206–12.

6. Booth, M., and D. A. P. Bundy. 1992. Comparative prevalences of *Ascaris lumbricoides, Trichuris trichiura* and hookworm infections and the prospects for combined control. *Parasitology* 105:151–57.

7. Bundy, D. A. P., and E. S. Cooper. 1989. *Trichuris* and trichuriasis in humans. In Baker, J. R., and R. Muller, eds. *Advances in parasitology* 28. London: Academic Press, 107–73.

8. Campbell, W. C. 1988. Trichinosis revisited—another look at modes of transmission. *Parasitol. Today* 4:83–86.

9. Centers for Disease Control. 1979. Trichinosis associated with meat from a grizzly bear—Alaska. *Morb. Mortal. Weekly Rep.* 30:115–16, 121.

10. Centers for Disease Control. 1991. Trichinosis surveillance, United States, 1987–1990. *Morb. Mortal. Weekly Rep.* 40(SS-3):35–42.

11. Chan, L., D. A. P. Bundy, and S. P. Kan. 1994. Genetic relatedness as a determinant of predisposition to *Ascaris lumbricoides* and *Trichuris trichiura* infection. *Parasitology* 108:77–80.

12. Choe, G., S. Lee, J. K. Seo, J. -Y. Chai, S. -H. Lee, K. S. Eom, and J. G. Chi. 1993. Hepatic capillariasis: First case report in the Republic of Korea. *Am. J. Trop. Med. Hyg.* 48:610–25.

13. Cook, G. C. 1990. Use of benzimidazole chemotherapy in human helminthiases: Indications and efficacy. *Parasitol. Today* 4:133–36.

14. Cooper, E. S., C. A. M. Whyte-Alleng, J. S. Finzi-Smith, and T. T. MacDonald. 1992. Intestinal nematode infections in children: The pathophysiological price paid. *Parasitology* 104:S91–S103.

15. Cooper, E. S., and D. A. P. Bundy. 1989. *Trichuris* is not trivial. *Parasitol. Today* 4:301–6.

16. Cross, J. H. 1990. Intestinal capillariasis. *Parasitol. Today* 6:26–28.

17. Despommier, D. D. 1990. *Trichinella spiralis:* The worm that would be virus. *Parasitol. Today* 6:193–96.

18. Despommier, D. D. 1993. *Trichinella spiralis* and the concept of niche. *J. Parasitol.* 79:472–82.

19. Despommier, D. D., A. M. Gold, S. W. Buck, V. Capo, and D. Silberstein. 1990. *Trichinella spiralis:* Secreted antigen of the infective L_1 larva localizes to the cytoplasm and nucleoplasm of infected host cells. *Exp. Parasitol.* 71:27–38.

20. Duckett, M. G., D. A. Denham, and G. S. Nelson. 1972. Immunity to *Trichinella spiralis.* V. Transfer of immunity against the intestinal phase from mother to baby mice. *J. Parasitol.* 58:550–54.

21. Fitzgerald, P. R., and T. B. S. Prakasam. 1978. Survival of *Trichinella spiralis* larvae in sewage sludge anaerobic digesters. *J. Parasitol.* 64:445–47.

22. Gardiner, C. H. 1976. Habitat and reproductive behavior of *Trichinella spiralis. J. Parasitol.* 62:865–70.

23. Hopkins, D. R. 1992. Homing in on helminths. *Am. J. Trop. Med. Hyg.* 46:626–34.

24. Kehayov, I., Ch. Tankov, S. Komandarev, and S. Kyurkchiev. 1991. Antigenic differences between *Trichinella spiralis* and *T. pseudospiralis* detected by monoclonal antibodies. *Parasitol. Res.* 77:72–76.

25. Kenney, M., and V. Yermakov. 1980. Infection of man with *Trichuris vulpis,* the whipworm of dogs. *Am. J. Trop. Med. Hyg.* 29:1205–8.

26. Ko, R. C., L. Fan, D. L. Lee, and H. Crompton. 1994. Changes in host muscle induced by excretory/secretory products of larval *Trichinella spiralis* and *Trichinella pseudospiralis*. *Parasitology* 108:195–205.

27. Lee, D. L., R. C. Ko, X. Y. Yi, and M. H. F. Yeung. 1991. *Trichinella spiralis:* Antigenic epitopes from the stichocytes detected in the hypertrophic nuclei and the cytoplasm of the parasitized muscle fibre (nurse cell) of the host. *Parasitology* 102:117–23.

28. Leiby, D. A., C. H. Duffy, K. D. Murrell, and G. A. Schad. 1990. *Trichinella spiralis* in an agricultural ecosystem: Transmission in the rat population. *J. Parasitol.* 76:360–64.

29. Levine, N. D. 1980. *Nematode parasites of domestic animals and of man,* 2d ed. Minneapolis:Burgess Publishing Co.

30. Mace, T. F., and R. C. Anderson. 1975. Development of the giant kidney worm, *Dioctophyma renale* (Goeze, 1782) (Nematoda: Dioctophymatoidea). *Can. J. Zool.* 53:1552–68.

31. Maynard, J. E., and F. P. Pauls. 1962. Trichinosis in Alaska. A review and report of two outbreaks due to bear meat, with observations of serodiagnosis and skin testing. *Am. J. Hyg.* 76:252–61.

32. Measures, L. N., and R. C. Anderson. 1985. Centrarchid fish as paratenic hosts of the giant kidney worm, *Dioctophyme renale* (Goeze, 1782), in Ontario, Canada. *J. Wildl. Dis.* 21:11–19.

33. Murrell, K. D. 1991. Trichinosis. In Strickland, G. T., ed. *Hunter's Tropical Medicine,* 7th ed. Philadelphia: W. B. Saunders Co., 756–61.

34. Nokes, C., S. M. Grantham-McGregor, A. W. Sawyer, E. S. Cooper, B. A. Robinson, and D. A. P. Bundy. 1992. Moderate to heavy infections of *Trichuris trichiura* affect cognitive function in Jamaican school children. *Parasitology* 104:539–47.

35. Olsen, O. W., and H. A. Robinson. 1958. Role of rats and mice in transmitting *Trichinella spiralis* through their feces. *J. Parasitol.* 44(sect. 2):35.

36. Pozio, E., G. La Rosa, P. Rossi, and K. D. Murrell. 1992. Biological characterization of *Trichinella* isolates from various host species and geographical regions. *J. Parasitol.* 78:647–53.

37. Pozio, E., G. La Rosa, K. D. Murrell, and J. R. Lichtenfels. 1992. Taxonomic revision of the genus *Trichinella*. *J. Parasitol.* 78:654–59.

38. Pozio, E., P. Varese, M. A. Gomez Morales, G. P. Croppo, D. Pelliccia, and F. Bruschi. 1993. Comparison of human trichinellosis caused by *Trichinella spiralis* and by *Trichinella britovi*. *Am. J. Trop. Med. Hyg.* 48:568–75.

39. Preston, C. M., and T. Jenkins. 1984. *Trichuris muris:* Structure and formation of the eggshell. *Parasitology* 89:263–73.

40. Rausch, R. 1953. Animal-borne diseases. *Publ. Health Rep.* 68:533.

41. Robertson, L. J., D. W. T. Crompton, D. E. Walters, M. C. Nesheim, D. Sanjur, and E. A. Walsh. 1989. Soil-transmitted helminth infections in school children from Cocle Province, Republic of Panama. *Parasitology* 99:287–92.

42. Romashov, B. V. 1983. Details of the life cycle of *Hepaticola hepatica* (Nematoda, Capillariidae). Moscow: Parazitologicheskie issledovaniya v Zapovednikakh, 49–58.

43. Sheffield, H. G. 1963. Electron microscopy of the bacillary band and stichosome of *Trichuris muris* and *T. vulpes*. *J. Parasitol.* 49:998–1009.

44. Shirazian, D., E. L. Schiller, C. A. Glaser, and S. L. Vonderfect. 1984. Pathology of larval *Eustrongylides* in the rabbit. *J. Parasitol.* 70:803–6.

45. Singleton, G. R., and H. I. McCallum. 1990. The potential of *Capillaria hepatica* to control mouse plagues. *Parasitol. Today* 6:190–93.

46. Smith, H. J. 1975. Trichinae in tail musculature of swine. *Can. J. Comp. Med.* 39:362–63.

47. Soulé , C., J. -P. Guillou, J. Dupouy-Camet, C. Vallet, and E. Pozio. 1993. Differentiation of *Trichinella* isolates by polymerase chain reaction. *Parasitol. Res.* 79:461–65.

48. Spalding, M. G., and D. J. Forrester. 1993. Pathogenesis of *Eustrongylides ignotus* (Nematoda: Dioctophymatoidea) in Ciconiiformes. *J. Wildl. Dis.* 29:250–60.

49. Sundman, P. O. 1970. *The flight of the Eagle.* New York: Random House, Inc.

50. Upatham, E. S., V. Viyanant, W. Y. Brockelman, S. Kurathong, P. Ardsungnoen, and U. Chindaphol. 1992. Predisposition to reinfection by intestinal helminths after chemotherapy in south Thailand. *Int. J. Parasitol.* 22:801–6.

51. Wright, K. A. 1968. Structure of the bacillary band of *Trichuris myocastoris*. *J. Parasitol.* 54:1106–10.

52. Zohar, A. S., and M. R. Rau. 1986. The role of muscle larvae of *Trichinella spiralis* in the behavioral alterations of the mouse host. *J. Parasitol.* 72:464–66.

Additional References

Bundy, D. A. P., and E. S. Cooper. 1989. *Trichuris* and trichuriasis in humans. In Baker, J. R., and R. Muller, eds. *Advances in parasitology* 28, London: Academic Press, 107–73.

Campbell, W. C., ed. 1983. *Trichinella* and trichinellosis. New York: Plenum Press.

Neafie, R. C., and D. H. Connor. 1976. Trichuriasis. In Binford, C. H., and D. H. Connor, eds. *Pathology of tropical and extraordinary diseases,* vol. 2, sect. 9. Washington, D.C.: Armed Forces Institute of Pathology.

Neafie, R. C., D. H. Connor, and J. H. Cross. 1976. Capillariasis. In Binford, C. H., and D. H. Connor, eds. *Pathology of tropical and extraordinary diseases,* vol. 2, sect. 9. Washington, D.C.: Armed Forces Institute of Pathology.

Stehr-Green, J. K., P. M. Schantz, and E. M. Chisolm. 1986. Trichinosis surveillance. 1984. *Morb. Mortal. Weekly Rep.* (CDC Surveillance Summaries) 35:1155–56.

Chapter 24

NEMATODES: RHABDITIDA, PIONEERING PARASITES

. . . occasionally parasitic females and males are passed in diarrheic stools. . . . The rare parasitic males are practically indistinguishable from the free-living males of the indirect cycle.

E. C. Faust, 1932

Although Faust continued to describe the parasitic male in his textbook as late as 1970, no other investigators were able to find such worms and the concept fell into disrepute.

D. I. Grove, describing the controversy over the existence of a parasitic male *Strongyloides stercoralis*

The tiny worms in the order Rhabditida appear to bridge the gap between free-living and parasitic modes of life, since several species alternate between free-living and parasitic generations. Most species inhabit decaying organic matter and are common in soil, foul water, decaying fruit, and so on. For this reason they often find their way into the bodies of larger animals; the digestive, reproductive, respiratory, and excretory tracts are particularly susceptible, as are open wounds. Once in such locations, they may become facultatively parasitic for a time or may simply pass through the body. Rarely they cause serious disease.[6]

Some of the behavioral and developmental adaptations of free-living species (*Rhabditis, Pelodera*) in this order were mentioned in Chapter 22 (dauer juveniles). It is not difficult to visualize how analogous adaptations for transfer to new food supplies in the ancestors of the frankly parasitic species could have been preadaptations to parasitism. The diversity in life cycles of parasites in Rhabditida further illustrates this evolutionary opportunism. Thus, within the order, one can observe completely free-living species, species with preadaptations for parasitism, facultative parasites, obligate parasites, and even species that seem to produce obligate free-living or obligate parasitic forms, depending on conditions (Rhabdiasidae, Strongyloididae).

FAMILIES STEINERNEMATIDAE AND HETERORHABDITIDAE

Members of these families belong to two genera, *Steinernema* and *Heterorhabditis*. Their hosts are insects, which they kill, as do parasitoid insects (p. 591), and they have considerable potential as biological agents of pest control.[10] Their life cycles are quite remarkable. The infective stage is third-stage, dauer juveniles (J_3), which can live a considerable period in the soil, depending on species, temperature, and moisture.

The J_3's enter the mouth, anus, or spiracle of a host and penetrate to the hemocoel. There they begin to feed, and in the process they release symbiotic bacteria (*Xenorhabdus* spp.) that were residing in their gut. The bacteria multiply rapidly and kill the host, but they release antibiotics that inhibit growth of other bacteria and thus putrefaction. The nematodes develop, feeding on the *Xenorhabdus* and contents of the dead host's hemocoel. They continue for several generations until the supply of nutrients becomes limiting, then they produce dauer J_3's again, which go back into the soil.

The relationship between the nematodes and bacteria is a true mutualism. The bacteria depend on the nematodes for transmission from insect to insect and cannot survive in the soil, and absence of the bacteria inhibits nematode reproduction.

FAMILY RHABDIASIDAE

Rhabdias bufonis and *R. ranae,* common parasites in the lungs of toads and frogs, also have very curious life cycles. The parasitic adult is a **protandrous hermaphrodite;** that is, an individual that is a functional male before it becomes a female. Sperm (chromosome number [N] = 5 or 6)[17] are produced in an early male phase and stored in a seminal receptacle. Then the gonad produces functional ova (N = 6) that are fertilized by the stored sperm. The resulting shelled zygotes (2N = 11 or 12) pass up the trachea of the host and then are swallowed, embryonating along the way. The juveniles hatch in the intestine of the frog, and the first-stage juveniles (J_1's) accumulate in the cloaca to be voided with the feces. The J_1's often are referred to as **rhabditiform** because the posterior end of their esophagus has a prominent **bulb** that is separated from the anterior portion (**corpus**) by a narrower region (**isthmus**) (Fig. 24.1).

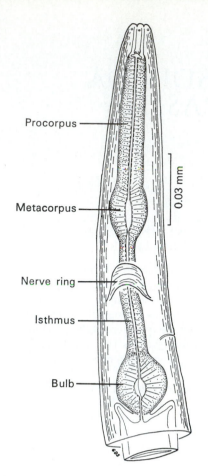

FIGURE 24.1

Typical rhabditiform esophagus.

From G. D. Schmidt and Robert E. Kuntz, "Nematode parasites of Oceania. XVIII, *Caenorhabditis avicoloa* sp. n. (Rhabditidae) found in a bird from Taiwan," in *Proc. Helm. Soc. Wash.* 39:189–191. Copyright © 1972. Reprinted with permission of the publisher.

These juveniles undergo four molts to produce a dioecious generation of free-living males (2N = 11) and females (2N = 12). This nonparasitic generation feeds on bacteria and other inhabitants of humusy soil. Its progeny hatch in utero and proceed to consume the internal organs of their mother, destroying her. "How sharper than a serpent's tooth!" By the time they escape from the female's body they have become infective J_3's. They are referred to as filariform at this stage because the esophagus has no terminal bulb or isthmus. The filariform juveniles undergo developmental arrest unless they penetrate the skin of a toad or frog. After penetration, they lodge in various tissues; those that reach the lungs mature into hermaphroditic adults, whereas the rest apparently expire. This type of life cycle is **heterogonic;** that is, a free-living generation is interspersed between parasitic generations.

Rhabdias fuscovenosa is a common parasite in the lungs of some kinds of aquatic snakes (Natricinae). The life cycle differs somewhat from those of *R. bufonis* and *R. ranae* in that most of the eggs from the parasitic forms yield filariform juveniles. Few free-living adults are found;[4] therefore, the life cycle in this species is predominately **homogonic,** and the worm is an obligate parasite.

FAMILY STRONGYLOIDIDAE

Strongyloides Spp.

Some species of this family are among the smallest nematode parasites of humans, the males being even smaller than *Trichinella*. *Strongyloides stercoralis* is the most common, widespread species; *S. fuelleborni* occurs commonly in parts of Africa.[14] *Strongyloides stercoralis* infects other primates besides humans, as well as dogs, cats, and some other mammals, and human races from various geographical areas vary in infectivity for different hosts.[13] *Strongyloides fuelleborni* has been reported from a variety of primates. Other species are parasites of other mammals, such as *S. ratti* in rats, *S. ransomi* in swine, and *S. papillosus* in sheep. They also are common in birds, amphibians, and reptiles. The species of *Strongyloides* are remarkable in their ability, at least in some cases, to maintain homogonic, parasitic life cycles or repeat free-living generations indefinitely, depending on conditions. The parasitic generation apparently consists only of parthenogenetic females, since sperm have not been found in the seminal receptacle of the females.[3] The free-living generation consists of both males and females. The following description pertains to *S. stercoralis* (Fig. 24.2).

• Morphology

Parthenogenetic females reach a length of 2.0 to 2.5 mm. The buccal capsule of both sexes is small, and the animals possess a long, cylindrical esophagus that lacks a posterior bulb. The vulva is in the posterior third of the body; the uteri are divergent and contain only a few eggs at a time. Both sexes of free-living adults have a rhabditiform esophagus. The male is up to 0.9 mm long and 40 to 50 µm wide. The male has two simple spicules and a gubernaculum; its pointed tail is curved ventrally. The female is stout and has a vulva that is about equatorial; the uteri generally contain more eggs than do those of the parasitic female.

• Biology

The parasitic females (Fig. 24.3) burrow their anterior ends into the submucosa of the small intestine. They are found occasionally in the respiratory, biliary, or pancreatic system. They produce several dozen, thin-shelled, partially embryonated eggs a day and release them into the gut lumen or submucosa. The eggs measure 50 to 58 µm by 30 to 34 µm. They hatch during passage through the gut or within the submucosa, and the juveniles escape to the lumen. These J_1's are 300 to 380 µm long, and they usually are passed with the feces. The juveniles go on either to develop into free-living adults or to become infective, filariform J_3's with a developmental arrest; they are now 490 to 630 µm long. The filariform juveniles develop no further unless they gain access to a new host by skin penetration or ingestion. If by skin penetration, the juveniles must undergo a tissue migration to reach the small intestine. It has been widely believed that the juveniles were carried by the

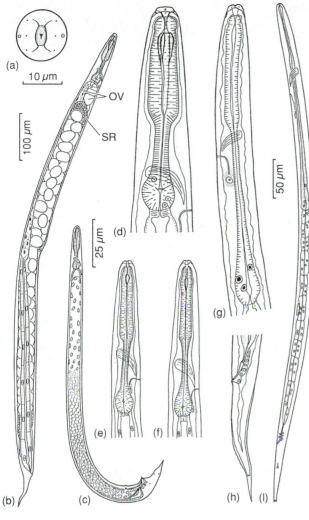

FIGURE 24.2

Strongyloides stercoralis. (*a*) Free-living female, en face view. (*b*) Free-living female, lateral view (**OV,** ovary; **SR,** seminal receptacle containing sperm). (*c*) Free-living male. (*d*) Anterior end of free-living female, showing details of esophagus. (*e*) Newly hatched J_1 obtained by duodenal aspiration from human. (*f*) J_1 from freshly passed feces of same patient as was juvenile in (*e*). (*g*) J_2 developing to the filariform (J_3) stage; cuticle is separating at anterior end. (*h*) Tail of same juvenile as shown in (*g*); notched tail is developing, and cuticle is separating at tip and in rectum. (*i*) Filariform J_3.

From M. D. Little, "Comparative morphology of six species of *Strongyloides* (Nematoda) and redefinition of the genus," in *J. Parasitol.* 52:69–84. Copyright © 1966. Reprinted with permission of the publisher.

blood to the lungs, where they emerged into the alveoli and gained access to the digestive system after being carried up the tracheae. However, the migration route of many *Strongyloides* juveniles includes the nasofrontal region of the host's head.[19,22] In rats very few juveniles of *S. ratti* migrate through the lungs and gain entry to the digestive tract via the trachea.[2] Given the lung symptoms sometimes seen in humans infected with *S. stercoralis,*[15] it appears that significant numbers of juveniles migrate through the lungs in that host.

The free-living adults can produce successive generations of free-living adults. Both parasitic and free-living females can produce juveniles that will become filariform, infective J_3's and juveniles that will mature into free-living adults. In other words the homogonic and heterogonic life cycles seem to be mixed in a random fashion. The mechanism that determines whether a given embryo will become a free-living male or female or a parasitic female is still unclear.

Still other permutations of the life cycle occur. If the juveniles have time to molt twice during their transit down the digestive tract, they may penetrate the lower gut mucosa or perianal skin, go through their migration, and mature. This process is called **autoinfection.** It appears that the host immune response slows down development of the juveniles in the gut because there is evidence for an "autoinfective burst" early in the infection, before the immune response becomes effective.[21] The capability of autoinfection means that *Strongyloides* is one of the few multicellular parasites that can multiply within its definitive host. Normally autoinfection is kept in check by the host immune response, but in immunocompromised hosts, hyperinfection can become severe and life threatening.[5,15]

Nevertheless, whether by maintenance of the original adult worms, by autoinfection, or otherwise, cases are known in which patients have had *Strongyloides* infections for up to 40 years. In 1984 Pelletier[16] reported on 142 American exprisoners of war who had worked on the Burma-Thailand Railroad in World War II. Of these, 52 had symptomatic, previously unrecognized strongyloidiasis. Such extreme longevity also has been reported in British and Australian exprisoners of war.

• Epidemiology

People typically become infected with strongyloidids by contacting the juveniles in contaminated soil or water. These primarily are parasites of tropical regions but extend well into temperate zones in several continents. Like most filthborne diseases, strongyloidiasis is most prevalent under conditions of low sanitation standards. Although traditionally a disease of poor, uneducated persons in depressed areas of the world, it is often found wherever conditions of filth prevail, such as some mental institutions. It could also become important among the affluent with "vacation hideways," with inadequate sewage disposal facilities, such as in mountain environments.

• Pathology

The effects of strongyloidiasis may be described in three stages: **invasion, pulmonary,** and **intestinal.**

Penetration of the skin by invasive juveniles results in slight hemorrhage and swelling, with intense itching at the site of entry. If pathogenic bacteria are introduced with the juveniles, inflammation may result as well.

During migration through the lungs, damage to lung tissues results in occasional pulmonary eosinophilic lung infiltrates or wheezing. A burning sensation in the chest, a nonproductive cough, and other symptoms of bronchial pneumonia may accompany this phase. The lung phase can be mistaken for asthma, which, if treated with corticosteroids, can result in rampant autoinfection.[5]

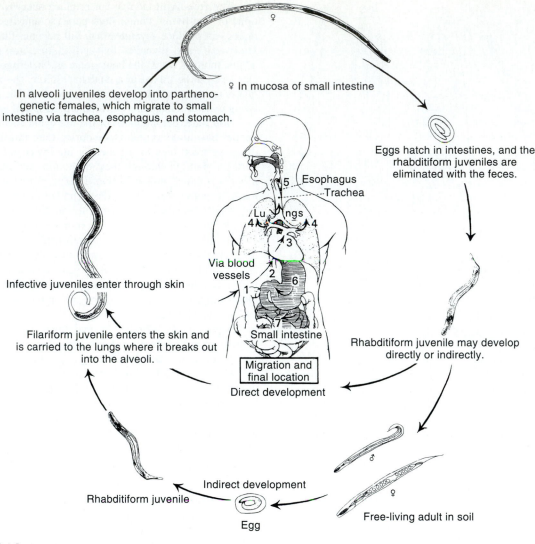

In alveoli juveniles develop into partheno-genetic females, which migrate to small intestine via trachea, esophagus, and stomach.

♀ In mucosa of small intestine

Eggs hatch in intestines, and the rhabditiform juveniles are eliminated with the feces.

Esophagus
Trachea

Lungs

Via blood vessels

Small intestine

Migration and final location

Direct development

Infective juveniles enter through skin

Filariform juvenile enters the skin and is carried to the lungs where it breaks out into the alveoli.

Rhabditiform juvenile may develop directly or indirectly.

Indirect development

Rhabditiform juvenile

Egg

Free-living adult in soil

FIGURE 24.3

Life cycle of *Strongyloides stercoralis* in humans.

Modified from *Medical Protozoology and Helminthology.* Copyright © 1959 U.S. Naval Medical School, Bethesda, MD. Reprinted with permission.

After the juvenile females enter the crypts of the intestinal mucosa, they rapidly mature and invade the tissues. They rarely penetrate deeper than the muscularis mucosae, but some cases of deeper penetration have been reported. The worms migrate randomly through the mucosa, depositing eggs. An intense, localized burning sensation or aching pain in the abdomen usually is felt at this time. Destruction of tissues by adult worms and juveniles results in sloughing of patches of mucosa, with fibrotic changes in chronic cases, sometimes with death resulting from septicemia (bacterial infection of the blood) following ulceration of the intestine.

Infection with *S. stercoralis* is most commonly asymptomatic, but cases are known in which hosts were asymptomatic carriers for years before developing serious disease.

Chronic strongyloidiasis can result in relapsing colitis.[1] In immunocompromised patients, the host-parasite relationship is altered in favor of the parasite, and fulminant, fatal hyperinfection can occur. Mortality has been observed in cases treated with high-dose steroid therapy and in patients with AIDS.[11,12,23] The course of infection in dogs and monkeys treated with immunosuppresive drugs has been studied.[8,9,20]

• Diagnosis and Treatment

Demonstration of rhabditiform (or occasionally filariform) juveniles in freshly passed stools is a sure means of diagnosis. A direct fecal smear is often effective in cases of massive infections, but in more moderate infections, finding juveniles may be very difficult.[15] Special isolation techniques will increase chances, but a more recently described technique of culturing fecal samples on nutrient agar is most effective.[7] Serodiagnosis by ELISA may be useful.[18] Rarely, after purgation or in severe diarrhea, embryonating eggs may be seen in the stool. These resemble hookworm eggs but are more rounded.

Difficulties in diagnosis arise because of day-to-day variability in numbers of juveniles in the feces. Furthermore, once autoinfection or hyperinfection becomes established, the number of juveniles passed in the feces may decrease. Duodenal aspiration is a very accurate technique but only applies to duodenal infections: Juveniles farther down the intestine cannot be obtained.

Once juveniles are obtained, problems of identification can occur. J_1's are similar to rhabditiform hookworm juveniles, which may be present if the stool was constipated or remained at room temperature long enough for hookworm eggs to hatch. Two morphological features can be useful to separate the two: *Strongyloides* has a short buccal cavity and a large genital primordium, whereas hookworm juveniles have a long buccal cavity and a tiny genital primordium (p. 412).

If the stool has been exposed to soil or water, species of *Rhabditis* or related genera may invade it, compounding the confusion. Furthermore, filariform juveniles appear in cases of constipation or autoinfection. These juveniles, however, are easily recognized by their notched tails.

No entirely satisfactory drug for strongyloidiasis is currently available. Thiabendazole is recommended, but side effects such as nausea, vomiting, malaise, dizziness, and smelly urine are common.[15] Therapy should be repeated after a week because of difficulty in confirming cure.

References

1. Berry, A. J., E. G. Long, J. H. Smith, W. K. Gourley, and D. P. Fine. 1983. Chronic relapsing colitis due to *Strongyloides stercoralis. Am. J. Trop. Med. Hyg.* 32:1289–93.

2. Bhopale, V. M., G. Smith, H. E. Jordan, and G. A. Schad. 1992. Developmental biology and migration of *Strongyloides ratti* in the rat. *J. Parasitol.* 78:861–68.

3. Bolla, R. I., and L. S. Roberts. 1968. Gametogenesis and chromosomal complement in *Strongyloides ratti* (Nematoda: Rhabdiasoidea). *J. Parasitol.* 54:849–55.

4. Chu, T. 1936. Studies on the life cycle of *Rhabdias fuscovenosa* var. *catanensis* (Rizzo, 1902). *J. Parasitol.* 22:140–60.

5. Dajer, T. 1994. Triumph by treachery. *Discover* 15(7):80–85.

6. Gardiner, C. H., D. S. Koh, and T. A. Cardella. 1981. *Micronema* in man: Third fatal infection. *Am. J. Trop. Med. Hyg.* 30:586–89.

7. Girard de Kaminsky, R. 1993. Evaluation of three methods for laboratory diagnosis of *Strongyloides stercoralis* infection. *J. Parasitol.* 79:277–80.

8. Grove, D. I., P. J. Heenan, and C. Northern. 1983. Persistent and disseminated infections with *Strongyloides stercoralis* in immunosuppressed dogs. *Int. J. Parasitol.* 13:483–90.

9. Harper, J. W. III, R. M. Genta, A. Gam, W. T. London, and F. A. Neva. 1984. Experimental disseminated strongyloidiasis in *Erythrocebus patas*. I. Pathology. *Am. J. Trop. Med. Hyg.* 33:431–43.

10. Hominick, W. M. 1990. Entomopathogenic rhabditid nematodes and pest control. *Parasitol. Today* 6:148–52.

11. Kalb, R. E., and M. E. Grossman. 1986. Periumbilical purpura in disseminated strongyloidiasis. *J.A.M.A.* 256:1170–71.

12. Katner, H. 1988. Personal communication.

13. Levine, N. D. 1980. *Nematode parasites of domestic animals and of man,* 2d ed. Minneapolis: Burgess Publishing Co.

14. Pampiglioni, S., and M. L. Ricciardi. 1971. The presence of *Strongyloides fülleborni* von Linstow, 1905, in man in central and east Africa. *Parassitologia* 13:257–69.

15. Pearson, R. D., and R. L. Guerrant. 1991. Intestinal nematodes that migrate through skin and lung. In Strickland, G. T., ed. *Hunter's tropical medicine,* 7th ed. Philadelphia: W. B. Saunders Company, 700–11.

16. Pelletier, L. L. Jr. 1984. Chronic strongyloidiasis in World War II. Far East ex-prisoners of war. *Am. J. Trop. Med. Hyg.* 33:55–61.

17. Runey, W. M., G. L. Runey, and F. H. Lauter. 1978. Gametogenesis and fertilization in *Rhabdias ranae* Walton 1929. I. The parasitic hermaphrodite. *J. Parasitol.* 64:1008–14.

18. Sato, Y., H. Toma, M. Takara, and Y. Shiroma. 1990. Application of enzyme-linked immunosorbent assay for mass examination of strongyloidiasis in Okinawa, Japan. *Int. J. Parasitol.* 20:1025–29.

19. Schad, G. A., L. M. Aikens, and G. Smith. 1989. *Strongyloides stercoralis:* Is there a canonical migratory route through the host? *J. Parasitol.* 75:740–49.

20. Schad, G. A., M. E. Hellman, and D. W. Muncey. 1984. *Strongyloides stercoralis:* Hyperinfection in immunosuppressed dogs. *Exp. Parasitol.* 57:289–96.

21. Schad, G. A., G. Smith, Z. Megyeri, V. M. Bhopale, S. Niamatali, and R. Maze. 1993. *Strongyloides stercoralis:* An initial autoinfective burst amplifies primary infection. *Am. J. Trop. Med. Hyg.* 48:716–25.

22. Tindall, N. R., and A.P.G. Wilson. 1988. Criteria for a proof of migration routes of immature parasites inside hosts exemplified by studies of *Strongyloides ratti* in the rat. *Parasitology* 96:551–63.

23. Vieyra-Herrera, G., et al. 1988. *Strongyloides stercoralis* hyperinfection in a patient with acquired immune deficiency syndrome. *Acta Cytol.* 32:277–78.

Additional References

Blunden, A. S., L. F. Khalil, and P. M. Webbon. 1987. *Halicephalobus deletrix* infection in a horse. *Equine Vet. J.* 19:255–60. A good historical account of *Micronema* in humans and horses.

Cable, R. M. 1971. Parthenogenesis in parasitic helminths. *Am. Zool.* 11:267–72. Contains a short discussion of parthenogenesis in nematodes, with emphasis on *Strongyloides*.

Grove, D. I., ed. 1989. Strongyloidiasis: A major roundworm infection of man. Basingstoke, Hants, U.K.: Taylor & Francis.

Little, M. D. 1966. Comparative morphology of six species of *Strongyloides* (Nematoda) and redefinition of the genus. *J. Parasitol.* 52:69–84.

Meyers, W. M., D. H. Connor, and R. C. Neafie. 1976. Strongyloidiasis. In Binford, C. H., and D. H. Connor, eds. *Pathology of tropical and extraordinary diseases,* vol. 2, sect. 9. Washington, D.C.: Armed Forces Institute of Pathology.

Popiel, I., and W. M. Hominick. 1992. Nematodes as biological control agents: Part II. In Baker, J. R., and R. Muller, eds. *Advances in parasitology* 31, London: Academic Press, 381–433.

Chapter 25

NEMATODES: STRONGYLIDA, BURSATE RHABDITIANS

The disease induced by the hookworm . . . was never suspected to be a disease at all. The people who had it were merely supposed to be lazy, and were therefore despised and made fun of, when they should have been pitied.

Mark Twain

The large order Strongylida is of great economic and medical importance. One feature of nearly all species is a broad copulatory bursa on the posterior end of the males. Most, but not all, species are parasites of the intestine of vertebrates and have a direct life cycle, requiring no intermediate hosts. We will illustrate the parasitological importance of the order chiefly with examples of importance to humans.

SUPERFAMILY ANCYLOSTOMATOIDEA

Family Ancylostomidae

Members of this family are commonly known as **hookworms.** They live in the intestine of their host, attaching to the mucosa and feeding on blood and tissue fluids sucked from it.

• Morphology

Much similarity of morphology and biology exists among the numerous species in this family, so we will first give them a general consideration. Most species are rather stout, and the anterior end is curved dorsally, giving the worm a hooklike appearance. The buccal capsule is large and heavily sclerotized and usually is armed with cutting plates, teeth, lancets, or a dorsal cone (Fig. 25.1). Lips are reduced or absent.

The esophagus is stout, with a swollen posterior end, giving it a club shape. It is mainly muscular, corresponding to its action as a powerful pump. The esophageal glands are extremely large and are mainly outside of the esophagus, extending posteriorly into the body cavity. Cervical papillae are present near the rear level of the nerve ring.

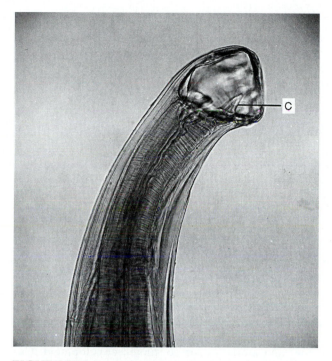

FIGURE 25.1

Lateral view of the anterior of *Bunostomum* sp., a hookworm of ruminants. Note the large buccal capsule and dorsal flexure, typical of hookworms. The dorsal cone (**C**) bears the duct of the dorsal esophageal gland, similar to that in *Necator americanus*.

Courtesy of Jay Georgi.

Males have a conspicuous copulatory bursa, consisting of two broad **lateral lobes** and a smaller **dorsal lobe,** all supported by fleshy rays (Fig. 25.2). These rays follow a common pattern in all species, varying only in relative size and point of origin; consequently, they are important taxonomic characters. The spicules are simple, needlelike, and similar. A gubernaculum is present.

Females have a simple, conical tail. The vulva is postequatorial, and two ovaries are present. About 5% of the daily output of eggs is found in the uteri at any one time; the total production is several thousand per day for as long as nine years.

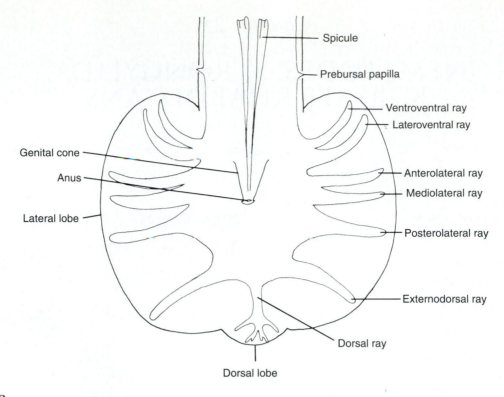

FIGURE 25.2

Ventral view of a typical strongyloid copulatory bursa. The basic pattern is found in all of the Strongylida.

• **Biology**

Hookworms mature and mate in the small intestine of their host (Fig. 25.3). The embryos develop into the two-, four-, or several-cell stage by the time they are passed with the feces (Fig. 25.4). The species infecting humans cannot be diagnosed reliably by the egg alone. The eggs require warmth, shade, and moisture for continued development. Coprophagous insects may mix the feces with soil and air, thus hastening embryonation, which may occur within 24 to 48 hours in ideal conditions (Fig. 25.5). Newly hatched J_1's have a rhabditiform esophagus (p. 400) with its characteristic constriction at the level of the nerve ring, such as that occurring in the rhabditiform J_1's of *Strongyloides* spp. In fact differentiation of hookworm juveniles from those of *Strongyloides* is difficult for the beginner.

The juveniles live in the feces, feeding on fecal matter, and molt their cuticle in two to three days. The J_2, which also has a rhabditiform esophagus, continues to feed and grow and after about five days, molts to the third stage, which is infective to the host. The second-stage cuticle may be retained as a loose-fitting sheath until penetration of a new host, or it may be lost earlier. The filariform J_3 has a **strongyliform esophagus;** that is, it has a posterior bulb, but the bulb is not separated from the corpus by an isthmus. The intestine is filled with food particles that sustain the worm through the nonfeeding J_3. The J_3 is similar to the filariform

J_3 of *Strongyloides* but can be distinguished from it by the tail, which is pointed in hookworms and notched in *Strongyloides* (p. 403).

The J_3's live in the upper few millimeters of soil, remaining in the capillary layer of water surrounding soil particles. Freezing or desiccation kills them quickly. There is a short, vertical migration in the soil, depending on the weather or time of day. When the surface of the ground is dry, they migrate a short distance into the soil, following the retreating water. When the surface of the ground is wet, after rain or morning dew, they move to the surface and extend themselves like snakes, waving back and forth in a manner that allows maximal opportunity for contact with a host. Thousands of juveniles often group together, crawling over each other and waving rhythmically in unison (see Fig. 22.34). They can live for several weeks under ideal conditions.

Infection usually occurs when J_3's contact the skin and burrow into it, and they resume feeding at about this time.[15] They usually shed the second-stage cuticle as they penetrate, but the presence of the cuticle does not preclude resumption of feeding.[19] Juveniles can penetrate any epidermis, although the parts most often in contact with the soil, such as hands, feet, and buttocks, are most often attacked. *Necator americanus* must penetrate the skin to infect humans, but *Ancylostoma duodenale* can penetrate skin and oral mucosa, be passed in mother's milk, and probably be acquired transplacentally.[12]

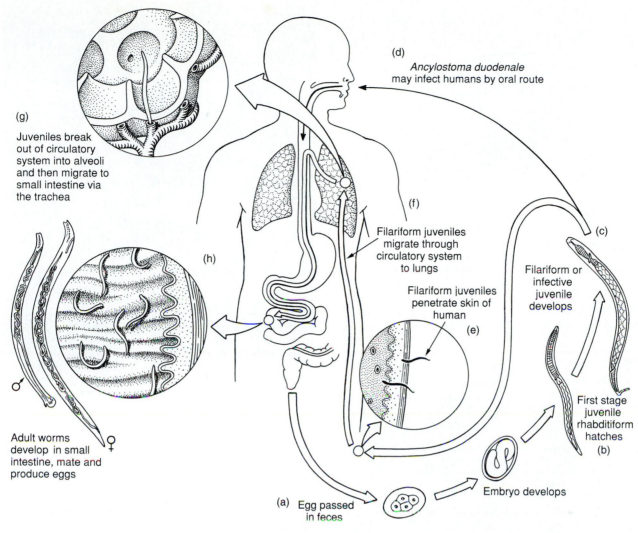

(d) *Ancylostoma duodenale* may infect humans by oral route

(g) Juveniles break out of circulatory system into alveoli and then migrate to small intestine via the trachea

(f) Filariform juveniles migrate through circulatory system to lungs

(c) Filariform or infective juvenile develops

Filariform juveniles penetrate skin of human

(e)

(h)

First stage juvenile rhabditiform hatches
(b)

Adult worms develop in small intestine, mate and produce eggs

(a) Egg passed in feces

Embryo develops

FIGURE 25.3

The life cycle of hookworms. (*a*) Shelled embryo passed in feces. (*b*) First-stage juvenile (rhabditiform) hatches. (*c*) Two molts ensue and then third-stage juvenile (infective, filariform) enters developmental arrest until it reaches a new host. (*d*) *Ancylostoma duodenale* may infect humans by oral route. (*e*) Filariform juveniles penetrate skin of humans. (*f*) Juveniles migrate through circulatory system to lungs. (*g*) Juveniles break out of circulatory system into alveoli and then migrate to small intestine via the trachea. (*h*) Adult worms develop in the small intestine, mate, and produce eggs.
Drawing by William Ober and Claire Garrison.

After gaining entry to a blood or lymph vessel, the juveniles are carried to the heart and then to the lungs. They break into the air spaces of the alveoli and are carried by ciliary action up the respiratory tree to the glottis. They are swallowed and finally arrive in the small intestine, there to attach to the mucosa, grow, and molt to the fourth stage, which has an enlarged buccal capsule. After further growth and a final molt, the worm becomes sexually mature. At least five weeks are required from the time of infection to the beginning of egg production. However, *A. duodenale* can undergo developmental arrest for up to 38 weeks, its maturation coinciding with seasonal return of environmental conditions favorable to transmission.[7]

Several genera and many species of hookworm plague humans and domestic and wild mammals. The following two species infect an estimated 900 million people.

Necator americanus. *Necator americanus,* the "American killer," was first discovered in Brazil and then Texas, but it was later found indigenous in Africa, India, southeast Asia, China, and the southwest Pacific islands. It probably came to the New World with the slave trade. The worm has had an important impact on the economic and cultural development of the southern United States, as well as other regions of the world in which it occurs.

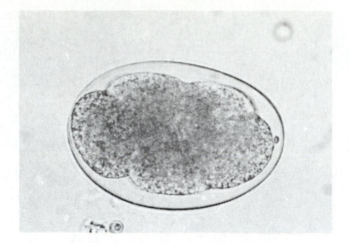

FIGURE. 25.4

Necator americanus egg in early cleavage stage, as normally passed in feces. Size is 65 to 75 by 36 to 40 μm. Eggs of *Ancylostoma duodenale,* 56 to 60 by 35 to 40 μm, are not distinguishable.

Photograph by Gerald D. Schmidt.

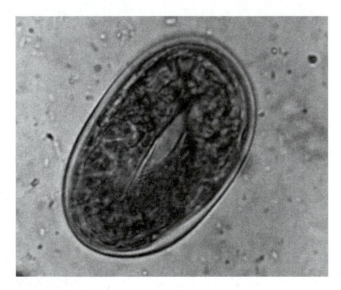

FIGURE 25.5

Hookworm egg containing fully developed J₁.

Courtesy of Robert E. Kuntz and J. Moore.

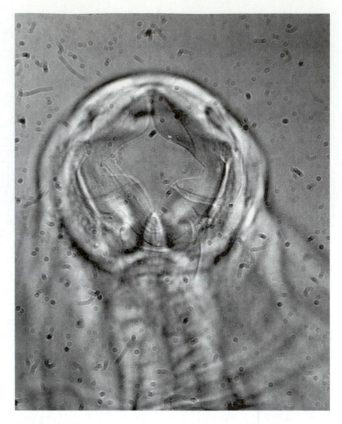

FIGURE 25.6

En face view of the mouth of *Necator americanus.* Note the two broad cutting plates in the ventrolateral margins (*top*).

Photograph by Larry S. Roberts.

Necator americanus has a pair of dorsal and a pair of ventral cutting plates surrounding the anterior margin of the buccal capsule (Fig. 25.6). In addition, a pair of subdorsal and a pair of subventral teeth are near the rear of the buccal capsule. The duct of the dorsal esophageal gland opens on a conspicuous cone that projects into the buccal cavity (see Fig. 25.1).

Males are 7 to 9 mm long and have a bursa diagnostic for the genus[24] (Fig. 25.7). The needlelike spicules have minute barbs at their tips and are fused distally. Females are 9 to 11 mm long and have the vulva located in about the middle of the body. Female worms produce about 9000 eggs per day.

Primarily a tropical parasite, *N. americanus* is the most common species in humans in most of the world. About 95% of the hookworms in the southern United States are this species. Adults survive in the gut up to 15 years.[7]

Ancylostoma duodenale. *Ancylostoma duodenale* is abundant in southern Europe, northern Africa, India, China, and southeastern Asia, as well as in other scattered locales, including small areas of the United States, Caribbean Islands, and South America. It is known in mines as far north as England and Belgium; since Lucretius, in the first century, it was known to cause a serious anemia in miners. Mines offer an ideal habitat for egg and juvenile development because of their constancy in temperature and humidity.

The anterior margin of the buccal capsule has two ventral plates, each with two large teeth that are fused at their bases (Fig. 25.8). A pair of small teeth is found in the depths of the capsule. The duct of the dorsal esophageal gland runs in a ridge in the dorsal wall of the buccal capsule and opens at the vertex of a deep notch on the dorsal margin of the capsule.

Adult males are 8 to 11 mm long and have a bursa characteristic for the species (Fig. 25.9). The needlelike spicules have simple tips and are never fused distally. Females are 10 to 13 mm long, with the vulva located about a third of the

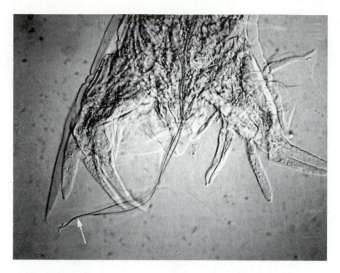

FIGURE 25.7

Copulatory bursa and spicules of *Necator americanus*. The spicules are fused at their distal ends (*arrow*) and form a characteristic hook.

Photograph by Larry S. Roberts.

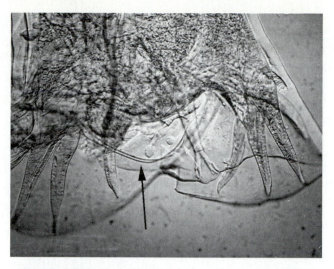

FIGURE 25.9

Copulatory bursa and spicules (*arrow*) of *Ancylostoma duodenale*. The tips of the spicules are not fused into a hook, as in *Necator americanus*.

Photograph by Larry S. Roberts.

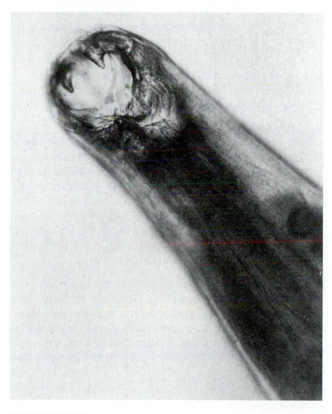

FIGURE 25.8

Ancylostoma duodenale: dorsal view. Notice the powerful ventral teeth.

AFIP neg. no. N-41730-2

body length from the posterior end. A single female can lay from 25,000 to 30,000 eggs per day. Adults may live up to five years.[7]

This is the first hookworm for which a life cycle was elucidated. In 1896 Arthur Looss, working in Egypt, was dropping cultures of *Ancylostoma* juveniles into the mouths of guinea pigs when he spilled some of the culture onto his hand. He noticed that it produced an itching and redness and wondered if infection would occur this way. He began examining his feces at intervals and after a few weeks, found that he was passing hookworm eggs. He next placed some juveniles on the leg of an Egyptian boy who was to have his leg amputated within an hour. Subsequent microscopic sections showed juveniles penetrating the skin. Looss' monograph on the morphology and life cycle of *A. duodenale* remains one of the most elegant of all works on helminthology.[26]

It is possible for swallowed juveniles to develop normally without a migration through the lungs, but this is probably a fairly rare means of infection.

• Other Hookworms Reported from Humans

Ancylostoma ceylanicum is normally a parasite of carnivores in Sri Lanka, southeast Asia, and the East Indies, but it has been reported from humans in the Philippines.[42] A very similar species, *A. braziliense,* is found in domestic and wild carnivores in most of the tropics. Although this species has been reported from humans in Brazil, Africa, India, Sri Lanka, Indonesia, and the Philippines, the infections probably were from *A. ceylanicum.*

Ancylostoma caninum is the most common hookworm of domestic dogs, especially in the Northern Hemisphere. It has been found in humans on at least five occasions, and the worm also is a common cause of creeping eruption. This hookworm is an important cause of eosinophilic enteritis (EE) in northeastern Australia and now reported in the United States.[11,27,35] EE is abdominal pain with peripheral blood eosinophilia, but with no eggs in the feces. Apparently development to maturity is inhibited, but the presence of even one immature worm can cause EE.

• Hookworm Disease

The distinction between hookworm infection and hookworm disease is important. Far more people are infected with the worm than exhibit symptoms of the disease. Presence and severity of disease depends strongly on three factors: the number of worms present, the species of hookworm, and the nutritional condition of the infected person. In general, fewer than 25 *N. americanus* in a person will cause no symptoms, 25 to 100 worms lead to light symptoms, 100 to 500 produce considerable damage and moderate symptoms, 500 to 1000 result in severe symptoms and grave damage, and more than 1000 worms causes very grave damage that may be accompanied by drastic and often fatal consequences. Because *Ancylostoma* spp. suck more blood, fewer worms cause greater disease; for example, 100 worms may cause severe symptoms. However, the clinical disease is intensified by the degree of malnutrition, corresponding impairment of the host's immune response, and other considerations. Infection stimulates little protective immunity.[7]

Epidemiology.   From the discussion on the biology of hookworms, it is obvious that the combination of poor sanitation and appropriate environmental conditions is necessary for high endemicity.

- ***Environmental Conditions.***   Environmental conditions that favor the development and survival of the juveniles promote transmission. The disease is restricted to warmer parts of the world (and to specialized habitats such as mines in more severe climates) because the juveniles will not develop to maturity at less than 17°C, with 23° to 30°C being optimal. Frost kills eggs and juveniles. Oxygen is necessary for hatching of eggs and juvenile development because their metabolism is aerobic. Thus, the juveniles will not develop in undiluted feces or in waterlogged soil. Therefore, a loose, humusy soil that has reasonable drainage and aeration is favorable. Both heavy clay and coarse sandy soils are unfavorable for the parasite, the latter because the juveniles are also sensitive to desiccation. Alternate drying and moistening are particularly damaging to the juveniles; hence, very sandy soils become noninfective after brief periods of frequent rainfall. However, the juveniles live in the film of water surrounding soil particles, and even apparently dry soil may have enough moisture to enable survival, particularly below the surface.

  The juveniles are quite sensitive to direct sunlight and survive best in shady locations, such as coffee, banana, or sugarcane plantations. Humans working in the plantations often have preferred defecation sites, not out in the open where the juveniles would be killed by the sun, of course, but in shady, cool, secluded spots beneficial for juvenile development. Repeated return of people to the defecation site exposes them to continual reinfection. Furthermore, the use of preferred defecation sites makes it possible for hookworms to be endemic in otherwise quite arid areas.

Juveniles develop best near a neutral pH, and acid or alkaline soils inhibit development, as does the acid pH of undiluted feces (pH 4.8 to 5.0). Chemical factors have an influence. Urine mixed with the feces is fatal to the eggs, and several strong chemicals that may be added to feces as disinfectants of fertilizers are lethal to the free-living stages. Salt in the water or soil inhibits hatching and is fatal to juveniles.

- ***Longevity of Juveniles and Adults.***   The longevity of the worms is important in transmission to new hosts, continuity of infection in a locality, and introduction to new areas. The juveniles can survive in reasonably good environmental conditions for about three weeks; in protected sites like mines, they can last for a year. There is some dispute about the life span of the adult, but a good estimate is 5 to 15 years. A person who moves from an endemic area loses the infection in about that time.

- ***Degree of Soil Contamination.***   Obviously a higher average number of worms per individual will seed the soil with more eggs. Promiscuous defecation, associated with poverty and ignorance, keeps soil contamination high. The use of nightsoil as fertilizer for crops is an especially important factor in the Orient.

- ***Soil Contact with Skin.***   Because the worm penetrates the skin, the habit of going barefoot in tropical countries makes an elemental contribution to transmission. The role of skin penetration presumably accounts for the general lack of high correlation of hookworm with *Ascaris* and *Trichuris* infections, which must be acquired by ingestion.[9] This finding has important implications for control strategies.

- ***Race.***   White people are about 10 times more susceptible to hookworm than are black persons. Prevalence and numbers of worms per individual tend to be higher among whites than blacks. The image of the lazy, apathetic poor whites ("poor white trash") in the Old South (see the Mark Twain quote at the beginning of the chapter) as contrasted with the more industrious blacks on comparable socioeconomic levels is generally thought due to the differential effects of hookworm.

- ***Paratenic Hosts.***   A new dimension in the epidemiology of hookworm disease was the discovery by Schad and coworkers[36] that juveniles will survive in the muscles of paratenic hosts. Thus, *A. duodenale*, at least, can be transmitted through ingestion of undercooked meat, including rabbit, lamb, beef, and pork.

Pathogenesis.   Hookworm disease manifests three main phases of pathogenesis: the cutaneous or invasion period, the migration or pulmonary phase, and the intestinal phase. When a juvenile enters an unsuitable host, it causes another pathogenic condition, which we will discuss separately.

- ***Cutaneous Phase.***   The cutaneous phase begins when juveniles penetrate the skin. They do little damage to the superficial layers, since they seem to slip through tiny cracks between skin scales or penetrate hair follicles. Once in the dermis, however, their attack on blood vessels initiates a tissue reaction that may isolate and kill the worms. If, as usually happens, pyogenic bacteria are introduced into the skin with the invading juvenile, a urticarial reaction will result, causing a condition known as **ground itch.**

- *Pulmonary Phase.*   The pulmonary phase occurs when the juveniles break out of the lung capillary bed into the alveoli and progress up the bronchi to the throat. Each site hemorrhages slightly, with serious consequences in massive infections; however, very large numbers of juveniles migrating through the lungs simultaneously is rare. The phase is usually asymptomatic, although there may be some dry coughing and sore throat. A pneumonitis may indicate severe infection.

- *Intestinal Phase.*   The intestinal phase is the most important period of pathogenesis. On reaching the small intestine, the young worm attaches to the mucosa with its strong buccal capsule and teeth, and it begins to feed on blood (Fig. 25.10). In heavy infections worms are found from the pyloric stomach to the ascending colon, but usually they are restricted to the anterior third of the small intestine. The worms move from place to place, but bleeding quickly ceases at a lesion that a worm has left, contrary to prevalent belief.[39] The worms pass substantially more blood through their digestive tracts than would appear necessary for their nutrition alone, but the reason for this is unknown. Blood loss per worm is about 0.03 ml per day for *Necator* and around 0.26 ml per day for *Ancylostoma.*

Patients with heavy infections may lose up to 200 ml of blood per day, but around 40% or so of the iron may be reabsorbed by the patient before it leaves the intestine.[22] Nevertheless, a moderate hookworm infection will gradually produce an iron-deficiency anemia as the body reserves of iron are used up. The severity of the anemia depends on the worm load and the dietary iron intake of the patient. Slight, intermittent abdominal pain, loss of normal appetite, and desire to eat soil (geophagy) are common symptoms of moderate hookworm disease. (Certain areas in the southern United States became locally famous for the quality of their clay soil, and people traveled for miles to eat it. In the early 1920s an enterprising person began a mail-order business, shipping clay to hookworm sufferers throughout the country!)

- *Heavy Infections.*   In very heavy infections, patients suffer severe protein deficiency, with dry skin and hair, edema, and potbelly in children and with delayed puberty, mental dullness, heart failure, and death. Intestinal malabsorption is not a marked feature of infection with hookworms, but hookworm disease is usually manifested in the presence of malnutrition and is often complicated by infection with other worms and/or malaria.

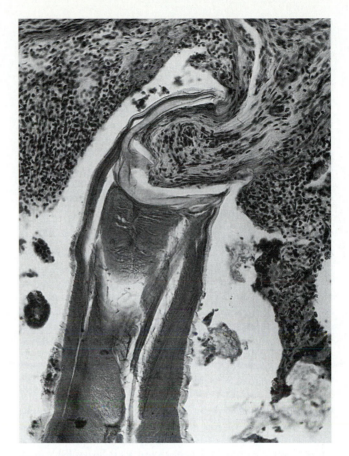

FIGURE 25.10

Hookworm attached to intestinal mucosa. Notice how the ventral tooth in the depth of the buccal capsule lacerates the host tissue.
AFIP neg. no. N-33818.

The drain of protein and iron is catastrophic to one who is subsisting on a minimal diet. Such chronic malnutrition, particularly in the young, often causes stunted growth and below-average intelligence, but treatment for the worms can significantly increase fitness, appetite, and growth.[20,40] Impairment in ability to produce IgG results in lowered antibody response to the hookworm as well as to other infectious agents. No living organism could be expected to live up to its potential under such conditions, and it is small wonder that development has been so difficult for many tropical countries.

Diagnosis.   Demonstration of hookworm eggs or the worms themselves in the feces is, as usual, the only definitive diagnosis of the disease. Demonstration of eggs in direct smears may be difficult, however, even in clinical cases, and one of the several concentration techniques should be used. If estimation of worm burden is necessary, techniques are available that give reliable data on egg counts.[28]

It is not necessary or possible to distinguish *Necator* eggs from those of *Ancylostoma*, but care should be taken to differentiate *Strongyloides* infections. This is not a problem unless some hours pass between time of defecation and time of examination of feces. Then the hookworm eggs may have hatched, and the juveniles must be distinguished from those of *Strongyloides.*

It is desirable, nevertheless, to distinguish *Necator* and *Ancylostoma* spp. in studies on the efficacy of various drugs or chemotherapeutic regimens because the two species are not equally sensitive to particular drugs. *Necator,* for example, has low sensitivity to ivermectin, in contrast with *Ancylostoma.*[8] Differentiation can be accomplished by recovery of adults after anthelmintic treatment or by culturing the juveniles from the feces. It may be possible to develop immunodiagnostic techniques.[33]

Treatment.   Mebendazole is the drug of choice for treatment, as it removes both species of hookworm and also any concurrent infection with *Ascaris lumbricoides.* Single-dose therapy is inexpensive, convenient, and effective.[1]

Treatment for hookworm disease should always include dietary supplementation. In many cases provision of an adequate diet alleviates the symptoms of the disease without worm removal, but treatment for the infection should be instituted, if only for public health reasons.

Control.   Control of hookworm disease depends on lowering worm burdens in a population to the extent that remaining worms, if any, can be sustained within the nutritional limitations of the people, without causing symptoms. Mass treatment campaigns do not eradicate the worms but certainly lower the "seeding" capacity of their hosts. Education and persuasion of the population in the sanitary disposal of feces are also vital. The economic dependence on nightsoil in family gardens remains one of the most persistent of all problems in parasitology.

Recognizing these factors, the American zoologist Charles W. Stiles persuaded John D. Rockefeller to donate $1 million in 1909 to establish the Rockefeller Sanitary Commission for the Eradication of Hookworm Disease.[2] (The activities of the Commission eventually led to the formation of the Rockefeller Foundation and then Rockefeller University.) Beginning state by state and then extending throughout the southeastern United States, the Commission first surveyed an area. Residents of the area were examined for infection and then treated with anthelmintics. Thousands of latrines were provided, together with instructions on how to use and maintain them. It is a study in human nature to note that many persons refused to use latrines and could be persuaded only with great difficulty. As the result of the efforts of this and other similar hygiene commissions, hookworm prevalence is now much lower than it used to be in the southern United States, most of the Caribbean, and a few other areas of the world. Nevertheless, hookworm remains one of the most important parasites of humans in the world today.

• **Creeping Eruption**

Also known as **cutaneous larva migrans,** creeping eruption is caused by invasive juvenile hookworms of species or strains normally maturing in animals other than humans. The juveniles manage to penetrate the skin of humans but are incapable of

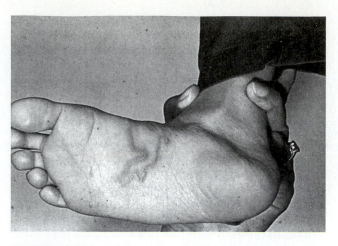

FIGURE 25.11

Creeping eruption caused by infection with *Ancylostoma* sp. juvenile.

H. Zaiman, editor, *A Pictorial Presentation of Parasites.* Photo courtesy of F. Battistine.

successfully completing migration to the intestine. However, before they are overcome by immune responses, they produce distressing and occasionally serious complications of the skin (Fig. 25.11). Possibly any species of hookworm can cause this condition, but those of cats, dogs, and other domestic animals are most likely to come into contact with people. *Ancylostoma braziliense* appears to be the most common agent, followed by *A. caninum.* Both are common parasites of domestic dogs and cats in some localities.

After entering the top layers of epithelium, the juveniles are usually incapable of penetrating the basal layer (stratum germinativum), so they begin an aimless wandering. As they tunnel through the skin, they leave a red, itchy wound that usually becomes infected by pyogenic bacteria. The worms may live for weeks or months. It is known that some can enter muscle fibers and become dormant.[25] The juveniles can attack any part of the skin, but people's feet and hands are more in contact with the ground and so are most often affected. Thiabendazole has revolutionized treatment of creeping eruption, and topical application of a thiabendazole ointment has supplanted all other forms of treatment.

SUPERFAMILY STRONGYLOIDEA

Family Strongylidae

Members of the Strongylidae occur in a variety of mammals, especially horses, where they are a serious veterinary problem. They are commonly recognized as large strongyles (several species of *Strongylus,* of which *S. vulgaris* is the most important) and small strongyles (mostly the numerous species of *Cyathostomum*).[16] Adults of both are found in the large intestine of equines. Eggs pass out in the feces, hatch, and develop in the soil to the infective J_3's. These crawl onto vegetation and are eaten by the grazing host. All undergo a migration and period of development in various tissues, the details of which vary with the species.

Developing juveniles of *S. vulgaris* migrate into the arteries, especially the anterior mesenteric artery, where they cause thrombosis and arteritis. Formerly, arterial stages of *S. vulgaris* showed 90% to 100% prevalence in horses in the United States, and it was the most feared equine parasite.[16] Small strongyles were essentially neglected. Now, because *S. vulgaris* is sensitive to the benzimidazole and ivermectin anthelmintics and the cyathostomes are relatively resistant, *S. vulgaris* is almost wiped out and *Cyathostomum* spp. are a much bigger problem.

Oesophagostomum spp. are the nodular worms of livestock. Adults live in the large intestine of ruminants, primates, and swine, and the developing juveniles form nodules in the walls of the small and large intestine. Infections in humans have been reported rarely, and they were considered zoonoses. However, *O. bifurcum* is common in humans in one small area in Africa, where it is probably not a zoonotic infection.[32]

Family Syngamidae

Syngamus trachea is the gapeworm of poultry, so called because adults live in the trachea of their hosts, causing gasping and gaping. The fowl coughs up the eggs, swallows them, and then passes them in the feces. The juveniles molt twice in the egg to become the infective J_3's. The eggs may or may not hatch, and a variety of terrestrial molluscs, earthworms, and arthropods can serve as paratenic hosts. These worms can survive several years in earthworms, and numerous wild bird species serve as reservoirs.[24] When swallowed by the definitive host, the J_3's penetrate the gut wall, are carried by the blood to the lungs where they break out into the alveoli, and then proceed up to the trachea. Young birds are most severely affected and may die with a heavy infection.

SUPERFAMILY TRICHOSTRONGYLOIDEA

Family Trichostrongylidae

Many genera and an immense number of species comprise the family Trichostrongylidae. They are parasites of the small intestine of all classes of vertebrates, causing great economic losses in domestic animals, especially ruminants, and in a few cases causing disease in humans.

Trichostrongylids (Fig. 25.12) are small, very slender worms with a rudimentary buccal cavity in most cases. Lips are reduced or absent, and teeth rarely are present. The cuticle of the head may be inflated. Males have a well-developed bursa, and the spicules vary from simple to complex, depending on the species. Females are considerably larger than males. The vulva is located anywhere from preequatorial to near the anus, according to the species. The worms lay thin-shelled eggs that are in the morula stage.

Life cycles are similar in all species. No intermediate host is required; the eggs hatch in soil or water and develop directly to infective J_3's. Some infections may occur through the skin, but as a rule the juveniles must be swallowed with contaminated food or water. Enormous numbers of juveniles may accumulate on heavily grazed pas-

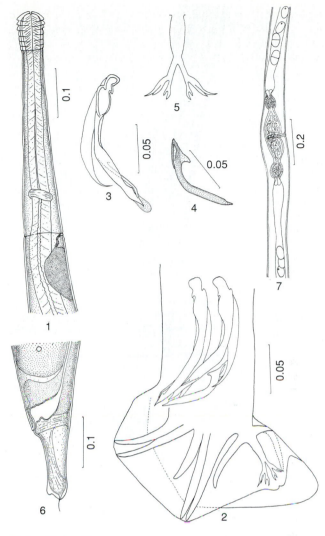

FIGURE 25.12

Molineus mustelae, showing the characters typical of the Trichostrongylidae. **1,** anterior end, lateral view. **2,** posterior end of male. **3,** complex spicules, lateral view. **4,** gubernaculum, lateral view. **5,** dorsal ray. **6,** posterior end of female, lateral view. **7,** midregion of female, showing ovijectors. (All scales are in millimeters.)

From G. D. Schmidt, "*Molineus mustelae* sp. n. (Nematoda: Trichostrongylidae) from the long-tailed weasel in Montana and *M. chabaudi* nom. n., with a key to the species of *Molineus,*" in *J. Parasitol.* 51:164–168. Copyright © 1965. Reprinted with permission of the publisher.

tures, causing serious or even fatal infections in ruminants and other grazers. A given host usually is infected with several species since their life cycles are similar, and severe pathogenesis results from the cumulative effects of all the worms.

Ivermectin has become a widely used anthelmintic for trichostrongyles, but several species in sheep and goats have become resistant to this drug.[37]

Haemonchus contortus. *Haemonchus contortus* lives in the "fourth stomach," or abomasum, of sheep, cattle, goats, and wild ruminants of many species. The species has been reported

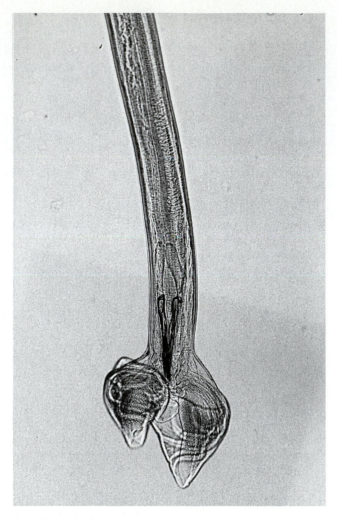

FIGURE 25.13

Haemonchus contortus: ventral view of male, showing asymmetrical copulatory bursa.

Courtesy of Jay Georgi.

in humans in Brazil and Australia. It is one of the most important nematodes of domestic animals, causing a severe anemia in heavy infections.

The small buccal cavity contains a single well-developed tooth that pierces the mucosa of the host. Blood is sucked from the wound, giving the transparent worms a reddish color. The large females have white ovaries wrapped around the red intestine, lending it a characteristic red and white appearance and leading to its common names: *twisted stomach worm* and *barber-pole worm.* Prominent cervical papillae are found near the anterior end. The male bursa is powerfully developed, with an asymmetrical dorsal ray (Fig. 25.13). The spicules are 450 to 500 μm long, each with a terminal barb. The vulva has a conspicuous anterior flap in many individuals but not in all. The frequency of occurrence of the vulvar flap seems to vary according to strain.

Infection occurs when the J₃, still wearing the loosely fitting second-stage cuticle, is eaten with forage. Exsheathment takes place in the forestomachs. Arriving in the abomasum or upper duodenum, the worm molts within 48 hours, becoming a J₄ with a small buccal capsule. It feeds on blood, which forms a clot around the anterior end of the worm. The worm molts for a final time in three days and begins egg production about 15 days later.

The anemia, emaciation, edema, and intestinal disturbances caused by these parasites result principally from loss of blood and injection of hemolytic proteins into the host's system. The host often dies with heavy infections, but those that survive usually develop an immunity and effect a self-cure, a result of the inflammatory response in the intestinal mucosa.

***Ostertagia* Species.**   *Ostertagia* spp. are similar to *H. contortus* in host and location but differ in color, being a dirty brown—hence their common name *brown stomach worm.* The buccal capsule is rudimentary and lacks a tooth. Cervical papillae are present. The male bursa is symmetrical. The vulva has a large anterior flap, and the tip of the female's tail bears several cuticular rings.

The life cycle is similar to that of *Haemonchus* except that the J₃ burrows into the abomasal mucosa, where it molts. Returning to the lumen, it feeds, molts, and begins producing eggs about 17 days after infection. *Ostertagia* spp. suck blood but not as much as does *Haemonchus.*

Some common species of *Ostertagia* are *O. circumcincta* in sheep, *O. ostertagi* in cattle and sheep, and *O. trifurcata* in sheep and goats. Economic losses in the cattle industry due to *O. ostertagi* probably exceed $600 million in the United States alone.[38]

***Trichostrongylus* Species.**   *Trichostrongylus* spp. are the smallest members of the family, seldom exceeding 7 mm in length. Many species exist parasitizing the small intestine of ruminants, rodents, pigs, horses, birds, and humans.

They are colorless, lack cervical papillae, and have a rudimentary, unarmed buccal cavity. The male bursa is symmetrical, with a poorly developed dorsal lobe. Spicules are brown and distinctive in size and shape in each species. The vulva lacks an anterior flap.

The life cycle is similar to that of *Ostertagia* except that the worms pass through the abomasum and burrow into the mucosa of the duodenum, where they molt. After returning to the lumen, they bury their heads in the mucosa and feed, grow, and molt for the last time. Egg production begins about 17 days after infection.

Common species of *Trichostrongylus* are *T. colubriformis* in sheep; *T. tenuis* in chickens and turkeys; *T. capricola, T. falcatus,* and *T. rugatus* in ruminants; *T. retortaeformis* and *T. calcaratus* in rabbits; and *T. axei* in a wide variety of mammals.

A number of species of *Trichostrongylus* have been reported in humans, with records from nearly every country of the world: There are nine species in Iran alone.[31] Rate of infection varies from very low to as high as 69% in southwest Iran[34] and 70% in a village in Egypt.[21]

Pathological conditions are identical in humans and other infected animals. Traumatic damage to the intestinal epithelium

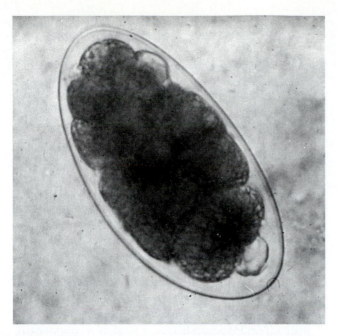

FIGURE 25.14

Trichostrongylus egg found in a human stool. Eggs of *Trichostrongylus* spp. resemble those of hookworms but are usually larger, about 81 to 104 by 40 to 48 μm.[24]

Courtesy of David Oetinger.

may be produced by burrowing juveniles and feeding adults. Systemic poisoning by metabolic wastes of the parasites and possible thyroid deficiency, hemorrhage, emaciation, and mild anemia may develop in severe infections.

Diagnosis can be made by finding the characteristic eggs (Fig. 25.14) in the feces or by culturing the juveniles in powdered charcoal. The juveniles are very similar to those of hookworms and *Strongyloides* spp., and careful differential diagnosis is required.

Treatment with thiabendazole or with pyrantel pamoate has proved effective. Cooking vegetables adequately will prevent many infections in humans.

Family Dictyocaulidae

Dictyocaulus filaria is an important parasite of sheep and goats. Adults live in the bronchi and bronchioles, where the females produce embryonated eggs. The eggs hatch while being carried toward the trachea by ciliary action. First-stage juveniles appear in the feces and develop to J_3's in contaminated soil, without feeding. The cuticles of both the first and second stages are retained by the third stage until the worm is eaten by a definitive host; then cuticles of all these stages are shed together. The J_4's penetrate the mucosa of the small intestine, enter the lymphatic system, mix with the blood, and are carried to the lungs, where they enter alveoli and migrate to the bronchioles. They commonly cause death to their host.

These worms are slender and quite long, males reaching 80 mm and females 100 mm. The bursa is small and symmetrical; the spicules are short and boot shaped in lateral view. The uterus is near the middle of the body. Other species in horses and cattle are similar to *D. filaria* in morphology and biology.

Other Trichostrongyloids

In addition to the species from ruminants already mentioned, *Cooperia curticei*, *Nematodirus spathiger*, and *N. filicollis* should be included in the group of trichostrongyles that often occur in the same host and cause so much damage. *Hyostrongylus rubidus* is a serious pathogen of swine and can cause death when present in large numbers. *Heligmosomoides polygyrus* (= *Nematospiroides dubius*) in mice and *Nippostrongylus brasiliensis* in rats are easily kept in the laboratory, and they serve as important tools for investigation of trichostrongyle nematodes.

SUPERFAMILY METASTRONGYOIDEA

Members of this superfamily live in the lungs of mammals. Most species for which the life cycles are known require an invertebrate intermediate host, and some also employ a vertebrate or invertebrate transport host. Most species mature in terrestrial mammals, although several species in numerous genera are important parasites of marine mammals. Taxonomy of the group is unsettled, with several schemes in use. The metastrongyles are fairly homogeneous morphologically, with buccal cavities reduced or absent and bursal lobes and rays reduced. Most are parasitic in the bronchioles, but some inhabit the pulmonary arteries, heart, muscles, and frontal sinuses.

Family Angiostrongylidae

Angiostrongylus cantonensis. *Angiostrongylus cantonensis* was first discovered in the pulmonary arteries and heart of domestic rats in China in 1935. Later the worm was found in many species of rats and bandicoots, and it may mature in other mammals throughout southeast Asia, the East Indies, Madagascar, and Oceanica, with infection rates as high as 88%. As a parasite of rats, it attracted little attention, but 10 years after its initial discovery, it was found in the spinal fluid of a 15-year-old boy in Taiwan. It has been discovered since in humans in Hawaii, Tahiti, the Marshall Islands, New Caledonia, Thailand, Vanuatu, the Loyalty Islands, and other places in the eastern hemisphere. It is now known to exist in Louisiana. This illustrates the value of basic research in parasitology to medicine, because when the medical importance of the parasite was realized, the reservoir of infection in rats already was known. Surveys of parasites endemic to the wild fauna of the world remain the first step in understanding the epidemiology of zoonotic diseases.

Angiostrongylus cantonensis is a delicate, slender worm with a simple mouth and no lips or buccal cavity. Males are 15.9 to 19.0 mm long, whereas females attain 21 to 25 mm. The bursa is small and lacks a dorsal lobe. Spicules are long, slender, and about equal in length and form. An inconspicuous gubernaculum is present. In the female the intertwining of the intestine and uterine tubules gives the worm a conspicuous barber-pole appearance. The vulva is about 0.2 mm in front of the anus. The eggs are thin shelled and unembryonated when laid.

- *Biology.* The eggs are laid in the pulmonary arteries, are carried to the capillaries, and break into the air spaces, where they hatch. The juveniles migrate up the trachea, are swallowed, and are expelled with the feces.

A number of types of molluscs serve as intermediate hosts, including slugs and aquatic and terrestrial snails. Terrestrial planarians, freshwater shrimp, land crabs, and coconut crabs serve as paratenic hosts. Frogs have been found naturally infected with infective juveniles.[5] Experimentally, Cheng[10] infected American oysters and clams, and Wallace and Rosen[43] succeeded in infecting crabs. All juveniles thus produced were infective to rats.

When eaten by a definitive host, the J$_3$'s undergo an obligatory migration to the brain, which they leave four weeks after ingestion. Maturation occurs in the pulmonary arteries in about six weeks. Many wander in the body and mature in other locations, primarily in the central nervous system, meninges, and eyes.

- *Epidemiology.*   Humans or other mammals become infected when they ingest J$_3$'s. There may be several avenues of human infection, depending on the food habits of particular peoples.[4,13] In Tahiti it is a common practice to catch and eat freshwater shrimp raw or to make sauce out of their raw juices. It is possible to eat slugs or snails accidentally with raw vegetables or fruit, so this route of infection should be considered. In Thailand and Taiwan, raw snails are often considered a delicacy. Infective juveniles escape from slugs and can be left behind in their mucous trail on vegetables over which they crawl.[17] Such juveniles were found on lettuce sold in a public market in Malaya. Thus, although the epidemiology of angiostrongyliasis is not completely known, ample opportunities for infection exist.

- *Pathology.*   For many years a disease of unknown cause was recognized in tropical Pacific islands and was named **eosinophilic meningoencephalitis.** Patients with this condition have high eosinophil counts in peripheral blood and spinal fluid in about 75% of the cases and increased lymphocytes in cerebrospinal fluid. Neural disorders commonly accompany these symptoms, occasionally followed by death. We now know that *A. cantonensis* is at least one cause of this condition.

   The presence of worms in blood vessels of the brain and meninges, as well as that of free-wandering worms in the brain tissue itself, results in serious damage. Some effects of such infection are severe headache, fever in some cases, paralysis of the fifth cranial nerve, stiff neck, coma, and death. Destruction of brain and spinal cord cells by trauma and immune responses evoked by dead worms results in vague symptoms for which the cause is most difficult to diagnose.

- *Diagnosis and Treatment.*   When the symptoms described appear in a patient in areas of the world where *A. cantonensis* exists, angiostrongyliasis should be suspected. It should be kept in mind that many of these symptoms can be produced by hydatids, cysticerci, flukes,

Strongyloides, Trichinella, various juvenile ascarids, and possibly other lungworms. Alicata[3] and Ash[6] differentiate the juveniles of several species of metastrongylids that could be confused with *A. cantonensis.*

Thiabendazole shows promise in treating early, invasive stages, but little is known of treating the adults. Dead worms in blood vessels and the central nervous system may be more dangerous than live ones. A spinal tap to relieve headache may be recommended.

Angiostrongylus costaricensis parasitizes the mesenteric arteries of rats in Central and South America, southern North America, and Cuba.[29] A number of cases have been diagnosed in southern Brazil.[14] The worms mature in the mesenteric arteries and their intramural branches. Most damage is to the wall of the intestine, especially the cecum and appendix, which become thickened and necrotic, with massive eosinophilic infiltration. Abdominal pain and high fever are the most evident symptoms. Evidently, these intestinal disorders are caused by blockage of the arterioles supplying the area by eggs and juveniles of the parasites. No symptoms of meningoencephalitis, typical of *A. cantonensis,* are noted. However, eosinophilic meningoencephalitis has been reported in patients infected with *A. cantonensis* in Cuba.[30]

Ubelaker[41] argued that the subgenus *Parastrongylus* should be a separate genus, containing *A. cantonensis* and *A. costaricensis* among other species. Cladistic analysis of this group of nematodes would be helpful.

Other Metastrongyloids

Protostrongylus rufescens parasitizes the bronchioles of ruminants in many parts of the world. Its intermediate hosts are terrestrial snails, in which it develops to the third stage, and the definitive host is infected when it eats the snail along with forage. Mountain sheep in America are seriously threatened by this and related species, which take a high toll of lambs every spring. Hibler and coworkers[18] demonstrated transplacental transmission of *Protostrongylus* spp. in bighorn sheep.

Muellerius capillaris lives in nodules in the parenchyma of the lungs of sheep and goats in most areas of the world. The life cycle is similar to that of *Protostrongylus rufescens,* involving a snail intermediate host.

Metastrongylus apri mainly infects swine, but infections in sheep and cattle and three cases in humans have been reported. Adults live in the bronchioles, and the eggs may hatch as in *Dictyocaulus,* or they may appear in the feces before hatching. An earthworm intermediate host is required for development to the infective third stage. This lungworm is also known to serve as a vector for the virus that causes **swine influenza.**[23] The nematodes serve as reservoirs for the disease, since they may live up to three years while carrying the virus within their bodies.

References

1. Abadi, K. 1985. Single dose mebendazole therapy for soil-transmitted nematodes. *Am. J. Trop. Med. Hyg.* 34:129–33.

2. Ackert, J. E. 1952. Some influences of the American hookworm. *Am. Midland Natur.* 47:749–62.

3. Alicata, J. E. 1963. Morphological and biological differences between the infective larvae of *Angiostrongylus cantonensis* and those of *Anafilaroides rostratus. Can. J. Zool.* 41:1179–83.

4. Alicata, J. E. 1991. The discovery of *Angiostrongylus cantonensis* as a cause of human eosinophilic meningitis. *Parasitol. Today* 7:151–53.

5. Ash, L. R. 1968. The occurrence of *Angiostrongylus cantonensis* in frogs of New Caledonia with observations on paratenic hosts of metastrongyles. *J. Parasitol.* 54:432–36.

6. Ash, L. R. 1970. Diagnostic morphology of the third-stage larvae of *Angiostrongylus cantonensis, Angiostrongylus vasorum, Aelurostrongylus abstrusus,* and *Anafilaroides rostratus* (Nematoda: Metastrongyloidea). *J. Parasitol.* 56:249–53.

7. Behnke, J. M. 1987. Do hookworms elicit protective immunity in man? *Parasitol. Today* 3:200–6.

8. Behnke, J. M., R. Rose, and P. Garside. 1993. Sensitivity to ivermectin and pyrantel of *Ancylostoma ceylanicum* and *Necator americanus. Int. J. Parasitol.* 23:945–52.

9. Booth, M., and D. A. P. Bundy. 1992. Comparative prevalences of *Ascaris lumbricoides, Trichuris trichiura* and hookworm infections and the prospects for combined control. *Parasitology* 105:151–57.

10. Cheng, T. C. 1965. The American oyster and clam as experimental intermediate hosts of *Angiostrongylus cantonensis. J. Parasitol.* 51:296.

11. Croese, J., A. Loukas, J. Opdebeeck, S. Fairley, and P. Prociv. 1994. Human enteric infection with canine hookworms. *Ann. Intern. Med.* 120:369–74.

12. Crompton, D. W. T. 1989. Hookworm disease: Current status and new directions. *Parasitol. Today* 5:1–2.

13. Cross, J. H. 1987. Public health importance of *Angiostrongylus cantonensis* and its relatives. *Parasitol. Today* 3:367–69.

14. Graeff-Teixeira, C., L. Camillo-Coura, and H. L. Lenzi. 1991. Histopathological criteria for the diagnosis of abdominal angiostrongyliasis. *Parasitol. Res.* 77:606–11.

15. Hawdon, J. M., S. W. Volk, R. Rose, D. I. Pritchard, J. M. Behnke, and G. A. Schad. 1993. Observations on the feeding behaviour of parasitic third-stage hookworm larvae. *Parasitology* 106:163–69.

16. Herd, R. P. 1990. The changing world of worms: The rise of the cyathostomes and the decline of *Strongylus vulgaris. Comp. Cont. Ed. Pr. Vet.* 12:732–36.

17. Heyneman, D., and B. L. Lim. 1967. *Angiostrongylus cantonensis:* Proof of direct transmission with its epidemiological implications. *Science* 158:1057–58.

18. Hibler, C. P., R. E. Lange, and C. J. Metzger. 1972. Transplacental transmission of *Protostrongylus* spp. in bighorn sheep. *J. Wildl. Dis.* 8:389.

19. Kumar, S., and D. I. Pritchard. 1994. Apparent feeding behaviour of ensheathed third-stage infective larvae of human hookworms. *Int. J. Parasitol.* 24:133–36.

20. Latham, M. C., L. S. Stephenson, K. M. Kurz, and S. N. Kinoti. 1990. Metrifonate or praziquantel treatment improves physical fitness and appetite of Kenyan schoolboys with *Schistosoma hematobium* and hookworm infections. *Am. J. Trop. Med. Hyg.* 43:170–79.

21. Lawless, D. K., R. E. Kuntz, and C. P. A. Strome. 1956. Intestinal parasites in an Egyptian village of the Nile Valley with emphasis on the protozoa. *Am. J. Trop. Med. Hyg.* 5:1010–14.

22. Layrisse, M., A. Paz, N. Blumenfeld, and M. Roche. 1961. Hookworm anemia: Iron metabolism and erythrokinetics. *Blood* 18:61–72.

23. Lee, D. L. 1971. Helminths as vectors of micoorganisms. In Fallis, A. M., ed. *Ecology and physiology of parasites.* Toronto: University of Toronto Press, 104–22.

24. Levine, N. D. 1980. *Nematode parasites of domestic animals and of man,* 2d ed. Minneapolis: Burgess Publishing Co.

25. Little, M. D., N. A. Halsey, B. L. Cline, and S. P. Katz. 1983. *Ancylostoma* larva in a muscle fiber of man following cutaneous larva migrans. *Am. J. Trop. Med. Hyg.* 32:1285–88.

26. Looss, A. 1898. Zur Lebensgeschichte des *Ankylostoma duodenale.* Cbt. Bakt. 24:441–49, 483–88.

27. Loukas, A., J. Opdebeeck, J. Croese, and P. Prociv. 1994. Immunologic incrimination of *Ancylostoma caninum* as a human enteric pathogen. *Am. J. Trop. Med. Hyg.* 50:69–77.

28. Markell, E. K. 1991. Examination of stool specimens. In Strickland, G. T., ed. *Hunter's tropical medicine,* 7th ed. Philadelphia: W. B. Saunders Company, 1075–84.

29. Morera, P. 1985. Abdominal angiostrongyliasis: A problem of public health. *Parasitol. Today* 1:173–75.

30. Pascual, J. E., R. P. Bouli, and H. Aguiar. 1981. Eosinophilic meningoencephalitis in Cuba, caused by *Angiostrongylus cantonensis. Am. J. Trop. Med. Hyg.* 30:960–62.

31. Pearson, R. D., and J. D. Schwartzman. 1991. Trichostrongyliasis. In Strickland, G. T., ed. *Hunter's tropical medicine,* 7th ed. Philadelphia: W. B. Saunders Company, 695–96.

32. Polderman, A. M., H. P. Krepel, S. Baeta, J. Blotkamp, and P. Gigase. 1991. Oesophagostomiasis, a common infection of man in northern Togo and Ghana. *Am. J. Trop. Med. Hyg.* 44:336–44.

33. Pritchard, D. I., P. G. McKean, and G. A. Schad. 1990. An immunological and biochemical comparison of hookworm species. *Parasitol. Today* 6:154–56.

34. Sabha, G. H., F. Arfaa, and H. Bijan. 1967. Intestinal helminthiasis in the rural area of Khuzestan, southwest Iran. *Ann. Trop. Med. Parasitol.* 61:352–57.

35. Schad, G. A. 1994. Hookworms: Pets to humans. *Ann. Intern. Med.* 120:434–35.

36. Schad, G. A., K. D. Murrell, R. Fayer, H. M. S. El Naggar, M. R. Page, P. K. Parrish, and T. B. Stewart. 1984. Paratenesis in *Ancylostoma duodenale* suggests possible meat-borne human infection. *Trans. R. Soc. Trop. Med. Hyg.* 78:203–4.

37. Shoop, W. L. 1993. Ivermectin resistance. *Parasitol. Today* 9:154–59.

38. Smith, G., and B. T. Granfell. 1985. The population biology of *Ostertagia ostertagi. Parasitol. Today* 1:76–81.

39. Spencer, H. 1973. Nematode diseases. I. In Spencer, H., ed. *Tropical pathology.* Berlin: Springer-Verlag, 457–509.

40. Stephenson, L. S., M. C. Latham, K. M. Kurz, and S. N. Kinoti. 1989. Single dose metrifonate or praziquantel treatment in Kenyan children. II. Effects on growth in relation to *Schistosoma haematobium* and hookworm egg counts. *Am. J. Trop. Med. Hyg.* 41:445–53.

41. Ubelaker, J. E. 1986. Systematics of species referred to the genus *Angiostrongylus. J. Parasitol.* 72:237–44.

42. Velasquez, C., and B. C. Cabrera. 1968. *Ancylostoma ceylanicum* (Looss), in a Filipino woman. *J. Parasitol.* 54:430–31.

43. Wallace, G. D., and L. Rosen. 1966. Studies on eosinophilic meningitis. 2. Experimental infection of shrimp and crabs with *Angiostrongylus cantonensis. Am. J. Epidemiol.* 84:120–41.

Additional References

Dooley, J. R., and R. C. Neafie. 1976. Angiostrongyliasis: *Angiostrongylus cantonensis* infections. In Binford, C. H., and D. H. Connor, eds. *Pathology of tropical and extraordinary diseases,* vol. 2, sect. 9. Washington, D.C.: Armed Forces Institute of Pathology.

Frenkel, J. K. 1976. Angiostrongyliasis: *Angiostrongylus costaricensis* infections. In Binford, C. H., and D. H. Connor, eds. *Pathology of tropical and extraordinary diseases,* vol. 2, sect. 9. Washington, D.C.: Armed Forces Institute of Pathology.

Meyers, W. M., and R. C. Neafie. 1976. Creeping eruption. In Binford, C. H., and D. H. Connor, eds. *Pathology of tropical and extraordinary diseases,* vol. 2, sect. 9. Washington, D.C.: Armed Forces Institute of Pathology.

Meyers, W. M., R. C. Neafie, and D. H. Connor. 1976. Ancylostomiasis. In Binford, C. H., and D. H. Connor, eds. *Pathology of tropical and extraordinary diseases,* vol. 2, sect. 9. Washington, D.C.: Armed Forces Institute of Pathology.

Pawlowski, Z. S., G. A. Schad, and G. J. Stott. 1991. *Hookworm infection and anaemia: Approaches to prevention and control.* Geneva: World Health Organization.

Schad, G. A., and K. S. Warren, eds. 1990. *Hookworm disease. Current status and new directions.* London: Taylor and Francis.

Stoll, N. R. 1972. The osmosis of research: Example of the Cort hookworm investigations. *Bull. N.Y. Acad. Med.* 48:1321–29.

Chapter 26

NEMATODES: ASCARIDIDA, INTESTINAL LARGE ROUNDWORMS

There are two things for which animals are to be envied: they know nothing of future evils, or of what people say about them.

Voltaire

The ascaridid worms are typically large, stout, intestinal parasites with three large lips. However, there are minute species with small or no lips, large species with two lips, and small species with well-defined lips. A preanal sucker is found on males of some. In one large group the esophagointestinal junction is highly specialized, with muscular or glandular appendages. Usually, however, the esophagus is simple and muscular. The life cycle is usually simple, lacking an intermediate host, although such a host is required in a few species.

Of the several superfamilies and families in this order, we will emphasize the Ascarididae in the Ascaridoidea, which has the most medical importance. We will discuss some members of other families briefly.

SUPERFAMILY ASCARIDOIDEA

Family Ascarididae

The ascaridids are among the largest of nematodes, some species achieving a length of 18 inches or more. Cervical, lateral, and caudal alae are absent, as are any esophageal ceca or ventriculi. Three large rounded or trapezoidal lips are present; interlabia are absent. Spicules are simple and equal. This family contains one of the oldest associates of humankind: *Ascaris,* the intestinal large roundworm.[19]

Ascaris lumbricoides* and *Ascaris suum. Because of their great size, abundance, and cosmopolitan distribution, these nematodes may well have been the first parasites known to humans. Certainly the ancient Greeks and the Romans were familiar with them, and they were mentioned in the Ebers Papyrus. It is probable that *A. lumbricoides* was originally a parasite of pigs that adapted to humans when swine were domesticated and began to live in close association with humans—or perhaps it was a human parasite that we gave to pigs. (The physiologies of people and swine are remarkably similar as, on occasion, are their eating and social habits.)

Today two populations of this parasite exist, one in humans and one in pigs. They show a strong host specificity, but the two forms are so close morphologically that they were long considered the same species. Sprent[43] pointed out slight differences in the tiny denticles on the dentigerous ridges along the inner edge of the lips. This difference seems consistent and is much clearer when the structures are viewed with the scanning electron microscope;[46] therefore, we can now consider the two as separate species: *A. suum* from pigs and *A. lumbricoides* from humans.

This seems to be a good example of evolution in action.[33] Each species may diverge even further with time, now that they have been reproductively isolated in many parts of the world where pigs no longer enjoy the homes of their masters. Considerable evidence indicates that some cases of human infection are due to *A. suum*.[29,42,45] Nevertheless, a study on mitochondrial DNA of *Ascaris* in Guatemala, where the worms are widely prevalent in both pigs and humans, showed that gene flow between the two populations was extremely limited.[1]

Aside from the host specificity and the characteristics of the denticles, there are few, if any, other differences in the two species, and the following remarks on morphology and biology apply to both equally.

- *Morphology.* In addition to their great size (Fig. 26.1), these species are characterized by having three prominent lips, each with a dentigerous ridge and no interlabia or alae. Lateral hypodermal cords are visible with the unaided eye.

Males are 15 to 31 cm long and 2 to 4 mm at greatest width. The posterior end is curved ventrally, and the tail is bluntly pointed. Spicules are simple, nearly equal, and measure 2.0 to 3.5 mm long. No gubernaculum is present.

Females are 20 to 49 cm long and 3 to 6 mm wide. The vulva is about one-third the body length from the anterior end. The ovaries are extensive, and the uteri may contain up to 27 million eggs at a time, with 200,000 being laid per day. Fertilized eggs (Fig. 26.2) are oval to round, 45 to 75 μm long by 35 to 50 μm wide, with a thick, lumpy outer shell (**mammillated,** uterine, or proteinaceous layer) that is contributed by the uterine wall. When the eggs are passed in the feces, the mammillated layer is bile stained to a golden brown. The embryos are usually uncleaved when laid and

FIGURE 26.1

Ascaris lumbricoides, males (*right*) and females (*left*). Females are up to 18 in long.

Courtesy of Ann Arbor Biological Center.

passed in the feces. A female before insemination or one in early stages of oviposition commonly deposits unfertilized eggs (Fig. 26.3) that are longer and narrower, measuring 88 to 94 μm long by 44 μm wide. Because formation of vitelline, chitinous, and lipid layers of the eggshell ensues only after sperm penetration of the oocyte (p. 373), only the proteinaceous layer can be clearly distinguished in unfertilized eggs.

- **Biology.**   A period of 9 to 13 days is the minimal time required for the embryo to develop into an active J_1. Although the embryo is extremely resistant to low

temperature, desiccation, and strong chemicals, embryogenesis is retarded by such factors. Sunlight and high temperatures are lethal in a short time. The juvenile molts to the second stage before hatching through an indistinct operculum (Fig. 26.4).

Infection occurs when shelled juveniles are swallowed with contaminated food and water (Fig. 26.5). They hatch in the duodenum, where the juveniles penetrate the mucosa and submucosa and enter lymphatics or venules. After passing through the right heart, they enter the pulmonary circulation and break out of capillaries into the air spaces. Many worms get lost during this migration and accumulate in almost every organ of the body, causing acute tissue reactions.

While in the lungs, the juveniles molt twice (or complete the second molt and then molt again) during a period of about 10 days, to a length of 1.4 to 1.8 mm. They then move up the respiratory tree to the pharynx, where they are swallowed. Many juveniles make this last step of their migration before molting to the fourth stage, but these cannot survive the gastric juices in the stomach. The J_4's are resistant to such a hostile environment and pass through the stomach to the small intestine, where they mature. Within 60 to 65 days of being swallowed, they begin producing eggs. It seems curious that the worms embark on such a hazardous migration only to end up where they began. One hypothesis to account for it suggests that the migration simulates an intermediate host, which normally would be required for the juvenile of the ancestral form to develop to the third stage. Another possibility is that the ancestor was a skin penetrator for which the migration was a developmental necessity.

- **Epidemiology.**   The dynamics of *Ascaris* infection are essentially the same as for *Trichuris*. Indiscriminate defecation, particularly near habitations, "seeds" the soil with eggs that remain viable for many months or even years. The resistance of *Ascaris* eggs to chemicals is almost legendary. They can embryonate successfully in 2% formalin, in potassium dichromate, and in 50% solutions of hydrochloric, nitric, acetic, and sulfuric acid, among other similar inhospitable substances.[41] This extraordinary chemical resistance is the result of the lipid layer of the eggshell, which contains the ascarosides.

The longevity of *Ascaris* eggs also contributes to the success of the parasite. Brudastov and coworkers[6] infected themselves with eggs kept for 10 years in soil at Samarkand, Russia. Of these eggs, 30% to 53% were still infective. Because of this longevity, it is impossible to prevent reinfection when houseyards have been liberally seeded with eggs, even when proper sanitation habits are initiated later.

Contamination, then, is the typical means of infection. Children are the most likely to become infected (or reinfected) by eating dirt or placing soiled fingers and toys in their mouths. In regions in which nightsoil is used as fertilizer, principally eastern Asia, Germany, and

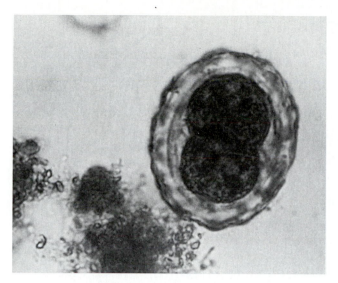

FIGURE 26.2

Fertilized egg of *Ascaris lumbricoides* from a human stool. Eggs of this species are 45 to 75 μm long.

Courtesy of Robert E. Kuntz.

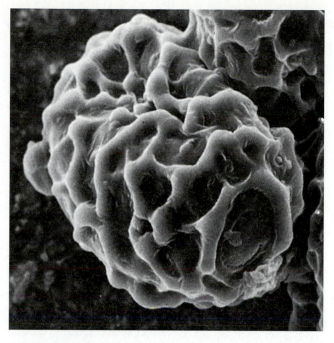

FIGURE 26.4

Scanning electron micrograph showing egg of *Ascaris lumbricoides*. An operculum is visible at one end.

Courtesy of John Ubelaker.

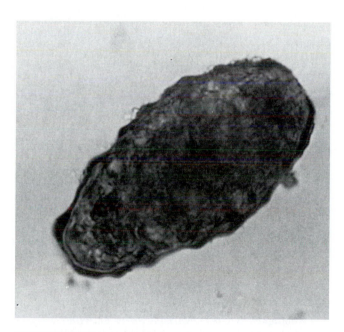

FIGURE 26.3

Unfertilized egg of *Ascaris lumbricoides* from a human stool. Such eggs are 88 to 94 μm long.

Courtesy of Robert E. Kuntz.

certain Mediterranean countries, uncooked vegetables become important vectors of *Ascaris* eggs. Experimental support for this hypothesis came from Mueller,[32] who seeded a strawberry plot with eggs; he and volunteers ate unwashed strawberries from this plot every year for six years and became infected each year. Cockroaches can carry and disseminate *Ascaris* eggs.[8] Even windborne dust can carry *Ascaris* eggs, when conditions permit. Bogojawlenski and Demidowa[5] found *Ascaris* eggs in nasal mucus of 3.2% of schoolchildren examined in the Soviet Union. From the nasal mucosa to the small intestine is a short trip in children. Dold and Themme[17] found *Ascaris* eggs on 20 German banknotes in actual circulation.

There can be little doubt that ascariasis is present in epidemic proportions in the southeastern United States.[14] Surveys in various states between 1956 and 1970 showed prevalences of 20% to 60% in the childhood population. Of 26,489 stool samples submitted to the South Carolina State Laboratory in 1969, 15% contained *A. lumbricoides.* Because no active campaign of eradication has been mounted since then, there is no reason to believe that the figure is any lower now.

Worldwide over 1 billion persons, almost one quarter of the world population, are infected.[12] The worms are commonly overdispersed in local populations, with a small number of people harboring infections of high intensity. These individuals seem to be predisposed to infection; when they are cured, they tend to become reinfected with large numbers of worms.[10,22] The reasons for predisposition may be social, behavioral, environmental, and genetic, either alone or in combination.[22] Some evidence suggests

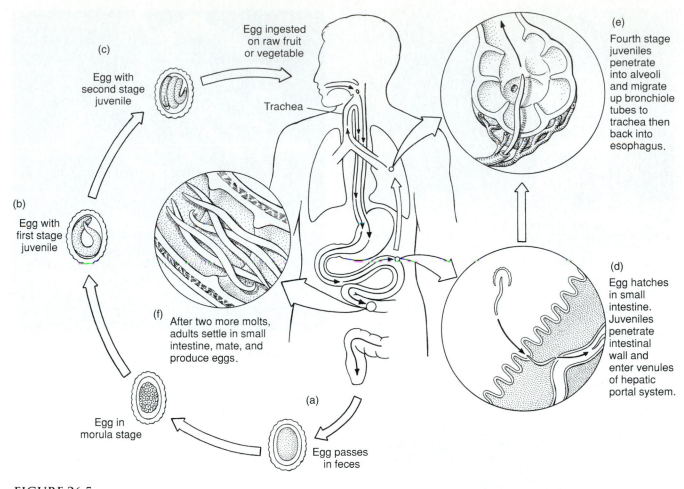

FIGURE 26.5

Life cycle of *Ascaris lumbricoides*. (*a*) Shelled embryo passes in feces. (*b*) Egg containing J$_1$. (*c*) Egg containing J$_2$ ingested on raw fruit or vegetable. (*d*) Egg hatches in small intestine, and juvenile penetrates intestinal wall and enters venules of hepatic portal system. Juveniles undergo further development during migration. (*e*) J$_4$ emerges into alveoli and migrates in bronchioles to trachea and then up trachea and into esophagus. (*f*) Worms reach small intestine again and develop into adults.

Drawing by William Ober and Claire Garrison.

genetic predisposition,[23] but other evidence suggests that genetic factors, if any, were overwhelmed by environmental or behavioral features of the families involved.[9]

- *Pathogenesis.* Little damage is caused by the penetration of intestinal mucosa by newly hatched worms. Juveniles that become lost and wander and die in anomalous locations, such as the spleen, liver, lymph nodes, or brain, often elicit an inflammatory response that may cause vague symptoms that are difficult to diagnose and may be confused with other diseases. Transplacental migration into a developing fetus is also known. Allergy and immunopathology of ascariasis was reviewed by Coles.[11]

When the juveniles break out of lung capillaries into the respiratory system, they cause a small hemorrhage at each site. Heavy infections will cause small pools of blood to accumulate, which then initiate edema with resultant clogging of air spaces. Accumulations of white blood cells and dead epithelium add to the congestion, which is known as *Ascaris* **pneumonitis** (Loeffler's pneu-

monia). Large areas of lung can become diseased, and when bacterial infections become superimposed, death can result. Once a perhaps unbalanced parasitology graduate student vented his ire on his roommates by "seeding" their breakfast with embryonated *Ascaris* eggs. They almost died before their malady was diagnosed.[2]

- *Pathogenesis from "Normal Worm Activities."* Although it is probable that *Ascaris* occasionally sucks blood from the intestinal wall, its main food is liquid contents of the intestinal lumen. In moderate and heavy infections the resulting theft of nourishment can cause malnutrition and underdevelopment in small children.[14,49] Abdominal pains and sensitization phenomena—including rashes, eye pain, asthma, insomnia, and restlessness—often result as allergic responses to metabolites produced by the worms.

A massive infection can cause fatal intestinal blockage[3] (Fig. 26.6). Why, in one case, do large numbers of worms cause no apparent problem, whereas in another, the worms knot together to form a mass that completely

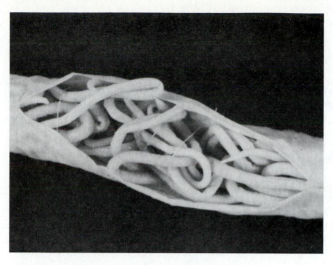

FIGURE 26.6

Intestine of a pig, nearly completely blocked by *Ascaris suum* (threads were inserted to hold worms in place). Such heavy infections are also fairly common with *A. lumbricoides* in humans.

Photograph by Larry S. Roberts.

blocks the intestine? Certain drugs, such as tetrachloroethylene, if used to treat hookworm, can aggravate *Ascaris* to knot up, but other factors are still unknown. Penetration of the intestine or appendix is not uncommon. The resulting peritonitis is usually quickly fatal. According to Louw,[30] 35.5% of all deaths in acute abdominal emergencies of children in Capetown were caused by *Ascaris.*

• *Wandering Worms.*   Wandering adult worms cause various conditions, some serious, some bizarre, all unpleasant. The tropism of a female to squirm through the coiled tail of a male causes her to wander if no males are present. A similar restlessness can be observed even in more highly evolved forms of animals. Overcrowding may also lead to wandering. A downstream wandering leads to the appendix, which can be clogged or penetrated, or to the anus, with an attendant surprise for the unsuspecting host. Upstream wandering leads to the pancreatic and bile ducts, possibly occluding them with grave results. Multiple liver abscesses have resulted from such invasion.[37] Worms reaching the stomach are aggravated by the acidity and writhe about, often causing nausea. The psychological trauma induced in one who vomits an 18-inch ascarid is difficult to quantify. Aspiration of the vomited worm can result in death.[11] Worms that reach the esophagus, usually while the host is asleep, may crawl into the trachea, causing suffocation or lung damage; they may crawl into the eustachian tubes and middle ears, causing extensive damage; or they may simply exit through the nose or mouth, causing a predictable consternation.

• *Diagnosis and Treatment.*   Accurate diagnosis of migrating juveniles is impossible at this time. Demonstration of juveniles in sputum is definitive, provided the technician can identify them. Most diagnoses are made by identifying the characteristic, mammillated eggs in the stool or by an appearance of the worm itself. So many eggs are laid each day by one worm that one or two direct fecal smears are usually sufficient to demonstrate at least one. *Ascaris* should be suspected when any of the previously listed pathogenic conditions are noted. Most light infections are asymptomatic, and presence of worms may be determined only by spontaneous elimination of spent individuals from the anus.

Mebendazole is the drug of choice, with pyrantel pamoate as an alternative. Mebendazole binds to tubulin in the worm's intestinal cells and body wall muscles.[7] No efficient treatment of migrating juveniles has been discovered.

Toxocara canis.   This species is a cosmopolitan parasite of domestic dogs and other canids, and it is the chief cause of visceral larva migrans, discussed later.

As the result of prenatal infections, one may expect even puppies in well-cared-for kennels to be infected at birth, and they are treated accordingly. It is not uncommon for 100% of puppies to be infected. The casual owner of a new puppy is likely to be startled by the vomiting by the pet of a number of large, active worms. Older dogs seem to develop strong resistance to further infection, and they harbor adult worms less often. The reported age resistance of dogs not previously exposed to *T. canis* is partly related to size of infective dose; a smaller number of eggs administered is more likely to lead to patent infection.[18]

Adults look basically like *Ascaris,* only they are much smaller. Three lips are present. Unlike *Ascaris,* however, *Toxocara* has prominent cervical alae in both sexes. Males are 4 to 6 cm, and females are 6.5 to more than 15.0 cm long. The brownish eggs are almost spherical, with surficial pits, and are unembryonated when laid.

• *Biology.*   Adult worms live in the small intestine of their host, producing prodigious numbers of eggs, which are passed with the host's feces (Fig. 26.7). Development of the J_2 takes five to six days under optimal conditions. The fate of ingested juveniles depends on the age and immunity of the host. If the puppy is young and has had no prior infection, the worms hatch and migrate through the portal system and lungs and back to the intestine, as in *A. lumbricoides.*

If the host is an older dog, the juveniles do not complete the lung migration. They wander through the body, eventually entering a developmental arrest for a long period. If a bitch becomes pregnant, the dormant juveniles apparently are activated by host hormones late in the pregnancy and reenter the circulatory system, where they are carried to the placentas. There they penetrate through to the fetal bloodstream, where they complete a lung migration en route to the intestine. Thus, a puppy can be born with an infection of *Toxocara,* even though the dam has shown no sign of patent infection. The puppy may become infected by the transmammary route, in the mother's milk, but this is probably less

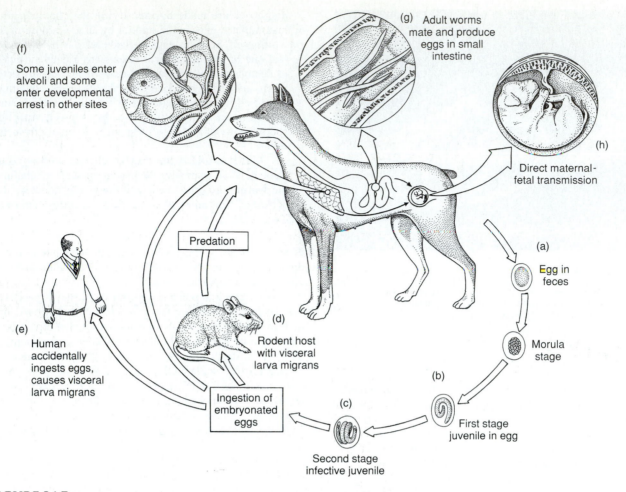

FIGURE 26.7

Life cycle of *Toxocara canis*. (*a*) Shelled embryo passed in feces. (*b*) J$_1$ in egg. (*c*) Infective J$_2$ in egg. (*d*) Eggs hatch in rodent host, and juveniles enter developmental arrest in viscera. (*e*) Eggs hatch in human, and juveniles cause visceral larva migrans. (*f*) After penetration of intestinal wall, some juveniles break out into alveoli, ascend trachea, and finally mature in small intestine. Other juveniles (especially in mature dogs) enter developmental arrest in other sites. (*g*) Adult worms mate and produce eggs in small intestine. (*h*) Direct maternal-fetal transmission.

Drawing by William Ober and Claire Garrison.

common.[20] If the lactating bitch ingests infective juveniles, they can complete the migration to the intestine and produce a patent infection.

Another option in the life cycle of *T. canis* is offered when a rodent or other mammal eats embryonated eggs. In this host the juvenile begins to migrate but then becomes dormant and continues its developmental arrest. If the rodent is eaten by a dog, the worms promptly migrate through the lungs to the intestine or into the tissues to continue their wait, depending on the age of the dog. Thus, the rodent is a paratenic host. Although this adaptability favors survival of the parasite, it bodes ill for certain accidental hosts, such as humans.

Visceral Larva Migrans.   When juveniles of several species of nematodes gain entry to an improper host, the juveniles begin the typical tissue migration. However, they do not complete the normal migration but undergo developmental arrest and begin an extended, random wandering through the body. The resulting disease entity is known as **visceral larva migrans,** in contrast to cutaneous larva migrans (p. 412). Visceral larva migrans can be caused by a variety of spirurid, strongylid, and other nematodes in addition to ascaridids. However, we will emphasize *Toxocara canis,* which is by far the most common species causing the condition in humans.

- *Epidemiology.*   Some years ago it was generally assumed that dog and cat worms could not infect humans or were not dangerous to them. We have come to realize since the early 1950s that the assumption is not true, particularly in the case of *T. canis*. Actually, very few cases of visceral larva migrans have been reported worldwide since then, but "most observers believe that this is the very small tip of a very large iceberg."[27] About 20% of adult dogs and 98% of puppies in the United

States are infected with *T. canis;* therefore, the risk of exposure is very high. Most cases are either unrecognized or unreported.

The development of a specific immunodiagnostic test, an ELISA using secretory-excretory antigens collected from cultured juveniles, has been a boon to epidemiological studies of visceral larva migrans.[40] This test can distinguish between *Ascaris* and *Toxocara.*[31] In the United States surveys have shown a seroprevalence among children 1 to 11 years old of 4.6% to 7.3%, but seroprevalence ranged up to 30% among black children of low socioeconomic status. Furthermore, seroprevalence among children in developing tropical countries was much higher: from 50% to 80%.

Dogs and cats defecating on the ground seed the area with eggs, which embryonate and become infective to any mammal eating them, including children. Considering the crawling-walking age of small children is a time when virtually every available object goes into the mouth for a taste, it is not surprising that the disease is most common in children between one and three years old. In the urban setting the dog owner looks on the city park as the perfect place to walk the dog, while the parent at the same time brings young children to play on the seeded grass. An especially unhappy fact in the epidemiology of larva migrans is the high risk to children by exposure to the environment of puppies.[40] Finally, a factor to contemplate in light of the foregoing is the durability and longevity of *Toxocara* eggs, which are comparable to those of *Ascaris* (discussed before).

- **Pathogenesis.** Juveniles provoke a delayed-type hypersensitivity reaction in paratenic hosts, and the degree and timing of the reaction depend on the infecting dose.[40] In experimental hosts most of the juveniles eventually end up in the brain; it is unclear whether this is because the juveniles have a predilection for the brain or because they are destroyed in other sites but remain in the brain. In sites other than the brain, the juveniles are encapsulated by a granulomatous reaction (Fig. 26.8).

  Characteristic symptoms of visceral larva migrans include fever, pulmonary symptoms, hepatomegaly, and eosinophilia. Extent of damage usually is related to the number of juveniles present and their ultimate homestead in the body. Various neurological symptoms have been reported, and deaths have occurred when juveniles were especially abundant in the brain. There seems little doubt that most cases result in rather minor, transient symptoms, which are undiagnosed or misdiagnosed. The most common site of larval invasion is the liver (Fig. 26.8), but no organ is exempt.

  Juveniles in the eye cause chronic inflammation of the inner chambers or retina or provoke dangerous granulomas of the retina. These reactions can lead to blindness in the affected eye. Ocular involvement has been reported in 245 patients with an average age of 7.5 years.[48] Other lesions destroy lung, liver, kidney, muscle, and nervous tissues. Generally, the ocular damage is the

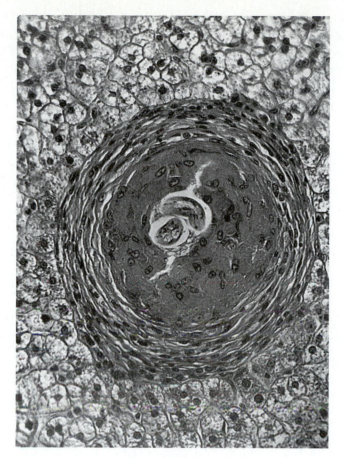

FIGURE 26.8

Toxocara canis juvenile in liver of a monkey at nine months' infection. The juvenile rests in a matrix of epithelioid cells surrounded by a fibrous capsule lacking intense inflammatory reaction.

From P. C. Beaver, "The nature of visceral larva migrans," in *J. Parasitol.* 55:3–12. Copyright © 1969.

result of the invasion of only a single juvenile.[40] It may be that, because heavy infections stimulate a much stronger immune response, juveniles survive longer in light infections, giving them more time to wander into the eye.

- **Diagnosis and Treatment.** The ELISA using secretory-excretory antigens has facilitated clinical diagnosis enormously. Previously diagnosis was difficult. A liver biopsy might demonstrate the characteristic granuloma surrounding the juvenile, but to obtain a biopsy containing a larva was a matter of luck. A high eosinophilia is suggestive, especially if the possibility of other parasitic infections can be eliminated.

  Usually only patients with severe symptoms are treated.[20] No dependably effective treatment is known, although several drugs have been tried.[27] Mebendazole has been used, along with systemic steroids to control inflammation.[20] Control consists of periodic worming of household pets, especially young animals, and proper disposal of the animal's feces. Dogs and cats should be restrained, if possible, from eating available transport hosts.

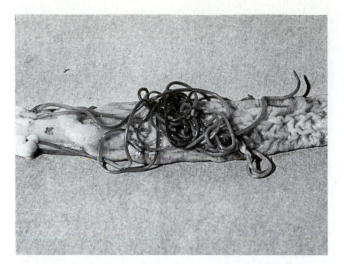

FIGURE 26.9

Intestine of a domestic cat, opened to show numerous *Toxocara cati*.
Courtesy of Robert E. Kuntz.

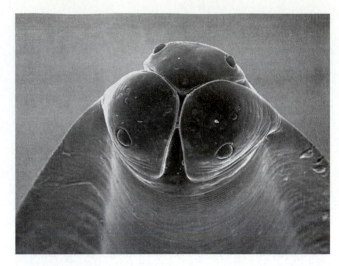

FIGURE 26.10

Scanning electron micrograph of *Toxocara cati:* en face view illustrating the three lips with sensory papillae; broad alae at each side.
Courtesy of John Ubelaker.

Other *Toxocara* Species. *Toxocara cati* is widely prevalent among domestic cats and other felids (Fig. 26.9). The cervical alae (Fig. 26.10) of *T. cati* are shorter and broader than those of *T. canis,* and the eggs of the two species have a slightly different appearance. The life cycles are similar, including the use of paratenic hosts, but kittens are infected with *T. cati* only by the transmammary route and not transplacentally.[20] *Toxocara cati* may be a cause of visceral larva migrans but is probably much less important than *T. canis.*

Toxocara vitulorum is the only ascaridid that occurs in cattle. Its life cycle is similar to that of *T. cati,* with the young being infected by the mother's milk.[35] Adult hosts are refractory to intestinal infection. Young calves may succumb to verminous pneumonia during the migratory stages of the parasites. Diarrhea or colic in later stages results in economic losses to the owner.

Toxocara pteropodis is a parasite of fruit bats (flying "foxes") of Australia. It is also passed by the translactational route. It was implicated at one point in hepatic disease in humans in southeastern Queensland but later exonerated. Prociv,[34] who reviewed the biology and life cycle of this fascinating form, postulated that translactational transmission in this genus is primitive for the group, and the transplacental route, as in *T. canis,* is probably the derived character state.

Parascaris equorum. This large nematode is the only ascaridid found in horses and other equids. *Parascaris equorum* is a cosmopolitan species. It is very similar in gross appearance to *A. lumbricoides* but is easily differentiated by its huge lips, which give it the appearance of having a large, round head.

The life cycle is similar to that of *A. lumbricoides,* involving a lung migration. Resulting pathogenesis is especially important in young animals, with pneumonia, bronchial hemorrhage, colic, and intestinal disturbances resulting in unthriftiness and morbidity. Intestinal perfora-

tion or obstruction is common. Older horses are usually immune to infection. Prenatal infection is not known to occur.

Baylisascaris procyonis. This is a very common intestinal parasite of raccoons in North America. Other, similar species are found in bears, skunks, badgers, and other carnivores. When embryonated eggs are swallowed by a raccoon, they will hatch in the small intestine and mature. Like *Toxocara* spp., *Baylisascaris* can be borne by paratenic hosts when the eggs are swallowed by a different vertebrate. Most commonly the paratenic host is a rodent, bird, or lagomorph.[26] In these animals the parasite juveniles wander, often invading the central nervous system. Resulting debilitation, sometimes confused with rabies, makes them vulnerable to attack by raccoons, in which the juveniles are freed by digestion to grow to maturity. Unfortunately the juveniles affect humans in the same way. Several proven and suspected cases of infection with *B. procyonis* have been reported, with fatal cases in children. Retinal involvement is common.

The epidemiology of infection involves close contact between humans and raccoons. Scavenging raccoons are bold animals that prowl near human dwellings and outbuildings. Their preferred defecation sites are dangerous sources of infection to humans and other animals. The eggs can remain infective for years under ideal conditions, so once an area is contaminated it is nearly impossible to decontaminate it.[25] Raccoons are popular pets, especially when young, but even very young animals are often infected. This is another way this dangerous parasite can be brought into contact with humans.

Other species of *Baylisascaris* may have similar pathogenicity, but most hosts are not as likely to come in close contact with humans. Pet skunks infected with *B. columnaris* are potential hazards, however.

Toxascaris leonina. *Toxascaris leonina* is a cosmopolitan parasite of dogs and cats and related canids and felids. It is similar in appearance to *Toxocara* spp., being recognized in

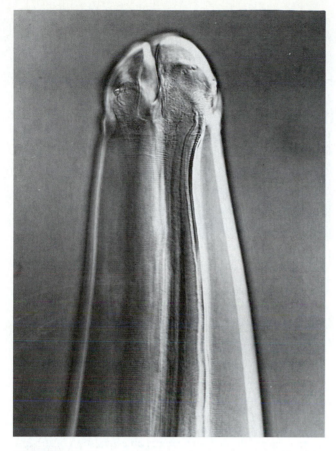

FIGURE 26.11

Anterior end of *Toxascaris leonina,* an intestinal parasite of dogs and cats. Note the narrow cervical alae as compared with the broad alae of *Toxocara cati.*

Courtesy of Jay Georgi.

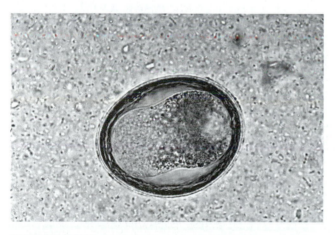

FIGURE 26.12

Egg of *Toxascaris leonina.* The size is 75 to 85 μm by 60 to 75 μm.

Courtesy of Jay Georgi.

the following ways: (1) the body tends to flex dorsally in *Toxascaris* and ventrally in *Toxocara;* (2) alae of *T. cati* are short and wide, whereas they are long and narrow in *T. canis* and *T. leonina* (Fig. 26.11); (3) the surface of the egg of

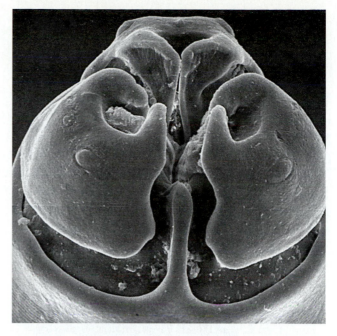

FIGURE 26.13

Lagochilascaris turgida. Note the prominent cleft in the tip of each lip, typical of the genus (Gr. *lagos,* hare + *cheilos,* lip).

Courtesy of John Sprent.

T. leonina is smooth (Fig. 26.12) but pitted in *Toxocara* spp.; and (4) the tail of male *Toxocara* is constricted abruptly behind the anus, whereas it gradually tapers in *Toxascaris.*

The life cycle of *T. leonina* is simple. Ingested eggs hatch in the small intestine, where the juveniles penetrate the mucosa. After a period of growth, they molt and return directly to the intestinal lumen, where they mature.

Although they are mildly pathogenic, their main importance is a diagnostic one: It may be useful to separate *T. leonina* from *Toxocara* spp. in identification procedures because *T. leonina* is considered of little importance as a source of visceral larva migrans.

Lagochilascaris minor. *Lagochilascaris minor* is evidently closely related to *Toxocara* and *Toxascaris,* differing only in minor morphological details (Fig. 26.13). Little is known about its biology in nature. It has been found in the stomach, pharynx, and trachea of various wild cats in South America and the Caribbean, with three related species in other felids and opossums in the same areas and in Africa and North America.

Lagochilascaris minor has been reported in humans at least eight times, usually in the tonsils, nose, or neck.[44,47] A fatal brain infection has been reported.[36] When present, the worms cause abscesses that may contain from one to more than 900 individuals (Fig. 26.14). They can mature in these locations, and they produce pitted eggs, much like those of *Toxocara.* It is common for human infections to last many years or to rapidly kill the infected person. *Lagochilascaris minor* epitomizes the zoonotic infection. How humans become infected is unknown. It seems probable that humans are unnatural, accidental hosts. The definitive host is not known.

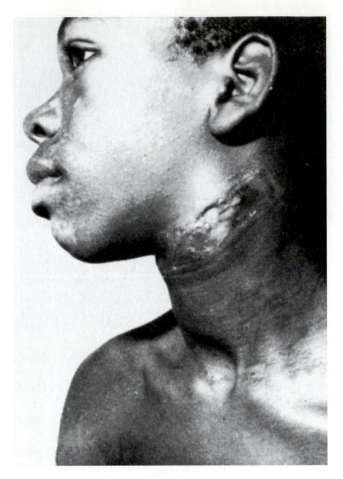

FIGURE 26.14

Abscess in the neck of a 15-year-old native of Surinam. It contained numerous adults, juveniles, and eggs of *Lagochilascaris* sp. After treatment with thiabendazole, the fistula closed, and the abscess healed, leaving only a small scar.

From B. F. J. Oostburg, "Thiabendazole therapy of *Lagochilascaris minor* infection in Surinam. Report of a case," in *Am. J. Trop. Med. Hyg.* 20:580–583. Copyright © 1971.

Family Anisakidae

Anisakis **Species.**   The many species in the genus *Anisakis* and related genera are parasites of the stomachs of marine fishes, birds, and mammals. The normal life cycles apparently involve passage of eggs in the feces of the definitive host, embryogenesis and hatching of the J_2, ingestion of the J_2 by a crustacean, development to the J_3 in the hemocoel of the crustacean, and then either (1) ingestion by the definitive host or (2) ingestion by a fish paratenic host, which is ultimately consumed by a definitive host.[15,38] The definitive hosts of *Anisakis* spp. are marine mammals. Two aspects of the situation are important to humans: esthetics and public health. The first relates to the disgust experienced by persons who find large, stout worms in the flesh of the meal they are preparing or eating (see *Frontispiece*). Many a finnan haddie has ended up in the garbage pail when *Anisakis* juveniles were discovered in it, a rather common occurrence!

FIGURE 26.15

Scanning electron micrograph of *Terranova* sp. juveniles (family Anisakidae) penetrating the stomach of a rat on day three postinfection. *Arrows* indicate acute lesions caused by juveniles. (Scale bar = 1 mm)

From T. L. Deardorff et al., "Histopathology induced by larval *Terranova* (Type HA) (Nematoda: Anisakinae) in experimentally infected rats," in *J. Parasitol.* 69:191–195. Copyright © 1983.

More importantly, *Anisakis* juveniles can produce pathological conditions in humans who eat them in raw, salted, or pickled fish. Such conditions may be asymptomatic, mild, or severe.[4] Symptoms generally commence when the juveniles begin to penetrate the stomach or intestine (Fig. 26.15), and this may happen from 1 to 12 hours after ingestion of the infected seafood (gastric) or up to 14 days in the case of intestinal penetration. There is severe epigastric or abdominal pain, usually nausea, and sometimes vomiting. Diagnosis of gastric anisakiasis by endoscope and removal of the worms with biopsy forceps is effective, although catching the lively worms with the forceps may be challenging.[15] In intestinal anisakiasis, or cases in which the worm has fully penetrated into the submucosa or migrated beyond the gastrointestinal tract, diagnosis is more problematic, and symptoms can mimic a number of other, more common conditions. In such cases immunodiagnosis can be very helpful.[39]

Most cases have been reported from Japan, Europe, and Scandinavia, where raw fish is relished. From several hundred to 1000 cases per year are reported from Japan, and the number of cases reported from the United States is increasing.[15,24] Fatalities due to peritonitis have been recorded.[4]

Cooking kills the juveniles, but the continued popularity of raw-fish dishes, such as sushi, sashimi, ceviche, and lomi-lomi, ensures a continued risk of human infection. Commercial blast-freezing causes little change in the texture or taste of fish, and the process renders the *Anisakis* juveniles harmless.[16]

SUPERFAMILY HETERAKOIDEA

Family Ascaridiidae

Ascaridia galli is a cosmopolitan parasite of the small intestine of domestic fowl. Males reach a length of 77 mm and females reach 115 mm.

FIGURE 26.16

Posterior end of *Heterakis variabilis,* a parasite of pheasants that is similar to *H. gallinarum.* Note the conspicuous preanal sucker.

From W. G. Ingles, et al., "Nematode parasites of Oceanica. XII. A review of *Heterakis* species, particularly from birds of Taiwan and Palawan," in *Rec. S. Aust. Mus.* 16:1–14. Copyright © 1971.

The J_2 hatches from the egg after it is ingested with contaminated food or water. The life cycle does not involve extensive migration. Instead, eight or nine days after infection, the juveniles molt to the third stage and begin to burrow into the mucosa, where they generally remain with their tails still in the intestinal lumen. After molting to the fourth stage at about 18 days, they return to the lumen, where they undergo their final molt and mature. Probably the majority of worms complete their three molts and attain maturity without ever leaving the lumen. Those that attack the mucosa, however, cause extensive damage, which may result in unthriftiness or even death. Adult chickens seem to be refractory to infection.

Family Heterakidae

Heterakis gallinarum is cosmopolitan in domestic chickens and related birds. It was probably brought to the United States in imported ringnecked pheasants. The worms live in the cecum, where they feed on its contents.

Three large lips and a bulbar esophageal swelling as well as lateral alae are found in this genus. Males are as long as 13 mm and possess wide caudal alae supported usually by 12 pairs of papillae (Fig. 26.16). The tail is sharply pointed, and there is a prominent preanal sucker. The spicules are strong and dissimilar, and a gubernaculum is not present. Females have the vulva near the middle of the body and a long, pointed tail.

Many species of *Heterakis* are known in birds, particularly in ground feeders, and one species, *H. spumosa,* is cosmopolitan in rodents.

- **Biology.** The eggs of *H. gallinarum* contain the zygote stage when laid. They develop to the infective stage in 12 to 14 days at 22°C and can remain infective for four years in soil. Infection is contaminative: When embryonated eggs are eaten, the second-stage juveniles hatch in the gizzard or duodenum and pass down to the ceca. Most complete their development in the lumen, but some penetrate the mucosa, where they remain for two to five days without further development. Returning to the lumen they mature, about 14 days after infection.

  If eaten by an earthworm, the juvenile may hatch and become dormant in the worm's tissues, remaining infective to chickens for at least a year. Since the nematodes do not develop further until eaten by a bird, the earthworm is a paratenic host.

- **Epidemiology.** As a result of the longevity of the eggs, it is difficult to eliminate *Heterakis* from a domestic flock. Thus, although adult chickens may effect a self-cure, infective eggs are still available the following spring, when new chicks hatch. Furthermore, as earthworms feed in contaminated soil, they accumulate large numbers of juveniles, which in turn causes massive infections in the unlucky birds that eat them.

- **Pathogenesis.** In heavy infections the cecal mucosa may thicken and bleed slightly. Generally speaking, *Heterakis* is not highly pathogenic in itself.

  However, a flagellate protozoan, *Histomonas meleagridis,* is transmitted between birds within eggs of *Heterakis gallinarum.*[28] This protozoan is the etiological agent of **histomoniasis,** a particularly serious disease in turkeys. The protozoan is eaten by the nematode and multiplies in the worm's intestinal cells, in the ovaries, and finally in the embryo within the egg (p. 90). Hatching of the worm within a new host releases *Histomonas.* Hence, we encounter the curious phenomenon of one parasite acting as a true intermediate host and vector of another.

- **Diagnosis and Treatment.** *Heterakis gallinarum* can be diagnosed by finding the eggs in the feces of its host. The worms are effectively eliminated with mebendazole. Usually a flock of birds routinely gets this or other drugs in its feed or water. This treatment, together with rearing the birds on hardware cloth, will eliminate the parasite from the flock. Birds that are allowed to roam the barnyard usually are infected.

References

1. Anderson, T. J. C., M. E. Romero-Abal, and J. Jaenike. 1993. Genetic structure and epidemiology of *Ascaris* populations: Patterns of host affiliation in Guatemala. *Parasitology* 107:319–34.

2. Anonymous. 1970. LIer [Long Islander] sought in roommates' poisoning. *Newsday* (27 February).

3. Baird, J. K., M. Mistrey, M. Pimsler, and D. H. Connor. 1986. Fatal human ascariasis following secondary massive infection. *Am. J. Trop. Med. Hyg.* 35:314–18.

4. Bier, J. W., T. L. Deardorff, G. J. Jackson, and R. B. Raybourne. 1987. Human anisakiasis. In Pawlowski, Z. S., ed. *Baillière's clinical tropical medicine and communicable diseases* 2(3). London: W. B. Saunders Company, 723–33.

5. Bogojawlenski, N. A., and A. Demidowa. 1928. Ueber den Nachweis von Parasiteneiern auf der menschlichen Nasenschleimhaut. *Russian J. Trop. Med.* 6:153–56.

6. Brudastov, A. N., V. R. Lemelev, S. K. Kholnukhanedov, and L. N. Krasnos. 1971. The clinical picture of the migration phase of ascariasis in self-infection. *Medskaya Parazitol.* 40:165–68.

7. Bughio, N. I., G. M. Faubert, and R. K. Prichard. 1994. Interaction of mebendazole with tubulin from body wall muscle, intestine, and reproductive system of *Ascaris suum*. *J. Parasitol.* 80:126–32.

8. Burgess, N. R. H. 1984. Hospital design and cockroach control. *Trans. R. Soc. Trop. Med. Hyg.* 78:293–94.

9. Chan, L., D. A. P. Bundy, and S. P. Kan. 1994. Genetic relatedness as a determinant of predisposition to *Ascaris lumbricoides* and *Trichuris trichiura* infection. *Parasitology* 108:77–80.

10. Chan, L., S. P. Kan, and D. A. P. Bundy. 1992. The effect of repeated chemotherapy on age-related predisposition to *Ascaris lumbricoides* and *Trichuris trichiura*. *Parasitology* 104:371–77.

11. Coles, G. C. 1985. Allergy and immunopathology of ascariasis. In Crompton, D. W. T., M. C. Nesheim, and Z. S. Pawlowski, eds. *Ascariasis and its public health importance.* London: Taylor and Francis.

12. Crompton, D. W. T. 1988. The prevalence of ascariasis. *Parasitol. Today* 4:162–69.

13. Crompton, D. W. T., and J. J. Tulley. 1987. How much ascariasis is there in Africa? *Parasitol. Today* 3:123–27.

14. Darby, C. P., and M. Westphal. 1972. The morbidity of human ascariasis. *J. S. C. Med. Assoc.* 68:104–8.

15. Deardorff, T. L., S. G. Kayes, and T. Fukumura. 1991. Human anisakiasis transmitted by marine food products. *Hawaii Med. J.* 50(1):9–16.

16. Deardorff, T. L., and R. Throm. 1988. Commercial blast-freezing of third-stage *Anisakis simplex* larvae encapsulated in salmon and rockfish. *J. Parasitol.* 74:600–3.

17. Dold, H., and H. Themme. 1949. Ueber die Möglichkeit der uebertragung der Askaridiasis durch Papiergeld. *Dtsch. Med. Wochenschr.* 74:409.

18. Dubey, J. P. 1978. Patent *Toxocara canis* infection in ascarid-naive dogs. *J. Parasitol.* 64:1021–23.

19. Faulkner, C. T. 1991. Prehistoric diet and parasitic infection in Tennessee: Evidence from the analysis of desiccated human paleofeces. *Am. Antiquity* 56:687–700.

20. Gillespie, S. H. 1988. The epidemiology of *Toxocara canis*. *Parasitol. Today* 4:180–82.

21. Holland, C. V., and S. O. Asaolu. 1990. Ascariasis in Nigeria. *Parasitol. Today* 6:143–47.

22. Holland, C. V., S. O. Asaolu, D. W. T. Crompton, R. C. Stoddart, R. Macdonald, and S. E. A. Torimiro. 1989. The epidemiology of *Ascaris lumbricoides* and other soil-transmitted helminths in primary school children from Ile-Ife, Nigeria. *Parasitology* 99:275–85.

23. Holland, C. V., D. W. T. Crompton, S. O. Asaolu, W. B. Crichton, S. E. A. Torimiro, and D. E. Walters. 1992. A possible genetic factor influencing protection from infection with *Ascaris lumbricoides* in Nigerian children. *J. Parasitol.* 78:915–16.

24. Kagei, N., and H. Isogaki. 1992. A case of abdominal syndrome caused by the presence of a large number of *Anisakis* larvae. *Int. J. Parasitol.* 22:251–53.

25. Kazacos, K. R. 1982. Contaminative ability of *Baylisascaris procyonis* infected raccoons in an outbreak of cerebrospinal nematodiasis. *Proc. Helm. Soc. Wash.* 49:155–57.

26. Kazacos, K. R. 1986. Raccoon ascarids as a cause of larva migrans. *Parasitol. Today* 2:253–55.

27. Levine, N. D. 1980. *Nematode parasites of domestic animals and of man,* 2d ed. Minneapolis: Burgess Publishing Co.

28. Long, P. L., W. L. Current, and G. P. Noblet. 1987. Parasites of the Christmas turkey. *Parasitol. Today* 3:360–66.

29. Lord, W. D., and W. L. Bullock. 1982. Swine *Ascaris* in humans. *N. Eng. J. Med.* 306:1113.

30. Louw, J. H. 1966. Abdominal complications of *Ascaris lumbricoides* infestation in children. *Br. J. Surg.* 53:510–21.

31. Lynch, N. R., I. Hagel, V. Vargas, A. Rotundo, M. C. Varela, M. C. Di Prisco, and A. N. Hodgen. 1993. Comparable seropositivity for ascariasis and toxocariasis in tropical slum children. *Parasitol. Res.* 79:547–50.

32. Mueller, G. 1953. Untersuchungen ueber die Lebensdauer von Askarideiern in Gartenerde. *Zentralbl. Bakt. I. Orig.* 159:377–79.

33. Nadler, S. A. 1987. Biochemical and immunological systematics of some ascaridoid nematodes: Genetic divergence between congeners. *J. Parasitol.* 73:811–16.

34. Prociv, P. 1989. *Toxocara pteropodis* and visceral larva migrans. *Parasitol. Today* 5:106–9.

35. Roberts, J. A. 1990. The life cycle of *Toxocara vitulorum* in Asian buffalo (*Bubalus bubalus*). *Int. J. Parasitol.* 20:833–40.

36. Rosemberg, S., M. B. S. Lopes, Z. Masuda, R. Campos, and M. C. R. Vieira Bressan. 1986. Fatal encephalopathy due to *Lagochilascaris minor* infection. *Am. J. Trop. Med. Hyg.* 35:575–78.

37. Rossi, M. A., and F. W. Bisson. 1983. Fatal case of multiple liver abscesses caused by adult *Ascaris lumbricoides*. *Am. J. Trop. Med. Hyg.* 32:523–25.

38. Sakanari, J. A. 1990. *Anisakis*—from the platter to the microfuge. *Parasitol. Today* 6:323–27.

39. Sakanari, J. A., H. M. Loinaz, T. L. Deardorff, R. B. Raybourne, J. H. McKerrow, and J. G. Frierson. 1988. Intestinal anisakiasis. A case diagnosed by morphologic and immunologic methods. *Am. J. Clin. Pathol.* 90:107–13.

40. Schantz, P. M. 1989. *Toxocara* larva migrans now. *Am. J. Trop. Med. Hyg.* 41(3, suppl.):21–34.

41. Schwartz, B. 1960. Evolution of knowledge concerning the roundworm *Ascaris lumbricoides*. *Smithsonian Report for 1959*. Washington, D.C.: The Smithsonian Institution, 465–81.

42. Shoemaker-Nawas, P., F. Frost, J. Kobayashi, and P. Jones. 1982. *Ascaris* infection in Washington. *Western J. Med.* 136:436–37.

43. Sprent, J. F. A. 1952. Anatomical distinction between human and pig strains of *Ascaris. Nature* 170:627–28.

44. Sprent, J. F. A. 1971. Speciation and development in the genus *Lagochilascaris. Parasitology* 62:71–112.

45. Taffs, L. F. 1985. *Ascaris* in man: A reply to Dr. Denham. *Trans. R. Soc. Trop. Med. Hyg.* 79:732.

46. Ubelaker, J. E., and V. F. Allison. 1972. Scanning electron microscopy of the denticles and eggs of *Ascaris lumbricoides* and *Ascaris suum*. In Arceneaux, C. J., ed. *Thirtieth Annual Proceedings of the Electron Microscopy Society of America*. Baton Rouge, La.: Claitor's Publishing Division.

47. Volcan, G., F. R. Ochoa, C. E. Medrano, and Y. de Valera. 1982. *Lagochilascaris minor* infection in Venezuela. *Am. J. Trop. Med. Hyg.* 31:1111–13.

48. Warren, K. S. 1974. Helminthic diseases endemic in the United States. *Am. J. Trop. Med. Hyg.* 23:723–30.

49. Willett, W. C., W. L. Kilama, and C. M. Kihamia. 1979. *Ascaris* and growth rates: A randomized trial of treatment. *Am. J. Public Health* 69:987–91.

Additional References

Chabaud, A. G. 1974. Keys to subclasses, orders and superfamilies. In Anderson, R. C., A. G. Chabaud, and S. Willmott, eds. *CIH keys to the nematode parasites of vertebrates*. Bucks, Eng.: Commonwealth Agricultural Bureaux, Farnham Royal.

Crompton, D. W. T., M. C. Nesheim, and Z. S. Pawlowski, eds. 1985. *Ascariasis and its public health significance*. London: Taylor and Francis.

Crompton, D .W. T., M. C. Nesheim, and Z. S. Pawlowski, eds. 1989. *Ascariasis and its prevention and control*. London: Taylor and Francis.

Hartwich, G. 1974. Keys to genera of the Ascaridoidea. In Anderson, R. C., A. G. Chabaud, and S. Willmott, eds. *CIH keys to the nematode parasites of vertebrates*. Bucks, Eng.: Commonwealth Agricultural Bureaux, Farnham Royal.

Neafie, R. C., and D. H. Connor. 1976. Visceral larva migrans: Ascariasis. In Binford, C. H., and D. H. Connor, eds. *Pathology of tropical and extraordinary diseases,* vol. 2, sect. 9. Washington, D.C.: Armed Forces Institute of Pathology.

Smith, J. W., and R. Wootten. 1978. *Anisakis* and anisakiasis. In Lumsden, W. H. R., R. Muller, and J. R. Baker, eds. *Advances in parasitology* 16, London: Academic Press, 93–163.

Sprent, J. F. A. 1983. Observations on the systematics of ascaridoid nematodes. In Stone, A. R., H. M. Platt, and L. F. Khalil, eds. *Concepts in nematode systematics*. London: Academic Press.

Chapter 27

NEMATODES: OXYURIDA, THE PINWORMS

In nature there are neither rewards nor punishments—there are consequences.

Robert G. Ingersoll

Members of the Oxyurida are called *pinworms* because they, especially the females, typically have slender, sharp-pointed tails. Along with the Ascaridida, Spirurida, and Strongylida, the Oxyurida is one of the four major groups of parasitic nematodes that apparently evolved independently from free-living, soil-dwelling ancestors.[2] It was the only one that radiated successfully in both vertebrate and invertebrate hosts. Members of the order were originally grouped together because they were parasites in the posterior gut of vertebrates and they had a posterior pharyngeal bulb (Fig. 27.1). However, the pharyngeal bulb is a primitive character and has no particular phylogenetic significance.[1] Three lips are present in some (Fig. 27.2), but lips are reduced or absent in the more evolved species.

Nevertheless, the Oxyurida are the most clearly defined order of nematodes, with the synapomorphies listed on page 380, and they are the only endoparasites with haplodiploidy. In haplodiploidy, males are haploid and develop parthenogenetically, and females are diploid, developing from fertilized eggs. Haplodiploidy is known among some rotifers, insects (for example, most Hymenoptera), and Acari (but not ticks). It has important implications for population dynamics.[1,2,3] For example, haplodiploid species generally have colonizing habits, have relatively low vagility (ability to migrate), and are divided into small, semi-isolated subpopulations of related individuals. There is a high level of inbreeding, tolerable because deleterious recessives are effectively screened out.

Pinworms (superfamily Oxyuroidea) are common in mammals, birds, reptiles, and amphibians but are rare in fish. Most domestic birds and mammals harbor pinworms, but curiously they are absent in dogs and cats. Insects and millipedes commonly are infected (superfamily Thelastomatoidea). Two species, *Enterobius vermicularis* and *E. gregorii,* are among the most common nematode parasites of humans.

FAMILY OXYURIDAE

Enterobius vermicularis and *E. gregorii*

Pinworms have infected *Homo sapiens* since the time of the species' origin in Africa.[9] In some ways they are paradoxical among the nematode parasites of humans. For one thing they

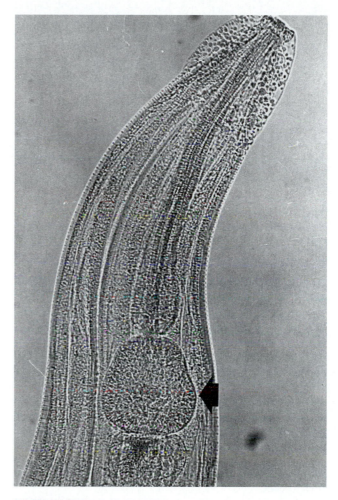

FIGURE 27.1

Anterior end of the pinworm *Enterobius vermicularis.* Note the large esophageal bulb (*arrow*) and the swollen cuticle at the head end, typical of this genus.

Courtesy of Warren Buss.

are not tropical in their distribution, thriving best in the temperate zones of the world. Furthermore, pinworms often are found in families at high socioeconomic levels, where after introduction into the premises by one member, they rapidly become a "family affair." It is fair to say, however, that the greatest pinworm problems are among institutionalized persons, such as those in orphanages and mental hospitals, where conditions facilitate transmission and reinfection.

433

FIGURE 27.2

Pharyngodon, a pinworm of reptiles, showing the primitive, three-lipped condition. Each lip (*arrow*) has a lateral notch.

Courtesy of John Ubelaker.

The fact that these worms inhabit at least 400 million persons[6] is perhaps less surprising than the fact that practically nothing is being done to eliminate the infection. At least part of the reason is simple and practical: Pinworms cause no obvious debilitating or disfiguring effects. Their presence is an embarrassment and an irritation, like acne or dandruff. Resources of democracies, kingdoms, and dictatorships could scarcely be expected to mobilize to combat such an innocuous foe, particularly when they seem to have so much trouble mounting efforts against more disabling infectious agents.

And yet, is enterobiasis so unimportant after all? Certainly it is important to the millions of persons who suffer the discomforts of infection. Furthermore, a great deal of money is spent in efforts to be rid of pinworms. The frantic efforts by persons to rid their households of the tiny worms often lead to what has been called a "pinworm neurosis." The mental stress and embarrassment suffered by families who know they harbor parasites are unmeasurable but very real consequences of infection, especially when multiplied by the vast number of persons involved. Finally, the pathogenesis of these worms may be greatly underrated.[14]

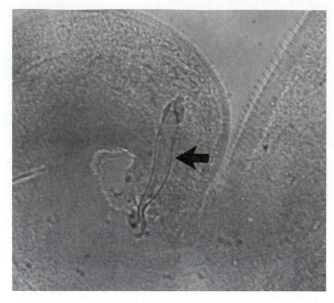

FIGURE 27.3

Posterior end of a male *Enterobius vermicularis,* illustrating the single spicule (*arrow*).

Courtesy of Warren Buss.

• Morphology

Both sexes have three lips surrounding the mouth, followed by a cuticular inflation of the head (Fig. 27.1). Females of *E. vermicularis* and *E. gregorii* are nearly identical. Males, however, are easily differentiated by the single spicule, which is 100 to 141 μm long in *E. vermicularis* (Fig. 27.3) and 68 to 80 μm long in *E. gregorii.* There are additional, more subtle differences.[7] Males of both species are 1 to 4 mm long and have the posterior ends strongly curved ventrally. The conspicuous caudal alae are supported by papillae.

Females measure 8 to 13 mm long and have the posterior end extended into a long, slender point (Fig. 27.4), giving pinworms their name. The vulva opens between the first and second thirds of the body. When gravid, the two uteri contain thousands of eggs, which are elongated-oval and flattened on one side (Fig. 27.5), measuring 50 to 60 μm by 20 to 30 μm.

• Biology

Adult worms congregate mainly in the ileocecal region of the intestine, but they commonly wander throughout the gastrointestinal tract from the stomach to the anus. They attach themselves to the mucosa where they presumably feed on epithelial cells and bacteria. Gravid females begin migrating within the lumen of the intestine, commonly passing out of the anus onto the perianal skin. As they crawl about, both within the bowel and on the outer skin, they leave a trail of eggs. One worm may deposit from 4600 to 16,000 eggs. Females die soon after oviposition, whereas males die soon after copulation. Consequently, it is usual to find many more females than males within a host.

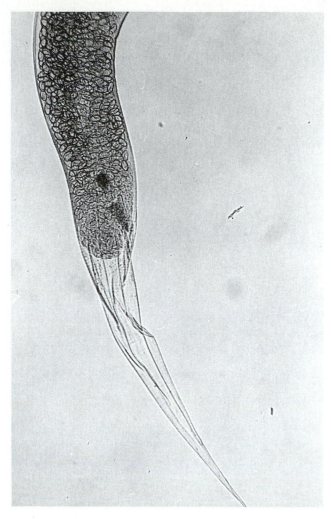

FIGURE 27.4

Posterior end of a gravid female *Enterobius vermicularis*. The long pointed tail lends this species the name *pinworm*.

Courtesy of Warren Buss.

When laid, each egg contains a partially developed juvenile, which can develop to infectivity within six hours at body temperature.[8] Shelled juveniles are resistant to putrefaction and disinfectants but succumb to dehydration in dry air within a day.

Reinfection occurs by two routes. Most often the eggs, containing J_3's, are swallowed, and they hatch in the duodenum. They slowly move down the small intestine, molting twice to become adults by the time they arrive at the ileocecal junction. Total time from ingestion of eggs to sexual maturity of the worms is 15 to 43 days.

If the perianal folds are unclean for long periods, the attached eggs may hatch and the juveniles may wander into the anus and hence to the intestine in a process known as **retrofection** (or **retroinfection**). Hatching of the eggs while they are still inside the intestine apparently does not occur, except, perhaps, during constipation.

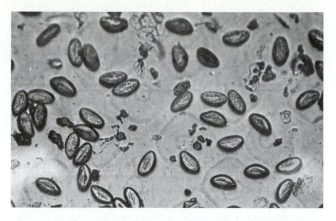

FIGURE 27.5

Eggs of *Enterobius vermicularis*. They are 50 to 60 μm long.

Courtesy of David Oetinger.

• Epidemiology

Clothing and bedding rapidly become seeded with eggs when an infection occurs. Even curtains, walls, and carpets become sources of subsequent infection (or reinfections). Eggs have been found in the dust of school rooms and school cafeterias,[5] providing a source of infection not only for children but also for teachers and other school personnel. The microscopic eggs are very light and are wafted about by the slightest air currents, depositing them throughout a building. The eggs remain viable in cool, moist conditions for up to a week.

The most common means of infection is through insertion of soiled fingers or other objects in the mouth, as well as through use of contaminated bedding, towels, and so on. Obviously, it becomes next to impossible to avoid contamination when eggs are abundant. Furthermore, it remains impossible to avoid reinfection when retrofection occurs.

Humans can inhale and subsequently swallow airborne eggs, or the eggs may remain in the nose until they hatch. This, together with nosepicking, accounts for the occasional case of pinworm in the nose. Contrary to popular belief, pinworms cannot be transmitted by dogs and cats because these animals are free of pinworms. They can, of course, pick up eggs from their environment on their fur and serve as a source of infection for a human who affectionately nuzzles them.

White people seem more susceptible to pinworms than black people.

• Pathogenesis

About one-third of infections are completely asymptomatic, and in many more, clinical symptoms are negligible. Nevertheless, very large numbers of worms may be present and lead to more serious consequences. Pathogenesis has two aspects: damage caused by worms within the intestine and damage resulting from egg deposition around the anus. Minute ulcerations of the intestinal mucosa from attachment of adults may lead to mild inflammation and bacterial infection.[13] Very rarely, pinworms will penetrate into the submucosa with fatal results. The movements of the females out of the anus to deposit eggs, especially when the patient is asleep, lead to a tickling sensation of the perianus, causing the patient to scratch. The subsequent vicious circle of bleeding, bacterial infection, and intensified itching can lead to a nightmare of discomfort.

It is common for worms to wander into the vulva where they remain for several days, causing a mild irritation. Cases have been reported where pinworms have wandered up the vagina, uterus, and oviducts into the coelom, to become encased by granulomatous tissue in the peritoneum. They have been known to become encapsulated within an ovarian follicle.[4] Granulomas caused by *E. vermicularis* have been reported in perianal tissue and even in the vulva.[10,14]

A variety of other symptoms have been ascribed to heavy pinworm infection in children: nervousness, restlessness, irritability, loss of appetite, nightmares, insomnia, bed wetting, grinding of the teeth, perianal pain, nausea, and vomiting.

• Diagnosis and Treatment

Positive diagnosis can be made only by finding eggs or worms on or in the patient. Ordinary fecal examinations are usually unproductive because few eggs are deposited within the intestine and passed in the feces. Heavy infections can be discovered by examining the perianus closely under bright light, during the night or early morning. Wandering worms glisten and can be seen easily.

When adults cannot be found, eggs often can be, as they are left behind in the perianal folds. A short piece of cellophane tape, held against a flat, wooden applicator or similar instrument, sticky side out, is pressed against the junction of the anal canal and the perianus. The tape is then reversed and stuck to a microscope slide for observation. If a drop of xylene or toluene is placed on the slide before the tape, it will dissolve the glue on the tape and clear away bubbles, simplifying the search for the characteristic, flat-sided eggs (Fig. 27.5). A physician can teach a parent how to prepare the slide, since it should be done just after awakening in the morning, certainly before bathing the child for a trip to the doctor's office.

The preferred drug is mebendazole (Vermox). Treatment should be repeated after about 10 days to kill worms acquired after the first dose, and sanitation procedures should be instituted concurrently. All members of the household should be treated simultaneously, regardless of whether the infection has been diagnosed in all.

Although diagnosis and cure of enterobiasis are easy, preventing reinfection is more difficult. Personal hygiene is most important. Completely sterilizing the household is a gratifyingly difficult activity but of limited usefulness. Nevertheless, at time of treatment, all bed linens, towels, and the like should be washed in hot water, and the household should be cleaned as well as possible to lower the prevalence of infective eggs in the environment. If all persons are undergoing chemotherapy while reasonable care is taken to avoid reinfection, the family infection can be eradicated—until the next time a child brings it home from school.

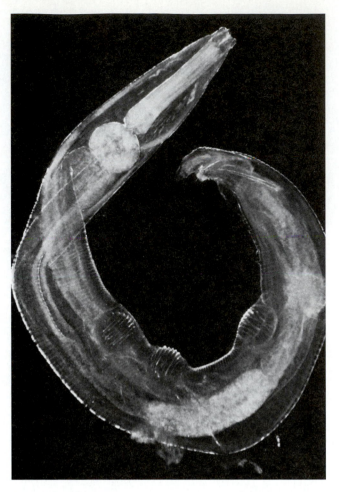

FIGURE 27.6

Syphacia, a pinworm of rodents. Note the three corrugated mamelons on the ventral surface of the male.

Courtesy of Warren Buss.

Syphacia Species

The tiny parasites of the genus *Syphacia* are rarely found in humans but are commonly encountered in their natural hosts: wild and domestic rodents. Laboratory rats and mice are frequently infected by *S. muris* and by *S. obvelata,* which lives in the cecum and has been reported from humans.[12]

Male *Syphacia* are easily recognized by their **mamelons,** two or three ventral, serrated projections (Fig. 27.6). Females are typical pinworms, with long, pointed tails. The eggs are operculated. The life cycle is direct: The worms mature in the cecum or large intestine.[11] No migration within the host is known.

References

1. Adamson, M. L. 1989. Evolutionary biology of the Oxyurida (Nematoda): Biofacies of a haplodiploid taxon. In Baker, J. R., and R. Muller, eds. *Advances in parasitology* 28. London: Academic Press, 175–228.

2. Adamson, M. L., A. Buck, and S. Noble. 1992. Transmission pattern and intraspecific competition as determinants of population structure in pinworms (Oxyurida: Nematoda). *J. Parasitol.* 78:420–26.

3. Adamson, M. L., and S. J. Noble. 1993. Interspecific and intraspecific competition among pinworms in the hindgut of *Periplaneta americana. J. Parasitol.* 79:50–56.

4. Beckman, E. N., and J. B. Holland. 1981. Ovarian enterobiasis—a proposed pathogenesis. *Am. J. Trop. Med. Hyg.* 30:74–76.

5. Brown, H. W., and F. A. Neva. 1983. *Basic clinical parasitology,* 5th ed. Norwalk, Conn.: Appleton-Century-Crofts, Inc.

6. Hopkins, D. R. 1992. Homing in on helminths. *Am. J. Trop. Med. Hyg.* 46:626–34.

7. Hugot, J. P., and C. Tourte-Schaffer. 1985. Etude morphologique des oxyures parasites de l'homme: *Enterobius vermicularis* et *E. gregorii. Ann. Parasitol. Hum. Comp.* 60:57–64.

8. Hulinskaca, D. 1968. The development of the female *Enterobius vermicularis* and the morphogenesis of its sexual organ. *Folia Parasitol.* 15:15–27.

9. Kliks, M. M. 1990. Helminths as heirlooms and souvenirs: A review of New World paleoparasitology. *Parasitol. Today* 6:93–100.

10. Mattia, A. R. 1992. Perianal mass and recurrent cellulitis due to *Enterobius vermicularis. Am. J. Trop. Med. Hyg.* 47:811–15.

11. Prince, M. J. R. 1950. Studies on the life cycle of *Syphacia obvelata,* a common nematode parasite of rats. *Science* 111:66–67.

12. Riley, W. A. 1920. A mouse oxyurid, *Syphacia obvelata,* as a parasite of man. *J. Parasitol.* 6:89–92.

13. Shubenko-Gabuzova, I. N. 1965. Appendicitis in enterobiasis. *Med. Parasitol. Dis.* 34:563–66.

14. Sun, T., N. S. Schwartz, C. Sewell, P. Lieberman, and S. Gross. 1991. Enterobius egg granuloma of the vulva and peritoneum: Review of the literature. *Am. J. Trop. Med. Hyg.* 45:249–53.

Additional References

Petter, A. J., and J. -C. Quentin. 1976. *CIH keys to the nematode parasites of vertebrates, No. 4. Keys to genera of the Oxyuroidea.* Bucks, Eng.: Commonwealth Agricultural Bureaux, Farnham Royal.

Skrjabin, K. I., N. P. Schikhobolova, and E. A. Lagodovskaya. 1960–1967. *Essentials of nematodology 8, 10, 13, 15, and 18. Oxyurata.* Moscow: Akademii Nauk SSSR. Indispensable reference works for the oxyurid taxonomist.

Skrjabin, K. I., N. P. Schikhobolova, and A. A. Mosgovoi. 1951. *Key to parasitic nematodes 2.* Oxyurata *and* Ascaridata. Moscow: Akademii Nauk SSSR. A useful key to genera, with lists of species.

Chapter 28
NEMATODES: SPIRURIDA, A POTPOURRI

We learn, as the thread plays out, that we belong

Less to what flatters us than to what scars. . . .

Stanley Kunitz

Spirurids are parasitic in all classes of vertebrates and employ an intermediate host in their development, usually an arthropod. They are a very large, heterogeneous group, with many species. The many variations of morphology in this order make generalization difficult, but most have two lateral lips, called **pseudolabia,** and an esophagus that is divided into anterior muscular and posterior glandular portions. The lips do not represent the fusion of primitive lips but are evolutionarily new structures that originate in anterior shifting of tissues from within the buccal walls. Although the esophagus usually has both muscular and glandular portions, there are species whose esophagus is primarily muscular and others in which it is mainly glandular. Some of these, however, are of uncertain taxonomic position. Spirurid spicules are usually dissimilar in size and shape.

Chabaud[5] arranged the order Spirurida into two suborders: the Camallanina and the Spirurina. The Camallanina contains two superfamilies, of which one contains a species (*Dracunculus medinensis*) that is an important parasite of humans. The suborder Spirurina includes 10 superfamilies. Members of nine of the superfamilies infect a variety of wild and domestic animals and occasionally humans as zoonoses. The remaining superfamily, the Filaroidea, includes members with vast public health significance. We will devote this chapter to brief descriptions of some interesting families of the Spirurina other than filaroids, the next chapter to the Filaroidea, and finally a chapter to the Camallanina.

FAMILY ACUARIIDAE

Nematodes of this family, all parasites of birds, exhibit very peculiar morphological structures at their head ends. Some have four grooves or ridges, called **cordons,** which begin two dorsally and two ventrally at the junctions of the lateral lips and proceed posteriorly for varying distances (Fig. 28.1). The cordons may be straight, sinuous, recurving, or even anastomosing in pairs. Other acuariids do not have cordons but instead possess four extravagant cuticular projections, sometimes simple, sometimes serrated or even feathered (Fig. 28.2). Both specializations, cordons and cuticular projections

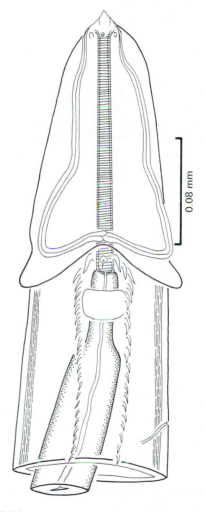

FIGURE 28.1

Cordonema venusta, from the stomach of an aquatic bird (dipper). Note the helmetlike inflation of the cuticle, which bears two cordons on each side. The cordons of each side join at their posterior ends in this genus.

From G. D. Schmidt and R. E. Kuntz, "Nematode parasites of Oceanica. XVI. *Cordonema venusta* gen. et sp. nov., and *Skryabinoclava* spp. (Acuariidae: Echinuriinae), from birds," in *Parasitology* 64:235–244. Copyright © 1972. Reprinted with permission of the publisher.

of the head, seem to correlate with the parasite's location within the host, the stomach. Most mature under the koilon, or gizzard lining, where they cause considerable damage to the underlying epithelium. How these anterior modifications aid the parasite is not known.

439

The members of only one genus, *Echinuria* spp., in ducks, geese, and swans are of economic importance; however, acuariids represent an interesting example of adaptive radiation.

FAMILY GNATHOSTOMATIDAE

Family Gnathostomatidae contains the genera *Tanqua* from reptiles, *Echinocephalus* from elasmobranchs, and *Gnathostoma* from the stomachs of carnivorous mammals (Fig. 28.3). These distinctive nematodes have two powerful, lateral lips, followed by a swollen "head," which is separated from the rest of the body by a constriction. Internally four peculiar, glandular cervical sacs, reminiscent of acanthocephalan lemnisci, hang into the pseudocoel from their attachments near the anterior end of the esophagus. The head bulb is divided internally into four hollow areas called **ballonets.** Each cervical sac has a central canal, which is continuous with a ballonet. The functions of these organs are unknown.

Gnathostoma spp. are particularly interesting because of their widespread distribution and peculiar biology and because they often cause disease in humans in some areas of the world. In the United States *G. procyonis* is common in the stomachs of raccoons and opossums, and *G. spinigerum* has been reported from a wide variety of carnivores in Asia. *Gnathostoma doloresi* is common in pigs in Asia (Fig. 28.4). Of the 20 or so species that have been described, *G. spinigerum* has most consistently been demonstrated as a cause of disease in humans, so we will examine this parasite in more detail.

Gnathostoma spinigerum

In 1836 Richard Owen, the famous British anatomist who established our basic definitions of homology and analogy, discovered *G. spinigerum* in the stomach wall of a tiger that had

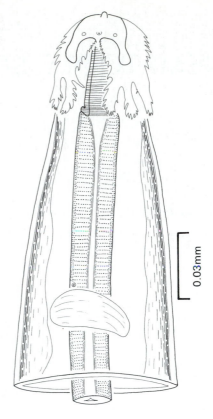

FIGURE 28.2

Sobolevicephalus chalcyonis from the stomach of a kingfisher. The head cuticle has four feathered projections.

From G. D. Schmidt and R. E. Kuntz, "Nematode parasites of Oceanica. XVII. Schistorophidae, Spiruridae, Physalopteridae, and Trichostrongylidae of birds," in *Parasitology* 64:269–278. Copyright © 1972. Reprinted with permission of the publisher.

FIGURE 28.3

Morphological comparison among six species of female *Gnathostoma; S, G. spinigerum; H, G. hispidum; T, G. turgidum; D, G. doloresi; N, G. nipponicum; P, G. procyonis*. This figure indicates the arrangement and shape of the cuticular spines and fresh fertilized uterine eggs, which at times may show various developmental stages when preserved.

From I. Miyazaki, "*Gnathostoma* and gnathostomiasis in Japan," in *Progress of Medical Parasitology in Japan*, 3:529–586. Edited by K. Morishita, et al. Copyright © 1966 Meguro Parasitological Museum, Tokyo. Reprinted with permission of the publisher.

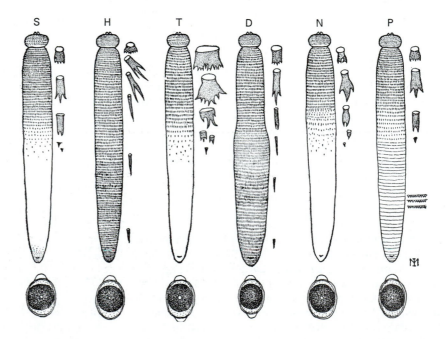

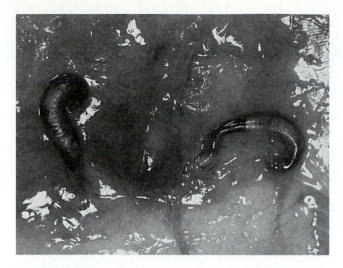

FIGURE 28.4

Gnathostoma doloresi attached to the stomach mucosa of a pig.
Courtesy of Robert E. Kuntz.

died in the London Zoo. Since then, it has been found in many kinds of mammals in several countries, although it is most common in southeast Asia.

• Morphology

The body is stout and pink in life. The swollen head bulb is covered with four circles of stout spines. The anterior half of the body is covered with transverse rows of flat, toothed spines, followed by a bare portion. The posterior tip of the body has numerous tiny cuticular spines.

Males are 11 to 31 mm long and have a bluntly rounded posterior end. The anus is surrounded by four pairs of stumpy papillae. The spicules are 1.1 mm and 0.4 mm long and are simple with blunt tips.

Females are 11 to 54 mm long and also have a blunt posterior end. The vulva is slightly postequatorial in position. The eggs are unembryonated when laid, 65 to 70 μm by 38 to 40 μm in size, and have a polar cap at only one end. The outer shell is pitted.

• Biology

The eggs complete embryonation and hatch in about a week at 27° to 31°C.[8] The actively swimming J_1 is eaten by a cyclopoid copepod, where it penetrates into the hemocoel and develops further into a J_2 in 7 to 10 days (Fig. 28.5). The J_2 already has a swollen head bulb covered with four transverse rows of spines.

When the infected crustacean is eaten by a vertebrate second intermediate host, the J_2 penetrates the intestine of its new host and migrates to muscle or connective tissue, where it molts to the third stage. The J_3 is infective to a definitive host. If it is eaten by the wrong host, it may wander in that

animal's tissues without further development. More than 35 species of paratenic hosts are known, among them crustaceans, freshwater fishes, amphibians, reptiles, birds, and mammals, including humans.[9]

Adult worms are found embedded in tumorlike growths in the stomach wall of the definitive host. They begin producing eggs about 100 days after infection.

• Epidemiology

Human infection results from eating a raw or undercooked intermediate or paratenic host containing third-stage juveniles. In Japan, this is most often fish, whereas in Thailand, domestic ducks and chickens are probably the most important vectors.[6] However, any amphibian, reptile, or bird may harbor juveniles and thereby contribute an infection if eaten raw. A number of cases of infection with *G. hispidum, G. doloresi,* and *G. nipponicum* also have been reported in Japan.[11]

• Pathology

In humans the J_3's usually migrate to the superficial layers of the skin, causing creeping eruption (cutaneous larva migrans). They may become dormant in abscessed pockets in the skin, or they may wander, leaving swollen red trails in the skin behind them.

If the worms remain in the skin with little wandering, they cause relatively little disease. Often they will erupt out of the skin spontaneously. However, erratic migration may take them into an eye, the brain, or the spinal cord, with serious results that even may cause death.

• Diagnosis and Treatment

Diagnosis depends on recovery and accurate identification of the worm. An intradermal test, using an antigen prepared from *G. spinigerum,* has been employed with success in Japan. Gnathostomiasis should be suspected in an endemic area when a localized edema is accompanied by leukocytosis with a high percentage of eosinophils.

Chemotherapy has not been effective against this zoonosis. The only effective treatment is surgical removal of the worm.

Prevention is the most realistic means of controlling this disease. Cooling by any means will kill the worms. In regions where ritualistic consumption of raw fish is an important tradition, the fish should be well-frozen before preparation, or marine fish should be used. Consumption of raw, previously unfrozen fish in any area of the world is dangerous for a variety of parasitological reasons.

FAMILY PHYSALOPTERIDAE

Members of the family Physalopteridae are mostly rather large, stout worms that live in the stomachs or intestines of all classes of vertebrates. All have two large, lateral pseudolabia, usually armed with teeth. The head papillae are on the pseudolabia. The cuticle at the base of the lips is

FIGURE 28.5

Life history of *G. spinigerum*. **1,** J₁ (averaging 0.27 mm long) swimming in the water; **2,** J₂ (averaging 0.5 mm long) parasitic in copepod; **3,** encysted J₃ (averaging 3 to 4 mm long) in the muscle of the second intermediate host. Double *arrow* in the figure indicates secondary infection. Humans are susceptible to the infection from a variety of second intermediate hosts (*wavy lines*) *Heavier wavy line,* most important source of human infection.

From I. Miyazaki, "*Gnathostoma* and gnathostomiasis in Japan," in *Progress of Medical Parasitology in Japan,* 3:529–586. Edited by K. Morishita, et al. Copyright © 1966 Meguro Parasitological Museum, Tokyo. Reprinted by permission of the publisher.

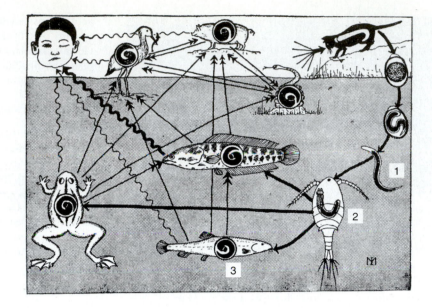

swollen into a "collar" in some genera. Caudal alae are well developed on males. Spicules are equal or unequal, and a gubernaculum is absent. This family has a tendency toward **polydelphy,** or having many ovaries and uteri. Of the several genera in this family, we will briefly consider *Physaloptera.*

Physaloptera Species

In the genus *Physaloptera* the triangular pseudolabia are armed with varying numbers of teeth, and a conspicuous cephalic collar is present. The male has numerous pedunculated caudal papillae and caudal alae that join anterior to the anus. In a few species the cuticle of the posterior end is inflated into a prepucelike sheath, which encloses the tail. Three species occur in Amphibia, around 45 species in reptiles, 24 in birds, and nearly 90 in mammals.

Physaloptera praeputialis (Fig. 28.6) lives in the stomachs of domestic and wild dogs and cats throughout the world except Europe. It is common in dogs, cats, coyotes, and foxes in the United States. A flap of cuticle covers the posterior ends of both sexes. Its life cycle is incompletely known, but development has been experimentally obtained in cockroaches.

Physaloptera rara is the most common physalopterid of carnivores in North America, to which it is apparently restricted. It is similar to *P. praeputialis* but lacks the posterior cuticular flap. The life cycle involves an insect intermediate host, usually a field cricket, in which it develops to the J₃. A paratenic host, such as a snake, is commonly necessary in the life cycle because of the feeding habits of the definitive hosts.

Physaloptera caucasica is the only species recorded from humans.[13] It is normally parasitic in African monkeys. Most recorded cases in humans were from Africa, although several records, some based only on eggs found in patients' feces, have been reported from South and Central America, India, and the Middle East. It is possible that some of these eggs were misidentified.

The life cycle is unknown but most likely involves insect intermediate hosts and a vertebrate paratenic host.

Symptoms in humans include vomiting, stomach pains, and eosinophilia. Tentative diagnosis can be made by demonstrating the eggs in a fecal sample or by obtaining an adult specimen for accurate identification.

FAMILY TETRAMERIDAE

Members of the family Tetrameridae are bizarre in their degree of sexual dimorphism. Whereas males exhibit typical nematoid shape and appearance, the females are greatly swollen and often colored bright red. The three genera in this family are all parasitic in the stomachs of birds. The genus *Geopetitia* is represented by five rare species that live in cysts on the outside of the proventriculus or gizzard of birds, where they communicate with the enteric lumen through a tiny pore. The other two genera in the family are very common parasites that live in the branched secretory glands of the proventriculus, although males can be found wandering throughout the organ.

Tetrameres (Fig. 28.7) is a large genus of about 50 species, mainly parasites of aquatic birds. A well-developed, sclerotized buccal capsule is present in both sexes. Males are typically nematoid in form, lacking caudal alae and possessing spicules that are vastly dissimilar in size. Lateral, longitudinal rows of spines are present on many species. Females, however, are greatly swollen, with only the front and back ends retaining the appearance of a nematode. In addition, females are blood-red with a black, saclike intestine. They are easily seen as reddish spots in the wall of the proventriculus, where they mature with the tail end near the lumen of that organ. The vulva is near the anus and thus is available to males who find it. The eggs are embryonated when laid. The intermediate hosts are crustaceans or insects. Definitive hosts are water birds, chickens, owls, and hawks.

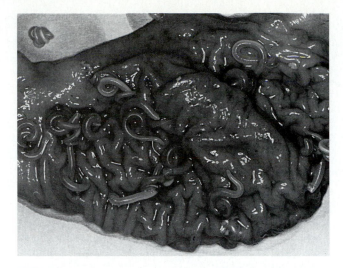

FIGURE 28.6

Physaloptera praeputialis in the stomach of a domestic cat.
Courtesy of Robert E. Kuntz.

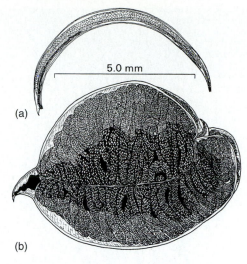

FIGURE 28.7

Tetrameres strigiphila from owls. (*a*) Male; (*b*) female.

From D. B. Pence et al., "*Tetrameres (Gynaecophila) strigiphila* sp. n. from the Florida barred owl, *Strix varia georgica,* with notes on the status of the subgenus *Gynaecophila* (Nematode: Tetrameridae)," in *J. Parasitol.* 61:494–498. Copyright © 1975. Reprinted with permission of the publisher.

The terrestrial counterpart of *Tetrameres* is *Microtetrameres* (Fig. 28.8). About 40 species have been described in this genus, all of which live in the proventricular glands of insectivorous birds. Females of this genus are also swollen and, in addition, are twisted into a spiral. Morphological and biological characteristics are otherwise similar to those of *Tetrameres,* with terrestrial crustaceans and insects serving as intermediate hosts.

The quaint little worms in this family are familiar to all who survey the parasites of birds, since they are common in many species of hosts. Even though a hundred or more females are commonly embedded in the proventriculus of a single duck, for example, they seem to have little effect on the overall health of their host.

FAMILY GONGYLONEMATIDAE

This family contains the single genus *Gongylonema,* which has several species that are found in the upper digestive tracts of birds and mammals. Morphologically, they resemble several spiruridans, except that the cuticle of the anterior end is covered with large bosses, or irregular scutes, arranged in eight longitudinal rows (Fig. 28.9). Cervical alae are present, as are cervical papillae. The posterior end of the male bears wide caudal alae, which are supported by numerous pedunculated papillae.

Of the 25 or so species in this genus, *G. pulchrum* is probably the best known. Primarily a parasite of ruminants and swine, the worm also has been reported from monkeys, hedgehogs, bears, and humans.[7,12] The life cycle involves an insect intermediate host, either a dung beetle or a cockroach. Although these would appear rather unpalatable fare for people, numerous cases of human gongylonemiasis have been reported. In normal hosts the worms invade the esophageal epithelium, where they burrow stitchlike in shallow tunnels. In an abnormal host, such as humans, they behave similarly but do not mature and seem to wander farther, being found

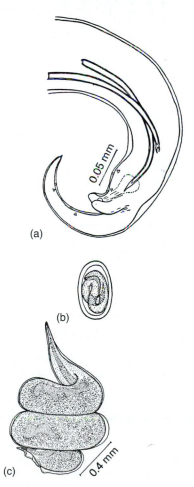

FIGURE 28.8

Microtetrameres aguila, a parasite of the proventriculus of eagles. (*a*) Posterior end of male, lateral view; (*b*) embryonated egg; (*c*) female.

From S. C. Schell, "Four new species of Microtetrameres (Nematodea: Spiruroidea) from North American Birds," in *Trans Am. Microsc. Soc.* 72:227–236. Copyright © 1953. Reprinted with permission of the publisher.

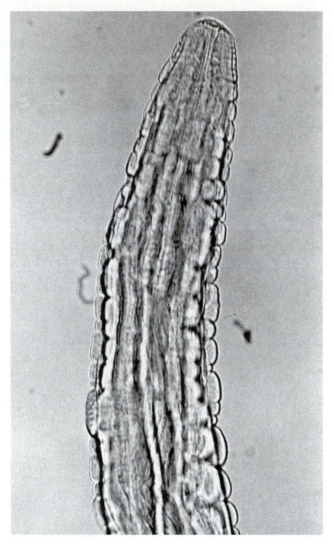

FIGURE 28.9

Anterior end of *Gongylonema*, demonstrating the cuticular bosses typical of the genus.

Courtesy of Warren Buss.

often in the epithelium of the tongue, gums, or buccal cavity. Their active movements, together with resulting irritation and bleeding, soon make their presence known. Treatment is surgical removal of worms that can be seen. Chemotherapy is seldom employed.

Gongylonema neoplasticum and *G. orientale* in domestic rats are thought to induce neoplastic tumors. They are not known to infect humans.

FAMILY SPIROCERCIDAE

Members of the family Spirocercidae are all parasites of mammals. Of these, *Spirocerca lupi* is the most interesting because of its complex life cycle and its relationship to esophageal cancer in dogs.[1,2]

Spirocerca lupi

This stout worm is bright pink to red when alive. The mouth is surrounded by six rudimentary lips, and the buccal capsule is well-developed, with thick walls. A short muscular portion of the esophagus is followed by a longer glandular portion. Males are 30 to 54 mm long, with a left spicule 2.45 to 2.80 mm long and a right spicule 475 to 750 μm long. Females are 50 to 80 mm long, with the vulva 2 to 4 mm from the anterior end. The eggs are cylindrical and embryonated when laid.

• Biology

Adults normally are found in clusters, entwined within the wall of the upper digestive tract of dogs, although they have been reported in other organs, mostly the dorsal aorta, and in a wide variety of carnivorous hosts. Hounds seem to be the breeds most frequently infected in the United States. This is probably related to their opportunities for exposure rather than breed susceptibility.

Shelled juveniles pass out of the host with its feces. Any of several species of scarabaeid dung beetles can serve as intermediate host. A wide variety of paratenic hosts is known, including birds, reptiles, and other mammals. Dogs can become infected by eating dung beetles or infected paratenic hosts. Domestic dogs are probably most often infected by eating the offal of chickens that have J_3's encysted in their crops.

Once in the stomach of a definitive host, the juveniles penetrate its wall and enter the wall of the gastric artery, migrating up to the dorsal aorta and forward to the area between the diaphragm and the aortic arch. They remain in the wall of the aorta for two and one-half to three months, after which they emerge and migrate to the nearby esophagus, which they penetrate. After establishing a passage into the lumen of the esophagus, they move back into the submucosa or muscularis where they complete their development about five to six months after infection. Eggs pass into the esophageal lumen through the tiny passage formed by the worm.

Many worms get lost during migration and may be found in abnormal locations, including lung, mediastinum, subcutaneous tissue, trachea, urinary bladder, and kidney.

• Epidemiology

Spirocerca lupi is most common in warm climates but has been found in Manchuria and northern regions of the Soviet Union. Many questions are still unanswered: Why is the parasite distributed so sporadically throughout the United States and the world? What factors influence the change in prevalence of the infection in a given area? The attractiveness of dog feces to susceptible beetle species is a factor, as is the application of pesticides in an endemic area. Certainly the successful transfer of J_3's from one paratenic host to another increases the parasite's chances for survival.

• Pathology

When the J_3 penetrates the mucosa of the stomach, it causes a small hemorrhage in the area. This irritation commonly causes the dog to vomit. The lesions in the aorta caused by the migrating worms are often severe, with hemorrhage

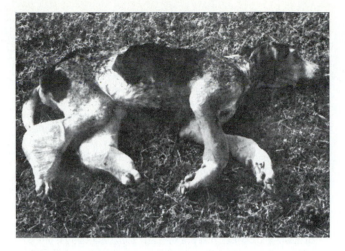

FIGURE 28.10

Hound with severe hypertrophic pulmonary osteoarthropathy associated with esophageal sarcoma.

From W. S. Bailey, "Parasites and cancer: sarcoma with *Spirocerca lupi,* in *Ann. N.Y. Acad. Sci.* 108(Art 3):890–923. Copyright © 1963.

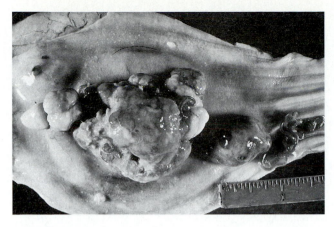

FIGURE 28.11

Esophageal sarcoma associated with *Spirocerca lupi* infection. Pedunculated masses protrude into the lumen; adult *S. lupi* are partially embedded in the neoplasm.

From W. S. Bailey, "*Spirocerca lupi:* a continuing inquiry," in *J. Parasitol.* 58:3–22. Copyright © 1972. Photo courtesy of the Department of Pathology and Parasitology, Auburn University School of Veterinary Medicine, Auburn, AL.

accounting for death in some dogs with heavy infections. Destruction of tissues in the wall of the aorta with subsequent scarring is typical of this disease and may lead to numerous aneurysms.

Worms that leave the aorta and migrate upward come in contact with the tissues surrounding the vertebral column, where they frequently cause, by a mechanism not yet elucidated, a condition known as **spondylosis.** This deformation may be so severe as to cause adjacent vertebrae to fuse. Hypertrophic pulmonary osteoarthropathy, with inflamed and swollen joints, is a common sequel to this disease (Fig. 28.10).

The most striking lesion associated with spirocercosis is in the wall of the esophagus, where the worms mature. Here their presence stimulates the formation of a **reactive granuloma,** made up of fibroblasts. The granulomas are rather more loosely organized than is usually the case in granulomatous reactions, and the cell structure is characteristic of incipient **neoplasia** (cancer). Some of the granulomas change to **sarcomas,** true cancerous growths (Fig. 28.11). The worms may continue to live inside these tumors for some time, or they may be extruded or compressed and killed by the rapidly growing tissue. The precise oncogenic factor responsible for stimulation of neoplasia is still unknown. Although several other helminths have been associated with malignancy, in no other instance is there as strong evidence for a cause-effect relationship as in canine spirocercosis.

• Diagnosis and Treatment

Diagnosis in dogs usually is by demonstration of the characteristic eggs, which measure 40 μm by 12 μm and have nearly parallel sides, in a fecal examination. At necropsy,

aortic scarring, aneurysms, and esophageal granuloma or sarcoma are considered diagnostic, even if worms are no longer present.

Disophenol is effective against *S. lupi,* but if extensive aneurysms or a sarcoma have already developed, the treatment will not affect these conditions.

FAMILY THELAZIIDAE

Members of the family Thelaziidae live on the surface of the eye in birds and mammals, usually remaining in the lacrimal ducts or conjunctival sacs or under the nictitating membrane. Most are parasites of wild animals, but two species of *Thelazia* have been reported from humans.[4]

These worms lack lips but show evidence of the primitive condition in having a hexagonal mouth (Fig. 28.12*a*). The buccal capsule is well-developed, with thick walls. Alae and cuticular ornamentations are absent, except for conspicuous transverse striations near the anterior end (Fig. 28.12*b*). These are deep, and their overlapping edges ostensibly aid movements across the smooth surface of the cornea.

Thelazia callipaeda is a parasite of dogs and other mammals in southeast Asia, China, and Korea, and *T. californiensis* parasitizes deer and other mammals in western North America. Both species have been reported from humans several times.[3]

The female worm releases J_1's onto the surface of the host eye, where they are picked up by face flies, such as *Musca autumnalis,* feeding on lacrimal secretions. After development in the fly, the J_3's emerge from the fly's labellum when it is feeding at the eye of another host.[10] In the United States, *T. skrjabini* and *T. gulosa* most commonly occur in the eyes of cattle, and *T. lacrymalis* is found in the eyes of horses.

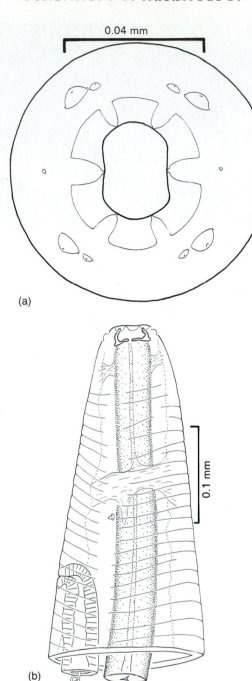

FIGURE. 28.12

Thelazia digiticauda from under the nictitating membrane of the eye of a kingfisher. (*a*) En face view; (*b*) female, lateral view of anterior end.

From G. D. Schmidt and R. E. Kuntz, "Nematode parasites of Oceanica. XIII. Subuluridae, Thelaziidae, and Acuariidae of birds," in *Parasitology* 63:91–99. Copyright © 1971. Reprinted with permission of the publisher.

References

1. Bailey, W. S. 1963. Parasites and cancer: Sarcoma in dogs associated with *Spirocerca lupi. Ann. N.Y. Acad. Sci.* 108:890–923.

2. Bailey, W. S. 1971. *Spirocerca lupi:* A continuing enquiry. *J. Parasitol.* 58:3–22.

3. Bhaibulaya, M., S. Prasertsilpa, and S. Vajrasthira. 1970. *Thelazia callipaeda* Railliet and Henry, 1910, in man and dog in Thailand. *Am. J. Trop. Med. Hyg.* 19:476–79.

4. Brown, H. W., and F. A. Neva. 1983. *Basic clinical parasitology.* Norwalk, Conn.: Appleton-Century-Crofts, Inc.

5. Chabaud, A. G. 1974. Class Nematoda. Keys to subclasses, orders and superfamilies. In Anderson, R. C., A. G. Chabaud, and S. Willmott, eds. *CIH keys to the nematode parasites of vertebrates,* no. 1. Bucks, Eng.: Commonwealth Agricultural Bureaux, Farnham Royal. 6–17.

6. Daengsvang, S., P. Thienprasitthi, and P. Chomcherngpat. 1966. Further investigations on natural and experimental hosts of larvae of *Gnathostoma spinigerum* in Thailand. *Am. J. Trop. Med. Hyg.* 15:727–29.

7. Feng, L. C., M. S. Tung, and S. C. Su. 1955. Two Chinese cases of *Gongylonema* infection. A morphological study of the parasite and clinical study of the case. *Chin. Med. J.* 73:149–62.

8. Miyazaki, I. 1954. Studies on *Gnathostoma* occurring in Japan (Nematoda: Gnathostomidae). II. Life history of *Gnathostoma* and morphological comparison of its larval forms. *Kyushu Mem. Med. Sci.* 5:123–40.

9. Miyazaki, I. 1966. Gnathostoma and gnathostomiasis in Japan. In Morishita, K., Y. Komiya, and H. Matsubayashi, eds. *Progress of medical parasitology in Japan* 3. Tokyo: Meguro Parasitological Museum, 529–86.

10. O'Hara, J. E., and M. J. Kennedy. 1991. Development of the nematode eyeworm, *Thelazia skrjabini* (Nematoda: Thelazoidea), in experimentally infected face flies, *Musca autumnalis* (Deptera: Muscidae). *J. Parasitol.* 77:417–25.

11. Sato, H., H. Kamiya, and K. Hanada. 1992. Five confirmed human cases of gnathostomiasis nipponica recently found in northern Japan. *J. Parasitol.* 78:1006–10.

12. Thomas, L. J. 1952. *Gongylonema pulchrum,* a spirurid nematode infecting man in Illinois, U.S.A. *Proc. Helm. Soc. Wash.* 19:124–26.

13. Vandepitte, G., J. Michaux, J. L. Fain, and F. Gatti. 1964. Premières observations congolaises de physaloptérose humaine. *Ann. Soc. Belg. Med. Trop.* 44:1067–76.

Additional References

Anderson, R. C., and O. Bain. 1976. Keys to genera of the order Spirurida, part 3. Diplotriaenoidea, Aproctoidea, and Filaroidea. In Anderson, R. C., A. G. Chabaud, and W. Willmott, eds. *CIH keys to the nematode parasites of vertebrates,* no. 3., Bucks, Eng.: Commonwealth Agricultural Bureaux, Farnham Royal.

Chabaud, A. G. 1954. Valeur des charactérs biologiques pour la systématique des nématodes spirurides. *Vie Milieu* 5:299–309.

Chabaud, A. G. 1975. Keys to the genera of the order Spirurida, part 1. Camallanoidea, Dracunculoidea, Gnathostomatoidea, Physalopteroidea, Rictularioidea, and Thelazioidea. In Anderson, R. C., A. G. Chabaud, and W. Willmott, eds. *CIH keys to the nematode parasites of vertebrates,* no. 3. Bucks, Eng.: Commonwealth Agricultural Bureaux, Farnham Royal.

Chabaud, A. G. 1975. Keys to the order Spirurida, part 2. Spiruroidea, Habronematoidea and Acuarioidea. In Anderson, R. C., A. G. Chabaud, and S. Willmott, eds. *CIH keys to the nematode parasites of vertebrates,* no. 3. Bucks, Eng.: Commonwealth Agricultural Bureaux, Farnham Royal.

Skrjabin, K. I., A. A. Sobolev, and V. M. Ivaskin. 1963–1967. *Essentials of nematodology, 11, 12, 14, 16, and 19. Spirurata of animals and man and the diseases they cause.* Moscow: Akademii Nauk SSSR. The most complete monographs on the subject.

Chapter 29

NEMATODES: FILAROIDEA, THE FILARIAL WORMS

Our eye-beams twisted, and did thread

our eyes, upon one double string . . .

John Donne

The filaroids are tissue-dwelling parasites. They are among the most highly evolved of the parasitic nematodes. All species employ arthropods as intermediate hosts, most of which deposit J_3's on the skin with their bite (Fig. 29.1). They are parasitic in all classes of vertebrates except fish. Generally speaking, filaroids are slender worms with reduced lips and buccal capsule. Most are parasites of wild animals, particularly of birds, but several are very important disease organisms of humans and domestic animals. The majority of these belong to the large family Onchocercidae.

FAMILY ONCHOCERCIDAE

Members of the family Onchocercidae live in the tissues of amphibians, reptiles, birds, and mammals. Most are of no known medical or economic importance, but a few cause some of the most tragic, horrifying, and debilitating diseases in the world today. Of these, species of *Wuchereria, Brugia, Onchocerca,* and *Loa* will be considered in some detail. Short mention will be made of others.

Wuchereria bancrofti

Perhaps the most striking disease of humans is the clinical entity known as **elephantiasis** (Fig. 29.2). The horribly swollen parts of the body afflicted with this condition have been known since antiquity. The ancient Greek and Roman writers likened the thickened and fissured skin of infected persons to that of the elephant, although they also confused leprosy with this condition. Actually *elephantiasis* is a non-sense word, since literally translated it means "a condition caused by elephants." The word is so deeply entrenched, however, that it is not likely ever to be abandoned. Classical elephantiasis is a rather rare consequence of infection by *Wuchereria bancrofti* and by at least two other species of filaroids.

Infection of the lymphatic system by filarial worms is best referred to as *lymphatic filariasis* and is much more common than elephantiasis. **Bancroftian filariasis** (*W. bancrofti*) is responsible for 90% of lymphatic filariasis,[24] extending throughout central Africa, the Nile Delta, Turkey, India and southeast Asia, the East Indies, the Philippine and Oceanic Islands, Australia, and parts of South America (in short, across a broad equatorial belt). It was probably brought to the New World by the slave trade.[27] A nidus of infection remained in the vicinity of Charleston, South Carolina, until it died out spontaneously in the 1920s.[8]

In the Pacific theater in World War II, lymphatic filariasis was a cause of great psychological concern to American armed forces, who visualized themselves returning home carrying their scrotum in a wheelbarrow. Although thousands of cases of filariasis were in fact contracted by American servicemen, no single case of classical elephantiasis resulted. Some persons experienced symptoms for as long as 16 years.[50] The evolution of knowledge of this disease and its cause remains one of the classics of medical history.[16]

• Morphology

Adult worms are long and slender with a smooth cuticle and bluntly rounded ends. The head is slightly swollen and bears two circles of well-defined papillae. The mouth is small; a buccal capsule is lacking.

The male is about 40 mm long and 100 μm wide. Its tail is fingerlike. The female is 6 to 10 cm long and 300 μm wide. The vulva is near the level of the middle of the esophagus.

• Biology

Adult *Wuchereria* live in the major lymphatic ducts of humans, tightly coiled into nodular masses. They are normally found in the afferent lymph channels near the major lymph glands in the lower half of the body. Rarely they invade a vein. The females are ovoviviparous, producing thousands of juveniles known as **microfilariae** (Fig. 29.3). Microfilariae are not as differentiated as are normal J_1's and are sometimes considered advanced embryos. The microfilariae of *W. bancrofti* retain the egg membrane as a "sheath" (not to be confused with the sheath of some strongyle J_3's, which is the second-stage cuticle). The sheath is rather delicate and close fitting but can be detected where it projects at the anterior and posterior ends of the microfilaria. When stained, several internal nuclei and primordia of organs can be seen in the microfilariae. The location of these and the presence or absence of a sheath help to identify the several species of microfilariae found in humans (Fig. 29.4).

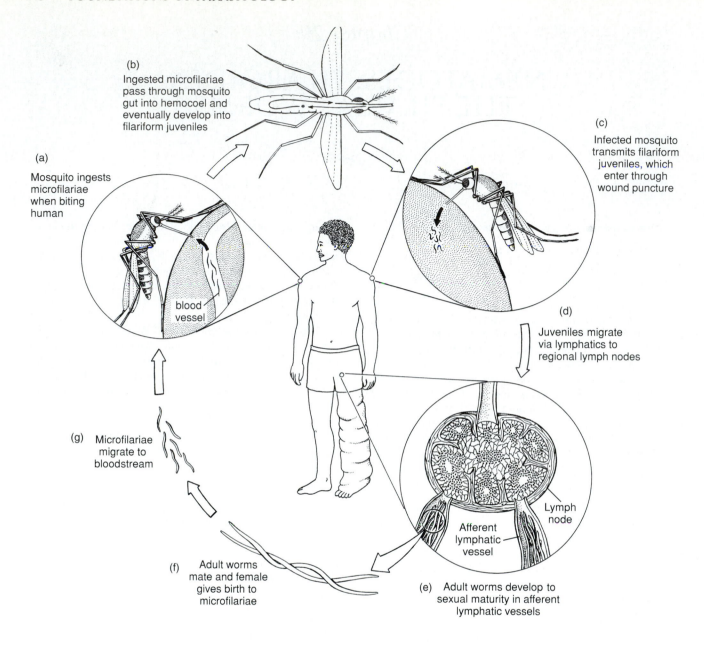

FIGURE 29.1

Life cycle of *Wuchereria bancrofti*. (*a*) Mosquito ingests microfilariae when biting human. (*b*) Microfilariae pass through mosquito gut and develop to filariform J$_3$. (*c*) Filariform juveniles escape from mosquito's proboscis when the insect is feeding and then penetrate wound structure. (*d*) Juveniles migrate via lymphatics to regional lymph nodes. (*e*) Worms develop to sexual maturity in afferent lymphatic vessels. (*f*) Adult worms mate, and female gives birth to microfilariae. (*g*) Microfilariae enter blood circulation.

Drawing by William Ober and Claire Garrison.

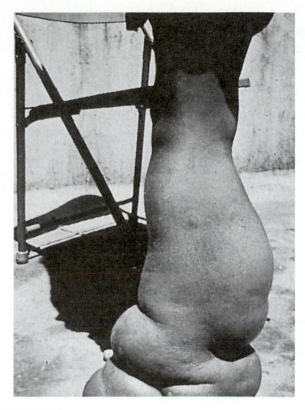

FIGURE 29.2

Elephantiasis of the leg caused by infection with *Wuchereria bancrofti* in India.

Courtesy of E. L. Schiller. AFIP neg. no. 74–6426–2.

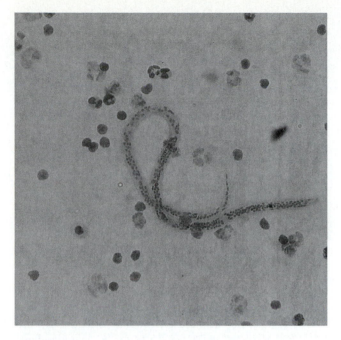

FIGURE 29.3

Microfilaria of *Wuchereria bancrofti*. A sheath is present, and the discrete nuclei do not reach the tip of the tail.

Photograph by Larry S. Roberts.

The female releases the microfilariae into the surrounding lymph. Some may wander into the adjacent tissues, but most are swept into the blood through the thoracic duct. Throughout much of the geographical distribution of this parasite, there is a marked **periodicity** of microfilariae in the peripheral blood; that is, they can be demonstrated at certain times of the day, whereas at other times they virtually disappear from the peripheral circulation. The maximal number usually can be found between 10 P.M. and 2 A.M. For this reason, night-feeding mosquitoes are the primary vectors of *Wuchereria* in areas where microfilarial periodicity occurs. During the day the microfilariae are concentrated in blood vessels of the deep tissues of the body, predominately in the pulmonary capillaries.[28] Causes of the periodicity remain obscure, but they apparently do not involve daily release of a new generation of progeny by the adult female. Stimuli such as arterial oxygen tension and body temperature probably are involved. Administration of pure oxygen to a patient during peak microfilaremia (the condition of having microfilariae in the blood) can cause the microfilariae to localize in the deep tissues. Reversal of the patient's sleep schedule causes reversal of periodicity so that microfilaremia becomes diurnal. The adaptive value of the periodicity is difficult to explain. Although it is clearly advantageous for the microfilariae to be present in the peripheral blood when the vector is likely to be feeding, what value is there in being absent when the vector is

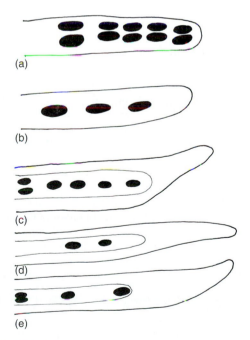

FIGURE 29.4

Presence or absence of a sheath and the arrangement of nuclei in the tail are useful criteria in identifying microfilariae. (*a*) *Mansonella perstans;* (*b*) *Mansonella ozzardi;* (*c*) *Loa loa;* (*d*) *Wuchereria bancrofti;* (*e*) *Brugia malayi.*

not feeding? Some strains of *Wuchereria* have a diurnal periodicity. The periodicity is unimportant clinically, but it has significant diagnostic and epidemiological implications.

In certain areas of the South Pacific, including Fiji, Samoa, the Philippines, and Tahiti, a strain of *Wuchereria* that lacks periodicity or shows diurnal periodicity is common. It is called **subperiodic.** The morphology of the adults is identical to that of *Wuchereria* producing periodic microfilariae; most investigators believe that only one species is involved, although some designate the subperiodic type as a separate species named *W. pacifica.* Daytime-feeding mosquitoes are the major vectors of the subperiodic strain.

The mosquito ingests microfilariae along with its blood meal. The microfilariae lose their sheath in the first two to six hours in the insect's stomach, after which they penetrate the gut of the host and reach the thoracic muscles. The first cuticular molt occurs about two days later. The J_2 is a short, sausage-shaped worm (**sausage stage**) (see Fig. 29.8 on p. 453) in which most of the organ systems are present. Within two weeks the second molt takes place. The juvenile is now an elongated, slender **filariform** J_3 (see Fig. 29.9 on p. 453), and development ceases. The filariform juveniles are 1.4 to 2.0 mm long and are infective to the definitive host. They migrate throughout the hemocoel, eventually reaching the labium, or proboscis sheath, from which they escape when the mosquito is feeding. They enter the skin through the wound made by the mosquito. After migrating through the peripheral lymphatics, the worms settle in the larger lymph vessels, where they mature.

• Epidemiology

Many mosquito vectors of *Wuchereria* have a preference for human blood and often breed near human habitation. At least 77 species and subspecies of mosquitoes in the genera *Anopheles, Aedes, Culex,* and *Mansonia* are known intermediate hosts for *Wuchereria.* In areas in which the periodic strain of *Wuchereria* is found, the mosquito vectors are primarily night feeders. The species of mosquito serving as vector in a particular area seems to depend more on coincidence (which species feeds when the juveniles within the definitive host are available) than on physiological determinants of host specificity. Nevertheless, the periodicity has practical epidemiological significance because it determines which mosquito species must be controlled and consequently what control measures must be applied.

Suitable breeding sites for mosquitoes abound in tropical areas. Some sites are difficult or impossible to control, such as tree holes and hollows at the bases of palm fronds, whereas others can be controlled with a degree of effort. Hollow coconuts, killed while still green by rats gnawing holes in them, fall, fill with rainwater, and become havens for developing mosquito larvae. These can be collected and burned. Even dugout canoes that are unused for a few days can partially fill with rainwater and become mosquito nurseries. Conditions for transmission of *Wuchereria* vary from locality to locality and country to country. The epidemiologist must consider each case independently within the framework of the biology of the vector and host, putting the economic and technical resources that are available to best advantage.

Humans seem to be the only animals naturally infected with *W. bancrofti;* there is no evidence for a reservoir.[4]

• Pathogenesis

Pathogenesis in lymphatic filariasis depends heavily on inflammatory and immune responses. Effects of infection with *W. bancrofti* display a wide spectrum from clinically silent infections, with no apparent inflammation or parasite damage, to mild-to-intense nongranulomatous chronic lymphatic inflammation, to a variety of granulomatous obstructive reactions.[24] Some investigators contend that these states represent a progression from initial infection, through inflammation, to obstructive disease.[5] Others maintain that progression from one clinical form to another is not inevitable and that there is a plasticity in the response.[38] After the infective bite of the mosquito, the worms require 6 to 12 months to mature and for the females to begin releasing microfilariae. Microfilariae can be released for 5 to 10 years in the absence of reinfection.[4]

Asymptomatic Phase.   Individuals with asymptomatic infections usually have high microfilaremias. It appears that in such persons the T_h1, inflammatory arm of the immune response is downregulated, and the T_h2 arm is stimulated.[29] The cytokine IL-4, which suppresses activation of T_h1 cells, is elevated, and IFN-γ is depressed (see p. 25). Eventually, after a period of perhaps several years, this tolerance (or hyporesponsiveness) often breaks down and inflammatory reactions rise.

Inflammatory (Acute) Phase.   Inflammatory responses are made to the adult worms, particularly the female; little or no disease is caused by the microfilariae. Intense lymphatic inflammation occurs, usually in the lower half of the body, with chills, fever, and toxemia. The area of the affected lymphatic is swollen and painful, and the overlying skin may be reddened. The attack usually subsides after a few days, but this and the other manifestations described further often recur at frequent intervals. Edema of the affected part may continue for some period after the inflammation has subsided.

Additional common symptoms in the acute stage of filariasis include **lymphadenitis** (inflammation of the lymph nodes), **lymphangitis** (inflammation of the lymph channels), **orchitis** (inflammation of the testes, usually with sudden enlargement and considerable pain), **hydrocele** (forcing of lymph into the tunica vaginalis of the testis or spermatic cord), and **epididymitis** (inflammation of the spermatic cord). Acute febrile episodes called **elephantoid fever** recur frequently. These are marked by sudden onset, rigors and sweating, and fever to 104°F (40°C) and endure from a few hours to several days. On the histological level, extensive proliferation of the lining cells occurs in the lymphatics, with much inflammatory cell infiltration, especially of polymorphonuclear leukocytes and eosinophils, around the lymphatics and adjacent veins. The most prominent cells in the infiltration become lymphocytes, plasma cells, and eosinophils, as the most acute phase subsides. Abscesses around dead worms may develop, with accompanying bacterial infection. Microfilariae may disappear from the peripheral blood during and after the acute phase, perhaps because the lymphatic vessel containing the female becomes blocked, or perhaps because they are now being destroyed by host inflammatory reactions.

Obstructive Phase.   The obstructive phase is marked by **lymph varices, lymph scrotum, hydrocele, chyluria,** and **elephantiasis.** Lymph varices are "varicose" lymph ducts, caused when lymph return is obstructed and the lymph "piles up," greatly dilating the affected duct. This causes chyluria, or lymph in the urine, a common symptom of lymphatic filariasis. The chyle gives the urine a milky appearance, and some blood is often present. A feature of the chronic obstructive phase is progressive infiltration of the affected areas with fibrous connective tissue, or "scar" formation, after inflammatory episodes. However, dead worms are sometimes calcified instead of absorbed, usually causing little further difficulty.

In a certain proportion of cases, thought to be associated with repeated attacks of acute lymphatic inflammation, the condition known as elephantiasis gradually develops. This is a chronic lymphedema with much fibrous infiltration and thickening of the skin. In men the organs most commonly afflicted with elephantiasis are scrotum, legs, and arms; in women the legs and arms are usually afflicted, with vulva and breasts being affected more rarely. Elephantoid organs are composed mainly of fibrous connective tissues, granulomatous tissue, and fat. The skin becomes thickened and cracked, and invasive bacteria and fungi further complicate the matter. Microfilariae usually are not present.

Elephantiasis is thus a result of complex immune responses of long duration. After worms die and are absorbed, the symptoms gradually disappear. Repeated superinfections over many years are usually necessary for elephantiasis to occur.

Immunotolerance and Microfilaremia.   Casual visitors into endemic areas may well become infected with the parasite and suffer from acute lymphatic inflammation, but they usually have no microfilaremia.[37] This condition may persist for many years, subsiding and recurring from time to time.[50]

One of the most perplexing problems of the disease has been determining why a person first infected as an adult so rarely shows a microfilaremia. In World War II, 10,431 U.S. naval personnel were infected with *W. bancrofti,* yet only 20 showed a microfilaremia.[3]

We now know that in endemic areas fetuses in pregnant women are exposed to substantial amounts of filarial antigens.[14,20] Exposure to antigens at this stage of development promotes tolerance by the immune system. Consequently, microfilariae may not be recognized as foreign. A child born to a microfilaremic mother is 2.4 to 2.9 times as likely to become microfilaremic as a child born to an amicrofilaremic mother.

• Diagnosis and Treatment

Demonstration of microfilariae in the blood involves a simple and fairly accurate diagnostic technique,[21] provided that thick blood smears are made during the period when the juveniles are in the peripheral blood. Concentration may be necessary (Knott technique).[28] The technician must be able

to distinguish this species from others that could be present. X-ray examinations can detect dead, calcified worms. The Quantitative Buffy Coat tube with acridine orange is a rapid and convenient method for detecting microfilariae.[17] Because many infected people are amicrofilaremic (for example, nonnative residents), immunodiagnosis is very helpful.[22]

The drug of choice for the past 40 years has been diethylcarbamazine (DEC, Hetrazan), which eliminates microfilariae from the blood and with careful administration usually kills the adults.[4] DEC suffers from serious disadvantages: It has general and local adverse reactions, and for best effect it must be administered over a course of 7 to 12 days.[9] Ivermectin is also microfilaricidal, and several research groups are investigating its use for lymphatic filariasis.[1,25,33,51]

Edematous limbs are sometimes successfully treated by applying pressure bandages, which force the lymph out of the swollen area. This may gradually reduce the size of the member to nearly normal. Any connective tissue proliferation that might have developed will not be affected, however. Surgical removal of elephantoid tissue is often possible.

Prevention primarily involves protection against the bite of mosquitoes in endemic areas. People temporarily visiting such places should use insect repellent, mosquito netting, and other preventive measures rigorously. Long-term protection requires mosquito control and mass chemotherapy of indigenous people to eliminate microfilariae from the circulating blood, where they are available to mosquitoes.

Brugia malayi

It was first noticed in 1927 that a microfilaria, different from that of *W. bancrofti,* occurred in the blood of natives of Celebes. It was not until 1940 that the adult form was found in India; a year later it was discovered in Indonesia. We now know that *Brugia malayi* parasitizes humans in China, Korea, Japan, southeast Asia, India, Sri Lanka, the East Indies, and the Philippines. Much of its distribution overlaps that of *Wuchereria,* but unlike *W. bancrofti, B. malayi* has not spread to Africa and the New World.[27]

The morphology of this parasite is very similar to that of *W. bancrofti,* although the male is only about half as large. The number of anal papillae of males differs slightly between the two species, and the left spicule of *B. malayi* is a little more complex than that of *W. bancrofti.* These are feeble differences on which to separate two genera, but because a large literature has accumulated on Malayan filariasis under the name of *Brugia,* we reluctantly follow common usage.

• Morphology

Males are 13.5 to 20.5 mm long and 70 to 80 μm wide. The tail is curved ventrally and bears three or four pairs of adanal and three or four pairs of postanal papillae. The spicules are unequal and dissimilar, and a small gubernaculum is present.

Females are 80 to 100 mm long by 240 to 300 μm wide. The fingerlike tail is covered with minute cuticular bosses. The vulva is near the level of the middle of the esophagus.

• Biology and Pathology

The life cycle of *B. malayi* is nearly identical to that of *W. bancrofti.* Mosquitoes of the genera *Mansonia, Aedes,* and *Culex* are intermediate hosts. Adults live in the lymphatics and cause the same disease symptoms as *W. bancrofti,* although elephantiasis, when it occurs, is more restricted to the distal extremities of the arms and legs.[4]

The microfilaria is somewhat similar to that of *Wuchereria* but can be differentiated from it by the presence of nuclei in the tail tip. There are both periodic and subperiodic strains. One strain is found in leaf-eating monkeys in some parts of Malaysia, where it is a zoonosis.[4]

Diagnosis and treatment are as for *Wuchereria.* Control is also primarily by mosquito eradication. *Mansonia* spp. are the major vectors in many areas. Their larvae pierce the stems of aquatic vegetation to tap air, so they do not need to reach the water surface regularly. For this reason, herbicides can be put to good use in mosquito control.

Another species of *Brugia, B. timori,* was first known from its distinctive microfilariae, and since then adults have been described.[39] It has been found only from the Lesser Sunda Islands of southeast Indonesia and can cause severe disease in affected populations. It shows nocturnal periodicity and is transmitted by *Anopheles barbirostris;* there is no known animal reservoir.[43]

Onchocerca volvulus

River blindness is a disease caused by this large filaroid worm in areas of Africa (where 30 million are infected), Arabia, Guatemala, Mexico, Venezuela, and Colombia. It probably has an even greater distribution than is currently known, since it is a cryptic disease that does not always manifest overt symptoms. When it does, the symptoms often are overlooked by health authorities, most of whom are busy with more pressing and immediate problems. An estimated 2000 infected persons lived as immigrants in London in 1975.[53]

Onchocerciasis, as the condition is also known, is not a fatal disease. However, it does cause disfigurement and blindness in many cases; in some small communities in Africa and Central America, most of the people middle aged and older are blind. Eradication of this disease from the earth would not result in the "parasitologist's dilemma," since it would not increase the birthrate or increase chances for infant survival. It would, instead, free hundreds of thousands of persons from a debilitating disease and thereby remove this economic burden from developing nations. It now appears that the use of ivermectin may eliminate the burden of blindness in endemic areas, although *O. volvulus* itself is unlikely to be eradicated in the foreseeable future.[7,11]

• Morphology

The morphology of *Onchocerca volvulus* is not very different from that of *W. bancrofti.* The worms characteristically are knotted together in pairs or groups in the subcutaneous tissues (Fig. 29.5). They are slender and blunt at both ends. Lips and a buccal capsule are absent, and two circles of four papillae each surround the mouth. The esophagus is not conspicuously divided.

Males are 19 to 42 cm long by 130 to 210 μm wide; females are 33.5 to 50.0 cm long by 270 to 400 μm wide, with the vulva just behind the posterior end of the esophagus. The tail of the male is curled ventrally and lacks alae; it bears four pairs of adanal and six or eight pairs of postanal papillae. The microfilariae are unsheathed.

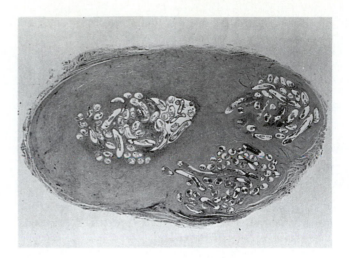

FIGURE 29.5

Cross section of a fibrous nodule (**onchocercoma**) removed from the chest of an African. It contained several worms bound together in a mass.

From D. H. Connor et al., "Onchocercal dermatitis, lymphadenitis, and elephantiasis in the Ubangi Territory," in *Human Pathol.* 1:553–579 Copyright © 1970. AFIP neg. no. 69–3625.

• Biology

Adult worms locate under the skin, where they become encapsulated by host reactions. If the site is over a bone, such as at a joint or over the skull, a prominent nodule appears (Fig. 29.6). The location of these nodules differs according to geographical area. In Africa most infections are below the waist, whereas in Central America they are usually above the waist. This correlates with the biting preferences of the insect vectors, since the microfilariae are concentrated in the areas where the insects prefer to bite. Perhaps the cause is simply that "bush country" Americans are more inclined to wear trousers than are their African counterparts.

The unsheathed microfilariae remain in the skin, where they can be ingested by the black fly intermediate hosts *Simulium* spp. (Fig. 29.7). These ubiquitous pests become infected when they take a blood meal. Their mouthparts are not adapted for deep piercing, so much of their food consists of tissue juices, which contain numerous microfilariae in infected persons. The J_1 migrates from the intestinal tract of the fly to its thoracic muscles. There it molts to the sausage stage (Fig. 29.8) and then molts again to the infective, filariform J_3 (Fig. 29.9). The J_3 moves to the labium of the fly and can infect a new host when the insect next feeds. Mature worms appear in the skin in less than a year.

Onchocerca volvulus was probably introduced to the Americas with African slaves. It became established in Central America and has since mutated sufficiently to cause different clinical symptoms in its definitive host and to differ in its infectivity to various vectors and laboratory animals. That the species has done this within about 400 years is an indication of the mutability of dioecious parasites with high reproductive capacity. Evidently the parasite is expanding its distribution northward through Mexico.[15] Humans appear to be the only natural definitive host for *O. volvulus.*

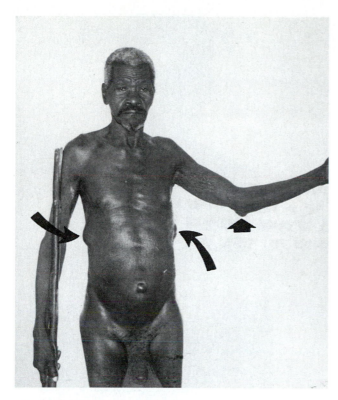

FIGURE 29.6

Several nodules (*arrows*) filled with *Onchocerca volvulus* are found in the skin of this man. Note also the elephantoid scrotum and the depigmentation and wrinkling of the skin of the upper arms, also symptoms of onchocerciasis.

From D. H. Connor et al., "Onchocercal dermatitis, lymphadenitis, and elephantiasis in the Ubangi Territory," in *Human Pathol.* 1:553–579 Copyright © 1970. AFIP neg. no. 68–10071–3.

FIGURE 29.7

A black fly, *Simulium damnosum,* biting the arm of a human. This insect is a major vector of *Onchocerca volvulus* in Africa.

From D. H. Connor et al., "Onchocercal dermatitis, lymphadenitis, and elephantiasis in the Ubangi Territory," in *Human Pathol.* 1:553–579 Copyright © 1970. AFIP neg. no. 68–2763–1.

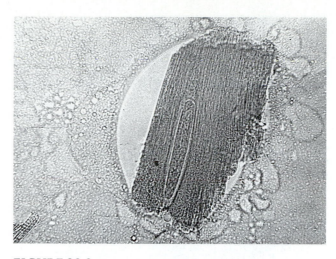

FIGURE 29.8

Piece of thoracic muscle from *Simulium damnosum.* Note the second, or "sausage-stage," juvenile of *Onchocerca volvulus.*

Courtesy of John Davies.

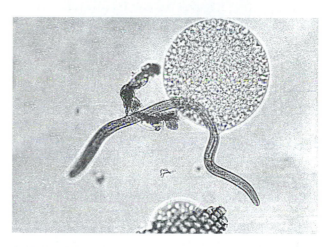

FIGURE 29.9

Third-stage, or filariform, juvenile of *Onchocerca volvulus,* dissected from the head of a *Simulium damnosum.*

Courtesy of John Davies.

• Epidemiology

Generally speaking, onchocerciasis is a model system for the landscape epidemiologist. *Simulium* spp. live their larval stages only in clear, fast-running streams. The adult flies survive only where there is high humidity and plenty of streamside vegetation. Some African indigenous people have long known that the disease is associated with rivers (and even with black flies, although it was not officially "discovered" until 1926), and they gave it the name *river blindness.* Anyone who intrudes into such a river area is viciously attacked by these insects. Wild-caught black flies are often infected with a variety of species of filaroids, most of which are still unidentified, but in areas endemic with *O. volvulus* the juveniles can often be recognized as that species. In hyperendemic areas more than 90% of the population can be microfilaremic.[46]

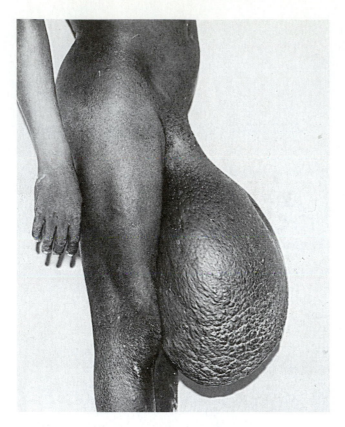

FIGURE. 29.10

Severe elephantoid scrotum on a native of Ubangi territory. It was removed surgically and a good cosmetic result was obtained. The scrotum weighed 20 kg and, when viewed microscopically, appeared as an edematous mass of interlacing collagen and smooth muscle fibers.

From D. H. Connor et al., "Onchocercal dermatitis, lymphadenitis, and elephantiasis in the Ubangi Territory," in *Human Pathol.* 1:553–579 Copyright © 1970. AFIP neg. no. 68–8582–9.

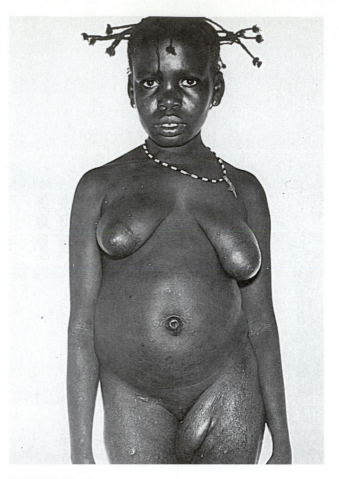

FIGURE 29.11

"Hanging groin," or adenolymphocele. The tissue was excised and contained a group of lymph nodes embedded in subcutaneous tissue. The nodes contained many microfilariae of *Onchocerca volvulus*.

From D. H. Connor et al., "Onchocercal dermatitis, lymphadenitis, and elephantiasis in the Ubangi Territory," in *Human Pathol.* 1:553–579 Copyright © 1970. AFIP neg. no. 68–10066–1.

Surprisingly, foci of onchocerciasis occur in the arid savanna of west Africa and the desert areas along the Nile near the Egypt-Sudan border. The epidemiology in these areas has not been thoroughly studied, but it is certain to depend on adaptations for survival of the black fly vectors.

• Pathogenesis

Two different elements contribute to the pathogenesis of onchocerciasis: the adult worms and the microfilariae, both as consequences of the host immune response. The adult is the least pathogenic, often causing no symptoms whatever and, at the worst, stimulating the growth of palpable subcutaneous nodules called **onchocercomas** (Fig. 29.6), especially over bony prominences. In the African strain these nodules are most frequent in the pelvic area, with a few along the spine, chest, and knees. The Venezuelan form is much like the African one, but in Central America the nodules are mostly above the waist, especially on the neck and head. These nodules are relatively benign, causing some disfigurement but no pain or ill health. The number of nodules may vary from one to well over a hundred. They consist mainly of collagen fibers surrounding one to several adult worms.[19] Rarely the nodule will degenerate to form an abscess, or the worm will become calcified.

True elephantiasis sometimes ensues (Fig. 29.10), and another condition, known as *hanging groin,* is common in some areas of Africa. A loss of elasticity of the skin causes a sagging of the groin into pendulous sacs, often containing lymph nodes. The testes and scrotum are not affected, and hydrocele does not accompany the condition. Females are similarly affected (Fig. 29.11). Leonine face is a rare complication in Central America and Africa. Onchocerciasis frequently causes hernias, especially femoral hernia, in Africa.

Live microfilariae elicit little inflammatory response,[26] but degenerating juveniles in the skin often result in a severe **dermatitis** caused by allergic responses. The first symptom is an intense itching, which may lead to secondary bacterial infection, often accompanied by abnormal pigmentation of the skin in small or extensive areas. This is followed by a thickening, discoloration, and cracking of the skin. The last stage of the skin lesion is characterized by loss of elasticity, which gives the patient a look of premature aging. **Depigmentation** (loss of pigment) is accentuated and may extend over large areas, especially of the legs (Fig. 29.12). Patients at this stage

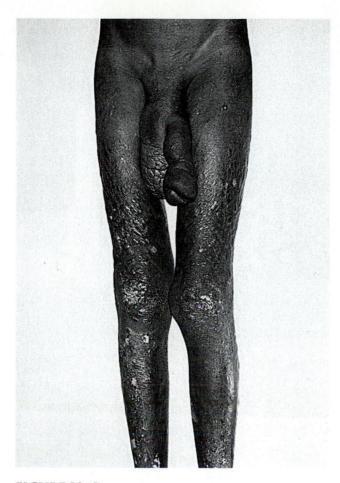

FIGURE 29.12

An 11-year-old boy with severe dermatitis characterized by depigmentation, wrinkling, and thickening of the skin. He also has elephantoid changes of the penis and scrotum and onchocercomas over the knees.

From D. H. Connor et al., "Onchocercal dermatitis, lymphadenitis, and elephantiasis in the Ubangi Territory," in *Human Pathol.* 1:553–579 Copyright © 1970. AFIP neg. no. 68–7912–1.

are often misdiagnosed as leprous. The distressing effects on the lifestyle of these persons can only be imagined.

Microfilariae in advanced cases often are located in the deeper part of the dermis and are not detected by skin-snip biopsy.

By far the most dreadful complications of onchocerciasis involve the eyes. The rate of impaired vision may reach 30% in some communities of Africa, where blindness affects in excess of 10% of the adult population. In these areas, and in similar areas of Guatemala, it is not unusual to see a child with good vision leading a string of blind adults to the local market. Ocular complications are less common in the rain forest areas of Africa but are frequent in the savanna. The reason for this difference remains one of the enigmas of parasitology. It is possible that vitamin A deficiency contributes to the condition in the savannas, where lesions due to deficiency in that vitamin are also more frequent.[47]

Lesions of the eye take many years to develop; most affected persons are over 40 years old. However, in Central

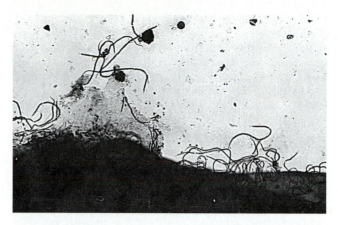

FIGURE 29.13

Skin snip from a patient with onchocerciasis. Note the emerging microfilariae.

Courtesy of Warren Buss.

America, with more worms concentrated on the head, young adults also show symptoms.

Live microfilariae invade many parts of the eye, but here again, they cause little reaction; their death leads to lesions.[12] Chronic inflammatory cells with eosinophils and neutrophils surround the worm, which is followed by fibroblast proliferation and chronic inflammatory infiltrates.[26] The major cause of blindness is **sclerosing keratitis,** a hardening inflammation of the cornea. In some cases there is a chronic, nongranulomatous chorioretinitis, consisting of infiltration with lymphocytes, plasma cells, and eosinophils, with degenerative changes in the retinal layers.

Microfilariae in the chambers of the eye are easily demonstrated in onchocerciasis. In fact often an ophthalmologist first diagnoses the disease during routine ocular inspection.

A bizarre manifestation of onchocerciasis in Uganda is **dwarfism.** At first thought to be pygmies, people afflicted with this condition are now known to derive from normal parents, and they are always infected with *O. volvulus.* The symptoms are pituitary deficiency, and there is little doubt that the pituitary has been damaged, either directly or indirectly, by microfilariae.

Adult worms may live as long as 16 years.

• Diagnosis and Treatment

The best method of diagnosis is the demonstration of microfilariae in bloodless skin snips. These are made by raising a small bit of skin with a needle and slicing it off with a razor or scissors. The bit of skin is then placed in saline on a slide and observed with a microscope for emerging microfilariae (Fig. 29.13). These must be differentiated from other species that might be present. Nodules may be aspirated, but no microfilariae will be found if only males or dead worms are present. Skin-snip biopsies can be taken anywhere, but if only a single snip is available, it should be taken from the buttock. If the snip is so deep as to draw blood, it might be contaminated with other species of filaroid. On the other hand, in old cases the microfilariae may be so deep as to elude the snip. A given case may show other, overt symptoms that obviate the need for demonstrating the microfilariae.

Treatment of onchocerciasis is of two types: surgical and chemotherapeutic. Excision of nodules, especially those around the head, may be effective in lowering both the rate of eye damage and the number of new infections within a population. It is a simple operation that can be performed under rather primitive conditions. Chemotherapy with ivermectin has been a major advance and a modern success story. Merck, Sharpe & Dohme developed and marketed the drug primarily for veterinary use. After its effectiveness in onchocerciasis became known, the company generously donated the drug for mass treatment in developing countries. Previously there had been no satisfactory drug available; the serious side effects of DEC and suramin precluded their use for mass treatment. Ivermectin is well-tolerated, and annual treatment significantly decreases incidence of infection in populations.[10,48,52] A single dose of ivermectin eliminates skin microfilariae and suppresses their release from adult females for a year or more.[44] Although usually considered only a microfilaricide, ivermectin in repeated doses slowly kills adult worms.[13]

Prevention of the disease by elimination of the vectors has been, over the years, very difficult. Application of DDT to swift-running streams destroys all simuliids,[18] but undesirable environmental effects of DDT are well-known. Other, more biodegradable insecticides are more expensive. The success of mass treatments with ivermectin may obviate costly fly-control programs. It is possible, however, that *O. volvulus* could develop resistance to ivermectin.[45]

Loa loa

Loa loa is the "eye worm" of Africa, which produces **loiasis** or **fugitive** or **Calabar swellings.** It is distributed in the rain forest areas of west Africa and equatorial Sudan. Although it was established for a short time in the West Indies, where it was first discovered during slavery, it no longer exists there.

The morphology of *L. loa* is typical of the family: a simple head with no lips and eight cephalic papillae; a long, slender body; and a blunt tail. The cuticle is covered with irregular, small bosses, except at the head and tail. Males are 20 to 34 mm long by 350 to 430 μm wide. The three pairs of preanal and five pairs of postanal papillae are often asymmetrical. The spicules are uneven and dissimilar, 123 and 88 μm long. Females are 20 to 70 mm long and about 425 μm wide. The vulva is about 2.5 mm from the anterior end, and the tail is about 265 to 300 μm long.

• Biology

Adults live in subcutaneous tissues (Fig. 29.14), including back, chest, axilla, groin, penis, scalp, and eyes in humans. Infections of deep tissues are also known, including fatal encephalitis.[35] The microfilariae (Fig. 29.4) appear in the peripheral blood in maximal numbers during daylight hours and concentrate in the lungs at night. The intermediate host is any of several species of deer fly, genus *Chrysops,* which feeds by slicing the skin and imbibing the blood as it wells into the wound. The worms develop to the filariform J_3's in the fat body of the fly, after which they migrate to the mouthparts. The prepatent period in humans is about a year, and adult

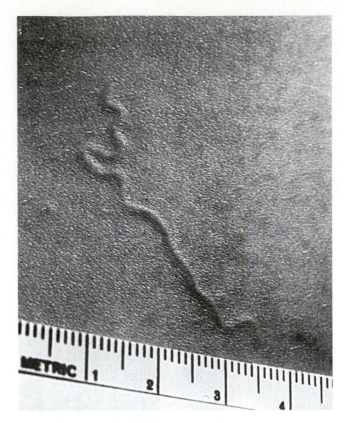

FIGURE 29.14

Adult female *Loa loa* visible under the skin of a patient.

From D. L. Price and H. C. Hopps, in R. A. Marcial-Rojas, editor, *Pathology of Protozoal and Helminthic Diseases, with Clinical Correlation,* © 1971 The Williams & Wilkins, Co. AFIP neg. no. 67-5366.

worms may live at least 15 years. There are probably no important reservoir hosts of the *L. loa* strain found in humans.[42]

• Pathogenesis

These worms have a tendency to wander through the subcutaneous connective tissues, provoking inflammatory responses as they go. When they remain in one spot for a short time, the host reaction results in localized Calabar swellings, especially in the wrists and ankles, which disappear when the worm moves on. Occasionally adult worms migrate through the conjunctiva and cornea (Fig. 29.15), with swelling of the orbit and psychosomatic results to the host. Intense pruritis (itching), arthralgia (joint pain), and fatigue are common, and serious complications can occur.[42]

As in *Wuchereria bancrofti,* many people indigenous to endemic areas can be asymptomatic but have a high microfilaremia, whereas newcomers may be amicrofilaremic but clinically and immunologically hyperresponsive.[37]

• Diagnosis and Treatment

Demonstration of typical microfilariae in the blood (Fig. 29.4) is ample proof of loiasis. The visual observation of a worm in the cornea or over the bridge of the nose is also

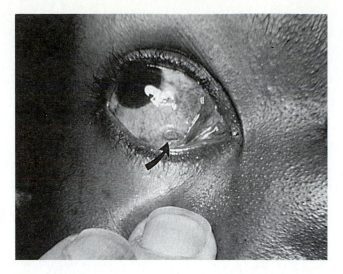

FIGURE 29.15

Adult female *Loa loa* coiled under the conjunctival epithelium (*arrow*) of the eye of an African from the Congo.

From D. L. Price and H. C Hopps, in R. A. Marcial-Rojas, editor, *Pathology of Protozoal and Helminthic Diseases, with Clinical Correlation,* © 1971 The Williams & Wilkins, Co. AFIP neg. no. 67-5368-1.

indicative of this species. Finally, transient swellings of the skin are suspect, although sparganosis or onchocerciasis may be confused with loiasis before the parasite is excised and examined. Surgical removal is simple and effective, providing the worm is properly located, but most of the worms are inapparent.

Conventional treatment with DEC has troublesome side effects and can lead to encephalitis, sometimes fatal.[6] Ivermectin seems to be effective,[30] but neither drug affects the adults.[42] Control of deer flies, which breed in swampy areas of the forest, is extremely difficult.

Other Filariids Found in Humans

Mansonella ozzardi is a filaroid parasite of the New World, with distribution that encompasses northern Argentina, the Amazon drainage, the northern coast of South America, Central America, and several islands of the West Indies. It has never been found in the Old World. Adults live in the body cavity, threaded among the mesenteries and peritoneum and in subcutaneous tissues. Its manifestations can mimic bancroftian filariasis, with polylymphadenitis, lymphedema, elephantiasis, and hepatomegaly.[23]

The species was redescribed by Orihel and Eberhard.[36] The intermediate hosts are species of *Culicoides* and *Simulium*.[49]

Mansonella perstans (formerly *Dipetalonema perstans*) exists in people in tropical Africa and South America. Several primates have been incriminated as reservoir hosts. Adult worms live in the coelom and produce unsheathed microfilariae (Fig. 29.16). Intermediate hosts are species of biting midges of the genus *Culicoides*. They appear to cause little pathological effect.

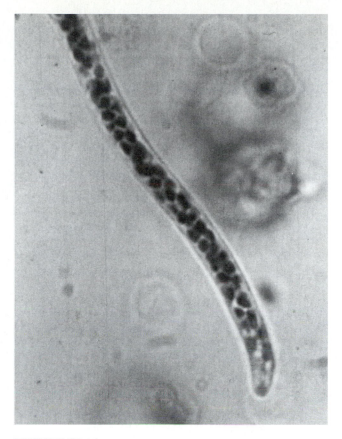

FIGURE 29.16

Tail end of microfilaria of *Mansonella perstans.* The infection was acquired in Nigeria.

From H. Zaiman, editor, *A Pictorial Presentation of Parasites.* Photo by M. G. Schultz.

Mansonella streptocerca (formerly *Dipetalonema streptocerca*) is a common parasite in the skin of humans in many of the rain forests of Africa.[32] It probably is a parasite of chimpanzees in nature.

Dirofilaria immitis

The heartworm *Dirofilaria immitis* (Fig. 29.17) is parasitic in the right side of the heart and pulmonary artery of dogs and other mammals throughout most of the world. It has been found in humans several times, including in the United States.

• Biology

Mosquitoes ingest microfilariae with their blood meal, and filariform J_3's develop in the malpighian tubules. The J_1's then migrate through the thorax and to the mouthparts of the mosquito, to drop onto the host's skin during the mosquito's next meal. After about five months the worms mature in the right heart and pulmonary arteries of the vertebrate host. Dogs and other canids, as well as cats, ferrets, and sea lions, are susceptible hosts.[2] Neither mature worms nor microfilariae have been detected in humans.[34]

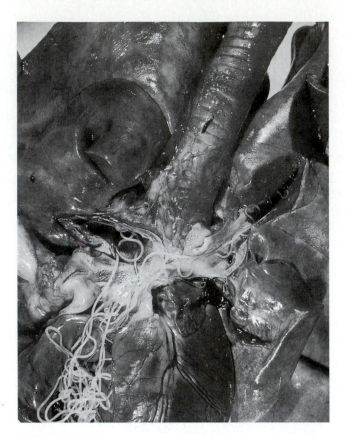

FIGURE 29.17

Dirofilaria immitis in right ventricle, extending up into right and left pulmonary arteries of an eight-year-old Irish setter.

Courtesy of Sharon Patton.

• Epidemiology

The worm is transmitted by many species of mosquitoes. The highest prevalence in the United States corresponds with highest mosquito density. Along the Atlantic and Gulf Coast states and northward along the Mississippi River throughout the midwestern states, prevalence among dogs is up to 45%.[2] Prevalence is much lower in the western United States but can range to 5% in some areas of California and Oregon.

No responsible pet owner should fail to provide chemoprophylaxis during mosquito season, which depends on locality. Transmission may occur whenever the daily temperature regularly exceeds 57°F (14°C), the threshold for development of juveniles in mosquitoes.[2]

• Pathogenesis

This worm constitutes a dangerous pathogen for dogs, and although prevalence and worm load are usually lower in cats, even a few worms can cause serious disease in cats.[31] Clinical signs include respiratory insufficiency, vomiting, chronic cough, and exercise intolerance.[41] Obviously, large worms extending through the openings of the tricuspid and semilunar valves will prevent efficient operation of those heart valves. The pulmonary arteries show thickening and inflammation of their inner walls. Death may occur from cardiopulmonary failure.

In humans the symptoms are vague and unpredictable. They may include chest pain, cough, coughing up blood, fever, and malaise.[34]

• Diagnosis and Treatment

Routine diagnosis is by finding microfilariae in a blood smear, but microfilariae of the nonpathogenic *Dipetalonema reconditum* must be distinguished those of *D. immitis*.[40] However, many infected animals do not have a detectable microfilaremia; therefore, immunodiagnosis (ELISA) should be employed.[2]

Dogs and other susceptible animals should be given prophylactic doses of ivermectin or milbemycin (a drug related to ivermectin) once a month or DEC once each day. These drugs do not affect adults, which must be killed with other drugs. Dying adult worms passing from the heart into the lungs clog arteries, lead to granuloma formation, and become life-threatening in severe cases. Animals require diligent care during treatment.

References

1. Addiss, D. G., M. L. Eberhard, P. L. Lammie, M. B. McNeeley, S. H. Lee, D. F. McNeeley, and H. C. Spencer. 1993. Comparative efficacy of clearing-dose and single high-dose ivermectin and diethylcarbamazine against *Wuchereria bancrofti* microfilaremia. *Am. J. Trop. Med. Hyg.* 48:178–85.

2. American Heartworm Society. 1992. *Recommended procedures for the diagnosis and management of heartworm* (Dirofilaria immitis) *infection.* Batavia, Ill.: American Heartworm Society (P.O. Box 667, 60510).

3. Beaver, P. C. 1970. Filariasis without microfilaremia. *Am. J. Trop. Med. Hyg.* 19:181–89.

4. Buck, A. A. 1991. Filariasis. In Strickland, G. T., ed. *Hunter's tropical medicine,* 7th ed. Philadelphia: W. B. Saunders Company, 713–27.

5. Bundy, D. A. P., B. T. Grenfell, and P. K. Rajagopalan. 1991. Immunoepidemiology of lymphatic filariasis: The relationship between infection and disease. *Parasitol. Today* 7(3):A71–A75.

6. Carme, B., J. Boulesteix, H. Boutes, and M. F. Puruehnce. 1991. Five cases of encephalitis during treatment of loiasis with diethylcarbamazine. *Am. J. Trop. Med. Hyg.* 44:684–90.

7. Centers for Disease Control. 1990. International task force for disease eradication. *Morb. Mortal. Weekly Rep.* 39:209–12.

8. Chernin, E. 1987. The disappearance of bancroftian filariasis from Charleston, South Carolina. *Am. J. Trop. Med. Hyg.* 37:111–14.

9. Coutinho, A. D., G. Dreyer, Z. Medeieros, E. Lopes, G. Machado, E. Galdino, J. A. Rizzo, L. D. Andrade, A. Rocha, I. Moura, J. Godoy, and E. A. Ottesen. 1994. Ivermectin treatment of bancroftian filariasis in Recife, Brazil. *Am. J. Trop. Med. Hyg.* 50:339–48.

10. Cupp, E. W. 1992. Treatment of onchocerciasis with ivermectin in Central America. *Parasitol. Today* 8:212–14.

11. Duke, B. O. L. 1990. Onchocerciasis (river blindness)—can it be eradicated? *Parasitol. Today* 6:82–84.

12. Duke, B. O. L., and H. R. Taylor. 1991. Onchocerciasis. In Strickland, G. T., ed. *Hunter's tropical medicine,* 7th ed. Philadelphia: W. B. Saunders Company, 729–44.

13. Duke, B. O. L., G. Zea-Flores, J. Castro, E. W. Cupp, and B. Muñoz. 1992. Effects of three-month doses of ivermectin on adult *Onchocerca volvulus. Am. J. Trop. Med. Hyg.* 46:189–94.

14. Eberhard, M. L., W. L. Hitch, D. F. McNeeley, and P. J. Lammie. 1993. Transplacental transmission of *Wuchereria bancrofti* in Haitian women. *J. Parasitol.* 79:62–66.

15. Editorial Board. 1986. Is New World onchocerciasis spreading? *Parasitol. Today* 2:131.

16. Foster, W. D. 1965. *A history of parasitology.* Edinburgh: E. & S. Livingston.

17. Freedman, D. O., and R. S. Berry. 1992. Rapid diagnosis of bancroftian filariasis by acridine orange staining of centrifuged parasites. *Am. J. Trop. Med. Hyg.* 47:787–93.

18. Garnham, P. C. G., and J. P. McMahon. 1947. The eradication of *Simulium neavei* Raubaud, from an onchocerciasis area in Kenya Colony. *Bull. Ent. Res.* 37:619–28.

19. George, G. H., J. R. Palmieri, and D. H. Connor. 1985. The onchocercal nodule: interrelationship of adult worms and blood vessels. *Am. J. Trop. Med. Hyg.* 34:1144–48.

20. Hightower, A. W., P. J. Lammie, and M. L. Eberhard. 1993. Maternal filarial infection—a persistent risk factor for microfilaremia in offspring? *Parasitol. Today* 9:418–21.

21. Hoegaerden, M. van, and B. Ivanoff. 1986. A rapid, simple method for isolation of viable microfilariae. *Am. J. Trop. Med. Hyg.* 35:148–51.

22. Hui-Jun, Z, T. Zheng-Hou, C. Weng-Feng, and W. F. Piessens. 1990. Comparison of dot-ELISA with sandwich-ELISA for the detection of circulating antigens in patients with bancroftian filariasis. *Am. J. Trop. Med. Hyg.* 42:546–49.

23. Jörg, M. E. 1983. Filariasis por *Mansonella ozzardi* (Manson 1897) Faust 1929 en la Argentina, con descripción de un caso grave. *Prensa Medica Argentina* 70:181–92.

24. Jungmann, P., J. Figueredo-Silva, and G. Dreyer. 1991. Bancroftian lymphadenopathy: A histopathologic study of fifty-eight cases from northeastern Brazil. *Am. J. Trop. Med. Hyg.* 45:325–31.

25. Kazura, J., J. Greenberg, R. Perry, G. Weil, K. Day, and M. Alpers. 1993. Comparison of single-dose diethylcarbamazine and ivermectin for treatment of bancroftian filariasis in Papua, New Guinea. *Am. J. Trop. Med. Hyg.* 49:804–11.

26. Kazura, J. W., T. B. Nutman, and B. M. Greene. 1993. Filariasis. In Warren, K. S., ed. *Immunology and molecular biology of parasitic infections,* 3d ed. Boston: Blackwell Scientific Publications, 473–95.

27. Laurence, B. R. 1989. The global dispersal of bancroftian filariasis. *Parasitol. Today* 5:260–64.

28. Levine, N. D. 1980. *Nematode parasites of domestic animals and of man,* 2d ed. Minneapolis: Burgess Publishing Co.

29. Maizels, R. M., and R. A. Lawrence. 1991. Immunological tolerance: The key feature in human filariasis? *Parasitol. Today* 7:272–76.

30. Martin-Prevel, Y., J. -Y. Cosnefroy, P. Tshipamba, P. Ngari, J. A. Chodakewitz, and M. Pinder. 1993. Tolerance and efficacy of single high-dose ivermectin for the treatment of loiasis. *Am. J. Trop. Med. Hyg.* 48:186–92.

31. McCracken, M. D., and S. Patton. 1993. Pulmonary arterial changes in feline dirofilariasis. *Vet. Pathol.* 30:64–69.

32. Meyers, W. M., and R. C. Neafie. 1991. Streptocerciasis. In Strickland, G. T., ed. *Hunter's tropical medicine,* 7th ed. Philadelphia: W. B. Saunders Company, 746–48.

33. Moulia-Pelat, J. -P., L. N. Nguyen, P. Glaziou, S. Chanteau, E. A. Ottesen, R. Cardines, P. -M. V. Martin, and J. -L. L. Cartel. 1994. Ivermectin plus diethylcarbamizine: An additive effect on early microfilarial clearance. *Am. J. Trop. Med. Hyg.* 50:206–9.

34. Neafie, R. C., and W. M. Meyers. 1991. Dirofilariasis. In Strickland, G. T., ed. *Hunter's tropical medicine,* 7th ed. Philadelphia: W. B. Saunders Company, 748–49.

35. Negesse, Y., L. O. Lanoie, R. C. Neafie, and D. H. Connor. 1985. Loiasis: "Calabar swellings" and involvement of deep organs. *Am. J. Trop. Med. Hyg.* 34:537–46.

36. Orihel, T. C., and M. L. Eberhard. 1982. *Mansonella ozzardi:* A redescription with comments on its taxonomic relationships. *Am. J. Trop. Med. Hyg.* 31:1142–47.

37. Ottesen, E. A. 1989. Filariasis now. *Am. J. Trop. Med. Hyg.* 41(3, suppl.):9–17.

38. Ottesen, E. A. 1992. Infection and disease in lymphatic filariasis: An immunological perspective. *Parasitology* 104:S71–S79.

39. Partono, F., Purnomo, D. T. Dennis, S. Atmosoedjono, S. Oemijati, and J. H. Cross. 1977. *Brugia timori* sp. n. (Nematoda: Filaroidea) from Flores Island, Indonesia. *J. Parasitol.* 63:540–46.

40. Patton, S., and C. T. Faulkner. 1992. Prevalence of *Dirofilaria immitis* and *Dipetalonema reconditum* infection in dogs: 805 cases (1980–1989). *J. Am. Vet. Med. Assoc.* 200:1533–34.

41. Patton, S., and M. D. McCracken. 1991. Prevalence of *Dirofilaria immitis* in cats and dogs in eastern Tennessee. *J. Vet. Diagn. Invest.* 3:79–80.

42. Pinder, M. 1988. *Loa Loa*—a neglected filaria. *Parasitol. Today* 4:279–84.

43. Purnomo, D. T. Dennis, and F. Partono. 1977. The microfilaria of *Brugia timori* Partono et al. 1977 (= Timor microfilaria, David and Edeson, 1964): Morphologic description with comparison to *Brugia malayi* of Indonesia. *J. Parasitol.* 63:1001–6.

44. Schulz-Key, H., P. T. Soboslay, and W. H. Hoffmann. 1992. Ivermectin-facilitated immunity. *Parasitol. Today* 8:152–53.

45. Shoop, W. L. 1993. Ivermectin resistance. *Parasitol. Today* 9:154–59.

46. Somo, R. M., P. A. Enyong, G. Fobi, J. S. Dinga, C. Lafleur, P. Agnamey, A. Ngosso, and E. M. Ngolle. 1993. A study of onchocerciasis with severe skin and eye lesions in a hyperendemic zone in the forest of southwestern Cameroon: Clinical, parasitologic, and entomologic findings. *Am. J. Trop. Med. Hyg.* 48:4–19.

47. Storey, D. M. 1993. Filariasis: Nutritional interactions in human and animal hosts. *Parasitology* 107:S147–S158.

48. Taylor, H. R., M. Pacqué, B. Muñoz, and B. M. Greene. 1990. Impact of mass treatment of onchocerciasis with ivermectin on the transmission of infection. *Science* 250:116–18.

49. Tidwell, M. A., and M. A. Tidwell. 1982. Development of *Mansonella ozzardi* in *Simulium amazonicum, S. argentiscutum,* and *Culicoides insinuatus* from Amazonas, Colombia. *Am. J. Trop. Med. Hyg.* 31:1137–41.

50. Trent, S. 1963. Reevaluation of World War II veterans with filariasis acquired in the South Pacific. *Am. J. Trop. Med. Hyg.* 12:877–87.

51. Vande Waa, E. A. 1991. Chemotherapy of filariases. *Parasitol. Today* 7:194–99.

52. Whitworth, J. 1992. Treatment of onchocerciasis with ivermectin in Sierra Leone. *Parasitol. Today* 8:138–40.

53. Woodhouse, D. F. 1975. Tropical eye diseases in Britain. *Practitioner* 214:646–53.

Additional References

Chabaud, A., and R. C. Anderson. 1959. Nouvel essai de classification des filaries (Superfamille des Filarioidea) II, 1959. *Ann. Parasitol.* 34:64–87. An accurate, easy-to-use key to the genera of Filaroidea.

Chernin, E. 1983. Sir Patrick Manson's studies on the transmission and biology of filariasis. *Rev. Infect. Dis.* 5:148–66.

Chernin, E. 1983. Sir Patrick Manson: An annotated bibliography and a note on a collected set of his writings. *Rev. Infect. Dis.* 5:353–86.

Connor, D. H., et al. 1970. Onchocerciasis, onchocercal dermatitis, lymphadenitis, and elephantiasis in the Ubangi Territory. *Hum. Pathol.* 1:553–79.

Duke, B. O. L. 1971. The ecology of onchocerciasis in man and animals. In Fallis, A. M., ed. *Ecology and physiology of parasites.* Toronto: University of Toronto Press, 213–22.

Khanna, N. N., and G. K. Joshi. 1971. Elephantiasis of female genitalia. A case report. *Plast. Reconstr. Surg.* 48:374–81.

Meyers, W. M., et al. 1976. Diseases caused by filarial nematodes. In Binford, C. H., and D. H. Connor, eds. *Pathology of tropical and extraordinary diseases,* vol. 2, sect. 8. Washington, D.C.: Armed Forces Institute of Pathology.

Nelson, G. S. 1970. Onchocerciasis. In Dawes, B., ed. *Advances in parasitology* 8. New York: Academic Press, Inc., 173–224.

Sasa, M. 1976. *Human filariasis. A global survey of epidemiology and control.* Baltimore: University Park Press.

Chapter 30

NEMATODES: CAMALLANINA, THE GUINEA WORMS AND OTHERS

And they journeyed from Mount Hor by the way of the Red Sea, to compass the land of Edom. . . .

And the Lord sent fiery serpents among the people, and they bit the people; and much people of Israel died. . . .

And the Lord said unto Moses, "Make thee a fiery serpent and set it upon a pole; and it shall come to pass that everyone that is bitten, when he looketh upon it, shall live."

Numbers 21:4–8

The suborder Camallanina of the order Spirurida differs in important ways from the suborder Spirurina. It has conspicuous phasmids with broad cavities and prominent pores. The esophageal glands are usually uninucleate, and the intermediate hosts are copepods. They are divided into the superfamily Camallanoidea, with a well-developed buccal cavity, and the Dracunculoidea, with a weakly developed buccal cavity. Two families, Camallanidae and Philometridae, are commonly encountered in fishes, and a third, Dracunculidae, has a species of great medical importance to humans. These three families will serve to illustrate the order.

FAMILY CAMALLANIDAE

Included in the family Camallanidae are several structurally similar genera that inhabit the intestines of fishes, amphibians, and reptiles. Their most conspicuous character is the head, in which the buccal capsule has been replaced with a pair of large, bilateral sclerotized valves (Fig. 30.1). The complex ornamentation of these valves (Fig. 30.2) is a useful taxonomic character.

The genus *Camallanus* is common in freshwater fishes and turtles in the United States. *Camallanus oxycephalus* is often seen as a bright red worm extending from the anus of a crappie (*Pomoxis*) or other warm-water panfish. The life cycles of all species that have been investigated involve a cyclopoid copepod crustacean as intermediate host. Development proceeds to maturity in the intestine of the vertebrate with no tissue migration.

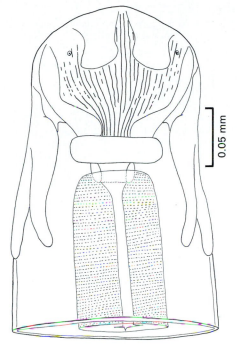

FIGURE 30.1

Head of *Camallanus marinus,* lateral view. In this genus, the buccal cavity is replaced by large, sclerotized valves with various markings.

From G. D. Schmidt and R. E. Kuntz, "Nematode parasites of Oceanica. V. Four new species from fishes of Palawan, P. I., with a proposal for *Oceanicucullanus* gen. nov.," in *Parasitology* 59:389–396. Copyright © 1969. Reprinted with permission of the publisher.

FAMILY PHILOMETRIDAE

Two common genera in the family Philometridae are *Philometra,* with a smooth cuticle, and *Philometroides,* which has a cuticle covered with bosses. Each is a tissue parasite of fishes. The mouth is small, there is no sclerotized buccal capsule, and the esophagus is short. Males of many species are unknown. Gravid females live under the skin, in the swim bladder, or in the coelom of fishes, where they release first-stage juveniles. After reaching the external environment, the juveniles develop further if they are eaten by a cyclopoid copepod. If the worm is to survive, a definitive host must eat the microcrustacean containing J_3's. Development to the adult is not well-known. Males and females mate

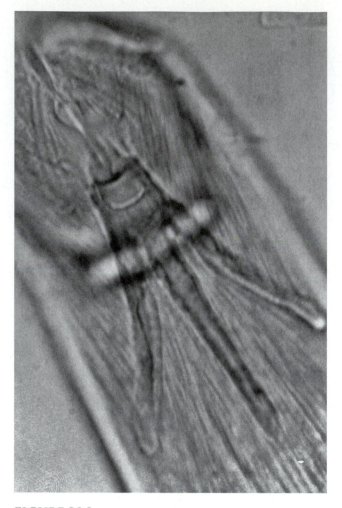

FIGURE 30.2

Dorsal view of the head of *Camallanus marinus,* showing the large, sclerotized trident characteristic of this genus.

Photograph by Gerald D. Schmidt.

in the deep tissues of the body, and the males die soon after. The females then migrate to their definitive site, where they release young. *Philometra oncorhynchi,* a parasite of salmon in the western United States and Canada, apparently passes out with the fish's eggs when it spawns, bursts in the fresh water, and thus releases its juveniles.[12]

Philometroides, under the skin of the head and fins of suckers (Catostomidae), are familiar sights to those who work with these fish in the United States (Fig. 30.3).

FAMILY DRACUNCULIDAE

Members of the family Dracunculidae are tissue-dwellers of reptiles, birds, and mammals. Morphological characteristics of the several genera and species are remarkably similar, with, for example, small differences between those in reptile hosts and those in mammals.

Several species of *Dracunculus* are known from snakes, and one is common in snapping turtles in the United States. The genus *Micropleura* is found in crocodilians and turtles in South America and India, whereas *Avioserpens* has species in aquatic birds.

FIGURE 30.3

Philometroides sp. in the skin of a fin of a white sucker, *Catastomus commersoni.*

Courtesy of John S. Mackiewicz.

The genus *Dracunculus* is also known from mammals. In the Americas a species known as *D. insignis* is common in muskrats, opossums, and raccoons and other carnivores, especially those occupying semiaquatic environments. *Dracunculus medinensis* is prevalent in circumscribed areas of Africa, India, and the Middle East. It has been reported from humans in the United States several times, but these cases may have been caused by *D. insignis*. In fact, *D. insignis* may well be *D. medinensis,* a strain perhaps not infective to humans. This might explain the scarcity of reports of it in humans in this country. It is infective to rhesus monkeys.[1]

However, *D. medinensis* is not scarce in humans in all countries of the world, so we will examine it in greater detail.

Dracunculus medinensis

Dracunculus medinensis has been known since antiquity, particularly in the Middle East and Africa, where it causes great suffering even today. Hopkins estimated its worldwide prevalence at 3 million cases in 1992.[7] Because of its large size and the conspicuous effects of infection, it is not surprising that the parasite was mentioned by classical authors. The Greek Agatharchidas of Cnidus, who was tutor to one of the sons of Ptolemy VII in the second century B.C., gave a lucid description of the disease: ". . . the people taken ill on the Red Sea suffered many strange and unheard of attacks, amongst other worms, little snakes, which came out upon them, gnawed away their legs and arms, and when touched retracted, coiled themselves in the muscles, and there gave rise to the most unsupportable pains."[4] The Greek and Roman writers Paulus Aegineta, Soranus, Aetius, Actuarus, Pliny, and Galen all described the disease, although most of them probably never saw an actual case. The Spanish and Arabian scholars Avicenna (Abu Ali al-Husein ibn Sina), Avenzoar, Rhazes, and Albucasis also discussed this parasite, probably from firsthand observations. In 1674 Velschius described winding the worm out on a stick as a cure. European parasitologists remained ignorant of this worm until about the beginning of the nineteenth century, when British army medical officers began serving in India. Information about *D. medinensis* slowly accumulated, but it remained for a young Russian traveler and scientist, Aleksej Fedchenko, to give the first detailed account of the

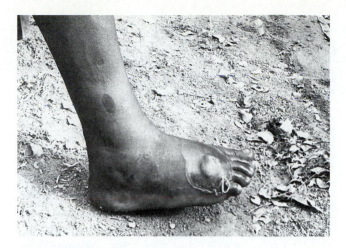

FIGURE 30.4

Blister, caused by a female *Dracunculus medinensis,* in the process of bursting. There has been an unusually severe tissue reaction resulting in a very large blister. A loop of the worm can be seen protruding through the skin.

From R. Muller, in B. Dawes, editor, *Advances in Parasitology, Vol. 9,* © 1971, Academic Press, Inc., (London), Ltd.

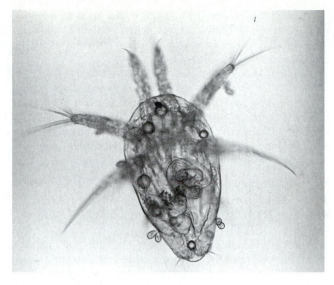

FIGURE 30.5

Living nauplius of *Cyclops vernalis* with a juvenile of *Dracunculus medinensis* in its hemocoel.

Courtesy of Ralph Muller.

morphology and life cycle of the worm between 1869 and 1870.[5] His discovery that humans become infected by swallowing infected *Cyclops* pointed the way to a means of prevention of dracunculiasis.

• Morphology

Dracunculus medinensis is one of the largest nematodes known. Adult females have been recorded up to 800 mm long, although the few males known do not exceed 40 mm. The mouth is small and triangular and is surrounded by a quadrangular, sclerotized plate. Lips are absent. Cephalic papillae are arranged in an outer circle of four double papillae at about the same level as the amphids and an inner circle of two double papillae, which are peculiar in that they are dorsal and ventral. The esophagus has a large glandular portion that protrudes and lies alongside the thin muscular portion.

In the female the vulva is about equatorial in young worms; it is atrophied and nonfunctional in adults. The gravid uterus has an anterior and a posterior branch, each of which is filled with hundreds of thousands of embryos. The intestine becomes squashed and nonfunctional as a result of the pressure of the uterus.

A major difficulty in the taxonomy of dracunculids is the sparsity of discovered males. The few specimens known range from 12 to 40 mm long; the spicules are unequal and 490 to 730 μm long. The gubernaculum ranges from 115 to 130 μm long. Genital papillae vary considerably in published descriptions. In fact, in monkeys, at least, males taken from a single animal have varying numbers of papillae. It is possible that more than one species is responsible for dracunculiasis, or there may be a complex of subspecies. Because of the technical difficulties of obtaining many specimens, it remains for an experimental approach to solve the taxonomy of the species.

• Biology

Dracunculus medinensis is **ovoviviparous.** When the parasite is gravid, the thousands of embryos in the uteri cause a high in-

ternal pressure. At this stage the female has migrated to the skin of the host. Usually the legs and feet are affected, but nearly any portion of the body is susceptible. Internal pressure and progressive senility cause the body wall and uterus of the parasite to burst, forcing a loop of the uterus through, freeing many juveniles. The juveniles cause a violent allergic reaction that causes a blister in the skin of the host (Fig. 30.4). This eventually ruptures, forming an exit for the young worms, which trickle out onto the surface of the skin. Sometimes, instead of the body wall rupturing, the uterus forces itself out of the mouth of the worm. Muscular contractions of the body wall force juveniles out in periodic spurts, with more than half a million ejected at a single time. These contractions are instigated by cool water, which causes the worm and its uterus to protrude through the wound. As portions of the uterus empty, they disintegrate, and adjacent portions move into the ulcer. Eventually all of the worm will be "used up," and the wound will heal.

After leaving its mother and host, the J_1 must enter directly into water to survive. It can live for four to seven days but is able to infect an intermediate host for only three days. To develop further, it must be eaten by a cyclopoid crustacean. Once in the intestine of their new host, the juveniles penetrate into the hemocoel, especially dorsally to the gut, where they develop to the infective J_3 in 12 to 14 days at 25°C (Fig. 30.5).

The definitive host is infected by swallowing infected copepods with drinking water. The released juveniles penetrate the duodenum, cross the abdominal mesenteries, pierce the abdominal muscles, and enter the subcutaneous connective tissues, where they migrate to the axillary and inguinal regions. The third molt occurs about 20 days after infection, and the final one at about 43 days. Females are fertilized by the third month after infection. Males die between the third and seventh months, become encysted, and degenerate. Gravid females migrate to the skin of the extremities between the eighth and tenth months, by which time the embryos are fully formed. Between 10 and 14 months after initial infection, the female causes a blister in the skin. The blister causes intense burning pain, which is alleviated by immersion in water.

Little is known about the physiology of this parasite, but the gut is often filled with a dark-brown material, suggesting that the worms feed on blood. Glycogen is stored in several tissues of the mature female. Glucose utilization and the rate of formation of lactic acid are not affected by the presence or absence of oxygen.[2] The blister formation in the definitive host is an immunological response to parasite antigens.

• Epidemiology

To become infected, a person must swallow a copepod that had been exposed to juveniles previously released from the skin of a definitive host. Thus, three conditions must be met before the parasite's life cycle can be completed: The skin of an infected individual must come in contact with water, the water must contain the appropriate species of microcrustacean, and the water must be used for drinking. There is circumstantial evidence that infection can be acquired by eating a fish paratenic host.[9]

It is curious that a parasite life cycle that is so dependent on water is most successfully completed under conditions of drought. In some areas of Africa, for instance, people depend on rivers for their water. During periods of normal river flow, few or no new cases of dracunculiasis occur. During the dry season, however, rivers are reduced to mere trickles with occasional deep pools, which are sometimes enlarged and deepened by those who depend on them as a water source (Fig. 30.6). Planktonic organisms flourish in this warm, semistagnant water, and a cyclopean population explosion occurs. At the same time any bathing, washing, and water drawing bring infected persons in contact with water, into which juveniles are shed. When such water is drunk, many infected copepods may be downed at a quaff.

In areas of India the step well (Fig. 30.7) is a time-honored method of exposing ground water. These wells, often centuries old, have steps leading into the water, on which water bearers enter the well to fill their jars and, incidentally, release juveniles into the water at the same time.

In many desert areas the populace depends on deep wells, which are crustacean free, during the dry season. Most villages also have one or more ponds that fill during the rainy season and become a source of infection with *Dracunculus*. Most villagers prefer the pond water because they have to pay for well water, and, moreover, the well water is usually saline.

Given these examples, it is no wonder that a parasite with an aquatic life cycle should thrive in a desert environment, since all animals, humans and beasts alike, depend on isolated waterholes for their existence. So does *D. medinensis*. On the other hand, the parasite's dependence on isolated waterholes exposes a weakness in its life cycle. Guinea worm is the only helminthic disease transmitted solely through drinking water.

• Pathogenesis

Dracunculiasis may result in three major disease conditions: emergent adult worms, secondary bacterial infection, and nonemergent worms.

FIGURE 30.6

Pond in the Mabauu area of Sudan, in the Sahel savannah zone. Conditions such as these favor the transmission of the guinea worm.

From R. Muller, in B. Dawes, editor, *Advances in Parasitology, Vol. 9,* © 1971 Academic Press, Inc., (London), Ltd. Photo by J. Bloss.

At the onset of migration to the skin, the female worm elicits an allergic reaction caused by the release of metabolic wastes into the host's system. The reaction may produce a rash, nausea, diarrhea, dizziness, and localized edema. The worms remain just under the skin for about a month before a reddish papule develops. This rapidly becomes a blister. The feet and legs are most often affected, although the blister may appear nearly anywhere on the surface of the body. On rupture of the blister, the allergic reactions usually subside. The site of the blister becomes abscessed, but this heals rapidly if serious secondary complications do not occur. A tiny hole remains, through which the worm protrudes. When the worm is removed or is expelled, healing is completed. Infection, however, does not confer immunity and a person may be reinfected many times.

Serious complications can result from the introduction of bacteria under the skin by the retreating worm. In parts of Africa this is the third most common mode of entry of tetanus spores.[10] Some other complications are abscesses, synovitis, arthritis, and bubo.

FIGURE 30.7

Step well at Kantarvos, near Kherwara, India, infected with *Dracunculus medinensis*.

From R. Muller, in B. Dawes, editor, *Advances in Parasitology, Vol. 9,* © 1971 Academic Press, Inc., (London), Ltd. Photo by A. Banks.

FIGURE 30.8

Ancient woodcut showing removal of guinea worm by winding it on a stick.

Velschius, 1674.

Worms that fail to reach the skin often cause complications in deeper tissues of the body, although many die and are absorbed or calcified, with no apparent effect on the host. Chronic arthritis, with a calcified worm in or alongside the joint, is common. More serious symptoms, such as paraplegia, result from a worm in the central nervous system. Adult worms have also been found in the heart and urogenital system.

Commonly, when worms do not emerge, they eventually begin to degenerate and release antigens. These cause aseptic abscesses, which also can lead to arthritis. The abscesses can be large, with up to half a liter of fluid containing leukocytes and frequently numerous embryos. Usually, however, the worms become calcified.

• Diagnosis and Treatment

The appearance of an itchy, red papule that rapidly transforms into a blister is the first strong symptom of dracunculiasis. On a few occasions, the patient can feel or see the worm in the skin before papule formation. After the blister ruptures, juveniles can be obtained by placing cold water on the wound; when mounted on a slide, they can be seen actively moving about under a low-power microscope.

When a part of the worm emerges, diagnosis is fairly evident, although the drying, disintegrating worm does not show the typical morphology of a nematode. An occasional sparganum may be diagnosed as dracunculiasis.

Pulling out guinea worms by winding them on a stick is a treatment used successfully since antiquity (Fig. 30.8). The biblical excerpt at the beginning of this chapter is a pretty fair account of dracunculiasis and its treatment. The burning pain of the blister could well be interpreted as the bite of a fiery serpent, and the serpent on a pole could easily represent the worm on a stick. Moses and his people were, at the time, near the Gulf of Akaba, where *Dracunculus* is still endemic. Also, the Israelites had for some time been in a drought area, existing on water where they could find it. This is consistent with the epidemiology of dracunculiasis.

The staff with serpent carried by Aesculapius, the Roman god of medicine, adopted today as the official symbol of medicine (and the double-serpent caduceus of the military), may also depict the removal of *Dracunculus* (Fig. 30.9). This form of cure is still widely used (Fig. 30.10). If cold water is applied to the worm, she will expel enough juveniles to allow about 5 cm of her body to be pulled out. The procedure is repeated once a day, complete removal requiring about three weeks. In some areas of India the worms are said to be sucked out by indigenous doctors using crude aspirators!

FIGURE 30.9

Seal of the American Medical Association and the double-serpent caduceus of the military medical profession. Might the serpent on a staff originally have depicted the removal of guinea worm?

Courtesy of the AMA.

An alternate method is removal of the complete worm by surgery. This is often successful when the entire worm is near the skin and also in the case of deep abscesses containing worms that failed to reach the skin. However, if the worm is threaded through a tendon or deep fascia or is broken into several pieces, it may be impossible to remove completely.

Several drugs have been used for treatment of dracunculiasis, but evidence for their effectiveness is dubious.[11]

Eradication Efforts.   This is perhaps the only parasite covered in this book that is potentially eradicable at the present time, and the World Health Assembly declared in 1991 its goal of eradicating dracunculiasis by 1995.[3] Although it is unlikely that the 1995 deadline will be met, major progress has been achieved. Annual incidence has been reduced from an estimated 10 million at the beginning of the campaign to an incidence of 2 million in 1993. The disease is nearly eradicated in Pakistan and India, and in Africa, Cameroon and Senegal are close to the goal.[6,8] Global success is eminently achievable, but serious obstacles are civil wars, apathy, and inadequate funding.

The most important strategies for the campaign are as follows:[11]

1. *Supply of safe drinking water.* Tubewells and handpumps are being installed wherever possible. Such sources preclude infected persons from entering the water. Field workers, however, persist in drinking from small ponds.

2. *Health education.* Recruiting and training workers from local populations is vital to this intervention. Once trained, the local health workers attempt to educate their fellow villagers to follow two procedures:(a) Boil or filter drinking water. Boiling is often not feasible because fuel is scarce. However, filtration through monofilament nylon nets is easy and removes copepods. The manufacturer, DuPont de Nemours, donated one million square meters of the mesh. (b) Do not enter water when a blister or worm is evident.

3. *Treat water sources.* Temephos (Abate; American Cyanamid) is a chemical that has low toxicity to mammals and fish, and it kills copepods for four to five weeks at a concentration of one part per million. The manufacturer has donated an amount of temephos estimated for total eradication requirements in Africa.

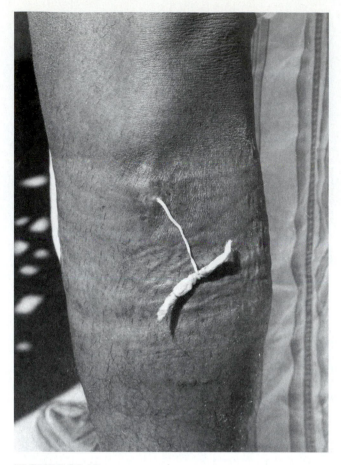

FIGURE 30.10

Uncomplicated case of dracunculiasis. The worm is being pulled out through a small hole left after the ulcer is mostly healed.

From R. Muller, In B. Dawes, editor, *Advances in Parasitology, Vol. 9,* © 1971 Academic Press, Inc. (London), Ltd.

References

1. Beverly-Burton, M., and V. F. J. Chrichton. 1973. Identification of guinea-worm species. *Trans. R. Soc. Trop. Med. Hyg.* 67:152.

2. Bueding, E., and J. Oliver-Gonzalez. 1950. Aerobic and anaerobic production of lactic acid by the filarial worm *Dracunculus insignis. Br. J. Pharmacol. Chemother.* 5:62–64.

3. Centers for Disease Control. 1993. Recommendations of the International Task Force for Disease Eradication. *Morb. Mortal. Weekly Rep.* 42(RR-16):1–38.

4. Cobbold, T. S. 1864. *Entozoa.* London: Groombridge.

5. Fedchenko, A. P. 1870. Concerning the structure and reproduction of the guinea worm (*Filaria medinensis*). *Proc. Imp. Soc. Friends Nat. Sci. Anthropol. Ethnograph.* 8:columns 71–81. Translation in *Am. J. Trop. Med. Hyg.* 20:511–23.

6. Greer, G., M. Dama, S. Graham, R. Migliani, M. Alami, and A. Sam-Abbenyi. 1994. Cameroon: An African model for final stages of guinea worm eradication. *Am. J. Trop. Med. Hyg.* 50:393–400.

7. Hopkins, D. R. 1992. Homing in on helminths. *Am. J. Trop. Med. Hyg.* 46:626–34.

8. Hopkins, D. R., E. Ruiz-Tiben, R. L. Kaiser, A. N. Agle, and P. C. Withers Jr. 1993. Dracunculiasis eradication: Beginning of the end. *Am. J. Trop. Med. Hyg.* 49:281–89.

9. Kobayashi, A., et al. 1986. Human case of dracunculiasis in Japan. *Am. J. Trop. Med. Hyg.* 35:159–61.

10. Lauckner, T. R., A. M. Rankin, and F. C. Adi. 1961. Analysis of medical admissions to University College Hospital, Ibadan. *W. Afr. Med. J.* 10:3.

11. Muller, R. 1992. Guinea worm eradication: Four more years to go. *Parasitol. Today* 8:387–90.

12. Platzer, E. G., and J. R. Adams. 1967. The life history of a dracunculoid *Philonema onchorhynchi,* in *Onchorhynchus nerka. Can. J. Zool.* 45:31–43.

Additional References

Chabaud, A. G. 1975. Keys to genera of the order Spirurida, part 1. In Anderson, R. C., A. G. Chabaud, and S. Willmott, eds. *CIH keys to the nematode parasites of vertebrates.* Bucks, Eng.: Commonwealth Agricultural Bureaux, Farnham Royal.

Ivashkin, V. N., A. A. Sobolev, and L. A. Hromova. 1971. *Essentials of nematodology, vol. 22. Camallanata of animals and man and the diseases they cause.* Moscow: Akademii Nauk SSSR. A most valuable reference to all species in this order.

Muller, R. 1971. *Dracunculus* and dracunculiasis. In Dawes, B., ed. *Advances in parasitology* 9. New York: Academic Press, Inc., 73–151.

Neafie, R. C., D. H. Connor, and W. M. Meyers. 1976. Dracunculiasis. In Binford, C. H., and D. H. Connor, eds. *Pathology of tropical and extraordinary diseases,* vol. 2, sect. 9. Washington, D.C.: Armed Forces Institute of Pathology.

Chapter 31

PHYLUM ACANTHOCEPHALA: THORNY-HEADED WORMS

Perhaps, indeed, the Echinorhynchi *are not simple, but double animals—in this fashion, that the proboscidial apparatus represents one, and the so-called sexual organ the other animal, whilst the body envelope is common to both. . . . Similar conditions, as is well known, occur among the* Bryozoa.

A. Schneider, 1871 (translated by W. S. Dallas)

Few zoologists and still fewer veterinarians and physicians ever encounter a thorny-headed worm. Compared with parasitic platyhelminths or nematodes, they are fairly rare. Still, representatives are to be found inhabiting the intestines of fishes, amphibians, reptiles (rarely), birds, and mammals, where they have established a parasitic relationship with their host and occasionally cause serious disease.

The first recognizable description of an acanthocephalan in the literature is that of Redi, who, in 1684, reported white worms with hooked, retractable proboscides in the intestines of eels. From the time of Linnaeus to the end of the nineteenth century, all acanthocephalan species were placed in the collective genus *Echinorhynchus* Zoega in Mueller, 1776, although Koelreuther is credited with naming the genus *Acanthocephalus* in 1771. Hamann[18] divided *Echinorhynchus,* which by then had become large and unwieldy, into *Gigantorhynchus, Neorhynchus,* and *Echinorhynchus,* thereby beginning the modern classification of the Acanthocephala.

Lankester[26] proposed elevating the order Acanthocephala, proposed by Rudolphi in 1808, to the level of phylum. This suggestion was not widely accepted until Van Cleave[50,51] convincingly argued in its favor. Today the Acanthocephala is widely accepted as a separate phylum.

FORM AND FUNCTION

The morphology of the acanthocephalans reflects an extensive adaptation to their parasitic mode of life and enteric habitat. There appears to have been an evolutionary reduction in muscular, nervous, circulatory, and excretory systems and a complete loss of a digestive system. The remaining animal seems little more than a pseudocoelomate bag of reproductive organs with a spiny holdfast at one end. The worms range in size from the tiny *Octospiniferoides chandleri,* only 0.92 to 2.40 mm long, to *Oligacanthorhynchus longissimus,* exceeding a meter in length. Extensive reviews of acanthocephalan physiology have been published.[7,37]

General Body Structure

Superficially the acanthocephalan body consists of an anterior proboscis, a neck, and a trunk (Fig. 31.1).

The proboscis is variable in shape, from spherical to cylindrical, depending on the species (Fig. 31.2). It is covered by a tegument and has a thin, muscular wall within which are embedded the roots of recurved, sclerotized hooks. The sizes, shapes, and numbers of these hooks are among the most useful characters in the taxonomy of the worms. The proboscis is hollow and fluid filled. Attached to its inner apex is a pair of muscles, called **proboscis inverter muscles,** which extend the length of the proboscis and neck and insert in the wall of a muscular sac called the **proboscis receptacle** (Figs. 31.1 and 31.3).

Proboscis receptacle morphology varies somewhat depending on the family, but generally speaking the receptacle consists of one or two layers of muscle fibers; it is attached to the inner wall of the proboscis. When the proboscis inverter muscles contract, the proboscis invaginates into the proboscis receptacle, with the hooks completely inside. When the proboscis receptacle contracts, it forces the proboscis to evaginate by a hydraulic system.[19] A nerve ganglion called the **brain** or **cerebral ganglion** is located within the receptacle.

The proboscis and its receptacle are sometimes referred to as the **presoma.** The neck is a smooth, unspined zone between the most posterior hooks of the proboscis and an infolding of the body wall. **Neck retractor muscles** attach this infolding of the body wall to the inner surface of the trunk. When the proboscis retractor and the neck retractor muscles contract, the entire anterior end is withdrawn into the trunk. Some species have a sensory pit on each side of the neck, and two similar pits are found on the tip of the proboscis of many species.

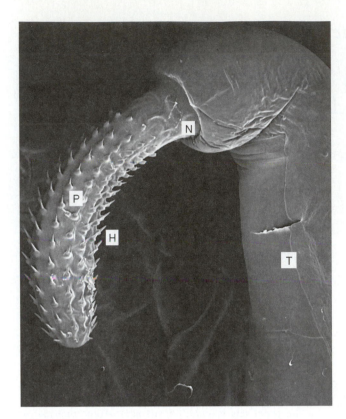
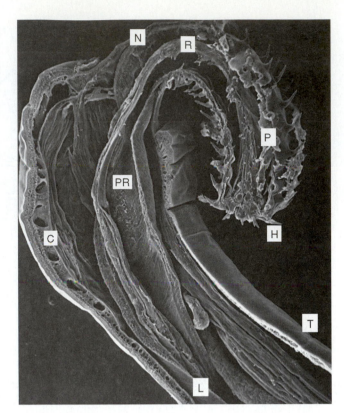

FIGURE 31.1

Scanning electron micrographs of *Leptorhynchoides thecatus* from a sunfish, showing some of the major anatomical features of acanthocephalans.
P, proboscis; **H,** hook; **N,** neck; **T,** trunk; **PR,** proboscis receptacle; **R,** proboscis retractor muscle; **L,** lemniscus; **C,** canals of the lacunar system.
Courtesy of Peter D. Olson.

The rest of the body, posterior to the neck, is called the **trunk,** or **metasoma.** Like the proboscis and neck, it is covered by a tegument and has muscular internal layers. Many species have simple, sclerotized spines embedded in the trunk wall that maintain close contact with the mucosa of the host's intestine. The trunk contains the reproductive system (Fig. 31.3) and also functions in absorbing and distributing nutrients from the host's intestinal contents. In the living worm the trunk is bilaterally flattened, usually with numerous transverse wrinkles, but when the worm is placed in a hypotonic solution, such as tap water, it swells and becomes turgid. This is desirable for ease of study of the specimen, since it places the internal organs in constant relationship with each other, and it usually forces the introverted proboscis to evaginate.

Body Wall

The body wall is a complex syncytium containing nuclear elements and a series of internal, interconnecting canals called the **lacunar system** (Figs. 31.4, 31.5, and 31.6). In some species the nuclei are gigantic but few in number. In others the nuclei fragment during larval development and are widely distributed throughout the trunk wall. When entire nuclei are present, their number is constant for each species, demonstrating the principle of **eutely,** or nuclear constancy. Development of the wall was described by Butterworth.[5]

• Tegument

The tegument has no true layers, but several regions differ in their construction. These are, beginning with the outermost, the (1) surface coat, (2) striped zone, (3) vesicular zone, (4) felt zone, (5) radial fiber zone, and (6) basement lamina (Fig. 31.4). Inside the tegument is a layer of irregular connective tissue, followed by circular and longitudinal muscle layers. Like that of the trematodes and cestodes, the tegument is syncytial, but unlike in those groups, the nuclei are in the basal region of the tegument, not in cytons separated from the distal cytoplasm. The **surface coat,** or **glycocalyx,** a filamentous material, was formerly known as the epicuticle. It is about 0.5 μm thick, for instance, on *Moniliformis moniliformis,* an acanthocephalan of rats and the most commonly investigated species in the laboratory. The surface coat is composed of acid mucopolysaccharides and neutral polysaccharides and/or glycoproteins.[56] The surface coat fits the definition of a glycocalyx, a carbohydrate-rich coat found on a variety of eukaryotic and prokaryotic cells. The stabilized system of polyelectrolytic filaments in the surface coat constitutes an extensive surface for molecular interactions, including those involved in transport functions and enzyme-substrate interactions.

Immediately beneath the surface coat and limited by the trilaminar outer membrane is the **striped zone.** This zone is 4 to 6 μm thick and is punctuated by a large number

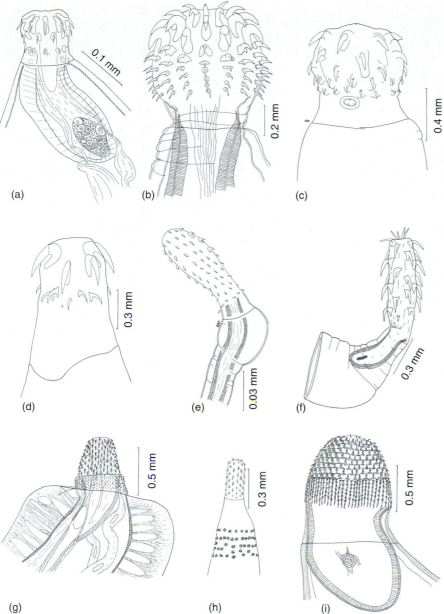

(a)   (b)   (c)

(d)   (e)   (f)

(g)   (h)   (i)

FIGURE 31.2

Examples of different types of acanthocephalan proboscides. (*a*) *Octospiniferoides australis;* (*b*) *Sphaerechinorhynchus serpenticola;* (*c*) *Oncicola spirula;* (*d*) *Acanthosentis acanthuri;* (*e*) *Pomphorhynchus yamagutii;* (*f*) *Paracanthocephalus rauschi;* (*g*) *Mediorhynchus wardae;* (*h*) *Palliolisentis polyonca;* (*i*) *Owilfordia olseni.*

(*a, e, h*) From G. D. Schmidt and E. H. Hugghins, "Acanthocephala of South American Fishes. Part I, Eoacanthocephala," in *J. Parasitol.* 59:829–835. Copyright © 1973. Reprinted with permission of the publisher. (*b*) From G. D. Schmidt and R. E. Kuntz, in "*Sphaerechinorhynchus serpenticola* sp. n. (Acanthocephala: Sphaerechinorhynchinae), a parasite of the Asian cobra, *Naja naja* (Cantor), in Borneo (Malaysia)," in *J. Parasitol.* 52:913–916. Copyright © 1966. Reprinted with permission of the publisher. (*c*) From G. D. Schmidt in *Pathology of Simian Primates,* 2:144–156. Edited by R. N. T. W. Fiennes. Copyright © 1972 S. Karger AG, Basel. Reprinted with permission. (*d*) From G. D. Schmidt, "Redescription of *Acanthosentis acanthuri* Cable et Quick 1954 (Acanthocephala: Quadrigyridae)," in *J. Parasitol.* 61:865–867. Copyright © 1975. Reprinted with permission of the publisher. (*f*) From G. D. Schmidt, "*Paracanthocephalus rauschi* sp. n. (Acanthocephala: Paracanthocephalidae) from grayling, *Thymallus articus* (Pallus) in Alaska," in *Canad. J. Zool.* 47:383–385. Copyright © 1969 National Research Council of Canada. Reprinted with permission of the publisher. (*g*) From G. D. Schmidt and A. G. Canaris, "Acanthocephala from Kenya with descriptions of two new species," in *J. Parasitol.* 53:634–637. Copyright © 1967. Reprinted with permission of the publisher. (*i*) From G. D. Schmidt and R. E. Kuntz, "Revision of the Porrorchinae (Acanthocephala: Plagiorhynchiddae) with descriptions of two new genera and three new species," in *J. Parasitol.* 53:130–141. Copyright © 1967. Reprinted with permission of the publisher.

of crypts about 2 to 4 μm deep that open to the surface by pores.[6] These crypts give this zone a striped appearance (*Streifenzone*) under the light microscope. The crypts increase the surface area of the worm by 44 times the area of a smooth surface. A filamentous molecular sieve is seen in the necks of the crypts, but particles of less than about 8.5 nm can gain access to the crypts and undergo pinocytosis by the crypt membrane.[6] The importance of pinocytosis in the acquisition of nutrients by the worms is unknown. In the deeper aspects of the striped zone, numerous lipid droplets, mitochondria, Golgi complexes, and lysosomes are found.

The striped zone grades into a region of numerous, closely packed, randomly arranged fibrils known as the **felt-fiber zone.** Mitochondria, numerous glycogen particles, vesicles, and occasionally lipid droplets and lysosomes also are found in the felt-fiber zone. The **radial fiber zone** is just within the felt-fiber zone and makes up about 80% of the thickness of the body wall. It contains large bundles of filaments that course radially through the cytoplasm, large lipid droplets, and nuclei of the body wall. Here, too, are many glycogen particles, mitochondria, Golgi complexes, and lysosomes. Rough endoplasmic reticulum is found in the perinuclear cytoplasm. The nuclei have numerous nucleoli. The lacunar canals course through the radial fiber zone.

The structure of the proboscis wall is similar to that of the trunk, except it has fewer crypts and a thinner radial zone, and it lacks a felt zone.

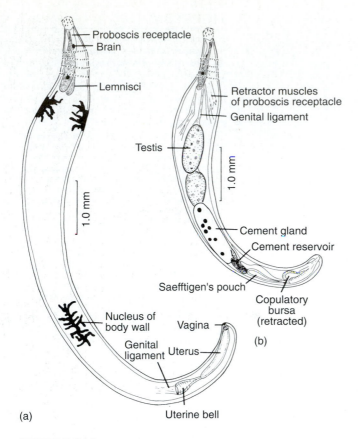

(a)

FIGURE 31.3

Quadrigyrus nickoli, illustrating basic acanthocephalan morphology. (*a*) Female; (*b*) male.

From G. D. Schmidt and E. H. Hugghins, "Acanthocephala of South American Fishes. Part I, Eoacanthocephala," in *J. Parasitol.* 59:829–835. Copyright © 1973. Reprinted with permission of the publisher.

• Lacunar System and Muscles

The system of fluid-filled channels in the body wall called the lacunar system has been long known, but its function has been enigmatic. A fascinating picture of the relationship of this curious system to the functioning of the body wall muscles has emerged, largely as a result of the efforts of Miller and Dunagan and their coworkers.[28,29,30,31,56]

The lacunar system is present in two parts, apparently unconnected to each other: that in the proboscis and neck and that in the trunk. The presomal lacunar system has channels that run into two structures called **lemnisci** that grow from the base of the neck into the pseudocoelom. Each lemniscus has a central canal that is continuous with the presomal lacunar system. The function of the lemnisci is unknown, although it may contribute to the hydraulics of the proboscis mechanism. The metasomal lacunar system consists of a complicated network of interconnecting canals. In most species there are two main longitudinal canals, either dorsal and ventral or lateral. These are connected by numerous irregular or regular transverse canals. The location and arrangement of the lacuni are used as taxonomic characters. In addition, at least in some species, there is a pair of medial longitudinal channels, each connected periodically by short radial canals to circular canals coursing between the dorsal and ventral longitudinal channels[28] (Fig. 31.5). The medial longitudinal channels lie on the pseudocoel side of the body-wall muscles, and the radial canals pierce the muscle layers to intercept the ring canals. Some of the ring canals give rise to branches that run throughout the radial fiber zone of the tegument.

The body-wall muscles are composed of a longitudinal layer surrounded by a circular muscle layer (Fig. 31.6). These muscles have a very curious structure. They are hollow, with tubelike cores and numerous, anastomosing interconnectives.[29] It has been found that the lumina of the muscles are continuous with the lacunar system; therefore circulation of the lacunar fluid may well bring nutrients to and remove wastes from the muscles. Although there is no heart or other circulatory organ, contraction of the circular muscles would force fluid into the longitudinal components and vice versa. Thus, the lacunar system seems to function as an effective fluid transport system and possibly a hydrostatic skeleton.[32]

Acanthocephalan muscles are peculiar in other respects. They are electrically inexcitable, have low membrane potentials, and are slow conductors.[20] They are characterized by rhythmic, spontaneous depolarizations. Although the muscles appear to be stimulated by acetylcholine, nervous control of contraction is at present unclear. It is believed that nerves initiate contractions via the **rete system,** which is a highly branched, anastomosing network of thin-walled tubules lying on the medial surface of the longitudinal muscles or between the longitudinal and circular muscle layers.[56] The rete system itself seems to be modified muscle cells.

Reproductive System

Acanthocephalans are dioecious and usually demonstrate some degree of sexual dimorphism in size, with the female being larger (Fig. 31.3). In both sexes one or two thin **ligament sacs** are attached to the posterior end of the proboscis receptacle and extend to near the distal genital pore. Within these sacs are the gonads and some accessory organs of the reproductive systems. In some species the ligament sacs are permanent; in others they break down as the worm matures.

• Male Reproductive System

Two testes normally occur in all species, and their location and size are somewhat constant for each species (Fig. 31.3). Spermiogenesis has been described.[55] Each testis has a vas efferens through which mature spermatozoa, which appear as slender, headless threads, travel to a common vas deferens and/or to a small penis. Several accessory organs also are present, the most obvious of which are the **cement glands.** These syncytial organs, numbering from one to eight, contain one or more giant nuclei or several nuclear fragments. In many species they are joined in places by slender bridges. They secrete a **copulatory cement** of tanned protein, which in some species is stored in a **cement reservoir** until copulation occurs. At that time the cement plugs the vagina after sperm transfer and rapidly hardens to form a **copulatory cap.** This remains attached to the posterior end of the female during subsequent development of the embryos within her body but eventually disintegrates.

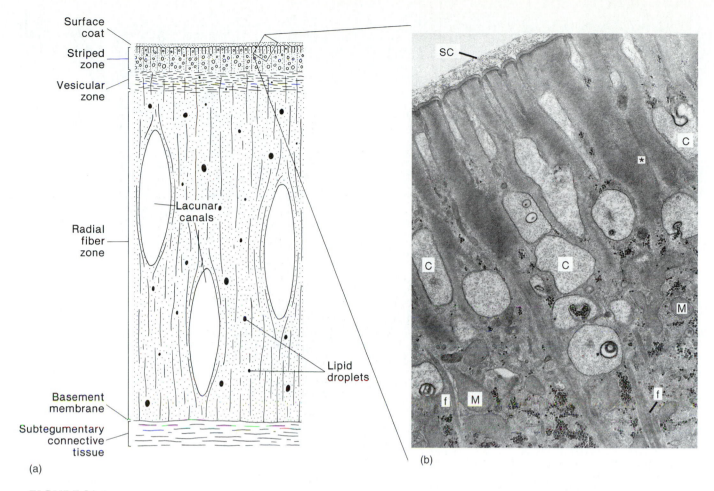

(a)

(b)

FIGURE 31.4

Tegument of *Moniliformis moniliformis*. (*a*) Diagram of transverse section to show layers. The vesicular zone is transitional between the striped zone and the felt-fiber zone; it contains many vesicles and mitochondria with poorly developed cristae. Lacunar canals are in the radial fiber zone. (*b*) Electron micrograph showing the major features of the striped zone. The worm is coated with a finely filamentous surface coat (**SC**). Numerous surface crypts (**C**) appear as large scattered vesicular structures with elements occasionally appearing to course to the surface of the helminth. The crypts are separated by patches of moderately electron-opaque material (*), giving the zone its striped appearance under the light microscope. Mitochondria (**M**), glycogen particles, microtubules, and other cytoplasmic details are evident in the inner portion of the striped zone. Bundles of fine cytoplasmic filaments (**f**) extend between this region and the deeper cytoplasm of the body wall. (×42,000.)

(*a*) Drawing by William Ober. (*b*) From J. E. Byram and F. M. Fisher, Jr., "The absorptive surface of *Moniliformis dubius* (Acanthocephala). I. Fine structure," in *Tiss. Cell* 5:553–579. Copyright © 1974.

Another male accessory sex organ is the **copulatory bursa** (Fig. 31.7), a bell-shaped specialization of the distal body wall that is invaginated into the posterior end of the body cavity except during copulation. A muscular sac, **Saefftigen's pouch,** is attached to the base of the bursa. When it contracts, fluid is forced into the lacunar system of the bursa, and it is everted by hydrostatic pressure. Many sensory papillae line the bursa; when it contacts the posterior end of a female, it clasps the female by muscular contraction, and sperm transfer is effected with a small penis.

• Female Reproductive System

The ovary of the female acanthocephalan is peculiar in that it fragments into **ovarian balls** early in the worm's life, often while it is still a juvenile in the intermediate host. These balls of oogonia float freely within the ligament sac, increasing slightly in size before insemination occurs. The posterior end of the ligament sac is attached to a muscular **uterine bell** (Fig. 31.8). This organ allows mature eggs to pass through into the uterus and vagina and out the genital pore, while returning immature eggs to the ligament sac.

After copulation, the spermatozoa migrate from the vagina, through the uterus and uterine bell, and into the ligament sac. There they begin fertilizing the oocytes of the ovarian balls. After the first few cleavages the embryos detach from the ovarian ball and float freely in the pseudocoelomic fluid. This exposes underlying oocytes for fertilization. Thus, several stages of early embryogenesis may be found in a single female. Eventually, from this one copulation, many thousands or even millions of embryonated eggs are produced and released by each female. These, when sufficiently mature, pass from the host in its feces, where they may become available to the proper intermediate host.

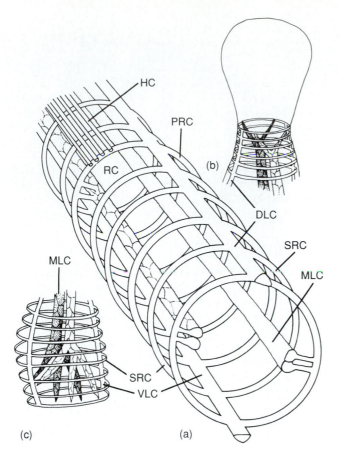

FIGURE 31.5

Organization of lacunar system in *Macracanthorhynchus hirudinaceus.* (*a*) Midmetasomal region; (*b*) region near neck, presomal lacunar system not indicated; (*c*) near posterior end of metasoma. **DLC,** dorsal longitudinal channel; **HC,** hypodermal canal (in radial fiber zone); **MLC,** medial longitudinal channel; **PRC,** primary ring canal; **RC,** radial canal; **SRC,** secondary ring canal; **VLC,** ventral longitudinal channel.

From D. M. Miller and T. T. Dunagan, "Body wall organization of the acanthocephalan, *Macracanthorhynchus hirudinaceus:* a reexamination of the lacunae system," in *Proc. Helm. Soc. Wash.* 43:99–106. Copyright © 1976. Reprinted with permission of the publisher.

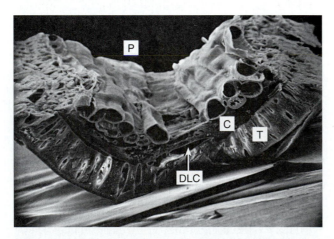

FIGURE 31.6

Scanning electron micrograph of body wall, *Oligacanthorhynchus tortuosa.* **C,** circular muscle; **P,** pseudocoel; **T,** tegument, showing hypodermal lacunar canals; **DLC,** dorsal longitudinal channel.

Courtesy of D. M. Miller and T. T. Dunagan.

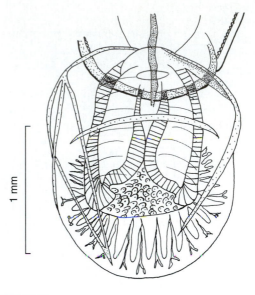

FIGURE 31.7

Extended copulatory bursa of *Owilfordia olseni.* Note the numerous sensory papillae.

From G. D. Schmidt and R. E. Kuntz, "Revision of the Porrorchinae (Acanthocephala: Plagiorhynchiddae) with descriptions of two new genera and three new species," in *J. Parasitol.* 53:130–141. Copyright © 1967. Reprinted with permission of the publisher.

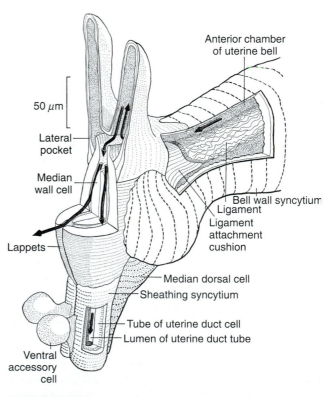

FIGURE 31.8

Stereogram of mature uterine bell, cut away to reveal complex internal luminal system. *Heavy arrows,* possible routes for egg translocation.

From J. P. Whitfield, "A histological description of the uterine bell of *Polymorphus minutus* (Acanthocephala), in *Parasitology.* 58:671–682. Copyright © 1968. Reprinted with permission of the publisher.

As the shelled embryos are pushed into the uterine bell by peristaltic action, two possible routes are available. They may pass back into the pseudocoelom through slits in the bell or on into the uterus. Fully developed embryos are slightly longer than immature ones and therefore cannot pass through the bell slits;[53] hence, they are passed on into the uterus. Immature eggs, however, are retained for further maturation. The efficiency of the sorting is quite high, and apparently no immature forms are passed into the uterus.

Excretory System

Excretion in most species appears to be effected by diffusion through the body wall. However, members of one family in the class Archiacanthocephala, the Oligacanthorhynchidae, are unique in possessing two **protonephridial excretory organs.** Each comprises many anucleate flame bulbs with tufts of flagella and may or may not be encapsulated, depending on the species.[13,25] In the male these organs are attached to the vas deferens and empty through it; in the female they are attached to the uterine bell and empty into the uterus.

Acanthocephalans show little ability to osmoregulate, swelling in hypotonic, balanced saline or sucrose solutions and becoming flaccid in hypertonic solutions. The osmotic pressure of their pseudocoelomic fluid is close to or somewhat above that of the intestinal contents. They take up sodium and potassium, swelling in hypertonic solutions of sodium chloride or potassium chloride at 37°C. In balanced saline they lose sodium and accumulate potassium against a concentration gradient. Their hexose transport mechanism is not sodium coupled.

Nervous System

The nervous system of acanthocephalans is simple. The cerebral ganglion consists of only 54 to 88 cells in the species studied; it lies in the proboscis receptacle.[12,33] Relatively few nerves issue from the ganglion, the largest of which are the anterior proboscis nerve and the lateral posterior nerves.[11] Nerves supply the two lateral sense organs and the apical sense organ, if present. A large multinucleate cell referred to as a *support cell* is located ventral and slightly anterior to the cerebral ganglion.[31] Processes from the support cell lead to the lateral and apical sensory organs, but these processes are not nerves. Their function is unknown, but they may be secretory and help explain the inflammatory reaction of the host to the worm's proboscis.

ACQUISITION AND USE OF NUTRIENTS

Because of the availability of a good laboratory subject (*M. moniliformis* in rats), investigators have been able to accumulate some knowledge of acanthocephalan metabolism. However, the problem of assessing the general applicability of observations on *M. moniliformis* and the few others reported is acute. (*Moniliformis dubius* is a junior synonym of *M. moniliformis,* and much literature accumulated on the physiology of the organism under that name.) Hatchery-reared rainbow trout are also good hosts for experimental work, and some recent work has been done on *Neoechinorhynchus rutili* in that host.

Uptake

Since the Cestoda and the Acanthocephala are both groups that must obtain all nutrient molecules through their body surfaces, comparisons between the two are quite interesting, particularly in light of their structural differences.[27]

Acanthocephalans can absorb at least some triglycerides, amino acids, nucleotides, and sugars. The presoma is the site of triglyceride uptake. In experiments using ^{3}H-labelled glyceroltrioleate and seven species of acanthocephalans, Taraschewski and Mackenstedt[45] showed that uptake started in the anterior half of the proboscis, but that the radioisotopes eventually accumulated in the lemnisci. Amino acids are absorbed, at least partially, by stereospecific membrane transport systems in *M. moniliformis* and *Macracanthorhynchus hirudinaceus.*[49] The surface of *Moniliformis moniliformis* contains peptidases, which can cleave several dipeptides, and the amino acid products are then absorbed by the worm.[48] In several other species lysine is absorbed across the metasomal tegument, especially the anterior portion, and accumulates in nuclei and the outer muscle belt.[46] Absorbed thymidine is incorporated into DNA in the perilacunar regions and into the nuclei of the ovarian balls and testes. Nuclei in the body wall are not labeled by radioactive thymidine; therefore, it is assumed that the DNA synthesized there is mitochondrial.

Like the tapeworm *Hymenolepis diminuta, M. moniliformis* has an absolute dependence on host dietary carbohydrate for growth and energy metabolism as an adult.[42,43] The worm can absorb glucose, mannose, fructose, and galactose, as well as several glucose analogs. In contrast to *H. diminuta, M. moniliformis* can grow and mature in the host fed a diet containing fructose as the sole carbohydrate source.[8,9,23] Absorption of glucose is through a single transport locus, whereas transport of mannose, fructose, and galactose is mediated both by the glucose locus and another site, referred to by Starling and Fisher[43] as the *fructose site.* Maltose and glucose-6-phosphate (G6P) are absorbed also, but first they are hydrolyzed to glucose by enzymes in or on the tegumental surface. But in sharp contrast to tapeworm and other glucose transport systems, glucose absorption by *M. moniliformis* is not coupled to cotransport of sodium. In *M. moniliformis,* glucose is rapidly phosphorylated, removing free glucose from the vicinity of the tegumental transport loci, thus theoretically forming a metabolic sink for the flow of additional hexose down its concentration gradient. However, substantial amounts of free glucose are found in the body wall.[44] Evidence suggests that the free glucose pool is not derived directly from absorbed glucose but instead from the nonreducing disaccharide trehalose. If so, the glucose may be deposited, perhaps by intervention of a membrane-bound trehalase, in an internal membranous compartment that by some means can resist the efflux of the glucose it contains. The scheme offers an interesting possible metabolic role for trehalose, perhaps similar to that in insects.

Metabolism

As in the other helminth parasites, the energy metabolism is adapted for facultative anaerobiosis. *Moniliformis moniliformis* can ferment the hexoses it absorbs. The tricarboxylic acid cycle apparently does not operate in *M. moniliformis* or *Macracanthorhynchus hirudinaceus,* although there is evidence for it in *Echinorhynchus gadi,* a parasite of cod. *Moniliformis* fixes carbon dioxide, and the principal enzyme of carbon dioxide fixation is phosphoenolpyruvate carboxykinase. Lactate and succinate are the main end products of glucose degradation in *Polymorphus minutus.* Interestingly the main end products of glycolysis in *M. moniliformis* are ethanol and carbon dioxide with a small amount of lactate and only traces of succinate, acetate, and butyrate. Even though PEP carboxykinase activity in *M. moniliformis* is high,[24] it must be regulated in such a way that the major end products are ethanol and lactate, rather than succinate.

Lipids apparently are not used as energy sources. Körting and Fairbairn[24] found that endogenous lipids were not metabolized during in vitro incubation of *M. dubius.* This was correlated with the fact that enzymes necessary for the beta-oxidation of lipids were low in activity, and one of them seemed to be completely absent.

Electron transport in acanthocephalans has been studied very little. Oxidation of both succinate and NADH leads to reduction of cytochrome *b.*[4] Two pathways for reoxidation of this compound have been postulated, the major one independent of cytochrome *c* and cytochrome oxidase. This one is somewhat similar to the branched-chain electron transport postulated for the cestode *Moniezia expansa.*

DEVELOPMENT AND LIFE CYCLES

Each species of Acanthocephala uses at least two hosts in its life cycle. The first is an insect, crustacean, or myriapod, and the arthropod must eat an egg that was voided with the feces of a definitive host. Development proceeds through a series of stages until the juvenile is infective to a definitive host. Many species, when eaten by a vertebrate that is an unsuitable definitive host, can penetrate the gut and encyst in some location where they survive without further development. This unsuitable vertebrate becomes a paratenic host, since if it is eaten by the proper definitive host, the parasite excysts, attaches to the intestinal mucosa, and matures. Such adaptability has survival value. For example, ecological gaps exist in the food chain between a microcrustacean and a large predaceous fish or between a grasshopper and an eagle. The paratenic host is one member in a food chain that bridges such a gap and incidentally ensures the survival of the parasite.

The manner of early embryogenesis is an unusual characteristic of the group. Early cleavage is spiral, although this pattern is somewhat distorted by the spindle shape of the eggshell. At about the 4- to 34-cell stage, the cell boundaries begin to disappear, and the entire organism becomes syncytial. Gastrulation occurs by migration of nuclei to the interior of the embryo.[40] The nuclei continue to divide but become smaller, until they form a dense core of tiny nuclei, the **inner nuclear mass.** These nuclei give rise to all internal organ

systems of the worm. In some species the uncondensed nuclei remaining in the peripheral area give rise to the tegument; in some the tegument is derived from a nucleus that separates from the inner mass; in others there are contributions from both.

The fully embryonated larva that is infective to the arthropod intermediate host is called the **acanthor.** The acanthor is an elongated organism that is usually armed at its anterior end with six or eight bladelike hooks. The hooks may be replaced by smaller spines in some species. The hooks or spines with their muscles are called the **aclid organ** or **rostellum.** The hooks aid in penetration of the gut of the intermediate host.[54] The acanthor is a resting, resistant stage and will undergo no further development until it reaches the intermediate host. Under normal environmental conditions, the acanthors may remain viable for months or longer. Acanthors of *Macracanthorhynchus hirudinaceus* can withstand subzero temperatures and desiccation, and they can remain viable for up to three and one-half years in the soil. The acanthors of some species completely penetrate the gut, coming to lie in the host's hemocoel, whereas others stop just under the serosa. In both cases the worm then becomes parasitic on the arthropod, absorbing nutrients and enlarging, thus initiating the developmental stage known as the **acanthella.** The end of the acanthor that bears the aclid organ apparently becomes the anterior end of the adult in some species, whereas others exhibit a curious 90-degree change in polarity, in which the anterior end of the adult develops from the side of the acanthor. During the acanthella stage, the organ systems develop from the central nuclear mass and the hypodermal nuclei of the acanthor.

At termination of this development, the juvenile is an infective stage called a **cystacanth.** In most species the anterior and posterior ends invaginate, and the entire cystacanth becomes encased in a hyaline envelope. The parasite then must be eaten by the definitive host before it can fulfill its potential. Obviously mortality is very high, since only a tiny fraction of the immense number of eggs produced may survive the numerous hazards involved in completion of the life cycle.

Complete life cycles are known for only about 20 species in the phylum, although we have partial information on several more. The following examples illustrate the pattern followed in the life histories of the three major groups.

Class Eoacanthocephala

Neoechinorhynchus saginatus is an eoacanthocephalan parasite of various species of suckers and of creek chubs, *Semotilus atromaculatus,* fish distributed from Maine to Montana. Its life cycle and embryology were described by Uglem and Larson[47] (Fig. 31.9). When the eggs are eaten by the common ostracod crustacean *Cypridopsis vidua,* they hatch within an hour and begin penetrating the gut within 36 hours. After penetration the unattached larva begins to enlarge and rearrange its nuclei, initiating the formation of internal organs. By 16 days after infection, the acanthella has developed into an infective cystacanth. Time required for maturation within the fish has not been determined.

Other eoacanthocephalan life cycles are similar, although paratenic hosts are known for *N. cylindratus* and *N. emydis.*[21,52]

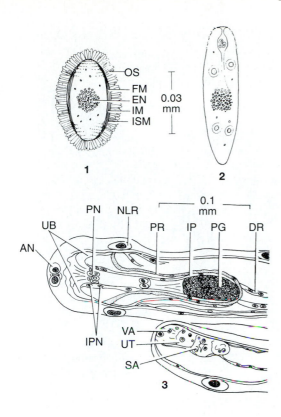

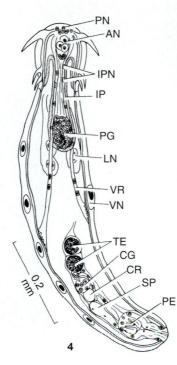

FIGURE 31.9

Stages in development of *Neoechinorhynchus saginatus*. **1**, Shelled acanthor from body cavity of adult female; **2**, acanthor from gut of ostracod, one hour after feeding; **3**, female acanthella, age 12 days (one lemniscal nucleus omitted for clarity); **4**, late male acanthella, age 14 days (neck retractors omitted); **AN**, apical nuclei; **CG**, cement gland; **CR**, cement reservoir; **DR**, dorsal retractor of proboscis receptacle; **EN**, condensed nuclear mass; **FM**, fertilization membrane; **IM**, inner membrane; **IP**, proboscis inverter; **IPN**, proboscis inverter nuclei; **ISM**, inner shell membrane; **LN**, lemniscal nucleus; **NLR**, lemniscal ring nuclei; **OS**, outer shell; **PE**, penis; **PG**, brain anlage; **PN**, proboscis nuclear ring; **PR**, proboscis receptacle muscle sheath; **SA**, selector apparatus; **SP**, Saefftigen's pouch; **TE**, testes; **UB**, uncinogenous bands; **UT**, uterus; **VA**, vagina; **VN**, giant nucleus of ventral trunk wall; **VR**, ventral retractor of proboscis receptacle.

From G. L. Uglem and O. R. Larson, "The Life history and larval development of *Neoechinorhynchus saginatus* Van Cleave and Bangham, 1949 (Acanthocephala: Neoechinorhynchidae), in *J. Parasitol.* 55:1212–1217. Copyright © 1969. Reprinted with permission of the publisher.

Class Palaeacanthocephala

Plagiorhynchus cylindraceus is a palaeacanthocephalan that is common in robins and other passerine birds in North America. Its life cycle and embryology were described by Schmidt and Olsen[41] (Fig. 31.10). When the eggs are eaten by the terrestrial isopod crustacean *Armadillidium vulgare,* they hatch in the midgut within 15 minutes to two hours. Active entrance of the acanthor into the gut wall occurs within 1 to 12 hours, and the acanthor lies within the tissues of the gut wall. After 15 to 25 days of apparent dormancy, it migrates to the outside of the gut, where it clings loosely to the serosa. Progressive changes follow in which the overall size increases, and the organs of the mature worm are delineated. The cystacanth (Fig. 31.10) appears fully developed in 30 to 40 days but is not infective to the definitive host until 60 to 65 days. On ingestion of an infected isopod by the definitive host, the proboscis of the cystacanth evaginates, pierces the cyst, and attaches to the gut wall, where it develops to maturity.

Nickol and Oetinger[38] found encapsulated *P. cylindraceus* in the mesenteries of a shrew. This observation illustrates how the interjection of a paratenic host into a life cycle may doom a parasite rather than serve it, because it is unlikely, although possible, that a robin would eat a shrew. (However, the parasite fauna of the American robin reveals the bird's diet to be more varied than most people realize.) But chances of the worm invading a hawk or owl are improved by paratenesis, and the parasite's host range could thus be extended if the worm survives in the large predatory birds.

Class Archiacanthocephala

Macracanthorhynchus hirudinaceus is a cosmopolitan parasite of pigs. Its life cycle has been known since 1868 and was more recently reported in detail by Kates.[22] When the eggs are eaten by white grubs (larvae of the beetle family Scarabaeidae), they hatch in the midgut within an hour and penetrate its lining. Within 5 to 20 days of infection, the developing acanthellas are found free in the hemocoel or attached to the outer surface of the serosa. By 60 to 90 days after infection, the cystacanth is infective to the definitive host. Pigs are infected by eating the grubs or the adult beetles, which have metamorphosed with their parasites intact.

Most archiacanthocephalans are parasites of predaceous birds and mammals, so paratenic hosts often are involved in life cycles within this class.

EFFECTS OF THE PARASITE ON ITS HOST

The behavior of an intermediate host is sometimes altered by infection, evidently increasing the probability of its being eaten by the definitive host.[34,35] Cockroaches (*Periplaneta americana* and *Blatella germanica*) infected with *Moniliformis moniliformis* move more slowly, travel shorter distances, and spend more time on horizontal surfaces than do uninfected controls.[16] However, a third cockroach species, *Supella longipalpa,* spends more time in the shade when infected with the same parasite.[36] Yet infection does not alter the behavior of a fourth cockroach species, *Diploptera punctata.*[1] If these behavioral changes do indeed increase the

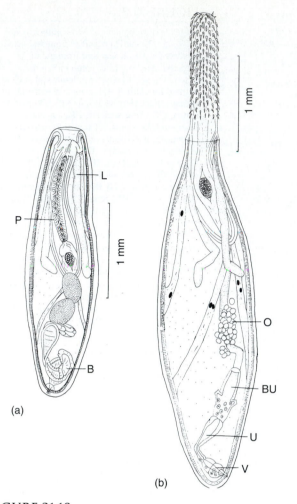

(a)

(b)

FIGURE 31.10

Cystacanths of *Plagiorhynchus cylindraceus*. (*a*) 37 days after infection of the pillbug intermediate hosts; (*b*) 60 days after infection; **B**, bursa; **L**, lemniscus; **P**, proboscis; **BU**, uterine bell; **O**, ovarian balls; **U**, uterus; **V**, vagina.

From G. D. Schmidt and O. W. Olsen, "Life cycle and development of *Prosthorynchus formosus* (Van Cleave 1918) Travassos, 1926, an acanthocephalan parasite of birds," in *J. Parasitol.* 50:721–730. Copyright © 1964. Reprinted with permission of the publisher.

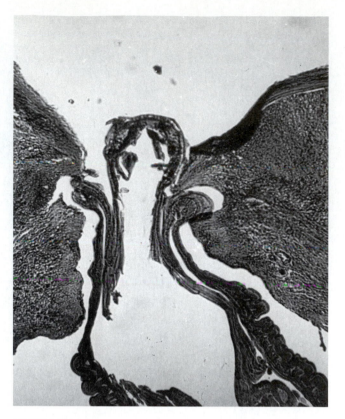

FIGURE 31.11

Complete perforation of the large intestine of a squirrel monkey by *Prosthenorchis elegans*.

From G. D. Schmidt, in R.N.T.W. Fiennes, editor, *Pathology of Simian Primates, Vol. 2,* 1972 S. Karger AG, Basel.

probability of transmission, then *M. moniliformis* may be one of the few species that has found a way to increase the utility of a cockroach.

Although the mechanism by which behavioral alterations occur is not known, the demonstration of these changes has stimulated additional work on parasite and intermediate host relationships. Infection with *Pomphorhynchus laevis,* a parasite of ducks, alters haemolymph proteins in the crustacean intermediate host, *Gammarus pulex.*[3] Furthermore, intermediate hosts are sometimes capable of defending themselves against infection. Zhao and Wang[58] showed that *Macracanthorhynchus hirudinaceus* elicited a hemocyte response in beetles, and some of the host cells penetrated the cystacanth cyst. Subsequently the cystacanths degenerated.

In a definitive host the nature of damage to intestinal mucosa is primarily traumatic, by penetration of the proboscis, and is compounded by the tendency of the worm to release its hold occasionally and reattach at another place. Complete

perforation of the gut sometimes occurs, and in mammals, at least, the results are often rapidly fatal (Fig. 31.11). Great pain accompanies this phase: Infected monkeys show evident distress, and Grassi and Calandruccio[17] recorded the symptoms of pain and delirium experienced by Calandruccio after he voluntarily infected himself with cystacanths of *Moniliformis moniliformis,* a common parasite of domestic rats.

It is suspected that secondary bacterial infection is responsible for localized and generalized peritonitis, hemorrhage, pericarditis, myocarditis, arteritis, cholangiolitis, and other complications.

In view of the invasive nature of the parasites, it is surprising that they elicit so little inflammatory response in many cases. The reaction seems mainly a result of the traumatic damage, with granulomatous infiltration and sometimes collagenous encapsulation around the proboscis (Fig. 31.12). Some species show evidence that antigens are released from the proboscis (as in *M. hirudinaceus*), followed by an intense inflammatory response. It is clear that the pathogenesis caused by acanthocephalans can be severe, but little consideration usually is given to the effects of this group of parasites as a controlling factor of wildlife populations.

Little chemotherapy has been developed for acanthocephalans. Various authors have proposed chenopodium and castor oil, calomel and santonin, carbon tetrachloride, and tetrachloroethylene for primates and pigs, with varying results. Oleoresin of aspidium has been used successfully

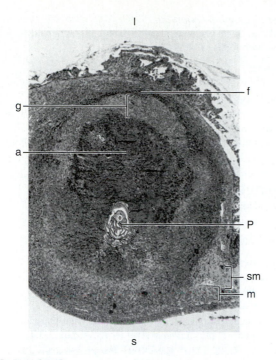

FIGURE 31.12

Cross section of a lesion produced by *Oligacanthorhynchus tortuosa* in an opossum; **p,** proboscis; **a,** necrotic abscess; **g,** granuloma; **f,** region of active fibrocyte proliferation; **m,** muscularis; **sm,** submucosa; **s,** serosal side of the intestinal wall; **l,** lumenal side.

Courtesy of Dennis Richardson.

in human cases but is not recommended for children. Mebendazole was used successfully in a 12-month-old child.[14] Control of intermediate hosts is helpful in preventing infection of domestic or captive animals.

ACANTHOCEPHALA IN HUMANS

Records of Acanthocephala in humans are few, no doubt because of the nature of the intermediate and paratenic hosts involved in the life cycles of the parasites. Few people of the world eat such animals as insects, microcrustaceans, toads, or lizards, at least without cooking them first. However, human infections with five different species have been reported.[39] *Macracanthorhynchus hirudinaceus* occasionally has been recognized as a parasite of humans from 1859 to the present.[10] Nine *M. hirudinaceus* were recovered from a one-year-old child in Austin, Texas, in 1983.[2] *Moniliformis moniliformis* has been found repeatedly in people. *Acanthocephalus rauschi* is known only from specimens taken from the peritoneum of an Alaskan Eskimo, an obvious case of accidental parasitism, since the proper host is undoubtedly a fish. The zest of Eskimos for raw fish probably commonly contributes to such zoonotic infections. *Corynosoma strumosum,* a common seal parasite, also has been found in humans. More puzzling is a case of *Acanthocephalus bufonis,* a toad parasite, in an Indonesian. In this instance it is probable that the man ate a raw paratenic host.[39]

Thus, it seems that the Acanthocephala do not pose much of a threat to human health. They are much more important as parasites of wild and captive animals, where sudden epizootics have been known to kill a great number of individuals in a short time.

PHYLOGENETIC RELATIONSHIPS

The thorny-headed worms are closely related to no known form. The presence of a pseudocoel and exterior "cuticle" has been used as evidence for an affinity with the nematodes and Nematomorpha. However, the tegument is quite different from the cuticle of nematodes, and the body wall structure (including lacunar system and tubular muscles), the eversible spined proboscis with its accompanying mechanisms, the nervous system, and the complete lack of a digestive system set these worms well apart from other phyla in the superphylum Aschelminthes. A further paradox is the fact that the Archiacanthocephala, parasites of the most highly evolved vertebrates, are in many ways more primitive than are the Eoacanthocephala, which are parasites of fishes. Golvan[15] discussed the subject in detail and proposed a hypothetical ancestor, which he named *Protacanthocephala.*

CLASSIFICATION OF PHYLUM ACANTHOCEPHALA

Class Archiacanthocephala

Main longitudinal lacunar canals dorsal and ventral or just dorsal; hypodermal nuclei few; giant nuclei present in lemnisci and cement glands; two ligament sacs persist in females; protonephridia present in one family; cement glands separate, pyriform; eggs oval, usually thick shelled; parasites of birds and mammals; intermediate hosts are insects or myriapods.

Order Moniliformida

Trunk usually pseudosegmented; proboscis cylindrical, with long, approximately straight rows of hooks; sensory papillae present; proboscis receptacle double walled, outer wall with muscle fibers usually arranged spirally; proboscis retractor muscles pierce posterior end of receptacle or are somewhat ventral; brain near posterior end or near middle of receptacle; protonephridial organs absent.

Family
Moniliformidae.

Order Gigantorhynchida

Trunk occasionally pseudosegmented; proboscis a truncated cone, with approximately longitudinal rows of rooted hooks on the anterior portion and rootless spines on the basal portion; sensory pits present on apex of proboscis and each side of neck; proboscis receptacle single walled with numerous accessory muscles, complex, thickest dorsally; proboscis retractor muscles pierce ventral wall of receptacle; brain near ventral, middle surface of receptacle; protonephridial organs absent.

Family
Gigantorhynchidae.

Order Oligacanthorhynchida

Trunk may be wrinkled but not pseudosegmented; proboscis subspherical, with short, approximately longitudinal rows of few hooks each; sensory papillae present on apex of proboscis and each side of neck; proboscis receptacle single walled, complex, thickest dorsally; proboscis retractor muscle pierces dorsal wall of receptacle; brain near ventral, middle surface of receptacle; protonephridial organs present.

Family
Oligacanthorhynchidae.

Order Apororhynchida

Trunk short, conical, may be curved ventrally; proboscis large, globular, with tiny spinelike hooks (which may not pierce the surface of the proboscis) arranged in several spiral rows; proboscis not retractable; neck absent or reduced; protonephridial organs absent.

Family
Apororhynchidae.

Class Palaeacanthocephala

Main longitudinal lacunar canals lateral; hypodermal nuclei fragmented, numerous, occasionally restricted to anterior half of trunk; nuclei of lemnisci and cement glands fragmented; spines present on trunk of some species; single ligament sac of female not persistent throughout life; protonephridia absent; cement glands separate, tubular to spheroid; eggs oval to elongated, sometimes with polar thickenings of second membrane; parasites of fishes, amphibians, reptiles, birds, and mammals.

Order Echinorhynchida

Trunk never pseudosegmented; proboscis cylindrical to spheroid, with longitudinal, regularly alternating rows of hooks; sensory papillae present or absent; proboscis receptacle double walled; proboscis retractor muscles pierce posterior end of receptacle; brain near middle or posterior end of receptacle; parasites of fishes and amphibians.

Families
Diplosentidae, Echinorhynchidae, Fessisentidae, Heteracanthocephalidae, Heterosentidae, Hypoechinorhynchidae, Illiosentidae, Pomporhynchidae, Rhadinorhynchidae, Cavisomidae, Arythmacanthidae.

Order Polymorphida

Proboscis spheroid to cylindrical, armed with numerous hooks in alternating longitudinal rows; proboscis receptacle double walled, with brain near center; parasites of reptiles, birds, and mammals.

Families
Centrorhynchidae, Plagiorhynchidae, Polymorphidae.

Class Eoacanthocephala

Main longitudinal lacunar canals dorsal and ventral, often no larger in diameter than irregular transverse commissures; hypodermal nuclei few, giant, sometimes ameboid; proboscis receptacle single walled; proboscis retractor muscle pierces posterior end of receptacle; brain near anterior or middle of receptacle; nuclei of lemnisci few, giant; two persistent ligament sacs in female; protonephridia absent; cement gland single, syncytial, with several nuclei, with cement reservoir appended; eggs variously shaped; parasites of fish, amphibians, and reptiles.

Order Gyracanthocephalida

Trunk small or medium size, spined; proboscis small, spheroid, with a few spiral rows of hooks.

Family
Quadrigyridae.

Order Neoechinorhynchida

Trunk small to large, unarmed; proboscis spheroid to elongated, with hooks arranged variously.

Families
Neoechinorhynchidae, Tenuisentidae, Dendronucleatidae.

Class Polyacanthocephala

Trunk spinose; hypodermic nuclei many and small; main longitudinal lacunar canals dorsal and ventral; many hooks, in longitudinal rows; proboscis receptacle single walled; cement glands elongated, with giant nuclei; protonephridia absent; parasites of fishes and (?) crocodilians.

Order Polyacanthorhynchida

With characters of the class.

Family
Polyacanthorhynchidae.

References

1. Alley, Z., J. Moore, and N. J. Gotelli. 1992. *Moniliformis moniliformis* infection has no effect on some behaviors of the cockroach *Diploptera punctata*. *J. Parasitol.* 78:524–26.

2. Associated Press. 1983. Austin tot suffers parasite usually found only in pigs. *Lubbock Avalanche Journal* (6, July).

3. Bentley, C. R., and H. Hurd. 1993. *Pomphorhynchus laevis* (Acanthocephala): Elevation of haemolymph protein concentrations in the intermediate host, *Gammarus pulex* (Crustacea: Amphipoda). *Parasitology* 107:193–98.

4. Bryant, C., and W. L. Nicholas. 1966. Studies on the oxidative metabolism of *Moniliformis dubius* (Acanthocephala). *Comp. Biochem. Physiol.* 17:825–40.

5. Butterworth, P. E. 1969. The development of the body wall of *Polymorphus minutus* (Acanthocephala) in the intermediate host, *Gammarus pulex*. *Parasitology* 59:373–88.

6. Byram, J. E., and F. M. Fisher Jr. 1973. The absorptive surface of *Moniliformis dubius* (Acanthocephala). I. Fine structure. *Tissue Cell* 5:553–79.

7. Crompton, D. W. T. 1970. *An ecological approach to acanthocephalan physiology*. Cambridge: Cambridge University Press.

8. Crompton, D. W. T., A. Keymer, A. Singhvi, and M. C. Nesheim. 1983. Rat-dietary fructose and the intestinal distribution and growth of *Moniliformis* (Acanthocephala). *Parasitology* 86:57–71.

9. Crompton, D. W. T., A. Singhvi, and A. Keymer. 1982. Effects of host dietary fructose on experimentally stunted *Moniliformis* (Acanthocephala). *Int. J. Parasitol.* 12:117–21.

10. Dingley, D., and P. C. Beaver. 1985. *Macracanthorhynchus ingens* from a child in Texas. *Am. J. Trop. Med. Hyg.* 34:918–20.

11. Dunagan, T. T., and D. M. Miller. 1970. Major nerves in the anterior nervous system of *Macracanthorhynchus hirudinaceus* (Acanthocephala). *Comp. Biochem. Physiol.* 37:235–42.

12. Dunagan, T. T., and D. M. Miller. 1975. Anatomy of the cerebral ganglion of the male acanthocephalan, *Moniliformis dubius*. *J. Comp. Neurol.* 164:483–94.

13. Dunagan, T. T., and D. M. Miller. 1986. A review of protonephridial excretory systems in Acanthocephala. *J. Parasitol.* 72:621–32.

14. Goldsmid, J. M., M. E. Smith, and F. Fleming. 1974. Human infections with *Moniliformis* sp. in Rhodesia. *Ann. Trop. Med. Parasitol.* 68:363–64.

15. Golvan, Y. J. 1958. Le phylum Acanthocephala. Première note. Sa place dans l'échelle zoologique. *Ann. Parasitol.* 33:539–602.

16. Gotelli, N. J., and J. Moore. 1992. Altered host behaviour in a cockroach-acanthocephalan association. *Animal Behaviour* 43:949–59.

17. Grassi, B., and S. Calandruccio. 1888. Ueber einen Echinorhynchus, welcher auch in Menschen parasitirt und dessen Zwischenwirt ein Blaps ist. *Zentralbl. Bakt. Parasitenkd. Orig.* 3:521–25.

18. Hamann, O. 1892. Das system der Acanthocephalen. *Zool. Anz.* 15:195–97.

19. Hammond, R. A. 1966. The proboscis mechanism of *Acanthocephalus ranae*. *J. Exp. Biol.* 45:203–13.

20. Hightower, K., D. M. Miller, and T. T. Dunagan. 1975. Physiology of the body wall muscles in an acanthocephalan. *Proc. Helm. Soc. Wash.* 42:71–80.

21. Hopp, W. B. 1954. Studies on the morphology and life cycle of *Neoechinorhynchus emydis* (Leidy), an acanthocephalan parasite of the map turtle, *Graptemys geographica* (La Sueur). *J. Parasitol.* 40:284–99.

22. Kates, K. C. 1943. Development of the swine thorn-headed worm, *Macracanthorhynchus hirudinaceus,* in its intermediate host. *Am. J. Vet. Res.* 4:173–81.

23. Keymer, A., D. W. T. Crompton, and D. E. Walters. 1983. Parasite population biology and host nutrition: Dietary fructose and *Moniliformis* (Acanthocephala). *Parasitology* 87:265–78.

24. Körting, W., and D. Fairbairn. 1972. Anaerobic energy metabolism in *Moniliformis dubius* (Acanthocephala). *J. Parasitol.* 58:45–50.

25. Krapf, K., and T. T. Dunagan. 1987. Structural features of the protonephridia in female *Macracanthorhynchus hirudinaceus* (Acanthocephala). *J. Parasitol.* 73:1176–81.

26. Lankester, R. 1900. *A treatise on zoology.* London: Adam & Charles Black.

27. Lumsden, R. D. 1975. Surface ultrastructure and cytochemistry of parasitic helminths. *Exp. Parasitol.* 37:267–339.

28. Miller, D. M., and T. T. Dunagan. 1976. Body wall organization of the acanthocephalan, *Macracanthorhynchus hirudinaceus:* A reexamination of the lacunar system. *Proc. Helm. Soc. Wash.* 43:99–106.

29. Miller, D. M., and T. T. Dunagan. 1977. The lacunar system and tubular muscles in Acanthocephala. *Proc. Helm. Soc. Wash.* 44:201–5.

30. Miller, D. M., and T. T. Dunagan. 1978. Organization of the lacunar system in the acanthocephalan, *Oligacanthorhynchus tortuosa*. *J. Parasitol.* 64:436–39.

31. Miller, D. M., and T. T. Dunagan. 1983. A support cell to the apical and lateral sensory organs in *Macracanthorhynchus hirudinaceus* (Acanthocephala). *J. Parasitol.* 69:534–38.

32. Miller, D. M., and T. T. Dunagan. 1985. New aspects of acanthocephalan lacunar system as revealed in anatomical modeling by corrosion cast method. *Proc. Helm. Soc. Wash.* 52:221–26.

33. Miller, D. M., T. T. Dunagan, and J. Richardson. 1973. Anatomy of the cerebral ganglion of the female acanthocephalan, *Macracanthorhynchus hirudinaceus*. *J. Comp. Neurol.* 152:403–15.

34. Moore, J. 1983. Responses of an avian predator and its isopod prey to an acanthocephalan parasite. *Ecology* 64:1000–15.

35. Moore, J. 1984. Altered behavioral responses in intermediate hosts—an acanthocephalan parasite strategy. *Am. Natural.* 123:572–77.

36. Moore, J., and N. J. Gotelli. 1992. *Moniliformis moniliformis* increases cryptic behaviors in the cockroach *Supella longipalpa*. *J. Parasitol.* 78:49–53.

37. Nicholas, W. L. 1973. The biology of the Acanthocephala. In Dawes, B., ed. *Advances in parasitology* 11. New York: Academic Press, Inc., 671–706.

38. Nickol, B. B., and D. F. Oetinger. 1968. *Prosthorhynchus formosus* from the short-tailed shrew (*Blarina brevicauda*) in New York State. *J. Parasitol.* 54:456.

39. Schmidt, G. D. 1971. Acanthocephalan infections of man, with two new records. *J. Parasitol.* 57:582–84.

40. Schmidt, G. D. 1973. Early embryology of the acanthocephalan *Mediorhynchus grandis* Van Cleave, 1916. *Trans. Am. Microsc. Soc.* 92:512–16.

41. Schmidt, G. D., and O. W. Olsen. 1964. Life cycle and development of *Prosthorhynchus formosus* (Van Cleave 1918) Travassos, 1926, an acanthocephalan parasite of birds. *J. Parasitol.* 50:721–30.

42. Starling, J. A. 1975. Tegumental carbohydrate transport in intestinal helminths: Correlation between mechanisms of membrane transport and the biochemical environment of absorptive surfaces. *Trans. Am. Microsc. Soc.* 94:508–23.

43. Starling, J. A., and F. M. Fisher Jr. 1975. Carbohydrate transport in *Moniliformis dubius* (Acanthocephala). I. The kinetics and specificity of hexose absorption. *J. Parasitol.* 61:977–90.

44. Starling, J. A., and F. M. Fisher Jr. 1979. Carbohydrate transport in *Moniliformis dubius* (Acanthocephala). III. Post-absorptive fate of fructose, mannose, and galactose. *J. Parasitol.* 65:8–13.

45. Taraschewski, H., and U. Mackenstedt. 1991. Autoradiographic and morphological studies on the uptake of the triglyceride [^{3}H]-glyceroltrioleate by acanthocephalans. *Parasitol. Res.* 77:247–54.

46. Taraschewski, H., and U. Mackenstedt. 1991. Autoradiographic and morphological investigations on the uptake and incorporation of tritiated lysin by acanthocephalans. *Parasitol. Res.* 77:536–41.

47. Uglem, G. L., and O. R. Larson. 1969. The life history and larval development of *Neochinorhynchus saginatus* Van Cleave and Bangham, 1949 (Acanthocephala: Neoechinorhynchidae). *J. Parasitol.* 55:1212–17.

48. Uglem, G. L., P. W. Pappas, and C. P. Read. 1973. Surface aminopeptidase in *Moniliformis dubius* and its relation to amino acid uptake. *Parasitology* 67:185–95.

49. Uglem, G. L., and C. P. Read. 1973. *Moniliformis dubius:* Uptake of leucine and alanine by adults. *Exp. Parasitol.* 34:148–53.

50. Van Cleave, H. J. 1941. Relationships of the Acanthocephala. *Am. Natural.* 75:31–47.

51. Van Cleave, H. J. 1948. Expanding horizons in the recognition of a phylum. *J. Parasitol.* 34:1–20.

52. Ward, H. L. 1940. Studies on the life-history of *Neoechinorhynchus cylindratus* (Van Cleave 1913) (Acanthocephala). *Trans. Am. Microsc. Soc.* 59:327–47.

53. Whitfield, P. J. 1970. The egg sorting function of the uterine bell of *Polymorphus minutus* (Acanthocephala). *Parasitology* 61:111–26.

54. Whitfield, P. J. 1971. The locomotion of the acanthor of *Moniliformis dubius* (Archiacanthocephala). *Parasitology* 62:35–47.

55. Whitfield, P. J. 1971. Spermiogenesis and spermatozoan ultrastructure in *Polymorphus minutus* (Acanthocephala). *Parasitology* 62:415–30.

56. Wong, B. S., D. M. Miller, and T. T. Dunagan. 1979. Electrophysiology of acanthocephalan body wall muscles. *J. Exp. Biol.* 82:273–80.

57. Wright, R. D., and R. D. Lumsden. 1968. Ultrastructural and histochemical properties of the acanthocephalan epicuticle. *J. Parasitol.* 54:1111–23.

58. Zhao, B., and M. X. Wang. 1992. Ultrastructural study of the defense reaction against the larvae of *Macracanthorhynchus hirudinaceus* in laboratory-infected beetles. *J. Parasitol.* 78:1098–1101.

Additional References

Amin, O. M. 1987. Key to the families and subfamilies of Acanthocephala, with the erection of a new class (Polyacanthocephala) and a new order (Polyacanthorhynchida). *J. Parasitol.* 73:1216–19.

Bullock, W. L. 1969. Morphological features as tools and as pitfalls in acanthocephalan systematics. In Schmidt, G. D., ed. *Problems in systematics of parasites*. Baltimore: University Park Press, 9–43. A useful, philosophical discussion of the subject, with recommended techniques for study.

Crompton, D. W. T. 1970. *An ecological approach to acanthocephalan physiology*. Cambridge: Cambridge University Press. An outstanding summation of the subject.

Crompton, D. W. T. 1975. *Relationships between Acanthocephala and their hosts. Symposium of the Society for Experimental Biology, vol. 29. Symbiosis.* Cambridge: Cambridge University Press, 467–504.

Crompton, D. W. T., and B. B. Nickol, eds. 1985. *Biology of the Acanthocephala*. Cambridge: Cambridge University Press.

Golvan, Y. J. 1969. Systématiques des Acanthocéphales (Acanthocéphala Rudolphi 1801). L'ordre des Palaeacanthocephala Meyer 1931. La super-famille des Echinorhynchoidea (Cobbold 1876) Golvan et Houin 1963. *Mem. Mus. Nat. Hist. Nat.* 47:1–373. An excellent account of this important superfamily. Besides descriptions of each species, it contains a key to genera and a host list.

Pappas, P. W., and C. P. Read. 1975. Membrane transport in helminth parasites: A review. *Exp. Parasitol.* 37:469–530.

Petrochenko, V. I. 1956, 1958. *Acanthocephala of domestic and wild animals,* 1 and 2. Moscow: Akademii Nauk SSSR. English translations: Israel Program for Scientific Translations, 1971. An indispensable resource for students of the phylum. Descriptions are given for nearly every species known at the time of writing.

Schmidt, G. D. 1969. Acanthocephala as agents of disease in wild mammals. *Wildl. Dis.* 53:1–10.

Schmidt, G. D. 1972. Revision of the class Archiacanthocephala Meyer, 1931 (Phylum Acanthocephala), with emphasis on Oligacanthorhynchidae Southwell and MacFie, 1925. *J. Parasitol.* 58:290–97. A modern classification of this difficult class.

Yamaguti, S. 1963. *Systema Helminthum, vol. 5. Acanthocephala.* New York: Interscience. In most regards a practical key to genera of Acanthocephala known to 1963. Lists of species and their hosts are included.

Chapter 32

PHYLUM PENTASTOMIDA: TONGUE WORMS

Pentastome work warms up all the time. . . . The day after Labor Day I take off for Africa and ultimately will visit Nigeria, Kenya, Bombay, Bangkok, Kuala Lumpur, Taiwan, and Okinawa returning home the 28th of September.

J. T. Self, in a letter to one of the authors, July 7, 1969

The pentastomids, or tongue worms, are wormlike parasites of the respiratory systems of vertebrates. About 100 species are known. As adults, most live in the respiratory system of reptiles, especially snakes, lizards, amphibians, and crocodilians, but one species lives in the air sacs of sea birds, and another inhabits the nasopharynx of canines and felines. The latter species is occasionally found as transient nymphs in the nasopharynx of humans; other species, in their nymphal stages, also parasitize humans. Thus, pentastomids are certainly of zoological interest and also are of some medical importance.[25]

The evolutionary relationships of pentastomids are obscured by their seemingly aberrant adult morphology. Certain similarities with the Annelida have been pointed out, but most modern taxonomists align them with the Arthropoda.[18,22] It is possible that the group reached its zenith in the Mesozoic age of reptiles and that today's few species are relicts derived from those ancestors. Wingstrand[28] proposed that the Pentastomida be regarded as an order of the crustacean class Branchiura (Chapter 34). Riley and coworkers[19] more or less agreed on the basis of embryogenesis, tegumental structure, and gametogenesis, concluding that pentastomids should be regarded as a subclass of Crustacea, closely allied to the Branchiura.

Wingstrand's[28] conclusion was based on a demonstration that the spermatozoa of the two groups are almost identical with regard to structure and development and that this type of spermatozoon represents a type of its own, not encountered in other animals. His argument is strengthened by the fact that each major crustacean group is characterized by its own type of spermatozoon, and if the Pentastomida and the Branchiura were unrelated, their sperm structure and development would represent a most extraordinary example of convergence in detail. Subsequently Storch and Jamieson[26] confirmed the findings of Wingstrand and Riley and his coworkers using morphological characters, and Abele and his coworkers[1] did the same using 18S ribosomal RNA nucleotide sequences. Only infrequently do great parasitological mysteries get solved, but

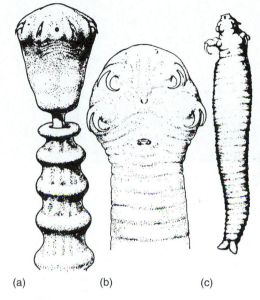

FIGURE 32.1

Examples of pentastome body types. (*a*) Anterior end of *Armillifer annulatus*; (*b*) head of *Leiperia gracilis*; (*c*) entire specimen of *Raillietiella mabuiae.*

Modified from R. Heymons, Pentastomida. Copyright © 1935. In Bronn's Klass. Ord. Tier. 5:4, book 1, in J. G. Baer, *Ecology of Animal Parasites.* Copyright © 1952 The University of Illinois Press, Urbana, IL.

this one—the taxonomic home of the pentastomids—appears to be nearing that point. We retain the Pentastomida as a separate phylum, however, because although the basic scientific observations to support inclusion in the Crustacea have been made, a revised classification has yet to be published in accordance with the rules of zoological nomenclature.

MORPHOLOGY

The body of a pentastomid (Fig. 32.1) is elongated, usually tapering toward the posterior end and often showing distinct segmentation, forming numerous **annuli.** It is indistinctly divided into an anterior **forebody** and a posterior **hindbody,** which is bifurcated at its tip in some species.

The exoskeleton contains chitin,[27] which is sclerotized around the mouth opening and accessory genitalia. A striking characteristic of all adult pentastomids is the presence of two pairs of sclerotized hooks in the mouth region (Fig. 32.2). These may be located at the ends of stumpy stalks or may be

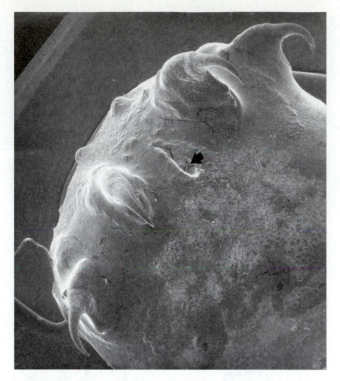

FIGURE 32.2

Anterior end of a pentastome. Note both the mouth (*arrow*) between the middle hooks and the apical sensory papillae.

Courtesy of John Ubelaker.

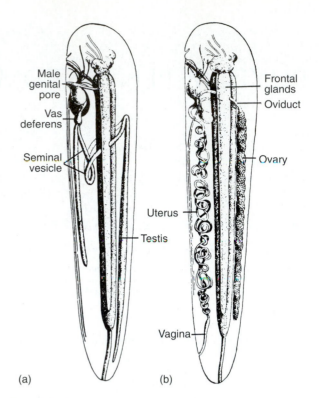

FIGURE 32.3

Reproductive systems of the pentastome *Waddycephalus teretiuscules.* (*a*) Male; (*b*) female.

Modified from W. B. Spencer, in R. Heymons, Pentastomida. Copyright © 1935. In Bronn's Klass. Ord. Tier. 5:4, book 1, in J. G. Baer, *Ecology of Animal Parasites.* Copyright © 1952 The University of Illinois Press, Urbana, IL.

nearly flush with the surface of the cephalothorax; in either case they can be withdrawn into cuticular pockets. The hooks are single in some species and double in others. The apparently double hooks are actually single, with an accessory hooklike protrusion of the cuticle. In some species the hook articulates against a basal fulcrum. The hooks are manipulated by powerful muscles and are used to tear and embed the mouth region into host tissues.

The body cuticle in some species also has circular rows of simple spines; the annuli may overlap enough to make the abdomen look serrated. There usually are transverse rows of cuticular glands, with conspicuous pores, which apparently function in regulation of the hydromineral balance in the hemolymph.[4]

The cuticle is similar to that of the arthropods, although it is thin and weak.[27] The muscles, too, are arthropodan in nature, being striated and segmentally arranged. The only sensory structures so far recognized are papillae, especially on the exoskeleton of the cephalothorax. The digestive system is simple and complete, with the anus opening at the posterior end of the abdomen. The mouth is permanently held open by its sclerotized lining, the **cadre,** which may be circular, oval, or U shaped and is an important taxonomic character. The nervous system is arthropodan in nature and has been described by Doucet.[6]

Pentastomids are dioecious and show sexual dimorphism in that males are usually smaller than females. The male has a single, tubular testis (two in *Linguatula*), which occupies one-third to one-half of the body cavity (Fig. 32.3*a*). It is con-

tinuous with a seminal vesicle, which in turn connects to a pair of ejaculatory organs. These each have a duct extending to a terminal penis that fits into a **dilator organ.** The dilator organ, which is usually sclerotized, serves as an intromittent organ in some species and as a dilator and guide for the penis in others. The male genital pore is midventral on the anterior abdominal segment, near the mouth.

In the female a single ovary extends nearly the length of the body cavity (Fig. 32.3*b*). It may bifurcate at its distal end to become two oviducts. These unite to form the uterus. The oviducts and uterus usually are extensively coiled within the body. One or more diverticulae of the uterus serve as seminal receptacles. The uterus terminates as a short vagina that opens through the female gonopore, at the anterior end of the abdomen in the order Cephalobaenida or at the posterior end in the order Porocephalida.[9] The female mates once; the male may be polygamous.

BIOLOGY

Adult pentastomids feed on tissue fluids and blood cells of their host. They appear to stimulate a strong host immune response, but because they are long lived, they must evade its consequences.[20] Their frontal and subparietal glands elaborate a lamellate secretion (Excretory/Secretory, E/S), which is poured over the entire surface of the cuticle, and this may protect vital areas of the parasite from antibody action.[11,20]

Host cells get caught up in this covering of E/S products, evidently reducing the inflammatory response that occurs each time the parasite molts. In experiments with *Porocephalus crotali* in mice, sustained cytotoxic responses killed parasites whose E/S coating had been removed.[3]

The female, depending on its size, may produce several million fully embryonated eggs, which pass up the trachea of its host, are swallowed, and then pass out with the feces. The intact egg appears to be surrounded by two shell membranes—an outer, thin membrane and an inner, thick one.[8] The inner layer, however, consists of three distinct layers. A characteristic of the pentastomid egg is the **facette,** a permanent, funnel-shaped opening through the inner membrane complex, with an inner opening extending toward the larva. In the embryo a gland called the **dorsal organ** (to be described further) secretes a mucoid substance that pours through the facette and ruptures the original outer membrane, which is lost. The mucoid material then flows over the inner membrane to form a new outer membrane, which is sticky when wet.[14] The viscid eggs cling together, sometimes resulting in massive infections in the intermediate host. The eggs can withstand drying for at least two weeks, in the case of *Porocephalus crotali,* and they can remain viable in water at refrigerator temperatures for about six months.[8]

The larva that hatches from the egg is an oval, tailed creature with four stumpy legs, each with one or two retractable claws. The claws are manipulated by a combination of muscle fibers and an inner hydraulic mechanism. A **penetration organ** is located at the anterior end of the body. This is composed of a median spear and two lateral, pointed forks; together with the clawed legs, these structures can tear through the tissues of the intermediate host. A pair of ducts open on either side of the median spear. Accessory spinelets are present around the penetration organ of many species.

Between the anterior legs is a simple mouth, surrounded by a U-shaped sclerotized cadre. An esophagus extends into the dorsal part of the body and expands into a blind sac; a thin hindgut is present in some species. Within the body cavity are a number of irregular giant cells, some with neutrophilic and others with eosinophilic granules. Their function is unknown. A consistent feature of the pentastomid embryo is the dorsal organ, referred to before. It consists of a number of gland cells surrounding a central hollow vesicle. The vesicle opens through the cuticle by a dorsal pore.

Complete life cycles are known for few species, but partial information is available for several more. With the exceptions of *Reighardia sternae* in birds and a few species of *Linguatula* in mammals, all pentastomids mature in reptiles. The intermediate hosts are various fishes, amphibians, reptiles, insects, and mammals. Typically, after ingestion by a poikilothermous vertebrate, the larva hatches and penetrates the intestine and migrates randomly in the body, finally becoming quiescent and metamorphosing into a nymph (Fig. 32.4). The nymph is infective to the definitive host; when eaten by the latter, it penetrates the host's intestine and bores into the lung, where it matures (Fig. 32.5). Each developmental stage undergoes one to several molts of the cuticle. The nymphal instars are difficult to differentiate. Some species even become sexually mature before completing the final ecdysis. A definitive host that eats the egg can also

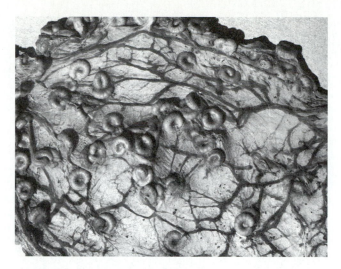

FIGURE 32.4

Nymphs of *Porocephalus* sp. in the mesenteries of a vervet, *Cercopithicus aethops.*

From J. T. Self, "Pentastomiasis: host responses to larval and nymphal infections," in *Trans. Am. Microsc. Soc.* 91:2–8. Copyright © 1972. Photograph by Robert E. Kuntz.

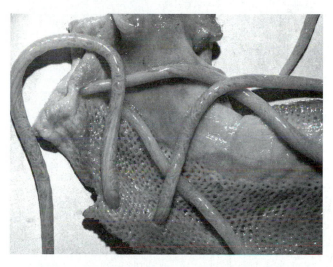

FIGURE 32.5

Kiricephalus pattoni in the lung of an Oriental rat snake, *Ptyas mucosus.* Courtesy of Robert E. Kuntz.

serve as intermediate host, similar to the case of *Trichinella.* The parasites, however, probably cannot migrate to the lung and mature. Whereas vertebrates are the intermediate hosts for the Porocephalida, cockroaches are used by some species of *Raillietiella,*[2,13] and *Reighardia sternae* in gulls has a direct life cycle. *Subtriquetra subtriquetra,* a parasite of the nasopharynx of South American crocodilians, differs in having a free-living larva, which somehow finds its fish intermediate host. Pentastomid reproductive biology was reviewed by Riley.[17]

Porocephalus crotali

The life cycle of *P. crotali* of crotalid snakes was experimentally demonstrated by Esslinger[7] (Fig. 32.6), using white mice as intermediate hosts; further elaboration was provided

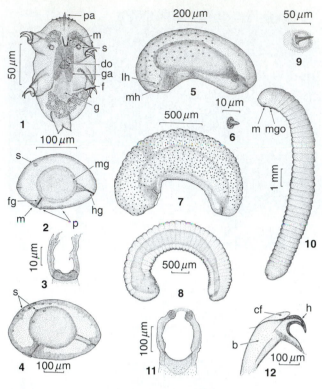

FIGURE 32.6

Developmental stages of *Porocephalus crotali* in experimental intermediate hosts (camera lucida drawings made from living specimens). **1,** primary larva (ventral view) after release from egg; **2,** first nymphal stage (nymph I) in left lateral view (all succeeding nymphs identically oriented); **3,** mouth ring of nymph I (en face view with anterior margin uppermost); **4,** nymph II; **5,** nymph III; **6,** lateral mouth hook of nymph III; **7,** nymph IV; **8,** nymph V (individual stigmata not shown); **9,** lateral mouth hook of nymph V; **10,** nymph VI (infective stage), male, removed from enveloping cuticle of nymph V; **11,** mouth ring of nymph VI; **12,** lateral mouth hook of nymph VI; **b,** base of mouth hook; **cf,** cuticular fold or auxiliary hook; **do,** dorsal organ; **f,** foot or leg; **fg,** foregut; **g,** gut; **ga,** ganglion; **h,** external clawlike portion of mouth hook; **hg,** hindgut; **lh,** lateral mouth hook; **m,** mouth ring; **mg,** midgut; **mgo,** male genital opening; **mh,** medial mouth hook; **p,** papilla; **pa,** penetrating apparatus; **s,** stigma.

From J. H. Esslinger, "Development of *Procephalus crotali* (Humboldt, 1808) (Pentastomida) in experimental intermediate hosts," in *J. Parasitol.* 48:452–456. Copyright © 1962. Reprinted with permission of the publisher.

by Riley.[16] On hatching, the larva penetrates the duodenal mucosa and works its way to the abdominal cavity. Complete penetration can be accomplished within an hour after the egg is swallowed. After wandering about for seven or eight days the larva molts and becomes lightly encapsulated in host tissue. The subsequent nymphal stages are devoid of the larval characteristics, having lost the legs, penetration apparatus, and tail. During the next 80 days or so, *P. crotali* molts five more times, gradually increasing in size and becoming segmented. The mouth hooks appear during the fourth nymphal instar and increase in size through subsequent ecdyses. The sexes can be differentiated after the fifth molt. After the sixth molt, the nymph becomes heavily encapsulated and dormant. When eaten by a snake, the nymph is activated; it quickly penetrates the snake's intestine and usually passes directly to

the lung, since the lung and intestine are adjacent to each other. It buries its forebody into lung tissues, feeds on blood and tissue fluids, and matures.

Linguatula serrata

Linguatula serrata (Fig. 32.7) is unusual among the Pentastomida in that the adults live in the nasopharyngeal region of mammals. Cats, dogs, foxes, and other carnivores are the normal hosts of this cosmopolitan parasite. Apparently almost any mammal is a potential intermediate host.

Adult *L. serrata* embed their forebody into the nasopharyngeal mucosa, feeding on blood and fluids. Females live at least two years and produce millions of eggs.[10] The eggs are about 90 by 70 μm, with an outer shell that wrinkles when dry. Eggs exit the host in nasal secretions or, if swallowed, with the feces. When swallowed by an intermediate host, the four-legged larva hatches in the small intestine, penetrates the intestinal wall, and lodges in tissues, particularly in lungs, liver, and lymph nodes. There the nymphal instars develop, with the infective stage becoming surrounded by host tissues. When eaten by a definitive host, the infective nymph either attaches in the upper digestive tract or quickly travels there from the stomach, eventually reaching the nasopharynx. Females begin egg production in about six months.

PATHOGENESIS

There are two types of **pentastomiasis** in humans. **Visceral pentastomiasis** results when eggs are eaten and nymphs develop in various internal organs, and **nasopharyngeal pentastomiasis** results when nymphs that are eaten locate in the nasopharynx. Both types are rather common in some parts of the world.

Visceral Pentastomiasis

Several species of pentastomids have been found encysted in humans. Probably the most commonly involved species is *Armillifer armillatus,* which has been reported from the liver, spleen, lungs, eyes, and mesenteries of people in, among other places, Africa, Malaysia, the Philippines, Java, and China.[5,24,25] Other reported species are *A. moniliformis, Pentastoma najae, L. serrata,* and *Porocephalus* sp.

Most infections cause few if any symptoms and therefore go undetected. In fact most recorded cases were found at autopsy, after death from other causes. However, infection of the spleen, liver, or other organs causes some tissue destruction. Ocular involvement may cause vision damage.[15] Prior visceral infection may sensitize a person, resulting in an allergy to subsequent infection.[12,23] The host response to nymphs is often highly inflammatory, although little pathological response is elicited in definitive reptilian hosts. Dead nymphs are often calcified and are sometimes detected in X-ray films. Others begin a slow deterioration, causing a mononuclear cell response, with a subsequent abscess and granuloma formation. Experimentally produced heavy infections in rodents may kill them, indicating that visceral pentastomiasis possibly may be more important in human medicine than usually is thought.[23]

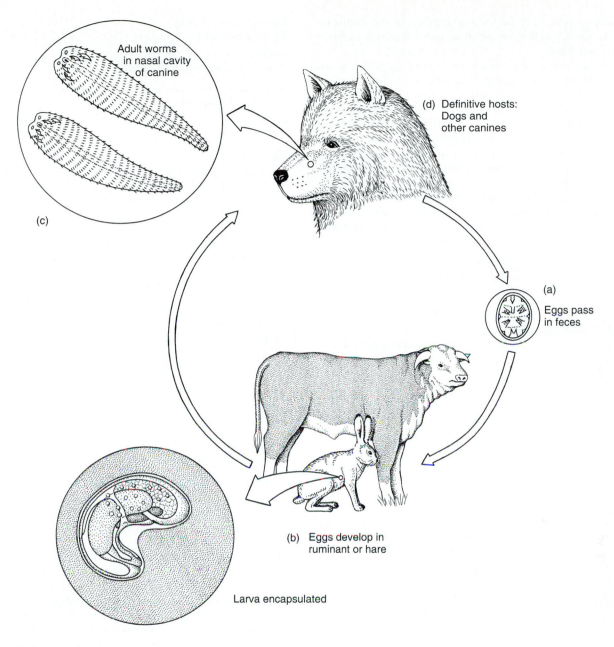

FIGURE 32.7

Life cycle of *Linguatula serrata*. Humans are dead end hosts in this case. (*a*) Eggs pass in feces. (*b*) Larvae develop in ruminant or hare. (*c*) Adult worms in nasal cavity of canine. (*d*) Definitive hosts: dogs and other canines.

Drawing by William Ober and Claire Garrison.

Nasopharyngeal Pentastomiasis

When nymphs of *L. serrata* invade the nasopharyngeal spaces of humans, they cause a condition usually called **halzoun** but also known as **marrara** or **nasopharyngeal linguatulosis.** According to Schacher and coworkers,[21] halzoun has been a clinically well recognized but etiologically obscure disease in the Levant since its original description by Khouri (1905); in the Sudan it is known as the marrara syndrome. In Lebanon the disease is linked in the popular mind with the eating of raw or undercooked

sheep or goat liver or lymph nodes; in the Sudan it is linked with the ingestion of various raw visceral organs of sheep, goats, cattle or camels. A few minutes to half an hour or more after eating, there is discomfort, and a prickling sensation deep in the throat; pain may later extend to the ears. Oedematous congestion of the fauces, tonsils, larynx, eustachian tubes, nasal passages, conjunctiva and lips is sometimes marked. Nasal and lachrymal discharges, episodic sneezing and coughing, dyspnoea, dysphagia, dysphonia and frontal headache are common.

Complications may include abscesses in the auditory canals, facial swelling or paralysis and sometimes asphyxiation and death.

At various times, this condition was suspected to be caused by the trematodes *Fasciola hepatica, Clinostomum complanatum,* and *Dicrocoelium dendriticum* and also by leeches. However, the recovery of *L. serrata* nymphs from the nasal passages and throats of patients in India, Turkey, Greece, Morocco, and Lebanon indicates that this species is the main cause of the condition in these areas. It is possible that the parasites can become mature if not removed or lost initially.

The epidemiology of this condition depends on cultural food patterns, in which nymphs are ingested when visceral organs, primarily liver or mesenteric lymph nodes of domestic herbivores, are consumed raw or undercooked.[21]

CLASSIFICATION OF PHYLUM PENTASTOMIDA

This classification follows that of Riley.[18]

Order Cephalobaenida
Mouth anterior to hooks; hooks lacking fulcrum; vulva at anterior end of abdomen.

Family Cephalobaenidae
Parasites of snakes, lizards, and amphibians.

Genera
Cephalobaena, Raillietiella.

Family Reighardiidae
Parasites of marine birds.

Genus
Reighardia.

Order Porocephalida
Mouth between or below level of anterior hooks; hooks with fulcrum; vulva near posterior end of body.

Family Sebekidae
Parasites of crocodilians and chelonians.

Genera
Sebekia, Alofia, Leiperia.

Family Subtriquetridae
Parasites of crocodilians.

Genus
Subtriquetra.

Family Sambonidae
Parasites of monitor lizards and snakes.

Genera
Sambonia, Elenia, Waddycephalus, Parasambonia.

Family Diesingidae
Parasites of chelonians.

Genus
Diesingia.

Family Porocephalidae
Parasites of snakes.

Genera
Porocephalus, Kiricephalus.

Family Armilliferidae
Parasites of snakes.

Genera
Armillifer, Cubirea, Gigliolella.

Family Linguatulidae
Parasites of mammals.

Genus
Linguatula.

References

1. Abele, L. G., W. Kim, and B. E. Felgenhauer. 1989. Molecular evidence for inclusion of the phylum Pentastomida in the Crustacea. *Mol. Biol. Evol.* 6:685–91.

2. Ali, J. H., and J. Riley. 1983. Experimental life-cycle studies of *Raillietiella gehyrae* Bovien, 1927 and *Raillietiella frenatus* Ali, Riley and Self, 1981: Pentastomid parasites of geckos utilizing insects as intermediate hosts. *Parasitology* 86:147–60.

3. Ambrose, N. C., and J. Riley. 1989. Further evidence for the protective role of sub-parietal cell membranous secretory product on the cuticle of a pentastomid arthropod parasite developing in its rodent intermediate host. *Tissue and Cell* 21:699–722.

4. Banaja, A. A., J. L. James, and J. Riley. 1977. Observations on the osmoregulatory system of pentastomids: The tegumental chloride cells. *Int. J. Parasitol.* 7:27–40.

5. Dönges, J. 1966. Parasitäre abdominalcysten bei Nigeriarern. *Z. Trop. Parasitol.* 17:252–56.

6. Doucet, J. 1965. Contribution à l'étude anatomique, histologique et histochimique des pentastomes (Pentastomida). *Mem. Office Rech. Sci. Tech. Outre-Mer Paris* 14:1–150.

7. Esslinger, J. H. 1962. Development of *Porocephalus crotali* (Humboldt, 1808) (Pentastomida) in experimental intermediate hosts. *J. Parasitol.* 48:452–56.

8. Esslinger, J. H. 1962. Morphology of the egg and larva of *Porocephalus crotali* (Pentastomida). *J. Parasitol.* 48:457–62.

9. Fain, A. 1961. Les pentastomides d'Afrique central. *Mus. R. Afr. Cent. Ann.* 92:1–115.

10. Hobmeier, A., and M. Hobmeier. 1940. On the life cycle of *Linguatula rhinaria. Am. J. Trop. Med.* 20:199–210.

11. Jones, D. A. C., R. J. Henderson, and J. Riley. 1992. Preliminary characterization of the lipid and protein components of the protective surface membranes of a pentastomid *Porocephalus crotali. Parasitology* 104:469–78.

12. Khalil, G. M., and J. F. Schacher. 1965. *Linguatula serrata* in relation to halzoun and the marrara syndrome. *Am. J. Trop. Med. Hyg.* 14:736–46.

13. Lavoippierre, M. M. J., and M. Lavoippierre. 1966. An arthropod intermediate host of a pentastomid. *Nature* 210:845–46.

14. Osche, G. 1963. Die systematische Stellung und Phylogenie der Pentastomids. Embryologische und vergleichendanatomische Studien an *Reighardia sternae. Z. Morphol. Ökol. Tiere* 52:487–596.

15. Rendtdorff, R. C., M. W. Deiwesse, and W. Murrah. 1962. The occurrence of *Linguatula serrata,* a pentastomid, within the human eye. *Am. J. Trop. Med. Hyg.* 11:762–64.

16. Riley, J. 1981. An experimental investigation of the development of *Porocephalus crotali* (Pentastomida: Porocephalida) in the western diamondback rattlesnake (*Crotalus atrox*). *Int. J. Parasitol.* 11:127–32.

17. Riley, J. 1983. Recent advances in our understanding of pentastomid reproductive biology. *Parasitology* 86:59–83.

18. Riley, J. 1986. The biology of pentastomids. In Baker, J. R., and R. Muller, eds. *Advances in parasitology* 25. New York: Academic Press, Inc. 45–128.

19. Riley, J., A. A. Banaja, and J. L. James. 1978. The phylogenetic relationships of the Pentastomida: The case for their inclusion within the Crustacea. *Int. J. Parasitol.* 8:245–54.

20. Riley, J., J. L. James, and A. A. Banaja. 1979. The possible role of the frontal and sub-parietal gland systems of the pentastomid *Reighardia sternae* (Diesing, 1864) in the evasion of the host immune response. *Parasitology* 78:53–66.

21. Schacher, J. F., S. Saab, R. Germanos, and N. Boustany. 1969. The aetiology of halzoun in Lebanon: Recovery of *Linguatula serrata* nymphs from two patients. *Trans. R. Soc. Trop. Med. Hyg.* 63:854–58.

22. Self, J. T. 1969. Biological relationships of the Pentastomida: A bibliography of the Pentastomida. *Exp. Parasitol.* 24:63–119.

23. Self, J. T. 1972. Pentastomiasis: Host responses to larval and nymphal infections. *Trans. Am. Microsc. Soc.* 91:2–8.

24. Self, J. T., H. C. Hopps, and A. O. Williams. 1972. Porocephaliasis in man and experimental mice. *Exp. Parasitol.* 32:117–26.

25. Self, J. T., H. C. Hopps, and A. O. Williams. 1975. Pentastomiasis in Africans, a review. *Trop. Geogr. Pathol.* 27:1–13.

26. Storch, V., and B. G. M. Jamieson. 1992. Further spermatological evidence for including the Pentastomida (tongue worms) in the Crustacea. *Int. J. Parasitol.* 22:95–108.

27. Trainer, J. E. Jr., J. T. Self, and K. H. Richter. 1975. Ultrastructure of *Porocephalus crotali* (Pentastomida) cuticle with phylogenetic implications. *J. Parasitol.* 61:753–58.

28. Wingstrand, K. G. 1972. Comparative spermatology of a pentastomid, *Raillietiella hemidactyli,* and a branchiuran crustacean, *Argulus foliaceus,* with a discussion of pentastomid relationships. *Kong. Danske Vidensk. Selsk. Biol. Skrift.* 19:1–72.

Chapter 33

PHYLUM ARTHROPODA: FORM, FUNCTION, AND CLASSIFICATION

Marvels indeed they are, and a feast to the eye and intellect of anyone interested in polymorphisms, local races, rare aberrations, teratological specimens, gynandromorphs, intersexes, and mosaics . . . and every aspect of mendelian genetics.

Cyril Clarke

The phylum Arthropoda includes an enormous assemblage of both fossil and extant species that far outnumbers all other known animals put together. Nearly a million species of insects have been described, and almost a quarter of these are beetles. There are over 50,000 species of arachnids, and another 30,000 of crustaceans. And yet the diversity of arthropods today may be lower than it was half a billion years ago. Recent analysis of Cambrian fossils reveals a number of arthropod body plans not represented among living groups.[9]

Arthropods are involved in virtually every kind of parasitic relationship. They serve as both definitive and intermediate hosts for protozoans, flatworms, nematodes, and even other arthropods. They also function as vectors, transmitting infective stages of parasites to vertebrates, including humans and domestic animals. Many arthropods, such as fleas, ticks, and some of the crustaceans, are highly adapted parasites in their own right. Arthropod life cycles, including those of vectors, are as varied as their structure, and thus they present us with a seemingly never-ending series of challenges in our attempts to control some parasitic infections. Their large populations and rapid reproductive rate contribute to their ability to evolve genetic resistance to pesticides. And there is plenty of evidence that arthropod-borne diseases have been the deciding factor in numerous military operations, thus accomplishing a defeat or ensuring a victory for which a human commander subsequently received blame or claimed credit.[34]

Traditionally the arthropods have composed a single phylum of metameric, coelomate animals. Arthropods share many features with annelids, such as metamerism and a nervous system consisting of supraesophageal ganglia, nerves encircling the esophagus, and a ventral series of segmental ganglia. Such similarities have led to claims that the two phyla are related, and that arthropods likely evolved from

annelidlike ancestors. But a century of comparative embryology, reanalysis of a remarkable set of Canadian fossils, and modern cladistic methods have all helped clarify the still murky picture of early arthropod evolution. At present some scientists consider the Hexapoda (insects), Myriapoda (centipedes and millipedes), and Onychophora (wormlike tropical and subtropical organisms) a monophyletic group, the Uniramia, probably descended from an ancestor in common with the annelids. The Crustacea and Chelicerata may not be very closely related to annelids, nor to the Uniramia, nor, for that matter, to each other, except to the extent that all are derived from spiral-cleaving, Cambrian or pre-Cambrian protostomes. And that last statement can be made for many present day phyla.[2,5,16,17,18,20]

Two features have contributed significantly to the success of arthropods: relatively small size and a chitinous exoskeleton. Although there are large species (lobsters, king crabs), the vast majority of arthropods are less than 1 cm in length. The planet provides many places that small organisms can occupy: spaces between sand grains, for example, or cracks in tree bark, and of course the bodies of animals. As a general rule, complex environments support relatively diverse faunas and floras, and on a small scale the earth is an exceedingly complex environment. Small species, especially parasitic ones, therefore have a rich supply of potential ecological niches. In a number of the previous chapters arthropods have been discussed mainly in their roles as intermediate hosts and vectors; in the following chapters they will be discussed as parasites, with the understanding that the parasitic lifestyle is what makes some of them important transmitters of infectious disease. This chapter is intended to be an easily accessible source of reference material on basic arthropod biology, for use in reviewing some aspect of that biology to place arthropod parasitism into its proper context.

GENERAL FORM AND FUNCTION

Arthropod Metamerism

Segmentation, or **metamerism,** is the overriding structural feature of arthropods. We believe that in the ancestral condition each segment bore a pair of jointed appendages, although most living arthropods have lost some of these appendages, particularly ones on the abdomen, during their evolutionary history. Each segment (or **somite**) of an arthropod's body is a

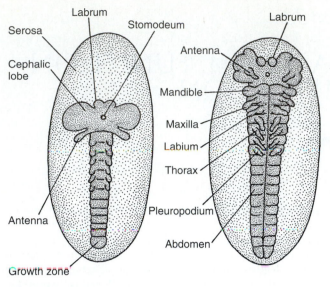

FIGURE 33.1

Two stages in the embryonic development of the grain beetle, *Tenebrio* sp., showing the undifferentiated mouth appendages and legs as relatively similar serial homologs.

From D. T. Anderson, *Embryology and Phylogeny in Annelids and Arthropods.* Copyright © 1973 Pergamon Press, Oxford. Reprinted with permission.

serial homolog of every other segment; that is, one segment can be thought of as a repeat expression of the genes that were expressed during construction of another segment. This basic plan is manifested in many arthropod embryos, which pass through a stage in which they consist of a linear series of relatively similar tissue blocks, each with generalized, and similar, appendage buds (Fig. 33.1).

Most adult arthropods are "modified" in the sense that segments are grouped into body regions, a feature known as **tagmatism** or **tagmatization.** The different body regions are called **tagmata** (singular **tagma**). As a result of grouping and subsequent development, individual skeletal plates may be shifted out of their embryonic positions or even fused together. Thus, the fundamental arthropod body architecture is not always obvious from adults, especially from their exterior. In this regard arthropod metamerism differs somewhat from that of the annelids, which also exhibit fusion and grouping of segments but not nearly to the extreme degree as do many arthropods (see Figs. 33.9 and 33.11 on pp. 500 and 501).

Metameres of two separate species, which correspond in relative position and embryological origin, are said to be **special homologs.** Much of the evolutionary history of arthropods is written in the variation, divergence in form and function, of both serial and special homologs. Evidence for the homology of metameres and their appendages comes mainly from a study of embryology and internal anatomy, especially nerve tracts.

Exoskeleton

The **cuticular exoskeleton** consists mostly of tanned proteins and chitin; is usually hard, virtually indigestible, and insoluble; may be impregnated with calcium salts or covered with wax; and provides physical as well as physiological protection, a place for muscle attachment, and in the case of insects, wings (Fig. 33.2). The body skeleton is constructed of plates, or **sclerites,** laid down as dorsal **tergites,** ventral **sternites,**

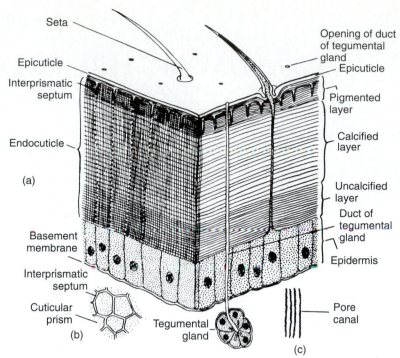

FIGURE 33.2

(*a*) Diagram showing structure of crustacean cuticle. All layers are secreted by the hypodermis (epidermis). The thin epicuticle is of sclerotized protein, and the procuticle (endocuticle) contains protein, chitin, and mineral salts. Protein in the uncalcified layer is unsclerotized. The procuticle of insects and arachnids is divided into the highly sclerotized exocuticle and the less sclerotized endocuticle. (*b*) Horizontal section through pigmented layer of endocuticle. (*c*) Pore canals as they appear in vertical sections.

From R. Dennell, in *The Physiology of Crustacea,* vol. 1. Edited by T. H. Waterman. Copyright © 1960 Academic Press, Inc., New York, NY. Reprinted with permission of the publisher.

and lateral **pleurites** (Fig. 33.3). Appendage segments are basically cylinders but like the body plates, may be so modified as to obscure their fundamental nature. Appendage joints are either hinges or pivots made from **chondyles** and **sockets.** Skeletal pieces are joined by **articular membranes** where the cuticle is very thin; therefore, they are flexible, allowing movement of the body and limbs. Muscles are attached to inner skeletal ridges (**apophyses**) or spines (**apodemes**). Arthropod locomotion is influenced greatly by the size, shape, and location of skeletal plates, as well as by the angles at which limbs are attached to the body and the origin and insertion of muscles.

The cuticle is made up of several layers containing protein, lipid, and polysaccharides, all secreted by the underlying **epidermis** (also called **hypodermis**) (Fig. 33.2). Much of the protein is stabilized (rendered relatively inert chemically) by processes of **sclerotization,** which involve crosslinking of amino acid chains in adjacent polypeptides (Fig. 33.4). The sclerotization reactions involve N-acetyldopamine, a tyrosine derivative.[1,12,19] In **quinone tanning,** the N-acetyldopamine is first oxidized by phenoloxidase to a quinone, which then spontaneously reacts with amine groups in adjacent polypeptides to form linkages between chains. In β-**sclerotization** the β-carbon of N-acetyldopamine is activated by an enzyme that

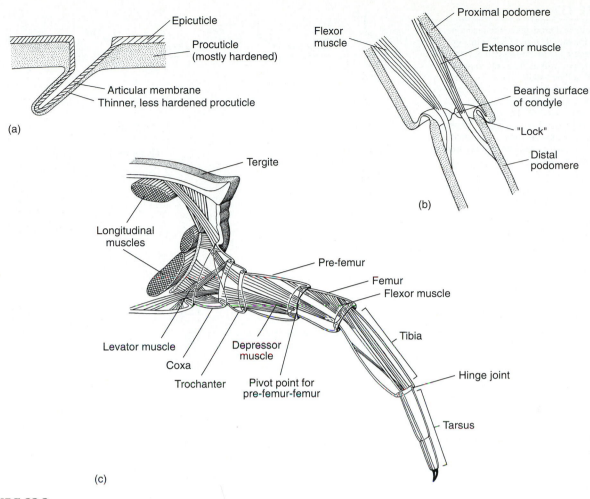

FIGURE 33.3

Diagram of articulation and musculature of arthropod trunk and limbs. (*a*) Flexibility is provided by thinner cuticle between sclerites. (*b*) Movement of a joint is by contraction of muscles inserted on opposite sides of the articulation, and the bearing surfaces of the joints are the condyles. (*c*) Muscles of a centipede leg. Levator muscles raise the leg; depressor muscles lower the leg; flexors bend the leg toward the body.

(*c*) From S. Manton, *The Arthropoda: Habits, Functional Morphology, and Evolution.* Copyright © 1977 Clarendon Press, Oxford, UK.

forms covalent bonds between the β-carbon and adjacent polypeptides (Fig. 33.4). So far, β-sclerotization is known to occur only in insects.[21] Protein stabilization by the formation of dityrosine and trityrosine crosslinks has also been reported. Some workers believe that sclerotization does not occur by covalent crosslinking of polypeptide chains but by controlled dehydration driven by quinones and other chemicals secreted into the cuticle.[31] In any case when the protein is stabilized, it is virtually insoluble except by vigorous chemical treatment.

The main carbohydrate component of cuticle is the polysaccharide **chitin,** a polymer of N-acetylglucosamine linked by 1,4-α-glycosidic bonds into long, unbranched molecules of high molecular weight. Chitin is flexible and contributes little to the rigidity of an arthropod's skeleton; the hardness is conferred by the proteins and by deposition of inorganic salts. Evidence also exists that the chitin is bonded to the protein, although the structural significance of this linkage is unclear.[21] These reactions are important evolutionary events that allowed arthropods to achieve their long history of success, including their present and relatively recent success as distributors of human misery.

The outermost layer of cuticle, the **epicuticle,** is thin and contains stabilized protein, sometimes called **cuticulin,** but no chitin. Insects and arachnids usually have a lipoidal layer covering, or perhaps lipid interspersed in the cuticulin, that functions to prevent water loss. Over the lipid is a "varnish" that protects the wax from abrasion. Beneath the epicuticle lies the thicker **procuticle,** which in insects and arachnids is further divided into **exocuticle** and **endocuticle,** the endocuticle being much less sclerotized than the exocuticle.

In Crustacea the waxy and varnish layers are absent, but the procuticle is often impregnated with calcium carbonate, calcium phosphate, and other inorganic salts. The entire procuticle of crustaceans is also called an endocuticle, which means that the term does not apply to the same structure as it does when one is referring to insects. In Crustacea the hardened layers containing salts and sclerotized proteins are the **pigmented** and **unpigmented calcified** layers. The unpigmented layers also contain chitin and protein, but the protein is unsclerotized and the layer is membranous and flexible.

FIGURE 33.4

(a) Substrate and products of the reaction sequence for stabilization of sclerotins by quinone tanning in insects and other arthropods and by β-sclerotization in insects (based on Richards[21] and on Riddiford and Truman[23]). (b) Disulfide crosslinkages of adjacent amino acid chains, as in keratin; probably also occurs in arthropod cuticle.[1] (c) The highly elastic protein, resilin, stabilized by dityrosine and trityrosine crosslinks, is sometimes found in insect cuticle, either mixed with other proteins or in almost pure form.[1]

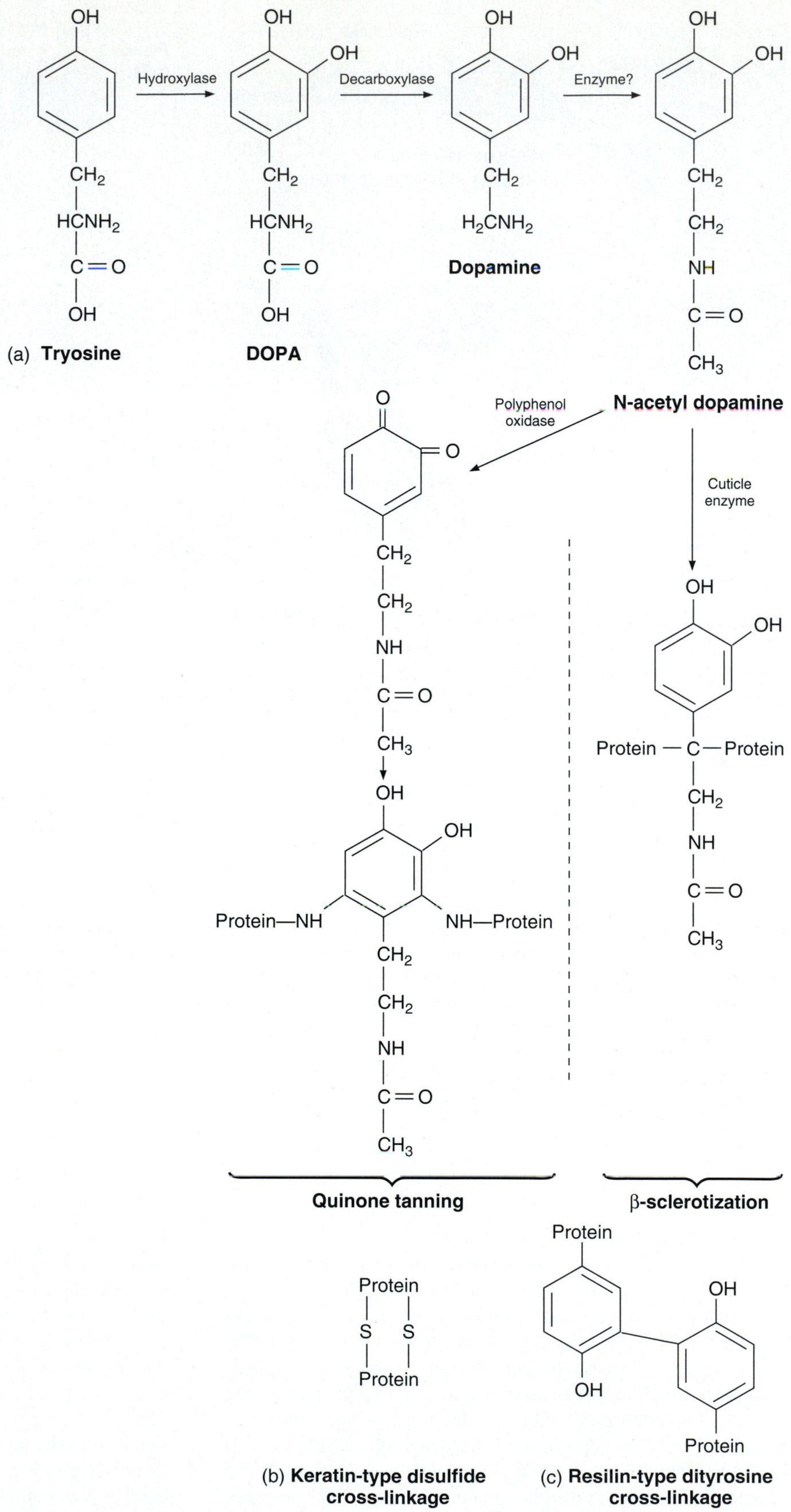

(a) **Tryosine**   **DOPA**   **Dopamine**   **N-acetyl dopamine**

Quinone tanning   **β-sclerotization**

(b) **Keratin-type disulfide cross-linkage**   (c) **Resilin-type dityrosine cross-linkage**

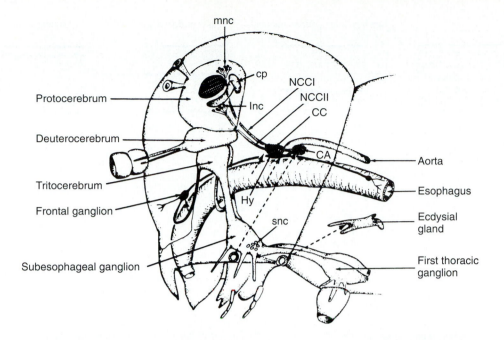

FIGURE 33.5

Generalized central nervous system of insect, showing sources of hormones in the head and prothorax. Neurosecretory hormones are produced by the median (**mnc**), lateral (**lnc**), and subesophageal (**snc**) neurosecretory cells and perhaps also by the corpora pedunculata (**cp**). Hormones from the neurosecretory centers (**NCCI, NCCII**) in the protocerebrum pass in two paired nerves to be stored in the corpora cardiaca (**CC**). Hormones are also secreted by the corpora allata (**CA**) and the prothoracic or ecdysial glands. The **dashed lines** mark the original embryonic origin of the organs indicated and their subsequent migration route during development. (**Hy**), hypocerebral ganglion.

From P. M. Jenkins, *Animal Hormones: A Comparative Survey,* part 1. Copyright © 1962 Pergamon Press Ltd., Oxford, UK.

Insertion of the muscles in the unyielding exoskeleton, coupled with fulcra provided by the condyles in the flexible joints, makes mechanically possible very fine control of movements. Increase in complexity of movements meant corresponding evolution of nervous elements to coordinate the movements. Furthermore, small changes in a given sclerite could substantially increase efficiency of a particular body part for a given function. Arthropods have many sclerites; the cuticle has evidently given evolutionary forces a great deal of raw material with which to work.

Molting

Although the arthropod cuticle confers many evolutionary opportunities, it presents some problems as well, most important of which is growth of an animal enclosed in a nonexpansible covering. The solution to this problem is a series of **molts,** or **ecdyses,** through which all arthropods go during their development. Much of the physiological activity of any arthropod is related to the molting cycle, and this relationship must have held true since the evolutionary origin of the phylum; a sizeable fraction of trilobite fossils, for example, consists of cast exoskeletons.

Growth in tissue mass occurs during an intercdysial period, and dimensional increase occurs immediately after molting, while the new cuticle is still soft. The stages of the animal between each molt are referred to as **instars.** The length of the intermolt phase depends on the species involved, the animal's age and stage of development, and any

interceding diapause (to be discussed later). The number of instars also varies with the species and in some cases is influenced by nutritional state, especially in acarines and insects.

In the groups in which the processes have been best studied—insects and malacostracan crustaceans—molting is controlled by hormones. In the Malacostraca order Decapoda, which includes familiar lobsters and crayfish, a **molt-inhibiting hormone (MIH)** is produced in the **X-organ,** which is composed of neurosecretory cells in the eyestalk; **molting hormone (MH)** is produced in the **Y-organs,** a pair of glands near the mandibular adductor muscles. Structures comparable to X-organs are found within the head of sessile-eyed Crustacea (those species whose eyes are not on stalks).

In insects at least three hormones are directly involved in the molting process: **prothoracicotropic hormone (PTTH), ecdysone,** and **bursicon.** PTTH is secreted by neurosecretory cells in the brain and released by organs called **corpora cardiaca** (Fig. 33.5). The hormone is carried in the hemolymph to the **prothoracic glands,** which are stimulated to produce ecdysone. The latter stimulates molting and is analogous, perhaps homologous, to MH of crustaceans. Bursicon is secreted from the organs associated with ventral nerve ganglia. It regulates postecdysial hardening of the cuticle.

As the level of MIH decreases and that of MH increases in the crustacean or as the level of ecdysone increases in the insect, the organism undergoes a series of changes preparatory for a molt, the **preecdysial** period. DNA synthesis in the hypodermal cells is stimulated and then RNA and protein

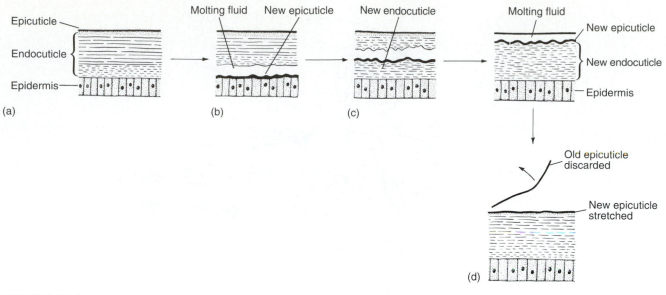

FIGURE 33.6

Cuticle secretion and resorption in preecdysis. (*a*) Interecdysis condition. (*b*) Old endocuticle separates from epidermis, which secretes new epicuticle. (*c*) As new endocuticle is secreted, molting fluid dissolves old endocuticle, and the solution products are resorbed. (*d*) At ecdysis, little more than the old epicuticle is left to discard. In postecdysis, new cuticle is stretched and unfolded, and more endocuticle is secreted.
Drawing by William Ober.

synthesis. The next effect of the ecdysial hormones is to cause the hypodermis to detach from the old procuticle **(apolysis)** and start secreting a new epicuticle (Fig. 33.6). At the same time beneath the old procuticle, enzymes (including chitinases and proteinases) begin to dissolve it. As the solution proceeds, the products of the reactions, including amino acids, *N*-acetylglucosamine, and calcium and other ions, are resorbed into the animal's body. These materials are thus salvaged and later are incorporated into the new cuticle.

Almost immediately after apolysis, the new epicuticle becomes limited in permeability. The new procuticle is thus protected from the enzymes dissolving the old cuticle above.[15,30,33] The old cuticle is not completely dissolved; in insects the epicuticle and sclerotized exocuticle remain, and in crustaceans the epicuticle and calcified regions remain, although some decalcification of these layers occurs. At the time of ecdysis, the old cuticle splits, normally along particular lines of weakness, or dehiscence, and the organism climbs out of its old clothes.

The old cuticle is split, and the new one stretched, by expansion of the body. Insects accomplish this feat by inhaling air and crustaceans by rapid imbibition of water, a process aided by the fact that the osmotic pressure in the tissues and blood has been increased before the molt by the calcium ions mobilized from the cuticle.[24] The increase in blood and tissue volume causes the small wrinkles in the still soft cuticle to smooth out, increasing the body dimensions, and the cuticle begins to harden again. In this postecdysial period, sclerotization of the protein and redeposition of calcium salts in the procuticle occur, and more procuticle is secreted. As in many biological phenomena, molting has both a benefit and a cost. The benefit is the growth allowed by the molt; the cost is the vulnerability to predation that comes with a temporarily soft cuticle.

• Molt Cycles

The length of time spent in an instar depends on the species involved, its age and stage of development, the season or annual cycle, and sometimes the species' nutritional state. Decapod crustaceans that molt on an annual cycle and have a long intermolt period are said to be **anecdysic.**[13] Species in which one ecdysial cycle grades rapidly into another are **diecdysic.** Some crabs reach maximal size and stop molting, undergoing *terminal anecdysis.* In crustaceans other than malacostracans, little is known of the hormonal control of molting. Barnacles, at least, seem to be in a permanent diecdysis.[4] Furthermore, many copepod parasites of fish cease molting when they reach the adult stage, although they continue to grow actively. For example, a female *Lernaeocera* is about 2 mm long after her last molt, but may attain an ultimate size of up to 60 mm *without molting.* What changes in the cuticle when the copepod reaches sexual maturity, and what causes the change? We do not know, but it is clear that the change permits continuous growth.

More is known about the mechanisms allowing such expansion in some parasitic insects and ticks than in crustaceans. The fourth-stage nymph of the bug *Rhodnius prolixus* takes a large blood meal that necessitates stretching its abdominal cuticle threefold.[32] Before this blood meal, its cuticle is stiff and inextensible, but this condition changes as the bug feeds. The change is mediated by neurosecretory axons running to the hypodermis, apparently stimulating an enzyme discharge, which affects the substance of the cuticle. After feeding, additional cuticle material is deposited, probably to provide protection and a template for the cuticle of the next instar.[10] The female cattle tick, *Boophilus microplus,* ingests 150 times its body weight in blood after molting to the adult, increasing in length from 2.5 mm to 11.0 mm. Filshie[8] reported that, before the last molt, the epicuticle is laid down

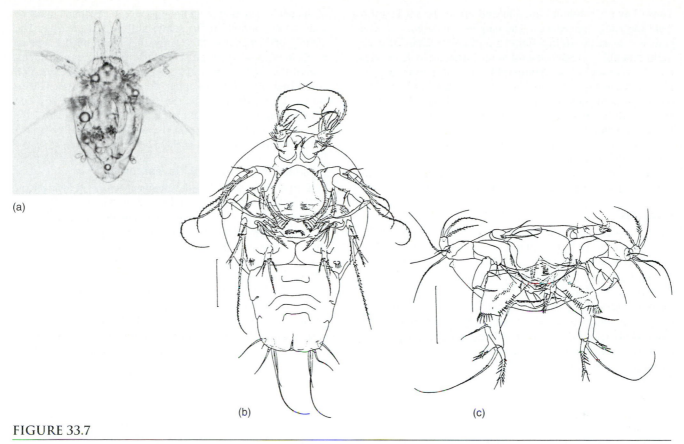

FIGURE 33.7

Examples of nauplius larvae. (*a*) Copepod nauplius. This particular nauplius also has a parasitic nematode in it, as well as some ectocommensal peritrich ciliates attached to the surface. (*b*) Late nauplius of *Tisbe cucumariae*. (*c*) An early nauplius of *Stenhelia palustris*. *Tisbe* and *Stenhelia* are both harpacticoid copepods.

(*a*) Courtesy of Ralph Muller. (*b*) and (*c*) From H.-U. Dahms, "Pictorial keys for the identification of crustacean nauplii from the marine meiobenthos," in *J. Crust. Biol.* 13:609–616. Copyright © 1993. Reprinted with permission of the publisher.

as a highly folded layer. The subsequent expansion is accommodated by unfolding of the inexpansible epicuticle and stretching and growth of the underlying procuticle.

Early Development and Embryology

In arthropods, the sexes are separate and fertilization is internal. However, the mating behaviors, methods of sperm transfer, and subsequent treatment of the eggs are all as varied as the phylum is structurally. Early embryology is discussed in fascinating detail, with wonderful pictures, in the book by D. T. Anderson.[2] Biologists have long used embryological development as evidence for evolutionary relationships, and Anderson provides a fairly accessible explanation of the way arthropod embryology has helped us understand the history of the phylum.

In most Crustacea and Insecta, cleavage is **intralecithal,** with the nuclei undergoing several divisions within the yolk mass and then migrating to the periphery to become the **blastoderm.** Yolk is concentrated in the interior of the embryo **(centrolecithal),** and differentiation proceeds in the superficial areas. A blastoderm is also formed in the chelicerates, but the initial cleavages are sometimes complete **(holoblastic).** Most arthropod embryos exhibit at least a remnant of the **acron,** a preoral (anatomically anterior to the mouth) segment without jointed appendages, and a varying number of postoral or trunk segments (somites), each with a

pair of jointed appendages. The segments are initially laid down as sometimes hollow mesoderm blocks; the cavities are the segmented coelom. Serially homologous metameres and appendages differentiate as the embryo grows. Comparative embryology allows us to establish special homologies by discovering, for example, the fate of corresponding appendages, in various groups of arthropods. Thus, we would not be able to claim that the walking legs of spiders correspond to the mouth parts of crayfish without reference to the embryological development of both animals.

Postembryonic Development

Arthropods differ in their postembryonic development. In most species the egg develops into a **larva** (because beetles and crustaceans make up the majority of known arthropods and both have larval stages!). A larva is a life cycle stage that is structurally distinct from the adult, normally occupies an ecological niche separate from the adult, is sexually immature, and must undergo a structural reorganization **(metamorphosis)** before becoming an adult.

• Crustaceans

The typical larva that hatches from the crustacean egg is called the **nauplius** (Fig. 33.7). The nauplius has only three pairs of appendages: antennules, antennae, and mandibles.

These have locomotor function and are different in form from the adult appendages. The nauplius undergoes several ecdyses, usually adding somites and appendages at each molt. Nauplii typically have several instars, and later ones may be referred to as **metanauplii.** Metamorphosis may be gradual, occurring over several instars, or more abrupt, from one instar to the next. But if a distinguishable larval stage occurs, development is said to be **indirect. Direct** development is that in which a juvenile, rather than a larva, hatches with segmentation and appendages complete. Juveniles, however, are sexually immature.

Crustacea may vary widely in their development patterns, even within the same class (see Chapter 34).

• Insects

Of the Insecta, only certain wingless orders (Collembola, Thysanura, and Diplura) have direct development. The winged orders, or Pterygota, are all metamorphic. In several orders (Dermaptera, Dictyoptera, Mallophaga, Anoplura, and Hemiptera) the larval instars are called **nymphs,** and they become gradually more like the adult (**imago**) with each ecdysis (**gradual metamorphosis** or **hemimetabolous** development) (see Fig. 36.5). The wing buds develop externally (**exopterygote**) and can be observed readily in the nymphal instars (except for bedbugs, which are wingless!). The nymphs of hemimetabolic insects have well-developed appendages, compound eyes, and the rudiments of external genitalia, and their habits are generally similar to those of the adult, although their microenvironments may differ slightly or, as in the case of dragonflies, significantly. Nymphal dragonflies are aquatic and the adults are terrestrial, but both stages are active predators.

In other orders (Neuroptera, Coleoptera, Strepsiptera, Siphonaptera, Diptera, Lepidoptera, and Hymenoptera) the larva bears little resemblance to the adult. After several larval instars, these insects enter a nonfeeding period, the **pupa** stage, in which the animal is completely reorganized (see Fig. 38.3) and the wing buds grow internally (**endopterygote**). This type of metamorphosis is **complete metamorphosis** or **holometabolic.** The larvae of holometabolic insects are extremely diverse and occupy ecological niches quite distinct from those of the adults. Consider mosquitoes, for example (Fig. 38.3); the larvae are aquatic and filter feed on microorganisms, while the adults are terrestrial and feed on plant juices. Adult females of most species suck blood. Holometabolic larvae occur in a variety of forms and exhibit a range of behaviors. Some are active predators with well-defined heads and thoracic appendages (**campodeiform** or **oligopod,** such as many beetles); some lack heads and appendages altogether (**vermiform** or **protopod,** such as Diptera and Hymenoptera); others are **polypod,** with abdominal **prolegs** (Lepidoptera); and still others are grublike, with swollen abdomens and lightly sclerotized body cuticle (**scarabaeiform,** such as some Coleopteran families) (Fig. 33.8).

• Ticks and Mites

Postembryonic development of acarines is characteristically direct, with the sexually immature **nymph,** a tiny facsimile of the adult, hatching from the egg. Some arachnids undergo one or two "larval" molts while still within the egg. The number of nymphal instars varies depending on the type of arachnid. A six-legged larva hatches from the egg in members of the order Acari and becomes an eight-legged nymph at the first molt. In most mites there are three nymphal instars: the **protonymph,** the **deuteronymph,** and the **tritonymph.** However, the hard ticks (family Ixodidae) have only one nymphal stage, and the soft ticks (family Argasidae) may have as many as eight.

• Endocrine Control of Development

Endocrine function during development is best known for insects, and it is essentially similar in both holometabolous and hemimetabolous forms. A **juvenile hormone (JH)** is produced by the **corpora allata** (Fig. 33.5). Although three different chemical forms of the hormone are present in varying proportions in different insects, all three forms produce similar effects.[23] The best known and documented of such actions is the so-called status quo effect, in which JH acts on insect tissues to maintain larval or nymphal characters (impedance of maturation). The level (titer) of JH in the blood of the insect decreases as development proceeds through the juvenile instars. Consequently the tissues and organs become progressively more adultlike. Finally, the titer drops to an undetectable level at about the beginning of the last nymphal instar in hemimetabolous insects, and the adult emerges at the next ecdysis. Disappearance of JH from the blood of holometabolous insects usually occurs about midway through the last larval instar; the next molt produces the pupa.

The status quo action of JH is believed to result from its direction of the kinds of RNA produced after ecdysone stimulation, but the mechanism by which JH does this is still not well-understood.[23] If a high titer of JH is present, the RNA produced will lead to larval proteins; if little or no JH is present in the blood, the RNA for synthesis of proteins with adult characteristics will be produced. Interestingly, although a shutdown of JH production by the corpora allata is necessary for maturation to the adult form, the corpora allata are reactivated in the adult of many insects and again secrete JH. The hormone is necessary for egg maturation in females and proper development of the sex accessory glands in males. Furthermore, the adult of almost all insects does not molt again, but ecdysone is once more secreted and also plays a role in reproductive function.

Diapause

Without a doubt, another factor contributing greatly to the evolutionary success of arthropods has been their ability to withstand adverse environmental conditions, such as subfreezing temperatures or extreme dryness, in which normal physiological function would be impossible. Under these conditions many arthropods can enter a period of developmental arrest known as **diapause.** Although diapause occurs in some crustaceans and arachnids, much more is known about it in insects than in those other groups. Knowledge of the role of diapause in the life of a particular species is sometimes of vital importance to our understanding of the biology of arthropod vectors of parasitic diseases.

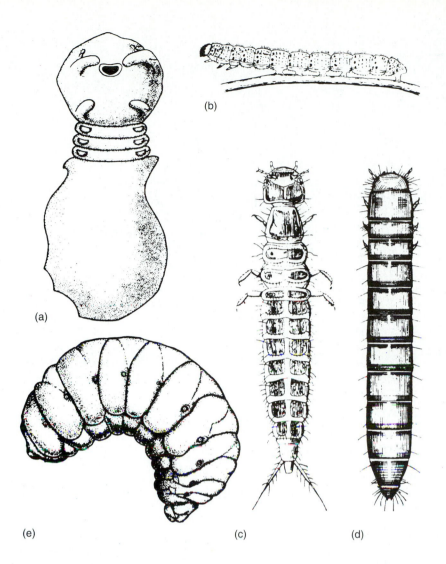

FIGURE 33.8

Types of endopterygote larvae. (*a*) Protopod type, found in some parasitic Diptera and Hymenoptera, has poorly defined segmentation, rudimentary appendages, and incompletely differentiated internal organs. (*b*) Polypod or eruciform, typical of Lepidoptera (butterflies and moths). (*c*), (*d*) Oligopod or campodeiform, as found in some beetles (Coleoptera). (*e*) Apodous type, found in Hymenoptera, Diptera, and some Coleoptera; usually lives among abundant food.

From O. W. Richards and R. G. Davies, *Imm's Outlines of Entomology,* 6th ed. Copyright © 1978 Chapman & Hall Ltd, London, England. Reprinted with permission.

Diapause in insects may occur in the egg, larva, pupa, or adult, depending on the species. In the immature stages it is characterized by a cessation of development and prolongation of that stage. In the adult, reproduction is inhibited. In addition other physiological processes more or less cease and the animal is quiescent. In all insects that have been studied, the mechanism of diapause initiation and termination is hormonal, but different hormones are involved, depending on the stage in the life cycle.[23]

EXTERNAL MORPHOLOGY

The detailed morphology and related physiology of a group as enormous and diverse as the arthropods are far beyond the scope of this book. However, parasitism as a phenomenon is inextricably intertwined with the lives of arthropods. From their roles as intermediate hosts, vectors of important human and veterinary diseases, and definitive hosts in their own right, to their often extraordinary adaptations to parasitic life, the arthropods constantly present us with evidence of the many forms that parasitism can take. Thus, a parasitology student needs to know a modicum of structure to identify parasitic arthropods and to understand host/parasite relationships. In addition to the brief discussion that follows, more specialized details of structure and biology will be given in the chapters to come.

Form of the Crustacea

In the Crustacea the head is usually not clearly set off from the trunk. One or more thoracic somites are commonly fused with the head, but some thoracic segments are distinct (Fig. 33.9). Thus, the normal tagmatization of crustaceans is a **cephalothorax,** a free **thorax,** and an **abdomen,** although the degree of

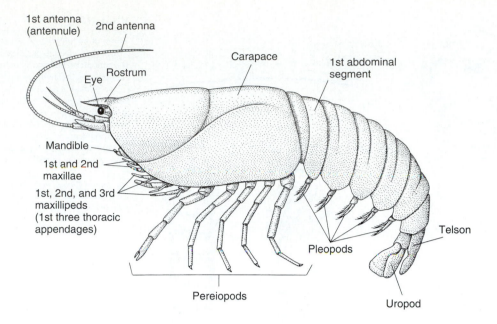

FIGURE 33.9

Lateral view of a generalized malacostracan crustacean.
Drawing by William Ober.

prominence and of fusion of the tagmata varies greatly from group to group. The head seems to have been formed by fusion of five somites.[27] The cephalothorax and sometimes even the entire body may be covered by a **carapace,** which arises as a fold from the posterior margin of the head (Fig. 33.9). The two types of eyes found in the crustaceans are **median eyes** and **compound eyes.** The median eye, consisting of three or four pigment-cup ocelli, is also called the **nauplius eye** because it is present in that larval stage. This eye sometimes persists into the adult, for example in copepods. Adults of most species have a pair of compound eyes; these may be sessile, or they may be mounted on stalks and be very convex, with an angle of vision of 180 degrees or more.

The anteriormost appendages of the head are the **antennules** (first antennae) followed by the **antennae** (second antennae). Crustaceans are the only arthropods with two pairs of antennae. The antennules and antennae are usually sensory, but in some forms they may be adapted for locomotion or prehension. Feeding appendages on the head are the **mandibles, maxillules** (first maxillae), and **maxillae** (second maxillae) (Fig. 33.9). One or more pairs of thoracic appendages may be incorporated into the mouthparts and are then called **maxillipeds.** Other thoracic appendages are the **pereiopods,** and the abdominal appendages are the **pleopods.** The pereiopods and pleopods may be variously modified for walking, swimming, or copulation. The number of pereiopods and pleopods varies from group to group, and in some groups pleopods are absent. The abdomen ends in a **telson,** which may be flanked by the posteriormost pleopods, called **uropods.**

The appendages of Crustacea were primitively **biramous** (having two branches) (Fig. 33.10), and this condition prevails in at least some appendages of all living species during their lives. The terminology applied by various workers to crustacean appendages has not been blessed with uniformity.

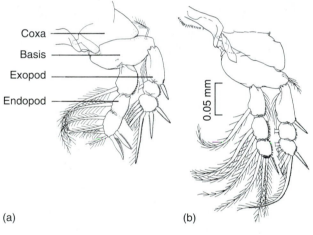

(a)   (b)

FIGURE 33.10

First (*a*) and second (*b*) thoracic appendages of *Ergasilus megaceros* (Copepoda), a parasite of the sucker *Catostomus commersoni.* The terminal segments of the first endopod are fused, the ancestral condition being indicated by the presence of vestigial condyles. The medial side of the coxa may be modified for food handling in some Crustacea and is called a gnathobase.

From L. S. Roberts, "*Ergasilus* (Copepoda: Cyclopoida): revision and key to species in North America," in *Trans. Am. Microsc. Soc.* 89:134–161. Copyright © 1970. Reprinted with permission of the publisher.

At least two systems are currently in wide use, and we have given the alternative term for each structure in parentheses. The lateral branch is the **exopod** (exopodite), and the medial one is the **endopod** (endopodite). Each of these branches may contain several segments, varying by appendage and according to species. The endopod and exopod are borne on a **basis** (basipodite), and the basis, in turn, is attached to the **coxa** (coxopodite); together they are referred to as the **protopod.**

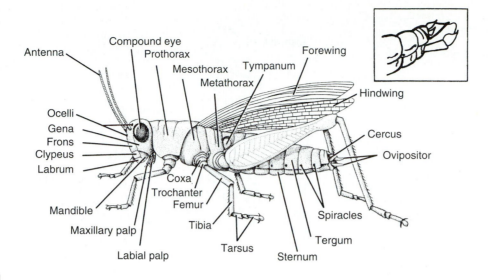

FIGURE 33.11

External features of a relatively generalized insect, the grasshopper *Romalea*. The terminal segment of a male with external genitalia is shown in inset.
From Cleveland P. Hickman, Jr. et al., *Integrated Principles of Zoology,* 8th edition. Copyright © 1988 by Mosby-Year Book Inc. Reprinted by permission of Times Mirror Higher Education Group, Inc., Dubuque, Iowa. All Rights Reserved.

Processes from the protopod are termed **endites** and **exites;** the exites may be called **epipods** (epipodites). The two branches of the legs may not be homologous through all crustacean classes.[25]

The Form of Pterygote (Winged) Insects

In all members of the class Insecta the tagmata are the **head, thorax,** and **abdomen.** The head is made up of six fused metameres, four of which bear appendages in modern insects. The bases of the freely movable, sensory **antennae** are above or between the eyes (Fig. 33.11). The **mandibles** are usually the primary feeding appendages and are borne ventrally, lateral to the mouth. Immediately posterior to the mandibles are the **maxillae** and following these, the **labium,** interpreted as a fused pair of appendages (Fig. 33.12). The maxillae and labium also may have **palps** with food handling and sensory functions. Anteriorly the mouth is covered by the **labrum,** or upper lip. A tonguelike lobe, the **hypopharynx** arises from the floor of the mouth in some insects, and a similar extension, the **epipharynx,** may emerge from the roof of the mouth in others. The labrum, epi-, and hypopharynx are not considered appendages, but they do function in feeding. Depending on the insect group, these mouthparts may be highly modified or secondarily lost altogether. In addition to the antennae and mouthparts, most insects' heads have a pair of **compound eyes** and one or more simple eyes, the **ocelli.**

The insect thorax consists of three segments—the **prothorax, mesothorax,** and **metathorax**—each of which bears a pair of legs. Each leg usually is divided into five segments, or **podomeres** (Fig. 33.11). The basal segment, or **coxa,** articulates with both the body and the **trochanter;** the latter is also fixed to the **femur,** the largest of the podomeres, which, in turn, articulates with the more slender **tibia.** Distal to the tibia is the **tarsus,** which is subdivided into two to five segments. The **pretarsus** consists of claws or other structures attached to the terminal tarsal segment.

Adult pterygote insects characteristically have wings, although some, such as fleas, lice, worker ants, and termites, have lost their wings during the course of evolution. Both the mesothorax and metathorax bear a pair of wings, but in the order Diptera, which includes the mosquitoes, tsetse flies, sand flies, and black flies, all vectors of parasitic diseases, the metathoracic wings are reduced to balancing organs called **halteres** (see Fig. 38.2). In male Strepsiptera, the mesothoracic wings are reduced to halteres, whereas the females are highly modified parasites with no wings at all. Wings develop from evaginations of thoracic epidermis and thus consist of a double layer of epidermis. The layers are penetrated by canals, called **lacunae,** and the lacunae contain nerves, tracheae, and blood.

The epidermal cells atrophy as the wing approaches full development; thus, the wing consists of two thin layers of cuticle secreted by the epidermis supported by the more heavily sclerotized **veins,** which are the remains of the lacunae. The pattern of wing venation is constant within a species and therefore is often of value in taxonomy. Entomologists have adopted a standard nomenclature for wing venation (Fig. 33.13). Anyone who regularly identifies insects using taxonomic keys quickly memorizes the terms for wing veins, but someone who tries such identification infrequently must relearn them time and again.

The abdomen of adult insects consists basically of 11 somites plus a terminal **telson,** but all these segments usually can be discerned only in the embryo. Abdominal segment appendages also occur in embryos, but except for those on the genital segments, the appendages are lost in the transformation of the embryo into an adult. Appendages of the genital segments are called **external genitalia;** these consist of the **penis** or **aedeagus,** on the ninth segment of males, and the **ovipositor,** formed of the eighth and ninth segment appendages of females. The eleventh segment may also have appendages, the **cerci,** which range from vestigial in size to longer with sensory function (Figs. 33.11 and 39.8).

FIGURE 33.12

Anterior view of the mouthparts of a
grasshopper.
Drawing by William Ober.

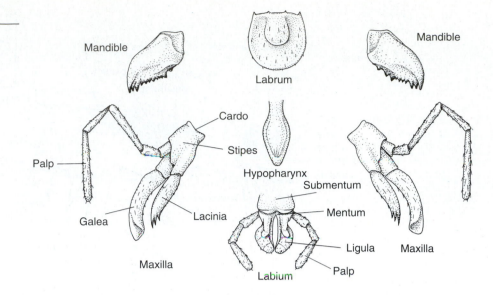

FIGURE 33.13

Diagram of typical venation of wing in modern
insects, showing standardized abbreviations for
the veins: **C** (*costa*) is the unbranched thickened
anterior margin of the wing. **Sc** (*subcosta*) is
typically branched. **R** (*radius*) is the second
major vein, connecting to the base of a sclerite.
The radius usually branches, and the cells
formed by these branches are important
taxonomic characters. **M** (*media*) also branches
into marginal cells which, like those of the
radius, are numbered anterior to posterior. **C**
(*cubitus*) articulates with a sclerite and is
posterior to the media. **1A–4A** (anal veins) form
a set. **jf** (jugal furrow) is a crease separating the
anal region or fold from the jugal fold, which is
the small area at the basal posterior corner of the
wing. *Crossveins* are named according to the
veins they connect.

From Herbert H. Ross, *Textbook of Entomology,*
3d ed. Copyright © 1965 John Wiley & Sons, Inc.
Reprinted with permission.

Most insects have **spiracles,** or openings into the respiratory system (Fig. 33.11). Although the vast majority of insects have only two pairs of thoracic spiracles, the mesothoracic and metathoracic, the former often migrates forward during embryological development and thus appears to be on the prothorax. Members of the order Diplura, primitively wingless forms typically found in leaf litter and rotting wood, are the only insects with true prothoracic spiracles. Adult insects typically have eight abdominal pairs of spiracles. Spiracles usually have closing mechanisms that function to reduce water loss.

Form of the Acari

The primary tagmata in the class Arachnida are a cephalothorax (**prosoma**) and an abdomen (**opisthosoma**). The somites of these tagmata are fused to a greater or lesser degree, depending on the order. In the spiders (order Araneae) the fusion is complete in almost all species, but the prosoma and the opisthosoma are distinct. In the other large order of arachnids, the Acari, even these tagmata are fused, and the opisthosoma is defined rather arbitrarily as the region posterior to the legs. This situation has given rise to a special nomenclature for the body regions applied only to the Acari, given by Savory[26] as follows:

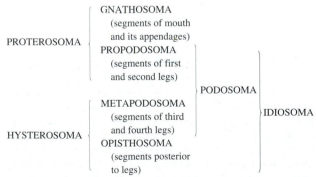

The **proterosoma** can usually be distinguished from the **hysterosoma** by a boundary between the second and third pairs of legs (Fig. 33.14). Dorsally the **idiosoma** is often covered by a single, sclerotized plate, the **carapace.** The

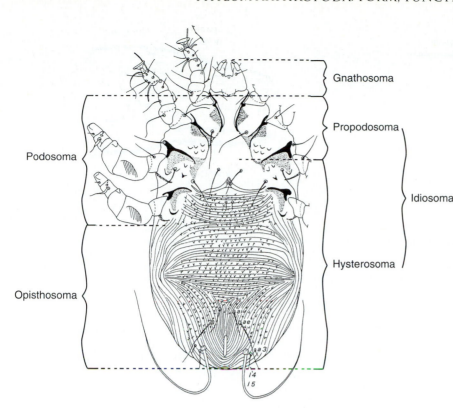

FIGURE 33.14

A representative mite, *Mycoptes neotomae* (female, ventral view), parasitizing woodrats and white-footed mice.

From A. Fain et al., "Two new Myocoptidae (Acari, Astigmata) from North American rodents," in *J. Parasitol.* 70:126–130. Copyright © 1970. Reprinted with permission of the publisher.

gnathosoma, or **capitulum,** is usually sharply set off from the idiosoma, and it carries the feeding appendages. These appendages are the **chelicerae,** usually with three podomeres, and the **pedipalps,** whose free segments may vary from one to five in different groups. The chelicerae may be **chelate** (pincerlike) in scavenging and predatory mites, but in parasitic mites they are usually modified to form stylets or bear teeth for piercing. The bases of the pedipalps are lateral and just posterior to the bases of the chelicerae. The pedipalps may be leglike or chelate, or they may be reduced in size and serve as sense organs. Ventrally the fused coxae of the pedipalps extend forward to form the **hypostome,** which, together with a labrum, makes up the **buccal cone** (Fig. 33.15). The dorsal part of the capitulum projects forward over the chelicerae as a **rostrum,** or **tectum.**

Acarines typically have four pairs of legs, as in other arachnids, but only one to three pairs may be present. The podomeres of the legs may vary from two to seven, but six is the usual number: **coxa, trochanter, femur, patella, tibia,** and **tarsus.** The tarsi of most acarines each have a pair of claws. Spiracles may or may not be present, and their position and existence are important criteria for distinguishing the suborders. The anus is near the posterior end of the body, but the location of the gonopore is more variable, being found as far forward as the first legs in some forms. Some male mites have an intromittent organ, or **aedeagus.** The gonopore commonly opens through a more heavily sclerotized area, the **genital plate.** Other plates or shields are found on the idiosoma; their location and form are of taxonomic value.

The body and legs of most ticks and mites are well supplied with sensory (tactile) setae, which may be simple and hairlike, plumose, or leaflike; movement of a seta stimulates nerve cells at its base. One or two pairs of simple eyes are found laterally on the propodosoma in members of most suborders. Some mites have paired **Claparedé organs,** or **urstigmata,** between the coxae of the first and second legs. The urstigmata are apparently humidity receptors. Ticks have a depression in the first tarsi called **Haller's organ,** which bears four different kinds of sensory setae.[3] Haller's organ is a humidity and olfactory receptor and is of considerable value to the tick in finding hosts.[14,29] More detailed anatomical information on ticks and mites can be found in Chapter 40.

Internal Structure

• Body Cavity and Circulation of Fluids

The arthropod coelom is greatly reduced, its remnants being found in the excretory organ or gonad spaces. The main body cavity of arthropods is thus a secondary space—the **hemocoel**—filled with blood (**hemolymph**) and a variety of cell types. Muscles, sometimes very large ones, are bathed in this blood, which is circulated through an open circulatory system by means of a dorsal tubular heart. Hemolymph enters the heart from the surrounding **pericardial sinus** through pairs of lateral openings, the **ostia.** The ostia are one-way valves; when the heart contracts, the ostia close, forcing the hemolymph anteriorly into the arteries and finally into a system of tissue spaces, or **sinuses.** Blood works its way back to the heart through these sinuses, often aided by body movements (Fig. 33.16).

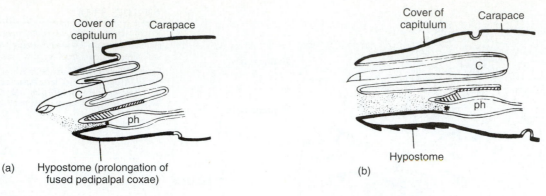

(a)   Hypostome (prolongation of fused pedipalpal coxae)

(b)

FIGURE 33.15

Diagrammatic longitudinal section through the capitulum of acarines. (*a*) Mite; (*b*) hard tick. The hypostome and the labrum (*crosshatched*) form the buccal cone, the anterior of which surrounds the preoral food canal (*shaded*). The mouth, designated by an *asterisk*, leads into the muscular pharynx (**ph**). The chelicerae (**C**) of the tick lie in a sheath and can be protracted and retracted.

From K. R. Snow, *The Arachnids: An Introduction.* Copyright 1970 Columbia University Press, New York, NY.

FIGURE 33.16

Insect circulatory system showing route of blood circulation. Although the blood flows from the arteries into the open hemocoel, its circulation through the body is assured by partitions. (*a*) Schematic of insect with fully developed circulatory system; (*b*) transverse section of **a**; (*c*) transverse section of abdomen; *arrows,* course of circulation; **a,** aorta; **apo,** accessory pulsatile organ of antenna; **d,** dorsal diaphragm with aliform muscles; **h,** heart; **n,** nerve cord; **o,** ostia; **pc,** pericardial sinus; **pn,** perineural sinus; **po,** mesothoracic and metathoracic pulsatile organs; **s,** septa dividing appendages; **v,** ventral diaphragm; **vs,** visceral sinus.

From V. B. Wigglesworth, *Principles of Insect Physiology,* 7th ed. Copyright © 1972 Chapman & Hall Ltd., London, England. Reprinted with permission of publisher.

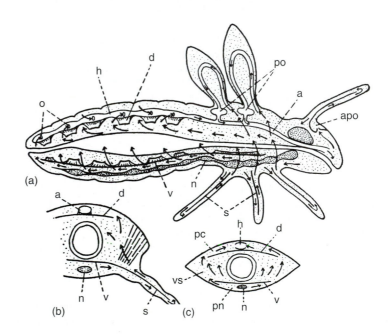

Formed elements of the blood are mostly **amebocytes.** Parasites, especially larval stages, may penetrate the gut and come to lie in the hemocoel, as in the case of acanthocephalan or tapeworm larvae. And malarial sporozoites escape from their oocyst on the gut and migrate through the hemocoel to the mosquito vector's salivary glands (Chapter 9).

• Respiratory System

Gas exchange takes place directly through the body wall in very small arthropods that may lack specialized respiratory organs and even a heart. Larger Crustacea have **gills,** which are extensive folds of the epidermis, covered with thin cuticle, through which hemolymph circulates. Most insects, as well as many Acari, have **tracheal systems,** a branching network of tubes. The tracheal system opens at the spiracle and ramifies through the body into a large number of very fine **tracheoles** (Figs. 33.17 and 33.18). The cuticle of the tra-

cheae, but not that of the tracheoles, is shed at ecdysis. Ventilation of the tracheal system is accomplished by pressure of body muscles on the walls of elastic tracheae, on tracheal air sacs, or both. Arachnid tracheal systems are thought to have evolved from **book lungs,** membranous folds inside a chamber that opens through a slit or spiracle. Book lungs occur in several arachnid orders but not in Acari.

• Nervous System

The arthropod central nervous system consists of a dorsal ganglionic mass, the **brain,** lying above the **stomodaeum** (anteriormost part of the digestive system); nerves that supply the cephalic sense organs; nerve trunks or **commissures** surrounding the esophagus and connecting the brain to a **subesophageal ganglion;** and a ventral nerve trunk that lies beneath the digestive tract. The ventral trunk consists of a double cord connecting the **segmental ganglia.** However, in

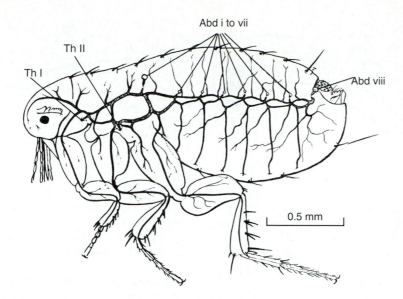

FIGURE 33.17

Half of the tracheal system of the flea *Xenopsylla.* The main tracheae and locations of the spiracles are shown.

From V. B. Wigglesworth, *Principles of Insect Physiology,* 7th ed. Copyright © 1972 Chapman & Hall Ltd., London, England. Reprinted with permission of the publisher.

many if not most arthropods, this fundamental structure is modified by postembryonic compression and shortening of the nerve trunk, fusion of ganglia, and lengthening of fibers to the posterior part of the animal.

The brain itself consists of three major regions: the **protocerebrum,** which may be homologous to part of the annelid prostomial ganglia and which supplies the optic nerves, the **deuterocerebrum,** which in crustacea supplies nerves to the first antennae, and the **tritocerebrum** (Fig. 33.5). On the basis of evidence from comparative anatomy and embryological studies, the tritocerebrum consists of segmental ganglia incorporated by fusion into the brain. Evidence for the special homology of arthropod anterior appendages is found in the fact that nerve centers of crustacean second antennae, the chelicerae of chelicerates, and the antennae of insects are all located in the tritocerebrum.[28]

The peripheral nervous system includes axons that innervate muscles and glands and bi- or multipolar **neurocytes,** their distal processes and axons. The sensory neurocytes are connected to a variety of sense organs, including tactile hairs and bristles and chemoreceptors.

• Digestive System

In crustaceans the digestive tract consists of a **foregut, midgut,** and **hindgut.** Part of the foregut may be enlarged into a **triturating stomach,** bearing calcareous ossicles, chitinous ridges, or denticles on its walls and functioning to grind up food. The midgut is often enlarged to form a stomach, and it usually bears one or more pairs of **ceca.** One pair of ceca may be modified to form a **digestive gland,** or **hepatopancreas,** which produces digestive enzymes. Absorption is confined to the midgut and tubules of the digestive gland.

The digestive system of insects is also divided into foregut, midgut, and hindgut. Distinct regions of the foregut are **esophagus, crop,** and **proventriculus** (Fig. 33.19). In insects that suck fluid meals from their hosts, the esophagus is a muscular **pharynx.** The crop is a storage chamber. The

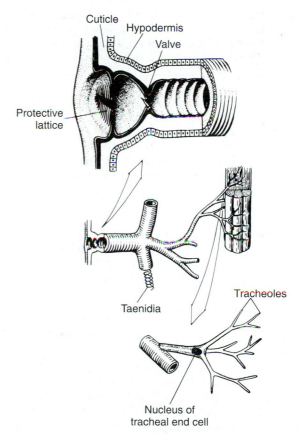

FIGURE 33.18

Diagram of trachea of an insect. The tracheae are virtually impermeable to liquids, but the finely branching tracheoles, leading into the tissues, are freely permeable, and their tips normally contain fluid. Oxygen primarily diffuses through the tracheolar walls, and elimination of carbon dioxide takes place more generally through the tracheal walls and body surface. Taenidia are chitinous bands that strengthen the tracheae.

From Cleveland P. Hickman, Jr. et al., *Integrated Principles of Zoology,* 8th edition. Copyright © 1988 by Mosby-Year Book Inc. Reprinted by permission of Times Mirror Higher Education Group, Inc., Dubuque, Iowa. All Rights Reserved.

FIGURE 33.19

Diagram of the digestive system of an insect. **An,** anus; **Ati,** anterior intestine; **Bu,** buccal cavity; **Cdv,** cardia valve; **Cm,** cecum; **Co,** crop; **Es,** esophagus; **Fg,** foregut; **Hg,** hindgut; **M,** mouth; **Mg,** midgut; **Mt,** malpighian tubules; **Phn,** pharynx; **Pti,** posterior intestine; **Pv,** proventriculus; **sd,** salivary duct.

From R. M. Fox and J. W. Fox, *Introduction to Comparative Entomology.* Copyright © 1966 Reinhold Publishing Co., New York, NY.

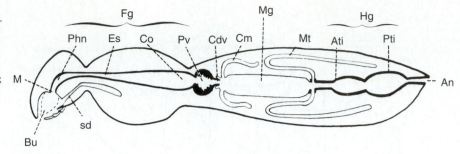

form and function of the proventriculus correspond to the type of food; in insects that eat solid food, the proventriculus is a gizzard, and in sucking insects, it is a valve regulating passage of food into the midgut. A pair of salivary glands usually lies beneath the midgut; the glands open into the buccal cavity by a common duct. Salivary secretions contain digestive enzymes and a variety of other substances, such as anticoagulants in bloodsucking species. The midgut is the principal site of digestive and absorptive function. In many insects the midgut secretes a thin, chitinous layer, the **peritrophic membrane,** which invests the food mass. The peritrophic membrane is permeable to enzymes and the products of digestion and probably protects the delicate epithelial lining of the midgut. Insects that live on liquid diets do not secrete a peritrophic membrane.

Gastric ceca, which increase the absorptive area of the midgut, are found near the anterior end of the midgut of most insects. The hindgut, divided into **intestine** and **rectum,** functions not only in the elimination of wastes but also in the regulation of water and ions.

In the Acari, the mouth leads into the muscular, sucking **pharynx,** which lies partly in the buccal cone. A slender esophagus proceeds posteriorly through the brain to the stomach, or **ventriculus.** A large pair of **salivary glands** lies above the ventriculus and esophagus and open by means of ducts into the **salivarium** in the buccal cone, over the labrum (Fig. 33.20). In bloodsucking forms, the salivary secretions contain anticoagulants and histolytic components. The ventriculus has up to five pairs of ceca, which contain secretory and absorptive cells. The hindgut may be a short tube leading from the midgut to the anus, or an enlarged portion, the **rectal sac,** may precede the anus. In ticks, chiggers, water mites, feather mites, and many parasitic forms, the ventriculus has lost its connection with the midgut and ends blindly. In some of these acarines, the indigestible food residues are removed from the body by a remarkable process called **schizeckenosy.** The residues are stored in gut cells that detach from the epithelial lining and move into the posterodorsal gut lobes. When one of the lobes fills with waste-laden cells, it breaks free from the ventriculus and is extruded through a split in the dorsal cuticle.

• **Excretory System**

Crustacean excretory organs are pairs of **antennal** and **maxillary glands,** opening to the outside on or near the bases of the antennae or maxillae, respectively. Both pairs are often present in larvae; adults normally retain only one or the other. The principal nitrogenous excretory products are ammonia with some amines and small amounts of urea and uric acid. Considerable excretion of ammonia also takes place across the gills.

Almost all insects have **malpighian tubules,** ranging in number from four to over 100 (Fig. 33.19). These thin-walled tubules are closed at their distal ends but open into the midgut near its junction with the hindgut. Uric acid is excreted, usually as an ammonium, potassium, or sodium salt. Water in the urine is reabsorbed by the proximal malpighian tubules or by the rectal wall, and the sodium and potassium are resorbed as bicarbonates, leaving virtually insoluble free uric acid as a precipitate. Thus, water and cations are recycled, part of the overall water conservation devices of insects. Bloodsucking forms, however, produce large amounts of fluid urine after a meal, an event that rids the animal of excess water.

Excretory **coxal glands** are found in some mites and other arachnids. These glands open to the outside at the bases of one or more pairs of appendages. Most ticks and mites also have malpighian tubules (Fig. 33.20). Waste from the hemocoel is taken up by the tubule walls and excreted into the lumen as guanine, the main excretory product. In those Prostigmata and Metastigmata whose ventriculus does not connect with the hindgut, an anteriorly directed excretory canal is joined to the hindgut, and guanine is excreted by this organ through the "anus" (uropore).

• **Reproductive Systems**

Most Crustacea are dioecious, and the gonopores open on a sternite or at the base of a trunk appendage. The male may have a penis, or appendages may be modified for copulation. Many crustaceans have nonflagellated, nonmobile sperm. In some groups the male places a packet of sperm (**spermatophore**) in the seminal receptacle or on the body surface of the female. Many Crustacea retain the fertilized eggs during embryonation, either in a brood chamber, attached to certain appendages, or within a sac formed during extrusion of the eggs.

Insects have a pair of testes. The vasa deferentia lead to a common, median **ejaculatory duct** that opens to the outside by the **aedeagus** (Fig. 33.21). **Accessory glands** join the ejaculatory duct and in many cases provide the material comprising the spermatophores. In females the paired ovaries are subdivided into **ovarioles** (Fig. 33.21). Each ovary usually has four to eight ovarioles, but some insects have more than 200; the viviparous Diptera, such

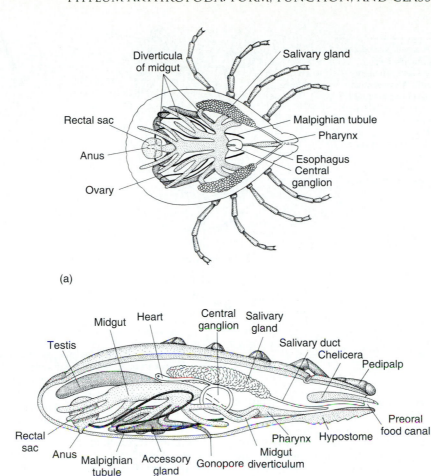

FIGURE 33.20

Internal anatomy of hard tick. (*a*) Dorsal view of female; (*b*) lateral view of male.

Drawings by William Ober.

as tsetse flies, however, have only one. The upper end of the ovariole produces oocytes and nurse cells. The developing oocytes become larger by the accumulation of yolk produced by the nurse cells and surrounding follicular cells. The **common oviduct** enlarges into a vagina, which opens to the exterior behind the eighth or ninth abdominal sternite. The **seminal receptacle** connects to the oviduct or vagina by a slender **spermathecal duct.** Accessory glands (**colleterial glands**) also open into the common oviduct or vagina, and these glands may produce a substance that cements the eggs together or to the substrate when they are laid or when they produce material for an egg capsule (**ootheca**).

CLASSIFICATION OF ARTHROPODAN TAXA WITH SYMBIOTIC MEMBERS

This classification of the Crustacea relies heavily on Kabata,[11] Marcotte,[18] and Bowman and Abele.[6] Classification of the Arachnida is according to Savory,[26] and diagnoses of the orders of pterygotes mainly follow Richards and Davies.[22]

Subphylum Crustacea
Head appendages consisting of two pairs of antennae, one pair of mandibles, and two pairs of maxillae; mostly aquatic; respiration usually with gills, sometimes through general body surface; head usually not clearly defined from trunk; cephalothorax usually with dorsal carapace; appendages, except first antennae (antennules), primitively biramous; sexes usually separate; development primitively with nauplius stage.

Class Maxillopoda
Typically with five cephalic, six thoracic, and four abdominal somites plus a telson, but reductions are common; no typical appendages on abdomen; naupliar eye (when present) is of unique structure and is referred to as maxillopodan eye.

Subclass Ostracoda
Body entirely enclosed in bivalve carapace; body unsegmented or indistinctly segmented; no more than two pairs of trunk appendages (only a few species recorded as gill parasites of marine teleosts and elasmobranchs).

FIGURE 33.21

General structure of insect reproductive organs. (*a*) Male: **AcGls,** accessory glands; **Dej,** ejaculatory duct; **Pen,** penis; **Gpr,** gonopore; **Tes,** testis; **Vd,** vasa deferentia; **Vsm,** seminal vesicle. (*b*) Female: **Lg,** ovarian ligament; **Ov,** ovary; **Ovl,** ovariole; **Clx,** calyx; **Odl,** lateral oviduct; **Odc,** common oviduct; **Gpr,** gonopore; **GC,** genital chamber; **AcGl,** accessory gland; **Spt,** spermatheca; **SptGl,** spermathecal gland.

From R. E. Snodgrass, *Principles of Insect Morphology.* Copyright © 1993 by Cornell University. Used by permission of the publisher, Cornell University Press.

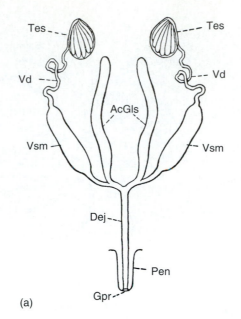
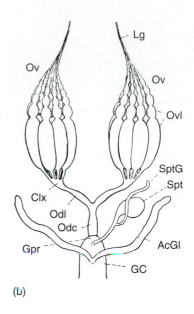

(a)

(b)

Family
Cypridinidae.

Subclass Copepoda

Typically with elongated, segmented body consisting of head, thorax, and abdomen; thorax with seven somites, of which first and sometimes second fused with head to form cephalothorax; thoracic appendages biramous, but maxillipeds and often fifth swimming legs uniramous; no appendages on abdomen except pair of rami on telson; no carapace; compound eyes absent, but median nauplius eye often present; gonopores on "genital segment," usually considered last somite of thorax; parasitic forms may not fit some or much of foregoing diagnosis and may be highly modified as adults and sometimes as juveniles.

Order Calanoida

No symbiotic members, but included because of ecological importance; antennules very long with 16 to 26 articles; buccal cavity open; antennae, mandibles, and maxillules biramous; mandibles gnathostomous; maxillae and maxilliped uniramous; first thoracic legs biramous, multiarticular, with plumose setae for swimming; last thoracic leg uniramous, modified, or missing; heart present in many; large order of important marine and freshwater planktonic organisms, never symbiotic.

Order Monstrilloida

Nauplii free swimming and then saclike endoparasites of marine polychaetes, prosobranch gastropods, and occasionally echinoderms; adults planktonic, without antennae, mouthparts, or functional gut; adult thoracic legs biramous for swimming.

Order Siphonostomatoida

Adult segmentation often reduced or lost; antennules reduced or elongated and multiarticulated; antennae may end in single massive claw for attachment to host; labrum and labium prolonged into siphon or tube, sometimes with some fusion; mandibles enclosed in buccal siphon, uniramous; maxillules ancestrally biramous, modified or reduced in derived forms; maxillae subchelate or brachiform (like human arm) for attachment to host; maxilliped subchelate or absent, sometimes absent in female only; adult thoracic limbs may be normal swimming appendages in some, variously modified and reduced in majority; adults ectoparasitic or endoparasitic on freshwater and marine fishes and on various invertebrates.

Families (representative)
Caligidae, Cecropidae, Dichelesthiidae, Lernaeopodidae, Pandaridae, Pennellidae, Sphyriidae.

Order Cyclopoida

Antennules short with 10 to 16 articles; buccal cavity open; antennae uniramous; mandibles and maxillules usually biramous; mandibles gnathostomous; free-living planktonic and benthic; commensal and ectoparasitic.

Families
Ascidicolidae, Enterocolidae, Lernaeidae, Notodelphyidae.

Order Poecilostomatoida

Adult segmentation often lost with copepodid metamorphosis; antennules often insignificant in size; buccal cavity slitlike; antennae often end in many small claws for attaching to host; mandibles with falcate (*falcatus* = sickle-shaped) gnathobase, rami missing; maxillules much reduced; maxillae reduced with denticulate inward pointing claw or slender, armed grasping claws; maxillipeds subchelate in males, often missing in females; adult thoracic limbs variously modified and reduced; adults parasitic on mostly marine invertebrates and fishes.

Families (representative)
Bomolochidae, Chondracanthidae, Clausiidae, Ergasilidae, Lichomolgidae, Philichthyidae, Sarcotacidae, Tuccidae.

Order Harpacticoida

Antennules short with fewer than 10 articles; buccal cavity open; antennae and mandibles biramous; mandibles gnathostomous; maxillules usually biramous; various degrees of fusion, reduction, and loss of rami in cephalic and thoracic appendages; heart absent; mostly free living, benthic, epibenthic, planktonic.

Subclass Tantulocarida

No recognizable cephalic appendages; solid median cephalic stylet; six free thoracic somites, each with pair of appendages, anterior five biramous; six abdominal somites; anterior five thoracic appendages with well-developed protopod and large endite arising from base of protopod; class of minute, copepodlike ectoparasites of other deep-sea benthic crustaceans; described in 1983;[7] examples: *Basipodella, Deoterthron.*

Subclass Branchiura

Body with head, thorax, and abdomen; head with flattened, bilobed, cephalic fold incompletely fused to first thoracic somite; thorax with four pairs of appendages, biramous, and with proximal extension of exopod of first and second legs; abdomen without appendages, unsegmented, bilobed; eyes compound; both pairs of antennae reduced; claws on antennules; maxillules often forming pair of suctorial discs; maxillae uniramous; gonopore at base of fourth leg; ectoparasites of marine and freshwater fishes and occasionally of amphibians.

Order Argulidea

Families

Argulidae, Dipteropeltidae.

Subclass Cirripedia

Sessile or parasitic as adults; head reduced and abdomen rudimentary; paired, compound eyes absent; body segmentation indistinct; usually hermaphroditic; in nonsymbiotic and epizoic forms, carapace becomes mantle, which secretes calcareous plates; antennules become organs of attachment; antennae disappear; young hatches as nauplius and develops to bivalved cypris larva; all marine.

Order Thoracica

With six pairs of thoracic appendages; alimentary canal; usually nonsymbiotic, although some epizoic and commensal on whales, fishes, sea turtles, and crabs; examples: *Chelonibia, Conchoderma, Coronula, Xenobalanus.*

Order Acrothoracica

Bores into mollusc shells or coral; females usually with four pairs of thoracic appendages; gut present, no abdomen; dioecious; males very small, without gut and appendages except antennules; parasitic on outside of mantle of female.

Order Ascothoracica

With segmented or unsegmented abdomen; usually six pairs of thoracic appendages; gut present; parasitic on echinoderms and soft corals; example: *Trypetesa.*

Order Rhizocephala

Adults with no segmentation, gut, or appendages; with rootlike absorptive processes through tissue of host; common parasites of decapod crustaceans.

Families

Lernaeodiscidae, Peltogastridae, Sacculinidae.

Class Malacostraca

Distinctly segmented bodies, typically with eight somites in the thorax and six somites plus the telson in the abdomen (except seven in Nebaliacea); all segments with appendages; antennules often biramous; first one to three thoracic appendages often maxillipeds; carapace covering head and part or all of thorax, a primitive character, but carapace lost in some orders; gills usually thoracic epipods; female gonopores on sixth thoracic segment; male gonopores on eighth thoracic segment; largest subclass is marine and freshwater, few terrestrial; many free living, but parasitic members relatively few, found in only three of the 10 to 12 extant orders commonly recognized.

SUPERORDER PERACARIDA

Without carapace or with carapace leaving at least four free thoracic somites; first thoracic somite fused with head; brood pouch in female (typically formed from modified thoracic epipods, the oostegites); several small, marine orders; the two large orders have parasitic members.

Order Amphipoda

No carapace; ventral brood pouch of oostegites; antennules often biramous; eyes usually sessile; gills on thoracic coxae; first thoracic limbs maxillipeds, second and third pairs usually prehensile (gnathopods); usually bilaterally compressed body form; marine, freshwater, and terrestrial; free living and symbiotic.

Suborder Hyperiidea

Head and eyes very large; only one thoracic somite fused with head; pelagic or symbiotic in medusae or tunicates.

Families

Hyperiidae, Phronimidae.

Suborder Caprellidea

So-called skeleton shrimp and whale lice. Two thoracic somites fused with head; abdomen much reduced, with vestigial appendages.

Families

Caprellidae, Cyamidae.

Order Isopoda

No carapace; ventral brood pouch of oostegites; antennules usually uniramous, sometimes vestigial; eyes sessile; gills on abdominal appendages; second and third appendages usually not prehensile; body usually dorsoventrally flattened.

Suborder Gnathiidea

Thorax much wider than abdomen; first and seventh thoracic somites reduced, seventh without appendages; larvae parasitic on marine fishes.

Family

Gnathiidae.

Suborder Flabellifera

Flattened body, with ventral coxal plates sometimes joined to body; telson fused with next abdominal somite, and other abdominal somites sometimes fused; uropods flattened, forming tail fan; marine, free living, and ectoparasitic on fishes.

Families with Parasitic Members

Aegidae, Crallanidae, Cymothoidae.

Suborder Epicaridea

Females greatly modified for parasitism; somites and appendages fused, reduced, or absent; mouthparts modified for sucking and mandible for piercing; maxillae reduced or absent; males small but less modified; marine parasites of Crustacea.

Families

Bopyridae, Cryptoniscidae, Dajidae, Entoniscidae, Phryxidae.

SUPERORDER EUCARIDA
All thoracic segments fused with and covered by carapace; no oostegites or brood pouch; eyes on stalks; usually with zoea larval stage.

Order Decapoda
First three pairs of thoracic appendages modified to maxillipeds (therefore appendages on remaining five thoracic somites equal 10 Gr. [*deka,* ten, + *podos,* foot]); includes crabs, lobsters, and shrimp.

Suborder Pleocyemata
Eggs carried by female and brooded on pleopods, hatch as zoeae.

INFRAORDER BRACHYURA
Carapace broad; abdomen reduced and tightly flexed beneath cephalothorax; first legs in form of heavy chelipeds; typical crabs.

Families with Symbiotic Members
Parthenopidae, Pinnotheridae.

Subphylum Uniramia
All appendages uniramous; head appendages consisting of one pair of antennae, one pair of mandibles, and one or two pairs of maxillae.

Class Insecta
Body with distinct head, thorax, and abdomen; one pair of antennae; thorax of three somites; abdomen with variable number, usually 11, of somites; thorax usually with two pairs of wings (sometimes one pair or none) and three pairs of jointed legs; separate sexes; usually oviparous; gradual or abrupt metamorphosis, few with direct development.

Subclass Apterygota
Primitively wingless insects; development direct or through slight metamorphosis; according to some recent authors, this subclass is paraphyletic (may not contain all the descendants of a common ancestor).

Orders
Collembola, Diplura, Protura, Thysanura.

Subclass Pterygota
Insects with wings (some secondarily wingless); all metamorphic; includes 97% of all insects; although members of all orders serve as hosts, what follow are only those with some medical or veterinary importance, in addition to orders that have appreciable numbers of symbiotic members.

Order Dermaptera
The earwigs. Forewings represented by small tegmina; hindwings large, membranous, and complexly folded; mouthparts for biting; ligula bilobed; body terminated by forceps; few ectoparasites of mammals (*Arixenia, Hemimerus*); some intermediate hosts of nematodes.

Order Dictyoptera
The cockroaches and mantids. Antennae nearly always filiform with many segments; mouthparts for biting; legs similar to each other or forelegs raptorial; tarsi with five segments; forewings more or less thickened into tegmina with marginal costal vein; many cerci segmented; ovipositor reduced and concealed; eggs contained in an ootheca; none symbiotic, but some implicated in mechanical transmission of human pathogens; some are intermediate hosts of Acanthocephala; examples: *Blatta, Blatella, Periplaneta, Supella.*

Order Mallophaga
The biting lice. Wingless; mouthparts are modified biting type; prothorax free; mesothorax and metathorax often im-

perfectly separated; tarsi of one or two segments, with one or two claws; cerci absent; metamorphosis slight; ectoparasitic in all stages on birds, less frequently on mammals; examples: *Menacanthus, Menopon, Piagetiella, Trichodectes.*

Order Anoplura
The sucking lice. Wingless; mouthparts modified for sucking and piercing; retracted when not in use; thoracic segments fused; tarsi unisegmented, claws single; cerci absent; metamorphosis slight; ectoparasitic in all stages on mammals; examples: *Haematopinus, Pediculus, Phthirus.*

Order Hemiptera
The true bugs, aphids, scale insects, etc. Wings variably developed with reduced or greatly reduced venation; forewings often more or less corneous; wingless forms frequent; mouthparts for piercing and sucking with mandibles and maxillae styletlike and lying in the projecting grooved labium, palps never evident; metamorphosis gradual with an incipient pupal instar sometimes present; many free living, some ectoparasites of birds and mammals; examples: *Cimex, Leptocimex, Rhodnius, Triatoma.*

Order Neuroptera
The alder flies, lacewings, ant lions, etc. Small to large, soft-bodied insects with two pairs of membranous wings without anal lobes; venation generally with many accessory branches and numerous costal veinlets; mouthparts for biting; antennae well-developed; cerci absent; complete metamorphosis; campodeiform larvae with biting or suctorial mouthparts; few are parasites of freshwater sponges and of spiders' egg cocoons.

Families
Mantispidae, Sisyridae (*Climacia, Sisyra*).

Order Coleoptera
The beetles. Minute to large insects whose forewings are modified to form elytra and abut down line of dorsum; hindwings membranous, folded beneath elytra, or absent; prothorax large; mouthparts for biting; metamorphosis complete, larvae of diverse types but never typically polypod; largest order of animals (more than 330,000 species); 1.5% are protelean parasites (immature stages parasitic) of insects; few are ectosymbionts of mammals.

Families
Leptinidae, Meloidae (some), Platypsyllidae, Rhipiphoridae, Staphylinidae (some).

Order Strepsiptera
Minute; males with branched antennae and degenerated biting mouthparts; forewings modified into small clublike processes; hindwings very large, plicately folded; females almost always extensively modified as internal parasites of other insects; larviform and devoid of wings, legs, eyes, and antennae; all protelean parasites of insects; examples: *Corioxenos, Elenchus, Eoxenos, Stylops.*

Order Siphonaptera
The fleas. Very small; wingless; laterally compressed body; mouthparts for piercing and sucking; complete metamorphosis with vermiform larvae; pupation in silk cocoons; adults all parasitic on warm-blooded animals; examples: *Pulex, Ctenocephalides, Xenopsylla, Tunga.*

Order Diptera
The flies and mosquitoes. Moderate size to very small; single pair of membranous wings (forewings), hindwings modified into halteres; mouthparts for sucking or for piercing as well and usually forming a proboscis; complete metamorphosis with vermiform larvae; many species of invertebrates and

vertebrate protelean parasites; vertebrate and insect ectoparasites; examples: *Aedes, Anopheles, Bombylius, Chrysops, Conops, Culex, Glossina, Hippobosca, Melophagus, Phlebotomus, Simulium, Stomoxys, Stylogaster, Tabanus.*

Order Lepidoptera
The butterflies and moths. Small to very large insects clothed with scales; mouthparts with galeae usually modified into a spirally coiled suctorial proboscis; mandibles rarely present; complete metamorphosis with larvae phytophagous, polypodous; large order with mostly free-living members; few insect protelean parasites and mammal ectoparasites; examples: *Bradypodicola, Calpe, Cyclotorna, Fulgoraecia.*

Order Hymenoptera
The sawflies, ants, bees, wasps, ichneumon flies, etc. Minute to moderate size; membranous wings, hindwings smaller and connected with forewings by hooklets, venation specialized by reduction; mouthparts for biting and licking; abdomen with first segment fused with thorax; sawing or piercing ovipositor present; complete metamorphosis with usually polypodous or apodous larvae; enormous insect order, about half of which are protelean parasites, mainly of other insects.

SUPERFAMILIES
Bethyloidea, Chalcidoidea (many), Cynipoidea (some), Evanioidea, Ichneumonoidea, Orussoidea, Proctorupoidea (Serphoidea), Trigonaloidea, Vespoidea (some).

Subphylum Chelicerata
Mostly terrestrial; respiration by gills, book lungs, or tracheae or through general body surface; first pair of appendages modified to form chelicerae; pair of pedipalps and usually four pairs of legs in adults; no antennae; tagmatization of prosoma (cephalothorax) and opisthosoma (abdomen), usually unsegmented.

Class Arachnida
Adult body fundamentally composed of 18 somites, divisible into six-unit prosoma and 12-unit opisthosoma, but segmentation often obscured in either or both of these tagmata; eyes, if present, simple (ocelli), not more than 12; chelicerae of two or three podomeres, chelate or unchelate; pedipalps of six podomeres, may be chelate or leglike, often with gnathobases; respiration through general body surface or by book lungs or tracheae (or both); sexes separate, with orifices on lower side of second opisthosomatic somite.

Orders
Acari, Amblypygi, Araneae, Opiliones, Palpigradi, Pseudoscorpiones, Ricinulei, Schizomida, Scorpiones, Solifugae, Uropygi.

Order Pseudoscorpiones
Prosoma undivided; opisthosoma with 12 distinguishable somites; chelicerae of two articles, chelate; pedipalps large, six articles, chelate; no pedicel; no telson; several pseudoscorpions symbiotic on mammals, prey on ectoparasites (lice, mites); examples: *Lasiochernes, Megachernes, Chiridiochernes.*

Order Acari
Highly specialized arachnids, in which modifications of segmentation divide body into proterosoma and hysterosoma, usually distinguishable as boundary between second and third pairs of legs; segments of mouth and its appendages borne on gnathosoma (capitulum), more or less sharply set off from rest of body (idiosoma); typically four pairs of legs but sometimes three, two, or one pair; often six podomeres in legs but may vary from two to seven; position of respiratory and genital openings variable; includes free-living suborders Notostigmata, Tetrastigmata.

Suborder Mesostigmata
Several sclerotized plates on dorsal and ventral surfaces; single pair of spiracular openings between second and fourth coxae; large group, many free living; parasitic examples: *Dermanyssus, Ornithonyssus, Sternostoma.*

Suborder Metastigmata
Large acarines (ticks); hypostome with recurved teeth, used as holdfast organ; sensory organ (Haller's organ) on tarsus of first leg with olfactory and hygroreceptor setae; single pair of spiracular openings close to coxae of fourth legs except in larvae; all parasitic; examples: *Amblyomma, Argas, Boophilus, Dermacentor, Ixodes, Ornithodoros, Otobius, Rhipicephalus.*

Suborder Prostigmata
Spiracular openings, when present, paired and located either between the chelicerae or on dorsum of anterior portion of hysterosoma; usually weakly sclerotized; chelicerae vary from strongly chelate to reduced; pedipalps simple, fanglike, or clawed; terrestrial and aquatic free-living, phytophagous, and parasitic forms; examples: *Demodex, Trombicula.*

Suborder Cryptostigmata
Oribatid or beetle mites (so called because of superficial resemblance to beetles); spiracles absent, although some have trachea associated with paired dorsal pseudostigmata and with bases of first and third legs; free living, but some are vectors of tapeworms (*Galumna, Oppia*).

Suborder Astigmata
Mostly slow moving and weakly sclerotized; no spiracles, respire through body surface; free-living and parasitic forms; examples: *Megninia, Otodectes, Sarcoptes.*

References

1. Andersen, S. O. 1976. Cuticular enzymes and sclerotization in insects. In Hepburn, H. R., ed. *The insect integument.* Amsterdam: Elsevier Scientific Publishing Co., 121–44.

2. Anderson, D. T. 1973. *Embryology and phylogeny in annelids and arthropods.* Oxford: Pergamon Press.

3. Arthur, D. R. 1956. The morphology of the British Prostriata with particular reference to *Ixodes hexagonus* Leach. III. *Parasitology* 46:261–307.

4. Barnes, H., and J. J. Gonor. 1958. Neurosecretory cells in the cirripede, *Pollicipes polymerus. J. Mar. Res.* 17:81–102.

5. Boudreaux, H. B. 1979. *Arthropod phylogeny with special reference to insects.* New York: John Wiley & Sons, Inc.

6. Bowman, T. E., and L. G. Abele. 1982. Classification of the recent Crustacea. In Abele, L. G., ed. *The biology of Crustacea, vol. 1. Systematics, the fossil record, and biogeography.* New York: Academic Press, Inc., 1–27.

7. Boxshall, G. A., and R. J. Lincoln. 1983. Tantulocarida, a new class of Crustacea ectoparasitic on other crustaceans. *J. Crust. Biol.* 3:1–16.

8. Filshic, B. K. 1976. The structure and deposition of the epicuticle of the adult female cattle tick (*Boophilus microplus*). In Hepburn, H. R., ed. *The insect integument.* Amsterdam: Elsevier Scientific Publishing Co., 193–206.

9. Gould, S. J. 1989. *Wonderful life.* New York: W. W. Norton.

10. Hillerton, J. E. 1979. Changes in the mechanical properties of the extensible cuticle of *Rhodnius* through the fifth larval instar. *J. Insect Physiol.* 25:73–77.

11. Kabata, Z. 1979. *Parasitic Copepoda of British fishes.* London: Ray Society.

12. Karlson, P., and E. C. Sekeris. 1976. Control of tyrosine metabolism and cuticle sclerotization by ecdysone. In Hepburn, H. R., ed. *The insect integument.* Amsterdam: Elsevier Scientific Publishing Co., 145–56.

13. Knowles, F. G. W., and D. B. Carlisle. 1956. Endocrine control in the Crustacea. *Biol. Rev.* 31:396–473.

14. Krantz, G. W. 1971. *A manual of acarology.* Corvallis, Ore.: Oregon State University Bookstores, Inc.

15. Krishnan, G. 1951. Phenolic tanning and pigmentation of the cuticle in *Carcinus meanas. Q. J. Microsc. Soc.* 92:333–44.

16. Manton, S. M. 1973. Arthropod phylogeny—a modern synthesis. *J. Zool.* 171:111–30.

17. Manton, S. M. 1977. *The Arthropoda: Habits, functional morphology, and evolution.* Oxford: Clarendon Press.

18. Marcotte, B. M. 1982. Evolution within the Crustacea, Part 2: Copepoda. In Abele, L. G., ed. *The biology of Crustacea, vol. 1. Systematics, the fossil record, and biogeography.* New York: Academic Press, Inc., 185–97.

19. Mordue, W., G. J. Goldsworthy, J. Brady, and W. M. Blaney. 1980. *Insect physiology.* New York: John Wiley and Sons, Inc.

20. Paulus, H. F. 1979. Eye structure and the monophyly of the Arthropoda. In Gupta, A. P., ed. *Arthropod phylogeny.* New York: Van Nostrand Reinhold Co., 299–383.

21. Richards, A. G. 1978. The chemistry of insect cuticle. In Rockstein, R., ed. *Biochemistry of insects.* New York: Academic Press, Inc., 205–32.

22. Richards. O. W., and R. G. Davies. 1978. *Imms' outlines of entomology,* 6th ed. London: Chapman & Hall, Ltd.

23. Riddiford, L. M., and J. W. Truman. 1978. Biochemistry of insect hormones and insect growth regulators. In Rockstein, R., ed. *Biochemistry of insects.* New York: Academic Press, Inc., 307–57.

24. Robertson, J. D. 1960. Ionic regulation in the crab, *Carcinus meanas* (L.) in relation to the moulting cycle. *Comp. Biochem. Physiol.* 1:183–212.

25. Sanders, H. L. 1957. The Cephalocarida and crustacean phylogeny. *Syst. Zool.* 6:112–28, 148.

26. Savory, T. 1977. *Arachnida,* 2d ed. New York: Academic Press, Inc.

27. Schram, F. R. 1986. *Crustacea.* New York: Oxford University Press.

28. Snodgrass, R. E. 1935. *Principles of insect morphology.* New York: McGraw-Hill Book Co.

29. Snow, K. R. 1970. *The arachnids: An introduction.* New York: Columbia University Press.

30. Travis, D. R. 1955. The moulting cycle of the spiny lobster, *Panulirus argus,* Latreille. II. Preecdysial histological and histochemical changes in the hepatopancreas and integumental tissues. *Biol. Bull.* 108:88–112.

31. Vincent, J. F. V., and J. E. Hillerton. 1979. The tanning of insect cuticle—a critical review and a revised mechanism. *J. Insect Physiol.* 25:653–58.

32. Wigglesworth, V. B. 1976. The distribution of lipid in the cuticle of *Rhodnius.* In Hepburn, H. R., ed. *The insect integument.* Amsterdam: Elsevier Scientific Publishing Co., 89–106.

33. Zacharuk, R. Y. 1976. Structural changes of the cuticle associated with moulting. In Hepburn, H. R., ed. *The insect integument.* Amsterdam: Elsevier Scientific Publishing Co., 299–321.

34. Zinsser, H. 1935. *Rats, lice and history.* New York: Little, Brown, and Company.

Additional References

Barrington, E. J. W. 1979. *Invertebrate structure and function,* 2d ed. New York: John Wiley & Sons, Inc. An excellent treatment from the functional standpoint.

Berenbaum, M. R. 1995. *Bugs in the system. Insects and their impact on human affairs.* Reading, Mass.: Addison-Wesley Publishing Co. An enjoyable account of the many ways in which insects are important to humans.

Brusca, R. C., and G. J. Brusca. 1990. *Invertebrates.* Sunderland, Mass.: Sinauer Associates, Inc., Publishers.

Burgess, N. R. H., and G. O. Cowan. 1993. *A colour atlas of medical entomology.* New York: Chapman and Hall.

Busvine, J. R. 1975. *Arthropod vectors of disease.* London: Edward Arnold.

Clarke, K. U. 1973. *The biology of the Arthropoda.* New York: American Elsevier Publishing Co., Inc.

Cloudsley-Thompson, J. L. 1976. *Insects and history.* London: Weidenfield and Nicolson.

Fox, R. M., and J. W. Fox. 1964. *Introduction to comparative entomology.* New York: Reinhold Publishing Corp. Covers chelicerates and insects but not Crustacea.

Gupta, A. P., ed. 1979. *Arthropod phylogeny.* New York: Van Nostrand Reinhold Company.

Gupta, A. P., ed. 1990. *Morphogenetic hormones of arthropods: Discoveries, synthesis, metabolism, evolution, modes of action, and techniques.* New Brunswick, N.J.: Rutgers University Press.

Harwood, R. F., and M. T. James. 1979. *Entomology in human and animal health,* 7th ed. New York: Macmillan Publishing Co., Inc. One of the best texts available in medical entomology.

Mordue, W., G. J. Goldsworthy, J. Brady, and W. M. Blaney. 1980. *Insect physiology.* New York: John Wiley and Sons, Inc.

Neville, A. C. 1975. *Biology of the arthropod cuticle.* New York: Springer-Verlag.

Service, M. W. 1986. *Blood-sucking insects: Vectors of disease.* London: Edward Arnold.

Snodgrass, R. E. 1935. *Principles of insect morphology.* New York: McGraw-Hill Book Co. This and the following are classics that remain valuable references on arthropod structure.

Snodgrass, R. E. 1965. *A textbook of arthropod anatomy.* (Facsimile of the 1952 edition.) New York: Hafner Publishing Co.

Strickland, G. T., ed. 1991. *Hunter's tropical medicine,* 7th ed. Philadelphia: W. B. Saunders Company.

U.S. Department of Health, Education, and Welfare. 1960 (1979 revision). *Introduction to arthropods of public health importance. HEW Pub. No. (CDC) 79-8139.* Washington, D.C.: U.S. Government Printing Office. A short, concise introduction to the subject, with a key to some common classes and orders of public health importance.

Wigglesworth, V. B. 1972. *Principles of insect physiology,* 7th ed. London: Chapman & Hall Ltd.

Chapter 34

PARASITIC CRUSTACEANS

Morphological details of small animals are often ignored by observers just because they are small. . . . [I] venture to suggest that, had the copepod been the size of a cow, the tip of its first antenna would have become a topic for exhaustive studies. One tends to forget that the dimensional scale does not influence the biological importance.

Z. Kabata

A fascinating array of adaptations for symbiosis can be found among crustaceans. In addition to being of academic interest, many parasitic crustaceans are of substantial economic importance. Nonetheless, they are often neglected in both parasitology and invertebrate zoology courses.

Some crustacean parasites have been known since antiquity, although the fact that they were crustaceans, or even arthropods, was not recognized until the early nineteenth century. Aristotle and Pliny recorded the affliction of tunny and swordfish by large parasites we would now recognize as penellid copepods. In 1554 Rondelet[61] figured a tunny with one of the copepods in place near the pectoral fin. In 1746 Linnaeus first established the genus *Lernaea*,[41] and in his 1758 edition of *Systema Naturae,* he called the species (from European carp) *Lernaea cyprinacea*.[42] Various other highly modified copepods were described in the latter half of the eighteenth and early nineteenth centuries, but they had so few obvious arthropod features that they were variously classified as worms, gastropod molluscs, cephalopod molluscs, and annelids. Finally, Oken[50] (1815–1816) associated these animals with other parasitic copepods that could be recognized as such. Based on Surriray's important observation that their young resemble those of *Cyclops,* de Blainville[6] (1822) firmly established these animals as crustaceans. Copepoda is not the only crustacean group with members so modified for parasitism as to be superficially unrecognizable as arthropods; some of these other groups will be discussed as well.

The status of higher taxa in the Crustacea, even the status of the taxon "Crustacea" itself, continues in a state of flux. Some recent texts have accepted the classification of Bowman and Abele,[7] and some have not.[10,23,62] Here we will accord the Crustacea the rank of subphylum, and for the remaining higher classification adopt that of Bowman and Abele.[7]

CLASS MAXILLOPODA

The class Maxillopoda includes a number of crustacean groups traditionally considered classes unto themselves. There is evidence that these groups descended from a common ancestor and thus form a clade within the Crustacea. They basically have five cephalic, six thoracic, and usually four abdominal somites plus a telson, but reductions are common. There are no typical appendages on the abdomen. The eye of the nauplius (when present) has a unique structure and is referred to as a **maxillopodan eye.**

Subclass Copepoda

The copepods are extremely important both as free-living organisms and as parasites. They display enormous evolutionary versatility in exploitation of symbiotic niches and by their spectrum of adaptations to symbiosis, ranging from the slight to the extreme. Although penellids are so bizarre that eighteenth century biologists did not recognize them as arthropods, many parasitic and commensal copepods are comparatively much less highly modified. In fact one can arrange examples of the various groups in an arbitrary series to demonstrate the progression from little to very high specialization.[34]

We shall cite but a few examples to illustrate the trends in adaptation to parasitism in copepods. Some of these trends are (1) reduction in locomotor appendages; (2) development of adaptations for adhesion, both by modification of appendages and by development of new structures; (3) increase in size and change in body proportions, caused by much greater growth of genital or reproductive regions; (4) fusion of body somites and loss of external evidence of segmentation; (5) reduction of sense organs; and (6) reduction in numbers of instars that are free living, both by passing more stages before hatching and by larval instars becoming parasitic. "Typical" or primitive copepod development can be regarded as gradual metamorphosis with a series of **copepodid**

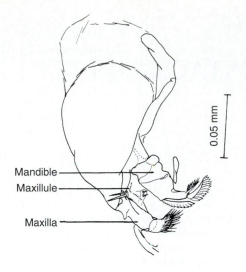

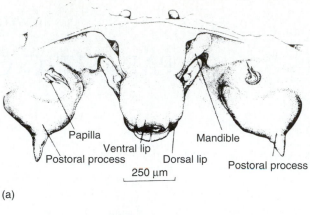

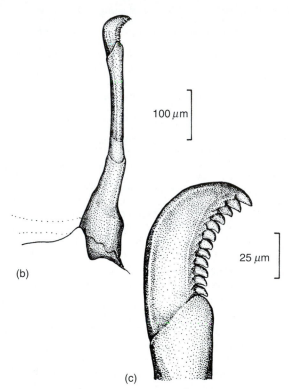

FIGURE 34.1

Poecilostome mouthparts (one side drawn) of *Ergasilus cerastes,* a parasite of catfish (*Ictalurus* spp.).

From L. S. Roberts "*Ergasilus cerastes* sp. n. (Copepoda: Cyclopoida) from North American catfishes," in *J. Parasitol.* 55:1266–1270. Copyright © 1969. Reprinted with permission of the publisher.

(subadult) instars succeeding the naupliar instars. Copepodid juveniles bear considerable similarity to the adults except in dramatically metamorphic families like Lernaeopodidae and Pennellidae.

Taxonomy of the Copepoda at the ordinal level is based partly on morphology of mouthparts:[35] **gnathostome, poecilostome,** and **siphonostome.** Gnathostomous mandibles are fairly short, broad, biting structures with teeth at their ends, and the buccal cavities are large and widely open. This is apparently the ancestral condition and is possessed by several copepod orders. Poecilostome mouths are rather similar, except that they are somewhat slitlike and have falcate (sickle-shaped) mandibles (Fig. 34.1). The siphonostome condition is characterized by a more or less elongated, conical, siphonlike mouth formed by the labrum and labium (Fig. 34.2*a*), and the mandibles are styletlike and enclosed within the siphon (Fig. 34.2*b*). The possession of poecilostome and of siphonostome mouths forms the basis for recognition of the orders Poecilostomatoida and Siphonostomatoida, respectively. According to Ho,[24] these two orders are sister groups.

• Order Cyclopoida

The order Cyclopoida is a large group of copepods, most species of which are free living. Free-living cyclopoids occupy important niches as primary consumers in many aquatic habitats, particularly freshwater. Several families of cyclopoids are parasites on invertebrates, including a highly modified one in the blood of a freshwater snail.[26] Lernaeidae is a highly specialized group of fish parasites.

Family Lernaeidae.   This is a relatively small family that parasitizes freshwater teleosts (bony fishes). Often quite large and conspicuous, some species, especially *Lernaea cyprinacea,* are serious pests of economically important fishes. Therefore, they are among the best known parasitic

FIGURE 34.2

Siphonostome mouthparts *Caligus curtus,* which parasitizes a variety of marine fishes. (*a*) The base of the mandible can be seen as it extends into the tube formed by the dorsal and ventral lips. (*b, c*) The mandible is a long, flat blade with teeth at its distal tip.

From R. R. Parker et al., "A review and description of *Caligus curtus* Miller, 1785 (Caligidae: Copepoda), type species of its genus." in *J. Fish. Res. Bd. Can.* 25:1923–1969. Copyright © 1968. Reprinted with permission.

copepods. *Lernaea cyprinacea* can infect a variety of fish hosts and even frog tadpoles.[66] The anterior of the parasite is embedded in the host's flesh and is anchored there by large processes that arise from the parasite's cephalothorax and thorax, hence the common name *anchor worm.* It causes damage to the scales, skin, and underlying muscle tissue. There may be considerable inflammation, ulceration, and secondary bacterial and fungous infection.[5] Fishes that are small relative to the parasite can easily be

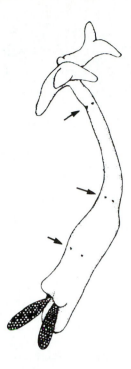

FIGURE 34.3

Lernaea cyprinacea, the "anchor worm," is a serious pest of a variety of fishes, including several of economic importance. The anterior holdfast ("horns") is embedded in the host's flesh, and the posterior part of the body projects to the exterior. The swimming legs (*arrows*) do not participate in the rapid, final growth of the adult female (up to 16 mm long) and so remain proportionately very tiny.

From Z. Kabata, "Crustacea as enemies of fishes," in *Diseases of Fish,* book I, edited by S. F. Snieszko and H. R. Axelrod. Copyright © 1970 T.F.H. Publications, Inc., Neptune City, NJ.

killed by infection with several individuals. A fully developed *L. cyprinacea* may be more than 12 mm long. Epizootics of this pest occur in wild fish populations, and it is a serious threat wherever fishes are raised in hatcheries.[54]

Lernaeids are among the most highly specialized copepods. Once the sexually mature female is fertilized, she embeds her anterior end beneath a scale, near a fin base or in the buccal cavity. At that point the parasite is less than 1.5 mm long and is superficially quite similar to *Cyclops* or other unspecialized cyclopoids. The female begins to grow rapidly, reaching "normal" size in little more than a week. The largest specimen recorded was 15.9 mm (22.0 mm including the egg sacs).[21] Interestingly the swimming legs and mouthparts remain in place but do not take part in this growth, so that they quickly become inconspicuous. At the same time, the large anchoring processes, two ventral and two dorsal (Fig. 34.3), grow into the fish's muscle. The body segmentation becomes blurred, the location of the somites being recognized only with location of the tiny legs. The result is an embryo-producing machine that bears practically no resemblance to an arthropod and that has its head permanently anchored in its food source. It is little wonder that the early taxonomists had such trouble correctly placing *Lernaea* in their system.

Nevertheless, the larvae can be recognized clearly as crustacean and are typical nauplii. The primitive series of

naupliar instars has been shortened to three. When the nauplii hatch, they contain enough yolk material within their bodies to eliminate the need for feeding in any of the three naupliar stages. The third nauplius molts to give rise to the first copepodid, and this marks the end of the free-living life of a *Lernaea.* Thus, the length of time spent as a free-living organism has been markedly shortened, compared with the primitive condition, and the free-living instars do not even feed.

• Order Poecilostomatoida

This order illustrates a progression from little specialized parasites (Ergasilidae) to some highly modified and bizarre forms (Philichthyidae, Sarcotacidae). Poecilostomes have been especially successful as symbionts of other invertebrates, particularly with cnidarian hosts. Of 1475 species of copepods known from invertebrates, 416 belong to this order, and 373 species of poecilostomes are associated with cnidarians.[28]

Family Ergasilidae. Ergasilids are among the most common copepod parasites of fishes. They have been a "thorn in the flesh for many valuable fisheries in the Old World" for a long time[34] and often frequent the gills of a variety of fishes in North America.[59] *Ergasilus* spp. are primarily parasites of freshwater hosts but are common on several marine fishes, especially the more euryhaline ones such as sticklebacks, killifish, and mullets.

Ergasilidae show primitive morphological characteristics reminiscent of free-living copepods, with few, but effective adaptations for parasitism. The antennules are sensory, but the antennae have become modified into powerful organs of prehension (Fig. 34.4). *Ergasilus* females usually are found clinging by their antennae to one of the fish's gill filaments. Each antenna ends in a sharp claw. The third segment and claw are opposable with the second (subchelate) (Fig. 34.5). Rather than depending on muscle and heavy sclerotization of the antennae, the antennal tips may be fused or locked so that the gill filament is completely encircled (*E. amplectens, E. tenax*) (Fig. 34.6).

When removed from their position on the gill, most *Ergasilus* can swim reasonably well; their pereiopods retain the flat copepod form, with setae and hairs well-adapted for swimming. The first legs, however, show adaptation for their feeding habit. These appendages are supplied with heavy, bladelike spines; in some species the second and third endopodal segments are fused, presumably lending greater rigidity to the leg. Such modifications increase the ability of the animal to rasp off mucus and tissue from the gill to which it is clinging (Fig. 34.7). The first legs dislodge epithelial and underlying cells in this manner and sweep them forward to the mouth[17] (Fig. 34.8). It is easy to see that a heavy infestation with *Ergasilus* could severely damage gill tissue, interfere with respiration, open the way to secondary infection, and lead to death. Epizootics of *Ergasilus* on mullet (*Mugil*) were recorded in Israel; in one case up to 50% of the stock in some ponds was lost, and hundreds of dead mullet were found daily.[63] Rogers and Hawke[60] found large numbers of *Ergasilus* infesting skin lesions of shad (*Dorosoma*) in Tennessee; they believed the copepods were the primary cause of the moribund condition of the fish.

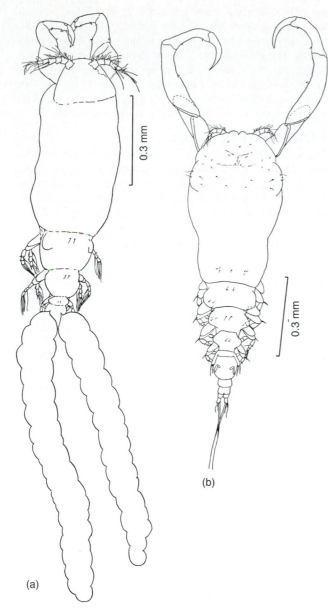

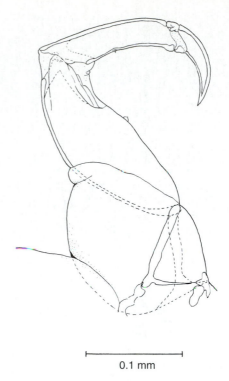

FIGURE 34.5

Antenna of *E. centrarchidarum,* a common parasite of members of the sunfish family (Centrarchidae). The antennae of *Ergasilus* are usually modified into a powerful organ used to grasp their host's gill filament, with the third and fourth joints opposable with the second.

From L. S. Roberts, "*Ergasilus* (Copepoda: Cyclopoida): revision and key to species in North America," in *Trans. Am. Microsc. Soc.* 89:134–161. Copyright © 1970. Reprinted with permission of the publisher.

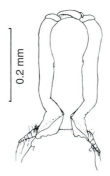

FIGURE 34.6

Tips of the antennae of *Ergasilus tenax* "lock" together, completely encircling the host's gill filament.

From L. S. Roberts, "*Ergasilus tenax* sp. n.. (Copepoda: Cyclopoda) from white crappie, *Pomoxis annularis* Rafinesque," in *J. Parasitol.* 51:987–989. Copyright © 1965. Reprinted with permission of the publisher.

FIGURE 34.4

Examples of *Ergasilus,* a common parasite of freshwater and some marine fishes. (*a*) *Ergasilus celestis,* from eels (*Anguilla rostrata*) and burbot (*Lota lota*), bearing egg sacs. (*b*) *Ergasilus arthrosis,* reported from several species of freshwater hosts, nonovigerous.

From L. S. Roberts, "*Ergasilus arthrosis* n. sp. (Copepoda: Cyclopoida) and the taxonomic status of *Ergasilus versicolor* Wilson, 1911, *Ergasilus elegans* Wilson, 1916, and *Ergasilus celestis* Mueller, 1936, from North American fishes," in *J. Fish Res. Bd. Can.* 26:997–1011. Copyright © 1969. Reprinted with permission of the publisher.

Ergasilus spp. have three naupliar and five copepodid stages, all free living.[67] The adult males are planktonic as well, and the female is fertilized before attaching to the fish host. Only the female has been found as a parasite. In one species even the females are planktonic as adults (*E. chautauquaensis,* which may be the only nonparasitic species in the genus), although females of several other species are sometimes encountered in the plankton.[9]

Family Lichomolgidae.   Lichomolgids are symbionts with a wide variety of marine invertebrates, including serpulid polychaetes, alcyonarian and madreporarian corals, ascidians, sea anemones, nudibranchs, holothurians, starfish, pelecypods, and sea urchins. It is evident that many species are involved, and many are yet to be discovered. The Lichomolgidae (family here broadly accepted) are divided by Humes and Stock[29] into five families, embracing 76 genera and 324 species.

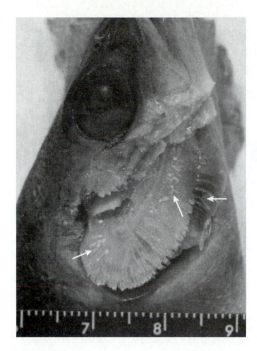

FIGURE 34.7

Ergasilus labracis in situ on gills of striped bass, *Morone saxatilis* (two specimens are indicated by *arrows*). The gill operculum has been removed. Note also that the fish is infected by an isopod, *Lironeca ovalis,* partly hidden under gill (*right*).

Photograph by Larry S. Roberts.

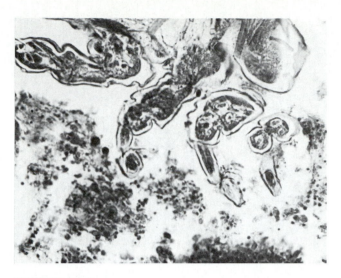

FIGURE 34.8

Section of *Ergasilus sieboldi* in situ showing damage to gills inflicted by thoracic appendages. The tissue is rasped off, and the parasite feeds on detached epithelial, mucous, and blood cells. The first legs (*top left*) are particularly important in directing dislodged tissue anteriorly toward the mouth. (× 200.)

From T. Einszporn, "Nutrition of *Ergasilus sieboldi* Nordmann. II. The uptake of food and the food material," in *Acta Parasitol. Polon.* 13:373–380. Copyright © 1965.

Lichomolgids are generally cyclopoid in body form, retaining segmentation and swimming legs (Fig. 34.9). Segments of the antennae are reduced to three or four, and they often end in one to three terminal claws. The antennae are apparently adapted for prehension in much the same manner as those of the ergasilids. Higher specialization is shown in some species, in which one or more swimming legs may be reduced or vestigial. The copepodids are often found parasitic on the same hosts as are the adults, and relatively little time is apparently spent in the free-living naupliar stages.

Families Philichthyidae, Sarcotacidae. Little is known of these families, but they deserve at least brief mention because of their great specialization for parasitism.

The general appearance of philichthyids is startling; unlikely looking processes emanate from their bodies (Fig. 34.10). This is a small group, completely endoparasitic in the subdermal canals of teleosts and elasmobranchs—that is, in the frontal mucous passages and sinuses and the lateral line canal. Some species retain external evidence of segmentation, but in others it is less apparent. Organs of attachment are reduced. The males are much smaller than the females and are less highly modified. Some philichthyids seem to prefer the left side of the host to the right for unknown reasons.[16]

Sarcotacids are also endoparasitic copepods and are probably the most highly specialized of any copepod parasite of a vertebrate (Fig. 34.11). They live in cysts in the muscle or abdominal cavity of their fish hosts. Their appendages are vestigial, and

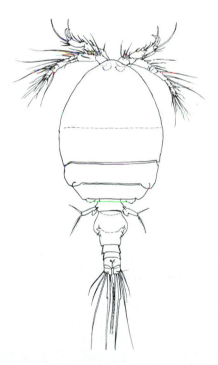

FIGURE 34.9

Typical lichomolgid, *Ascidioxynus jamaicensis,* from the branchial sac of an ascidian, *Ascidia atra* (dorsal view of female).

Reprinted from *Smithsonian Contributions to Zoology,* no. 127, from the chapter entitled "A Revision of the Family Lichomolgidae Kossman, 1877" (Washington, DC: Smithsonian Institution Press), page 143, by permission of the publisher. Copyright © 1973.

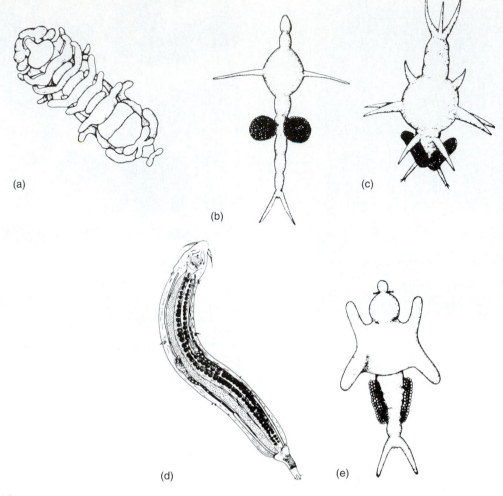

FIGURE 34.10

Philichthyids, parasites in subdermal canals of fish. (*a*) *Philichthys xiphiae;* (*b*) *Sphaerifer leydigi;* (*c*) *Colobomatus sciaenae;* (*d*) *Lerneascus nematoxys;* (*e*) *Colobomatus muraenae.*

From Z. Kabata, "Crustacea as enemies of fishes," in *Diseases of Fish,* book I, edited by S. F. Snieszko and H. R. Axelrod. Copyright © 1970 T.F.H. Publications, Inc., Neptune City, NJ.

they appear to feed on blood from the vascular wall of the cyst. The adult female is little more than a reproductive bag within the cyst and may reach several centimeters in size. Males are much smaller, and one lives in each cyst, mashed between the wall of the cyst and the huge body of its mate.

Nothing is known of the development and many other aspects of the biology of sarcotacids and philichthyids.

• Order Siphonostomatoida

The members of this large group are mostly parasites of fishes. Only 31 species have been recorded from cnidarians, but there are probably many more.[28] Although even the most primitive siphonostomes show some adaptations to parasitism, like the poecilostomes they show an array from generalized to extremely modified and bizarre. The majority of siphonostomes are parasites of marine fishes, and with increased aquaculture of marine fishes, siphonostomes will have an increasing economic impact. A variety of pathogenic effects have been demonstrated (Fig. 34.12). In western Japan, where yellowtail (*Seriola*) are cultured intensively in small bays, *Caligus spinosus* can inflict considerable damage.[31] In Ireland, wrasse are used as cleaners (fish that eat ectoparasites of other fish) against caligid parasites of cultured Atlantic salmon.[16]

Family Caligidae. Adult caligids are obviously arthropods, although they have departed from the "typical" free-living copepod plan. They have, at least, some adaptations for prehension, tend to be larger than most free-living groups, and have some dorsoventral flattening for closer adhesion to the host surface. Some tend to be more sedentary, being mostly confined to the fish's branchial chamber, but the adults of many species can move rapidly over the host's surface (fins, gills, and mouth). Adults can swim and change hosts.

The usual caligid body form shows a fusion of the ancestral body somites: a large, flat cephalothorax followed by one to three free thoracic segments, a large genital segment, and a smaller unsegmented abdomen. Three segments between the cephalothorax and genital somite, as in *Dissonus* (Fig. 34.13), is a primitive character, and reduction to a single segment, as in *Caligus* (Fig. 34.14), is derived.[51] *Caligus curtus'* principal appendages for prehension are the antennae and maxillipeds. It has two **lunules** on the anterior margin of the cephalothorax that function as accessory organs of adhesion. The cephalothorax is roughly disc shaped and bears a flexible, membranous margin. The posterior portion of the disc is not formed by the cephalothorax itself but by the greatly enlarged, fused protopod of the third thoracic legs

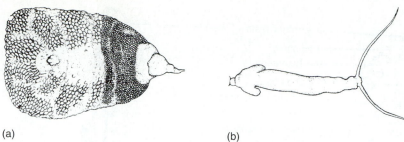

(a)

(b)

FIGURE 34.11

Sarcotacids may be the most highly specialized copepod parasites of vertebrates. (*a*) *Sarcotaces* sp., female; (*b*) *Sarcotaces* sp., male; (*c*) *Ichthyotaces pteroisicola.*

From Z. Kabata, "Crustacea as enemies of fishes," in *Diseases of Fish,* book I, edited by S. F. Snieszko and H. R. Axelrod. Copyright © 1970 T.F.H. Publications, Inc., Neptune City, NJ.

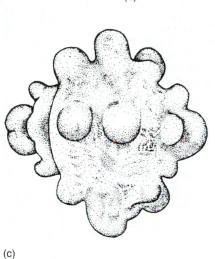

(c)

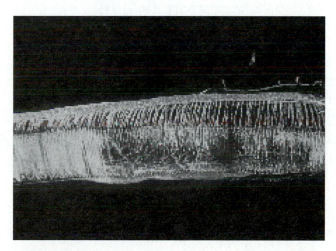

FIGURE 34.12

Gill of a coral trout (*Plectropomus leopardi*) perfused with microfil yellow. The photograph shows that the perfusate (and thus blood in the living fish) could not penetrate gill filaments damaged by the siphonostomatoid copepod *Hatschekia plectropomi.*

Courtesy of R. J. G. Lester.

FIGURE 34.13

Dissonus nudiventris, a caligid with the primitive character of three segments between the cephalothorax and genital somite.

From Z. Kabata, "Crustacea as enemies of fishes," in *Diseases of Fish,* book I, edited by S. F. Snieszko and H. R. Axelrod. Copyright © 1970 T.F.H. Publications, Inc., Neptune City, NJ.

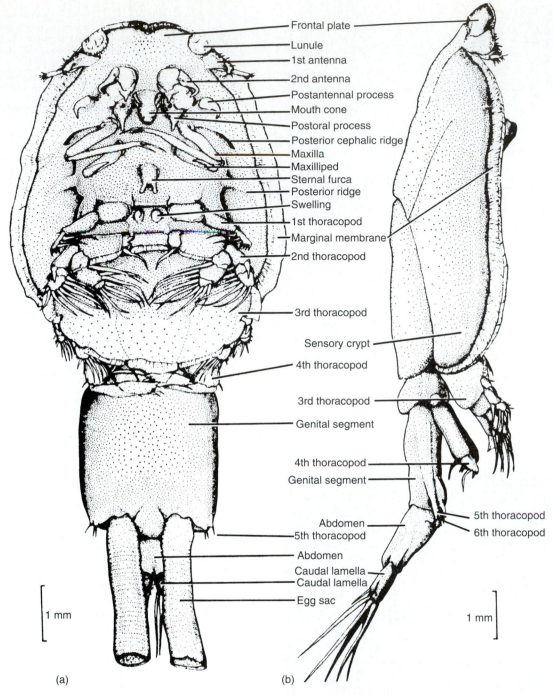

Frontal plate
Lunule
1st antenna
2nd antenna
Postantennal process
Mouth cone
Postoral process
Posterior cephalic ridge
Maxilla
Maxilliped
Sternal furca
Posterior ridge
Swelling
1st thoracopod
Marginal membrane
2nd thoracopod

3rd thoracopod

Sensory crypt

4th thoracopod

3rd thoracopod

Genital segment

4th thoracopod
Genital segment

5th thoracopod
6th thoracopod

Abdomen
5th thoracopod

Abdomen

Caudal lamella
Caudal lamella

Egg sac

1 mm

1 mm

(a)   (b)

FIGURE 34.14

Caligus curtus, a caligid with a derived character of only one segment between the cephalothorax and genital somite. (*a*) Female, ventral view; (*b*) male, lateral view.

From R. R. Parker et al., "A review and description of *Caligus curtus* Miller, 1785 (Caligidae: Copepoda), type species of its genus," in *J. Fish. Res. Bd. Can.* 25:1923–1969. Copyright © 1968. Reprinted with permission of the publisher.

(Fig. 34.15). Membranes on the margin of the protopods match those on the cephalothorax, and the arrangement forms an efficient suction disc when the cephalothorax is applied to the fish's surface and is arched.

The feeding apparatus of *C. curtus* is a good example of the tubular mouth type of the siphonostomes. The mouth tube is carried in a folded position parallel to the body axis, but it can be erected so that its tip can be applied directly to the host surface. The tip of the tube bears flexible membranes analogous to

those on the margin of the cephalothorax, again increasing the efficiency of the organ as a suction device. The bases of the mandibles are lateral to and outside the mouth tube (Fig. 34.2). They enter the buccal cavity through longitudinal canals so that their tips lie within the opening of the cone. The mandibular tips bear a sharp cutting blade on one side and a row of teeth on the other. Thus, the mandibles can work back and forth like little pistons in their canals, piercing and tearing off bits of host tissue to be sucked up by the muscular action of the mouth tube.

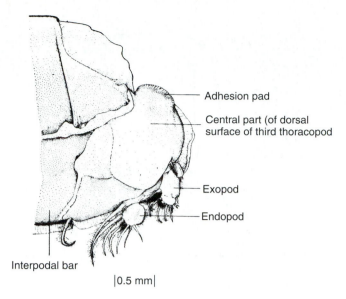

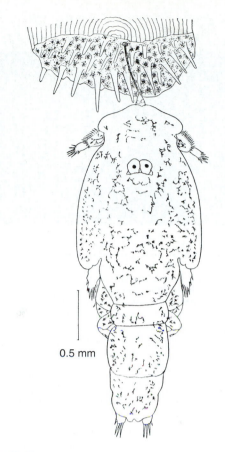

Adhesion pad

Central part (of dorsal
surface of third thoracopod

Exopod

Endopod

Interpodal bar

|0.5 mm|

FIGURE 34.15

Third thoracic leg of *Caligus curtus.* Note greatly enlarged, fused
protopod with flexible marginal membrane.

From R. R. Parker et al., "A review and description of *Caligus curtus* Miller,
1785 (Caligidae: Copepoda), type species of its genus," in *J. Fish. Res. Bd. Can.*
25:1923–1969. Copyright © 1968. Reprinted with permission of the publisher.

0.5 mm

FIGURE 34.16

Chalimus larva of *Caligus rapax.*

From C. B. Wilson, "North American parasitic copepods belonging to the
family Caligidae. Part 1, The Caligidae," in *Proc. U.S. Nat. Mus.* Copyright
© 1905.

Caligus and *Lepeophtheirus* have only two naupliar
stages, and these appear not to feed.[31,40] The second nauplius
molts to produce the first copepodid. The first copepodid must
find a host or perish. If it finds its host, the copepodid clings to
the fish with its prehensile antennae and molts to produce the
specialized type of copepodid called the **chalimus** (Fig.
34.16). Three more chalimus instars follow, all of them attached to the host by the frontal filament. The actual attachment process of a caligid is unknown, but it is probably similar
to that of the lernaeopodid. The chalimus backs off from its
point of attachment, thus pulling more filament out of the
frontal organ, while stroking the filament with its maxillae.[36]
The four chalimus instars are followed by two preadult stages
in all caligids. The preadults are detached from the frontal filament, and they, as well as the adults, have the capacity for free
movement over the host's body. Males are parasitic and not
much smaller than females.

Family Lernaeopodidae.   The lernaeopodids are common,
widespread parasites of fish that frequently occur on freshwater hosts. *Salmincola,* a parasite of salmonids, has caused
great damage to hatchery stocks in North America.[34] Lernaeopodids are substantially more modified away from the
ancestral copepod form than are the Caligidae. Virtually all
external signs of segmentation have disappeared in the adult
(Fig. 34.17), as is the case with the Lernaeidae and Pennellidae. Similarly, the adult females are permanently anchored in
one place on the host. However, in contrast to these families,
lernaeopodid females are attached almost completely outside
the host; the anchor, or **bulla,** is nonliving and is formed
from head and maxillary gland secretions. The maxillae
themselves are fused to the bulla, and they are often huge.
Occasionally the maxillae are very short, as in *Clavella* (Fig.
34.18); however, in these cases a very long, mobile
cephalothorax provides a "grazing range" similar in extent to
that possible with longer maxillae.

The maxillipeds are modified to form powerful grasping
structures; although they were primitively posterior to the
maxillae, in most species they are now located and function
more anteriorly. The bases of the maxillae mark the approximate posterior limit of the cephalothorax, and the rest of the
body is the trunk, or fused thoracic and genital segments.
The abdomen and swimming legs are absent or vestigial.
There is extreme sexual dimorphism. The males are pygmies
and are free to move around in search of females after the
last chalimus stage. Both the maxillae and maxillipeds of the
males are used as powerful grasping organs. The males, however, do not use a bulla to anchor themselves.

Kabata and Cousens[36] give a fascinating account of lernaeopodid development (Fig. 34.19). *Salmincola californiensis*
hatches from the egg as a nauplius and molts simultaneously to
the copepodid. After the cuticle hardens, the copepodid must
find a host within about 24 hours, or it dies. It attaches to the
host with prehensile hooks on the antennae and the powerful
claws of its maxillae, and then it must find a suitable position
on the fish for placement of the frontal filament. It wanders
over the host's skin until it finds a solid structure, such as a
bone or fin ray, close to the surface. The maxillipeds excavate a
small cavity at that position and press the anterior end of the
cephalothorax into the cavity. The terminal plug of the frontal
filament detaches and is fixed to the underlying host structure
by a rapidly hardening cement produced by the frontal gland.

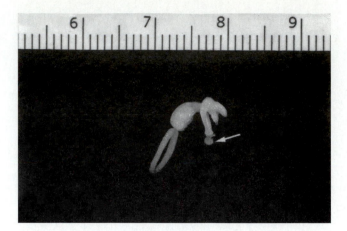

FIGURE 34.17

Salmincola inermis, a lernaeopodid parasite of whitefish, *Coregonus* spp. The huge maxillae are fused to the bulla (*arrow*), which is embedded in the host's flesh, anchoring the female to that site. The powerful maxillipeds can be seen anterior to the maxillae, and the mouth is at the tip of the anteriormost conelike projection.

Photograph by Larry S. Roberts.

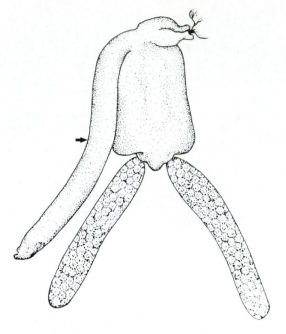

FIGURE 34.18

The maxillae are very short in *Clavella,* but the cephalothorax (*arrow*) is long and mobile, providing an extended "grazing range."

From Z. Kabata, "Crustacea as enemies of fishes," in *Diseases of Fish,* book I, edited by S. F. Snieszko and H. R. Axelrod. Copyright © 1970 T.F.H. Publications, Inc., Neptune City, NJ.

The copepodid moves backward, pulling the filament out of the frontal gland, and if the attachment site is favorable and the copepodid has not been too much damaged by detachment of the frontal filament, it soon molts to the first chalimus stage.

These hazards destroy many copepodids, but even after the copepodid is safely attached to a host, it must pass through four chalimus instars. Each chalimus molt involves a complicated series of maneuvers in which the frontal filament is detached by the maxillae and then reattached when the molt is completed (Fig. 34.20). The fourth chalimus finally breaks free of the frontal filament.

After this chalimus is free, it must find a suitable location for its permanent residence. With its antennae and mouth appendages, it rasps out a site for the bulla, now developing in the frontal organ. The maxillipeds cannot be used because they are the principal means of prehension. After molting, the bulla is everted, placed in the excavation, and detached from the anterior end. These processes are again dangerous for the parasite, which loses considerable body fluid, causing substantial mortality. Finally the linked tips of the maxillae must find the opening in the implanted bulla, where they connect with small ducts and secrete cement from the maxillary glands. If and when this last maneuver is successful, the parasite is permanently attached to its host and can graze at will on the surface epithelium. It is no surprise that many copepods fail in this complicated series of developmental events; it is amazing that so many succeed.

Family Pennellidae. The Pennellidae (formerly Lernaeoceridae) are widespread and conspicuous parasites of marine fish and mammals. They carry the evolutionary tendencies mentioned earlier to the extreme. Even the small ones are usually large by free-living standards, and the large ones are the mammoths of the copepod world. *Pennella balaenopterae* from whales may be more than 30 cm long! Their loss of external segmentation, obscuration of swimming appendages in the adult, and invasion of host tissue by their anterior ends are rem-

iniscent of the cyclopoid family Lernaeidae. However, pennellids tend to be more invasive of the circulatory system, sense organs, and viscera than are lernaeids. Each species usually has a characteristic site into which the anterior end grows and feeds. Several species, including all *Lernaeocera* spp., invade particular parts of the circulatory system, normally a large blood vessel. (The large trunk, bearing the reproductive organs and ovisacs, is external to the fish surface.) Common sites are the heart, branchial vessels, and ventral aorta. On the Atlantic cod *Gadus morhua, L. branchialis* (Fig. 34.21) invades the bulbus arteriosus. The parasite generally attaches in the branchial area, and the cephalothorax may have to grow into and follow the ventral aorta for some distance. The associated pathogenesis is severe and probably impacts commerical fisheries. Two or more mature parasites on haddock (*Melanogrammus aeglefinus*) can cause the fish to be as much as 29% underweight, have less than half the normal amount of liver fat, and lose half the hemoglobin content of the blood.[33,38] Concurrent infections with a trypanosome can compound the damage.[37]

The form of adult females is wonderfully grotesque. Anchoring processes, sometimes referred to as antlers, emanate from the anterior end. These are often more elaborate than those found in lernaeids. The greatest development of the antlers seems to be in *Phrixocephalus,* where many branches are found (Fig. 34.22). *Lernaeolophus* and *Pennella* have curious, branched outgrowths at the posterior part of the trunk, the function of which is unknown. As in the lernaeids, the appendages do not participate in the metamorphosis undergone by the rest of the female body; they are so small compared with the rest of the body as to be hardly discernible.

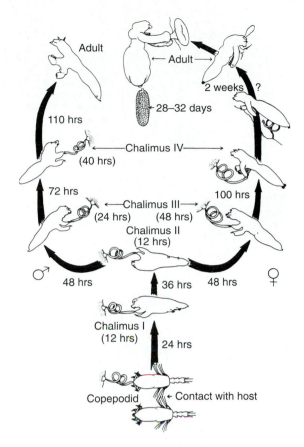

FIGURE 34.19

Life cycle of *Salmincola californiensis.* The time periods in parentheses refer to the duration of each stage, whereas those without parentheses denote time from first contact with host.

From Z. Kabata and B. Cousens, "Life cycle of *Salmincola californiensis* (Dana, 1852) (Copepoda: Lernaeopodidae)," in *J. Fish. Res. Bd. Can.* 30:881–903. Copyright © 1973. Reprinted with permission of the publisher.

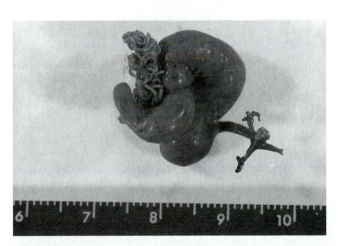

FIGURE 34.21

Lernaeocera branchialis from the Atlantic cod, *Gadus morhua.* The voluminous trunk of the organism, containing the reproductive organs, along with the coiled egg sacs, protrudes externally from the host in the region of the gills. The anterior end (*right*) extends into the flesh of the host, and the antlers are embedded in the wall of the bulbus arteriosus, which is severely damaged. The antlers rarely penetrate the lumen of the bulbus, since this would lead to thrombus formation and death of both the parasite and host.

Photograph by Larry S. Roberts.

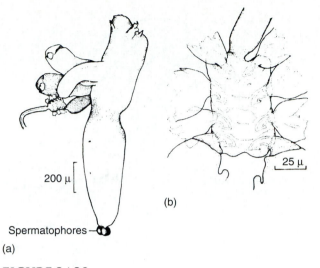

FIGURE 34.20

Early fourth chalimus of female *Salmincola californiensis,* (*a*) showing maxillae embedded in frontal filament. (*b*) Enlarged end of frontal filament, showing tips of maxillae embedded in it (at bottom), along with the molted cuticle of maxillae tips from earlier chalimus stages.

From Z. Kabata and B. Cousens, "Life cycle of *Salmincola californiensis* (Dana, 1852) (Copepoda: Lernaeopodidae)," in *J. Fish Res. Bd. Can.* 30:881–903. Copyright 1973.

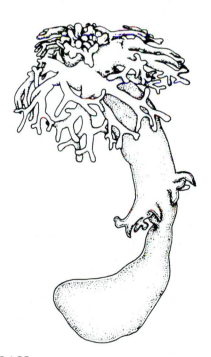

FIGURE 34.22

Phrixocephalus longicollum, a lernaeocerid whose antlers proliferate into a luxuriant, intertwining growth.

From Z. Kabata, "Crustacea as enemies of fishes," in *Diseases of Fish,* book I, edited by S. F. Snieszko and H. R. Axelrod. Copyright © 1970 T.F.H. Publications, Inc., Neptune City, NJ.

The life cycles of pennellids are unique among the copepods in that they often require an intermediate host. Usually the intermediate host is another species of fish, but it may be an invertebrate. *Lernaeocera branchialis* apparently has only one nTnaupliar stage, which leads a brief pelagic existence. The copepodid infects any of several different species of fishes[35]

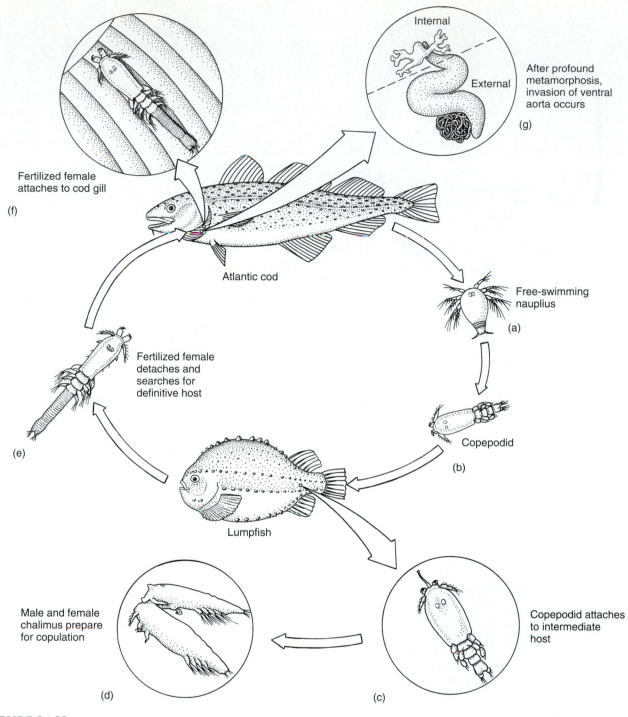

Internal

External

After profound
metamorphosis,
invasion of ventral
aorta occurs

(g)

Fertilized female
attaches to cod gill

(f)

Atlantic cod

Free-swimming
nauplius

(a)

Fertilized female
detaches and
searches for
definitive host

(e)

Copepodid

(b)

Lumpfish

Male and female
chalimus prepare
for copulation

(d)

Copepodid attaches
to intermediate
host

(c)

FIGURE 34.23

Life cycle of *Lernaeocera branchialis*. (*a*) Free-swimming nauplius. (*b*) Copepodid. (*c*) Copepodid attaches to intermediate host—in this example, the lumpfish (*Cyclopterus lumpus*). Flounders (Pleuronectidae) and sculpins (Cottidae) can also serve. (*d*) Male and female chalami preparing for copulation. (*e*) Fertilized female detaches and searches for a definitive host. (*f*) Fertilized female attaches to gill of a cod or other Gadidae (cod family). (*g*) After profound metamorphosis, invasion of ventral aorta occurs.

Drawing by William Ober and Claire Garrison.

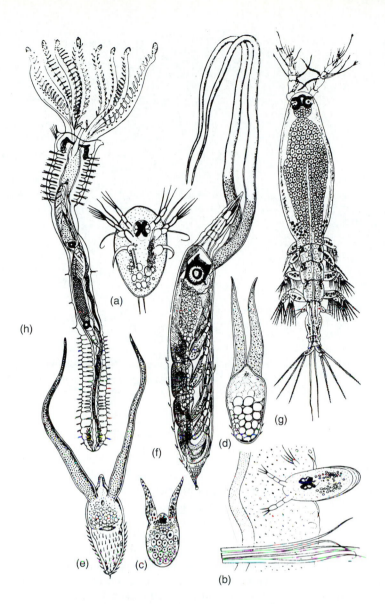

FIGURE 34.24

Haemocera danae, a monstrilloid parasite of polychaete annelids. (*a*) Nauplius. (*b*) Nauplius penetrating integument of host. (*c–e*) Successive larval stages, showing development of absorptive appendages. (*f*) Fully developed copepodid within spiny sheath. (*g*) Adult female. (*h*) Polychaete containing two copepodids in coelom.

From J. G. Baer, *Ecology of Animal Parasites.* Copyright © 1951 by the Board of Trustees of the University of Illinois. Used with permission of the University of Illinois Press.

and undergoes several chalimus instars (Fig. 34.23). The female is fertilized as a late chalimus while on the intermediate host and then detaches from the frontal filament. She undergoes another pelagic phase to search out the definitive host, usually a species of gadid (cod family). The copepod attaches in the gill cavity; the anterior end burrows into the host tissue, aided by the strong antennae; and the dramatic metamorphosis begins. At the time she leaves the intermediate host, the female is only 2 to 3 mm long and is copepodan in appearance. In her metamorphosis she loses all semblance of external segmentation and grows to 40 mm or more.

Adult females of *Cardiodectes medusaeus* are found on lanternfishes, with their anterior ends embedded in the bulbus arteriosus of the heart.[52] The intermediate hosts are not fish but are thecosomate gastropods. Both *Lernaeocera* and *Cardiodectes* feed on blood. *Cardiodectes* completely di-

gests the hemoglobin and stores the waste iron as ferritin crystals; the manner in which *Lernaeocera* disposes of the excess iron is unknown.[53]

• Order Monstrilloida

The Monstrilloida have the distinction of being parasitic only during their larval stages. The adult is the free-living dispersal agent. They have the further distinction among the Crustacea of having only one pair of antennae. Neither monstrilloid larvae nor adults have a mouth or functional gut. The nauplius penetrates its host, which is either a polychaete or a prosobranch gastropod, depending on the species. It molts to become a rather undifferentiated larva with one to three pairs of apparently absorptive appendages (Fig. 34.24). Progressive differentiation and copepodid stages ensue, and the adult finally breaks out of the host to reproduce. Thus, only the adult

FIGURE 34.25

Ventral view of *Argulus viridis,* female. Note suctorial proboscis, modification of maxillules into sucking discs, and lateral expansion of carapace into alae.

Drawing by William Ober.

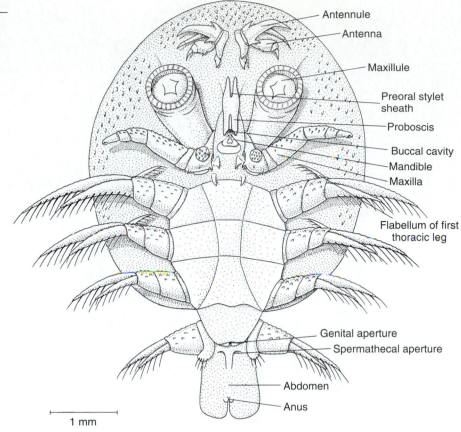

Antennule

Antenna

Maxillule

Preoral stylet sheath

Proboscis

Buccal cavity

Mandible

Maxilla

Flabellum of first thoracic leg

Genital aperture

Spermathecal aperture

Abdomen

Anus

1 mm

and the nauplius are free living, and the intermediate stages absorb food in a manner analogous to that of a tapeworm. Adults may be found in the plankton, and hosts of the juvenile stages are often unknown.[65]

Forms Not Assignable to Orders.   There are a number of strange copepods whose adults are so modified, and of whose developmental stages we are so ignorant, that we cannot even place them in an order. Males and females of *Coelotrophus nudus* in the coelomic cavity of a sipunculan do not have a mouth.[25] The females have no appendages, and the males have only one pair, with which they grasp the female.

Female *Ischnochitonika* spp. live in the branchial (pallial) groove of chitons, with long (presumably absorptive) processes extending into the host viscera. They have no appendages or segmentation, and several pygmy males live on the body of each female.[18,47]

Pectenophilus ornatus females attach to a gill of a scallop, where they consume host blood. They lack segmentation and appendages, and pygmy males live in a special chamber in the female adjacent to the brood pouch.[46] They are a serious pest in the Japanese scallop industry.[48]

Xenocoeloma spp. adults are essentially gonads grafted into the body wall of their hosts, polychaete annelids. Long thought hermaphroditic, they apparently are an example of cryptogonochorism (p. 528); all that remains of the male is a testis producing spermatozoa.[55]

Subclass Branchiura

The subclass Branchiura is relatively small in numbers of species but great in its destructive potential in fish culture. All species are ectoparasites of fishes, although some can use frogs and tadpoles as hosts, too. They are dorsoventrally flattened, reminiscent of caligid copepods with which they are sometimes confused, and can adhere closely to the host's surface. Some species are moderately large, up to 12 mm or so. The most common cosmopolitan genus is *Argulus* (Fig. 34.25). *Argulus* spp. can swim well as adults; the females must leave their hosts to deposit eggs on the substrate. Many *Argulus* spp. are not host specific and so have been recorded from a large number of fish species.

The branchiuran carapace expands laterally to form respiratory alae. The parasites have two pairs of antennae. Homologies of the remaining head appendages have been disputed, but the best evidence suggests that the only appendages in the suctorial proboscis, or mouth tube, are the mandibles.[44] The large, prominent sucking discs are modified maxillules. Immediately posterior to the maxillular discs are the large maxillae, apparently used to maintain the animal's position on its host and to clean the other appendages. *Argulus* has four pairs of thoracic swimming legs of the typically crustacean biramous form. The exopods of the first two pairs often bear an odd, recurved process, the **flabellum,** thought by some to indicate affinities with the subclass Branchiopoda (Fig. 34.26). An unsegmented abdomen follows the four segments of the thorax.

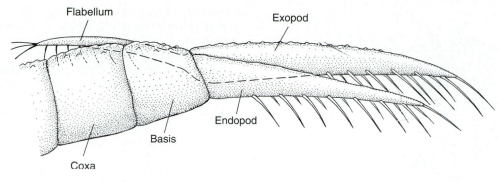

FIGURE 34.26

First thoracic appendage of *Argulus viridis,* showing flabellum.
Drawing by William Ober.

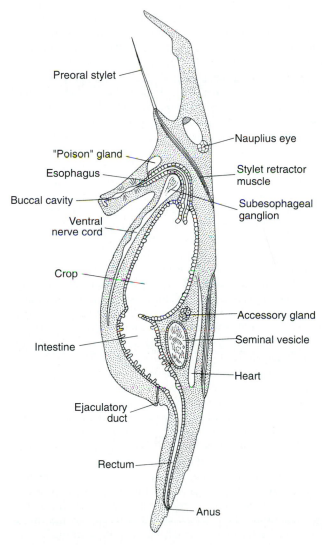

FIGURE 34.27

Median longitudinal section of male *Argulus viridis,* semidiagrammatic,
with preoral stylet extruded.
Drawing by William Ober.

The Branchiura were long associated taxonomically with the Copepoda, but present knowledge does not justify this. Branchiurans have, among other characteristics that differ from the copepods, a carapace, compound eyes, and an unsegmented abdomen behind the genital apertures, and no thoracic segments are completely fused with the head. Another feature present in most branchiurans, but not in other Crustacea, is the piercing stylet, or "sting." It is located on the midventral line, just posterior to the antennae (Fig. 34.27). The function of this curious organ is unknown.

The development of *Argulus japonicus* and *Chonopeltis brevis* and various developmental stages of other species have been described.[19] Here again is a difference between the Branchiura and Copepoda. Whereas the development of copepods is metamorphic, that of most branchiurans is usually direct, although in some the first instar may be a modified nauplius.[20] As noted, the eggs are laid on the substrate (no ovisacs, as in copepods), and the first instar is usually a juvenile.

Subclass Cirripedia

The most familiar cirripedes belong to the order Thoracica, the barnacles. They are important members of the littoral and sublittoral benthic fauna and are economically important as fouling organisms. Some members of the Thoracica are commonly found growing on other animals. Interestingly *Conchoderma virgatum* is often found on *Pennella,* a good example of hyperparasitism. (Numerous species of parasitic copepods frequently have epizooic suctorians, hydroids, algae, and so on growing on them, encouraged by the fact that the copepod is in terminal anecdysis; that is, it does not molt further.) However, other orders of cirripedes contain some fascinating organisms that are among the most highly specialized parasites known. These are parasites of other invertebrates, and space will permit consideration only of the most important order, the Rhizocephala.

• Order Rhizocephala

Members of the order Rhizocephala are highly specialized parasites of decapod malacostracans. The decapods include the animals most of us know as crabs, crayfish, lobsters, and shrimp. The Sacculinidae are primarily parasites of a variety

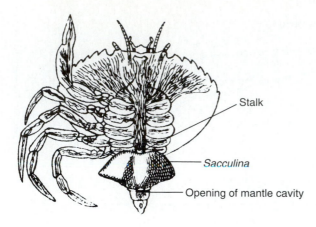

FIGURE 34.28

Shore crab, *Carcinus,* infected with a mature rhizocephalan, *Sacculina.*
From L. A. Borradaile et al. (editors), *The Invertebrata: A Manual for Students,* 2d ed. Copyright 1956 Cambridge University Press, New York, NY. Reprinted with permission of the publisher.

of brachyurans ("true" crabs), the Peltogastridae are found on hermit crabs (anomurans), and the Lernaeodiscidae prefer the anomuran families Galatheidae and Porcellanidae as hosts.

As adults, the rhizocephalans resemble arthropods even less than do lernaeids and pennellids. They have no gut or appendages, not even reduced ones, but get nutrients by means of rootlike processes ramifying through the tissues of the crab host (Fig. 34.28). They start life much as do many other crustaceans, with a nauplius larva, but the nauplius has no mouth or gut. The nauplius undergoes four molts, and the fifth larval instar is referred to as a **cyprid** or **cypris** (Fig. 34.29) because of its resemblance to the free-living ostracod *Cypris.*

Thus far the rhizocephalan life cycle is not unlike that of a normal, thoracican barnacle. The cypris of a barnacle, however, would attach to a suitable spot on the substrate by its antennules and metamorphose to the adult form. The halves of the carapace become the mantle and secrete the calcareous covering plates. However, the female cypris of *Sacculina* and related rhizocephalans attaches to a decapod with its antennules.

Most of the differentiated structures, including swimming legs and their muscles, are shed from between the two valves of the carapace. The remaining, largely undifferentiated cell mass forms a peculiar larva called the **kentrogon.** The kentrogon may be likened to a living hypodermic syringe. The cell mass within is actually injected into the hemocoel of the crab at the base of a seta or other vulnerable spot where the cuticle is thin. The mass of cells migrates to a site just dorsal to the ventral nerve cord and begins to grow (Fig. 34.29). As the absorptive processes grow out into the crab's tissue, the central mass also begins to enlarge. This mass contains the developing gonads of the female. As it grows larger, it appears to press against the host hypodermis in the ventral cephalothorax and thereby prevents cuticle secretion. Finally the weakened cuticle overlying the parasite breaks open, and the gonadal mass of *Sacculina* becomes external (now called the **externa**).[14] After the sacculinid becomes externalized, it inhibits molting by the host;[49] therefore further development of the host essentially ceases. The

externa attracts one or more male cyprids. These extrude a mass of cells that become a spine-covered larva called a **trichogon** and come to lie within male cell receptacles in the female and undergo spermatogenesis.

Formerly, these cells from the male cyprids in the females were interpreted as testes, and *Sacculina* was considered hermaphroditic. However, we now know that sexuality in rhizocephalans is an example of **cryptogonochorism** (Gr. *kryptos,* hidden, + NL *gonas,* reproductive organ, + Gr. *choris,* apart; gonochoristic, gonochorism = dioecious, sexes in separate individuals).[55]

The life histories of peltogastrids and lernaeodiscids seem to be similar to those of sacculinids in many respects, although further ecdyses of the host are not hindered.[49] However, the life cycle of *Clistosaccus* shows interesting differences.[27] The functional female cyprids inject the parasitic cells directly through an antennule (without transforming into a kentrogon) and develop within the host near the site of original injection. The functional male cyprids likewise inject spermatogonia through an antennule into an internal parasite (before she becomes external). In *Thompsonia* there is no nauplius, and the first free instar is the cyprid.[43]

In light of the invasiveness of rhizocephalans, it is not surprising that there is a range of pathogenic effects on their hosts, including damage to the hepatic, blood, and connective tissues and to the thoracic nerve ganglion of infected crabs.[57] However, some of the most interesting effects are on the hormonal and reproductive processes of the host, through so-called *parasitic castration.* Crabs exhibit some degree of sexual dimorphism, and morphological differences between the sexes are especially pronounced in the Brachyura. In the normal sequence of ecdyses the secondary sexual characteristics of the respective sexes become increasingly apparent as the crab approaches maturity. When the young male crab is infected with *Sacculina,* various degrees of "feminization" are exhibited in the subsequent instars. The manifestations vary, depending on the species of host and its degree of development when infected, but they may include a broader, more completely segmented abdomen and alteration of the pleopods toward the female type. In female crabs the effects seem to be more complex, involving some aspects of both hyperfeminization and hypofeminization. Somewhat similar effects of parasitic castration have been reported in the hosts of the Peltogastridae and Lernaeodiscidae. The mechanism of host castration has yet to be explained satisfactorily. It may be a result of copious nutrient withdrawal by the large parasites, interference with the host endocrine system, or both.[4]

Whatever the mechanism of host castration, however, the combined results of the parasite structure and the castration lead to an astonishing diversion of host behavior to promote parasite survival. The externa of the parasite (in many species) is in the same position and is the same size as the egg mass of the crab, since it would be carried by the crab's abdomen.[58] Reacting as though the parasite were in fact its egg mass, the crab protects, grooms, and ventilates the parasite. If the grooming legs of the crab are removed artificially, the externa of the parasite soon becomes fouled and necrotic. At the time the parasite begins to release its larvae, the crab performs spawning behavior. The parasites apparently release pheromones that elicit larval-release behavior in the

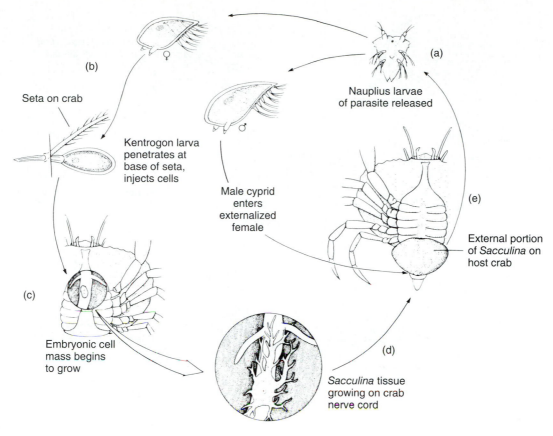

FIGURE 34.29

Life cycle of *Sacculina,* a rhizocephalan parasite of crabs. (*a*) Nauplius larvae of parasite released. (*b*) Kentrogon larva penetrates at base of seta and injects cells. (*c*) Embryonic cell mass attaches and begins to grow. (*d*) *Sacculina* tissue growing on crab midgut. (*e*) External portion of *Sacculina* on host crab.

From Cleveland P. Hickman, Jr. et al., *Integrated Principles of Zoology,* 8th edition. Copyright © 1988 by Mosby-Year Book Inc. Reprinted by permission of Times Mirror Higher Education Group, Inc., Dubuque, Iowa. All Rights Reserved.

host crab.[15] The crab comes from its normal hiding place, stands high on its legs, and waves its abdomen back and forth. Thus, the nauplii of the parasite are released into the current created by the host. And a final note: Because they are castrated and feminized, male crabs also display appropriate maternal behavior!

Subclass Tantulocarida

This subclass was discovered and described only recently.[8] Its members are bizarre, minute, ectoparasites of other crustaceans. They are most "uncrustacean" crustaceans, with no molts and no cephalic appendages except one pair of antennae on the sexual female (Fig. 34.30). These are characters apparently unique to this group, and they apparently undergo a parthenogenetic phase, which is unusual in Crustacea.[30] The odd little larva, called a **tantulus,** attaches directly to the host by its median cephalic stylet. Immediately behind the larval head, a saclike trunk forms, and the larval trunk is sloughed off. Apparently, within the expanded trunk sac either (1) a brood of tantulus larvae develops parthenogenetically; (2) an adult male differentiates and breaks out of the trunk to swim free; or (3) a sexual female differentiates to be fertilized by the male during her free-swimming existence. Neither the free-swimming male or female feeds after liberation from the trunk sac.

CLASS MALACOSTRACA

Although the malacostracans constitute the largest class of Crustacea, with members widespread and abundant in marine and freshwater habitats, comparatively few are symbiotic. Those that are, by and large, are confined to the peracaridan orders Amphipoda and Isopoda. The isopods have been particularly diverse in this regard, and some of them have become highly modified for parasitism. The eucaridan order Decapoda is the largest order of crustaceans, but few of its members are symbiotic.

Order Amphipoda

Free-living amphipods are widely prevalent aquatic organisms, often abundant along the seashore. Not many symbiotic species have been described, but some of the ones that have are quite common. Some Hyperiidae are frequent parasites of jellyfish (*Aurelia, Cyanea*) and Phronimidae are found in the tunic of planktonic ascidians (*Salpa*), apparently killing the tunicate itself and taking over its gelatinous case. *Laphystius* spp. (suborder Gammaridea) are relatively unmodified amphipods, parasitizing a variety of marine fishes.[22] The most interesting symbiotic and most unlikely looking amphipods are among the Caprellidea. The Cyamidae are curious ectoparasites of whales (Fig. 34.31). The abdomen is vestigial.

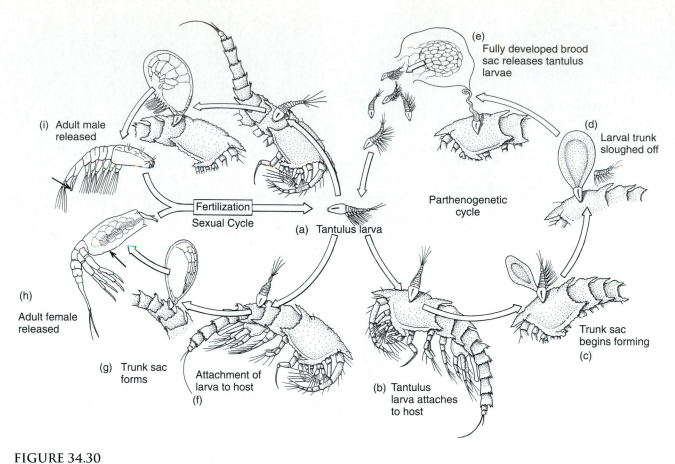

FIGURE 34.30

Presumed life cycle of tantulocarids. Positions of genital apertures of free-swimming sexual stages marked with *small arrows*. Parthenogenetic cycle: (*a*) Tantulus larva; (*b*) tantulus larva attaches to host; (*c*) trunk sac begins forming; (*d*) larval trunk is sloughed off; (*e*) fully developed brood sac releases tantulus larvae. Sexual cycle: (*f*) tantulus larva attaches to host; (*g*) trunk sac forms; (*h*) adult female is released; (*i*) adult male is released.

Source: Based on R. Huys et al., "The tantulocaridan life cycle: the circle closed?" in *J. Crustacean Biol.* 13:432–442. Drawing by William Ober and Claire Garrison.

In contrast with most amphipods, cyamids are dorsoventrally flattened, a clearly adaptive characteristic in their ectoparasitic habitat. The second and the fifth through seventh legs are strongly modified adhesive organs.

Order Isopoda

Members of the order Isopoda have exploited terrestrial environments more than any other group of crustaceans, limited though that may be, and they are abundant in a variety of marine and freshwater habitats. Furthermore, they have invaded parasitic niches more than any other malacostracans.

The gnathiidean and flabelliferan families parasitic on marine fishes have relatively few modifications. *Gnathia* is a parasite only as a larva, the **praniza,** which was originally described as a separate genus before its true identity was recognized. The praniza stage (Fig. 34.32) attaches to a fish host and feeds on blood until its gut is hugely distended. It then leaves its host and molts to become an adult. The adults are benthic and do not feed. Some of the Cymothoidae are of economic importance as fish parasites. The young of *Lironeca amurensis* and some other species burrow under a scale on their host. As the isopod grows, the underlying skin stretches to accommodate it; finally the enveloped crustacean communicates to the exterior only by a small hole. *Lironeca*

ovalis (Fig. 34.33) is a common parasite of a variety of teleosts in the Atlantic Ocean and has been reported along the United States coast from Texas to Massachusetts.[64] The parasite is usually found beneath the gill operculum, where, on small host individuals, it causes a marked pressure atrophy of the adjacent gills (Fig. 34.34). Juveniles of *L. vulgaris* locate their host by slowing their swimming activity in the presence of fish mucus, and white color (either paper or fish skin) plus mucus induces a settling response.[45]

The epicaridean isopods are highly specialized parasites of other Crustacea. The adult females of some species are comparable in loss of external segmentation and appendages to the most specialized copepods and the Rhizocephala. Portions of the appendages that are not lost, however, are the oostegites forming the brood pouch. These may become enormously developed, whereas most of the other appendages disappear or become vestigial.

Examples of the family Entoniscidae are *Pinnotherion vermiforme* (Fig. 34.35), a parasite of a brachyuran crab, *Pinnotheres pisum* (itself a parasite of the mussel *Mytilus edulis*), and *Portunion conformis,* a parasite of the shore crab *Hemigrapsus oregonensis.* The larval isopod (cryptoniscus) enters the crab's hemocoel, apparently through the gills. In the gonads or alongside skeletal apodemes, it molts to an

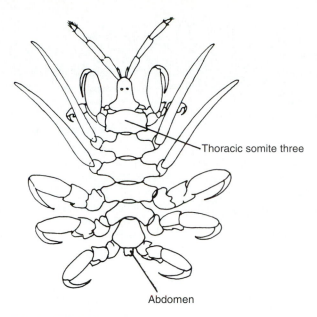

Thoracic somite three

Abdomen

FIGURE 34.31

Paracyamus, an amphipod parasite of whales. Cyamids are ectoparasites and are dorsoventrally flattened with several pairs of legs modified for clinging to their hosts.

From G. O. Sars, from W. T. Calman, "Crustacea." Copyright © 1909. In *A Treatise on Zoology,* vol. 8, edited by R. Lankester. Adams & Charles Black, London.

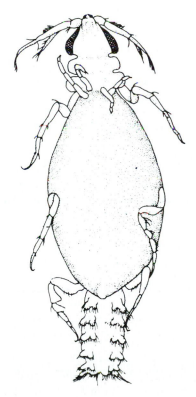

FIGURE 34.32

Praniza larva of the isopod *Gnathia.* The praniza is the only parasitic stage of this isopod. The gut becomes greatly distended with blood from its fish host.

From Z. Kabata, "Crustacea as enemies of fishes," in *Diseases of Fish,* book I, edited by S. F. Snieszko and H. R. Axelrod. Copyright © 1970 T.F.H. Publications, Inc., Neptune City, NJ.

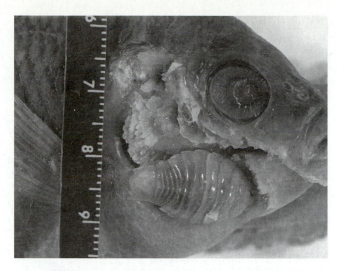

FIGURE 34.33

Lironeca ovalis is a common parasite of fish along the Atlantic coast of the United States. Here it is found on a *Lepomis gibbosus* (operculum removed) from Chesapeake Bay.

Photograph by Larry S. Roberts.

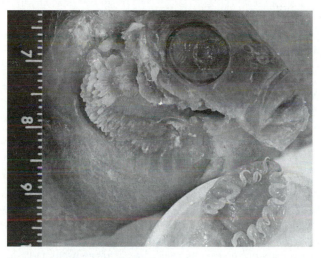

FIGURE 34.34

Lironeca ovalis, same specimen as in Figure 34.33, removed from its site on the gills. The gills show pressure atrophy and traumatic damage.

Photograph by Larry S. Roberts.

apodous juvenile.[39] The juvenile becomes invested with a cellular covering produced by the host. The female isopod becomes quite large and occupies the space normally filled by a host ovary, but the males remain small and live on the surface of the female. The female's oostegites are produced into extensive, thin lamellae that, together with the host-produced sheath, form a brood chamber. Extending from the sides of the abdomen are highly vascularized pleural lamellae, apparently of respiratory function, but the pleopods are vestigial. The short esophagus leads into a peculiar "cephalogaster," a contractile organ for sucking blood that apparently has absorptive function as well.[2] Upon hatching,

FIGURE 34.35

Young female *Pinnotherion vermiforme,* an entoniscid isopod parasite of a crab, *Pinnotheres pisum.* The parasite develops in a closely investing, thin layer of host origin, that communicates with the branchial chamber by a small opening. The opening into the host's branchial chamber is surrounded by a somewhat thickened ring of cuticle (*enlarged at left*). The vascularized pleural lamellae extending from the abdomen are prominent. Note the vestigial nature of the appendages and presence of the peculiar contractile "cephalogaster."

Redrawn from D. Atkins, "*Pinnotherion vermiforme* Giard and Bonnier, an entoniscid infecting *Pinnotheres pisum*," in *Proc. Zool. Soc. Lond.* [pp. 319–363]. Copyright © 1933 The Zoological Society of London. (Atkins noted that this specimen probably had not yet spawned; the respiratory (?) folds and pleural lamellae had not yet reached their final stage of complexity; and the cephalogaster was abnormal in that the two lobes were unequal in size.)

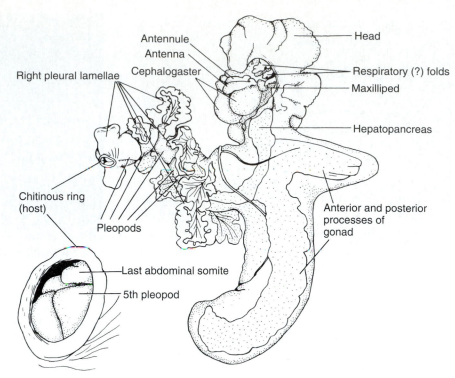

the larvae remain in the brood chamber until they exit through a pore formed between the sheath and the thin cuticle of the host's gill cavity.

In the Cryptoniscidae, morphological modification appears even more extreme. After infection of the definitive host, the female *Ancyroniscus bonnieri* feeds heavily, and the gorged isopod begins to produce eggs. During this process most of the internal organs, including the digestive and nervous system, disappear, and the animal becomes increasingly distended with eggs. Finally all that is left is a large, pulsatile sac of eggs that ruptures, freeing the eggs.[11]

The life histories of epicarideans are of great interest. The larva that hatches is an **epicaridium,** quite isopodlike in appearance. The epicaridium has bloodsucking mouthparts and attaches to a free-swimming, calanoid copepod. There it feeds, grows rapidly,[1] and molts to become a **microniscus** larva in three to four days.[13] In a few more days the microniscus molts several times and develops into a **cryptoniscus,** which again is free swimming and must find a suitable definitive host. In the Bopyridae, Entoniscidae, and Dajidae, the first isopod to infect the definitive host becomes a female, and subsequent cryptoniscus larvae become small males, sometimes living as parasites within the female brood sac. In the bopyrid, *Stegophryxus hyptius,* the cryptoniscus actually has to enter the female's brood pouch to become a male; a masculinizing substance derived from feeding on the female may be responsible.[56] It seems clear that sex determination in a number of epicarideans is epigamic, depending on circumstances other than the chromosomal complement of the gametes. If the female partner of a cymothoid pair dies, the androgenic glands of the male degenerate, and he becomes a she.[55]

Effects of epicarideans on their hosts are similar to those described for rhizocephalans, including parasitic castration.

Secondary sexual characteristics of the male host are lost, the host becomes feminized, and the gonads of both males and females are suppressed or atrophied.[4] If the parasite dies (perhaps killed by host defense mechanisms), reproductive capacity of the crab may return.[39]

Feminization of brachyuran males by epicarideans is not as striking as that produced by rhizocephalans.[57] It is interesting that some rhizocephalans are themselves hyperparasitized by epicarideans, and these, in turn, induce castration of their rhizocephalan hosts!

Order Decapoda

The Decapoda contains a relatively small number of symbiotic species, although some of them are quite common. They are of interest because of the slight, but definite, modifications for parasitism that they illustrate. Pinnotherids are frequent commensals with polychaetes, in their tubes or burrows, or are parasites in the mantle cavity of pelecypods. *Pinnotheres pisum* has been mentioned, and *P. ostreum* is found in the commercially important oyster *Crassostrea virginica* (Fig. 34.36). *Pinnotheres ostreum* interferes with the feeding of its host and damages its gills sufficiently to cause female oysters to become males.[3] The same sex change can be produced experimentally by starving the oysters.

Pinnotherids are modified relatively little from the typical, free-living brachyuran. The adult females tend to be white or cream colored with thin, soft cuticle in the carapace and with reduced eyes and chelae. Younger stages and males have hard carapaces and more well-developed eyes and chelae.[12]

A few species of symbiotic decapods belonging to other infraorders, such as Caridea and Anomura, are known. They

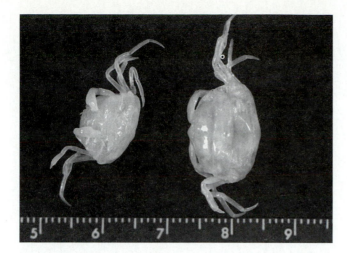

FIGURE 34.36

Pinnotheres ostreum damages the gills of its host, the commercial oyster (*Crassostrea virginica*). The carapace is soft, and the eyes and chelae are reduced.

Photograph by Larry S. Roberts.

live in such places as in the mantle cavities of clams, in the tubes of polychaetes, and on the stems of sea pens (Octocorallia) and are modified for symbiosis to about the same extent as are pinnotherids. The supposed rarity of these forms is probably a result of the failure of collectors to look for them.[32] A number of different decapods are involved in cleaning associations (Chapter 1).

References

1. Anderson, C. 1975. Larval metabolism of the epicaridian isopod parasite *Probopyrus pandalicola* and metabolic effects of *P. pandalicola* on its copepod intermediate host *Acartia tonsa*. *Comp. Biochem. Physiol.* 50A:747–51.

2. Atkins, D. 1933. *Pinnotherion vermiforme* Giard and Bonnier, an entoniscid infecting *Pinnotheres pisum*. *Proc. Zool. Soc. Lond.* 319–63.

3. Awaiti, P. R., and H. S. Rai. 1931. *Ostrea cucullata. Indian Zool. Mem.* 3:1–107.

4. Beck, J. T. 1980. The effects of an isopod castrator, *Probopyrus pandalicola,* on the sex characters of one of its caridean shrimp hosts, *Palaemonetes paludosus. Biol. Bull.* 158:1–15.

5. Berry, C. R. Jr., G. J. Babey, and T. Shrader. 1991. Effect of *Lernaea cyprinacea* (Crustacea: Copepoda) on stocked rainbow trout (*Oncorhynchus mykiss*). *J. Wildl. Dis.* 27:206–13.

6. de Blainville, M. H. D. 1822. Mémoire sur les Lernées (Lernaea, Linn.). *J. Physiol. (Paris)* 95:372–80, 437–47.

7. Bowman, T. E., and L. G. Abele. 1982. Classification of the recent Crustacea. In Abele, L. G., ed. *The biology of Crustacea, vol. 1. Systematics, the fossil record, and biogeography.* New York: Academic Press, Inc., 1–27.

8. Boxshall, G. A., and R. J. Lincoln. 1983. Tantulocarida, a new class of Crustacea ectoparasitic on other crustaceans. *J. Crust. Biol.* 3:1–16.

9. Bricker, K. S., J. E. Gannon, L. S. Roberts, and B. G. Torke. 1978. Observations on the ecology and distribution of free-living Ergasilidae (Copepoda, Cyclopoida). *Crustaceana* 35:313–17.

10. Brusca, R. C., and G. J. Brusca. 1990. *Invertebrates.* Sunderland, Mass.: Sinauer Associates, Inc., Publishers.

11. Caullery, M., and F. Mesnil. 1920. *Ancyroniscus bonnieri* C. et M., epicaride parasite d'un sphéromide (*Dynamene bidentata* Mont.). *Bull. Sci. Fr. Belg.* 34:1–36.

12. Christensen, A. M., and J. J. McDermott. 1958. Life-history and biology of the oyster crab, *Pinnotheres ostreum* Say. *Biol. Bull.* 114:146–79.

13. Dale, W. E., and G. Anderson. 1982. Comparison of morphologies of *Probopyrus bithynis, P. floridensis,* and *P. pandalicola* larvae reared in culture (Isopoda, Epicaridea). *J. Crust. Biol.* 2:392–409.

14. Day, J. H. 1935. The life history of *Sacculina. Q. J. Microsc. Sci.* 77:549–583.

15. DeVries, M. C., D. Rittschof, and R. B. Forward Jr. 1989. Response by rhizocephalan-parasitized crabs to analogues of crab larval-release pheromones. *J. Crust. Biol.* 9:517–24.

16. Donnelly, R. E., and J. D. Reynolds. 1994. Occurrence and distribution of the parasitic copepod *Leposphilus labrei* on corkwing wrasse (*Crenilabrus melops*) from Mulroy Bay, Ireland. *J. Parasitol.* 80:331–32.

17. Einszporn, T. 1965. Nutrition of *Ergasilus sieboldi* Nordmann. II. The uptake of food and the food material. *Acta Parasitol. Pol.* 13:373–80.

18. Franz, C. J., and R. C. Bullock. 1990. *Ischnochitonika lasalliana,* new genus, new species (Copepoda), a parasite of tropical western Atlantic chitons (Polyplacophora: Ischnochitonidae). *J. Crust. Biol.* 10:544–49.

19. Fryer, G. 1961. Larval development in the genus *Chonopeltis* (Crustacea: Branchiura). *Proc. Zool. Soc. Lond.* 137:61–69.

20. Fryer, G. 1968. The parasitic Crustacea of African freshwater fishes; their biology and distribution. *J. Zool. Lond.* 156:45–95.

21. Grabda, J. 1963. Life cycle and morphogenesis of *Lernaea cyprinacea* L. *Acta. Parasitol. Pol.* 11:169–99.

22. Grant, D. 1993. A parasite for *Prionotus. Underwater Naturalist* 22(1):32–34.

23. Hickman, C. P. Jr., and L. S. Roberts. 1994. *Biology of animals,* 6th ed. Dubuque, Iowa: Wm. C. Brown Publishers.

24. Ho, J. -S. 1990. Phylogenetic analysis of copepod orders. *J. Crust. Biol.* 10:528–36.

25. Ho, J. -S., F. Katsumi, and Y. Honma. 1981. *Coelotrophus nudus* gen. et sp. nov., an endoparasitic copepod causing sterility in a sipunculan *Phascolosoma scolops* (Selenka and De Man) from Sado Island, Japan. *Parasitology* 82:481–88.

26. Ho, J. -S., and V. E. Thatcher. 1989. A new family of cyclopoid copepods (Ozmanidae) parasitic in the hemocoel of a snail from the Brazilian Amazon. *J. Nat. Hist.* 23:903–11.

27. Høeg, J. T. 1990. "Akentrogonid" host invasion and an entirely new type of life cycle in the rhizocephalan parasite *Clistosaccus paguri* (Thecostraca: Cirripedia). *J. Crust. Biol.* 10:37–52.

28. Humes, A. G. 1985. Cnidarians and copepods: A success story. *Trans. Am. Microsc. Soc.* 104:313–20.

29. Humes, A. G., and J. H. Stock. 1973. A revision of the Family Lichomolgidae Kossmann, 1877, cyclopoid copepods mainly associated with marine invertebrates. *Smithsonian Contrib. Zool.* no. 127.

30. Huys, R., G. A. Boxshall, and R. J. Lincoln. 1993. The tantulocaridan life cycle: The circle closed? *J. Crust. Biol.* 13:432–42.

31. Izawa, K. 1969. Life history of *Caligus spinosus* Yamaguti, 1939 obtained from cultured yellowtail, *Seriola quinqueradiata* T. and S. (Crustacea: Caligoida). *Report of Faculty of Fisheries, Perfectural University of Mie* 6:127–57.

32. Johnson, D. S. 1967. On some commensal decapod crustaceans from Singapore (Palaemoidae and Porcellanidae). *J. Zool.* 153:499–526.

33. Kabata, Z. 1958. *Lernaeocera obtusa* n. sp.; its biology and its effects on the haddock. *Mar. Res. Scot.* 1–26.

34. Kabata, Z. 1970. *Diseases of fishes, book 1: Crustacea as enemies of fishes.* Snieszko, S. F., and H. R. Axelrod, eds. Neptune City, N.J.: T. F. H. Publications, Inc.

35. Kabata, Z. 1979. *Parasitic Copepoda of British fishes.* London: Ray Society.

36. Kabata, Z., and B. Cousens. 1973. Life cycle of *Salmincola californiensis* (Dana, 1852) (Copepoda: Lernaeopodidae). *J. Fish. Res. Bd. Can.* 30:881–903.

37. Khan, R. A. 1988. Experimental transmission, development, and effects of a parasitic copepod, *Lernaeocera branchialis,* on Atlantic cod, *Gadus morhua. J. Parasitol.* 74:586–99.

38. Khan, R. A., E. M. Lee, and D. Barker. 1990. *Lernaeocera branchialis:* A potential pathogen to cod ranching. *J. Parasitol.* 76:913–17.

39. Kuris, A. M., G. O. Poinar, and R. T. Hess. 1980. Post-larval mortality of the endoparasitic isopod castrator *Portunion conformis* (Epicaridea: Entoniscidae) in the shore crab, *Hemigrapsus oregonensis,* with a description of the host response. *Parasitology* 80:211–32.

40. Lewis, A. G. 1963. Life history of the caligid copepod *Lepeophtheirus dissimulatus* Wilson, 1905 (Crustacea: Caligoida). *Pacific Sci.* 17:195–242.

41. Linnaeus, C. 1746. *Fauna suecica sistems animalia suecica regni,* 1st ed. Stockholm.

42. Linnaeus, C. 1758. *Systema naturae,* 10th ed. Stockholm.

43. Lützen, J. 1992. Morphology of *Thompsonia reinhardi,* new species (Cirripedia: Rhizocephala), parasitic on the northeast Pacific hermit crab *Discorsopagurus schmitti* (Stevens). *J. Crust. Biol.* 12:83–93.

44. Martin, M. F. 1932. On the morphology and classification of *Argulus* (Crustacea). *Proc. Zool. Soc. Lond.* 771–806.

45. Moser, M., and J. Sakanari. 1985. Aspects of host location in the juvenile isopod *Lironeca vulgaris* (Stimpson, 1857). *J. Parasitol.* 71:464–68.

46. Nagasawa, K., J. Bresciani, and J. Lützen. 1988. Morphology of *Pectenophilus ornatus,* new genus, new species, a copepod parasite of the Japanese scallop *Patinopecten yessoensis. J. Crust. Biol.* 8:31–42.

47. Nagasawa, K., J. Bresciani, and J. Lützen. 1991. *Ischnochitonika japonica,* new species (Copepoda), a parasite on *Ischnochiton* (*Ischnoradsia*) *hakodadensis* (Pilsbry) (Polyplacophora: Ischnochitonidae) from the Sea of Japan. *J. Crust. Biol.* 11:315–21.

48. Nagasawa, K., and M. Nagata. 1992. Effects of *Pectenophilus ornatus* (Copepoda) on the biomass of culture Japanese scallop *Patinopecten yessoensis. J. Parasitol.* 78:552–54.

49. O'Brien, J. J., and D. M. Skinner. 1990. Overriding of the molt-inducing stimulus of multiple limb autotomy in the mud crab *Rhithropanopeus harrisii* by parasitization with a rhizocephalan. *J. Crust. Biol.* 10:440–45.

50. Oken, L. 1816. *Lehrbuch der Naturgeschichte, vols. 1 and 2. Dritter Theil, Zoologie.* Jena.

51. Parker, R. R., Z. Kabata, L. Margolis, and M. D. Dean. 1968. A review and description of *Caligus curtus* Miller, 1785, type species of its genus. *J. Fish. Res. Bd. Can.* 25:1923–69.

52. Perkins, P. S. 1983. The life history of *Cardiodectes medusaeus* (Wilson), a copepod parasite of lanternfishes (Myctophidae). *J. Crust. Biol.* 3:70–87.

53. Perkins, P. S. 1985. Iron crystals in the attachment organ of the erythrophagous copepod *Cardiodectes medusaeus* (Penneliidae). *J. Crust. Biol.* 5:581–605.

54. Putz, R. E., and J. T. Bowen. 1968. *Parasites of freshwater fishes. IV. Miscellaneous, the anchor worm* (Lernaea cyprinacea) *and related species.* U.S. Department of Interior, Bureau of Sport Fisheries and Wildlife, Division of Fisheries Research FDL-12.

55. Raibaut, A., and J. P. Trilles. 1993. The sexuality of parasitic crustaceans. In Baker, J. R., and R. Muller, eds. *Advances in parasitology* 32. London: Academic Press, 367–444.

56. Reinhard, E. G. 1949. Experiments on the determination and differentiation of sex in the bopyrid *Stegophryxus hyptius* Thompson. *Biol. Bull.* 96:17–31.

57. Reinhard, E. G. 1956. Parasitic castration of Crustacea. *Exp. Parasitol.* 5(1):79–107.

58. Ritchie, L. E., and J. T. Høeg. 1981. The life history of *Lernaeodiscus porcellanae* (Cirripedia: Rhizocephala) and coevolution with its porcellanid host. *J. Crust. Biol.* 1:334–47.

59. Roberts, L. S. 1970. *Ergasilus* (Copepoda: Cyclopoida): Revision and key to species in North America. *Trans. Am. Microsc. Soc.* 39:134–61.

60. Rogers, W. A., and J. P. Hawke. 1978. The parasitic copepod *Ergasilus* from the skin of the gizzard shad *Dorosoma cepedianum. Trans. Am. Microsc. Soc.* 97:244.

61. Rondelet, G. 1554. *Libri de piscibus marinus* 1. Lyon.

62. Ruppert, E. E., and R. D. Barnes. 1994. *Invertebrate zoology,* 6th ed. Philadelphia: Saunders College Publishing.

63. Sarig, S. 1971. *Diseases of fishes, book 3: The prevention and treatment of diseases of warmwater fishes under subtropical conditions, with special emphasis on fish farming.* Snieszko, S. F., and H. R. Axelrod, eds. Neptune City, N.J.: T. F. H. Publications, Inc.

64. Sindermann, C. J. 1970. *Principal diseases of marine fish and shellfish.* New York: Academic Press, Inc.

65. Suárez-Morales, E. 1993. Two new monstrilloids (Copepoda: Monstrilloida) from the coastal area of the Mexican Caribbean Sea. *J. Crust. Biol.* 13:349–56.

66. Tidd, W. M. 1962. Experimental infestations of frog tadpoles by *Lernaea cyprinacea. J. Parasitol.* 48:870.

67. Zmerzlaya, E. I. 1972. *Ergasilus sieboldii* Nordmann, 1832, its development biology and epizootic significance. (English summary.) *Izv. Gos-NIORKh* 80:132–77.

Additional References

Bliss, D. E., editor in chief. 1982–1985. *The biology of Crustacea* 1–10. New York: Academic Press, Inc. This series is a standard reference for all aspects of crustacean biology.

Harrison, R. W., and A. G. Humes, eds. 1992. *Microscopic anatomy of invertebrates. Vol 9. Crustacea.* New York: Wiley-Liss.

Raibaut, A., and J. P. Trilles. 1993. The sexuality of parasitic crustaceans. In Baker, J. R., and R. Muller, eds. *Advances in parasitology* 32. London: Academic Press, 367–444.

Chapter 35

PARASITIC INSECTS: MALLOPHAGA AND ANOPLURA, THE LICE

Riddle: What we caught we threw away;

what we didn't we kept.

Homer

Until recently in human history, lice and fleas were such common companions of *Homo sapiens* that they were considered one of life's inevitable nuisances for rich and poor, royalty and beggar alike. Not so long ago, the education of a French princess included instructions that "it was bad manners to scratch when one did it by habit and not by necessity, and that it was improper to take lice or fleas or other vermin by the neck to kill them in company, except in the most intimate circles."[36] Although lice remain widely prevalent, mostly among the very poor and in developing countries, we in modern, industrialized countries tend to think of them as pests of the past. Hence, many upper- and middle-class parents in the United States today are astonished to receive notes from the school nurse that their offspring must seek treatment for head lice. The astonishment usually evolves quickly into strong negative emotional reactions; in one psychological study of kindergarten drawings of infested people, unhappy faces, sometimes without mouths, were thought to reflect the universal negative reaction of parents and physicians, regardless of the cleanliness of the child's hair.[21]

Lice are part of our cultural heritage, giving rise to regularly used words and phrases, the origins of which most people are unaware. Do students, when complaining that their professor is a "lousy" teacher realize they are literally stating that the august personage is infested with lice? Few people appreciate the original meanings of the phrases "nit-picking" and "going over with a fine-toothed comb," which refer to the removal of louse eggs (nits) from a companion's hair. "Getting down to the nitty-gritty" and "nitwit" may thus assume new dimensions for some readers. And Robert Burns's poem "To a Louse" is the source of an insightful, louse-related quotation: "Oh wad some power the giftie gie us/To see oursels as ithers see us!"

Lice traditionally are assigned to two orders, the Mallophaga and the Anoplura, both of which were probably derived from an ancestor in common with the free-living book lice (Psocoptera). Some workers consider the Mallophaga not a natural phyletic unit and so join the Anoplura and Mallophaga in a single order, Phthiraptera.[11] We adopt the conventional arrangement, as have other authors. These insects are wingless, are dorsoventrally flattened, and have reduced or no eyes; their tarsal claws are often enlarged, an adaptation for clinging to hair and feathers. Development is hemimetabolous. The eggs are cemented to the feathers or hairs of their hosts (Fig. 35.1), and there are three nymphal instars. The most critical difference between the two orders is the structure of the mouthparts, modified for chewing in the Mallophaga and for sucking in the Anoplura.

Both orders are highly adapted for parasitism: They have no free-living stages and soon die when separated from their host. Birds and mammals are structurally complex environments with many microenvironments on their surfaces, so one would expect lice to evolve rapidly. Molecular studies show that this expectation is met, at least in part and in some cases.[24] For example, pocket gophers have a rich louse fauna; five species of lice from Central American pocket gophers of the genus *Orthogeomys* form a clade distinct from those on North American gopher genera (*Geomys, Thomomys,* and *Cratogeomys*). The enzyme studies in this case generally confirmed earlier conclusions based on morphology.[24] On the other hand, similar studies of the lice of wallabies revealed much host switching and evident replacement of some parasite species by others, suggesting more convoluted evolutionary histories.[3] And there is some evidence that lice evolve faster than their hosts: Human head and body lice may be evolutionary divergences related to the loss of hair during human evolution.[5]

ORDER MALLOPHAGA

Mallophagans commonly are referred to as the *biting lice,* but the term *chewing* is more appropriate, since anoplurans bite, too. About 3000 species parasitize various birds and mammals. None is of direct medical importance, but some species are vectors of filarial nematodes, and others may become significant pests on domestic animals. They feed primarily on feathers and hair, but some eat sebaceous secretions, mucus, and sloughed epidermis; one study of cleared museum specimens showed gut contents consisting of eggs and nymphs of the same species as well as mites.[25] Most will

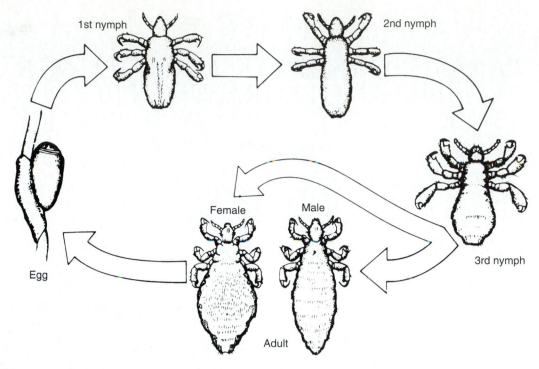

1st nymph

2nd nymph

Female   Male

3rd nymph

Egg

Adult

FIGURE 35.1

Life cycle of the head louse, *Pediculus humanus capitis.* The eggs (nits) are cemented to hairs and require 5 to 10 days to hatch. The life cycle requires about 21 days from egg to egg.

Source: H. D. Pratt and K. S. Littig, *Lice of Public Health Importance and Their Control.* Department of Health, Education, and Welfare, Pub. No. (CDC) 77–8265, 1973. U.S. Government Printing Office, Washington, D.C.

eat blood, if available, such as that resulting from scratching by the host. *Menacanthus stramineus,* a louse of chickens and turkeys, chews into developing quills to feed on blood, and some species on small birds pierce the skin to do so.

Morphology

Most lice are small, from one to a few millimeters in length. *Laemobothrion circi* is virtually a giant among lice at almost a centimeter long.[2] The head of mallophagans usually is broader than the prothorax and lacks ocelli. The short antennae have three to five segments, and the tarsi have one or two segments. Among the parasitic insects, the mouthparts of the Mallophaga are the most similar to the primitive chewing apparatus of the free-living forms (Fig. 33.12). The mandibles are the most conspicuous of these appendages, whereas the maxillae and labium are reduced. The mandibles cut off pieces of feather or hair, and the labrum pushes them into the mouth.[2]

Three suborders of Mallophaga are recognized: Amblycera, Ischnocera, and Rhynchophthirina. The Amblycera are the most generalized and least host specific. They have maxillary palps, which have been lost in the other two suborders, and their antennae are carried in grooves in the head (Fig. 35.2). The filiform, easily seen antennae of the Ischnocera (Fig. 35.3) distinguish them from the Amblycera. The Ischnocera are more specialized, more host specific, and more limited in food preferences. That is, their food is more confined to hairs and feathers (keratin) than that of others in the order. The Ischnocera of birds are so specialized that they are usually limited to a particular region of their host's body or to a certain part of a feather. The Rhynchophthirina is a

much smaller suborder than the other two, comprising only two species: *Haematomyzus elephantis* on African and Indian elephants (Fig. 35.4) and *H. hopkinsi* on warthogs. Although of the chewing type, the mouthparts of these lice are carried at the end of a projecting structure, and the insects feed on blood.

Biology of Some Representative Species

Since the majority of Mallophaga are parasites of birds, it is not surprising that domestic fowls host a number of species. The most common amblycerans are *Menopon gallinae,* the shaft louse of fowl, and *Menacanthus stramineus,* the yellow body louse of chickens and turkeys. The shaft louse is about 2 mm in length and usually causes little economic loss. *Menacanthus stramineus,* however, is about 3 mm long and may occur in large numbers, up to 35,000 per bird.[13] It frequents lightly feathered areas such as the breast, the thigh, and around the anus, gnawing through the skin to reach the quills of the pinfeathers. This irritation can cause restlessness and disrupt the bird's feeding. The birds often become unthrifty, with reduced egg production and retarded development.

Important ischnoceran parasites of fowl include *Goniocotes gallinae,* called the fluff louse because it is found in the fluff at the base of the feather (Fig. 35.3); *Goniodes dissimilis,* the brown chicken louse; *Lipeurus caponis,* the wing louse; *Cuclotogaster heterographus,* the chicken head louse; *Chelopistes meleagridis,* the large turkey louse; *Oxylipeurus polytrapzius,* the slender turkey louse; *Columbicola columbae,* the slender pigeon louse (Fig. 35.5); and *Anaticola crassicornis* and *A. anseris,* duck lice.

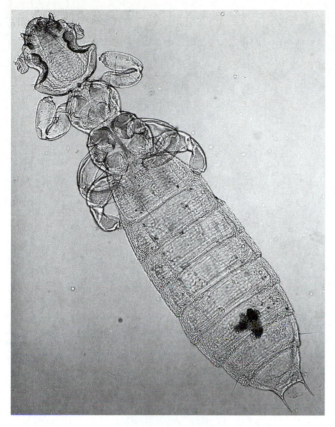

FIGURE 35.2

Gliricola porcelli (Mallophaga, Amblycera), a chewing louse of guinea pigs. Antennae are normally held in the deep grooves on the sides of the head.

Courtesy of Jay Georgi.

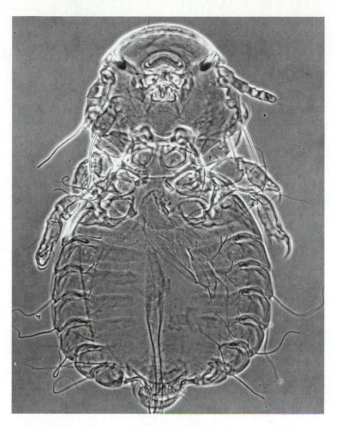

FIGURE 35.3

Goniocotes gallinae (Mallophaga, Ischnocera), the fluff louse of fowl. The antennae are clearly visible and do not lie in grooves on the head.

Courtesy of Jay Georgi.

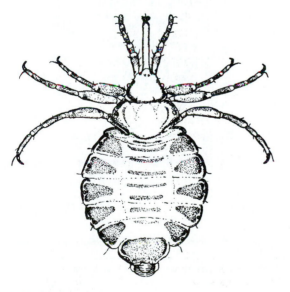

FIGURE 35.4

Haematomyzus elephantis (Mallophaga, Rhynchophthirina), a parasite of Indian and African elephants. The chewing mouthparts are at the end of a long proboscis.

From G. Lapage, *Veterinary Parasitology.* Copyright © 1956 Oliver & Boyd Ltd., Essex, UK. Reprinted with permission of the publisher.

Amblyceran parasites of mammals include *Gyropus ovalis* and *Gliricola porcelli* (Fig. 35.2) of guinea pigs and *Heterodoxus spiniger* on dogs. The guinea pig parasites are important because of the wide use of their hosts in laboratory experiments. *Heterodoxus spiniger* is common on dogs in warmer parts of the world. Several ischnocerans are pests of other domestic mammals, including *Bovicola bovis* on cattle; *B. equi* on horses, mules, and donkeys; *B. ovis* on sheep; *B. caprae* on goats; *Trichodectes canis* (Fig. 35.6) on dogs; and *Felicola subrostratus* on cats. The species of *Bovicola,* when abundant, cause considerable irritation to their hosts, although the lice are only 1.5 to 1.8 mm long. *Bovicola bovis* causes cattle to rub against solid objects and bite at their skin in an attempt to alleviate the irritation, with consequent abrasions and hair loss. The reddish brown color of *B. bovis* distinguishes it from the sucking lice commonly found on cattle. Irritation caused by *T. canis* can become severe, especially on puppies. *Trichodectes canis* is an important intermediate host, along with dog and cat fleas, of the tapeworm *Dipylidium caninum,* which also can develop in humans who accidentally ingest the insects while playing with or petting their pets (p. 343).

Chewing lice have been implicated as intermediate hosts for several other endoparasites.[13] Several bird species,

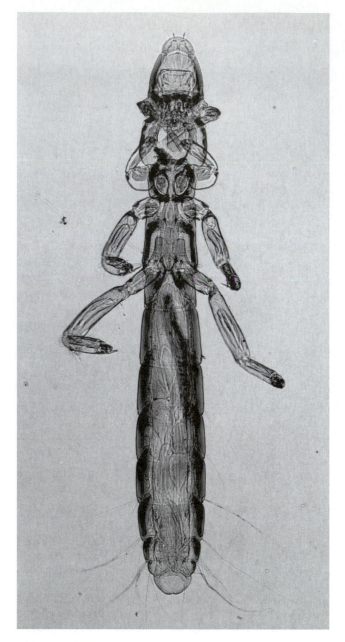

FIGURE 35.5

Columbicola columbae (Mallophaga, Ischnocera), the slender pigeon louse.

Courtesy of Jay Georgi.

especially coots, grebes, and parrots, have filarial nematodes in their legs and ankles; the worms are transmitted by mallophagans, and Bartlett and Anderson[4] believe louse host specificity is responsible for evolutionary isolation of the worms. Even thick skin is no protection from lice; elephants may suffer a severe dermatitis caused by the rhynchophthirean *Haematomyzus elephantis* (Fig. 35.4).[28]

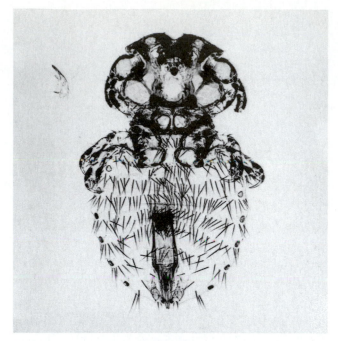

(a)

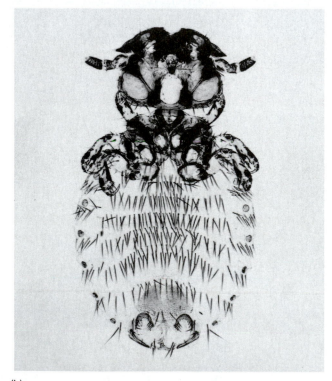

(b)

FIGURE 35.6

Trichodectes canis (Mallophaga, Ischnocera), the chewing louse of dogs. (*a*) Male; (*b*) female.

Courtesy of Jay Georgi.

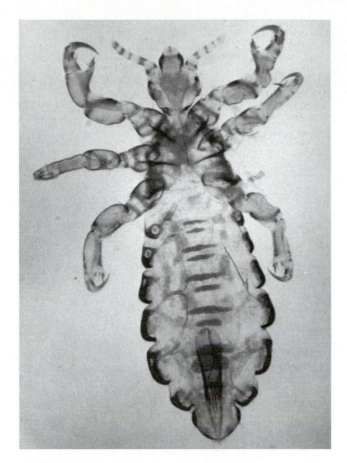

FIGURE 35.7

Pediculus humanus (Anoplura), the head and body louse of humans.
Courtesy of Warren Buss.

ORDER ANOPLURA

With fewer than 500 species,[15] the sucking lice are a much smaller group than are the Mallophaga, parasitizing only mammals. Morphologically they are more specialized than the chewing lice, but medically their importance and impact on human history are infinitely greater. Two species parasitize humans, *Pediculus humanus* (Fig. 35.7) and *Phthirus pubis* (Fig. 35.8), of which *P. humanus* is the more important. The several species on domestic mammals are of considerable veterinary significance.

Morphology

The anoplurans superficially resemble the chewing lice, with their small, wingless, flattened bodies, but the anopluran head is narrower than the prothorax. The sucking mouthparts are retracted into the head when the animal is not feeding (Fig. 35.9*a*). Each leg has a single tarsal segment with a large claw, an adaptation for clinging to the hairs of the host. The first legs, with their terminal claws, are often smaller than the other legs, and the third legs and their claws

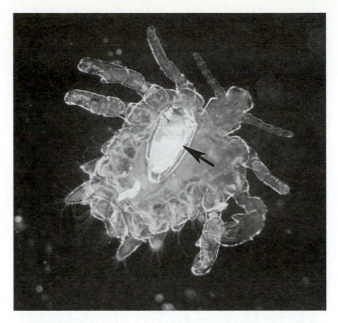

FIGURE 35.8

Phthirus pubis (Anoplura), the pubic, or crab, louse of humans. *Arrow,* developing egg.
Courtesy of Warren Buss.

are usually largest. Eyes, if present, are small, and there are no ocelli. The antennae are short, clearly visible, and composed of a scape, a pedicel, and a flagellum that is divided into three subsegments. All three subsegments of the flagellum bear tactile hairs, and subsegments two and three bear chemoreceptors.[31]

Mode of Feeding

Lavoipierre[16] distinguished two distinct feeding methods used by bloodsucking arthropods. One of these he termed **solenophage** (from the Greek *pipe + eating*) for arthropods that introduce their mouthparts directly into a blood vessel to withdraw blood; the other he called **telmophage** (from the Greek *pool + eating*) for those whose mouthparts cut through the skin and vessels to produce and feed from a small pool of blood. Anoplurans are true solenophages.[17] Their proboscis is formed from the maxillae, hypopharynx, and labium, which are produced into long, thin **stylets** (Fig. 35.9*b*). The maxillae are flattened and rolled transversely to form the food canal, and the salivary duct passes down the hypopharynx. The third member of the stylet bundle, or **fascicle,** is the labium, which bears three serrated lobes at its tip. Mandibles are absent in most adult anoplurans.

Lavoipierre[17] observed the feeding of *Haematopinus suis;* the process in other anoplurans is likely to be quite similar. The anterior tip of the head is formed by the labrum, which bears in its interior several strong, recurved teeth (Fig. 35.10). When the louse begins to feed, it places the tip of the labrum on the skin of its host and begins to evert the structure, including the buccal teeth. The teeth serve to cut

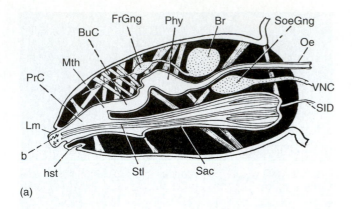

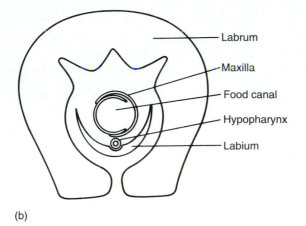

(b)

FIGURE 35.9

(*a*) Piercing and sucking apparatus of Anoplura. At rest, the buccal teeth (**b**) are within the labrum (**Lm**), but when the louse bites its host, the labrum is everted, and the buccal teeth cut into the epidermis of the host. **PrC,** preoral cavity, or "buccal funnel"; **Mth,** mouth; **BuC,** buccal cavity, first chamber of the sucking pump; **FrGng,** frontal ganglion; **Phy,** pharynx, second chamber of the sucking pump; **Br,** brain; **SoeGng,** subesophageal ganglion; **Oe,** esophagus; **VNC,** ventral nerve cord; **SID,** salivary duct; **Sac,** inverted sac holding the fascicle; **Stl,** stylet bundle, or fascicle; **hst,** hypostome. (*b*) transverse section through the mouthparts of a sucking louse.

(*a*) From R. E. Snodgrass, *Principles of Insect Morphology.* Copyright © 1935 McGraw-Hill Book Co., New York, NY.; (*b*) From R. R. Askew, *Parasitic Insects.* Copyright © 1971 American Elsevier Publishing Co., New York, NY.

through the horny outer layer of the skin and when the labrum is fully everted, are oriented so that their cutting edges point away from the central axis of the labrum (Fig. 35.10). During feeding, the everted labrum and its teeth anchor the louse in place. The stylets evert and probe the tissues until they penetrate a blood vessel, usually a venule, whereupon the louse begins to suck blood. The louse sucks by means of a two-chambered pump in its head, the first chamber comprising the buccal cavity and the second the pharynx (Fig. 35.9*a*). Contraction of muscles inserted on the walls of these structures and on the inner surface of the head cuticle serves to dilate the chambers.

How can the louse know, when it is probing the tissue with its fascicle, that it has penetrated a venule and can begin to suck blood? The stimulus for a number of hematophagous insects, presumably including lice, is the detection by chemoreceptors of adenine nucleotides, particularly ADP and

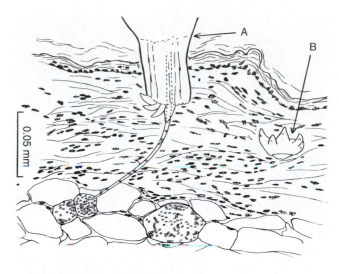

FIGURE 35.10

Schematic diagram illustrating the feeding mechanism of *Haemotopinus suis,* drawn from a histologic section of the insect's mouthparts embedded in a mouse's skin. The labrum (**A**) is anchored in the dermis by the everted buccal teeth, and the stylets are inserted in a venule. (**B**) Lateral view of everted buccal teeth.

From M. M. J. Lavoipierre, "Feeding Mechanism of *Haematopinus suis,* on the transilluminated mouse ear," in *Exp. Parasitol.* 20:303–311. Copyright © 1967.

ATP.[12] The nucleotides are released by platelets that aggregate in the bitten region as a result of the damage done to the blood vessel by the probing fascicle.

The ability of lice (and fleas) to transmit prokaryotic pathogens may be due to the way in which they digest blood meals.[34] In contrast to mosquitoes, lice hemolyze erythrocytes rapidly, their blood meals remain liquid, and they lack peritrophic membranes. Clotting and peritrophic membranes presumably inhibit prokaryote transmission but allow eukaryotic parasite (e.g. *Plasmodium* species) transmission.[34]

Pediculus humanus

Two distinct forms of *P. humanus* (Fig. 35.7) parasitize humans: the body louse (*P. humanus humanus*) and the head louse (*P. humanus capitis.*) The body louse also has been called *P. humanus corporis* and *P. humanus vestimenti,* although laypersons may know them by such common names as *cooties, graybacks, mechanized dandruff,* or even more vulgar appellations. The two subspecies are difficult to distinguish morphologically, although they have slight differences. The subspecies will interbreed and are only slightly interfertile.[2] A very convincing argument for separate species status for the head louse (*Pediculus capitus*) and the body louse (*Pediculus humanus*) was given by Busvine.[10] It seems likely that the body louse descended from a head louse ancestor after humans began wearing clothes. Body lice are much more common in cooler parts of the world; in tropical areas persons who wear few clothes usually have only head lice.[26] This makes typhus (discussed further) a disease of the cooler climates because only body lice are vectors. Curiously, however, head lice can serve as hosts for the typhus organism and have a high potential for transmitting it.[23] Body lice are extremely unusual among the Anoplura in that they spend most of their time in their host's clothing, visiting

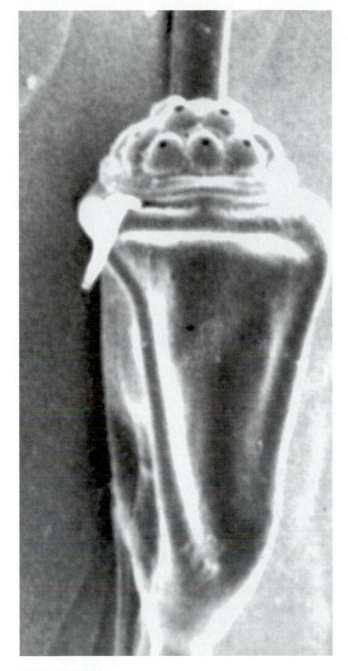

FIGURE 35.11

Nit of *Phthirus pubis* cemented to a hair. The nits of *Pediculus* spp. are essentially similar. Note the operculum with pores.

Courtesy of John Ubelaker.

the host's body only during feeding. They nevertheless stay close to the body and are most commonly found in areas where the clothing is in close contact.

The eggs (nits) of body lice are cemented to fibers in the clothes. They have a cap at one end to admit air and to facilitate hatching (Fig. 35.11). The eggs hatch in about a week, and the combined three nymphal stages usually require eight to nine days to mature when they are close to the host's body. Lower temperature lengthens the time of the complete cycle; for example, if the clothing is removed at night, the life cycle will require two to four weeks. If the clothing is not worn for several days, the lice will die. The female can lay 9 or 10 eggs per day, up to a total of about 300 eggs in her lifetime; therefore she has a high reproductive potential. Fortunately, the potential is usually not realized. It is typical to find no more than 10 lice per host, although as many as a thousand have been removed from the clothes of one person.[27]

Body lice normally do not leave their host voluntarily, but their temperature preferences are rather strict. They will depart when the host's body cools after death or if the person has a high fever. Nevertheless, they travel from one host to another fairly easily, and a person may acquire them by contact with infested people in crowded areas such as buses and trains. Of course they may be acquired easily by donning infested clothing or occupying bedding recently vacated by a person with lice. Potential for transmission is highest when people are in crowded, institutionalized conditions or in war or prison camps, where sanitation is bad and clothing cannot be changed often.

Head lice tend to be somewhat smaller than body lice: 1.0 to 1.5 mm for males and 1.8 to 2.0 mm for females, contrasted with 2 to 3 mm and 2 to 4 mm for male and female body lice, respectively.[27] The nits of both are about 0.8 mm by 0.3 mm. The nits of head lice are cemented to hairs. The lice are usually most prevalent on the back of the neck and behind the ears, and they do not infest the eyebrows and eyelashes. They are easily transmitted by physical contact and stray hairs, even under good sanitary conditions. Accordingly, they sometimes occur among schoolchildren. As in the case of body lice, however, the heaviest infestations are associated with crowded conditions and poor sanitation.[19]

Infestation with lice (**pediculosis**) is not life threatening, unless the lice carry a disease organism, but it can subject the host to considerable discomfort. The bites cause a red papule to develop, which may continue to exude lymph. The intense pruritis induces scratching, which frequently leads to dermatitis and secondary infection. Symptoms may persist for many days in sensitized persons. Years of infestation lead to a darkened, thickened skin, a condition called **vagabond's disease.** In untreated cases of head lice the hair becomes matted together from exudate, a fungus grows, and the mass develops a fetid odor. This condition is known as **plica polonica.** Large numbers of lice are found under the mat of hair.

Phthirus pubis

The origin of the common name of this insect, **crab louse** or, more popularly, **crabs,** is evident from its appearance (Fig. 35.8). The lice are 1.5 to 2.0 mm long and nearly as broad as long, and the grasping tarsi on the two larger pairs of their legs are reminiscent of crabs' pincers (Fig. 35.12). *Phthirus pubis* dwells primarily in the pubic region, but it may also be found in the armpits and rarely in the beard, mustache, eyebrows, and eyelashes. *Phthirus* is less active than is *Pediculus,* and it may remain in the same position for some time with its mouthparts inserted in the skin. The bites can cause an intense pruritis but fortunately do not seem to transmit disease organisms.

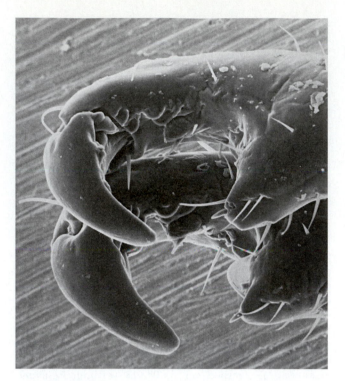

FIGURE 35.12

Scanning electron micrograph of the tarsi of the second and third legs of *Phthirus pubis,* with the terminal claw, nicely illustrating the adaptation for grasping the hairs of its host.

Courtesy of John Ubelaker.

FIGURE 35.13

Haematopinus suis, the pig louse.

Courtesy of Warren Buss.

The nits are cemented to hair, and the complete life cycle requires less than a month. The female deposits only about 30 eggs during her life. Infection can occur through contact with bedding or other objects, especially in crowded situations, but transmission is characteristically venereal and often a surprise to the new host. A few years ago one of the authors received an envelope, addressed to the "Department of Microscopic Analysis, University of Massachusetts," along with the rather urgent instruction, "Please microscope these specimens immediately. They were taken from a living organism." The distress and embarrassment of the sender were apparent.

Other Anoplurans of Note

Like many of the Mallophaga, the Anoplura tend to be host specific. Interestingly *Pediculus humanus* can also live and breed on pigs,[2] and *Haematopinus suis* of swine will readily feed on humans when it is hungry.[13] The principal effects on the host are irritation, weight loss, and anemia in heavy infestations. The USDA estimated that the combined effects of chewing and sucking lice amounted to a $47 million loss each in the cattle and sheep industries in 1965.[33] *Haematopinus suis* (Fig. 35.13) on swine is considered their most serious infection after hog cholera.[33]

Other species of *Haematopinus* infest cattle: *H. eurysternus,* the short-nosed cattle louse; *H. quadripertusus,* the cattle tail louse; and *H. tuberculatus,* primarily a parasite of water buffalo. The most serious economic losses resulting from lice on cattle are caused by *H. eurysternus. Haematopinus asini* is a parasite of horses, mules, and donkeys. *Haematopinus* spp. are large, blind lice; female *H. suis* are as

long as 6 mm. Species of the genus *Linognathus* parasitize cattle, sheep, goats, and dogs, and *Solenopotes capillatus* is found on cattle. *Pediculus mjobergi* is found on New World monkeys, sometimes becoming a problem in zoos; it may be a subspecies of *P. humanus.*[13] Like *Pediculus,* different members of the genus *Linognathus* may specialize on different regions of the body: *L. pedalis* is found on the legs of sheep, whereas *L. ovillus* predominates on the head.

Polyplax spinulosa (Fig. 35.14) of *Rattus* can transmit *Rickettsia typhi,* the causative agent of murine typhus, carried also by fleas (Chapter 37).

LICE AS VECTORS OF HUMAN DISEASE

Three important human diseases are transmitted by *Pediculus humanus humanus:* epidemic, or louse-borne, typhus; trench fever; and relapsing fever.

Epidemic, or Louse-Borne, Typhus

Typhus is caused by a rickettsial organism, *Rickettsia prowazekii.* Rickettsias are bacteria that usually are obligate intracellular parasites. Various species can inflict vertebrate and/or invertebrate hosts with effects ranging from symptomless to severe. Epidemic typhus has had an enormous impact on human history, detailed in Zinsser's classic book *Rats, Lice and History.*[36] Typhus epidemics tend to coincide with conditions favoring heavy and widely prevalent infestations of body lice, such as pre- and postwar situations and crowding, stress, poverty, and mass migration. Mortality rates during epidemics may approach 100%. It is not certain which or how many of the great epidemics of earlier human history

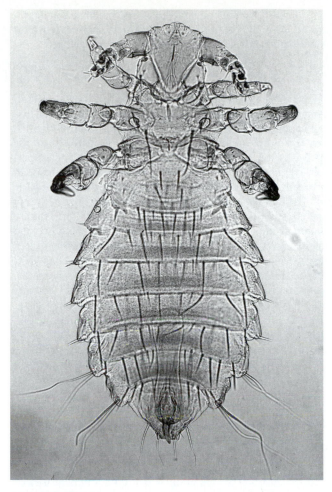

FIGURE 35.14

Polyplax spinulosa (Anoplura), parasitic on brown and black rats. This louse transmits murine typhus from rat to rat, although not from rats to humans. Another species of *Polyplax, P. serrata,* parasitizes mice.
Courtesy of Jay Georgi.

were caused by typhus, but in historical accounts of the decimation of the Christian and Moorish armies in Spain during 1489 and 1490 the role of typhus is clear. In 1528 typhus reduced the French army besieging Naples from 25,000 to 4000, leading to its defeat, to the crowning of Charles V of Spain as Holy Roman Emperor, and to the dominance of Spain among European powers for more than a century. The Thirty Years' War can be divided epidemiologically into two periods: 1618 to 1630, when the chief scourge was typhus, and 1630 to 1648, when the major epidemic was plague. Zinsser contends that between 1917 and 1921, there "were no less and probably more than twenty-five million cases of typhus in the territories controlled by the Soviet Republic, with from two and one-half to three million deaths."[36] Numerous other examples could be cited. As recently as 1971, 10,272 cases and 106 deaths were reported throughout the world.[13]

The disease starts with a high fever (39.5° to 40.0° C), which continues for about two weeks, and backache, intense headache, and often bronchitis and bronchopneumonia. There is malaise, vertigo, and loss of appetite, and the face becomes flushed. A petechial rash appears by the fifth or sixth day, first in the axilla and on the flanks and then extending to the chest, abdomen, back, and extremities. The palms, soles, and face are rarely affected.[7] After about the second week, the fever drops, and profuse sweating begins. At this point, stupor ends with clearing consciousness, which is followed either by convalescence or by an increased involvement of the central nervous system and death. The rash often remains after death, and subdermal hemorrhagic areas frequently appear. The disease can be treated effectively by broad-spectrum antibiotics of the tetracycline group and chloramphenicol. Also although prior vaccination with killed *R. prowazekii* does not result in complete protection, the severity of the disease is greatly ameliorated in persons who have been vaccinated.

Curiously typhus is a fatal disease for lice. When the louse picks up the rickettsia along with the blood meal from the human host, the organism invades the louse's gut epithelial cells and multiplies so plentifully that the cells become distended and rupture. After about 10 days, so much damage has been done to the insect's gut that the louse dies. For several days before its demise, however, the louse's feces contain large numbers of the rickettsias. Scratching the louse bites, the human is inoculated with the typhus organism from the louse feces or when the offending creature is crushed. The louse's strong preference for normal body temperature causes it to leave the febrile patient and search for a new host, thus aiding in the rapid spread of the disease in epidemics. A person can also become infected with typhus by inhaling dried louse feces or getting them in the eye. *Rickettsia prowazekii* can remain viable in dried louse feces for as long as 60 days at room temperature.[13]

Because the infection is fatal to lice, transovarial transmission cannot occur, and humans are an important reservoir host. After surviving the acute phase of the disease, humans can be asymptomatic but capable of infecting lice for many years. The disease can recrudesce and produce a mild form known as **Brill-Zinsser disease.** Flying squirrels (*Glaucomys volans*) also can be a reservoir host, with the infection transmitted by lice (*Neohaematopinus sciuropteri*) and fleas (*Orchopeas howardii*).[6,32] Some recent cases in the United States are probably caused by contact with such animals.[20] The human and possibly the animal reservoirs could provide the source to begin a new epidemic in the event of a war, famine, or other disaster. As Harwood and James[13] point out, "Current standards of living in well-developed countries have largely eliminated the disease there, but its cause lies smoldering, ready to erupt quickly and violently under conditions favorable to it."

Both Ricketts and Prowazek, the pioneers of typhus research, became infected with typhus and died in the course of their work.

Trench Fever

Trench fever is a nonfatal but very debilitating disease caused by another rickettsia, *Rochalimaea quintana,* transmitted by *Pediculus humanus humanus.* Epidemics occurred in Europe during World Wars I and II, and foci have since been discovered in Egypt, Algeria, Ethiopia, Burundi, Japan, China, Mexico, and Bolivia. In the louse the rickettsia multiplies in the lumen of the gut. Infection of humans occurs by contamination of abraded skin with louse feces or a crushed louse or by inhalation of louse feces. The organism is not pathogenic for the louse; thus, the vector remains infective for the duration of its life.

A latent infection period lasts about 10 to 30 days, toward the end of which the person may experience headache, body pain, and malaise. The temperature then rises rapidly to 39.5° to 40.0° C, accompanied by headache, pain in the back and legs (especially in the shins), dizziness, and postorbital pain in movement of the eyes. A typhuslike rash appears, usually early in the attack, on the chest, back, and abdomen, but it disappears within 24 hours. The fever continues for as long as a week and occasionally for several weeks. Convalescence is often slow, and the initial attack is followed in about half the cases by a regularly or irregularly relapsing fever curve. Tetracyclines are effective in treatment.

Humans are the primary reservoirs. *Rochalimaea quintana* has been recovered from the blood of convalescents as long as eight years after the initial attack.[9]

Relapsing Fever

The third important disease of humans transmitted by body lice is epidemic relapsing fever, which is caused by a spirochete, *Borrelia recurrentis*. Mortality is usually low, but the fatality rate can reach more than 50% in groups of undernourished people.[27] The louse picks up the bacterium along with the blood meal, and the spirochete penetrates the insect's gut to reach the hemocoel. It multiples in the hemolymph but does not invade the salivary glands, gonads, or Malpighian tubules. Therefore, transmission is accomplished only when the louse is crushed by host scratching, which releases the spirochetes in the hemolymph. Hence, the infectious organisms gain entrance through abraded skin, but evidence also indicates that they can penetrate unbroken skin.[8] Louse-borne relapsing fever apparently has disappeared from the United States, but scattered foci are in South America, Europe, Africa, and Asia. Ethiopia had 4700 cases and 29 deaths in 1971.[13] Frequent epidemics occurred in Europe during the eighteenth and nineteenth centuries, and major epidemics befell Russia, central Europe, and North Africa during and after World Wars I and II. During the war in Vietnam an epidemic occurred in the Democratic People's Republic of Vietnam.[26]

Clinically, louse-borne relapsing fever is indistinguishable from the tick-borne relapsing fevers that are caused by other species of *Borrelia* (Chapter 40). After an incubation period of 2 to 10 days, the victim is struck rather suddenly by headache, dizziness, muscle pain, and a fever that develops rapidly. Transitory rash is common, especially around the neck and shoulders and then extending to the chest and abdomen. The patient is severely ill for four to five days, when the temperature suddenly falls, accompanied by profuse sweating. Considerable improvement is seen for 3 to 10 days, and then another acute attack occurs. The cycle may be repeated several times in untreated cases. Antibiotic treatment is effective but complicated in this disease by serious systemic reactions to the drugs.

Humans are the only reservoirs, and epidemics are associated with the same kind of conditions connected with louse-borne typhus. The diseases often occur together.

CONTROL OF LICE

For complete information on control of lice on humans, consult Pratt and Littig.[27] A variety of commercial preparations containing insecticides effective against lice are available. Within the last few years no fewer than six brands were found on the shelves of a supermarket in Lubbock, Texas. Insecticides (permethrin) may also be incorporated into hair care products. In one study of 38,160 patients who used a permethrin creme rinse for 47,578 treatments, the delousing product proved both safe and effective.[1] But in a similar study in Israel, 14 different antilouse shampoos varied in their ability to kill both lice and eggs.[22] Good personal hygiene with ordinary laundering of garments, including dry cleaning of woolens, will control body lice. Devices for large-scale treatment of civilian populations, troops, and prisoners of war, which blow insecticide dust into clothing, are effective and have controlled or prevented typhus epidemics.

Lice on pets and domestic animals can be controlled by insecticidal dusts and dips. Ear tags impregnated with cypermethrin (a synthetic pyrethroid)[14] and slow-release moxidectin injected subcutaneously[35] have both been used on livestock. However, acquired resistance to cypermethrin has been demonstrated in laboratory studies.[18] Experimental work with mice has shown that it is possible for a host to be partially immune to lice, and injection of soluble antigens resulted in some protection.[29]

Normal, healthy mammals and birds usually apply some natural louse control by grooming and preening themselves. Poorly nourished or sick animals that do not exhibit normal grooming behavior often are heavily infested with lice. Many species of passerine birds show an interesting behavior known as *anting* that may represent another natural method of louse control. The bird settles on the ground near a colony of ants, allowing the ants to crawl into its plumage or even picking up ants and applying them to the feathers. However, the bird uses only ant species whose workers exude or spray toxic substances in attack and defense but do not sting. Ants in two subfamilies of Formicidae either spray formic acid or exude droplets of a repugnatorial fluid from their anus.[30] The worker ants liberally anoint the feathers with noxious fluids. Numbers of dead and dying lice have been found in the plumage of birds immediately after anting.

References

1. Andrews, E. B., M. C. Joseph, M. J. Magenheim, H. H. Tilson, P. A. Doi, and M. W. Scultz. 1992. Postmarketing surveillance study of permethrin creme rinse. *Am. J. Public Health* 82:857–61.

2. Askew, R. R. 1971. *Parasitic insects.* New York: American Elsevier Publishing Co., Inc.

3. Barker, S. C., D. A. Briscoe, and R. L. Close. 1992. Phylogeny inferred from allozymes in the *Heterodoxus octoseriatus* group of species (Phthiraptera: Boopiidae). *Aust. J. Zool.* 40:411–22.

4. Bartlett, C. M., and R. C. Anderson. 1989. Mallophagan vectors and the avian filaroids: New subspecies of *Pelecitus fulicaeatrae* (Nematoda: Filaroidea) in sympatric North American hosts, with development, epizootiology, and pathogenesis of the parasite in *Fulica americana. Can. J. Zool.* 67:2821–33.

5. Beaucornu, J. C. 1993. Lice and fleas: True or false indicators? *Bull. Soc. Zool. de Fr. Evol. et Zool.* 118:11–24.

6. Bozeman, F. M., S. A. Masiello, M. S. Williams, and B. L. Elisberg. 1975. Epidemic typhus rickettsiae isolated from flying squirrels. *Nature* 255:545–47.

7. Burgdorfer, W. 1976. Epidemic (louse-borne) typhus. In Hunter, G. W., J. C. Swartzwelder, and D. F. Clyde. *Tropical medicine,* 5th ed. Philadelphia: W. B. Saunders Company.

8. Burgdorfer, W. 1976. The relapsing fevers. In Hunter, G. W., J. C. Swartzwelder, and D. F. Clyde. *Tropical medicine,* 5th ed. Philadelphia: W. B. Saunders Company.

9. Burgdorfer, W. 1976. Trench fever. In Hunter, G. W., J. C. Swartzwelder, and D. F. Clyde. *Tropical medicine,* 5th ed. Philadelphia: W. B. Saunders Company.

10. Busvine, J. R. 1978. Evidence from double infestations for the specific status of human head lice and body lice (Anoplura). *Systematic Entomology* 3:1–8.

11. Clay, T. 1970. The Amblycera (Phthiraptera: Insecta). *Bull. Br. Mus. Nat. Hist. (Entomol.)* 25:73–98.

12. Galun, R. 1977. The physiology of hematophagous insect/animal host relationships. In White, D., ed. *Proceedings of the Fifteenth International Congress of Entomology.* College Park, Md.: Entomological Society of America, 257–65.

13. Harwood, R. F., and M. T. James. 1979. *Entomology in human and animal health,* 7th ed. New York: Macmillan Publishing Co., Inc.

14. James, P. J., P. Erkerlenz, and R. J. Meade. 1990. Evaluation of ear tags impregnated with cypermethrin for the control of sheep body lice (*Damalinia ovis*). *Aust. Vet. J.* 67:128–31.

15. Kim, K. C., and H. W. Ludwig. 1978. The family classification of the Anoplura. *Systematic Entomology* 3:249–84.

16. Lavoipierre, M. M. J. 1965. Feeding mechanisms of blood-sucking arthropods. *Nature* 208:302–3.

17. Lavoipierre, M. M. J. 1967. Feeding mechanism of *Haematopinus suis,* on the transilluminated mouse ear. *Exp. Parasitol.* 20:303–11.

18. Levot, G. W., and P. B. Hughes. 1990. Laboratory studies on resistance to cypermethrin in *Damalinia ovis* (Schrank) (Phthiraptera: Trichodectidae). *J. Aust. Ent. Soc.* 29:257–59.

19. Lindsey, S. W. 1993. 200 years of lice in Glasgow: An index of social deprivation. *Parasitol. Today* 9:412–17.

20. McDade, J. E., C. C. Shepard, M. A. Redus, V. F. Newhouse, and J. D. Smith. 1980. Evidence of *Rickettsia prowazekii* infections in the United States. *Am. J. Trop. Med. Hyg.* 29:277–84.

21. Mumcuoglu, K. Y. 1991. Head lice in drawings of kindergarten children. *Israel J. Psychiatry and Rel. Sci.* 28:25–32.

22. Mumcuoglu, K. Y., and J. Miller. 1991. The efficacy of pediculicides in Israel. *Israel J. Med. Sci.* 27:562–65.

23. Murray, E. S., and S. B. Torrey. 1975. Virulence of *Rickettsia prowazekii* for head lice. *Ann. N.Y. Acad. Sci.* 266:25–34.

24. Nadler, S. A., and M. S. Hafner. 1993. Systematic relationships among pocket gopher chewing lice (Phthiraptera: Trichodectidae) inferred from electrophoretic data. *Int. J. Parasitol.* 23:191–201.

25. Oniki, Y., and J. F. Butler. 1989. The presence of mites and insects in the gut of two species of chewing lice (*Myrsidea* sp. and *Philopterus* sp., Mallophaga): Accident or predation? *Rev. Brasil. Biol.* 49:1013–16.

26. Pan American Health Organization. 1973. *Proceedings of the International Symposium on the Control of Lice and Louse-borne Diseases. PAHO Scientific Pub. No. 263.* Washington, D.C.

27. Pratt, H. D., and K. S. Littig. 1973. *Lice of public health importance and their control. U.S. Department of Health, Education, and Welfare Pub. No. (CDC) 77-8265.* Washington, D.C.: U.S. Government Printing Office.

28. Raghavan, R. S., K. R. Reddy, and G. A. Khan. 1968. Dermatitis in elephants caused by the louse *Haematomyzus elephantis* (Piaget, 1869). *Indian Vet. J.* 45:700–1.

29. Ratzlaff, R. E., and S. K. Wikel. 1990. Murine immune responses and immunization against *Polyplax serrata* (Anoplura: Polyplacidae). *J. Med. Ent.* 27:1002–7.

30. Simmons, K. E. L. 1966. Anting and the problem of self-stimulation. *J. Zool.* 149:145–62.

31. Slifer, E. H., and S. S. Sekhon. 1980. Sense organs on the antennal flagellum of the human louse, *Pediculus humanus* (Anoplura). *J. Morphol.* 164:161–66.

32. Sonenshine, D. E., F. M. Bozeman, M. S. Williams, S. A. Masiello, D. P. Chadwick, N. I. Stocks, D. M. Lauer, and B. L. Elisberg. 1978. Epizootiology of epidemic typhus (*Rickettsia prowazekii*) in flying squirrels. *Am. J. Trop. Med. Hyg.* 27:339–49.

33. Steelman, C. D. 1976. Effects of external and internal arthropod parasites on domestic livestock production. *Ann. Rev. Entomol.* 21:155–78.

34. Vaughan, J. A., and A. F. Azad. 1993. Patterns of erythrocyte digestion by bloodsucking insects: Constraints on vector competence. *J. Med. Ent.* 30:214–16.

35. Webb, J. D., J. G. Burg, and F. W. Knapp. 1991. Moxidectin evaluation against *Solenoptes capillatus* (Anoplura: Linognathidae), *Bovicola bovis* (Mallophaga: trichodectidae), and *Musca autumnalis* (Diptera: Muscidae) on cattle. *J. Econ. Entomol.* 84:1266–69.

36. Zinsser, H. 1934. *Rats, lice and history.* Boston: Little, Brown, and Co.

Additional References

Kim, K. C., H. D. Pratt, and C. J. Stojanovich. 1986. *The sucking lice of North America: An illustrated manual for identification.* University Park, Pa.: Pennsylvania State University Press.

Rothschild, M., and T. Clay. 1952. *Fleas, flukes and cuckoos.* New York: Philosophical Library.

Chapter 36

PARASITIC INSECTS: HEMIPTERA, THE BUGS

Snug as a bug in a rug.

Benjamin Franklin

Bug off!

familiar slang

Although to the layperson all insects, and sometimes even bacteria and viruses (including computer viruses), are *bugs,* to the zoologist the only true bugs are members of the order Hemiptera. It is one of the larger insect orders, containing more than 55,000 species, but relatively few (about 100 species) are ectoparasites of mammals and birds.[3] Two suborders are commonly recognized: the Homoptera and the Heteroptera. The Homoptera (considered a separate order by some authors) include such insects as aphids, scale insects, leaf hoppers, and cicadas. Because the Homoptera feed on the juices of plants, they are not of public health or veterinary importance, but in many cases they are very important agricultural pests. Both the Homoptera and the Heteroptera have sucking mouthparts, which are folded back beneath the head and thorax when at rest (Fig. 36.1). The members of the suborders with wings are easily distinguished: Both pairs of wings of homopterans are wholly membranous, whereas the forewings (**hemielytra**) of heteropterans are divided into a heavier, leathery proximal portion and a membranous distal portion.

Hemipterans are exopterygotes with hemimetabolous development; consequently, the nymphs have habitat and feeding preferences similar to those of the adults. Like the Homoptera, most heteropterans suck the juices of plants, but many are predatory, using their mouthparts to suck the body fluids of smaller arthropods. Some are quite cannibalistic, at least in culture, and can acquire parasites (especially trypanosomatid flagellates) from their prey.[26] Although some normally plant-feeding forms suck blood in rare instances,[30] relatively few heteropterans obtain most or all of their nutrition from the blood of birds or mammals. Most of the ones who do are found in the families Cimicidae and Reduviidae.

MOUTHPARTS AND FEEDING

Regardless of whether they feed on plant or animal juices, homopterans and heteropterans have similar basic mouthparts.[9] The labrum is short and inconspicuous, but the labium is elongated and forms a tube containing the mandibles and maxillae (Fig. 36.2). The maxillae enclose a food canal,

through which the fluid food is drawn, and a salivary canal, through which the saliva is injected. The mandibles run alongside the maxillae in the labial tube and together with the maxillae comprise the **fascicle.** The tips of the mandibles and maxillae may be barbed or spined.

The operation of the mouthparts in the bedbug *Cimex* and in the reduviid *Rhodnius* has been studied.[5] Both are solenophages. *Rhodnius* applies the tip of its labial tube to the skin of the host, anchors itself with the barbed tips of the mandibles, and slides each maxilla against the other, piercing the dermal tissue (Fig. 36.3). The route of the maxillary penetration curves because the tip of one maxilla is hooked and the other is spiny. In *Cimex* the entire fascicle penetrates, and the labium folds at its joints (Fig. 36.4). In both cases feeding begins when a blood vessel is pierced. Some, perhaps all, bloodsucking bugs have sensory receptors on the tips of the mandibles and maxillae,[24] which are undoubtedly essential to the success of the feeding process.

Like the lice, both the bloodsucking reduviids (Triatominae) and the Cimicidae depend for part of their nutrition on endosymbiotic bacteria. The bacteria are contained in the epithelial cells of the triatomine gut, and the cimicids bear theirs in two disc-shaped **mycetomes** in the abdomen beside the gonads (see Fig. 36.6 on p. 550). *Triatoma infestans* has symbiotic bacteria throughout many of its tissues, including muscles, gonads, and nervous system. There is no evidence that they do the bug any harm.[10] The mutualistic bacteria of the gut, at least, are essential for the growth and maturation of the bugs; aposymbiotic triatomines, those that were artificially "cured" of their bacteria, can reach only the second, third, or fourth instars before death, depending on the species.[12]

FAMILY CIMICIDAE

The cimicids include a variety of small, wingless bugs that feed on the blood of warm-blooded animals, primarily birds and bats (Fig. 36.5). *Cimex lectularius, C. hemipterus,* and *Leptocimex boueti* are known as bedbugs and attack humans. *Cimex lectularius* is cosmopolitan but is found primarily in the temperate zones, whereas *C. hemipterus* tends to be more tropical, and *L. boueti* is confined to West Africa. Of the 22 genera of cimicids, 12 are parasites of bats. The bird feeders of the group most often attack birds that commonly nest in caves, although swallows, especially the colonial cliff swallow that builds structurally complex mud nests, can support large populations of the swallow bug *Oeciacus vicarius.*

FIGURE 36.1

Diagram of typical bug (Hemiptera). The labium forms a tube within which lies the fascicle (maxillae and mandibles.)

Drawing by Ian Grant.

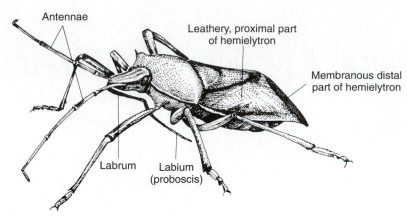

FIGURE 36.2

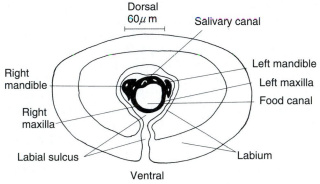

Cross section of the proboscis of the nonfeeding *Rhodnius prolixus* close to the base of the labium.

From M. M. J. Lavoipierre et al., "Studies on the methods of feeding of blood-sucking arthropods. I. The manner in which triatomine bugs obtain their blood-meal, as observed in the tissues of the living rodent, with some remarks on the effects of the bite on human volunteers," in *Ann. Trop. Parasitol.* 53:235–250. Copyright © 1959.

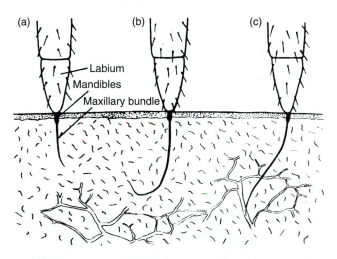

FIGURE 36.3

Schematic diagram showing successive stages in the introduction of the fascicle of *Rhodnius prolixus* into the skin of a rodent. (*a*) The maxillary bundle is being thrust into the tissues. (*b*) Probing has commenced, and the flexible maxillary bundle is shown bending sharply. (*c*) The tip of the maxillary bundle has entered the lumen of a vessel. The barbed mandibles act as anchors to the fascicle while the maxillae are projected deep into the tissues.

From M. M. J. Lavoipierre et al., "Studies on the methods of feeding of blood-sucking arthropods. I. The manner in which triatomine bugs obtain their blood-meal, as observed in the tissues of the living rodent, with some remarks on the effects of the bite on human volunteers," in *Ann. Trop. Parasitol.* 53:235–250. Copyright © 1959.

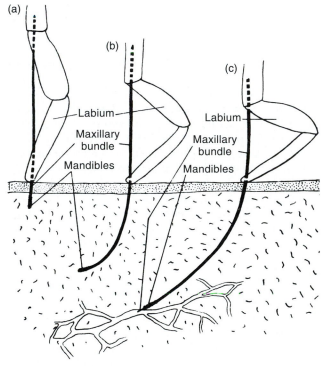

FIGURE 36.4

Schematic diagram showing successive stages in the introduction of the fascicle of *Cimex lectularius* into the ear of a rodent. (*a*) The fascicle (mandibles and maxillae) is being thrust into the tissues. (*b*) Probing has commenced, and the flexible fascicle is shown bending in the tissues. (*c*) The tip of the maxillary bundle has entered the lumen of a vessel, but the mandibles remain outside. Both the mandibles and the maxillae enter and probe the tissues of the host as a compact bundle (fascicle), and the labium is progressively bent to allow the fascicle to be projected deep into the tissues.

From G. Dickerson and M. M. J. Lavoipierre, "Studies on the methods of feeding of blood-sucking arthropods. II. The method of feeding adopted by the bed-bug (*Cimex lectularius*) when obtaining a blood-meal from a mammalian host," in *Ann. Trop. Med. Parasitol.* 53:347–357. Copyright © 1959.

Brown and Brown[4] reported up to 700 bugs per nest in large swallow colonies, and blood loss due to ectoparasitism significantly reduced the weight of nestlings.

It is believed that caves were probably the home of the ancestors of the Cimicidae,[3] and it is not unreasonable to assume that humans acquired bedbugs during their cave-dwelling period. All three species that parasitize humans will also feed on bats. In addition *Cimex* spp. will feed on

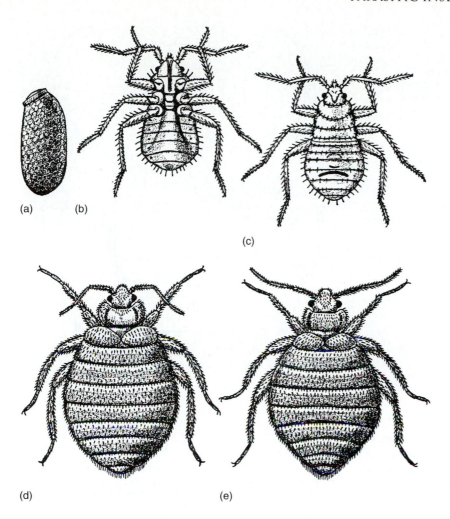

FIGURE 36.5

(*a–d*) Developmental stages of *Cimex lectularius.* (*a*) Egg; (*b*) newly hatched nymph, ventral view; (*c*) newly hatched nymph, dorsal view; (*d*) adult female; (*e*) *Cimex hemipterus,* adult female.

From D. L. Belding, *Textbook of Clinical Parasitology,* 2d ed. Copyright © 1952 Appleton-Century-Crofts, Inc., East Norwalk, CT. Reprinted with permission of the publisher.

(a)   (b)

(c)

(d)   (e)

chickens, and *C. lectularius* readily attacks rodents and some other domestic animals.

Although bedbugs are not known to carry any human disease, they can be extremely annoying. Persons who are chronically infested by bedbugs suffer loss of sleep, sores from infected bites, iron and hemoglobin deficiencies, and rarely mechanical transmission of hepatitis B virus.[19] *Cimex hemipterus* has been experimentally infected with Human Immunodeficiency Virus (HIV), which remained viable in the bug for as long as eight days.[31] However, mechanical transmission could not be demonstrated and there was no virus in the insect feces.

For an excellent treatment of the taxonomy, ecology, morphology, reproduction, and control of cimicids, consult Usinger's monograph.[29]

Morphology

Cimicids are reddish brown bugs, up to about 8 mm long (Fig. 36.5). Their appendages are not particularly adapted for clinging to their hosts, but they enable the bugs to run rather rapidly, perhaps because they do not stay on their hosts for any longer than the 5 to 10 minutes of feeding. They are flattened dorsoventrally. The adults have no wings, although rudimentary wing pads form in the nymphs. Doubtless this is an adaptation for inhabiting the narrow crevices in which they reside between feedings. Their antennae have four segments, and the distal two are much more slender than are the medial ones. Cimicids have conspicuous compound eyes but

no ocelli, and the first thoracic somite forms a rim around the posterior portion of the head. The tergum of the first thoracic somite (**pronotum**) of *C. lectularius* is about two and one-half times broader than long, the pronotum of *C. hemipterus* is just more than twice as broad as long (Fig. 36.5), and the pronotum of *L. boueti* is only a bit wider than the head. Scent glands opening on the ventral side of the third thoracic somite produce an oily secretion and give bedbug-infested dwellings a disagreeable odor. The secretion is probably a defense against predators.

Biology

Bedbugs are nocturnal, emerging from their daytime hiding places to feed on their resting hosts during the night. The peak activity of *C. lectularius* is just before dawn.[18] The bites cause little reaction in some people, whereas in others they cause considerable inflammation as a result of allergic reactions to the bug's saliva. The annoyance may disturb sleep, and persistent feeding may reduce the person's hemoglobin count significantly.[9]

Bedbugs can survive long periods of starvation. Adults commonly live without food for more than four months and can survive to at least 18 months.[3] Cannibalism is common.[6]

Cimicids practice a rather startling type of copulation known as traumatic insemination, employed among other insects only by the related families Anthocoridae and Polyctenidae.[3] The male has a copulatory appendage, the **paramere,** which curves strongly to the left; the right paramere has

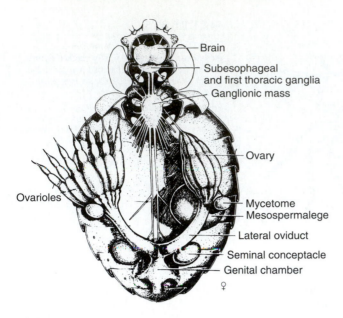

FIGURE 36.6

Internal anatomy of *Cimex lectularius*. Female reproductive organs, mycetomes, and parts of nervous system.

From R. L. Usinger, drawings by G. P. Catts, *Monograph of Cimicidae* (Hemiptera-Heteroptera). Copyright © 1966 Entomological Society of America, College Park, MD.

been lost. The aedeagus is small and lies immediately above the base of the paramere. During copulation the paramere of the male *Cimex* stabs into a notch (paragenital sinus) near the right side of the posterior border of the female's fifth abdominal sternite[29] (Fig. 36.7). The sperm enter a pocket (**spermalege**) from which they emerge into the hemocoel and make their way to organs at the base of the oviducts, the **seminal conceptacles** (Figs. 36.6 and 36.7). The seminal conceptacles are analogous, but not homologous, to the spermathecae of other insects. From the conceptacles, the sperm travel by minute ducts in the walls of the oviducts to the ovarioles and ova. Male cimicids mate with females repeatedly, and homosexual behavior is common. In the laboratory, male *C. hemipterus* and female *C. lectularius* interbreed rather freely; the mating produces sterile eggs, however, and in mixed infestations *C. lectularius* females may lay mostly infertile eggs.[20]

The female lays from 200 to 500 eggs in batches of 10 to 50. The eggs hatch in about 10 days, and the five nymphal instars must each have at least one blood meal. A blood meal is also necessary before the males will mate and the females oviposit. The time from egg to maturity is between 37 and 128 days, but this time is lengthened in periods of starvation.

Epidemiology and Control

Their shape makes it possible for bedbugs to insinuate themselves into a variety of tight places. Their daytime sites can be seams of mattresses and box springs, wooden bedsteads, cracks in the wall, and the space behind loose wallpaper.[11] These sites may be some distance from their host, which they find by following a temperature, and perhaps a carbon dioxide, gradient. They can be transported from one dwelling to another in secondhand furniture, suitcases, bedding, laundry, and other items. Only a single female is required to create the nucleus of a new infestation. Transmission in public conveyances and gathering places occurs frequently.[9]

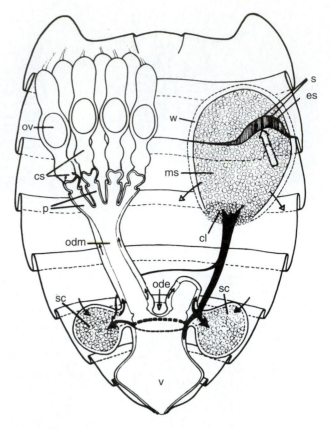

FIGURE 36.7

Diagram of paragenital system and process of insemination in Cimicidae, based principally on *Cimex* type. Right ovary and nearly all of corresponding lateral oviduct omitted, and spermalege shown farther forward than is actually the case in *Cimex. Broad white arrow,* course followed by paramere of male in reaching ectospermalege (hatched and crossed by three black bands representing scars of copulation). *Black arrows,* normal routes of migration of spermatozoa from mesospermalege to bases of ovarioles. *Small arrows with white points,* migratory routes never or rarely used in *Cimex* but seen in other Cimicidae. **cl,** conductor lobe; **cs,** syncitial body; **es,** ectospermalege; **ms,** mesospermalege; **ode,** paired ectodermal oviduct; **odm,** paired mesodermal oviduct; **ov,** oocyte; **p,** pedicel; **s,** scars or traces of copulation; **sc,** seminal conceptacle; **v,** vagina; **w,** wall of mesospermalege.

From R. L. Usinger, *Monograph of Cimicidae* (Hemiptera-Heteroptera). Copyright © 1966 Entomological Society of America, College Park, MD.

Control of bedbugs by application of residual insecticides to the areas of likely hiding places is usually effective, although resistance to some insecticides has been encountered. A high level of domestic cleanliness certainly helps in control.

FAMILY REDUVIIDAE

Most reduviids are predators on other insects and are commonly called *assassin bugs* for this reason. They often are valuable for their predation on pest species; *Reduvius personatus* may even enter houses and feed on bedbugs! Most of these can, but usually do not, bite humans. The bite is quite painful. One of the authors has vivid memories of a college field trip on which he carelessly picked up an unidentified reduviid with his fingers—the reduviid remained unidentified because the author released it quite abruptly when it bit him.

One subfamily of the Reduviidae, the Triatominae, is of great public health significance because its members are vectors for *Trypanosoma cruzi,* the causative agent of Chagas' disease (Chapter 5). The prevalence of Chagas' disease is currently estimated at more than 10 million cases in South and Central America.[9] The Triatominae characteristically feed on blood of various vertebrates. In contrast to the assassin bug's bite, the bite of Triatominae, as might be expected of forms that must suck blood several minutes unnoticed by the host, is essentially painless. They are called *kissing bugs* because they often bite the lips of sleeping persons.

Morphology

Triatomines (Fig. 36.8) are relatively large bugs, up to 34 mm in length. They usually have wings, which are held in a concavity on top of the abdomen. The head is narrow, and large eyes are located midway or far back on the sides of the head. Two ocelli may be present behind the eyes. The antennae are slender and in four segments. The apparently three-segmented labial tube folds backward at rest into a groove between the forelegs. The bugs can make a squeaking sound by rubbing the labium against ridges in the groove (stridulation).

Biology

The various reduviids characteristically frequent different sites; for example, some species are normally found on the ground, some in trees, and some in human dwellings. The eggs, numbering from a few dozen to a thousand depending on species, are deposited in the normal habitat of the adult. There are usually five nymphal instars. Triatomines do not seem to be very choosy about their food sources; whatever vertebrate is available in their habitat is apparently acceptable. Triatomines that inhabit human dwellings feed on humans, dogs, cats, and rats; other species depend more on wild animals.

Research on trypanosomiasis cruzi, as well as on xenodiagnosis (Chapter 5), demands a supply of lab-reared bugs, and much effort has gone into the development of rearing techniques, especially feeding stimuli. Temperature seems to be a stimulus for *Triatoma infestans;* in experimental feeders mammalian body temperatures evoked a feeding response, although crop filling did not depend on blood temperature.[16] Evidently *Triatoma infestans* can detect objects that are near mammalian body temperatures and orient toward them when seeking food.[17] *Triatoma infestans* is also somewhat particular in what it eats, being partial to citrated over heparinized blood (sodium citrate and heparin are both anticoagulants), but mouse odor added to the feeder does not make the blood more appetizing.[22]

Mating in triatomines can involve a fairly complex set of behaviors. In one laboratory study involving *Triatoma mazzottii,* nine steps were identified, including vigilance on the part of the male, female advancement, gyrations, copulation, and separation, all happening in about 10 minutes.[25] Only about 1 in 10 of the matings was completed, however, due to a combination of nonreceptiveness on the part of females and indifference on the part of males.[25]

Egg laying follows a circadian rhythm in *Rhodnius prolixus,* and lab-reared populations can be made to lay more or less synchronously using light-dark cycles.[1] The restricted timing of egg laying persists when the insects are transferred to total darkness, suggesting environmental control of population level ovulation and oviposition. Both egg laying and

FIGURE 36.8

Specimen of *Triatoma dimidiata* discovered feeding on a parasitologist. Courtesy of Warren Buss.

feeding in *Rhodnius prolixus* are likely under hormonal control; serotonin is evidently secreted from tissues associated with the abdominal nerves and builds up in the hemolymph during feeding.[14] Blood meals are essential to egg production, but adults may lay eggs without feeding, provided the nymphs have fed well.[21]

Epidemiology and Control

We discussed epidemiology of trypanosomiasis cruzi in Chapter 5. Apparently all species of triatomines are suitable hosts for *T. cruzi,* but species differ in their susceptibility and presumably in their ability to serve as vectors. In experiments using infected mice and 11 species of triatomes, *Dipetalogaster maximus* and *Triatoma rubrovaria* passed the most trypomastigotes in their feces, and *T. vitticeps* passed the fewest, but all were infectable.[27] The importance of a particular species depends on its domesticity (**synanthropism**). The most common vectors are *Panstrongylus megistus, Triatoma infestans, T. dimidiata,* and *Rhodnius prolixus.* The relative importance of each varies with locality (Fig. 5.13). The insects are nocturnal and hide by day in cracks, crevices, and roof thatching. Poorly constructed houses are thus a significant epidemiological factor. *Triatoma infestans* does not have to be alive to transmit *T. cruzi* infections. Live trypomastigotes infective for mice were found in dead bugs for up to two weeks after one spraying campaign.[2]

Dogs, cats, and rats are important reservoir hosts around human habitations, and there is a wide variety of sylvatic reservoirs, the most important of which is the opossum, *Didelphis marsupialis.* The opossum is a common and successful marsupial occurring from the northern United States to Argentina. Other important reservoirs include armadillos, bats, squirrels, wild rats and mice, guinea pigs, and sloths.

The number of triatomines in a house increases with the number of people living there,[23] but the triatomine population can be reduced by reducing the number of hiding places for the bugs—that is, by improving construction and altering nearby environments. For example, removal of stacked firewood from near houses and replacement of dirt floors with concrete nearly eliminates *Triatoma dimidiata* infestation.[32] Replacement of thatched roofs with sheet metal is of great aid, and even whitewashing of mud walls helps; but in one case in Brazil even repeated plastering of a house did not permanently reduce the population of *Triatoma infestans* because the owner would not replace the roof tiles.[8] Reducing the number of other food sources for the bugs around the dwelling, such as dogs, birds, and rats, also is of value in controlling triatomines.

As with bedbugs, residual insecticides around the potential hiding places are effective in control, and paint with insecticide (chlorpyrifos) has been used with some success for control of triatomes, especially on wood interiors.[13] However, nutritional state affects the susceptibility of *T. infestans* to insecticide. Starved nymphs were nearly 200 times as resistant as well-fed ones to DDT.[7] Precocene II, a natural product extracted from the plant *Ageratum* sp., shows promise as a fumigant against triatomines. It is cytotoxic to the corpora allata, preventing production of juvenile hormone. Precocene blocks oogenesis in adult females and causes immatures to molt precociously to sterile adults.[28]

Triatomine bugs occur in the United States from New England to California. Similarly *T. cruzi* has been found from coast to coast in wild mammals, including wood rats, raccoons, opossums, and skunks. Several cases of human infection were diagnosed in Arizona.

References

1. Ampleford, E. J., and K. G. Davey. 1989. Egg laying in the insect *Rhodnius prolixus* is timed in a circadian fashion. *J. Insect Physiol.* 35:183–88.

2. Asin, S. N., and S. S. Catala. 1991. Are dead *Triatoma infestans* a competent vector of *Trypanosoma cruzi? Mem. Inst. Oswaldo Cruz* 86:301–6.

3. Askew, R. R. 1971. *Parasitic insects.* New York: American Elsevier Publishing Co., Inc.

4. Brown, C. R., and M. B. Brown. 1986. Ectoparasitism as a cost of coloniality in cliff swallows. *Ecology* 67:1206–81.

5. Dickerson, G., and M. M. J. Lavoipierre. 1959. Studies on the methods of feeding of blood-sucking arthropods. II. The method of feeding adopted by the bedbug (*Cimex lectularius*) when obtaining a blood-meal from the mammalian host. *Ann. Trop. Med. Parasitol.* 53:347–57.

6. Durden, L. A. 1987. Predator-prey interactions between ectoparasites. *Parasitol. Today* 3:306–8.

7. Fontan, A., and E. Zerba. 1992. Influence of the nutritional state of *Triatoma infestans* over the insecticidal activity of DDT. *Comp. Biochem. Physiol. (C) Comp. Pharmacol. Toxicol.* 101:589–91.

8. Garcia-Zapata, M. T. A., P. D. Marsden, V. A. Soares, and C. N. Castro. 1992. The effect of plastering in a house persistently infested with *Triatoma infestans* (Klug) 1934. *J. Trop. Med. Hyg.* 95:420–23.

9. Harwood, R. F., and M. T. James. 1979. *Entomology in human and animal health,* 7th ed. New York: Macmillan Publishing Co., Inc.

10. Hypsa, V. 1993. Endocytobionts of *Triatoma infestans:* Distribution and transmission. *J. Invertebr. Pathol.* 61:32–38.

11. King, F., I. Dick, and P. Evans. 1989. Bed bugs in Britain. *Parasitol. Today* 5:100–2.

12. Koch, A. 1967. Insects and their endosymbionts. In Henry, S. M., ed. *Symbiosis* 2. New York: Academic Press, Inc., 1–106.

13. Lacey, L. A., A. D'Alessandro, and M. Barreto. 1989. Evaluation of a chlorpyrifos-based paint for the control of Triatominae. *Bull. Soc. Vect. Ecol.* 14:81–86.

14. Lange, A. B., I. Orchard, and F. M. Barrett. 1989. Changes in hemolymph serotonin levels associated with feeding in the blood-sucking bug, *Rhodnius prolixus. J. Insect Physiol.* 35:393–400.

15. Lavoipierre, M. M. J., G. Dickerson, and R. M. Gordon. 1959. Studies on the methods of feeding of blood-sucking arthropods.

I. The manner in which triatomine bugs obtain their blood-meal, as observed in the tissues of the living rodent, with some remarks on the effects of the bite on human volunteers. *Ann. Trop. Med. Parasitol.* 53:235–50.

16. Lazzari, C. R., and J. A. Núñez. 1989. Blood temperature and feeding behavior in *Triatoma infestans* (Heteroptera: Reduviidae). *Entomologia Generalis* 14:183–88.

17. Lazzari, C. R., and J. A. Núñez. 1989. The response to radiant heat and the estimation of the temperature of distant sources in *Triatoma infestans. J. Insect Physiol.* 35:525–29.

18. Mellanby, K. 1939. The physiology and activity of the bedbug (*Cimex lectularius* L.) in a natural infestation. *Parasitology* 31:200–11.

19. Newberry, K., and E. J. Jansen. 1986. The common bedbug *Cimex lectularius* in African huts. *Trans. R. Soc. Trop. Med. Hyg.* 80:653–58.

20. Newberry, K. 1989. The effects on domestic infestations of *Cimex lectularius* bedbugs of interspecific mating with *Cimex hemipterus. Med. and Vet. Entomol.* 3:407–14.

21. Noriega, F. G. 1992. Autogeny in three species of Triatominae: *Rhodnius prolixus, Triatoma rubrovaria,* and *Triatoma infestans* (Hemiptera: Reduviidae). *J. Med. Ent.* 29:273–77.

22. Núñez, J. A., and C. R. Lazzari. 1990. Rearing of *Triatoma infestans* Klug (Heteroptera: Reduviidae) in the absence of a live host: I. Some factors affecting the artificial feeding. *J. Applied Entomol.* 109:87–92.

23. Piesman, J., I. A. Sherlock, and H. A. Christensen. 1983. Host availability limits population density of *Panstrongylus megistus. Am. J. Trop. Med. Hyg.* 32:1445–50.

24. Pinet, J. M., J. Bernard, and J. Boistel. 1969. Etude électrophysiologique des récepteurs des stylets chez une punaise hématophage: *Triatoma infestans. C. R. Séances Soc. Biol.* 163:1939–46.

25. Rojas, J. C., E. A. Malo, A. Gutierrez-Martinez, and R. N. Ondarza. 1990. Mating behavior of *Triatoma mazzottii* Usinger (Hemiptera: Reduviidae) under laboratory conditions. *Ann. Ent. Soc. Amer.* 83:598–602.

26. Schaub, G. A., C. A. Boeker, C. Jensen, and D. Reduth. 1989. Cannibalism and coprophagy are modes of transmission of *Blastocrithidia triatomae* (Trypanosomatidae) between triatomines. *J. Protozool.* 36:171–75.

27. Silva, I. G. D., and H. H. G. D. Silva. 1993. Differences in susceptibility to "Y" strain of *Trypanosoma cruzi* (Kinetoplastida: Trypanosomatidae) among 11 species Triatomines (Hemiptera, Reduviidae). *Rev. Brasil. Entomol.* 37:459–63.

28. Tarrant, C., E. W. Cupp, and W. S. Bowers. 1982. The effects of precocene II on reproduction and development of triatomine bugs (Reduviidae: Triatominae). *Am. J. Trop. Med. Hyg.* 31:416–20.

29. Usinger, R. L. 1966. *Monograph of Cimicidae (Hemiptera-Heteroptera).* College Park, Md.: Entomological Society of America.

30. Usinger, R. L., and J. G. Myers. 1929. Facultative bloodsucking in phytophagous Hemiptera. *Parasitology* 21:472–80.

31. Webb, P. A., C. M. Happ, G. O. Maupin, B. J. B. Johnson, C. Y. Ou, and T. P. Monath. 1989. Potential for insect transmission of HIV: Experimental exposure of *Cimex hemipterus* and *Toxorhynchites amboinensis* to human immunodeficiency virus. *J. Infect. Dis.* 160:970–77.

32. Zeledón, R., and L. G. Vargas. 1984. The role of dirt floors and of firewood in rural dwellings in the epidemiology of Chagas' disease in Costa Rica. *Am. J. Trop. Med. Hyg.* 33:232–35.

Chapter 37

PARASITIC INSECTS: SIPHONAPTERA, THE FLEAS

The combined effects of Nero and Kubla Khan, of Napoleon and Hitler, all the Popes, all the Pharaohs, and all the incumbents of the Ottoman throne are as a puff of smoke against the typhoon blast of fleas' ravages through the ages.

B. Lehane

How can this be? Ravages of *fleas?* It no longer seems an exaggeration when we realize that fleas transmit the dreaded plague, the killer of millions of people from the dawn of civilization through the beginning of the twentieth century. Of all of humanity's major diseases and important insect pests, the combination of the flea and plague bacillus has had the greatest impact on human history. We will return to the subject of plague later in this chapter.

The approximately 2500 species of fleas are small insects, from just less than a millimeter to a few millimeters long. Most are parasites of mammals, but approximately a hundred species regularly occur on birds. They are rather heavily sclerotized, bilaterally flattened, and secondarily wingless. Evolutionary loss of wings is a condition commonly found in parasitic insects. Some fleas are tan or yellow, but they are commonly reddish brown to black. Their mouthparts are of the piercing-sucking type; the adults feed exclusively on blood. The larvae usually are not parasitic but feed on debris and materials associated with the nest or surroundings of the host, especially the feces of the adult fleas. Strong evidence suggests that fleas descended from a winged ancestor much like the present-day scorpion flies (Mecoptera). In fact several features of the jumping mechanism, which is well-developed in most fleas, seem to be homologous with flight structures of flying insects.[26]

Flea taxonomy still is a relatively unsettled discipline, and various authors disagree mainly on numbers of families and the status of subspecies.[21]

MORPHOLOGY

The head is broadly joined to the thorax and often bears a **genal ctenidium** (Fig. 37.1). Ctenidia are series of rather stout, peglike spines often found on the posterior margin of the first thoracic tergite (**pronotal ctenidium**) as well as on the head. Many species lack ctenidia, however. The ctenidia and the backwardly directed setae on the body are adaptations that help the flea retain itself among the fur or feathers of its host. The width of the space between adjacent spine tips in the ctenidia of a particular flea species is correlated with the diameter of the hairs of its usual host, being slightly wider than the maximal diameter of its host's hairs.[11] Thus, in backward movement, the hairs tend to catch between the ctenidial spines. Because of the resistance to being dragged backward, removal of the flea by host grooming or preening is much more difficult. Obviously these structures do not impede forward progress of the flea between the hairs; neither do the antennae, which fold back into grooves on the sides of the head. The antennae appear trisegmented, but the apparent terminal segment of the antenna actually consists of 9 or 10 segments. When present, the eyes are simple and have only a single, small lens. Ocelli are absent. Fleas bear a peculiar sensory organ near their posterior end called the **pygidium** that apparently functions to detect air currents.[2]

The legs are strong, and the hind legs are commonly much larger than are the other two pairs and are modified for jumping.

Jumping Mechanism

Many fleas are champion jumpers. The oriental rat flea, *Xenopsylla cheopis,* can jump more than a hundred times its body length, and cat and human fleas are capable of a standing leap of 33 cm high.[29] In terms of proportionate body length, this would be the equivalent of a 6-foot human executing a standing high jump of almost 800 feet! It is not at all obvious how fleas accomplish this remarkable feat. *Xenopsylla* reaches an acceleration of $140 \times g$ in a little more than a millisecond, yet the fastest single muscle twitches (as in the locust, for example) require 15 milliseconds to reach peak force. The answer lies in the use of the flight structures that the flea inherited from its ancestors. Fleas have **resilin** in the **pleural arch,** an area between the internal ridges of the metapleuron and metanotum (Fig. 37.1). The pleural arch is homologous to the wing-hinge ligament in dragonflies, locusts, and scorpion flies.[29] Resilin is a stabilized protein with highly unusual elastic properties. It is a better "rubber" than rubber, releasing 97% of its stored energy on returning from a stretched position, compared with only 85% in most commercial rubber.

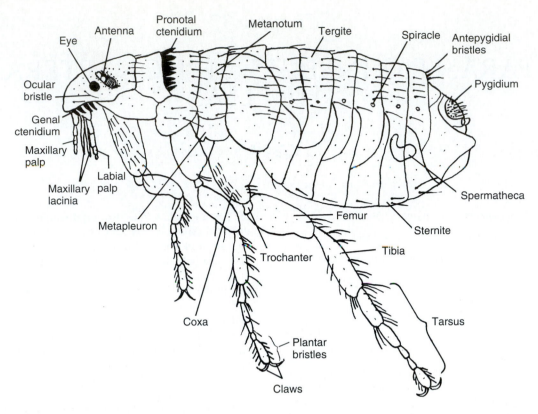

FIGURE 37.1

Diagram of a flea.
Drawing by Larry S. Roberts.

When the flea prepares to jump, it rotates its hind femurs up so that they lie almost parallel to the coxae, and the flea rests on the tarsi of its front two pairs of legs and on the trochanters and tarsi of its hind legs. The resilin pad is compressed and is maintained in that condition by catch structures on certain sclerites (two on each side of the flea). In effect the flea has cocked itself; to take off it must exert the relatively small muscular action to unhook the catches, allowing the resilin to expand. Then the flea rotates its femurs down toward the substrate and pushes off with its hind tarsi. By using resilin, the fleas have circumvented two major limitations of muscle: the relatively slow rate of contraction and relaxation and the poor performance at low temperature. Because resilin does not become deformed under prolonged strain, once the jumping mechanism is cocked, little energy is required to retain this state. The flea can lie in waiting, ready to hop aboard a host in a fraction of an instant. Interestingly, flea species whose preferred hosts are small or relatively accessible tend to have reduced pleural arches, whereas those that prefer the largest hosts (such as deer, sheep, cats, and humans) have the largest pleural arches and are the best jumpers.[29]

Mouthparts and Mode of Feeding

Like the Anoplura and Hemiptera, the Siphonaptera have piercing-sucking mouthparts, but the structure is different from that of lice and bugs. The broad maxillae bear conspicuous, segmented palps (Fig. 37.2), as does the slender labium. The piercing fascicle comprises two elongated maxillary lobes (**laciniae**) and a median, unpaired **epipharynx.** The laciniae lie closely on each side of the epipharynx, and the fascicle is held in a channel formed by grooves in the inner side of the labial palps. The hypopharynx cannot be demonstrated (Fig. 37.3), and the labium is rudimentary. Lavoipierre and Hamachi[19] reported that several species of fleas are solenophages, but Rothschild and her coworkers believed that *Spilopsyllus cuniculi* is primarily a telmophage.[9,28] The laciniae are the cutting organs, and piercing is achieved by the back-and-forth cutting action of these structures. The tip of the epipharynx, but not the laciniae, enters a small blood vessel. Saliva is ejected into the area near the puncture of the vessel by the laciniae, but it does not enter the vessel lumen. Very little damage is done by the penetrating fascicle, and after withdrawal from the skin, hemorrhage is scant or absent.

DEVELOPMENT

Fleas undergo holometabolous development, and the larvae thus are quite different in form and habits from the adults. Although the adult female usually oviposits while on the host, the eggs are not sticky and so drop off the host's body. This often happens in the host's nest or lair, where there is a supply of detritus and flea feces on which the larvae feed. The eggs are relatively large (about 0.5 mm) (Fig. 37.4), providing many essential nutrients to the larvae.

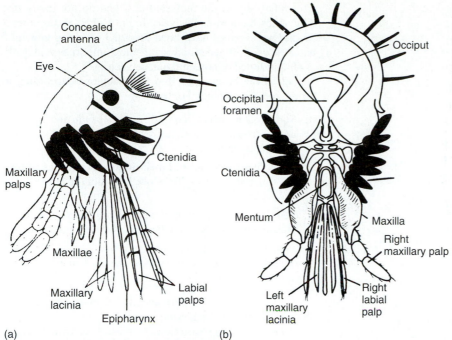

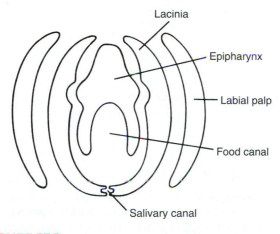

FIGURE 37.2

Head and mouthparts of a flea. (*a*) Side view; (*b*) ventral view.

From R. F. Harwood and M. T. James, *Entomology in Human and Animal Health,* 7th ed. Copyright © 1979 Macmillan Publishing Company. Reprinted with permission of Simon & Schuster, Inc.

(a)   (b)

FIGURE 37.3

Diagrammatic representation of transverse section through flea mouthparts.

From R. R. Askew, *Parasitic Insects.* Copyright © 1971 American Elsevier Publishing Co., New York, NY.

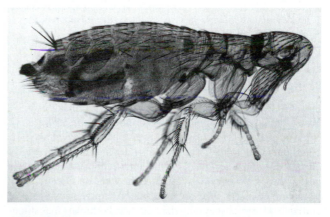

FIGURE 37.4

Leptopsyllus segnis, the European mouse flea, which is also common in parts of the United States. Note the large size of the eggs visible within this cleared specimen.

Courtesy of Jay Georgi.

Under favorable conditions, the larvae hatch within 2 to 21 days; the larval instars (usually three) require 9 to 15 days; and the pupa completes development in as short a time as a week. Low temperatures and those as high as the host's body temperature retard development. Low temperatures can extend the larval period to more than 200 days and the pupal stage to nearly a year. Because the larvae cannot close their spiracles, they are sensitive to low humidity. High humidity tends to favor egg laying in the adults and of course is apt to be the prevalent condition in nests and burrows. The larvae are white, legless, and eyeless, resembling the maggots of some Diptera (Fig. 37.5). They have chewing mouthparts and stout body hairs. The pupa spins a loose, silken cocoon from its salivary secretions, often picking up debris from the surroundings in the cocoon.

In common with other lair parasites such as bedbugs and in contrast with lice, fleas can survive long periods as adults without food, particularly under conditions of high humidity. Unfed *Pulex irritans* have survived 125 days at 7° to 10° C and *Xenopsylla cheopis* for 38 days; periodically fed *P. irritans* may live up to 513 days and *X. cheopis* to 100 days.[9] Periodically fed *Ctenophthalmus wladimiri* have survived for more than three years at 7° to 10° C and 100% relative humidity. Such longevity has clear epidemiological importance because it allows flea-transmitted pathogens to survive long periods when vertebrate hosts are absent. Cases are known in which long survival occurs even in the face of highly adverse conditions: *Glaciopsyllus* larvae, pupae, and some adults can withstand freezing in their host's nest and being covered with ice for nine months out of the year.[26]

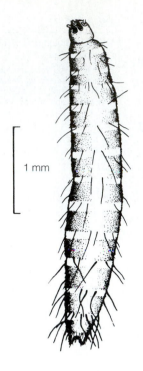

FIGURE 37.5

Third instar larva of *Spilopsyllus cuniculi,* the European rabbit flea, showing mostly the ventral surface.

From R. R. Askew, *Parasitic Insects.* Copyright © 1971 American Elsevier Publishing Co., New York, NY.

Two genera of the subfamily Spilopsyllinae, *Spilopsyllus* and *Cediopsylla,* are unusual in that their reproduction is closely controlled by their host's hormones. *Spilopsyllus cuniculi,* the **European rabbit flea,** is a relatively sedentary flea, commonly attaching itself to its host's ears for long periods. However, it does not breed on the adult rabbit. About 10 days before the pregnant doe gives birth, the fleas on the doe begin to mature sexually. The pregnant rabbit experiences a rise in cortisol and corticosterone, which are the hormones that stimulate flea maturation. The hormone levels also are high in the newborn rabbit. By the time the rabbit's young are born, the flea's eggs are ripe, and the fleas detach from the doe's ear and move onto her face. As she tends her young, the fleas hop onto the newborn rabbits and feed voraciously, mate, and lay eggs. After about 12 days, the fleas leave the young and return to the doe. Good evidence indicates that the growth hormone somatotropin, present in young rabbits, constitutes the stimulus for fleas to mate and lay eggs. Reproductive control of *C. simplex* by host hormones is essentially similar to that of *S. cuniculi.*[28] This remarkable coordination of the flea's reproduction with that of the host assures that the flea's eggs will be ripe at just the right moment to be deposited into the host's nest and assures the larvae of a plentiful supply of food.[25]

HOST SPECIFICITY

In general, fleas are not very host specific, although they have preferred hosts. Most can transfer from one of their hosts to another or to a host of a different species. Their common names (for example, rat flea, chicken flea, and human flea) refer only to the preferred host and do not imply that they attack the host exclusively. In the United States at least 19 different species have been recorded as biting humans,[9] but well over 50 genera are of medical importance globally as demonstrated or potential plague vectors.[21]

Fleas can be grouped into four categories according to the degree of attachment to the host:

1. Some rodent fleas, such as *Conorhinopsylla* and *Megarthroglossus* spp., are seldom on the host but occur abundantly in its nest.

2. Most fleas spend most of their time on the host as adults but can transfer easily from one host individual to another.

3. Females of the sticktight flea, *Echidnophaga gallinacea,* attach permanently to the host skin by their mouthparts.

4. Females of the **chigoe,** *Tunga penetrans,* burrow beneath the host skin and become stationary, intracutaneous, and subcutaneous parasites.

Those that become permanently attached play little or no role in disease transmission.

There are some variations to these four categories. A species is known (*Uropsylla tasmanica,* on Tasmanian devils) in which the larvae burrow beneath the skin and live as endoparasites. The larvae of *Hoplopsyllus* on the arctic hare live as ectoparasites in the fur of the host. Bird fleas also exhibit some rather remarkable adaptations to their hosts. The southern fulmar, an oceanic bird that spends vast amounts of time at sea, is the major host for *Glaciopsyllus antarcticus.* Live *G. antarcticus* do not occur in the nest material, suggesting that the fleas survive the South Pole winter by going to sea with their hosts.[4,35]

FAMILIES CERATOPHYLLIDAE AND LEPTOPSYLLIDAE

The **northern rat flea,** *Nosopsyllus fasciatus,* is a common parasite of domestic rats and mice (*Rattus* and *Mus* spp.) throughout Europe and North America, and it has been recorded on many other hosts, including humans. Although it may be of some importance in transmission of plague from rat to rat, it is not regarded as an important plague vector because it usually does not bite humans, and it is widespread in temperate climates where plague is not commonly a problem.[24] The **ground squirrel flea,** *Diamanus montanus,* is found in western North America from Nebraska and Texas to the Pacific coast and may be of some importance in transmission of plague in wild rodents.

Ceratophyllus niger and *C. gallinae* are bird fleas, although both will bite humans. *Ceratophyllus niger* is the **western chicken flea.** It can be distinguished easily from another common chicken flea, the sticktight (*E. gallinacea*), in that *C. niger* is larger and does not attach permanently. The **European chicken flea,** *C. gallinae,* commonly parasitizes a wide variety of other birds, especially passerines.

Leptopsyllus segnis is the **European mouse flea** (Fig. 37.4), but it is common throughout the Gulf states and in parts of California. It is more common on *Rattus* than on *Mus.* Although it can be infected with plague, it is not an important vector because it does not readily bite humans.

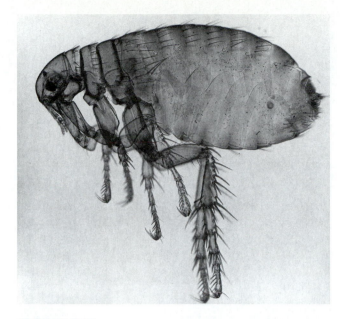

FIGURE 37.7

Female *Pulex irritans.*
Courtesy of Jay Georgi.

FIGURE 37.6

Male *Pulex irritans,* the human flea. The copulatory apparatus is visible in the abdomen. Rothschild[25] has called the genital organs of male fleas the most elaborate in the animal kingdom.
Courtesy of Jay Georgi.

FAMILY PULICIDAE

Pulex irritans, the **human flea** (Figs. 37.6 and 37.7), and other species of medical and veterinary importance belong to this family. *Pulex irritans* has been recorded from pigs, dogs, coyotes, prairie dogs, ground squirrels, and burrowing owls, but some of the records may refer to another species, *P. simulans,* which occurs in the central and southwestern United States and in Central and South America.[31] Both species lack genal and pronotal ctenidia, and their metacoxae have a row or patch of short spinelets on the inner side of the podomere. However, the maxillary laciniae of *P. irritans* extend only about half the length of the forecoxae, whereas those of *P. simulans* extend about three-fourths the distance. *Pulex irritans* can transmit plague, and it has been implicated in the transmission from person to person in some epidemics. Kalkofen[14] found in a survey in Georgia that more than 80% of fleas on dogs were *P. irritans* and that dogs seemed to be the preferred hosts. Since dogs are susceptible to plague (see Kalkofen[14] for references), this has important public health implications.

Echidnophaga gallinacea is an important poultry pest, but it also attacks cats, dogs, horses, rabbits, humans, and other animals. It is called the **sticktight flea** because it buries its fascicle in the skin of the host and remains in place. The maxillary laciniae are broad and coarsely serrated (Fig. 37.8). This flea is widespread in tropical and subtropical regions and also occurs in the United States as far north as Kansas and Virginia.[6] Like those of *Tunga penetrans,* its thoracic segments are reduced, being shorter together than the head or the first abdominal segment; however, *E. gallinacea* has a patch of spinelets on the inner side of its metacoxa (absent in *Tunga*). It prefers to attach in areas with few feathers, such as the comb, wattles, around the eyes, and around the anus of the host. The infestation causes ulcers, into which the female deposits the eggs. The larvae hatch in the ulcers but then drop to the ground to develop off the host, as in most other fleas. Heavy infestations may kill the chickens.

Ctenocephalides canis (Fig. 37.9) and *C. felis* are the **dog** and **cat fleas.** They can be distinguished from the other common fleas by the presence of a genal ctenidium with more than five teeth. In spite of the names, both species attack cats and dogs, as well as humans, other mammals, and occasionally chickens.[9] *Ctenocephalides felis* is more common than is *C. canis* on dogs in North America. They can be very annoying pests of humans, particularly when cats and dogs are kept on the premises. Oddly they do not occur in the mid– to north–Rocky Mountain area. Cat fleas deposit most of their eggs at times their hosts are least active (midnight to very early morning) and most likely to be in their resting areas.[15] Flea feces (larval food) make up more than 40% of the debris that falls off infested cats, and so are also concentrated in these areas.[16] The effect of this coordination is a localization of larval development sites.

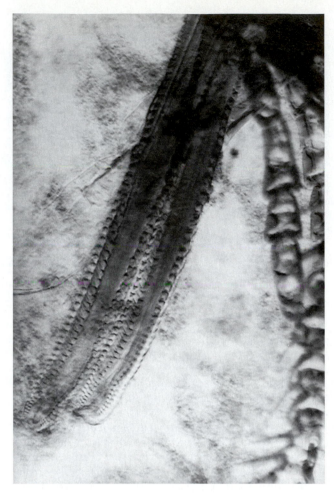

FIGURE 37.8

Maxillary laciniae of *Echidnophaga gallinacea,* the sticktight flea. The laciniae are broad and coarsely serrated, and the thoracic somites are much reduced compared with most other fleas. Maxillary palps are to the right in this photograph.

Photograph by Larry S. Roberts.

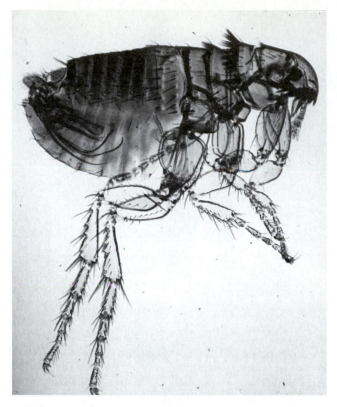

FIGURE 37.9

Ctenocephalides canis, the dog flea. This species and *C. felis* bite humans frequently and are the source of much annoyance. They are the most common intermediate hosts of the tapeworm, *Dipylidium caninum,* of dogs and cats.

Courtesy of Jay Georgi.

Like many insects, cat fleas and their feces are allergenic and may contribute to the allergenicity of house dust.[34] At least 15 different proteins from *C. felis* are allergenic.[8]

Xenopsylla cheopis (Fig. 37.10) is called the **oriental** or **tropical rat flea,** although it is almost cosmopolitan on *Rattus* spp., except in cold climates. In the United States it ranges as far north as New Hampshire, Minnesota, and Washington.[24] Like *Pulex irritans, X. cheopis* lacks both genal and pronotal ctenidia. It can be distinguished by the location of the ocular bristle, which originates in front of the eye in *X. cheopis* and beneath the eye in *P. irritans.* Females can be recognized easily by the dark-colored spermatheca, since this is the only species in the United States with a pigmented spermatheca (compare Figs. 37.7 and 37.10).[24] *Xenopsylla cheopis* has enormous public health significance because it is the most important vector of plague and of murine typhus. *Xenopsylla brasiliensis,* an African species that has become established in South America and India, appears to be a more important plague vector in Kenya and Uganda. Some other species of *Xenopsylla* are implicated or suspected as vectors in various plague outbreaks.

FIGURE 37.10

Xenopsylla cheopis, the oriental, or tropical, rat flea. This flea is the most common vector of plague and murine typhus. The spermatheca is darkly pigmented and clearly visible.

Courtesy of Jay Georgi.

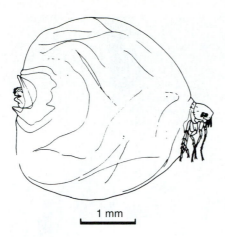

1 mm

FIGURE 37.11

Engorged female *Tunga penetrans,* the chigoe, or jigger. This stage is found in a subcutaneous sinus that communicates with the outside by a small pore, through which the larvae escape. The legs are degenerating by the time the female is engorged.

From R. R. Askew, *Parasitic Insects.* Copyright © 1971 American Elsevier Publishing Co., New York, NY.

FAMILY TUNGIDAE

Tunga penetrans is called the **chigoe, jigger, chigger, chique,** and **sand flea** (Fig. 37.11). Some of these names are said to result from the irritation the flea causes, prompting the host to "jig" about. This flea is apparently a native of Central and South America and the West Indies, from which it was introduced to Africa in the seventeenth century and again in the nineteenth century. It spread all over tropical Africa and then to India. *Tunga penetrans* attacks several other mammals besides humans, particularly swine.

The female chigoe penetrates the skin, most commonly around the nail bases of the hands and feet or between the toes. Only a small aperture through the skin is left to communicate with the outside world. The male does not penetrate host skin, but it copulates with the female after she has reached her final position. When she enters the skin, she is barely 1 mm long, but she gradually expands to about the size of a pea. Her body is enclosed in a sinus, into which she lays her eggs. After hatching, the larvae exit through the aperture and develop on the ground. The presence of the female causes extreme itching, pain, inflammation, and often secondary infection (Fig. 37.12). Tetanus and gangrene occasionally are complications. Autoamputation has been attributed to results of infection with this flea and its secondary infection in Angola.[6] Surgical removal of the flea with careful sterilization and dressing of the wound is the recommended remedy.

FLEAS AS VECTORS

Plague

Plague, also known as pest and black death, is caused by a bacterium, *Yersinia pestis* (formerly *Pasturella pestis*). The bacterium releases two potent toxins that have identical serious effects.[13] Some animals, such as rats and mice, are more sensitive to the toxins than are others (rabbits, dogs, monkeys, and chimpanzees). Evidence indicates that the toxins act on the mitochondrial membranes of susceptible animals, inhibiting ion uptake and interfering with normal functioning of the respiratory chain.[13]

The great pandemics of plague have probably had more profound effect on human history than any other single infection. The pandemic of the fourteenth century alone, for example, took 25 million lives, or a fourth of the population of Europe, and has been called the worst disaster that has ever befallen humanity.[18] Adequate treatment of this subject in relation to its importance is far beyond the scope of this book. Interested readers may avail themselves of numerous sources on plague, including highly readable accounts of plague and fleas in history by Lehane[20] and Pollitzer.[23] A fascinating picture of human behavior during the terror of universal catastrophe is given by Langer.[18] The last pandemic began in the interior of China toward the end of the nineteenth century, reached Hong Kong and Canton by 1895 and Bombay and Calcutta in 1896, and then spread throughout the world, including numerous port cities in the United States. In the period between 1898 and 1908, more than 548,000 people per year died from plague in India.[23]

Between 1900 and 1972 there were 992 cases in the United States (416 of them in Hawaii), of which 720 were fatal.[24] The disease has decreased in incidence and severity in recent years: Between 1958 and 1972 there were 51 cases in the United States, of which only nine were fatal. A similar decline occured globally.[21] It is not clear if this improvement results entirely from better medical care.

The disease was formerly centered in seaports, from which it spread out in epidemics, recent cases in the United States have been virtually all rural (**campestral** or **sylvatic**); that is, they were contracted after contacts with wild rodents in the countryside rather than with *Rattus* in the cities. *Yersinia pestis* is widely distributed in rodents and occurs across broad areas of virtually every continent.[21] It is unknown whether the urban plague spread into the wild rodents in the United States or whether it was there before 1900.

Plague is essentially a disease of rodents, from which it is contracted by humans through the bites of fleas, particularly *Xenopsylla cheopis*. The bacteria are consumed by a flea along with its blood meal, and the organisms multiply in the flea's gut, often to the extent that passage of food through the proventricular teeth is blocked. When the flea next feeds, the new blood meal cannot pass the obstruction, but is contaminated by the bacteria and then regurgitated back into the bite wound. The propensity of a particular flea species to have its gut blocked by growth of *Yersinia pestis* is an important determinant of its efficacy as a vector. *Xenopsylla cheopis* is a good vector because it becomes blocked easily, feeds readily both on infected rodents and humans, and is abundant near human habitations.[24] Although the black, or roof, rat, *Rattus rattus,* is common and abundant around human habitations, it appears not to be a reservoir of infection.[30]

The three forms of plague are **bubonic, primary pneumonic,** and **primary septicemic.** Bubonic plague, which is the most common type in epidemics, demonstrates definite bubo formation in about 75% of cases. **Buboes** are swollen lymph nodes in the groins or armpits; they sometimes reach the size of a hen's egg (Fig. 37.13) and may rupture to the outside. They are hard, tender, and filled with

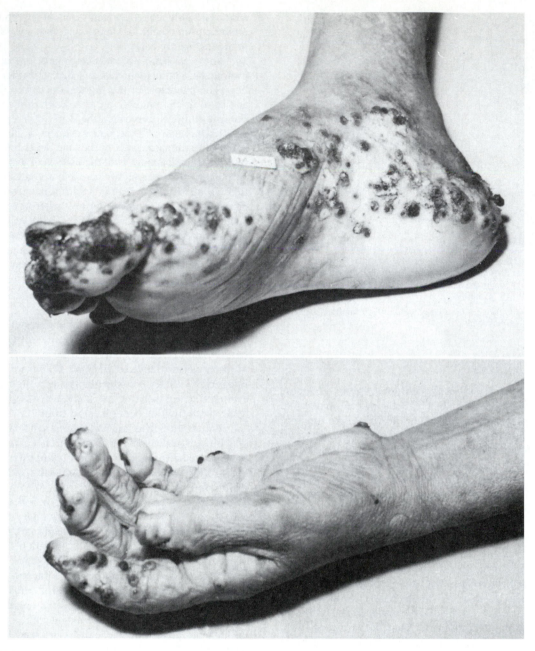

FIGURE 37.12

Lesions caused by *Tunga penetrans.*

From G. W. Hunter, J. C. Swartzweider, and D. F. Clyde, *Tropical Medicine,* 5th ed., 1976 W. B. Saunders, Co. Photo by Rodolfo Cespedes, Hospital San Juan de Dios, San Jose, Costa Rica.

bacteria. Pneumonic plague is a condition in which the lungs are most heavily involved, producing a pneumonialike disease that is highly contagious to other persons. Primary septicemic plague is a generalized blood infection with little or no prior lymph node swelling, apparently because the blood is invaded too rapidly for the typical nodal inflammation to develop. However, secondary septicemic plague usually occurs in pneumonic and sometimes in bubonic plague.[3] Bubonic plague is fatal in about 25% to 50% of untreated cases, and pneumonic and septicemic plague are usually fatal.

The incubation period after the bite of the flea is usually two to four days, followed by a chill and rapidly rising temperature to 39.5° and 40° C. The lymph nodes draining the site of the infection swell, becoming hemorrhagic and often necrotic. Damage to vascular and lymphatic endothelium in many parts of the body leads to petechial and diffuse hemorrhages. At first there is mental dullness, followed by anxiety or excitement and then delirium or lethargy and coma. If untreated, the disease may cause death within five days. The patient usually dies within three days of the onset of primary

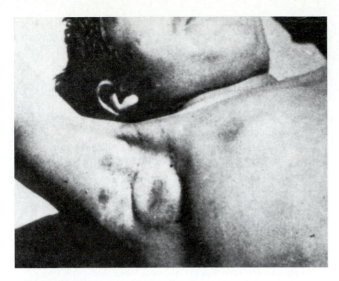

FIGURE 37.13

Plague bubo in right axilla of human.
AFIP neg. no. ACC 219900–7–B.

septicemic plague. If the patient is to recover, the fever begins to drop in two to five days. Treatment with antibiotic and antitoxin is usually effective. Persons traveling into an area in which plague is known to occur with any regularity should be vaccinated.[3]

Conditions conducive to high rat and flea populations contribute to plague outbreaks. The disease may exist in rodent populations in acute, subacute, and chronic forms. Epidemics among humans usually closely follow epizootics, with high mortality among rats. When the rat dies, its fleas depart and seek greener pastures. Meteorological conditions are important, doubtless because of their effects on the flea population. Plague is usually seen in temperate climates during summer and autumn and in the tropics during the cool months. Extreme heat and dryness inhibit spread of plague. Epidemics of primary pneumonic plague occur only during conditions of low temperatures and high humidity, which are conditions that favor survival of the bacilli in sputum droplets.

Campestral plague (so called because it is associated with animals of open areas rather than wooded or sylvatic country) is widespread among wild rodents and rabbits in the United States west of the 100th meridian. Cases among humans are reported sporadically, usually after a human has contacted wild rodents and their fleas. New Mexico has had the highest case rate. One Californian contracted bubonic plague with secondary pneumonic plague after hunting ground squirrels and was the source of 13 cases of primary pneumonic plague in other persons, with 12 deaths.[24] A number of cases have been associated with skinning, cooking, and eating wild rabbits and hares; but the human victims may have been bitten by the rabbit's fleas.[1]

Throughout recorded history, plague has been cyclical, smoldering in endemic foci and then giving rise to great outbreaks. The world seems to be in a remission phase at present, but a vast and easily accessible supply of *Y. pestis* is still present in nature. Given the widespread distribution of rodents and their fleas, the plague bacillus seems to persist regardless of control measures; a decade-long study accompanying control efforts in Tanzania implicated dogs as reservoirs, with over 6% seropositive, and reported insecticide resistance in *Pulex irritans.*[17]

Murine Typhus

Murine typhus, also called endemic or flea-borne typhus, is caused by *Rickettsia mooseri* (= *R. typhi*) which is morphologically indistinguishable from *R. prowazekii* (p. 542). It occurs in warmer climates throughout the world. It can infect a wide range of small mammals, including the opossum, *Didelphis marsupialis,* but the most important reservoir is *Rattus norvegicus,* in which it causes slight disease symptoms. Murine typhus can be transmitted from rat to rat by *Xenopsylla cheopis; Nosopsyllus fasciatus; Leptopsyllus segnis;* the rat louse, *Polyplax spinulosa;* and the tropical rat mite, *Ornithonyssus bacoti.* In humans the disease is a rather mild, febrile illness of about 14 days' duration, with chills, severe headaches, body pains, and rash. It tends to be more severe in elderly persons. *Xenopsylla cheopis* is considered the primary vector transmitting the disease to humans, either through the bite or through contamination of skin abrasions with flea feces by scratching. Ingestion of infected fleas and their feces also can produce infection in rats.[7] The rickettsias proliferate in the midgut cells of the flea but do not kill it. Rupture of the midgut cells releases the organisms into the gut of the flea.

Before 1945 the incidence of murine typhus was high in the United States and reached a peak of 5401 cases in 1945. After the institution of a rat control program, use of DDT, and increasing use of antibiotics, the reported incidence dropped dramatically, ranging between 18 and 36 cases per year between 1969 and 1972. However, opossums are proliferating in many urban and suburban areas, creating a possible resurgence in the number of cases. In one Los Angeles study, city opossums were heavily infested with the cat flea, *Ctenocephalides felis* (mean flea count = 104.7/animal), a species that readily bites humans. In that report the authors concluded that the biology of this focus differed substantially from the classical transmission cycle, and that cats, opossums, and *C. felis* may play an important role in the occurrence of human cases.[33]

Myxomatosis

The myxoma virus causes a disease in rabbits and is transmitted by several bloodsucking arthropods, including mosquitoes, fleas, and mites.[32] The principal vector in England is *Spilopsyllus cuniculi,* and the disease causes considerable losses in the domestic rabbit industry. The virus was apparently introduced from South America, where the rabbits are relatively resistant to myxomatosis.[25] It was intentionally introduced into Australia to control the abundant rabbits there. However, the principal vectors in Australia are mosquitoes, which are not ideal vectors for rabbit control. The mosquitoes confer a selective advantage on an attenuated "field strain" of the virus and are most abundant during the warm months, when the rabbits have the best chance of surviving the disease.[32] Consequently, resistance to the virus in rabbit populations was unintentionally selected for. Introduction of *Spilopsyllus cuniculi* subsequently offered more hope of better rabbit control, along with reintroduction of virulent virus strains.[32]

Other Parasites

Nosopsyllus fasciatus is a vector for the nonpathogenic *Trypanosoma lewisi* of rats (p. 67). *Ctenocephalides canis, C. felis,* and *Pulex irritans* serve as intermediate hosts of *Dipylidium caninum,* a common tapeworm of cats and dogs (p. 343). *Nosopsyllus fasciatus* and *Xenopsylla cheopis* can serve as vectors for the rat tapeworm, *Hymenolepis diminuta;* the mouse tapeworm, *Vampirolepis nana,* can develop in *X. cheopis, C. felis, C. canis,* and *P. irritans* (pp. 341 and 342). All of these fleas acquire the tapeworms as larvae when they consume the eggs passed in the feces of the vertebrate host, retaining the cysticercoids in their hemocoel through metamorphosis to the adult. All three species can be transmitted to humans if the person inadvertently ingests an infected flea.

A filarial worm of dogs, *Dipetalonema reconditum,* which lives in the subcutaneous, connective, and perirenal tissues, is transmitted by *C. canis* and *C. felis.* The microfilariae are picked up by the fleas in their blood meal, develop to the infective stage in the flea's fat body in about six days, migrate to the head, and then pass to the wound when the flea next feeds. *Dipetalonema reconditum* is of slight or no pathogenicity, but its microfilaria may be easily confused with those of the serious pathogen *Dirofilaria immitis* (p. 457). For techniques to distinguish the two, see Ivens, Mark, and Levine.[12]

CONTROL OF FLEAS

Many of us occasionally need to control fleas around our homes or on our pets. It is sometimes extremely important for public health reasons to control rat fleas and more importantly their hosts. For more complete instructions about and techniques for flea and rat control, consult Pratt and Stark.[24]

Within habitations one should keep debris that harbors larval fleas to a minimum, such as under carpets, in floor crevices, and in pet bedding. Some persistent insecticides may be used indoors. These tend to act especially on eggs and larvae. For example, diflubenzuron inhibits *C. felis* development in carpets for up to a year, but it breaks down when used in the yard.[10] Pyriproxyfen and methoprene make fleas lay infertile eggs.[22] More subtle control attempts include light traps with yellow-green filters to which fleas respond positively. One such device collected 86% of the live fleas released into a carpeted room, attracting them from as far away as 8 m.[5]

It is important to keep premises where livestock are maintained as free from debris, manure, and other litter as possible. Various insecticidal flea powders for use on dogs and cats are available, but repeated application may be necessary because they easily pick up more fleas in outdoor areas not treated with insecticide. In recent years flea collars with slow-release vapors have proven very effective.

Personal protection may be achieved by use of insect repellants. In areas in which *Tunga penetrans* is found, it is important to wear shoes.

References

1. Anonymous. 1984. 1983 sets high mark for plague. *Lubbock Avalanche-Journal.* (24 March).

2. Askew, R. R. 1971. *Parasitic insects.* New York: American Elsevier Publishing Co., Inc.

3. Butler, T. 1991. Plague. In Strickland, T., ed. *Hunter's tropical medicine,* 7th ed. Philadelphia: W. B. Saunders Company, 408–16.

4. Chastel, O., and J. C. Beaucournu. 1992. Notes sur la spécificité et l'éco-éthologie des puces d'oiseaux aux Iles Kerguelen (*Insecta; Siphonaptera*). *Ann. Parasitol. Hum. Comp.* 67:213–220.

5. Dryden, M. W., and A. B. Broce. 1993. Development of a trap for collecting newly emerged *Ctenocephalides felis* (Siphonaptera: Pulicidae) in homes. *J. Med. Ent.* 30:901–6.

6. Grothaus, R. H., and D. E. Weidhaas. 1976. Class Insecta (Hexapoda). In Hunter, G. W., J. C. Swartzwelder, and D. F. Clyde. *Tropical medicine,* 5th ed. Philadelphia: W. B. Saunders Company.

7. Farhang Azad, A., and R. Traub. 1985. Transmission of murine typhus rickettsiae by *Xenopsylla cheopis,* with notes on experimental infection and effects of temperature. *Am. J. Trop. Med. Hyg.* 34:555–63.

8. Greene, W. K., R. L. Carnegie, S. E. Shaw, R. C. A. Thompson, and W. J. Penhale. 1993. Characterization of allergens of the cat flea, *Ctenocephalides felis:* Detection and frequency of IgE antibodies in canine sera. *Parasite Immunol.* 15:69–74.

9. Harwood, R. F., and M. T. James. 1979. *Entomology in human and animal health,* 7th ed. New York: Macmillan Publishing Co., Inc.

10. Henderson, G., and L. D. Foil. 1993. Efficacy of diflubenzuron in simulated household and yard conditions against the cat flea *Ctenocephalides felis* (Bouche) (Siphonaptera: Pulicidae). *J. Med. Ent.* 30:619–21.

11. Humphries, D. A. 1967. The function of combs in fleas. *Entomol. Mon. Mag.* 102:232–36.

12. Ivens, V. R., D. L. Mark, and N. D. Levine. 1978. *Principal parasites of domestic animals in the United States. Special Pub. No. 52.* Urbana, Ill.: Colleges of Agriculture and Veterinary Medicine, University of Illinois.

13. Kadis, S., T. C. Montie, and S. J. Ajl. 1969. Plague toxin. *Sci. Am.* 220(3):93–100.

14. Kalkofen, U. P. 1974. Public health implications of *Pulex irritans* infestations of dogs. *J. Am. Vet. Med. Assoc.* 165:903–5.

15. Kern, W. H. Jr., P. G. Koehler, and R. S. Patterson. 1992. Diel patterns of cat flea (Siphonaptera: Pulicidae) egg and fecal deposition. *J. Med. Ent.* 29:203–6.

16. Kern, W. H. Jr., P. G. Koehler, R. S. Patterson, and R. W. Wadleigh. 1992. Spectrophotometric method of quantifying adult cat flea (Siphonaptera: Pulicidae) feces. *J. Med. Ent.* 29:221–25.

17. Kilonzo, B. S., R. H. Makundi, and T. J. Mbise. 1992. A decade of plague epidemiology and control in the Western Usambara mountains, northeast Tanzania. *Acta Tropica* 50:323–29.

18. Langer, W. L. 1964. The black death. *Sci. Am.* 210(2):114–21.

19. Lavoipierre, M. M. J., and M. Hamachi. 1961. An apparatus for observations on the feeding mechanism of the flea. *Nature* 192:998–99.

20. Lehane, B. 1969. *The compleat flea.* New York: The Viking Press.

21. Lewis, R. E. 1993. Fleas (Siphonaptera). In Lane, R. P., and R. W. Crosskey, eds. *Medical insects and arachnids.* London: Chapman and Hall, Ltd. 529–75.

22. Palma, K. G., S. M. Meola, and R. W. Meola. 1993. Mode of action of pyriproxyfen and methoprene on eggs of *Ctenocephalides felis* (Siphonaptera: Pulicidae). *J. Med. Ent.* 30:421–26.

23. Pollitzer, R. 1954. *Plague.* Geneva: World Health Organization.

24. Pratt, H. D., and H. E. Stark. 1973. *Fleas of public health significance and their control. U.S. Department of Health, Education, and Welfare Pub. No. (CDC) 75-8267.* Washington, D.C.: U.S. Government Printing Office.

25. Rothschild, M. 1965. Fleas. *Sci. Am.* 213(6):44–53.

26. Rothschild, M. 1975. Recent advances in our knowledge of the order Siphonaptera. *Ann. Rev. Entomol.* 20:241–59.

27. Rothschild, M., and B. Ford. 1972. Breeding cycle of the flea *Cediopsylla simplex* is controlled by breeding cycle of host. *Science* 178:625–26.

28. Rothschild, M., B. Ford, and M. Hughes. 1970. Maturation of the male rabbit flea (*Spilopsyllus cuniculi*) and the Oriental rat flea (*Xenopsylla cheopis*): Some effect of mammalian hormones on development and impregnation. *Trans. Zool. Soc. Lond.* 32:105–88.

29. Rothschild, M., Y. Schlein, K. Parker, C. Neville, and S. Sternberg. 1973. The flying leap of the flea. *Sci. Am.* 229(5):92–100.

30. Schwan, T. G., D. Thompson, and B. C. Nelson. 1985. Fleas on roof rats in six areas of Los Angeles County, California: Their potential role in the transmission of plague and murine typhus to humans. *Am. J. Trop. Med. Hyg.* 34:372–79.

31. Smit, F. G. A. M. 1958. A preliminary note on the occurrence of *Pulex irritans* L. and *Pulex simulans* Baker in North America. *J. Parasitol.* 44:523–26.

32. Sobey, W. R., and D. Conolly. 1971. Myxomatosis: The introduction of the European rabbit flea *Spilopsyllus cuniculi* (Dale) into wild rabbit populations in Australia. *J. Hyg.* 69:331–46.

33. Sorvillo, F. J., B. Gondo, R. Emmons, P. Ryan, S. H. Waterman, A. Tilzer, E. M. Andersen, R. A. Murray, and A. R. Barr. 1993. A suburban focus of endemic typhus in Los Angeles county: Association with seropositive domestic cats and opossums. *Am. J. Trop. Med. Hyg.* 48:269–73.

34. Trudeau, W. L., E. Caldas-Fernandez, R. W. Fox, R. Brenner, G. A. Bucholtz, and R. F. Lockey. 1993. Allergenicity of the cat flea (*Ctenocephalides felis felis*). *Clinical and Exp. Allergy* 23:377–83.

35. Whitehead, M. D., H. R. Burton, P. J. Bell, J. P. Y. Arnould, and D. E. Rounsevell. 1991. A further contribution on the biology of the Antarctic flea, *Glaciopsyllus antarcticus* (Siphonaptera: Ceratophyllidae). *Polar Biology* 11:379–84.

Chapter 38

PARASITIC INSECTS: DIPTERA, THE FLIES

Time is fun and we're having flies.

Kermit the Frog

Among all the orders of insects, the Diptera stands out as by far the most medically important, directly or indirectly causing each year more than a million human deaths.[20] Moreover, various flies contribute to disfiguring, debilitating diseases of many kinds, either as vectors of pathogenic organisms or as active parasites in their own right. And who could not imagine the vexation caused by ravenous hoards of mosquitoes, black flies, midges, deer flies, and others? The order is so vast, with approximately 120,000 species in up to 140 families (depending on source), that we can do no more than introduce the subject in this text. Published information on mosquitoes alone would fill a small library.

The name *fly* applies to insects that have a pair of wings on the mesothorax and a reduced pair, the **halteres,** on the metathorax. Halteres are knoblike appendages that function as gyroscopic balance organs. Of course some parasitic flies secondarily have lost all wings. Flies are holometabolous, with obtect, coarctate, or puparious pupae. Dragonflies and mayflies and other insects whose common names are written as one word are not actually flies. The names of true flies such as the house fly and horse fly are written as two words.

Because the order is large and its members vary considerably in feeding habits, both as larvae and adults, the structure of mouthparts is diverse. We can divide them into five general types, differing in the structures and functions of the palps and labium, mandibles, maxillae, and hypopharynx: mosquito, horse fly, house fly, stable fly, and louse fly.[23] The mouthparts of some larvae are of medical interest.

Historically the order Diptera was divided into three suborders: Nematocera, Brachycera, and Cyclorrhapha. However, major recent works combine the last two and make the Cyclorrhapha an infraorder, Muscomorpha, a practice we follow here.[7] Taxonomic characters most used in the Diptera are the mouthparts, head sutures, ocelli, antennae, wing venations, tarsi, and placement of bristles (**chaetotaxy**). Male genitalia offer useful taxonomic characters at the genus and species levels.

SUBORDER NEMATOCERA

The antennae of species in this group have many segments and are filamentous. They may be plumose, especially in the males, but basically they are simple and longer than the head. The wings have many veins, which is a primitive character. The larvae are active, with a well-developed head capsule, and the pupae often are free swimming. Most larval and pupal stages are aquatic, although some develop in bogs or wet soil.

Family Psychodidae

Two subfamilies are of medical importance: The Psychodinae contain the **moth flies,** which are of little serious importance; the Phlebotominae consist of the **sand flies,** of great importance in many parts of the world.

• Subfamily Psychodinae

The body and the ovoid wings are densely covered with hairs. This character, plus the rooflike position of the wings when at rest, suggests the appearance of a tiny moth; hence, the common name *moth fly.*

Psychodids breed in substrates rich in organic decomposition. *Psychoda alternata,* the **trickling filter fly,** is often found in incredible numbers in sewage disposal plants; it and other species are common in cesspools, washbasins, and drains where the larvae develop in the gelatinous linings of pipes. Emerging adults may be so numerous as to constitute a genuine annoyance to householders. *Psychoda alternata* is a cosmopolitan cause of such situations; in addition to its being a pest, its larvae have been reported in pseudomyiasis, and it can be involved in mechanical transmission of nematodes of livestock.[59]

• Subfamily Phlebotominae

In contrast to the moth flies, sand flies are not so hairy and hold their wings at rest 60 degrees from the body, not rooflike. More important, the mouthparts are of the horse fly subtype, with cutting mandibles. Females feed on plant fluids and on blood by telmophagy, whereas males feed mainly on plant juices and never on blood. The fascicle structures of mouthparts in this group vary somewhat depending on the type of usual host, related to its skin structure.[38] Many species feed on reptiles or amphibians, whereas others feed on birds and mammals, including humans.

Adult sand flies (see Fig. 5.16) usually feed at night or at twilight and early morning, although some are day feeders. They are weak flyers, capable of navigating only short distances, and are inactive when any wind blows. Because of their soft, delicate exoskeleton, their survival depends on avoiding hot, dry places. However there are desert species that thrive by hiding in dry mud cracks, burrows, and so on, emerging only during the few humid night hours. These hours often coincide with the timing of some human activities, such as water fetching.

There are two Old World genera of sand flies, *Phlebotomus* and *Sergentomyia,* and American form, *Lutzomyia. Lutzomyia* spp. occur as far north as Canada, but medically important sand flies are primarily south of Texas. *Lutzomyia diabolica* and *L. shannoni* are the only known anthropophilic sand flies in the United States. Both have been shown experimentally to transmit *Leishmania mexicana.* The other North American species feed mainly on reptiles and rodents.

To breed, phlebotomines require a combination of darkness, high humidity, organic debris on which the larvae feed, and possibly a sleeping host. These requirements are met in animal burrows, crevices, and hollow trees and under logs and among dead leaves. Lek behavior has been observed in *Lutzomyia longipalpis,* with males setting up territories on anesthetized mice, maintaining these territories with wing fanning and aggressive behavior, and competing for females.[27] Females are chemically attracted to dead plant material and feces of both small mammals and larval sand flies.[14] They lay several eggs at a time. The tiny, white larvae feed on such matter as animal feces, decaying vegetation, and fungi. They have simple, chewing mandibles. The four larval instars require 2 to 10 weeks before pupation. The pupa develops in about 10 days.

Sand flies are good vectors of disease, which is surprising considering their weakness and fragility. They are the vectors of the leishmaniases, bartonellosis, and some viral diseases.

The visceral and cutaneous leishmaniases of humans are discussed in Chapter 5. The many species of *Leishmania* in other vertebrates are beyond the scope of this book, but all are transmitted by phlebotomines.

The bacterium *Bartonella bacilliformis* causes a disease known as **Carrión's disease,** with two clinical forms, **Oroya fever** and **verruga peruana.** It is found in Ecuador, southern Colombia, and the Andean region of Peru, being transmitted by *Lutzomyia verrucarum* and probably *L. colombiana.* Oroya fever is a sometimes fatal, visceral form of the disease, accompanied by bone, joint, and muscle pains; anemia; and jaundice. Verruga peruana is a mild, nonfatal cutaneous form. The disease is named after Daniel Carrión, who inoculated himself with organisms obtained from a verruga patient and subsequently developed Oroya fever. Before he died of it, he recognized that the two entities were actually expressions of the same disease.

Sand fly fever is transmitted by *Phlebotomus papatasi, P. sergenti,* and other flies in much of the Old World. Also known as **papatasi fever** and **three-day fever,** it occurs in the Mediterranean region, eastward to central Asia, southern China, and India. Sand fly fever is a nonfatal, febrile, viral disease of short duration but with a long convalescence period. Sand flies acquire the virus when they feed, but because males also can be infected, it seems probable that transovarial transmission occurs. The fly, then, also is a reservoir. Sporadic epidemics occur, such as in Yugoslavia in 1948 when three-fourths of the population (1.2 million persons) acquired it.[23]

Family Culicidae

Mosquitoes are the most important insect vectors of human disease and the most common blood-sucking arthropods. They feed on amphibians, reptiles, birds, mammals, intrepid explorers, and homemakers. Some species exhibit considerable host specificity, while others have more catholic tastes. Mosquitoes have greatly affected the course of human events and continue to do so even today when we have an arsenal of insecticides at our disposal and a vast amount of knowledge about these insects and the diseases they carry. More than a million people die every year from malaria, and other mosquito-borne diseases cause incalculable misery, poverty, and debilitation. The annoyance of hordes of ravenous mosquitoes is in itself enough to affect real estate values, tourist industries, and outdoor activities. There is considerable concern as to whether mosquitoes can transmit the AIDS virus. At this writing transmission has not been demonstrated, but the potential should not be ignored. And humans are not alone in their mosquito-borne misery; significant agricultural losses occur as a result of attacks on domestic animals.[55]

The world's mosquito fauna is rich and diverse, and populations are often enormous. Approximately 3500 species have been described, at least 150 of these in North America. Extensive treatises have been published on all stages of the mosquito life cycle.[55]

Morphologically mosquitoes are fairly simple insects (Fig. 38.1). They are readily differentiated from the superficially similar Dixidae, Chaoboridae, and Chironomidae by the combination of slender wings with scales on the veins and margins and elongated mouthparts that form a proboscis. The fascicle consists of six stylets in the mosquito subtype. The two mandibles, two maxillae, hypopharynx, and labrum-epipharynx are loosely ensheathed in the elongated labium, which has a lobelike tip, the **labella.** The mosquito inserts the fascicle into a blood vessel and pumps blood into the food channel formed by the labrum-epipharynx and hypopharynx. Males lack mandibular stylets and do not feed on blood. Mosquito antennae are long and filamentous with 14 or 15 segments. Whorls of hairs on the antennae are quite plumose in males of most species. Male terminalia are taxonomic characters useful to experts in differentiating species.

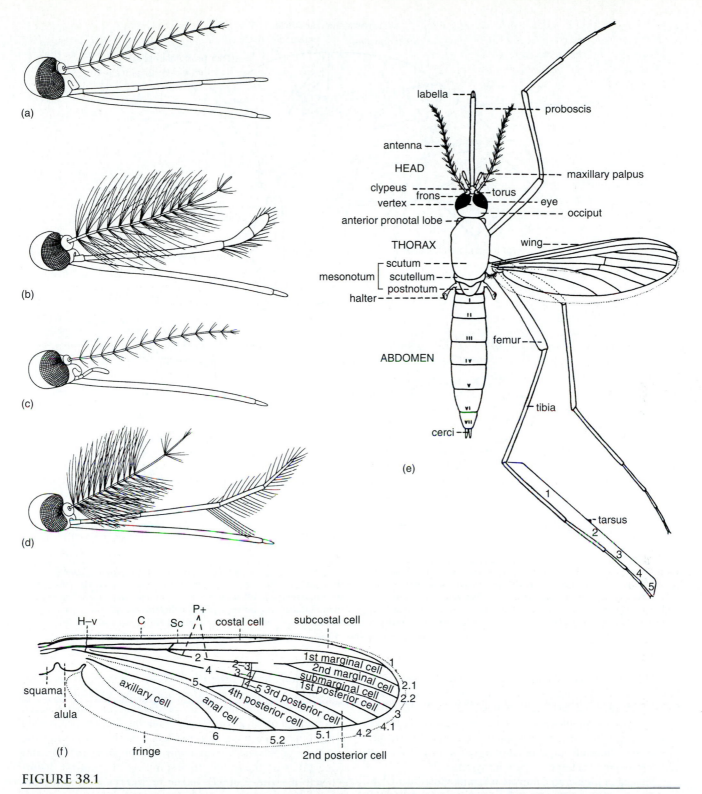

FIGURE 38.1

Mosquito anatomy. (*a–d*) Heads and appendages: (*a*) female *Anopheles;* (*b*) male *Anopheles;* (*c*) female culicine; (*d*) male culicine; (*e*) female culicine; (*f*) wing with veins and cells labeled: **C,** costa; **Sc,** subcosta; **numbers** are those of the longitudinal veins; **Pt,** petiole of vein 2.

From Stanley Carpenter and Walter LaCasse, *Mosquitos of North America (North of Mexico).* Copyright © 1974 University of California Press, Berkeley, CA. Reprinted by permission.

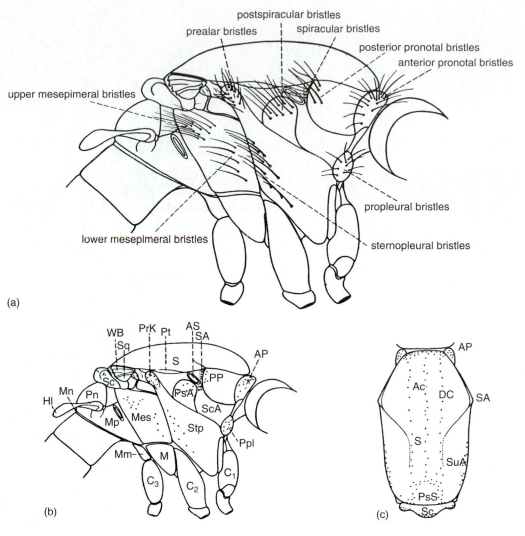

FIGURE 38.2

Thoracic anatomy of a mosquito. (*a*) Chaetotaxy, with bristles named. Spiracular and postspiracular bristles are important identification characters at the generic level. (*b*) Thoracic sclerites: **AP,** anterior pronotum; **AS,** anterior spiracle; **C₁, C₂, C₃,** first, second, and third coxae respectively; **Hl,** haltere; **M,** meron; **Mes,** mesepimeron; **Mm,** metameron; **Mn,** metanotum; **Mp,** metapleuron; **Pn,** postnotum; **PP,** posterior pronotum; **Ppl,** propleuron; **Prk,** prealar knob; **PsA,** postspiracular area; **Stp,** sternopleuron; **WB,** wing base; (*c*) dorsal view of thorax with bristle locations indicated: **Ac,** acrostichal bristles; **AP,** anterior pronotal lobe; **DC,** dorsocentral bristles; **PsS,** prescutellar space; **S,** scutum; **SA,** scutal angle; **Sc,** scutellum; **SuA,** supra-alar bristles.

From Stanley Carpenter and Walter LaCasse, *Mosquitos of North America (North of Mexico)*. Copyright © 1974 University of California Press, Berkeley, CA. Reprinted by permission.

Most species, particularly females, are fairly easily identified. Anyone studying mosquitoes will soon become familiar with thoracic sclerites and bristles (Fig. 38.2). Members of some species complexes can be differentiated only by advanced techniques such as crossmating, electron microscopy of sensilla, study of polytene chromosomes, or comparison of electrophoretic patterns of enzyme systems.

Mosquitoes undergo complete metamorphosis, with egg, larval, pupal, and adult stages (Fig. 38.3). Larval and pupal stages can develop only in water. Adults deposit eggs singly on water or soil or in rafts of eggs on water. They either hatch quickly or, in the case of those on soil, after a period of drought followed by flooding. Most floodwater mosquitoes hatch after the first flooding, but some remain for subsequent floodings, in some cases for up to four years.[66]

Most mosquito larvae hang suspended by surface tension at the surface of the water by a prominent breathing siphon, or **air tube** (Fig. 38.4). The majority are filter feeders or browse on microorganisms inhabiting solid substrates. Some larvae are predaceous on other insects, including other mosquitoes. Four larval instars precede the pupa. The pupa is remarkably active, breaking the surface of the water with a pair of trumpet-shaped breathing tubes on the thorax to respire. At the faintest disturbance it swims quickly to the bottom in a tumbling action. The pupation period is short, usually two to three days. When fully developed the skin on the thorax splits, and the adult quickly emerges to fly away. Adult females live for four to five months, especially if they undergo a period of hibernation. During hot summer months of greatest activity, females live only about two weeks. Males live about a week, but under optimal conditions of food and humidity, their life span may extend to more than a month.

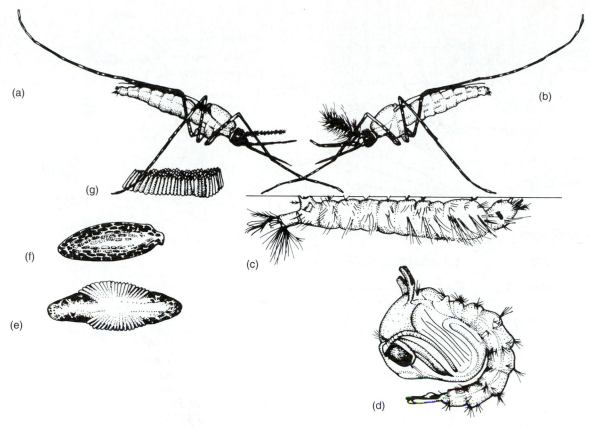

FIGURE 38.3

Life history stages of *Anopheles*. (*a*) Adult female; (*b*) adult male; (*c*) larva; (*d*) pupa; (*e*) egg; (*f*) egg of *Aedes* sp.; (*g*) egg raft of *Culex* sp.; (*f*) and (*g*) are included for comparison.

Drawings by Ian Grant.

• Subfamily Culicinae

Adult members of the Culicinae have a scutellum with a trilobed posterior margin, in dorsal view. The abdomen is densely covered with scales. They lay eggs in rafts or singly on soil. Larvae have a prominent air tube and hang by it nearly perpendicular to the surface of the water. About 3000 species, in more than 30 genera, are placed in this subfamily. Most species are in the genera *Culex* and *Aedes*.

Genus *Culex*.   *Culex* females have rounded tips on their abdomens, and their palps are less than half as long as the proboscis. They have no thoracic spiracular or postspiracular bristles. The larva has a long, slender, air tube bearing many hair tufts. Most *Culex* spp. are bird feeders but do not have a narrow host specificity. They overwinter as inseminated females. Several species are important vectors of bird malaria parasites and arboviruses.

Culex tarsalis, a robust, handsome mosquito, is widespread and common in the semiarid western United States and in the southern states as far northwest as Indiana. Its coloration is distinctive: nearly black with a white band on the *lower half* of each leg joint and a prominent white band in the middle of the proboscis. *Culex tarsalis* breeds in water in almost any sunny location. It is a bird feeder, most active at night, but is not reluctant to feed on humans and other mammals; hence, it is the main vector of **western**

equine encephalitis (WEE) and also transmits **St. Louis encephalitis** virus. WEE is normally a bird disease, with no apparent symptoms, but it can be acquired by other hosts. Horses are particularly susceptible, with a high rate of mortality. Humans also can be infected; it is not as commonly fatal in humans as in horses but can be severe in children. In adults it results in fever and drowsiness; hence, it is sometimes called **sleeping sickness.** Rarely, following a coma, a person may have reduced physical capabilities.

Culex pipiens, the **house mosquito** (Fig. 38.5), is nearly worldwide in distribution. It is a plain, brown insect that breeds freely around human habitation, laying egg rafts in tin cans, tires, cisterns, clogged rain gutters, and any other receptacle of water. It enters houses readily and is a night feeder, causing consternation in many a bedroom. *Culex pipiens* actually is a complex of species with slight physiological differences, only some of which are understood. The members of this complex are important not only for their annoyance factor but because they are major vectors of the filarial worms *Wuchereria bancrofti* and *Dirofilaria immitis* (Chapter 29). They also transmit bird malaria, avian pox, and arbovirus encephalitides.

Genus *Aedes*.   The posterior end of the *Aedes* female abdomen is rather pointed; postspiracular bristles are present on the thorax; the female claws are toothed; and pulvilli are

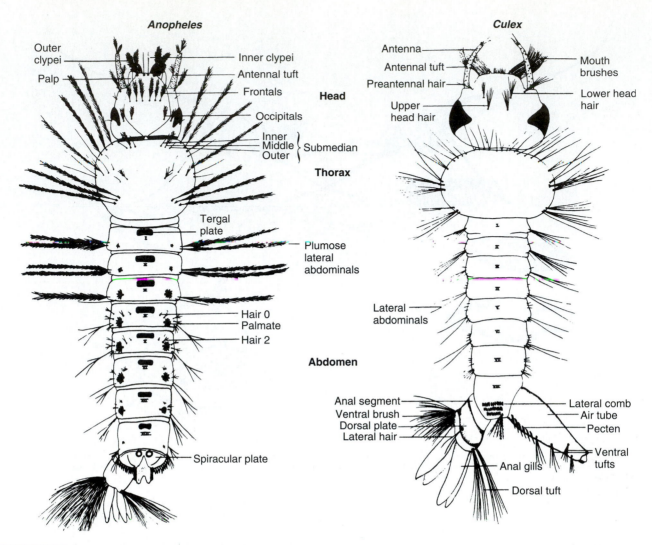

FIGURE 38.4

Mosquito larvae, showing basic taxonomic characters.

Courtesy of Communicable Disease Center, 1953, U.S. Public Health Service, Washington, D.C.

absent or hairlike. Larvae have siphons bearing only one pair of posteroventral hair tufts. Because nearly half of North American mosquitoes are *Aedes* spp. and many of the rest are *Culex,* the pointed abdomen of the female usually is all one needs for field identification of this genus.

Species of *Aedes* are notable for their ferocity. Most are diurnal or crepuscular in their activities, as contrasted with the night-biting *Culex* spp. They lay their eggs singly on water, mud, or soil in places likely to be flooded. Mosquitoes of this genus are not only among the most obnoxious of bloodsucking insects but also are extremely important medically because of the diseases they transmit.

Two species, *Aedes dorsalis* and *A. vexans* (Fig. 38.5), are scourgemates in the western United States. Both are fierce daytime biters; at a single swat a person may kill half a dozen of each species. *Aedes dorsalis* has a wide range, including most of the Holarctic region, North

Africa, and Taiwan. It breeds in salt marshes as well as fresh water. *Aedes vexans,* aptly named, overlaps the range of *A. dorsalis* and includes South Africa and the Pacific Islands. *Aedes dorsalis* is easily recognized as a straw-colored, medium-sized mosquito of great beauty but the utmost persistence. *Aedes vexans* is brown to black with white bands encompassing *both halves* of each leg joint. *Aedes sollicitans* (Fig. 38.5) is a floodwater mosquito found throughout most of the eastern two-thirds of the United States and southern Canada, where it makes life miserable for biologists and others who frequent the marshes. Last but certainly not least, *Aedes taeniorhynchus,* the black salt marsh mosquito, gives up nothing in the biter reputation category. In the words of one experienced parasitologist, it "considers deet a trivial impediment not to be taken seriously." Deet is the major insect repellant used by the United States military.

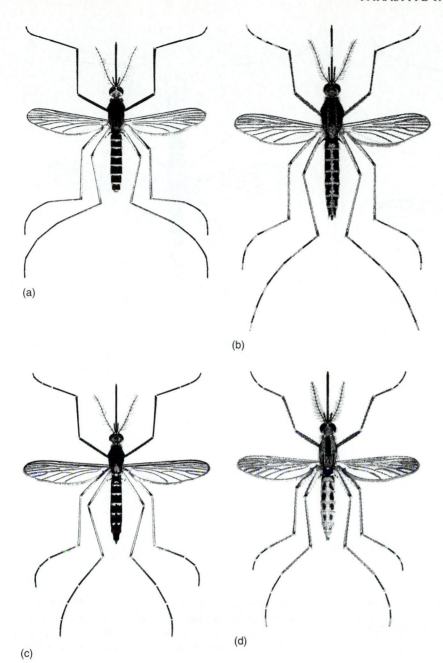

(a)

(b)

(c)

(d)

FIGURE 38.5

Some important mosquitoes: (*a*) *Culex pipiens;* (*b*) *Aedes sollicitans;* (*c*) *Aedes vexans;* (*d*) *Aedes dorsalis.*

From Stanley Carpenter and Walter LaCasse, *Mosquitos of North America* (*North of Mexico*). Copyright © 1974 University of California Press, Berkeley, CA. Reprinted by permission.

Among the many other species of *Aedes* are the snow-water mosquitoes of the far north and western mountains. A difficult complex of species, they are all characterized by their immense numbers and ferocious appetites. Usually there is only one generation per year: The females lay eggs singly in low-lying areas destined to become flooded by melting snow water the following year. Although snow-water mosquitoes transmit no known diseases to humans and domestic animals, their presence in large numbers precludes carefree sport by those who venture into their domain.

Several species of *Aedes* are tree-hole breeders. Species that have adapted to breeding in small containers or leaf axils appear to have derived from tree-hole breeders, such as *A. aegypti* (Fig. 38.6). *Aedes triseriatus* is a widespread tree-hole breeder east of the Rocky Mountains. It is similar in appearance to *A. vexans* but lacks white rings on the tarsi. It is an important vector of California (La Crosse) encephalitis virus. The most common tree-hole breeder in the United States is *A. hendersoni*, which is very similar to *A. triseriatus*.

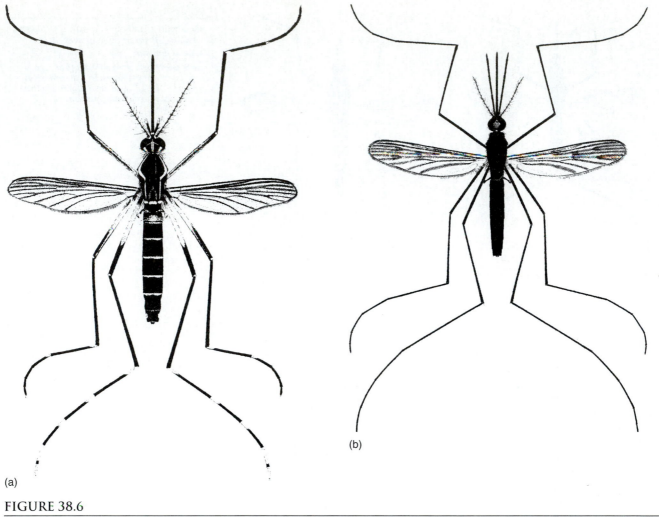

FIGURE 38.6

More important mosquitoes: (*a*) *Aedes aegypti;* (*b*) *Anopheles quadrimaculatus.*

From Stanley Carpenter and Walter LaCasse, *Mosquitos of North America (North of Mexico)*. Copyright © 1974 University of California Press, Berkeley, CA. Reprinted by permission.

Aedes albopictus, the Asian tiger mosquito, was discovered in Houston, Texas in 1985. Apparently it arrived in this country in a shipload of used tires. It now is found in most areas east of the Rocky Mountains. Experimentally it is a good vector for dengue, equine encephalitis, yellow fever, and La Crosse virus.[24]

Many species of *Aedes* are vectors of a variety of virus diseases. The topic is too extensive to explore adequately here, but one species, *A. aegypti,* the **yellow fever mosquito,** is of extreme importance and wide distribution. It is found within a belt from 40° N to 40° S latitude, except in hot, dry locations. It is common in much of the southern United States. It is a beautiful mosquito, jet black or brown with silvery white or golden stripes on the abdomen and legs; the last tarsal segment is white. A lyre-shaped pattern covers the dorsal surface of the thorax. *Aedes aegypti* is a tree-breeding species in sylvatic situations, but when associated with human habitation, it breeds freely in containers, cisterns, and other water storage units. As many as 140 eggs are laid singly at or near the waterline, and they can withstand desiccation for up to a year.

Aedes aegypti originated in Africa, whence it was widely distributed by the slave trade. It was transported to much of the world via water barrels in ships. With the mosquito went a *Flavivirus,* which causes yellow fever, a devastating disease that has wrought havoc wherever it has emerged. After establishing in the New World, *A. aegypti* caused many epidemics. For example, the British army lost 20,000 of 27,000 men who attempted to conquer Mexico in 1741; the French lost 29,000 of 33,000 men trying to acquire Haiti and the Mississippi Valley. It was largely because of the presence of yellow fever in Louisiana and parts north that France was willing to negotiate the Louisiana Purchase. Many outbreaks hit coastal cities in the United States, such as Charleston, New Orleans, and Philadelphia, and gold-rush settlements in California. Yellow fever and malaria forced France to abandon the completion of the Panama Canal and prevented the job from being attempted again until William Gorgas developed a program of mosquito control in Havana and then applied it to Panama. Strangely yellow fever has never established in Asia.

Urban yellow fever is transmitted only by *A. aegypti,* but a sylvatic form existing in monkeys, both in Africa and South America, is transmitted by other *Aedes* and *Haemogogus* mosquitoes.

Another *Flavivirus* disease transmitted by *A. aegypti* is **dengue,** also called **breakbone fever** and **epidemic hemorrhagic fever.** The four distinct serotypes of dengue virus cannot be differentiated by symptoms. In uncomplicated cases the patient has fever, severe headaches, and pains in the muscles and joints; weakness and temporary prostration are common. Recovery is rapid. A hemorrhagic complication occurs occasionally, especially in indigenous Asian children of three to six years old. This condition ranges from a rash and mottled skin to severe hemorrhaging in the lungs, digestive tract, and skin. The mortality rate is up to 7% in those who are hospitalized; unhospitalized cases must have a higher rate. Other *Aedes* spp. can transmit the virus, but *A. aegypti* is the principal vector, being implicated in all serious epidemics. Dengue occurs from eastern Europe through most of Asia; North, Central, and South America; and the Caribbean.

Other Culicine Genera. Anyone who studies mosquitoes seriously discovers a rich fauna that presents many challenges for a biologist. *Toxorhynchites* spp., for example, are predatory treehole breeders and candidates for biological control agents focused on species of medical importance. In *Mansonia* and *Coquillettidia* spp., larval air tubes are sharply pointed, enabling them to pierce stems of aquatic plants to obtain air. This has significance in control: Simply coating the water with oil will not prevent these insects from obtaining oxygen. Species in this complex are important vectors of **brugian filariasis** (Chapter 29).

The genus *Culiseta* contains eight North American species and subspecies. They are large, brownish mosquitoes, some restricted to feeding on birds and other mammals. However, *C. inornata* and *C. melanura* are involved in the transmission of **western** and **eastern encephalitis viruses.** *Culiseta inornata* is the most widespread of the two, being found in southern Canada and the conterminous United States, whereas *C. melanura* is restricted to the eastern and central United States.

Other genera of Culicinae are marginally important in the transmission of arboviruses.

• Subfamily Anophelinae

Adult anophelines have a scutellum that is rounded or straight but never trilobed (except slightly in *Chagasia* spp.) in dorsal view. The abdominal sternites largely lack scales. The palpi of both sexes are almost as long as the proboscis (except in *Bironella* spp.). The larva lacks an air tube, and its dorsal surface bears branching hairs (Fig. 38.4).

The subfamily contains three genera: *Bironella,* with seven species in New Guinea and Melanesia; *Chagasia,* with four species in tropical America; and *Anopheles,* with about 400 species, including 15 in North America. Resting and feeding postures are distinctive for *Anopheles* spp. When the mosquito is at rest, the head, proboscis, and abdomen are almost in a straight line; while feeding, the body is inclined at a sharp angle from the surface of the host. Because the genera *Bironella* and *Chagasia* are of no medical importance, we will consider only *Anopheles* spp. in this chapter.

Genus *Anopheles.* Female *Anopheles* (Figs. 38.3 and 38.6) lay up to a thousand eggs, depositing them singly on the water. The eggs have useful taxonomic characters, such as the presence or absence of lateral floats, which are characteristically marked, and a lateral frill. The eggs must remain in contact with water to survive. Usually they hatch within two to six days and develop through four larval instars in about two weeks, followed by a three-day pupal stage. Development from egg to adult takes from three weeks to one month.

Preferred breeding sites vary tremendously among the species of *Anopheles,* a factor that must be understood before effective control of malaria vectors can be undertaken. Thus, some species breed most efficiently in stagnant mangrove swamps; others, in sunny, partly shaded pools; and still others, along the edges of trickling streams. A few are tree-hole breeders.

Taxonomy of the genus *Anopheles* is complicated by the existence of several species complexes. For example, the species initially known as *A. maculipennis* we now know to consist of at least seven subspecies (or perhaps sibling species), differing slightly in host preferences and egg characteristics. American representatives of this complex are *A. quadrimaculatus* (Fig. 38.6), *A. freeborni, A. aztecus, A. earlei,* and *A. occidentalis.* What was formerly known as *A. gambiae* in Africa comprises six species, including some freshwater and some saltwater breeders. There are reproductive isolation in and genetic barriers between all six species.[41]

Of all diseases transmitted to humans by insects, that caused by *Plasmodium falciparum* takes more lives and causes more suffering than the others put together. This and the other malaria parasites of humans (*P. ovale, P. malariae,* and *P. vivax*) are all transmitted by species of *Anopheles.* We do not need to repeat our discussion of malaria in Chapter 9 except to say that historically the primary vectors of the disease in North America were *A. quadrimaculatus* and *A. freeborni.* Both are still common on the continent, as is *Aedes aegypti,* but like yellow fever, endemic malaria has been eradicated. The conquest of these two diseases in the United States stands among the greatest triumphs of this country.

That victory, however, is due more to elevated standards of living than it is to elimination of mosquitoes. From experiences in many parts of the world, it is now obvious that insecticide resistance is one of the more predictable results of insecticide use, a realization that has inspired a broad research effort to find alternate ways of controlling not only mosquitoes, but also black flies, tsetse flies, and bot flies. Some of these approaches being tested are ones that 50 years ago would have been considered "far out" or frivolous; others that have been tried for almost a century now are getting a modern reexamination.

Predatory fish are being used in attempts to control mosquito larvae in many parts of the world, in settings ranging from rice fields to cisterns.[57] The infamous *Gambusia affinis* has been spread far and wide as a result and such

FIGURE 38.7

Simulium sp. (*a*) Larva; (*b*) head. Note the short, cutting-type mandibles.

(*a*) Courtesy of Warren Buss; (*b*) Courtesy of Jay Georgi.

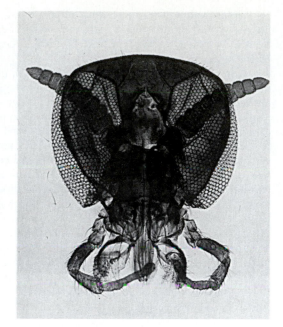

(b)

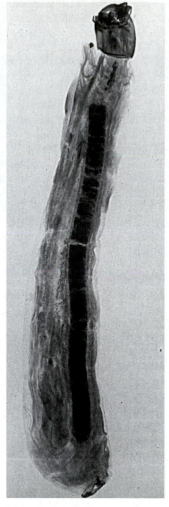

(a)

human-aided dispersal has the potential for reducing the numbers of competing native fishes.[40] Australian and South Pacific copepods of the genus *Mesocyclops* are also being tested as larval predators; these crustaceans have the advantage of occurring naturally in habitats ranging from lakes and streams to wells and tree holes.[3] Species of *Toxorhynchites* have been tested as controlling predators on their fellow tree-hole breeder mosquitoes.[60] Biological control with nematodes was discussed by Platzer.[47] Other methods and devices that have been tried in recent years include planarians,[39] specially designed lids for water jars,[31] sustained release insecticide pellets,[34] runnelling (digging of shallow channels to help flushing of tidal marshes),[10] use of copper linings in cemetery flower vases,[45] root and bark extracts of mangroves,[58] strains of the bacterium *Bacillus sphaericus,* and genetically engineered cyanobacteria containing *B. sphaericus* genes.[29,67] Obviously humanity has a long way to go before it finds the perfect solution to problems of malaria, filariasis, and encephalitis transmission.

Family Simuliidae

Simuliids (Fig. 29.7) are commonly called **black flies,** although many species are gray or tan. They are small, 1 to 5 mm long. The prescutum of their mesonotum is reduced, giving them a humpbacked appearance, which explains their other common name, **buffalo gnat.** The wings are broad and iridescent, with strongly developed anterior veins. The antennae are filiform, usually with 11 segments. The eyes of the female are separated, whereas those of the male are contiguous above the antennae. There are no ocelli. Mouthparts are of the horse fly subtype, although delicate (Fig. 38.7*b*). Serrated teeth on the edges of the mandibles are cutting structures, whereas recurved teeth on the maxillary lacinia serve to anchor the mouthparts during feeding.[12]

Black flies are found worldwide but are most abundant in northern temperate and subarctic zones. Females of most species feed on blood as well as nectar, but males feed only on plant juices. Mating occurs in flight, when females fly into swarms of hovering males. A female simuliid produces

200 to 800 eggs, laying them on the surface of the water, where they rapidly sink. In some species the female lands at the water's edge, crawls down a rock or plant to deposit the eggs underwater, and then crawls back out of the water and flies away.

Larval development can occur only in running, well-oxygenated water. Hence, black flies are most numerous near rivers and streams, although they are known to travel several miles when aided by winds. On hatching, the larva (Fig. 38.7a), with its modified salivary glands, spins a silken mat on some underwater object. It attaches itself to this mat with a hooked sucker at the posterior end of its abdomen. Thus, its head hangs downstream and with the fanlike projections around the mouth, filters protozoa, algae, and other small organisms and organic detritus from the passing water. Often the larval numbers are so great that they form a solid covering on a favorable location, such as a cement spillway or the downstream side of a rock or log. A larva is capable of changing locations rapidly by stretching out, spinning a new mat and clinging to it with its mandibles, and then releasing the old mat and hooking onto the new one. The six or seven larval instars require 7 to 12 days under ideal temperature conditions and food availability, but this time may be greatly extended. Some species overwinter as larvae.

Before pupation, the larva spins a flimsy cocoon around itself. After molting, the pupa remains nearly immobile, respiring through long filamentous gills on its anterior end. The number and arrangement of these filaments are of taxonomic importance. The pupal stage lasts from a few days to three or four weeks. To emerge, the imago first cuts a T-shaped slit in the pupal thorax and crawls through it. It quickly fills its air sacs with air extracted from the water, forming an internal balloon; releases its hold on the substrate; and shoots to the surface. One to six generations may mature per year, depending on the locality.

Classification and identification of simuliids are often difficult because of numerous complexes of sibling species. The most important genus is *Simulium*, with more than 1200 species. Also important medically are *Prosimulium* and *Cnephia* in North America and *Austrosimulium* in Australia and New Zealand. Black flies are fairly host specific. Few species will bite humans, but those that do are extremely vexatious.

In North America, *Prosimulium mixtum* can be very annoying, and *Cnephia pecuarum,* the southern buffalo gnat, has been known to ravage and kill entire herds of livestock. *Simulium vittatum* is widespread in the United States and is particularly irritating to livestock. *Simulium meridionale,* the turkey gnat, torments poultry, biting them on the combs and wattles.

All fishermen and campers in the northern United States and Canada are familiar with the attacks of *S. venustum,* which often occur in such numbers as to ruin a vacation. Vast numbers of *S. arcticum* killed more than a thousand cattle in western Canada annually from 1944 to 1948. *Simulium colombaschense,* of central and southern Europe, killed 16,000 cattle, horses, and mules in 1923 and 13,900 in

FIGURE 38.8

Fossil simuliid pupa, *Simulimima grandis,* from the middle Jurassic. This pupa is virtually indistinguishable from that of a modern member of the genus *Prosimulium.*

From R. W. Crosskey, "The fossil pupa *Simulimima* and the evidence it provides for the Jurassic origin of the Simuliidae (Diptera)," in *Syst. Entomol.* 16:401–406. Copyright © 1992.

1934.[23] But simuliids probably have been biting vertebrates for well over 100 million years; a Jurassic fossil pupa indistinguishable from that of a modern *Prosimulium* (Fig. 38.8) has been found.[6]

Individuals react differently to the bites of black flies. Few people have little or no reaction; most develop local reactions in the form of reddened, itching wheals. **Black fly fever,** a combination of nausea, headache, fever, and swollen limbs, occurs in particularly sensitive persons. Deet (N,N-diethyl-3-methylbenzamide) and an extended-duration repellant formulation (EDRF) of deet are the repellants of choice against a variety of insects and arachnids, including black flies.[48]

Black flies are the vectors of *Onchocerca volvulus,* the cause of human onchocerciasis, discussed in detail in Chapter 29. The most common vector in Africa is *S. damnosum,* although *S. neavei* also is important. In the New World *S. ochraceum, S. metallicum, S. callidum,* and *S. exiguum* are the most efficient vectors because of their preference for humans and the timing of their activity, coinciding with that of humans. *Onchocerca gutterosa* commonly is transmitted to cattle by *S. ornatum* in Europe. In Australia *Simulium* and *Culicoides* spp. infect cattle with *Onchocerca gibsoni,* causing considerable loss to flesh and hides. Mathematical models of onchocerciasis transmission suggest some fascinating interactions between humans and simuliid flies. Davies,[9] using data from a single village, predicted that a 99% effective vector control program would have to continue for 18 years to eradicate the worms and that immigrant infected flies posed a greater threat for reinfection than did infected resident humans.

The malarialike bird disease caused by *Leucocytozoon* spp. is transmitted by various species of *Simulium* (Chapter 9). Virtually any species of bird is likely to be infected with a species of *Leucocytozoon; L. simondi,* a severe pathogen of ducks and other anatids, is transmitted by *S. rugglesi, S. anatinum,* and other species.

Family Ceratopogonidae

This family comprises the **biting midges,** also called **punkies, no-see-ums,** and "sand flies." They are very small, usually less than 1 mm long, but what they lack in size they make up for in ferocity. The majority are daytime feeders that cannot cope with blowing winds so they are most pesky on hot, still days. Their small size enables them to crawl through ordinary window screening; some species, particularly in the tropics, enter houses freely.

Most of the 60 or more genera feed on insects, a few feed on poikilothermic vertebrates, and only four genera feed on mammals: *Culicoides* (Fig. 38.9), *Forcipomyia, Austroconops,* and *Leptoconops.* All four have species that happily feed on humans. Only females feed on blood. Biting midges are recognized by, in addition to their small size, their narrow wings, which have few veins, often are distinctly spotted, and are folded over the abdomen when at rest.

These flies breed in a wide variety of situations, a factor no doubt contributing to their worldwide range. Larvae are aquatic or subaquatic or develop in moist soil, tree holes, decaying vegetation, and cattle dung. *Leptoconops* larvae have been found as deep as 3 feet in the soil. Some species breed readily in salt or brackish water, notably mangrove swamps and salt marshes. The life cycle is completed in between six months and three years.

Of the 1000 or so species of *Culicoides,* the several that bite humans may be so annoying as to affect tourism in infested areas, such as beaches, mountain lodges, and resorts. Farm workers often are intensely annoyed, and domestic livestock are plagued by these tiny flies. An estimated 10,000 *Culicoides* have been witnessed on a single cow.[43]

Culicoides furens is common and widespread from Massachusetts to Brazil, throughout the West Indies, and on the Pacific coast from Mexico to Ecuador. Species of *Leptoconops* abound in south temperate and tropical areas of the world. *Leptoconops torrens* and *L. carteri* are common pests in the western United States. The taxonomy of North American species of ceratopogonids has been treated by Wirth and Atchley.[65]

Three apparently nonpathogenic filariid nematodes are transmitted to humans by ceratopogonids: *Mansonella per-*

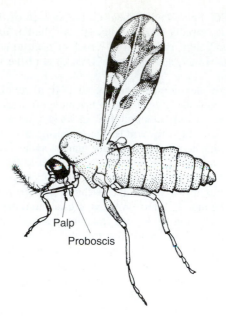

FIGURE 38.9

Female *Culicoides* sp.
Drawing by Ian Grant.

stans and *M. streptocerca* in Africa and *Mansonella ozzardi* in South and Central America. We discussed these briefly in Chapter 29. *Onchocerca cervicalis* of horses and *O. gibsoni* of cattle are transmitted by *Culicoides* spp., as are other filariids of domestic and wild animals.

Blood-dwelling protozoan parasites also use *Culicoides* spp. as vectors. *Hepatocystis* in monkeys and other arboreal mammals, some species of *Haemoproteus* in birds, and various species of *Leucocytozoon* are transmitted by ceratopogonids (see Chapter 9).

These midges can transmit viruses through their bites. *Orbivirus,* the etiological agent of **bluetongue,** is spread by *C. variipenis* in North America and by other species of *Culicoides* in Africa and Asia Minor. Bluetongue is a hemorrhagic disease of sheep, cattle, bison, deer, and other ruminants. It causes some mortality, but possibly of more importance, it is responsible for unthriftiness of infected animals, with loss of flesh, wool, and breeding. Transmission of encephalitis viruses, bovine ephemeral fever, and African horse sickness has also been implicated with the bites of these tiny midges. Like the simuliids, *Culicoides* has been making life miserable for large animals ever since the Mesozoic.[54]

SUBORDER BRACHYCERA

In this group of mostly robust flies the antennae are reduced to three apparent segments, the terminal one being drawn into a sharp point, or **style.** A flagellumlike **arista** may be present. Wing venation is reduced. The larvae are active, usually predaceous; their heads may be incomplete, retractable (can be pulled into the thorax), or vestigial. Life cycles are aquatic or semiaquatic. Only the Tabanidae and Rhagionidae among the Brachycera have bloodsucking habits as adults; larvae of the calliphorid genus *Protocalliphora* feed on blood.

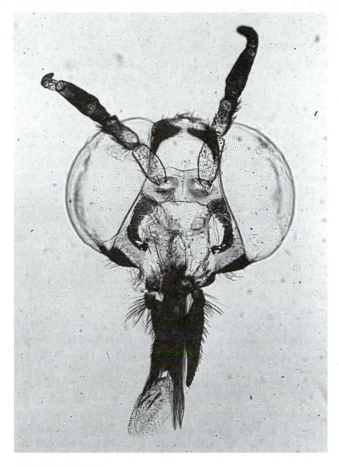

FIGURE 38.10

Head of *Chrysops* sp., showing the tips of the mandibles.
Courtesy of Warren Buss.

Infraorder Tabanomorpha

Tabanomorph flies have a bulbous adult face and a re-tractable larval head.

• Family Tabanidae

Horse flies and deer flies are widely distributed in the world, and their fierce questing for blood has earned them universal animosity. They are large, powerful flies, from 6 to 25 mm long. Tabanids are mainly daytime feeders. Only the female feeds on blood; the male lacks mandibles and eats only plant juices. The eyes are widely separated in females and are contiguous in males. About 4000 species are divided into 30 to 80 genera, depending on the authority.

The mouthparts of tabanids (Fig. 38.10) are of the horse fly type. They are similar to those of the Cerato-pogonidae and Simuliidae but are stouter and stronger. The fascicle consists of six piercing organs: two flattened, bladelike mandibles with toothlike serrations; two more narrow maxillae, also serrated; a median hypopharynx; and a median labrum-epipharynx. In biting, the mandibles cut in a scissorslike motion, whereas the maxillae pierce and rend the tissues, rupturing blood vessels. The fly feeds on the pool of blood that wells into the wound (tel-mophagy). The hypopharynx and labrum-epipharynx form

the food canal. Some fierce-looking species with long mouthparts, such as those in Pangoniinae, actually are not blood feeders.

Tabanids usually breed in aquatic or near aquatic environments, although some complete larval development in soil. Generally the female lays from a hundred to a thousand eggs at water's edge or on overhanging vegetation, rocks, and so on. At egg-laying time, such locations may be swarming with ovipositing flies. On hatching, the larva falls or crawls into the water or burrows into the mud. Many feed on organic debris, but others are voracious predators on insect larvae, worms, and other soft-bodied animals, including other horse fly larvae and even toads.

The larva has a small retractable head provided with powerful, sharp mandibles capable of inflicting a painful wound on the unwary. The body has 12 segments and a tracheal siphon that retracts into the posterior end of the body.

In temperate zones the fly requires about a year to develop to pupation. At this time the larva crawls into drier earth and pupates. The pupa is obtect and requires from five days to two weeks to complete metamorphosis. The adult escapes the pupal case by cutting a T-shaped opening in the dorsal thorax; it then crawls out and makes its way to the surface. In the tropics two or more generations per year may occur.

Several species of horse flies are serious pests of humans and livestock in the United States. *Tabanus quinquevittatus* and *T. nigrovittatus* are the large, gray, "greenhead" horse flies of most of the United States. *Tabanus atratus* (Fig. 38.11) is a huge, uniformly black horse fly of eastern North America; *T. lineola* and *T. similis* are smaller, striped flies also in the eastern states. In the western states *T. punctifer* is a very large, black horse fly with a yellow thorax. Such species as *T. atratus* and *T. punctifer* seldom bite humans; they buzz so loudly when approaching that they seldom are allowed to land. *Haematopota americana*, *Hybomitra* spp., and *Silvius* spp. are also common western pests. *Diachlorus* spp. are common, aggressive pests in Central America, where they are called **doctor flies.** Unlike most tabanids they freely enter houses.

The name **deer fly** is applied to the genus *Chrysops*, of which there are about 80 North American species. Deer flies (Fig. 38.12) usually are smaller than horse flies and have brown-spotted wings. Their flight is not as noisy as that of most horse flies, so they bite humans more commonly.

The medical importance of Tabanidae is two sided: the annoyance and blood loss occasioned by the bite and infections transmitted mechanically and biologically by the flies.

Because of the large size of the mouthparts, tabanid bites are quite painful. Most people have little or no allergic reaction to them, although such sequellae are known to occur. Their annoyance factor may seriously interfere with the use of recreational areas, and field and timber workers may have lowered productivity as a result of harassment by these flies.

A serious problem is blood loss and aggravated behavior in livestock. Beef and milk production in the United States were reduced an estimated $40 million in 1965 as a result of interrupted grazing, energy consumed in trying to escape the insects, and blood loss. Tethered or caged animals particularly suffer from these flies because they are unable to escape their tormentors, even briefly. One well-known parasitologist

FIGURE 38.11

Tabanus atratus.
Courtesy of Warren Buss.

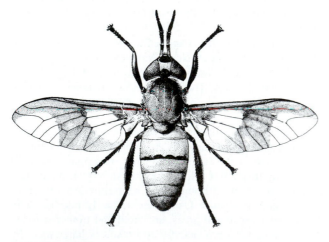

FIGURE 38.12

Fly of the genus *Chrysops.*

From Richard P. Lane and Roger W. Crosskey, *Medically Important Insects and Arachnids.* Copyright © 1992 Chapman & Hall, London. Reprinted with permission.

reported seeing a caged mule deer simultaneously being fed on by a dozen or more *Tabanus punctifer* and dozens of open, freely bleeding wounds covering the wretched animal's face. No tabanids are strictly host specific, but most have host preferences. Thus, *Haematopota,* with at least 300 species, feeds mainly on cattle and antelope, with which it may have evolved. Birds are attacked uncommonly; amphibians and reptiles may provide nourishment for these parasites.

Certain adaptations of tabanids enhance their capabilities to transmit pathogens:[35] (1) **anautogeny,** the necessity of a blood meal for development of eggs, stimulating host-seeking behavior; (2) **telmophagy,** where blood-dwelling pathogens can enter the pools from which the fly sucks blood; (3) **relatively large blood meals,** enhancing the possibility that pathogens will be imbibed; (4) **long engorgement time,** enabling the pathogens to infect the fly's tissues; and (5) **intermittent feeding behavior,** increasing the chances for mechanical transmission of pathogens.

Tabanids are involved in the transmission of protozoan, helminthic, bacterial, and viral diseases of animals and humans. Among the diseases caused by protozoa, two species of *Trypanosoma* are transmitted mechanically by tabanids. *Trypanosoma evansi,* the causative agent of **surra** in many wild and domestic animals, is spread by species of *Tabanus* (Chapter 5). Other vectors, such as stable flies, other genera of horse flies, and vampire bats, also can be involved; but *Tabanus* spp. appear to be the most effective vectors of this trypanosome. *Trypanosoma theileri* is a cosmopolitan parasite of cattle and antelopes. Cyclopropagative development occurs in the insect gut, so the tabanid species involved are actually true intermediate hosts. Examples of such species are *Haematopota pluvialis, Tabanus striatus,* and *T. glaucopis.*[35]

The African eye worm, *Loa loa,* is transmitted by species of *Chrysops* (Chapter 29). There appear to be two strains of this filariid, one in monkeys in the forest canopy and one in humans. Night-feeding *Chrysops langi* and *C. centurionis* transmit the former, and diurnal *C. silaceus* and *C. dimidiatus* transmit the latter.

The arterial filariid *Elaeophora schneideri* lives in vessels of the head and neck of American species of deer, elk, moose, and domestic sheep in the western states. It is symptomless in deer but in other hosts causes much distress, such as blindness, nervous dysfunction, necrosis, and deformity of the head. Species of *Tabanus* and *Hybomitra* are the vectors of this worm.[25]

Bacterial infections known to be mechanically transmitted by tabanids are anaplasmosis and anthrax. Similarly, hog cholera and equine infectious anemia (swamp fever) viruses use horse flies as vectors.

• Family Rhagionidae

Rhagionids, known as **snipe flies,** are predominantly non-bloodsucking, but the several species that feed on blood are considerably annoying. Dominant among the snipe flies that bite humans is the genus *Symphoromyia.* The behavior of its species is like that of *Chrysops* spp., landing suddenly and quickly piercing the skin. The mouthparts and mode of feeding are also like those of the Tabanidae.

The biology of Rhagionidae is poorly known. Larvae develop in boggy or mossy stream banks and similar places. It is not known if snipe flies transmit any diseases, but because *Symphoromyia* species and others take a considerable amount of blood at a meal and feed several times during their lives, they certainly must be regarded as potential vectors. As far as is known, rhagionids are important primarily as pests whose biting proclivities may doom an otherwise pleasant outing.[61]

Infraorder Muscomorpha

Except for the mosquitoes in Nematocera, members of this group are the most important flies in veterinary and human medicine. The antennae are short and pendulous, usually with a conspicuous arista on the second segment. Three prominent ocelli are arranged in a triangle on the vertex and frons. The compound eyes are very large, separated in females and close together in males. The arrangement of bristles on the head and thorax is important in the taxonomy of flies. At the anal angle of the wing is a prominent lobe, the **squama,** or **calypter,** in the medically important families (members of the subsection Calyptratae).

Larval calyptrates are **maggots,** with an elongated, simple body, usually tapered toward the head end. A true head is absent; the vertically biting mandibles are part of a conspicuous cephalopharyngeal skeleton. These sclerotized structures are important in the taxonomy of larvae. Two spiracles are found on the posterior end. Each species has distinctive markings on the spiracular plates that are useful in identification.

When fully developed, the third-stage larva undergoes **pupariation,** resulting in a pupalike surrounding case (**puparium**) made of the hardened third-stage larval tegument. After internal reorganization, the adult emerges through a circular hole in the puparium. It pushes the operculum off the puparium by inflating a balloonlike organ, the **ptilinum,** in the head. The ptilinum extends from the frontal suture of the head, pushes off the operculum, and then withdraws back into the head; the frontal suture immediately heals.

• Family Chloropidae

Looking very much like tiny house flies, the **eye gnats** are of considerable medical importance. Unlike those of the gnats discussed in the Nematocera, the antennae of this family are aristate. The antennae resemble those of Drosophilidae (fruit flies) except that the arista is nearly smooth, whereas it is distinctly feathered in fruit flies.

The most important genus in this family is *Hippolates.* These flies are very small, 1.5 to 2.5 mm long, and are acalypterate. They are called eye gnats because they are attracted to eye secretions, as well as to other body secretions and free blood. They do not bite but feed like house flies, sponging up liquid food. Also like house flies, they vomit liquid stomach contents onto their food, to be reeaten. For this and other reasons, eye gnats are important vectors of disease. In some species the labellum is provided with tiny spines that scarify the skin of a host, also leading to infection by pathogens. *Hippolates* spp. are very persistent when hungry and hence may become intensely irritating. They are strong flyers, able to fly against weak winds.

Hippolates spp. occur only in North America. *Hippolates pusio* and several other related species constitute the most important group.[49] *Siphunculina funicola* plays a similar role in Southeast Asia.

The life cycles all seem to be similar. About 50 eggs are laid on the surface of or slightly under the soil, which must be loose and have an abundance of well-aerated organic matter, either animal or vegetable. Larval development consumes 7 to 12 days under optimal conditions; the pupal period requires about six days, and the adult ages seven days before oviposition.[19]

Aside from the annoyance caused by these flies, which can be considerable, they are important vectors of disease. Although they are not biting insects, they congregate at wounds caused by others, such as tabanids and stable flies, thereby further contaminating the host.

Hippolates spp. probably are a factor in transmission of the bacillum causing human **pinkeye,** or **bacterial conjunctivitis,** although this has not been proved. Similarly, the spirochete *Treponema pertenue,* the etiological agent of **yaws,** most likely is transmitted mechanically by eye gnats, although this also has not been proved. *Hippolates* flies feeding at the tips of the teats spread a bacterial disease, **bovine mastitis,** from cow to cow.

Control of eye gnats is difficult. The best system in use so far is a combination of attractant baits with a pesticide and efficient soil management.

• Family Glossinidae

These are the infamous **tsetse flies.** The family contains a single genus, *Glossina,* and occurs only in Africa and two localities on the Arabian peninsula. It was once more widespread, however; four species have been found in the Oligocene shales of Colorado.

***Glossina* Species.**   Tsetse flies (Fig. 5.4) are 7.5 to 14.0 mm long and brownish gray. When at rest, the wings cross like scissors. The palpi are almost as long as the proboscis, which protrudes from the front of the head. The mouthparts, and thus feeding habits, are much like those of stable flies. The base of the proboscis is swollen into a characteristic bulb. Tsetse are daytime feeders and are visually attracted to moving objects. Both sexes feed exclusively on the blood of a wide variety of animals, including humans, and are particularly attracted to the Suidae.

Tsetse flies are larviparous and pupiparous, giving birth to a single, completely developed larva at intervals, producing from 8 to 20 in all. While in the oviduct, the larva feeds on secretions from specialized milk glands. The larvae is deposited on loose, dry soil, usually under shelter of some type. The larva has no locomotor structures but by contraction and extension buries itself under a few centimeters of loose soil. Hardening of the integument to form a puparium occurs within an hour of larviposition. The integument darkens to brownish black; the pupa is barrel shaped and has two prominent posterior lobes. The adult emerges within two to four weeks. The biology and influence of tsetses are beautifully illustrated by Gerster.[18]

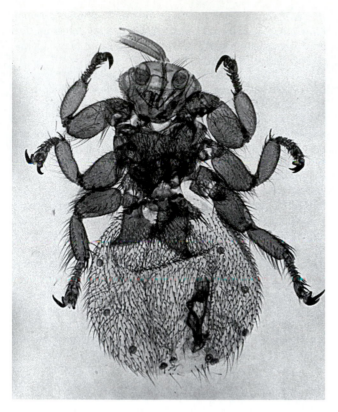

FIGURE 38.13

Sheep ked, *Melophagus ovinus.*
Courtesy of Warren Buss.

Tsetse flies have chemoreceptors on their tarsi, among them sensillae that stimulate feeding when exposed to uric acid, leucine, valine, and lactic acid (stimulatory ingredients in human sweat).[63] *Glossina morsitans* females also respond to chemicals released by larvae before pupation by depositing larvae in moist sand where there are already larvae.[37] The result is aggregation in shady, relatively moist areas, especially during dry seasons.

Twenty-three species of *Glossina* are usually recognized and can be identified with the use of the key in Lane and Crosskey.[28] All but three have been found capable of transmitting trypanosomes of mammals. Six of these are of outstanding medical importance: *Glossina palpalis, G. fuscipes,* and *G. tachinoides* are found along rivers; *G. morsitans, G. swynnertoni,* and *G. pallidipes* are savannah species. The first three are the primary vectors of Gambian sleeping sickness; the last three principally transmit Rhodesian sleeping sickness. Probably all of these, as well as several other species, can transmit nagana to cattle. We discussed African trypanosomiases in detail in Chapter 5. Vale and coworkers discussed the use of traps to control tsetses.[62]

• Family Hippoboscidae

The **louse flies** look neither like lice nor flies but rather like six-legged ticks. In most species the males are winged and the females wingless, although in some, such as *Hippobosca* spp., both sexes are winged. Both sexes are bloodsuckers, with some species parasitizing mammals and others, birds.

This is another pupiparous family: The larvae are retained within the female, feeding on secretions from special glands, and when born are ready to pupariate.

The **sheep ked,** *Melophagus ovinus* (Fig. 38.13), is distributed worldwide except in the tropics. Its puparium is glued to the wool of its host at any season of the year. Each female produces from 10 to 12 young. The entire life of the ked is spent on the host; when the ked is removed, it dies in about four days. A heavy infestation causes emaciation, anemia, and general unthriftiness of sheep. The skin may be so scarred by bites as to lose its market value. Hippoboscids are not loath to bite humans, and sheep shearers particularly are vulnerable to their attacks. The bite is said to be as painful as a yellow jacket wasp sting.

Other genera and species are found on mammals and birds in various parts of the world. *Olfersia coriacea* has been observed to bite humans in Panama. Its painless bite is a potential disease vector to humans.[22] In the western United States *Neolipoptena ferrisi* is common on mule deer. The **pigeon fly,** *Pseudolynchia canariensis,* is common on pigeons throughout most of the warm, temperate regions of the world. Both sexes are winged. Besides its importance as a bloodsucking parasite of pigeons, it also is the intermediate host of the malarialike *Haemoproteus columbae,* discussed in Chapter 9. The pigeon fly is willing to bite people, with painful results.

• Families Streblidae and Nycteribiidae

These two small, poorly known families are the **bat flies,** parasitic only on bats. The streblids may be winged or wingless, or they may have reduced wings. Compound eyes are small or absent. Six species are found in North America; most are associated with New World tropical bats. Nycteribiids are called **bat spider flies** (Fig. 38.14) because of their superficial resemblance to spiders. They are wingless, with the head folded back into a groove in the dorsum of the thorax. Five species are found in North America; most species feed on Old World bats. Both families are pupiparous.

• Family Fanniidae

This family contains over 260 species, the large majority of which are in the genus *Fannia.* Their calypteres are small and approximately equal in size, and ends of their anal wing veins (see Fig. 33.13) lie close together.

Genus *Fannia.* About 220 species are known in this genus, two of which will be mentioned briefly. *Fannia canicularis,* the **lesser house fly,** looks much like a smaller version of *M. domestica.* It is more slender and dark, and it has three brown longitudinal stripes on the thorax, rather than four as in the house fly. The biology of the lesser house fly parallels that of *M. domestica,* with the life cycle completed in 15 to 30 days. The larva is unlike that of *M. domestica,* since it is covered with long, slender projections. *Fannia canicularis* will enter houses freely; in fact it often is the dominant species at any given time. Unlike *M. domestica,* however, *F. canicularis* is not particularly attracted to food and so is not as efficient a vector as is the house fly.

The **latrine fly,** *F. scalaris,* is very similar in appearance to *F. canicularis* but seldom enters houses. It breeds in

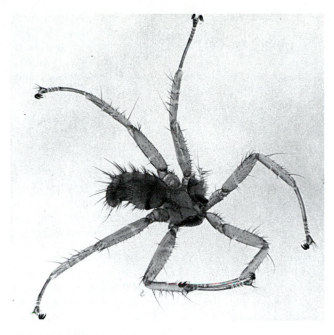

FIGURE 38.14

Bat spider fly, family Nycteribiidae.
Courtesy of Warren Buss.

fresh dung, particularly that of swine. The larva of the two species also are similar, but the lateral processes of *F. scalaris* are distinctly feathered.

Both species of *Fannia* are known to cause accidental myiasis of the rectum of humans, presumably when they lay eggs on the anus and the larvae crawl into the rectum and begin developing. No medical problem is caused by this beyond consternation in a person who discovers maggots in his or her stool.

• Family Muscidae

Members of this family are often **synanthropic;** that is, they live closely with humans. Many freely enter houses and readily avail themselves of whatever food and drink they may find there. They are small- to medium-sized flies, usually dull colored, with well-developed squamae and mouthparts.

Musca domestica.   The house fly is the most familiar of all flies, as well as being one of the most medically important. It is gray, 6 to 9 mm long, and has four conspicuous, dark, longitudinal stripes on the top of the thorax. Its distribution is nearly cosmopolitan but has changed markedly in societies with increased sanitation and decreased dependence on the horse.

House flies breed in all types of organic wastes except decaying flesh. Feces of any kind are preferred, although decaying milk around dairy barns, silage, slops around hog troughs, rotten fruit and vegetables, and so on are all stock in trade for house flies. Garbage cans are a favorite breeding place, and rotting garbage in the tropics can become transformed into an apparently equal mass of maggots, seemingly overnight.

Under ideal conditions, the egg can develop to adult in 10 days. One female deposits 120 to 150 eggs in each of at least six lots in its short lifetime. You can easily calculate that if all offspring of a pair of flies in April lived and reproduced, as did their succeeding generations, by August there would be 191,010,000,000,000,000,000 flies, which would cover the earth to a depth of 47 feet. This illustrates how a depleted population of flies can recover in a very short time.

The house fly is an efficient disease carrier for three primary reasons:

1. Its construction favors carrying bacteria. The multitude of tiny hairs covering most of the body readily collect bacteria, spores, and helminth eggs; and the mouthparts and six feet also have sticky pads that collect such matter.

2. It relishes human food and excrement alike. While walking on food and utensils, it not only leaves a trail of bacteria, but also while feeding, it defecates and vomits the remains of its last meal. Helminth eggs, protozoan cysts, and bacteria survive the intestinal tract of the fly and thus can be widely distributed from the site of their initial deposition.

3. Because of their synanthropy and powerful flight, house flies move about freely between indoor and outdoor attractions. Thus, it would appear that house flies achieve the ideal of mechanical transmission of disease.

The list of diseases known to be transmitted by house flies is too long to be repeated here; the interested reader can consult any current text in medical entomology. Briefly, most are enteric diseases occasioned by fecal contamination, such as typhoid fever, cholera, polio, hepatitis, shigellosis, salmonellosis, and other dysenteries. Other diseases are yaws, leprosy, anthrax, trachoma, tuberculosis, and those caused by various worms, such as *Ascaris*. Several diseases of domestic animals also are transmitted by these pests; in one study 70% of the flies from a goat yard contained coccidian oocysts in their gut.[11]

Other Species of *Musca*.   About 60 species have been placed in the genus *Musca*.[8] In Australia the bush fly, *M. vetustissima,* occurs in incredible numbers. Its importance as a vector is much less than that of *M. domestica,* partly because it is not so willing to enter human habitation and partly because of the widely scattered human population in much of that continent. However, the propensity for *M. vetustissima* to walk on people's faces is a genuine nuisance. It is not unusual for a person walking through the central Australian desert to have at least a thousand of these flies on his or her back.

The face fly, *M. autumnalis,* is a native of Africa, Asia, and Europe and was introduced into the United States in 1950. It now is found from coast to coast and well into Canada. It is a little larger than the house fly. The sides of the abdomen of the female are black, and those of the male are orangish. Larval development is in cow dung. Adults feed on secretions around the eyes of cattle and other large ruminants. They will enter houses and can annoy people. They serve as vectors of the eye worm, *Thelazia* (Chapter 28). Annual losses in the United States are estimated at $60 million.

Stomoxys calcitrans.   The **stable fly** is the only member of the genus *Stomoxys* that occurs in North America; it is a cosmopolitan species and there is molecular evidence that gene flow among *S. calcitrans* populations is high, suggesting equally high intercontinental motility.[32] The stable fly is similar in appearance to *M. domestica* and is often mistaken for it. The gray abdomen is rather checked, and the long slender proboscis protrudes in front of the head. A valuable reference to this fly is Zumpt.[69]

Stable flies are daytime biters. Both sexes feed on blood. The labella are equipped with rows of teeth that can readily pierce the skin and underlying tissues. The flies then sponge up blood that wells into the wound. Stable flies avidly bite humans and other animals, especially cattle and horses. They prefer to breed in decaying vegetation rather than manure but are adaptable.

When the insect occurs in great numbers, its attacks on humans are intolerable, affecting tourist industries in some areas. The fly is one of the most important pests of livestock, causing great losses annually. Loss of weight and milk production cost the United States $142 million in 1965. Hides can be damaged by the bites, and adult cattle can be killed if bitten enough times. It has been calculated that 25 flies a day per cow is the economic threshold; more flies cause a recognizable loss.[55] A thousand flies have been observed on a cow at one time.

Several diseases are known or suspected to be transmitted by stable flies. Among these is *Trypanosoma evansi*, the agent of surra (Chapter 5). Mechanical transmission of the *T. brucei* complex also occurs. Other diseases transmitted by stable flies are epidemic relapsing fever, anthrax, brucellosis, swine erysipelas, equine swamp fever, African horse sickness, and fowl pox. Also, the stable fly is the intermediate host of the horse stomach worm, *Habronema microstoma*, which infects the horse when an infected fly is swallowed. Feeding habits of *S. calcitrans* allow for potential mechanical transmission of HIV and herpes simplex virus (HSV). Experimentally, both HIV and HSV remained viable in blood spontaneously regurgitated within a few minutes of feeding, but the epidemiological significance of this observation cannot yet be assessed.[2]

In addition to insecticides, control measures include release of parasitoid wasps (*Muscidifurax raptor*) and solar-powered electrocuting traps designed to attract both house and stable flies.[46,52] As in the case of *Lucilia* infections, vaccine development shows some long-term promise. In one study, *S. calcitrans* fed on rabbits that had been immunized with fly-derived antigens exhibited higher mortality and lower fecundity than control flies.[64]

Haematobia irritans.   The horn fly (also known as *Hydrotaea irritans*) is found in the Americas, Europe, Asia Minor, and Africa. It closely resembles the stable fly but is more slender. It feeds with its head toward the ground, whereas the stable fly feeds with its head up. Horn flies breed in fresh cow manure. They will bite humans but are not as active fliers as are stable flies. They may be vectors of bovine mastitis.[26]

The importance of horn flies is mainly veterinary, with loss to livestock approaching that caused by stable flies. However, we know of no diseases transmitted by this insect, with the exception of the filarial nematode of cattle, *Stephanofilaria stilesi*. This skin parasite causes thickening and scabbing of the epidermis, especially around the navel, thus attracting more horn flies.

Horn flies are among the ectoparasite targets of insecticides incorporated into cattle ear tags, but such use produces resistant horn fly populations within a few weeks.[4, 53]

• Family Calliphoridae

This is the large family of blow flies, most of which help destroy carcasses. A few, however, are of great importance in causing myiasis. The common blow flies are usually metallic green, blue, or copper color, although some are nonmetallic.

Common Species.   *Calliphora vomitoria* is a large, metallic blue fly, probably the "blue-tailed fly" of song. It is conspicuous because of its large size and loud buzz as it flies. It can cause pseudomyiasis. Other species of *Calliphora* are facultative parasites.

Phaenicia and *Lucilia* spp. are metallic green with copper iridescence. *Phaenicia sericata* will breed in carrion, as well as excrement and garbage. Along with *L. cuprina*, it is important in **wool strike** in Australia. *Strike* is the term for the action of a fly laying its eggs or larvae on an animal. In wool strike the eggs are laid on the soiled wool on the rump of a sheep. The maggots feed on feces and bacteria; their activities cause a great deal of irritation to the sheep, which leads to other complications. This is why the tail is docked from lambs soon after they are born.

Sheep tend to develop some resistance to *L. cuprina*, but without continuous exposure, that resistance tends to be short lived.[50] However, inflammatory responses in sheep bred for resistance to *L. cuprina* infection were more intense than those of susceptible strains.[44] Efforts to develop vaccines against myiasis and fly attacks on livestock have been partially successful. In one study, *Lucilia cuprina* larvae grown on sheep immunized with fly peritrophic membrane antigens were half the size of those grown on control sheep.[13] On the other hand, immune suppression by *L. cuprina* excretory/secretory products has also been reported.[30]

Laboratory-reared *P. sericata* were used in World War I to clean wounds in servicemen; some species of calliphorid blow flies limit infections in wounds.[15] It has been reported as a facultative parasite in the ear canal and in open wounds.

Other blow flies that are facultative parasites are species of *Phormia*, *Cochliomyia*, and *Chrysomyia*.

Cochliomyia hominivorax.   This is the **primary screwworm** (Fig. 38.15) which is the most important cause of myiasis in the world. It is an obligate parasite, occurring throughout the Neotropical region. It causes dermal myiasis in nearly any mammal, as well as nasopharyngeal myiasis in humans.

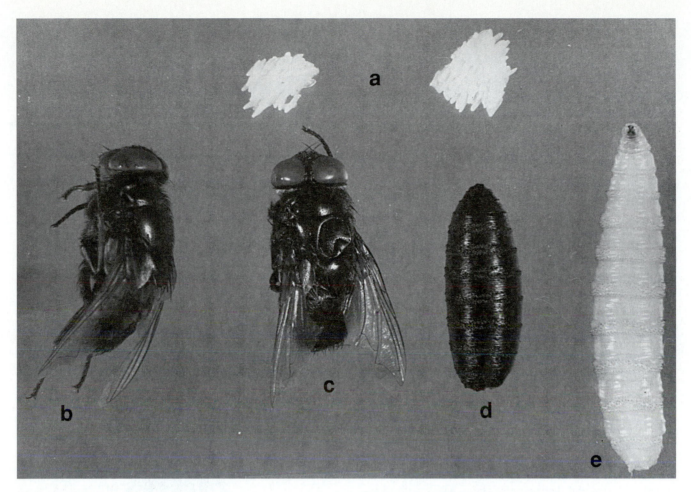

FIGURE 38.15

Life history stages of the primary screwworm, *Cochliomyia hominivorax*. (*a*) Two egg clusters; (*b*) male; (*c*) female; (*d*) puparium; (*e*) larva.

The adult fly is a deep, greenish-blue metallic color with a yellow, orange, or reddish face and three dark stripes on the thorax. It is difficult to differentiate *C. hominivorax* from the **lesser screwworm,** *C. macellaria,* which is not an important myiasis-causing insect.

Screwworm larvae cannot penetrate intact skin, although mucous membranes of the face and genitalia are susceptible to their attack. Usually a preexisting wound, however small, attracts the fly. Cuts from barbed wire or needle grass, castration and dehorning of calves, and insect and tick bites are all examples of sources of entry for screwworm maggots. Wounds from dog fights are commonly attacked. A noted parasitologist once removed more than a hundred screwworms from around the ear of an embattled dog in Trinidad.

Screwworms in humans are not uncommon. Generally the more abundant they are in livestock, the greater the chances of human infection. Infection in the head can be fatal, and urogenital infection can be grossly deforming.

The primary screwworm cannot survive winter in cold climates, but summer migrations have brought it as far north as Montana and Minnesota. Its normal range is from Mexico to Chile and Argentina. Severe epizootics have occurred in Texas cattle. More than 1.2 million cases were recorded in 1935 in that state alone.

The best control of this pest so far developed involves raising the flies in the laboratory, sterilizing the males, and freeing them to mate with wild females. Since females of this species usually mate only once, they thus cannot produce offspring after a sterile mating. *Cochliomyia hominivorax* has been eradicated from the United States and in fact is present in North America only in the Mexican state of Yucatan.[33]

The **Old World screwworm,** *Chrysomyia megacephala* (Fig. 38.16*a*), occupies the same niche in Africa, India, the Philippines, and the East Indies. It also attacks humans. Its biology and pathogenesis parallel those of *C. hominivorax.* It now is known to be endemic in South America.[36]

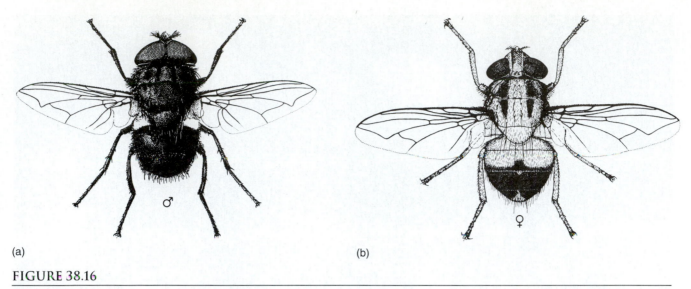

(a)

(b)

FIGURE 38.16

Two species of Calliphoridae of medical importance: (*a*) male of the Old World screwworm, *Chrysomya megacephala;* (*b*) female of the tumbu fly of Africa, *Cordylobia anthropophaga.*

From Richard P. Lane and Roger W. Crosskey, *Medically Important Insects and Arachnids.* Copyright © 1992 Chapman & Hall, London. Reprinted with permission.

• *Cordylobia anthropophaga*

The **tumbu fly** (Fig. 38.16*b*) is an African calliphorid restricted to south of the Sahara. The adult is yellowish, as contrasted with the calliphorids we are accustomed to in the northern hemisphere. It is stimulated to lay its eggs on soil that has been contaminated with urine. When the first-stage larva contacts mammalian skin, it penetrates and begins to grow, causing furuncular myiasis. Reports of this parasite from other parts of the world probably reflect infection acquired in Africa and detected elsewhere. Many wild mammals are reservoirs.

• *Auchmeromyia luteola*

The **Congo floor maggot** is a bloodsucking species found south of the Sahara. It is the only dipterous larva known to suck the blood of humans. Eggs are laid on floor mats, dry soil, or crevices in huts. The larvae are quite resistant to desiccation. They feed like bedbugs: When a person is asleep on the floor or on a mat, the maggots come out of hiding and pierce the skin with their powerful mouth hooks. Feeding is completed in 15 to 20 minutes, after which the larvae return to hiding. They are not known to transmit any disease.

Bloodsucking maggots of birds are common. In the northern hemisphere *Protocalliphora* spp. sometimes destroy entire broods of young birds.

• Family Sarcophagidae

These ubiquitous insects are known as **flesh flies** (Fig. 38.17). They are closely related to Calliphoridae, but instead of being metallic, the abdomen is checkered gray and black. Most are parasites of invertebrates, including insects and snails, and some are carrion breeders, but others are parasitic in the skin of vertebrates, including humans.

Sarcophaga hemorrhoidalis is widespread in the northern hemisphere and well into the tropics. It looks like a large house fly, but the tip of the male abdomen is red. Most sarcophagids are larviparous and normally breed in carrion, but

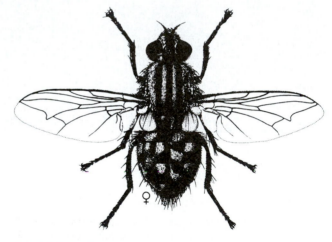

FIGURE 38.17

A member of the genus *Sarcophaga,* illustrating the typical three-striped thorax and checkered abdomen.

From Richard P. Lane and Roger W. Crosskey, *Medically Important Insects and Arachnids.* Copyright © 1992 Chapman & Hall, London. Reprinted with permission.

the female will deposit larvae in open wounds to become facultative parasites. Similarly, *Wohlfartia magnifica* is a facultative parasite of mammals in the warmer zones of the Palearctic region. Fatal human cases have been recorded.

Cutaneous furuncular myiasis is caused by *W. vigil* in Canada and the northern United States and by *W. opaca* in the western United States. The larvae are deposited on unbroken skin, which they quickly penetrate. Human infections usually occur in infants left unattended outdoors, although sleeping adults have been infected indoors. In the northern United States *W. vigil* is a serious pathogen of mink and fox kits in fur farms where newborns are often struck and soon die of the infection. Rodents and rabbits probably are reservoirs, as are carnivores.[16]

FIGURE 38.18

Rice rat, *Oryzomys capito,* with larvae of *Cuterebra* sp. in the skin.
Courtesy of C.O.R. Everard.

• Family Oestridae

According to recent classifications, this family now contains several subfamilies that were formerly given family status.[7] Students seeking additional information about the oestrids, especially in the older literature, should expect to find it under the subfamily names, with the *-inae* endings given as the family level *-idae* instead.

Subfamily Cuterebrinae. The cuterebrids are the **skin bot flies.** The common genus *Cuterebra* is a large black or blue fly about the size of a bumblebee. Found from the north temperate to tropical zones of the New World, species of this genus parasitize rodents, lagomorphs, and marsupials (Fig. 38.18). Eggs are laid on or near natural orifices. After hatching, the larvae enter the body, tunnel under the skin, cut an air hole in it, and begin to feed. The larvae are densely covered with thick spines and in some cases grow as large as a human adult's thumb. When located in the scrotum, *Cuterebra* spp. often will castrate the rodent host. Human cases are rare, with entry made through the anus, nose, mouth, or eye.

Often the host is disproportionately small, and it seems incredible that it can survive parasitism by such a large bot (Fig. 38.18). But it is possible that our impressions of "survival" are rather naive, because bot flies have been postulated as major participants in evolutionary and ecological phenomena. For example, among three chipmunk species in the Colorado Rockies, the most aggressive one lives at the highest elevation but suffers severe pathological effects of *Cuterebra fontinella* infections at lower elevations, while the less agressive, smaller species survives at lower elevations and is resistant to myiasis.[1] Previous explanations of the habitat separation were based on the higher species aggression, assumed to exclude the smaller species. Parasite avoidance has also been suggested as a basis for postcalving reindeer migrations.[17]

Dermatobia hominis is the **human skin bot** (Fig. 38.19*a*). It is common from Mexico through most of South America. A forest-inhabiting fly, it develops in the skin of almost any warm-blooded animals, including birds. Adults resemble bluebottles. Unlike any other myiasis-causing fly, *D. hominis* does not lay its eggs directly on the host. Instead it catches another insect, such as a mosquito, and glues its eggs to the side of it, with the operculated anterior end hanging down. At least 48 species of flies and one species of tick are carriers.[21]

When the carrier insect lands on warm skin, the eggs immediately hatch, and the larva drops onto the new host, penetrating unbroken skin. It bores into the dermis and remains there without further wandering. Development to pupal stage requires about six weeks; pupation is in the soil.

This fly commonly parasitizes humans, in whom it causes painful lesions. An infected traveler may have returned home to a distant place before noting the infection and

FIGURE 38.19

Representative life cycle stages of bot flies.
(*a*) Third-stage larva of *Dermatobia hominis;*
(*b–e*) *Oestrus ovis;* (*b*) ventral view of third-
stage larva; (*c*) first-stage larva; (*d*) mouthparts
of first-stage larva, lateral view; (*e*) posterior
spiracles of third-stage larva; (*f*) posterior
spiracle slits of third-stage *Cuterebra
emasculator* larva.

From Richard P. Lane and Roger W. Crosskey,
Medically Important Insects and Arachnids.
Copyright © 1992 Chapman & Hall, London.
Reprinted with permission.

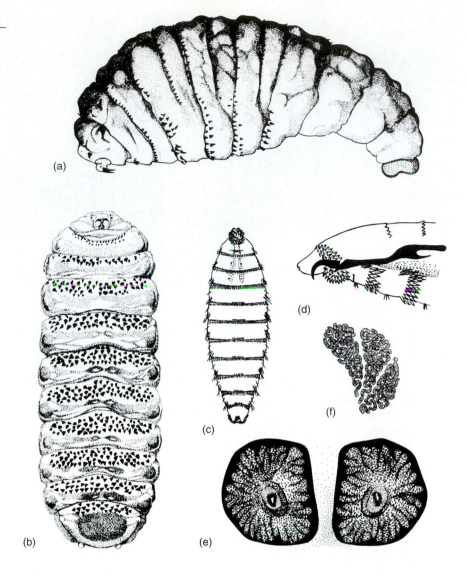

having it diagnosed. A small incision in the skin allows the larva to be expressed. It is readily recognized by two cauliflowerlike projections at the posterior end.

Subfamily Oestrinae.   The **head maggots** are about the same size and shape as honeybees and do not feed as adults. The larvae develop within the sinuses and nasal passages of hoofed animals.

The **sheep bot,** *Oestrus ovis* (Fig. 38.19*b–e*), is a cosmopolitan parasite of domestic sheep and goats and related wild species. The female deposits active larvae in the nostrils of the host during summer or early autumn. The larvae rapidly crawl up into the sinuses, where they attach to the mucosa and feed. Often they are in great numbers, causing considerable damage and pain to the host. By spring the larvae are developed and crawl back down to the nostrils, where they fall or are sneezed out. Pupation in the soil lasts from three to six weeks. Heavy infections can be fatal, but usually the host is only tormented, showing evidence of

great distress by sneezing, shaking of the head, loss of appetite, and a purulent discharge from the nose.

Other head maggots are *Rhinoestrus purpureus* in horses of Europe, Asia, and Africa; *Gedoelstia* spp. in African antelopes; and *Cephenemyia* spp. in Old and New World deer.

Ophthalmomyiasis occasionally occurs in humans, usually because of strike by *O. ovis* or *R. purpureus.* The larva cannot develop beyond first stage and usually does not last long. Inflammation and conjunctivitis may result. Head maggot strike in humans is most common in shepherds and others who work closely with sheep or horses.

Subfamily Hypodermatinae.   Variously known as **cattle grubs, ox warbles,** and **heel flies,** these skin parasites are found in most of the northern hemisphere. They primarily infect cattle and Old World deer, including reindeer, but they have been known to parasitize horses and humans as well.

Hypoderma lineatum and *H. bovis* are the two species that infect cattle. The former is common in Asia, Europe, and the United States, whereas the latter is slightly more northern in its distribution. Both look much like small bumblebees, with light and dark bands on the bodies.

The life cycles of the two species are similar. Both flies strike the hair of cattle, mainly on the hind legs. Although this is painless, the cattle become agitated and even terrified and gallop back and forth to avoid them. This action is called **gadding** and gives rise to the term *gadfly,* sometimes applied to people. On hatching within a week, the larva penetrates the skin and makes a remarkable migration, first to the front end of its relatively huge host and then back to the lumbar region, where it develops until pupation. All aspects of the migration are not yet known, but *H. bovis* reaches the spinal cord, usually in the neck, and burrows posteriorly between the periosteum and the dura mater for a distance and then completes its journey through tissues to the back. *Hypoderma lineatum* rests for a time in the wall of the esophagus and appears not to invade the spinal cord. Both species, on arriving at the lumbar skin, cut a hole in it, reverse position, apply the spiracles to the hole, and begin to feed. When ready to pupate, the grub cuts its way out, falls to the ground, and buries itself. The entire life cycle requires about a year.

These flies cause considerable damage to their hosts. In 1965 the USDA cited a $192 million loss, resulting primarily from the loss of weight, reduced milk production, and damage to hides. Warble fly has been nearly eradicated in Britain.[56]

Numerous cases of *Hypoderma* spp. in humans are recorded, mostly in people who have a close association with cattle. Unlike those of *Gasterophilus* spp., the larvae of these flies can successfully migrate and develop in the human body. Usually they surface in the neck region, probably because of the upright position of humans. Results of the migration can be dire, including partial or total paralysis of the legs. Ocular myiasis can occur, with loss of an eye.

Other species of warbles infect sheep and goats in Africa, Asia, and Europe. The **reindeer warble,** *Oedemagena tarandi,* is distributed over the range of wild and domestic reindeer, causing considerable loss in young animals. Some Eskimo tribes consider the fresh live grubs a delicacy to be eaten immediately upon the slaughter of caribou.

Subfamily Gasterophilinae. This family comprises the **stomach bots** of equids, elephants, and rhinoceroses. The adult flies are similar to honeybees in size and appearance and are strong fliers. The ovipositor is long and protuberant. The larvae of these species cause true enteric myiasis, attaching to the mucosa of the host's stomach.

Three species have been introduced into the United States from the Old World. They are parasites of horses, asses, and mules.

Gasterophilus intestinalis is called the **horse bot fly.** It is very common in North America and throughout most of the world. The female attaches approximately a thousand eggs to the hair of the horse, mainly on the knees. When the animal licks its hair, the warmth and moisture stimulate hatching. The first-stage larva immediately penetrates the tongue epithelium and tunnels its way down to the stomach, where it emerges and attaches with powerful mouth hooks.

FIGURE 38.20

Third-stage larvae of the horse stomach bot, *Gasterophilus intestinalis.*
Courtesy of Warren Buss.

Feeding on blood, it grows through two ecdyses to the third stage (Fig. 38.20). All instars have circles of strong spines on all but the last few segments. They remain attached until the following spring and early summer, when they detach and pass out with the feces. Pupation takes place in loose earth, and after three to five weeks the adults emerge. Cogley and coauthors well illustrate migration in the oral cavity.[5]

Gastrophilus nasalis, the **throat bot fly,** has a similar life cycle except that the eggs are attached to hairs under the jaw. The larvae hatch in four to five days without the need of moisture, crawl along the jaw, and enter between the lips.

The **nose bot fly,** *G. haemorrhoidalis,* strikes the horse on the lips. The remainder of its life cycle is similar to that of *G. intestinalis* except that the third instar larvae attach inside the anus for a short time before passing out.

Other genera are *Cobboldia, Platycobboldia,* and *Rodhainomyia* in elephants and *Gyrostigma* in rhinoceroses.

A few stomach bots cause little or no problems in horses, but a heavy infection may cause enough damage to the mucosa of the stomach and to the intestine during migration to kill the animal. Blockage of the pylorus also can occur.

Occasionally first instars will penetrate human skin and cause creeping myiasis, but they cannot mature or move deeper into the tissues.

• Myiasis

Myiasis is the name given to infection by fly maggots. There are several categories of myiasis, as in other forms of parasitism. In **obligatory myiasis** the insect depends on a period of parasitism before it can complete its life cycle. In **facultative myiasis** a normally free-living maggot becomes parasitic when it accidentally gains entrance into a host. Other terms are useful in describing the location of the larva: **gastric, intestinal,** or **rectal** for invasion of the digestive system; **nasopharyngeal** for nose, sinuses, and pharynx; **cutaneous,** either **creeping,** when the larva burrows through the skin, or **furuncular,** if it remains in a boil-like lesion; **urinary** or **urogenital; auricular** for the ears; and **ophthalmic** for the eyes. Some larvae are intermittent bloodsuckers and are loosely included under the term *myiasis.*

Accidental myiasis is usually enteric, occurring when eggs or larvae are eaten with contaminated food or drink. Such cases also are called **pseudomyiasis**.[68] Pseudomyiasis usually involves muscoid flies, such as *M. domestica* and *Fannia* spp.

References

1. Bergstrom, B. J. 1992. Parapatry and encounter competition between chipmunk (*Tamias*) species and the hypothesized role of parasitism. *Am. Mid. Nat.* 128:168–79.

2. Brandner, G., W. J. Kloft, C. Schlager-Vollmer, E. Platten, and P. Neumann-Opitz. 1992. Preservation of HIV infectivity during uptake and regurgitation by the stable fly, *Stomoxys calcitrans* L. *AIDS-Forschung* 7:253–56.

3. Brown, M. D., B. H. Kay, and J. K. Hendrikz. 1991. Evaluation of Australian *Mesocyclops* (Cyclopoida: Cyclopidae) for mosquito control. *J. Med. Ent.* 28:618–23.

4. Cilek, J. E., and F. W. Knapp. 1993. Influence of insecticide mixtures impregnated in cattle ear tags for management of pyrethroid-resistant horn flies (Diptera: Muscidae). *J. Econ. Ent.* 86:444–49.

5. Cogley, T. P., J. R. Anderson, and L. J. Cogley. 1982. Migration of *Gasterophilus intestinalis* larvae (Diptera: Gasterophilidae) in the equine oral cavity. *Int. J. Parasitol.* 12:473–80.

6. Crosskey, R. W. 1992. The fossil pupa *Simulimima* and the evidence it provides for the Jurassic origin of the Simuliidae (Diptera). *Systematic Entomology* 16:401–6.

7. Crosskey, R. W. 1993. Introduction to the Diptera. In Lane, R. P., and R. W. Crosskey, eds. 1993. *Medical insects and arachnids.* London: Chapman and Hall, Ltd., 51–77.

8. Crosskey, R. W. 1993. Stable-flies and horn-flies (bloodsucking Muscidae). In Lane, R. P., and R. W. Crosskey, eds. 1993. *Medical insects and arachnids.* London: Chapman and Hall, Ltd., 389–428.

9. Davies, J. B. 1993. Description of a computer model of forest onchocerciasis transmission and its application to field scenarios of vector control and chemotherapy. *Ann. Trop. Med. Parasitol.* 87:41–63.

10. Dale, P. E. R., P. T. Dale, K. Hulsman, and B. H. Kay. 1993. Runnelling to control saltmarsh mosquitoes: Long-term efficacy and environmental impacts. *J. Am. Mosq. Control Assoc.* 9:174–81.

11. Dipeolu, O. O., and V. Kahn. 1991. Studies on the dissemination of coccidia oocysts by *Musca* Linne, 1758 species. *Insect Sci. Appl.* 12:395–400.

12. Downes, J. A. 1971. The ecology of blood-sucking Diptera: An evolutionary perspective. In A. M. Fallis, ed. *Ecology and physiology of parasites.* Toronto: University of Toronto Press.

13. East, I. J., C. J. Fitzgerald, R. D. Pearson, R. A. Donaldson, T. Vucolo, L. C. Cadogan, R. L. Tellam, and C. H. Eisemann. 1993. *Lucilia cuprina:* Inhibition of larval growth induced by immunization of host sheep with extracts of larval peritrophic membrane. *Int. J. Parasitol.* 23:221–29.

14. Elnaiem, D. E. A., and R. D. Ward. 1992. Oviposition attractants and stimulants for the sand fly *Lutzomyia longipalpis* (Diptera: Psychodidae). *J. Med. Ent.* 29:5–12.

15. Erdmann, G. R. 1987. Antibacterial action of myiasis-causing flies. *Parasitol. Today* 3:214–16.

16. Eschle, J. L., and G. R. DeFoliart. 1965. Rearing and biology of *Wohlfartia vigil* (Diptera: Sarcophagidae). *Ann. Entomol. Soc. Am.* 58:849–55.

17. Folstad, I., A. C. Nilssen, O. Halvorsen, and J. Anderson. 1991. Parasite avoidance: The cause of post-calving migrations in *Rangifer? Can. J. Zool.* 69:2423–29.

18. Gerster, G. 1986. Tsetse—the deadly fly. *Natl. Geogr.* 170:814–33.

19. Greenberg, B. 1971. *Flies and disease, vol. 1. Ecology, classification, and biotic associations.* Princeton, N.J.: Princeton University Press.

20. Greenberg, B. 1973. *Flies and disease, vol. 2. Biology and disease transmission.* Princeton N.J.: Princeton University Press.

21. Guimarães, J. H., and N. Papavero. 1966. A tentative annotated bibliography of *Dermatobia hominis* (Linnaeus, 1781) (Diptera, Cuterebridae). *Arq. Zool.* 14:223–94.

22. Harlan, H. J., and B. N. Chaniotes. 1983. Report of *Olfersia coriacea* (Diptera: Hippoboscidae) feeding on a human in Panama. *J. Parasitol.* 69:1026.

23. Harwood, R. F., and M. T. James. 1979. *Entomology in human and animal health,* 7th ed. New York: Macmillan Publishing Co., Inc.

24. Hawley, W. A., P. Reiter, R. S. Copeland, C. B. Pumpuni, and G. B. Craig Jr. 1987. *Aedes albopictus* in North America: Probable introduction in used tires from northern Asia. *Science* 236:1114–16.

25. Hibler, C. P., and J. L. Adcock. 1971. Elaeophorosis. In Davis, J. W., and R. C. Anderson, eds. *Parasitic diseases of wild mammals.* Ames, Iowa: Iowa State University Press.

26. Hillerton, J. E. 1987. Summer mastitis: Vector transmission or not? *Parasitol. Today* 3:121–23.

27. Jarvis, E. K., and L. C. Rutledge. 1992. Laboratory observations on mating and lek-like aggregations in *Lutzomyia longipalpis* (Diptera: Psychodidae). *J. Med. Ent.* 29:171–77.

28. Jordan, A. M. 1993. Tsetse-flies (Glossinidae). In Lane, R. P., and R. W. Crosskey, eds. 1993. *Medical insects and arachnids.* London: Chapman and Hall, Ltd., 333–88.

29. Karch, S., N. Asidi, Z. M. Manzambi, and J. J. Salaun. 1992. Efficacy of *Bacillus sphaericus* against the malaria vector *Anopheles gambiae* and other mosquitoes in swamps and rice fields in Zaire. *J. Am. Mosq. Control Assoc.* 8:376–80.

30. Kerlin, R. L., and I. J. East. 1992. Potent immunosuppression by secretory/excretory products of larvae from the sheep blowfly *Lucilia cuprina. Parasite Immunol.* 14:595–604.

31. Kittayapong, P., and D. Strickman. 1993. Three simple devices for preventing development of *Aedes aegypti* larvae in water jars. *Am. J. Trop. Med. Hyg.* 49:158–65.

32. Krafsur, E. S. 1993. Allozyme variation in stable flies (Diptera: Muscidae). *Biochem. Gen.* 31:231–40.

33. Krafsur, E. S., C. J. Whitten, and J. E. Novy. 1987. Screwworm eradication in North and Central America. *Parasitol. Today* 3:131–37.

34. Kramer, V. L., E. R. Carper, and C. Beesley. 1993. Control of *Aedes dorsalis* with sustained-release methoprene pellets in a saltwater marsh. *J. Am. Mosq. Control Assoc.* 9:127–30.

35. Krinsky, W. L. 1976. Animal disease agents transmitted by horse flies and deer flies (Diptera: Tabanidae). *J. Med. Ent.* 13:225–75.

36. Lawrence, B. R. 1986. Old World blowflies in the New World. *Parasitol. Today* 2:77–79.

37. Leonard, D. E., and R. K. Saini. 1993. Semiochemicals from anal exudate of larvae of tsetse flies *Glossina morsitans morsitans* Westwood and *Glossina morsitans centralis* Machado attract gravid females. *J. Chem. Ecol.* 19:2039–46.

38. Lewis, D. J. 1975. Functional morphology of the mouthparts in New World phlebotomine sand flies (Diptera: Psychodidae). *Trans. R. Soc. Entomol.* 126:493–532.

39. Loh, P. Y., H. H. Yap, N. L. Chong, and S. C. Ho. 1992. Laboratory studies on the predatory activity of a turbellarian, *Dugesia* sp. (Penang) on *Aedes aegypti, Anopheles maculatus, Culex quinquefasciatus* and *Mansonia uniformis. Mosq. Borne Dis. Bull.* 9:55–59.

40. Lynch, J. D. 1988. Introduction, establishment, and dispersal of western mosquitofish in Nebraska (Actinopterygii: Poeciliidae). *Prairie Nat.* 20:203–16.

41. Mattingly, P. F. 1977. Names for the *Anopheles gambiae* complex. *Mosq. Syst.* 9:323–28.

42. Mulligan, H. W., and W. H. Potts. 1970. *The African trypanosomiases.* London: Allen & Unwin.

43. Nielsen, B. O., and O. Christensen. 1975. A mass attack by the biting midge *Culicoides nubeculosus* (Mq.) (Diptera: Ceratopogonidae) on grazing cattle in Denmark. A new aspect of sewage discharge. *Nord. Vet. Med.* 27:365–72.

44. O'Meara, T. J., M. Nesa, H. W. Baadsma, D. G. Saville, and R. M. Sandeman. 1992. Variation in skin inflammatory responses between sheep bred for resistance or susceptibility to fleece rot and blowfly strike. *Res. Vet. Sci.* 52:205–10.

45. O'Meara, G. F., L. F. Evans Jr., and A. C. Gettman. 1992. Reduced mosquito production in cemetery vases with copper liners. *J. Am. Mosq. Control Assoc.* 8:419–20.

46. Pickens, L. G., and G. D. Mills Jr. 1993. Solar-powered electrocuting trap for controlling house flies and stable flies (Diptera: Muscidae). *J. Econ. Entomol.* 30:872–77.

47. Platzer, E. G. 1981. Biological control of mosquitoes with mermithids. *J. Nematol.* 13:257–62.

48. Robert, L. L., R. E. Coleman, D. A. LaPointe, P. J. S. Martin, R. Kelly, and J. D. Edman. 1992. Laboratory and field evaluation of five repellents against the black flies *Prosimulium mixtum* and *Prosimulium fuscum* (Diptera: Simuliidae). *J. Econ. Entomol.* 29:267–72.

49. Sabrosky, C. W. 1941. The *Hippolates* flies or eye gnats: Preliminary notes. *Can. Entomol.* 73:23–27.

50. Sandeman, R. M., R. A. Chandler, B. J. Collins, and T. J. O'Meara. 1992. Hypersensitivity responses and repeated infections with *Lucilia cuprina,* the sheep blowfly. *Int. J. Parasitol.* 22:1175–77.

51. Service, M. W. 1993. Mosquitoes (Culicidae). In Lane, R. P., and R. W. Crosskey, eds. 1993. *Medical insects and arachnids.* London: Chapman & Hall, Ltd., 120–240.

52. Seymour, R. C., and J. B. Campbell. 1993. Predators and parasitoids of house flies and stable flies (Diptera: Muscidae) in cattle confinements in West Central Nebraska. *Env. Entomol.* 22:212–19.

53. Sheppard, D. C., and J. A. Joyce. 1992. High levels of pyrethroid resistance in horn flies (Diptera: Muscidae) selected with cyhalothrin. *J. Econ. Entomol.* 85:1587–93.

54. Szadziewski, R., and T. Schulueter. 1992. Biting midges (Diptera: Ceratopogonidae) from Upper Cretaceous (Cenomanian) amber of France. *Ann. Soc. Entomologique de Fr.* 28:73–81.

55. Steelman, C. D. 1976. Effects of external and internal arthropod parasites on domestic livestock production. *Ann. Rev. Entomol.* 21:155–78.

56. Tarry, D. W. 1986. Progress in warble fly eradication. *Parasitol. Today* 2:111–16.

57. Taylor, S. D., S. A. Ritchie, and E. Johnson. 1992. The killifish *Rivulus marmoratus:* A potential biocontrol agent for *Aedes taeniorhynchus* and brackish water *Culex. J. Am. Mosq. Control Assoc.* 8:80–83.

58. Thangam, T. S., and K. Kathiresan. 1992. Mosquito larvicidal activity of mangrove plant extracts against *Aedes aegypti. Int. Pest Control* 34:116–19.

59. Tod, M. E., D. E. Jacobs, and A. M. Dunn. 1971. Mechanisms for the dispersal of parasitic nematode larvae. I. Psychodid flies as transport hosts. *Helminthology* 45:133–37.

60. Toma, T., and I. Miyaga. 1992. Laboratory evaluation of *Toxorhynchites splendens* (Diptera: Culicidae) for predation of *Aedes albopictus* mosquito larvae. *Med. Vet. Entomol.* 6:281–89.

61. Turner, W. N. 1978. A case of severe human allergic reaction to bites of *Symphoromyia* (Diptera: Rhagionidae). *J. Med. Ent.* 15:138–39.

62. Vale, G. A., E. Bursell, and J. W. Hargrove. 1985. Catching-out the tsetse fly. *Parasitol. Today* 1:106–10.

63. van der Goes van Naters, W. M., and T. H. N. Rinkes. 1993. Taste stimuli for tsetse flies on the human skin. *Chem. Senses* 18:437–44.

64. Webster, K. A., M. Rankin, N. Goddard, D. W. Tarry, and G. C. Coles. 1993. Immunological and feeding studies on antigens derived from the biting fly, *Stomoxys calcitrans. Vet. Parasitol.* 44:143–50.

65. Wirth, W. W., and W. R. Atchley. 1973. A review of the North America *Leptoconops* (Diptera: Ceratopogonidae). *Grad. Stud. Texas Tech. Univ.* 5:1–57.

66. Woodard, D. B., and H. C. Chapman. 1970. Hatching of flood-water mosquitoes in screened and unscreened enclosures exposed to natural flooding of Louisiana salt marshes. *Mosq. News* 30:545–50.

67. Xudong, X., K. Renqiu, and H. Yuxiang. 1993. High larvicidal activity of intact recombinant cyanobacterium *Anabaena* sp. PCC 7120 expressing Gene 51 and Gene 42 of *Bacillus sphaericus* sp. 2297. *FEMS Microbiol. Letters* 107:247–50.

68. Zumpt, F. 1965. *Myiasis in man and animals in the Old World.* London: Butterworth & Co. (Publishers) Ltd.

69. Zumpt, F. 1973. *The stomoxyine biting flies of the world.* Stuttgart: Gustav Fischer Verlag.

Chapter 39

PARASITIC INSECTS: STREPSIPTERA, HYMENOPTERA, AND OTHERS

. . . weirdness feeds on itself. . . .

Hunter S. Thompson

Thus far we have considered the orders of insects containing members of medical or veterinary significance. Remaining are several orders that are not covered often in parasitology texts but that are of biological interest and particularly in the case of the Hymenoptera, have considerable impact on human welfare. Half or more of hymenopterans are parasites of other insects, and many are extremely important natural controls of agricultural and forestry pests. Some of these species are prime candidates for use in integrated pest management schemes (see p. 599). The diversity of hymenopterans alone confirms the assertion that parasitism in its many forms is the most common way of life on earth. In a typical study, for example, Butler[7] found eight families and 74 species of parasitic arthropods in 46 species of caterpillars from a West Virginia deciduous forest. And interest in biological control is not limited to agricultural pests. An encyrtid wasp was recently discovered parasitizing *Ixodes dammini*, the vector of Lyme disease (see Chapter 40).[20]

Many of the insects discussed in this chapter are parasitic as larvae, growing inside or outside the host and eventually killing it. Such insects would seem to fit the conventional definition of parasite when they are small, but because they kill their host during or at the completion of development, they seem to become predators at this stage. Consequently, they are often referred to as **parasitoids,** rather than parasites. Other authors prefer to use the term **protelean** to describe all insects whose immature stages are parasitic and whose adults are free living. This designation would include some members of most orders covered in this chapter as well as certain Diptera, such as *Gasterophilus, Dermatobia,* and *Hypoderma* (see Chapter 38). The adults of protelean parasites are generally the dispersal mechanism and host finders. Because of their potential agricultural importance, the host finding behavior of parasitoids is a major research focus among ecologists.

Doutt[15] pointed out several ways in which parasitoids differ from typical parasites, including (1) the host is usually of the same taxonomic class, Insecta; (2) the development of an individual destroys its host; (3) parasitoids are relatively large in size compared with their hosts; and (4) they are parasitic as larvae only, the adults being free living. Nonetheless, Doutt used the term *parasite* interchangeably with *parasitoid,* and we concur, adopting the broad definition of parasitoids as parasites.

Although of less economic significance than the Hymenoptera, all Strepsiptera are parasitic and demonstrate very interesting adaptations to parasitism. We will consider the Hymenoptera and Strepsiptera in some detail and briefly will mention some members of other orders that are parasitic.

ORDERS WITH FEW PARASITIC SPECIES

Order Dermaptera (The Earwigs)

The comparatively small order Dermaptera (1100 species) has hemimetabolous development and is composed primarily of the free-living earwigs. Only one genus, *Hemimerus,* can be considered parasitic in the same sense as other animals discussed in this text, and some investigators place members of this genus in their own separate order.[31] The abdomen of adult free-living earwigs terminates in a pair of unsegmented, pincerlike cerci, whereas the cerci of *Hemimerus* are short and threadlike. *Hemimerus* are parasites of pouched rats in tropical Africa. They are eyeless and wingless, and they feed on their host's epidermis and a fungus that grows thereon.[21,32]

Order Neuroptera (The Lacewings)

Members of the Neuroptera are holometabolous, as are those of the remaining orders to be considered. The Neuroptera is an order of moderate size (5000 species), and only about 190 species can be considered protelean parasites. Most neuropterans are predators as larvae and sometimes as adults on other insects and mites. They are considered beneficial to humans because they help control numerous pests. Most larvae are terrestrial, although members of the Sisyridae have aquatic larvae. The larvae of *Sisyra* and *Climacia* spp. parasitize freshwater sponges, a food source apparently distasteful to most other animals. The first instar larvae float in the water until they encounter a sponge, enter ostia, and begin to feed. Larvae of the family Mantispidae parasitize the egg cocoons of several families of spiders. When fully developed, mantispid larvae are a contrast with the active, predaceous larvae of other families. They have a small head, large abdomen, and small legs, characteristics correlated with the fact that their food source is abundant and does not have to be caught and killed.

FIGURE 39.1

Eight *Lobocraspis griseifusa* (Lepidoptera, Noctuidae) suck tears from the eye of a banteng, *Bos banteng,* in northern Thailand. Note the proboscis of each moth extended to feed at the eye perimeter.

Courtesy of Hans Bänzinger.

Order Lepidoptera (Butterflies and Moths)

Members of the large order Lepidoptera (120,000 species) are familiar to most people. The adults have generally rather large wings covered with tiny scales, and they have a long suctorial proboscis formed from parts of the maxillae, the mandibles being absent or rudimentary. They are adapted for feeding on nectar or other extruded plant juices. The polypodous larvae are different in form, with strong mandibles adapted for chewing. The adults and larvae are almost entirely phytophagous. A few species are ectoparasites of mammals as adults, and a few are protelean parasites of insects. Although small in number of species, the eye-feeding moths in Southeast Asia demonstrate an interesting progression from nectar feeding on plants to blood-sucking from animals. Members of several families occasionally visit the eyes of domestic ungulates (Fig. 39.1), where they feed on lacrimation; those of a different group visit the eyes fairly frequently, feeding on lacrimation and fluid running down the host's cheeks; and those of another group (for example, *Arcyophora* and *Lobocraspis*) feed only at the eyelid on lacrimation and pus and sometimes penetrate the conjunctiva to feed on blood.[8] Finally, a species of noctuid, *Calpe eustrigata,* has developed the ability to pierce vertebrate skin and feed on blood[3] (Fig. 39.2).

Cyclotorna monocentra in Australia has a very curious life history.[14] Its first instar larva finds, attaches to, and feeds on but does not kill nymphs or adults of a species of leafhopper (Hemiptera, Homoptera). When *Cyclotorna* molts to the second instar, it somehow induces an ant (*Irdomyrmex*) to pick it up and carry it back to its colony. The caterpillar then ingratiates itself by providing the ants with a sweet secretion and by running its mouthparts over the bodies of the ants—activities that the ants seem to enjoy. The caterpillar exacts payment for these services by feeding on ant larvae. Finally, the lepidopteran larva emerges from the ant colony and pupates on the trunk of a tree.

FIGURE 39.2

A noctuid moth, *Calpe eustrigata,* piercing the skin and sucking the blood of a Malayan tapir. This is the only known bloodsucking moth.

Courtesy of Hans Bänzinger.

Order Coleoptera (The Beetles)

Judged by its number of described species, the order Coleoptera is the most successful order of animals on earth (300,000 species). Less than 2% are parasites, but both mammalian ectoparasites and protelean parasites of insects are represented. As to be expected in such a large and speciose order, the Coleoptera exhibit a wide range of morphological form, habit, and adaptation. In most species the mandibles are well-developed for biting and chewing, but in some they are adapted for piercing and sucking. Although most beetles can fly well, the forewings are hardened into sheathlike elytra. Most coleopterans that are protelean parasites are **hypermetamorphic,** leaving the host-finding task to the first instar larva.

Hypermetamorphosis is a condition in which different larval instars have dissimilar forms, and it is found among the mantispid neuropterans, the Strepsiptera, and several families of Diptera and Hymenoptera, as well as in some Coleoptera. In all of these the first instar is rather heavily sclerotized and quite active. In the Neuroptera, Strepsiptera, and Coleoptera, the larva is a campodeiform oligopod and is called a **triungulin** or **triungulinid** (Fig. 39.3*c*). The active dipteran and hymenopteran larvae are apodous, but they move about with the aid of thoracic and caudal setae; the larva is called a **planidium** (Fig. 39.3*a,b*). Subsequent instars of both types are typically much less sclerotized and have a smaller head and much reduced means of locomotion. Planidia and triungulins are interesting instances of convergent evolution.

Examples of protelean parasitic beetles include the aleocharine staphylinids, whose triungulins seek out the puparia of Diptera, penetrate them, and feed on the pupa within. The larvae of some Meloidae feed on grasshopper eggs, and others feed on the eggs of solitary bees and then on the honey

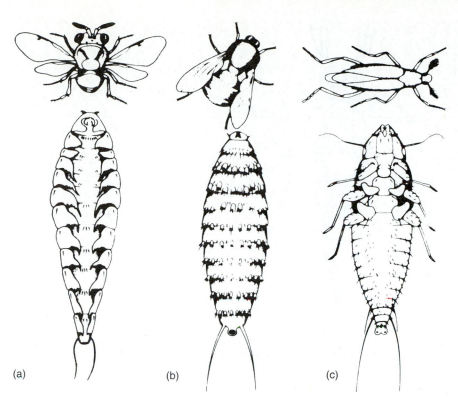

FIGURE 39.3

First instar larvae with corresponding adults (*not* drawn to same scale). The larvae are active and must find the host. (*a*) Perilampidae (Hymenoptera), and (*b*) Acroceridae (Diptera), exemplify the planidium type of larva. (*c*) A triungulin larva (with well-developed legs) of Rhipiphoridae (Coleoptera).

From R. R. Askew, *Parasitic Insects.* Copyright © 1971 American Elsevier Publishing Co., New York, NY.

(a)   (b)   (c)

stored in the cell in which the egg is laid. Triungulins of the latter may congregate on flowers, lying motionless until the flower is visited by a bee. They quickly attach themselves to the bee, hang on until the bee returns to its nest, and then drop off into the cell in which the bee deposits its egg.

A few beetles are symbiotic as adults. *Platypsyllus castoris* (Platypsyllidae) is an ectoparasite of beavers in both the Palearctic and Nearctic. It is a blind, obligate parasite and an ectoparasite in both the adult and larval states, feeding on skin debris. Some species of Staphylinidae are apparently mutuals of marsupials and certain rodents, feeding on fleas and mites.[16] The beetles are rather large (5 to 16 mm) but are well-tolerated by the host even as they cling to its hair in areas such as the base of the tail where a more noxious passenger would excite grooming attention.

ORDER STREPSIPTERA (THE STYLOPS)

In numbers of species (only about 300), the Strepsiptera is a very small order. It is of great interest biologically, however, and demonstrates some of the most extreme adaptations to protelean parasitism of any insects. Commonly known as **stylops,** the insects show certain similarities to the Coleoptera and are considered by some investigators a superfamily (Stylopoidea) of the Coleoptera.[11] The two orders have important differences, including the fact that the last larval instar of the most primitive strepsipteran family, Mengeidae, has a primitive character: compound eyes. This is in striking contrast to the condition in all other endopterygote larvae, which have only simple, ocelluslike stemmata.

Extant strepsipteran species have been recovered from Oligocene amber, strongly suggesting that their remarkable life cycles evolved during the age of dinosaurs.[22]

Morphology

The strepsipterans exhibit extreme sexual dimorphism. The males are small, robust, beetlelike insects about 1.5 to 4.0 mm long (Fig. 39.4). The forewings, believed by Crowson[11] to represent reduced elytra, are small halteres bearing numerous sensory endings. The hindwings are large, membranous, and fan shaped and are borne on a large, third thoracic somite. The insects are various shades of brown to black. The compound eyes are large and protuberant. One or more segments of the antennae have lateral processes so that the antennae appear branched (Fig. 39.4). Mandibles are present but are simple and sickle shaped; other mouthparts are reduced or absent. These characteristics reflect the fact that the adult male spends its very short existence agitatedly seeking females to mate with; thus, it has more use for well-developed sense organs than for feeding appendages.

Except in the Mengeidae, the adult female is parasitic. It is vermiform, up to 20 to 30 mm long according to species, and has no wings, eyes, legs, or antennae (Fig. 39.5). The mouth and anus are tiny and nonfunctional, and the gut has no lumen. The head and thorax are fused to form a cephalothorax and are rather heavily sclerotized. The animal breathes through a spiracle on each side of the cephalothorax, which protrudes from the host between two sclerites, usually between tergites of the host abdomen. The abdomen of the female stylops is large and soft, lying within the host body and

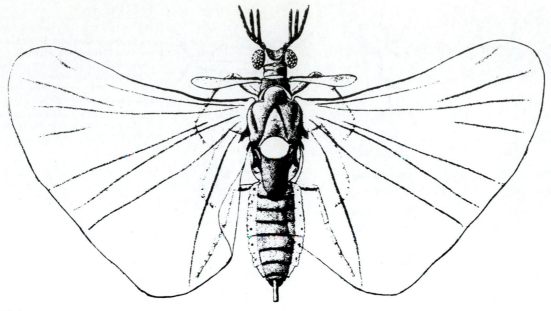

FIGURE 39.4

Eoxenos laboulbenei, adult male (Strepsiptera). The mesothoracic wings have been reduced to halteres, and the large, membranous metathoracic wings are borne on the very large metathorax.

From H. L. Parker and H. D. Smith, "Further notes on *Eoxenos laboulbenei* peyerimoff with a description of the male," in *Ann. Entomol. Soc. Am.* 27:468–479. Copyright © 1934.

FIGURE 39.5

Diagram of female strepsipteran. (*a*) Ventral view; (*b*) longitudinal section. The more heavily sclerotized cephalothorax protrudes from the host body between abdominal tergites, whereas the soft abdomen lies within the host's hemocoel and is covered by the cuticle of the last larval instar. Initially closed, the brood canal is pierced by the male during copulation, and the triungulins subsequently exit through that opening.

From R. R. Askew, *Parasitic Insects.* Copyright © 1971 American Elsevier Publishing Co., New York, NY.

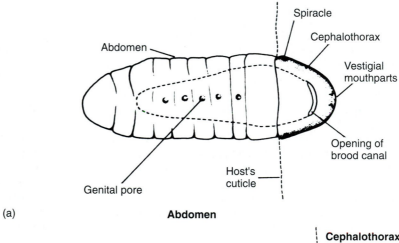

(a)

Abdomen

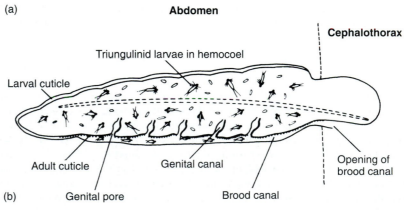

(b)

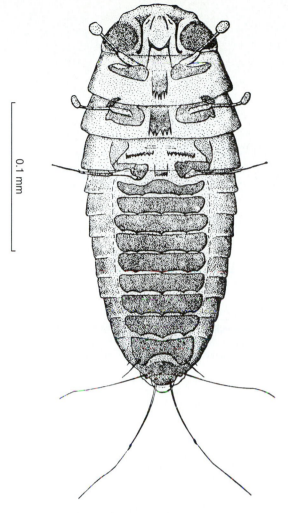

FIGURE 39.6

First instar larva (triungulin) of the strepsipteran, *Corioxenos antestiae.*
The triungulin characteristically lies motionless, resting on its hind legs
and the central caudal bristles. It is stimulated to leap in the air by
movements of nearby black and orange objects, such as its host, a
pentatomid bug. Adhesive pads on its forelegs help it stay on the host
when it makes contact.

From T. W. Kirkpatrick, "Studies on the ecology of coffee plantations in east
Africa. II. The autecology of *Antestia* spp. (Pentatomidae) with a particular
account of a Strepsipterous parasite," in *Trans. R. Entomol. Soc. Lond.*
86:247–343. Copyright © 1937.

Development

The embryos develop and hatch within the hemocoel of the
female to produce triungulin larvae, which exit through the
genital pores and brood canal. One female can produce 2000
or more triungulins, and polyembryony has been reported.
The triungulins are only about 0.2 mm long, but they have a
well-sclerotized cuticle, well-developed legs, and one or two
pairs of long caudal filaments (Fig. 39.6). They die in a short
time if they do not reach a new host.

Although the vast majority of triungulins will perish,
they possess some adaptations to increase their chances of
survival. For example, the triungulins of *Corioxenos antestiae* rest motionless, with the anterior part of the body raised,
on the hind legs and the central caudal bristles, which are
bent forward beneath the abdomen. A movement in their
vicinity, particularly if the moving object is black and orange
(the colors of their host, a pentatomid bug),[24] stimulates the
triungulin to leap up to 10 mm vertically and 25 mm hori-
zontally, and adhesive pads on its front two pairs of tarsi help
maintain contact if it hits a host (Fig. 39.6). Although the triungulin will attach to a wide variety of insects, it soon detaches unless the insect is the correct host. Other species are
known, however, that will penetrate an unsuitable host and
die therein.

Although parasites of Hemiptera and those of social
members of Hymenoptera reach their hosts directly, stylops
parasitic in solitary Hymenoptera have special problems because the host larva is hidden in a cell, and they must be
transported to that site. A bee infected with *Halictoxenos
jonesi,* for example, drags her abdomen among the stamens
of flowers, depositing triungulins.[4] When another bee visits
the flower, a triungulin attaches to it, is transported by the
adult bee back to the cell containing its larva, and transfers to
the larva while the adult bee is feeding its young.

Once on its host, the triungulin penetrates the hemocoel through an intersegmental membrane and begins to
feed. The second instar, however, differs vastly from the
triungulin. It is a grublike organism without mouthparts and
legs, and it feeds by absorbing nutrients through its cuticle.
There are normally six larval instars, including the triungulin.[17] The sixth instar regains its mandibles and chews its
way through the intersegmental membrane between the
sclerites of the host abdomen. If it is a male, it emerges
completely and pupates within the cuticle of the last larval
instar. If it is a female, only the cephalothorax emerges
with no obvious pupation.

In contrast to the situation in most families, in the
Mengeidae females retain several primitive characters: They
emerge completely, pupate, and become free living. Although wingless, they have functional eyes, legs, and antennae. Mengeids are parasites of Thysanura (the silverfish), an
order of insects also with numerous primitive characters, but
they probably colonized their hosts rather than coevolved
with them because the Strepsiptera are of more recent origin
than are the Thysanura.[1]

ensheathed in the cuticle of the last larval instar. The space
between the abdomen and its larval cuticle forms a brood
canal, which opens to the outside beneath the cephalothorax.
Copulation is accomplished by insertion of the male's aedeagus into the brood canal. The sperm make their way through
two to five genital openings in the female's abdomen and
thence to her hemocoel. There the sperm fertilize the eggs,
which are lying free in the hemocoel. This type of reproductive system is found in no other insects.

Stylops differ from virtually all other protelean parasites of insects in that their host usually lives approximately a normal life span. Even so, the strepsipterans may cause the reproductive death of the host through parasitic castration.[27] Although there is some variation, depending on whether the host is infected in an early or late instar, secondary sexual characteristics tend to take on an intersex appearance because effects on the gonads may be profound. Williams[39] reported that the aedeagus and parameres were often reduced or absent in leafhoppers (*Dicranotropis muiri*) infected with *Elenchus templetoni* in Mauritius, and the length of the ovipositors and sheath was reduced in female hosts. The gonads of hosts with advanced larvae or extruded forms were much reduced or could not be found at all.

Strepsipterans are particularly attractive as potential biological control agents for the fire ant now spreading across much of the southern United States.[23]

ORDER HYMENOPTERA (ANTS, BEES, AND WASPS)

The Hymenoptera is the largest order of insects after the Coleoptera, estimated to include more than 200,000 species; at least half of these are insect protelean parasites. Free-living hymenopterans are familiar to everyone as ants, bees, and wasps. Although feared and avoided by most people because of their sometimes potent sting, they are nevertheless extremely important pollinators of flowering plants. Most of our vegetable and fruit plants require the services of honeybees. At least as important to humans, if not more so, is the role played by the parasitic Hymenoptera in population control of other insects.

Morphology

Hymenopterans usually have a fairly heavily sclerotized cuticle and show a vast range of body form, size, and color. They include the smallest insect known (*Alaptus*, 0.21 mm long), and insects up to 115 mm (some ichneumons, including the ovipositor). They have two pairs of membranous wings, of which the forewings are largest (Fig. 39.7) except in the Symphyta. The hindwings are attached to the forewings by a row of small hooks on the leading edge of the hindwings. Wing venation tends to be reduced and may be absent in some minute species. Some forms are wingless. The first abdominal segment is fused to the thorax and is called the **propodeum.** The second abdominal segment in most Hymenoptera is constricted to a waistlike **pedicel,** or **petiole.** The head is remarkably free, with a small neck and large compound eyes. Antennae usually have 12 segments in males and 13 in females. The mandibles, maxillae, and labium are present and variously modified for different feeding habits. A **glossa** (considered the hypopharynx by some authors) is present, and it and some other mouthparts may be lengthened to form a tongue or proboscis for gathering nectar from flowers.

The **ovipositor** is modified to form a stinging organ in some species, but it is important to the biology of the parasitic forms as well. In common with many other insects, the ovipositor of hymenopterans has been derived from pairs of segmental appendages on the abdomen (Fig. 39.8). Its basal portions represent the coxae of abdominal segments eight

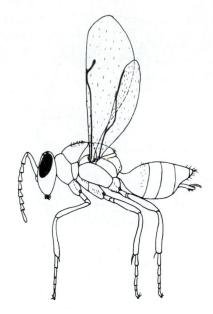

FIGURE 39.7

Chalcidoid hymenopteran (Encyrtidae). Members of this family commonly are parasites of scale insects (Hemiptera, Homoptera), and several species have been used successfully in biological control of important pests.[34]

Figure from *An Introduction to the Study of Insects,* Third Edition by Donald J. Borror and Dwight M. DeLong, Copyright © 1974 by Saunders College Publishing. Reproduced by permission of the publisher.

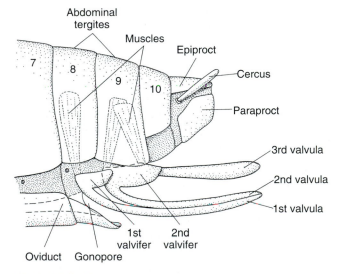

FIGURE 39.8

Diagram of generalized appendiculate ovipositor of a pterygote insect. Some internal muscles and the position of the oviduct are shown. The valvulae and valvifers are all paired, although the second valvulae are fused in the Hymenoptera. They join with the first valvulae to form the ovipositor tube (terebra), and the third valvulae form the ovipositor sheath.

Drawing by William Ober.

and nine and are called the **valvifers.**[35] The coxal endites have been lengthened to become the body of the ovipositor and are called **valvulae.** Another process from the second valvifer (third valvula) has developed to become the ovipositor sheath. In Hymenoptera the pair of second valvulae are fused and have longitudinal ridges that interlock with grooves on the first valvulae. This arrangement of parts forms the tube down which the egg must pass, and the functional unit is called the **terebra.** The much broader third valvulae form a sheath on each side of the terebra; they are well-endowed with sensory endings on their outer surfaces and are much less rigid than are the terebra. Cutting ridges on the end of the first and second valvulae enable the insect to drill through host cuticle, through cells that contain the host, or even through hard vegetable matter, before oviposition. Since the lumen of the terebra is much smaller in diameter than the egg, the eggshell must be elastic enough to travel down the ovipositor to gain entrance to the host body.

Development

Parthenogenesis occurs throughout the order Hymenoptera. Three types are recognized: thelyotoky, deuterotoky, and arrhenotoky. In **thelyotoky** all individuals are uniparental (parthenogenetic), and virtually no males are produced. Some sawflies and parasitic Hymenoptera in several families are in this category. Some males are produced in **deuterotoky,** but all individuals are nevertheless uniparental. The most common condition is **arrhenotokous parthenogenesis,** in which only the males are uniparental and are haploid. The females come from fertilized eggs and are diploid. By some still obscure means, the mated, egg-laying female can influence whether a given egg will be fertilized. Needless to say, all of the foregoing types of parthenogenesis may lead to sex ratios that diverge strongly from the 50/50 that we normally expect. An interesting discussion of parthenogenesis in Hymenoptera is given by Doutt.[15]

Some parasitic hymenoptera, especially members of the Encyrtidae, exhibit polyembryony. Within the embryo, a series of cell divisions results in cell aggregations called morulae, each of which is contained in a membrane.[1,29] The parasite may become a chain or branched sac filled with larvae. This chain later breaks up as larval development continues. All embryos from a single egg are of the same sex; female embryos develop from fertilized eggs and males from unfertilized eggs. In some species, up to 3000 embryos may be formed from a single egg and subsequent morphogenesis is controlled by host hormones, primarily ecdysone.[2]

A few hymenopterans are hypermetamorphic and have a planidium larva. An example is *Perilampus hyalinus,* an American chalcid, which lays its eggs on foliage.[1] The planidia search for and penetrate caterpillars such as the fall webworm (*Hyphantria*). Once inside, they must find and penetrate the larvae of another parasite of *Hyphantria,* the tachinid dipteran, *Ernestia.* Thus, they are hyperparasites, a common condition among Hymenoptera. Hypermetamorphosis is found among the Perilampidae, Echaritidae, and Ichneumonidae, but of course not all species are hyperparasites.

Classification and Examples

The Hymenoptera are divided into two suborders, the Symphyta and Apocrita, easily distinguished because the Symphyta do not have a pedicel. The latter are mostly phytophagous, but the small family Orussoidea parasitizes larvae of cerambycid and buprestid beetles.

Two divisions of the Apocrita, the Aculeata and the Parasitica (also called Terebrántes), are commonly recognized, but the latter is possibly polyphyletic. Some of the Aculeata are parasitic, and some of the Parasitica are not. In the Aculeata the eighth and ninth tergites are reduced and are retracted into the seventh, so that the ovipositor (sting) seems to issue from the apex of the abdomen. In the Parasitica the eighth segment is not retracted into the seventh, and the ovipositor is exposed almost to its base. Superfamilies in the Aculeata with parasitic forms are the Bethyloidea and Vespoidea (some). In the Parasitica, parasitic forms are the Evanioidea, Trigonaloidea, Ichneumonoidea, Proctotrupoidea (Serphoidea), Chalcidoidea (many), and Cynipoidea (some).[1] We will discuss a few examples of the biology of this group.

The arbitrary nature of the distinction between parasitism and predation in hymenopterans is well-illustrated in the Aculeata, which contains the ants, solitary and social wasps, and bees. A number of the wasps sting their prey (host) to paralyze it and then lay their eggs on the host, which provides food for the young. Some entomologists distinguish parasite from predator on the basis of whether the adult wasp constructs a cell to house the paralyzed host. An intermediate position is occupied by members of the Bethylidae, which drag the host to a sheltered position after paralyzing it and then oviposit on it. The female, who stands guard while the larvae develop as ectoparasites of the host, sometimes bites the host and feeds on its body fluids. Additional females may lay eggs on the same host individual, and the insects guard the young cooperatively. Such behavior may demonstrate an early stage in the evolution of the maternal care practiced by the social wasps. The various vespoid families exhibit the spectrum of maternal behavior, ranging from the typical parasitoid practice of laying an egg on the prey and then abandoning it, to the building of a cell to house the prey and wasp young, and finally to the maternal care of the social wasps.

The Ichneumonoidea comprises the Ichneumonidae, the Braconidae, and the Aphidiidae (sometimes considered a subfamily of braconids). The Ichneumonidae is a very large family with more than 3000 described species. They are all parasitic, sometimes on other ichneumonids, and most are endoparasitic. They tend to have a slender abdomen and often have a very long ovipositor that is permanently extruded. The largest ichneumonids in the United States may exceed 40 mm, and their ovipositors may be twice that length. Such insects attack the larvae of horntails, wood wasps, and wood-boring beetles, somehow detecting the location of the host within its tunnel, which may be several centimeters below the surface of the wood. In Britain *Rhyssa persuasoria* is attracted by a substance produced by a fungus in the frass (feces and wood pulp) of the tunnels of the host, a wood wasp, *Sirex.* When the location of the host has been determined, apparently by the antennae, the ichneumonid raises its abdomen and places the tip of its ovipositor exactly in the location indicated by its antennae. Aided by sharp cutting ridges at the end of the ovipositor, the ichneumonid forces the ovipositor through the wood by pressure and twisting motions of its abdomen[1] (Fig. 39.9).

It is astonishing that the apparently delicate ovipositor can penetrate the wood to reach the host. *Pseudorhyssa alpestris,* a parasite of the alder wood wasp, *Xiphydria camelus,* has solved the problem of reaching the host in another way. *Xiphydria* is also parasitized by *Rhysella*

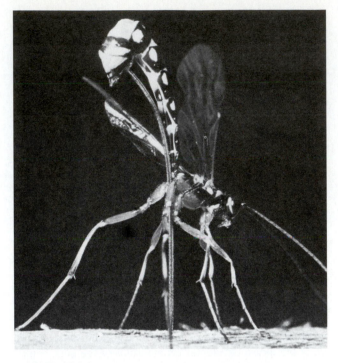

FIGURE 39.9

Ichneumonid with the end of her abdomen raised to thrust her ovipositor through the wood to the host, a wood-boring, larval beetle. The coxae of the hind legs help steady and support the ovipositor as it is thrust into the wood.

Courtesy of L. L. Rue, III.

FIGURE 39.10

Larva of a tomato hornworm (Lepidoptera, Sphingidae) parasitized by the braconid *Apanteles* sp. The white objects on the dorsum of the caterpillar are the pupal cocoons of the wasp.

Courtesy of O. W. Olsen.

curvipes, which locates its host and oviposits much like *R. persuasoria. Pseudorhyssa,* however, simply locates the oviposition holes of *Rhysella,* inserts its ovipositor, and lays an egg in *Rhysella's* host. When the *Pseudorhyssa* hatches, it kills the larva of *Rhysella* and takes over the host![37]

Members of the Braconidae are similar to those of Ichneumonidae, although they are generally smaller (15 mm or less in length) and tend to be more heavy bodied. They also differ in that they pupate in silken cocoons on the outside surface of the host, rather than within its body (Fig. 39.10). A number of species of ichneumonid and braconid adults are known to feed on the host, as well as to oviposit within it. *Polysphincta* stings and paralyzes its spider host and then lays an egg on the spider's opisthosoma and feeds on its body fluids.[12] It is known that some species cannot mature their eggs or achieve full reproductive potential without first feeding on host body fluids,[6,28] and it is possible that many other species have this requirement. Larvae of the ichneumonid, *Hyposoter* spp., can parasitize the tussock moth only if the host has been previously parasitized by a braconid, *Cotesia melanoscela.*[18] The first parasite apparently modifies the host defense system in such a way that the second can survive.

The developing larvae may modify their host environment to the advantage of the parasite. No more than one larva of the aphidiid parasites of pea aphids, *Aphidius smithi,* can develop in a single host; if more than one egg is deposited, the supernumerary parasites are somehow eliminated. However, superparasitized pea aphids have a greater food incorporation efficiency and growth rate than singly

parasitized hosts; thus, the early presence of more than one parasite larva modifies the host physiology by some unknown mechanism.[10] Some investigators have proposed that the dominant egg releases substances that suppress the development of supernumerary parasites.[33]

The Chalcidoidea is a large superfamily that contains 18 families. Its members are very small (less than 5 mm), metallic green or black wasps with few wing veins. Some of them induce the growth of plant galls. The parasitic chalcidoids attack a wide variety of hosts, the majority of which are in the Lepidoptera, Coleoptera, Hymenoptera, Diptera, and Hemiptera. A few are parasites of ticks, mites, and egg cocoons of spiders. The Trichogrammatidae and Mymaridae are parasites of insect eggs. The mymarids include the smallest insects known (0.2 mm); they are called **fairyflies** and are smaller than some protozoa (Fig. 39.11).

Several Encyrtidae have been very valuable in control of scale insects (Hemiptera, Homoptera). Encyrtids also illustrate the rather remarkable complexity of parasitoid evolution. *Ooencyrtus nezarae,* for example, prefers previously parasitized *Megacopta* (Hemiptera) over unparasitized ones, even though larval survival is lower in hosts that already contain *O. nezarae* larvae.[36] Evidently the effort of boring a new hole in the host is a more expensive physiological burden to the parasite than the lowered survival of its offspring.

Most of the Cynipoidea are phytophagous, and they induce the formation of galls in the tissue on which they feed. However, a number are parasitic on various other insects. *Charips* is a hyperparasite of a braconid primary parasite of aphids (Fig. 39.12).

The Evanioidea, Trigonaloidea, and Proctotrupoidea are less common parasites of a variety of insects. Trigonaloids parasitize larvae of social wasps (Vespoidea), and some are hyperparasites of ichneumonid and tachinid dipteran parasites

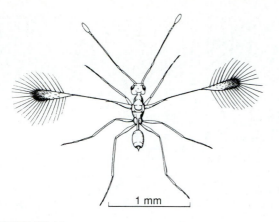

FIGURE 39.11

Fairyfly, *Mymar pulchellus,* female (Chalcidoidea, Mymaridae). This is one of the larger species in the family, commonly parasitic in the eggs of other insects.

From R. R. Askew, *Parasitic Insects.* Copyright © 1971 American Elsevier Publishing Co., New York, NY.

of lepidopterans. The trigonalid female lays many eggs, not on its prospective host but on vegetation. The eggs do not hatch until they are eaten by a caterpillar, but they can remain viable for several months.[1] Once eaten by the caterpillar, they hatch, penetrate the gut to the hemocoel, and then try to find a larval primary parasite in which to develop further. The Evanioidea contains several families, one of which (Evaniidae) contains parasites of cockroach eggs. The evaniid egg is laid within one of the cockroach eggs. After hatching, the evaniid larva consumes the roach egg and then may eat the other eggs in the egg case (ootheca) of the cockroach. The Proctotrupoidea is a somewhat larger group, including seven families.[1] They parasitize various insects, including homopterans, dipterans, neuropterans, and coleopterans. Pelecinidae (Fig. 39.13) have a long, attenuated abdomen, which may compensate for the short ovipositor when the female burrows through the ground in search of her host, the larvae (grubs) of May beetles.

BIOLOGICAL CONTROL

The examples presented in this chapter are but a few of the rapidly growing list of parasitoid/host combinations now being studied as potential biological control agents along with viral, bacterial, nematode, and other pathogens and pheromones. The major stimuli for this explosion of research effort are (1) pesticide resistance and (2) the cost of agricultural chemicals. We have learned, through what amounts to a massive experiment in evolution, that insecticides quickly select for resistant pest populations; thus, new chemicals continually must be developed. The cost of agricultural chemicals includes not only research, development, patenting, and associated legal expenses, but also fertilizer, the hydrocarbon stocks from which new compounds are synthesized, and transportation. Finally, chemical insecticides are toxic in some degree to other fauna, including humans; that is, they are nonspecific. In addition, they are often quite toxic to beneficial species, such as pollinators and natural predators of pest insects.

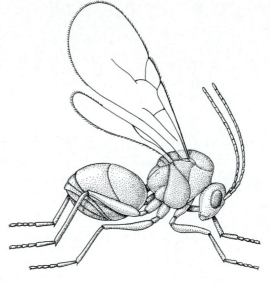

FIGURE 39.12

Female *Charips victrix* (Cynipoidea, Cynipidae), a hyperparasite on a braconid primary parasite of aphids.

Drawing by William Ober.

By contrast, at least in theory, once a successful biological control agent has been introduced, no additional cost accrues except occasionally for reintroduction of the parasite. Furthermore, a good biological control agent should keep the host population to a low but tolerable level in equilibrium. Evolution of greater resistance in the host population to the parasite is likely to be matched by adaptations in the parasite population to restore the equilibrium. However, the initial costs of developing biological controls for practical use may be quite high because extensive research is required. Careful ecological investigations of both the parasite and pest must be conducted before a foreign organism can be introduced into an area. Sibling species or different biological races of a parasite may have different host preferences or biological characteristics that determine success or failure. Parasites that are successful in biological control usually have the following qualities (modified from Askew):[1]

1. They must have a high host-searching capacity.

2. They must have a very limited range of hosts but be able to use a few other host species in addition to the target; that is, when the pest population is reduced, the parasite should be able to maintain itself on alternative hosts. Thus, a high enough parasite population is available to counteract surges in the pest population.

3. Their life cycle must be substantially shorter than that of the pest if the pest population consists of overlapping generations, or their life cycle must be synchronized with the pest life cycle if the pest population is composed of a single developmental stage at any time.

4. They must be able to survive in all habitats occupied by the pest.

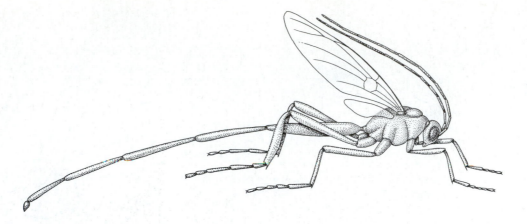

FIGURE 39.13

Female *Pelecinus polyturator* (Proctotrupoidea, Pelecinidae). This striking insect is 2 in. or longer and shining black. The rare males are about an inch long and have a swollen abdomen. They are parasites of the larvae (grubs) of May beetles (Scarabaeidae), which burrow in the soil.
Drawing by William Ober.

5. They must be easily cultured in the laboratory so that large enough numbers are available for introductions.

6. They must control the pest population rapidly (some workers have suggested that control must occur within three years of the time of introduction).[9]

   These conditions mean that although biological control is theoretically rather straightforward, in practice it is often a rather difficult goal to achieve. The exploration and experimentation needed to find, domesticate, and develop wild insects with the listed properties can easily consume one's career. Hokkanen[19] estimates that only about one out of seven biological control introductions results in positive economic results. Nevertheless, the approximately 300 instances of successful introductions have returned about 30 times their original investments.[19]

   A massive body of literature on biological control agents of all kinds, many of them obscure insects a lay person would never recognize by name or by sight, is accumulating rapidly. For a wondrous eye-opening tour through this world, see the paper by Waage and Hassell,[38] the major three-volume set edited by Pimentel,[30] the theoretical work on integrated pest management edited by Kogan,[26] the volume on plants, parasitoids, and insect predators edited by Boethel and Eikenbarry,[5] and the second edition of DeBach and Rosen[13] (particularly useful for students).

References

1. Askew, R. R. 1971. *Parasitic insects.* New York: American Elsevier Publishing Co., Inc.

2. Baehrecke, E. H., J. M. Aiken, B. A. Dover, and M. R. Strand. 1993. Ecdysteroid induction of embryonic morphogenesis in a parasitic wasp. *Dev. Biol.* 158:275–87.

3. Bänziger, H. 1968. Preliminary observations on a skin-piercing blood-sucking moth (*Calyptra eustrigata* Hmps.) (Lep., Noctuidae) in Malaya. *Bull. Entomol. Res.* 58:159–63.

4. Batra, S. W. T. 1965. Organisms associated with *Lasioglossum zyphyrum* (Hymenoptera: Halictidae). *J. Kans. Entomol. Soc.* 38:367–89.

5. Boethel, D. J., and R. D. Eikenbarry, eds. 1986. *Interactions of plant resistance and parasitoids and predators of insects.* Chichester, England: Ellis Horwood, Ltd.

6. Bracken, G. K. 1965. Effects of dietary components on fecundity of the parasitoid *Exeristes comstockii* (Cress.) (Hymenoptera: Ichneumonidae). *Can. Entomol.* 97:1037–41.

7. Butler, L. 1993. Parasitoids associated with the macrolepidoptera community at Cooper's Rock State Forest, West Virginia: A baseline study. *Proc. Entomol. Soc. Wash.* 95:504–10.

8. Büttiker, W. 1967. Biological notes on eye-frequenting moths from N. Thailand. *Mitt. Schweiz. Entomol. Ges.* 39(1966):151–79.

9. Clausen, C. P. 1951. The time factor in biological control. *J. Econ. Entomol.* 44:1–9.

10. Cloutier, C., and M. Mackauer. 1980. The effect of superparasitism by *Aphidius smithi* (Hymenoptera: Aphidiidae) on the food budget of the pea aphid, *Acyrthosiphon pisum* (Homoptera: Aphidiidae). *Can. J. Zool.* 58:241–44.

11. Crowson, R. A. 1955. *The natural classification of the families of Coleoptera.* London: Nathaniel Lloyd.

12. Cushman, R. A. 1926. Some types of parasitism among the Ichneumonidae. *Proc. Entomol. Soc. Wash.* 28:5–6.

13. DeBach, P., and D. Rosen. 1991. *Biological control by natural enemies,* 2d ed. Cambridge: Cambridge University Press.

14. Dodd, F. P. 1912. Some remarkable ant friend Lepidoptera of Queensland. *Trans. R. Entomol. Soc. Lond.* 59(1911):577–90.

15. Doutt, R. L. 1959. The biology of parasitic Hymenoptera. *Ann. Rev. Entomol.* 4:161–82.

16. Durden, L. A. 1987. Predator-prey interactions between ectoparasites. *Parasitol. Today* 3:306–8.

17. Greathead, D. J. 1968. Further descriptions of *Halictophagus pontilifex* Fox and *H. regina* Fox (Strepsiptera: Halictophagidae) from Uganda. *Proc. R. Entomol. Soc. Lond.* (B)37:91–97.

18. Guzo, D., and D. B. Stoltz. 1985. Obligatory multiparasitism in the tussock moth, *Orgyia leucostigma. Parasitology* 90:1–10.

19. Hokkanen, H. M. T. 1991. New approaches in biological control. In Pimentel, D., ed. *CRC handbook of pest management in agriculture.* Boca Raton, Fl.: CRC Press, 185–98.

20. Hu, R., K. E. Hyland, and T. N. Mather. 1993. Occurrence and distribution in Rhode Island of *Hunterellus hookeri* (Hymenoptera: Encyrtidae), a wasp parasitoid of *Ixodes dammini*. *J. Med. Ent.* 30:277–80.

21. Jordan, K. 1910. Notes on the anatomy of *Hemimerus talpoides*. *Novit. Zool.* 16:327–30.

22. Kathirithamby, J., and D. Grimaldi. 1993. Remarkable stasis in some Lower Tertiary parasitoids: Descriptions, new records, and review of Strepsiptera in the Oligo-Miocene amber of the Dominican Republic. *Entomol. Scandinavica* 24:31–41.

23. Kathirithamby, J., and J. S. Johnston. 1992. Stylopization of *Solenopsis invicta* (Hymenoptera: Formicidae) by *Caenocholax fenyesi* (Strepsiptera: Myrmecolacidae) in Texas. *Ann. Entomol. Soc. Amer.* 85:293–97.

24. Kirkpatrick, T. W. 1937. Colour vision in the triungulin larva of a strepsipteron (*Corioxenos antestiae* Blair). *Proc. R. Entomol. Soc. Lond.* (A)12:40–44.

25. Kirkpatrick, T. W. 1937. Studies on the ecology of coffee plantations in East Africa. II. The autecology of *Antestia* spp. (Pentatomidae) with a particular account of a strepsipterous parasite. *Trans. R. Entomol. Soc. Lond.* 86:247–343.

26. Kogan, M., ed. 1986. *Ecological theory and integrated pest management practice.* New York: John Wiley & Sons, Inc.

27. Kuris, A. M. 1974. Trophic interactions: Similarity of parasitic castrators to parasitoids. *Q. Rev. Biol.* 49:129–48.

28. Leius, K. 1961. Influence of food on fecundity and longevity of adults of *Itoplectis conquisitor* (Say) (Hymenoptera: Ichneumonidae). *Can. Entomol.* 93:771–800.

29. Nikol'skaya, M. N. 1952. *The chalcid fauna of the U.S.S.R.* Moskva: Izdatel'stvo Akademii Nauk SSSR. Birron, A., and Z. S. Cole, trans. 1963. Israel Program for Scientific Translations. Washington, D.C.: U.S. Dept. Commerce.

30. Pimentel, D., ed. 1990–1991. *CRC Handbook of pest management in agriculture.* I, II, III, 2d ed. Boca Raton, Fl.: CRC Press, Inc.

31. Popham, E. J. 1961. On the systematic position of *Hemimerus* Walker—a case for ordinal status. *Proc. R. Entomol. Soc. Lond.* (B)30:19–25.

32. Popham, E. J. 1962. The anatomy related to the feeding habits of *Arixenia* and *Hemimerus* (Dermaptera). *Proc. Zool. Soc. Lond.* 139:429–50.

33. Silvers, M. J., and A. J. Nappi. 1986. *In vitro* study of physiological suppression of supernumerary parasites by the endoparasitic wasp *Leptopilina heterotoma*. *J. Parasitol.* 72:405–9.

34. Simmonds, F. J., and F. D. Bennett. 1977. Biological control of agricultural pests. In White, D., ed. *Proceedings of the Fifteenth International Congress of Entomology.* College Park, Md.: Entomological Society of America, 464–72.

35. Snodgrass, R. E. 1935. *Principles of insect morphology.* New York: McGraw-Hill Book Co.

36. Takasu, K., and Y. Hirose. 1991. The parasitoid *Ooencyrtus nezarae* (Hymenoptera: Encyrtidae) prefers hosts parasitized by conspecifics over unparasitized hosts. *Oecologia* 87:319–23.

37. Thompson, G. H., and E. R. Skinner. 1961. The alder wood wasp and its insect enemies. Film, Commonwealth Forestry Institute. Quoted in Askew, R. R. 1971. *Parasitic insects.* New York: American Elsevier Publishing Co., Inc.

38. Waage, J. K., and M. P. Hassell. 1982. Parasitoids as biological control agents—a fundamental approach. In Anderson, R. M., and E. U. Canning, eds. Parasites as biological control agents. Symposia of the British Society for Parasitology, vol. 19. *Parasitology* 84(4):241–68.

39. Williams, J. R. 1957. The sugar-cane Delphacidae and their natural enemies in Mauritius. *Trans. R. Entomol. Soc. Lond.* 109:65–110.

Chapter 40

PARASITIC ARACHNIDS: SUBCLASS ACARI, THE TICKS AND MITES

There is a degree of originality about being an arachnologist, about being detectably superior to those who cannot distinguish a Pholcus *from a* Phalangium.

Theodore Savory

Ticks and mites are immensely important in human and veterinary medicine, many by causing diseases themselves and others by acting as vectors of serious pathogens. A large body of literature exists on all aspects of acarine biology, yet vast areas remain relatively unexplored. For a more extensive treatment of the subjects covered in this chapter, see volumes written or edited by Dusbábek and Bukva,[7] Evans,[9] Houck,[18] Sauer and Hair,[39] Schuster and Murphy,[40] and Sonenshine.[47]

All ticks are epidermal parasites during their larval, nymphal, and adult instars; and many mites are parasites on or in the skin or in the respiratory system or other organs of their hosts. Some mites, although not actually parasites of vertebrates, stimulate allergic reactions when they or their remains come into contact with a susceptible individual. Ticks are found in nearly every country of the world; mites are even more ubiquitous, thriving on land, in fresh water, and in the oceans. Many millions of dollars are spent annually throughout the world in attempts to control these pests and the diseases they transmit.

The general morphology of the Acari, described in detail in Chapter 33, can be summarized as follows. Segmentation is reduced externally, having been obscured by fusion. Tagmatization has resulted in two body regions: an anterior **gnathosoma,** or **capitulum,** bearing the mouthparts, and a single **idiosoma,** containing most internal organs and bearing the legs. Most adult Acari have eight legs but some mites have only one to three pairs. The idiosoma is further divided into regions (Fig. 33.14) as follows: The portion bearing the legs is the **podosoma,** the first and second pairs of legs are on the **propodosoma,** and the third and fourth pairs are on the **metapodosoma.** The portion of the body posterior to the legs is the **opisthosoma.** The gnathosoma and propodosoma together comprise the **proterosoma,** and the metapodosoma and opisthosoma together are the **hysterosoma.** These terms may seem confusing at first, but they are very useful in describing and therefore identifying acarines.

The capitulum mainly is made up of feeding appendages surrounding the mouth. On each side of the mouth is a **chelicera,** which functions in piercing, tearing, or gripping host tissues. The form of the chelicerae varies greatly in different families; thus, they are useful taxonomic features. Lateral to the chelicerae is a pair of segmental **pedipalps,** which also vary greatly in form and function related to feeding. Ventrally the coxae of the pedipalps are fused to form a **hypostome;** a **rostrum,** or **tectum,** extends dorsally over the mouth. Some or all of these structures can be retracted in some acarines.

Ticks and mites, although basically similar, have distinct differences, as follows:

Ticks	Mites
• hypostome toothed, exposed	• hypostome unarmed, hidden
• large, easily macroscopic	• small, usually microscopic
• Haller's organ (Fig. 40.1) present on first tarsi	• Haller's organ absent
• peritreme absent	• peritreme present in Mesostigmata

The mouthparts of the Acari are modified for specialized feeding. In ticks the pedipalps grasp a fold of skin while the chelicerae cut through it. As cutting progresses, the hypostome is thrust into the wound, and its teeth help anchor the tick to its host. Blood and lymph from lacerated tissues well into the wound and are sucked up. Soft ticks feed rapidly, leaving their host after engorging, whereas hard ticks remain attached for several days. Some hard ticks, particularly those with short mouthparts, secrete a cementing substance that hardens, further securing them to the host.

Mites feed on vertebrates in much the same way, but because their mouthparts are so small, most feed on lymph or other secretions rather than on blood. There is no toothed hypostome, and the chelicerae vary from chelate cutters to hooks or stylets. Variations in feeding behavior of mites are discussed in the sections on their respective groups.

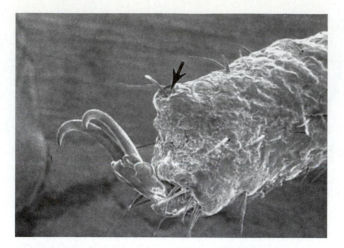

FIGURE 40.1

Haller's organ (**arrow**).
Courtesy of Tyler Woolley.

Classification of the Arachnida and Acari

Acarology is a highly active discipline in which biologists the world over are still describing many species annually while pursuing sophisticated work in chemical ecology, immunology, biological control, and epidemiology. Phylogenetic research is an indispensable part of this effort, but the taxonomic database on acarines is a massive one that grows monthly. Evans[9] elevates the Acari to subclass level (cl. Arachnida) and recognizes two superorders and seven orders but admits that other schemes are equally acceptable, given the fluid state of arachnid systematics. Krantz[25] proposes two orders: Acariiformes, with medically important members in the suborders Acaridida and Actinedida (all mites), and Parasitiformes, with medically important species in the suborders Mesostigmata (mites) and Ixodida (ticks). Both orders also contain nonparasitic species. Varma[54] agrees with Krantz.[25] The major difference between classifications is the level (order, suborder) assigned to the taxa. The system used in this book is based on Evans[9] mainly because her book gives a detailed morphological and historical presentation of the subclass, and thus it provides a reader with the information needed to make decisions about the hierarchical levels presented by various authors.

ORDER IXODIDA: THE TICKS

Because of their large size and pesky habits, ticks have been recognized for centuries. Both Homer and Aristotle referred to them in their writings, but Linnaeus in 1746 was the first to attempt to classify them among other animals. Evans[9] recognizes three superfamilies: Ixodoidea, the hard ticks, Argasoidea, the soft ticks, and Nuttallielloidea. The latter contains a single African species, *Nuttalliella namaqua,* of which only the females are known, and will not be considered further here. Ticks' importance as agents and vectors of disease also has long been recognized. Pathogenesis attributable to these parasites appears in several ways:

1. **Anemia.** Blood loss in heavy infections can be considerable; as much as 200 pounds of blood can be lost from a single large host in one season.[19]

2. **Dermatosis.** Inflammation, swelling, ulcerations, and itching can result from a tick bite. These reactions often are caused by pieces of mouthparts remaining in the wound after the tick is forcibly removed, but constituents of the tick's saliva and secondary infection by bacteria probably are also involved.

3. **Paralysis.** A condition known as tick paralysis is common in humans, dogs, cattle, and other mammals when they are bitten near the base of the skull. This paralysis seems to result from toxic secretions by the tick and is quickly reversed when the parasite is removed.[55] Mechanisms were reviewed by Gothe, Kunze, and Hoogstraal.[13]

4. **Otoacariasis.** Infestation of the ear canal by ticks causes a serious irritation to the host, sometimes accompanied by severe infection.

5. **Infections.** In addition to common pyogenic infections, ticks can transmit viruses, bacteria, rickettsias, spirochetes, protozoa, and filariae. These will be discussed further with the appropriate vectors.

Livestock losses from all arthropod infestations were estimated at $2.8 billion annually in the mid-1970s.[34] In 1965 the annual losses caused by ticks to cattle production were estimated at $60 million and for sheep production, at $4.7 million.[40] The loss to beef cattle in Australia is estimated at $25 million annually.[56] Control efforts are worth their expense, however; prior to 1906 *Babesia bigemina* transmitted by *Boophilus microplus* caused annual losses of $100 million in the United States alone.[34]

Biology

All ticks undergo four basic stages in their life cycles—egg, larva, nymph, and adult—all of which might require from six weeks to three years to complete. Ixodids have only one nymphal instar, but argasids have as many as five. Copulation nearly always occurs on the host. The male tick produces a spermatophore, which it places under the genital operculum of the female. A blood meal is usually required for egg production, although exceptions are known. Also, some ticks are parthenogenetic. The engorged female drops to the ground, where she deposits her eggs in the soil or humus. A six-legged larva (Fig. 40.2) hatches from each of the 100 to 18,000 eggs and climbs onto low vegetation, where it quests for a host. On finding one, it feeds and then molts to an eight-legged nymph. If molting through all instars occurs on the same host animal, the tick is called a **one-host tick,** such as *Boophilus.* If the nymph drops off, molts to adult, and attaches to another host, the tick is a **two-host tick.** Confinement of instars to one or two hosts is an adaptation to feeding on wide-ranging hosts.[17.] Most ixodids are **three-host ticks,** whereas argasids, with their multiple nymphal stages, are **many-host ticks.** Clearly the use of a series of hosts increases the opportunities for transmission of pathogens.

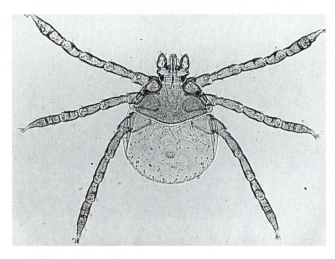

FIGURE 40.2

Six-legged larva of *Ixodes* sp.
Courtesy of Jay Georgi.

Some ticks are rather host specific, but most are opportunists that will feed on a variety of hosts. Ticks are hardy and can withstand periods of starvation as long as 16 years. Some may have a life span of up to 21 years.

Ticks exhibit a surprising amount of complex behavior, much of which is under chemical control. The tick pheromones that have been recognized influence aggregation, attachment, and reproduction, especially mate recognition and subsequent courtship. This research has been reviewed by Sonenshine.[46]

The major pheromones, and the behaviors they influence, include guanine (aggregation), *o*-nitrophenol (searching and aggregation), methyl salicylate and pelargonic acid (clasping and attachment during mating), 2,6-dichlorophenol (attraction and potential mate recognition in males), cholesteryl oleate (mounting), and 20-hydroxyecdysone (copulation), although a particular behavior may depend on mixtures of phermones as well as on the developmental stage and nutritional state of the tick. Species also differ, as might be expected, in their aggregation behaviors, mating rituals, and responses to pheromones. The chemicals themselves emanate from the anus, coxal glands, and female genital aperture.

Mating behavior is typically specific, complex, and controlled at several stages by chemicals.[46] In *Dermacentor variabilis,* 2,6-dichlorophenol excites feeding males, which then detach, move toward females, and mount them. The male then begins probing for the female gonopore. Unless the male encounters a genital sex pheromone in the vulva, using sensillae on his chelicerae, he will stop the mating attempt. The species-specific aspect of this behavior occurs at this last stage. Thus, a *Dermacentor andersoni* male cannot distinguish a female *D. variabilis* from one of his own species until he probes the genital opening. In camel ticks the decision-making events occur at the beginning rather than at the end of the mating ritual. *Hyalomma dromedarii* and *H. anatolicum* males respond to the different amounts of 2,6-dichlorophenol produced by females of their respective species. Complex behavior does not necessarily end with copulation. Male *Ixodes dammini,* for example, mate preferentially with feeding females, repel competing males, and may remain attached to (thus protective of) the female after she finishes feeding.[59]

The overall picture of tick chemical ecology is that of highly evolved systems with significant population level consequences.

Family Ixodidae

The family of **hard ticks** is divided into three subfamilies: Ixodinae, with the single genus *Ixodes;* Amblyominae, containing *Amblyomma, Haemaphysalis, Aponomma,* and *Dermacentor;* and Rhipicephalinae, with *Rhipicephalus, Anocentor, Hyalomma, Boophilus,* and *Margaropus.* We will discuss these genera individually.

Hard ticks are easily recognized as such, since the capitulum is terminal and can be seen in dorsal view. By contrast, in soft ticks the capitulum is subterminal and cannot be seen in dorsal view. Following are other characters of the Ixodidae: (1) a large anterodorsal sclerite, the **scutum,** is present; (2) eyes, when present, are on the scutum; (3) pedipalps are rigid, not leglike; (4) marked sexual dimorphism occurs in size and often coloration; (5) the female has **porose areas** (regions with many small pits) on the basis capituli; (6) the coxae usually have spurs; (7) a pulvillus is present on the tarsus; (8) the stigmatal plates are behind the fourth pair of legs; (9) the posterior margin of the opisthosoma usually is subdivided into sclerites called **festoons;** and (10) there is only one nymphal instar.

• *Ixodes* Species

Ixodes is the largest genus of hard ticks, with over 200 species; nearly 40 species are known from North America. Most parasitize small mammals and are so small themselves that they are easily overlooked. Phylogenetically *Ixodes* is a unique, highly specialized group; however, the absence of eyes is considered a primitive character. The pronounced sexual dimorphism of the mouthparts, which are longer in the female, is a condition unknown in any other genus. Festoons are absent, and the anal groove is anterior to the anus. *Ixodes* spp. are three-host ticks.

Ixodes scapularis, the **blacklegged tick** (Fig. 40.3), is common in the eastern and south-central United States. It feeds on a wide variety of hosts and can be a major pest on dogs. It bites humans freely, commonly resulting in a strong reaction with pain at the site and generalized malaise for a short time. *Ixodes pacificus* is found along the west coast of California, Oregon, and Washington on deer, cattle, and other mammals. It also welcomes a human meal and is a vector of Lyme disease, caused by the spirochaete *Borrelia burgdorferi.* Transovarial and transstadial passage of Lyme disease has been reported.[26] *Ixodes holocyclus* is the major cause of tick paralysis in Australia. Other species in the genus are also common agents of paralysis in several parts of the world. Tick-borne encephalitis in Europe and Asia can be transmitted by the bite of *I. ricinus, I. persulcatus,* and *I. pavlovskyi.*

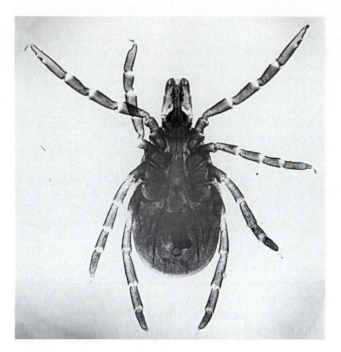

FIGURE 40.3

Ixodes scapularis, the black-legged tick.
Courtesy of Jay Georgi.

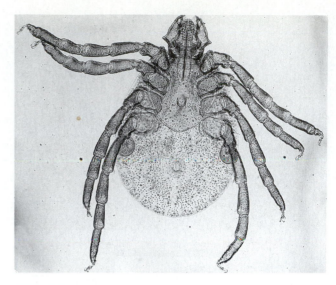

FIGURE 40.4

Haemaphysalis leporispalustris, the rabbit tick.
Courtesy of Jay Georgi.

Lyme disease takes its name from the town of Old Lyme, Connecticut, where an epidemic of arthritis occurred during the mid-1970s.[20] Many of those afflicted had noticed a skin erythema (red swelling and rash) at the site of a tick bite prior to the onset of joint pains. The rash sometimes spread, and it was often accompanied by fever and headaches. We now know that neurological symptoms, facial paralysis, meningitis, and occasionally cardiac disease occur in some cases. *Ixodes dammini* is the primary vector in the northeastern United States, but seven other species of *Ixodes* are competent to transmit the disease.[20] Half of all arthropod-transmitted infections in the United States are with *Borrelia burgdorferi,* making it the leading arthropod-borne disease in the country.

In the northeastern United States, between 25% and 50% of *I. dammini* are infected with the Lyme disease spirochaete.[20] This high prevalence is attributed to the relatively unselective feeding habits of the tick, especially the nymphal stages, and the participation of rodent and deer reservoirs. Deer, cats, raccoons, opossums, chickens, and lizards all serve as hosts, but the ticks vary in feeding and egg production, depending on their hosts.[21,33,58] Mice of the genus *Peromyscus* are the primary reservoirs; deer serve mainly as a large source of blood to maintain a high tick population. *Borrelia burgdorferi* is widely distributed in the northern hemisphere,[20] and the spirochaete has also been isolated from the ticks *Amblyomma americanum* and *A. maculatum* and the flea *Ctenocephalides felis.*[52]

• *Haemaphysalis* Species

Haemaphysalis spp. are easily recognized by the second segments of the pedipalps, which are produced laterally into spurs. These small ticks often are overlooked unless they are engorged. About 150 species are known worldwide, with only two found in North America. Most are three-host ticks. There is little sexual dimorphism, and both sexes have festoons.

Haemaphysalis leporispalustris, the **rabbit tick** (Fig. 40.4), is common on rabbits from Alaska to Argentina. It occasionally feeds on domestic animals but rarely bites humans. Its main importance is as a vector of tularemia and Rocky Mountain spotted fever among wild mammals. *Haemaphysalis cordeilis,* the **bird tick,** is common on turkeys, quail, pheasants, and related game birds. It may vector diseases among these hosts. It seldom is found on mammals and is not a pest on humans. Most species of *Haemaphysalis* occur in Asia, Africa, and the East Indies.

• *Dermacentor* Species

Among the most medically important of all the ticks, particularly in North America, this genus contains about 30 species. At least seven of these are found in the United States. These ticks are ornate, with punctations and colored markings, and usually show sexual dimorphism of color. Both sexes have festoons. The eyes are well-developed, and the stigmatal plates have numerous shallow depressions called **goblets.** The sides of the basis capituli are parallel. Most species are three-host ticks, but a few are one-host.

Dermacentor andersoni (Fig. 40.5) is the **Rocky Mountain wood tick.** It is distributed throughout most of the western United States west of the Great Plains and is most prevalent in mountainous, brushy terrain. The larvae feed on small mammals, especially chipmunks, ground squirrels, and rabbits; nymphs also feed on these hosts, as well as on marmots, porcupines, and other medium-sized mammals. Adults may feed on any of these animals but seem to prefer larger hosts such as deer, sheep, cattle, and coyotes, and they are not reluctant to bite humans. When hosts are plentiful, the entire life cycle may take a year, but if long waits occur between meals, the life cycle may extend up to three years. All stages can survive about a year without feeding. This species becomes most

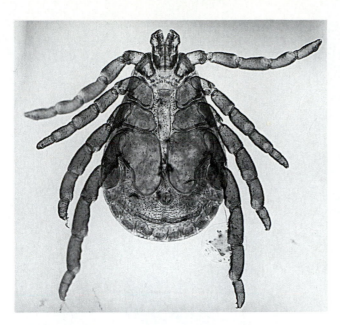

FIGURE 40.5

Dermacentor andersoni, the Rocky Mountain wood tick.
Courtesy of Jay Georgi.

active in early spring, as soon as the snow cover is off, and actively continues to seek hosts until about the beginning of July, when the ticks become dormant.

Dermacentor andersoni is the vector for several diseases that afflict humans: tick paralysis, Powassan encephalitis virus (mainly transmitted by *Ixodes spinipalpis*), Colorado tick fever virus, tularemia, Rocky Mountain spotted fever, and some nonpathogenic viruses. These infections and others, such as anaplasmosis, are transmitted to other mammals by *D. andersoni.*

Dermacentor variabilis, the **American dog tick,** is common throughout the eastern United States and is extending its range rapidly. One focus has been established in the Pacific Northwest, mainly along river valleys in Washington, Oregon, and Idaho. It prefers to feed on dogs but will attack horses and other mammals, including humans. This tick appears to occupy the same niche in the eastern states that *D. andersoni* occupies in the western states, although *D. variabilis* is more urban. Isolated foci of *D. variabilis* are known in western Oregon.

The female lays 4000 to 6500 eggs on the ground, which hatch in about 35 days. The six-legged larva feeds on small rodents, especially voles and deer mice, and then drops off to molt. Nymphs feed again on these hosts for about a week, dropping off again to molt to adult. Adults prefer larger hosts. They often accumulate at the edges of trails and roadways where they have access to passersby.

Dermacentor variabilis is the principal vector of Rocky Mountain spotted fever in the central and eastern United States. Oddly enough, by far the majority of cases of this poorly named disease occur in the eastern half of the country. This tick also causes paralysis in dogs and humans and transmits tularemia.

Dermacentor occidentalis, the **Pacific Coast tick,** is very similar in morphology and biology to *D. andersoni.* Adults are found on many species of large mammals, including humans, in California and Oregon. *Dermacentor occidentalis* transmits Colorado tick fever virus, Rocky Mountain spotted fever, tularemia, anaplasmosis, and chlamydial abortion of cattle.

Dermacentor albipictus is called the **horse tick** or because it does not feed during the summer months, **winter tick.** This one-host tick is widely distributed in the northern United States and Canada, where it feeds on elk, moose, horse, and deer. It rarely attacks cattle or humans. The eggs are laid in the spring. The larvae hatch in three to six weeks and become dormant until the cold weather of autumn stimulates them to seek a host. Once aboard, the tick remains through its instars and subsequent mating, until it drops off in the spring. Infection can be so heavy as to kill the host. Because it is a one-host tick, *D. albipictus* is an inefficient vector for microorganisms.

Other species of *Dermacentor* throughout the world are vectors of several diseases caused by viruses and rickettsias.

• *Amblyomma* Species

The hundred or so species in this genus mostly are restricted to the tropics. The only exceptions are a few species in the southern United States. All appear to be one-host ticks and to have a one-year life cycle. *Amblyomma* spp. are fairly long, highly ornate ticks with long mouthparts. The second segment of the pedipalp is longer than the others, resulting in the mouthparts, including the hypostome, being longer than the basis capituli.

Larvae and nymphs are common on birds, although adults usually feed only on large animals. Immature stages will feed on almost any terrestrial vertebrate and also can be found alongside adults on the final host. Because all three stages will readily bite humans, which is unusual among hard ticks, *Amblyomma* is exceptionally annoying.

Amblyomma americanum is the lone star tick (Fig. 40.6). Both sexes are dark brown with a bright silver spot at the posterior margin of the scutum. Males may have more than one spot. The tick ranges throughout much of the southern United States and well into Mexico.[41] It has a wide variety of hosts, including livestock and humans, and is a vector for Rocky Mountain spotted fever and tularemia.

• *Aponomma* Species

The few species of *Aponomma* parasitize reptiles, particularly in Asia and Africa. *Aponomma elaphensis* is known from rat snakes in Texas, and four species have been described from South America and Haiti. They are not known to be of medical or economic importance, although it has been speculated that they may transmit haemogregarines among reptiles.[16] They are eyeless.

• *Rhipicephalus* Species

Continental Africa appears to be the place of origin and center of distribution of rhipicephalid ticks.[16] The approximately 60 species and subspecies in this genus usually are ecologically distributed, being restricted to forests, mountains, and semidesert regions or at least those with certain limits of rainfall. Most species show little host specificity, biting

FIGURE 40.6

Amblyomma americanum, the lone star tick.
Courtesy of Warren Buss.

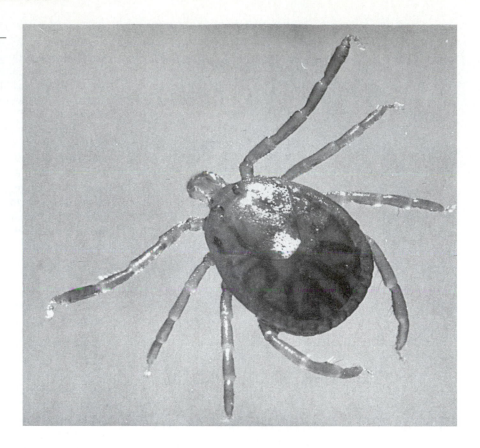

many mammals and even reptiles and ground-dwelling birds. Both two-host and three-host species are known. The genus is easily recognized by the combination of festoons and spurs on each side of the basis capituli. The only other genus with a similar basis capituli is *Boophilus,* which lacks festoons.

Rhipicephalus sanguineus (Fig. 40.7), the **brown dog tick** or **kennel tick,** is the most widely distributed of all ticks, being found in practically all countries between 50° N and 35° S, including most of North America. It quickly becomes established whenever it is introduced.[17] *Rhipicephalus sanguineus* is a three-host tick, since it leaves the host to molt. All three stages feed mainly on dogs, mostly between their toes, in their ears, and behind the neck. The host range is very wide and occasionally includes human attacks, but the tick has a definite taste for dogs.

In some areas of Europe and Africa *R. sanguineus* is a major vector of boutenneuse fever (*Rickettsia connorii*), which usually is acquired by crushing the tick against the skin. In Mexico it transmits Rocky Mountain spotted fever. Other diseases of animals are transmitted by this tick: *Borrelia theileri,* a spirochaete of sheep, goats, horses, and cattle that is carried by this arthropod in some areas; a highly fatal canine rickettsiosis caused by *Rickettsia canis;* and malignant jaundice, caused by *Babesia canis.* The protozoan *Hepatozoon canis* infects dogs when they swallow infected ticks. For a list of successful and unsuccessful experimental transmissions of diseases by *R. sanguineus* see Hoogstraal.[16]

By far the most important disease transmitted by any species of *Rhipicephalus* is **East Coast fever,** a protozoan disease of the red blood cells of cattle (Chapter 9). This highly malignant infection mainly kills adult cattle, with mortality ranging from an average of 80% to 100% of infected animals. The causative agent, *Theileria parva,* is transmitted by several ticks, but the most important vector seems to be *R. appendiculatus.* An infected tick can transmit the disease only during the stadium (period between molts) following the infected meal.

• *Anocentor* Species

Only one species, *A. nitens,* is known in this genus. The **tropical horse tick** is found through South America up to Texas, Georgia, and Florida. It is very similar to *Dermacentor* but has seven festoons rather than 11, its eyes are poorly developed, and it is inornate. It feeds primarily on horses and is not known to attack humans. It transmits *Babesia caballi,* a protozoan blood parasite in horses.

• *Hyalomma* Species

Few, if any, ticks are as difficult to identify as are species of *Hyalomma* (Fig. 40.8). This difficulty results partly from a natural genetic variation but also from a tendency toward hybridization. Furthermore, extrinsic factors, such as periods of starvation and climatic conditions, will cause morphological variations. This group probably originated in Iran or the

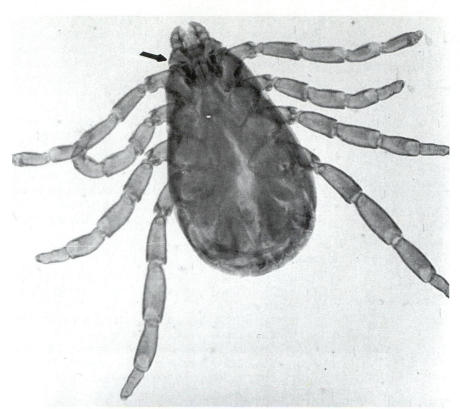

FIGURE 40.7

Rhipicephalus sanguineus, the brown tick. Note the spurs on the basis capituli (*arrow*).
Courtesy of Warren Buss.

These are fairly large ticks with no ornamentation. The legs are banded. Eyes are present, and the festoons are indistinct. They are very active, and it has been reported that in Africa and Arabian deserts these ticks will come rushing from beneath every shrub when persons or other animals stop by.[31]

Hyalommas must be the hardiest of ticks, since they are found in desert conditions where there is little shelter away from hosts, where there are few small mammals available for larvae and nymphs to feed on, and where large mammals are far ranging. They usually are the only ticks existing in such places.

Species of adult *Hyalomma* usually feed only on domestic animals. Occasionally they will bite humans, and because they transmit serious pathogens, they are among the most dangerous of ticks. Immature forms often feed on birds, rodents, and hares that are the reservoirs of viruses and rickettsias. For instance, **Crimean-Congo hemorrhagic fever** is carried between Africa and Europe by *H. marginatum* on migrating birds. Other viruses isolated from this species of tick are Dugbe virus and West Nile virus, and *H. anotolicum* harbors Thogoto virus and a swine poxvirus. Rickettsial diseases that can be transmitted by *Hyalomma* include Siberian tick typhus, boutonneuse fever, Q fever, and those that are caused by *Ehrlichia* spp. Malignant jaundice of dogs, caused by the protozoan *Babesia canis,* is transmitted by *H. marginatum* and *H. plumbeum* in Russia. Another protozoan, *Theileria annulata,* is transmitted to cattle by the bite of *H. anatolicum* in Eurasia.

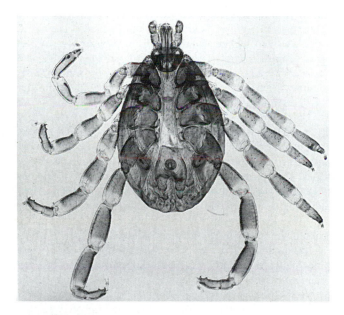

FIGURE 40.8

Hyalomma sp.
Courtesy of Jay Georgi.

southern portion of the former Soviet Union and has radiated into Asia, the Middle East, southern Europe, and Africa.[16] Frequently nymphs are carried from Africa to Europe by migrating birds.

• *Boophilus* Species

Boophilus ticks resemble *Rhipicephalus* spp. in that the basis capituli is bilaterally produced into points. They differ, however, in lacking festoons and an anal groove. Because unengorged specimens are quite small and easily overlooked, they have spread to many parts of the world when cattle are imported from an endemic zone. Two hypotheses on their place of origin have been suggested: Either they came from the Indian subcontinent attached to Brahman cattle (zebu), or they originally were parasites of American bison or deer, they adapted to cattle, and they thence were exported to other places.[16]

Taxonomy of this genus has been confused, but at least three species are clearly recognized. *Boophilus annulatus* is the most widely distributed of these. It is often called the **American cattle tick,** since it was once widespread in the southern United States and is still common in Mexico, Central America, and some Caribbean islands. It is also known in Africa; the species known as *B. calcaratus,* from the Near East and Mediterranean region, is actually *B. annulatus.*[16]

This tick has been eradicated from the United States but appears sporadically along the Mexican border, as cattle and deer carry it across.

Boophilus microplus is similar in biology to *B. annulatus* and also has been eradicated from the United States. It still is found in Mexico and Africa, as well as Australia, Central and South America, Madagascar, and Taiwan. The high incidence of parthenogenesis in this species aids its survival when harsh conditions restrict the size of a population.[50] Cattle are the primary hosts of this tick; but sheep, goats, horses, and other animals may be infested.

The **blue tick,** *B. decoloratus,* occurs widely in continental Africa. Mainly it attacks cattle, but it bites many other animals, including humans.

Boophilus spp. are all one-host ticks. Larval, nymphal, and adult stages are all spent on the same host animal, a rarity among ticks. Engorged females drop off and lay 2000 to 4000 eggs during the next 12 to 14 days. Newly hatched larvae are quite active, crawling to the tips of grasses and other plants, where they often accumulate in great numbers. After reaching a host, they remain until after breeding and feeding. Obviating the finding of two or three hosts during its life has obvious survival value to these ticks.

Control, however, is greatly aided by the fact that all stages are to be found on the same animal. Dipping kills all parasitic stages at once. Unfed larvae die in about 65 days, so a pasture becomes tick free if cattle are dipped and kept off for this duration. Larvae, though, can be windblown for considerable distances.[30]

Species of *Boophilus* have been implicated as vectors of Crimean-Congo hemorrhagic fever and Ganjon viruses, as well as Bhanja virus in Nigeria and Thogoto virus in Kenya. The rickettsia *Anaplasma marginale* is transmitted to cattle by all three species of *Boophilus* in Africa. Mortality ranges from 30% to 50% in infected animals. Experimentally, *B. decoloratus* can transmit *Trypanosoma theileri* among cattle.[2]

By far the most important disease transmitted by a species of *Boophilus* is **Texas cattle fever,** also called **redwater fever.** The agent of this disease is a piroplasm,

Babesia bigemina, discussed in detail in Chapter 9. To transmit most of the aforementioned diseases, ticks must change hosts, usually under crowded conditions in pens, railroad cars, and so on. However, because of transovarial transmission of *Babesia bigemina,* newly hatched ticks are already infected and capable of passing the disease on to cattle. Redwater fever was eradicated in the United States, along with the tick vectors, in 1939, but it persists in Central and South America, Africa, South Europe, Mexico, and the Philippines.

• *Margaropus* Species

The four rare species in this genus, including *M. winthemi* and *M. reidi,* are found in East Africa and Sudan. They are parasites of giraffes and occasionally horses but are not known to be medically or otherwise economically important. For a review see Hoogstraal.[16]

Family Argasidae

The family of soft ticks includes only five genera, *Argas, Ornithodoros, Otobius, Nothoaspis,* and *Antricola,* with a total of about 160 species. *Antricola* and *Nothoaspis* infest cave-dwelling bats in North and Central America and will not be considered further here.

Soft ticks are easily distinguished from hard ticks by a number of characters, most obviously a subterminal capitulum in nymphs and adults (but not larvae) that cannot be seen in dorsal view. The capitulum lies within a groove or depression called the **camerostome.** The dorsal wall of the camerostome, which extends over the capitulum, is called the **hood.** In addition, (1) there are no festoons or scutum; (2) sexual dimorphism is slight; (3) the pedipalps are freely articulated and leglike; (4) there are no porose areas on the basis capituli; (5) eyes are on the supracoxal fold; (6) the coxae lack spurs; (7) pulvilli are absent on the tarsi; (8) the stigmatal plates are behind the third leg; and (9) there may be two to eight nymphal stages.

In general, argasid ticks inhabit localities of extremely low relative humidity. Those that occur in wet climates seek dry microhabitats in which to live. Unlike ixodid ticks, most argasids feed repeatedly, resting away from the host between meals. This makes them difficult to collect, since they hide in loose soil, crevices, birds' nests, and the like. Examination, including sifting, of the soil or detritus of burrows, rodent nests, big game resting and rolling places, caves, and other animal lairs is usually required to find them.

Adult females lay eggs in their hiding places several times between feedings. Even so, the total number of eggs produced is small, usually fewer than 500. Although larvae of some species remain dormant and molt to the first nymphal stage before feeding, most feed actively in this stage. Likewise, first-stage nymphs of a few species molt to second stage without feeding, although most feed first. Larvae usually remain on the host until molting, but nymphs, like the adult, leave the host. Exceptions to this are found in the genera *Otobius* and *Argas,* discussed later.

Hosts may be few and far between in desert habitats, and argasid ticks have adapted to potentially long periods without meals. They are capable of estivating for months or even many years without food. A blood meal not only provides

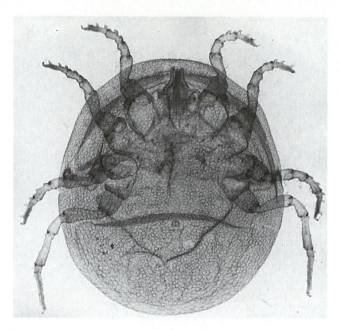

FIGURE 40.9

Ornithodoros sp.
Courtesy of Jay Georgi.

nutrition for developing eggs but also triggers a ganglion in the brain to release a hormone that instigates egg production in the ovaries.[43]

• *Ornithodoros* Species

Ornithodoros ticks are thick, leathery, and rounded (Fig. 40.9). The tegument in nonengorged specimens is densely wrinkled in fairly consistent patterns, allowing them great distention when feeding. Over 100 species of *Ornithodoros* parasitize mammals, including bats. It is unusual for them to feed on birds or reptiles, although *O. capensis* is found on marine birds in North America. Some species in this genus are very important in that they are vectors for relapsing fever spirochaetes, and the bite of several species is itself highly toxic and painful.

Ornithodoros hermsi is found throughout the Rocky Mountains and Pacific Coast states. It is an important vector of *Borrelia recurrentis,* the etiological agent of relapsing fever, which was first reported in North America from gold miners near Denver in 1915. Basically the tick is a rodent parasite. Its life cycle is typical for most species of *Ornithodoros.* The female lays up to 200 eggs in crannies and crevices like those in which adults hide. The larvae actively seek hosts and feed for about 12 to 15 minutes. After molting, the two nymphal instars each feed again and then molt to adult. The life cycle under laboratory conditions takes about four months but may be greatly prolonged in the absence of food.

Ornithodoros cariaeceus occurs from Mexico to southern Oregon, hiding in the soil of bedding areas of large mammals, such as deer and cattle. It is greatly feared by many people because of its painful bite and the venomous aftereffects of its attack. Like many bloodsucking arthropods, it is attracted to carbon dioxide and thus can be trapped using dry ice for bait.

Ornithodoros moubata is an eyeless argasid found in widely dispersed arid regions of Africa. Closely related species are *O. compactus* in South Africa, *O. aperatus* in mideastern Africa, and *O. porcinus* from middle to southern Africa. *Ornithodoros porcinus* is primarily a parasite of the burrowing warthog but readily invades human habitation. It feeds on many mammals and birds and can survive starvation for at least five years. The larvae of these species do not feed but molt directly into the first nymphal instar. Some populations are parthenogenetic. An interesting physiological adaptation in this species complex is the absence of a passage between the midgut and hindgut, with the result that all waste matter must remain within the intestinal diverticuli during the life of the tick.[8] A great deal is known about the biology, control, and other aspects of this group,[17] members of which are all important vectors of relapsing fever.

Ornithodoros savignyi is similar in appearance to *O. moubata,* except that it has eyes. It is found in arid regions of North, East, and southern Africa; the Near East; India; and Ceylon. Basically it is a parasite of camels but will bite practically any mammal and fowl. It does not invade human habitation, as does *O. porcinus,* but buries itself in shallow soil, awaiting its prey. It is quite bold, attacking across wide open areas if the ground is not too hot, and feeds quickly. Its bite is quite painful, but *O. savignyi* apparently does not transmit diseases in nature.

• *Otobius* Species

These argasids are called **spinose ear ticks** because the nymphs have a spiny tegument and usually feed within the folds of the external auditory canal. *Otobius megnini* (Fig. 40.10) is widely distributed in the warmer parts of the United States and in British Columbia. It also has been introduced into India, South Africa, and South America. It feeds mainly on cattle but attacks many domestic and wild mammals, as well as humans.

Adult *O. megnini* do not feed. The adult capitulum is submarginal, but the hypostome is vestigial; its tegument is not spiny. The capitulum of the larva and nymph is marginal; the hypostome is well-developed; and the tegument of the nymphs, especially the second stage, is spiny. Eggs are laid on the soil. On hatching, the larvae contact a host, wander upward toward the head, and attach in the ear. There they remain through two molts, detaching and dropping to the ground for a final molt to the adult stage. Heavy infections can have serious, even fatal, effects on livestock.

Otobius lagophilus has a similar life cycle, except that the larvae and nymphs feed on the faces of rabbits in western North America.

• *Argas* Species

Species of *Argas* are almost exclusively parasites of birds and bats. However, most of them have been known to bite humans, although they apparently do not transmit diseases to them. *Argas* spp. (Fig. 40.11) are superficially similar to those of *Ornithodoros* but are flatter and retain a lateral ridge even when engorged. Furthermore, the peripheral tegument of the body is typically sculptured in *Argas* spp., whereas

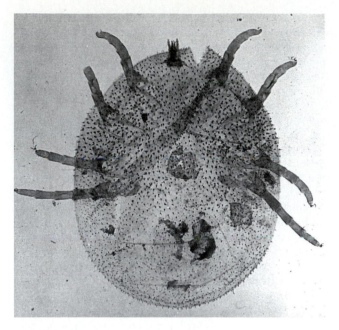

FIGURE 40.10

Otobius megnini, the spinose ear tick: nymph.
Courtesy of Warren Buss.

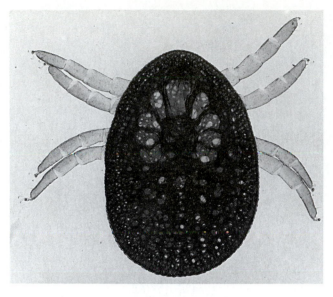

FIGURE 40.11

Argas sp.
Courtesy of Jay Georgi.

that of *Ornithodoros,* although minutely wrinkled, does not show an obvious pattern without very high magnification. Eyes are absent.

Argas ticks have bedbuglike habits, feeding briefly by night and hiding by day in crevices and under litter. Eggs are laid in these hiding places, and the hatched larvae eagerly seek a host. They usually remain attached, feeding for a few days before dropping off to molt to the first-stage nymph. The two nymphal stages feed in the same manner as do the adults, engorging in less than an hour and then leaving the host to hide and digest the meal.

Argas persicus, the **fowl tick,** is primarily an Old World species, although it does exist in the New World, along with the similar *A. miniatus, A. sanchezi,* and *A. radiatus,* all parasites of domestic fowl and other birds. Under favorable conditions, these ticks may build up a huge population in a henhouse, and their nocturnal depredations can exhaust a flock or even kill individuals. Vagabonds or others who try to spend the night in a deserted chicken house are sometimes surprised by masses of ravenous fowl ticks that literally come out of the woodwork to attack them. The bite is painful, often with toxic aftereffects, but such attacks on humans are rare.

The **pigeon tick,** *A. reflexus,* is a Near and Middle Eastern pest that has spread northward through Europe and Russia and eastward to India and other Asian localities. It has been reported in North and South America but probably was misidentified. *Argas reflexus* mainly attacks domestic pigeons, but because these birds are closely associated with human habitation, this tick bites people more often than does *A. persicus.*

Argas cooleyi is commonly associated with cliff swallows and other birds in the United States. *Argas vespertillionis* is widely distributed among Old World bats and occasionally

bites humans. The largest of all ticks is *A. brumpti,* a parasite inhabiting dens of the hyrax and some rodents in Africa. It is 15 to 20 mm long by 10 mm wide.

Immunity to Ticks

Mammals can develop resistance to hard ticks.[27] Some strains of host species are naturally more resistant than others; in these species inflammatory lesions occur at attachment sites.[28] Rabbits, at least two species of domestic cattle, and goats all exhibit such reactions. These observations suggest vaccination may be a practical means of controlling economic losses due to acarine infestations (see also Chapter 38 on immunity to myiasis).

Acquired resistance often is manifested as failure of the parasite, especially larvae and nymphs, to fully engorge.[12] The antigens responsible for immune reactions are of both salivary gland and gut origin.[22,29] Some of the salivary gland antigens are shared by *Ixodes dammini, Dermacentor variabilis,* and *Amblyomma americanum.*[22] Adult *Rhipicephalus sanguineus, Dermacentor variabilis,* and *Amblyomma maculatum* elicit stronger immune reactions in rabbits than either larvae or nymphs. The resistance is nonspecific and is passively transferrable with immune serum.[6] Unless the rabbits are continually exposed, however, the resistance is lost after about three months. Experimental vaccines for cattle, developed from *Boophilus* midgut membrane antigens, reduce egg production in adult ticks whose larvae have fed on the vaccinated cattle.[29] Thus, recent research seems to hold promise for eventual successful tick vaccine development.

ORDER MESOSTIGMATA

The mesostigmatid mites (Fig. 40.12) have a pair of respiratory spiracles, the **stigmata,** that are located just behind and lateral to the third coxae. Usually extending anteriorly from each stigma is a tracheal trunk, the **peritreme,** which makes

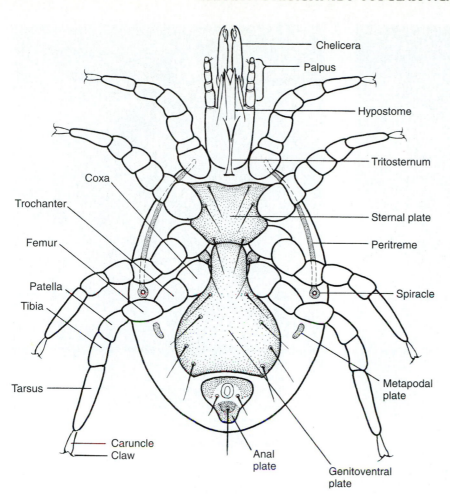

FIGURE 40.12

Generalized mesostigmatid mite, ventral view.

Source: Courtesy of Communicable Disease Center, 1949, U.S. Public Health Service, Washington, D.C.

it easy to recognize specimens belonging to this suborder. The gnathosoma forms a tube surrounding the mouthparts. A **tectum** is present above the mouth, and a ventral bristlelike organ, the **tritosternum,** usually is present immediately behind the gnathosoma. The palpal tarsus has a forked tine at its base. The dorsum of adults usually has one or two sclerites called **shields** or **dorsal plates.**

Family Laelaptidae

The cosmopolitan family Laelaptidae includes a large number of diverse genera. They are the most common ectoparasites of mammals, and some species parasitize invertebrates. Most species have pretarsi, caruncles, and claws on all legs. The dorsal shield is undivided. The second coxa has a toothlike projection from the anterior border.

The **common rat mite,** *Echinolaelaps echidinus* (Fig. 40.13), transmits the protozoan *Hepatozoon muris* from rat to rat. Although laelaptids are not known to transmit diseases to humans, they are suspected of causing dermatitis. The virus of epidemic hemorrhagic fever has been demonstrated in several species collected in rodent burrows in the Far East.[4]

Family Halarachnidae

Closely related to the laelaptid mites, the halarachnids are parasites of the respiratory systems of mammals. They are easily recognized by a combination of morphological features: The dorsal shield is undivided and reduced; the sternal plate is reduced and has three pairs of setae (if much

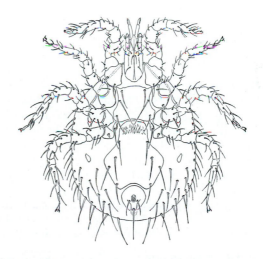

FIGURE 40.13

Echinolaelaps echidinus, the common rat mite. Ventral view of female.

From S. Hirst, "Mites injurious to domestic animals (with an appendix on the acarine disease of hive bees)," in *Brit. Mus.* (*Nat. Hist.*) *Econ. Ser.* 13:1–107. Copyright © 1922.

reduced, some setae may be displaced from the plate); the female genital sclerite, the epigynial plate, is rudimentary; the male genital opening is in the anterior margin of the sternal plate; the tritosternum is absent; and the movable digit of the chelicera is more strongly developed than is the fixed digit.

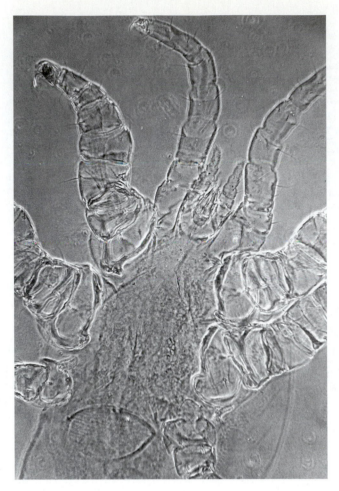

FIGURE 40.14

Orthohalarachne attenuata from the nasal passages of a northern fur seal.
Courtesy of Warren Buss.

FIGURE 40.15

Lung of a baboon, *Papio cynocephalus,* with nodules caused by the mite *Pneumonyssus* sp.
Courtesy of Robert Kuntz.

The genus *Halarachne* is found only in the respiratory system of seals of the family Phocidae, and *Orthohalarachne attenuata* parasitizes other families of Pinnipedia (Fig. 40.14). Several species of *Pneumonyssus* are found in primates, although infection in humans is unknown. Respiratory problems caused by these mites in captive monkeys and baboons are common[53] (Fig. 40.15) *Pneumonyssus caninum* inhabits the nasal passages and sinuses of dogs and may cause central nervous system disorders.[3] *Raillietea auris* is the **cattle ear mite.** Not known to be pathogenic, apparently it feeds on secretions and dead cells of the external auditory meatus.

Family Dermanyssidae

Dermanyssids are parasites on vertebrates and are of considerable economic and medical importance. The dorsal plate in the female either is undivided or is divided with a very small posterior part. The sternal plate has three pairs of setae, and the metasternal plates are reduced and lateral to the genital plates. A tritosternum is present. The chelicerae may be normal, with reduced chelae, or they may be quite elongated and needlelike. All legs have pretarsi, caruncles, and claws.

Dermanyssus gallinae, the **chicken mite** (Fig. 40.16), attacks domestic fowl, particularly chickens and pigeons, throughout the world. The mites hide by day in crevices near the roosting places of birds, emerging at night to feed. Their numbers may be so great as to kill the birds. Setting hens may abandon their nests, and young chicks may rapidly perish. This mite readily attacks humans, especially children, causing a severe dermatitis. Roosting pigeons may bring *D. gallinae* into proximity with human habitation, where wandering mites may encounter a mammalian meal.[42] They are attracted to warm objects and so tend to accumulate in electric clocks and around fireplaces, water pipes, and so on.

The viruses of western and St. Louis equine encephalitis have been isolated from *D. gallinae.* It is unlikely that these mites play an important role in transmission of these diseases to mammals, but they may help keep up the reservoir of infection among birds.[45] Natural and experimental transmissions of fowl poxvirus have been demonstrated.[44] Experiments have shown that *D. gallinae* also can transmit Q fever and fowl spirochaetosis.

Liponyssus sanguineus, the **house mouse mite,** prefers to feed on that host but will readily attack humans. It is important in that this mite can transmit the rickettsialpox pathogen to humans. Rickettsialpox is a mild, febrile condition with a vesicular rash commencing three to four days after the onset of fever. A scab develops at the site of the bite, and healing is slow. Besides fever, the patient has chills,

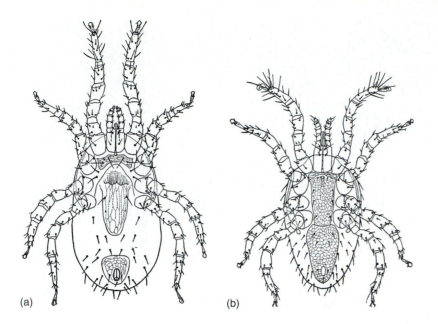

(a)   (b)

FIGURE 40.16

Ventral views of *Dermanyssus gallinae,* the chicken mite. (*a*) Female; (*b*) male.

From S. Hirst, "Mites injurious to domestic animals (with an appendix on the acarine disease of hive bees)," in *Brit. Mus.* (*Nat. Hist.*) *Econ. Ser.* 13:1–107. Copyright © 1922.

sweating, backache, and muscle pains. Patients recover in one to two weeks; no fatalities are known. Q fever has been experimentally transmitted by this mite. The biology of the house mouse mite is summarized by Baker and coworkers.[1]

The **tropical rat mite,** *Ornithonyssus bacoti* (Fig. 40.17), is found worldwide, in both temperate and tropical climates, where, as its name implies, it normally infests rats. It is a serious pathogen of laboratory mouse colonies, where it can retard growth and eventually kill young mice. When rats are killed or abandon their nests, the mites can migrate considerable distances to enter human habitation. They cause a sharp, itching pain at the time of their bite, and skin-sensitive persons may develop a severe dermatitis. A nonfeeding larva hatches and rapidly molts to a bloodsucking **protonymph.** This molts to become a nonfeeding **deutonymph,** which in turn becomes a feeding adult. Parthenogenesis is common, producing only males. The life cycle from egg to egg can be completed in 13 days. Adult females live about 60 days and produce approximately a hundred eggs.

These mites apparently do not transmit any pathogens to humans, although experimentally they have been shown to transmit plague, rickettsialpox, Q fever, and murine typhus. *Ornithonyssus bacoti* is the intermediate host of the filarial nematode *Litomosoides carinii,* a parasite of the cotton rat, *Sigmodon hispidus.*[57] Because rats, mites, and worms can easily be maintained in the laboratory, this host-parasite system has been used as a model for studies on filariasis, including drug testing.

The **northern fowl mite,** *Ornithonyssus sylviarum,* is widespread in northern temperate climes and has been reported from Australia and New Zealand. It will bite humans and can be a nuisance to egg processors. It does not appear to be particularly pathogenic to fowl. *Ornithonyssus bursa* is the **tropical fowl mite.** It is ectoparasitic on chickens, turkeys, and some wild birds, including the English sparrow.[15] It can be pathogenic to poultry, causing them to be listless and poorly developed. It bites humans but causes only a slight irritation.

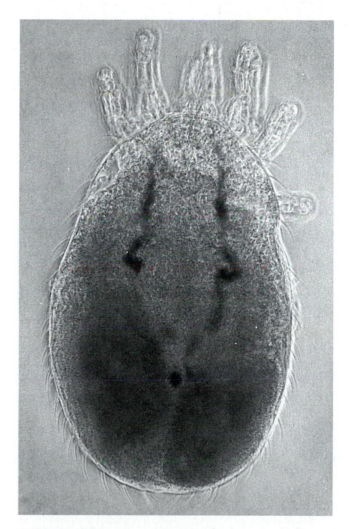

FIGURE 40.17

Ornithonyssus bacoti, the tropical rat mite.
Courtesy of Warren Buss.

FIGURE 40.18

Female *Sternostoma tracheacolum*, the canary lung mite. (*a*) Dorsal view; (*b*) ventral view.

From D. B. Pence, "Keys, species and host list, and bibliography for nasal mites of North American birds (Acarina: Phinonyssinae, Turbinoptina, Speleognathinae, and Cytoditidae)," in *Spec. Pub. Mus. Texas Tech. Univ.* 8:1–148. Copyright © 1975.

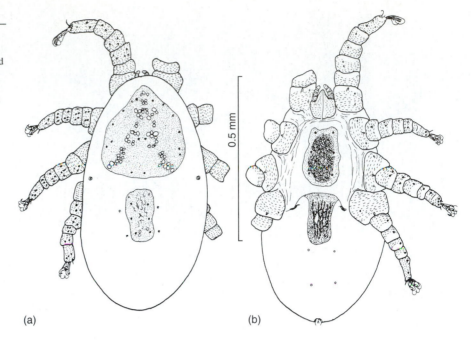

(a)   (b)

Family Rhinonyssidae

This family is considered by some authorities to be a subfamily of Dermanyssidae. All members are parasitic in the respiratory tracts of birds. Rhinonyssids are oval in shape and have weakly sclerotized plates. All tarsi have pretarsi, caruncles, and claws. Stigmata are present with or without short dorsal peritremes. The tritosterum is absent.

These mites are viviparous, producing larvae in which the protonymph is already developed. Nearly every species of bird examined has nasal mites; many species have been described, and many more species undoubtedly are yet to be discovered.[35] Because of their blood- or tissue-feeding habits, these mites may be regarded as significant disease agents in wild bird populations. The **canary lung mite,** *Sternostoma tracheacolum* (Fig. 40.18), can sicken and kill captive canaries and finches.

ORDER PROSTIGMATA

In this order the spiracles are located either between the chelicerae or on the dorsum of the hysterosoma. These mites usually are weakly sclerotized. The chelicerae vary from strongly chelate to reduced. Pedipalps are simple, fanglike, or clawed. There are phytophagous, terrestrial and aquatic free-living, and parasitic forms.

Family Cheyletidae

The cheyletid mites are small, measuring 0.2 to 0.8 mm long. Most are yellowish or reddish, oval, and plump, except for the feather-inhabiting species, which are elongated. The propodosoma and hysterosoma are clearly delineated and usually have one or more dorsal shields. Eyes are present or absent. Strong peritremes, which usually surround the gnathostoma, are present. The chelicerae are short and styletlike; the palpi are large and pincerlike.

In *Cheyletiella parasitivorax* the male genital opening is dorsal, a rare occurrence in arthropods. These are common denizens of fur coats, possibly predaceous on other mites. *Cheyletiella yasguri* of dogs and *C. blakei* of cats cause a mange dermatitis on their normal hosts. They also will feed on humans, although temporarily.

Family Pyemotidae

Pyemotid mites mainly parasitize insects that, in turn, infest cereal crops. They are brought into contact with humans when people harvest or work with stored grains or sleep on straw mattresses. When these mites bite, they leave a small, itching vesicle that may become inflamed rapidly and cause considerable discomfort. Itching, headache, nausea, and internal pains may accompany severe attacks.

These are soft-bodied mites with tiny chelicerae and pedipalps. A wide space occurs between the third and fourth pairs of legs. Sexual dimorphism is marked. The female becomes enormously swollen when gravid.

Pyemotes tritici, the **straw itch mite,** normally is a parasite of various stored grain beetles, but it readily attacks humans. The male can be seen by the unaided eye only with difficulty, but the gravid female reaches nearly a millimeter in length. Her body contains 200 to 300 large eggs, which hatch internally. The developing mites complete all larval instars before being born. The few males emerge first and cluster around the genital pore of the mother, copulating with the females as they emerge.

The **grain itch mite,** *Pyemotes ventricosus,* is similar in biology and pathogenicity. It normally infests boring beetle larvae, grain moths, and numerous other insects.

Family Psorergatidae

These are small- to medium-sized mites that are unarmored and have striated skin and peritremes. Because they are soft, they are susceptible to desiccation and are less numerous

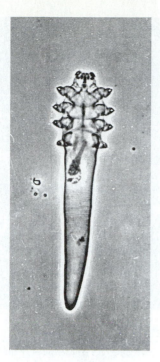

FIGURE 40.19

Demodex folliculorum, the human follicle mite.

From C. Desch and W. B. Nutting, "*Demodex folliculorum* (Simon) and *D. brevis* Akbulatova of man: redescription and reevaluation," in *J. Parasitol.* 58:169–177. Copyright © 1972.

during dry periods. The chelicerae are minute and styletlike. The pedipalps are simple and minute and are not used for grasping. The first pair of legs is modified for grasping hairs.

Psorergates ovis is the itch mite of sheep and is a serious pest in sheep-raising countries, including the United States, New Zealand, and Australia. It causes skin injury and fleece derangement. *Psorergates simplex* is found on laboratory mice, and *P. bos* is known from cattle in the western United States.[23]

Family Demodicidae

These minute, cigar-shaped parasites are known as the **follicle mites.** They range in length from 100 to 400 μm and have short, stumpy, five-segmented legs on the anterior half of the body. The opisthosoma is transversely striated. Species of *Demodex* live in hair follicles and sebaceous glands of many species of mammals. Although numerous species have been described, it is probable that many more exist, especially in wild mammals. There seems to be rigid host specificity.

Humans serve as hosts to two species. *Demodex folliculorum* (Fig. 40.19) lives in hair follicles, whereas *D. brevis,* a stubbier species, inhabits sebaceous glands.[5] Both exist mainly on the face, particularly around the nose and eyes. All life stages may be found in a single follicle. These mites may penetrate the skin and lodge in various internal organs, where they elicit a granulomatous response.

The prevalence of these mites in humans is very high, from about 20% in persons 20 years of age or younger to nearly 100% in the aged. Infection usually is benign, although rarely there may be loss of eyelashes or granulomatous skin eruptions.[14] It is suspected that follicle mites may be involved in introducing acne-causing bacteria into the skin follicles of susceptible individuals. An easy means of diagnosing both species in humans is to examine microscopically some oil expressed from the side of the nose.

Much more pathogenic is the **dog follicle mite,** *Demodex canis.* This species, together with some form of the bacterium *Staphylococcus pyogenes,* causes **red mange,** or **canine demodectic mange.** Infection in young dogs can be serious, even fatal. There is hair loss on the muzzle, around the eyes, and on the forefeet. The skin develops reddish pimples and pustules, becoming hot, thickened, and covered with a foul-smelling reddish-yellow exudate. Exact diagnosis depends on demonstrating a mite in skin scrapings. Treatment is difficult, and severely infected puppies may have to be killed. Symptoms may disappear gradually, and older, although perhaps infected dogs show no further signs of disease, probably because of acquired immunity.

Other demodicids of importance are the **cattle follicle mite,** *D. bovis,* the **horse follicle mite,** *D. equi,* and the **hog follicle mite,** *D. phylloides.* All three cause a pustular dermatitis with nodules and loss of hair. Holes in the skin caused by these mites may reduce the value of hides.

Family Trombiculidae

This family contains the infamous **chigger mites,** which are all too common in most tropical and temperate countries of the world. They are unique among mites that attack humans in that only the larval stage is parasitic; the nymphs and adults are predators on small terrestrial invertebrates or their eggs. Over 1200 species have been described, many from the larva only; in numerous cases the nymphs and adults are unknown or have not been correlated with the larvae.

In a **generalized chigger life cycle** the egg hatches in about a week, releasing the developing larva, or **deutovum.** It rapidly completes differentiation into a six-legged larva (Fig. 40.20), which finds a host and feeds. The engorged larva leaves its host and becomes a quiescent **prenymph** or **nymphochrysalis.** Emerging nymphs feed on insect eggs or soft-bodied invertebrates and pass into a quiescent **imagochrysalis** before molting into an adult. Males deposit onto the substrate a stalked spermatophore, which the female inserts into her genital pore. Most chiggers show little host specificity.

The taxonomy of trombiculids is based mainly on the larvae. The larval body is rounded and usually is red, although it may be colorless. It bears a dorsal plate, or scutum, at the level of the anterior two pairs of legs; usually two pairs of eyes are near the lateral margins of the scutum. The scutum bears a pair of sensillae and three to seven setae. The chelicerae have two segments: the basal segment is stout and muscular, whereas the distal segment is a curved blade with or without teeth. The pedipalps consist of five segments. The fifth, or tarsus, bears several setae and opposes a tibial claw like a thumb. Distributed on the body are numerous plumose setae.

Adult chiggers are among the largest of mites, reaching a millimeter or more in length. There is a conspicuous constriction between the propodosoma and hysterosoma. Eyes may be present or absent. Both sexes are clothed

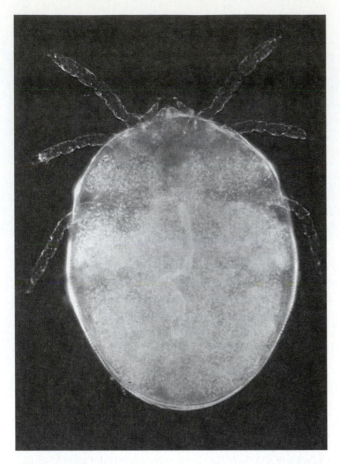

FIGURE 40.20

A chigger larva.
Courtesy of Mark Pope.

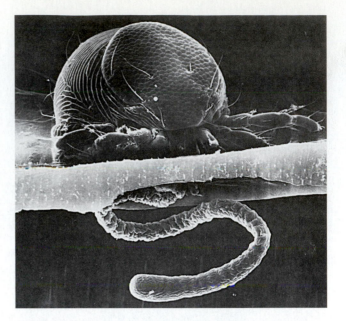

FIGURE 40.21

Larval *Arrenurus* sp. feeding on a damselfly. Note the long stylostome penetrating the insect's body wall. The internal organs of the damselfly have been removed.

From B. L. Redmond and J. Hochberg, "The stylostome of *Arrenurus* spp. (Acari: Parasitengona) studied with the scanning electron microscope," in *J. Parasitol.* 67:308–313. Copyright © 1981.

with a dense covering of plumose setae, which gives them the appearance of velvet. Commonly they are bright red or yellow.

There are two medical aspects of chigger bite: **chigger dermatitis** and transmission of pathogens. These are considered separately.

Larval chiggers do not burrow into the skin, as is commonly thought. After the mouthparts penetrate the epidermis, the mite injects salivary secretions. These are proteolytic, killing and digesting host cells, which the parasite then sucks, along with interstitial fluids. Simultaneously host cells harden under the influence of other salivary secretions to become a tube, the **stylostome** (Fig. 40.21). The mite retains its mouthparts in the stylostome, using it like a drinking straw, until engorged and then drops off. Not all chiggers cause an itching reaction; those that do usually have detached before a host reaction begins. Some people are immune to their bites, whereas others may incur a violent reaction.

Most chiggers of medical importance in North America are of the genus *Trombicula*. Of these, *T. alfreddugesi* is the most common species, ranging throughout the United States except in the western mountain states. It is most abundant in

disturbed forest that has been overgrown with shrubs, vines, and similar second-growth vegetation. It feeds on most terrestrial vertebrates. *Trombicula splendens* is the most abundant chigger in the southeastern United States, especially in moist areas such as swamps and bogs. The two species overlap their ranges in many areas but are active in different seasons.

Other species, some representing other genera, are found in the United States and elsewhere. Many bite humans, and others are important pests of livestock. The turkey chigger, *Neoschoengastia americana,* causes discoloring of the skin and loss of feathers of turkeys and related birds, rendering them less fit for market.[10] *Euschoengastia latchmani* causes a mangelike dermatitis on horses in California.

Several species of *Leptotrombidium* are vectors of the rickettsial disease in humans called **scrub typhus** (tsutsugamushi disease). The microorganism *Rickettsia tsutsugamushi* is transovarially transmitted among mites; wild rodents, particularly species of *Rattus,* are reservoirs. This disease was first described from Japan and now is known from Southeast Asia; adjacent islands of the Indian Ocean and the southwest Pacific; and coastal north Queensland, Australia. A primary lesion appears at the site of the chigger bite. It slowly enlarges to 8 to 12 mm and becomes necrotic in the center. By the fifth to eighth day, a red rash appears on the trunk and may spread to the extremities. Other symptoms are enlarged spleen, delirium and other nervous disturbances, prostration, and possibly deafness. Mortality rates range from 6% to 60%. Early treatment with broad-spectrum antibiotics usually is successful.

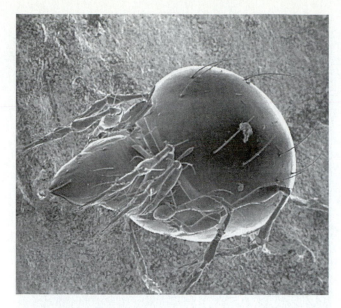

FIGURE 40.22

Oppia coloradensis, a beetle mite.
Courtesy of Tyler Woolley.

ORDER ORIBATIDA

In these mites the exoskeleton usually is strongly sclerotized or leathery and often deep brown (Fig. 40.22). Stigmata and tracheae are usually present, opening into a **porose area** (pitted region). The mouthparts are withdrawn into a tube, the **camerostome,** which may have a hoodlike sclerite over it.

A complex of families within this group is called the Oribatei. All feed on organic detritus and as such are among the dominant fauna of humus. They are of no direct medical importance, but many serve as intermediate hosts of *Moniezia expansa* (p. 344) and other anoplocephalid tapeworms. Members of this group are also intermediate hosts for *Bertiella studeri,* the primate cestode that sometimes infects humans (p. 344).

ORDER ASTIGMATA

Mites of the order Astigmata totally lack tracheal systems; they respire through the tegument, which is soft and thin. They lack claws, which are replaced with suckerlike structures on their pretarsi (Fig. 40.23). Some of the most medically and economically important mites belong to this order.

Family Psoroptidae

Members of this family are very similar to those of the family Sarcoptidae and are easily confused with them. However, unlike the sarcoptids, they do not burrow into the skin; instead, psoroptids pierce the skin at the bases of hairs, causing an inflammation that can become severe. Furthermore, the Psoroptidae lack propodosoma vertical setae, which are present on the Sarcoptidae.

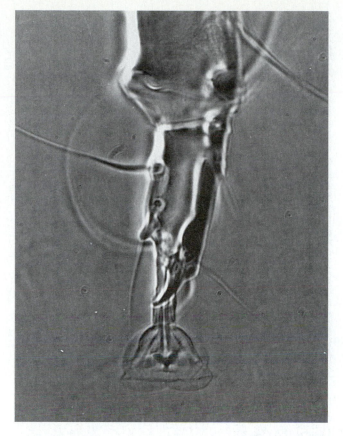

FIGURE 40.23

Pretarsus of *Chorioptes* sp., showing suckerlike modification.
Courtesy of Jay Georgi.

Chorioptic mange is a condition of domestic animals caused by mites of the genus *Chorioptes* (Fig. 40.24). Formerly each host species of *Chorioptes* was thought to harbor a distinct species of parasite. However, experiments have shown that only a single species, *C. bovis,* is involved.[51] These mites usually inhabit the feet and lower hind legs of cattle and horses. In sheep, chorioptic mange of the scrotum is well-known to cause seminal degeneration. In fact surveys of sheep in the United States have shown *C. bovis* to be the most common arthropod parasite, although pathogenic results usually are rare.

Psoroptic mange is caused by several species of mites on several species of hosts. *Psoroptes* spp. are distinguished by long legs that extend beyond the body and by the pedicel of the caruncle, which is segmented. They pierce the skin and suck exudates. These fluids congeal to form scabs that provide a protective cover for the parasites; then the *Psoroptes* spp. can reproduce under ideal conditions, increasing their numbers into millions in only a few days. Most domestic and a number of wild animals suffer from psoroptic mange. Wool production may be greatly inhibited by *P. ovis* in sheep. Unlike sarcoptic mange, *P. ovis* infests parts of the body most densely covered with wool. The species was reviewed by Kirkwood.[24]

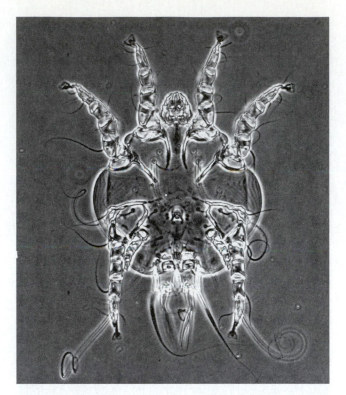

FIGURE 40.24

Chorioptes sp., male.
Courtesy of Jay Georgi.

Mites very closely related to *Psoroptes* spp. are in the genus *Otodectes.* Fairly common in cats, dogs, foxes, and ferrets, *O. cynotis* (Fig. 40.25) is usually found in the ears, although other parts of the head may be infested. Thousands of these mites swarming in the ears of a luckless host can cause desperate distress, with scabby, flowing ears and fit-like behavior.

Family Sarcoptidae

Sarcoptic mange, or **scabies,** may result from an infestation of the itch mite, *Sarcoptes scabiei* (Fig. 40.26). Although separate *Sarcoptes* spp. have been described from a wide variety of domestic animals as well as humans, they are morphologically indistinguishable and probably represent physiological races or perhaps sibling species. Thus, *S. scabiei* var. *equi* has a predilection for horses but will readily bite the rider as well.

The sarcoptids are skin parasites of homoiotherms. The body is rounded without a constriction separating the propodosoma from the hysterosoma. A propodosomal shield may be present or absent; either way, a pair of vertical setae projects from the dorsal propodosoma. The tegument has fine striae arranged in fields interrupted by scales, spines, or setae. The legs are very short and may or may not have claws or caruncles.

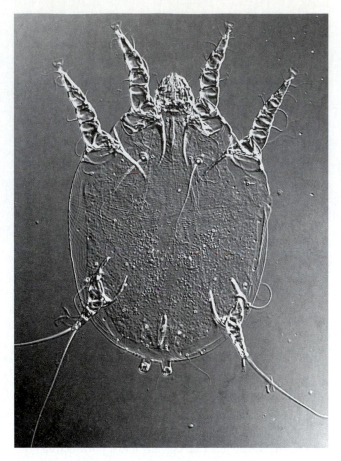

FIGURE 40.25

Otodectes cynotis nymph.
Courtesy of Jay Georgi.

Scabies mites mate on the skin of the host, the male inseminating immature females. The immature females move rapidly over the skin and probably are at this stage transmissible between hosts. Males do not burrow into the skin but remain on the surface, along with nymphs. The mature female uses the long bristles on the posterior legs to lift her back end up until she is nearly vertical. She cuts rapidly with her mouthparts and claws, becoming completely embedded in two and one-half minutes. She remains within the horny layer of the skin, forming tortuous tunnels for about two months. Scattered along the burrows are eggs, hatched larvae, ecdysed cuticles, and excrement. Eggs hatch in three to eight days, and the larvae and nymphs emerge to wander on the surface of the skin.

The tunneling and the secretory and excretory products produce an intense itching sensation in most infected persons. Usually a person does not notice any symptoms until the case is well-advanced. A rash begins to show, and vesicles and crusts may begin to form in some cases. This disease has several names, such as **seven-year itch, Norwegian itch,** or simply **scabies.** The skin between the fingers, the

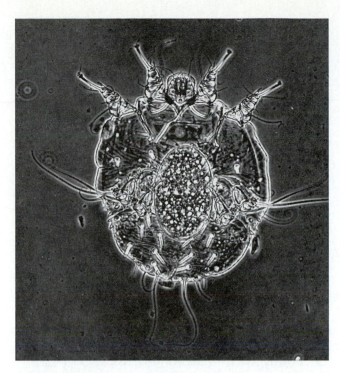

FIGURE 40.26

Sarcoptes scabiei, the itch mite.
Courtesy of Jay Georgi.

breasts, and the shoulder blades and around the penis and in the creases of the knees and elbows is most often infected. Scratching can cause bleeding and secondary infection. Transmission occurs primarily by physical contact between persons. Scabies was well-reviewed by Robinson.[38]

A 17- to 20-year cycle of resurgence of scabies infection in the world apparently occurs for unknown reasons, possibly because of a changing immunity in the human population.[48]

Sarcoptic mange in domestic and wild mammals is basically the same as that in humans. Hairless or short-haired regions of the body are most affected. Secondary infection by bacteria is more common in animals other than humans, resulting in severe weight loss or failure to gain weight, loss of hair, and pruritic dermatitis.[36] Infection is readily passed on to humans in contact with mangy animals.

Cats may develop mange caused by *Notoedres cati. Notoedres* is very similar in appearance to *Sarcoptes* but is smaller and more circular. It affects rodents and dogs but apparently not humans. **Notoedric mange** usually begins at the tips of the ears and spreads down over the head, sometimes onto the body.

Family Knemidokoptidae

Knemidokoptid mites are very similar in morphology and biology to the Sarcoptidae, with which they have been aligned in the past. However, some authorities accord them distinct family status.[11] They all are parasites of birds.

FIGURE 40.27

Dermatophagoides sp., a house dust mite.
From G. W. Wharton, "Mites and commercial extracts of house dust," in *Science* 167:1382–1383. © 1970 by the AAAS.

Knemidokoptes mutans causes a condition known as **scaly leg** in chickens and other small domestic birds. The mites burrow into the skin and under the scales of the feet and lower legs, which become distorted and covered with thick, nodular, spongy crusts. Infection may be so severe as to kill the birds, either directly or by secondary sensitization, which can involve internal organs. Scaly leg is highly contagious. Closely related to *K. mutans* is the **depluming mite,** *K. gallinae.* It embeds in the skin at the bases of the quills; infected birds pluck out their feathers in an attempt to alleviate the itching, or the feathers may fall out by themselves. Usually, large patches of deplumation extend over the body. *Knemidokoptes jamaicensis* is a parasite of canaries. *K. pilae* infests the face and legs of budgerigars. For a discussion of control of these pests in cage birds see Rickards.[37]

Family Pyroglyphidae

Important mites in this family are species of *Dermatophagoides* (Fig. 40.27). Most are not parasitic, but *D. scheremetewskyi* can cause a severe dermatitis on the scalp, face, and ears of humans. Probably it is a normal parasite of sparrows, bats, or other house-inhabiting animals.

More importantly, mites in this family are responsible for **house dust allergy.** Several species, especially those in *Dermatophagoides,* are abundant denizens of house dust. When whole mites or their parts or excrement are inhaled, as happens to everyone every day, they can stimulate an allergic reaction in sensitive individuals.

References

1. Baker, E. W., et al. 1956. *A manual of parasitic mites of medical and economic importance.* New York: National Pest Control Association.

2. Burgdorfer, W., M. C. Schmidt, and H. Hoogstraal. 1973. Detection of *Trypanosoma theileri* in Ethiopian cattle ticks. *Acta Tropica* 30:340–46.

3. Christensen, H. A., and C. Rehbinder. 1971. *Pneumonyssus caninum,* a mite in the nasal cavities and sinuses of the dog. *Nord. Vet. Med.* 23:499–505.

4. Chumakov, M. P. 1957. *Etiology, epidemiology and prophylaxis of hemorrhagic fevers. U.S. Public Health Service Monograph No. 50.* Washington, D.C.: U.S. Government Printing Office, 19–25.

5. Desch, C., and W. B. Nutting. 1972. *Demodex folliculorum* (Simon) and *D. brevis* Akbulatova of man: Redescription and reevaluation. *J. Parasitol.* 58:169–77.

6. Dipeolu, O. O. 1990. Expression and quantification of degrees of resistance by rabbits to infestation with *Rhipicephalus sanguineus* (L), *Dermacentor variabilis* (Say) and *Amblyomma maculatum* Koch (Acari, Ixodidae). *Insect Sci. Appl.* 11:235–44.

7. Dusbábek, F., and V. Bukva. 1990–1991. *Modern acarology 1 and 2.* Prague: Academia.

8. Enigk, K., and I. Grittner, 1952. Die Excretion der Zecken. *Z. Tropenmed. Parasitol.* 4:77–94.

9. Evans, G. O. 1992. *Principles of acarology.* Wallingford, England: C. A. B. International.

10. Everitt, R. E., M. A. Price, and S. E. Kunz. 1973. Biology of the chigger *Neoschöngastia americana* (Acarina: Trombiculidae). *Ann. Entomol. Soc. Am.* 66:429–35.

11. Fain, A., and P. Elsen. 1967. Mites of the family Knemidokoptidae that cause mange in birds (a revision with notes). *Acta Zool. Pathol. Antverp.* no. 45.

12. Fivaz, B. H., J. P. Nurton, and T. N. Petney. 1991. Resistance of restrained *Bos taurus* dairy bull calves to the bont tick *Amblyomma hebraeum* (Acarina: Ixodidae). *Vet. Parasitol.* 38:299–316.

13. Gothe, R., K. Kunze, and H. Hoogstraal. 1979. The mechanisms of pathogenicity in the tick paralyses. *J. Med. Ent.* 16:357–69.

14. Grosshans, E. M., M. Dremer, and J. Maleville. 1974. *Demodex folliculorum* and the histogenesis of granulomatous acne rosacea. *Hautartz* 25:166–77.

15. Hadani, A., K. Rauchbach, Y. Weissman, and R. Bock. 1975. The occurrence of the tropical fowl mite, *Ornithonyssus bursa* (Berlese, 1888), Dermanyssidae, on turkeys in Israel. *Refuah Vet.* 31:111–13.

16. Hoogstraal, H. 1956. *African Ixodoidea, vol. 1. Ticks of the Sudan.* Washington, D.C.: U.S. Department of the Navy. U.S. Government Printing Office.

17. Hoogstraal, H. 1972. The influence of human activity on tick distribution, density, and diseases. *Wiad. Parazytol.* 18:501–11.

18. Houck, M. A., ed. 1994. *Mites: Ecological and evolutionary analysis of life-history patterns.* New York: Chapman & Hall, Ltd.

19. Hunter, W. D., and W. A. Hooker. 1907. Information concerning the North American fever tick. *USDA Bur. Entomol. Bull.* 72:1–87.

20. Jaenson, T. G. T. 1991. The epidemiology of Lyme borreliosis. *Parasitol. Today* 7:39–45.

21. James, A. M., and J. H. Oliver Jr. 1990. Feeding and host preference of immature *Ixodes dammini, Ixodes scapularis,* and *Ixodes pacificus* (Acari: Ixodidae). *J. Med. Ent.* 27:324–30.

22. Jaworski, D. C., M. T. Muller, F. A. Simmen, and G. R. Needham. 1990. *Amblyomma americanum:* Identification of tick salivary gland antigens from unfed and early feeding females with comparisons to *Ixodes dammini* and *Dermacentor variabilis. Exp. Parasitol.* 70:217–26.

23. Johnston, D. E. 1964. *Psorergates bos,* a new mite parasite of domestic cattle (Acari, Psorergatidae). *Ohio Agric. Exp. Sta. Res. Circ.* 129:1–7.

24. Kirkwood, A. C. 1986. History, biology and control of sheep scab. *Parasitol. Today* 2:302–7.

25. Krantz, G. W. 1978. *A manual of acarology.* 2d ed. Corvallis, Ore.: Oregon State University Bookstore.

26. Lane, R. S., and W. Burgdorfer. 1987. Transovarial and transtadial passage of *Borrelia burgdorferi* in the western black-legged tick, *Ixodes pacificus* (Acari: Ixodidae). *Am. J. Trop. Med. Hyg.* 37:188–92.

27. Latif, A. A., R. M. Newson, and T. S. Dhadialla. 1988. Feeding performance of *Amblyomma variegatum* (Acarina: Ixodidae) fed repeatedly on rabbits. *Exp. Appl. Acarol.* 5:83–92.

28. Latif, A. A., D. K. Punyua, P. B. Capstick, S. Nokoe, A. R. Walker, and J. D. Fletcher. 1991. Histopathology of attachment sites of *Amblyomma variegatum* and *Rhipicephalus appendiculatus* on Zebu cattle of varying resistance to ticks. *Vet. Parasitol.* 38:205–14.

29. Lee, R. P., and J. P. Opdebeeck. 1991. Isolation of protective antigens from the gut of *Boophilus microplus* using monoclonal antibodies. *Immunology* 72:121–26.

30. Lewis, I. J. 1970. Observations on the dispersal of larvae of the cattle tick *Boophilus microplus* (Can.). *Bull. Entomol. Res.* 59:595–604.

31. Mann, W. M. 1915. A cursorial tick. *Psyche* 22:60.

32. Mather, T. N., J. M. C. Ribeiro, and A. Spielman. 1987. Lyme disease and babesiosis: Acaricide focused on potentially infected ticks. *Am. J. Trop. Med. Hyg.* 36:609–14.

33. Mather, T. N., S. R. Telford III, S. I. Moore, and A. Spielman. 1990. *Borrelia burgdorferi* and *Babesia microti:* Efficiency of transmission from reservoirs to vector ticks (*Ixodes dammini*). *Exp. Parasitol.* 70:55–61.

34. Morgan, N. O. 1991. Potential impact of alien arthropod pests and vectors of animal diseases on the U.S. livestock industry. In Pimentel, D., ed. *CRC Handbook of pest management in agriculture,* 2d ed. Boca Raton, Fl.: CRC Press, Inc., 99–105.

35. Pence, D. B. 1975. Keys, species and host list, and bibliography for nasal mites of North American birds. (Acarina: Rhinonyssinae, Turbinoptinae, Speleognathinae, and Cytoditidae). *Spec. Pub. Mus. Texas Tech. Univ.* 8:1–148.

36. Pence, D. B., L. A. Windberg, B. C. Pence, and R. Sprowls. 1983. The epizootiology and pathology of sarcoptic mange in coyotes, *Canis latrans,* from south Texas. *J. Parasitol.* 69:1101–15.

37. Rickards, D. A. 1975. Cnemidocoptic mange in parakeets. *Vet. Med. Small Anim. Clin.* 70:729–31.

38. Robinson, R. 1985. Fight the mite and ditch the itch. *Parasitol. Today* 1:140–42.

39. Sauer, J. R., and J. A. Hair, eds. 1986. *Morphology, physiology, and behavioral biology of ticks.* Chichester, England: Ellis Horwood, Ltd.

40. Schuster, R., and P. W. Murphy, eds. 1991. *The Acari: Reproduction, development and life history strategies.* London: Chapman & Hall, Ltd.

41. Semtner, P. J., and J. A. Hair. 1973. The ecology and behavior of the lone star tick (Acarina: Ixodidae). V. Abundance and seasonal distribution in different habitat types. *J. Med. Entomol.* 10:618–28.

42. Sexton, D. J., and B. Hayes. 1975. Bird-mite infestation in a university hospital. *Lancet* 1(7904):445.

43. Shanbaky, N. M., and G. M. Khalil. 1975. The subgenus *Persicargas* (Ixodoidea: Argasidae: *Argus*). 22. The effect of feeding on hormonal control of egg development in *Argas* (*Persicargas*) *arboreus. Exp. Parasitol.* 37:361–66.

44. Shirinov, F. B., A. I. Ibragimova, and Z. G. Misirov. 1968. The dissemination of the virus of fowl-pox by the mite *D. gallinae. Veterinariia* 4:48–49.

45. Smith, M. G., R. J. Blattner, F. M. Heys, and A. Miller. 1948. Experiments on the role of the chicken mite *Dermanyssus gallinae* and the mosquito in the epidemiology of St. Louis encephalitis. *J. Exp. Med.* 87:119–38.

46. Sonenshine, D. E. 1986. Tick phermones: An overview. In Sauer, J. R., and J. A. Hair, eds. *Morphology, physiology, and behavioral biology of ticks.* Chickester, England: Ellis Horwood, Ltd.

47. Sonenshine, D. E. 1993. *Biology of ticks.* New York: Oxford University Press.

48. Sönnichsen, N., and H. Barthelmes. 1976. Epidemiological and immunological investigations on human scabies. *Agnew. Parasitol.* 17:65–70.

49. Steelman, C. D. 1976. Effects of external and internal arthropod parasites on domestic livestock production. *Ann. Rev. Entomol.* 21:155–78.

50. Stone, B. F. 1963. Parthenogenesis in the cattle tick, *Boophilus microplus. Nature* 200:1233.

51. Sweatman, G. K. 1957. Life history, non-specificity, and revision of the genus *Chorioptes,* a parasitic mite of herbivores. *Can J. Zool.* 35:641–89.

52. Teltow, G. J., P. V. Fournier, and J. A. Rawlings. 1991. Isolation of *Borrelia burgdorferi* from arthropods collected in Texas (USA). *Am. J. Trop. Med. Hyg.* 44:469–74.

53. Testi, B., and F. de Michelis. 1972. An interesting parasitological discovery in monkeys imported into Italy for laboratory use. *Zooprofilassi* 27:353–69.

54. Varma, M. R. G. 1993. Ticks and mites (Acari). In Lane, R. P., and R. W. Crosskey, eds. *Medical insects and arachnids.* London: Chapman & Hall, Ltd.

55. Viljoen, G. J., et al. 1986. Isolation of a neurotoxin from the salivary glands of female *Rhipicephalus evertsi evertsi. J. Parasitol.* 72:865–74.

56. Wharton, R. H., K. L. S. Harley, P. R. Wilkinson, K. B. Utech, and B. M. Kelly. 1969. A comparison of cattle tick control by pasture spelling, planned dipping, and tick-resistant cattle. *Aust. J. Agric. Res.* 20:783–97.

57. Williams, R. W., and H. W. Brown. 1945. The development of *Litomosoides carinii* filariid parasite of the cotton rat in the tropical rat mite. *Science* 102:482–83.

58. Wilson, M. L., T. S. Litwin, T. A. Gavin, M. C. Capkanis, D. C. MacLean, and A. Spielman. 1990. Host-dependent differences in feeding and reproduction of *Ixodes dammini* (Acari: Ixodidae). *J. Med. Ent.* 27:945–54.

59. Yuval, B., R. D. Deblinger, and A. Spielman. 1990. Mating behavior of male deer ticks *Ixodes dammini* (Acari: Ixodidae). *J. Insect Behavior* 3:765–72.

Additional References

Arthur, D. R. 1962. *Ticks and disease.* Oxford: Pergamon Press.

Baker, E. W., and G. W. Wharton. 1952. *An introduction to acarology.* New York: The Macmillan Co. An excellent reference for the beginner and professional alike, although a bit out of date.

Brennan, J. M., and E. K. Jones. 1959. Keys to the chiggers of North America with synonymic notes and descriptions of two new genera (Acarina: Trombiculidae). *Ann. Entomol. Soc. Am.* 52:7–16.

Lang, J. D., L. D. Charlet, and M. S. Mulla. 1976. Bibliography (1864 to 1974) of house-dust mites *Dermatophagoides* spp. (Acarina: Pyroglyphidae), and human allergy. *Sci. Biol. J.* 2:62–83.

McDaniel, B. 1979. *How to know the ticks and mites.* Dubuque, Iowa: William C. Brown Publishers. Useful, well-illustrated keys to genera and higher categories of ticks and mites in the United States.

Piesman, J. 1987. Emerging tick-borne diseases in temperate climates. *Parasitol. Today* 3:197–99.

Uilenberg, G. 1986. Highlights in recent research on tick-borne diseases of domestic animals. *J. Parasitol.* 72:485–91.

GLOSSARY

A

abscess (AB-ses)   Tissue necrosis in a localized area with increase in hydrostatic pressure from pus accumulation.

acanthella (ā-kan-THEL-ə)   Developing acanthocephalan larva, between an acanthor and a cystacanth, in which the definitive organ systems are developed.

acanthor (ā-KANTH-ər)   Acanthocephalan larva that hatches from the egg.

accidental myiasis (mī-Ī-ə-səs)   Presence within a host of a fly not normally parasitic. Also called **pseudomyiasis.**

accidental parasite   Parasite found in other than its normal host. Also called an **incidental parasite.**

acetabulum (a-set-TAB-ū-ləm)   Sucker: the ventral sucker of a fluke; a sucker on the scolex of a tapeworm.

acquired immunity   Immunity arising from a specific immune response, stimulated by antigen in the host's body (active) or in the body of another individual with the antibodies or lymphocytes transferred to the host (passive).

adhesive disc   Suckerlike circular organ or organelle used for attachment.

adoptive immunity   Immune state conferred by inoculation of lymphocytes, not antibodies, from an immune animal rather than by exposure to the antigen itself.

aedeagus (ə-DĒ-ə-gəs)   Copulatory organ or penis in insects and acarines.

aggregated   Describes the concentration of most parasites of a single species in a minority of hosts. Also called **overdispersed.**

ala (pl. alae; Ā-lə; A-lē)   Term often applied to winglike structure on plants or animals: the lateral winglike expansions of the branchiuran carapace to form respiratory alae, cuticular winglike expansions of nematodes, and others.

allograft (AL-lə-graft)   Graft of a piece of tissue or organ from one individual to another of the same species.

alveoli (al-VĒ-ə-lī)   Pockets or spaces bounded by membrane or epithelium.

amastigote (ā-MAST-i-gōt)   Form of Trypanosomatidae that lacks a long flagellum. Also called a **Leishman-Donovan (L-D) body,** as in *Leishmania.*

ameboma (a-mē-BŌ-mə)   Granuloma containing active trophozoites, occasionally resulting from a chronic amebic ulcer; rare except in Central and South America.

amebula (a-MĒ-byu-lə)   Daughter cell resulting from mitosis and cytokinesis of an encysted ameba.

bat / āpe / ärmadillo / herring / fēmale / finch / līce / crocodile / crōw / duck / ūnicorn / ə indicates unaccented vowel sound "uh" as in mammal, fishes, cardinal, heron, vulture / stress as in bi-OL-o-gy, BI-o-LOG-i-cal

amphid (AM-fid)   Sensory organ on each side of the "head" of nematodes.

amphistome (AM-fi-stōm)   Fluke with the ventral sucker located at the posterior end.

anamnestic response (an-əm-ES-tik)   Immune response to a challenge or secondary antigen inoculation, marked by more rapid and stronger manifestation of the immune reaction (specifically, antibody titer) than after the primary immunizing dose.

anapolysis (AN-ə-POL-ə-sis)   Detachment of a senile proglottid after it has shed its eggs.

anautogeny (AN-ä-TOJ-ə-nē)   In some Diptera the necessity of a blood meal before eggs can develop within the female.

androgenic gland (AN-drə-JEN-ik)   Gland located near the vas deferens in many Crustacea. Its secretions are responsible for development of male secondary sexual characteristics.

anecdysis (AN-ek-DĪ-səs)   Ecdysis in which successive molts are separated by quite long intermolt phases; referred to as *terminal anecdysis* when maximal size is reached and no more ecdyses occur.

anisogametes (AN-īs-ə-GAM-ēts)   Outwardly dissimilar male and female gametes. **Anisogamy** (AN-ī-SOG-ə-mē) is the condition of having dissimilar male and female gametes.

annuli (AN-yū-lī)   Rings on the body of a parasite; not necessarily indicative of internal segmentation.

antennae (second antennae of crustaceans) (an-TEN-ē)   Second pair of appendages in Crustacea, with bases usually immediately posterior to antennules; primarily sensory but sometimes adapted for other functions; derived from appendages on primitive third preoral somite; no homologous appendage in insects.

antennules (first antennae) (an-TEN-ūls)   Anteriormost pair of appendages of Crustacea; primarily sensory but often adapted for additional or other functions in particular species; derived from appendages on primitive second preoral somite; homologous to **antennae** of insects.

anterior station   Development of a protozoan in the middle or anterior intestinal portions of its insect host, such as the section *Salivaria* of Trypanosomatidae.

antibody (AN-tē-BOD-ē)   Immunoglobulin protein, produced by B cells (or plasma cells derived from B cells), that binds with a specific antigen.

antibody titer (TI-tər)   Measure of the amount of antibody present, usually given in units per milliliter of serum.

anticoagulant (an-tē-kō-AG-ū-lənt)   Substance that prevents blood clotting.

antigen (AN-tə-jən)   Any substance that will stimulate an immune response.

antigen challenge   Dose or inoculation with an antigen given to an animal some time after primary immunization with that antigen has been achieved.

antigenic determinant (AN-tə-JEN-ik dē-TERM-ə-nənt)   Area on an antigen molecule that binds with antibody or specific receptor sites on the sensitized lymphocyte; it "determines" the specificity of the antibody or lymphocyte. See **epitope.**

apical complex (Ā-pē-kal KOM-plex)   Dense ring and conelike structure, along with associated microtubules, micronemes, and rhoptries, at the anterior end of an apicomplexan sporozoite. (apical = at the apex.)

apical organ (Ā-pē-kal OR-gən)   Organ of unknown function at the apex of a cestode's scolex.

apodeme (AP-ō-dēm)   Spinelike inward projection of the cuticle in arthropods on which a muscle inserts; a ridgelike projection is an **apophysis** (a-POF-ə-sis).

apodous larva (a-PŌD-us LARV-ə)   Larva with no legs and with reduced head; usual in Hymenoptera, Diptera, some Coleoptera; requires maternal care or deposition in or on food source.

apolysis (a-POL-ə-sis)   Disintegration or detachment of a gravid tapeworm segment; also, the detachment of the hypodermis from the old procuticle in arthropods before molting.

apomorphic (ap-ō-MOR-fik)   Adjective that refers to the form of characters, in particular characters whose form differs from that of the same character in an outgroup.

arista (ə-RIS-tə)   Flagellumlike appendage on the antenna of a fly of the suborder Brachycera and some members of the Nematocera.

arrhenotoky (ə-REN-ō-TŌK-kē)   Parthenogenetic production of males. See **haplodiploidy.**

arthropodization (är-thrō-PÄD-i-ZĀ-shən)   Evolutionary development of the combination of characteristics associated with the Arthropoda, including a firm cuticular exoskeleton containing chitin.

ascaridine (əs-KAR-ə-den)   Protein of unknown function in the sperm of *Ascaris.*

ascaroside (əs-KAR-ə-sīd)   Glycoside found in *Ascaris,* made of the sugar *ascarylose* and a series of secondary monol and diol alcohols.

ascites (ə-SĪT-ēz)   Edema, or accumulation of tissue fluid, in the mesenteries and abdominal cavity.

autoimmunity (Ä-tō-im-MŪN-i-tē)   Immune response to ones' own proteins or other antigens.

autoinfection (AW-tō-in-FEK-shun)   Reinfection by a parasite juvenile without its leaving the host.

autotrophic nutrition (AW-tō-TRŌF-ik)   Feeding that does not require preformed organic molecules as nutritive substances.

axial cells (AX-ē-əl)   Central cells of a dicyemid mesozoan.

axoneme (AX-ō-nēm)   Core of a cilium or flagellum, comprising microtubules.

axostyle (AX-ō-stīl)   Tubelike organelle in some flagellate protozoa, extending from the area of the kinetosomes to the posterior end, where it often protrudes.

B

B cell   Type of lymphocyte that gives rise to plasma cells that liberate antibody to the antigen; so called because in birds they are processed through a lymphoid organ called the *bursa of Fabricius;* of primary importance in humoral immune response.

bacillary bands (BAS-ə-la-rē)   Lateral zones in the body wall of some nematodes, consisting of glandular and nonglandular cells of unknown function.

Baer's disc (BĀ-ərz)   Large, ventral sucker of an aspidogastrean trematode.

ballonets (bal-ō-NETS)   Four inflated areas within the "head" of nematodes of the family Gnathostomatidae; each is connected to an internal cervical sac of unknown function.

basal body (BĀ-səl)   Centriole from which an axoneme arises; also called a **kinetosome** or *blepharoplast.*

basis (basipodite) (BĀ-sis; bā-SIP-ə-dīt)   Joint of a crustacean appendage from which the exopod and the endopod originate; that is, the joint between the coxa and the exopod and endopod.

basophil (BĀ-sō-fil)   Least numerous of polymorphonuclear leukocytes, so called because it stains with basic stains.

bilharziasis (bil-härz-Ī-ə-sis)   Disease caused by *Schistosoma* spp. Also called *schistosomiasis.*

biological vector (VEK-tər)   Vector in which a disease organism lives or develops. Contrast with **mechanical vector.**

biramous appendage (bī-RĀM-əs ə-PEN-dij)   Appendage with two main branches from a common basal joint, characteristic of Crustacea, although not all appendages of a crustacean may be biramous.

black fly fever   Combination of symptoms resulting from sensitization to bites by black flies (Simuliidae).

blackhead   Disease of turkeys caused by the protozoan *Histomonas meleagridis.* Also called *histomoniasis* or *infectious enterohepatitis.*

bladderworm (BLA-dər-wərm)   See **cysticercus.**

blastocyst (BLAST-ō-sist)   In cestodes, posterior portion of plerocercus metacestode into which the body can withdraw.

blastoderm (BLAST-ō-derm)   "Primary epithelium" formed in early embryonic development of many arthropods when the nuclei migrate to the periphery and undergo superficial cleavage; usually encloses the central yolk mass.

bluetongue   Virus disease of ruminants transmitted by biting midges (Ceratopogonidae).

bothridium (Bäth-RID-ē-əm)   Muscular lappet on the dorsal or ventral side of the scolex of a tapeworm. Bothridia are often highly specialized, with many types of adaptations for adhesion.

bothrium (bÄTH-rē-əm)   Dorsal or ventral groove, which may be variously modified, on the scolex of a cestode.

bradyzoite (brā-dē-ZŌ-īt)   Small stage in various coccidia of the *Isospora* group that develops in a zoitocyst; similar to a **merozoite.**

breakbone fever   Another name for **dengue,** a virus disease transmitted by mosquitoes.

bubo (BŪ-bō)   Swollen lymph node.

buccal cone (BUK-kəl)   Portion of the mouthparts of acarines composed of hypostome and labrum.

bulla (BUL-ə)   Nonliving structure serving as an anchor to which the maxillae are permanently attached; secreted by head and maxillary glands of female copepods in the family Lernaeopodidae.

C

cadre (KAD-rē)   Sclerotized mouth lining of a pentastomid.

calabar swelling (KAL-ə-bär)   Transient subcutaneous nodule, provoked by the filarial nematode *Loa loa.*

calotte (kə-LOT)   "Head" end of a dicyemid mesozoan.

calypter (kə-LIP-tər)   Squama or lobe in the anal angle of a dipteran wing.

camerostome (kə-MER-ə-stōm)   Ventral groove in the propodosoma of soft ticks wherein lies the capitulum.

campestral (kəm-PES-trəl)   Characteristic of rural locations, especially open country and grasslands.

campodeiform (kam-pō-DĒ-ə-form)   Insect larva with well-defined head and thoracic appendages; typically predatory.

capitulum (kə-PIT-ū-ləm)   Anterior of two basic body regions of a mite or tick. Also called a **gnathosoma.**

bat / āpe / ärmadillo / herring / fēmale / finch / līce / crocodile / crōw / duck / ūnicorn / ə indicates unaccented vowel sound "uh" as in mammal, fishes, cardinal, heron, vulture / stress as in bi-OL-o-gy, BI-o-LOG-i-cal

capsule (KAP-sūl)   In reference to the eggshell of flatworms, that portion composed of sclerotin, with the precursors principally contributed by vitelline cells. Contrast with **coat.**

carapace (KAR-ə-pās)   Structure formed by posterior and lateral extension of dorsal sclerites of the head in many Crustacea, usually covering and/or fusing with one or more thoracic somites; considered as arising from a fold of head exoskeleton. Also a dorsal sclerotized plate often covering the idiosoma of acarines.

Carrion's disease (KAR-ē-onz də-zēz)   Bacterial disease transmitted by sand flies. See also **Oroya fever** and **verruga peruana.**

cecum (SĒ-kum)   Blind pouch or diverticulum of an intestine.

cell-mediated immunity (CMI)   Immunity in which antigen is bound to receptor sites on the surface of sensitized T lymphocytes that have been produced in response to prior immunizing experience with that antigen and in which manifestation is through macrophage response with no intervention of antibody.

cellular immune response (SEL-ū-lər)   Binding of antigen with receptor sites on sensitized T lymphocytes to cause release of lymphokines that affect macrophages, a direct response with no intervention of antibody. Also, the entire process by which the body responds to an antigen, resulting in a condition of cell-mediated immunity.

cement glands (SĒ-ment)   Glands in a male acanthocephalan that produce secretions sealing the female reproductive tract after copulation.

centrolecithal egg (cen-trō-LES-ə-thəl)   Type of egg found in many arthropods, in which the nucleus is located centrally in a small amount of nonyolky cytoplasm, surrounded by a large mass of yolk. After fertilization and some nuclear divisions, the nuclei migrate to the periphery to proceed with superficial cleavage, the yolk remaining central.

cephalogaster (sef-AL-ə-gas-tər)   Contractile organ in adult epicaridean isopods that functions in sucking blood and perhaps in respiration.

cercaria (ser-KAR-ē-ə)   Juvenile digenetic trematode, produced by asexual reproduction within a sporocyst or redia.

cerci (SER-sē)   Appendages on the 11th abdominal somite of some insects; usually sensory.

cercomer (SER-kō-mer)   Posterior, knoblike attachment on a procercoid or cysticercoid. It usually bears the hooks of the oncosphere.

chaetotaxy (KE-tō-tax-ē)   Taxonomic study of the location and arrangement of bristles on an insect. Especially important in the order Diptera.

Chagas' disease (SHÄ-gəs)   Disease of humans and other mammals caused by *Trypanosoma cruzi.*

chagoma (shä-GŌ-mə)   Reddish nodule that forms at the site of entrance of *Trypanosoma cruzi* into the skin.

chalimus (KAL-ə-məs)   Specialized, parasitic copepodid, found in the copepod order Siphonostomatoida; attached to its host by an anterior "frontal filament" that is secreted by the frontal gland.

chelate (KĒ-lāt)   Condition of an arthropod appendage in which the subterminal podomere bears a distal process to form a pincer with the terminal podomere; sometimes (incorrectly) used to describe the subchelate condition.

chelicerae (kə-LIS-ər-ē)   Anteriormost pair of appendages in the chelicerate arthropods, which include spiders, ticks, and mites; generally the most important feeding appendages in these groups.

chigger (CHIG-ər)   Mite of the family Trombiculidae. Also, sometimes applied to *Tunga penetrans,* the chigoe flea.

chitin (KĪ-tin)   High molecular weight polymer of N-acetyl glucosamine linked by 1,4-β-glycosidic bonds.

choanomastigote (kō-an-ō-MAST-ə-gōt)   Like a promastigote but with the flagellum emerging from a collarlike process, as in *Crithidia* spp.

chorioptic mange (kōr-ē-OP-tik mānj)   Disease caused by mites of the genus *Chorioptes.*

chromatoid bar (KRŌ-ma-toid)   Masses of RNA, visible with light microscopy, in young cysts of *Entamoeba* spp.

chyluria (ky-LŪR-ē-ə)   Lymph in the urine, characterized by a milky color.

ciliary organelles (SIL-ē-ar-ē or-gən-ELZ)   Organelles of specialized function formed by the fusion of cilia.

cirri (SER-rī)   Fused tufts of cilia in some protozoa that function like tiny legs. Also, plural for **cirrus.**

cirrus (SER-əs)   Penis or copulatory organ of a flatworm.

cladistics (kla-DIS-tiks)   General method of recovering hypothesized evolutionary histories using shared apomorphic characters as criteria for grouping.

clamp   Complex set of sclerotized bars, forming a "pinching" organ on the opisthaptor of a monogenetic trematode.

Claparedé organs (klap-er-ə-DĀ)   See **urstigmata.**

coarctate pupa (kō-ARK-tāt PŪ-pə)   Pupa in which the last larval cuticle is retained as a puparium.

coat   In reference to the eggshell of many cestodes, the portion contributed by the outer envelope, derived from embryonic blastomeres.

coelozoic (sēl-ə-ZŌ-ik)   Living in the lumen of a hollow organ, such as the intestine.

coenurus (sē-NŪR-əs)   Tapeworm metacestode in the family Taeniidae, in which several scolices bud from an internal germinative membrane; not enclosed in an internal secondary cyst.

colleterial glands (kō-lə-TER-ē-əl)   Female accessory glands in insects that produce a substance to cement eggs together or material for an ootheca.

commensalism (kō-MEN-səl-izm)   Kind of symbiosis in which one symbiont, the commensal, benefits, and the other symbiont, the host, is neither harmed nor helped by the association.

complement (KOMP-lə-mənt)   Collective name for a series of proteins that bind in a complex series of reactions to antibody (either IgM or IgG) when the antibody is itself bound to an antigen; produces lysis of cells if the antibody is bound to antigens on the cell surface.

complement fixation test   Immunological method used to detect presence of antibodies that bind (or fix) complement; a standard diagnostic test for many infections.

concomitant immunity (kon-KOM-ə-tənt)   Premunition.

condyles (KON-dīlz)   Bearing surfaces between arthropod joints that provide the fulcra on which the joints move.

conoid (KŌ-noid)   Truncated cone of spiral fibrils located within the polar rings of the suborder Eimeriina.

contagious (kon-TĀJ-əs)   Capable of being transmitted through direct contact. Also used to describe population distributions that are **aggregated,** such as in an area.

contaminative antigen (kon-TAM-in-ə-tiv)   Antigen borne by the parasite that is common to both the host and the parasite but that genetically is of host origin.

copepodid (ko-PEP-ə-did)   Juvenile stage that succeeds the naupliar stages in copepods, often quite similar in body form to the adult.

coracidium (kōr-ə-SID-ē-əm)   Larva with a ciliated epithelium, hatching from the egg of certain cestodes; a ciliated oncosphere.

costa (KOS-tə)   Prominent striated rod in some flagellate protozoa that courses from one of the kinetosomes along the cell surface beneath the recurrent flagellum and undulating membrane.

cotylocidia (kot-ə-lō-SID-ē-ə)   Larva of Aspidobothria.

coxa (KOX-a)   Most proximal podomere of an arthropod limb, sometimes called *coxopodite* in crustaceans.

coxal glands   Excretory organs of arachnids, consisting of a sac, tubule, and opening on the coxa.

crabs Infestation with the crab louse, *Phthirus pubis.*

creeping eruption Skin condition caused by hookworm larvae not able to mature in a given host.

crura (KRŪ-rə) Branches of the intestine of a flatworm.

cryptogonochorism (KRIP-tō-gō-nō-KŌR-izm) Separate sexes joined or associated to form the appearance of hermaphroditism.

cryptoniscus (krip-tō-NIS-kə s) Intermediate, free-swimming larval stage of the isopod suborder Epicaridea, developing after microniscus; attaches to definitive host.

cryptozoite (krip-tō-ZŌ-īt) Preerythrocytic schizont of *Plasmodium* spp.

ctenidium (tē-NID-ē-ə m) Series of stout, peglike spines on the head (genal ctenidium) and first thoracic tergite (pronotal ctenidium) of many fleas.

cutaneous (kū-TĀN-ē-ə s) Pertaining to the skin (e.g., a skin infection).

cuticulin (kū-TIK-ū-lin) Protein component of arthropod exoskeletons.

cypris (SĪ-prə s) Postnaupliar larva of barnacles (crustacean subclass Cirripedia) in which the carapace largely envelops the body; so called because of its resemblance to the ostracod genus *Cypris.*

cystacanth (SIS-tə-kanth) Juvenile acanthocephalan that is infective to its definitive host.

cysticercoid (sis-tə-SER-koid) Metacestode developing from the oncosphere in most Cyclophyllidea. It usually has a "tail" and a well-formed scolex.

cysticercosis (sis-tə-ser-KŌ-sis) Infection with one or more cysticerci.

cysticercus (sis-tə-SER-kə s) Metacestode with a fluid-filled bladder as the "tail."

cystogenic cells (sis-tō-JEN-ik) Secretory cells in a cercaria that produce a metacercarial cyst.

cytokine (SĪT-ə-kīn) Protein hormone produced by one cell type that modifies the physiological condition of the cells that produce it and/or other cells.

cyton (SĪ-ton) Cell body; contains nucleus and some other organelles but excludes processes extending from cell. For example, the neurocyton is the nerve cell body, excluding the axon and dendrites.

cytophaneres (SĪ-tō-fan-ER-ēz) Fibers radiating out from a zoitocyst into surrounding muscle; found in some species of Sarcocystidae.

D

dauer juvenile (DOW-er JŪ-və n-ə l) Nematode juvenile in which development is arrested during unsuitable conditions and resumed when conditions improve.

decacanth (DEK-ə-kanth) Ten-hooked larva that hatches from the egg of a cestodarian tapeworm. Also called a **lycophora.**

definitive host (də-FIN-ə-tiv) Host in which a parasite achieves sexual maturity. If there is no sexual reproduction in the life of the parasite, the host most important to humans is the definitive host.

deirid (DAR-id) Sensory papilla on each side near the anterior end of some nematodes.

bat / āpe / ärmadillo / herring / fēmale / finch / līce / crocodile / crōw / duck / ūnicorn / ə indicates unaccented vowel sound "uh" as in mammal, fishes, cardinal, heron, vulture / stress as in bi-OL-o-gy, BI-o-LOG-i-cal

delayed type hypersensitivity (DTH) Manifestation of cell-mediated immunity, distinguished from immediate hypersensitivity in that maximal response is reached about 24 hours or more after intradermal injection of the antigen. The lesion site is infiltrated primarily by monocytes and macrophages.

dengue (DEN-gē) Virus disease transmitted by mosquitoes.

denticles (denticulate) (DENT-ə-klz; den-TIK-ū-lāt) Small, toothlike projections.

dermatitis (derm-ə-TĪ-tis) Infection or inflammation of the skin.

deuterotoky (DŪT-ər-ō-tō-kē) Type of parthenogenesis in which all individuals are uniparental but in which both males and females occur.

deutomerite (dū-TŌM-er-īt) Posterior half of a cephaline gregarine protozoan.

deutonymph (DŪT-ō-nimf) In the life cycle of some mesostigmatid mites, a nonfeeding stage that molts into the adult.

deutovum (dū-TŌV-ə m) Incompletely developed larva that hatches from the egg of a chigger mite.

diapause (DĪ-ə-pawz) Quiescent phase in arthropods in which most physiological processes are suspended.

diapolar cells (DĪ-ə-PŌL-ə r) Ciliated somatodermal cells located between the parapolar and uropolar cells of a mesozoan.

diecdysis (dī-ek-DĪ-sis) Condition in which ecdysis processes are going on continuously and one ecdysis cycle grades rapidly into another.

dioecious (dī-Ē-sh ə s) Separate sexes; males and females are different individuals.

diplostomulum (dip-lō-STŌM-ū-l ə m) Strigeoid metacercaria in the family Diplostomatidae.

diporpa (dī-PŌRP-ə) Larval stage in the life cycle of the monogenean *Diplozoon.*

direct development In arthropods refers to development in which a juvenile hatches from the egg, and the juvenile is not distinctly different from adult except in size and maturity.

distal cytoplasm (DIS-tə l SĪT-ō-plazm) Distal cytoplasmic layer in the tegument of Monogenea, Digenea, and Cestoidea.

distome (DĪ-stōm) Fluke with two suckers, oral and ventral.

dorsal plate Dorsal plate on the body of a mesostigmatid mite.

duo-adhesive gland (DU-o-ad-HES-iv) Platyhelminth tegument gland with two types of cells, one producing an adhesive substance and the other a releasing substance.

dourine (DOW-rēn) Disease of horses and other equids caused by *Trypanosoma equiperdum.*

dyspnea (DISP-nē-ə) Difficult or labored breathing.

E

ecdysis (ek-DĪ-sis) Molting or discarding of inexpansible portions of cuticle, after which there is an increase in physical dimensions of the animal's body before the newly secreted cuticle hardens.

eclipsed antigen (ē-KLIPST) Antigen borne by the parasite that is common to both the host and the parasite but that genetically is of parasite origin.

ectocommensal (ek-tō-kō-MEN-sə l) Commensal symbiont that lives on the outer surface of its host.

ectoparasite (ek-tō-PAR-ə-sīt) Parasite that lives on the outer surface of its host.

ectopic (ek-TOP-ik) Infection in a location other than normal or expected.

edema (ē-DĒM-ə) Accumulation of more than normal amounts of tissue fluid, or lymph, in the intercellular spaces, resulting in localized swelling of the area.

elephantiasis (el-ə-fan-TĪ-ə-sis)   Permanently swollen body parts, usually limbs and scrotum, resulting from a lengthy filarial nematode infection.

enzyme-linked immunosorbent assay (ELISA) (ē-LĪS-ə)   Immunodiagnostic test designed to detect the presence of fixed antibody through linkage with an enzymatic reaction.

embryophore (EM-br ē-ə-for)   In reference to the eggshell of many cestodes, that portion contributed by the inner envelope, derived from embryonic blastomeres.

encephalitis (en-cef-ə-LĪ-tis)   Infection of the brain, especially by viruses or amebas.

endemic (en-DEM-ik)   Normally present in a certain geographic area or part of an area.

endemnicity (en-dem-NIS-ə-tē)   Amount or severity of a disease in a particular geographic area.

endite (EN-dīt)   Medial process from the protopod.

endocommensal (EN-dō-kō-MEN-s ə l)   Commensal symbiont that lives inside its host.

endocytosis (EN-dō-si-TŌ-sis)   Ingestion of particulate matter or fluid by phagocytosis or pinocytosis; that is, bringing material into a cell by invagination of its surface membrane and then pinching off the invaginated portion of a vacuole.

endodyogeny (EN-dō-dī-OJ-ə-nē)   Same as endopolyogeny except that only two daughter cells are formed.

endoparasite   Parasite that lives inside its host.

endopod (endopodite) (END-ō-pod; en-DOP-ə-dit)   Medial branch of a biramous appendage.

endopolyogeny (EN-dō-pol-ē-OJ-ə-nē)   Formation of daughter cells, each surrounded by its own membrane, while still in the mother cell.

endopterygote (EN-dop-TER-ə-gōt)   Condition of internal wing bud development in an insect. Also, an insect in which the wing buds develop externally or any insect secondarily wingless but derived from such an ancestor; associated with holometabolous insects.

endosome (END-ə-sōm)   Nucleoluslike organelle that does not disappear during mitosis.

eosinophil (ē-ō-SIN-ō-fil)   Type of polymorphonuclear leukocyte very important in many parasitic infections; so called because it stains with acidic stains such as eosin.

eosinophilia (ē-ō-SIN-ō-FIL-ē-ə)   Elevated eosinophil count in the circulating blood; commonly associated with chronic parasite infections.

epicaridium (ep-ē-kar-ID-ē-ə m)   First larval stage of the isopod suborder Epicaridea; attaches to a free-living copepod.

epicuticle (ep-ē-KŪT-i-k ə l)   Thin, outermost layer of arthropod cuticle; contains sclerotin but not chitin.

epidemic (ep-ē-DEM-ik)   Sharp rise in the incidence of an infection or disease.

epidemic hemorrhagic fever (him-ō-RAJ-ik)   Virus disease transmitted by mosquitoes. Also called **dengue.**

epidemiology (ep-ē-dem-ē-OL-ō-jē)   Study concerned with all ecological aspects of a disease to explain its transmission, distribution, prevalence, and incidence.

epimastigote (ep-ē-MAST-ə-gōt)   Trypanosomatid flagellate similar to a promastigote but with a short undulating membrane, such as in *Blastocrithidia.*

epipod (epipodite) (EP-ē-pod; e-PIP-ə-dīt)   Lateral process, from the protopod, usually with one or more joints. May be called an **exite.**

epitope (EP-ē-tōp)   Antigenic determinant; the portion of the antigen molecule displayed on the surface of an antigen-presenting cell (APC).

epizootic (ep-ē-zō-OT-ik)   Massive infection rate among animals other than humans; identical to an epidemic in humans.

espundia (es-PŪN-dē-ə)   Disease caused by *Leishmania braziliensis.* Also called *chiclero ulcer, uta, pian bois,* or *mucocutaneous leishmaniasis.*

eutely (Ū-te-lē)   Cell or nuclear constancy; the adult has the same number of nuclei or cells as the first-stage juvenile. Eutely may exist in tissues, organs, or entire animals.

exflagellation (EX-flaj-el-Ā-sh ə n)   Rapid formation of microgametes from a microgametocyte of *Plasmodium* and related genera.

exite (EX-īt)   Lateral process or joint from the protopod, sometimes referred to as an **epipod.**

exopod (exopodite) (EX-ō-pod; ex-OP-ə-dīt)   Lateral branch of a biramous appendage.

exopterygote (ex-op-TER-ə-gōt)   Condition of external wing bud development in an insect. Also, any insect in which the wing buds develop externally; associated with hemimetabolous insects.

F

facette (fa-SET)   Funnel-shaped opening through the inner membrane complex of the egg of a pentastomid. It receives the product of the dorsal organ.

facultative symbiont   Opportunistic symbiont, establishing a relationship with a host only if the opportunity presents itself but not physiologically dependent on doing so.

fascicle (FAS-i-c ə l)   Stylet bundle or combination of mouthparts used to pierce the skin in a blood-feeding arthropod. Composition of a fascicle varies according to group.

femur (FĒ-m ə r)   Podomere of an insect or acarine leg fixed to the trochanter proximally and articulating with the tibia distally in insects and with the patella in acarines.

festoons (fes-TOONS)   Sclerites on the posterior margin of the opisthosoma of certain hard ticks.

flabellum (fla-BEL-ə m)   Recurved process often found on the first two thoracic exopods of branchiuran crustaceans.

flagellar pocket (fla-JEL-ə r)   Depression, sometimes long and deep, from which a flagellum arises.

G

gametocyst (g ə-MĒT-ō-sist)   Cyst produced by some apicomplexan parasites. Sexual reproduction and spore formation occurs within this cyst.

gamont (GA-mont)   Apicomplexan life cycle stage that is committed to undergoing gametogenesis.

gena (JE-n ə)   Anterioventral portion of an insect head. For example, genal ctenidium is a row of heavy spines on the gena of a flea.

genital atrium (JEN-ə-t ə l ĀT-rē-ə m)   Cavity in the body wall of a flatworm into which male and female genital ducts open.

genitointestinal canal (JEN-ə-tō-in-TES-tin-ə l)   Duct connecting the oviduct and intestine of some polyopisthocotylean Monogenea.

gid (GID)   Disorientation caused by cysticercia in the brain; usually manifested by staggering or whirling.

glossa (GLOS-ə)   Tonguelike mouthpart in Hymenoptera (considered a hypopharynx by some authors).

glycocalyx (GLĪ-kō-KĀL-ix)   Finely filamentous layer containing carbohydrate, found on the outer surface of many cells, from 7.5 to 200.0 nm thick.

glycosomes (GLĪ-kō-sōmz)   Organelles found in *Trypanosoma* that contain enzymes of glycolysis and for oxidizing reduced NAD.

glyoxylate cycle (glī-OX-ə-lāt)   Metabolic pathway that functions to convert fatty acids or acetate to carbohydrate.

gnathopod (NATH-ə-pod)   Prehensile appendages of some Crustacea, such as the second and third thoracic legs of Amphipoda and the first thoracic legs of some Isopoda.

gnathosoma (nath-ə-SŌM-ə)   Anterior of two basic regions of the body of a mite or tick. Also called a **capitulum.**

goblets (GOB-lets)   Markings on the stigmatal plates of certain hard ticks.

gonotyl (GŌN-ō-til)   Muscular sucker or other perigenital specialization surrounding or associated with the genital atrium of a digenetic trematode.

granuloma, granulomatous tissue (gran-ū-LŌM-ə; gran-ū-LŌM-ə-təs)   Repaired area of a body marked by fibrous connective tissue (fibrosis). Also, fibrous connective tissue surrounding an antigen source.

ground itch   Skin rash caused by bacteria introduced by invasive hookworm larvae.

gynandry (JEN-an-drē)   In a hermaphroditic organism, maturation first of the female gonads and then of the male organs. Also called *protogyny.*

gynocophoral canal (gin-ə-KOF-ōr-əl)   Longitudinal groove in the ventral surface of a male schistosome fluke.

H

Haller's organ (HAL-erz)   Depression on the first tarsi of ticks; functions as an olfactory and humidity receptor.

haltere (HAL-tər)   Vestigial wing on the metathorax of a fly of the order Diptera; necessary for balance during flight.

halzoun (hal-ZŪN)   Disease resulting from blockage of the nasopharynx by a parasite. Also called **marrara.**

hamuli (HAM-ū-lī)   Large hooks on the opisthaptor of a monogenetic trematode; referred to as *anchors* by American authors.

haplodiploidy (HAP-lō-DIP-loi-dē)   Reproduction in which the males are haploid (parthenogenetically produced) and the females are diploid (from a fertilized egg).

haptens (HAP-tenz)   Molecules of small molecular weight (usually) that are immunogenic only when attached to carrier molecules, usually proteins.

hemelytra (HIM-ē-LIT-rə)   The front wing of an insect of the order Hemiptera.

hemimetabolous metamorphosis (HIM-ē-mə-TAB-ō-ləs met-ə-MORF-ə-sis)   In insects gradual metamorphosis in which the nymphs are generally similar in body form to the adults and become more like the adults with each instar.

hemocoel (HĒM-ə-sēl)   Main body cavity of arthropods, the embryonic development of which differs from that of a true coelom but that includes a vestige of a true coelom.

hemoglobinuria (HĒM-ə-glōb-in-ŪR-ē-ə)   Bloody urine.

hemolymph (HĒM-ə-limf)   Fluid within the hemocoel of arthropods. Also, the pseudocoelomic fluid of nematodes.

hepatosplenomegaly (hē-PAT-ō-SPLEN-ō-meg-ə-lē)   Swollen liver and spleen.

hermaphroditism (her-MAF-rō-di-tizm)   Possession of gonads of both sexes by a single (monoecious) individual.

heterogonic life cycle (het-ər-ə-GŌN-ik)   Life cycle involving alternation of parasitic and free-living generations.

heterophile reaction (HET-er-ə-fīl)   Antigen-antibody reaction in which the antibody was not specifically elicited by the antigen to which it binds.

heteroxenous (het-ər-ə-ZĒN-əs)   Describes a parasite that lives within more than one host during its life cycle.

hexacanth (HEX-ə-kanth)   Oncosphere; a six-hooked larva hatching from the egg of a eucestode.

histozoic (HIS-tə-ZŌ-ik)   Dwelling within the tissues of a host.

bat / āpe / ärmadillo / herring / fēmale / finch / līce / crocodile / crōw / duck / ūnicorn / ə indicates unaccented vowel sound "uh" as in mammal, fishes, cardinal, heron, vulture / stress as in bi-OL-o-gy, BI-o-LOG-i-cal

holoblastic cleavage (HŌ-lō-BLAS-tik)   Each nuclear division in an early embryo that is accompanied or closely followed by complete cytokinesis, the nuclei being separated by cell membranes.

holometabolous metamorphosis (HŌ-lō-mə-TAB-ə-ləs met-ə-MORF-ə-sis)   Metamorphosis in an insect with a larva, pupa, and adult.

holophytic nutrition (HŌ-lō-FIT-ik)   Formation of carbohydrates by chloroplasts.

holozoic nutrition (HŌ-lō-ZŌ-ik)   Feeding by active ingestion of organisms or particles.

homogonic life cycle (HŌ-mō-GŌN-ik)   Life cycle in which all generations are parasitic or all are free living. There is no (or little) alternation of the two.

homothetogenic fission (HŌ-mō-thet-ə-JEN-ik FISH-shən)   Mitotic fission across the rows of cilia of a protozoan.

hood   Dorsal wall of the camerostome that extends over the capitulum.

host specificity   Degree to which a parasite is able to mature in more than one host species.

humoral immune response (HŪM-er-əl)   Binding of antigen with soluble antibody in blood serum. Also, the entire process by which the body responds to an antigen by producing antibody to that antigen.

hydatid cyst (hī-DAT-id sist)   Metacestode of the cyclophyllidean cestode genus *Echinococcus,* with many protoscolices, some budding inside secondary brood cysts.

hydatid sand   Sediment in a hydatid cyst formed by free protoscolices.

hydrogenosomes (hī-drə-JEN-ə-sōmz)   Small organelles in certain anaerobic protozoa that produce molecular hydrogen as an end product of energy metabolism.

hyperapolysis (HĪ-pər-ap-ō-LĪ-sis *or* HĪ-per-ə-POL-ə-sis)   Detachment of a tapeworm proglottid while still immature, before eggs are formed.

hyperendemic (HĪ-pər-en-DEM-ik)   Condition in which a disease or infection has high, usually seasonal, transmission in a certain geographic area.

hyperinfection (HĪ-pər-in-FEK-shən)   Condition in *Strongyloides* infections in which filariform juveniles repenetrate mucosa of the small intestine and proceed with migration.

hypermetamorphosis (HĪ-pər-met-ə-MORF-ə-sis)   Type of metamorphic development in which different larval instars have markedly dissimilar body forms.

hyperparasitism (HĪ-pər-PAR-ə-sit-izm)   Condition in which an organism is a parasite of another parasite.

hypnozoite (HIP-nō-ZŌ-it)   Dormant exoerythrocytic form found in certain *Plasmodium* species.

hypodermis (HĪ-pō-DER-mis)   Syncytial layer that secretes the cuticle in nematodes.

hypopharynx (HĪ-pō-FAR-inx)   Tonguelike lobe arising from the floor of the mouth in insects; variously modified for feeding in many groups.

hypostome (HĪ-pō-stōm)   Portion of the mouthparts of acarines; composed of fused coxae of pedipalps.

hysterosoma (HIST-er-ō-SŌM-ə)   Combination of the metapodosoma and opisthosoma of the body of a tick or mite.

I

ick (ik)   Serious disease of freshwater fishes, caused by the ciliate protozoan *Ichthyophthirius multifiliis.*

icterus (jaundice) (IK-tər-əs; JÄN-dis)   Yellowing of the skin and other organs because of bile pigments in the blood.

idiosoma (ID-ē-ō-SŌM-ə)   Posterior of the two basic parts of the body of a mite or tick, bearing the legs and most internal organs.

imago (i-MAG-ō)   Adult or final instar in the development of an insect.

imagochrysalis (i-MAG-ō-KRIS-ə-lis)   Quiescent stage between the nymph and adult in the life cycle of a chigger mite.

immediate hypersensitivity (hī-per-sen-sə-TIV-ə-tē)   Biological manifestation of an antigen-antibody reaction in which the maximal response is reached in a few minutes or hours. Intradermal injection of the antigen produces local swelling and redness with heavy infiltration of polymorphonuclear leukocytes. Intravenous injection may produce anaphylactic shock and death.

immune cross reaction (im-MŪN)   Binding of an antibody or cell receptor site with an antigen other than the one that would provide an exact "fit"; that is, an antigen-antibody reaction in which the antigen is not the same one that stimulated the production of that antibody.

immunity (im-MŪN-ə-tē)   State in which a host is more or less resistant to an infective agent; preferably used in reference to resistance arising from tissues that are capable of recognizing and protecting the animal against "nonself."

immunogenic (IM-ū-nō-JEN-ik)   Refers to any substance that is antigenic; that is, stimulates production of antibody or cell-mediated immunity.

immunoglobulin (IM-ū-nō-GLOB-u-lin) Any one of five classes of proteins in blood serum that function as antibodies; abbreviated IgM, IgG, IgA, IgD, and IgE.

incidence (IN-sə-dens)   In epidemiology, the number of new cases of a disease per unit time; that is, a rate measurement. Contrast with **prevalence.**

incidental parasite (in-se-DEN-tal)   Accidental parasite.

indirect development   In arthropods, refers to development in which larva or nymph hatches from an egg and is distinctly different in body form from the adult; that is, development with metamorphosis.

inflammation (in-flə-MĀ-shən)   Defense process of body including congestion of blood vessels, escape of plasma to interstitial tissue space, swelling, and warmth.

infraciliature (IN-frə-SIL-ē-ə-tūr)   All cilia, basal bodies, and their associated fibrils in a ciliate protozoan.

infrapopulation (IN-frə-POP-ū-lā-shən)   All individuals of a single parasite species in one host.

infusoriform larva (IN-fū-SŌR-ə-form)   Ciliated larva produced by an infusorigen within a dicyemid mesozoan.

infusorigen (IN-fū-SŌR-ə-jen)   Mass of reproductive cells within a rhombogen.

ingroup (IN-grūp)   Taxon being studied in a cladistic analysis of evolutionary history.

instar (IN-star)   Molt stage in the life of an arthropod.

intensity (in-TEN-sə-tē)   Number of parasites in an infected host (**infrapopulation**). Mean intensity is the average number of parasites per infected host.

interleukin (IN-ter-LŪ-kin)   Cytokines produced by white blood cells and mediating their own activities or those of other white blood cells.

intermediate host (IN-ter-MĒD-ē-ət)   Host in which a parasite develops to some extent but not to sexual maturity.

intermittent parasite (IN-ter-MIT-tent)   Temporary parasite.

internuncial processes (in-ter-NUN-sē-əl)   Cytoplasmic channels that connect one part of a cell to another, such as those linking the distal cytoplasm to the tegumental cytons in many flatworms.

intralecithal cleavage (IN-tra-LES-ə-thal CLĒ-vaj)   Cleavage in which the nuclei undergo several divisions within the yolk mass without concurrent cytokinesis; common in arthropods.

iodinophilous vacuole (ī-ō-din-OF-ə-lus VAK-ū-ōl)   Vacuole within a protozoan that stains readily with iodine.

isogametes (Ī-sō-GAM-ēts)   Outwardly similar male and female gametes.

J

jacket cells (JAK-et)   Ciliated somatoderm of an orthonectid mesozoan.

K

kala-azar (ka-lə-Ā-zar)   Disease caused by *Leishmania donovani.* Also called *Dumdum fever* or *visceral leishmaniasis.*

kentrogon (KEN-trō-gən)   Larva in the crustacean order Rhizocephala that is attached to its host crab; formed after the cypris larva molts and its appendages and carapace are discarded.

kinetid (kī-NET-id)   Axoneme of a cilium of flagellum together with its basal fibrils and organelles. Also called a **mastigont.**

kinetodesmose (kinetodesmata) (kī-net-ō-DES-mōs; kī-net-ō-des-MÄ-tə)   Compound fiber joining cilia into rows.

kinetoplast (kī-NET-ō-plast)   Conspicuous part of a mitochondrion in a trypanosome; usually found near the kinetosome.

kinetosome (kī-NET-ō-sōm)   Centriole from which an axoneme arises. Also called a **basal body** or *blepharoplast.*

kinety (kī-NET-tē)   Row of cilia basal bodies and their kinetodesmose. All kineties and kinetodesmata in the organism are its infraciliature.

Koch's blue bodies (KŌKS)   Schizonts of *Theileria parva* in circulating lymphocytes.

K-strategist   Species of organism that uses a survival and reproductive "strategy" characterized by low fecundity, low mortality, and longer life and with populations approaching the carrying capacity of the environment, controlled by density-dependent factors.

Kupffer cells (KŪP-fer)   Phagocytic epithelial cells lining the sinusoids of the liver.

L

labellum (la-BEL-əm)   Expanded tip of the labium in an insect.

labium (LĀB-ē-əm)   Mouthpart in insects composed of fused second maxillae; homologous to second maxillae of crustaceans.

labrum (LĀ-brəm)   Sclerite forming the anterior closure of the mouth in arthropods; specifically the free lobe overhanging the mouth.

lacunae (lak-KUN-ē)   Channels making up the lacunar system in Acanthocephala. Also, in developing wings of insects, canals that contain nerves, tracheae, and hemolymph.

lacunar system (lak-KŪN-ər)   System of canals in the body wall of an acanthocephalan, functioning as a circulatory system.

landscape epidemiology   Approach to epidemiology that employs all ecological aspects of a nidus. By recognizing certain physical conditions, the epidemiologist can anticipate whether a disease can be expected to exist.

larva (LAR-və)   Progeny of any animal that is markedly different in body form from the adult.

Laurer's canal (LÄ-rərz)   Usually blind canal extending from the base of the seminal receptacle of a digenetic trematode. It probably represents a vestigial vagina.

Leishman-Donovan (L-D) body (LĪSH-mən DON-ə-vən) Amastigote in the Trypanosomatidae.

leishmaniasis (LĪSH-mən-Ī-ə-sis)   Infection by a species of *Leishmania.*

lemniscus (lem-NIS-cəs)   Structure occurring in pairs attached to the inner, posterior margin of the neck of an acanthocephalan, extending into the trunk cavity. Its function is unknown.

leukorrhea (lu-kō-RĒ-ə)   White, puslike discharge resulting from infection.

liposome (LIP-ə-sōm)   Artificial lipoid particle used to deliver antiparasitic drugs directly to macrophages (which eat the particles).

loculi (LOK-ū-lī)   Shallow, suckerlike depressions in an adhesive organ of a flatworm.

lumen (LŪ-men)   Space within any hollow organ.

lunules (LŪN-ūls)   Small, suckerlike discs on the anterior margin of some copepods in the family Caligidae, functioning as organs of adhesion.

lycophora (lī-KOF-ōr-ə *or* lī-kə-FOR-ə)   Ten-hooked larva that hatches from the egg of a cestodarian tapeworm. Also called a **decacanth.**

lymph varices (limf VER-ə-sēz)   Dilated lymph ducts.

lymphadenitis (limf-FAD-ən-ī-tis)   Inflamed lymph node.

lymphocyte (LIMF-ō-sīt)   Type of leukocyte vital in immune response. Several different types are known. See **B cell** and **T cell.**

lymphokine (LIM-fə-kīn)   Cytokine released by a lymphocyte.

lysosome (LĪ-sə-sōm)   Intracellular vacuole or vesicle containing digestive enzymes (lysozymes).

M

macrogamete (MA-krō-GA-mēt)   Large, quiescent, "female" anisogamete.

macrogametocyte (MA-krō-gə-MĒT-ə-cīt)   Cell giving rise to a macrogamete.

macroparasite (MA-krə-PAR-ə-sīt)   Large parasite that does not multiply in the host of interest. Examples are cestodes, trematodes, and most nematodes in their definitive hosts.

macrophage (MA-krə-fāj)   Important phagocytic cell and antigen-presenting cell, derived from monocyte.

macrophage migration inhibitory factor (MIF)   Cytokine released by sensitized lymphocytes that tends to inhibit migration of macrophages in the immediate vicinity, thus contributing to accumulation of larger numbers of macrophages close to the site of MIF release.

Malpighian tubule (mal-PIG-ē-ən)   Blind tubules opening into the hindgut of nearly all insects and some myriapods and arachnids and functioning primarily as excretory organs.

mamelon (MA-mə-lon)   Ventral, serrated projection on the ventral surface of a male nematode of the family Syphaciidae. Its function is unknown.

mandibles   Third pair of appendages from the anterior in Crustacea; second pair in Insecta; primarily function in feeding; derived from appendages on primitive fourth (first postoral) somite.

mange (mānj)   Dermatitis caused by species of mites, often designated with the causative organism. For example, *Sarcoptes* causes sarcoptic mange.

marginal bodies   Sensory pits or short tentacles between the marginal loculi of the opisthaptor of an aspidogastrean trematode.

marrara (mə-RA-rə)   Nasopharyngeal blockage by a parasite. Also called **halzoun.**

mast cell   Type of cell in various tissues that releases pharmacologically active substances with a role in inflammation.

mastigont (MAS-tə-gont)   Axoneme of a cilium or flagellum together with its basal fibrils and organelles.

mastitis (mas-TĪ-təs)   Infection of the udder of cattle.

bat / āpe / ärmadillo / herring / fēmale / finch / līce / crocodile / crōw / duck / ūnicorn / ə indicates unaccented vowel sound "uh" as in mammal, fishes, cardinal, heron, vulture / stress as in bi-OL-o-gy, BI-o-LOG-i-cal

Maurer's clefts (MÄR-rərz)   Blotches on the surface of an erythrocyte infected with *Plasmodium falciparum.*

maxillae (second maxillae) (MAX-ə-lē *or* max-IL-ē)   Fifth pair of appendages in Crustacea, primarily feeding in function, derived from appendages on primitive sixth (third postoral) somite; homologous to **labium** in insects. The maxillae of insects are the third pair of head appendages, homologous to **maxillules** of Crustacea.

maxillipeds (max-IL-ə-pedz)   One or more pairs of head appendages originating posterior to maxillae in Crustacea; derived from appendages on somites that were primitively posterior to gnathocephalon; usually function in feeding but sometimes adapted for other functions, such as prehension, in parasitic forms.

maxillopodan eye (max-ə-LOP-ə-dən)   Naupliar eye of crustacean class Maxillopoda; has a *tapetum* (crystalline reflective layer).

maxillules (first maxillae) (MAX-ə-lulz)   Fourth pair of appendages in Crustacea, primarily feeding in function; derived from appendages on primitive fifth (second postoral) somite; homologous to **maxillae** in insects.

mechanical vector   Vector that transmits disease organism by mechanical means only. Contrast with **biological vector.**

megacolon   Flabby distended colon caused by chronic Chagas' disease.

megaesophagus   Distended esophagus caused by chronic Chagas' disease.

Mehlis' glands (MĀ-ləs)   Unicellular mucous and serous glands surrounding the ootype of a flatworm.

membranelle (mem-brən-EL)   Short, transverse rows of cilia, fused at their bases, serving to move food particles toward the oral groove of a protozoan.

merogony (mer-OG-ə-nē)   Multiple fission to produce merozoites; schizogony.

merozoite (mer-ə-ZŌ-īt)   Daughter cell resulting from schizogony.

mesocercaria (mez-ə-ser-KAR-ē-ə)   Juvenile stage of the digenetic trematode *Alaria*. It is an unencysted form between the cercaria and the metacercaria.

metacercaria (met-ə-ser-KAR-ē-ə)   Stage between the cercaria and adult in the life cycle of most digenetic trematodes; usually encysted and quiescent.

metacestode (met-ə-SES-tōd)   Developmental stage of a cestode after metamorphosis of the oncosphere; a juvenile cestode.

metacryptozoite (met-ə-krip-tə-ZŌ-īt)   Merozoite developed from a cryptozoite.

metacyclic   Stage in the life cycle of a parasite that is infective to its definitive host.

metacyst (MET-ə-sist)   Cystic stage of a parasite that is infective to a host.

metamere (MET-ə-mer)   One of the segments in a metameric animal.

metamerism (met-AM-ər-izm)   Division of the body along the anteroposterior axis into a serial succession of segments, each of which contains identical or similar representatives of all the organ systems of the body; primitively in arthropods, including externally a pair of appendages and internally a pair of nerve ganglia, a pair of nephridia, a pair of gonads, paired blood vessels and nerves, and a portion of the digestive and muscular systems.

metamorphosis (met-ə-MORF-ə-sis)   Type of development in which one or more juvenile types differ markedly in body form from the adult; occurs in numerous animal phyla. Also applies to the actual process of changing from larval to adult form.

metanauplius (met-ə-NÄ-plē-əs)   Later naupliar larvae of some crustaceans; that is, occurring after several naupliar stages but before another larval type or preadult in the developmental sequence.

metapodosoma (MET-ə-PŌD-ə-sō-mə)   Portion of the podosoma that bears the third and fourth pairs of legs of a tick or mite.

metapolar cells   Posterior tier of cells in the calotte of a dicyemid mesozoan.

metasome (MET-ə-sōm)   Portion of the body anterior to the major point of body flexion in many copepods; usually includes the cephalothorax and several free thoracic segments.

metraterm (MET-rə-tərm)   Muscular, distended termination of the uterus of a digenetic trematode.

microfilaria (MĪK-rə-fi-LAR-ē-ə)   First-stage juvenile of any filariid nematode that is ovoviviparous; usually found in the blood or tissue fluids of the definitive host.

microgamete (MĪK-rə-GAM-ēt)   Slender, active "male" anisogamete.

microgametocyte (MĪK-rə-gam-ET-ə-sīt)   Cell that gives rise to microgametes.

microparasite (MĪK-rə-par-ə-sīt)   Small (or very small) parasite that multiplies within the host of interest. Examples are protistan and prokaryotic parasites.

micronemes (MĪK-rə-nēmz)   Slender, convoluted bodies that join a duct system with the rhoptries, opening at the tip of a sporozoite or merozoite.

microniscus (MĪK-rə-NIS-kəs)   Intermediate larval stages of the isopod suborder Epicaridea, parasitic on free-living copepods.

micropredator   Temporary parasite.

micropyle (MĪK-rə-pīl)   Pore in the oocyst of some coccidia and in the egg of an insect.

microthrix (microtriches) (MĪK-rə-thrix; MIK-rə-trich-ēz)   Minute projections of the tegument of a cestode.

Miescher's tubules (MESH-ərz)   Sarcocysts; tissue cysts of *Sarcocystis*.

miracidium (mir-ə-SID-ē-əm)   First larval stage of a digenetic trematode; ciliated and often free swimming.

monoecious (mən-Ē-shəs)   Hermaphroditic; an individual that contains reproductive systems of both sexes.

monostome (MON-ə-stōm)   Fluke that lacks a ventral sucker.

monoxenous (MON-ə-ZĒN-əs *or* MON-ox-ĒN-əs)   Living within a single host during a parasite's life cycle.

monozoic (MON-ə-ZŌ-ik)   Tapeworm whose "strobila" consists of a single proglottid.

mucron (MŪ-krən)   Apical anchoring device on an acephaline gregarine protozoan.

muscularis mucosae (məs-kū-LAR-is mū-KŌ-sē)   Smooth muscle fibers around the mucosa of the gut wall, surrounding the lamina propria and surrounded by the submucosa.

mutualism (MŪ-chū-əl-izm)   Type of symbiosis in which both host and symbiont benefit from the association.

mycetome (MĪ-sē-tōm)   Specialized organ in some insects that bears mutualistic bacteria.

myiasis (mī-Ī-ə-sis)   Infection by fly maggots.

myzorhynchus (MĪ-zō-RINK-əs)   Apical stalked, suckerlike organ on the scolex of some tetraphyllidean cestodes.

N

nagana (nə-GA-nə)   Disease of ruminants caused by *Trypanosoma brucei brucei* or *T. congolense*.

nauplius (NÄ-plē-əs)   Typically the earliest larval stage(s) of crustaceans; has only three pairs of appendages: antennules, antennae, and mandibles—all primarily of locomotive function.

neascus (nē-AS-kəs)   Strigeoid metacercaria with a spoon-shaped forebody.

necrosis (nə-KRŌS-is)   Cell or tissue death.

nematogen (nə-MAT-ə-jən)   State in the life cycle of a dicyemid mesozoan.

neoteny (nē-OT-ə-nē)   The attainment of sexual maturity in the larval condition. Also, the retention of larval characters into adulthood.

neutrophil (NU-trə-fil)   Most abundant of polymorphonuclear leukocytes; an important phagocyte; so called because it stains with both acidic and basic stains.

nidus (NĪ-dəs)   Specific locality of a given disease; result of a unique combination of ecological factors that favors the maintenance and transmission of the disease organism.

nymphochrysalis (NIM-fə-KRIS-ə-lis)   Nonfeeding, prenymph stage in the life cycle of a chigger mite.

nymphs (nimfs)   Juvenile instars in insects with hemimetabolous metamorphosis. Also, juvenile instars of mites and ticks with a full complement of legs.

O

obligate symbiont   Organism that is physiologically dependent on establishing a symbiotic relationship with another.

obtect pupa (OB-tekt PŪ-pə)   Pupa with wings and legs tightly appressed to its body and covered by an external cuticle.

oligopod larva (ə-LIG-ə-pod)   Usual larva in Coleoptera and Neuroptera, with a well-developed head and thoracic legs.

onchocercoma (ON-kō-sər-KŌ-mə)   Subcutaneous nodule containing masses of the nematode *Onchocerca volvulus*.

oncomiracidium (ON-kō-mir-ə-SID-ē-əm)   Ciliated larva of a monogenetic trematode.

oocyst (Ō-ə-sist)   Cystic form in the Apicomplexa, resulting from sporogony; the oocyst may be covered by a hard, resistant membrane (as in *Eimeria*), or it may not (as in *Plasmodium*).

oocyst residuum (re-ZI-jū-əm)   Cytoplasmic material not incorporated into the sporocyst within an oocyst; seen as an amorphous mass within an oocyst.

oogenotop (ō-ə-GEN-ə-tōp)   Female genital complex of a flatworm, including oviduct, ootype, Mehlis' glands, common vitelline duct, and upper uterus.

ookinete (ō-ə-KĪN-ēt)   Motile, elongated zygote of a *Plasmodium* or related organism.

oostegites (ō-OS-tə-gīts)   Modified thoracic epipods in females of the crustacean superorder Peracarida. They form a pouch for brooding embryos.

ootheca (ō-ə-THĒK-ə)   Egg packet secreted by some insects; may be covered with sclerotin.

ootype (Ō-ə-tīp)   Expansion of the flatworm female duct, surrounded by Mehlis' glands, where, in some flatworms, ducts from a seminal receptacle and vitelline reservoir join.

operculum (ō-PER-kū-ləm)   Lidlike specialization of a parasite eggshell through which the larva escapes.

opisthaptor (Ō-pist-HAP-tər)   Posterior attachment organ of a monogenetic trematode.

opisthomastigote (ō-PIS-thə-MAS-ti-gōt)   A form of Trypanosomatidae with the kinetoplast at the posterior end. The flagellum runs through a long reservoir to emerge at the anterior. There is no undulating membrane. An example is *Herpetomonas*.

opisthosoma (ō-PIS-thə-SŌ-mə)   Portion of the body posterior to the legs in a tick or mite.

opsonization (OP-sən-i-ZĀ-shən)   Modification of the surface characteristics of an invading particle or organism by binding with antibody or a nonspecific molecule in such a manner as to facilitate phagocytosis by host cells.

orchitis (or-KĪT-is)   Inflammation of the testis.

oriental sore   Disease caused by *Leishmania tropica*. Also called *Jericho boil*, *Delhi boil*, *Aleppo boil*, or *cutaneous leishmaniasis*.

Oroya fever   Clinical form of Carrion's disease, caused by the bacterium *Bartonella bacilliformis* and transmitted by sand flies.

otoacariasis (OT-ō-ak-ər-Ī-ə-sis)   Infestation of the external ear canal by ticks or mites.

outgroup   In cladistic analysis, a taxon chosen that is related to the taxa in the ingroup and has ancestral (plesiomorphic) characters in common with the ingroup.

overdispersion   Nonrandom dispersion of individuals in a habitat, such as when a minority of host individuals bears a majority of parasites.

ovicapt (Ō-vi-kapt)   Sphincter on the oviduct of a flatworm.

ovipositor (Ō-vē-PAZ-əd-ər)   Structure on a female animal modified for deposition of eggs. In many insects it is derived from segmental appendages of the abdomen.

ovisac   External sac attached to the somite that bears openings of gonoducts in females of many Copepoda. Fertilized eggs pass into the ovisacs for embryonation.

ovovitellarium (Ō-vō-vit-ə-LAR-ē-əm)   Mixed mass of ova and vitelline cells; found in the monogenean genus *Gyrodactylus* and in a few tapeworms.

ovoviviparous (Ō-vō-vī-VI-par-əs)   Describes reproduction in which embryos develop within the maternal body without additional nourishment from the parent and hatch within the parent or immediately after emerging.

P

paedogenesis (pē-dō-JEN-ə-sis)   Reproduction by immature or larval animals caused by acceleration of maturation.

pandemic   Very widely distributed epidemic.

pansporoblast (pan-SPŌR-ə-blast)   Myxosporidean sporoblast that gives rise to more than one spore. Also called a sporoblast mother cell.

Papatasi fever   Virus disease transmitted by sand flies. Also called *sand fly fever.*

parabasal body   Golgi body located near the basal body (kinetosome) of some flagellate protozoa, from which the parabasal filament runs to the basal body.

parabasal filament   Fibril, with periodicity visible in electron micrographs, that courses between the parabasal body and a kinetosome.

paramastigote (PA-rə-MAS-ti-gōt)   Form of trypanosomatid in which the kinetosome and kinetoplast are beside the nucleus.

paramere (PA-rə-mər)   Copulatory appendage in male cimicid bugs.

parapolar cells   Cells making up the ciliated somatoderm immediately behind the calotte of a mesozoan.

parasite   *Raison d'étre* for parasitologists.

parasitic castration   Condition in which a parasite causes retardation in development or atrophy of host gonads, often accompanied by failure of secondary sexual characteristics to develop.

parasitism   Symbiosis in which the symbiont benefits from the association while the host is harmed in some way.

parasitoid   Organism that is a typical parasite early in its development but that finally kills the host during or at the completion of development; often used in reference to many insect parasites of other insects.

parasitologist   Quaint person who seeks truth in strange places; a person who sits on one stool, staring at another.

parasitophorous vacuole (PAR-ə-sit-OF-ər-əs VAK-ū-ōl)   Vacuole within a host cell that contains a parasite.

paratenic host (par-ə-TĒN-ik)   Host in which a parasite survives without undergoing further development. Also known as a **transport host.**

paraxial (crystalline) rod (par-AX-ē-əl)   Rod that runs alongside the axoneme in the flagellum of a kinetoplastid flagellate.

parenchyma (pə-REN-kə-mə)   Spongy mass of vacuolated mesenchymal cells filling spaces between viscera, muscles, or epithelia. In some flatworms the cells are cell bodies of muscle cells. Also, the specialized tissue of an organ as distinguished from the supporting connective tissue.

pars prostatica (parz prə-STAT-i-kə)   Dilation of the ejaculatory duct of a flatworm, surrounded by unicellular prostate cells.

parthenogenesis (PAR-thə-nō-JEN-ə-sis)   Development of an unfertilized egg into a new individual.

paruterine organ (par-ŪT-ər-in)   Fibromuscular organ in some cestodes that replaces the uterus.

passive immunization   Immune state in an animal created by inoculation with serum (containing antibodies) or lymphocytes from an immune animal, rather than by exposure to the antigen.

patent (PĀ-tent)   Stage in an infection at which infectious agents produce evidence of their presence, such as eggs or cysts. Contrast with **prepatent.**

pathogenesis (PA-thə-JEN-ə-sis)   Production and development of disease.

pathogenicity   Capability of an agent to produce disease.

pedicel (petiole) (PED-ə-sel; PĒT-ē-ōl)   Slender, second abdominal segment that forms a "waist" in most Hymenoptera.

pedipalps (PĒD-ə-palps)   Second pair of appendages in chelicerate arthropods, modified variously in different groups.

peduncle (PĒ-dun-kəl)   Stalk. The tapering posterior part of the body of a monogenean just anterior to the opisthaptor.

pellicle (PEL-i-kəl)   Thin, translucent, secreted envelope coving many protozoa.

pereiopod (pə-RĪ-ə-pod)   Thoracic appendage of a crustacean.

perikaryon (perikarya) (pe-ri-KAR-yən; pe-ri-KAR-yə)   Portion of the cell that contains the nucleus (karyon); sometimes called the **cyton** or cell body; used in reference to cells that have processes extending some distance away from the area of the nucleus, such as nerve axons or tegumental cells of cestodes and trematodes.

peritreme (PE-rə-trēm)   Elongated sclerite extending forward from the stigma of certain mites, mainly in the suborder Mesostigmata.

peritrophic membrane (pe-ri-TRŌ-fik)   Noncellular, delicate membrane lining an insect's midgut.

permanent parasite   Parasite that lives its entire adult life within or on a host.

peroxisomes (pe-ROKS-ə-sōmz)   Small organelles containing enzymes of the glyoxylate cycle, catalase, and peroxidases.

Peyer's patches (PĪ-yərz)   Lymphoid tissue in the wall of the intestine; not circumscribed by a tissue capsule.

phagocytosis (FA-gə-sī-TŌ-səs)   Endocytosis of a particle by a cell.

phagolysosome (FA-gə-LĪ-sə-sōm)   Vacuole in a cell in which a phagocytosed particle is digested.

phagosome (FA-gə-sōm)   Vacuole in a cell containing phagocytosed particle.

phasmid (FAZ-məd)   Sensory pit on each side near the end of the tail of nematodes of the class Phasmidea.

pheromone (FE-rə-mōn)   Substance produced by one animal that affects the physiological state or the behavior of another individual.

phoresis (fō-RĒ-səs)   Form of symbiosis when the symbiont, the phoront, is mechanically carried about by its host. Neither is physiologically dependent on the other.

phyletics (fī-LET-iks)   Phylogenetic systematics, cladistics.

phylogenetic systematics (FĪ-lō-jən-ET-ik)   Cladistics, an analytical method to infer evolutionary histories.

pinkeye   Bacterial conjunctivitis, sometimes transmitted by flies of the genus *Hippolates.*

pipestem fibrosis (fī-BRŌ-səs)   Thickening of the walls of a bile duct as the result of the irritating presence of a parasite.

piroplasm (PI-rə-plazm)   Any of the class Piroplasmea, while in a circulating erythrocyte.

planidium (pla-NID-ē-əm)   First instar of hypermetamorphic, parasitic Diptera and Hymenoptera, which is apodous but moves actively by means of thoracic and caudal setae.

plasmotomy (plaz-MOT-ə-mē)   Division of a multinucleate cell into multinucleate daughter cells, without accompanying mitosis.

pleopods (PLĒ-ə-podz)   Abdominal appendages of Crustacea.

plerocercoid (PLĒ-rə-SER-koid)   Metacestode that develops from a procercoid. It usually shows little differentiation.

plerocercus (PLE-rə-SER-kəs)   Tapeworm metacestode in the order Trypanorhyncha in which the posterior forms a bladder, the blastocyst, into which the rest of the body withdraws.

plesiomorphic (PLEZ-ē-ə-MORF-ik)   Ancestral characters; characters possessed by members of both ingroup and outgroup.

plica polonica (PLĒ-kə pə-LŌN-i-kə)   Develops in untreated head louse (*Pediculus humanus capitis*) infection, hair matted together with exudate, fungal growth, fetid odor.

pleurite (PLU-rīt)   Lateral sclerite of a somite in an arthropod.

podomere (PŌ-də-mer)   More or less cylindrical segment of a limb of an arthropod, generally articulated at both ends.

podosoma (pō-də-SŌ-mə)   Portion of the body of a tick or mite that bears the legs.

poecilostome (pē-SIL-ə-stōm)   Describes mouthparts borne by members of the copepod order Poecilostomata (buccal cavity large, somewhat slitlike, with sickle-shaped mandibles); also, a member of the order Poecilostomata.

polar capsule   Compartment bearing the polar filaments in myxozoans.

polar filament   Threadlike organelles in Myxozoa and Microspora.

polar granule   Refractile granule within a coccidian oocyst.

polar ring   Electron-dense organelles of unknown function, located under the cell membrane at the anterior tip of sporozoites and merozoites.

polaroplast (pō-LA-rə-plast)   Organelle, apparently a vacuole, near the polar filament of a microsporidean.

polyembryony (po-lē-EM-brē-ə-nē)   Development of a single zygote into more than one offspring.

polykinetid (PO-lē-kī-NE-təd)   Rows or fields of kinetids in ciliates linked by fibrous networks.

polypod larva (PO-lē-pod)   Caterpillar type of larva found in Lepidoptera and some Hymenoptera. It has thoracic appendages and abdominal locomotory processes (prolegs). Also called *cruciform.*

polyzoic (po-lē-ZŌ-ik)   Strobila, when consisting of more than one proglottid.

porose area (PŌ-rōs *or* PŌ-rəs)   Sunken areas on the basis capituli of certain mites and ticks.

posterior station   Development of a protozoan in the hindgut or posterior midgut of its insect host, such as in the section Stercoraria of the Trypanosomatidae.

post-kala-azar dermal leishmanoid   Disfiguring dermal condition developing about one to two years after inadequate treatment of kala-azar.

praniza (prə-NĒ-zə)   Parasitic larva of the isopod suborder Gnathiidea. It parasitizes fishes and feeds on blood.

predation   Animal interaction in which the predator kills the prey outright. It does not subsist on the prey while the prey is alive.

prepatent (pre-PĀ-tənt)   Developmental stage in an infection before agents produce evidence of their presence.

premunition (prē-mū-NI-shən)   Resistance to reinfection or superinfection, conferred by a still existing infection, that does not destroy the organisms of the infection already present.

prenymph (PRE-nimf)   Nonfeeding, quiescent stage in the life cycle of a chigger mite.

presoma (PRĒ-sō-mə)   The proboscis, neck, and attached muscles and organs of an acanthocephalan.

prevalance (PRE-və-ləns)   In epidemiology, the number of cases of a disease at a given time; that is, a static measurement. Contrast with **incidence.**

primite (PRĪ-mīt)   Anterior member of a pair of gregarines in syzygy.

procercoid (prō-SER-koid)   Cestode metacestode developing from a coracidium in some orders. It usually has a posterior cercomer.

procuticle (PRŌ-kū-tə-kəl)   Thicker layer beneath the epicuticle of arthropods that lends mass and strength to the cuticle. It contains chitin, sclerotin, and also inorganic salts in Crustacea. The layers within the procuticle vary in structure and composition.

proglottid (prō-GLO-təd)   One set of reproductive organs in a tapeworm strobila. It usually corresponds to a segment.

prohaptor (PRŌ-hap-tər)   Collective adhesive and feeding organs at the anterior end of a monogenetic trematode.

prokaryote (prō-KA-rē-ət)   Organism in which the chromosomes are not contained within membrane-bound nuclei.

prolegs (PRŌ-legz)   Unjointed abdominal appendage in the larva of Lepidoptera and some other insects.

promastigote (prō-MAS-tə-gōt)   Form of Trypanosomatidae with the free flagellum and the kinetoplast anterior to the nucleus, as in *Leptomonas.*

propodeum (prō-PŌ-dē-əm)   First abdominal segment of hymenopterans, fused to the thorax.

propodosoma (PRŌ-pō-də-SŌ-mə)   Portion of the podosoma that bears the first and second pairs of legs of a tick or mite.

propolar cells (PRŌ-po-lər)   Anterior tier of cells in the calotte of a dicyemid mesozoan.

prosoma (PRŌ-sō-mə)   Anterior tagma of arachnids, consisting of cephalothorax; fused imperceptibly to opisthosoma in Acari.

protandry (prō-TAN-drē)   Maturation first of the male gonads and then of the female organs within a hermaphroditic individual. Also called *androgyny.*

protelean parasite (prō-TĒL-ē-ən)   Organism parasitic during its larval or juvenile stages and free living as an adult, usually changing form with each stage.

proterosoma (PRŌ-te-rə-SŌ-mə)   Combination of the gnathosoma and propodosoma of the body of a tick or mite.

protomerite (prō-TOM-ə-rīt)   Anterior half of a cephaline gregarine protozoan.

protonymph (PRŌ-tə-nimf)   Early, bloodsucking stage in the life cycle of some mesostigmatid mites.

protopod (protopodite) (PRŌ-tə-pod; prō-TOP-ə-dīt)   Coxa and basis together.

protopod larva   Larva found in some parasitic Hymenoptera and Diptera; limbs are rudimentary or absent; internal organs are incompletely differentiated; requires highly nutritive and sheltered environment for further development.

protoscolex (PRŌ-tə-SKŌ-leks)   Juvenile scolex budded within a coenurus or a hydatid metacestode of a taeniid cestode.

pseudocyst (SU-də-sist)   Pocket of protozoa within a host cell but not surrounded by a cyst wall of parasite origin.

pseudolabia (SU-də-LĀ-bē-ə)   Bilateral lips around the mouth of many nematodes of the order Spirurata; they are not homologous to the lips of most other nematodes but develop from the inner wall of the buccal cavity.

pseudomyiasis (SU-də-mī-Ī-ə-sis)   Presence within a host of a fly not normally parasitic.

ptilinum (ti-LĪ-nəm)   Balloonlike organ in the head of teneral dipterans that pushes off the operculum of the puparium.

pupariation (pū-pa-rē-Ā-shən)   Formation of a puparium by the third-stage larvae of certain families of Diptera.

puparium (pū-PA-rē-əm)   Pupal stage of certain families of Diptera.

pygidium (pī-JID-ē-əm)   Sensory organ on a posterior tergite of fleas, which apparently detects air currents.

pyogenic (PĪ-ə-JEN-ik)   Pus producing.

pyrogenic (PĪ-rə-JEN-ik)   Substance that causes a rise in body temperature; causes fever.

Q

quartan malaria (KUAR-tən)   Malaria with fevers recurring every 72 hours. Caused by *Plasmodium malariae.*

quotidian malaria (kuo-TID-ē-ən)   Malaria with fevers recurring every 24 hours. Found in cases of overlapping infections.

R

rachis (RĀ-kis)   Central, longitudinal, supporting structure in the ovary of some nematodes.

redia (RĒ-dē-ə)   Larval, digenetic trematode, produced by asexual reproduction within a miracidium, sporocyst, or mother redia.

reservoir (REZ-ər-vuar)   Living or (rarely) nonliving means of maintaining an infectious agent in nature that can serve as source of infection for humans or domestic animals.

resilin (rə-ZIL-in)   Elastic protein that releases up to 97% of stored energy upon release from stretched condition.

rete system (RĒ-tē)   Highly branched system of tubules in acanthocephalans, lying on longitudinal muscles or between longitudinal and circular muscles; thought to assist in contraction stimuli.

reticuloendothelial (RE) system (re-TIK-ū-lō-en-də-THĒ-lē-əl)   Total complement of fixed macrophages in the body, especially reticular connective tissue and the lining epithelium of the blood vascular system. Some authorities also include the phagocytic white blood cells.

retrofection (RE-trə-FEK-shən)   Process of reinfection, whereby juvenile nematodes hatch on the skin and reenter the body before molting to third-stage larvae.

rhombogen (ROM-bə-jən)   Stage in the life cycle of a dicyemid mesozoan.

rhoptries (RŌP-trēs)   Elongated, electron-dense bodies extending within the polar rings of an apicomplexan.

Romaña's sign (rō-MÄN-yəz sīn)   Symptoms of recent infection by *Trypanosoma cruzi,* consisting of edema of the orbit and swelling of the preauricular lymph node.

Romanovsky stain (RŌ-mən-OV-skē)   Complex stain, based on methylene blue and eosin, used to stain blood cells and hemoparasites. Wright's and Giemsa's stains are two common examples.

rostrum (tectum) (RÄS-trəm; TEK-təm)   Dorsal part of capitulum projecting over chelicerae in acarines.

r-strategist   Species of organism that uses a survival and reproductive "strategy" characterized by high fecundity, high mortality, short longevity. Populations are controlled by density-independent factors.

ruffles   Slender projections of the exterior surface of a dicyemid mesozoan.

S

Saefftigen's pouch (SĀF-ti-gənz)   Internal, muscular sac near the posterior end of a male acanthocephalan. It contains fluid that aids in manipulating the copulatory bursa.

salivarium (sa-li-VA-rē-əm)   Chamber in buccal cone of acarines into which salivary ducts open.

sand fly   Member of the dipteran subfamily Phlebotominae, family Psychodidae; sometimes also applied to Simuliidae (New Zealand) and Ceratopogonidae (Caribbean).

saprophytic (SAP-rə-FIT-ik)   Plant living on dead organic matter.

saprozoic nutrition (SAP-rə-ZŌ-ik)   Nutrition of an animal by absorption of dissolved salts and simple organic nutrients from surrounding medium. Also refers to feeding on decaying organic matter.

sarcocystin (SÄR-kə-SIS-tin)   Powerful toxin produced by zoitocysts of *Sarcocystis.*

sarcoptic mange (sär-KOP-tik mānj)   Disease caused by mites of the genus *Sarcoptes.* Also called **scabies.**

satellite   Posterior member of a pair of gregarines in syzygy.

scabies (SKĀ-bēz)   Disease caused by mites of the genus *Sarcoptes.* Also called **sarcoptic mange.**

scarabaeiform (SKA-rə-BĒ-ə-form)   Grublike larvae with lightly sclerotized cuticle; found in some coleopteran families.

schistosomule (shis-tə-SOM-ūl)   Juvenile stage of a blood fluke, between a cercaria and an adult; a migrating form taking the place of a metacercaria in the life cycle.

schizeckenosy (shiz-ə-KEN-ə-sē)   System of waste elimination found in some mites with a blindly ending midgut; the lobe breaks free from the ventriculus and is expelled through a split in the posterodorsal cuticle.

schizogony (shiz-ÄG-ə-nē *or* skiz-ÄG-ə-nē)   Form of asexual reproduction in which multiple mitoses take place, followed by simultaneous cytokineses, resulting in many daughter cells at once.

schizont (SHIZ-änt *or* SKIZ-änt)   Cell undergoing schizogony, in which nuclear divisions have occurred but cytokinesis is not completed; in its late phase sometimes called a *segmenter.*

Schüffner's dots (SHŪF-nerz)   Small surface invaginations that appear as stippling on the membrane of an erythrocyte infected with *Plasmodium vivax* after Romanovsky staining.

sclerite (SKLER-īt)   Any well-defined, sclerotized area of arthropod cuticle limited by suture lines or flexible, membranous portions of cuticle.

sclerotin (SKLER-ə-tən)   Highly resistant and insoluble protein occurring in the cuticle of arthropods; also thought to occur in structures secreted by various other animals, such as in the eggshells of some trematodes, in which stabilization of the protein is achieved by orthoquinone crosslinks between free imino or amino groups of the protein molecules.

scolex (SKŌ-leks)   "Head" or holdfast organ of a tapeworm.

scoliosis (skō-lē-ŌS-əs)   Lateral curvature of the spine.

scrub typhus (TĪ-fəs)   Rickettsial disease transmitted by certain chigger mites.

scutum (SKU-təm)   Large, anteriodorsal sclerite on a tick or mite.

septicemia (SEP-ti-SĒM-ē-ə)   Systemic infection where a pathogen is present in the circulating blood.

serial homolog   Series of segments in which each repeats the genetic expression of the genes in the segment (somite) before it.

sleeping sickness   African trypanosomiasis and mosquito-borne, virus-induced encephalitis.

slime ball   Mass of mucus-covered cercariae of dicrocoeliid flukes, released from land snails. Also a term of derogation applied to really disgusting persons.

solenophage (sō-LEN-ə-fāj)   Blood-feeding arthropod that introduces its mouthparts directly into a blood vessel to feed.

---

bat / āpe / ärmadillo / herring / fēmale / finch / līce / crocodile / crōw / duck / ūnicorn / ə indicates unaccented vowel sound "uh" as in mammal, fishes, cardinal, heron, vulture / stress as in bi-OL-o-gy, BI-o-LOG-i-cal

somite (SŌ-mīt)   Body segment or metamere, usually used in reference to arthropods.

sparganum (spär-GA-nəm)   Cestode plerocercoid of unknown identity.

spermalege (SPER-mə-lēj)   Organ that receives the sperm in the female cimicid bug during copulation.

spermatodactyl (spər-MAT-ə-DAK-təl)   Modification in some Acari of chelicera, which functions in transfer of sperm from male's gonopore to copulatory receptacles between third and fourth coxae of female.

spermatophore (spər-MAT-ə-for)   Formed "container" or packet of sperm that is placed in or on the body of a female, in contrast to the sperm in copulation which are conducted directly from male reproductive structures into the female's body.

spiracle (SPI-rə-kəl)   Opening into the respiratory system in various arthropods.

spondylosis (spon-də-LŌ-səs)   Degeneration of a vertebra.

sporadin (SPŌR-ə-dən)   Mature trophozoite of a gregarine protozoan.

sporoblast (SPŌR-ə-blast)   Cell mass that will differentiate into a sporocyst within an oocyst.

sporocyst (SPŌR-ə-sist)   Stage of development of a sporozoan protozoan, usually with an enclosing membrane, the oocyst. Also an asexual stage of development in some trematodes.

sporocyst residuum (SPŌR-ə-sist rē-ZID-ū-əm)   Cytoplasmic material "left over" within a sporocyst after sporozoite formation; seen as an amorphous mass.

sporogony (spōr-ÄG-ə-nē)   Multiple fission of a zygote; such a cell also is called a **sporont.**

sporont (SPŌR-ənt)   Undifferentiated cell mass within an unsporulated oocyst.

sporoplasm (SPŌR-ə-pla-zəm)   Amebalike portion of a microsporan or myxosporan cyst that is infective to the next host.

sporozoite (SPŌR-ə-ZŌ-īt)   Daughter cell resulting from sporogony.

squama (SKUA-mə)   Prominent lobe in the anal angle of a dipteran wing.

sternite (STER-nīt)   Main ventral sclerite of a somite of an arthropod.

stichosome (STIK-ə-sōm)   Column of large, rectangular cells called stichocytes, supporting and secreting into the esophagus of most nematodes of the family Trichuridae.

Stieda body (STĒ-də)   Plug in the inner wall of one end of a coccidian oocyst.

stigma (pl. **stigmata;** STIG-mə; stig-MÄ-tə)   Operculumlike area of an eggshell through which the miracidium of a schistosome fluke hatches. Also an arthropod spiracle.

strike   Deposition of fly eggs or larvae on a living host.

strobilation (STRŌ-bə-LĀ-shən)   Formation of a chain of zoids by budding, as in the strobila of a tapeworm.

strobilocercoid (STRŌ-bə-lō-SER-koid)   Cysticercoid that undergoes some strobilation; found only in *Schistotaenia.*

strobilocercus (STRŌ-bə-lō-SER-kəs)   Simple cysticercus with some evident strobilation.

style   Terminal segment of the antenna of a brachyceran dipteran. It is drawn into a sharp point.

stylops (STĪ-lops)   Member of the insect order Strepsiptera.

stylostome (STĪ-lə-stōm)   Hardened, tubelike structure secreted by a feeding chigger mite.

subchelate (səb-KĒL-āt)   Condition of an arthropod appendage in which the terminal podomere can fold back like a pincer against the subterminal podomere.

substiedal body (səb-STĒ-dəl)   Additional plug material underlying a Stieda body.

suprapopulation of parasites   All individuals of a single parasite species at all stages in the life cycle in all hosts in an ecosystem.

surra (SU-rə)   Disease of large mammals caused by *Trypanosoma evansi.*

swarmer   Daughter trophozoites resulting from multiple fissions of *Ichythophthirius multifiliis* and a few other protozoa.

sylvatic (sil-VA-tik)   Existing normally in the wild, not in the human environment.

symbiology (SIM-bi-OL-ə-gē)   Study of symbioses.

symbiont (SIM-bī-änt *or* SIM-bē-änt)   Any organism involved in a symbiotic relationship with another organism, the host.

symbiosis (SIM-bī-OS-əs *or* SIM-bē-OS-əs)   Interaction among organisms in which one organism lives with, in, or on the body of another.

symmetrogenic fission (si-ME-trə-JEN-ik)   Mitotic fission between the rows of flagella of protozoa.

synanthropism (si-NAN-thrə-pizm)   Habit of an organism of living in or around human dwellings.

synapomorphy (si-NAP-ə-mor-fē)   Shared derived characters that set a taxon apart from related taxa.

syngamy (SIN-gə-mē)   Sexual reproduction by fusion of gametes.

syzygy (SIZ-ə-jē)   Stage during sexual reproduction of some gregarines in which two or more sporadins connect end to end.

T

T cell   Type of lymphocyte with a vital regulatory role in immune response; so called because they are processed through the thymus. Subsets of T cells may be stimulatory or inhibitory. They communicate with other cells by protein hormones called cytokines.

tachyzoite (TAK-ē-ZŌ-īt)   Small, merozoitelike stages of *Toxoplasma.* They develop in the host cells' parasitophorous vacuole by endodyogeny.

tagmatization (TAG-mə-tə-ZĀ-shən)   Specialization of metameres in animals, particularly arthropods, into distinct body regions, each known as a tagma (pl. tagmata).

tarsus (TÄR-səs)   Most distal podomere of the insect or acarine limb; articulates proximally with the tibia and usually is subdivided into two to five subsegments in insects.

tectum (TEK-təm)   Dorsal extension over the mouth of a crustacean or acarine. Also called the **rostrum.**

tegument (TEG-ū-mənt)   Surficial covering of a multicellular organism, an integument.

telmophage (TEL-mə-fāj)   Blood-feeding arthropod that cuts through skin and blood vessels to cause a small hemorrhage of blood from which it feeds.

temporary parasite   Parasite that contacts its host only to feed and then leaves. Also called an **intermittent parasite** or **micropredator.**

teneral (TEN-ə-rəl)   Newly emerged adult arthropod that is soft and weak.

terebra (tə-RĒ-brə)   Functional unit of a hymenopteran ovipositor, formed from first and second valvulae.

tergite (TER-gīt)   Main dorsal sclerite of a somite of an arthropod.

tertian malaria (TER-shən)   Malaria in which fevers recur every 48 hours. Caused by *Plasmodium vivax, P. ovale,* and *P. falciparum.*

tetracotyle (TET-rə-CÄ-təl)   Strigeoid metacercaria in the family Strigeidae.

tetrathyridium (TET-rə-thī-RI-dē-əm)   Only metacestode form known in the tapeworm cyclophyllidean genus *Mesocestoides.* A large, solid-bodied cysticercoid.

theileriosis (TĪ-lər-ē-OS-əs)   Disease of cattle and other ruminants, caused by *Theileria parva.* Also called *East Coast fever.*

thelyotoky, thelytoky (THĒ-lē-ō-TŌ-kē, THĒ-lē-TŌ-kē)   Type of parthenogenesis in which all individuals are uniparental and essentially no males are produced.

thrombus   Blood clot in a blood vessel or in one of the cavities of the heart.

tibia (TIB-ē-ə)   Podomere of an insect or acarine leg that articulates proximally with the femur in insects and patella in acarines and distally with the tarsus in insects or with the metatarsus or tarsus in acarines.

titer (TĪ-tər)   Concentration of a substance in a solution as determined by titration.

trabecula (trə-BEK-ū-lə)   In general anatomical usage, a septum extending from an envelope through an enclosed substance, which, together with other trabeculae, forms part of the framework of various organs; here referring specifically to the cell processes connecting the perikarya of cestode and trematode tegumental cells with the distal cytoplasm. Also called **internuncial process.**

tracheal system (TRĀ-kē-əl)   System of cuticle-lined tubes in many insects and acarines that functions in respiration; opens to outside through spiracles.

transport host   Paratenic host.

triactinomyxon (TRĪ-ak-TIN-ə-MIX-ən)   Stage in the life cycle of a myxozoan, formerly assigned to a separate class.

tribocytic organ (TRI-bə-SI-tik)   Glandular, padlike organ behind the acetabulum of a strigeoid trematode.

trichogon (TRĪK-ə-gän)   Spiny male larva of a rhizocephalan cirripede that comes to lie within a special receptacle in the female.

tritonymph (TRĪ-tə-nimf)   Third nymphal stage in most acarines.

tritosternum (TRĪ-tə-STER-nəm)   Ventral, bristlelike sensory organ just behind the gnathosoma of a mesostigmatid mite.

triungulin (triungulinid) (trī-UN-gū-lən; trī-UN-gū-LI-nəd)   First instar larva of some parasitic, hypermetamorphic Neuroptera and Coleoptera and of the Strepsiptera, which is an active, campodeiform oligopod.

trochanter (TRŌ-kan-tər)   Podomere of an insect or acarine leg that articulates basally with the coxa and distally with the femur; usually fixed to the femur in insects.

trophont (TRŌ-fənt)   Stage in the life cycle of gregarines.

trophozoite (TRŌ-fə-ZŌ-īt)   Active, feeding stage of a protozoan, in contrast to a cyst. Also called the *vegetative stage.*

trypomastigote (TRI-pə-MAS-tə-gōt)   Form of Trypanosomatidae with an undulating membrane and the kinetoplast located posterior to the nucleus. An example is *Trypanosoma.*

tsetse fly (TSET-sē *or* SET-sē)   Bloodsucking fly of the genus *Glossina.*

U

ulcer (UL-sər)   Area of inflammation that opens out to the skin or a mucous surface.

ultrastructure   Structure of an organism or cell at the electron microscopic level.

undulating membrane (UN-dū-LĀT-ing)   Name applied to two quite different structures in protozoa. In some Mastigophora it is a finlike ridge across the surface of a cell, with the axoneme of a flagellum near its surface. In some ciliates it is a line of cilia that are fused at their bases, usually beating to force food particles toward the gullet.

undulating ridges   Undulatory waves in the surface of some protozoa, probably aided by subpellicular microtubules; the means of locomotion in some species.

uniramous appendage (Ū-nē-RĀ-məs)   Arthropod appendage that is unbranched, characteristic of living arthropods other than Crustacea, although some crustacean appendages are uniramous.

urban   Peculiar to the human environment, as contrasted with that found normally around wild animals.

urn   Region near the center of an infusoriform larva of a dicyemid mesozoan.

uropolar cells (Ū-rə-PŌ-lər)   Somatoderm cells at the posterior end of the trunk of a dicyemid mesozoan.

urosome (Ū-rə-sōm)   Portion of the body posterior to the major point of body flexion in many copepods; usually includes one or more free thoracic segments and abdomen.

urstigmata (UR-stig-MÄ-tə)   Sense organs between the coxae of the first and second pairs of legs on some mites. Apparently they are humidity receptors. Also called **Claparedé organs.**

uterine bell (ŪT-ər-ən)   Structure in female acanthocephalans that allows fully developed, shelled embryos to pass out of the body and that retains undeveloped ones.

V

vagabond's disease   Darkened, thickened skin caused by years of infestation with body lice, *Pediculus humanus humanus.*

valvifers (VALV-i-fərz)   Basal portions of the ovipositor in Hymenoptera, derived from coxae of segmental appendages.

valvulae (VALV-ū-lē)   Processes from the valvifers to form the body of the ovipositor (terebra) and the ovipositor sheath (third valvulae) in Hymenoptera.

variant antigen type (VAT)   Applied to certain trypanosomes, any one of numerous antigenic types expressed on the surface of the organisms and "seen" by the immune system of the host. See **variant-specific surface glycoprotein.**

variant-specific surface glycoprotein (VSG)   Glycoprotein on the surface of certain trypanosomes recognized by the host's immune system. Each VSG is responsible for one VAT.

vector (VEK-tər)   Any agent, such as water, wind, or insect, that transmits a disease organism.

veins (vānz)   Blood vessels conducting blood toward the heart in any animal. Also more heavily sclerotized portions of wings of insects, which are remains of lacunae.

vermicle (VER-mə-kəl)   Infective stage of *Babesia* in a tick.

vermiform (VER-mə-form)   Wormlike.

verruga peruana (vər-UG-ə PE-ru-ÄN-ə)   Clinical form of Carrion's disease, caused by the bacterium *Bartonella bacilliformis* and transmitted by sand flies.

vesicular disease (ve-SIK-ū-lər)   Any disease of the urinary bladder, such as vesicular schistosomiasis.

vestibulum (ves-TI-bū-ləm)   Cavity leading into another cavity or passage, such as in the ciliate order Vestibulifera.

virulence (VIR-ə-ləns)   Degree of pathogenicity of an agent; how much damage the agent can cause.

---

bat / āpe / ärmadillo / herring / fēmale / finch / līce / crocodile / crōw / duck / ūnicorn / ə indicates unaccented vowel sound "uh" as in mammal, fishes, cardinal, heron, vulture / stress as in bi-OL-o-gy, BI-o-LOG-i-cal

W

whirling disease   Disease of fishes, caused by the protozoan *Myxobolus cerebralis.*

weir (WĒ-ər)   Filtration apparatus in flame-cell protonephridia of flatworms.

Winterbottom's sign   Swollen lymph nodes at the base of the skull, symptomatic of African sleeping sickness.

X

xenodiagnosis (ZĒN-ə-dī-əg-NŌ-səs)   Diagnosis of a disease by infecting a test animal.

xenograft (ZĒN-ə-graft)   Graft of a piece of tissue or organ from one individual to another of a different species.

xenosomes (ZĒN-ə-sōmz)   Body or organelle living within a cell that contains its own DNA and is capable of reproducing itself, once having functioned as a free-living organism. Examples are zooxanthellae and zoochlorellae.

xiphidiocercaria (zī-FID-ē-ə-ser-KA-rē-ə)   Cercaria with a stylet in the anterior rim of its oral sucker.

Y

yaws   Bacterial disease caused by the spirochete *Treponema pertenue,* often transmitted by flies.

yellow fever   Virus disease transmitted by the mosquito *Aedes aegypti.*

Z

zoid (ZŌ-id)   Member of a colonial organism.

zoitocyst (zō-ĪT-ə-sist)   Tissue phase in some of the coccidia of the *Isospora* group. They usually have internal septae and contain thousands of bradyzoites. Also called a *sarcocyst* or **Miescher's tubule.**

zoonosis (ZŌ-ə-NŌ-sis)   Disease of animals that is transmissible to humans. Some authors subdivide the concept into zooanthroponosis, an infection humans can acquire from animals, and anthropozoonosis, a disease of humans transmissible to other animals.

zooxanthellae (ZŌ-ə-zan-THEL-ē)   Dinoflagellate protistans living mutualistically in the cells of certain marine animals. Among other benefits, their hosts derive nutrients from the photosynthetic reactions of the zooxanthellae.

INDEX

A

Abcess, 28
Abdomen, 499
Ablastin, 68
Abortion
    toxoplasmosis, 127
    Trichomonas foetus, 89–90
    Trypanosoma lewisi, 67
Acanthamoeba, 109–10
Acanthella, 476
Acanthobothrium spp., 347
Acanthobothrium coronatum, 305
Acanthobothrium urolophi, 348
Acanthocephala (phylum)
    body structure, 469–70
    body wall, 460
    development and life cycles, 476–77
    digestive system, 475–76
    effects of on host, 477–78
    excretory system, 475
    in humans, 479
    metabolism, 476
    muscular system, 472
    nervous system, 475
    reproductive system, 472–75
    taxonomy, 479–80
    tegument, 470–71
Acanthocephalus spp., 13, 479
Acanthocotyle lobianchi, 284
Acanthopodina (suborder), 47
Acanthor, 476
Acanthosentis acanthuri, 471
Acanthostomum brauni, 222
Acari (order), 511
    Astigmata, 619–21
    form of, 502–3
    Ixodida, 604–12
    Mesostigmata, 612–16
    Orbatida, 619
    Prostigmata, 616–18
    taxonomic characteristics of, 603, 604
Accessory duct, 220
Accessory filament, 86
Accessory glands, 506
Accessory piece, 288
Accessory sclerites, 283
Accidental myiasis, 588
Accidental parasites, 6
Accommodation, immunological, 31
Acephaline gregarines, 114
Acetabula, 299
Acetylcholine, 211, 364
Aclid organ, 476
Acoelomorpha (superclass), 189
Acoels, 192

Acquired immunity, 21, 23–28
Acraspedote, 298
Acron, 497
Acrothoracica (order), 509
Actinocephalus carrilynnae, 17
Actinomyxida (order), 49
Actinopoda (superclass), 48
Actinosphaerium, 36
Actinosporea (class), 49
Acuariidae (family), 439–40
Acute infections, 126
Acute phase, of schistosomiasis, 244
Adeleorina (subclass), 48, 117–18
Adenophorea (class), 379
Adhesive disc, 82
Adhesive organ, 233
Adoral zone of membranelles (AZM), 40, 177
Adult stem nematogen, 179
Aedeagus, 501, 503, 506
Aedes (genus), 569–73
Aedes spp., 570, 571, 572
Aedes aegypti, 571, 572
Africa, politics and public health, 61. *See also* Third World
Agamete, 179
Agamococcidiorida (order), 48
Aggregated populations, 11
Agriculture. *See also* Cattle; Horses; Poultry; Swine
    disease control and, 14
    parasites of domestic animals, 3–4
AIDS (acquired immunodeficiency syndrome), 4, 27. *See also* Human immunodeficiency virus (HIV)
    cryptosporidiosis, 130
    Encephalitozoon cuniculi, 170
    Isospora belli, 122
    mosquitoes and transmission of, 566
    Pneumocystis carinii, 132
    public awareness of parasitic disease, 2
    strongyloidiasis, 402
Air tube, 568
Akiba (genus), 154
Alae, 357
Alaria spp., 216, 221
Alaria americana, 233–34, 235
Alaria marcianae, 234
Albendazole, 387, 388
Aleppo boil, 71
Algid malaria, 147
Allergies, 28
Allocorrigia filiformis, 221
Alloeocoels, 194–95
Alloglossidium hirudicola, 206
Allograft, 22

Allopatric speciation, 342
Allopuranol, 45
Alofia (genus), 488
Alveococcus (genus), 340
Alveolar hydatid, 313, 340
Alveoli, 35, 197
Amastigote, 55
Amblycera, 536
Amblyomma spp., 606, 607, 608
Amblyospora, 169
Amebastomes, 109
Amebiasis, 44, 99–105
Amebocytes, 504
Ameboma, 103
American Type Culture Collection (Rockville, Maryland), 19
Ameson michaelis, 170
Amino acids, 315, 378
Ammonotelic excretion, 45
Amoeba spp., 36
Amoeba proteus, 5
Amoebida, 47, 99–108
Amphibdellatinea (suborder), 294
Amphidial nerves, 362
Amphids, 362
Amphilina foliacea, 350
Amphilinidea (subcohort), 192, 318, 349
Amphimictic reproduction, 42
Amphiphos squamata, 182
Amphipoda (order), 509, 529–30
Amphiscela macrocephala, 193
Amphiscolops, 188
Amphistome, 205
Amphotericin B, 109
Anamnestic response, 26, 60
Anaphylaxis, 28, 339
Anaplasma marginale, 610
Anapolysis, 298
Anaticola spp., 536
Anatrichosoma spp., 389
Anatrichosomatidae (family), 389
Anautogeny, 578
Anchors, 283
Anchor worm, 514
Ancylostoma spp., 369, 370, 410
Ancylostoma braziliense, 409, 412
Ancylostoma caninum, 409, 412
Ancylostoma ceylanicum, 409
Ancylostoma duodenale, 406, 408–9
Ancylostoma tubaeforme, 377
Ancylostomatoidea (superfamily), 405–12
Ancylostomidae (family), 405–12
Ancyroniscus bonnieri, 532
Androgyny, 307
Anecdysis, 496

Owls, 7
Oxylipeurus polytrapzus, 536
Oxymonadida (order), 46
Oxyurida (order), 380
Oxyuridae (family), 433–36
Oyster, 532. *See also* Molluscs

P

Palaeacanthocephala (class), 477, 480
Palliolisentis polyonca, 471
Palps, 501
Pamaquine, 148
Pandemics, of plague, 559
Panopistus, 221
Pansporoblasts, 164
Panstrongylus megistus, 551
Papatasi fever, 566
Papillary nerves, 362
Parabasal body, 36, 40
Parabasal filament, 40
Paracanthocephalus rauschi, 471
Paracostal granules, 87
Paracyamus, 531
Paradujardinia, 368
Paragonimiasis, 270–71
Paragonimus spp., 267–71
Paragonimus kellicotti, 212, 213, 271
Paragonimus westermani, 222, 267, 268, 270
Parahaemoproteus, 154
Parahistomonas wenrichi, 92
Paralysis, tick, 604
Paramastigote, 55
Paramere, 549
Paramphistomidae (family), 259
Paramphistomiformes (order), 227
Paramphistomoidea (superfamily), 259–60
Paramphistomum cervi, 226, 259
Paramyxea (class), 49
Paramyxida (order), 49
Paranoplocephala mamillana, 300
Parasambonia (genus), 488
Parascaris spp., 373
Parascaris equorum, 426
Parasite species assemblage, 12
Parasitic castration, 528
Parasitic crustaceans, 513–29
Parasitism, definition of, 6
Parasitoids, 6, 591
Parasitology
    basic definitions, 4–7
    careers in, 4
    echinostomatids as models in
        experimental, 253
    human welfare and, 2–3
    parasites of domestic and wild animals,
        3–4
    relationship to other sciences, 1–2
Parasitophorus vacuole, 45, 55
Parasomal sac, 40
Paratenic host, 6–7
Paraxial rod, 40, 53
Paraxostylar granules, 87
Parenchyma, 187, 304
Pars prostatica, 213
Parthenogenesis, 215
Paruterine organs, 309
Paryphostomum surfrartyfex, 253
Patella, 503
Pathology, pathogenesis of parasitic
        infections, 30–31. *See also*
        Epidemiology

Pectenophilus ornatus, 526
Pedicel, 596
Pediculosis, 541
Pediculus humanus, 536, 539, 540–41
Pediculus mjobergi, 542
Pedipalps, 503, 603
Pelecinus polyturator, 600
Pellicle, 35
Pellicular microtubules, 35–36, 55
Pelmatosphaera (genus), 185
Pelmatosphaeridae (family), 185
Pelobiontida (order), 48
Pelodera strongyloides, 375
Pelta, 87
Penarchigetes oklensis, 330
Penetration glands, 216, 220
Penetration organ, 485
Penis, 288, 501
Pennella spp., 522
Pennellidae (family), 522–25
Pentastoma majae, 486
Pentastomiasis, 486–88
Pentastomida
    biology, 484–86
    morphology, 483–84
    pathogenesis, 486–88
    taxonomy, 483, 488
Pentatrichomonas hominis, 36, 89
Pentavalent antimonials, 70
Pentostam, 70
Peracarida (superorder), 509
Pereiopods, 500
Pericardial sinus, 503
Perilampus hyalinus, 597
Peritonitis, 103
Peritreme, 612
Peritrichia (subclass), 51, 177
Peritrichida (order), 177
Peritrophic membrane, 506
Perkinsasida (class), 48
Perkinsorida (order), 48
Perkinsus spp., 113
Permanent parasites, 6
Peroxisomes, 37
Phaenicia sericata, 582
Phagocytosis, 21–22, 44–45
Pharyngodon, 434
Pharynx, 188, 197, 287, 505, 506
Phasmidea (subclass), 379
Phasmids, 364
Pheromone sex attractants, 372, 605
Philichthyidae (family), 517
Philichthys xiphae, 518
Philometra spp., 461
Philometra oncorhynchi, 462
Philometridae (family), 461–62
Philometroides, 461, 462
Philophthalmus megalurus, 213
Phlebotominae, 565–66
Phlebotomus spp., 566
Phoresis, 5
Phoronts, 43
Phosphofructokinase (PFK), 226
Photoautotrophic feeding, 44
Phrixocephalus spp., 522
Phrixocephalus longicollum, 523
Phthirus pubis, 539, 541–42
Phyletics, 16
Phyllobothrium spp., 300
Phyllobothrium kingae, 347
Phylogenetic systematics, 16
Phylogeny
    of Aspidobothrea, 203
    definition, 1

of digenetic trematodes, 226–29
hosts of parasitic copepods, 19
of Mesozoa, 182–83
of Monogenea, 294–95
of Nematoda, 355
Physaloptera spp., 442, 443
Physalopteridae (family), 441–42
Phytomastigophorea (class), 46
Phytomonas (genus), 77
Pian bois, 76
Pigeon fly, 580
Pigeon tick, 612
Pilot fish, 6
Pinkeye, 579
Pinnotheres spp., 532, 533
Pinnotherion vermiforme, 530, 532
Pinocytosis, 44–45
Pinworms
    Dientamoeba fragilis, 93
    infection in humans, 2, 433–36
Piroplasmasina (subclass), 49
Piroplasmea (subclass), 155–58
Plagiorchiata (suborder), 263–66
Plagiorchiformes (order), 229, 263–73
Plagiorchiidae (family), 266
Plagiorchis maculosus, 266, 267
Plagiorchis nobeli, 266
Plagiorhynchus cylindraceus, 477, 478
Plagiotomidae (family), 178
Plague, 559–61
Planidium, 592
Planorbula armigera, 265
Plasma cells, 26
Plasma membrane, 35, 37
Plasmodiophorida (order), 48
Plasmodium (genus). *See* Malaria
Plasmodium spp.. *See also* Malaria
    hosts of, 6
    hyperparasitism, 7
    life cycle and general morphology of,
        140–43
    metabolism, 45, 151–53
    taxonomy, 143–46
Plasmodium berghei, 143
Plasmodium brasilianum, 146
Plasmodium cathemerium, 42, 44
Plasmodium falciparum, 145, 147, 573
Plasmodium gallinaceum, 143, 144
Plasmodium inui, 151
Plasmodium lophurae, 144
Plasmodium malariae, 145–46
Plasmodium ovale, 146
Plasmodium vivax, 140, 144–45
Plasmotomy, 42
Platycobboldia, 587
Platyhelminthes. *See also* Trematoda
    taxonomy, 19, 188–92
    tubellarians, 192–95
Platymyarian muscle cell, 359
Platypsyllus castoris, 593
Platysporina (suborder), 49
Pleistophora, 170
Pleistophoridida (order), 49
Pleocyemata (suborder), 510
Pleodicyema (genus), 184
Pleomorphy, 58
Pleopods, 500
Plerocercoid, 311
Plerocercoid growth factor (PGF), 318
Plerocercus, 312
Plesiomorphic species, 16
Pleural arch, 553
Pleurites, 492
Pleurolophocercous cercaria, 275